FOURTH EDITION

Elements of Engineering Electromagnetics

Nannapaneni Narayana Rao

Professor of Electrical and Computer Engineering
University of Illinois at Urbana–Champaign

PRENTICE HALL, Englewood Cliffs, New Jersey 07632

Library of Congress Cataloging-in-Publication Data

Narayana Rao, Nannapaneni,
Elements of engineering elctromagnetics / Nannapaneni Narayana Rao. -- 4th ed.
p. cm.
Includes index.
ISBN: 0-13-948746-8
1. Electromagnetic theory. I. Title.
QC670.N3 1994
621.3'01--dc20
93-27009
CIP

Acquisitions editor: **ALAN APT**
Production editor: **RICHARD DeLORENZO**
Cover design: **WANDA LUBELSKA DESIGN**
Production coordinator: **LINDA BEHRENS**
Editorial assistant: **SHIRLEY McGUIRE**
Supplements editor: **ALICE DWORKIN**

A Paramount Communications Company
Englewood Cliffs, New Jersey 07632

Printed in the United States of America

10 9 8 7 6 5 4 3 2

ISBN 0-13-948746-8

Prentice-Hall International (UK) Limited, London
Prentice-Hall of Australia Pty. Limited, Sydney
Prentice-Hall Canada Inc., Toronto
Prentice-Hall Hispanoamericana, S.A., Mexico
Prentice-Hall of India Private Limited, New Delhi
Prentice-Hall of Japan, Inc., Tokyo
Simon & Schuster Asia Pte. Ltd., Singapore
Editora Prentice-Hall do Brasil, Ltda., Rio de Janeiro

From the Upanishads—
Matrudevo bhava: *Revere the mother as God!*
Pitrudevo bhava: *Revere the father as God!*
Acharyadevo bhava: *Revere the preceptor as God!*
Atidhidevo bhava: *Revere the guest as God!*
—as enunciated by Sri Satya Sai

The guiding *essence*, cherished and followed,
in communicating the *science* that follows.

Contents

6
Uniform Plane Waves 296

7
Transmission Lines 1. Time-Domain Analysis 358

8
Transmission Lines 2. Sinusoidal Steady-State Analysis 435

9
Metallic Waveguides and Resonators 506

Preface

Electromagnetics textbooks and this edition

The existing introductory textbooks on engineering electromagnetics can be classified broadly into three categories:

1. One-semester textbooks based on the historical approach, covering essentially electrostatics and magnetostatics and culminating in Maxwell's equations with some discussion of their applications.
2. Two-semester textbooks, with the first half or more covering electrostatics and magnetostatics as in category 1 and the remainder devoted to topics associated with electromagnetic waves.
3. One- or two-semester textbooks that have deviated from the historical approach with the degree and nature of deviation depending on the author.

This book belongs to category 3, and the deviation from the historical approach originating with the first edition, a one-semester textbook, has been preserved in the subsequent editions and expanded for one- or two-semester usage to provide flexibility and include PC programs, among other teaching aids. The substantial changes leading to the fourth edition have been prompted by the increasing need for coverage of material at the introductory level for application beyond the microwave region into the optical regime of the electromagnetic spectrum with the advent of the era of photonics overlapping with that of electronics. Thus Chapter 10 of the third edition on antennas has been converted to Chapter 11 and a new Chapter 10 has been created by shifting the material on total internal reflection and dielectric slab waveguides in the previous Chapter 9 and adding new material on topics of interest to photonics. Sections on cylindrical metallic waveguides and losses in waveguides and resonators have been added to Chapter 9, and a section on aperture antennas has been added to Chapter 11. The 18 PC programs written in BASIC for the IBM PC have been retained and one new program has been included.

As in the second and third editions, this edition incorporates flexibility to facilitate its adoption according to the following options:

1. For a three-credit one-semester course or for a four-credit one-quarter course based on coverage of a combination of chapters, depending on the background preparation of the students and the needs of the curriculum. Some examples are
   (a) Chapters 1 through 6
   (b) Chapters 3 through 6 plus parts of Chapters 7, 8, 9, and 10
   (c) Chapters 6 through 11
2. For a two-semester or two-quarter sequence covering the entire book
3. As a text or supplementary text for a course emphasizing PC-assisted instruction

Thread of development of material

The thread of development of the material, evident from a reading of the table of contents, is essentially along the lines of the second and third editions. Some of the salient features are the following:

1. Discusses materials following the presentation of electric and magnetic field concepts and prior to the study of Maxwell's equations
2. Introduces collectively Maxwell's equations for time-varying fields, first in integral form and then in differential form
3. Considers boundary conditions following Maxwell's equations in integral form, and potential functions and associated equations following Maxwell's equations in differential form
4. Devotes a chapter to the development of selected topics in static and quasistatic fields in addition to the coverage of static fields in earlier chapters
5. Obtains uniform plane wave solutions by considering the infinite plane current sheet source first in free space and then in a material medium
6. Develops time-domain analysis of transmission lines in a progressive manner beginning with the case of a resistive load and culminating in the discussion of interconnections between logic gates
7. Presents sinusoidal-steady-state analysis of transmission lines comprising the topics of standing waves, resonance, power transfer, and matching with emphasis on computer and graphical solutions
8. Discusses metallic waveguides by first introducing the parallel-plate waveguide by considering the superposition of two obliquely propagating uniform plane waves between two perfect conductors and then extending to rectangular and cylindrical waveguides
9. Devotes a chapter for electromagnetic principles for photonics, building up on the coverage of wave phenomena in earlier chapters
10. Introduces radiation by obtaining the complete field solution to the Hertzian dipole field through the magnetic vector potential, and then developing the basic concepts of antennas

Teaching and learning aids

All the teaching and learning aids employed in the previous editions have been retained: (1) examples distributed throughout the text, (2) discussion of practical applications of field concepts and phenomena interspersed among presentations of basic subject matter, (3) descriptions of brief experimental demonstrations suitable for presentation in the classroom, (4) summary of material and review questions for each chapter, (5) inclusion of drill problems (**D**) with answers at the end of each section, (6) marginal notes, and (7) key words (**K**) at the end of each section. Answers are provided for about 40 percent of the end-of-chapter problems. The comprehensive, user-interactive software package, extending the PC programs in the book and including additional topics made available with the third edition, has been updated for this edition and is again available free of charge. For information, write to the author, c/o Department of Electrical and Computer Engineering, University of Illinois at Urbana–Champaign, William L. Everitt Laboratory, 1406 West Green St., Urbana, IL 61801.

Acknowledgements

I wish to express my appreciation to many colleagues at the University of Illinois who have taught from the previous editions of the book during the period from 1977 to 1993. Listed in alphabetical order are M. T. Birand, D. J. Brady, K. Y. Cheng, S. L. Chuang, D. H. Cooper, C. Daft, T. A. DeTemple, S. J. Franke, L. A. Frizzell, S. D. Gedney, R. Gilbert, S. Gnanalingam, K. C. Hsieh, J. Joseph, K. Kim, P. W. Klock, J. Kolodzey, E. Kudeki, J. P. Leburton, S. W. Lee, C. H. Liu, R. L. Magin, K. Mahadevan, P. E. Mast, E. A. Mechtly, H. Merkelo, K. L. Miller, R. Mittra, H. Morkoc, G. C. Papen, A. F. Peterson, P. L. Ransom, U. Ravaioli, J. Schutt-Aine, C. F. Sechrist, Jr., L. G. Smith, A. Steinbach, D. R. Tanner, R. J. Turnbull, A. W. Wernik, K. Whites, and K. C. Yeh. Thanks are also due to the numerous users of the book at other schools. The evolution of this book and the associated software would not have been possible without the many opportunities provided to me by my Department Heads since joining the University of Illinois in 1965, the late E. C. Jordan, followed by G. W. Swenson, Jr., E. W. Ernst, and T. N. Trick. Many individuals in the department have provided support over the years; I am particularly appreciative of Mrs. Lilian H. Beck, Editor, Publications, in this regard. Sheryle Carpenter and Laurie Oakes performed the office duties in an admirable manner that ensured smooth functioning of the office at all times during my tenure as Associate Head of the Department since 1987. The typing of the new portions of the manuscript was done by Kelly Voyles in a prompt and skillful manner. As always, I am deeply indebted to my wife Sarojini for her continued understanding and patience.

N. Narayana Rao
Urbana, Illinois

1

Vectors and Fields

Electromagnetics deals with the study of electric and magnetic fields. It is at once apparent that we need to familiarize ourselves with the concept of a field, and in particular with electric and magnetic fields. These fields are vector quantities and their behavior is governed by a set of laws known as Maxwell's equations. The mathematical formulation of Maxwell's equations and their subsequent application in our study of the elements of engineering electromagnetics require that we first learn the basic rules pertinent to mathematical manipulations involving vector quantities. With this goal in mind, we devote this chapter to vectors and fields.

We first study certain simple rules of vector algebra without the implication of a coordinate system and then introduce the Cartesian, cylindrical, and spherical coordinate systems. After learning the vector algebraic rules, we turn our attention to a discussion of scalar and vector fields, static as well as time-varying, by means of some familiar examples. We devote particular attention to sinusoidally time-varying fields, scalar as well as vector, and to the phasor technique of dealing with sinusoidally time-varying quantities. With this general introduction to vectors and fields, in Chapter 2 we study the concepts of electric and magnetic fields, from considerations of the experimental laws of Coulomb and Ampère.

1.1 VECTOR ALGEBRA

Vectors versus scalars

In the study of elementary physics we come across several quantities such as mass, temperature, velocity, acceleration, force, and charge. Some of these quantities have associated with them not only a magnitude but also a direction in space whereas others are characterized by magnitude only. The former class of quantities are known as *vectors* and the latter class of quantities are known as *scalars*.

Mass, temperature, and charge are scalars, whereas velocity, acceleration, and force are vectors. Other examples are voltage and current for scalars and electric and magnetic fields for vectors.

Vector quantities are represented by symbols in boldface roman type (e.g., **A**), to distinguish them from scalar quantities, which are represented by symbols in lightface italic type (e.g., A). Graphically, a vector, say **A**, is represented by a straight line with an arrowhead pointing in the direction of **A** and having a length proportional to the magnitude of **A**, denoted $|\mathbf{A}|$ or simply A. Figure 1.1 shows four vectors drawn to the same scale. If the top of the page represents north, then vectors **A** and **B** are directed eastward, with the magnitude of **B** being twice that of **A**. Vector **C** is directed toward the northeast and has a magnitude three times that of **A**. Vector **D** is directed toward the southwest and has a magnitude equal to that of **C**. Since **C** and **D** are equal in magnitude but opposite in direction, one is the negative of the other.

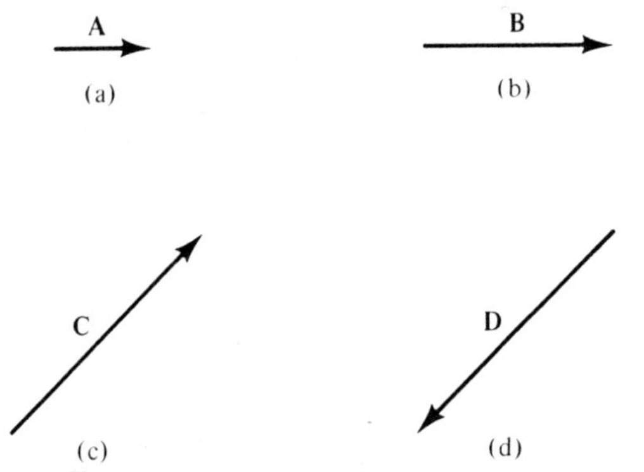

Figure 1.1. Graphical representation of vectors.

Unit vector defined

Since a vector may have in general an arbitrary orientation in three dimensions, we need to define a set of three reference directions at each and every point in space in terms of which we can describe vectors drawn at that point. It is convenient to choose these three reference directions to be mutually orthogonal, as, for example, east, north, and upward, or the three contiguous edges of a rectangular room. Thus let us consider three mutually orthogonal reference directions and direct *unit vectors* along the three directions as shown, for example, in Fig. 1.2(a). A unit vector has magnitude *unity*. We shall represent a unit vector by the symbol **i** and use a subscript to denote its direction. We shall denote the three directions by subscripts 1, 2, and 3. We note that for a fixed orientation of $\mathbf{i}_1$, two combinations are possible for the orientations of $\mathbf{i}_2$ and $\mathbf{i}_3$, as shown in Fig. 1.2(a) and (b). If we take a right-hand screw and turn it from $\mathbf{i}_1$ to $\mathbf{i}_2$ through the 90° angle, it progresses in the direction of $\mathbf{i}_3$ in Fig. 1.2(a) but opposite to the direction of $\mathbf{i}_3$ in Fig.1.2(b). Alternatively, a left-hand screw when turned from $\mathbf{i}_1$ to $\mathbf{i}_2$ in Fig. 1.2(b) will progress in the direction of $\mathbf{i}_3$. Hence the set of unit vectors in Fig. 1.2(a) corresponds to a right-handed system, whereas the set in Fig. 1.2(b) corresponds to a left-handed system. We shall work consistently with the right-handed system.

A vector of magnitude different from unity along any of the reference directions can be represented in terms of the unit vector along that direction. Thus $4\mathbf{i}_1$ represents a vector of magnitude 4 units in the direction of $\mathbf{i}_1$, $6\mathbf{i}_2$ represents a

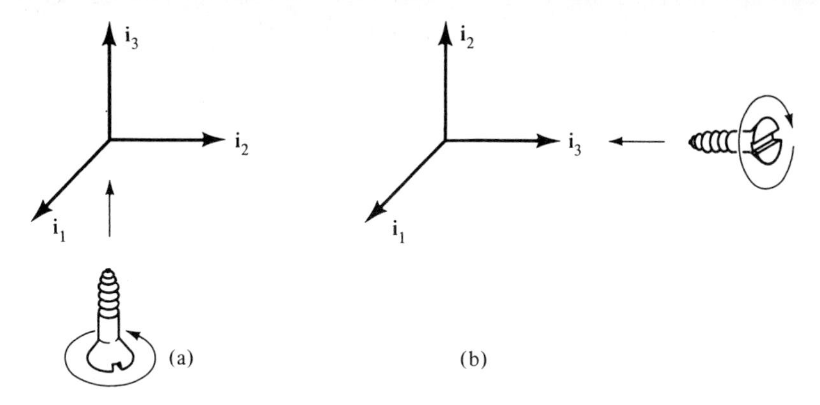

Figure 1.2. (a) Set of three orthogonal unit vectors in a right-handed system. (b) Set of three orthogonal unit vectors in a left-handed system.

vector of magnitude 6 units in the direction of $\mathbf{i}_2$, and $-2\mathbf{i}_3$ represents a vector of magnitude 2 units in the direction opposite to that of $\mathbf{i}_3$, as shown in Fig. 1.3. Two vectors are added by placing the beginning of the second vector at the tip of the first vector and then drawing the sum vector from the beginning of the first vector to the tip of the second vector. Thus to add $4\mathbf{i}_1$ and $6\mathbf{i}_2$, we simply slide $6\mathbf{i}_2$ without changing its direction until its beginning coincides with the tip of $4\mathbf{i}_1$ and then draw the vector $(4\mathbf{i}_1 + 6\mathbf{i}_2)$ from the beginning of $4\mathbf{i}_1$ to the tip of $6\mathbf{i}_2$, as shown in Fig. 1.3. To see this, imagine that on the floor of an empty rectangular room you are going from one corner to the opposite corner. Then to reach the destination, you can first walk along one edge and then along the second edge. Alternatively, you can go straight to the destination along the diagonal. By adding $-2\mathbf{i}_3$ to the vector $(4\mathbf{i}_1 + 6\mathbf{i}_2)$ in a similar manner, we obtain the vector $(4\mathbf{i}_1 + 6\mathbf{i}_2 - 2\mathbf{i}_3)$, as shown in Fig. 1.3. We note that the magnitude of $(4\mathbf{i}_1 + 6\mathbf{i}_2)$ is $\sqrt{4^2 + 6^2}$ or 7.211 and that the magnitude of $(4\mathbf{i}_1 + 6\mathbf{i}_2 - 2\mathbf{i}_3)$ is $\sqrt{4^2 + 6^2 + 2^2}$ or 7.483. Conversely to the foregoing discussion, a vector $\mathbf{A}$ at a given point is simply the superposition of three vectors $A_1\mathbf{i}_1$, $A_2\mathbf{i}_2$, and $A_3\mathbf{i}_3$ which are the projections of $\mathbf{A}$ onto the reference directions at that point. A_1, A_2, and A_3 are known as the *components* of $\mathbf{A}$ along the 1, 2, and 3 directions, respectively, Thus

$$\boxed{\mathbf{A} = A_1\mathbf{i}_1 + A_2\mathbf{i}_2 + A_3\mathbf{i}_3} \tag{1.1}$$

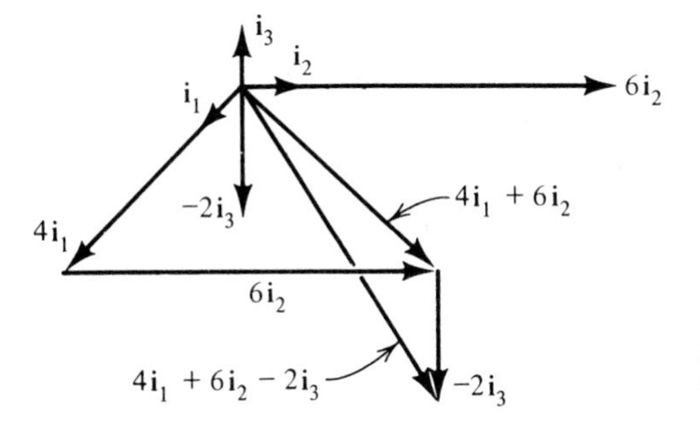

Figure 1.3. Graphical addition of vectors.

We now consider three vectors, **A**, **B**, and **C** given by

$$\mathbf{A} = A_1\mathbf{i}_1 + A_2\mathbf{i}_2 + A_3\mathbf{i}_3 \tag{1.2a}$$

$$\mathbf{B} = B_1\mathbf{i}_1 + B_2\mathbf{i}_2 + B_3\mathbf{i}_3 \tag{1.2b}$$

$$\mathbf{C} = C_1\mathbf{i}_1 + C_2\mathbf{i}_2 + C_3\mathbf{i}_3 \tag{1.2c}$$

at a point and discuss several algebraic operations involving vectors as follows:

Vector addition and subtraction. Since a given pair of like components of two vectors are parallel, addition of two vectors consists simply of adding the three pairs of like components of the vectors. Thus

$$\begin{aligned}\mathbf{A} + \mathbf{B} &= (A_1\mathbf{i}_1 + A_2\mathbf{i}_2 + A_3\mathbf{i}_3) + (B_1\mathbf{i}_1 + B_2\mathbf{i}_2 + B_3\mathbf{i}_3)\\ &= (A_1 + B_1)\mathbf{i}_1 + (A_2 + B_2)\mathbf{i}_2 + (A_3 + B_3)\mathbf{i}_3\end{aligned} \tag{1.3}$$

Vector subtraction is a special case of addition. Thus

$$\begin{aligned}\mathbf{B} - \mathbf{C} &= \mathbf{B} + (-\mathbf{C})\\ &= (B_1\mathbf{i}_1 + B_2\mathbf{i}_2 + B_3\mathbf{i}_3) + (-C_1\mathbf{i}_1 - C_2\mathbf{i}_2 - C_3\mathbf{i}_3)\\ &= (B_1 - C_1)\mathbf{i}_1 + (B_2 - C_2)\mathbf{i}_2 + (B_3 - C_3)\mathbf{i}_3\end{aligned} \tag{1.4}$$

Multiplication and division by a scalar. Multiplication of a vector **A** by a scalar m is the same as repeated addition of the vector. Thus

$$m\mathbf{A} = m(A_1\mathbf{i}_1 + A_2\mathbf{i}_2 + A_3\mathbf{i}_3) = mA_1\mathbf{i}_1 + mA_2\mathbf{i}_2 + mA_3\mathbf{i}_3 \tag{1.5}$$

Division by a scalar is a special case of multiplication by a scalar. Thus

$$\frac{\mathbf{B}}{n} = \frac{1}{n}(\mathbf{B}) = \frac{B_1}{n}\mathbf{i}_1 + \frac{B_2}{n}\mathbf{i}_2 + \frac{B_3}{n}\mathbf{i}_3 \tag{1.6}$$

Magnitude of a vector. From the construction of Fig. 1.3 and the associated discussion, we have

$$\boxed{|\mathbf{A}| = |A_1\mathbf{i}_1 + A_2\mathbf{i}_2 + A_3\mathbf{i}_3| = \sqrt{A_1^2 + A_2^2 + A_3^2}} \tag{1.7}$$

Unit vector along A. The unit vector $\mathbf{i}_A$ has a magnitude equal to unity, but its direction is the same as that of **A**. Hence

$$\boxed{\mathbf{i}_A = \frac{\mathbf{A}}{|\mathbf{A}|} = \frac{A_1}{|\mathbf{A}|}\mathbf{i}_1 + \frac{A_2}{|\mathbf{A}|}\mathbf{i}_2 + \frac{A_3}{|\mathbf{A}|}\mathbf{i}_3} \tag{1.8}$$

Dot product

Scalar or dot product of two vectors. The scalar or dot product of two vectors **A** and **B** is a scalar quantity equal to the product of the magnitudes of **A** and **B** and the cosine of the angle between **A** and **B**. It is represented by a boldface dot between **A** and **B**. Thus if α is the angle between **A** and **B**, then

$$\boxed{\mathbf{A} \cdot \mathbf{B} = |\mathbf{A}||\mathbf{B}| \cos \alpha = AB \cos \alpha} \tag{1.9}$$

For the unit vectors $\mathbf{i}_1$, $\mathbf{i}_2$, $\mathbf{i}_3$, we have

$$\mathbf{i}_1 \cdot \mathbf{i}_1 = 1 \qquad \mathbf{i}_1 \cdot \mathbf{i}_2 = 0 \qquad \mathbf{i}_1 \cdot \mathbf{i}_3 = 0 \tag{1.10a}$$

$$\mathbf{i}_2 \cdot \mathbf{i}_1 = 0 \qquad \mathbf{i}_2 \cdot \mathbf{i}_2 = 1 \qquad \mathbf{i}_2 \cdot \mathbf{i}_3 = 0 \tag{1.10b}$$

$$\mathbf{i}_3 \cdot \mathbf{i}_1 = 0 \qquad \mathbf{i}_3 \cdot \mathbf{i}_2 = 0 \qquad \mathbf{i}_3 \cdot \mathbf{i}_3 = 1 \tag{1.10c}$$

By noting that $\mathbf{A} \cdot \mathbf{B} = A(B \cos \alpha) = B(A \cos \alpha)$, we observe that the dot product operation consists of multiplying the magnitude of one vector by the scalar obtained by projecting the second vector onto the first vector as shown in Fig. 1.4(a) and (b). The dot product operation is commutative since

$$\mathbf{B} \cdot \mathbf{A} = BA \cos \alpha = AB \cos \alpha = \mathbf{A} \cdot \mathbf{B} \tag{1.11}$$

The distributive property also holds for the dot product as can be seen from the construction of Fig. 1.4(c), which illustrates that the projection of $\mathbf{B} + \mathbf{C}$ onto $\mathbf{A}$ is equal to the sum of the projections of $\mathbf{B}$ and $\mathbf{C}$ onto $\mathbf{A}$. Thus

$$\mathbf{A} \cdot (\mathbf{B} + \mathbf{C}) = \mathbf{A} \cdot \mathbf{B} + \mathbf{A} \cdot \mathbf{C} \tag{1.12}$$

Using this property, we have

$$\begin{aligned}
\mathbf{A} \cdot \mathbf{B} &= (A_1\mathbf{i}_1 + A_2\mathbf{i}_2 + A_3\mathbf{i}_3) \cdot (B_1\mathbf{i}_1 + B_2\mathbf{i}_2 + B_3\mathbf{i}_3) \\
&= A_1\mathbf{i}_1 \cdot B_1\mathbf{i}_1 + A_1\mathbf{i}_1 \cdot B_2\mathbf{i}_2 + A_1\mathbf{i}_1 \cdot B_3\mathbf{i}_3 \\
&\quad + A_2\mathbf{i}_2 \cdot B_1\mathbf{i}_1 + A_2\mathbf{i}_2 \cdot B_2\mathbf{i}_2 + A_2\mathbf{i}_2 \cdot B_3\mathbf{i}_3 \\
&\quad + A_3\mathbf{i}_3 \cdot B_1\mathbf{i}_1 + A_3\mathbf{i}_3 \cdot B_2\mathbf{i}_2 + A_3\mathbf{i}_3 \cdot B_3\mathbf{i}_3
\end{aligned}$$

Then using the relationships (1.10a)–(1.10c), we obtain

$$\boxed{\mathbf{A} \cdot \mathbf{B} = A_1B_1 + A_2B_2 + A_3B_3} \tag{1.13}$$

Thus the dot product of two vectors is the sum of the products of the like components of the two vectors.

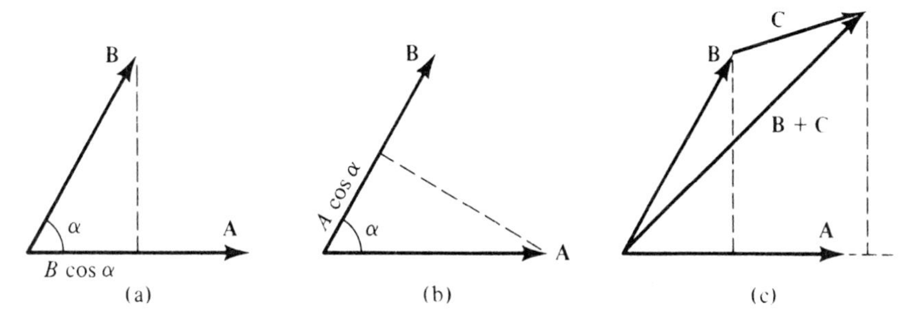

Figure 1.4. (a) and (b) For showing that the dot product of two vectors **A** and **B** is the product of the magnitude of one vector and the projection of the second vector onto the first vector. (c) For proving the distributive property of the dot product operation.

Finding angle between two vectors

From (1.9) and (1.13), we note that the angle between the vectors **A** and **B** is given by

$$\alpha = \cos^{-1}\frac{\mathbf{A}\cdot\mathbf{B}}{AB} = \cos^{-1}\frac{A_1B_1 + A_2B_2 + A_3B_3}{AB} \tag{1.14}$$

Thus the dot product operation is useful for finding the angle between two vectors. In particular, the two vectors are perpendicular if $\mathbf{A}\cdot\mathbf{B} = A_1B_1 + A_2B_2 + A_3B_3 = 0$.

Cross product

Vector or cross product of two vectors. The vector or cross product of two vectors **A** and **B** is a vector quantity whose magnitude is equal to the product of the magnitudes of **A** and **B** and the sine of the smaller angle α between **A** and **B** and whose direction is normal to the plane containing **A** and **B** and toward the side of advance of a right-hand screw as it is turned from **A** to **B** through the angle α, as shown in Fig. 1.5. It is represented by a boldface cross between **A** and **B**. Thus if $\mathbf{i}_N$ is the unit vector in the direction of advance of the right-hand screw, then

$$\boxed{\mathbf{A}\times\mathbf{B} = |\mathbf{A}||\mathbf{B}|\sin\alpha\ \mathbf{i}_N = AB\sin\alpha\ \mathbf{i}_N} \tag{1.15}$$

For the unit vectors $\mathbf{i}_1$, $\mathbf{i}_2$, $\mathbf{i}_3$, we have

$$\mathbf{i}_1\times\mathbf{i}_1 = \mathbf{0} \qquad \mathbf{i}_1\times\mathbf{i}_2 = \mathbf{i}_3 \qquad \mathbf{i}_1\times\mathbf{i}_3 = -\mathbf{i}_2 \tag{1.16a}$$

$$\mathbf{i}_2\times\mathbf{i}_1 = -\mathbf{i}_3 \qquad \mathbf{i}_2\times\mathbf{i}_2 = \mathbf{0} \qquad \mathbf{i}_2\times\mathbf{i}_3 = \mathbf{i}_1 \tag{1.16b}$$

$$\mathbf{i}_3\times\mathbf{i}_1 = \mathbf{i}_2 \qquad \mathbf{i}_3\times\mathbf{i}_2 = -\mathbf{i}_1 \qquad \mathbf{i}_3\times\mathbf{i}_3 = \mathbf{0} \tag{1.16c}$$

Note that the cross product of two identical unit vectors is the null vector **0**, that is, a vector whose components are all zero. If we arrange the unit vectors in the manner $\mathbf{i}_1\mathbf{i}_2\mathbf{i}_3\mathbf{i}_1\mathbf{i}_2$, then going to the right the cross product of any two successive unit vectors is the following unit vector, whereas going to the left the cross product of any two successive unit vectors is the negative of the following unit vector.

The cross product operation is not commutative since

$$\mathbf{B}\times\mathbf{A} = |\mathbf{B}||\mathbf{A}|\sin\alpha\,(-\mathbf{i}_N) = -AB\sin\alpha\ \mathbf{i}_N = -\mathbf{A}\times\mathbf{B} \tag{1.17}$$

The distributive property holds for the cross product (we shall prove this later in this section) so that

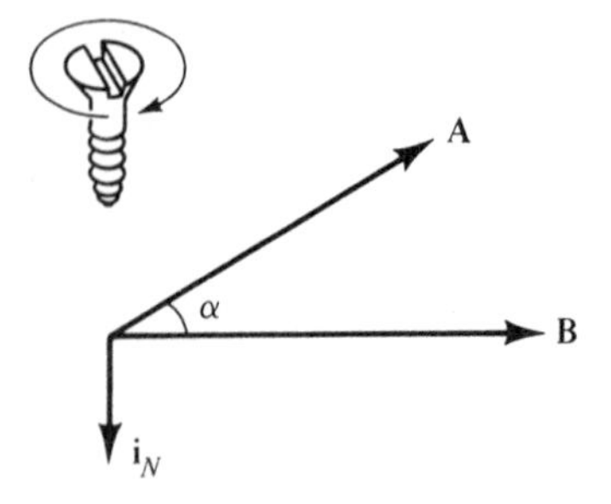

Figure 1.5. Cross product operation **A** × **B**.

$$\mathbf{A} \times (\mathbf{B} + \mathbf{C}) = \mathbf{A} \times \mathbf{B} + \mathbf{A} \times \mathbf{C} \tag{1.18}$$

Using this property and the relationships (1.16a)–(1.16c), we obtain

$$\begin{aligned}
\mathbf{A} \times \mathbf{B} &= (A_1\mathbf{i}_1 + A_2\mathbf{i}_2 + A_3\mathbf{i}_3) \times (B_1\mathbf{i}_1 + B_2\mathbf{i}_2 + B_3\mathbf{i}_3) \\
&= A_1\mathbf{i}_1 \times B_1\mathbf{i}_1 + A_1\mathbf{i}_1 \times B_2\mathbf{i}_2 + A_1\mathbf{i}_1 \times B_3\mathbf{i}_3 \\
&\quad + A_2\mathbf{i}_2 \times B_1\mathbf{i}_1 + A_2\mathbf{i}_2 \times B_2\mathbf{i}_2 + A_2\mathbf{i}_2 \times B_3\mathbf{i}_3 \\
&\quad + A_3\mathbf{i}_3 \times B_1\mathbf{i}_1 + A_3\mathbf{i}_3 \times B_2\mathbf{i}_2 + A_3\mathbf{i}_3 \times B_3\mathbf{i}_3 \\
&= A_1B_2\mathbf{i}_3 - A_1B_3\mathbf{i}_2 - A_2B_1\mathbf{i}_3 + A_2B_3\mathbf{i}_1 \\
&\quad + A_3B_1\mathbf{i}_2 - A_3B_2\mathbf{i}_1 \\
&= (A_2B_3 - A_3B_2)\mathbf{i}_1 + (A_3B_1 - A_1B_3)\mathbf{i}_2 \\
&\quad + (A_1B_2 - A_2B_1)\mathbf{i}_3
\end{aligned}$$

This can be expressed in determinant form in the manner

$$\mathbf{A} \times \mathbf{B} = \begin{vmatrix} \mathbf{i}_1 & \mathbf{i}_2 & \mathbf{i}_3 \\ A_1 & A_2 & A_3 \\ B_1 & B_2 & B_3 \end{vmatrix} \tag{1.19}$$

Finding unit vector normal to two vectors

The cross product operation is useful for obtaining the unit vector normal to two given vectors at a point. This can be seen by rearranging (1.15) in the manner

$$\mathbf{i}_N = \frac{\mathbf{A} \times \mathbf{B}}{AB \sin \alpha} = \frac{\mathbf{A} \times \mathbf{B}}{|\mathbf{A} \times \mathbf{B}|} \tag{1.20}$$

Triple cross product. A triple cross product involves three vectors in two cross product operations. Caution must be exercised in evaluating a triple cross product since the order of evaluation is important; that is, $\mathbf{A} \times (\mathbf{B} \times \mathbf{C})$ is not in general equal to $(\mathbf{A} \times \mathbf{B}) \times \mathbf{C}$. This can be illustrated by means of a simple example involving unit vectors. Thus if $\mathbf{A} = \mathbf{i}_1$, $\mathbf{B} = \mathbf{i}_1$, and $\mathbf{C} = \mathbf{i}_2$, then

$$\mathbf{A} \times (\mathbf{B} \times \mathbf{C}) = \mathbf{i}_1 \times (\mathbf{i}_1 \times \mathbf{i}_2) = \mathbf{i}_1 \times \mathbf{i}_3 = -\mathbf{i}_2$$

whereas

$$(\mathbf{A} \times \mathbf{B}) \times \mathbf{C} = (\mathbf{i}_1 \times \mathbf{i}_1) \times \mathbf{i}_2 = \mathbf{0} \times \mathbf{i}_2 = \mathbf{0}$$

Scalar triple product. The scalar triple product involves three vectors in a dot product operation and a cross product operation as, for example, $\mathbf{A} \cdot \mathbf{B} \times \mathbf{C}$. It is not necessary to include parentheses since this quantity can be evaluated in only one manner, that is, by evaluating $\mathbf{B} \times \mathbf{C}$ first and then dotting the resulting vector with $\mathbf{A}$. It is meaningless to try to evaluate the dot product first since it results in a scalar quantity, and hence we cannot proceed any further. From (1.19) we have

$$\mathbf{A}\cdot\mathbf{B}\times\mathbf{C} = (A_1\mathbf{i}_1 + A_2\mathbf{i}_2 + A_3\mathbf{i}_3)\cdot\begin{vmatrix}\mathbf{i}_1 & \mathbf{i}_2 & \mathbf{i}_3\\ B_1 & B_2 & B_3\\ C_1 & C_2 & C_3\end{vmatrix}$$

From (1.13), we then have

$$\boxed{\mathbf{A}\cdot\mathbf{B}\times\mathbf{C} = \begin{vmatrix}A_1 & A_2 & A_3\\ B_1 & B_2 & B_3\\ C_1 & C_2 & C_3\end{vmatrix}} \tag{1.21}$$

Since the value of the determinant on the right side of (1.21) remains unchanged if the rows are interchanged in a cyclical manner,

$$\mathbf{A}\cdot\mathbf{B}\times\mathbf{C} = \mathbf{B}\cdot\mathbf{C}\times\mathbf{A} = \mathbf{C}\cdot\mathbf{A}\times\mathbf{B} \tag{1.22}$$

The scalar triple product has the geometrical meaning that its absolute value is the volume of the parallelepiped having the three vectors as three of its contiguous edges, as will be shown in Section 1.2.

We shall now show that the distributive law holds for the cross product operation by using (1.22). Thus let us consider $\mathbf{A}\times(\mathbf{B}+\mathbf{C})$. Then if $\mathbf{D}$ is any arbitrary vector, we have

$$\begin{aligned}\mathbf{D}\cdot\mathbf{A}\times(\mathbf{B}+\mathbf{C}) &= (\mathbf{B}+\mathbf{C})\cdot(\mathbf{D}\times\mathbf{A}) = \mathbf{B}\cdot(\mathbf{D}\times\mathbf{A}) + \mathbf{C}\cdot(\mathbf{D}\times\mathbf{A})\\ &= \mathbf{D}\cdot\mathbf{A}\times\mathbf{B} + \mathbf{D}\cdot\mathbf{A}\times\mathbf{C} = \mathbf{D}\cdot(\mathbf{A}\times\mathbf{B} + \mathbf{A}\times\mathbf{C})\end{aligned}$$

where we have used the distributive property of the dot product operation. Since this equality holds for any $\mathbf{D}$, it follows that

$$\mathbf{A}\times(\mathbf{B}+\mathbf{C}) = \mathbf{A}\times\mathbf{B} + \mathbf{A}\times\mathbf{C}$$

Example 1.1

Given three vectors

$$\mathbf{A} = \mathbf{i}_1 + \mathbf{i}_2$$

$$\mathbf{B} = \mathbf{i}_1 + 2\mathbf{i}_2 - 2\mathbf{i}_3$$

$$\mathbf{C} = \mathbf{i}_2 + 2\mathbf{i}_3$$

let us carry out several of the vector algebraic operations:

(a) $\mathbf{A}+\mathbf{B} = (\mathbf{i}_1+\mathbf{i}_2) + (\mathbf{i}_1 + 2\mathbf{i}_2 - 2\mathbf{i}_3) = 2\mathbf{i}_1 + 3\mathbf{i}_2 - 2\mathbf{i}_3$

(b) $\mathbf{B}-\mathbf{C} = (\mathbf{i}_1 + 2\mathbf{i}_2 - 2\mathbf{i}_3) - (\mathbf{i}_2 + 2\mathbf{i}_3) = \mathbf{i}_1 + \mathbf{i}_2 - 4\mathbf{i}_3$

(c) $4\mathbf{C} = 4(\mathbf{i}_2 + 2\mathbf{i}_3) = 4\mathbf{i}_2 + 8\mathbf{i}_3$

(d) $|\mathbf{B}| = |\mathbf{i}_1 + 2\mathbf{i}_2 - 2\mathbf{i}_3| = \sqrt{(1)^2 + (2)^2 + (-2)^2} = 3$

(e) $\mathbf{i}_B = \dfrac{\mathbf{B}}{|\mathbf{B}|} = \dfrac{\mathbf{i}_1 + 2\mathbf{i}_2 - 2\mathbf{i}_3}{3} = \dfrac{1}{3}\mathbf{i}_1 + \dfrac{2}{3}\mathbf{i}_2 - \dfrac{2}{3}\mathbf{i}_3$

(f) $\mathbf{A}\cdot\mathbf{B} = (\mathbf{i}_1+\mathbf{i}_2)\cdot(\mathbf{i}_1 + 2\mathbf{i}_2 - 2\mathbf{i}_3) = (1)(1) + (1)(2) + (0)(-2) = 3$

(g) Angle between $\mathbf{A}$ and $\mathbf{B}$ $= \cos^{-1}\dfrac{\mathbf{A}\cdot\mathbf{B}}{AB} = \cos^{-1}\dfrac{3}{(\sqrt{2})(3)} = 45°$

(h) $\mathbf{A} \times \mathbf{B} = \begin{vmatrix} \mathbf{i}_1 & \mathbf{i}_2 & \mathbf{i}_3 \\ 1 & 1 & 0 \\ 1 & 2 & -2 \end{vmatrix} = (-2 - 0)\mathbf{i}_1 + (0 + 2)\mathbf{i}_2 + (2 - 1)\mathbf{i}_3$

$= -2\mathbf{i}_1 + 2\mathbf{i}_2 + \mathbf{i}_3$

(i) Unit vector normal to $\mathbf{A}$ and $\mathbf{B} = \dfrac{\mathbf{A} \times \mathbf{B}}{|\mathbf{A} \times \mathbf{B}|} = -\dfrac{2}{3}\mathbf{i}_1 + \dfrac{2}{3}\mathbf{i}_2 + \dfrac{1}{3}\mathbf{i}_3$

(j) $(\mathbf{A} \times \mathbf{B}) \times \mathbf{C} = \begin{vmatrix} \mathbf{i}_1 & \mathbf{i}_2 & \mathbf{i}_3 \\ -2 & 2 & 1 \\ 0 & 1 & 2 \end{vmatrix} = 3\mathbf{i}_1 + 4\mathbf{i}_2 - 2\mathbf{i}_3$

(k) $\mathbf{A} \cdot \mathbf{B} \times \mathbf{C} = \begin{vmatrix} 1 & 1 & 0 \\ 1 & 2 & -2 \\ 0 & 1 & 2 \end{vmatrix} = (1)(6) + (1)(-2) + (0)(1) = 4$

K1.1. Scalars; Vectors; Unit vectors; Right-handed system; Components of a vector; Vector addition; Multiplication of vector by a scalar; Magnitude of a vector; Dot product; Cross product; Triple cross product; Scalar triple product.

D1.1. Vector **A** has a magnitude of 4 units and is directed toward north. Vector **B** has a magnitude of 3 units and is directed toward east. Vector **C** has a magnitude of 4 units and is directed 30° toward south from west. Find the following: **(a)** $\mathbf{A} + \mathbf{C}$ **(b)** $|\mathbf{A} - \mathbf{B}|$; **(c)** $3\mathbf{A} + 4\mathbf{B} + 3\mathbf{C}$; **(d)** $\mathbf{B} \cdot (\mathbf{A} - \mathbf{C})$; and **(e)** $\mathbf{A} \times (\mathbf{B} \times \mathbf{C})$.
Ans. **(a)** 4 units directed 60° west of north; **(b)** 5; **(c)** 6.212 units directed 15° east of north; **(d)** 10.392; **(e)** 24 units directed westward

D1.2. Given three vectors

$$\mathbf{A} = 3\mathbf{i}_1 + 2\mathbf{i}_2 + \mathbf{i}_3$$

$$\mathbf{B} = \mathbf{i}_1 + \mathbf{i}_2 - \mathbf{i}_3$$

$$\mathbf{C} = \mathbf{i}_1 + 2\mathbf{i}_2 + 3\mathbf{i}_3$$

Find the following: **(a)** $|\mathbf{A} + \mathbf{B} - 4\mathbf{C}|$; **(b)** unit vector along $(\mathbf{A} + 2\mathbf{B} - \mathbf{C})$; **(c)** $\mathbf{A} \cdot \mathbf{C}$; **(d)** $\mathbf{B} \times \mathbf{C}$; and **(e)** $\mathbf{A} \cdot \mathbf{B} \times \mathbf{C}$.
Ans. **(a)** 13; **(b)** $(2\mathbf{i}_1 + \mathbf{i}_2 - 2\mathbf{i}_3)/3$; **(c)** 10; **(d)** $5\mathbf{i}_1 - 4\mathbf{i}_2 + \mathbf{i}_3$; **(e)** 8

D1.3. Three vectors **A**, **B**, and **C** are given by

$$\mathbf{A} = \mathbf{i}_1 + 2\mathbf{i}_2 + 2\mathbf{i}_3$$

$$\mathbf{B} = 2\mathbf{i}_1 + \mathbf{i}_2 - 2\mathbf{i}_3$$

$$\mathbf{C} = \mathbf{i}_1 - \mathbf{i}_2 + \mathbf{i}_3$$

Find the following: **(a)** $\mathbf{A} \times (\mathbf{B} \times \mathbf{C})$; **(b)** $\mathbf{B} \times (\mathbf{C} \times \mathbf{A})$; and **(c)** $\mathbf{C} \times (\mathbf{A} \times \mathbf{B})$.
Ans. **(a)** $2\mathbf{i}_1 + \mathbf{i}_2 - 2\mathbf{i}_3$; **(b)** $\mathbf{i}_1 + 2\mathbf{i}_2 + 2\mathbf{i}_3$; **(c)** $-3\mathbf{i}_1 - 3\mathbf{i}_2$

1.2 CARTESIAN COORDINATE SYSTEM

In the preceding section we introduced the technique of expressing a vector at a point in space in terms of its component vectors along a set of three mutually orthogonal directions defined by three mutually orthogonal unit vectors at that point. Now to relate vectors at one point in space to vectors at another point in space, we must define the set of three reference directions at each and every point in space. To do this in a systematic manner, we need to use a coordinate system. Although there are several different coordinate systems, we shall be concerned with only three of those, namely, the Cartesian, cylindrical, and spherical coordinate systems. The *Cartesian coordinate system,* also known as the *rectangular coordinate system,* is the simplest of the three since it permits the geometry to be simple, yet sufficient to study many of the elements of engineering electromagnetics. We introduce the Cartesian coordinate system in this section and devote the next section to the cylindrical and spherical coordinate systems.

The Cartesian coordinate system is defined by a set of three mutually orthogonal planes as shown in Fig. 1.6(a). The point at which the three planes intersect is known as the origin O. The origin is the reference point relative to which we locate any other point in space. Each pair of planes intersects in a straight line. Hence the three planes define a set of three straight lines that form the coordinate axes. These coordinate axes are denoted as the x-, y-, and z-axes. Values of x, y, and z are measured from the origin; hence the coordinates of the origin are (0, 0, 0), that is, $x = 0$, $y = 0$, and $z = 0$. Directions in which values of x, y, and z increase along the respective coordinate axes are indicated by arrowheads. The same set of three directions is used to erect a set of three unit vectors, denoted $\mathbf{i}_x$, $\mathbf{i}_y$, and $\mathbf{i}_z$, as shown in Fig. 1.6(a), for the purpose of describing vectors drawn at the origin. Note that the positive x-, y-, and z-directions are chosen such that they form a right-handed system, that is, a system for which $\mathbf{i}_x \times \mathbf{i}_y = \mathbf{i}_z$.

On one of the three planes, namely, the yz-plane, the value of x is constant and equal to zero, its value at the origin, since movement on this plane does not require any movement in the x-direction. Similarly, on the zx-plane the value of y is constant and equal to zero, and on the xy-plane the value of z is constant and equal to zero. Any point other than the origin is now given by the intersection of three planes

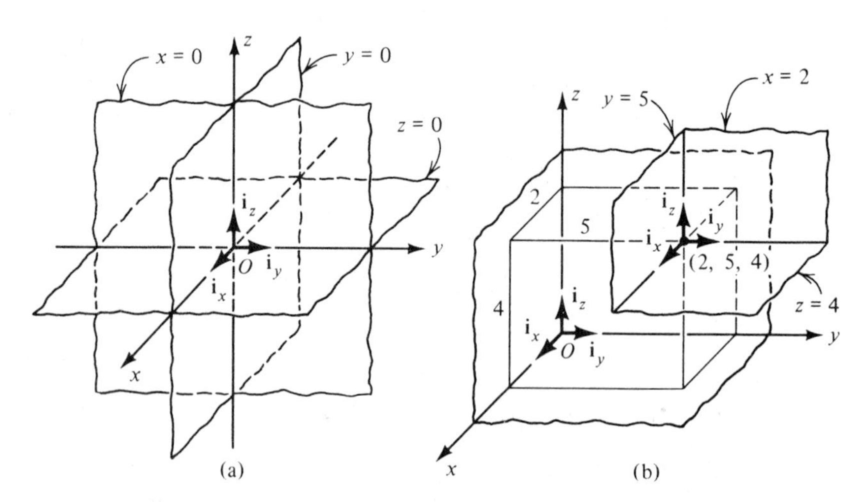

Figure 1.6. Cartesian coordinate system. (a) The three orthogonal planes defining the coordinate system. (b) To show that the unit vectors in the Cartesian coordinate system are uniform.

$$
\boxed{\begin{aligned} x &= \text{constant} \\ y &= \text{constant} \\ z &= \text{constant} \end{aligned}} \tag{1.23}
$$

obtained by incrementing the values of the coordinates by appropriate amounts. For example, by displacing the $x = 0$ plane by 2 units in the positive x-direction, the $y = 0$ plane by 5 units in the positive y-direction, and the $z = 0$ plane by 4 units in the positive z-direction, we obtain the planes $x = 2$, $y = 5$, and $z = 4$, respectively, which intersect at point (2, 5, 4) as shown in Fig. 1.6(b). The intersections of pairs of these planes define three straight lines along which we can erect the unit vectors $\mathbf{i}_x$, $\mathbf{i}_y$, and $\mathbf{i}_z$ toward the directions of increasing values of x, y, and z, respectively, for the purpose of describing vectors drawn at that point. These unit vectors are parallel to the corresponding unit vectors drawn at the origin, as can be seen from Fig. 1.6(b). The same is true for any point in space in the Cartesian coordinate system. Thus each one of the three unit vectors in the Cartesian coordinate system has the same direction at all points, and hence it is uniform. This behavior does not, however, hold for all unit vectors in the cylindrical and spherical coordinate systems, as we shall see in the next section.

Expression for vector joining two points

It is now a simple matter to apply what we have learned in Section 1.1 to vectors in Cartesian coordinates. All we need to do is to replace the subscripts 1, 2, and 3 for the unit vectors and the components along the unit vectors by the subscripts x, y, and z, respectively, and also utilize the property that $\mathbf{i}_x$, $\mathbf{i}_y$, and $\mathbf{i}_z$ are uniform vectors. Thus let us, for example, obtain the expression for the vector $\mathbf{R}_{12}$ drawn from point $P_1(x_1, y_1, z_1)$ to point $P_2(x_2, y_2, z_2)$, as shown in Fig. 1.7. To do this, we note that the position vector $\mathbf{r}_1$ drawn from the origin to the point P_1 is given by

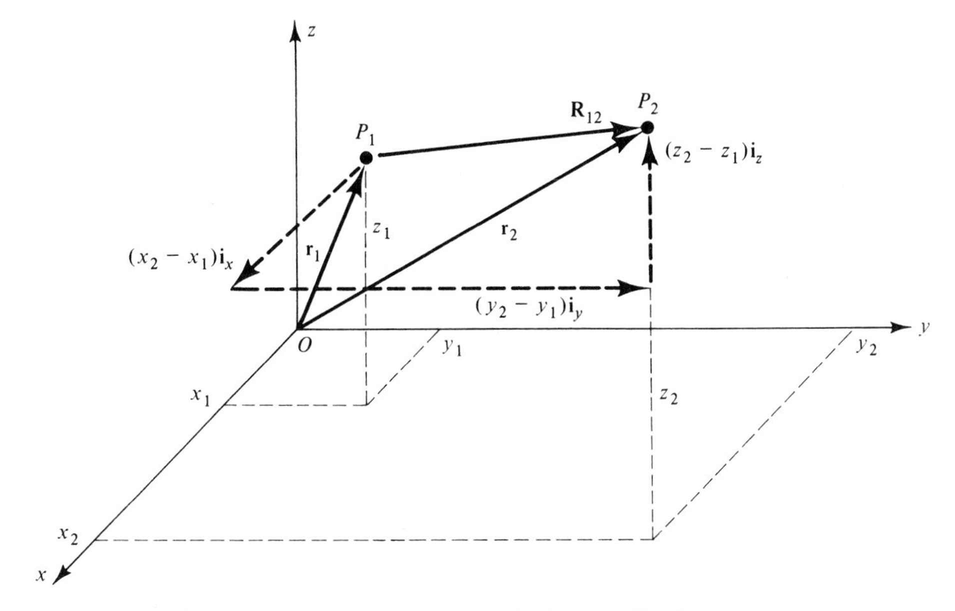

Figure 1.7. For obtaining the expression for the vector $\mathbf{R}_{12}$ from, $P_1(x_1, y_1, z_1)$ to $P_2(x_2, y_2, z_2)$.

$$\mathbf{r}_1 = x_1\mathbf{i}_x + y_1\mathbf{i}_y + z_1\mathbf{i}_z \qquad (1.24a)$$

The *position vector* is so termed because it defines the position of the point in space relative to the origin. Similarly, the position vector $\mathbf{r}_2$ drawn from the origin to the point P_2 is given by

$$\mathbf{r}_2 = x_2\mathbf{i}_x + y_2\mathbf{i}_y + z_2\mathbf{i}_z \qquad (1.24b)$$

Since, from the rule for vector addition, $\mathbf{r}_1 + \mathbf{R}_{12} = \mathbf{r}_2$, we obtain the vector $\mathbf{R}_{12}$ to be

$$\boxed{\begin{aligned}\mathbf{R}_{12} &= \mathbf{r}_2 - \mathbf{r}_1 \\ &= (x_2 - x_1)\mathbf{i}_x + (y_2 - y_1)\mathbf{i}_y + (z_2 - z_1)\mathbf{i}_z\end{aligned}} \qquad (1.25)$$

Thus to find the components of the vector drawn from one point to another in the Cartesian coordinate system, we simply subtract the coordinates of the initial point from the corresponding coordinates of the final point. These components are just the distances one has to travel along the x-, y-, and z-directions, respectively, if one chooses to go from P_1 to P_2 by traveling parallel to the coordinate axes instead of traveling along the direct straight-line path.

Proceeding further, we can obtain the unit vector along the line drawn from P_1 to P_2 to be

$$\boxed{\mathbf{i}_{12} = \frac{\mathbf{R}_{12}}{R_{12}} = \frac{(x_2 - x_1)\mathbf{i}_x + (y_2 - y_1)\mathbf{i}_y + (z_2 - z_1)\mathbf{i}_z}{[(x_2 - x_1)^2 + (y_2 - y_1)^2 + (z_2 - z_1)^2]^{1/2}}} \qquad (1.26)$$

For a numerical example, if P_1 is $(1, -2, 0)$ and P_2 is $(4, 2, 5)$, then

$$\mathbf{R}_{12} = 3\mathbf{i}_x + 4\mathbf{i}_y + 5\mathbf{i}_z$$

$$\mathbf{i}_{12} = \frac{1}{5\sqrt{2}}(3\mathbf{i}_x + 4\mathbf{i}_y + 5\mathbf{i}_z)$$

In our study of electromagnetic fields, we have to work with line, surface, and volume integrals. These involve differential lengths, surfaces, and volumes, obtained by incrementing the coordinates by infinitesimal amounts. Since in the Cartesian coordinate system the three coordinates represent lengths, the differential length elements obtained by incrementing one coordinate at a time, keeping the other two coordinates constant, are $dx\,\mathbf{i}_x$, $dy\,\mathbf{i}_y$, and $dz\,\mathbf{i}_z$ for the x-, y-, and z-coordinates, respectively.

Differential length vector. The differential length vector $d\mathbf{l}$ is the vector drawn from a point $P(x, y, z)$ to a neighboring point $Q(x + dx, y + dy, z + dz)$ obtained by incrementing the coordinates of P by infinitesimal amounts. Thus it is the vector sum of the three differential length elements, as shown in Fig. 1.8, and given by

$$\boxed{d\mathbf{l} = dx\,\mathbf{i}_x + dy\,\mathbf{i}_y + dz\,\mathbf{i}_z} \qquad (1.27)$$

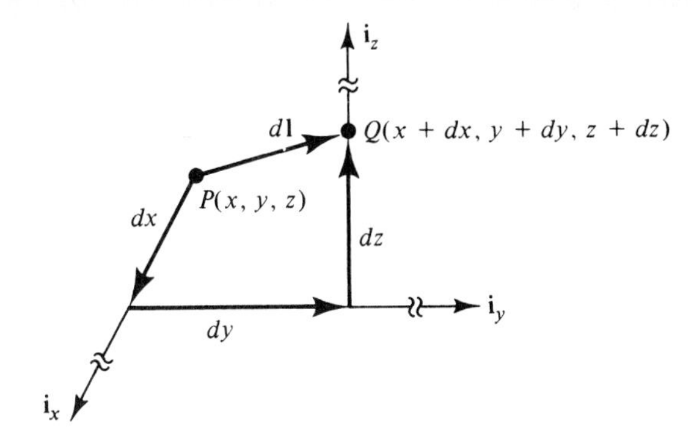

Figure 1.8. Differential length vector $d\mathbf{l}$.

The differential lengths dx, dy, and dz in (1.27) are, however, not independent of each other since in the evaluation of line integrals, the integration is performed along a specified path on which the points P and Q lie. We shall illustrate this by means of an example.

Example 1.2

Finding differential length vector along a curve

Let us consider the curve $x = y = z^2$ and obtain the expression for the differential length vector $d\mathbf{l}$ along the curve at the point (1, 1, 1) and having the projection dz on the z-axis.

The geometry pertinent to the problem is shown in Fig. 1.9. From elementary calculus, we know that for $x = y = z^2$, $dx = dy = 2z\,dz$. In particular, at the point (1, 1, 1), $dx = dy = 2\,dz$. Thus

$$\begin{aligned} d\mathbf{l} &= dx\,\mathbf{i}_x + dy\,\mathbf{i}_y + dz\,\mathbf{i}_z \\ &= 2\,dz\,\mathbf{i}_x + 2\,dz\,\mathbf{i}_y + dz\,\mathbf{i}_z \\ &= (2\mathbf{i}_x + 2\mathbf{i}_y + \mathbf{i}_z)\,dz \end{aligned}$$

Note that the z-component of the $d\mathbf{l}$ vector found is dz, thereby satisfying the requirement of projection dz on the z-axis specified in the problem.

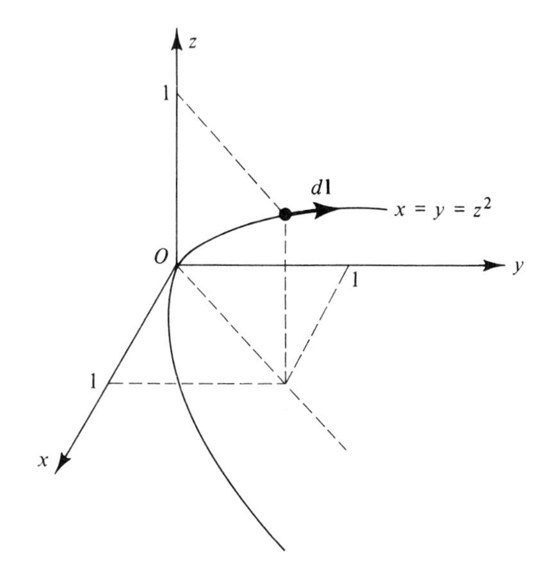

Figure 1.9. For finding the differential length vector along the curve $x = y = z^2$.

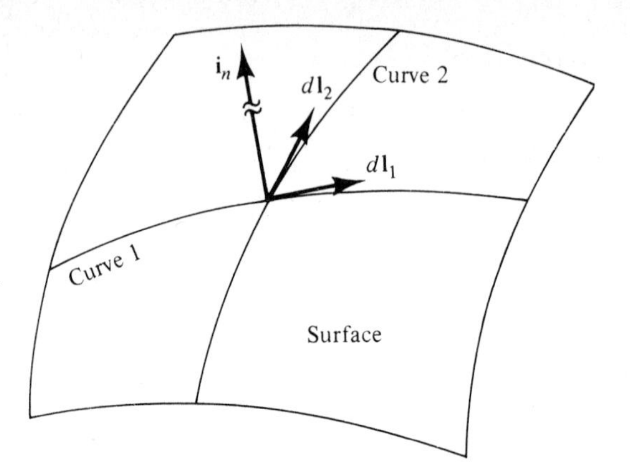

Figure 1.10. Finding the unit vector normal to a surface by using differential length vectors.

Differential length vectors are useful for finding the unit vector normal to a surface at a point on that surface. This is done by considering two differential length vectors at the point under consideration and tangential to two curves on the surface and then using (1.20). Thus with reference to Fig. 1.10, we have

$$\boxed{\mathbf{i}_n = \frac{d\mathbf{l}_1 \times d\mathbf{l}_2}{|d\mathbf{l}_1 \times d\mathbf{l}_2|}} \tag{1.28}$$

Let us consider an example.

Example 1.3

Finding unit normal vector at a point on a surface

Find the unit vector normal to the surface $2x^2 + y^2 = 6$ at the point (1, 2, 0).

With reference to the construction shown in Fig. 1.11, we consider two differential length vectors $d\mathbf{l}_1$ and $d\mathbf{l}_2$ at the point (1, 2, 0). The vector $d\mathbf{l}_1$ is along the straight line $x = 1$, $y = 2$, whereas the vector $d\mathbf{l}_2$ is tangential to the curve $2x^2 + y^2 = 6$, $z = 0$. For $x = 1$ and $y = 2$, $dx = dy = 0$. Hence

$$d\mathbf{l}_1 = dz\,\mathbf{i}_z$$

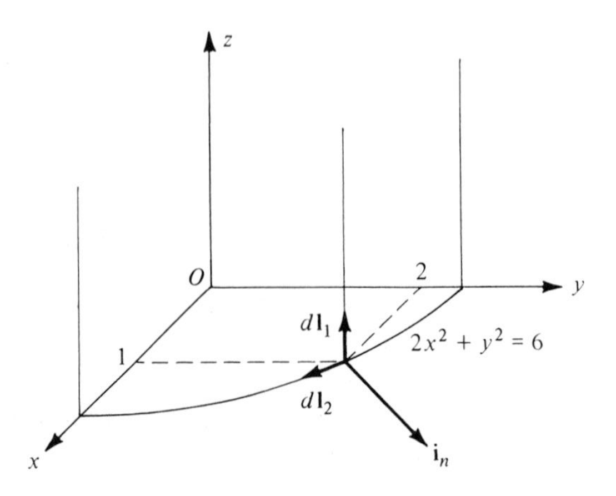

Figure 1.11. Example of finding the unit vector normal to a surface.

For $2x^2 + y^2 = 6$ and $z = 0$, $4x\,dx + 2y\,dy = 0$ and $dz = 0$. Specifically, at the point (1, 2, 0), $dy = -dx$ and $dz = 0$. Hence

$$d\mathbf{l}_2 = dx\,\mathbf{i}_x - dx\,\mathbf{i}_y = dx\,(\mathbf{i}_x - \mathbf{i}_y)$$

The unit normal vector is then given by

$$\mathbf{i}_n = \frac{dz\,\mathbf{i}_z \times dx\,(\mathbf{i}_x - \mathbf{i}_y)}{|dz\,\mathbf{i}_z \times dx\,(\mathbf{i}_x - \mathbf{i}_y)|}$$

$$= \frac{1}{\sqrt{2}}(\mathbf{i}_x + \mathbf{i}_y)$$

Differential surface vector. Two differential length vectors $d\mathbf{l}_1$ and $d\mathbf{l}_2$ originating at a point define a differential surface whose area dS is that of the parallelogram having $d\mathbf{l}_1$ and $d\mathbf{l}_2$ as two of its adjacent sides, as shown in Fig. 1.12(a). From simple geometry and the definition of the cross product of two vectors, it can be seen that

$$dS = dl_1\,dl_2 \sin\alpha = |d\mathbf{l}_1 \times d\mathbf{l}_2| \tag{1.29}$$

In the evaluation of surface integrals, it is convenient to define a differential surface vector $d\mathbf{S}$ whose magnitude is the area dS and whose direction is normal to the differential surface. Thus recognizing that the normal vector can be directed to either side of a surface, we can write

$$d\mathbf{S} = \pm dS\,\mathbf{i}_n = \pm|d\mathbf{l}_1 \times d\mathbf{l}_2|\mathbf{i}_n$$

or

$$\boxed{d\mathbf{S} = \pm d\mathbf{l}_1 \times d\mathbf{l}_2} \tag{1.30}$$

Applying (1.30) to pairs of three differential length elements $dx\,\mathbf{i}_x$, $dy\,\mathbf{i}_y$, and $dz\,\mathbf{i}_z$, we obtain the differential surface vectors

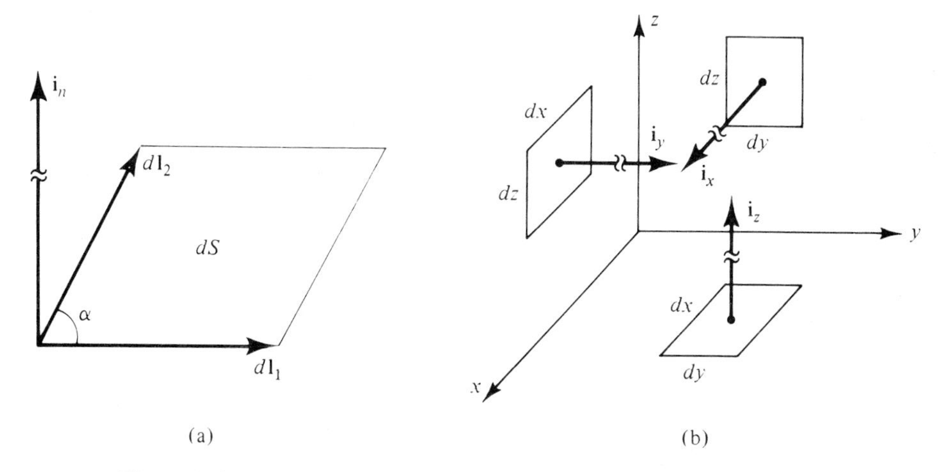

Figure 1.12. (a) For illustrating the differential surface vector concept. (b) Differential surface vectors in the Cartesian coordinate system.

$$\pm dy\, \mathbf{i}_y \times dz\, \mathbf{i}_z = \pm dy\, dz\, \mathbf{i}_x \tag{1.31a}$$

$$\pm dz\, \mathbf{i}_z \times dx\, \mathbf{i}_x = \pm dz\, dx\, \mathbf{i}_y \tag{1.31b}$$

$$\pm dx\, \mathbf{i}_x \times dy\, \mathbf{i}_y = \pm dx\, dy\, \mathbf{i}_z \tag{1.31c}$$

associated with the planes x = constant, y = constant, and z = constant, respectively. These are shown in Fig. 1.12(b) for the plus signs in (1.31a)–(1.31c).

Differential volume. Three differential length vectors $d\mathbf{l}_1$, $d\mathbf{l}_2$, and $d\mathbf{l}_3$ originating at a point define a differential volume dv which is that of the parallelepiped having $d\mathbf{l}_1$, $d\mathbf{l}_2$, and $d\mathbf{l}_3$ as three of its contiguous edges, as shown in Fig. 1.13(a). From simple geometry and the definitions of cross and dot products, it can be seen that

$$\begin{aligned} dv &= \text{area of the base of the parallelepiped} \times \text{height of the parallelepiped} \\ &= |d\mathbf{l}_1 \times d\mathbf{l}_2||d\mathbf{l}_3 \cdot \mathbf{i}_n| \\ &= |d\mathbf{l}_1 \times d\mathbf{l}_2|\left|\frac{d\mathbf{l}_3 \cdot d\mathbf{l}_1 \times d\mathbf{l}_2}{|d\mathbf{l}_1 \times d\mathbf{l}_2|}\right| \\ &= |d\mathbf{l}_3 \cdot d\mathbf{l}_1 \times d\mathbf{l}_2| \end{aligned}$$

or

$$dv = d\mathbf{l}_1 \cdot d\mathbf{l}_2 \times d\mathbf{l}_3 \tag{1.32}$$

For the three differential length elements $dx\, \mathbf{i}_x$, $dy\, \mathbf{i}_y$, and $dz\, \mathbf{i}_z$ associated with the Cartesian coordinate system, we obtain the differential volume to be

$$dv = dx\, dy\, dz \tag{1.33}$$

which is that of the rectangular parallelepiped shown in Fig. 1.13(b).

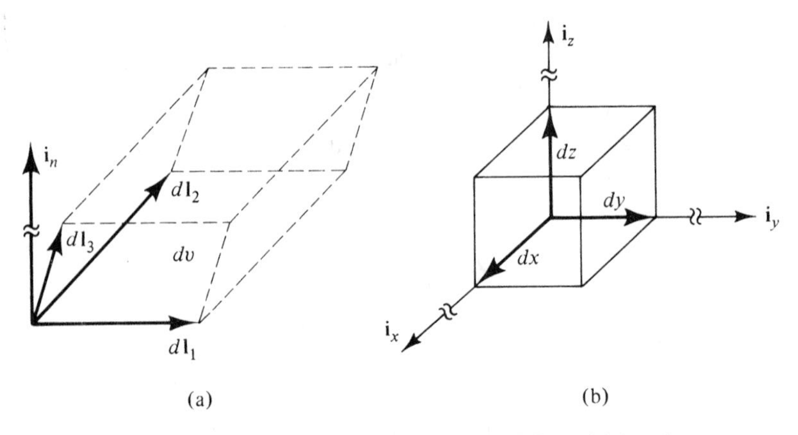

Figure 1.13. (a) Parallelepiped defined by three differential length vectors originating at a point. (b) Differential volume in the Cartesian coordinate system.

We shall conclude this section with a brief review of some elementary analytic geometrical details that will be useful in our study of electromagnetics. An arbitrary surface is defined by an equation of the form

$$f(x, y, z) = 0 \tag{1.34}$$

In particular, the equation for a plane surface making intercepts a, b, and c on the x-, y-, and z-axes, respectively, is given by

$$\frac{x}{a} + \frac{y}{b} + \frac{z}{c} - 1 = 0 \tag{1.35}$$

Since a curve is the intersection of two surfaces, an arbitrary curve is defined by a pair of equations

$$f(x, y, z) = 0 \quad \text{and} \quad g(x, y, z) = 0 \tag{1.36}$$

Alternatively, a curve is specified by a set of three parametric equations

$$x = x(t), \qquad y = y(t), \qquad z = z(t) \tag{1.37}$$

where t is an independent parameter. For example, a straight line passing through the origin and making equal angles with the positive x-, y-, and z-axes is given by the pair of equations $y = x$ and $z = x$, or by the set of three parametric equations $x = t$, $y = t$, and $z = t$.

K1.2. Cartesian or rectangular coordinate system; Orthogonal surfaces; Unit vectors; Position vector; Vector joining two points; Differential length vector; Differential surface vector; Differential volume.

D1.4. Three points P_1, P_2, and P_3 are given by $(1, -2, 2)$, $(3, 1, 0)$, and $(5, 2, -2)$, respectively. Obtain the following: **(a)** the vector drawn from P_1 to P_2; **(b)** the straight line distance from P_2 to P_3; and **(c)** the unit vector along the line from P_1 to P_3.
Ans. **(a)** $(2\mathbf{i}_x + 3\mathbf{i}_y - 2\mathbf{i}_z)$; **(b)** 3; **(c)** $(\mathbf{i}_x + \mathbf{i}_y - \mathbf{i}_z)/\sqrt{3}$

D1.5. For each of the following straight lines, find the differential length vector along the line and having the projection dz on the z-axis: **(a)** $x = 3$, $y = -4$; **(b)** $x + y = 0$, $y + z = 1$; and **(c)** the line passing through the points $(2, 0, 0)$ and $(0, 0, 1)$.
Ans. **(a)** $dz\,\mathbf{i}_z$; **(b)** $(\mathbf{i}_x - \mathbf{i}_y + \mathbf{i}_z)\,dz$; **(c)** $(-2\mathbf{i}_x + \mathbf{i}_z)\,dz$

D1.6. For each of the following pairs of points, obtain the equation for the straight line passing through the points: **(a)** $(1, 2, 0)$ and $(3, 4, 0)$; **(b)** $(0, 0, 0)$ and $(2, 2, -1)$; and **(c)** $(1, 1, 1)$ and $(3, -2, 4)$.
Ans. **(a)** $y = x + 1$, $z = 0$; **(b)** $x = y = -2z$; **(c)** $3x + 2y = 5$, $3x - 2z = 1$

1.3 CYLINDRICAL AND SPHERICAL COORDINATE SYSTEMS

Cylindrical coordinate system

In the preceding section we learned that the Cartesian coordinate system is defined by a set of three mutually orthogonal surfaces, all of which are planes. The cylindrical and spherical coordinate systems also involve sets of three mutually orthogonal surfaces. For the cylindrical coordinate system, the three surfaces are a cylinder and two planes, as shown in Fig. 1.14(a). One of these planes is the same as the z = constant plane in the Cartesian coordinate system. The second

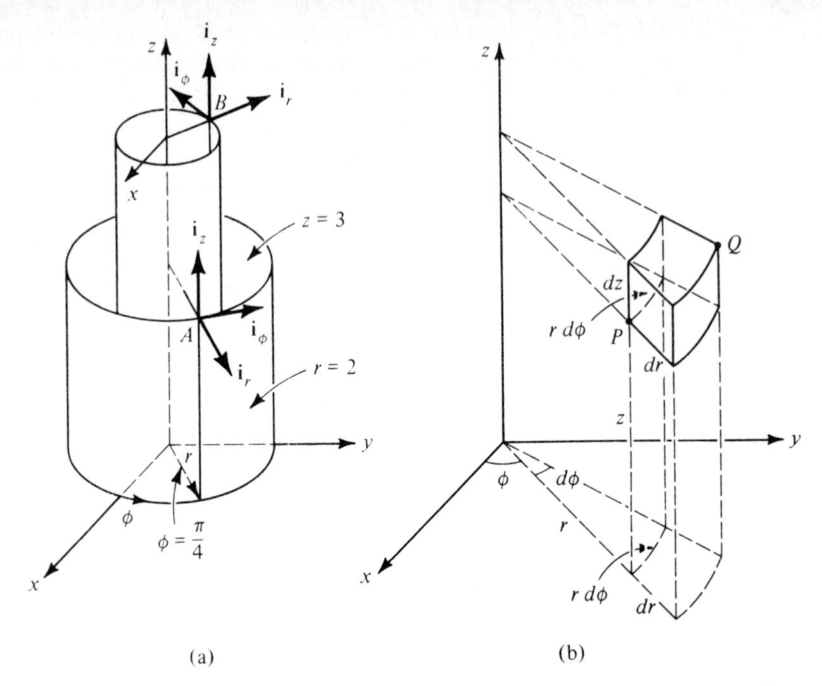

Figure 1.14. Cylindrical coordinate system. (a) Orthogonal surfaces and unit vectors. (b) Differential volume formed by incrementing the coordinates.

plane contains the z-axis and makes an angle ϕ with a reference plane, conveniently chosen to be the xz-plane of the Cartesian coordinate system. This plane is therefore defined by ϕ = constant. The cylindrical surface has the z-axis as its axis. Since the radial distance r from the z-axis to points on the cylindrical surface is a constant, this surface is defined by r = constant. Thus the three orthogonal surfaces defining the cylindrical coordinates of a point are

$$\boxed{\begin{aligned} r &= \text{constant} \\ \phi &= \text{constant} \\ z &= \text{constant} \end{aligned}} \tag{1.38}$$

Only two of these coordinates (r and z) are distances; the third coordinate (ϕ) is an angle. We note that the entire space is spanned by varying r from 0 to ∞, ϕ from 0 to 2π, and z from $-\infty$ to ∞.

The origin is given by $r = 0$, $\phi = 0$, and $z = 0$. Any other point in space is given by the intersection of three mutually orthogonal surfaces obtained by incrementing the coordinates by appropriate amounts. For example, the intersection of the three surfaces $r = 2$, $\phi = \pi/4$, and $z = 3$ defines the point $A(2, \pi/4, 3)$, as shown in Fig. 1.14(a). These three orthogonal surfaces define three curves that are mutually perpendicular. Two of these are straight lines and the third is a circle. We draw unit vectors, $\mathbf{i}_r$, $\mathbf{i}_\phi$, and $\mathbf{i}_z$ tangential to these curves at the point A and directed toward increasing values of r, ϕ, and z, respectively. These three unit vectors form a set of mutually orthogonal unit vectors in terms of which vectors drawn at A can be described. In a similar manner, we can draw unit vectors at any other point in the cylindrical coordinate system, as shown, for example, for point $B(1, 3\pi/4, 5)$ in Fig. 1.14(a). It can now be seen that the unit vectors $\mathbf{i}_r$ and $\mathbf{i}_\phi$ at point B are not parallel to the corresponding unit vectors at point A. Thus unlike in the Cartesian coordinate system, the unit vectors $\mathbf{i}_r$ and $\mathbf{i}_\phi$ in the cylindrical coordinate system do not have the same directions everywhere; that is, they are not uniform. Only the unit vector $\mathbf{i}_z$, which is the same as in the Cartesian coordinate system, is uniform. Finally, we note that for the choice of ϕ as in Fig. 1.14(a), that is, increasing from the positive x-axis toward the positive y-axis, the coordinate system is right-handed, that is, $\mathbf{i}_r \times \mathbf{i}_\phi = \mathbf{i}_z$.

To obtain expressions for the differential lengths, surfaces, and volumes in the cylindrical coordinate system, we now consider two points $P(r, \phi, z)$ and $Q(r + dr, \phi + d\phi, z + dz)$ where Q is obtained by incrementing infinitesimally each coordinate from its value at P, as shown in Fig. 1.14(b). The three orthogonal surfaces intersecting at P, and the three orthogonal surfaces intersecting at Q define a box which can be considered to be rectangular since dr, $d\phi$, and dz are infinitesimally small. The three differential length elements forming the contiguous sides of this box are $dr\,\mathbf{i}_r$, $r\,d\phi\,\mathbf{i}_\phi$, and $dz\,\mathbf{i}_z$. The differential length vector $d\mathbf{l}$ from P to Q is thus given by

$$\boxed{d\mathbf{l} = dr\,\mathbf{i}_r + r\,d\phi\,\mathbf{i}_\phi + dz\,\mathbf{i}_z} \tag{1.39}$$

The differential surface vectors defined by pairs of the differential length elements are

$$\pm r\,d\phi\,\mathbf{i}_\phi \times dz\,\mathbf{i}_z = \pm r\,d\phi\,dz\,\mathbf{i}_r \tag{1.40a}$$

$$\pm dz\,\mathbf{i}_z \times dr\,\mathbf{i}_r = \pm dr\,dz\,\mathbf{i}_\phi \tag{1.40b}$$

$$\pm dr\,\mathbf{i}_r \times r\,d\phi\,\mathbf{i}_\phi = \pm r\,dr\,d\phi\,\mathbf{i}_z \tag{1.40c}$$

These are associated with the r = constant, ϕ = constant, and z = constant surfaces, respectively. Finally, the differential volume dv formed by the three differential lengths is simply the volume of the box; that is

$$\boxed{dv = (dr)(r\,d\phi)(dz) = r\,dr\,d\phi\,dz} \tag{1.41}$$

Spherical coordinate system

For the spherical coordinate system, the three mutually orthogonal surfaces are a sphere, a cone, and a plane, as shown in Fig. 1.15(a). The plane is the same as the ϕ = constant plane in the cylindrical coordinate system. The sphere has the origin as its center. Since the radial distance r from the origin to points on the spherical surface is a constant, this surface is defined by r = constant. The spherical coordinate r should not be confused with the cylindrical coordinate r. When these two coordinates appear in the same expression, we shall use the subscripts c and s to distinguish between cylindrical and spherical. The cone has its vertex at the origin and its surface is symmetrical about the z-axis. Since the angle θ is the angle that the conical surface makes with the z-axis, this surface is defined by

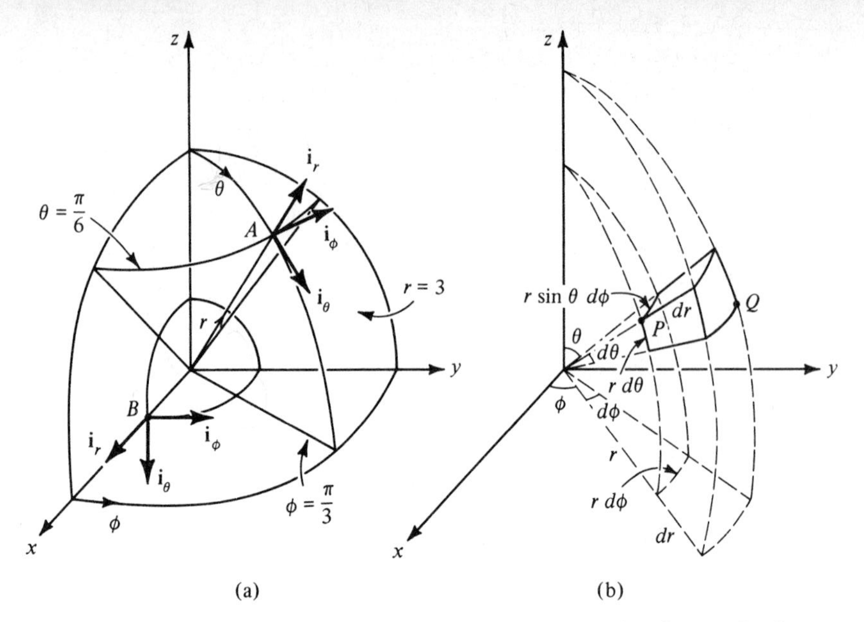

Figure 1.15. Spherical coordinate system. (a) Orthogonal surfaces and unit vectors. (b) Differential volume formed by incrementing the coordinates.

θ = constant. Thus the three orthogonal surfaces defining the spherical coordinates of a point are

$$\boxed{\begin{aligned} r &= \text{constant} \\ \theta &= \text{constant} \\ \phi &= \text{constant} \end{aligned}} \tag{1.42}$$

Only one of these coordinates (r) is distance; the other two coordinates (θ and ϕ) are angles. We note that the entire space is spanned by varying r from 0 to ∞, θ from 0 to π, and ϕ from 0 to 2π.

The origin is given by $r = 0$, $\theta = 0$, and $\phi = 0$. Any other point in space is given by the intersection of three mutually orthogonal surfaces obtained by incrementing the coordinates by appropriate amounts. For example, the intersection of the three surfaces $r = 3$, $\theta = \pi/6$, and $\phi = \pi/3$ defines the point $A(3, \pi/6, \pi/3)$, as shown in Fig. 1.15(a). These three orthogonal surfaces define three curves that are mutually perpendicular. One of these is a straight line and the other two are circles. We draw unit vectors $\mathbf{i}_r$, $\mathbf{i}_\theta$, and $\mathbf{i}_\phi$ tangential to these curves at point A and directed toward increasing values of r, θ, and ϕ, respectively. These three unit vectors form a set of mutually orthogonal unit vectors in terms of which vectors drawn at A can be described. In a similar manner, we can draw unit vectors at any other point in the spherical coordinate system, as shown, for example, for point $B(1, \pi/2, 0)$ in Fig. 1.15(a). It can now be seen that these unit vectors at point B are not parallel to the corresponding unit vectors at point A. Thus in the spherical coordinate system all three unit vectors $\mathbf{i}_r$, $\mathbf{i}_\theta$, and $\mathbf{i}_\phi$ do not have the same directions everywhere; that is, they are not uniform. Finally, we note that for the choice of θ as in Fig. 1.15(a), that is, increasing from the positive z-axis toward the xy-plane, the coordinate system is right-handed, that is, $\mathbf{i}_r \times \mathbf{i}_\theta = \mathbf{i}_\phi$.

To obtain expressions for the differential lengths, surfaces, and volume in the spherical coordinate system, we now consider two points $P(r, \theta, \phi)$ and $Q(r + dr, \theta + d\theta, \phi + d\phi)$ where Q is obtained by incrementing infinitesimally each coordinate from its value at P, as shown in Fig. 1.15(b). The three orthogonal surfaces intersecting at P and the three orthogonal surfaces intersecting at Q define a box that can be considered to be rectangular since dr, $d\theta$, and $d\phi$ are infinitesimally small. The three differential length elements forming the contiguous sides of this box are $dr\ \mathbf{i}_r$, $r\ d\theta\ \mathbf{i}_\theta$, and $r \sin\theta\ d\phi\ \mathbf{i}_\phi$. The differential length vector $d\mathbf{l}$ from P to Q is thus given by

$$d\mathbf{l} = dr\ \mathbf{i}_r + r\ d\theta\ \mathbf{i}_\theta + r \sin\theta\ d\phi\ \mathbf{i}_\phi \tag{1.43}$$

The differential surface vectors defined by pairs of the differential length elements are

$$\pm r\ d\theta\ \mathbf{i}_\theta \times r \sin\theta\ d\phi\ \mathbf{i}_\phi = \pm r^2 \sin\theta\ d\theta\ d\phi\ \mathbf{i}_r \tag{1.44a}$$

$$\pm r \sin\theta\ d\phi\ \mathbf{i}_\phi \times dr\ \mathbf{i}_r = \pm r \sin\theta\ dr\ d\phi\ \mathbf{i}_\theta \tag{1.44b}$$

$$\pm dr\ \mathbf{i}_r \times r\ d\theta\ \mathbf{i}_\theta = \pm r\ dr\ d\theta\ \mathbf{i}_\phi \tag{1.44c}$$

These are associated with the r = constant, θ = constant, and ϕ = constant surfaces, respectively. Finally, the differential volume dv formed by the three differential lengths is simply the volume of the box, that is,

$$dv = (dr)(r\ d\theta)(r \sin\theta\ d\phi) = r^2 \sin\theta\ dr\ d\theta\ d\phi \tag{1.45}$$

Conversions between coordinate systems

In the study of electromagnetics it is useful to be able to convert the coordinates of a point and vectors drawn at a point from one coordinate system to another, particularly from the cylindrical system to the Cartesian system and vice versa, and from the spherical system to the Cartesian system and vice versa. To derive first the relationships for the conversion of the coordinates, let us consider Fig. 1.16(a), which illustrates the geometry pertinent to the coordinates of a point P in the three different coordinate systems. Thus from simple geometrical considerations, we have

$$x = r_c \cos\phi \qquad y = r_c \sin\phi \qquad z = z \tag{1.46a}$$

$$x = r_s \sin\theta \cos\phi \qquad y = r_s \sin\theta \sin\phi \qquad z = r_s \cos\theta \tag{1.46b}$$

Conversely, we have

$$r_c = \sqrt{x^2 + y^2} \qquad \phi = \tan^{-1}\frac{y}{x} \qquad z = z \tag{1.47a}$$

$$r_s = \sqrt{x^2 + y^2 + z^2} \qquad \theta = \tan^{-1}\frac{\sqrt{x^2 + y^2}}{z} \qquad \phi = \tan^{-1}\frac{y}{x} \tag{1.47b}$$

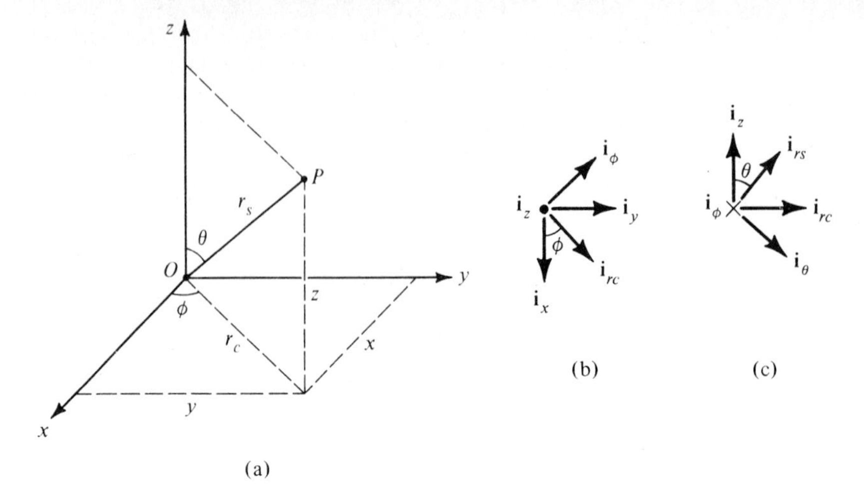

Figure 1.16. (a) For conversion of coordinates of a point from one coordinate system to another. (b) and (c) For expressing unit vectors in cylindrical and spherical coordinate systems, respectively, in terms of unit vectors in the Cartesian coordinate system.

Relationships (1.46a) and (1.47a) correspond to conversion from cylindrical coordinates to Cartesian coordinates, and vice versa. Relationships (1.46b) and (1.47b) correspond to conversion from spherical coordinates to Cartesian coordinates, and vice versa. It should be noted that in computing ϕ from y and x, consideration should be given to the quadrant of the xy-plane in which the projection of the point P onto the xy-plane lies.

Considering next the conversion of vectors from one coordinate system to another, we note that to do this, we need to express each of the unit vectors of the first coordinate system in terms of its components along the unit vectors in the second coordinate system. From the definition of the dot product of two vectors, the component of a unit vector along another unit vector, that is, the cosine of the angle between the unit vectors, is simply the dot product of the two unit vectors. Thus considering the sets of unit vectors in the cylindrical and Cartesian coordinate systems, we have with the aid of Fig. 1.16(b),

$$\mathbf{i}_{rc} \cdot \mathbf{i}_x = \cos\phi \qquad \mathbf{i}_{rc} \cdot \mathbf{i}_y = \sin\phi \qquad \mathbf{i}_{rc} \cdot \mathbf{i}_z = 0 \tag{1.48a}$$

$$\mathbf{i}_\phi \cdot \mathbf{i}_x = -\sin\phi \qquad \mathbf{i}_\phi \cdot \mathbf{i}_y = \cos\phi \qquad \mathbf{i}_\phi \cdot \mathbf{i}_z = 0 \tag{1.48b}$$

$$\mathbf{i}_z \cdot \mathbf{i}_x = 0 \qquad \mathbf{i}_z \cdot \mathbf{i}_y = 0 \qquad \mathbf{i}_z \cdot \mathbf{i}_z = 1 \tag{1.48c}$$

Similarly, for the sets of unit vectors in the spherical and Cartesian coordinate systems, we obtain with the aid of Fig. 1.16(b) and (c),

$$\mathbf{i}_{rs} \cdot \mathbf{i}_x = \sin\theta\cos\phi \qquad \mathbf{i}_{rs} \cdot \mathbf{i}_y = \sin\theta\sin\phi \qquad \mathbf{i}_{rs} \cdot \mathbf{i}_z = \cos\theta \tag{1.49a}$$

$$\mathbf{i}_\theta \cdot \mathbf{i}_x = \cos\theta\cos\phi \qquad \mathbf{i}_\theta \cdot \mathbf{i}_y = \cos\theta\sin\phi \qquad \mathbf{i}_\theta \cdot \mathbf{i}_z = -\sin\theta \tag{1.49b}$$

$$\mathbf{i}_\phi \cdot \mathbf{i}_x = -\sin\phi \qquad \mathbf{i}_\phi \cdot \mathbf{i}_y = \cos\phi \qquad \mathbf{i}_\phi \cdot \mathbf{i}_z = 0 \tag{1.49c}$$

We shall now illustrate the use of these relationships by means of an example.

Example 1.4

Let us consider the vector $3\mathbf{i}_x + 4\mathbf{i}_y + 5\mathbf{i}_z$ at the point (3, 4, 5) and convert it to one in spherical coordinates.

First, from the relationships (1.47b), we obtain the spherical coordinates of the point (3, 4, 5) to be

$$r_s = \sqrt{3^2 + 4^2 + 5^2} = 5\sqrt{2}$$

$$\theta = \tan^{-1}\frac{\sqrt{3^2 + 4^2}}{5} = \tan^{-1} 1 = 45°$$

$$\phi = \tan^{-1}\tfrac{4}{3} = 53.13°$$

Then noting from the relationships (1.49) that at the point under consideration,

$$\begin{aligned}\mathbf{i}_x &= \sin\theta\cos\phi\,\mathbf{i}_{rs} + \cos\theta\cos\phi\,\mathbf{i}_\theta - \sin\phi\,\mathbf{i}_\phi \\ &= 0.3\sqrt{2}\,\mathbf{i}_{rs} + 0.3\sqrt{2}\,\mathbf{i}_\theta - 0.8\mathbf{i}_\phi \\ \mathbf{i}_y &= \sin\theta\sin\phi\,\mathbf{i}_{rs} + \cos\theta\sin\phi\,\mathbf{i}_\theta + \cos\phi\,\mathbf{i}_\phi \\ &= 0.4\sqrt{2}\,\mathbf{i}_{rs} + 0.4\sqrt{2}\,\mathbf{i}_\theta + 0.6\mathbf{i}_\phi \\ \mathbf{i}_z &= \cos\theta\,\mathbf{i}_{rs} - \sin\theta\,\mathbf{i}_\theta = 0.5\sqrt{2}\,\mathbf{i}_{rs} - 0.5\sqrt{2}\,\mathbf{i}_\theta\end{aligned}$$

we obtain

$$\begin{aligned}3\mathbf{i}_x + 4\mathbf{i}_y + 5\mathbf{i}_z &= (0.9\sqrt{2} + 1.6\sqrt{2} + 2.5\sqrt{2})\mathbf{i}_{rs} \\ &\quad + (0.9\sqrt{2} + 1.6\sqrt{2} - 2.5\sqrt{2})\mathbf{i}_\theta + (-2.4 + 2.4)\mathbf{i}_\phi = 5\sqrt{2}\,\mathbf{i}_{rs}\end{aligned}$$

This result is to be expected since the given vector has components equal to the coordinates of the point at which it is specified. Hence its magnitude is equal to the distance of the point from the origin, that is, the spherical coordinate r_s of the point, and its direction is along the line drawn from the origin to the point, that is, along the unit vector $\mathbf{i}_{rs}$ at that point. In fact, the given vector is a particular case of the position vector $x\mathbf{i}_x + y\mathbf{i}_y + z\mathbf{i}_z = r_s\mathbf{i}_{rs}$, which is the vector drawn from the origin to the point (x, y, z).

The listing of a computer program written in the BASIC language for the IBM PC microcomputer to perform the conversions considered in this example and the output from a run of the program for the vector of this example are reproduced as PL 1.1.

PL 1.1. Program listing and sample output for conversion of vector in Cartesian coordinates to one in spherical coordinates.

```
100 '*****************************************************
110 '* CONVERSION OF VECTOR IN CARTESIAN COORDINATES TO *
120 '* ONE IN SPHERICAL COORDINATES                     *
130 '*****************************************************
140 PI=3.141593:RD=180/PI
150 DEF FN TRD(ARG)=INT(ARG*10000+.5)/10000:'* ROUNDS ARG
160 '  TO FOUR DECIMAL PLACES *
170 CLS:PRINT "ENTER CARTESIAN COORDINATES OF POINT:"
180 INPUT "X = ",X
190 INPUT "Y = ",Y
200 INPUT "Z = ",Z:PRINT
210 PRINT "ENTER COMPONENTS OF VECTOR IN CARTESIAN"
220 PRINT "COORDINATES:"
230 INPUT "X COMPONENT = ",VX
240 INPUT "Y COMPONENT = ",VY
```

PL1.1 (continued)

```
250 INPUT "Z COMPONENT = ",VZ
260 '* COMPUTE SPHERICAL COORDINATES OF POINT *
270 RC=SQR(X*X+Y*Y):RS=SQR(RC*RC+Z*Z)
280 IF Z=0 THEN THETA=SGN(RS)*PI/2:GOTO 310
290 THETA=ATN(RC/Z)
300 IF Z<0 THEN THETA=THETA+PI
310 IF X=0 THEN PHI=SGN(Y)*PI/2:GOTO 360
320 PHI=ATN(Y/X)
330 IF X<0 THEN PHI=PHI+PI
340 '* COMPUTE COMPONENTS OF VECTOR IN SPHERICAL
350 '  COORDINATES *
360 IF X=0 AND Y=0 THEN VPHI=0:GOTO 430:'* SPECIAL CASE
370 '  OF POINT ON Z-AXIS *
380 ST=SIN(THETA):CT=COS(THETA):SP=SIN(PHI):CP=COS(PHI)
390 VRS=VX*ST*CP+VY*ST*SP+VZ*CT
400 VTHETA=VX*CT*CP+VY*CT*SP-VZ*ST
410 VPHI=-VX*SP+VY*CP
420 GOTO 520
430 IF Z<>O THEN 470
440 VRS=SQR(VX*VX+VY*VY+VZ*VZ):VTHETA=0:'* SPECIAL CASE
450 '  OF POINT AT ORIGIN *
460 GOTO 520
470 IF VX=0 THEN PHI=SGN(VY)*PI/2:GOTO 500
480 PHI=ATN(VY/VX)
490 IF VX<0 THEN PHI=PHI+PI
500 VRS=SGN(Z)*VZ
510 VTHETA=SGN(Z)*SQR(VX*VX+VY*VY)
520 PRINT:PRINT
530 PRINT "SPHERICAL COORDINATES OF POINT ARE:"
540 PRINT "RS =";FN TRD(RS)
550 PRINT "THETA =";FN TRD(THETA*RD);"DEG"
560 PRINT "PHI =";FN TRD(PHI*RD);"DEG":PRINT
570 PRINT "COMPONENTS OF THE VECTOR IN SPHERICAL"
580 PRINT "COORDINATES ARE:"
590 PRINT "RS COMPONENT =";FN TRD(VRS)
600 PRINT "THETA COMPONENT =";FN TRD(VTHETA)
610 PRINT "PHI COMPONENT =";FN TRD(VPHI)
620 PRINT:PRINT "STRIKE ANY KEY TO CONTINUE";:C$=INPUT$(1)
630 GOTO 170
640 END
```

```
RUN
ENTER CARTESIAN COORDINATES OF POINT:
X = 3
Y = 4
Z = 5

ENTER COMPONENTS OF VECTOR IN CARTESIAN
COORDINATES:
X COMPONENT = 3
Y COMPONENT = 4
Z COMPONENT = 5


SPHERICAL COORDINATES OF POINT ARE:
RS = 7.0711
THETA = 45 DEG
PHI = 53.1301 DEG

COMPONENTS OF THE VECTOR IN SPHERICAL
COORDINATES ARE:
RS COMPONENT = 7.0711
```

PL1.1 (continued)

```
THETA COMPONENT = 0
PHI COMPONENT = 0

STRIKE ANY KEY TO CONTINUE
```

K1.3. Cylindrical coordinate system; Orthogonal surfaces; Unit vectors; Differential lengths, surfaces, and volume; Spherical coordinate system; Orthogonal surfaces; Unit vectors; Differential lengths, surfaces, and volume; Conversions between coordinate systems.

D1.7. Convert into Cartesian coordinates each of the following points: **(a)** $(2, 5\pi/6, 3)$ in cylindrical coordinates; **(b)** $(4, 4\pi/3, -1)$ in cylindrical coordinates; **(c)** $(4, 2\pi/3, \pi/6)$ in spherical coordinates; and **(d)** $(\sqrt{8}, \pi/4, \pi/3)$ in spherical coordinates.
Ans. **(a)** $(-\sqrt{3}, 1, 3)$; **(b)** $(-2, -2\sqrt{3}, -1)$; **(c)** $(3, \sqrt{3}, -2)$; **(d)** $(1, \sqrt{3}, 2)$

D1.8. Convert into cylindrical coordinates the following points specified in Cartesian coordinates: **(a)** $(-2, 0, 1)$; **(b)** $(1, -\sqrt{3}, -1)$; and **(c)** $(-\sqrt{2}, -\sqrt{2}, 3)$.
Ans. **(a)** $(2, \pi, 1)$; **(b)** $(2, 5\pi/3, -1)$; **(c)** $(2, 5\pi/4, 3)$

D1.9. Convert into spherical coordinates the following points specified in Cartesian coordinates: **(a)** $(0, -2, 0)$; **(b)** $(-3, \sqrt{3}, 2)$; and **(c)** $(-\sqrt{2}, 0, -\sqrt{2})$.
Ans. **(a)** $(2, \pi/2, 3\pi/2)$; **(b)** $(4, \pi/3, 5\pi/6)$; **(c)** $(2, 3\pi/4, \pi)$

1.4 SCALAR AND VECTOR FIELDS

Before we take up the task of studying electromagnetic fields, we must understand what is meant by a *field*. A field is associated with a region in space, and we say that a field exists in the region if there is a physical phenomenon associated with points in that region. For example, in everyday life we are familiar with the earth's gravitational field. We do not "see" the field in the same manner as we see light rays, but we know of its existence in the sense that objects are acted upon by the gravitational force of the earth. In a broader context, we can talk of the field of any physical quantity as being a description, mathematical or graphical, of how the quantity varies from one point to another in the region of the field and with time. We can talk of scalar or vector fields depending on whether the quantity of interest is a scalar or a vector. We can talk of static or time-varying fields depending on whether the quantity of interest is independent of time or changing with it.

Scalar fields

We shall begin our discussion of fields with some simple examples of scalar fields. Thus let us consider the case of the conical pyramid shown in Fig. 1.17(a). A description of the height of the pyramidal surface versus position on its base is an example of a scalar field involving two variables. Choosing the origin to be the projection of the vertex of the cone onto the base and setting up an xy-coordinate system to locate points on the base, we obtain the height field as a function of x and y to be

$$h(x, y) = 6 - 2\sqrt{x^2 + y^2} \tag{1.50}$$

Although we are able to depict the height variation of points on the conical surface graphically by using the third coordinate for h, we will have to be content with the visualization of the height field by a set of constant-height contours on

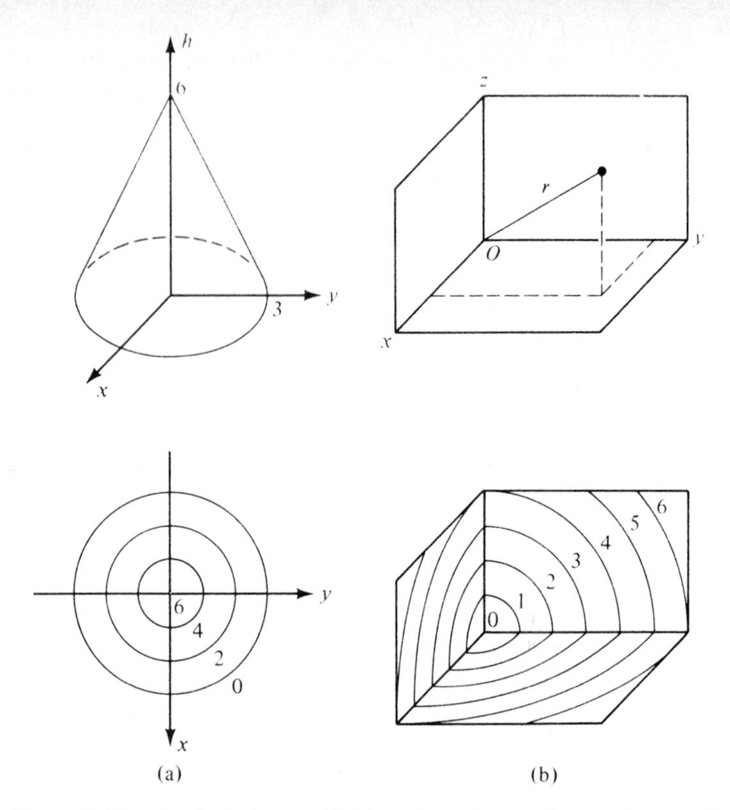

Figure 1.17. (a) Conical pyramid lying above the xy-plane and a set of constant-height contours for the conical surface. (b) Rectangular room and a set of constant-distance surfaces depicting the distance field of points in the room from one corner of the room.

the xy-plane if only two coordinates were available, as in the case of a two-dimensional space. For the field under consideration, the constant-height contours are circles in the xy-plane centered at the origin and equally spaced for equal increments of the height value, as shown in Fig. 1.17(a).

For an example of a scalar field in three dimensions, let us consider a rectangular room and the distance field of points in the room from one corner of the room, as shown in Fig. 1.17(b). For convenience, we choose this corner to be the origin O and set up a Cartesian coordinate system with the three contiguous edges meeting at that point as the coordinate axes. Each point in the room is defined by a set of values for the three coordinates x, y, and z. The distance r from the origin to that point is $\sqrt{x^2 + y^2 + z^2}$. Thus the distance field of points in the room from the origin is given by

$$r(x, y, z) = \sqrt{x^2 + y^2 + z^2} \tag{1.51}$$

Since the three coordinates are already used up for defining the points in the field region, we have to visualize the distance field by means of a set of constant-distance surfaces. A constant-distance surface is a surface for which points on it correspond to a particular constant value of r. For the case under consideration, the constant-distance surfaces are spherical surfaces centered at the origin and are equally spaced for equal increments in the value of the distance, as shown in Fig. 1.17(b).

The fields we have discussed thus far are static fields. A simple example of a time-varying scalar field is provided by the temperature field associated with points in a room, especially when it is being heated or cooled. Just as in the case of the distance field of Fig. 1.17(b), we set up a three-dimensional coordinate system and to each set of three coordinates corresponding to the location of a point in the room, we assign a number to represent the temperature T at that point. Since the temperature at that point, however, varies with time t, this number is a function of time. Thus we describe mathematically the time-varying temperature field in the room by a function $T(x, y, z, t)$. For any given instant of time, we can visualize a set of constant-temperature or isothermal surfaces corresponding to particular values of T as representing the temperature field for that value of time. For a different instant of time, we will have a different set of isothermal surfaces for the same values of T. Thus we can visualize the time-varying temperature field in the room by a set of isothermal surfaces continuously changing their shapes as though in a motion picture.

Vector fields

The foregoing discussion of scalar fields may now be extended to vector fields by recalling that a vector quantity has associated with it a direction in space in addition to magnitude. Hence to describe a vector field we attribute to each point in the field region a vector that represents the magnitude and direction of the physical quantity under consideration at that point. Since a vector at a given point can be expressed as the sum of its components along the set of unit vectors at that point, a mathematical description of the vector field involves simply the descriptions of the three component scalar fields. Thus for a vector field $\mathbf{F}$ in the Cartesian coordinate system, we have

$$\boxed{\mathbf{F}(x, y, z, t) = F_x(x, y, z, t)\mathbf{i}_x + F_y(x, y, z, t)\mathbf{i}_y + F_z(x, y, z, t)\mathbf{i}_z} \tag{1.52}$$

Similar expressions in cylindrical and spherical coordinate systems are as follows:

$$\mathbf{F}(r, \phi, z, t) = F_r(r, \phi, z, t)\mathbf{i}_r + F_\phi(r, \phi, z, t)\mathbf{i}_\phi + F_z(r, \phi, z, t)\mathbf{i}_z \tag{1.53a}$$

$$\mathbf{F}(r, \theta, \phi, t) = F_r(r, \theta, \phi, t)\mathbf{i}_r + F_\theta(r, \theta, \phi, t)\mathbf{i}_\theta + F_\phi(r, \theta, \phi, t)\mathbf{i}_\phi \tag{1.53b}$$

We should however recall that the unit vectors $\mathbf{i}_r$ and $\mathbf{i}_\phi$ in (1.53a) and all three unit vectors in (1.53b) are themselves nonuniform but known functions of the coordinates.

Finding equations for direction lines of a vector field

A vector field is described by a set of *direction lines,* also known as *stream lines* and *flux lines*. A direction line is a curve constructed such that the field is tangential to the curve for all points on the curve. To find the equations for the direction lines for a specified vector field $\mathbf{F}$, we consider the differential length vector $d\mathbf{l}$ tangential to the curve. Then since $\mathbf{F}$ and $d\mathbf{l}$ are parallel, their components must be in the same ratio. Thus in the Cartesian coordinate system, we obtain the differential equation

$$\boxed{\frac{dx}{F_x} = \frac{dy}{F_y} = \frac{dz}{F_z}} \tag{1.54}$$

which upon integration gives the required algebraic equation. Similar expressions in cylindrical and spherical coordinate systems are as follows:

$$\frac{dr}{F_r} = \frac{r\,d\phi}{F_\phi} = \frac{dz}{F_z} \tag{1.55}$$

$$\frac{dr}{F_r} = \frac{r\,d\theta}{F_\theta} = \frac{r\sin\theta\,d\phi}{F_\phi} \tag{1.56}$$

We shall illustrate the concept of direction lines and the use of (1.54)–(1.56) to obtain the equations for the direction lines by means of an example.

Example 1.5

Consider a circular disk of radius a rotating with constant angular velocity ω about an axis normal to the disk and passing through its center. We wish to describe the linear velocity vector field associated with points on the rotating disk.

We choose the center of the disk to be the origin and set up a two-dimensional coordinate system, as shown in Fig. 1.18(a). Note that we have a choice of the coordinates (x, y) or the coordinates (r, ϕ). We know that the magnitude of the linear velocity of a point on the disk is then equal to the product of the angular velocity ω and the radial distance r of the point from the center of the disk. The direction of the linear velocity is tangential to the circle drawn through that point and concentric with the disk. Hence we may depict the linear velocity field by drawing at several points on the disk vectors that are tangential to the circles concentric with the disk and passing through those points, and whose lengths are pro-

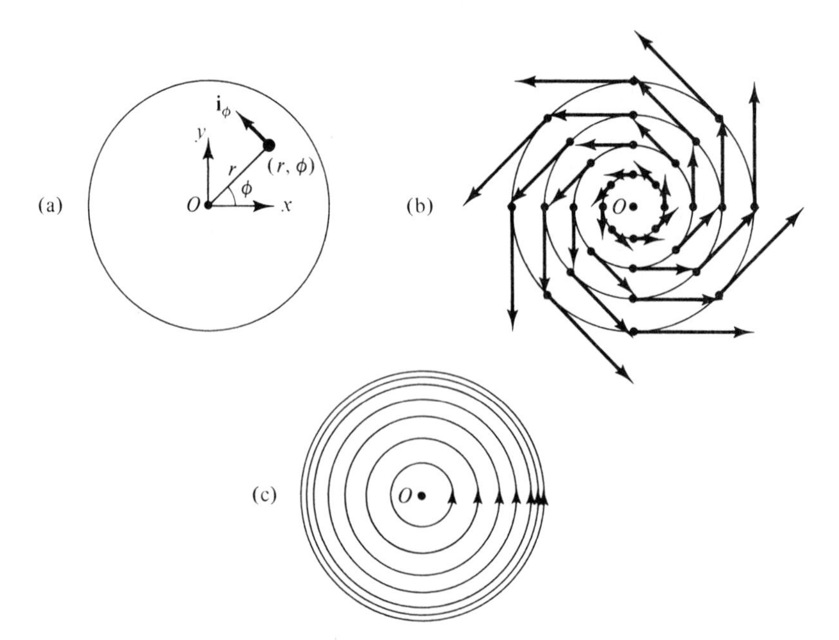

Figure 1.18. (a) Rotating disk. (b) Linear velocity vector field associated with points on the rotating disk. (c) Same as (b) except that the vectors are omitted and the density of direction lines is used to indicate the magnitude variation.

portional to the radii of the circles, as shown in Fig. 1.18(b), where the points are carefully selected to reveal the circular symmetry of the field with respect to the center of the disk. We find, however, that this method of representation of the vector field results in a congested sketch of vectors. Hence we may simplify the sketch by omitting the vectors and simply placing arrowheads along the circles thereby obtaining a set of direction lines. We note that for the field under consideration the direction lines are also contours of constant magnitude of the velocity, and hence by increasing the density of the direction lines as r increases, we can indicate the magnitude variation, as shown in Fig. 1.18(c).

For this simple example, we have been able to obtain the direction lines without resorting to the use of mathematics. We shall now consider the mathematical determination of the direction lines and show that the same result is obtained. To do this, we note that the linear velocity vector field is given by

$$\mathbf{v}(r, \phi) = \omega r \mathbf{i}_\phi$$

Then considering that the geometry associated with the problem is two-dimensional and using (1.55), we have

$$\frac{dr}{0} = \frac{r\,d\phi}{\omega r}$$

or

$$dr = 0$$

$$r = \text{constant}$$

which represents circles centered at the origin, as in Fig. 1.18(c).

If we wish to obtain the equations for the direction lines using Cartesian coordinates, we first write

$$\begin{aligned}\mathbf{v}(x, y) &= \omega r(\mathbf{i}_\phi \cdot \mathbf{i}_x)\mathbf{i}_x + \omega r(\mathbf{i}_\phi \cdot \mathbf{i}_y)\mathbf{i}_y \\ &= \omega r(-\sin\phi\,\mathbf{i}_x + \cos\phi\,\mathbf{i}_y) \\ &= \omega(-y\mathbf{i}_x + x\mathbf{i}_y)\end{aligned}$$

Then from (1.54), we have

$$\frac{dx}{-y} = \frac{dy}{x}$$

or

$$x\,dx + y\,dy = 0$$

$$x^2 + y^2 = \text{constant}$$

which again represents circles centered at the origin.

K1.4. Field; Static field, Time-varying field; Scalar field; Constant magnitude contours and surfaces; Vector field; Direction lines.

D1.10. The time-varying temperture field in a certain region of space is given by

$$T(x, y, z, t) = T_0\{[x(1 + \sin \pi r)]^2 + [2y(1 - \cos \pi t)]^2 + 4z^2]\}$$

where T_0 is a constant. Find the shapes of the constant-temperature surfaces for each of the following values of t: **(a)** $t = 0$; **(b)** $t = 0.5$ s; and **(c)** $t = 1$ s.
Ans. **(a)** elliptic cylinders; **(b)** spheres; **(c)** ellipsoids

D1.11. For the vector field $\mathbf{F} = (3x - y)\mathbf{i}_x + (x + z)\mathbf{i}_y + (2y - z)\mathbf{i}_z$, find the following: **(a)** the magnitude of $\mathbf{F}$ and the unit vector along $\mathbf{F}$ at the point (1, 1, 0); **(b)** the point at which the magnitude of $\mathbf{F}$ is 3 and the direction of $\mathbf{F}$ is along the unit vector $\frac{1}{3}(2\mathbf{i}_x + 2\mathbf{i}_y + \mathbf{i}_z)$; and **(c)** the point at which the magnitude of $\mathbf{F}$ is 3 and the direction of $\mathbf{F}$ is along the unit vector $\mathbf{i}_z$.
Ans. **(a)** $3, \frac{1}{3}(2\mathbf{i}_z + \mathbf{i}_y + 2\mathbf{i}_z)$; **(b)** (1, 1, 1) **(c)** (0.6, 1.8, −0.6)

D1.12. A vector field is given in cylindrical coordinates by

$$\mathbf{F} = \frac{1}{r^2}(\cos\phi\, \mathbf{i}_r + \sin\phi\, \mathbf{i}_\phi)$$

Express the vector $\mathbf{F}$ in Cartesian coordinates at each of the following points specified in Cartesian coordinates: **(a)** (1, 0, 0); **(b)** (1, −1, −3); and **(c)** $(1, \sqrt{3}, -4)$.

Ans. **(a)** $\mathbf{i}_x$; **(b)** $-\frac{1}{2}\mathbf{i}_y$; **(c)** $\frac{1}{8}(-\mathbf{i}_x + \sqrt{3}\mathbf{i}_y)$

1.5 SINUSOIDALLY TIME-VARYING FIELDS

In our study of electromagnetics, we will be particularly interested in fields that vary sinusoidally with time. Sinusoidally time-varying fields are important because of their natural occurrence and ease of generation. For example, when we speak, we emit sine waves; when we tune our radio dial to a broadcast station, we receive sine waves; and so on. Also, any function for which the time variation is arbitrary can be expressed in terms of sinusoidally time-varying functions having a discrete or continuous spectrum of frequencies, depending upon whether the function is periodic or aperiodic. Thus if the response of a system to a sinusoidal excitation is known, its response for a nonsinusoidal excitation can be found.

Let us first consider a scalar sinusoidal function of time. Such a function is given by an expression of the form $A \cos(\omega t + \phi)$. In this expression, A is the amplitude of the sinusoidal variation and $(\omega t + \phi)$ is the phase. In particular, the phase of the function at $t = 0$ is ϕ. The quantity $\omega = 2\pi f = 2\pi/T$ is the rate of change of phase with time and is known as the radian frequency, having the units radians per second. The quantity $f = \omega/2\pi$ is the number of times the phase changes by 2π radians in 1 second and is known as the linear frequency or simply frequency, having the units hertz, abbreviated Hz. The quantity $T = 1/f = 2\pi/\omega$ is the period, that is, the time interval in which the phase changes by 2π radians. A plot of the function versus t shown in Fig. 1.19 illustrates how the function changes periodically between positive and negative values. Note that the value of the function at $t = 0$ is $A \cos\phi$ and that a positive maximum occurs for $\omega t + \phi = 0$ or $t = -\phi/\omega$. For a numerical example, let us consider a sinusoidal function having the period 10^{-3} s, amplitude 5 units, and a positive maximum at $t = 2 \times 10^{-4}$ s. Then the expression for the function is given by $5 \cos(2\pi \times 10^3 t - 0.4\pi)$.

If we now have a sinusoidally time-varying scalar field, we can visualize the field quantity varying sinusoidally with time at each point in the field region with the amplitude and phase governed by the spatial dependence of the field quantity. Thus, for example, the field $Ae^{-\alpha z} \cos(\omega t - \beta z)$, where A, α, and β are positive constants, is characterized by sinusoidal time variations with amplitude decreasing exponentially with z and the phase at any given time decreasing linearly with z.

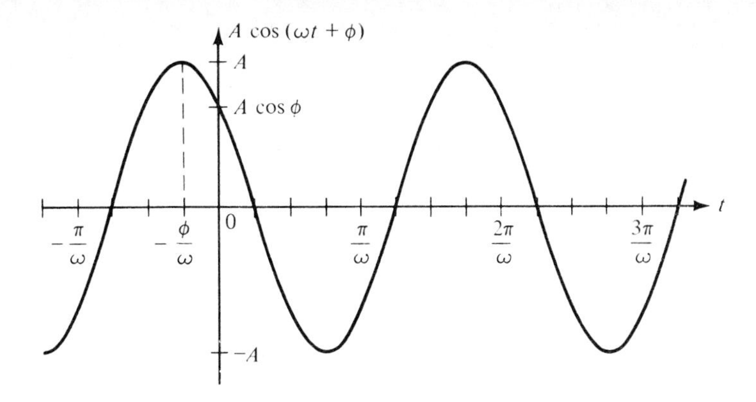

Figure 1.19. Sinusoidally time-varying scalar function $A \cos(\omega t + \phi)$.

For a sinusoidally time-varying vector field, the behavior of each component of the field may be visualized in the manner just discussed. If we now fix our attention on a particular point in the field region, we can visualize the sinusoidal variation with time of a particular component at that point by a vector changing its magnitude and direction as shown, for example, for the x-component in Fig. 1.20(a). Since the tip of the vector simply moves back and forth along a line, which in this case is parallel to the x-axis, the component vector is said to be *linearly polarized* in the x-direction. Similarly, the sinusoidal variation with time of the y-component of the field can be visualized by a vector changing its magnitude and direction as shown in Fig. 1.20(b), not necessarily with the same amplitude and phase as those of the x-component. Since the tip of the vector moves back and forth parallel to the y-axis, the y-component is said to be linearly polarized in the y-direction. In the same manner, the z-component is linearly polarized in the z-direction.

Polarization of sinusoidally time-varying fields

When two component sinusoidally time-varying vectors at a point are added, the polarization of the resulting sinusoidally time-varying vector can be linear, circular, or elliptical, that is, the tip of the vector can describe a straight

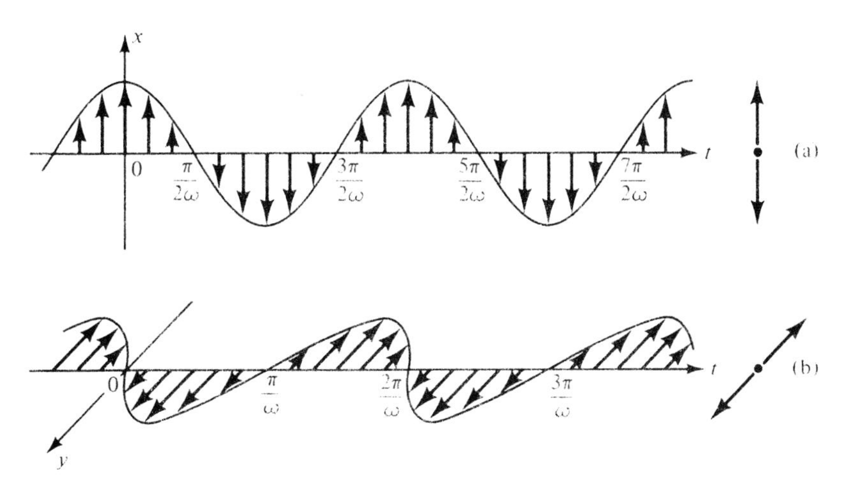

Figure 1.20. (a) Time variation of a linearly polarized vector in the x-direction. (b) Time variation of a linearly polarized vector in the y-direction.

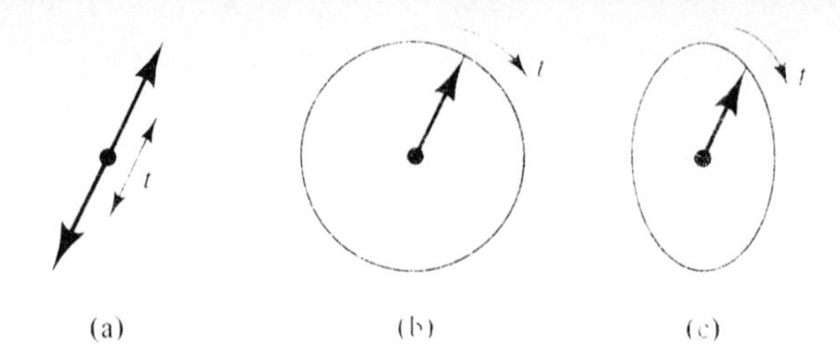

Figure 1.21. (a) Linear, (b) circular, and (c) elliptical polarizations.

line, a circle, or an ellipse with time, as shown in Fig. 1.21, depending on the relative amplitudes and phase angles of the component vectors.

Note that in the case of linear polarization, the direction of the vector remains along a straight line, but its magnitude changes with time. For circular polarization, the magnitude remains constant, but its direction changes with time. Elliptical polarization is characterized by both magnitude and direction of the vector changing with time.

Linear polarization

If two component sinusoidally time-varying vectors have arbitrary amplitudes but are in phase as, for example,

$$\mathbf{F}_1 = F_1 \cos(\omega t + \phi)\, \mathbf{i}_x \tag{1.57a}$$

$$\mathbf{F}_2 = F_2 \cos(\omega t + \phi)\, \mathbf{i}_y \tag{1.57b}$$

then the sum vector $\mathbf{F} = \mathbf{F}_1 + \mathbf{F}_2$ is linearly polarized in a direction making an angle

$$\alpha = \tan^{-1}\frac{F_y}{F_x} = \tan^{-1}\frac{F_2}{F_1} \tag{1.58}$$

with the x-direction as shown in the series of sketches in Fig. 1.22 illustrating the time history of the magnitude and direction of $\mathbf{F}$ over an interval of one period. The reasoning can be extended to two (or more) linearly polarized vectors not necessarily along the coordinate axes but all of them *in phase*. Thus the sum vector of any number of linearly polarized vectors having different directions and amplitudes but in phase is also a linearly polarized vector.

Circular polarization

If two component sinusoidally time-varying vectors have equal amplitudes, differ in direction by 90°, and differ in phase by $\pi/2$, as, for example,

$$\mathbf{F}_1 = F_0 \cos(\omega t + \phi)\, \mathbf{i}_x \tag{1.59a}$$

$$\mathbf{F}_2 = F_0 \sin(\omega t + \phi)\, \mathbf{i}_y \tag{1.59b}$$

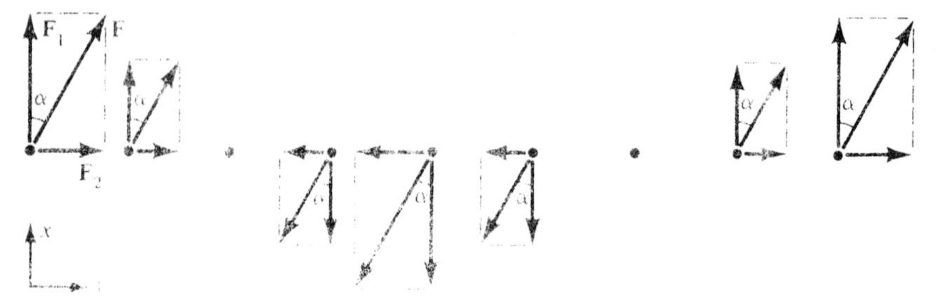

Figure 1.22. The sum vector of two linearly polarized vectors in phase is a linearly polarized vector.

then, to determine the *polarization* of the sum vector $\mathbf{F} = \mathbf{F}_1 + \mathbf{F}_2$, we note that the magnitude of $\mathbf{F}$ is given by

$$|\mathbf{F}| = |F_0 \cos(\omega t + \phi)\,\mathbf{i}_x + F_0 \sin(\omega t + \phi)\,\mathbf{i}_y| = F_0 \tag{1.60}$$

and that the angle α which $\mathbf{F}$ makes with $\mathbf{i}_x$ is given by

$$\alpha = \tan^{-1}\frac{F_y}{F_x} = \tan^{-1}\frac{F_0 \sin(\omega t + \phi)}{F_0 \cos(\omega t + \phi)} = \omega t + \phi \tag{1.61}$$

Thus the sum vector rotates with constant magnitude F_0 and at a rate of ω rad/s so that its tip describes a circle. The sum vector is then said to be *circularly polarized*. The series of sketches in Fig. 1.23 illustrates the time history of the magnitude and direction of $\mathbf{F}$ over an interval of one period.

The reasoning can be generalized to two linearly polarized vectors not necessarily along the coordinate axes. Thus if two linearly polarized vectors satisfy the three conditions of (1) equal amplitudes, (2) perpendicularity in direction, and (3) phase difference of 90°, then their sum vector is circularly polarized. We shall illustrate this by means of an example.

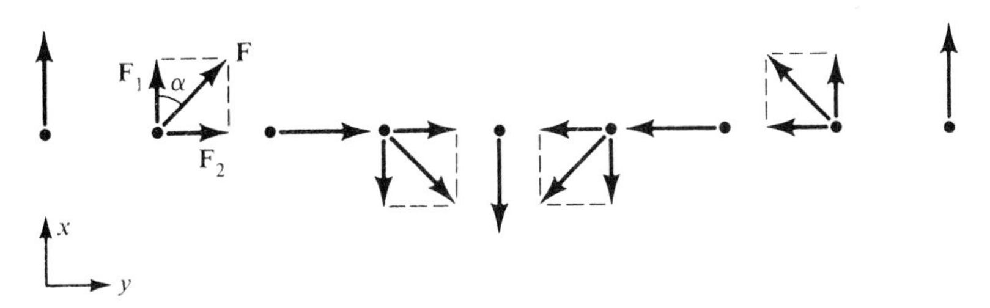

Figure 1.23. Circular polarization.

Example 1.6

Given two vectors

$$\mathbf{F}_1 = (3\mathbf{i}_x - 4\mathbf{i}_z)\cos 2\pi \times 10^6 t$$

$$\mathbf{F}_2 = 5\mathbf{i}_y \sin 2\pi \times 10^6 t$$

at a point. Note that the vector $\mathbf{F}_1$ consists of two components which are in phase. Hence it is linearly polarized but along the direction of $(3\mathbf{i}_x - 4\mathbf{i}_z)$. The vector $\mathbf{F}_2$ is linearly polarized along the y-direction. We wish to determine the polarization of the vector $\mathbf{F} = \mathbf{F}_1 + \mathbf{F}_2$.

Since the two linearly polarized vectors $\mathbf{F}_1$ and $\mathbf{F}_2$ are not in phase, we rule out the possibility of $\mathbf{F}$ being linearly polarized. In fact, since $\mathbf{F}_1$ varies with time in a cosine manner whereas $\mathbf{F}_2$ varies in a sine manner, we note that $\mathbf{F}_1$ and $\mathbf{F}_2$ differ in phase by 90°. The amplitude of $\mathbf{F}_1$ is $\sqrt{3^2 + (-4)^2}$, or 5, which is equal to that of $\mathbf{F}_2$. Also,

$$\mathbf{F}_1 \cdot \mathbf{F}_2 = (3\mathbf{i}_x - 4\mathbf{i}_z) \cdot 5\mathbf{i}_y = 0$$

so that $\mathbf{F}_1$ and $\mathbf{F}_2$ are perpendicular. Thus $\mathbf{F}_1$ and $\mathbf{F}_2$ satisfy all three conditions for the sum of two linearly polarized vectors to be circularly polarized. Therefore, $\mathbf{F}$ is circularly polarized.

Alternatively, we observe that

$$
\begin{aligned}
|\mathbf{F}| &= |\mathbf{F}_1 + \mathbf{F}_2| \\
&= |3 \cos 2\pi \times 10^6 t \ \mathbf{i}_x + 5 \sin 2\pi \times 10^6 t \ \mathbf{i}_y - 4 \cos 2\pi \times 10^6 t \ \mathbf{i}_z| \\
&= (25 \cos^2 2\pi \times 10^6 t + 25 \sin^2 2\pi \times 10^6 t)^{1/2} \\
&= \sqrt{25} = 5, \text{ a constant with time}
\end{aligned}
$$

Hence **F** is circularly polarized.

Elliptical polarization

For the general case in which the conditions for the sum vector to be linearly polarized or circularly polarized are not satisfied, the sum vector is *elliptically polarized*; that is, its tip describes an ellipse. Thus linear and circular polarizations are special cases of elliptical polarization. The listing of a PC program for demonstrating linear, circular, and elliptical polarizations as resulting from the superposition of two sinusoidally time-varying, orthogonal component vectors $\mathbf{F}_1 = A \cos(\omega t + \phi)\, \mathbf{i}_x$ and $\mathbf{F}_2 = B \cos(\omega t + \theta)\, \mathbf{i}_y$ of the same frequency under different conditions pertaining to their amplitudes and phase angles is reproduced in PL 1.2. The ellipse obtained from a run of the program for $A = 40$, $B = 60$, $\phi = 60°$, and $\theta = 105°$ is shown in Fig. 1.24.

PL 1.2. Program listing for demonstration pertinent to polarization of sinusoidally time-varying vector fields.

```
100 '*****************************************************
110 '* POLARIZATION OF SINUSOIDALLY TIME-VARYING VECTOR *
120 '* FIELDS                                            *
130 '*****************************************************
140 SC=1.2:'* SCALE FACTOR TO EQUALIZE VERTICAL AND
150 '  HORIZONTAL SCALES *
160 PI=3.1416:DR=PI/180:XC=80:YC=160
170 CLS:SCREEN 1:COLOR 0,1
180 PRINT "ENTER VALUES OF A, B, PHI, AND THETA:"
190 PRINT:INPUT "A = ",AV:'* AMPLITUDE OF VERTICAL
200 '  COMPONENT *
210 PRINT:INPUT "B = ",AH:'* AMPLITUDE OF HORIZONTAL
220 '  COMPONENT *
230 PRINT:INPUT "PHI IN DEG = ",PV:'* PHASE ANGLE OF
240 '  VERTICAL COMPONENT *
250 PRINT:INPUT "THETA IN DEG = ",PH:'* PHASE ANGLE OF
260 '  HORIZONTAL COMPONENT *
270 PRINT:PRINT "PRESS ANY KEY TO CONTINUE":C$=INPUT$(1)
280 '* PRINT VALUES OF INPUT PARAMETERS *
290 CLS:LOCATE 21,1:PRINT "A =";AV;" PHI =";PV;"DEG"
300 PRINT "B =";AH;" THETA =";PH;"DEG"
310 '* SCALE A AND B SO THAT THE LARGEST OF A AND B IS
320 '  EQUAL TO 60 *
330 IF AV>AH THEN AX=60:AY=60*AH/AV:GOTO 350
340 AY=60:AX=60*AV/AH
350 AY=AY*SC:C=0:D=0:CC=XC:DD=YC
360 '* DRAW COMPONENT AND RESULTANT VECTORS, AND PLOT
370 '  POINTS ALONG TRAJECTORIES OF TIPS OF VECTORS, AT
380 '  INTERVALS OF ONE-HUNDRENTHS OF A PERIOD *
390 FOR I=0 TO 100
400 A=AX*COS(PI*I/50+PV*DR)
410 IF ABS(A)<ABS(C) THEN COLR=0 ELSE COLR=3
420 LINE (YC,CC)-(YC,XC-A),COLR:PSET (YC,CC)
430 B=AY*COS(PI*I/50+PH*DR)
440 IF ABS(B)<ABS(D) THEN COLR=0 ELSE COLR=3
450 LINE (DD,XC)-(YC+B,XC),COLR:PSET (DD,XC)
```

PL1.2 (continued)

```
460 LINE (YC,XC)-(DD,CC),0:PSET (DD,CC)
470 C=A:D=B:CC=XC-C:DD=YC+D
480 LINE (YC,XC)-(YC,CC)
490 LINE (YC,XC)-(DD,XC)
500 LINE (YC,XC)-(DD,CC)
510 NEXT
520 '* PLOTTING COMPLETED *
530 LOCATE 23,1:PRINT "STRIKE ANY KEY TO CONTINUE"
540 C$=INPUT$(1)
550 GOTO 170
560 END
```

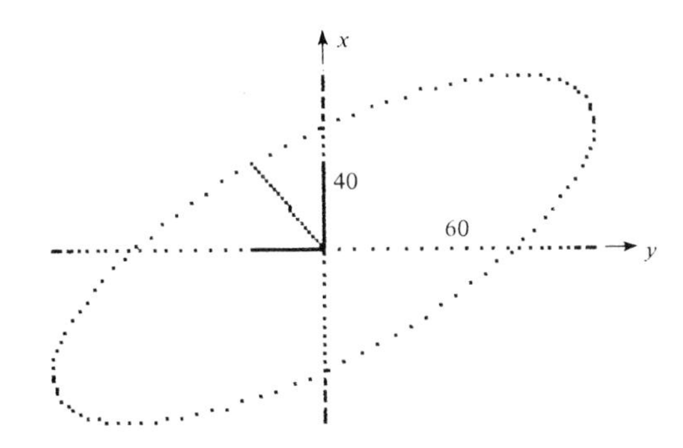

Figure 1.24. Ellipse obtained from a run of the program of PL 1.2 for values of $A = 40$, $B = 60$, $\phi = 60°$, and $\theta = 105°$.

Relevance of polarization in reception of radio waves

An example in which polarization is relevant is in the reception of radio waves. Here when we talk about polarization, we generally refer to the direction of the electric field of the wave. If the incoming signal is linearly polarized, then for *maximum* voltage to be induced in a *linear* receiving antenna, the antenna must be oriented parallel to the direction of polarization of the signal. Any other orientation of the antenna will result in a smaller induced voltage since the antenna "sees" only that component of the electric field parallel to itself. In particular, if the antenna is in the plane perpendicular to the direction of polarization of the incoming signal, *no* voltage is induced. On the other hand, if the incoming signal is circularly or elliptically polarized, a voltage is induced in the antenna except for one orientation which is along the line perpendicular to the plane of the circle or the ellipse.

K1.5. Sinusoidal function; Sinusoidally time-varying scalar field; Sinusoidally time-varying vector field; Polarization of sinusoidally time-varying vector field; Linear polarization; Circular polarization; Elliptical polarization.

D1.13. Write the expression for each of the sinusoidal functions of time having the specified characteristics: **(a)** period equal to 1 s, amplitude equal to 10 units, and a positive maximum occurring at $t = 0$; **(b)** frequency equal to 10^6 Hz, value at $t = 0$ equal to 5 units and a positive maximum occurring at $t = \frac{1}{6} \times 10^{-6}$ s; and **(c)** period equal to 2 s, amplitude equal to 5 units, and a negative maximum occurring at $\frac{3}{4}$ s.
Ans. **(a)** $10 \cos 2\pi t$; **(b)** $10 \cos (2\pi \times 10^6 t - \pi/3)$; **(c)** $5 \cos (\pi t + \pi/4)$

D1.14. Two sinusoidally time-varying vector fields are given by

$$\mathbf{F}_1 = F_0 \cos (2\pi \times 10^8 t - 2\pi z)\, \mathbf{i}_x$$

$$\mathbf{F}_2 = F_0 \cos (2\pi \times 10^8 t - 3\pi z)\, \mathbf{i}_y$$

Find the polarization of $\mathbf{F}_1 + \mathbf{F}_2$ at each of the following points: **(a)** (3, 4, 0); **(b)** (3, −2, 0.5); **(c)** (−2, 1, 1); and **(d)** (−1, −3, 0.2).
Ans. **(a)** Linear; **(b)** circular; **(c)** linear; **(d)** elliptical

D1.15. A sinusoidally time-varying vector field is given at a point by $\mathbf{F} = 1 \cos(\omega t + 60°)\,\mathbf{i}_x + 1 \cos(\omega t + \alpha)\,\mathbf{i}_y$. Find the value(s) of α between 0° and 360° for each of the following cases: **(a)** $\mathbf{F}$ is linearly polarized along a line lying in the second and fourth quadrants; **(b)** $\mathbf{F}$ is circularly polarized with the sense of rotation from the $+x$-direction toward the $+y$-direction with time; and **(c)** $\mathbf{F}$ is circularly polarized with the sense of rotation from the $+y$-direction toward the $+x$-direction with time.
Ans. **(a)** 240°; **(b)** 330°; **(c)** 150°

1.6 COMPLEX NUMBERS AND PHASOR TECHNIQUE

In this section we discuss a mathematical technique known as the phasor technique pertinent to operations involving sinusoidally time-varying quantities. The technique simplifies the solution of a differential equation in which the steady-state response for a sinusoidally time-varying excitation is to be determined, by reducing the differential equation to an algebraic equation involving phasors. A phasor is a complex number or a complex variable. We first review complex numbers and associated operations.

A complex number has two parts: a real part and an imaginary part. Imaginary numbers are square roots of negative real numbers. To introduce the concept of an imaginary number, we define

$$\boxed{\sqrt{-1} = j} \tag{1.62a}$$

or

$$\boxed{(\pm j)^2 = -1} \tag{1.62b}$$

Thus, $j5$ is the positive square root of -25, $-j10$ is the negative square root of -100, and so on. A complex number is written in the form $a + jb$, where a is the real part and b is the imaginary part. Examples are

$$3 + j4,\ -4 + j1,\ -2 - j2,\ 2 - j3,\ \ldots$$

Rectangular form

A complex number is represented graphically in a complex plane by using two orthogonal axes, corresponding to the real and imaginary parts, as shown in Fig. 1.25, in which are plotted the numbers just listed. Since the set of orthogonal axes resembles the rectangular coordinate axes, the representation $(a + jb)$ is known as the rectangular form.

Exponential and polar forms

An alternative form of representation of a complex number is the exponential form $Ae^{j\phi}$, where A is the magnitude and ϕ is the phase angle. To convert from one form to another, we first recall that

$$e^x = 1 + x + \frac{x^2}{2!} + \frac{x^3}{3!} + \cdots \tag{1.63}$$

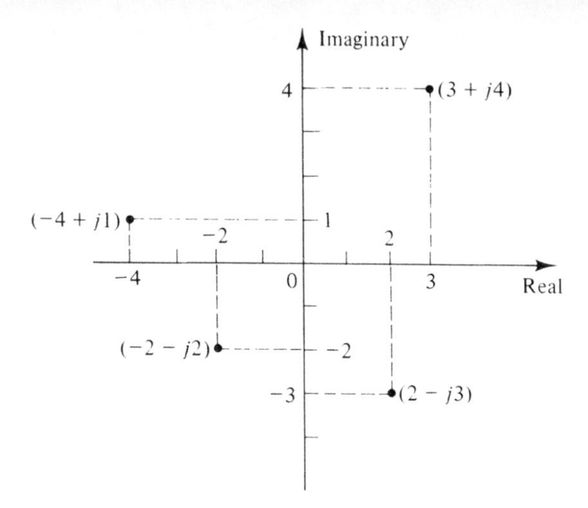

Figure 1.25. Graphical representation of complex numbers in rectangular form.

Substituting $x = j\phi$, we have

$$
\begin{aligned}
e^{j\phi} &= 1 + j\phi + \frac{(j\phi)^2}{2!} + \frac{(j\phi)^3}{3!} + \cdots \\
&= 1 + j\phi - \frac{\phi^2}{2!} - j\frac{\phi^3}{3!} + \cdots \\
&= \left(1 - \frac{\phi^2}{2!} + \cdots\right) + j\left(\phi - \frac{\phi^3}{3!} + \cdots\right) \\
&= \cos\phi + j\sin\phi
\end{aligned} \tag{1.64}
$$

This is the so-called *Euler's identity*. Thus

$$
\boxed{\begin{aligned} Ae^{j\phi} &= A(\cos\phi + j\sin\phi) \\ &= A\cos\phi + jA\sin\phi \end{aligned}} \tag{1.65}
$$

Now, equating the two forms of the complex numbers, we have

$$a + jb = A\cos\phi + jA\sin\phi$$

or

$$a = A\cos\phi \tag{1.66a}$$

$$b = A\sin\phi \tag{1.66b}$$

These expressions enable us to convert from exponential form to rectangular form. To convert from rectangular form to exponential form, we note that

$$a^2 + b^2 = A^2$$

$$\cos\phi = \frac{a}{A}, \quad \sin\phi = \frac{b}{A}, \quad \text{and} \quad \tan\phi = \frac{b}{a}$$

Thus

$$A = \sqrt{a^2 + b^2} \tag{1.67a}$$

$$\phi = \tan^{-1} \frac{b}{a} \tag{1.67b}$$

Note that in the determination of ϕ, the signs of $\cos \phi$ and $\sin \phi$ should be considered to see if it is necessary to add π to the angle obtained by taking the inverse tangent of b/a.

In terms of graphical representation, A is simply the distance from the origin of the complex plane to the point under consideration, and ϕ is the angle measured counterclockwise from the positive real axis ($\phi = 0$) to the line drawn from the origin to the complex number, as shown in Fig. 1.26. Since this representation is akin to the polar coordinate representation of points in two-dimensional space, the complex number is also written as $A\underline{/\phi}$, the polar form.

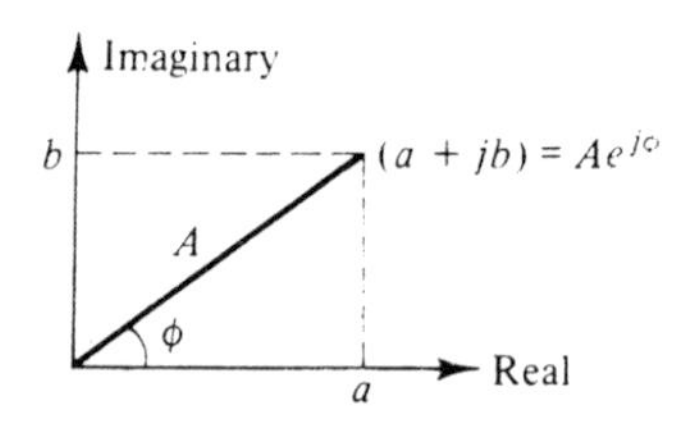

Figure 1.26. Graphical representation of a complex number in exponential form or polar form.

Turning now to the Euler's identity, we see that for $\phi = \pm\pi/2$, $Ae^{\pm j\pi/2} = A \cos \pi/2 \pm jA \sin \pi/2 = \pm jA$. Thus purely imaginary numbers correspond to $\phi = \pm\pi/2$. This justifies why the vertical axis, which is orthogonal to the real (horizontal) axis, is the imaginary axis.

Conversion from rectangular to exponential or polar form

The complex numbers in rectangular form plotted in Fig. 1.25 may now be converted to exponential form (or polar form):

$$3 + j4 = \sqrt{3^2 + 4^2}\, e^{j \tan^{-1}(4/3)} = 5e^{j0.295\pi} = 5\underline{/53.13^\circ}$$

$$-4 + j1 = \sqrt{4^2 + 1^2}\, e^{j[\tan^{-1}(-1/4)+\pi]} = 4.12e^{j0.922\pi} = 4.12\underline{/165.96^\circ}$$

$$-2 - j2 = \sqrt{2^2 + 2^2}\, e^{j[\tan^{-1}(1)+\pi]} = 2.83e^{j1.25\pi} = 2.83\underline{/225^\circ}$$

$$2 - j3 = \sqrt{2^2 + 3^2}\, e^{j \tan^{-1}(-3/2)} = 3.61e^{-j0.313\pi} = 3.61\underline{/-56.31^\circ}$$

These are shown plotted in Fig. 1.27. It can be noted that in converting from rectangular to exponential (or polar) form, the angle ϕ can be correctly determined if the number is first plotted in the complex plane to see in which quadrant it lies. Also note that angles traversed in the clockwise sense from the positive real axis are negative angles. Furthermore, adding or subtracting an integer multiple of 2π to the angle does not change the complex number.

Arithmetic of complex numbers

Complex numbers are added (or subtracted) by simply adding (or subtracting) their real and imaginary parts separately as follows:

$$(3 + j4) + (2 - j3) = 5 + j1$$

$$(2 - j3) - (-4 + j1) = 6 - j4$$

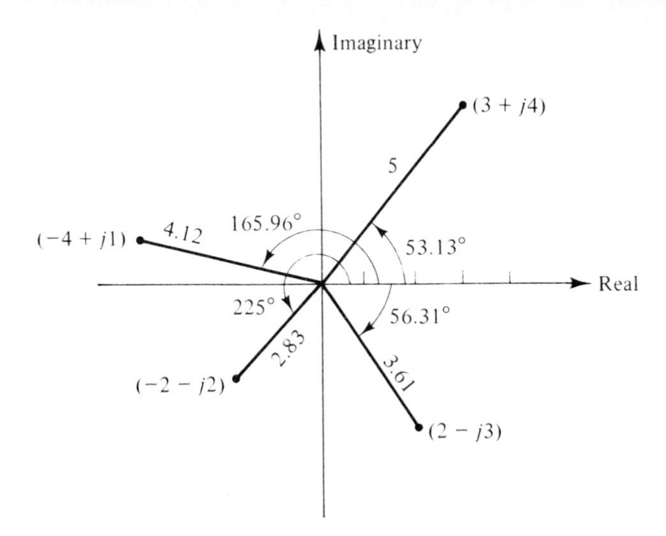

Figure 1.27. Polar form representation of the complex numbers of Fig. 1.25.

Graphically, this procedure is identical to the parallelogram law of addition (or subtraction) of two vectors.

Two complex numbers are multiplied by multiplying each part of one complex number by each part of the second complex number and adding the four products according to the rule of addition as follows:

$$\begin{aligned}(3 + j4)(2 - j3) &= 6 - j9 + j8 - j^2(12)\\ &= 6 - j9 + j8 + 12\\ &= 18 - j1\end{aligned}$$

Two complex numbers whose real parts are equal but whose imaginary parts are the negative of each other are known as complex conjugates. Thus $(a - jb)$ is the complex conjugate of $(a + jb)$, and vice versa. The product of two complex conjugates is a real number:

$$(a + jb)(a - jb) = a^2 - jab + jba + b^2 = a^2 + b^2 \tag{1.68}$$

This property is used in division of one complex number by another by multiplying both the numerator and the denominator by the complex conjugate of the denominator and then performing the division by real number. For example,

$$\frac{(3 + j4)}{(2 - j3)} = \frac{(3 + j4)(2 + j3)}{(2 - j3)(2 + j3)} = \frac{-6 + j17}{13} = -0.46 + j1.31$$

The exponential form is particularly useful for multiplication, division, and other operations, such as raising to the power, since the rules associated with exponential functions are applicable. Thus

$$(A_1e^{j\phi_1})(A_2e^{j\phi_2}) = A_1A_2e^{j(\phi_1+\phi_2)} \tag{1.69a}$$

$$\frac{A_1e^{j\phi_1}}{A_2e^{j\phi_2}} = \frac{A_1}{A_2}e^{j(\phi_1-\phi_2)} \tag{1.69b}$$

$$(Ae^{j\phi})^n = A^ne^{jn\phi} \tag{1.69c}$$

Let us consider some numerical examples:

(a) $(5e^{j0.295\pi})(3.61e^{-j0.313\pi}) = 18.05e^{-j0.018\pi}$

(b) $\dfrac{5e^{j0.295\pi}}{3.61e^{-j0.313\pi}} = 1.39e^{j0.608\pi}$

(c) $(2.83e^{j1.25\pi})^4 = 64.14e^{j5\pi} = 64.14e^{j\pi}$

(d) $\sqrt{4.12e^{j0.922\pi}} = [4.12e^{j(0.922\pi+2k\pi)}]^{1/2}, \quad k = 0, 1, 2, \ldots$

$$= \sqrt{4.12}\, e^{j(0.461\pi+k\pi)}, \quad k = 0, 1$$

$$= 2.03e^{j0.461\pi} \quad \text{or} \quad 2.03e^{j1.461\pi}$$

Note that in evaluating the square roots, although k can assume an infinite number of integer values, only the first two need to be considered since the numbers repeat themselves for higher values of integers. Similar considerations apply for cube roots, and so on.

Phasor defined

Having reviewed complex numbers, we are now ready to discuss the phasor technique. The basis behind the phasor technique lies in the fact that since

$$\boxed{Ae^{jx} = A\cos x + jA\sin x} \tag{1.70}$$

we can write

$$\boxed{A\cos x = \text{Re}[Ae^{jx}]} \tag{1.71}$$

where Re stands for "real part of." In particular, if $x = \omega t + \theta$, then we have

$$\begin{aligned} A\cos(\omega t + \theta) &= \text{Re}[Ae^{j(\omega t+\theta)}] \\ &= \text{Re}[Ae^{j\theta}e^{j\omega t}] \\ &= \text{Re}[\bar{A}e^{j\omega t}] \end{aligned} \tag{1.72}$$

where $\bar{A} = Ae^{j\theta}$ is known as the phasor (the overbar denotes that $\bar{A}$ is complex) corresponding to $A\cos(\omega t + \theta)$. Thus the phasor corresponding to a cosinusoidally time-varying function is a complex number having magnitude same as the amplitude of the cosine function and phase angle equal to the phase of the cosine function for $t = 0$. To find the phasor corresponding to a sine function, we first convert it into a cosine function and proceed as in (1.72). Thus

$$\begin{aligned} B\sin(\omega t + \phi) &= B\cos(\omega t + \phi - \pi/2) \\ &= \text{Re}[Be^{j(\omega t+\phi-\pi/2)}] \\ &= \text{Re}[Be^{j(\phi-\pi/2)}e^{j\omega t}] \end{aligned} \tag{1.73}$$

Hence, the phasor corresponding to $B\sin(\omega t + \phi)$ is $Be^{j(\phi-\pi/2)}$, or $Be^{j\phi}e^{-j\pi/2}$, or $-jBe^{j\phi}$.

Addition of two sine functions

Let us now consider the addition of two sinusoidally time-varying functions (of the same frequency), for example, $5\cos\omega t$ and $10\sin(\omega t - 30°)$, by using the phasor technique. To do this, we proceed as follows:

$$
\begin{aligned}
5\cos\omega t + 10\sin(\omega t - 30°) &= 5\cos\omega t + 10\cos(\omega t - 120°) \\
&= \mathrm{Re}[5e^{j\omega t}] + \mathrm{Re}[10e^{j(\omega t - 2\pi/3)}] \\
&= \mathrm{Re}[5e^{j0}e^{j\omega t}] + \mathrm{Re}[10e^{-j2\pi/3}e^{j\omega t}] \\
&= \mathrm{Re}[5e^{j0}e^{j\omega t} + 10e^{-j2\pi/3}e^{j\omega t}] \\
&= \mathrm{Re}[(5e^{j0} + 10e^{-j2\pi/3})e^{j\omega t}] \\
&= \mathrm{Re}\{[(5 + j0) + (-5 - j8.66)]e^{j\omega t}\} \\
&= \mathrm{Re}[(0 - j8.66)e^{j\omega t}] \\
&= \mathrm{Re}[8.66e^{-j\pi/2}e^{j\omega t}] \\
&= \mathrm{Re}[8.66e^{j(\omega t - \pi/2)}] \\
&= 8.66\cos(\omega t - 90°)
\end{aligned}
\tag{1.74}
$$

In practice, we need not write all the steps just shown. First, we express all functions in their cosine forms and then recognize the phasor corresponding to each function. For the foregoing example, the complex numbers $5e^{j0}$ and $10e^{-j2\pi/3}$ are the phasors corresponding to $5\cos\omega t$ and $10\sin(\omega t - 30°)$, respectively. Then we add the phasors and from the sum phasor write the required cosine function as one having the amplitude same as the magnitude of the sum phasor and argument equal to ωt plus the phase angle of the sum phasor. Thus the steps involved are as shown in the block diagram of Fig. 1.28.

Solution of differential equation

We shall now discuss the solution of a differential equation for sinusoidal steady-state response by using the phasor technique. To do this, let us consider the problem of finding the steady-state solution for the current $I(t)$ in the simple RL series circuit driven by the voltage source $V(t) = V_m \cos(\omega t + \phi)$, as shown in Fig. 1.29. From Kirchhoff's voltage law, we then have

$$
\boxed{RI(t) + L\frac{dI(t)}{dt} = V_m\cos(\omega t + \phi)}
\tag{1.75}
$$

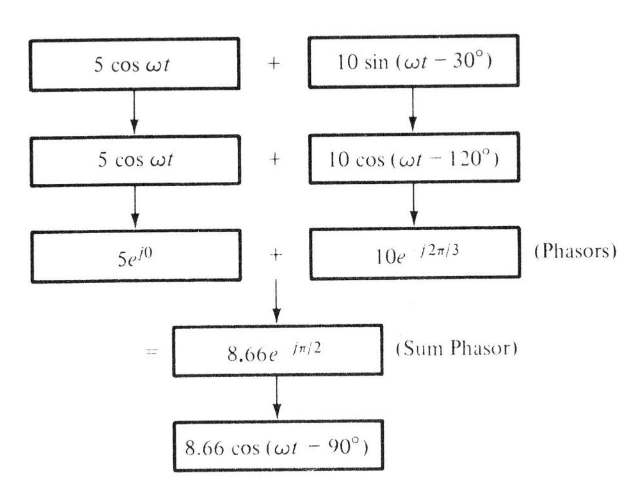

Figure 1.28. Block diagram of steps involved in the application of phasor technique to the addition of two sinusoidally time-varying functions.

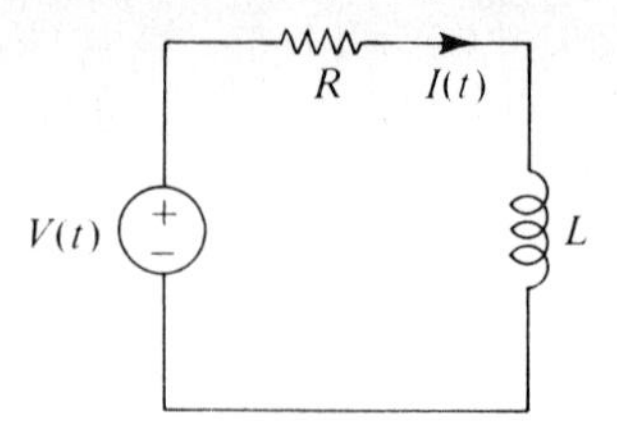

Figure 1.29. *RL* series circuit.

We know that the steady-state solution for the current must also be a cosine function of time having the same frequency as that of the voltage source but not necessarily in phase with it. Hence let us assume

$$I(t) = I_m \cos(\omega t + \theta) \tag{1.76}$$

The problem now consists of finding I_m and θ.

Using the phasor concept, we write

$$\begin{aligned} V_m \cos(\omega t + \phi) &= \text{Re}[V_m e^{j(\omega t + \phi)}] \\ &= \text{Re}[V_m e^{j\phi} e^{j\omega t}] \\ &= \text{Re}[\bar{V} e^{j\omega t}] \end{aligned} \tag{1.77a}$$

$$\begin{aligned} I_m \cos(\omega t + \theta) &= \text{Re}[I_m e^{j(\omega t + \theta)}] \\ &= \text{Re}[I_m e^{j\theta} e^{j\omega t}] \\ &= \text{Re}[\bar{I} e^{j\omega t}] \end{aligned} \tag{1.77b}$$

where $\bar{V} = V_m e^{j\phi}$ and $\bar{I} = I_m e^{j\theta}$ are the phasors corresponding to $V(t) = V_m \cos(\omega t + \phi)$ and $I(t) = I_m \cos(\omega t + \theta)$, respectively. Substituting these into the differential equation, we have

$$R\{\text{Re}[\bar{I} e^{j\omega t}]\} + L \frac{d}{dt} \{\text{Re}[\bar{I} e^{j\omega t}]\} = \text{Re}[\bar{V} e^{j\omega t}] \tag{1.78}$$

Since R and L are constants, and since d/dt and Re can be interchanged, we can simplify this equation in accordance with the following steps:

$$\text{Re}[R\bar{I} e^{j\omega t}] + \text{Re}\left[L \frac{d}{dt} (\bar{I} e^{j\omega t})\right] = \text{Re}[\bar{V} e^{j\omega t}]$$

$$\text{Re}[R\bar{I} e^{j\omega t}] + \text{Re}[j\omega L\bar{I} e^{j\omega t}] = \text{Re}[\bar{V} e^{j\omega t}]$$

$$\text{Re}[(R\bar{I} + j\omega L\bar{I}) e^{j\omega t}] = \text{Re}[\bar{V} e^{j\omega t}] \tag{1.79}$$

Let us now consider two values of ωt, say $\omega t = 0$ and $\omega t = \pi/2$. For $\omega t = 0$, we obtain

$$\text{Re}(R\bar{I} + j\omega L\bar{I}) = \text{Re}(\bar{V}) \tag{1.80}$$

For $\omega t = \pi/2$, we obtain

$$\text{Re}[j(R\bar{I} + j\omega L\bar{I})] = \text{Re}[j\bar{V}]$$

or

$$\text{Im}(R\bar{I} + j\omega L\bar{I}) = \text{Im}(\bar{V}) \tag{1.81}$$

where Im stands for "imaginary part of." Now, since the real parts as well as the imaginary parts of $(R\bar{I} + j\omega L\bar{I})$ and $\bar{V}$ are equal, it follows that the two complex numbers are equal. Thus

$$\boxed{R\bar{I} + j\omega L\bar{I} = \bar{V}} \tag{1.82}$$

By solving this equation, we obtain $\bar{I}$ and hence I_m and θ. Note that by using the phasor technique, we have reduced the problem of solving the differential equation (1.75) to one of solving the phasor (algebraic) equation (1.82). In fact, the phasor equation can be written directly from the differential equation without the necessity of the intermediate steps by recognizing that the time functions $I(t)$ and $V(t)$ are replaced by their phasors $\bar{I}$ and $\bar{V}$, respectively, and d/dt is replaced by $j\omega$. We have here included the intermediate steps merely to illustrate the basis behind the phasor technique. We shall now consider an example.

Example 1.7

For the circuit of Fig. 1.29, let us assume that $R = 1\ \Omega$, $L = 10^{-3}$H, and $V(t) = 10\cos(1000t + 30°)$ V and obtain the steady-state solution for $I(t)$.

The differential equation for $I(t)$ is given by

$$I + 10^{-3}\frac{dI}{dt} = 10\cos(1000t + 30°)$$

Replacing the current and voltage by their phasors $\bar{I}$ and $10e^{j\pi/6}$, respectively, and d/dt by $j\omega = j1000$, we obtain the phasor equation

$$\bar{I} + 10^{-3}(j1000\bar{I}) = 10e^{j\pi/6}$$

or

$$\bar{I}(1 + j1) = 10e^{j\pi/6}$$

$$\bar{I} = \frac{10e^{j\pi/6}}{1 + j1} = \frac{10e^{j\pi/6}}{\sqrt{2}\,e^{j\pi/4}}$$

$$= 7.07e^{-j\pi/12}$$

Having determined the value of $\bar{I}$, we now find the required solution to be

$$I(t) = \text{Re}[\bar{I}e^{j\omega t}]$$

$$= \text{Re}[7.07e^{-j\pi/12}e^{j1000t}]$$

$$= 7.07\cos(1000t - 15°)\ \text{A}$$

K1.6. Complex number; Rectangular form, exponential form, polar form, and conversions; Complex number arithmetic; Phasor; Phasor technique.

D1.16. Express the following complex numbers in exponential form: **(a)** $5 + j12$; **(b)** $-24 - j7$; and **(c)** $-1 + j\sqrt{3}$.
Ans. **(a)** $13e^{j0.374\pi}$; **(b)** $25e^{j1.09\pi}$; **(c)** $2e^{j0.667\pi}$

D1.17. Find: **(a)** $5e^{j0.205\pi} + (1 + j2)$ in exponential form; **(b)** $(-\sqrt{3} + j1)^5$ in rectangular form; and **(c)** $(3 + j1)(1 - j3)/(4 + j3)$ in polar form.
Ans. **(a)** $7.07e^{j0.25\pi}$; **(b)** $(16\sqrt{3} + j16)$; **(c)** $2\underline{/-90°}$

D1.18. Using the phasor technique, express each of the following functions as a single cosinusoidal function of time: **(a)** $4 \cos(\omega t + 30°) - 3 \cos(\omega t + 120°)$; **(b)** $5 \sin(\omega t + 60°) - 5 \cos(\omega t + 30°) - 3 \sin \omega t$.
Ans. **(a)** $5 \cos(\omega t - 6.87°)$; **(b)** $2 \cos(\omega t - 90°)$

1.7 SUMMARY

We first learned in this chapter several rules of vector algebra that are necessary for our study of the elements of engineering electromagnetics by considering vectors expressed in terms of their components along three mutually orthogonal directions. To carry out the manipulations involving vectors at different points in space in a systematic manner, we then introduced the Cartesian coordinate system and discussed the application of the vector algebraic rules to vectors in the Cartesian coordinate system. To summarize these rules, we consider three vectors

$$\mathbf{A} = A_x\mathbf{i}_x + A_y\mathbf{i}_y + A_z\mathbf{i}_z$$

$$\mathbf{B} = B_x\mathbf{i}_x + B_y\mathbf{i}_y + B_z\mathbf{i}_z$$

$$\mathbf{C} = C_x\mathbf{i}_x + C_y\mathbf{i}_y + C_z\mathbf{i}_z$$

in a right-handed Cartesian coordinate system, that is, with $\mathbf{i}_x \times \mathbf{i}_y = \mathbf{i}_z$. We then have

$$\mathbf{A} + \mathbf{B} = (A_x + B_x)\mathbf{i}_x + (A_y + B_y)\mathbf{i}_y + (A_z + B_z)\mathbf{i}_z$$

$$\mathbf{B} - \mathbf{C} = (B_x - C_x)\mathbf{i}_x + (B_y - C_y)\mathbf{i}_y + (B_z - C_z)\mathbf{i}_z$$

$$m\mathbf{A} = mA_x\mathbf{i}_x + mA_y\mathbf{i}_y + mA_z\mathbf{i}_z$$

$$\frac{\mathbf{B}}{n} = \frac{B_x}{n}\mathbf{i}_x + \frac{B_y}{n}\mathbf{i}_y + \frac{B_z}{n}\mathbf{i}_z$$

$$|\mathbf{A}| = \sqrt{A_x^2 + A_y^2 + A_z^2}$$

$$\mathbf{i}_A = \frac{A_x}{\sqrt{A_x^2 + A_y^2 + A_z^2}}\mathbf{i}_x + \frac{A_y}{\sqrt{A_x^2 + A_y^2 + A_z^2}}\mathbf{i}_y + \frac{A_z}{\sqrt{A_x^2 + A_y^2 + A_z^2}}\mathbf{i}_z$$

$$\mathbf{A} \cdot \mathbf{B} = A_xB_x + A_yB_y + A_zB_z$$

$$\mathbf{A} \times \mathbf{B} = \begin{vmatrix} \mathbf{i}_x & \mathbf{i}_y & \mathbf{i}_z \\ A_x & A_y & A_z \\ B_x & B_y & B_z \end{vmatrix}$$

$$\mathbf{A} \cdot \mathbf{B} \times \mathbf{C} = \begin{vmatrix} A_x & A_y & A_z \\ B_x & B_y & B_z \\ C_x & C_y & C_z \end{vmatrix}$$

Other useful expressions are

$$d\mathbf{l} = dx\,\mathbf{i}_x + dy\,\mathbf{i}_y + dz\,\mathbf{i}_z$$

$$d\mathbf{S} = \pm dy\,dz\,\mathbf{i}_x,\ \pm dz\,dx\,\mathbf{i}_y,\ \pm dx\,dy\,\mathbf{i}_z$$

$$dv = dx\,dy\,dz$$

We then discussed the cylindrical and spherical coordinate systems, and conversions between these coordinate systems and the Cartesian coordinate system. Relationships for carrying out the coordinate conversions are as follows:

Cylindrical to Cartesian, and vice versa:

$$x = r \cos \phi \qquad y = r \sin \phi \qquad z = z$$

$$r = \sqrt{x^2 + y^2} \qquad \phi = \tan^{-1} \frac{y}{x} \qquad z = z$$

Spherical to Cartesian, and vice versa:

$$x = r \sin \theta \cos \phi \qquad y = r \sin \theta \sin \phi \qquad z = r \cos \theta$$

$$r = \sqrt{x^2 + y^2 + z^2} \qquad \theta = \tan^{-1} \frac{\sqrt{x^2 + y^2}}{z} \qquad \phi = \tan^{-1} \frac{y}{x}$$

Other useful expressions are as follows:

Cylindrical:

$$d\mathbf{l} = dr\, \mathbf{i}_r + r\, d\phi\, \mathbf{i}_\phi + dz\, \mathbf{i}_z$$

$$d\mathbf{S} = \pm r\, d\phi\, dz\, \mathbf{i}_r,\ \pm dr\, dz\, \mathbf{i}_\phi,\ \pm r\, dr\, d\phi\, \mathbf{i}_z$$

$$dv = r\, dr\, d\phi\, dz$$

Spherical:

$$d\mathbf{l} = dr\, \mathbf{i}_r + r\, d\theta\, \mathbf{i}_\theta + r \sin \theta\, d\phi\, \mathbf{i}_\phi$$

$$d\mathbf{S} = \pm r^2 \sin \theta\, d\theta\, d\phi\, \mathbf{i}_r,\ \pm r \sin \theta\, dr\, d\phi\, \mathbf{i}_\theta,\ \pm r\, dr\, d\theta\, \mathbf{i}_\phi$$

$$dv = r^2 \sin \theta\, dr\, d\theta\, d\phi$$

Next we discussed the concepts of scalar and vector fields, static and time-varying, by means of some simple examples such as the height of points on a conical surface above its base, the temperature field of points in a room, and the velocity vector field associated with points on a disk rotating about its center. We learned about the visualization of fields by means of constant-magnitude contours or surfaces and in addition by means of direction lines in the case of vector fields. We also discussed the mathematical technique of obtaining the equations for the direction lines of a vector field.

Particular attention was then devoted to sinusoidally time-varying fields, in view of their importance in our study of electromagnetics. Polarization of sinusoidally time-varying vector fields was then considered. In the general case, the polarization is elliptical; that is, the tip of the vector at a point describes an ellipse with time. Linear and circular polarizations are special cases.

Finally, a review of complex numbers and associated operations was presented as a prelude to the discussion of phasor technique. The phasor technique was illustrated by considering two examples: (1) addition of two sinusoidal functions of time and (2) solution of a differential equation for the steady-state response due to a sinusoidal excitation.

REVIEW QUESTIONS

R1.1. Give some examples of scalars.

R1.2. Give some examples of vectors.

R1.3. Is it necessary for the reference vectors $\mathbf{i}_1$, $\mathbf{i}_2$, and $\mathbf{i}_3$ to be an orthogonal set?

R1.4. State whether $\mathbf{i}_1$, $\mathbf{i}_2$, and $\mathbf{i}_3$ directed westward, northward, and downward, respectively, is a right-handed or a left-handed set.

R1.5. State all conditions for which $\mathbf{A} \cdot \mathbf{B} = 0$.

R1.6. State all conditions for which $\mathbf{A} \times \mathbf{B} = \mathbf{0}$.

R1.7. What is the significance of $\mathbf{A} \cdot \mathbf{B} \times \mathbf{C} = 0$?

R1.8. What is the significance of $\mathbf{A} \times (\mathbf{B} \times \mathbf{C}) = \mathbf{0}$?

R1.9. What is the particular advantageous characteristic associated with the unit vectors in the Cartesian coordinate system?

R1.10. What is the position vector?

R1.11. What is the total distance around the circumference of a circle of radius 1 m? What is the total vector distance around the circle?

R1.12. Discuss the application of differential length vectors to find a unit vector normal to a surface at a point on the surface.

R1.13. Discuss the concept of a differential surface vector.

R1.14. What is the total surface area of a cube of sides 1 m? Assuming the normals to the surfaces to be directed outward of the cubical volume, what is the total vector surface area of the cube?

R1.15. Describe the three orthogonal surfaces that define the cylindrical coordinates of a point.

R1.16. Which of the unit vectors in the cylindrical coordinate system are not uniform? Explain.

R1.17. Discuss the conversion from the cylindrical coordinates of a point to its Cartesian coordinates, and vice versa.

R1.18. Describe the three orthogonal surfaces that define the spherical coordinates of a point.

R1.19. Discuss the nonuniformity of the unit vectors in the spherical coordinate system.

R1.20. Discuss the conversion from the spherical coordinates of a point to its Cartesian coordinates, and vice versa.

R1.21. Describe briefly your concept of a scalar field and illustrate with an example.

R1.22. Describe briefly your concept of a vector field and illustrate with an example.

R1.23. How do you depict pictorially the gravitational field of the earth?

R1.24. Discuss the procedure for obtaining the equations for the direction lines of a vector field.

R1.25. Discuss the parameters associated with a sinusoidal function of time.

R1.26. A sinusoidally time-varying vector is expressed in terms of its components along the x-, y-, and z-axes. What is the polarization of each of the components?

R1.27. What are the conditions for the sum of two linearly polarized sinusoidally time-varying vectors to be circularly polarized?

R1.28. What is the polarization for the general case of the sum of two sinusoidally time-varying linearly polarized vectors having arbitrary amplitudes, phase angles, and directions?

R1.29. Considering the second hand on your analog watch to be a vector, state its polarization. What is the frequency?

R1.30. Discuss the relevance of polarization in the reception of radio waves.

R1.31. Discuss the conversion of a complex number from rectangular form to exponential (or polar) form.

R1.32. How do you perform division of one complex number by another using their rectangular forms?

R1.33. What is a phasor?

R1.34. Is there any relationship between a phasor and a vector? Explain.

R1.35. Describe the phasor technique of adding two sinusoidal functions of time.

R1.36. Describe the phasor technique of solving a differential equation for the sinusoidal steady-state solution.

PROBLEMS

P1.1. A bug starts at a point and travels in successive segments of 1 m, $\frac{1}{2}$ m, $\frac{1}{4}$ m, $\frac{1}{8}$ m, and so on, with the initial segment in the northern direction, and making a 120° turn to the right at the end of each segment and halving the distance. **(a)** What is the total distance traveled by the bug? **(b)** Find the final position of the bug relative to its starting location in terms of distances to north and east. **(c)** Find the final position of the bug relative to its starting location in terms of straight-line distance and azimuthal angle, that is, angle measured from the north toward east.

P1.2. Three vectors satisfy the equations

$$\mathbf{A} + \mathbf{B} + \mathbf{C} = 4\mathbf{i}_1 + 2\mathbf{i}_2 + 5\mathbf{i}_3$$

$$\mathbf{A} + 2\mathbf{B} + 3\mathbf{C} = 9\mathbf{i}_1 + 2\mathbf{i}_2 + 10\mathbf{i}_3$$

$$2\mathbf{A} - \mathbf{B} + \mathbf{C} = 3\mathbf{i}_1 - \mathbf{i}_2$$

By writing a matrix equation for the 3 × 3 matrix

$$\begin{bmatrix} A_1 & A_2 & A_3 \\ B_1 & B_2 & B_3 \\ C_1 & C_2 & C_3 \end{bmatrix}$$

and solving it, obtain the vectors **A**, **B**, and **C**.

P1.3. Two vectors **A** and **B** originate from a common point. **(a)** Show that the area of the triangle having **A** and **B** as its sides is equal to $\frac{1}{2}|\mathbf{A} \times \mathbf{B}|$. **(b)** If $\mathbf{C} = \mathbf{B} - \mathbf{A}$ comprises the third side of the triangle, obtain using $\mathbf{C} \cdot \mathbf{C} = (\mathbf{B} - \mathbf{A}) \cdot (\mathbf{B} - \mathbf{A})$ the law of cosines relating **C** to **A**, **B**, and the angle α between **A** and **B**.

P1.4. Three vectors **A**, **B**, and **C** originating from a common point, lie in a plane. Obtain expressions for the following: **(a)** the vector from the common point to the midpoint of the line joining the tips of **A** and **B** and **(b)** the vector distance along **C** from the common point to the line joining the tips of **A** and **B**.

P1.5. Four vectors originating from a common point are given as follows:

$$\mathbf{A} = \mathbf{i}_1 + m\mathbf{i}_2 - 2\mathbf{i}_3$$

$$\mathbf{B} = m\mathbf{i}_1 - 2\mathbf{i}_2 + \mathbf{i}_3$$

$$\mathbf{C} = 2\mathbf{i}_1 + 8\mathbf{i}_2 - m\mathbf{i}_3$$

$$\mathbf{D} = m\mathbf{i}_1 - 2\mathbf{i}_2 + m\mathbf{i}_3$$

Find, if any, the value(s) of m for each of the following cases: **(a)** $\mathbf{A}$ is perpendicular to $\mathbf{B}$; **(b)** $\mathbf{A}$ is parallel to $\mathbf{C}$; **(c)** $\mathbf{B}$, $\mathbf{C}$, and $\mathbf{D}$ lie in the same plane; and **(d)** $\mathbf{B}$ is perpendicular to both $\mathbf{A}$ and $\mathbf{C}$.

P1.6. The tips of three vectors $\mathbf{A}$, $\mathbf{B}$, and $\mathbf{C}$, originating from a common point, define a plane. **(a)** Show that a fourth vector $\mathbf{D}$ is parallel to the plane if $\mathbf{D} \cdot (\mathbf{A} \times \mathbf{B} + \mathbf{B} \times \mathbf{C} + \mathbf{C} \times \mathbf{A}) = 0$ and perpendicular to the plane if $\mathbf{D} \times (\mathbf{A} \times \mathbf{B} + \mathbf{B} \times \mathbf{C} + \mathbf{C} \times \mathbf{A}) = \mathbf{0}$. **(b)** If $\mathbf{A} = \mathbf{i}_1$, $\mathbf{B} = 2\mathbf{i}_2$, and $\mathbf{C} = 2\mathbf{i}_3$, determine if $\mathbf{D}$ is parallel to the plane, perpendicular to the plane, or neither for each of the following values of $\mathbf{D}$: (i) $\mathbf{i}_1 + 2\mathbf{i}_2 + 2\mathbf{i}_3$; (ii) $2\mathbf{i}_1 + \mathbf{i}_2 + \mathbf{i}_3$; and (iii) $\mathbf{i}_1 - \mathbf{i}_2 - \mathbf{i}_3$.

P1.7. **(a)** Show that

$$\mathbf{A} \times (\mathbf{B} \times \mathbf{C}) = (\mathbf{A} \cdot \mathbf{C})\mathbf{B} - (\mathbf{A} \cdot \mathbf{B})\mathbf{C}$$

(b) Using the result of part **(a)**, show the following:

(i) $\mathbf{A} \times (\mathbf{B} \times \mathbf{C}) + \mathbf{B} \times (\mathbf{C} \times \mathbf{A}) + \mathbf{C} \times (\mathbf{A} \times \mathbf{B}) = \mathbf{0}$

(ii) $(\mathbf{A} \times \mathbf{B}) \cdot (\mathbf{B} \times \mathbf{C}) \times (\mathbf{C} \times \mathbf{A}) = (\mathbf{A} \times \mathbf{B} \cdot \mathbf{C})^2$

P1.8. Consider four points (x_1, y_1, z_1), (x_2, y_2, z_2), (x_3, y_3, z_3), and (x_4, y_4, z_4). Show that the center point (x_0, y_0, z_0) of the sphere passing through these points is given by the solution of the equation

$$2\begin{bmatrix} x_2 - x_1 & y_2 - y_1 & z_2 - z_1 \\ x_3 - x_1 & y_3 - y_1 & z_3 - z_1 \\ x_4 - x_1 & y_4 - y_1 & z_4 - z_1 \end{bmatrix}\begin{bmatrix} x_0 \\ y_0 \\ z_0 \end{bmatrix} = \begin{bmatrix} (x_2^2 + y_2^2 + z_2^2) - (x_1^2 + y_1^2 + z_1^2) \\ (x_3^2 + y_3^2 + z_3^2) - (x_1^2 + y_1^2 + z_1^2) \\ (x_4^2 + y_4^2 + z_4^2) - (x_1^2 + y_1^2 + z_1^2) \end{bmatrix}$$

Then find the center point of the sphere and its radius if the four points are (1, 1, 4), (3, 3, 2), (2, 3, 3), and (3, 2, 3).

P1.9. Three points $\mathbf{A}$, $\mathbf{B}$, and $\mathbf{C}$ are given by (1, 1, 1), (2, 3, 2), and (−1, 2, 3), respectively. **(a)** Find the perpendicular distance from $\mathbf{A}$ to the line from $\mathbf{B}$ to $\mathbf{C}$. **(b)** Determine if the point D (4, −3, −4) lies on the plane formed by $\mathbf{A}$, $\mathbf{B}$, and $\mathbf{C}$.

P1.10. Consider the plane surface given by

$$\frac{x}{a} + \frac{y}{b} + \frac{z}{c} = 1$$

Show that the distance from an arbitrary point (x_0, y_0, z_0) to the plane is given by

$$\frac{|bcx_0 + cay_0 + abz_0 - abc|}{\sqrt{b^2c^2 + c^2a^2 + a^2b^2}}$$

Then find the distance from the origin to the plane $x + y + z = 1$.

P1.11. Find an expression for the unit vector tangential to the curve given by the parametric equations $x = 2\cos t$, $y = \sin t$, and $z = t$ at an arbitrary point on the curve. Then obtain the unit vectors tangential to the curve at the following points: **(a)** (2, 0, 0); **(b)** $(1, \sqrt{3}/2, \pi/3)$; and **(c)** $(0, 1, \pi/2)$.

P1.12. Find the expression for the unit vector normal to the curve $x = y = z^2$ at the point (1, 1, −1) and having no component tangential to the curve $xy = 1$, $z = -1$.

P1.13. By considering two differential length vectors tangential to the surface $x^2 - y^2 = 9$ at the point (5, 4, 1), find the unit vector normal to the surface.

P1.14. Consider the differential surface lying on the plane $2x + y = 2$ and having the projection on the xz-plane the rectangular differential surface of sides dx and dz in the x- and z-directions, respectively. Obtain the expression for the vector $d\mathbf{S}$ associated with that surface.

P1.15. Three points are given by $A(1, \sqrt{3}, 1)$ in Cartesian coordinates, $B(2, 5\pi/6, 0)$ in cylindrical coordinates, and $C(\sqrt{3}, \pi/2, 2.5)$ in cylindrical coordinates. **(a)** Find the volume of the parallelepiped having the lines from the origin to the three points as one set of its contiguous edges. **(b)** Determine if the point $D(4, 2\pi/3, \pi/6)$ in spherical coordinates lies in the plane containing A, B, and C.

P1.16. In Fig. 1.30, a point of observation T on the surface of the earth is defined by a spherical coordinate system with the origin at the center of the earth. The spherical coordinates of T are its distance (r_0) from the center of the earth, its colatitude θ_T (90° minus the latitude, with south latitudes being negative), and its east longitude ϕ_T. N is the north pole. A point P is now defined by a coordinate system with origin at the point of observation T. The coordinates of P in this new coordinate system are the azimuthal angle α, which is the angle between the great circle paths TN and TR, where R is the projection of P onto the earth's surface, the elevation angle Δ in the plane TPR, and the range S. The colatitude and east longitude of R are θ_R and ϕ_R, respectively.

(a) Show that

$$\cos \eta = \sin \theta_T \sin \theta_R \cos (\phi_T - \phi_R) + \cos \theta_T \cos \theta_R$$

$$\cos \alpha = \frac{\cos \theta_R - \cos \eta \cos \theta_T}{\sin \eta \sin \theta_T}$$

(b) Find α, Δ, and S if T is at Urbana, Illinois (40.069° N latitude, 88.225° W longitude) and P represents a geostationary satellite parked above the equator at 50° W longitude. The earth radius r_0 is equal to 6370 km and the height h of the geostationary satellite above the earth's surface is equal to 35,800 km.

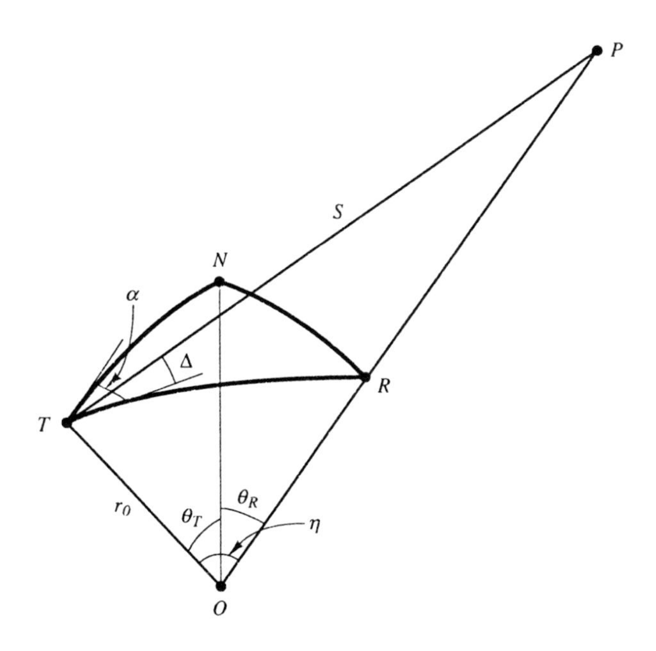

Figure 1.30. For Problem P1.16.

P1.17. Three unit vectors are given in cylindrical coordinates as follows: $\mathbf{A} = \mathbf{i}_r$ at $(1, 3\pi/2, 0)$, $\mathbf{B} = \mathbf{i}_\phi$ at $(2, \pi/3, -1)$, and $\mathbf{C} = \mathbf{i}_\phi$ at $(1, \pi/4, 2)$. Find: **(a)** $\mathbf{A} \cdot \mathbf{B}$; **(b)** $\mathbf{A} \cdot \mathbf{C}$; and **(c)** $\mathbf{A} \times \mathbf{B}$.

P1.18. Three unit vectors are given in spherical coordinates as follows: $\mathbf{A} = \mathbf{i}_r$ at $(1, \pi/6, \pi/2)$, $\mathbf{B} = \mathbf{i}_\theta$ at $(2, \pi/2, 3\pi/4)$, and $\mathbf{C} = \mathbf{i}_\phi$ at $(1, 3\pi/4, \pi/2)$. Find **(a)** $\mathbf{A} \cdot \mathbf{B}$; **(b)** $\mathbf{B} \cdot \mathbf{C}$; and **(c)** $\mathbf{A} \cdot \mathbf{B} \times \mathbf{C}$.

P1.19. Three vectors are given in spherical coordinates as follows: $\mathbf{A} = (3\mathbf{i}_r + \sqrt{3}\mathbf{i}_\theta - 2\mathbf{i}_\phi)$ at $(1, \pi/3, \pi/6)$, $\mathbf{B} = (3\mathbf{i}_r + \sqrt{3}\mathbf{i}_\theta - 2\mathbf{i}_\phi)$ at $(5.4, \pi/6, \pi/3)$, and $\mathbf{C} = (\mathbf{i}_r + \sqrt{3}\mathbf{i}_\theta - 2\sqrt{3}\mathbf{i}_\phi)$ at $(5.4, \pi/6, \pi/3)$. **(a)** Determine which of the vectors are equal in magnitude. **(b)** Determine which of the vectors are equal.

P1.20. For each of the following cases, find the unit vector tangential to the curve at the point specified: **(a)** $r^2 \sin 2\phi = 1$, $z = 0$ at $(1, \pi/4, 0)$ in cylindrical coordinates and **(b)** $r = \sqrt{1 + t^2}$, $\theta = \cot^{-1} t$, $\phi = \pi t$, $-\infty < t < \infty$, at $(\sqrt{2}, \pi/4, \pi)$ in spherical coordinates.

P1.21. An otherwise hemispherical dome of radius 2 m has a symmetrically situated hemispherical trough of radius 1 m, as shown by the cross-sectional view in Fig. 1.31. Assuming the origin to be at the center of the base of the dome, obtain the expression for the two-dimensional scalar field describing the height h of the dome, in each of the two coordinate systems: **(a)** rectangular (x, y) and **(b)** polar (r, ϕ).

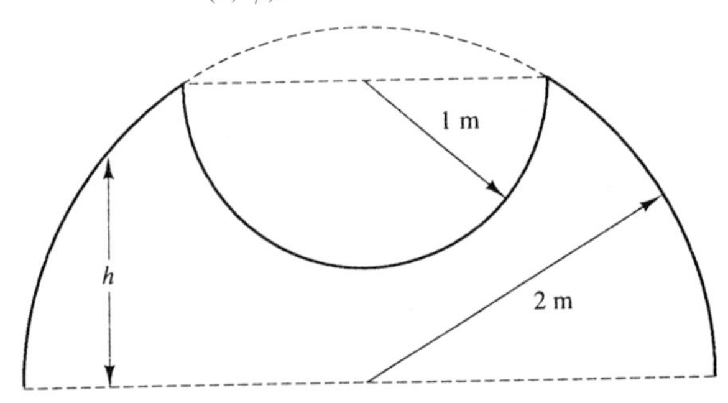

Figure 1.31. For Problem Pl.21.

P1.22. Assuming the origin to be the center of the earth and the z-axis to be passing through the poles, write vector functions for the following: **(a)** the force experienced by a mass m in the gravitational field of the earth (mass M) in the Cartesian coordinate system and **(b)** the linear velocity of points inside the earth due to its spin motion in the spherical coordinate system. Describe the constant magnitude surfaces and the direction lines for each case.

P1.23. Obtain the equations for the direction lines for the following vector fields: **(a)** $x\mathbf{i}_x + y\mathbf{i}_y + z\mathbf{i}_z$ and **(b)** $2xy\mathbf{i}_x + x^2\mathbf{i}_y$.

P1.24. Obtain the equations for the direction lines for the following vector fields in cylindrical coordinates: **(a)** $r^2\mathbf{i}_r$ and **(b)** $(\cos 2\phi\, \mathbf{i}_r - \sin 2\phi\, \mathbf{i}_\phi)$.

P1.25. Obtain the equations for the direction lines for the vector field $(2 \cos \theta\, \mathbf{i}_r - \sin \theta\, \mathbf{i}_\theta)$ in spherical coordinates.

P1.26. Given $f(z, t) = 10 \cos (3\pi \times 10^7 t + 0.1\pi z)$, draw on the same graph sketches of f versus z for $t = 0, \frac{1}{6} \times 10^{-7}$ s, and $\frac{1}{3} \times 10^{-7}$ s. Discuss the nature of the function.

P1.27. Given $f(z, t) = 10 \cos 0.1\pi z \cos 3\pi \times 10^7 t$, draw on the same graph sketches of f versus z for $t = 0, \frac{1}{12} \times 10^{-7}$ s, $\frac{1}{6} \times 10^{-7}$ s, $\frac{1}{4} \times 10^{-7}$ s, and $\frac{1}{3} \times 10^{-7}$ s. Discuss the nature of the function.

P1.28. Three sinusoidally time-varying, linearly polarized vector fields are given at a point by

$$\mathbf{F}_1 = (3\mathbf{i}_x + 2\mathbf{i}_y) \cos (2\pi \times 10^6 t + 30°)$$

$$\mathbf{F}_2 = (2\mathbf{i}_y + 5\mathbf{i}_z) \cos (2\pi \times 10^6 t + 30°)$$

$$\mathbf{F}_3 = (3\mathbf{i}_x + 4\mathbf{i}_y - 5\mathbf{i}_z) \cos (2\pi \times 10^6 t - 60°)$$

Determine the polarization of **(a)** $\mathbf{F}_1 - \mathbf{F}_2$; **(b)** $\mathbf{F}_1 + \mathbf{F}_2 + \mathbf{F}_3$; and **(c)** $\mathbf{F}_1 - \mathbf{F}_2 + \mathbf{F}_3$.

P1.29. Two sinusoidally time-varying, linearly polarized vector fields are given at a point by

$$\mathbf{F}_1 = (C\mathbf{i}_x - \mathbf{i}_y + C\mathbf{i}_z) \sin 2\pi \times 10^6 t$$

$$\mathbf{F}_2 = (C\mathbf{i}_x + C\mathbf{i}_y - \mathbf{i}_z) \cos 2\pi \times 10^6 t$$

where C is a constant. **(a)** What is the polarization of $(\mathbf{F}_1 + \mathbf{F}_2)$ for $C = 1$? **(b)** Find the value(s) of C for which the tip of $(\mathbf{F}_1 + \mathbf{F}_2)$ traces a straight line with time. **(c)** Find the value(s) of C for which the tip of $(\mathbf{F}_1 + \mathbf{F}_2)$ traces a circle with time.

P1.30. Consider an analog watch that keeps accurate time and assume the origin to be at the center of the dial, the x-axis to be passing through the "12" mark, and the y-axis to be passing through the "3" mark. **(a)** Write the expression for the time-varying unit vector directed along the hour hand of the watch. **(b)** Write the expression for the time-varying unit vector directed along the minute hand of the watch. **(c)** Obtain the specific expressions for these unit vectors when the hour hand is between the "7" and "8" marks and the minute hand is perpendicular to the hour hand.

P1.31. The components of a sinusoidally time-varying vector field are given at a point by

$$\mathbf{F}_x = 1 \cos \omega t$$

$$\mathbf{F}_y = 1 \cos (\omega t + 60°)$$

Show that the field is elliptically polarized in the xy-plane with the equation of the ellipse given by $x^2 - xy + y^2 = 3/4$. Further show that the axial ratio (ratio of the major axis to the minor axis) of the ellipse is $\sqrt{3}$ and the tilt angle (angle made by the major axis with the x-axis) is 45°.

P1.32. Find

$$\sqrt{\frac{(7 - j24)(5e^{j0.3\pi})}{(-3 + j4)} - 25\underline{/153.39°}}$$

and express in rectangular form.

P1.33. Find all values of $(-8 + j8\sqrt{3})^{3/4}$ and express in rectangular form.

P1.34. Find the value of $\ln(j10)$ having the smallest magnitude and express in polar form.

P1.35. Find the steady state solution for the differential equation

$$3 \times 10^{-6} \frac{dI}{dt} + 4I = 4 \cos^3 (10^6 t + 30°)$$

Hint: Expand $\cos^3 (10^6 t + 30°)$ in terms of cosine functions.

P1.36. Find the steady state solution for the integrodifferential equation

$$\frac{dI}{dt} + I + 5\int I\, dt = 10 \sin 3t$$

PC EXERCISES

PC1.1. Consider four points A, B, C, and D where the coordinates of each point are specified in any of the three coordinate systems. Write a program to compute and print: **(a)** the component of the vector from A to B along the vector from C to D; **(b)** the area of the triangle formed by A, B, and C; **(c)** the volume of the parallelepiped having AB, AC, and AD as three of its contiguous edges; and **(d)** the perpendicular distance from A to the plane containing B, C, and D, provided the three points do not lie along a straight line. Consider as input to the program the coordinates of each point and a code for each set of coordinates to specify the coordinate system.

PC1.2. Consider four vectors $\mathbf{A}_1$, $\mathbf{B}_1$, $\mathbf{C}_1$, and $\mathbf{D}_1$ originating from a common point P_1 specified in spherical coordinates and the vectors specified in terms of the unit vectors in spherical coordinates at P_1. Write a program to compute and print: **(a)** the spherical coordinates of the center point P_2 of the spherical surface passing through the tips of $\mathbf{A}_1$, $\mathbf{B}_1$, $\mathbf{C}_1$, and $\mathbf{D}_1$; and **(b)** the vectors $\mathbf{A}_2$, $\mathbf{B}_2$, $\mathbf{C}_2$, and $\mathbf{D}_2$ from P_2 to the tips of $\mathbf{A}_1$, $\mathbf{B}_1$, $\mathbf{C}_1$, and $\mathbf{D}_1$, respectively, in terms of the unit vectors in spherical coordinates at P_2. The program is to check if the tips of $\mathbf{A}_1$, $\mathbf{B}_1$, $\mathbf{C}_1$, and $\mathbf{D}_1$ do not lie in a plane and then only compute the quantities required.

PC1.3. Extend the program of PL1.2 for the case of elliptical polarization to compute and print: **(a)** the axial ratio (ratio of the major axis to the minor axis) of the ellipse; and **(b)** the tilt angle (angle made by the major axis with the x-axis) of the ellipse.

PC1.4. Consider the plotting of $F(x)$, the sum of the first $(n + 1)$ terms in the Fourier series

$$f(x) = a_0 + \sum_{m=0,1,2,\ldots}^{\infty} a_m \cos (m\pi x + \phi_m)$$

in the range $-1 \leq x \leq 1$, so that the input to the program are the values of a_0, n, the amplitudes a_m, $m = 1, 2, 3, \ldots, n$, the phase angles ϕ_m, $m = 1, 2, 3, \ldots, n$, and an odd number for p, the number of points to be plotted. Write a program to compute $F(x_i)$ for $x_i = -1 + 2i/(p - 1)$, $i = 0, 1, 2, \ldots, p - 1$, and plot $F(x_i)/|F(x_i)|_{\max}$ versus x.

PC1.5. Consider the sinusoidal steady-state solution of the integrodifferential equation

$$Ri + L\frac{di}{dt} + \frac{1}{C}\int i\,dt = V_0 \cos^n 2\pi ft$$

where n is a positive odd integer. Assuming the values of R, L, C, V_0, f, and n to be specified as input, write a program to compute and print for each frequency component of $i(t)$ in the steady state the value of the frequency, the amplitude of $i(t)$, and the phase angle of $i(t)$. Proceeding further, extend the program to plot on the same graph $\cos^n 2\pi ft_j$ and $i(t_j)/|i(t_j)|_{\max}$ versus t for $t_j = j/2fp$, $j = -p, -p + 1, \ldots, 0, 1, 2, \ldots, p - 1, p$, where p is a specified integer.

2

Fields and Materials

In Chapter 1 we provided a general introduction to vectors and fields. Basic to our study of the elements of engineering electromagnetics is an understanding of the concepts of electric and magnetic fields. Hence in this chapter we first study these concepts from considerations of the experimental laws of Coulomb and Ampère. We will learn how to compute the electric and magnetic fields due to charge and current distributions, respectively. Combining the electric and magnetic field concepts, we introduce the Lorentz force equation and use it to discuss charged particle motion in electric and magnetic fields. We devote the remainder of the chapter to materials.

Materials contain charged particles that respond to applied electric and magnetic fields to produce secondary fields. We will learn that there are three basic phenomena resulting from the interaction of the charged particles with the electric and magnetic fields. These are conduction, polarization, and magnetization. Although a given material may exhibit all three properties, it is classified as a conductor (including semiconductor), a dielectric, or a magnetic material, depending on whether conduction, polarization, or magnetization is the predominant phenomenon. Thus we introduce these materials one at a time and develop a set of relations known as the constitutive relations which enable us to avoid the necessity of explicitly taking into account the interaction of the charged particles with the fields.

2.1 THE ELECTRIC FIELD

From our study of Newton's law of gravitation in elementary physics, we are familiar with the gravitational force field associated with material bodies by virtue of their physical property known as *mass*. Newton's experiments showed that the gravitational force of attraction between two bodies of masses m_1 and m_2 separated by a distance R, which is very large compared to their sizes, is equal to $m_1 m_2 G/R^2$, where G is the universal constant of gravitation. In a similar manner, a force field known as the *electric field* is associated with bodies that are *charged*. A material body may be charged positively or negatively or may possess no net charge. In the International System of Units that we use throughout this book, the unit of charge is the coulomb, abbreviated C. The charge of an electron is -1.60219×10^{-19} C. Alternatively, approximately 6.24×10^{18} electrons represent a charge of one negative coulomb.

Coulomb's law

Experiments conducted by Coulomb showed that the following hold for two charged bodies that are very small in size compared to their separation so that they can be considered as *point charges*:

1. The magnitude of the force is proportional to the product of the magnitudes of the charges.
2. The magnitude of the force is inversely proportional to the square of the distance between the charges.
3. The magnitude of the force depends on the medium.
4. The direction of the force is along the line joining the charges.
5. Like charges repel; unlike charges attract.

For free space, the constant of proportionality is $1/4\pi\varepsilon_0$, where ε_0 is known as the permittivity of free space, having a value 8.854×10^{-12} or approximately equal to $10^{-9}/36\pi$. (For convenience, we shall use throughout this book a value of $10^{-9}/36\pi$ for ε_0.) Thus if we consider two point charges Q_1 C and Q_2 C separated R m in free space, as shown in Fig. 2.1, then the forces $\mathbf{F}_1$ and $\mathbf{F}_2$ experienced by Q_1 and Q_2, respectively, are given by

$$\boxed{\mathbf{F}_1 = \frac{Q_1 Q_2}{4\pi\varepsilon_0 R^2}\mathbf{i}_{21}} \tag{2.1a}$$

and

$$\boxed{\mathbf{F}_2 = \frac{Q_2 Q_1}{4\pi\varepsilon_0 R^2}\mathbf{i}_{12}} \tag{2.1b}$$

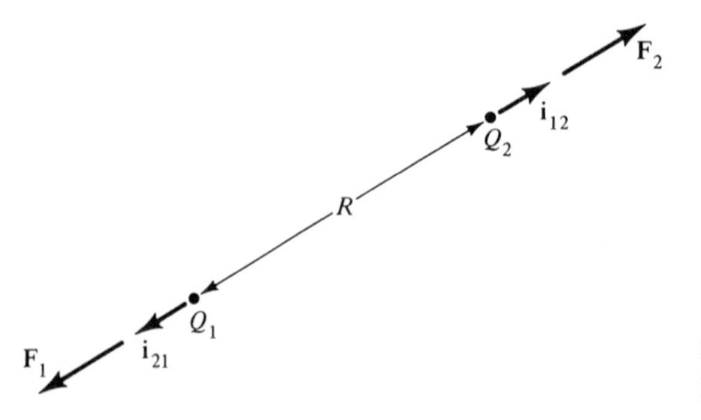

Figure 2.1. Forces experienced by two point charges Q_1 and Q_2.

where $\mathbf{i}_{21}$ and $\mathbf{i}_{12}$ are unit vectors along the line joining Q_1 and Q_2 as shown in Fig. 2.1. Equations (2.1a) and (2.1b) represent Coulomb's law. Since the units of force are newtons, we note that ε_0 has the units (coulomb)^2 per (newton-meter^2). These are commonly known as *farads per meter,* where a farad is (coulomb)^2 per newton-meter.

Electric field defined

In the case of the gravitational field of a material body, we define the gravitational field intensity as the force per unit mass experienced by a small test mass placed in that field. In a similar manner, the force per unit charge experienced by a small test charge placed in an electric field is known as the *electric field intensity,* denoted by the symbol **E**. Alternatively, if in a region of space, a test charge q experiences a force **F**, then the region is said to be characterized by an electric field of intensity **E** given by

$$\mathbf{E} = \frac{\mathbf{F}}{q} \tag{2.2}$$

The unit of electric field intensity is newton per coulomb, or more commonly volt per meter, where a volt is newton-meter per coulomb. The test charge should be so small that it does not alter the electric field in which it is placed. Ideally, **E** is defined in the limit that q tends to zero; that is,

$$\boxed{\mathbf{E} = \lim_{q \to 0} \frac{\mathbf{F}}{q}} \tag{2.3}$$

Equation (2.3) is the defining equation for the electric field intensity irrespective of the source of the electric field. Just as one body by virtue of its mass is the source of a gravitational field acting on other bodies by virtue of their masses, a charged body is the source of an electric field acting on other charged bodies. We will, however, learn in Chapter 3 that there exists another source for the electric field, namely, a time-varying magnetic field.

Electrostatic separation of minerals

Equation (2.2) or (2.3) tells us that the force experienced by a charged particle placed at a point in an external electric field is in the same direction as that of the electric field if the charge is positive, but opposite to that of the electric field if the charge is negative, as shown in Fig. 2.2. This phenomenon is the basis behind *electrostatic separation,* a process widely used in industry to separate minerals.[1] An example is illustrated in Fig. 2.3. Phosphate ore composed of granules of quartz and phosphate rock is dropped through a hopper onto a vibrating feeder. The friction between the two types of granules resulting from the vibration causes the quartz particles to be positively charged and the phosphate particles to be neg-

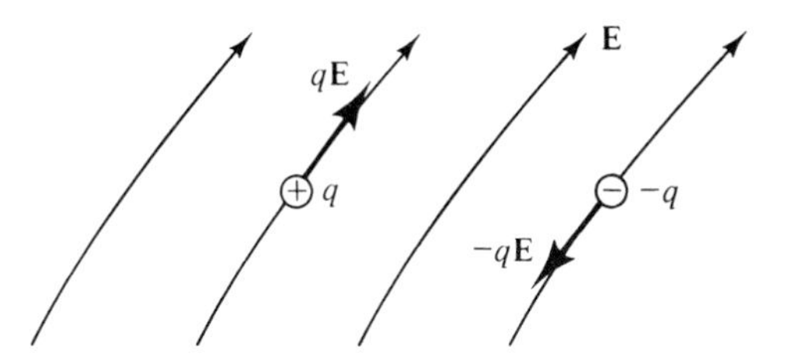

Figure 2.2. Forces experienced by positive and negative charges in an electric field.

[1] See, for example, A. D. Moore, ed., *Electrostatics and Its Applications* (New York: John Wiley & Sons, Inc., 1973), Chap. 10.

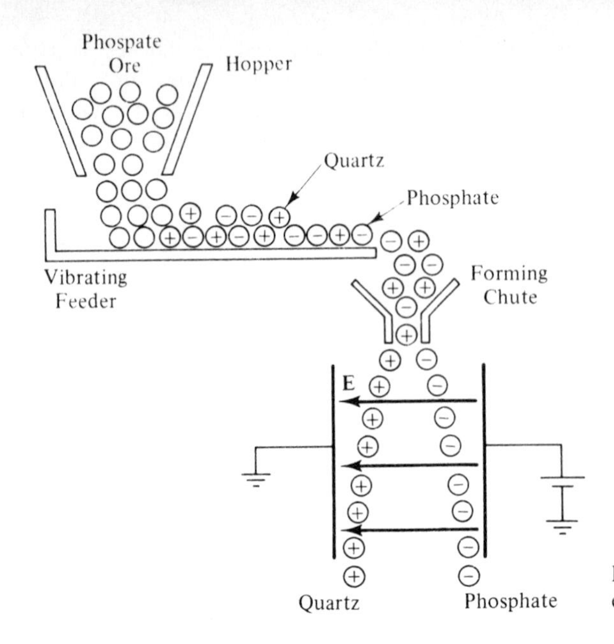

Figure 2.3. Example for illustrating electrostatic separation of minerals.

atively charged. The oppositely charged particles are then passed through a chute into the electric field region between two parallel plates, where they are separated and subsequently collected separately.

Cathode ray tube

There are many other devices based upon the electric force on a charged particle. We shall, however, discuss only one other application, the cathode ray tube, which is widely used in oscilloscopes, TV receivers, computer display terminals, and so on. The schematic of a cathode ray tube is shown in Fig. 2.4. Electrons are emitted from the heated cathode and are accelerated toward the anode by an electric field directed from the anode toward the cathode. After passing through the anode, they enter a region between two orthogonal pairs of parallel plates, one pair being horizontal and the other vertical. A voltage applied to the horizontal set of plates produces an electric field between the plates directed vertically, thereby deflecting the electrons vertically and imparting to them a verti-

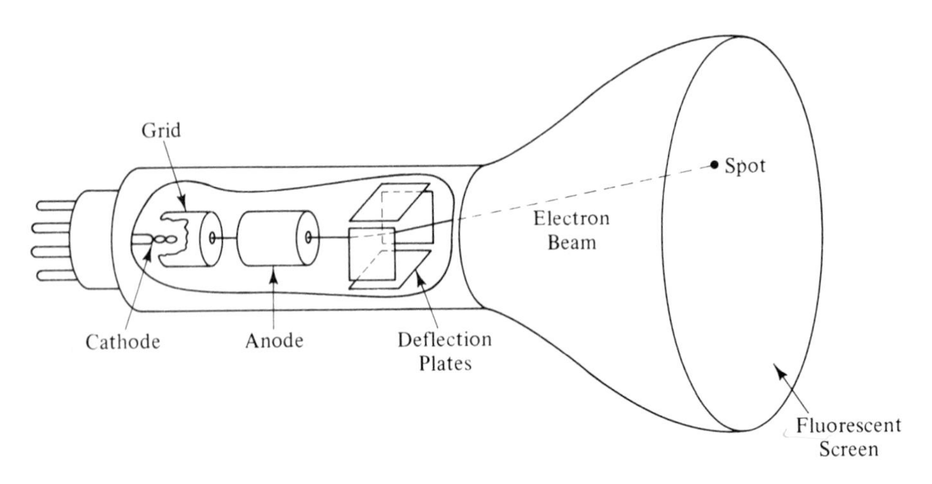

Figure 2.4. Schematic diagram of a cathode ray tube.

cal component of velocity as they leave the region between the plates. Similarly, a voltage applied to the vertical set of plates deflects the electrons horizontally sideways and imparts to them a sideways component of velocity as they leave the region between the plates. Thus, by varying the voltages applied to the two sets of plates, the electron beam can be made to strike the fluorescent screen and produce a bright spot at any point on the screen.

Electric field due to a point charge

Returning now to Coulomb's law and letting one of the two charges in Fig. 2.1 say, Q_2, be a small test charge q, we have

$$\mathbf{F}_2 = \frac{Q_1 q}{4\pi\varepsilon_0 R^2}\mathbf{i}_{12} \tag{2.4}$$

The electric field intensity $\mathbf{E}_2$ at the test charge due to the point charge Q_1 is then given by

$$\mathbf{E}_2 = \frac{\mathbf{F}_2}{q} = \frac{Q_1}{4\pi\varepsilon_0 R^2}\mathbf{i}_{12} \tag{2.5}$$

Generalizing this result by making R a variable, that is, by moving the test charge around in the medium, writing the expression for the force experienced by it, and dividing the force by the test charge, we obtain the electric field intensity $\mathbf{E}$ due to a point charge Q to be

$$\boxed{\mathbf{E} = \frac{Q}{4\pi\varepsilon_0 R^2}\mathbf{i}_R} \tag{2.6}$$

where R is the distance from the point charge to the point at which the field intensity is to be computed and $\mathbf{i}_R$ is the unit vector along the line joining the two points under consideration and directed away from the point charge. The electric field intensity due to a point charge is thus directed everywhere radially away from the point charge and its constant-magnitude surfaces are spherical surfaces centered at the point charge, as shown by the cross-sectional view in Fig. 2.5.

Using (2.6) in conjunction with (1.25) and (1.26), we can obtain the expression for the electric field intensity at a point $P(x, y, z)$ due to a point charge Q located at a point $P'(x', y', z')$. Thus noting that the vector $\mathbf{R}$ from P' to P is given by $[(x - x')\mathbf{i}_x + (y - y')\mathbf{i}_y + (z - z')\mathbf{i}_z]$ and the unit vector $\mathbf{i}_R$ is equal to $\mathbf{R}/R$, we obtain

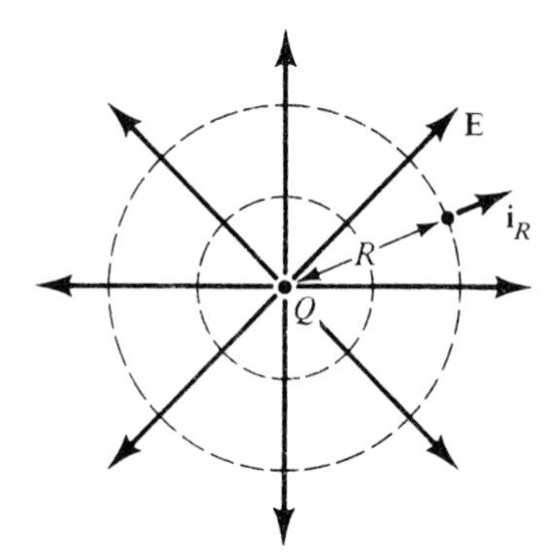

Figure 2.5. Direction lines and cross sections of constant-magnitude surfaces of electric field due to a point charge.

$$\boxed{\begin{aligned}\mathbf{E} &= \frac{Q\mathbf{R}}{4\pi\varepsilon_0 R^3}\\ &= \frac{Q}{4\pi\varepsilon_0}\frac{(x-x')\mathbf{i}_x + (y-y')\mathbf{i}_y + (z-z')\mathbf{i}_z}{[(x-x')^2 + (y-y')^2 + (z-z')^2]^{3/2}}\end{aligned}} \tag{2.7}$$

For a numerical example, if P and P' are (3, 1, 1) and (1, −1, 0), respectively, then

$$\mathbf{E} = \frac{Q}{108\pi\varepsilon_0}(2\mathbf{i}_x + 2\mathbf{i}_y + \mathbf{i}_z)$$

If we now have several point charges Q_1, Q_2, . . . , as shown in Fig. 2.6, the force experienced by a test charge situated at a point P is the vector sum of the forces experienced by the test charge due to the individual charges. It then follows that the electric field intensity at point P is the superposition of the electric field intensities due to the individual charges; that is,

$$\boxed{\mathbf{E} = \frac{Q_1}{4\pi\varepsilon_0 R_1^2}\mathbf{i}_{R_1} + \frac{Q_2}{4\pi\varepsilon_0 R_2^2}\mathbf{i}_{R_2} + \cdots + \frac{Q_n}{4\pi\varepsilon_0 R_n^2}\mathbf{i}_{R_n}} \tag{2.8}$$

We shall illustrate the application of (2.8) by means of an example involving two point charges.

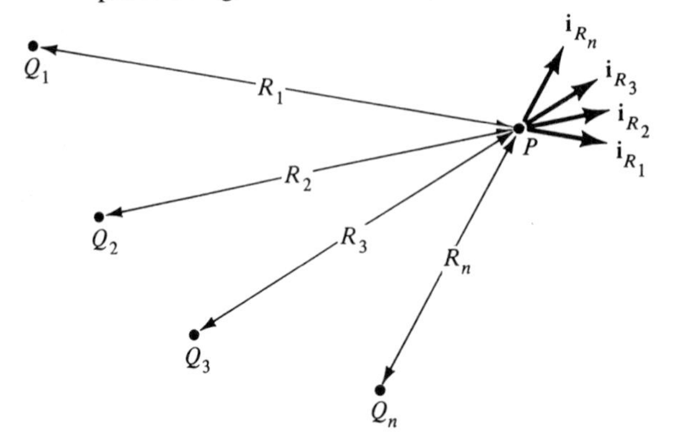

Figure 2.6. Collection of point charges and unit vectors along the directions of their electric fields at a point P.

Example 2.1

Let us consider two point charges $Q_1 = 8\pi\varepsilon_0$ C and $Q_2 = -4\pi\varepsilon_0$ C situated at (−1, 0, 0) and (1, 0, 0), respectively. We wish to (a) find the electric field intensity at the point (0, 0, 1) and (b) discuss computer generation of the direction line of **E** passing through that point.

(a) Using (2.8) and (2.7) in conjunction with the geometry in Fig. 2.7(a), we obtain

$$\begin{aligned}[\mathbf{E}]_{(0,0,1)} &= [\mathbf{E}_1]_{(0,0,1)} + [\mathbf{E}_2]_{(0,0,1)}\\ &= \frac{8\pi\varepsilon_0}{4\pi\varepsilon_0}\frac{(\mathbf{i}_x + \mathbf{i}_z)}{2^{3/2}} - \frac{4\pi\varepsilon_0}{4\pi\varepsilon_0}\frac{(-\mathbf{i}_x + \mathbf{i}_z)}{2^{3/2}} \\ &= 1.118\left(\frac{3\mathbf{i}_x + \mathbf{i}_z}{\sqrt{10}}\right)\end{aligned} \tag{2.9}$$

Note that the direction of **E** is given by the unit vector $(3\mathbf{i}_x + \mathbf{i}_z)/\sqrt{10}$ pointing away from the positive charge Q_1. The field vectors $\mathbf{E}_1$ and $\mathbf{E}_2$, and the resultant field vector **E**, are shown in Fig. 2.7(a).

(b) To discuss the computer generation of the direction line of **E**, we recall that a direction line is a curve such that at any given point on the curve, the field is tangential to the curve. For the case of the electric field, it is also the path followed by an infinitesimal test charge when released at a point on the curve. To obtain the direction line through the point (0, 0, 1), we go by an incremental distance from (0, 0, 1) along the direction of the electric field vector at that point to reach a new point, compute the field at the new point, and continue the process. Thus choosing for the purpose of illustration an incremental distance of 0.1 m and going along the unit vector $(3\mathbf{i}_x + \mathbf{i}_z)/\sqrt{10}$ from (0, 0, 1), we obtain the new point to be (0.095, 0, 1.032), as shown in Fig. 2.7(b). The electric field at this point is

$$[\mathbf{E}]_{(0.095,0,1.032)} = \frac{8\pi\varepsilon_0}{4\pi\varepsilon_0}\frac{(1.095\mathbf{i}_x + 1.032\mathbf{i}_z)}{(1.095^2 + 1.032^2)^{3/2}} - \frac{4\pi\varepsilon_0}{4\pi\varepsilon_0}\frac{(-0.905\mathbf{i}_x + 1.032\mathbf{i}_z)}{(0.905^2 + 1.032^2)^{3/2}}$$

$$= 1.015\left(\frac{4.8\mathbf{i}_x + \mathbf{i}_z}{4.9}\right) \tag{2.10}$$

Note that the direction of this electric field, which is along the unit vector $(4.8\mathbf{i}_x + \mathbf{i}_z)/4.9$, is slanted more toward the negative charge Q_2 than that of the electric field at the point (0, 0, 1), as shown in Fig. 2.7(b), indicating the swing of the direction line toward Q_2. The procedure is continued by going the incremental distance of 0.1 m from (0.095, 0, 1.032) along the unit vector $(4.8\mathbf{i}_x + \mathbf{i}_z)/4.9$ to the new point (0.193, 0, 1.052) and computing the field vector at that point, and so on, until the direction line is terminated close to the point charge Q_2. The same can be done to obtain the portion of the direction line from (0, 0, 1) toward the point charge Q_1, by moving opposite to **E**.

The listing of a PC program for generating the direction line in the manner just discussed and the output from a run of the program are given as PL 2.1. It can be seen from the output of the computer program that to compute the direction line, the test charge takes 17 steps toward Q_2 but only 14 steps back toward Q_1.

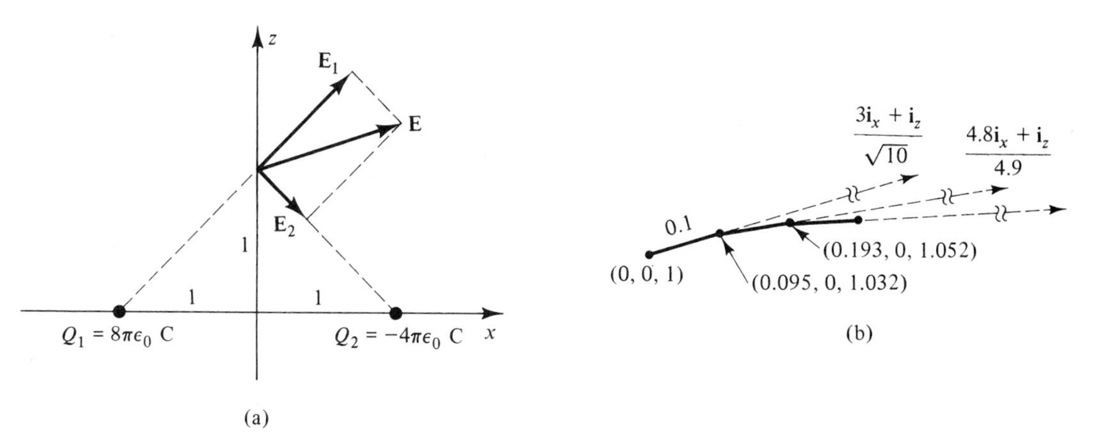

Figure 2.7. (a) Computation of the resultant electric field due to two point charges. (b) Generation of direction line of the electric field of (a).

PL 2.1. Program listing for generating the direction line of electric field due to the two point charges of Fig. 2.7(a) and the output from a run of the program.

```
100 '****************************************************
110 '* DIRECTION LINE OF ELECTRIC FIELD DUE TO A PAIR OF *
120 '* POSITIVE AND NEGATIVE POINT CHARGES               *
130 '****************************************************
140 K1=2:K2=-1:'* VALUES OF POINT CHARGES IN MULTIPLES OF
150 '   4*PI*PERMITTIVITY *
160 '* COMPUTE DIRECTION LINE TOWARD THE NEGATIVE CHARGE *
170 X=0:Z=1:DL=.1:N=0
180 GOSUB 290:N=N+1
190 IF ((X-1)^2+Z*Z)<DL*DL THEN 210
200 GOTO 180
210 GOSUB 290:PRINT "NUMBER OF STEPS =";N
220 '* COMPUTE DIRECTION LINE TOWARD THE POSITIVE CHARGE *
230 PRINT:X=0:Z=1:DL=-.1:N=0
240 GOSUB 290:N=N+1
250 IF ((X+1)^2+Z*Z)<DL*DL THEN 270
260 GOTO 240
270 GOSUB 290:PRINT "NUMBER OF STEPS =";N
280 END
290 '* SUBPROGRAM TO COMPUTE EX AND EZ AND STEP ALONG THE
300 '   DIRECTION LINE *
310 D1=((X+1)^2+Z*Z)^1.5:D2=((X-1)^2+Z*Z)^1.5
320 EX=K1*(X+1)/D1+K2*(X-1)/D2:EZ=(K1/D1+K2/D2)*Z
330 E=SQR(EX*EX+EZ*EZ):'* MAGNITUDE OF E *
340 UX=EX/E:UZ=EZ/E:'* COMPONENTS OF UNIT VECTOR ALONG THE
350 '   FIELD *
360 PRINT USING "X=#.###";X;
370 PRINT USING "  Z=#.###";Z;
380 PRINT USING "  E=###.###";E;
390 PRINT USING "  UX=##.###";UX;
400 PRINT USING "  UZ=##.###";UZ
410 X=X+DL*UX:Z=Z+DL*UZ
420 RETURN

RUN
X=0.000  Z=1.000  E=  1.118  UX= 0.949  UZ= 0.316
X=0.095  Z=1.032  E=  1.015  UX= 0.979  UZ= 0.204
X=0.193  Z=1.052  E=  0.942  UX= 0.997  UZ= 0.076
X=0.292  Z=1.060  E=  0.898  UX= 0.998  UZ=-0.065
X=0.392  Z=1.053  E=  0.882  UX= 0.977  UZ=-0.215
X=0.490  Z=1.032  E=  0.898  UX= 0.930  UZ=-0.368
X=0.583  Z=0.995  E=  0.951  UX= 0.858  UZ=-0.513
X=0.669  Z=0.944  E=  1.051  UX= 0.766  UZ=-0.643
X=0.745  Z=0.879  E=  1.212  UX= 0.660  UZ=-0.751
X=0.811  Z=0.804  E=  1.459  UX= 0.548  UZ=-0.836
X=0.866  Z=0.721  E=  1.837  UX= 0.439  UZ=-0.899
X=0.910  Z=0.631  E=  2.426  UX= 0.337  UZ=-0.942
X=0.944  Z=0.536  E=  3.391  UX= 0.246  UZ=-0.969
X=0.968  Z=0.440  E=  5.100  UX= 0.167  UZ=-0.986
X=0.985  Z=0.341  E=  8.537  UX= 0.101  UZ=-0.995
X=0.995  Z=0.241  E= 17.101  UX= 0.049  UZ=-0.999
X=1.000  Z=0.142  E= 49.846  UX= 0.010  UZ=-1.000
X=1.001  Z=0.042  E=577.540  UX=-0.023  UZ=-1.000
NUMBER OF STEPS = 17

X=0.000  Z=1.000  E=  1.118  UX= 0.949  UZ= 0.316
X=-.095  Z=0.968  E=  1.243  UX= 0.908  UZ= 0.420
X=-.186  Z=0.926  E=  1.411  UX= 0.862  UZ= 0.507
X=-.272  Z=0.876  E=  1.634  UX= 0.815  UZ= 0.580
X=-.353  Z=0.818  E=  1.931  UX= 0.768  UZ= 0.640
X=-.430  Z=0.754  E=  2.333  UX= 0.724  UZ= 0.689
```

PL2.1 (continued)

```
X=-.503   Z=0.685   E=   2.888   UX= 0.684   UZ= 0.730
X=-.571   Z=0.612   E=   3.681   UX= 0.648   UZ= 0.762
X=-.636   Z=0.536   E=   4.871   UX= 0.616   UZ= 0.788
X=-.697   Z=0.457   E=   6.769   UX= 0.590   UZ= 0.808
X=-.756   Z=0.376   E= 10.074   UX= 0.568   UZ= 0.823
X=-.813   Z=0.294   E= 16.616   UX= 0.551   UZ= 0.835
X=-.868   Z=0.210   E= 32.588   UX= 0.538   UZ= 0.843
X=-.922   Z=0.126   E= 91.176   UX= 0.529   UZ= 0.849
X=-.975   Z=0.041   E=860.610   UX= 0.522   UZ= 0.853
NUMBER OF STEPS = 14
```

In the simple procedure employed in Example 2.1, there is a (cumulative) error associated with each step. This error can be reduced by employing a modified procedure as follows: Instead of moving the test charge by 0.1 m from its current location, say point A, to a new location, say point B, along the direction of the electric field at point A, it is moved by 0.1 m to a point C along a direction that bisects the directions of the fields at points A and B. Computer-plotted field maps in the xz-plane for pairs of point charges Q_1 and Q_2 located at $(-1, 0)$ and $(1, 0)$, respectively, by using this modified procedure are shown in Fig. 2.8. In each map, plotting of a direction line begins at one of the point charges and terminates when the line reaches to within a distance of 0.1 m from the second point charge, or if it goes beyond a specified rectangular region. In

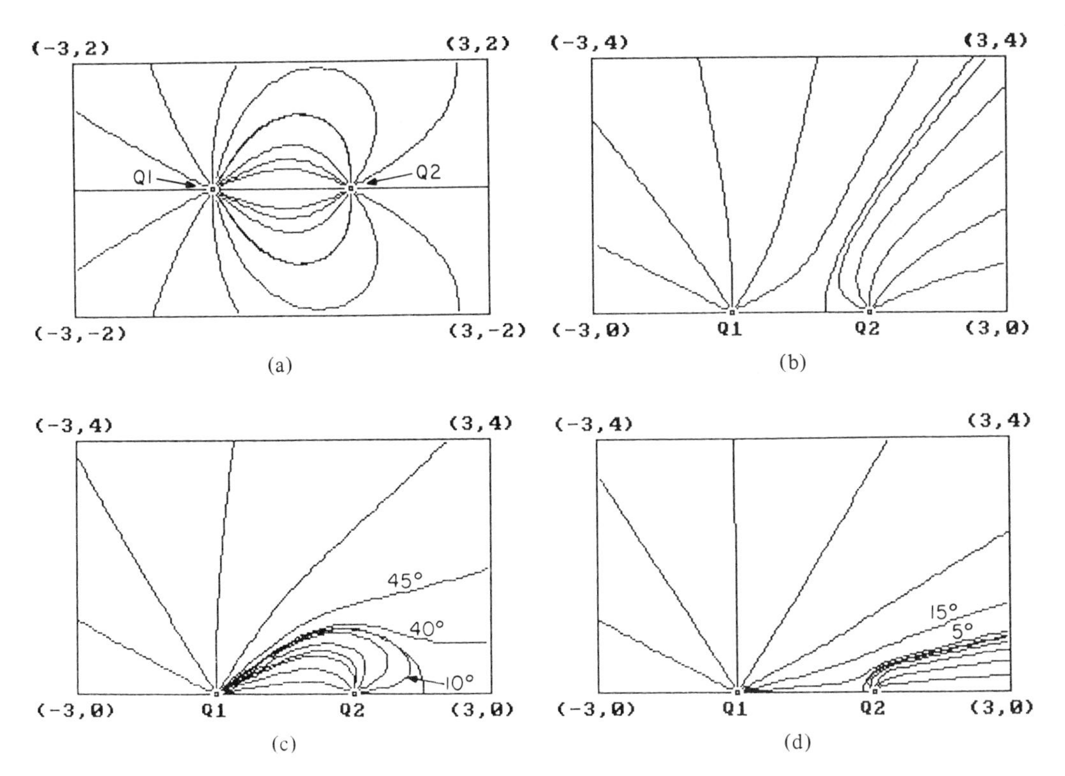

Figure 2.8. Computer-generated maps of direction lines of electric field for pairs of point charges Q_1 and Q_2 at $(-1, 0)$ and $(1, 0)$, respectively, in the xz-plane. (a) $Q_1 = 2Q$, $Q_2 = -Q$; (b) $Q_1 = 4Q$, $Q_2 = Q$; (c) $Q_1 = 9Q$, $Q_2 = -Q$; and (d) $Q_1 = 81Q$, $Q_2 = Q$.

this manner, direction lines beginning at points around each point charge and at 30° intervals on a circle of radius 0.1 m are plotted, with the 0° angle corresponding to the $+x$-direction.

For Fig. 2.8(a), $Q_1 = 2Q$ at $(-1, 0)$ and $Q_2 = -Q$ at $(1, 0)$. The rectangular region is one having corners at $(-3, 2)$, $(3, 2)$, $(3, -2)$, and $(-3, -2)$. The direction lines beginning at each point charge either end on the second charge or go out of the boundary of the rectangular region. For Fig. 2.8(b)–(d), region of map is rectangle having corners at $(-3, 4)$, $(3, 4)$, $(3, 0)$, and $(-3, 0)$, taking advantage of the symmetry of the field map about the axis through the charges, illustrated in Fig. 2.8(a). For Fig. 2.8(b), $Q_1 = 4Q$ at $(-1, 0)$ and $Q_2 = Q$ at $(1, 0)$. A zero field point exists within the region at $(\frac{1}{3}, 0)$, between the two charges. For direction lines passing through this point, the test charge gets trapped at that point and the procedure used is to untrap it by displacing it by 0.01 perpendicular to the axis and continue plotting of the line until it terminates at a point on the boundary of the region. For Fig. 2.8(c), $Q_1 = 9Q$ at $(-1, 0)$ and $Q_2 = -Q$ at $(1, 0)$. A zero field point exists within the region at $(2, 0)$, to the right of Q_2. Also, three additional field lines are shown plotted. Two of these are from Q_1 at angles of 40 and 45° and the third is from Q_2 at 10°. For Fig. 2.8(d), $Q_1 = 81Q$ at $(-1, 0)$ and $Q_2 = Q$ at $(1, 0)$. A zero field point exists just to the left of Q_2 between the two charges. The map also includes two additional field lines originating from Q_1 at 5 and 15° angles.

Types of charge distributions

The foregoing illustration of the computation of the electric field intensity due to two point charges can be extended to the computation of the field intensity due to continuous charge distributions. Continuous charge distributions are of three types: line charges, surface charges, and volume charges, depending on whether the charge is distributed along a line like chalk powder along a thin line drawn on the blackboard, on a surface like chalk powder on the erasing surface of a blackboard eraser, or in a volume like chalk powder in the chalk itself. The corresponding charge densities are the line charge density ρ_L, the surface charge density ρ_S, and the volume charge density ρ, having the units of charge per unit length (coulombs per meter), charge per unit area (coulombs per meter squared), and charge per unit volume (coulombs per meter cubed), respectively. The technique of finding the electric field intensity due to a given charge distribution consists of dividing the region of the charge distribution into a number of differential lengths, surfaces, or volumes, depending on the type of the distribution, considering the charge in each differential element to be a point charge, and using superposition. We shall illustrate the procedure by means of two examples.

Example 2.2

Ring charge

Charge Q C is distributed with uniform density along a circular ring of radius a lying in the xy-plane and having its center at the origin, as shown in Fig. 2.9. We wish to find the electric field intensity at a point on the z-axis.

Let us divide the ring into a large number of segments so that the charge in each segment can be considered to be a point charge located at the center of the segment. Let the segments be of equal length and numbered $1, 2, 3, \ldots, 2n$, as shown in Fig. 2.9. Then the electric field intensity at the point $(0, 0, z)$ due to the charge in the jth segment is given by

$$\mathbf{E}_j = \frac{Q_j}{4\pi\varepsilon_0 R_j^2}\mathbf{i}_{R_j}$$

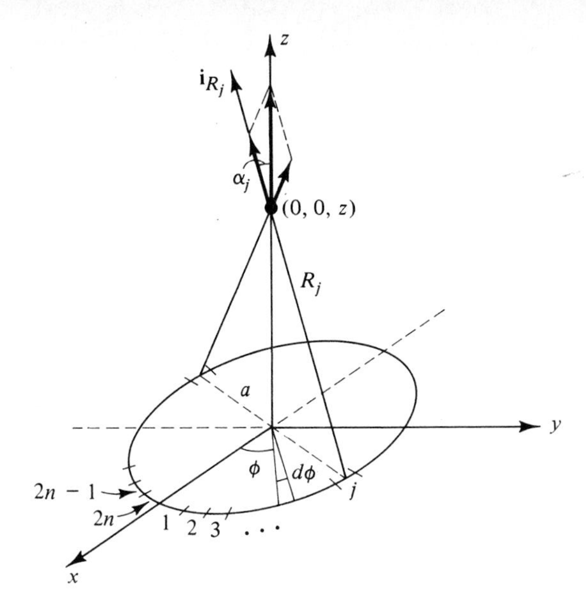

Figure 2.9. Determination of electric field due to a circular ring of charge of uniform density.

where Q_j is the charge in the jth segment and R_j and $\mathbf{i}_{R_j}$ are as shown in the figure. Since the charge is uniformly distributed, Q_j is the same for all j and is equal to the charge density times the length of the segment. Thus

$$Q_j = \left(\frac{Q}{2\pi a}\right)\left(\frac{2\pi a}{2n}\right) = \frac{Q}{2n}$$

Furthermore, since the point (0, 0, z) is along the axis of the ring, it is equidistant from all segments so that R_j is the same for all j. Hence

$$R_j = \sqrt{z^2 + a^2}$$

Now, from symmetry considerations, we note that for every segment 1, 2, 3, . . . , n, there is a corresponding segment diametrically opposite to it in the other half of the ring such that the electric field intensity due to the two segments together is directed along the z-axis, as illustrated for segment j in Fig. 2.9. Hence to find $\mathbf{E}$ due to the entire ring charge, it is sufficient if we consider the z-component of $\mathbf{E}_j$, multiply it by 2, and sum from $j = 1$ to $j = n$. Thus we obtain the required electric field intensity to be

$$\begin{aligned}
[\mathbf{E}]_{(0,0,z)} &= \sum_{j=1}^{n} \frac{2Q_j}{4\pi\varepsilon_0 R_j^2}(\mathbf{i}_{R_j} \cdot \mathbf{i}_z)\mathbf{i}_z \\
&= \sum_{j=1}^{n} \frac{Q_j}{2\pi\varepsilon_0 R_j^2} \cos\alpha_j \mathbf{i}_z \\
&= \sum_{j=1}^{n} \frac{Q_j z}{2\pi\varepsilon_0 R_j^3}\mathbf{i}_z \qquad (2.11) \\
&= \sum_{j=1}^{n} \frac{Qz}{4\pi\varepsilon_0 n(z^2 + a^2)^{3/2}}\mathbf{i}_z \\
&= \frac{Qz}{4\pi\varepsilon_0 (z^2 + a^2)^{3/2}}\mathbf{i}_z
\end{aligned}$$

Note that $[\mathbf{E}]_{(0,0,z)}$ is directed in the $+z$-direction above the origin ($z > 0$) and in the $-z$-direction below the origin ($z < 0$), as to be expected.

Alternative to the summation procedure just employed, we can obtain E_z at $(0, 0, z)$ by setting up an integral expression and evaluating it. Thus considering a differential length $a\, d\phi$ of the ring charge at the point $(a, \phi, 0)$, as shown in Fig. 2.9, and making use of symmetry considerations as discussed in connection with the summation procedure, we obtain

$$
\begin{aligned}
[E_z]_{(0,0,z)} &= \int_{\phi=0}^{\pi} \frac{2(Q/2\pi a)a\, d\phi}{4\pi\varepsilon_0(a^2 + z^2)} \frac{z}{(a^2 + z^2)^{1/2}} \\
&= \frac{Qz}{4\pi^2\varepsilon_0(a^2 + z^2)^{3/2}} \int_{\phi=0}^{\pi} d\phi \qquad (2.12) \\
&= \frac{Qz}{4\pi\varepsilon_0(a^2 + z^2)^{3/2}}
\end{aligned}
$$

For this example, the two results given by (2.11) and (2.12) are identical. In general, however, the summation procedure gives an approximate result for any finite value of n, and the integral gives the exact result, provided it can be evaluated in closed form. The summation procedure is, however, more illuminating as to the application of superposition and is convenient for computer solution.

Example 2.3

Infinite plane sheet of charge

Let us consider an infinite plane sheet of charge in the xy-plane with uniform surface charge density ρ_{S0} C/m² and find the electric field intensity due to it everywhere.

Let us first consider a point $(0, 0, z)$ on the z-axis, as shown in Fig. 2.10(a). Then the solution can be carried out by dividing the sheet into a number of infinitesimal surfaces in Cartesian coordinates and using superposition. An alternate procedure consists of using the result of Example 2.2 by dividing the sheet into concentric rings centered at the origin and each having infinitesimal width dr in the radial direction. One such ring having the arbitrary radius r and width dr is shown in

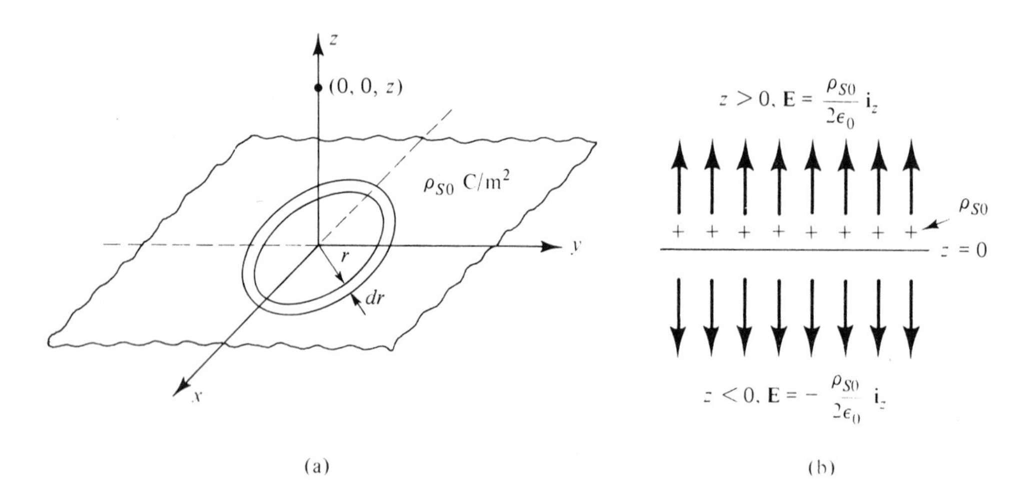

Figure 2.10. (a) Determination of electric field due to an infinite plane sheet of uniform surface charge density ρ_{S0} C/m². (b) Electric field due to the infinite plane sheet of charge of (a).

Fig. 2.10(a). The charge in that ring is equal to $\rho_{S0}(2\pi r\,dr)$, the product of the uniform surface charge density and the area of the ring. According to the result obtained in Example 2.2, the electric field intensity at (0, 0, z) due to this ring charge is given by

$$[d\mathbf{E}]_{(0,0,z)} = \frac{(\rho_{S0}2\pi r\,dr)z}{4\pi\varepsilon_0(r^2 + z^2)^{3/2}}\mathbf{i}_z$$

The electric field intensity due to the entire sheet of charge is then given by

$$\begin{aligned}[\mathbf{E}]_{(0,0,z)} &= \int_{r=0}^{\infty} [d\mathbf{E}]_{(0,0,z)} \\ &= \int_{r=0}^{\infty} \frac{\rho_{S0}rz\,dr}{2\varepsilon_0(r^2 + z^2)^{3/2}}\mathbf{i}_z \\ &= \frac{\rho_{S0}z}{2\varepsilon_0}\left[-\frac{1}{\sqrt{r^2 + z^2}}\right]_{r=0}^{\infty} \\ &= \frac{\rho_{S0}z}{2\varepsilon_0|z|}\mathbf{i}_z\end{aligned}$$

Finally, since the charge density is uniform and the origin of the coordinate system can be chosen anywhere on the infinite sheet without changing the geometry, this result is valid everywhere. Thus the required electric field intensity is

$$\mathbf{E} = \pm\frac{\rho_{S0}}{2\varepsilon_0}\mathbf{i}_z \qquad \text{for} \quad z \gtrless 0 \tag{2.13}$$

which has the magnitude $\rho_{S0}/2\varepsilon_0$ everywhere and directed normally away from the sheet, as shown by the cross-sectional view in Fig. 2.10(b). Defining $\mathbf{i}_n$ to be the unit normal vector directed away from the sheet, that is,

$$\mathbf{i}_n = \pm\mathbf{i}_z \qquad \text{for} \quad z \gtrless 0$$

we have

$$\boxed{\mathbf{E} = \frac{\rho_{S0}}{2\varepsilon_0}\mathbf{i}_n} \tag{2.14}$$

K2.1. Coulomb's law; Electric field intensity; **E** due to a point charge; Computation of **E** due to charge distributions; **E** due to an infinite plane sheet of charge of uniform density.

D2.1. Point charges, each of value $\sqrt{4\pi\varepsilon_0}$ C, are located at the vertices of an n-sided regular polygon circumscribed by a circle of radius a. Find the electric force on each charge for **(a)** $n = 3$; **(b)** $n = 4$; and **(c)** $n = 6$.
Ans. **(a)** $0.577/a^2$ N; **(b)** $0.957/a^2$ N; **(c)** $1.827/a^2$ N; all directed away from the center of the polygon.

D2.2. In Fig. 2.7, let the point charges be $Q_1 = 8\pi\varepsilon_0$ C at $(-1, 0, 0)$ and $Q_2 = 4\pi\varepsilon_0$ C at (1, 0, 0). Find the following: **(a)** **E** at (0, 0, 1); **(b)** the coordinates of point at the end of the second step; and **(c)** unit vector along **E** at the point computed in **(b)**.
Ans. **(a)** $(0.353\mathbf{i}_x + 1.061\mathbf{i}_z)$; **(b)** (0.060, 0, 1.191) **(c)** $(0.264\mathbf{i}_x + 0.965\mathbf{i}_z)$

D2.3. In Fig. 2.9, let there be a second ring of charge $-Q$, uniformly distributed along a circle of radius a, having its center at (0, 0, $2a$) and lying parallel to the xy-

plane. Find **E** due to the two rings of charge together at each of the following points: **(a)** (0, 0, 0); **(b)** (0, 0, a); and **(c)** (0, 0, $3a$).
Ans. **(a)** $(0.0142Q/\varepsilon_0 a^2)\mathbf{i}_z$; **(b)** $(0.0563Q/\varepsilon_0 a^2)\mathbf{i}_z$; **(c)** $(-0.0206Q/\varepsilon_0 a^2)\mathbf{i}_z$

D2.4. Infinite plane sheets of charge lie in the $z = 0$, $z = 2$, and $z = 4$ planes with uniform surface charge densities ρ_{S1}, ρ_{S2}, and ρ_{S3}, respectively. Given that the resulting electric field intensities at the points (3, 5, 1), (1, −2, 3), and (3, 4, 5) are **0**, $6\mathbf{i}_z$, and $4\mathbf{i}_z$ V/m, respectively, find the following: **(a)** ρ_{S1}; **(b)** ρ_{S2}; **(c)** ρ_{S3}; and **(d)** **E** at (−2, 1, −6).
Ans. **(a)** $4\varepsilon_0$ C/m²; **(b)** $6\varepsilon_0$ C/m²; **(c)** $-2\varepsilon_0$ C/m²; **(d)** $-4\mathbf{i}_z$ V/m

2.2 THE MAGNETIC FIELD

In the preceding section we presented an experimental law known as Coulomb's law having to do with the electric force associated with two charged bodies, and we introduced the electric field intensity vector as the force per unit charge experienced by a test charge placed in the electric field. In this section we present another experimental law known as *Ampère's law of force,* analogous to Coulomb's law, and use it to introduce the magnetic field concept.

Ampère's law of force

Ampère's law of force is concerned with magnetic forces associated with two loops of wire carrying currents by virtue of motion of charges in the loops. Figure 2.11 shows two loops of wire carrying currents I_1 and I_2 and each of which is divided into a large number of elements having infinitesimal lengths. The total force experienced by a loop is the vector sum of forces experienced by the infinitesimal current elements comprising the loop. The force experienced by each of these current elements is the vector sum of the forces exerted on it by the infinitesimal current elements comprising the second loop. If the number of elements in loop 1 is m and the number of elements in loop 2 is n, then there are $m \times n$ pairs of elements. A pair of magnetic forces is associated with each pair of these elements just as a pair of electric forces is associated with a pair of point charges. Thus if we consider an element $d\mathbf{l}_1$ in loop 1 and an element $d\mathbf{l}_2$ in loop 2, then the forces $d\mathbf{F}_1$ and $d\mathbf{F}_2$ experienced by the elements $d\mathbf{l}_1$ and $d\mathbf{l}_2$, respectively, are given by

$$d\mathbf{F}_1 = I_1\, d\mathbf{l}_1 \times \left(\frac{kI_2\, d\mathbf{l}_2 \times \mathbf{i}_{21}}{R^2}\right) \tag{2.15a}$$

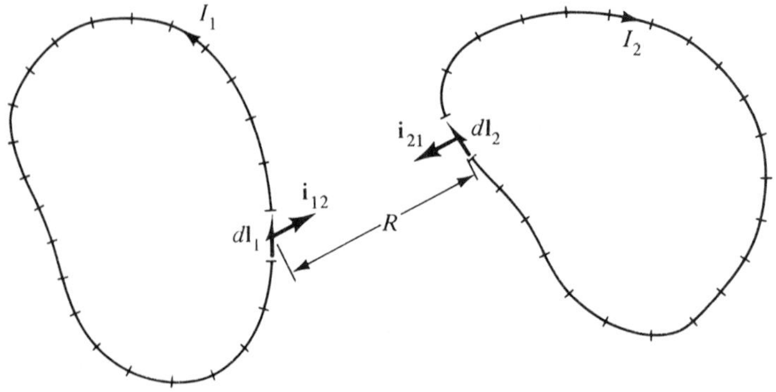

Figure 2.11. Two loops of wire carrying currents I_1 and I_2.

$$d\mathbf{F}_2 = I_2\, d\mathbf{l}_2 \times \left(\frac{kI_1\, d\mathbf{l}_1 \times \mathbf{i}_{12}}{R^2}\right) \tag{2.15b}$$

where $\mathbf{i}_{21}$ and $\mathbf{i}_{12}$ are unit vectors along the line joining the two current elements, R is the distance between them, and k is a constant of proportionality that depends on the medium. For free space, k is equal to $\mu_0/4\pi$ where μ_0 is known as the permeability of free space, having a value $4\pi \times 10^{-7}$. From (2.15a) or (2.15b), we note that the units of μ_0 are newtons per ampere squared. These are commonly known as *henrys per meter,* where a henry is a newton-meter per ampere squared.

Equations (2.15a) and (2.15b) represent Ampère's force law as applied to a pair of current elements. Some of the features evident from these equations are as follows:

1. The magnitude of the force is proportional to the product of the two currents and to the product of the lengths of the two current elements.
2. The magnitude of the force is inversely proportional to the square of the distance between the current elements.
3. To determine the direction of the force acting on the current element $d\mathbf{l}_1$, we first find the cross product $d\mathbf{l}_2 \times \mathbf{i}_{21}$ and then cross $d\mathbf{l}_1$ into the resulting vector. Similarly, to determine the direction of the force acting on the current element $d\mathbf{l}_2$, we first find the cross product $d\mathbf{l}_1 \times \mathbf{i}_{12}$ and then cross $d\mathbf{l}_2$ into the resulting vector. For the general case of arbitrary orientations of $d\mathbf{l}_1$ and $d\mathbf{l}_2$, these operations yield $d\mathbf{F}_{12}$ and $d\mathbf{F}_{21}$ which are not equal and opposite. To illustrate by means of an example, let us consider $I_1\, d\mathbf{l}_1 = I_1\, dx\, \mathbf{i}_x$ at (1, 0, 0) and $I_2\, d\mathbf{l}_2 = I_2\, dy\, \mathbf{i}_y$ at (0, 1, 0). Then

$$\mathbf{i}_{12} = -\mathbf{i}_{21} = \frac{1}{\sqrt{2}}(-\mathbf{i}_x + \mathbf{i}_y);\; R = \sqrt{2}$$

$$I_2\, d\mathbf{l}_2 \times \mathbf{i}_{21} = (I_2\, dy\, \mathbf{i}_y) \times \frac{1}{\sqrt{2}}(\mathbf{i}_x - \mathbf{i}_y) = -\frac{I_2}{\sqrt{2}}\, dy\, \mathbf{i}_z$$

$$d\mathbf{F}_1 = (I_1\, dx\, \mathbf{i}_x) \times \left(\frac{-kI_2\, dy\, \mathbf{i}_z}{2\sqrt{2}}\right) = \frac{kI_1I_2}{2\sqrt{2}}\, dx\, dy\, \mathbf{i}_y$$

$$I_1\, d\mathbf{l}_1 \times \mathbf{i}_{12} = (I_1\, dx\, \mathbf{i}_x) \times \frac{1}{\sqrt{2}}(-\mathbf{i}_x + \mathbf{i}_y) = \frac{I_1}{\sqrt{2}}\, dx\, \mathbf{i}_z$$

$$d\mathbf{F}_2 = (I_2\, dy\, \mathbf{i}_y) \times \left(\frac{kI_1\, dx\, \mathbf{i}_z}{2\sqrt{2}}\right) = \frac{kI_1I_2}{2\sqrt{2}}\, dx\, dy\, \mathbf{i}_x$$

Thus $d\mathbf{F}_2 \neq -d\mathbf{F}_1$. This is not a violation of Newton's third law since isolated current elements do not exist without sources and sinks of charges at their ends. Newton's third law, however, must and does hold for complete current loops.

Magnetic flux density

The forms of (2.15a) and (2.15b) suggest that each current element is acted upon by a field which is due to the other current element. By definition, this field is the magnetic field and is characterized by a quantity known as the *magnetic flux density vector,* denoted by the symbol $\mathbf{B}$. Thus we note from (2.15b) that the magnetic flux density at the element $d\mathbf{l}_2$ due to the element $d\mathbf{l}_1$ is given by

$$\mathbf{B}_1 = \frac{\mu_0}{4\pi} \frac{I_1\, d\mathbf{l}_1 \times \mathbf{i}_{12}}{R^2} \tag{2.16}$$

and that this flux density acting upon $d\mathbf{l}_2$ results in a force on it given by

$$d\mathbf{F}_2 = I_2\, d\mathbf{l}_2 \times \mathbf{B}_1 \tag{2.17}$$

Similarly, we note from (2.15a) that the magnetic flux density at the element $d\mathbf{l}_1$ due to the element $d\mathbf{l}_2$ is given by

$$\mathbf{B}_2 = \frac{\mu_0}{4\pi} \frac{I_2\, d\mathbf{l}_2 \times \mathbf{i}_{21}}{R^2} \tag{2.18}$$

and that this flux density acting upon $d\mathbf{l}_1$ results in a force on it given by

$$d\mathbf{F}_1 = I_1\, d\mathbf{l}_1 \times \mathbf{B}_2 \tag{2.19}$$

From (2.17) and (2.19), we see that the units of **B** are newtons per ampere-meter, commonly known as *webers per meter squared* (or tesla), where a weber is a newton-meter per ampere. The units of webers per unit area give the character of flux density to the quantity **B**.

Generalizing (2.17) and (2.19), we say that an infinitesimal current element of length $d\mathbf{l}$ and current I placed in a magnetic field of flux density **B** experiences a force $d\mathbf{F}$ given by

$$\boxed{d\mathbf{F} = I\, d\mathbf{l} \times \mathbf{B}} \tag{2.20}$$

as shown in Fig. 2.12. Alternatively, if a current element experiences a force in a region of space, then the region is said to be characterized by a magnetic field.

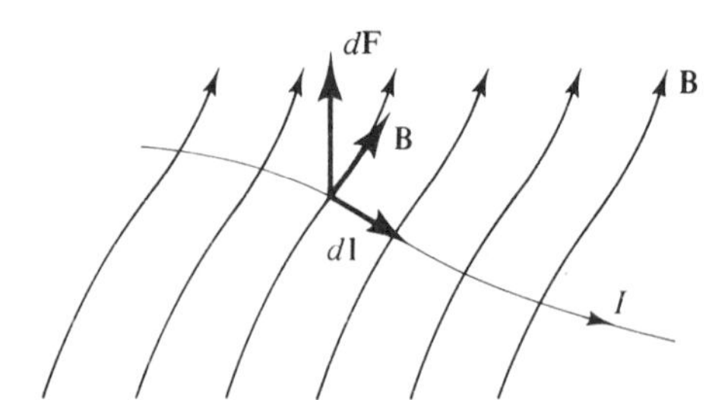

Figure 2.12. Force experienced by a current element in a magnetic field.

Principle of loudspeaker

There are many devices using the principle of magnetic force on a current carrying wire. One such device in everyday life is the loudspeaker. As shown by the cross-sectional view in Fig. 2.13, the loudspeaker consists of a permanent magnet between the poles of which is a coil wound around a cylinder attached to the apex of a movable cone-shaped diaphragm. Current through the coil varies in accordance with the audio signal from the output stage of the hi-fi amplifier or radio receiver. A magnetic force is thus exerted on the coil, vibrating it back and forth in step with the changes in the current. Since the coil assembly is attached to the cone, the cone also vibrates, thereby producing sound waves in the air.

Magnetic field due to current element

Returning now to (2.16) and (2.18) and generalizing, we obtain the magnetic flux density due to an infinitesimal current element of length $d\mathbf{l}$ and carrying current I to be

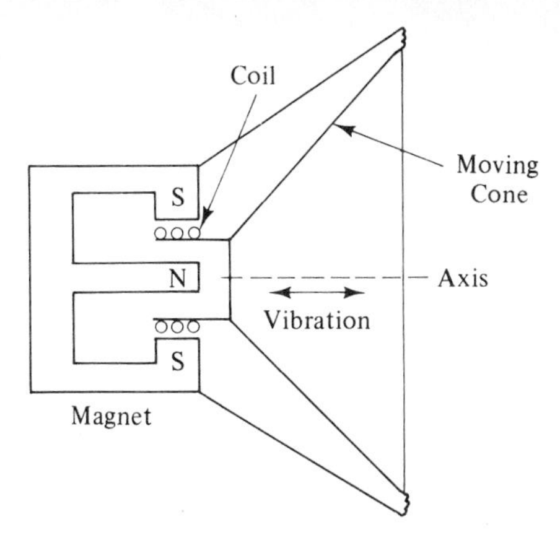

Figure 2.13. Cross-sectional view of a loudspeaker.

$$\boxed{\mathbf{B} = \frac{\mu_0}{4\pi} \frac{I\, d\mathbf{l} \times \mathbf{i}_R}{R^2}} \tag{2.21}$$

where R is the distance from the current element to the point at which the flux density is to be computed and $\mathbf{i}_R$ is the unit vector along the line joining the current element and the point under consideration and directed away from the current element, as shown in Fig. 2.14. Equation (2.21) is known as the *Biot–Savart law* and is analogous to the expression for the electric field intensity due to a point charge. The Biot–Savart law tells us that the magnitude of **B** at a point P is proportional to the current I, the element length dl, and the sine of the angle α between the current element and the line joining it to the point P, and is inversely proportional to the square of the distance from the current element to the point P. Hence the magnetic flux density is zero at points along the axis of the current element and increases in magnitude as the point P is moved away from the axis on a spherical surface centered at the current element, becoming a maximum for α equal to 90°. This is in contrast to the behavior of the electric field intensity due to a point charge which remains the same in magnitude at points on a spherical surface centered at the point charge. The direction of **B** at point P is normal to the plane containing the current element and the line joining the current element to P as given by the cross product operation $d\mathbf{l} \times \mathbf{i}_R$, that is, right circular to the axis of the wire. Thus the direction lines of the magnetic flux density due to a

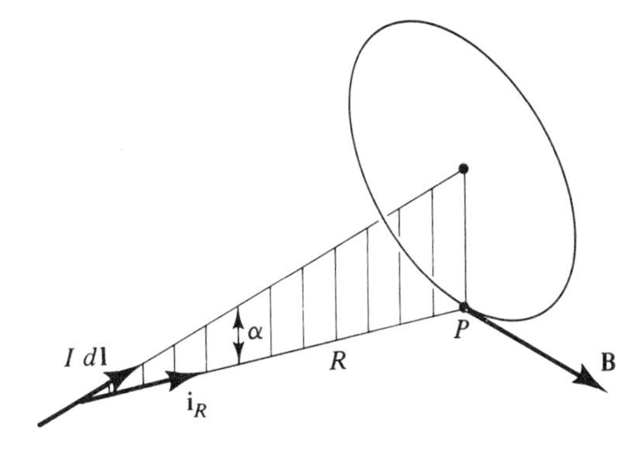

Figure 2.14. Magnetic flux density due to an infinitesimal current element.

current element are circles centered at points on the axis of the current element and lying in planes normal to the axis. This is in contrast to the direction lines of the electric field intensity due to a point charge which are radial lines emanating from the point charge.

Example 2.4

Let us consider an infinitesimal length 10^{-3} m of wire located at the point (1, 0, 0) and carrying current 2 A in the direction of the unit vector $\mathbf{i}_x$. We wish to find the magnetic flux density due to the current element at the point (0, 2, 2).

Noting that the current element is given by

$$I\,d\mathbf{l} = (2)(10^{-3})\mathbf{i}_x = 0.002\mathbf{i}_x$$

and the vector $\mathbf{R}$ from the location (1, 0, 0) of the current element to the point (0, 2, 2) is given by

$$\begin{aligned}\mathbf{R} &= (0 - 1)\mathbf{i}_x + (2 - 0)\mathbf{i}_y + (2 - 0)\mathbf{i}_z \\ &= -\mathbf{i}_x + 2\mathbf{i}_y + 2\mathbf{i}_z\end{aligned}$$

and using Biot–Savart law, we obtain

$$\begin{aligned}[\mathbf{B}]_{(0,2,2)} &= \frac{\mu_0}{4\pi}\frac{I\,d\mathbf{l} \times \mathbf{i}_R}{R^2} \\ &= \frac{\mu_0}{4\pi}\frac{I\,d\mathbf{l} \times \mathbf{R}}{R^3} \\ &= \frac{\mu_0}{4\pi}\frac{0.002\mathbf{i}_x \times (-\mathbf{i}_x + 2\mathbf{i}_y + 2\mathbf{i}_z)}{27} \\ &= \frac{0.001\mu_0}{27\pi}(-\mathbf{i}_y + \mathbf{i}_z) \text{ Wb/m}^2\end{aligned}$$

The Biot–Savart law can be used to find the magnetic flux density due to a current carrying filamentary wire of any length and shape by dividing the wire into a number of infinitesimal elements and using superposition. We shall illustrate the procedure by means of an example.

Example 2.5

Infinitely long, straight wire

Let us consider an infinitely long, straight wire situated along the z-axis and carrying current I A in the $+z$-direction. We wish to find the magnetic flux density everywhere.

Let us consider a point on the xy-plane specified by the cylindrical coordinates $(r, \phi, 0)$, as shown in Fig. 2.15(a). Then the solution for the magnetic flux density at $(r, \phi, 0)$ can be obtained by considering a differential length dz of the wire at the point $(0, 0, z)$ and using superposition. Applying Biot–Savart law (2.21) to the geometry in Fig. 2.15(a), we obtain the magnetic flux density at $(r, \phi, 0)$ due to the current element $I\,dz\,\mathbf{i}_z$ at $(0, 0, z)$ to be

$$\begin{aligned}[d\mathbf{B}]_{(r,\phi,0)} &= \frac{\mu_0}{4\pi}\frac{I\,dz\,\mathbf{i}_z \times \mathbf{i}_R}{R^2} \\ &= \frac{\mu_0 I\,dz}{4\pi}\frac{\sin\alpha}{R^2}\mathbf{i}_\phi\end{aligned}$$

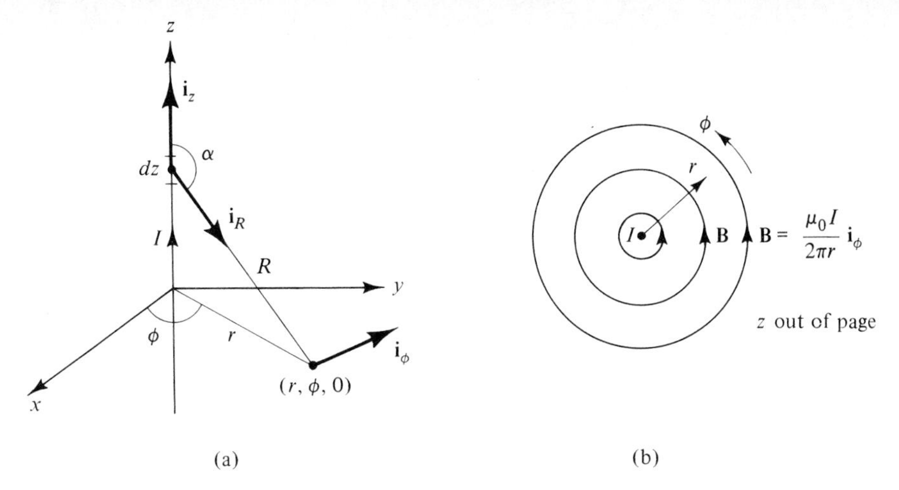

Figure 2.15. (a) Determination of magnetic field due to an infinitely long, straight wire of current I A. (b) Magnetic field due to the wire of (a).

$$= \frac{\mu_0 I\, dz}{4\pi}\frac{r}{R^3}\mathbf{i}_\phi$$

$$= \frac{\mu_0 Ir\, dz}{4\pi(z^2 + r^2)^{3/2}}\mathbf{i}_\phi$$

The magnetic flux density due to the entire wire is then given by

$$[\mathbf{B}]_{(r,\phi,0)} = \int_{z=-\infty}^{\infty} d\mathbf{B}$$

$$= \int_{z=-\infty}^{\infty} \frac{\mu_0 Ir}{4\pi(z^2 + r^2)^{3/2}}\, dz\ \mathbf{i}_\phi$$

$$= \frac{\mu_0 Ir}{4\pi}\left[\frac{z}{r^2\sqrt{z^2 + r^2}}\right]_{z=-\infty}^{\infty}\mathbf{i}_\phi$$

$$= \frac{\mu_0 I}{2\pi r}\mathbf{i}_\phi$$

Now, since the origin can be chosen to be anywhere on the wire without changing the geometry, this result is valid everywhere. Thus the required magnetic flux density is

$$\mathbf{B} = \frac{\mu_0 I}{2\pi r}\mathbf{i}_\phi \tag{2.22}$$

which has the magnitude $\mu_0 I/2\pi r$ and surrounds the wire, as shown by the cross-sectional view in Fig. 2.15(b).

Types of current distributions

The magnetic field computation illustrated in Example 2.5 can be extended to current distributions. Current distributions are of two types: surface currents and volume currents, depending upon whether current flows on a surface like rain water flowing down a smooth wall or in a volume like rain water flowing down a gutter downspout. The corresponding current densities are the surface current density $\mathbf{J}_S$ and the volume current density, or simply the current density $\mathbf{J}$, having the units of current crossing unit length (amperes per meter) and current crossing unit area (amperes per meter squared), respectively. Note that the current densities are vector quantities since flow is involved. Assuming for simplicity surface current of uniform density flowing on a plane sheet as shown in Fig. 2.16(a), one obtains the current I on the sheet by multiplying the magnitude of $\mathbf{J}_S$ by the dimension w of the sheet normal to the direction of $\mathbf{J}_S$. Similarly for volume current of uniform density flowing in a straight wire as shown in Fig. 2.16(b), the current I in the wire is given by the product of the magnitude of $\mathbf{J}$ and the area of cross section A of the wire normal to the direction of $\mathbf{J}$. If the current density is nonuniform, the current can be obtained by performing an appropriate integration along the width of the sheet or over the cross section of the wire, depending on the case. We shall illustrate the determination of the magnetic field due to a current distribution by means of an example.

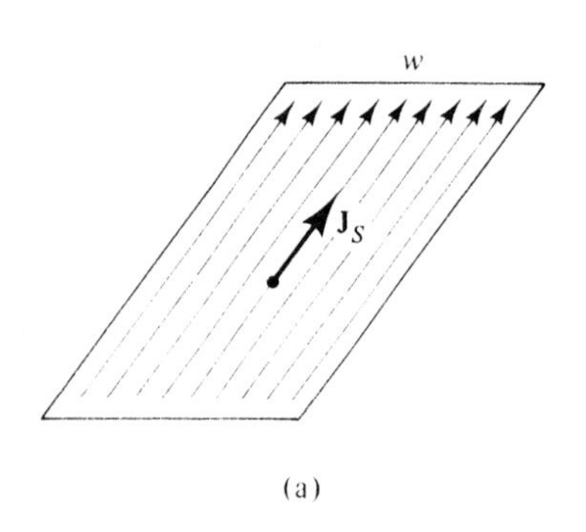

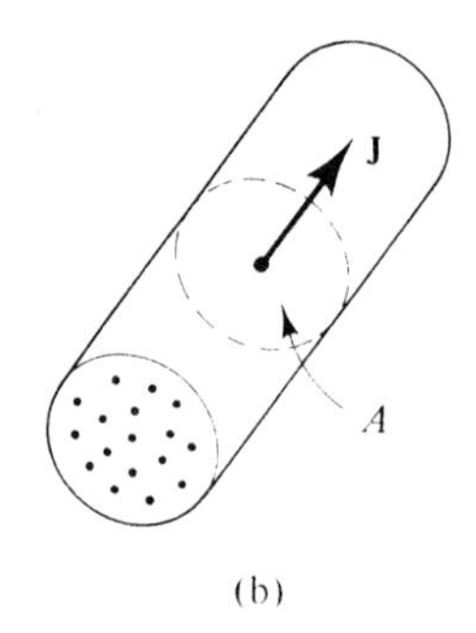

Figure 2.16. Determination of currents due to (a) surface current and (b) volume current distributions of uniform densities.

Example 2.6

Infinite plane sheet of current

Let us consider an infinite plane sheet of current in the xz-plane with uniform surface current density $\mathbf{J}_S = J_{S0}\mathbf{i}_z$ A/m and find the magnetic flux density everywhere.

Let us first consider a point $(0, y, 0)$ on the positive y-axis, as shown in Fig. 2.17(a). Then the solution can be carried out by dividing the sheet into a number of thin vertical strips and using superposition. Two such strips, which are on either side of the z-axis and equidistant from it, are shown in Fig. 2.17(a). Each strip is an infinitely long filamentary wire of current $J_{S0}\,dx$. Then applying the result of Example 2.5 to each strip and noting that the resultant magnetic flux density at $(0, y, 0)$ due to the two strips together has only an x-component, we obtain

$$\begin{aligned} d\mathbf{B} &= d\mathbf{B}_1 + d\mathbf{B}_2 \\ &= -2\,dB_1 \cos\alpha\,\mathbf{i}_x \\ &= -2\frac{\mu_0 J_{S0}\,dx}{2\pi\sqrt{x^2+y^2}}\frac{y}{\sqrt{x^2+y^2}}\mathbf{i}_x \\ &= -\frac{\mu_0 J_{S0} y\,dx}{\pi(x^2+y^2)}\mathbf{i}_x \end{aligned}$$

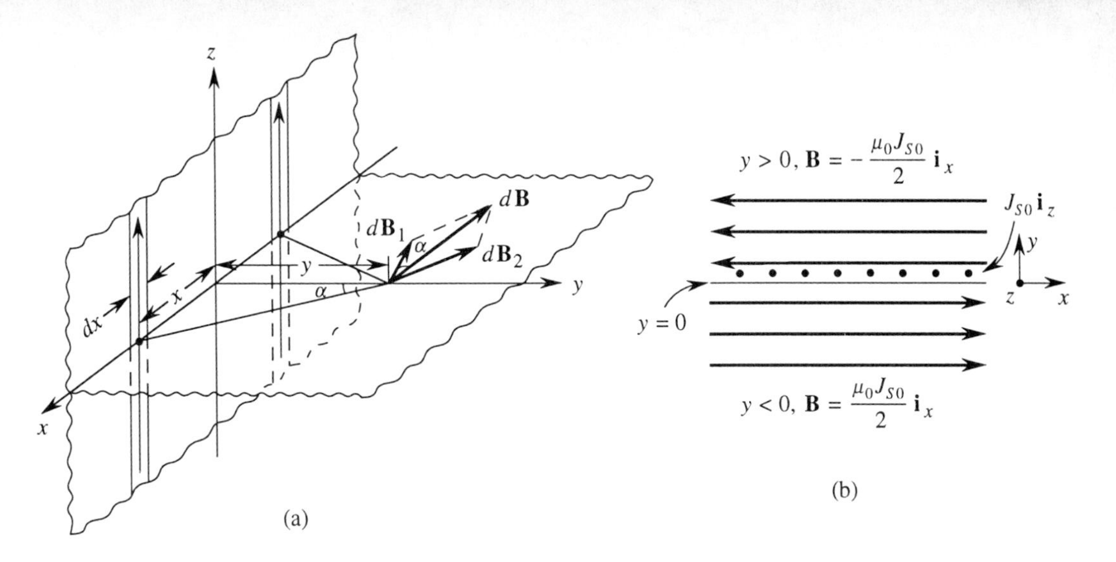

Figure 2.17. (a) Determination of magnetic field due to an infinite plane sheet of current density $J_{S0}\mathbf{i}_z$ A/m. (b) Magnetic field due to the current sheet of (a).

The magnetic flux density due to the entire sheet is then given by

$$
\begin{aligned}
[\mathbf{B}]_{(0,y,0)} &= \int_{x=0}^{\infty} d\mathbf{B} \\
&= -\int_{x=0}^{\infty} \frac{\mu_0 J_{S0} y}{\pi(x^2 + y^2)}\, dx\, \mathbf{i}_x \\
&= -\frac{\mu_0 J_{S0} y}{\pi}\left[\frac{1}{y}\tan^{-1}\frac{x}{y}\right]_{x=0}^{\infty} \mathbf{i}_x \\
&= -\frac{\mu_0 J_{S0}}{2}\mathbf{i}_x \qquad \text{for} \quad y > 0
\end{aligned}
$$

Since the magnetic field due to each strip is circular to that strip, a similar result applies for a point on the negative y-axis except for $+x$-direction for the field. Thus

$$[\mathbf{B}]_{(0,y,0)} = \frac{\mu_0 J_{S0}}{2}\mathbf{i}_x \qquad \text{for} \quad y < 0$$

Now, since the origin can be chosen to be anywhere on the sheet without changing the geometry, the foregoing results are valid everywhere in the respective regions. Thus the required magnetic flux density is

$$\mathbf{B} = \mp\frac{\mu_0 J_{S0}}{2}\mathbf{i}_x \qquad \text{for} \quad y \gtrless 0 \tag{2.23}$$

which has the magnitude $\mu_0 J_{S0}/2$ everywhere and is directed in the $\mp\mathbf{i}_x$ direction for $y \gtrless 0$, as shown in Fig. 2.17(b). Defining $\mathbf{i}_n$ to be the unit normal vector directed away from the sheet, that is,

$$\mathbf{i}_n = \pm\mathbf{i}_y \qquad \text{for} \quad y \gtrless 0$$

and noting that

$$\mathbf{B} = \frac{\mu_0}{2}(J_{S0}\mathbf{i}_z) \times (\pm\mathbf{i}_y) \qquad \text{for} \quad y \gtrless 0$$

we can write

$$\boxed{\mathbf{B} = \frac{\mu_0}{2}\mathbf{J}_S \times \mathbf{i}_n} \tag{2.24}$$

Magnetic force on charge

Returning now to (2.20), we can formulate the magnetic force in terms of moving charge, since current is due to flow of charges. Thus if the time taken by the charge dq contained in the length $d\mathbf{l}$ of the current element to flow with a velocity $\mathbf{v}$ across the infinitesimal cross-sectional area of the element is dt, then $I = dq/dt$, and $d\mathbf{l} = \mathbf{v}\,dt$ so that

$$d\mathbf{F} = \frac{dq}{dt}\mathbf{v}\,dt \times \mathbf{B} = dq\,\mathbf{v} \times \mathbf{B} \tag{2.25}$$

It then follows that the force $\mathbf{F}$ experienced by a test charge q moving with a velocity $\mathbf{v}$ in a magnetic field of flux density $\mathbf{B}$ is given by

$$\boxed{\mathbf{F} = q\mathbf{v} \times \mathbf{B}} \tag{2.26}$$

We may now obtain a defining equation for $\mathbf{B}$ in terms of the moving test charge. To do this, we note from (2.26) that the magnetic force is directed normally to both $\mathbf{v}$ and $\mathbf{B}$ as shown in Fig. 2.18 and that its magnitude is equal to $qvB \sin\delta$ where δ is the angle between $\mathbf{v}$ and $\mathbf{B}$. A knowledge of the force $\mathbf{F}$ acting on a test charge moving with an arbitrary velocity $\mathbf{v}$ provides only the value of $B \sin\delta$. To find $\mathbf{B}$, we must determine the maximum force qvB that occurs for δ equal to 90° by trying out several directions of $\mathbf{v}$, keeping its magnitude constant. Thus if this maximum force is $\mathbf{F}_m$ and it occurs for a velocity $v\mathbf{i}_m$, then

$$\mathbf{B} = \frac{\mathbf{F}_m \times \mathbf{i}_m}{qv} \tag{2.27}$$

As in the case of defining the electric field intensity, we assume that the test charge does not alter the magnetic field in which it is placed. Ideally, $\mathbf{B}$ is defined in the limit that qv tends to zero; that is,

$$\boxed{\mathbf{B} = \lim_{qv \to 0} \frac{\mathbf{F}_m \times \mathbf{i}_m}{qv}} \tag{2.28}$$

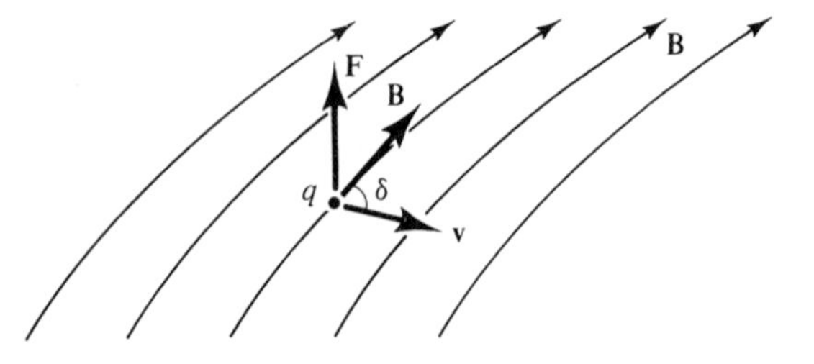

Figure 2.18. Force experienced by a test charge q moving with a velocity $\mathbf{v}$ in a magnetic field $\mathbf{B}$.

Equation (2.28) is the defining equation for the magnetic flux density irrespective of the source of the magnetic field. We have learned in this section that an electric current or a charge in motion is a source of the magnetic field. We will learn in Chapter 3 that there exists another source for the magnetic field, namely, a time-varying electric field.

Charge motion in uniform magnetic field

There are many devices based on the magnetic force on a moving charge. Of particular interest is the motion of a charged particle in a uniform magnetic field, as shown in Fig. 2.19. In this figure, a particle of mass m and charge q entering the magnetic field region with velocity $\mathbf{v}$ perpendicular to $\mathbf{B}$ experiences a force qvB perpendicular to $\mathbf{v}$. Hence the particle describes a circular path of radius R, equal to mv/qB, obtained by equating the centripetal force mv^2/R to the magnetic force qvB. The fact that the radius is equal to mv/qB is used in several different applications. In the mass spectrograph, the mass-to-charge ratio of the particles is obtained by measuring the radius of the circular orbit for known values of v and B. For ions of the same charge but of different masses, the radii of the circular paths are directly proportional to their masses and to their velocities. This forms the basis for electromagnetic separation of isotopes, two or more forms of a chemical element having the same chemical properties and the same atomic number but different atomic weights. In the cyclotron, a particle accelerator, the particle undergoes a series of semicircular orbits of successively increasing velocities and hence radii before it exits the field region with high energy.

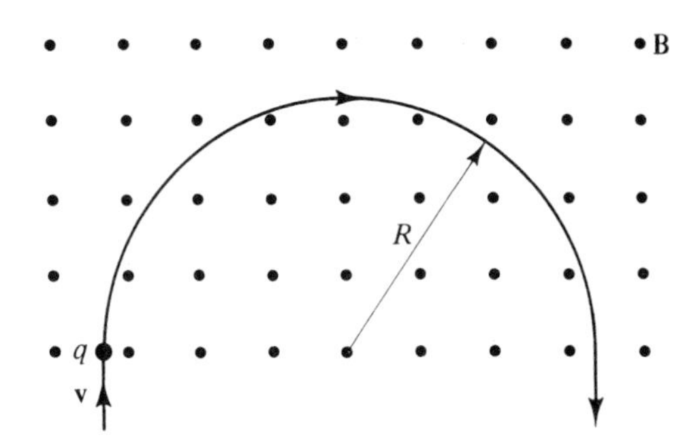

Figure 2.19. Circular motion of a charged particle entering a uniform magnetic field region.

K2.2. Ampère's law of force; Magnetic flux density; Biot–Savart law; Computation of **B** due to current distributions; **B** due to an infinitely long, straight wire; **B** due to an infinite plane sheet of current of uniform density.

D2.5. For $I_1\,d\mathbf{l}_1 = I_1\,dy\,\mathbf{i}_y$ located at (1, 0, 0) and $I_2\,d\mathbf{l}_2 = I_2\,dx\,\mathbf{i}_x$ located at (0, 1, 0), find: **(a)** $d\mathbf{F}_1$ and **(b)** $d\mathbf{F}_2$.
Ans. **(a)** $-(\mu_0 I_1 I_2/8\sqrt{2}\pi)\,dx\,dy\,\mathbf{i}_x$; **(b)** $-(\mu_0 I_1 I_2/8\sqrt{2}\pi)\,dx\,dy\,\mathbf{i}_y$

D2.6. A current I flows in a wire along the curve $x = 2y = z^2 + 2$ and in the direction of increasing z. If the wire is situated in a magnetic field $\mathbf{B} = (y\mathbf{i}_x - x\mathbf{i}_y)/(x^2 + y^2)$, find the magnetic force acting on an infinitesimal length of the wire having the projection dz on the z-axis at each of the following points: **(a)** (2, 1, 0); **(b)** (3, 1.5, 1); and **(c)** (6, 3, 2).
Ans. **(a)** $I\,dz\,(2\mathbf{i}_x + \mathbf{i}_y)/5$; **(b)** $I\,dz\,(2\mathbf{i}_x + \mathbf{i}_y - 5\mathbf{i}_z)/7.5$; **(c)** $I\,dz\,(2\mathbf{i}_x + \mathbf{i}_y - 10\mathbf{i}_z)/15$

D2.7. Given $\mathbf{B} = (B_0/3)(2\mathbf{i}_x + 2\mathbf{i}_y - \mathbf{i}_z)$, find the magnitude of the magnetic force acting on a test charge q moving with velocity v_0 at the point (2, 2, −1) for each of the following paths of the test charge: **(a)** $x = y = -2z$; **(b)** $4x = 4y = z + 9$; and **(c)** $x = y = 2z^2$.
Ans. **(a)** 0; **(b)** qv_0B_0; **(c)** $0.1641qv_0B_0$

D2.8. Infinite plane sheets of current lie in the $x = 0$, $y = 0$, and $z = 0$ planes with uniform surface current densities $J_{S0}\mathbf{i}_z$, $2J_{S0}\mathbf{i}_x$, and $-J_{S0}\mathbf{i}_x$ A/m, respectively. Find the resulting magnetic flux densities at the following points: **(a)** (1, 2, 2); **(b)** (2, −2, − 1); and **(c)** (−2, 1, −2).
Ans. **(a)** $\mu_0 J_{S0}(\mathbf{i}_y + \mathbf{i}_z)$; **(b)** $-\mu_0 J_{S0}\mathbf{i}_z$; **(c)** $\mu_0 J_{S0}(-\mathbf{i}_y + \mathbf{i}_z)$

2.3 LORENTZ FORCE EQUATION

In Section 2.1 we learned that a test charge q placed in an electric field of intensity $\mathbf{E}$ experiences a force

$$\boxed{\mathbf{F}_E = q\mathbf{E}} \tag{2.29}$$

and in Section 2.2 we learned that a test charge q moving with a velocity $\mathbf{v}$ in a magnetic field of flux density $\mathbf{B}$ experiences a force

$$\boxed{\mathbf{F}_M = q\mathbf{v} \times \mathbf{B}} \tag{2.30}$$

Combining (2.29) and (2.30), we can write the expression for the total force acting on a test charge q moving with velocity $\mathbf{v}$ in a region characterized by electric field of intensity $\mathbf{E}$ and magnetic field of flux density $\mathbf{B}$ to be

$$\boxed{\mathbf{F} = \mathbf{F}_E + \mathbf{F}_M = q(\mathbf{E} + \mathbf{v} \times \mathbf{B})} \tag{2.31}$$

Equation (2.31) is known as the *Lorentz force equation*.

Determination of electric and magnetic fields from forces on a test charge

We observe from (2.31) that the electric and magnetic fields at a point can be determined from a knowledge of the forces experienced by a test charge at that point for several different velocities. For a given $\mathbf{B}$, $\mathbf{E}$ can be found from the force for one velocity, since $\mathbf{F}_E$ acts in the direction of $\mathbf{E}$. For a given $\mathbf{E}$, $\mathbf{B}$ can be found from two forces for two noncollinear velocities, since $\mathbf{F}_M$ acts perpendicular to both $\mathbf{v}$ and $\mathbf{B}$. Thus to find both $\mathbf{E}$ and $\mathbf{B}$, the knowledge of a minimum of three forces is necessary. We shall illustrate the determination of $\mathbf{E}$ and $\mathbf{B}$ from three forces by means of an example.

Example 2.7

The forces experienced by a test charge q for three different velocities at a point in a region of electric and magnetic fields are given by

$$\mathbf{F}_1 = qE_0\mathbf{i}_x \qquad \text{for} \quad \mathbf{v}_1 = v_0\mathbf{i}_x$$

$$\mathbf{F}_2 = qE_0(2\mathbf{i}_x + \mathbf{i}_y) \qquad \text{for} \quad \mathbf{v}_2 = v_0\mathbf{i}_y$$

$$\mathbf{F}_3 = qE_0(\mathbf{i}_x + \mathbf{i}_y) \qquad \text{for} \quad \mathbf{v}_3 = v_0\mathbf{i}_z$$

where v_0 and E_0 are constants. We wish to find $\mathbf{E}$ and $\mathbf{B}$ at that point.

From the Lorentz force equation, we have

$$q\mathbf{E} + qv_0\mathbf{i}_x \times \mathbf{B} = qE_0\mathbf{i}_x \tag{2.32a}$$

$$q\mathbf{E} + qv_0\mathbf{i}_y \times \mathbf{B} = q(2E_0\mathbf{i}_x + E_0\mathbf{i}_y) \tag{2.32b}$$

$$q\mathbf{E} + qv_0\mathbf{i}_z \times \mathbf{B} = q(E_0\mathbf{i}_x + E_0\mathbf{i}_y) \tag{2.32c}$$

Eliminating **E** by subtracting (2.32a) from (2.32b) and (2.32c) from (2.32b), we obtain

$$v_0(\mathbf{i}_y - \mathbf{i}_x) \times \mathbf{B} = E_0(\mathbf{i}_x + \mathbf{i}_y) \tag{2.33a}$$

$$v_0(\mathbf{i}_y - \mathbf{i}_z) \times \mathbf{B} = E_0\mathbf{i}_x \tag{2.33b}$$

Since the cross product of two vectors is perpendicular to the two vectors, it follows from (2.33a) that $(\mathbf{i}_x + \mathbf{i}_y)$ is perpendicular to **B** and from (2.33b) that $\mathbf{i}_x$ is perpendicular to **B**. Thus **B** is perpendicular to both $(\mathbf{i}_x + \mathbf{i}_y)$ and $\mathbf{i}_x$. But the cross product of $(\mathbf{i}_x + \mathbf{i}_y)$ and $\mathbf{i}_x$ is perpendicular to both of them. Therefore, **B** must be directed parallel to $(\mathbf{i}_x + \mathbf{i}_y) \times \mathbf{i}_x$. Thus we can write

$$\mathbf{B} = C(\mathbf{i}_x + \mathbf{i}_y) \times \mathbf{i}_x = -C\mathbf{i}_z \tag{2.34}$$

where C is a proportionality constant to be determined. To do this, we substitute (2.34) into (2.33b) to obtain

$$v_0(\mathbf{i}_y - \mathbf{i}_z) \times (-C\mathbf{i}_z) = E_0\mathbf{i}_x$$

$$-v_0 C\mathbf{i}_x = E_0\mathbf{i}_x$$

or $C = -E_0/v_0$. Thus we get

$$\mathbf{B} = \frac{E_0}{v_0}\mathbf{i}_z$$

Alternatively, we can obtain this result by assuming $\mathbf{B} = B_x\mathbf{i}_x + B_y\mathbf{i}_y + B_z\mathbf{i}_z$, substituting in (2.33a) and (2.33b), equating the like components, and solving the resulting algebraic equations. Thus substituting in (2.33a), we have

$$v_0\begin{vmatrix} \mathbf{i}_x & \mathbf{i}_y & \mathbf{i}_z \\ -1 & 1 & 0 \\ B_x & B_y & B_z \end{vmatrix} = E_0(\mathbf{i}_x + \mathbf{i}_y)$$

or

$$v_0[B_z\mathbf{i}_x + B_z\mathbf{i}_y - (B_y + B_x)\mathbf{i}_z] = E_0\mathbf{i}_x + E_0\mathbf{i}_y$$

$$B_z = \frac{E_0}{v_0} \qquad \text{and} \qquad (B_y + B_x) = 0$$

Substituting in (2.33b), we have

$$v_0\begin{vmatrix} \mathbf{i}_x & \mathbf{i}_y & \mathbf{i}_z \\ 0 & 1 & -1 \\ B_x & B_y & B_z \end{vmatrix} = E_0\mathbf{i}_x$$

or

$$v_0[(B_z + B_y)\mathbf{i}_x - B_x\mathbf{i}_y - B_x\mathbf{i}_z] = E_0\mathbf{i}_x$$

$$B_z + B_y = \frac{E_0}{v_0} \qquad \text{and} \qquad B_x = 0$$

Thus we obtain $B_z = E_0/v_0$, $B_x = 0$, $B_y = 0$, and hence

$$\mathbf{B} = \frac{E_0}{v_0}\mathbf{i}_z$$

Finally, we can find **E** by substituting the result obtained for **B** in any one of the three equations (2.32a)–(2.32c). Thus substituting $\mathbf{B} = (E_0/v_0)\mathbf{i}_z$ in (2.32c), we obtain

$$\mathbf{E} = E_0(\mathbf{i}_x + \mathbf{i}_y)$$

Lorentz force applications

The Lorentz force equation is a fundamental equation in electromagnetics. Together with the pertinent laws of mechanics, it constitutes the starting point for the study of charged particle motion in electric and/or magnetic fields. Devices based on charged particle motion in fields are abundant in practice. Examples, some of which we discussed in Sections 2.1 and 2.2, are cathode ray tubes, ink-jet printers, electron microscopes, mass spectrographs, particle accelerators, and microwave tubes such as klystrons, magnetrons, and traveling wave tubes. Interaction between charged particles and fields is the basis for the study of the electromagnetic properties of materials and for the study of radio-wave propagation in gaseous media such as the earth's ionosphere, in which the constituent gasses are partially ionized by the solar radiation.

Tracing of charged particle motion in electric and magnetic fields

Tracing the path of a charged particle in a region of electric and magnetic fields involves setting the mechanical force, as given by the product of the mass of the test charge and its acceleration, equal to the electromagnetic force, as given by the Lorentz force equation, and solving the resulting differential equation(s) subject to initial condition(s). For simplicity, we shall consider a two-dimensional situation in which the motion is confined to the xy-plane in a region of uniform, crossed electric and magnetic fields, $\mathbf{E} = E_0\mathbf{i}_y$ and $\mathbf{B} = B_0\mathbf{i}_z$, as shown in Fig. 2.20, where E_0 and B_0 are constants. We shall assume that a test charge q having mass m starts at $t = 0$ at the point $(x_0, y_0, 0)$ with initial velocity $\mathbf{v} = v_{x0}\mathbf{i}_x + v_{y0}\mathbf{i}_y$.

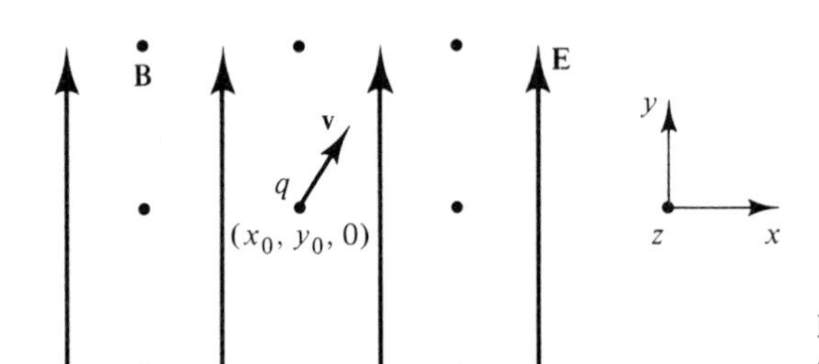

Figure 2.20. Test charge q in a region of crossed electric and magnetic fields.

From the Lorentz force equation (2.31) the force exerted by the crossed electric and magnetic fields on the test charge is given by

$$\begin{aligned}\mathbf{F} &= q(\mathbf{E} + \mathbf{v} \times \mathbf{B}) \\ &= qE_0\mathbf{i}_y + q(v_x\mathbf{i}_x + v_y\mathbf{i}_y + v_z\mathbf{i}_z) \times B_0\mathbf{i}_z \\ &= qB_0v_y\mathbf{i}_x + (qE_0 - qB_0v_x)\mathbf{i}_y\end{aligned} \tag{2.35}$$

The equations of motion of the test charge can then be written as

$$\frac{dv_x}{dt} = \frac{qB_0}{m}v_y \tag{2.36a}$$

$$\frac{dv_y}{dt} = \frac{qE_0}{m} - \frac{qB_0}{m}v_x \tag{2.36b}$$

$$\frac{dv_z}{dt} = 0 \tag{2.36c}$$

Equation (2.36c) together with the initial conditions $v_z = 0$ and $z = 0$ at $t = 0$ simply tells us that the path of the test charge is confined to the $z = 0$ plane.

Eliminating v_y from (2.36a) and (2.36b), we obtain

$$\frac{d^2 v_x}{dt^2} + \left(\frac{qB_0}{m}\right)^2 v_x = \left(\frac{q}{m}\right)^2 B_0 E_0 \tag{2.37}$$

the solution for which is

$$v_x = \frac{E_0}{B_0} + C_1 \cos \omega_c t + C_2 \sin \omega_c t \tag{2.38a}$$

where C_1 and C_2 are constants to be determined from the initial conditions and $\omega_c = qB_0/m$. From (2.36a), the solution for v_y is then given by

$$v_y = -C_1 \sin \omega_c t + C_2 \cos \omega_c t \tag{2.38b}$$

Using initial conditions $v_x = v_{x0}$ and $v_y = v_{y0}$ at $t = 0$ to evaluate C_1 and C_2 in (2.38a) and (2.38b), we obtain

$$v_x = \frac{E_0}{B_0} + \left(v_{x0} - \frac{E_0}{B_0}\right) \cos \omega_c t + v_{y0} \sin \omega_c t \tag{2.39a}$$

$$v_y = -\left(v_{x0} - \frac{E_0}{B_0}\right) \sin \omega_c t + v_{y0} \cos \omega_c t \tag{2.39b}$$

Integrating (2.39a) and (2.39b) with respect to t and using initial conditions $x = x_0$ and $y = y_0$ at $t = 0$ to evaluate the constants of integration, we then obtain

$$x = x_0 + \frac{E_0}{B_0} t + \frac{1}{\omega_c}\left(v_{x0} - \frac{E_0}{B_0}\right) \sin \omega_c t + \frac{v_{y0}}{\omega_c}(1 - \cos \omega_c t) \tag{2.40a}$$

$$y = y_0 - \frac{1}{\omega_c}\left(v_{x0} - \frac{E_0}{B_0}\right)(1 - \cos \omega_c t) + \frac{v_{y0}}{\omega_c} \sin \omega_c t \tag{2.40b}$$

Equations (2.40a) and (2.40b) give the position of the test charge versus time, whereas (2.39a) and (2.39b) give the corresponding velocity components. For $B_0 = 0$, $\omega_c \to 0$, and the solutions reduce to

$$x = x_0 + v_{x0} t \tag{2.41a}$$

$$y = y_0 + v_{y0} t + \frac{1}{2}\frac{qE_0}{m} t^2 \tag{2.41b}$$

$$v_x = v_{x0} \tag{2.41c}$$

$$v_y = v_{y0} + \frac{qE_0}{m} t \tag{2.41d}$$

These can also be obtained directly from (2.36a) and (2.36b) with B_0 set equal to zero.

We may now trace the path of a test charge in the crossed electric and magnetic fields by using (2.40a) and (2.40b) for B_0 not equal to zero and (2.41a) and (2.41b) for B_0 equal to zero. To illustrate by means of an example, we shall consider the test charge to be an electron so that $q/m = -1.7578 \times 10^{11}$ C/kg and the field components to be given by $E_0 = -k_1 \times 10^3$ V/m and $B_0 = -k_2 \times 10^{-4}$ Wb/m^2. We shall assume the initial position (x_0, y_0) of the electron to be (0, 0) at time $t = 0$ and the initial velocity components to be $v_{x0} = k_3 \times 10^7$ m/s and $v_{y0} = k_4 \times 10^7$ m/s and compute the positions of the electron using a time increment $\Delta t = k_5 \times 10^{-9}$ s. The listing of a PC program for tracing the path of the electron in this manner is given as PL 2.2.

The plot obtained by running the program with values of $k_1 = 1$, $k_2 = 1$, $k_3 = 1$, $k_4 = 3$, and $k_5 = 5$ is shown in Fig. 2.21. With minor changes, the program can also be used to trace approximately the path of the electron for the case in which E_0 and B_0 are functions of x and y, so long as a time increment is used such that the field components do not vary appreciably between two successive points on the plot.

PL 2.2. Program listing for tracing the path of an electron in a region of crossed electric and magnetic fields.

```
100 '*******************************************************
110 '* MOTION OF ELECTRON IN A REGION OF CROSSED ELECTRIC *
120 '* AND MAGNETIC FIELDS                                *
130 '*******************************************************
140 SC=1.2:'* SCALE FACTOR TO EQUALIZE VERTICAL AND
150 '   HORIZONTAL SCALES *
160 QM=-1.7578*10^11:PI=3.1416
170 '* DRAW BOUNDARY OF REGION, AXES, AND SCALE MARKS *
180 CLS:SCREEN 1:COLOR 0,1:X1=52:X2=X1+180*SC
190 LINE (X1,0)-(X2,120),3,B:LINE (X1,60)-(X2,60)
200 FOR I=1 TO 5
210 IC=I*30*SC
220 LINE (X1+IC,0)-(X1+IC,2)
230 LINE (X1+IC,59)-(X1+IC,61)
240 LINE (X1+IC,118)-(X1+IC,120)
250 NEXT
260 LINE (X1,30)-(X1+2,30)
270 LINE (X1,90)-(X1+2,90)
280 LINE (X2-2,30)-(X2,30)
290 LINE (X2-2,90)-(X2,90)
300 '* ENTER AND PRINT VALUES OF INPUT PARAMETERS *
310 LOCATE 20,1:PRINT "ENTER VALUES OF K1, K2, K3, K4, AND
    K5":INPUT K1,K2,K3,K4,K5
320 LOCATE 20,1:PRINT "
    "
330 PRINT "K1 =";K1;" K2 =";K2;" K3 =";K3;" K4 =";K4;" K5 =
    ";K5
340 '* PLOT PATH OF ELECTRON FOR SPECIFIED VALUES OF K1,
350 '  K2, K3, K4, AND K5 *
360 EO=-K1*1000:BO=-K2*.0001:VX=K3*1E+07:VY=K4*1E+07
370 DT=K5*10^-9:OT=0:T=0
380 IF BO THEN EB=EO/BO:OC=QM*BO:VB=VX-EB:GOTO 410
390 T=T+DT:X=VX*T:Y=VY*T+.5*QM*EO*T*T:GOTO 480:'* SPECIAL
400 '  CASE OF K2 EQUAL TO ZERO *
410 T=T+DT:OT=OC*T
420 IF EO THEN 450
430 IF OT>2*PI THEN 530:'* IF K1 IS ZERO, THEN STOP TRACING
440 '  CIRCULAR PATH OF ELECTRON AFTER ONE REVOLUTION *
```

PL2.2. (continued)

```
450 ST=SIN(OT):CT=COS(OT)
460 X=EB*T+VB*ST/OC+VY*(1-CT)/OC
470 Y=-VB*(1-CT)/OC+VY*ST/OC
480 IF X<0 OR X>12 OR ABS(Y)>4 THEN 530:'* CHECK IF
490 '  ELECTRON GOES OUT OF THE REGION *
500 PSET (X1+15*SC*X,60-15*Y)
510 IF BO THEN 410
520 GOTO 390
530 LOCATE 23,1:PRINT "STRIKE ANY KEY TO CONTINUE"
540 C$=INPUT$(1):GOTO 180
550 END
```

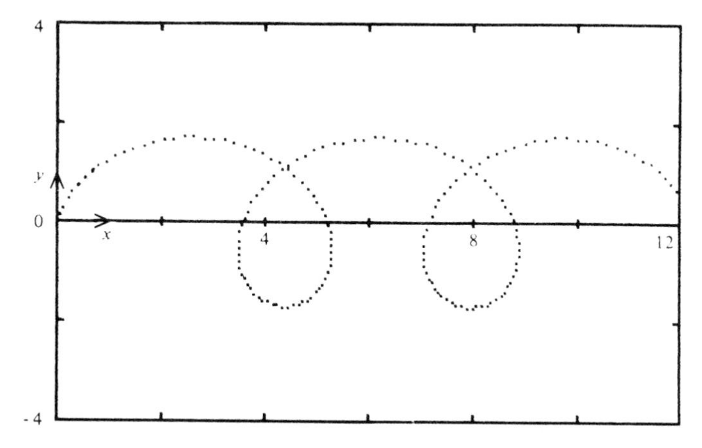

Figure 2.21. Specific example of computer plotting of the path of an electron in crossed electric and magnetic fields using the program listed in PL 2.2.

K2.3. Lorentz force equation; Determination of **E** and **B** from forces on a test charge; Charged-particle motion in electric and magnetic fields.

D2.9. A magnetic field $\mathbf{B} = \frac{B_0}{3}(\mathbf{i}_x + 2\mathbf{i}_y - 2\mathbf{i}_z)$ exists at a point. For each of the following velocities of a test charge q, find the electric field **E** at that point for which the acceleration experienced by the test charge is zero: **(a)** $v_0(\mathbf{i}_x - \mathbf{i}_y + \mathbf{i}_z)$; **(b)** $v_0(2\mathbf{i}_x + \mathbf{i}_y + 2\mathbf{i}_z)$; and **(c)** v_0 along the line $y = -z = 2x$.
Ans. **(a)** $-v_0 B_0(\mathbf{i}_y + \mathbf{i}_z)$; **(b)** $v_0 B_0(2\mathbf{i}_x - 2\mathbf{i}_y - \mathbf{i}_z)$; **(c)** **0**

D2.10. In a region of uniform electric and magnetic fields $\mathbf{E} = E_0\mathbf{i}_y$ and $\mathbf{B} = B_0\mathbf{i}_z$, respectively, a test charge q of mass m moves in the manner

$$x = \frac{E_0}{\omega_c B_0}(\omega_c t - \sin \omega_c t)$$

$$y = \frac{E_0}{\omega_c B_0}(1 - \cos \omega_c t)$$

$$z = 0$$

where $\omega_c = qB_0/m$. Find the forces acting on the test charge for the following values of t: **(a)** $t = 0$; **(b)** $t = \pi/2\omega_c$; and **(c)** $t = \pi/\omega_c$.
Ans. **(a)** $qE_0\mathbf{i}_y$; **(b)** $qE_0\mathbf{i}_x$; **(c)** $-qE_0\mathbf{i}_y$

2.4 CONDUCTORS AND SEMICONDUCTORS

Conduction

In our discussion of electric and magnetic fields thus far, we considered the medium to be free space. We shall now introduce materials. Materials contain charged particles that respond to applied electric and magnetic fields. Depending on their response to an applied electric field, they may be classified as conductors, semiconductors, or dielectrics. According to the classical model, an atom consists of a tightly bound, positively charged nucleus surrounded by a diffuse electron cloud having an equal and opposite charge to the nucleus. While the electrons for the most part are less tightly bound, the majority of them are associated with the nucleus and are known as *bound* electrons. These bound electrons can be displaced but not removed from the influence of the nucleus upon application of an electric field. Not taking part in this bonding mechanism are the *free* or *conduction* electrons. These electrons are constantly under thermal agitation, being released from the parent atom at one point and recaptured at another point. In the absence of an applied electric field, their motion is completely random; that is, the average thermal velocity on a macroscopic scale is zero so that there is no net current and the electron cloud maintains a fixed position. When an electric field is applied, an additional velocity is superimposed on the random velocities, thereby causing a *drift* of the average position of the electrons along the direction opposite to that of the electric field. This process is known as *conduction*. In certain materials, a large number of electrons may take part in this process. These materials are known as *conductors*. In certain other materials, only very few or a negligible number of electrons may participate in conduction. These materials are known as *dielectrics* or insulators. A class of materials for which conduction occurs not only by electrons but also by another type of carriers known as *holes*—vacancies created by detachment of electrons due to breaking of covalent bonds with other atoms—is intermediate to that of conductors and dielectrics. These materials are called *semiconductors*.

The quantum theory describes the motion of the current carriers in terms of energy levels. According to this theory, the electrons in an atom can have associated with them only certain discrete values of energy. When a large number of atoms are packed together, as in a crystalline solid, each energy level in the individual atom splits into a number of levels with slightly different energies, with the degree of splitting governed by the interatomic spacing, thereby giving rise to allowed bands of energy levels which may be widely separated, may be close together, or may even overlap. Four possible energy band diagrams are shown in Fig. 2.22, in which a forbidden band consists of energy levels which no electron in any atom of the solid can occupy. For case (a), the lower allowed band is only partially filled at the temperature of absolute zero. At higher temperatures, the electron population in the band spreads out somewhat, but only very few electrons reach higher energy levels. Thus, since there are many unfilled levels in the same band, it is possible to increase the energy of the system by moving the electrons to these unoccupied levels very easily by the application of an electric field, thereby resulting in drift of the electrons. The material is then classified as a conductor. For cases (b) and (c), the lower band is completely filled, whereas the next-higher band is completely empty at the temperature of absolute zero. If the width of the forbidden band is very large as in (b), the situation at normal tem-

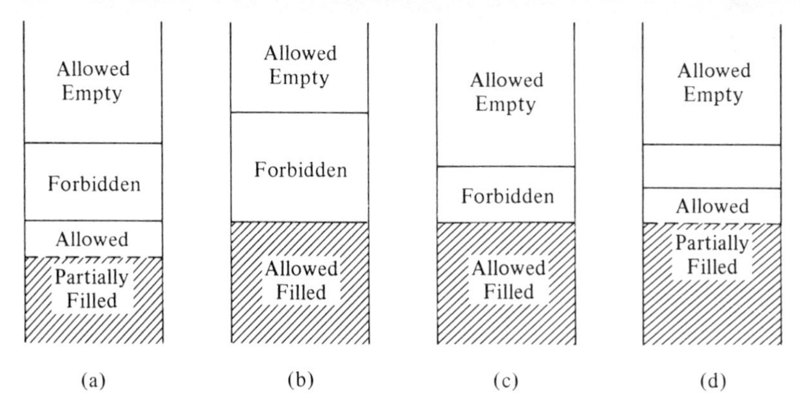

Figure 2.22. Energy band diagrams for different cases: (a) and (d) conductor; (b) dielectric; and (c) semiconductor.

peratures is essentially the same as at absolute zero, and hence there are no neighboring empty energy levels for the electrons to move. The only way for conduction to take place is for the electrons in the filled band to get excited and move to the next higher band. But this is very difficult to achieve with reasonable electric fields and the material is then classified as a dielectric. Only by supplying a very large amount of energy can an electron be excited to move from the lower band to the higher band where it has available neighboring empty levels for causing conduction. The dielectric is said to break down under such conditions. If, on the other hand, the width of the forbidden band in which the Fermi level lies is not too large, as in (c), some of the electrons in the lower band move into the upper band at normal temperatures so that conduction can take place under the influence of an electric field, not only in the upper band but also in the lower band because of the vacancies (holes) left by the electrons which moved into the upper band. The material is then classified as a semiconductor. A semiconductor crystal in pure form is known as an intrinsic semiconductor. The properties of an intrinsic crystal can be altered by introducing impurities into it. The crystal is then said to be an extrinsic semiconductor. For case (d), two allowed bands overlap; one or both of the bands is only partially filled and the situation corresponds to a conductor.

In the foregoing discussion we classified materials on the basis of their ability to permit conduction of electrons under the application of an external electric field. For conductors, we are interested in knowing about the relationship between the *drift velocity* of the electrons and the applied electric field, since the predominant process is conduction. But for collisions with the atomic lattice, the electric field continuously accelerates the electrons in the direction opposite to it as they move about at random. Collisions with the atomic lattice, however, provide the frictional mechanism by means of which the electrons lose some of the momentum gained between collisions. The net effect is as though the electrons drift with an average drift velocity $\mathbf{v}_d$, under the influence of the force exerted by the applied electric field and an opposing force due to the frictional mechanism. This opposing force is proportional to the momentum of the electron and inversely proportional to the average time τ between collisions. Thus the equation of motion of an electron is given by

$$m\frac{d\mathbf{v}_d}{dt} = e\mathbf{E} - \frac{m\mathbf{v}_d}{\tau} \tag{2.42}$$

where e and m are the charge and mass of an electron.

Rearranging (2.42), we have

$$m\frac{d\mathbf{v}_d}{dt} + \frac{m}{\tau}\mathbf{v}_d = e\mathbf{E} \tag{2.43}$$

For the sudden application of a constant electric field $\mathbf{E}_0$ at $t = 0$, the solution for (2.43) is given by

$$\mathbf{v}_d = \frac{e\tau}{m}\mathbf{E}_0 - \frac{e\tau}{m}\mathbf{E}_0 e^{-t/\tau} \tag{2.44}$$

where we have evaluated the arbitrary constant of integration by using the initial condition that $\mathbf{v}_d = \mathbf{0}$ at $t = 0$. The values of τ for typical conductors such as copper are of the order of 10^{-14} s so that the exponential term on the right side of (2.44) decays to a negligible value in a time much shorter than that of practical interest. Thus, neglecting this term, we have

$$\mathbf{v}_d = \frac{e\tau}{m}\mathbf{E}_0 \tag{2.45}$$

and the drift velocity is proportional in magnitude and opposite in direction to the applied electric field since the value of e is negative.

In fact, since we can represent a time-varying field as a superposition of step functions starting at appropriate times, the exponential term in (2.44) may be neglected as long as the electric field varies slowly compared to τ. For fields varying sinusoidally with time, this means that as long as the period T of the sinusoidal variation is several times the value of τ, or the radian frequency $\omega \ll 2\pi/\tau$, the drift velocity follows the variations in the electric field. Since $1/\tau \approx 10^{14}$, this condition is satisfied even at frequencies up to several hundred gigahertz, where a gigahertz is 10^9 Hz. Thus, for all practical purposes, we can assume that

$$\mathbf{v}_d = \frac{e\tau}{m}\mathbf{E} \tag{2.46}$$

Mobility

Now, we define the *mobility*, μ_e, of the electron as the ratio of the magnitudes of the drift velocity and the applied electric field. Then we have

$$\boxed{\mu_e = \frac{|\mathbf{v}_d|}{|\mathbf{E}|} = \frac{|e|\tau}{m}} \tag{2.47}$$

and

$$\mathbf{v}_d = -\mu_e\mathbf{E} \quad \text{for electrons} \tag{2.48a}$$

For values of τ typically of the order of 10^{-14} s, we note by substituting for $|e|$ and m on the right side of (2.47) that the electron mobilities are of the order of 10^{-3} C-s/kg. Alternative units for the mobility are square meters per volt-second.

In semiconductors, conduction is due not only to the movement of electrons but also to the movement of holes. We can define the mobility μ_h of a hole similarly to μ_e as the ratio of the drift velocity of the hole to the applied electric field. Thus we have

$$\mathbf{v}_d = \mu_h \mathbf{E} \quad \text{for holes} \tag{2.48b}$$

Note from (2.48b) that conduction of a hole takes place along the direction of the applied electric field since a hole is a vacancy created by the removal of an electron and hence a hole movement is equivalent to the movement of a positive charge of value equal to the magnitude of the charge of an electron. In general, the mobility of holes is lower than the mobility of electrons for a particular semiconductor. For example, for silicon, the values of μ_e and μ_h are 0.135 m^2/V-s and 0.048 m^2/V-s, respectively.

Conduction current

The drift of electrons in a conductor and that of electrons and holes in a semiconductor is equivalent to a current flow. This current is known as the *conduction current*. The conduction current density may be obtained in the following manner. If there are N_e free electrons per cubic meter of the material, then the amount of charge ΔQ passing through an infinitesimal area ΔS normal to the drift velocity at a point in the material in a time Δt is given by

$$\Delta Q = N_e e(\Delta S)(v_d \, \Delta t) \tag{2.49}$$

The current ΔI flowing across ΔS is given by

$$\Delta I = \frac{|\Delta Q|}{\Delta t} = N_e |e| v_d \, \Delta S \tag{2.50}$$

The magnitude of the current density at the point is the ratio of ΔI to ΔS in the limit ΔS tends to zero, and the direction is opposite to that of $\mathbf{v}_d$. Thus the conduction current density $\mathbf{J}_c$ resulting from the drift of electrons in the conductor is given by

$$\mathbf{J}_c = -N_e |e| \mathbf{v}_d \tag{2.51}$$

Substituting for $\mathbf{v}_d$ from (2.48a), we have

$$\mathbf{J}_c = \mu_e N_e |e| \mathbf{E} \tag{2.52}$$

Conductivity

Defining a quantity σ as

$$\boxed{\sigma = \mu_e N_e |e|} \tag{2.53}$$

we obtain the simple and important relationship between $\mathbf{J}_c$ and $\mathbf{E}$:

$$\boxed{\mathbf{J}_c = \sigma \mathbf{E}} \tag{2.54}$$

The quantity σ is known as the electrical conductivity of the material, and (2.54) is known as Ohm's law valid at a point. We shall show later that the well-known Ohm's law in circuit theory follows from it. In a semiconductor, the current density is the sum of the contributions due to the drifts of electrons and holes. If the densities of holes and electrons are N_h and N_e, respectively, the conduction current density is given by

$$\boxed{\mathbf{J}_c = (\mu_h N_h|e| + \mu_e N_e|e|)\mathbf{E}} \tag{2.55}$$

Thus the conductivity of a semiconducting material is given by

$$\boxed{\sigma = \mu_h N_h|e| + \mu_e N_e|e|} \tag{2.56a}$$

For an intrinsic semiconductor, $N_h = N_e$ so that (2.56a) reduces to

$$\boxed{\sigma = (\mu_h + \mu_e)N_e|e|} \tag{2.56b}$$

The units of conductivity are (meter^2/volt-second)(coulomb/meter^3) or ampere/volt-meter, also commonly known as siemens per meter (S/m), where a siemen is an ampere per volt. The ranges of conductivities for conductors, semiconductors, and dielectrics are shown in Fig. 2.23. Values of conductivities for a few materials are listed in Table 2.1. The constant values of conductivities do not imply that the conduction current density is proportional to the applied electric field intensity for all values of current density and field intensity. However, the

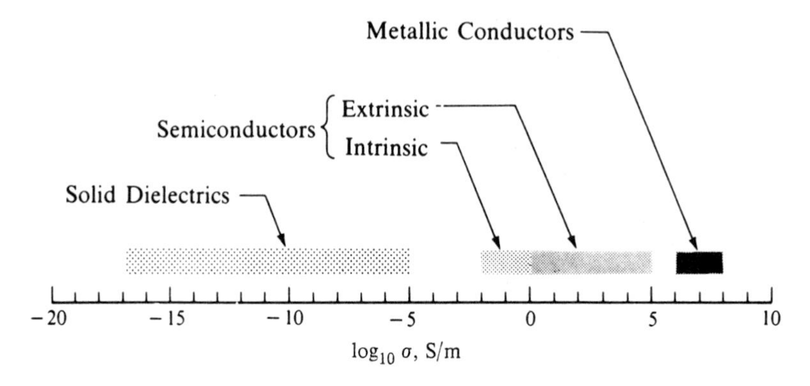

Figure 2.23. Ranges of conductivities for conductors, semiconductors, and dielectrics.

TABLE 2.1 CONDUCTIVITIES OF SOME MATERIALS

Material	Conductivity (S/m)	Material	Conductivity (S/m)
Silver	6.1×10^7	Seawater	4
Copper	5.8×10^7	Intrinsic germanium	2.2
Gold	4.1×10^7	Intrinsic silicon	1.6×10^{-3}
Aluminum	3.5×10^7	Fresh water	10^{-3}
Tungsten	1.8×10^7	Distilled water	2×10^{-4}
Brass	1.5×10^7	Dry earth	10^{-5}
Solder	7.0×10^6	Bakelite	10^{-9}
Lead	4.8×10^6	Glass	$10^{-10} - 10^{-14}$
Constantin	2.0×10^6	Mica	$10^{-11} - 10^{-15}$
Mercury	1.0×10^6	Fused quartz	0.4×10^{-17}

range of current densities for which the material is linear, that is, for which the conductivity is a constant, is very large for conductors.

Conductor in a static electric field

Let us now consider a conductor placed in a static electric field, as shown in Fig. 2.24(a). The free electrons in the conductor move opposite to the direction lines of the electric field. If there is a way by means of which the flow of electrons can be continued to form a closed circuit, then a continuous flow of current takes place. Since the conductor is bounded by free space, the electrons are held at the boundary from moving further. Thus a negative surface charge forms on the boundary, accompanied by an equal amount of positive surface charge, as shown in Fig. 2.24(b), since the conductor as a whole is neutral. The surface charge distribution formed in this manner produces a secondary electric field which, together with the applied electric field, makes the field inside the conductor zero. We shall illustrate the computation of the surface charge densities by means of a simple example.

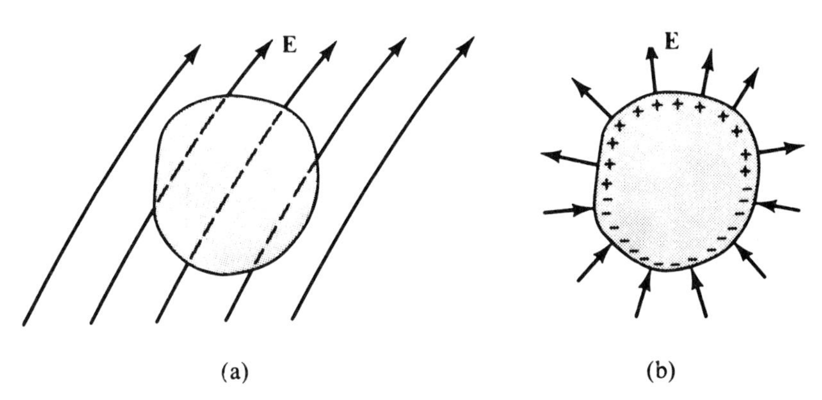

Figure 2.24. For illustrating the surface charge formation at the boundary of a conductor placed in a static electric field.

Example 2.8

Let us consider an infinite plane conducting slab of thickness d occupying the region between $z = 0$ and $z = d$ and in a uniform electric field $\mathbf{E} = E_0\mathbf{i}_z$ produced by two infinite plane sheets of equal and opposite uniform charge densities on either side of the slab, as shown in Fig. 2.25(a). We wish to find the charge densities induced on the surfaces of the slab.

Since the applied electric field is uniform and is directed along the z-direction, a negative charge of uniform density forms on the surface $z = 0$ due to the accumulation of free electrons at that surface. A positive charge of equal and opposite uniform density forms on the surface $z = d$ due to a deficiency of electrons at that surface. Let these surface charge densities be $-\rho_{S0}$ and ρ_{S0}, respectively. To satisfy the property that the field in the interior of the conductor is zero, the secondary field produced by the surface charges must be equal and opposite to the applied field; that is, it must be equal to $-E_0\mathbf{i}_z$. Now, each surface charge produces a field intensity directed normally from it and having a magnitude $1/2\varepsilon_0$ times the charge density so that the field due to the two surface charges together is equal to $-(\rho_{S0}/\varepsilon_0)\mathbf{i}_z$ inside the conductor and zero outside the conductor, as shown in Fig. 2.24(b). Thus, for zero field inside the conductor,

$$-\frac{\rho_{S0}}{\varepsilon_0}\mathbf{i}_z = -E_0\mathbf{i}_z$$

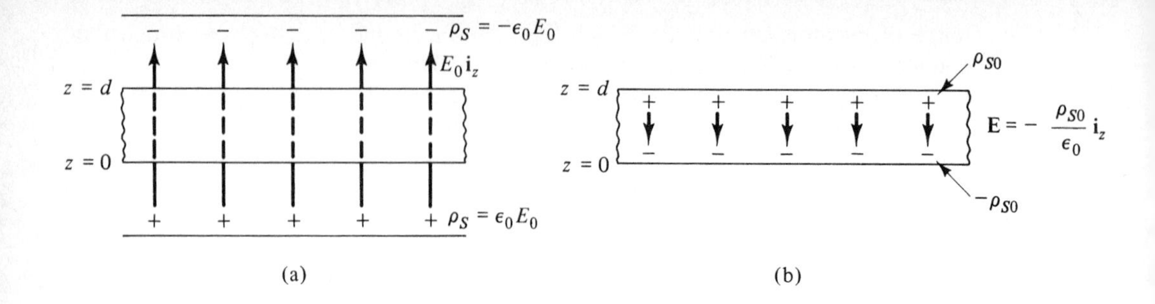

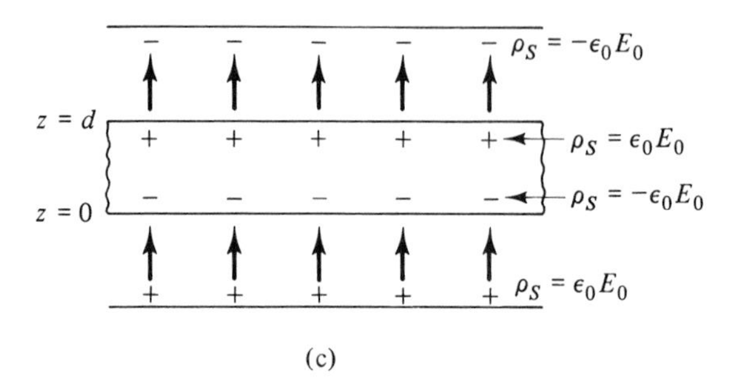

Figure 2.25. (a) Infinite plane slab conductor in a uniform applied field. (b) Induced surface charge at the boundaries of the conductor and the secondary field. (c) Sum of the applied and the secondary fields.

or

$$\rho_{S0} = \varepsilon_0 E_0$$

The field outside the conductor remains the same as the applied field since the secondary field in that region due to the surface charges is zero. The induced surface charge distribution and the fields inside and outside the conductor are shown in Fig. 2.25(c). In the general case, the induced surface charge produces a secondary field outside the conductor also, thereby changing the applied field.

Ohm's law, resistance

Returning now to (2.54), we shall show that the well-known Ohm's law in circuit theory follows from it. To do this, let us consider a bar of conducting material of conductivity σ, length l, and uniform cross-sectional area A and between the ends of which a voltage V is applied, as shown in Fig. 2.26. The voltage sets up an electric field directed along the length of the conductor, thereby giving rise to conduction current. We shall define voltage rigorously in Section 3.1, but for the purpose of discussion here, the voltage between the two ends of the conductor is simply the electric field intensity times the length of the conductor, that is,

$$V = El \tag{2.57}$$

Then from (2.54) and (2.57), the conduction current density magnitude is given by

$$J_c = \sigma E = \frac{\sigma V}{l} \tag{2.58}$$

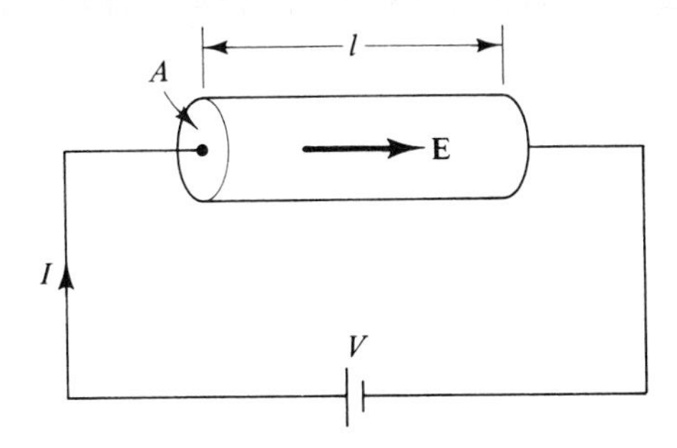

Figure 2.26. For the derivation of Ohm's law in circuit theory.

Assuming uniformity of the field and hence of the conduction current density in the cross-sectional area of the conductor, we then obtain the conduction current to be

$$I = J_c A = \frac{\sigma A}{l} V \tag{2.59}$$

Upon rearrangement, we get

$$V = I \frac{l}{\sigma A} \tag{2.60}$$

which is in the form of the familiar Ohm's law,

$$\boxed{V = IR} \tag{2.61}$$

From (2.60), the resistance R of the conducting bar can now be identified as

$$\boxed{R = \frac{l}{\sigma A}} \tag{2.62}$$

the units of R being ohms.

Hall effect

We shall conclude this section with a discussion of the Hall effect, an important phenomenon employed in the determination of charge densities in conducting and semiconducting materials, as well as in other techniques such as the measurement of fluid flow using electromagnetic flow meters. Let us consider the p-type semiconducting material in the form of a rectangular bar shown in Fig. 2.27, in which holes drift in the x-direction with a velocity $\mathbf{v} = v_x \mathbf{i}_x$ due to an applied voltage between the two ends of the bar. If a magnetic field $\mathbf{B} = B_z \mathbf{i}_z$ is applied in a perpendicular direction, then the drifting holes will experience a magnetic force $\mathbf{F}_m$ that deflects them in the $\mathbf{i}_x \times \mathbf{i}_z$ or $-\mathbf{i}_y$-direction. This deflection of holes toward the $-y$-direction establishes an electric field $\mathbf{E}_H = E_y \mathbf{i}_y$ in the material, resulting in the development of a voltage between the two sides of the bar. This phenomenon is known as the *Hall effect*, and the voltage developed is known as the *Hall voltage*. If it were not for the establishment of the Hall electric field, the holes will continually deflect toward the $-y$-direction as they drift in the x-direction. The Hall electric field exerts force $\mathbf{F}_H$ on the holes in the $+y$-direction, which in the steady-state balances exactly the magnetic force $\mathbf{F}_m$ in the

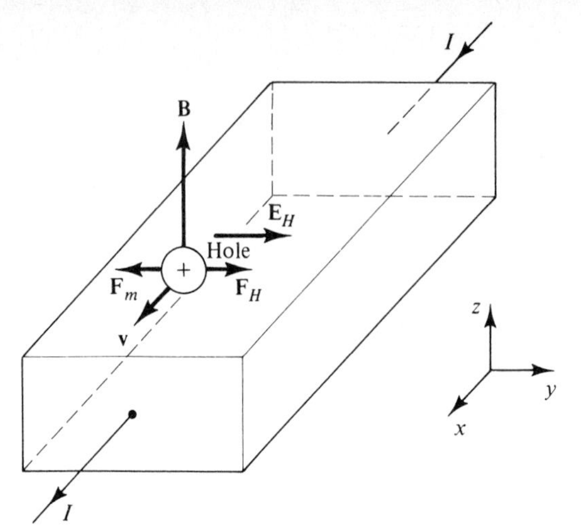

Figure 2.27. For illustrating the Hall effect phenomenon.

$-y$-direction so that the net y-directed force is zero. According to the Lorentz force equation (2.31), the Hall electric field which achieves this balance is given by

$$\begin{aligned} q(\mathbf{E}_H + \mathbf{v} \times \mathbf{B}) &= q(E_y\mathbf{i}_y + v_x\mathbf{i}_x \times B_z\mathbf{i}_z) \\ &= q(E_y - v_xB_z)\mathbf{i}_y = \mathbf{0} \end{aligned} \tag{2.63}$$

or $E_y = v_xB_z$. Using this result, the hole density can be computed from a measurement of the Hall voltage for known values of the magnetic field B_z, the current I, and the cross-sectional dimensions of the bar. If the material is n-type instead of p-type, then the charge carriers are electrons, and $\mathbf{v}$ would be in the $-x$-direction. The deflection of the charge carriers will still be toward the $-y$-direction since the charge is negative. This results in an electric field in the $-y$-direction and hence in a Hall voltage of opposite polarity to that in the case of the p-type material. Thus the polarity of the Hall voltage can be used to determine if the charge carriers are holes or electrons.

K2.4. Conduction; Conduction current density; Conductivity; Ohm's law; Conductor in a static electric field; Resistance; Hall effect.

D2.11. Find the magnitude of the electric field intensity required to establish the flow of a conduction current of 0.1 A across an area of 1 cm^2 normal to the field for each of the following cases: **(a)** in copper; **(b)** in an intrinsic semiconductor material with electron and hole mobilities of 3600 cm^2/V-s and 1700 cm^2/V-s, respectively, and electron and hole densities of 2.5×10^{13} cm^{-3}; and **(c)** in a metallic wire of circular cross section of radius 1 mm, length 1 m, and resistance 1 ohm.
Ans. **(a)** 17.24 μV/m; **(b)** 471.1 V/m; **(c)** 3.14 mV/m

D2.12. An infinite plane conducting slab lies between and parallel to two infinite plane sheets of charge of uniform surface charge densities ρ_{SA} and ρ_{SB}, as shown by the cross-sectional view in Fig. 2.28. Find the surface charge densities on the two surfaces of the slab: **(a)** ρ_{S1} and **(b)** ρ_{S2}.
Ans. **(a)** $\frac{1}{2}(\rho_{SB} - \rho_{SA})$; **(b)** $\frac{1}{2}(\rho_{SA} - \rho_{SB})$

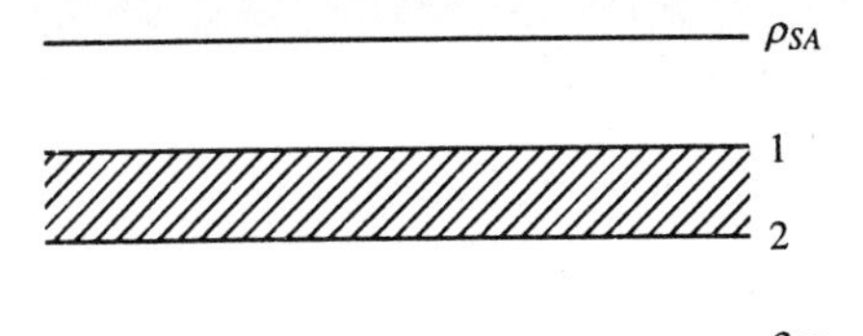

Figure 2.28. For Problem D2.12

2.5 DIELECTRICS

Polarization, electric dipole

In the preceding section we learned that conductors are characterized by an abundance of *conduction* or *free* electrons that give rise to conduction current under the influence of an applied electric field. In this section we turn our attention to dielectric materials in which the *bound* electrons are predominant. Under the application of an external electric field, the bound electrons of an atom are displaced such that the centroid of the electron cloud is separated from the centroid of the nucleus. The atom is then said to be *polarized,* thereby creating an *electric dipole,* as shown in Fig. 2.29(a). This kind of polarization is called *electronic polarization*. The schematic representation of an electric dipole is shown in Fig. 2.29(b). The strength of the dipole is defined by the electric dipole moment $\mathbf{p}$ given by

$$\boxed{\mathbf{p} = Q\mathbf{d}} \tag{2.64}$$

where $\mathbf{d}$ is the vector displacement between the centroids of the positive and negative charges, each of magnitude Q coulombs.

In certain dielectric materials, polarization may exist in the molecular structure of the material even under the application of no external electric field. The polarization of individual atoms and molecules, however, is randomly oriented, and hence the net polarization on a *macroscopic* scale is zero. The application of an external field results in torques acting on the *microscopic* dipoles, as shown in Fig. 2.30, to convert the initially random polarization into a partially coherent one along the field, on a macroscopic scale. This kind of polarization is known as *orientational polarization.* A third kind of polarization known as *ionic polarization* results from the separation of positive and negative ions in molecules formed by the transfer of electrons from one atom to another in the molecule. Certain materials exhibit permanent polarization, that is, polarization even in the absence of an applied electric field. Electrets, when allowed to solidify in the applied elec-

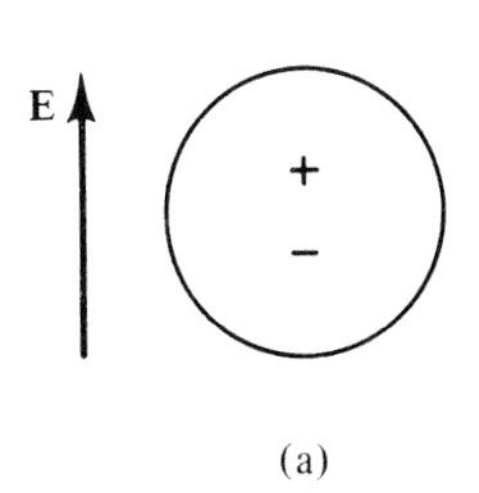

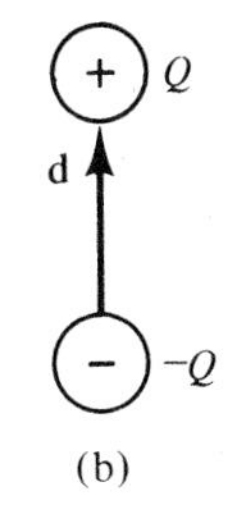

Figure 2.29. (a) Electric dipole. (b) Schematic representation of an electric dipole.

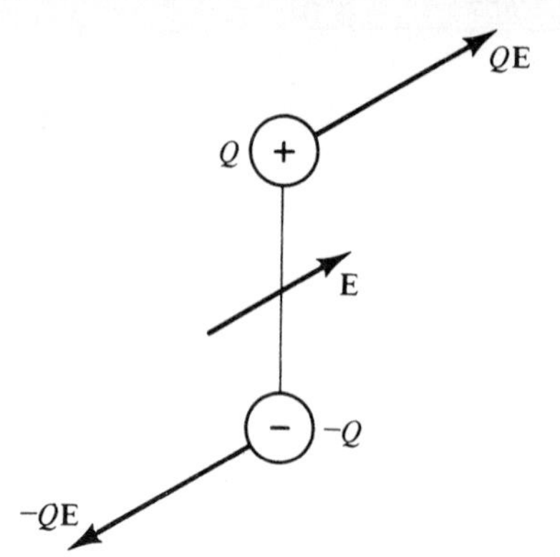

Figure 2.30. Torque acting on an electric dipole in an external electric field.

tric field, become permanently polarized, and ferroelectric materials exhibit spontaneous, permanent polarization.

On a macroscopic scale, we define a vector **P**, called the *polarization vector*, as the *electric dipole moment per unit volume*. Thus if N denotes the number of molecules per unit volume of the material, then there are $N\,\Delta v$ molecules in a volume Δv and

$$\boxed{\mathbf{P} = \frac{1}{\Delta v}\sum_{j=1}^{N\,\Delta v} \mathbf{p}_j = N\mathbf{p}} \tag{2.65}$$

where **p** is the average dipole moment per molecule. The units of **P** are coulomb-meter/meter^3 or coulombs per square meter. It is found that for many dielectric materials the polarization vector is related to the electric-field **E** in the dielectric in the simple manner given by

$$\boxed{\mathbf{P} = \varepsilon_0 \chi_e \mathbf{E}} \tag{2.66}$$

where χ_e, a dimensionless parameter, is known as the *electric susceptibility*. The quantity χ_e is a measure of the ability of the material to become polarized and differs from one dielectric to another.

Dielectric in an electric field

When a dielectric material is placed in an electric field, the induced dipoles produce a secondary electric field such that the resultant field, that is, the sum of the originally applied field and the secondary field, and the polarization vector satisfy (2.66). We shall illustrate this by means of a simple example.

Example 2.9

Let us consider an infinite plane dielectric slab of thickness d sandwiched between two infinite plane sheets of equal and opposite uniform charge densities ρ_{S0} and $-\rho_{S0}$ in the $z = 0$ and $z = d$ planes, respectively, as shown in Fig. 2.31(a). We wish to investigate the effect of polarization in the dielectric.

In the absence of the dielectric, the electric field between the sheets of charge is given by

$$\mathbf{E}_a = \frac{\rho_{S0}}{\varepsilon_0}\mathbf{i}_z \tag{2.67}$$

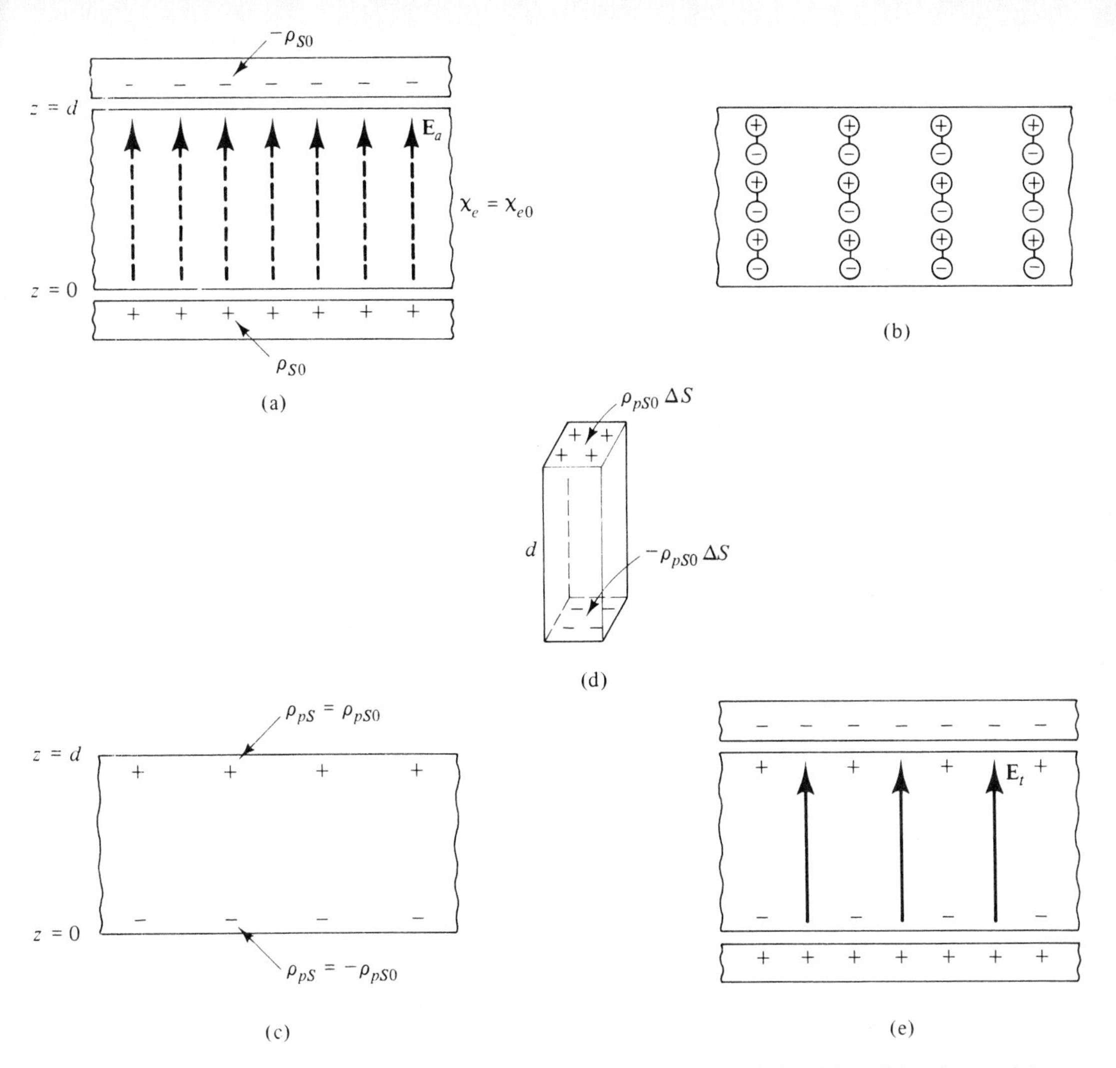

Figure 2.31. For investigating the effect of polarization induced in a dielectric material sandwiched between two infinite plane sheets of charge.

In the presence of the dielectric, this field acts as the applied electric field inducing dipole moments in the dielectric with the negative charges separated from the positive charges and pulled away from the direction of the field. Since the electric field and the electric susceptibility are uniform, the density of the induced dipole moments, that is, the polarization vector **P**, is uniform as shown in Fig. 2.31(b). Such a distribution results in exact neutralization of all the charges except at the boundaries of the dielectric since, for each positive (or negative) charge not on the surface, there is the same amount of negative (or positive) charge associated with the dipole adjacent to it, thereby canceling its effect. Thus the net result is the formation of a positive surface charge at the boundary $z = d$ and a negative surface charge at the boundary $z = 0$ as shown in Fig. 2.31(c). These surface charges are known as polarization surface charges since they are due to the polarization in the dielectric. In view of the uniform density of the dipole moments, the surface charge densities are uniform. Also, in the absence of a net charge in the interior of the dielectric, the surface charge densities must be equal in magnitude to preserve the charge neutrality of the dielectric.

Let us therefore denote the surface charge densities as

$$\rho_{pS} = \begin{cases} \rho_{pS0} & \text{for} \quad z = d \\ -\rho_{pS0} & \text{for} \quad z = 0 \end{cases} \tag{2.68}$$

where the subscript p in addition to the other subscripts stands for polarization. If we now consider a vertical column of infinitesimal rectangular cross-sectional area ΔS cut out from the dielectric as shown in Fig. 2.31(d), the equal and opposite surface charges make the column appear as a dipole of moment $(\rho_{pS0}\,\Delta S)d\mathbf{i}_z$. On the other hand, writing

$$\mathbf{P} = P_0\mathbf{i}_z \tag{2.69}$$

where P_0 is a constant in view of the uniformity of the induced polarization, the dipole moment of the column is equal to $\mathbf{P}$ times the volume of the column, or $P_0(d\,\Delta S)\mathbf{i}_z$. Equating the dipole moments computed in the two different ways, we have

$$\rho_{pS0} = P_0 \tag{2.70}$$

Thus we have related the surface charge density to the magnitude of the polarization vector. Now, the surface charge distribution produces a secondary field $\mathbf{E}_s$ given by

$$\mathbf{E}_s = \begin{cases} -\dfrac{\rho_{pS0}}{\varepsilon_0}\mathbf{i}_z = -\dfrac{P_0}{\varepsilon_0}\mathbf{i}_z & \text{for} \quad 0 < z < d \\ \mathbf{0} & \text{otherwise} \end{cases} \tag{2.71}$$

Denoting the total field in the dielectric to be $\mathbf{E}_t$, we have

$$\mathbf{E}_t = \mathbf{E}_a + \mathbf{E}_s = \frac{\rho_{S0}}{\varepsilon_0}\mathbf{i}_z - \frac{P_0}{\varepsilon_0}\mathbf{i}_z \tag{2.72}$$

But from (2.66),

$$\mathbf{P} = \varepsilon_0\chi_{e0}\mathbf{E}_t \tag{2.73}$$

Substituting (2.69) and (2.72) into (2.73), we obtain

$$P_0 = \chi_{e0}(\rho_{S0} - P_0)$$

or

$$P_0 = \frac{\chi_{e0}\rho_{S0}}{1 + \chi_{e0}} \tag{2.74}$$

Thus the polarization surface charge densities are given by

$$\rho_{pS} = \begin{cases} \dfrac{\chi_{e0}\rho_{S0}}{1 + \chi_{e0}} & \text{for} \quad z = d \\ -\dfrac{\chi_{e0}\rho_{S0}}{1 + \chi_{e0}} & \text{for} \quad z = 0 \end{cases} \tag{2.75}$$

and the electric field intensity in the dielectric is

$$\mathbf{E}_t = \frac{\rho_{S0}}{\varepsilon_0(1 + \chi_{e0})}\mathbf{i}_z \tag{2.76}$$

as shown in Fig. 2.31(e).

Displacement flux density, permittivity

We have just learned that the electric field in a dielectric material is the superposition of an applied field $\mathbf{E}_a$ and a secondary field $\mathbf{E}_s$ which results from the polarization $\mathbf{P}$, which in turn is induced by the total field $(\mathbf{E}_a + \mathbf{E}_s)$. Thus we have

$$\mathbf{P} = \varepsilon_0 \chi_e(\mathbf{E}_a + \mathbf{E}_s) \tag{2.77a}$$

$$\mathbf{E}_s = f(\mathbf{P}) \tag{2.77b}$$

where $f(\mathbf{P})$ denotes a function of $\mathbf{P}$. Determination of the secondary field $\mathbf{E}_s$ and hence the total field $(\mathbf{E}_a + \mathbf{E}_s)$ for a given applied field $\mathbf{E}_a$ requires a simultaneous solution of (2.77a) and (2.77b). To eliminate the need for the explicit determination of $\mathbf{P}$, we now define a new vector field $\mathbf{D}$, known as the displacement flux density as

$$\boxed{\mathbf{D} = \varepsilon_0 \mathbf{E} + \mathbf{P}} \tag{2.78}$$

Note that the units of $\mathbf{D}$ are the same as those of $\mathbf{P}$, that is, coulombs per square meter.

Substituting for $\mathbf{P}$ in (2.78) by using (2.66), we obtain

$$\begin{aligned} \mathbf{D} &= \varepsilon_0 \mathbf{E} + \varepsilon_0 \chi_e \mathbf{E} \\ &= \varepsilon_0(1 + \chi_e)\mathbf{E} \\ &= \varepsilon_0 \varepsilon_r \mathbf{E} \end{aligned} \tag{2.79}$$

or

$$\boxed{\mathbf{D} = \varepsilon \mathbf{E}} \tag{2.80}$$

where we define

$$\boxed{\varepsilon_r = 1 + \chi_e} \tag{2.81}$$

and

$$\boxed{\varepsilon = \varepsilon_0 \varepsilon_r} \tag{2.82}$$

The quantity ε_r is known as the *relative permittivity* or *dielectric constant* of the dielectric, and ε is the *permittivity* of the dielectric. The permittivity ε takes into account the effects of polarization, and there is no need to consider them when we use ε for ε_0! The relative permittivity is an experimentally measurable parameter,

TABLE 2.2 RELATIVE PERMITTIVITIES OF SOME MATERIALS

Material	Relative permittivity	Material	Relative permittivity
Air	1.0006	Dry earth	5
Paper	2.0–3.0	Mica	6
Teflon	2.1	Neoprene	6.7
Polystyrene	2.56	Wet earth	10
Plexiglass	2.6–3.5	Ethyl alcohol	24.3
Nylon	3.5	Glycerol	42.5
Fused quartz	3.8	Distilled water	81
Bakelite	4.9	Titanium dioxide	100

as we shall discuss in Section 3.3. Its values for several dielectric materials are listed in Table 2.2.

Returning now to Example 2.9, we observe that in the absence of the dielectric between the sheets of charge,

$$\mathbf{E} = \mathbf{E}_a = \frac{\rho_{S0}}{\varepsilon_0}\mathbf{i}_z \tag{2.83a}$$

$$\mathbf{D} = \varepsilon_0 \mathbf{E}_a = \rho_{S0}\mathbf{i}_z \tag{2.83b}$$

since **P** is equal to zero. In the presence of the dielectric between the sheets of charge,

$$\mathbf{E} = \mathbf{E}_t = \frac{\rho_{S0}}{\varepsilon_0(1 + \chi_{e0})}\mathbf{i}_z = \frac{\rho_{S0}}{\varepsilon}\mathbf{i}_z \tag{2.84a}$$

$$\mathbf{D} = \varepsilon\mathbf{E} = \rho_{S0}\mathbf{i}_z \tag{2.84b}$$

Thus the **D** fields are the same in both cases whereas the expressions for the **E** fields differ in the permittivities, that is, with ε_0 replaced by ε. The situation in general is however not so simple because the dielectric alters the original field distribution. In the case of Example 2.9, the geometry is such that the original field distribution is not altered by the dielectric. Also in the general case of a nonuniform susceptibility, the situation is equivalent to having a polarization volume charge inside the dielectric in addition to polarization surface charges on its boundaries. Furthermore, for time-varying fields, the electric dipoles oscillate with time, creating the equivalent of a polarization current in the dielectric. However, all these are implicitly taken into account by the permittivity ε of the dielectric.

Anisotropic dielectric materials

The nature of (2.54), which is characteristic of conductors, and of (2.80), which is characteristic of dielectrics, implies that $\mathbf{J}_c$ in the case of conductors and **D** in the case of dielectrics are in the same direction as that of **E**. Such materials are said to be *isotropic* materials. For *anisotropic* materials, this is not necessarily the case. To explain, we shall consider *anisotropic dielectric materials*. Then **D** is not in general in the same direction as that of **E**. This arises because the induced polarization is such that the polarization vector **P** is not necessarily in the same direction as that of **E**. In fact, the angle between the directions of the applied **E** and the resulting **P** depends on the direction of **E**. The relationship between **D** and $\mathbf{E}$ is then expressed in the form of a matrix equation as

$$\begin{bmatrix} D_x \\ D_y \\ D_z \end{bmatrix} = \begin{bmatrix} \varepsilon_{xx} & \varepsilon_{xy} & \varepsilon_{xz} \\ \varepsilon_{yx} & \varepsilon_{yy} & \varepsilon_{yz} \\ \varepsilon_{zx} & \varepsilon_{zy} & \varepsilon_{zz} \end{bmatrix} \begin{bmatrix} E_x \\ E_y \\ E_z \end{bmatrix} \tag{2.85}$$

Thus each component of $\mathbf{D}$ is in general dependent upon each component of $\mathbf{E}$. The square matrix in (2.85) is known as the *permittivity tensor* of the anisotropic dielectric.

Although $\mathbf{D}$ is not in general parallel to $\mathbf{E}$ for anisotropic dielectrics, there are certain polarizations of $\mathbf{E}$ for which $\mathbf{D}$ is parallel to $\mathbf{E}$. These are said to correspond to the characteristic polarizations, where the word *polarization* here refers to the direction of the field, not to the creation of electric dipoles. We shall consider an example to investigate the characteristic polarizations.

Example 2.10

An anisotropic dielectric material is characterized by the permittivity tensor

$$[\varepsilon] = \varepsilon_0 \begin{bmatrix} 7 & 2 & 0 \\ 2 & 4 & 0 \\ 0 & 0 & 3 \end{bmatrix}$$

Let us find $\mathbf{D}$ for several cases of $\mathbf{E}$.

Substituting the given permittivity matrix into (2.85), we obtain

$$D_x = 7\varepsilon_0 E_x + 2\varepsilon_0 E_y$$

$$D_y = 2\varepsilon_0 E_x + 4\varepsilon_0 E_y$$

$$D_z = 3\varepsilon_0 E_z$$

For $\mathbf{E} = E_0\mathbf{i}_z$, $\mathbf{D} = 3\varepsilon_0 E_0\mathbf{i}_z = 3\varepsilon_0\mathbf{E}$; $\mathbf{D}$ is parallel to $\mathbf{E}$.
For $\mathbf{E} = E_0\mathbf{i}_x$, $\mathbf{D} = 7\varepsilon_0 E_0\mathbf{i}_x + 2\varepsilon_0 E_0\mathbf{i}_y$; $\mathbf{D}$ is not parallel to $\mathbf{E}$.
For $\mathbf{E} = E_0\mathbf{i}_y$, $\mathbf{D} = 2\varepsilon_0 E_0\mathbf{i}_x + 4\varepsilon_0 E_0\mathbf{i}_y$; $\mathbf{D}$ is not parallel to $\mathbf{E}$.
For $\mathbf{E} = E_0(\mathbf{i}_x + 2\mathbf{i}_y)$, $\mathbf{D} = 11\varepsilon_0 E_0\mathbf{i}_x + 10\varepsilon_0 E_0\mathbf{i}_y$; $\mathbf{D}$ is not parallel to $\mathbf{E}$.
For $\mathbf{E} = E_0(2\mathbf{i}_x + \mathbf{i}_y)$, $\mathbf{D} = 16\varepsilon_0 E_0\mathbf{i}_x + 8\varepsilon_0 E_0\mathbf{i}_y = 8\varepsilon_0 E_0(2\mathbf{i}_x + \mathbf{i}_y) = 8\varepsilon_0\mathbf{E}$; $\mathbf{D}$ is parallel to $\mathbf{E}$.

When $\mathbf{D}$ is parallel to $\mathbf{E}$, that is, for the characteristic polarizations of $\mathbf{E}$, one can define an *effective permittivity* as the ratio of $\mathbf{D}$ to $\mathbf{E}$. Thus for the case of $\mathbf{E} = E_0\mathbf{i}_z$, the effective permittivity is $3\varepsilon_0$, and for the case of $\mathbf{E} = E_0(2\mathbf{i}_x + \mathbf{i}_y)$, the effective permittivity is $8\varepsilon_0$. For the characteristic polarizations, the anisotropic material behaves effectively as an isotropic dielectric having the permittivity equal to the corresponding effective permittivity.

K2.5. Polarization; Electric dipole; Polarization vector; Displacement flux density; Permittivity; Relative permittivity; Anisotropic dielectric; Characteristic polarizations; Effective permittivity.

D2.13. Infinite plane sheets of uniform charge densities 1 μC/m^2 and -1 μC/m^2 occupy the planes $z = 0$ and $z = d$, respectively. The region $0 < z < d$ is a dielectric of permittivity $4\varepsilon_0$. Find the values of **(a)** $\mathbf{D}$, **(b)** $\mathbf{E}$, and **(c)** $\mathbf{P}$, in the region $0 < z < d$.
Ans. **(a)** $10^{-6}\mathbf{i}_z$ C/m^2; **(b)** $9000\pi\,\mathbf{i}_z$ V/m; **(c)** $0.75 \times 10^{-6}\mathbf{i}_z$ C/m^2

D2.14. For an anisotropic dielectric material characterized by the **D** to **E** relationship

$$\begin{bmatrix} D_x \\ D_y \\ D_z \end{bmatrix} = \varepsilon_0 \begin{bmatrix} 8 & 2 & 0 \\ 2 & 5 & 0 \\ 0 & 0 & 9 \end{bmatrix} \begin{bmatrix} E_x \\ E_y \\ E_z \end{bmatrix}$$

find the value of the effective relative permittivity for each of the following electric field intensities corresponding to the characteristic polarizations: **(a)** $\mathbf{E} = E_0\mathbf{i}_z$; **(b)** $\mathbf{E} = E_0(\mathbf{i}_x - 2\mathbf{i}_y)$; and **(c)** $\mathbf{E} = E_0(2\mathbf{i}_x + \mathbf{i}_y)$.
Ans. **(a)** 9; **(b)** 4; **(c)** 9

2.6 MAGNETIC MATERIALS

Magnetization, magnetic dipole

In the preceding two sections we have been concerned with the response of materials to electric fields. We now turn our attention to materials known as magnetic materials which, as the name implies, are classified according to their magnetic behavior. According to a simplified atomic model, the electrons associated with a particular nucleus orbit around the nucleus in circular paths while spinning about themselves. In addition, the nucleus itself has a spin motion associated with it. Since the movement of charge constitutes a current, these orbital and spin motions are equivalent to current loops of atomic dimensions. A current loop is the magnetic analog of the electric dipole. Thus each atom can be characterized by a superposition of magnetic dipole moments corresponding to the electron orbital motions, electron spin motions, and the nuclear spin. However, owing to the heavy mass of the nucleus, the angular velocity of the nuclear spin is much smaller than that of an electron spin and hence the equivalent current associated with the nuclear spin is much smaller than the equivalent current associated with an electron spin. The dipole moment due to the nuclear spin can therefore be neglected in comparison with the other two effects. The schematic representations of a magnetic dipole as seen from along its axis and from a point in its plane are shown in Fig. 2.32(a) and (b), respectively. The strength of the dipole is defined by the magnetic dipole moment **m** given by

$$\boxed{\mathbf{m} = IA\,\mathbf{i}_n} \tag{2.86}$$

where A is the area enclosed by the current loop and $\mathbf{i}_n$ is the unit vector normal to the plane of the loop and directed in the right-hand sense.

In many materials the net magnetic moment of each atom is zero; that is, on the average, the magnetic dipole moments corresponding to the various electronic

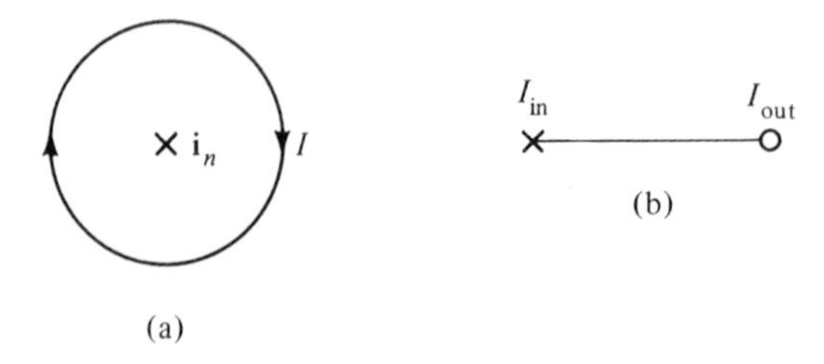

Figure 2.32. Schematic representation of a magnetic dipole as seen from (a) along its axis and (b) a point in its plane.

orbital and spin motions add up to zero. An external magnetic field has the effect of inducing a net dipole moment by changing the angular velocities of the electronic orbits, thereby magnetizing the material. This kind of magnetization, known as *diamagnetism,* is in fact prevalent in all materials. In certain materials known as *paramagnetic materials,* the individual atoms possess net nonzero magnetic moments even in the absence of an external magnetic field. These *permanent* magnetic moments of the individual atoms are, however, randomly oriented so that the net magnetization on a macroscopic scale is zero. An applied magnetic field has the effect of exerting torques on the individual permanent dipoles as shown in Fig. 2.33 to convert, on a macroscopic scale, the initially random alignment into a partially coherent one along the magnetic field, that is, with the normal to the current loop directed along the magnetic field. This kind of magnetization is known as *paramagnetism.* Certain materials known as *ferromagnetic, antiferromagnetic,* and *ferrimagnetic* materials exhibit permanent magnetization, that is, magnetization even in the absence of an applied magnetic field.

On a macroscopic scale we define a vector **M,** called the *magnetization vector,* as the *magnetic dipole moment per unit volume*. Thus if N denotes the number of molecules per unit volume of the material, then there are $N\,\Delta v$ molecules in a volume Δv and

$$\boxed{\mathbf{M} = \frac{1}{\Delta v}\sum_{j=1}^{N\Delta v} \mathbf{m}_j = N\mathbf{m}} \tag{2.87}$$

where **m** is the average dipole moment per molecule. The units of **M** are ampere-meter^2/meter^3 or amperes per meter. It is found that for many magnetic materials, the magnetization vector is related to the magnetic field **B** in the material in the simple manner given by

$$\boxed{\mathbf{M} = \frac{\chi_m}{1 + \chi_m}\frac{\mathbf{B}}{\mu_0}} \tag{2.88}$$

where χ_m, a dimensionless parameter, is known as the *magnetic susceptibility*. The quantity χ_m is a measure of the ability of the material to become magnetized and differs from one magnetic material to another.

Magnetic material in a magnetic field

When a magnetic material is placed in a magnetic field, the induced dipoles produce a secondary magnetic field such that the resultant field, that is, the sum of the originally applied field and the secondary field, and the magnetization vector satisfy (2.88). We shall illustrate this by means of an example.

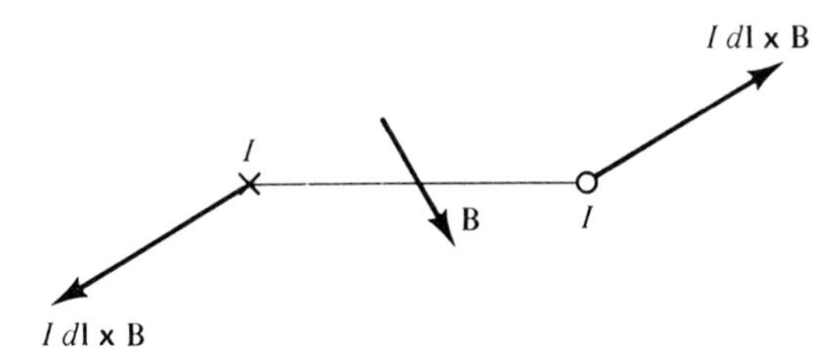

Figure 2.33. Torque acting on a magnetic dipole in an external magnetic field.

Example 2.11

Let us consider an infinite plane magnetic material slab of thickness d sandwiched between two infinite plane sheets of equal and opposite uniform current densities $J_{S0}\mathbf{i}_y$ and $-J_{S0}\mathbf{i}_y$ in the $z = 0$ and $z = d$ planes, respectively, as shown in Fig. 2.34(a). We wish to investigate the effect of magnetization in the magnetic material.

In the absence of the magnetic material, the magnetic field between the sheets of current is given by

$$\begin{aligned}\mathbf{B}_a &= \mu_0 J_{S0}\mathbf{i}_y \times \mathbf{i}_z \\ &= \mu_0 J_{S0}\mathbf{i}_x \end{aligned} \tag{2.89}$$

In the presence of the magnetic material, this field acts as the applied magnetic field resulting in magnetic dipole moments in the material which are oriented along the

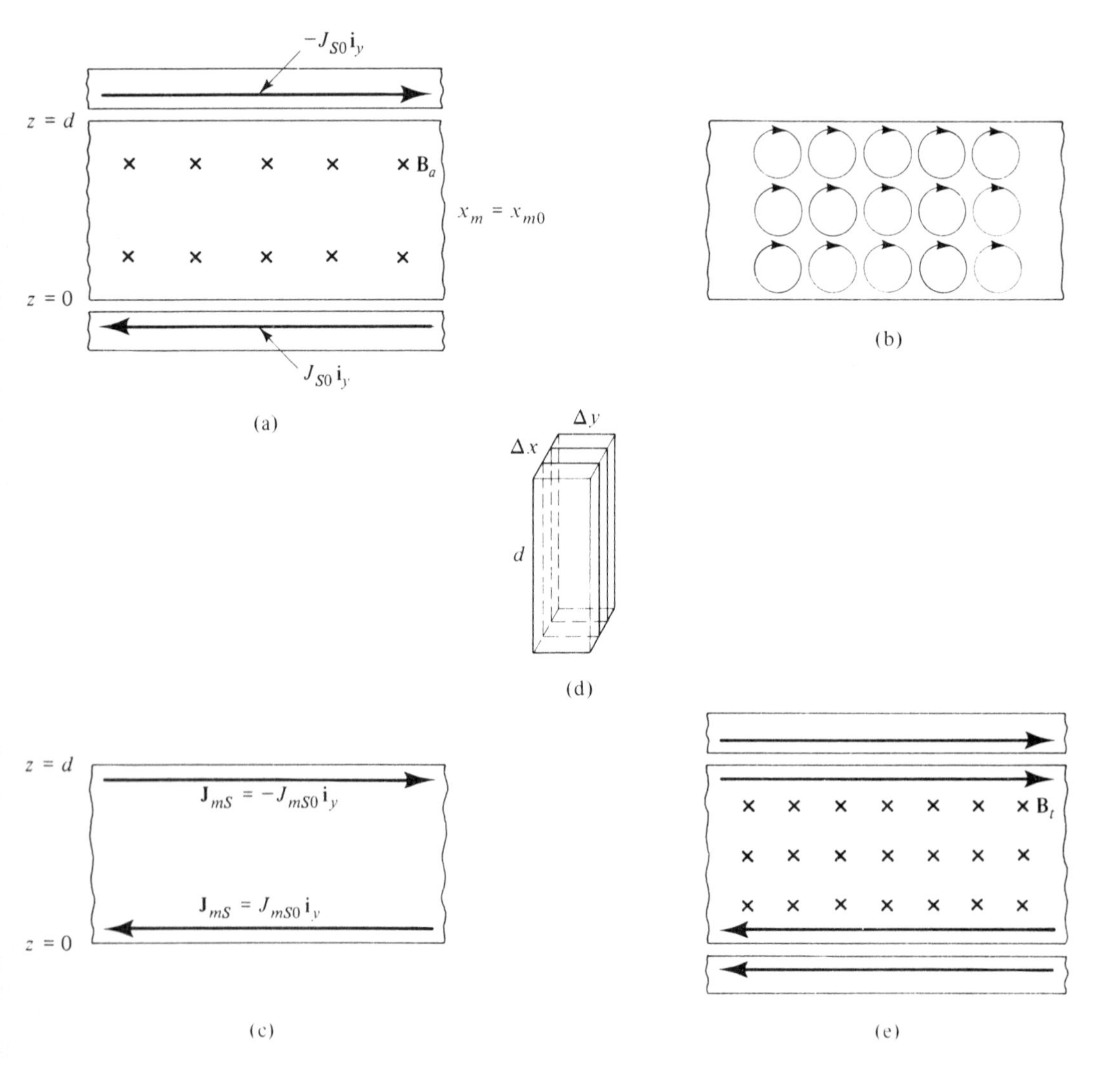

Figure 2.34. For investigating the effect of magnetization induced in a magnetic material sandwiched between two infinite plane sheets of current.

field. Since the magnetic field and the magnetic susceptibility are uniform, the density of the dipole moments, that is, the magnetization vector **M**, is uniform as shown in Fig. 2.34(b). Such a distribution results in exact cancellation of currents everywhere except at the boundaries of the material since, for each current segment not on the surface, there is a current segment associated with the dipole adjacent to it and carrying the same amount of current in the opposite direction, thereby canceling its effect. Thus the net result is the formation of a negative y-directed surface current at the boundary $z = d$ and a positive y-directed surface current at the boundary $z = 0$ as shown in Fig. 2.34(c). These surface currents are known as magnetization surface currents since they are due to the magnetization in the material. In view of the uniform density of the dipole moments, the surface current densities are uniform. Also, in the absence of a net current in the interior of the magnetic material, the surface current densities must be equal in magnitude so that whatever current flows on one surface returns via the other surface.

Let us therefore denote the surface current densities as

$$\mathbf{J}_{mS} = \begin{cases} J_{mS0}\mathbf{i}_y & \text{for} \quad z = 0 \\ -J_{mS0}\mathbf{i}_y & \text{for} \quad z = d \end{cases} \tag{2.90}$$

where the subscript m in addition to the other subscripts stands for magnetization. If we now consider a vertical column of infinitesimal rectangular cross-sectional area $\Delta S = (\Delta x)(\Delta y)$ cut out from the magnetic material as shown in Fig. 2.34(d), the rectangular current loop of width Δx makes the column appear as a dipole of moment $(J_{mS0}\,\Delta x)(d\,\Delta y)\mathbf{i}_x$. On the other hand, writing

$$\mathbf{M} = M_0\mathbf{i}_x \tag{2.91}$$

where M_0 is a constant in view of the uniformity of the magnetization, the dipole moment of the column is equal to $\mathbf{M}$ times the volume of the column, or M_0 $(d\,\Delta x\,\Delta y)\mathbf{i}_x$. Equating the dipole moments computed in two different ways, we have

$$J_{mS0} = M_0 \tag{2.92}$$

Thus we have related the surface current density to the magnitude of the magnetization vector. Now, the surface current distribution produces a secondary field $\mathbf{B}_s$ given by

$$\mathbf{B}_s = \begin{cases} \mu_0 J_{mS0}\mathbf{i}_x = \mu_0 M_0\mathbf{i}_x & \text{for} \quad 0 < z < d \\ \mathbf{0} & \text{otherwise} \end{cases} \tag{2.93}$$

Denoting the total field inside the magnetic material to be $\mathbf{B}_t$, we have

$$\begin{aligned} \mathbf{B}_t &= \mathbf{B}_a + \mathbf{B}_s = \mu_0 J_{S0}\mathbf{i}_x + \mu_0 M_0\mathbf{i}_x \\ &= \mu_0(J_{S0} + M_0)\mathbf{i}_x \end{aligned} \tag{2.94}$$

But, from (2.88),

$$\mathbf{M} = \frac{\chi_{m0}}{1 + \chi_{m0}}\frac{\mathbf{B}_t}{\mu_0} \tag{2.95}$$

Substituting (2.91) and (2.94) into (2.95), we have

$$M_0 = \frac{\chi_{m0}}{1 + \chi_{m0}}(J_{S0} + M_0)$$

or

$$M_0 = \chi_{m0} J_{S0} \tag{2.96}$$

Thus the magnetization surface current densities are given by

$$\mathbf{J}_{mS} = \begin{cases} \chi_{m0} J_{S0} \mathbf{i}_y & \text{for} \quad z = 0 \\ -\chi_{m0} J_{S0} \mathbf{i}_y & \text{for} \quad z = d \end{cases} \tag{2.97}$$

and the magnetic flux density in the magnetic material is

$$\mathbf{B}_t = \mu_0(1 + \chi_{m0}) J_{S0} \mathbf{i}_x \tag{2.98}$$

as shown in Fig. 2.34(e).

Magnetic field intensity, permeability

We have just learned that the magnetic field in a magnetic material is the superposition of an applied field $\mathbf{B}_a$ and a secondary field $\mathbf{B}_s$ which results from the polarization $\mathbf{M}$, which in turn is induced by the total field $(\mathbf{B}_a + \mathbf{B}_s)$. Thus we have

$$\mathbf{M} = \frac{\chi_m}{1 + \chi_m} \frac{\mathbf{B}_a + \mathbf{B}_s}{\mu_0} \tag{2.99a}$$

$$\mathbf{B}_s = f(\mathbf{M}) \tag{2.99b}$$

where $f(\mathbf{M})$ denotes a function of $\mathbf{M}$. Determination of the secondary field $\mathbf{B}_s$ and hence the total field $(\mathbf{B}_a + \mathbf{B}_s)$ for a given applied field $\mathbf{B}_a$ requires a simultaneous solution of (2.99a) and (2.99b). To eliminate the need for the explicit determination of $\mathbf{M}$, we now define a new vector field $\mathbf{H}$, known as the *magnetic field intensity*, as

$$\boxed{\mathbf{H} = \frac{\mathbf{B}}{\mu_0} - \mathbf{M}} \tag{2.100}$$

Note that the units of $\mathbf{H}$ are the same as those of $\mathbf{M}$, that is, amperes per meter.

Substituting for $\mathbf{M}$ by using (2.88), we obtain

$$\begin{aligned} \mathbf{H} &= \frac{\mathbf{B}}{\mu_0} - \frac{\chi_m}{1 + \chi_m} \frac{\mathbf{B}}{\mu_0} \\ &= \frac{\mathbf{B}}{\mu_0(1 + \chi_m)} \\ &= \frac{\mathbf{B}}{\mu_0 \mu_r} \end{aligned} \tag{2.101}$$

or

$$\boxed{\mathbf{H} = \frac{\mathbf{B}}{\mu}} \tag{2.102}$$

where we define

$$\boxed{\mu_r = 1 + \chi_m} \tag{2.103}$$

and

$$\boxed{\mu = \mu_0 \mu_r} \tag{2.104}$$

The quantity μ_r is known as the *relative permeability* of the magnetic material and μ is the *permeability* of the magnetic material. The permeability μ takes into account the effects of magnetization, and there is no need to consider them when we use μ for μ_0!

Returning now to Example 2.11, we observe that in the absence of the magnetic material between the sheets of current,

$$\mathbf{B} = \mathbf{B}_a = \mu_0 J_{S0} \mathbf{i}_x \tag{2.105a}$$

$$\mathbf{H} = \frac{\mathbf{B}}{\mu_0} = J_{S0} \mathbf{i}_x \tag{2.105b}$$

since $\mathbf{M}$ is equal to zero. In the presence of the magnetic material between the sheets of current,

$$\mathbf{B} = \mathbf{B}_t = \mu_0(1 + \chi_m) J_{S0} \mathbf{i}_x = \mu J_{S0} \mathbf{i}_x \tag{2.106a}$$

$$\mathbf{H} = \frac{\mathbf{B}}{\mu} = J_{S0} \mathbf{i}_x \tag{2.106b}$$

Thus the **H** fields are the same in both cases whereas the expressions for the **B** fields differ in the permeabilities, that is, with μ_0 replaced by μ. The situation in general is however not so simple because the magnetic material alters the original field distribution. In the case of Example 2.11, the geometry is such that the original field distribution is not altered by the magnetic material. Also in the general case of a nonuniform susceptibility, the situation is equivalent to having a magnetization volume current inside the material in addition to the surface current at the boundaries. However, all of these are implicitly taken into account by the permeability μ of the magnetic material. For anisotropic magnetic materials, **H** is not in general parallel to **B** and the relationship between the two quantities is expressed in the form of a matrix equation as given by

$$\begin{bmatrix} B_x \\ B_y \\ B_z \end{bmatrix} = \begin{bmatrix} \mu_{xx} & \mu_{xy} & \mu_{xz} \\ \mu_{yx} & \mu_{yy} & \mu_{yz} \\ \mu_{zx} & \mu_{zy} & \mu_{zz} \end{bmatrix} \begin{bmatrix} H_x \\ H_y \\ H_z \end{bmatrix} \tag{2.107}$$

just as in the case of the relationship between **D** and **E** for anistropic dielectric materials.

Ferromagnetic materials

For many materials for which the relationship between **H** and **B** is linear, the relative permeability does not differ appreciably from unity, unlike the case of linear dielectric materials, for which the relative permittivity can be very large, as shown in Table 2.2. In fact, for diamagnetic materials, the magnetic susceptibility χ_m is a small negative number of the order -10^{-4} to -10^{-8} whereas for paramagnetic materials, χ_m is a small positive number of the order 10^{-3} to 10^{-7}. Ferromagnetic materials, however, possess large values of relative permeability on the order of several hundreds, thousands, or more. The relationship between **B** and **H** for these materials is nonlinear, resulting in a nonunique value of μ_r, for a given material. In fact, these materials are characterized by hysteresis, that is, the relationship between **B** and **H** dependent on the past history of the material.

Ferromagnetic materials possess strong dipole moments owing to the predominance of the electron spin moments over the electron orbital moments. The theory of ferromagnetism is based on the concept of magnetic *domains*, as formulated by Weiss in 1907. A magnetic domain is a small region in the material in which the atomic dipole moments are all aligned in one direction, due to strong interaction fields arising from the neighboring dipoles. In the absence of an external magnetic field, although each domain is magnetized to saturation, the magnetizations in various domains are randomly oriented as shown in Fig. 2.35(a) for a single crystal specimen. The random orientation results from minimization of the associated energy. The net magnetization is therefore zero on a macroscopic scale. With the application of a weak external magnetic field, the volumes of the domains in which the original magnetizations are favorably oriented relative to the applied field grow at the expense of the volumes of the other domains, as shown in Fig. 2.35(b). This feature is known as domain wall motion. Upon removal of the applied field, the domain wall motion reverses, bringing the material close to its original state of magnetization. With the application of stronger external fields, the domain wall motion continues to such an extent that it becomes irreversible; that is, the material does not return to its original unmagnetized state on a macroscopic scale upon removal of the field. With the application of still stronger fields, the domain wall motion is accompanied by domain rotation, that is, alignment of the magnetizations in the individual domains with the applied field as shown in Fig. 2.35(c), thereby magnetizing the material to saturation. The material retains some magnetization along the direction of the applied field even after removal of the field. In fact, an external field opposite to the original direction has to be applied to bring the net magnetization back to zero.

Hysteresis curve

We may now discuss the relationship between **B** and **H** for a ferromagnetic material, which is depicted graphically as shown by a typical curve in Fig. 2.36. This curve is known as the hysteresis curve or the B–H curve. To trace the development of the hysteresis effect, we start with an unmagnetized sample of ferromagnetic material in which both **B** and **H** are initially zero, corresponding to point a on the curve. As H is increased, the magnetization builds up, thereby increasing B gradually along the curve ab and finally to saturation at b, according to the following sequence of events as discussed earlier: (1) reversible motion of domain walls, (2) irreversible motion of domain walls, and (3) domain rotation. The

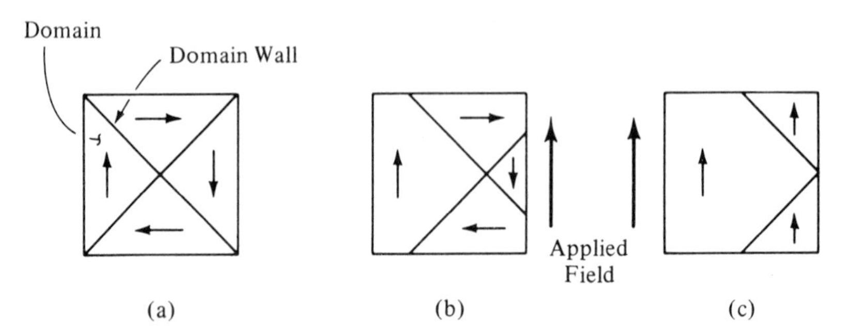

Figure 2.35. For illustrating the different steps in the magnetization of a ferromagnetic specimen: (a) unmagnetized state; (b) domain wall motion; and (c) domain rotation.

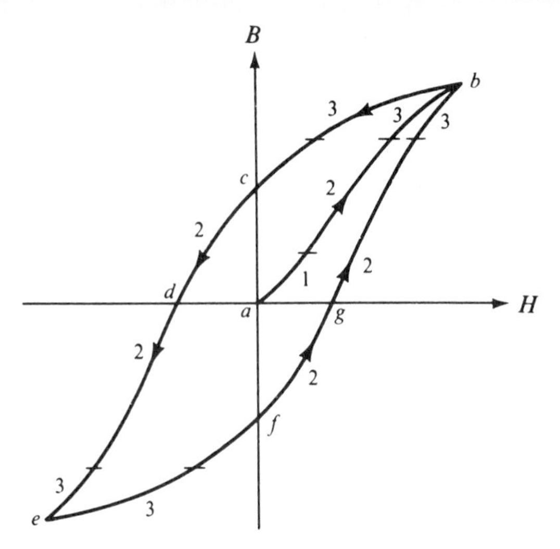

Figure 2.36. Hysteresis curve for a ferromagnetic material.

regions corresponding to these events along the curve *ab* as well as other portions of the hysteresis curve are shown marked 1, 2, and 3, respectively, in Fig. 2.36. If the value of H is now decreased to zero, the value of B does not retrace the curve *ab* backward but instead follows the curve *bc*, which indicates that a certain amount of magnetization remains in the material even after the magnetizing field is completely removed. In fact, it requires a magnetic field intensity in the opposite direction to bring B back to zero as shown by the portion *cd* of the curve. The value of B at the point c is known as the *remanence* or *retentivity,* whereas the value of H at d is known as the *coercivity* of the material. Further increase in $\mathbf{H}$ in this direction results in the saturation of $\mathbf{B}$ in the direction opposite to that corresponding to b as shown by the portion *de* of the curve. If $\mathbf{H}$ is now decreased to zero, reversed in direction, and increased, the resulting variation of $\mathbf{B}$ occurs in accordance with the curve *efgb*, thereby completing the hysteresis loop.

The nature of the hysteresis curve suggests that the hysteresis phenomenon can be used to distinguish between two states, for example, "1" and "0" in a binary number magnetic memory system. There are several kinds of magnetic memories. Although differing in details, all these are based on the principles of storing and retrieving information in regions on a magnetic medium. In disk, drum, and tape memories, the magnetic medium moves, whereas in bubble and core memories, the medium is stationary. We shall here briefly discuss only the floppy disk or diskette, commonly used as secondary memory in personal computers.[2]

Floppy disk

The floppy disk consists of a coating of ferrite material applied over a thin flexible nonmagnetic substrate for physical support. Ferrites are a class of magnetic materials characterized by almost rectangular-shaped hysteresis loops so that the two remanent states are well defined. The disk is divided into many circular tracks, and each track is subdivided into regions called sectors, as shown in Fig. 2.37. To access a sector, an electromagnetic read/write head moves across the spinning disk to the appropriate track and waits for the correct sector to rotate beneath it. The head consists of a ferrite core around which a coil is wound and

[2] See, for example, Robert M. White, "Disk-Storage Technology," *Scientific American,* August 1980, pp. 138–148.

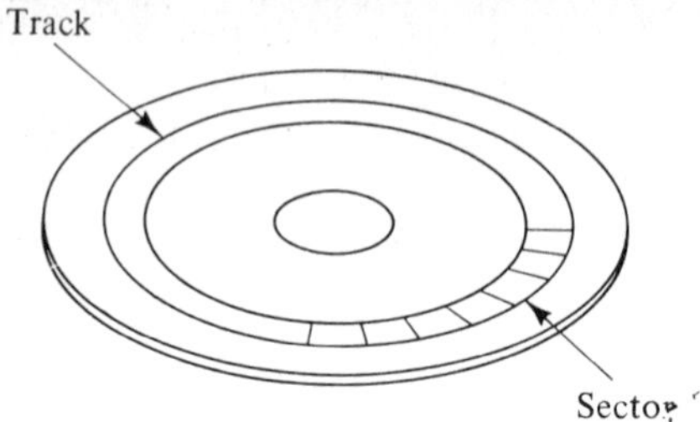

Figure 2.37. Arrangement of sectors on a floppy disk.

with a gap at the bottom, as shown in Fig. 2.38. Writing data on the disk is done by passing current through the coil. The current generates a magnetic field which in the core confines essentially to the material but in the air gap spreads out into the magnetic medium below it, thereby magnetizing the region to represent the 0 state. To store the 1 state in a region, the current in the coil is reversed to magnetize the medium in the reverse direction. Reading of data from the disk is accomplished by the changing magnetic field from the magnetized regions on the disk inducing a voltage in the coil of the head, as the disk rotates under the head. The voltage is induced in accordance with Faraday's law (which we shall study in Section 3.2) whenever there is a change in magnetic flux linked by the coil. We have here only discussed the basic principles behind storing data on the disk and retrieving data from it. There are a number of ways in which bits can be encoded on the disk. We shall however not pursue the topic here.

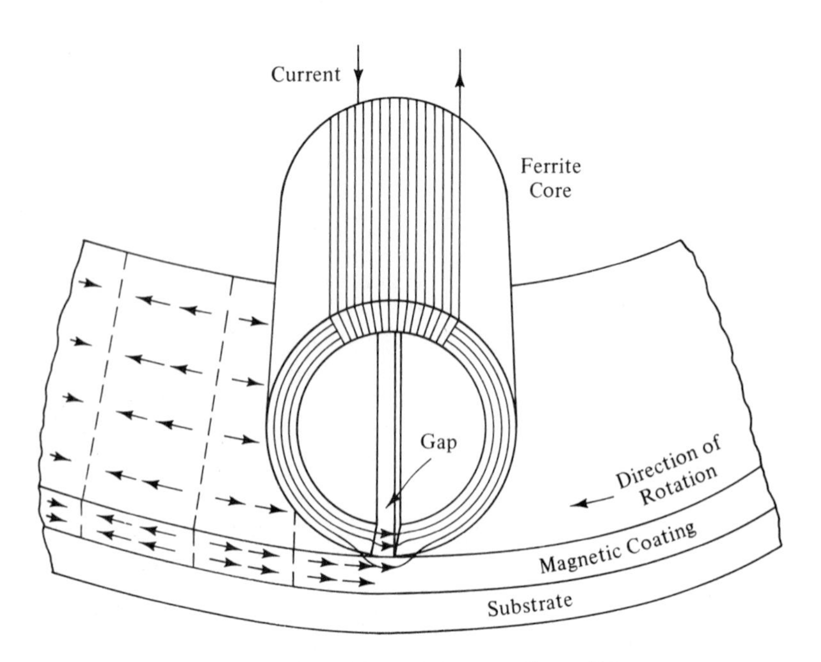

Figure 2.38. Writing of data on a floppy disk.

K2.6. Magnetization; Magnetic dipole; Magnetization vector; Magnetic field intensity; Permeability; Relative permeability; Ferromagnetic materials; Hysteresis.

D2.15. Find the magnetic dipole moment for each of the following cases: **(a)** 1 μC of charge in a circular orbit of radius $1/\sqrt{\pi}$ mm in the xy-plane around the z-axis in the sense of increasing ϕ with angular velocity of 1 revolution per millisecond; **(b)** a square current loop having the vertices at the points $A(10^{-3}, 0, 0)$, $B(0, 10^{-3}, 0)$, $C(-10^{-3}, 0, 0)$, and $D(0, -10^{-3}, 0)$ with current 0.1 A flowing in the sense $ABCDA$; and **(c)** an equilateral triangular current loop having vertices at the points $A(10^{-3}, 0, 0)$, $B(0, 10^{-3}, 0)$, and $C(0, 0, 10^{-3})$ with current 0.1 A flowing in the sense $ABCA$.
Ans. **(a)** $10^{-9}\mathbf{i}_z$ A-m^2; **(b)** $2 \times 10^{-7}\mathbf{i}_z$ A-m^2; **(c)** $5 \times 10^{-8}(\mathbf{i}_x + \mathbf{i}_y + \mathbf{i}_z)$ A-m^2

D2.16. Infinite plane sheets of current densities $0.1\mathbf{i}_y$ A/m and $-0.1\mathbf{i}_y$ A/m occupy the planes $z = 0$ and $z = d$, respectively. The region $0 < z < d$ is a magnetic material of permeability $100\mu_0$. Find **(a) H**, **(b) B**, and **(c) M**, in the region $0 < z < d$.
Ans. **(a)** $0.1\mathbf{i}_x$ A/m; **(b)** $4\pi \times 10^{-6}\mathbf{i}_x$ Wb/m^2; **(c)** $9.9\mathbf{i}_x$ A/m

2.7 SUMMARY

In this chapter we first introduced the electric field concept from consideration of an experimental law known as Coulomb's law, having to do with the electric forces between two charges. We learned that electric force acts on charges merely by virtue of the property of charge. The electric force acting on a test charge q at a point in the field region is given by

$$\mathbf{F} = q\mathbf{E}$$

where **E** is the electric field intensity at that point. The electric field intensity due to a point charge Q in free space is given by

$$\mathbf{E} = \frac{Q}{4\pi\varepsilon_0 R^2}\mathbf{i}_R$$

where ε_0 is the permittivity of free space, R is the distance from the point charge to the point at which the field intensity is to be computed, and $\mathbf{i}_R$ is the unit vector along the line joining the two points and directed away from the point charge. Using superposition in conjunction with the electric field due to a point charge, we discussed the computation of the electric field due to two point charges and the computer generation of the direction lines of the electric field. We then extended the determination of electric field intensity to continuous charge distributions.

Next we introduced the magnetic field concept from considerations of Ampère's law of force, having to do with the magnetic forces between two current loops. We learned that the magnetic field exerts force only on moving charges. The magnetic force acting on a test charge q moving with a velocity **v** at a point in the field region is given by

$$\mathbf{F} = q\mathbf{v} \times \mathbf{B}$$

where **B** is the magnetic flux density at that point. In terms of current flowing in a wire, the magnetic force acting on a current element of length $d\mathbf{l}$ and current I at a point in the field region is given by

$$\mathbf{F} = I\,d\mathbf{l} \times \mathbf{B}$$

The magnetic flux density due to a current element $I\,d\mathbf{l}$ in free space is given by the Biot–Savart law

$$\mathbf{B} = \frac{\mu_0}{4\pi}\frac{I\,d\mathbf{l} \times \mathbf{i}_R}{R^2}$$

where μ_0 is the permeability of free space and R and $\mathbf{i}_R$ have the same meanings as in the expression for $\mathbf{E}$ due to a point charge. Using superposition in conjunction with Biot–Savart law, we discussed the computation of the magnetic field due to current distributions.

Combining the electric and magnetic field concepts, we then introduced the Lorentz force equation

$$\mathbf{F} = q(\mathbf{E} + \mathbf{v} \times \mathbf{B})$$

which gives the force acting on a test charge q moving with velocity $\mathbf{v}$ at a point in a region characterized by electric field of intensity $\mathbf{E}$ and magnetic field of flux density $\mathbf{B}$. We used the Lorentz force equation to discuss (1) the determination of $\mathbf{E}$ and $\mathbf{B}$ at a point from a knowledge of forces acting on a test charge at that point for three different velocities and (2) the tracing of charged particle motion in a region of crossed electric and magnetic fields.

We devoted the rest of the chapter to introduce materials. We learned that materials can be classified as (1) conductors, (2) semiconductors, (3) dielectrics, and (4) magnetic materials, depending on the nature of the response of the charged particles in the materials to applied fields. Conductors are characterized by conduction, which is the phenomenon of steady drift of free electrons under the influence of an applied electric field, thereby resulting in a conduction current. In semiconductors, also characterized by conduction, the charge carriers are not only electrons but also holes. We learned that the conduction current density is related to the electric field intensity in the manner

$$\mathbf{J}_c = \sigma\mathbf{E} \tag{2.108}$$

where σ is the conductivity of the material. We discussed (1) the formation of surface charge at the boundaries of a conductor placed in a static electric field, (2) the derivation of Ohm's law in circuit theory, and (3) the Hall effect.

Dielectrics are characterized by polarization, which is the phenomenon of the creation and net alignment of electric dipoles, formed by the displacement of the centroids of the electron clouds from the centroids of the nucleii of the atoms, along the direction of an applied electric field. Magnetic materials are characterized by magnetization, which is the phenomenon of net alignment of the axes of the magnetic dipoles, formed by the electron orbital and spin motion around the nucleii of the atoms, along the direction of an applied magnetic field. To eliminate the need for explicitly taking into account the effects of polarization and magnetization, we defined two new vector fields known as the displacement flux density and the magnetic field intensity, given by

$$\mathbf{D} = \varepsilon_0\mathbf{E} + \mathbf{P}$$

$$\mathbf{H} = \frac{\mathbf{B}}{\mu_0} - \mathbf{M}$$

respectively, where **P** is the polarization vector and **M** is the magnetization vector. We learned that for isotropic materials, these expressions simplify to

$$\mathbf{D} = \varepsilon \mathbf{E} \tag{2.109}$$

$$\mathbf{H} = \frac{\mathbf{B}}{\mu} \tag{2.110}$$

where

$$\varepsilon = \varepsilon_0 \varepsilon_r$$

$$\mu = \mu_0 \mu_r$$

are the permittivity and the permeability, respectively, of the material and the quantities ε_r and μ_r are the relative permittivity and the relative permeability, respectively, which take into account implicitly the effects of polarization and magnetization, respectively. Equations (2.108), (2.109), and (2.110) are known as the constitutive relations.. Finally, we discussed the hysteresis phenomenon associated with ferromagnetic materials and discussed an application based on the use of the hysteresis curve.

REVIEW QUESTIONS

R2.1. State Coulomb's law. To what law in mechanics is Coulomb's law analogous?

R2.2. What is the value of the permittivity of free space? What are its units?

R2.3. What is the definition of electric field intensity? What are its units?

R2.4. Discuss two applications based on the electric force on a charged particle.

R2.5. Describe the electric field due to a point charge.

R2.6. Discuss the computer generation of the direction lines of the electric field due to two point charges.

R2.7. Discuss the different types of charge distributions. How do you determine the electric field intensity due to a charge distribution?

R2.8. Describe the electric field due to an infinite plane sheet of uniform surface charge density.

R2.9. State Ampère's force law as applied to current elements. Why is it not necessary for Newton's third law to hold for current elements?

R2.10. What are the units of magnetic flux density? How is magnetic flux density defined in terms of force on a current element?

R2.11. What is the value of the permeability of free space? What are its units?

R2.12. Describe the magnetic field due to a current element.

R2.13. Discuss the different types of current distributions. How do you determine the magnetic flux density due to a current distribution?

R2.14. Describe the magnetic field due to an infinite plane sheet of uniform surface current density.

R2.15. How is magnetic flux density defined in terms of force on a moving charge?

R2.16. Discuss two applications based on the magnetic force on a current carrying wire or on a moving charge.

R2.17. State the Lorentz force equation.

R2.18. Discuss the determination of **E** and **B** at a point from the knowledge of forces experienced by a test charge at that point for several velocities. What is the minimum required number of forces?

R2.19. Give some examples of devices based upon charged particle motion in electric and magnetic fields.

R2.20. Discuss the tracing of the path of a charged particle in a region of crossed electric and magnetic fields.

R2.21. Distinguish between bound electrons and free electrons in an atom and briefly describe the phenomenon of conduction.

R2.22. Discuss the classification of a material as a conductor, semiconductor, or dielectric with the aid of energy band diagrams.

R2.23. What is mobility? Give typical values of mobilities for electrons and holes.

R2.24. State Ohm's law valid at a point, defining the conductivities for conductors and semiconductors.

R2.25. Discuss the formation of surface charge at the boundaries of a conductor placed in a static electric field.

R2.26. Discuss the derivation of Ohm's law in circuit theory from the Ohm's law valid at a point.

R2.27. Discuss the Hall effect.

R2.28. Briefly describe the phenomenon of polarization in a dielectric material. What are the different kinds of polarization?

R2.29. What is an electric dipole? How is its strength defined?

R2.30. What is a polarization vector? How is it related to the electric field intensity?

R2.31. Discuss the effect of polarization in a dielectric material using the example of polarization surface charge.

R2.32. Discuss the definition of displacement flux density and the permittivity concept.

R2.33. What is an anisotropic dielectric material? When can an effective permittivity be defined for an anisotropic dielectric material?

R2.34. Briefly describe the phenomenon of magnetization in a magnetic material. What are the different kinds of magnetic materials?

R2.35. What is a magnetic dipole? How is its strength defined?

R2.36. What is the magnetization vector? How is it related to the magnetic flux density?

R2.37. Discuss the effect of magnetization in a magnetic material using the example of magnetization surface current.

R2.38. Discuss the definition of magnetic field intensity and the permeability concept.

R2.39. Discuss the phenomenon of hysteresis associated with ferromagnetic materials.

R2.40. Discuss the principles behind storing data on a floppy disk and retrieving the data from it.

PROBLEMS

P2.1. Point charges, each of value Q, are situated at the corners of a regular tetrahedron of edge length L. Find the electric force on each point charge.

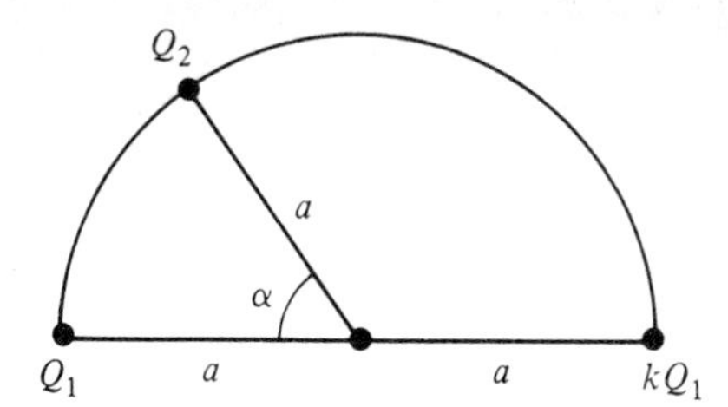

Figure 2.39. For Problem P2.2.

P2.2. Consider the arrangement of three point charges Q_1, kQ_1 $(k > 0)$, and Q_2 shown in Fig. 2.39, where Q_1 and kQ_1, are fixed and Q_2 is constrained to move on the semicircle. **(a)** Find the value of α in terms of k for which Q_2 is in equilibrium. **(b)** Find the numerical value of α for $k = 8$.

P2.3. Four point charges, each of value Q, are located at $(d, 0, 0)$, $(-d, 0, 0)$, $(0, d, 0)$, and $(0, -d, 0)$. An electron (charge e and mass m) is constrained to move along the z-axis without friction. The electron situated at the origin is displaced by a distance $z_0 \ll d$ and released at $t = 0$. **(a)** Obtain the approximate differential equation for the motion of the electron. **(b)** Show that the solution to the differential equation corresponds to simple harmonic motion and find the frequency of oscillation. Ignore the gravitational force between the charges.

P2.4. Two equal and opposite point charges Q and $-Q$ are located at $(0, 0, d/2)$ and $(0, 0, -d/2)$, respectively. Such an arrangement is known as the electric dipole. Show that the electric field intensity due to the electric dipole at very large distances from the origin compared to the spacing d is given approximately by $(Qd/4\pi\varepsilon_0 r^3)(2\cos\theta\,\mathbf{i}_r + \sin\theta\,\mathbf{i}_\theta)$.

P2.5. Assuming that the circular ring of Example 2.2 is coated with charge such that the charge density is given by $\rho_L = \rho_{L0}\sin\phi$ C/m, find the electric field intensity at a point on the z-axis by setting up the integral expression and evaluating it.

P2.6. Consider charge distributed uniformly with density ρ_{L0} C/m along the line between $(0, 0, -a)$ and $(0, 0, a)$. Obtain the expression for the electric field intensity at $(r, \phi, 0)$ in cylindrical coordinates by considering a differential length element along the line charge, setting up the field as an integral, and evaluating it. By letting $a \to \infty$, find the expression for the electric field intensity due to an infinitely long line charge of uniform density ρ_{L0} C/m.

P2.7. Consider a circular disk of radius a lying in the xy-plane with its center at the origin and carrying charge of density $4\pi\varepsilon_0 r$ C/m^2. Obtain the expression for the electric field intensity at the point $(0, 0, z)$.

P2.8. Consider volume charge distributed uniformly with density ρ_0 C/m^3 between the planes $z = -a$ and $z = a$. Using superposition in conjunction with the result of Example 2.3, show that the electric field intensity due to the slab of charge is given by

$$\mathbf{E} = \begin{cases} -(\rho_0 a/\varepsilon_0)\mathbf{i}_z & \text{for } z < -a \\ (\rho_0 z/\varepsilon_0)\mathbf{i}_z & \text{for } -a < z < a \\ (\rho_0 a/\varepsilon_0)\mathbf{i}_z & \text{for } z > a \end{cases}$$

P2.9. Six identical current elements $I\,dz\,\mathbf{i}_z$ are located at equally spaced points on a circle of radius unity centered at the origin and lying in the xy-plane. The first point is $(1, 0, 0)$. Find the magnetic force on each current element.

P2.10. An infinitesimal current element is situated at the point $(2, 2, 1)$ and along the line $x = y = 2z$. For each of the three points $A(3, 0, 3)$, $B(4, 4, 2)$, and $C(2, 2, 4)$ lying on a spherical surface of radius 3 m and centered at the current element, find the ratio of the magnitude of the magnetic flux density at the point to its maximum value on the spherical surface.

P2.11. A circular loop of wire of radius a is situated in the xy-plane with its center at the origin. It carries a current I in the clockwise sense as seen along the positive z-axis, that is, in the sense of increasing values of ϕ. Obtain the expression for $\mathbf{B}$ due to the current loop at a point on the z-axis.

P2.12. A straight wire along the z-axis carries current I in the positive z-direction. Consider the portion of the wire between $(0, 0, a_1)$ and $(0, 0, a_2)$, where $a_2 > a_1$. Show that the magnetic flux density at an arbitrary point $P(r, \phi, z)$ due to this portion of the wire is given by

$$\mathbf{B} = \frac{\mu_0 I}{4\pi r}(\cos\alpha_1 - \cos\alpha_2)\mathbf{i}_\phi$$

where α_1 and α_2 are the angles subtended by the lines from P to $(0, 0, a_1)$ and $(0, 0, a_2)$, respectively, with the z-axis. Verify your result in the limit $a_1 \to \infty$ and $a_2 \to -\infty$.

P2.13. A triangular loop of wire lies in the $x + y + z = 3$ plane with its corners at $(3, 0, 0)$, $(0, 3, 0)$, and $(0, 0, 3)$. A current of 1 A flows in the loop in the sense defined by connecting the specified points in succession. Applying the result of Problem P2.12 to each side of the loop, find the magnetic flux densities at three points: **(a)** $(0, 0, 0)$; **(b)** $(1, 1, 1)$; and **(c)** $(3, 3, 0)$.

P2.14. Three infinite plane current sheets, each of a uniform current density, exist in the coordinte planes of a Cartesian coordinate system. The magnetic flux densities due to these current sheets are given at three points: at $(2, 1, 4)$, $\mathbf{B} = B_0(2\mathbf{i}_x + 3\mathbf{i}_z)$; at $(-3, 5, 2)$, $\mathbf{B} = B_0(2\mathbf{i}_x + 2\mathbf{i}_y + \mathbf{i}_z)$; and at $(8, -4, 2)$, $\mathbf{B} = B_0(2\mathbf{i}_x + 2\mathbf{i}_y - 3\mathbf{i}_z)$. Find the magnetic flux densities at the following points: **(a)** $(2, -3, -5)$; **(b)** $(4, 8, -6)$; and **(c)** $(-1, -2, 3)$.

P2.15. Consider current flowing with uniform density $J_0\mathbf{i}_z$ A/m^2 in the volume between the planes $y = -a$ and $y = a$. Using superposition in conjunction with the result of Example 2.6, show the magnetic flux density due to the slab of current is given by

$$\mathbf{B} = \begin{cases} \mu_0 J_0 a\,\mathbf{i}_x & \text{for } y < -a \\ -\mu_0 J_0 y\,\mathbf{i}_x & \text{for } -a < y < a \\ -\mu_0 J_0 a\,\mathbf{i}_x & \text{for } y > a \end{cases}$$

P2.16. Show that the ratio of the radii of orbits of two charged particles of the same charge but different masses entering a region of uniform magnetic field perpendicular to the field and with equal kinetic energies is equal to the ratio of the square roots of their masses.

P2.17. Show that in a region of uniform, crossed electric and magnetic fields $\mathbf{E}$ and $\mathbf{B}$, respectively, a test charge released at a point in the field region with initial velocity $\mathbf{v} = (\mathbf{E} \times \mathbf{B})/B^2$ moves with constant velocity equal to the initial value. Compute $\mathbf{v}$ for $\mathbf{E}$ and $\mathbf{B}$ equal to $E_0(2\mathbf{i}_x + 2\mathbf{i}_y + \mathbf{i}_z)$ and $B_0(\mathbf{i}_x - 2\mathbf{i}_y + 2\mathbf{i}_z)$, respectively.

P2.18. The forces experienced by a test charge q at a point in a region of electric and magnetic fields $\mathbf{E}$ and $\mathbf{B}$, respectively, are given as follows for three different velocities of the test charge, where v_0 and B_0 are constants.

$$\mathbf{F}_1 = qE_0(\mathbf{i}_x + \mathbf{i}_y + \mathbf{i}_z) \quad \text{for } \mathbf{v}_1 = v_0\mathbf{i}_x$$
$$\mathbf{F}_2 = qE_0(\mathbf{i}_x + \mathbf{i}_y + \mathbf{i}_z) \quad \text{for } \mathbf{v}_2 = v_0\mathbf{i}_y$$
$$\mathbf{F}_3 = \mathbf{0} \quad \text{for } \mathbf{v}_3 = v_0\mathbf{i}_z$$

Find **E** and **B** at that point.

P2.19. Three forces experienced by a test charge q at a point in a region of electric and magnetic fields for three different velocities of the test charge are given as follows:

$$\mathbf{F}_1 = \mathbf{0} \quad \text{for } \mathbf{v}_1 = v_0\mathbf{i}_x$$
$$\mathbf{F}_2 = \mathbf{0} \quad \text{for } \mathbf{v}_2 = v_0\mathbf{i}_y$$
$$\mathbf{F}_3 = qE_0\mathbf{i}_z \quad \text{for } \mathbf{v}_3 = v_0(\mathbf{i}_x - 2\mathbf{i}_y)$$

Without explicitly evaluating the electric and magnetic fields, find the forces $\mathbf{F}_4$, $\mathbf{F}_5$, and $\mathbf{F}_6$ experienced by the test charge for three other velocities: **(a)** $\mathbf{v}_4 = \mathbf{0}$; **(b)** $\mathbf{v}_5 = v_0(\mathbf{i}_x + \mathbf{i}_y)$; and **(c)** $\mathbf{v}_6 = v_0(4\mathbf{i}_x - \mathbf{i}_y)$.

P2.20. Uniform electric and magnetic fields exist in a region of space. A test charge released at a point in the region with an initial velocity $\mathbf{v}_1$ or $\mathbf{v}_2$ moves with constant velocity equal to the initial value. Show that the test charge moves with constant velocity equal to the initial value when released with an initial velocity $(m\mathbf{v}_1 + n\mathbf{v}_2)/(m + n)$ for any nonzero $(m + n)$.

P2.21. Consider two electrons moving under thermal agitation with velocities equal in magnitude and opposite in direction. A uniform electric field is applied along the direction of motion of one of the electrons. Show that the gain in kinetic energy by the accelerating electron is greater than the loss in kinetic energy by the decelerating electron.

P2.22. **(a)** For a sinusoidally time-varying electric field $\mathbf{E} = \mathbf{E}_0 \cos \omega t$, where $\mathbf{E}_0$ is a constant, show that the steady-state solution to (2.43) is given by

$$\mathbf{v}_d = \frac{\tau e}{m\sqrt{1 + \omega^2\tau^2}}\mathbf{E}_0 \cos(\omega t - \tan^{-1}\omega\tau)$$

(b) Based on the assumption of one free electron per atom, the free electron density N_e in silver is 5.86×10^{28} m^{-3}. Using the conductivity for silver given in Table 2.1, find the frequency at which the drift velocity lags the applied field by $\pi/4$. What is the ratio of the mobility at this frequency to the mobility at zero frequency?

P2.23. Two infinite plane conducting slabs A and B, shown by the cross-sectional view in Fig. 2.40, carry surface charges of net uniform densities ρ_{SA} and ρ_{SB}, respectively. Find the surface charge densities on all four surfaces of the slabs.

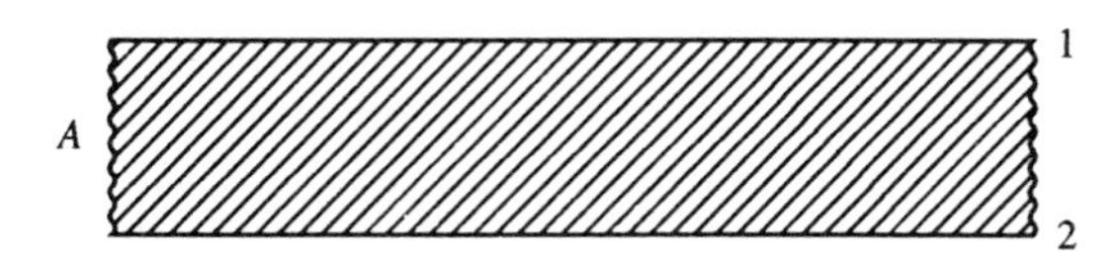

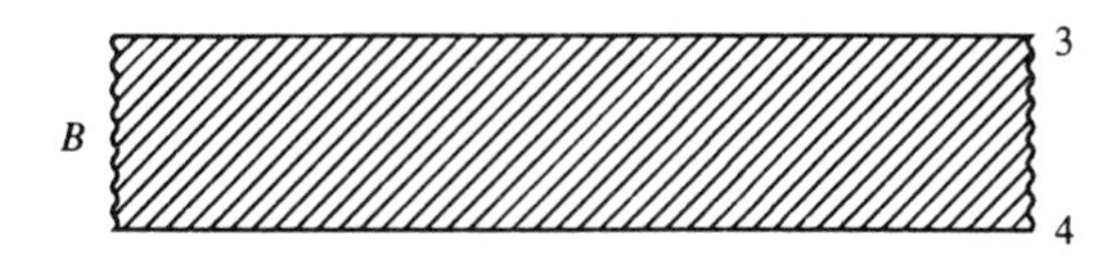

Figure 2.40. For Problem P2.23.

P2.24. The region $z < 0$ is occupied by a conductor. A point charge Q is situated at the point $(0, 0, d)$. From the secondary field required to make the total electric field inside the conductor equal to zero and from symmetry considerations as illustrated by the cross-sectional view in Fig. 2.41, show that the field outside the conductor is the same as the field due to the point charge Q at $(0, 0, d)$ and an "image" point charge $-Q$ situated at $(0, 0, -d)$. Find the expression for the field outside the conductor.

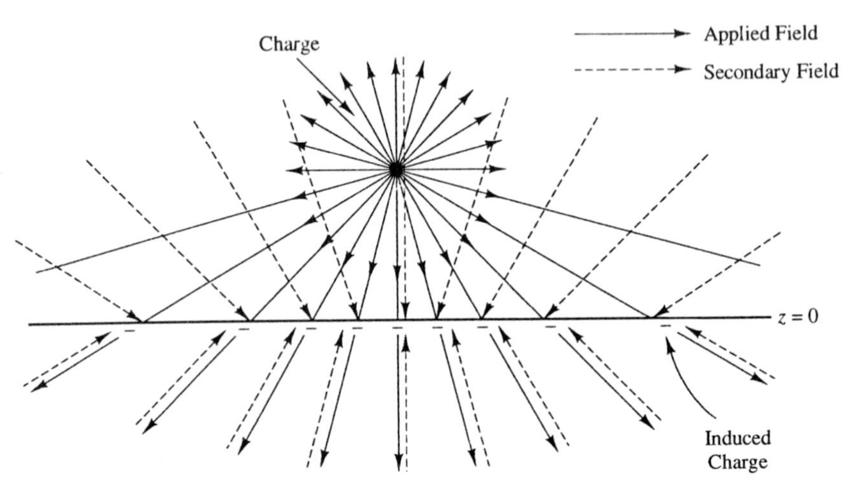

Figure 2.41. For Problem P2.24.

P2.25. Show that the torque acting on an electric dipole of moment $\mathbf{p}$ due to an applied electric field $\mathbf{E}$ is $\mathbf{p} \times \mathbf{E}$. Compute the torque for a dipole consisting of 1 μC of charge at $(0, 0, 10^{-3})$ and -1 μC of charge at $(0, 0, -10^{-3})$ in an electric field $\mathbf{E} = 10^3(\mathbf{i}_x - 2\mathbf{i}_y + \mathbf{i}_z)$ V/m.

P2.26. A point charge Q is situated at the origin surounded by a spherical dielectric shell of uniform permittivity $4\varepsilon_0$ and having inner and outer radii a and b, respectively. Find the following: **(a)** the $\mathbf{D}$ and $\mathbf{E}$ fields in the three regions $0 < r < a$, $a < r < b$, and $r > b$ and **(b)** the polarization vector inside the dielectric shell.

P2.27. An anisotropic dielectric material is characterized by the $\mathbf{D}$ to $\mathbf{E}$ relationship

$$\begin{bmatrix} D_x \\ D_y \\ D_z \end{bmatrix} = \varepsilon_0 \begin{bmatrix} 3 & 1 & 1 \\ 1 & 3 & 1 \\ 1 & 1 & 3 \end{bmatrix} \begin{bmatrix} E_x \\ E_y \\ E_z \end{bmatrix}$$

(a) Find $\mathbf{D}$ for $\mathbf{E} = E_0(\mathbf{i}_x + \mathbf{i}_y)$. **(b)** Find $\mathbf{D}$ for $\mathbf{E} = E_0(\mathbf{i}_x - \mathbf{i}_y)$. **(c)** Find $\mathbf{E}$ for $\mathbf{D} = D_0(\mathbf{i}_x + \mathbf{i}_y - 2\mathbf{i}_z)$. Comment on your result for each case.

P2.28. An anisotropic material is characterized by the $\mathbf{D}$ to $\mathbf{E}$ relationship

$$\begin{bmatrix} D_x \\ D_y \\ D_z \end{bmatrix} = \begin{bmatrix} \varepsilon_{xx} & \varepsilon_{xy} & 0 \\ \varepsilon_{yx} & \varepsilon_{yy} & 0 \\ 0 & 0 & \varepsilon_{zz} \end{bmatrix} \begin{bmatrix} E_x \\ E_y \\ E_z \end{bmatrix}$$

For $\mathbf{E} = E_x\mathbf{i}_x + E_y\mathbf{i}_y$, find the value(s) of E_y/E_x for which $\mathbf{D}$ is parallel to $\mathbf{E}$. Find the effective permittivity for each case.

P2.29. Find the magnetic dipole moment of an electron in circular orbit of radius a normal to a uniform magnetic field of flux density B_0. Compute its value for $a = 10^{-3}$ m and $B_0 = 5 \times 10^{-5}$ Wb/m^2.

P2.30. Considering for simplicity a rectangular current loop in the xy-plane, show that the torque acting on a magnetic dipole of moment $\mathbf{m}$ due to an applied magnetic field $\mathbf{B}$ is $\mathbf{m} \times \mathbf{B}$. Then find the torque acting on a circular current loop of radius 1 mm, in the xy-plane, centered at the origin and with current 0.1 A flowing in the sense of increasing ϕ in a magnetic field $\mathbf{B} = 10^{-5}(2\mathbf{i}_x - 2\mathbf{i}_y + \mathbf{i}_z)$ Wb/m^2.

P2.31. A portion of the B-H curve for a ferromagnetic material can be approximated by the analytical expression

$$\mathbf{B} = \mu_0 kH\mathbf{H}$$

where k is a constant having units of meter per ampere. Find μ, μ_r, χ_m, and $\mathbf{M}$.

P2.32. An anisotropic magnetic material is characterized by the $\mathbf{B}$ to $\mathbf{H}$ relationship

$$\begin{bmatrix} B_x \\ B_y \\ B_z \end{bmatrix} = k\mu_0 \begin{bmatrix} 7 & 6 & 0 \\ 6 & 12 & 0 \\ 0 & 0 & 3 \end{bmatrix} \begin{bmatrix} H_x \\ H_y \\ H_z \end{bmatrix}$$

where k is a constant. Find the effective permeability for $\mathbf{H} = H_0(3\mathbf{i}_x - 2\mathbf{i}_y)$.

PC EXERCISES

PC2.1. Consider four point charges $Q_1 = 4\pi\varepsilon_0$ C at (2, 0, 1), $Q_2 = -4\pi\varepsilon_0$ C at $(-2, 0, 1)$, $Q_3 = 4\pi\varepsilon_0$ C at $(-2, 0, -1)$, and $Q_4 = -4\pi\varepsilon_0$ C at $(2, 0, -1)$. A positive test charge released in the vicinity of Q_1 will move along the direction line of the electric field originating at that charge. Consider the plotting of these direction lines within the rectangular region having corners at (0, 0, 0), (5, 0, 0), (5, 0, 4) and (0, 0, 4). Write a program to plot a direction line originating on a circle of radius 0.1 m centered at Q_1 and at an angle to the $+x$-direction specified by your instructor. Use the modified procedure of tracing the path of the test charge as described following Example 2.1, that is, instead of moving the test charge by 0.1 m from its current location, say, point A to a new location, say, point B, along the direction of the electric field at point A, move it by 0.1 m to a point C along the direction which bisects the directions of the fields at points A and B.

PC2.2. Write a program for computing the x- and z-components of $\varepsilon_0\mathbf{E}$ for the ring charge of Example 2.2 at an arbitrary point in the xz-plane and not on the ring, by dividing the ring into $2n$ segments as in Fig. 2.8 and using superposition. Assume that $a = 1$ m and $Q = 4\pi\varepsilon_0$ C. Run the program for values of n and the x- and z-coordinates of the point specified by your instructor.

PC2.3. Consider a finite sheet of charge of density $4\pi\varepsilon_0 f(x,y)$ C/m^2 lying in the xy-plane between $x = \pm 1$ and $y = \pm 1$. Write a program for computing the components of $\varepsilon_0\mathbf{E}$ in Cartesian coordinates due to the sheet charge at an arbitrary point not on the sheet, by dividing the sheet into a $2n \times 2n$ grid of squares and using superposition. Run the program for the function $f(x, y)$ and values of n and the Cartesian coordinates of the point specified by your instructor.

PC2.4. Consider a circular loop of wire of radius 1 m situated in the xy-plane with its center at the origin and with current 1 A flowing in the sense of increasing ϕ. Write a program to compute the x- and z-components of $\mathbf{B}/\mu_0$ due to the wire at an arbitrary point in the xz-plane and not on the wire, by dividing the wire into $2n$ segments (as for the ring charge in Fig. 2.9) and applying superposition. Run the program for values of n and the x- and z-coordinates of the point specified by your instructor.

PC2.5. Consider a square loop of wire of sides 1 m situated in the xy-plane with its center at the origin and its sides parallel to the x- and y-axes. A current of 1 A flows in the loop in the clockwise sense as seen looking along the positive z-axis. Write a program to compute the components of $\mathbf{B}/\mu_0$ in Cartesian coordinates due to the current loop at an arbitrary point not on the loop, by applying the result of Problem P2.12 to each side of the loop. Run the program for values of n and the Cartesian coordinates of the point specified by your instructor.

PC2.6. Modify the program of PL 2.2 for nonuniform electric and magnetic fields of the form $E_0 = -k_1(1 + m_1x)10^3$ and $B_0 = -k_2(1 + m_2x)10^{-4}$, where m_1 and m_2 are additional input parameters. Run the program with values of $k_1 = 1$, $k_2 = 1$, $k_3 = 1$, $k_4 = 3$, $k_5 = 5$, and m_1 and m_2 varying from -0.05 to $+0.05$ in steps of 0.05.

3

Maxwell's Equations in Integral Form and Boundary Conditions

In Chapter 1 we learned the simple rules of vector algebra and familiarized ourselves with the basic concepts of fields in general. In Chapter 2 we introduced electric and magnetic fields in terms of forces on charged particles and extended the discussion to fields in materials. We now have the necessary background to introduce the additional tools required for the understanding of the various quantities associated with Maxwell's equations and then discuss Maxwell's equations. In particular, our goal in this chapter is to learn Maxwell's equations in integral form as a prerequisite to the derivation of their differential forms in the next chapter. Maxwell's equations in integral form govern the interdependence of certain field and source quantities associated with regions in space, that is, contours, surfaces, and volumes. The differential forms of Maxwell's equations, however, relate the characteristics of the field vectors at a given point to one another and to the source densities at that point.

Maxwell's equations in integral form are a set of four laws resulting from several experimental findings and a purely mathematical contribution. We shall, however, consider them as postulates and learn to understand their physical significance as well as their mathematical formulation. The source quantities involved in their formulation are charges and currents. The field quantities have to do with the line and surface integrals of the electric and magnetic field vectors. We shall therefore first introduce line and surface integrals and then consider successively the four Maxwell's equations in integral form.

3.1 LINE AND SURFACE INTEGRALS

Line integral To introduce the line integral, let us consider in a region of electric field **E** the movement of a test charge q from the point A to the point B along the path C as shown in Fig. 3.1(a). At each and every point along the path the electric field exerts a force on the test charge and hence does a certain amount of work in moving the charge to another point an infinitesimal distance away. To find the total amount of work done from A to B, we divide the path into a number of infinitesimal segments $\Delta\mathbf{l}_1, \Delta\mathbf{l}_2, \Delta\mathbf{l}_3, \ldots \Delta\mathbf{l}_n$, as shown in Fig. 3.1(b), find the infinitesimal amount of work done for each segment, and then add up the contributions from all the segments. Since the segments are infinitesimal in length, we can consider each of them to be straight and the electric field at all points within a segment to be the same and equal to its value at the start of the segment.

If we now consider one segment, say the jth segment, and take the component of the electric field for that segment along the length of that segment, we obtain the result $E_j \cos \alpha_j$, where α_j is the angle between the direction of the electric field vector $\mathbf{E}_j$ at the start of that segment and the direction of that segment. Since the electric field intensity has the meaning of force per unit charge, the electric force along the direction of the jth segment is then equal to $qE_j \cos \alpha_j$. To obtain the work done in carrying the test charge along the length of the jth segment, we then multiply this electric force component by the length Δl_j of that segment. Thus for the jth segment, we obtain the result for the work done by the electric field as

$$\Delta W_j = qE_j \cos \alpha_j \, \Delta l_j \tag{3.1}$$

If we do this for all the infinitesimal segments and add up all the contributions, we get the total work done by the electric field in moving the test charge from A to B along the path C to be

$$\begin{aligned} W_{AB} &= \Delta W_1 + \Delta W_2 + \Delta W_3 + \cdots + \Delta W_n \\ &= qE_1 \cos \alpha_1 \, \Delta l_1 + qE_2 \cos \alpha_2 \, \Delta l_2 + qE_3 \cos \alpha_3 \, \Delta l_3 + \cdots \\ &\quad + qE_n \cos \alpha_n \, \Delta l_n \\ &= q \sum_{j=1}^{n} E_j \cos \alpha_j \, \Delta l_j \\ &= q \sum_{j=1}^{n} (E_j)(\Delta l_j) \cos \alpha_j \end{aligned} \tag{3.2}$$

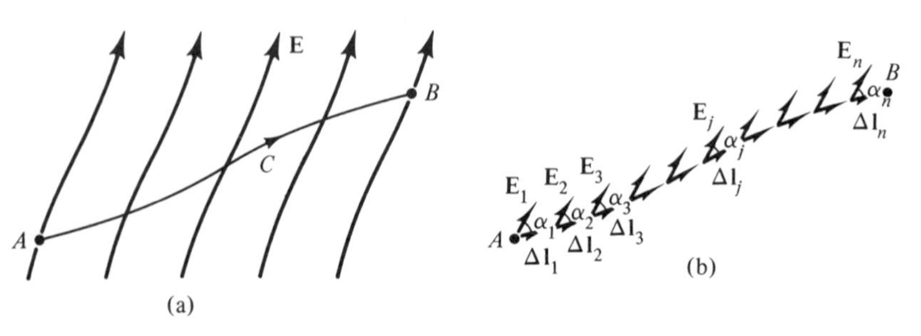

Figure 3.1. For evaluating the total amount of work done in moving a test charge along a path C from point A to point B in a region of electric field.

Using the dot product operation between two vectors, we obtain

$$\boxed{W_{AB} = q \sum_{j=1}^{n} \mathbf{E}_j \cdot \Delta\mathbf{l}_j} \tag{3.3}$$

For a numerical example, let us consider the electric field given by

$$\mathbf{E} = y\mathbf{i}_y$$

and determine the work done by the field in the movement of 3 μC of charge from the point $A(0, 0, 0)$ to the point $B(1, 1, 0)$ along the parabolic path $y = x^2$, $z = 0$ shown in Fig. 3.2(a).

For convenience, we shall divide the path into 10 segments having equal projections along the x-axis, as shown in Fig. 3.2(a). We shall number the segments 1, 2, 3, . . . , 10. The coordinates of the starting and ending points of the jth segment are as shown in Fig. 3.2(b). The electric field at the start of the jth segment is given by

$$\mathbf{E}_j = (j - 1)^2 0.01\mathbf{i}_y$$

The length vector corresponding to the jth segment, approximated as a straight line connecting its starting and ending points, is

$$\begin{aligned}\Delta\mathbf{l}_j &= 0.1\mathbf{i}_x + [j^2 - (j - 1)^2]0.01\mathbf{i}_y \\ &= 0.1\mathbf{i}_x + (2j - 1)0.01\mathbf{i}_y\end{aligned}$$

The required work is then given by

$$\begin{aligned}W_{AB} &= 3 \times 10^{-6} \sum_{j=1}^{10} \mathbf{E}_j \cdot \Delta\mathbf{l}_j \\ &= 3 \times 10^{-6} \sum_{j=1}^{10} [(j - 1)^2 0.01\mathbf{i}_y] \cdot [0.1\mathbf{i}_x + (2j - 1)0.01\mathbf{i}_y]\end{aligned}$$

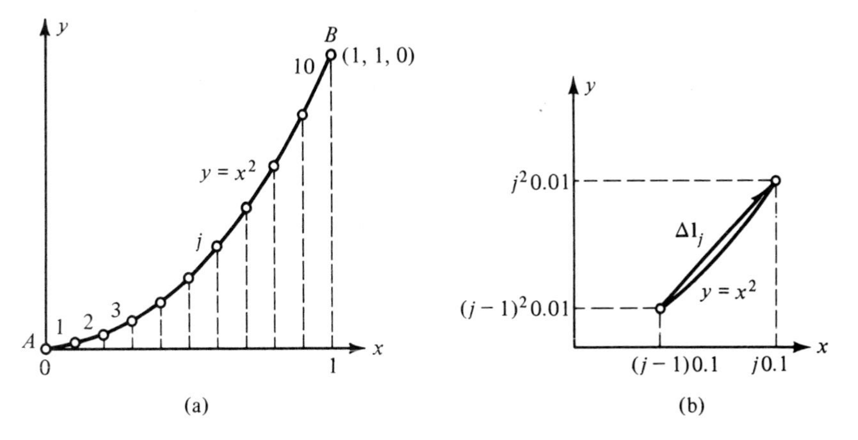

Figure 3.2. (a) Division of the path $y = x^2$ from $A(0, 0, 0)$ to $B(1, 1, 0)$ into 10 segments. (b) Length vector corresponding to the jth segment of part (a) approximated as a straight line.

$$= 3 \times 10^{-10} \sum_{j=1}^{10} (j - 1)^2(2j - 1)$$

$$= 3 \times 10^{-10}[0 + 3 + 20 + 63 + 144 + 275 + 468 + 735 + 1088 + 1539]$$

$$= 3 \times 10^{-10} \times 4335 \text{ J} = 1.3005 \ \mu\text{J}$$

The result that we have just obtained for W_{AB} is approximate since we divided the path from A to B into a finite number of segments. By dividing it into larger and larger numbers of segments, we can obtain more and more accurate results. In the limit that $n \to \infty$, the result converges to the exact value. The summation in (3.3) then becomes an integral, which represents exactly the work done by the field and is given by

$$\boxed{W_{AB} = q \int_A^B \mathbf{E} \cdot d\mathbf{l}} \qquad (3.4)$$

The integral on the right side of (3.4) is known as the *line integral of* **E** *from A to B*, along the specified path.

Evaluation of line integral

We shall illustrate the evaluation of the line integral by computing the exact value of the work done by the electric field in the movement of the 3 μC charge for the path in Fig. 3.2(a). To do this, we note that at any arbitrary point on the curve $y = x^2$, $z = 0$,

$$dy = 2x\,dx, \qquad dz = 0$$

so that the differential length vector tangential to the curve is given by

$$\begin{aligned} d\mathbf{l} &= dx\,\mathbf{i}_x + dy\,\mathbf{i}_y + dz\,\mathbf{i}_z \\ &= dx\,\mathbf{i}_x + 2x\,dx\,\mathbf{i}_y \end{aligned}$$

The value of $\mathbf{E} \cdot d\mathbf{l}$ at the point is

$$\begin{aligned} \mathbf{E} \cdot d\mathbf{l} &= y\mathbf{i}_y \cdot (dx\,\mathbf{i}_x + 2x\,dx\,\mathbf{i}_y) \\ &= x^2\mathbf{i}_y \cdot (dx\,\mathbf{i}_x + 2x\,dx\,\mathbf{i}_y) \\ &= 2x^3\,dx \end{aligned}$$

Thus the required work is given by

$$\begin{aligned} W_{AB} &= q \int_{(0,0,0)}^{(1,1,0)} \mathbf{E} \cdot d\mathbf{l} = 3 \times 10^{-6} \int_0^1 2x^3\,dx \\ &= 3 \times 10^{-6} \left[\frac{2x^4}{4}\right] = 1.5 \ \mu\text{J} \end{aligned}$$

Note that we have evaluated the line integral by using x as the variable of integration. Alternatively, using y as the variable of integration, we obtain

$$\begin{aligned} \mathbf{E} \cdot d\mathbf{l} &= y\mathbf{i}_y \cdot (dx\,\mathbf{i}_x + dy\,\mathbf{i}_y) \\ &= y\,dy \end{aligned}$$

$$W_{AB} = q\int_{(0,0,0)}^{(1,1,0)} \mathbf{E}\cdot d\mathbf{l} = 3\times 10^{-6}\int_0^1 y\,dy$$

$$= 3\times 10^{-6}\left[\frac{y^2}{2}\right] = 1.5\ \mu\text{J}$$

Thus the integration can be performed with respect to x or y (or z in the three-dimensional case). What is important, however, is that the integrand must be expressed as a function of the variable of integration and the limits appropriate to that variable must be employed.

Voltage defined

Returning now to (3.4) and dividing both sides by q, we note that the line integral of $\mathbf{E}$ from A to B has the physical meaning of work per unit charge done by the field in moving the test charge from A to B. This quantity is known as the *voltage between A and B* along the specified path and is denoted by the symbol V_{AB}, having the units of volts. Thus

$$\boxed{V_{AB} = \int_A^B \mathbf{E}\cdot d\mathbf{l}} \tag{3.5}$$

When the path under consideration is a closed path, that is, one which has no beginning or ending such as a rubber band, as shown in Fig. 3.3, the line integral is written with a circle associated with the integral sign in the manner $\oint_C \mathbf{E}\cdot d\mathbf{l}$. The line integral of a vector around a closed path is known as the *circulation* of that vector. In particular, the line integral of $\mathbf{E}$ around a closed path is the work per unit charge done by the field in moving a test charge around the closed path. It is the voltage around the closed path and is also known as the *electromotive force*. We shall now consider an example of evaluating the line integral of a vector around a closed path.

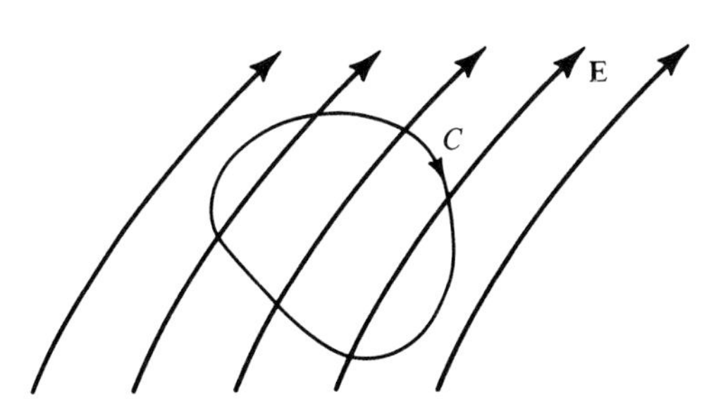

Figure 3.3. Closed path C in a region of electric field.

Example 3.1

Let us consider the force field

$$\mathbf{F} = x\mathbf{i}_y$$

and evaluate $\oint_C \mathbf{F}\cdot d\mathbf{l}$ where C is the closed path $ABCDA$ shown in Fig. 3.4.

Noting that

$$\oint_{ABCDA} \mathbf{F}\cdot d\mathbf{l} = \int_A^B \mathbf{F}\cdot d\mathbf{l} + \int_B^C \mathbf{F}\cdot d\mathbf{l} + \int_C^D \mathbf{F}\cdot d\mathbf{l} + \int_D^A \mathbf{F}\cdot d\mathbf{l}$$

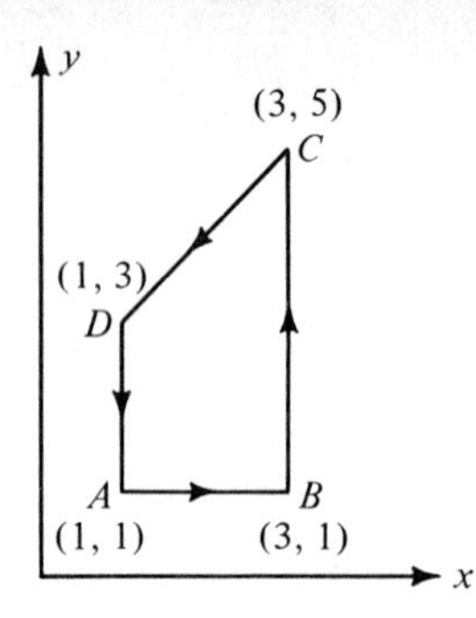

Figure 3.4. For evaluating the line integral of a vector field around a closed path.

we simply evaluate each of the line integrals on the right side and add them up to obtain the required quantity.

First we observe that since the entire closed path lies in the $z = 0$ plane, $dz = 0$ and $d\mathbf{l} = dx\,\mathbf{i}_x + dy\,\mathbf{i}_y$ for all four straight lines. Then for the side AB,

$$y = 1, \qquad dy = 0, \qquad d\mathbf{l} = dx\,\mathbf{i}_x + (0)\mathbf{i}_y = dx\,\mathbf{i}_x$$

$$\mathbf{F} \cdot d\mathbf{l} = (x\mathbf{i}_y) \cdot (dx\,\mathbf{i}_x) = 0$$

$$\int_A^B \mathbf{F} \cdot d\mathbf{l} = 0$$

For the side BC,

$$x = 3, \qquad dx = 0, \qquad d\mathbf{l} = (0)\mathbf{i}_x + dy\,\mathbf{i}_y = dy\,\mathbf{i}_y$$

$$\mathbf{F} \cdot d\mathbf{l} = (3\mathbf{i}_y) \cdot (dy\,\mathbf{i}_y) = 3\,dy$$

$$\int_B^C \mathbf{F} \cdot d\mathbf{l} = \int_1^5 3\,dy = 12$$

For the side CD,

$$y = 2 + x, \qquad dy = dx, \qquad d\mathbf{l} = dx\,\mathbf{i}_x + dx\,\mathbf{i}_y$$

$$\mathbf{F} \cdot d\mathbf{l} = (x\mathbf{i}_y) \cdot (dx\,\mathbf{i}_x + dx\,\mathbf{i}_y) = x\,dx$$

$$\int_C^D \mathbf{F} \cdot d\mathbf{l} = \int_3^1 x\,dx = -4$$

For the side DA,

$$x = 1, \qquad dx = 0, \qquad d\mathbf{l} = (0)\mathbf{i}_x + dy\,\mathbf{i}_y$$

$$\mathbf{F} \cdot d\mathbf{l} = (\mathbf{i}_y) \cdot (dy\,\mathbf{i}_y) = dy$$

$$\int_D^A \mathbf{F} \cdot d\mathbf{l} = \int_3^1 dy = -2$$

Finally,

$$\oint_{ABCDA} \mathbf{F} \cdot d\mathbf{l} = 0 + 12 - 4 - 2 = 6$$

Conservative vs. nonconservative fields

In this example, we found that the line integral of $\mathbf{F}$ around the closed path C is nonzero. The field is then said to be a *nonconservative field*. For a nonconservative field, the line integral between two points, say A and B, is dependent on

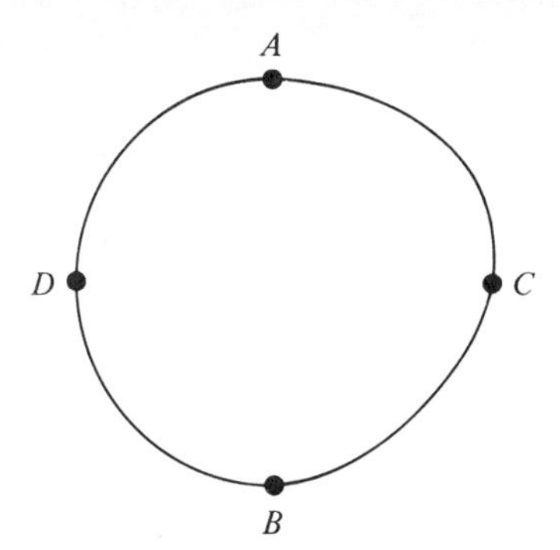

Figure 3.5. Two different paths from point A to point B.

the path followed from A to B. To show this, let us consider the two paths ACB and ADB as in Fig. 3.5. Then we can write

$$\begin{aligned}\oint_{ACBDA} \mathbf{F} \cdot d\mathbf{l} &= \int_{ACB} \mathbf{F} \cdot d\mathbf{l} + \int_{BDA} \mathbf{F} \cdot d\mathbf{l} \\ &= \int_{ACB} \mathbf{F} \cdot d\mathbf{l} - \int_{ADB} \mathbf{F} \cdot d\mathbf{l}\end{aligned} \tag{3.6}$$

It can be easily seen that if $\oint_{ACBDA} \mathbf{F} \cdot d\mathbf{l}$ is not equal to zero, then $\int_{ACB} \mathbf{F} \cdot d\mathbf{l}$ is not equal to $\int_{ADB} \mathbf{F} \cdot d\mathbf{l}$. The two integrals are equal only if $\oint_{ACBDA} \mathbf{F} \cdot d\mathbf{l}$ is equal to zero, which is the case for *conservative fields*. Examples of conservative fields are the earth's gravitational field and the static electric field. An example of nonconservative fields is the time-varying electric field. Thus in a time-varying electric field, the voltage between two points A and B is dependent on the path followed to evaluate the line integral of $\mathbf{E}$ from A to B, whereas in a static electric field, the voltage, more commonly known as the *potential difference,* between two points A and B is uniquely defined because the line integral of $\mathbf{E}$ from A to B is independent of the path followed from A to B.

Surface integral

To introduce the surface integral, let us consider a region of magnetic field and an infinitesimal surface at a point in that region. Since the surface is infinitesimal, we can assume the magnetic flux density to be uniform on the surface, although it may be nonuniform over a wider region. If the surface is oriented normal to the magnetic field lines, as shown in Fig. 3.6(a), then the magnetic

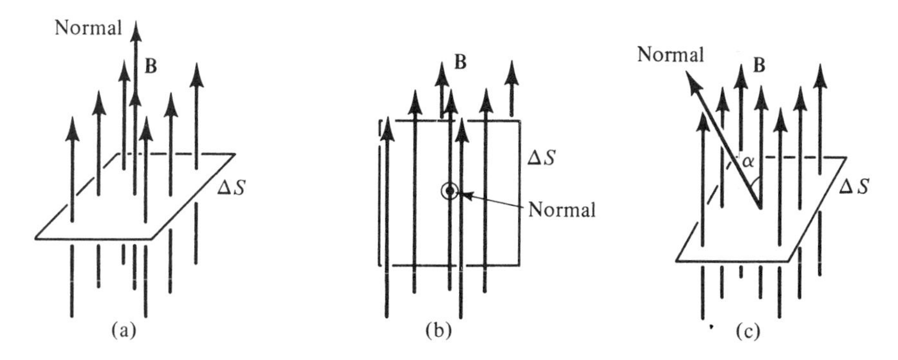

Figure 3.6. Infinitesimal surface ΔS in a magnetic field $\mathbf{B}$ oriented (a) normal to the field, (b) parallel to the field, and (c) with its normal making an angle α to the field.

flux (webers) crossing the surface is simply given by the product of the surface area (meters squared) and the magnetic flux density (Wb/m^2) on the surface, that is, $B\ \Delta S$. If, however, the surface is oriented parallel to the magnetic field lines, as shown in Fig. 3.6(b), there is no magnetic flux crossing the surface. If the surface is oriented in such a manner that the normal to the surface makes an angle α with the magnetic field lines as shown in Fig. 3.6(c), then the amount of magnetic flux crossing the surface can be determined by considering that the component of **B** normal to the surface is $B \cos \alpha$ and the component tangential to the surface is $B \sin \alpha$. The component of **B** normal to the surface results in a flux of $(B \cos \alpha)\ \Delta S$ crossing the surface whereas the component tangential to the surface does not contribute at all to the flux crossing the surface. Thus the magnetic flux crossing the surface in this case is $(B \cos \alpha)\ \Delta S$. We can obtain this result alternatively by noting that the projection of the surface onto the plane normal to the magnetic field lines is $\Delta S \cos \alpha$. Hence the magnetic flux crossing the surface ΔS is the same as that crossing normal to the area $\Delta S \cos \alpha$, that is, $B(\Delta S \cos \alpha)$ or $(B \cos \alpha)\ \Delta S$.

To aid further in the understanding of this concept, imagine raindrops falling vertically downward uniformly. If you hold a rectangular loop horizontally, the number of drops falling through the loop is simply equal to the area of the loop multipled by the density (number of drops per unit area) of the drops. If the loop is held vertically, no rain falls through the loop. If the loop is held at some angle to the horizontal, the number of drops falling through the loop is the same as that which would fall through another (smaller) loop, which is the projection of the slanted loop on to the horizontal plane.

Let us now consider a large surface S in the magnetic field region, as shown in Fig. 3.7. The magnetic flux crossing this surface can be found by dividing the surface into a number of infinitesimal surfaces $\Delta S_1, \Delta S_2, \Delta S_3, \ldots, \Delta S_n$ and applying the result just obtained for each infinitesimal surface and adding up the contributions from all the surfaces. To obtain the contribution from the jth surface, we draw the normal vector to that surface and find the angle α_j between the normal vector and the magnetic flux density vector $\mathbf{B}_j$ associated with that surface. Since the surface is infinitesimal, we can assume $\mathbf{B}_j$ to be the value of **B** at the centroid of the surface, and we can also erect the normal vector at that point. The contribution to the total magnetic flux from the jth infinitesimal surface is then given by

$$\Delta\psi_j = B_j \cos \alpha_j\ \Delta S_j \tag{3.7}$$

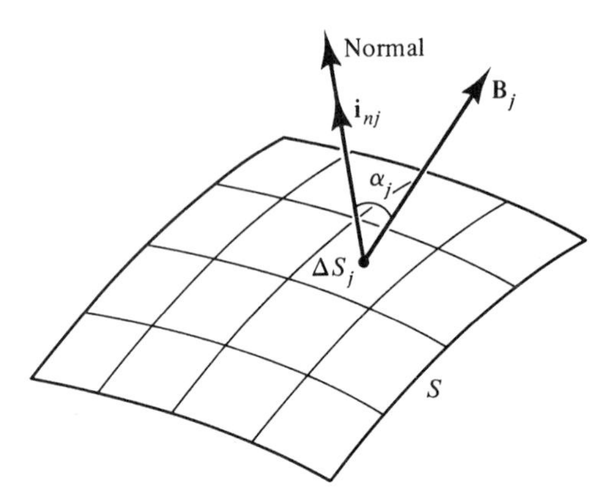

Figure 3.7. Division of a large surface S in a magnetic field region into a number of infinitesimal surfaces.

where the symbol ψ represents magnetic flux. The total magnetic flux crossing the surface S is then given by

$$
\begin{aligned}
[\psi]_S &= \Delta\psi_1 + \Delta\psi_2 + \Delta\psi_3 + \cdots + \Delta\psi_n \\
&= B_1 \cos\alpha_1 \, \Delta S_1 + B_2 \cos\alpha_2 \, \Delta S_2 + B_3 \cos\alpha_3 \, \Delta S_3 + \cdots \\
&\quad + B_n \cos\alpha_n \, \Delta S_n \\
&= \sum_{j=1}^{n} B_j \cos\alpha_j \, \Delta S_j \qquad (3.8) \\
&= \sum_{j=1}^{n} B_j(\Delta S_j) \cos\alpha_j
\end{aligned}
$$

Using the dot product operation between two vectors, we obtain

$$[\psi]_S = \sum_{j=1}^{n} \mathbf{B}_j \cdot \Delta S_j \, \mathbf{i}_{nj} \qquad (3.9)$$

where $\mathbf{i}_{nj}$ is the unit vector normal to the surface ΔS_j. Furthermore, by using the concept of an infinitesimal surface vector as one having magnitude equal to the area of the surface and direction normal to the surface, that is,

$$\Delta \mathbf{S}_j = \Delta S_j \, \mathbf{i}_{nj} \qquad (3.10)$$

we can write (3.9) as

$$\boxed{[\psi]_S = \sum_{j=1}^{n} \mathbf{B}_j \cdot \Delta \mathbf{S}_j} \qquad (3.11)$$

For a numerical example, let us consider the magnetic field given by

$$\mathbf{B} = 3xy^2 \mathbf{i}_z \text{ Wb/m}^2$$

and determine the magnetic flux crossing the portion of the xy-plane lying between $x = 0$, $x = 1$, $y = 0$, and $y = 1$. For convenience, we shall divide the surface into 25 equal areas as shown in Fig. 3.8(a). We shall designate the

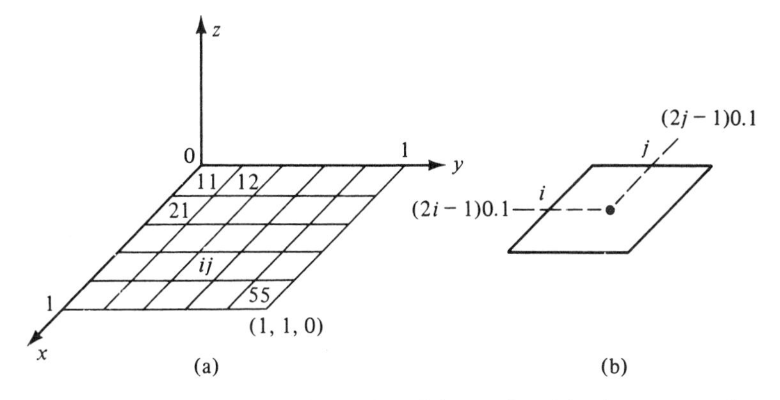

Figure 3.8. (a) Division of the portion of the xy-plane lying between $x = 0$, $x = 1$, $y = 0$, and $y = 1$ into 25 squares. (b) Area corresponding to the ijth square.

squares as 11, 12, . . . , 15, 21, 22, . . . , 55 where the first digit represents the number of the square in the x-direction and the second digit represents the number of the square in the y-direction. The x- and y-coordinates of the midpoint of the ijth square are $(2i - 1)0.1$ and $(2j - 1)0.1$, respectively, as shown in Fig. 3.8(b). The magnetic field at the center of the ijth square is then given by

$$\mathbf{B}_{ij} = 3(2i - 1)(2j - 1)^2 0.001\mathbf{i}_z$$

Since we have divided the surface into equal areas and since all areas are in the xy-plane,

$$\Delta\mathbf{S}_{ij} = 0.04\mathbf{i}_z \quad \text{for all } i \text{ and } j$$

The required magnetic flux is then given by

$$\begin{aligned}
[\psi]_S &= \sum_{i=1}^{5}\sum_{j=1}^{5} \mathbf{B}_{ij} \cdot \Delta\mathbf{S}_{ij} \\
&= \sum_{i=1}^{5}\sum_{j=1}^{5} 3(2i - 1)(2j - 1)^2 0.001\mathbf{i}_z \cdot 0.04\mathbf{i}_z \\
&= 0.00012 \sum_{i=1}^{5}\sum_{j=1}^{5} (2i - 1)(2j - 1)^2 \\
&= 0.00012(1 + 3 + 5 + 7 + 9)(1 + 9 + 25 + 49 + 81) \\
&= 0.495 \text{ Wb}
\end{aligned}$$

The result that we have just obtained for $[\psi]_S$ is approximate since we have divided the surface S into a finite number of areas. By dividing it into larger and larger numbers of squares, we can obtain more and more accurate results. In the limit that $n \to \infty$, the result converges to the exact value. The summation in (3.11) then becomes an integral that represents exactly the magnetic flux crossing the surface and is given by

$$\boxed{[\psi]_S = \int_S \mathbf{B} \cdot d\mathbf{S}} \tag{3.12}$$

where the symbol S associated with the integral sign denotes that the integration is performed over the surface S. The integral on the right side of (3.12) is known as the *surface integral of* **B** *over* S. The surface integral is a double integral since dS is equal to the product of two differential lengths.

Evaluation of surface integral

We shall illustrate the evaluation of the surface integral by computing the exact value of the magnetic flux crossing the surface in Fig. 3.8(a). To do this, we note that at any arbitrary point on the surface, the differential surface vector is given by

$$d\mathbf{S} = dx\, dy\, \mathbf{i}_z$$

The value of $\mathbf{B} \cdot d\mathbf{S}$ at the point is

$$\begin{aligned}
\mathbf{B} \cdot d\mathbf{S} &= 3xy^2\mathbf{i}_z \cdot dx\, dy\, \mathbf{i}_z \\
&= 3xy^2\, dx\, dy
\end{aligned}$$

Thus the required magnetic flux is given by

$$[\psi]_S = \int_S \mathbf{B} \cdot d\mathbf{S}$$

$$= \int_{x=0}^{1} \int_{y=0}^{1} 3xy^2 \, dx \, dy = 0.5 \text{ Wb}$$

When the surface under consideration is a closed surface, the surface integral is written with a circle associated with the integral sign in the manner $\oint_S \mathbf{B} \cdot d\mathbf{S}$. A closed surface is one which encloses a volume. Hence if you are anywhere in that volume, you can get out of it only by making a hole in the surface, and vice versa. A simple example is the surface of a balloon inflated and tied up at the mouth. The surface integral of **B** over the closed surface S is simply the magnetic flux *emanating* from the volume bounded by the surface. Thus whenever a closed surface integral is evaluated, the unit vectors normal to the differential surfaces are chosen to be pointing out of the volume, so as to give the outward flux of the vector field, unless specified otherwise. We shall now consider an example of evaluating $\oint_S \mathbf{B} \cdot d\mathbf{S}$.

Example 3.2

Let us consider the magnetic field

$$\mathbf{B} = (x + 2)\mathbf{i}_x + (1 - 3y)\mathbf{i}_y + 2z\mathbf{i}_z$$

and evaluate $\oint_S \mathbf{B} \cdot d\mathbf{S}$ where S is the surface of the cubical box bounded by the planes

$$x = 0, \qquad x = 1$$

$$y = 0, \qquad y = 1$$

$$z = 0, \qquad z = 1$$

as shown in Fig. 3.9.

Noting that

$$\oint_S \mathbf{B} \cdot d\mathbf{S} = \int_{abcd} \mathbf{B} \cdot d\mathbf{S} + \int_{efgh} \mathbf{B} \cdot d\mathbf{S} + \int_{adhe} \mathbf{B} \cdot d\mathbf{S} + \int_{bcgf} \mathbf{B} \cdot d\mathbf{S}$$
$$+ \int_{aefb} \mathbf{B} \cdot d\mathbf{S} + \int_{dhgc} \mathbf{B} \cdot d\mathbf{S}$$

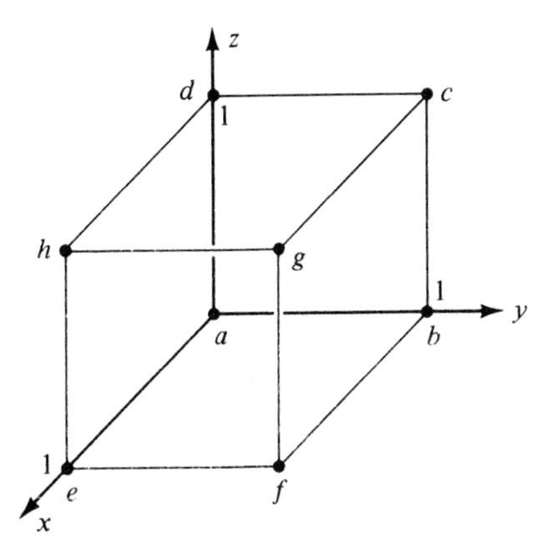

Figure 3.9. For evaluating the surface integral of a vector field over a closed surface.

we simply evaluate each of the surface integrals on the right side and add them up to obtain the required quantity. In doing so, we recognize that since the quantity we want is the magnetic flux out of the box, we should direct the unit normal vectors toward the outside of the box. Thus for the surface *abcd*,

$$x = 0, \qquad \mathbf{B} = 2\mathbf{i}_x + (1 - 3y)\mathbf{i}_y + 2z\mathbf{i}_z, \qquad d\mathbf{S} = -dy\,dz\,\mathbf{i}_x$$

$$\mathbf{B} \cdot d\mathbf{S} = -2\,dy\,dz$$

$$\int_{abcd} \mathbf{B} \cdot d\mathbf{S} = \int_{z=0}^{1} \int_{y=0}^{1} (-2)\,dy\,dz = -2$$

For the surface *efgh*,

$$x = 1, \qquad \mathbf{B} = 3\mathbf{i}_x + (1 - 3y)\mathbf{i}_y + 2z\mathbf{i}_z, \qquad d\mathbf{S} = dy\,dz\,\mathbf{i}_x$$

$$\mathbf{B} \cdot d\mathbf{S} = 3\,dy\,dz$$

$$\int_{efgh} \mathbf{B} \cdot d\mathbf{S} = \int_{z=0}^{1} \int_{y=0}^{1} 3\,dy\,dz = 3$$

For the surface *adhe*,

$$y = 0, \qquad \mathbf{B} = (x + 2)\mathbf{i}_x + 1\mathbf{i}_y + 2z\mathbf{i}_z, \qquad d\mathbf{S} = -dz\,dx\,\mathbf{i}_y$$

$$\mathbf{B} \cdot d\mathbf{S} = -dz\,dx$$

$$\int_{aehd} \mathbf{B} \cdot d\mathbf{S} = \int_{x=0}^{1} \int_{z=0}^{1} (-1)\,dz\,dx = -1$$

For the surface *bcgf*,

$$y = 1, \qquad \mathbf{B} = (x + 2)\mathbf{i}_x - 2\mathbf{i}_y + 2z\mathbf{i}_z, \qquad d\mathbf{S} = dz\,dx\,\mathbf{i}_y$$

$$\mathbf{B} \cdot d\mathbf{S} = -2\,dz\,dx$$

$$\int_{bfgc} \mathbf{B} \cdot d\mathbf{S} = \int_{x=0}^{1} \int_{z=0}^{1} (-2)\,dz\,dx = -2$$

For the surface *aefb*,

$$z = 0, \qquad \mathbf{B} = (x + 2)\mathbf{i}_x + (1 - 3y)\mathbf{i}_y + 0\mathbf{i}_z, \qquad d\mathbf{S} = -dx\,dy\,\mathbf{i}_z$$

$$\mathbf{B} \cdot d\mathbf{S} = 0$$

$$\int_{aefb} \mathbf{B} \cdot d\mathbf{S} = 0$$

For the surface *dhgc*,

$$z = 1, \qquad \mathbf{B} = (x + 2)\mathbf{i}_x + (1 - 3y)\mathbf{i}_y + 2\mathbf{i}_z, \qquad d\mathbf{S} = dx\,dy\,\mathbf{i}_z$$

$$\mathbf{B} \cdot d\mathbf{S} = 2\,dx\,dy$$

$$\int_{dhgc} \mathbf{B} \cdot d\mathbf{S} = \int_{y=0}^{1} \int_{x=0}^{1} 2\,dx\,dy = 2$$

Finally,

$$\oint_S \mathbf{B} \cdot d\mathbf{S} = -2 + 3 - 1 - 2 + 0 + 2 = 0$$

K3.1. Line integral; Line integral of **E**; Voltage; Surface integral; Surface integral of **B**; Magnetic flux.

D3.1. For each of the curves **(a)** $y = x^2, z = 0$, **(b)** $x^2 + y^2 = 2, z = 0$, and **(c)** $y = \sin 0.5\pi x, z = 0$ in a region of electric field $\mathbf{E} = y\mathbf{i}_x + x\mathbf{i}_y$, find the approximate value of the work done by the field in carrying a charge of 1 μC from the point (1, 1, 0) to the neighboring point on the curve, whose x coordinate is 1.1, by evaluating $\mathbf{E} \cdot \Delta\mathbf{l}$ along a straight line path.
Ans. **(a)** 0.31 μJ; **(b)** -0.0112 μJ; **(c)** 0.0877 μJ

D3.2. For $\mathbf{F} = y(\mathbf{i}_x + \mathbf{i}_y)$, find $\int \mathbf{F} \cdot d\mathbf{l}$ for the straight line paths between the following pairs of points from the first point to the second point: **(a)** (0, 0, 0) to (2, 0, 0); **(b)** (0, 2, 0) to (2, 2, 0); and **(c)** (2, 0, 0) to (2, 2, 0).
Ans. **(a)** 0; **(b)** 4; **(c)** 2

D3.3. Given $\mathbf{B} = (y\mathbf{i}_x - x\mathbf{i}_y)$ Wb/m^2, find by evaluating $\mathbf{B} \cdot \Delta\mathbf{S}$ the approximate absolute value of the magnetic flux crossing from one side to the other side of an infinitesimal surface of area 0.001 m^2 at the point (1, 2, 1) for each of the following orientations of the surface: **(a)** in the $x = 1$ plane; **(b)** on the surface $2x^2 + y^2 = 6$; and **(c)** normal to the unit vector $\frac{1}{3}(2\mathbf{i}_x + \mathbf{i}_y + 2\mathbf{i}_z)$.
Ans. **(a)** 2×10^{-3} Wb; **(b)** $\frac{1}{\sqrt{2}} \times 10^{-3}$ Wb; **(c)** 10^{-3} Wb

D3.4. For the vector field $\mathbf{A} = x(\mathbf{i}_x + \mathbf{i}_y)$, find the absolute value of $\int \mathbf{A} \cdot d\mathbf{S}$ over the following plane surfaces: **(a)** square having the vertices at (0, 0, 0), (0, 2, 0), (0, 2, 2), and (0, 0, 2); **(b)** square having the vertices at (2, 0, 0), (2, 2, 0), (2, 2, 2), and (2, 0, 2); **(c)** square having the vertices at (0, 0, 0), (2, 0, 0), (2, 0, 2), and (0, 0, 2); and **(d)** triangle having the vertices at (0, 0, 0), (2, 0, 0), and (0, 0, 2).
Ans. **(a)** 0; **(b)** 8; **(c)** 4; **(d)** $\frac{4}{3}$

3.2 FARADAY'S LAW

In the preceding section we introduced the line and surface integrals. We are now ready to consider Maxwell's equations in integral form. The first equation, which we shall discuss in this section, is a consequence of an experimental finding by Michael Faraday in 1831 that time-varying magnetic fields give rise to electric fields and hence it is known as *Faraday's law*. Faraday discovered that when the magnetic flux enclosed by a loop of wire changes with time, a current is produced in the loop, indicating that a voltage or an *electromotive force,* abbreviated as emf, is induced around the loop. The variation of the magnetic flux can result from the time variation of the magnetic flux enclosed by a fixed loop or from a moving loop in a static magnetic field or from a combination of the two, that is, a moving loop in a time-varying magnetic field.

Statement of Faraday's law

In mathematical form, Faraday's law is given by

$$\boxed{\oint_C \mathbf{E} \cdot d\mathbf{l} = -\frac{d}{dt}\int_S \mathbf{B} \cdot d\mathbf{S}} \tag{3.13}$$

where S is a surface bounded by the closed path C, as shown in Fig. 3.10. In words, Faraday's law states that *the electromotive force around a closed path is*

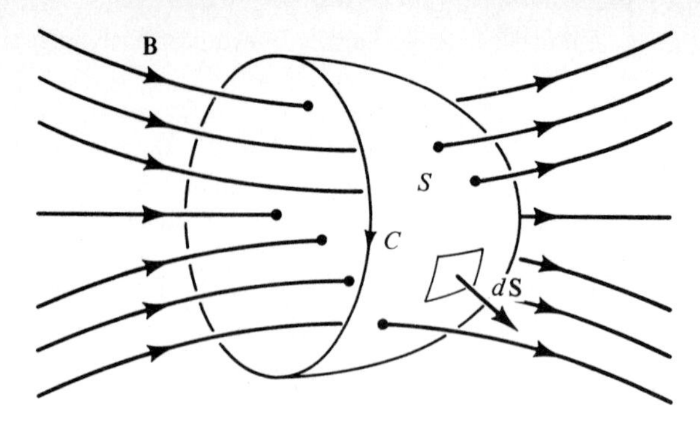

Figure 3.10. For illustrating Faraday's law.

equal to the negative of the time rate of change of the magnetic flux enclosed by that path. There are certain procedures and observations of interest pertinent to the application of (3.13). We shall discuss these next.

Right-hand screw rule

1. The magnetic flux on the right side is to be evaluated in accordance with the *right-hand screw rule* (R.H.S. rule), a convention that is applied consistently for all electromagnetic field laws involving integration over surfaces bounded by closed paths. The right-hand screw rule consists of imagining a right-hand screw being turned around the closed path, as illustrated in Fig. 3.11 for two opposing senses of paths, and using the resulting direction of advance of the screw to evaluate the surface integral. The application of this rule to the geometry of Fig. 3.10 means that in evaluating the surface integral of **B** over S, the normal vector to the differential surface dS should be directed as shown in that figure.
2. In evaluating the surface integral of **B**, any surface S *bounded by* C can be employed. For example, if the loop C is a planar loop, it is not necessary to consider the plane surface having the loop as its perimeter. One can consider a curved surface bounded by C or any combination of plane (or plane and curved) surfaces which together are bounded by C, and which is sometimes a more desirable choice. To illustrate this point, consider the planar loop $PQRP$ in Fig. 3.12(a). The most obvious surface bounded by this loop is the plane surface PQR inclined to the coordinate planes. Now imagine this plane surface to be an elastic sheet glued to the perimeter and pushed in toward the origin so as to conform to the coordinate planes. Then we obtain the combination of the plane surfaces OPQ, OQR, and ORP as shown in Fig. 3.12(b), which together constitute a surface also bounded by the loop.

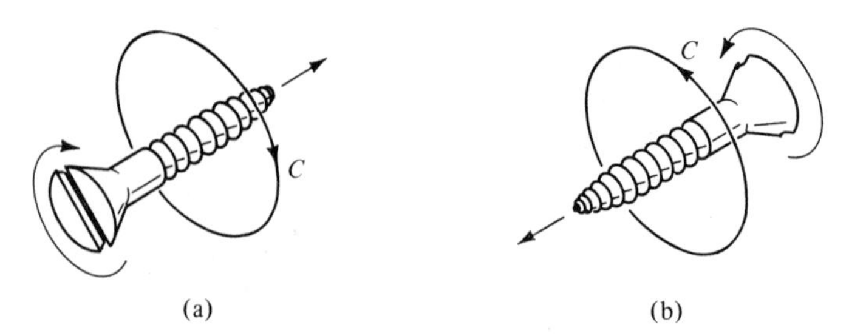

Figure 3.11. Right-hand screw rule convention employed in the formulation of electomagnetic field laws.

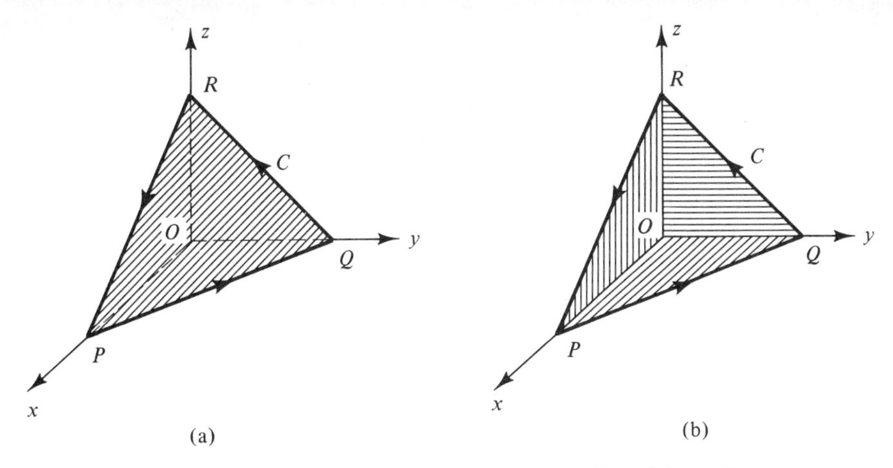

Figure 3.12. (a) A plane surface and (b) a combination of three plane surfaces, bounded by the closed path C.

To evaluate the surface integral of $\mathbf{B}$ for the surface in Fig. 3.12(a), we need to make use of the $d\mathbf{S}$ vector on that slant surface. On the other hand, for the geometry in Fig. 3.12(b), we can use the (simpler) $d\mathbf{S}$ vectors associated with the coordinate planes. The fact that any surface S bounded by a closed path C can be employed to evaluate the magnetic flux enclosed by C implies that the magnetic flux through all such surfaces is the same in order for the emf around C to be unique. As we shall learn in Section 3.4, it is a fundamental property of the magnetic field that the magnetic flux is the same through all surfaces bounded by a given closed path.

3. The closed path C on the left side need not represent a loop of wire but can be an imaginary contour. It means that the time-varying magnetic flux induces an electric field in the region and this results in an emf around the closed path. If a wire is placed in the position occupied by the closed path, the emf will produce a current in the loop simply because the charges in the wire are constrained to move along the wire.

Lenz's law

4. The minus sign on the right side together with the right-hand screw rule ensures that *Lenz's law* is always satisfied. Lenz's law states that *the sense of the induced emf is such that any current it produces tends to* oppose the change *in the magnetic flux producing it*. It is important to note that the induced emf acts to oppose the change in the flux and not the flux itself. To clarify this, let us consider that the flux is into the paper and increasing with time. Then the induced emf acts to produce flux out of the paper. On the other hand, if the same flux is decreasing with time, then the induced emf acts to produce flux into the paper.

Faraday's law for N*-turn coil*

5. If the loop C contains more than one turn, such as in an N-turn coil, then the surface S bounded by the periphery of the loop takes the shape of a spiral ramp, as shown in Fig. 3.13(a) for N equal to 2. This surface can be visualized by taking two paper plates, cutting each of them along a radius, as shown in Fig. 3.13(b) and (c), and joining the edge BO of plate in (c) to the edge $A'O$ of plate in (b). For a tightly wound coil, this is equivalent to the situation in which N separate, identical, single-turn loops are stacked so that the emf induced in the N-turn coil is N times that induced in one turn. Thus

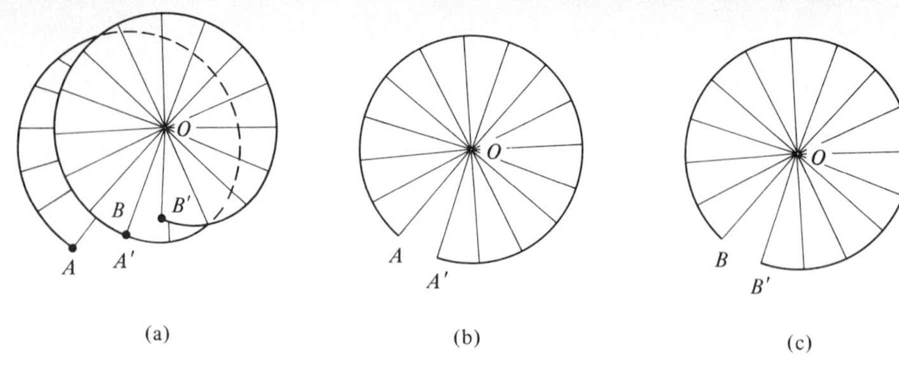

Figure 3.13. For illustrating the surface bounded by a loop containing two turns.

for an N-turn coil,

$$\boxed{\text{emf} = -N\frac{d\psi}{dt}} \tag{3.14}$$

where ψ is the magnetic flux computed as though the coil is a one-turn coil.

We shall now consider two examples to illustrate the determination of induced emf using Faraday's law, the first involving a stationary loop in a time-varying magnetic field and the second involving a moving conductor in a static magnetic field.

Example 3.3

Stationary loop in a time-varying magnetic field

A time-varying magnetic field is given by

$$\mathbf{B} = B_0 \cos \omega t \ \mathbf{i}_y$$

where B_0 is a constant. It is desired to find the induced emf around the rectangular loop C in the xz-plane bounded by the lines $x = 0$, $x = a$, $z = 0$, and $z = b$, as shown in Fig. 3.14.

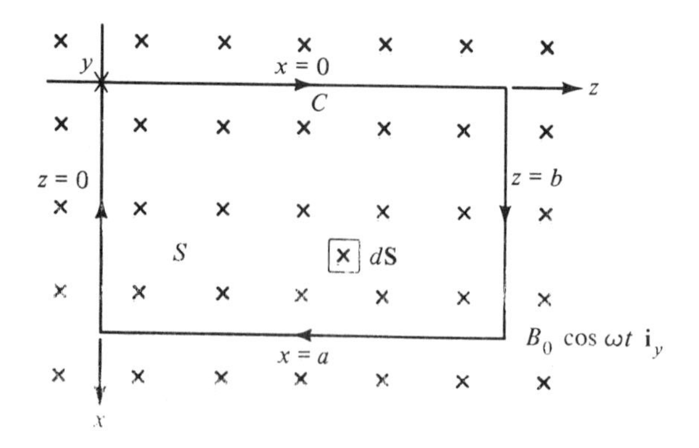

Figure 3.14. Rectangular loop in the xz-plane situated in a time-varying magnetic field.

Choosing $d\mathbf{S} = dx\, dz\, \mathbf{i}_y$ in accordance with the right-hand screw rule and using the plane surface S bounded by the loop, we obtain the magnetic flux enclosed by the loop to be

$$\psi = \int_S \mathbf{B} \cdot d\mathbf{S} = \int_{z=0}^{b} \int_{x=0}^{a} B_0 \cos \omega t\, \mathbf{i}_y \cdot dx\, dz\, \mathbf{i}_y$$

$$= B_0 \cos \omega t \int_{z=0}^{b} \int_{x=0}^{a} dx\, dz = abB_0 \cos \omega t$$

Note that since the magnetic flux density is uniform and normal to the plane of the loop, this result could have been obtained by simply multiplying the area ab of the loop by the component $B_0 \cos \omega t$ of the flux density vector. The induced emf around the loop is then given by

$$\oint_C \mathbf{E} \cdot d\mathbf{l} = -\frac{d}{dt}\int_S \mathbf{B} \cdot d\mathbf{S}$$

$$= -\frac{d}{dt}[abB_0 \cos \omega t] = abB_0\omega \sin \omega t$$

The time variations of the magnetic flux enclosed by the loop and the induced emf around the loop are shown in Fig. 3.15. It can be seen that when the magnetic flux enclosed by the loop into the paper is *decreasing* with time, the induced emf is positive, thereby producing a clockwise current as if the loop were a wire. This polarity of the current gives rise to a magnetic field directed into the paper inside the loop and hence acts to *oppose the decrease* of the magnetic flux enclosed by the loop. When the magnetic flux enclosed by the loop into the paper is *increasing* with time, the induced emf is negative, thereby producing a counterclockwise current around the loop. This polarity of the current gives rise to a magnetic field directed out of the paper inside the loop and hence acts to *oppose the increase* of the magnetic flux enclosed by the loop. These observations are consistent with Lenz's law.

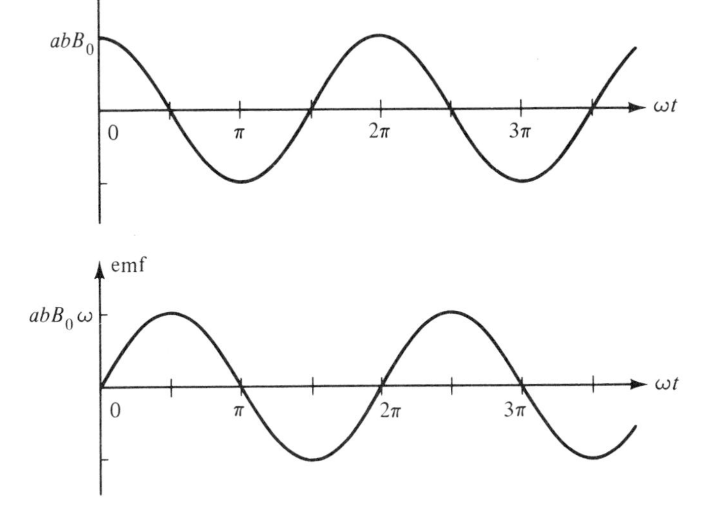

Figure 3.15. Time variations of magnetic flux ψ enclosed by the loop of Fig. 3.14 and the resulting induced emf around the loop.

Example 3.4

Moving conductor in a static magnetic field

A rectangular loop of wire with three sides fixed and the fourth side movable is situated in a plane perpendicular to a uniform magnetic field $\mathbf{B} = B_0\mathbf{i}_z$, as illustrated in Fig. 3.16. The movable side consists of a conducting bar moving with a velocity v_0 in the y-direction. It is desired to find the emf induced around the closed path C of the loop.

Letting the position of the movable side at any time t be $y_0 + v_0 t$ and considering $d\mathbf{S} = dx\,dy\,\mathbf{i}_z$ in accordance with the right-hand screw rule and using the plane surface S bounded by the loop, we obtain the magnetic flux enclosed by the loop to be

$$\begin{aligned}\int_S \mathbf{B} \cdot d\mathbf{S} &= \int_S B_0\mathbf{i}_z \cdot dx\,dy\,\mathbf{i}_z \\ &= \int_{x=0}^{l} \int_{y=0}^{y_0+v_0t} B_0\,dx\,dy \\ &= B_0 l(y_0 + v_0 t)\end{aligned}$$

Note that this result could also have been obtained as the product of the area of the loop $l(y_0 + v_0t)$ and the flux density B_0 because of the uniformity of the flux density within the area of the loop and its perpendicularity to the plane of the loop. The emf induced around C is given by

$$\begin{aligned}\oint_C \mathbf{E} \cdot d\mathbf{l} &= -\frac{d}{dt}\int_S \mathbf{B} \cdot d\mathbf{S} \\ &= -\frac{d}{dt}[B_0 l(y_0 + v_0 t)] \\ &= -B_0 l v_0\end{aligned}$$

Note that if the bar is moving to the right, the induced emf is negative and produces a current in the sense opposite to that of C. This polarity of the current is such that it gives rise to a magnetic field directed out of the paper inside the loop. The flux of this magnetic field is in opposition to the flux of the original magnetic field and hence tends to *oppose the increase* in the magnetic flux enclosed by the loop. On the other hand, if the bar is moving to the left, v_0 is negative, the induced emf is positive, and produces current in the same sense as that of C. This polarity of current is such that it gives rise to a magnetic field directed into the paper inside the loop. The flux of this magnetic field is in augmentation to the flux of the original magnetic field and hence tends to *oppose the decrease* in the magnetic flux enclosed by the loop. These observations are once again consistent with Lenz's law.

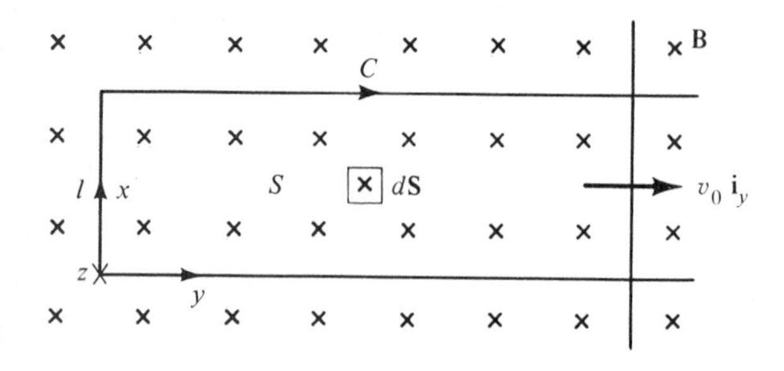

Figure 3.16. Rectangular loop of wire with a movable side situated in a uniform magnetic field.

It is also of interest to note that the induced emf can also be interpreted as being due to the electric field induced in the moving bar by virtue of its motion perpendicular to the magnetic field. Thus a charge Q in the bar experiences a force $\mathbf{F} = Q\mathbf{v} \times \mathbf{B}$ or $Qv_0\mathbf{i}_y \times B_0\mathbf{i}_z = Qv_0B_0\mathbf{i}_x$. To an observer moving with the bar, this force appears as an electric force due to an electric field $\mathbf{F}/Q = v_0B_0\mathbf{i}_x$. Viewed from inside the loop, this electric field is in the counterclockwise sense. Hence the induced emf, which is the line integral of $\mathbf{E}$ along the bar, is given by

$$\int_{x=0}^{l} v_0B_0\mathbf{i}_x \cdot dx\,\mathbf{i}_x = \int_{x=0}^{l} v_0B_0\,dx = v_0B_0l$$

in the counterclockwise sense (i.e., opposite to C), consistent with the result deduced from Faraday's law. This concept of induced emf is known as the *motional emf concept,* which is employed widely in the study of electromechanics.

In the two examples we just discussed, we have implicitly illustrated the principles behind two of the practical applications of Faraday's law. These are pertinent to the reception of radio and TV signals using a loop antenna and electromechanical energy conversion.

Principle of loop antenna

That the arrangement considered in Example 3.3 illustrates the principle of a loop antenna can be seen by noting that if the loop C were in the xy-plane or in the yz-plane, no emf would be induced in it since the magnetic flux density is then parallel to the plane of the loop and no flux is enclosed by the loop. In fact, for any arbitrary orientation of the loop, only that component of $\mathbf{B}$ normal to the plane of the loop contributes to the magnetic flux enclosed by the loop and hence to the emf induced in the loop. Thus for a given magnetic field, the voltage induced in the loop varies as the orientation of the loop is changed, with the maximum occurring when the loop is in the plane perpendicular to the magnetic field. Pocket AM radios generally employ inside them a type of loop antenna consisting of many turns of wire wound around a bar of magnetic material, and TV receivers generally employ a single-turn circular loop for UHF channels. Thus for maximum signals to be received, the AM radios and the TV loop antennas need to be oriented appropriately. Another point of interest evident from Example 3.3 is that the induced emf is proportional to ω, the radian frequency of the source of the magnetic field. Hence for the same voltage to be induced for a given amplitude B_0 of the magnetic flux density, the area of the loop times the number of turns is inversely proportional to the frequency.

Locating a radio transmitter

What is undesirable for one purpose can sometimes be used to advantage for another purpose. The fact that no voltage is induced in the loop antenna when the magnetic field is parallel to the plane of the loop is useful for locating the transmitter of a radio wave. Since the magnetic field of an incoming radio wave is perpendicular to its direction of propagation, no voltage is induced in the loop when its axis is along the direction of the transmitter. For a transmitter on the earth's surface, it is then sufficient to use two spaced vertical, loop antennas and find their orientations for which no signals are received. By then producing backward along the axes of the two loop antennas, as shown by the top view in Fig. 3.17, the location of the transmitter can be determined.

Energy conversion

That the arrangement considered in Example 3.4 is a simple example of an electromechanical energy converter can be seen by recognizing that in view of the current flow in the moving bar, the bar is acted upon by a magnetic force.

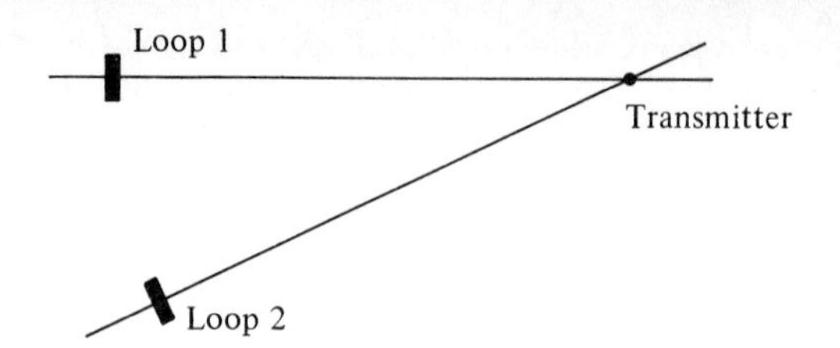

Figure 3.17. Top view of an arrangement consisting of two loop antennas for locating a transmitter of radio waves.

Since for positive v_0 the current flows in the loop in the sense opposite to that of C and hence in the positive x-direction in the moving bar, and since the magnetic field is in the z-direction, the magnetic force is exerted in the $\mathbf{i}_x \times \mathbf{i}_z$ or $-\mathbf{i}_y$-direction. Thus to keep the bar moving, an external force must be exerted in the $+\mathbf{i}_y$-direction, thereby requiring mechanical work to be done by an external agent. It is this mechanical work that is converted into electrical energy in the loop.

Principle of rotating generator

What we have just discussed is the principle of generation of electric power by linear motion of conductor in a magnetic field. Practical electric generators are of the rotating type. The principle of a rotating generator can be illustrated by considering a rectangular loop of wire situated symmetrically about the z-axis and rotating with angular velocity ω around the z-axis in a constant magnetic field $\mathbf{B} = B_0\mathbf{i}_x$, as shown in Fig. 3.18(a). Then noting from the view in Fig. 3.18(b) that the magnetic flux ψ enclosed by the loop at any arbitrary value of time is the same as that enclosed by its projection onto the yz-plane at that time, we obtain $\psi = B_0 A \cos \omega t$, where A is the area of the loop and the situation shown in Fig. 3.18(a) is assumed for $t = 0$. The emf induced in the loop is $-d\psi/dt$, or $\omega B_0 A \sin \omega t$. Thus the rotating loop in the constant magnetic field produces an alternating voltage. The same result can be achieved by a stationary loop in a rotating magnetic field. In fact, in most generators, a stationary member, or stator, carries the coils in which the voltage is induced, and a rotating member, or rotor, provides the magnetic field. As in the case of the arrangement of Example 3.4, a certain amount of mechanical work must be done to keep the loop rotating. It is this mechanical work, which is supplied by the prime mover (such as a turbine in the case of a hydroelectric generator or the engine of an automobile in the case of its alternator) turning the rotor, that is converted into electrical energy.

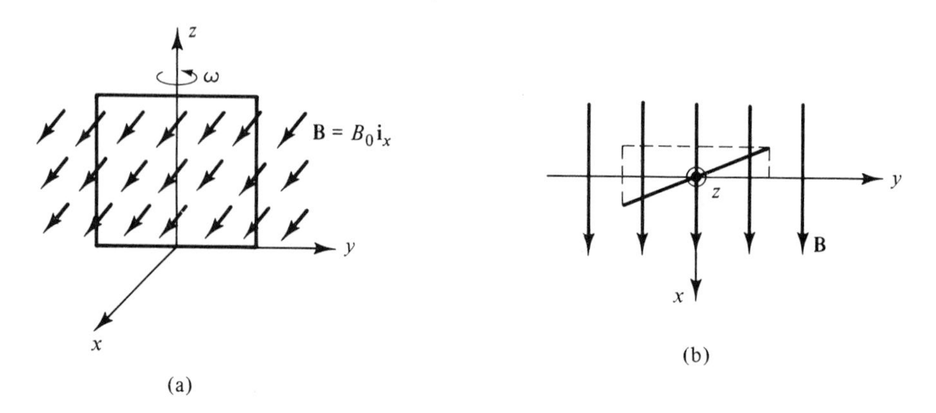

Figure 3.18. For illustrating the principle of a rotating generator.

Magnetic levitation

There are numerous other applications of Faraday's law, but we shall discuss only one more before we conclude this section. This is the phenomenon of *magnetic levitation,* a basis for rapid transit systems employing trains that hover over their guideways and do not touch the rail, among other applications. Magnetic levitation arises from a combination of Faraday's law and Ampère's force law. It can be explained and demonstrated through a series of simple experiments, culminating in a current-carrying coil lifting up above a metallic plate, as described in the following:

1. Consider a pair of coils (30 to 50 turns of No. 26 wire of about 4 in. diameter) attached to nails on a piece of wood, as shown in Fig. 3.19. Set to zero the output of a variable power supply obtained by connecting a variac to the 110 V ac mains. Connect one output terminal (A) of the variac to the beginnings (C_1 and C_2) of both coils and the second output terminal (B) to the ends (D_1 and D_2) of both coils so that currents flow in the two coils in the same sense. Apply some voltage to the coils by turning up the variac and note the attraction between the coils. Repeat the experiment by connecting A to C_1 and D_2 and B to C_2 and D_1 so that currents in the two coils flow in opposite senses, and note repulsion this time. What we have just described is Ampère's force law at work. If the currents flow in the same sense, the magnetic force is one of attraction, and if the currents flow in opposite senses, it is one of repulsion, as shown in Fig. 3.20(a) and (b), respectively, for straight wires, for the sake of simplicity.[1]
2. Connect coil No. 2 to variac and coil No. 1 to an oscilloscope to observe induced voltage in coil No. 1, thereby demonstrating Faraday's law. Note

[1] See L. Pearce Williams, "André-Marie Ampère," *Scientific American,* January 1989, pp. 90–97, for an interesting account of Ampère's experiments involving helical and spiral coils.

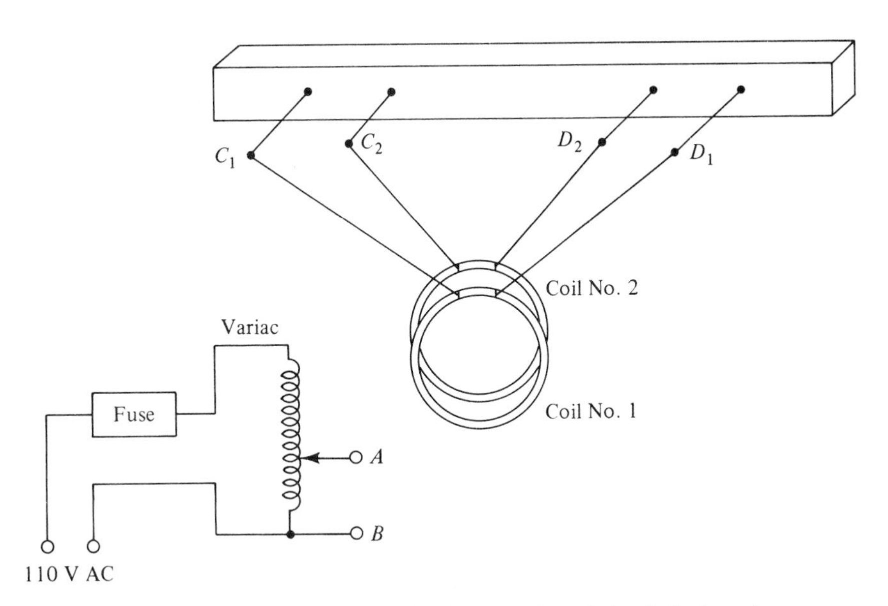

Figure 3.19. Experimental setup for demonstration of Ampère's force law, Faraday's law, and the principle of magnetic levitation.

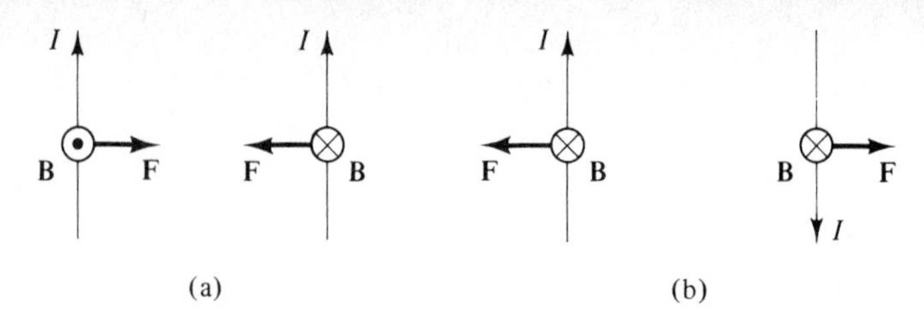

Figure 3.20. For explaining (a) force of attraction for currents flowing in the same sense and (b) force of repulsion for currents flowing in opposite senses.

change in induced voltage as variac voltage is changed. Also note change in induced voltage by keeping variac voltage constant and moving coil No. 1 away from coil No. 2 and/or turning it about the vertical.

3. Connect coil No. 2 to variac and leave coil No. 1 open circuited. Observe that no action takes place as the variac voltage is applied to coil No. 2. This is because although a voltage is induced in coil No. 1, no current flows in it. Now short circuit coil No. 1 and repeat the experiment to note repulsion. This is due to the induced voltage in coil No. 1 causing a current flow in it in the sense opposite to that in coil No. 2, and hence is a result of the combination of Faraday's law and Ampère's force law. That the force is one of repulsion can be deduced by writing circuit equations and showing that the current in the short-circuited coil does indeed flow in the sense opposite to that in the excited coil. However, it can be explained with the aid of physical reasoning as follows. When both coils are excited in the same sense in part (1) of the demonstration, the magnetic flux linking each coil is the sum of two fluxes in the same sense, due to the two currents. When the two coils are excited in opposite senses, the magnetic flux linking each coil is the algebraic sum of two fluxes in opposing senses, due to the two currents. Therefore, for the same source voltage and for the same pair of coils, the currents that flow in the coils in the second case have to be greater than those in the first case, for the induced voltage in each coil to equal the applied voltage. Thus the force of repulsion in the second case is greater than the force of attraction in the first case. Consider now the case of one of the coils excited by source voltage, say, V_g, and the other short circuited. Then the situation can be thought of as the first coil excited by $V_g/2$ and $V_g/2$ in series, and the second coil excited by $V_g/2$ and $-V_g/2$ in series, thereby resulting in a force of attraction and a force of repulsion. Since the force of repulsion is greater than the force of attraction, the net force, according to superposition, is one of repulsion.
4. Now to demonstrate actual levitation, place a smaller coil (about 30 turns of No. 28 wire of about 2 in. diameter) on a heavy aluminum plate (5 in. × 5 in. × $\frac{1}{2}$ in.) as shown in Fig. 3.21. Applying only the minimum necessary voltage and turning the variac only momentarily to avoid overheating, pass current through the coil from the variac to see the coil levitate. This levitation is due to the repulsive action between the current in the coil and the induced currents in the metallic plate. Since the plate is heavy and cannot move, the alternative is for the coil to lift up.

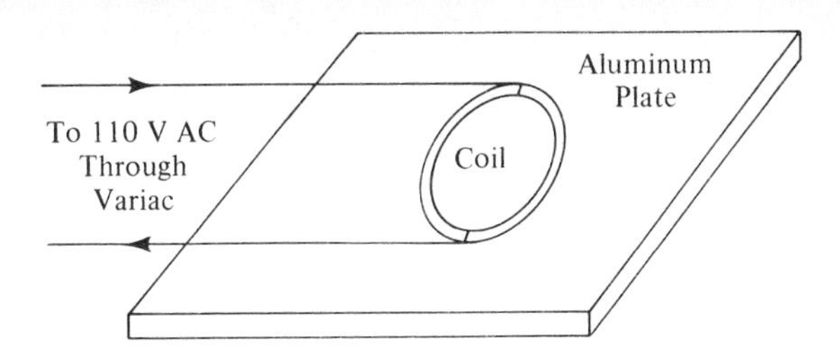

Figure 3.21. Setup for demonstrating magnetic levitation.

K3.2. Faraday's law; Right-hand screw rule; Lenz's law; Faraday's law for *N*-turn coil.

D3.5. Given $\mathbf{B} = B_0(\sin \omega t\ \mathbf{i}_x - \cos \omega t\ \mathbf{i}_y)$ Wb/m^2, find the induced emf around each of the following closed paths: **(a)** the rectangular path from (0, 0, 0) to (0, 1, 0) to (0, 1, 1) to (0, 0, 1) to (0, 0, 0); **(b)** the triangular path from (1, 0, 0) to (0, 1, 0) to (0, 0, 1) to (1, 0, 0); and **(c)** the rectangular path from (0, 0, 0) to (1, 1, 0) to (1, 1, 1) to (0, 0, 1) to (0, 0, 0).

Ans. **(a)** $-\omega B_0 \cos \omega t$ V; **(b)** $-\dfrac{\omega B_0}{\sqrt{2}} \cos (\omega t - \pi/4)$ V;

(c) $-\sqrt{2}\,\omega B_0 \sin (\omega t + \pi/4)$ V

D3.6. A square loop lies in the *xy*-plane forming the closed path *C* connecting the points (0, 0, 0), (1, 0, 0), (1, 1, 0), (0, 1, 0), and (0, 0, 0), in that order. A magnetic field **B** exists in the region. From considerations of Lenz's law, determine whether the induced emf around the closed path *C* at $t = 0$ is positive, negative, or zero for each of the following magnetic fields, where B_0 is a positive constant: **(a)** $\mathbf{B} = B_0 t\,\mathbf{i}_z$; **(b)** $\mathbf{B} = B_0 \cos (2\pi t + 60°)\,\mathbf{i}_z$; and **(c)** $\mathbf{B} = B_0 e^{-t^2}\mathbf{i}_z$.

Ans. **(a)** negative; **(b)** positive; **(c)** zero

D3.7. For $\mathbf{B} = B_0 \cos \omega t\ \mathbf{i}_z$ Wb/m^2, find the induced emf around the following closed paths: **(a)** the closed path comprising the straight lines successively connecting the points (0, 0, 0), (1, 0, 0), (1, 1, 0), (0, 1, 0), (0, 0, 0.001), and (0, 0, 0); **(b)** the closed path comprising the straight lines successively connecting the points (0, 0, 0), (1, 0, 0), (1, 1, 0), (0, 1, 0), (0, 0, 0.001), (1, 0, 0.001), (1, 1, 0.001), (0, 1, 0.001), (0, 0, 0.002), and (0, 0, 0) with a slight kink in the straight line at the point (0, 0, 0.001) to avoid touching the point; and **(c)** the closed path comprising the helical path $r = 1/\sqrt{\pi}$, $\phi = 1000\pi z$ from $(1/\sqrt{\pi}, 0, 0)$ to $(1/\sqrt{\pi}, 0, 0.01)$ and the straight line path from $(1/\sqrt{\pi}, 0, 0.01)$ to $(1/\sqrt{\pi}, 0, 0)$ with slight kinks to avoid touching the helical path.

Ans. **(a)** $\omega B_0 \sin \omega t$ V; **(b)** $2\omega B_0 \sin \omega t$ V; **(c)** $5\omega B_0 \sin \omega t$ V

3.3 AMPÈRE'S CIRCUITAL LAW

In the preceding section we introduced Faraday's law, one of Maxwell's equations, in integral form. In this section we introduce another of Maxwell's equations in integral form. This equation, known as *Ampère's circuital law,* is a combination of an experimental finding of Oersted that electric currents generate magnetic fields and a mathematical contribution of Maxwell that time-varying electric fields give rise to magnetic fields. It is this contribution of Maxwell that led to the prediction of electromagnetic wave propagation even before the phenomenon was discovered experimentally. In mathematical form, Ampere's circuital law is analogous to Faraday's law and is given by

$$\oint_C \mathbf{H} \cdot d\mathbf{l} = [I_c]_S + \frac{d}{dt}\int_S \mathbf{D} \cdot d\mathbf{S} \tag{3.15}$$

where S is a surface bounded by C.

The quantity $\oint_C \mathbf{H} \cdot d\mathbf{l}$ on the left side of (3.15) is the line integral of the vector field $\mathbf{H}$ around the closed path C. We learned in Section 3.1 that the quantity $\oint_C \mathbf{E} \cdot d\mathbf{l}$ has the physical meaning of work per unit charge associated with the movement of a test charge around the closed path C. The quantity $\oint_C \mathbf{H} \cdot d\mathbf{l}$ does not have a similar physical meaning. This is because the magnetic force on a moving charge is directed perpendicular to the direction of motion of the charge as well as to the direction of the magnetic field and hence does not do work in the movement of the charge. By recalling that $\mathbf{H}$ has the units of amperes per meter, we obtain the units of current (A) for $\oint_C \mathbf{H} \cdot d\mathbf{l}$. However, by analogy with the name *electromotive force* for $\oint_C \mathbf{E} \cdot d\mathbf{l}$, the quantity $\oint_C \mathbf{H} \cdot d\mathbf{l}$ is known as the *magnetomotive force,* abbreviated as mmf.

The quantity $[I_c]_S$ on the right side of (3.15) is the current due to flow of free charges crossing the surface S. It can be a convection current such as due to motion of a charged cloud in space or a conduction current due to motion of charges in a conductor. It cannot be due to the polarization and magnetization phenomena in a material because the effects of these phenomena are taken into account implicitly in the definitions of $\mathbf{D}$ and $\mathbf{H}$, respectively. While $[I_c]_S$ can be filamentary current, surface current, or volume current, or a combination of these, it is formulated in terms of the volume current density vector, $\mathbf{J}$, in the manner

$$[I_c]_S = \int_S \mathbf{J} \cdot d\mathbf{S} \tag{3.16}$$

Just as the surface integral of the magnetic flux density vector $\mathbf{B}$ (Wb/m^2) over a surface S gives the magnetic flux (Wb) crossing that surface, the surface integral of $\mathbf{J}$ (A/m^2) over a surface S gives the current (A) crossing that surface.

Displacement current

The quantity $\int_S \mathbf{D} \cdot d\mathbf{S}$ on the right side of (3.15) is the flux of the vector field $\mathbf{D}$ crossing S. Hence it is the displacement flux, or the electric flux, crossing the surface S. By recalling that $\mathbf{D}$ has the units of coulombs per meter squared, we obtain the units of charge (C) for the displacement flux. Hence the quantity $d/dt\,(\int_S \mathbf{D} \cdot d\mathbf{S})$ has the units of d/dt (charge) or current and is known as the *displacement current*. While in materials it includes currents resulting from the polarization phenomena, in free space it is physically not a current in the sense that it does not represent the flow of charges, but it gives rise to the same effect as a current does, namely, producing a magnetic field.

Statement of Ampère's circuital law

We may now write (3.15) in the manner

$$\oint_C \mathbf{H} \cdot d\mathbf{l} = \int_S \mathbf{J} \cdot d\mathbf{S} + \frac{d}{dt}\int_S \mathbf{D} \cdot d\mathbf{S} \tag{3.17}$$

where we recall that

$$\boxed{\begin{aligned}\mathbf{D} &= \varepsilon\mathbf{E} \qquad (3.18a)\\ \mathbf{H} &= \frac{\mathbf{B}}{\mu} \qquad (3.18b)\end{aligned}}$$

and it is understood that $\int_S \mathbf{J} \cdot d\mathbf{S}$, although formulated in terms of the volume current density vector $\mathbf{J}$, represents the algebraic sum of all currents due to flow of charges crossing the surface S. If the current is due to conduction, then

$$\boxed{\mathbf{J} = \mathbf{J}_c = \sigma\mathbf{E}} \qquad (3.18c)$$

In words, (3.17) states that *the magnetomotive force around a closed path C is equal to the algebraic sum of the current due to flow of charges and the displacement current bounded by C.* The situation is illustrated in Fig. 3.22.

As in the case of Faraday's law, there are certain procedures and observations pertinent to the application of (3.17). These are as follows.

1. The surface integrals on the right side of (3.17) are to be evaluated in accordance with the R.H.S. rule, which means that for the geometry of Fig. 3.22, the normal vector to the differential surface dS should be directed as shown in the figure.
2. In evaluating the surface integrals, *any surface S bounded by C* can be employed. However, *the same surface* must be employed for the two surface integrals. It is not correct to consider two different surfaces to evaluate the two surface integrals, although both surfaces may be bounded by C.

Observation 2 implies that for the mmf around C to be unique, the sum of the two currents (current due to flow of charges and displacement current) through all possible surfaces bounded by C is the same. Let us now consider two surfaces S_1 and S_2 bounded by the closed paths C_1 and C_2, respectively, as shown in Fig. 3.23, where C_1 and C_2 are traversed in opposite senses and touch each

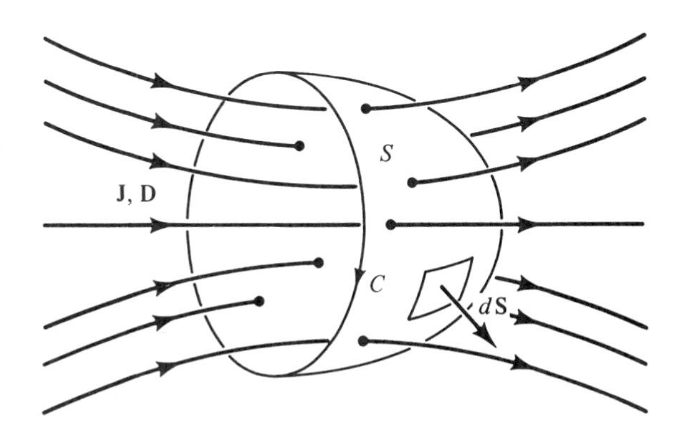

Figure 3.22. For illustrating Ampère's circuital law.

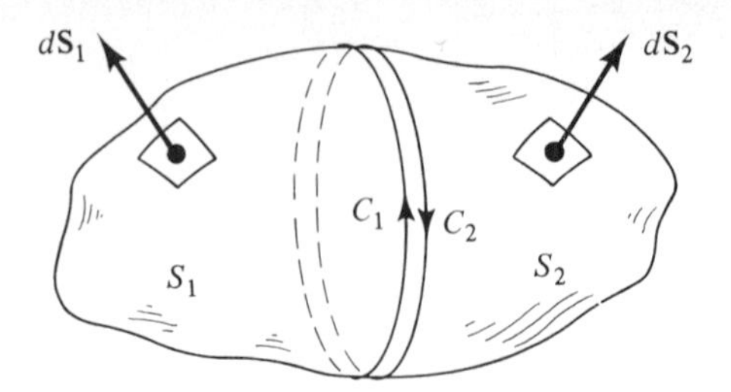

Figure 3.23. Two closed paths C_1 and C_2 touching each other and bounding the surfaces S_1 and S_2, respectively, which together form a closed surface.

other so that S_1 and S_2 together form a closed surface. The situation may be imagined by considering the closed surface to be that of a potato and C_1 and C_2 to be two rubber bands around the potato.

Applying Ampère's circuital law to C_1 and S_1 and noting that $d\mathbf{S}_1$ is chosen in accordance with the R.H.S. rule, we have

$$\oint_{C_1} \mathbf{H} \cdot d\mathbf{l} = \int_{S_1} \mathbf{J} \cdot d\mathbf{S}_1 + \frac{d}{dt}\int_{S_1} \mathbf{D} \cdot d\mathbf{S}_1 \tag{3.19a}$$

Similarly, applying Ampère's circuital law to C_2 and S_2 and noting again that $d\mathbf{S}_2$ is chosen in accordance with the R.H.S. rule, we have

$$\oint_{C_2} \mathbf{H} \cdot d\mathbf{l} = \int_{S_1} \mathbf{J} \cdot d\mathbf{S}_2 + \frac{d}{dt}\int_{S_2} \mathbf{D} \cdot d\mathbf{S}_2 \tag{3.19b}$$

Now adding (3.19a) and (3.19b), we obtain

$$0 = \oint_{S_1+S_2} \mathbf{J} \cdot d\mathbf{S} + \frac{d}{dt}\oint_{S_1+S_2} \mathbf{D} \cdot d\mathbf{S} \tag{3.20}$$

where the left side results from the fact that C_1 and C_2 are actually the same path but traversed in opposite senses, so that the two line integrals are the negatives of each other.

Since the closed surface $S_1 + S_2$ can be of any size and shape, we can generalize (3.20) to write

$$\oint_S \mathbf{J} \cdot d\mathbf{S} + \frac{d}{dt}\oint_S \mathbf{D} \cdot d\mathbf{S} = 0$$

or

$$\boxed{\frac{d}{dt}\oint_S \mathbf{D} \cdot d\mathbf{S} = -\oint_S \mathbf{J} \cdot d\mathbf{S}} \tag{3.21}$$

Thus the displacement current emanating from a closed surface is equal to the current due to charges flowing into the volume bounded by that closed surface.

Capacitor circuit

An important example of the property given by (3.21) at work is in a capacitor circuit, as shown in Fig. 3.24. In this circuit, the time-varying voltage source sets up a time-varying electric field between the plates of the capacitor and directed from one plate to the other. Therefore, one can talk about displacement current crossing a surface between the plates. According to (3.21) applied to a

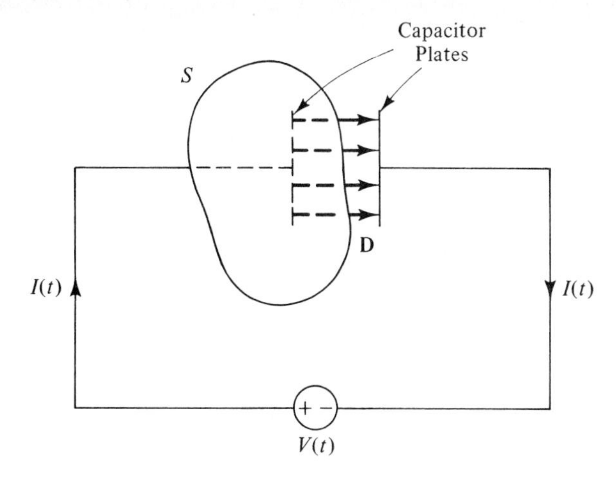

Figure 3.24. Capacitor circuit for illustrating that the displacement current from one plate to the other is equal to the wire current.

closed surface S enclosing one of the plates as shown in the figure,

$$\frac{d}{dt}\oint_S \mathbf{D} \cdot d\mathbf{S} = I(t) \tag{3.22}$$

where $I(t)$ is the current (due to flow of charges in the wire) drawn from the voltage source. Neglecting fringing effects and assuming that the electric field is normal to the plates and uniform, we have, from (3.22),

$$\frac{d}{dt}\oint_S \mathbf{D} \cdot d\mathbf{S} = \frac{d}{dt}(DA) = I(t) \tag{3.23}$$

where A is the area of each plate. Thus, where the wire current ends on one of the plates, the displacement current takes over and completes the circuit to the second plate.

Measurement of dielectric permittivity

Since for a given electric field intensity the displacement flux density between the plates of the capacitor varies with the permittivity of the dielectric, (3.23) can be used as a basis for the measurement of the relative permittivity of a dielectric material. We shall discuss the details of a simple experimental demonstration of this technique in the following.

A schematic of the experimental setup is shown in Fig. 3.25. It consists of a parallel-plate capacitor arrangement, made up of two 12 in. × 12 in. aluminum sheets attached to wooden blocks spaced about $\frac{1}{2}$ in. apart and driven by an ac voltage generator in series with a resistor. By observing the voltage across the resistor, the current through the circuit can be monitored. With no dielectric between the plates, $\mathbf{D} = \varepsilon_0 \mathbf{E}$, and

$$I(t) = \frac{d}{dt}\int_A \varepsilon_0 \mathbf{E} \cdot d\mathbf{S} = \varepsilon_0 A \frac{dE}{dt}$$

With a dielectric between the plates, $\mathbf{D} = \varepsilon \mathbf{E} = \varepsilon_0 \varepsilon_r \mathbf{E}$, and

$$I(t) = \frac{d}{dt}\int_A \varepsilon_0 \varepsilon_r \mathbf{E} \cdot d\mathbf{S} = \varepsilon_0 \varepsilon_r A \frac{dE}{dt}$$

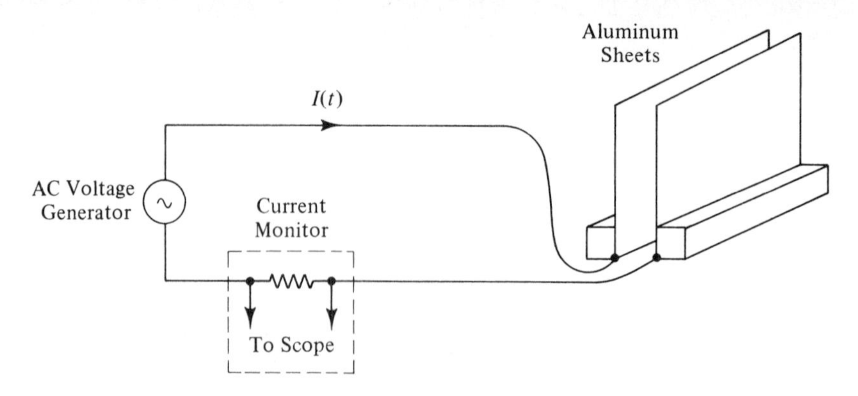

Figure 3.25. Experimental setup for demonstration of measurement of the relative permittivity of a dielectric material.

Thus if the electric field between the plates is the same in both cases,

$$\varepsilon_r = \frac{I(t) \text{ with the dielectric}}{I(t) \text{ without the dielectric}}$$

To maintain the same electric field between the plates, the voltage ($= \int \mathbf{E} \cdot d\mathbf{l}$) across the plates is ensured to be the same in both cases. The procedure therefore consists of the following steps:

1. With the dielectric sample of the same size as the plates inserted between the plates and the voltage source applied, measure the voltage across the resistor and the voltage across the capacitor plates.
2. Without changing the spacing between the plates, pull the dielectric sample out. Adjust the voltage source value such that the voltage across the capacitor plates is the same as in step 1. This can be omitted if the value of the resistor is chosen such that the voltage drop across it is only a small fraction of the source voltage. Measure the voltage across the resistor.
3. Compute the ratio of the voltage across the resistor measured in step 1 to that measured in step 2. This ratio is the relative permittivity ε_r of the dielectric sample, at the frequency of the voltage source and to within the assumptions made concerning the field between the plates.

In the experimental demonstration we just discussed, we maintained the electric field intensity **E** between the plates constant and made use of the variation of **D** and hence of $I(t)$ with ε. Alternatively, if ε remains the same and **E** changes, once again **D** and hence $I(t)$ changes. This is the basis behind the operation of a condenser microphone.

Condenser microphone

A condenser microphone consists of a tightly stretched but movable metallic-plate diaphragm electrically insulated from a second fixed plate, as shown in Fig. 3.26. A dc voltage is applied between the plates via a resistor. Sound waves impinging on the diaphragm cause it to vibrate, thereby changing the spacing between the two plates. This in turn changes the electric field between the plates, resulting in a change in the displacement current and hence in the wire current through the resistor. The voltage fluctuations across the resistor are picked up and amplified.

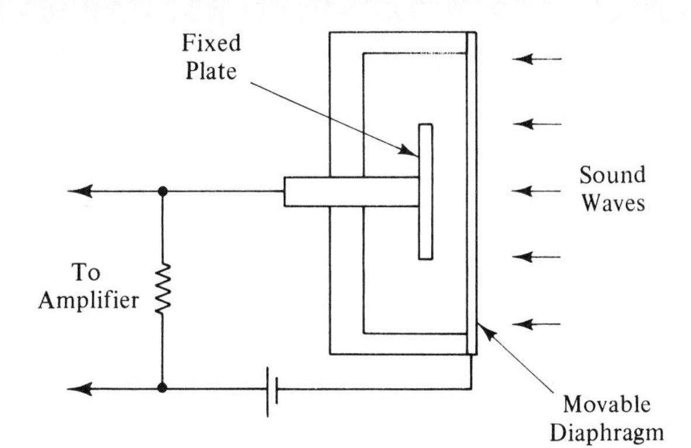

Figure 3.26. Schematic of a condenser microphone.

Radiation from antenna

Let us now return to Ampère's circuital law (3.17) and examine it together with Faraday's law (3.13). To do this, we repeat the two laws

$$\oint_C \mathbf{E} \cdot d\mathbf{l} = -\frac{d}{dt}\int_S \mathbf{B} \cdot d\mathbf{S} \tag{3.24}$$

$$\oint_C \mathbf{H} \cdot d\mathbf{l} = \int_S \mathbf{J} \cdot d\mathbf{S} + \frac{d}{dt}\int_S \mathbf{D} \cdot d\mathbf{S} \tag{3.25}$$

and observe that time-varying electric and magnetic fields are interdependent, since according to Faraday's law (3.24), a time-varying magnetic field produces an electric field, whereas according to Ampère's circuital law (3.25), a time-varying electric field gives rise to a magnetic field. In addition, Ampère's circuital law tells us that an electric current generates a magnetic field. These properties form the basis for the phenomena of radiation and propagation of electromagnetic waves. To provide a simplified, qualitative explanation of radiation from an antenna, we begin with a piece of wire carrying a time-varying current, $I(t)$, as shown in Fig. 3.27. Then, the time-varying current generates a time-varying magnetic field $\mathbf{H}(t)$, which surrounds the wire. Time-varying electric and magnetic fields, $\mathbf{E}(t)$ and $\mathbf{H}(t)$, are then produced in succession, as shown by two views in Fig. 3.27, thereby giving rise to electromagnetic waves. Thus just as water waves are produced when a rock is thrown in a pool of water, electromagnetic waves are radiated when a piece of wire in space is excited by a time-varying current.

K3.3. Ampère's circuital law; Displacement current; Capacitor circuit.

D3.8. For $\mathbf{E} = E_0 t e^{-t^2}\mathbf{i}_z$ in free space, find the displacement current crossing an area of 0.1 m^2 in the xy-plane from the $-z$-side to the $+z$-side for each of the following values of t: **(a)** $t = 0$; **(b)** $t = 1/\sqrt{2}$ s; and **(c)** $t = 1$ s.
Ans. **(a)** $0.1\varepsilon_0 E_0$ A; **(b)** 0; **(c)** $-0.1e^{-1}\varepsilon_0 E_0$ A

D3.9. Three point charges $Q_1(t)$, $Q_2(t)$, and $Q_3(t)$ situated at the corners of an equilateral triangle of sides 1 m are connected to each other by wires along the sides of the triangle. Currents of I A and $3I$ A flow from Q_1 to Q_2 and Q_1 to Q_3, respectively. The displacement current emanating from a spherical surface of radius 0.1 m and centered at Q_2 is $-2I$ A. Find the following: **(a)** the current flowing from Q_2 to Q_3; **(b)** the displacement current emanating from the spherical surface of radius

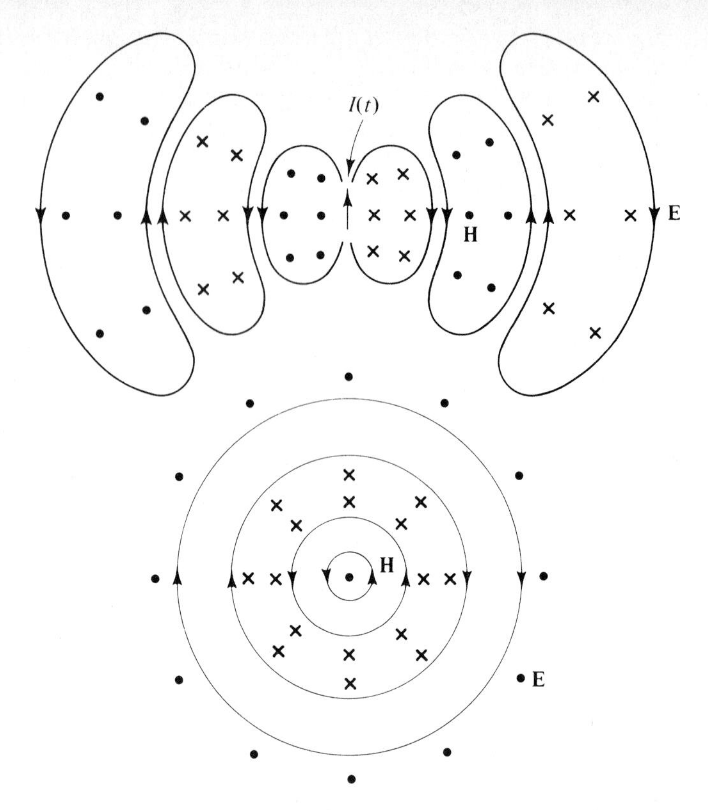

Figure 3.27. Two views of a simplified depiction of electromagnetic wave radiation from a piece of wire carrying a time-varying current.

0.1 m and centered at Q_1; and **(c)** the displacement current emanating from the spherical surface of radius 0.1 m and centered at Q_3.
Ans. **(a)** $3I$ A; **(b)** $-4I$ A; **(c)** $6I$ A

3.4 GAUSS' LAWS AND THE LAW OF CONSERVATION OF CHARGE

In the previous two sections we learned two of the four Maxwell's equations. These two equations have to do with the line integrals of the electric and magnetic fields around closed paths. The remaining two Maxwell's equations are pertinent to the surface integrals of the electric and magnetic fields over closed surfaces. These are known as *Gauss' laws.*

Gauss' law for the electric field

Gauss' law for the electric field states that *the displacement flux emanating from a closed surface S is equal to the charge contained within the volume V bounded by that surface*. This statement, although familiarly known as Gauss' law, has its origin in experiments conducted by Faraday. In mathematical form, it is given by

$$\boxed{\oint_S \mathbf{D} \cdot d\mathbf{S} = [Q]_V} \tag{3.26}$$

The quantity $[Q]_V$ is the free charge contained within the volume V bounded by S. It can be free charge such as in a charged cloud or in a conductor. It cannot be due to the polarization phenomenon in a dielectric material since the effect of this phenomenon is taken into account implicitly in the definition of **D.** While $[Q]_V$ can be a point charge, surface charge, or volume charge, or a combination of these, it is formulated as the volume integral of the volume charge density ρ, that is, in the manner

$$[Q]_V = \int_V \rho \, dv \tag{3.27}$$

Evaluation of volume integral

The volume integral is a triple integral since dv is the product of three differential lengths. For an illustration of the evaluation of a volume integral, let us consider

$$\rho = (x + y + z) \text{ C/m}^3$$

and the cubical volume V bounded by the planes $x = 0$, $x = 1$, $y = 0$, $y = 1$, $z = 0$, and $z = 1$. Then the charge Q contained within the cubical volume is given by

$$\begin{aligned}
Q = \int_V \rho \, dv &= \int_{x=0}^{1} \int_{y=0}^{1} \int_{z=0}^{1} (x + y + z) \, dx \, dy \, dz \\
&= \int_{x=0}^{1} \int_{y=0}^{1} \left[xz + yz + \frac{z^2}{2} \right]_{z=0}^{1} dx \, dy \\
&= \int_{x=0}^{1} \int_{y=0}^{1} \left(x + y + \frac{1}{2} \right) dx \, dy \\
&= \int_{x=0}^{1} \left[xy + \frac{y^2}{2} + \frac{y}{2} \right]_{y=0}^{1} dx \\
&= \int_{x=0}^{1} (x + 1) \, dx \\
&= \left[\frac{x^2}{2} + x \right]_{x=0}^{1} \\
&= \frac{3}{2} \text{ C}
\end{aligned}$$

We may now write Gauss' law for the electric field (3.26) in the manner

$$\boxed{\oint_S \mathbf{D} \cdot d\mathbf{S} = \int_V \rho \, dv} \tag{3.28}$$

where we recall that

$$\mathbf{D} = \varepsilon \mathbf{E}$$

and it is understood that $\int_V \rho \, dv$, although formulated in terms of the volume charge density ρ, represents the algebraic sum of all free charges contained within V. The situation is illustrated in Fig. 3.28.

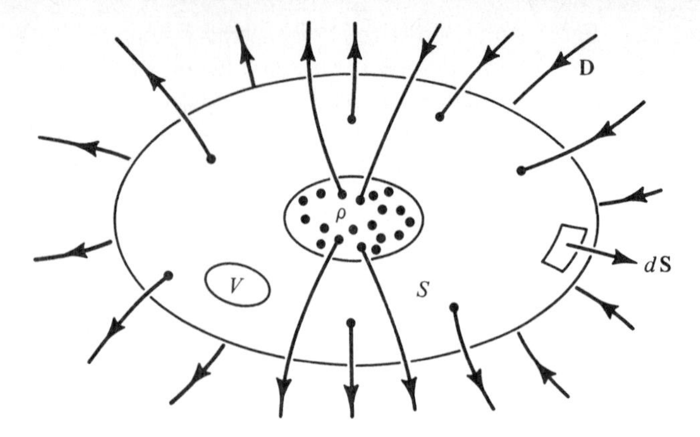

Figure 3.28. For illustrating Gauss' law for the electric field.

Gauss' law for the magnetic field

Gauss' law for the magnetic field is analogous to Gauss' law for the electric field and is given by

$$\boxed{\oint_S \mathbf{B} \cdot d\mathbf{S} = 0} \tag{3.29}$$

In words, (3.29) states that *the magnetic flux emanating from a closed surface is equal to zero*. In physical terms, (3.29) signifies that magnetic charges do not exist and magnetic flux lines are closed. Whatever magnetic flux enters (or leaves) a certain part of a closed surface must leave (or enter) through the remainder of the closed surface, as illustrated in Fig. 3.29.

This property of the magnetic field is sometimes useful in the computation of magnetic flux crossing a given surface (which is not closed). For example, to find the magnetic flux crossing the slanted plane surface S_1 in Fig. 3.30, it is not necessary to evaluate formally the surface integral of **B** over that surface. Since the slant surface S_1 and the three surfaces S_2, S_3, and S_4 in the coordinate planes together form a closed surface, the required flux is the same as the net flux crossing the surfaces S_2, S_3, and S_4. In fact, the net flux crossing the surfaces S_2, S_3, and S_4 is the same as that crossing any nonplanar surface having the same periphery as that of S_1. Thus as already pointed out in Sec. 3.2, it is a fundamental property of the magnetic field that the magnetic flux is the same through all sur-

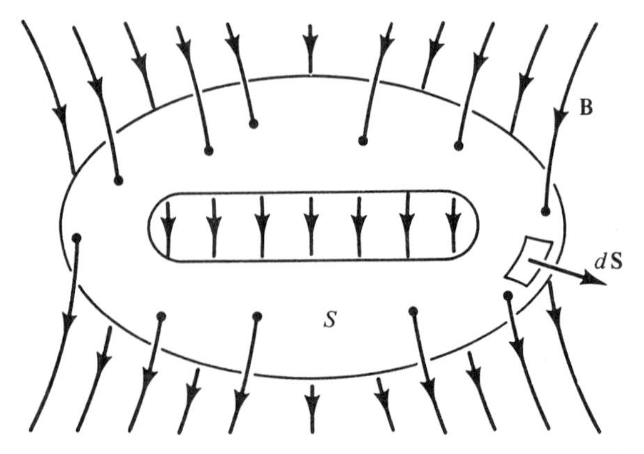

Figure 3.29. For illustrating Gauss' law for the magnetic field.

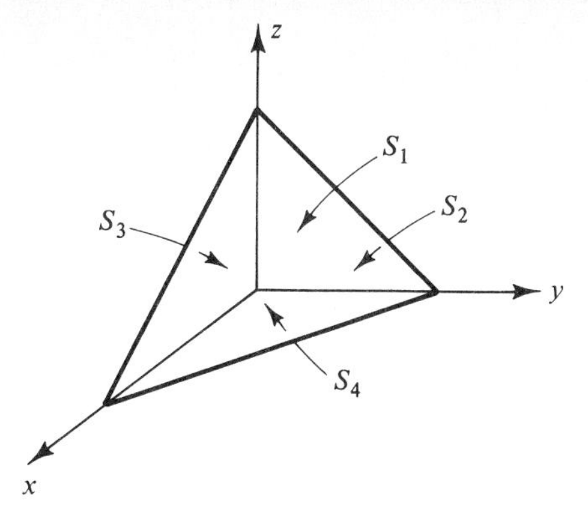

Figure 3.30. Slanted plane surface S_1 and surfaces S_2, S_3, and S_4 in the coordinate planes.

faces bounded by a closed path, and hence *any surface S bounded by closed path C can be used in Faraday's law.*

In view of the foregoing discussion, it can be seen that Gauss' law for the magnetic field is not independent of Faraday's law. To show this mathematically, we consider the geometry shown in Fig. 3.23 and apply Faraday's law to the two closed paths to write

$$\oint_{C_1} \mathbf{E} \cdot d\mathbf{l} = -\frac{d}{dt}\int_{S_1} \mathbf{B} \cdot d\mathbf{S}_1$$

$$\oint_{C_2} \mathbf{E} \cdot d\mathbf{l} = -\frac{d}{dt}\int_{S_2} \mathbf{B} \cdot d\mathbf{S}_2$$

Adding the two equations, we obtain

$$0 = -\frac{d}{dt}\oint_{S_1+S_2} \mathbf{B} \cdot d\mathbf{S}$$

or

$$\oint_{S_1+S_2} \mathbf{B} \cdot d\mathbf{S} = \text{constant with time} \tag{3.30}$$

Since there is no experimental evidence that the right side of (3.30) is nonzero, it follows that

$$\oint_S \mathbf{B} \cdot d\mathbf{S} = 0$$

where we have replaced $S_1 + S_2$ by S.

Law of conservation of charge

In a similar manner, Gauss' law for the electric field is not independent of Ampère's circuital law in view of the *law of conservation of charge*. The law of conservation of charge states that *the net current due to flow of charges emanating from a closed surface S is equal to the time rate of decrease of the charge within the volume V bounded by S*. It is given in mathematical form by

$$\boxed{\oint_S \mathbf{J} \cdot d\mathbf{S} = -\frac{d}{dt}\int_V \rho\, dv} \tag{3.31}$$

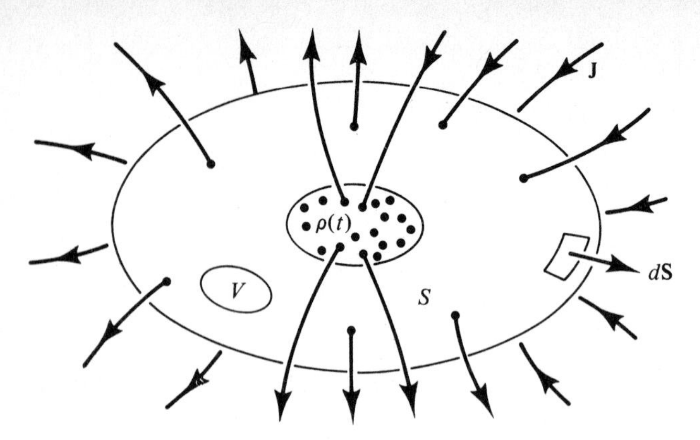

Figure 3.31. For illustrating the law of conservation of charge.

As illustrated in Fig. 3.31, this law follows from the property that electric charge is conserved. If the charge in a given volume is decreasing with time at a certain rate, there must be a net outflow of the charge at the same rate. Since current is defined to be the rate of flow of charge, (3.31) then follows. As in the case of (3.17), it is understood that $\oint_S \mathbf{J} \cdot d\mathbf{S}$ in (3.31), although formulated in terms of $\mathbf{J}$, represents the algebraic sum of all currents due to flow of charges crossing S.

Comparing (3.21) and (3.31), we obtain

$$\frac{d}{dt}\oint_S \mathbf{D} \cdot d\mathbf{S} = \frac{d}{dt}\int_V \rho\, dv$$

$$\frac{d}{dt}\left(\oint_S \mathbf{D} \cdot d\mathbf{S} - \int_V \rho\, dv\right) = 0$$

$$\int_S \mathbf{D} \cdot d\mathbf{S} - \int_V \rho\, dv = \text{constant with time} \tag{3.32}$$

Since there is no experimental evidence that the right side of (3.32) is nonzero, it follows that

$$\oint_S \mathbf{D} \cdot d\mathbf{S} = \int_V \rho\, dv$$

Thus since (3.21) follows from Ampère's circuital law, Gauss' law for the electric field follows from Ampère's circuital law with the aid of the law of conservation of charge.

We shall now illustrate the combined application of Gauss' law for the electric field, the law of conservation of charge, and Ampère's circuital law by means of an example.

Example 3.5

Let us consider current I A flowing from a point charge $Q(t)$ at the origin to infinity along a semi-infinitely long, straight wire occupying the positive z-axis, and find $\oint_C \mathbf{H} \cdot d\mathbf{l}$ where C is a circular path of radius a lying in the xy-plane and centered at the point charge, as shown in Fig. 3.32.

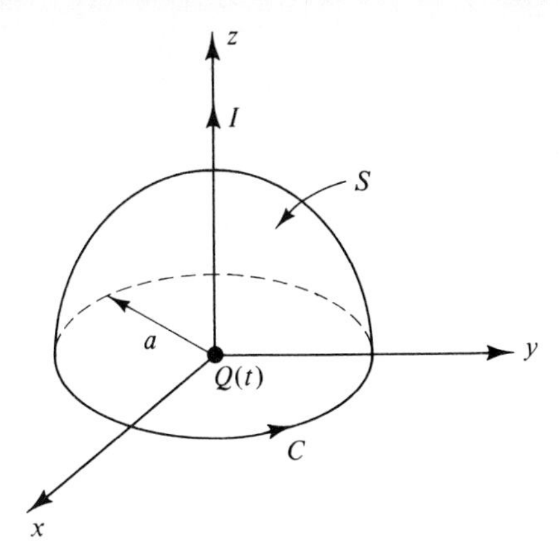

Figure 3.32. Semi-infinitely long wire of current I, with a point charge $Q(t)$ at the origin.

Considering the hemispherical surface S bounded by C, and above the xy-plane, as shown in Fig. 3.32, and applying Ampère's circuital law, we obtain

$$\oint_C \mathbf{H} \cdot d\mathbf{l} = I + \frac{d}{dt} \int_S \mathbf{D} \cdot d\mathbf{S} \qquad (3.33)$$

From Gauss' law for the electric field, the displacement flux emanating from a spherical surface centered at the point charge is equal to Q. In view of the spherical symmetry of the electric field about the point charge, half of the flux goes through the hemispherical surface. Thus

$$\int_S \mathbf{D} \cdot d\mathbf{S} = \frac{Q}{2} \qquad (3.34)$$

From the law of conservation of charge applied to a spherical surface centered at the point charge,

$$I = -\frac{dQ}{dt} \qquad (3.35)$$

Substituting (3.34) into (3.33) and then using (3.35), we obtain

$$\begin{aligned}\oint_C \mathbf{H} \cdot d\mathbf{l} &= I + \frac{d}{dt}\left(\frac{Q}{2}\right) \\ &= I + \frac{1}{2}\frac{dQ}{dt} \\ &= I + \frac{1}{2}(-I) \\ &= \frac{I}{2}\end{aligned}$$

K3.4. Volume integral; Gauss' law for the electric field; Gauss' law for the magnetic field; Law of conservation of charge.

D3.10. Several types of charge are located, in Cartesian coordinates, as follows: a point charge of 1 μC at $(1, 1, -1.5)$, a line charge of uniform density 2 μC/m along the

straight line from $(-1, -1, -1)$ to $(3, 3, 3)$, and a surface charge of uniform density $-1\ \mu\text{C/m}^2$ on that part of the plane $x = 0$ between $z = -1$ and $z = 1$. Find the displacement flux emanating from each of the following closed surfaces: **(a)** surface of the cubical box bounded by the planes $x = \pm 2$, $y = \pm 2$, and $z = \pm 2$; **(b)** surface of the cylindrical box of radius 2 m, having the z-axis as its axis and lying between $z = -2$ and $z = 2$; and **(c)** surface of the octahedron having its vertices at $(3, 0, 0)$, $(-3, 0, 0)$, $(0, 3, 0)$, $(0, -3, 0)$, $(0, 0, 3)$, and $(0, 0, -3)$.
Ans. **(a)** $3.3923\ \mu\text{C}$; **(b)** $1.3631\ \mu\text{C}$; **(c)** $-3.0718\ \mu\text{C}$

D3.11. Magnetic fluxes of absolute values ψ_1, ψ_2, and ψ_3 cross three surfaces S_1, S_2, and S_3, respectively, comprising a closed surface S. If $\psi_1 + \psi_2 + \psi_3 = \psi_0$, find the smallest of ψ_1, ψ_2, and ψ_3 for each of the following cases: **(a)** ψ_1, ψ_2, and ψ_3 are in arithmetic progression; **(b)** $1/\psi_1$, $1/\psi_2$, and $1/\psi_3$ are in arithmetic progression; and **(c)** $\ln \psi_1$, $\ln \psi_2$, and $\ln \psi_3$ are in arithmetic progression.
Ans. **(a)** $\frac{1}{6}\psi_0$; **(b)** $\frac{1}{2 + 2\sqrt{2}}\psi_0$; **(c)** $\frac{1}{3 + \sqrt{5}}\psi_0$

D3.12. Three point charges $Q_1(t)$, $Q_2(t)$, and $Q_3(t)$ are situated at the vertices of a triangle and are connected by means of wires carrying currents. A current I A flows from Q_1 to Q_2 and $3I$ A flows from Q_2 to Q_3. The charge Q_3 is increasing with time at the rate of $5I$ C/s. Find the following: **(a)** $\frac{dQ_1}{dt}$; **(b)** $\frac{dQ_2}{dt}$; and **(c)** the current flowing from Q_1 *to* Q_3.
Ans. **(a)** $-3I$ C/s; **(b)** $-2I$ C/s; **(c)** $2I$ A

3.5 APPLICATION TO STATIC FIELDS

Maxwell's equations for static fields

For static fields, that is, for $d/dt = 0$, Maxwell's equations in integral form become

$$\oint_C \mathbf{E} \cdot d\mathbf{l} = 0 \tag{3.36a}$$

$$\oint_C \mathbf{H} \cdot d\mathbf{l} = \int_S \mathbf{J} \cdot d\mathbf{S} \tag{3.36b}$$

$$\oint_S \mathbf{D} \cdot d\mathbf{S} = \int_V \rho\, dv \tag{3.36c}$$

$$\oint_S \mathbf{B} \cdot d\mathbf{S} = 0 \tag{3.36d}$$

whereas the law of conservation of charge becomes

$$\oint_S \mathbf{J} \cdot d\mathbf{S} = 0 \tag{3.37}$$

It can be immediately seen from (3.36a)–(3.36d) that the interdependence between the electric and magnetic fields no longer exists. Equation (3.36a) tells us simply that the static electric field is a conservative field. Similarly, (3.36d) tells us that the magnetic flux is the same through all surfaces bounded by a closed path. On the other hand, (3.36c) and (3.36b) enable us to find the static electric and magnetic fields for certain time-invariant charge and current distributions, respectively. These distributions must be such that the resulting electric and magnetic fields possess symmetry to be able to replace the integrals on the left sides of (3.36c) and (3.36b) by algebraic expressions involving the components of electric and magnetic fields, respectively.

In addition, in the case of (3.36b) the current on the right side must be uniquely given for a given closed path C, which property is ensured by (3.37). An example in which this current is uniquely given is that of the infinitely long wire in Fig. 3.33(a). This is because the current crossing all possible surfaces bounded by the closed path C is equal to I since the wire, being infinitely long, pierces through all such surfaces. This can also be seen in a different manner by imagining the closed path to be a rigid loop and visualizing that the loop cannot be moved to one side of the wire without cutting the wire. On the other hand, if the wire is finitely long, as shown in Fig. 3.33(b), it can be seen that for some surfaces bounded by C, the wire pierces through the surface, whereas for some other surfaces, it does not. Alternatively, a rigid loop occupying the closed path can be moved to one side of the wire without cutting the wire. Thus for this case, there is no unique value of the wire current enclosed by C and hence (3.36b) cannot be used to determine $\mathbf{H}$. The problem here is that (3.37) is not satisfied since for current to flow in the finitely long wire, there must be time-varying charges at the two ends, thereby giving rise to time-varying electric field. Hence a displacement current exists in addition to the wire current such that the algebraic sum of the two currents crossing all surfaces bounded by C is the same and requires the use of (3.17).

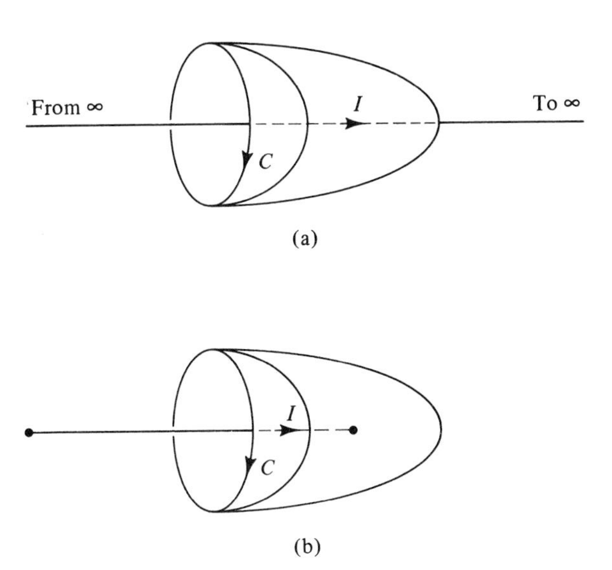

Figure 3.33. For illustrating that the current enclosed by a closed path C is uniquely given in (a) but not in (b).

We shall now illustrate the application of (3.36c) and (3.36b) by means of some examples.

Example 3.6

D *due to a line charge*

Let us consider charge distributed uniformly with density ρ_{L0} C/m along the z-axis and find the electric field due to the infinitely long line charge using (3.36c).

Let us consider the closed surface S of a cylinder of radius r, with the line charge as its axis and extending from $z = 0$ to $z = l$, as shown in Fig. 3.34. Then according to (3.36c),

$$\oint_S \mathbf{D} \cdot d\mathbf{S} = \rho_{L0} l \tag{3.38}$$

While this result is valid for any closed surface enclosing the portion of the line charge from $z = 0$ to $z = l$, we have chosen the particular surface in Fig. 3.34 to be able to reduce the surface integral of **D** in (3.36c) and hence in (3.38) to an algebraic quantity. To do this, we note the following:

(a) In view of the uniform charge density, the entire line charge can be thought of as the superposition of pairs of equal point charges located at equal distances above and below any given point on the z-axis. Hence the field due to the entire line charge has only a radial component independent of ϕ and z.

(b) In view of (a), the contribution to the closed surface integral from the top and bottom surfaces of the cylindrical box is zero.

Thus we have

$$\mathbf{D} = D_r(r)\mathbf{i}_r$$

and

$$\begin{aligned}\oint_S \mathbf{D} \cdot d\mathbf{S} &= \int_{\phi=0}^{2\pi} \int_{z=0}^{l} D_r(r)\mathbf{i}_r \cdot r\, d\phi\, dz\, \mathbf{i}_r \\ &= 2\pi r l D_r(r)\end{aligned} \tag{3.39}$$

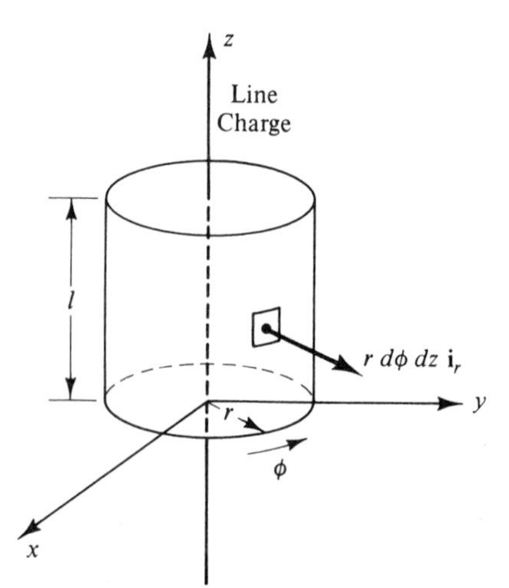

Figure 3.34. For the determination of electric field due to an infinitely long line charge of uniform density ρ_{L0} C/m.

Comparing (3.38) and (3.39), we obtain

$$2\pi r l D_r(r) = \rho_{L0} l$$

$$D_r(r) = \frac{\rho_{L0}}{2\pi r}$$

$$\boxed{\mathbf{D} = \frac{\rho_{L0}}{2\pi r}\mathbf{i}_r} \tag{3.40}$$

The field varies inversely with the radial distance away from the line charge.

Example 3.7

D *due to a spherical volume charge*

Let us consider charge distributed uniformly with density ρ_0 C/m^3 in the spherical region $r \leq a$, as shown by the cross-sectional view in Fig. 3.35, and find the electric field due to the spherical charge by using (3.36c).

As in Example 3.6, we shall once again choose a surface S that enables the replacement of the surface integral in (3.36c) by an algebraic quantity. To do this, we note from considerations of symmetry, and of the spherical charge as a superposition of point charges, that **D** possesses only an r-component dependent upon r only. Thus

$$\mathbf{D} = D_r(r)\mathbf{i}_r$$

Choosing then a spherical surface of radius r centered at the origin, we obtain

$$\begin{aligned}\oint_S \mathbf{D} \cdot d\mathbf{S} &= \int_{\theta=0}^{\pi} \int_{\phi=0}^{2\pi} D_r(r)\mathbf{i}_r \cdot r^2 \sin^2 \theta \, d\theta \, d\phi \, \mathbf{i}_r \\ &= 4\pi r^2 D_r(r)\end{aligned} \tag{3.41}$$

Noting that the charge exists only for $r < a$, and with uniform density, we obtain the charge enclosed by the spherical surface to be

$$\int_V \rho \, dv = \begin{cases} \frac{4}{3}\pi r^3 \rho_0 & \text{for} \quad r \leq a \\ \frac{4}{3}\pi a^3 \rho_0 & \text{for} \quad r \geq a \end{cases} \tag{3.42}$$

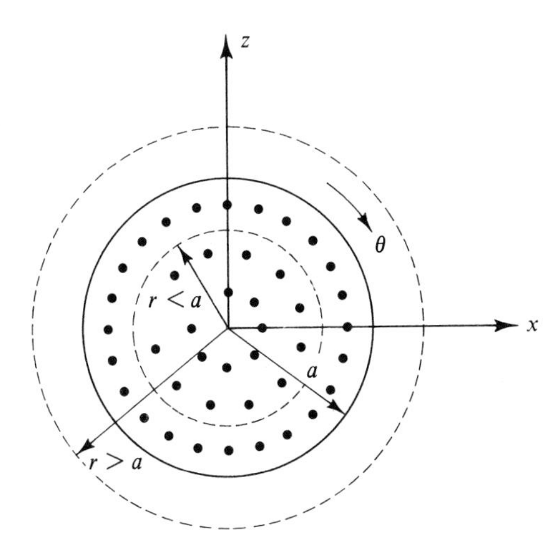

Figure 3.35. For the determination of electric field due to a spherical charge of uniform density ρ_0 C/m^3.

Substituting (3.41) and (3.42) into (3.36c), we get

$$4\pi r^2 D_r(r) = \begin{cases} \frac{4}{3}\pi r^3 \rho_0 & \text{for} \quad r \leq a \\ \frac{4}{3}\pi a^3 \rho_0 & \text{for} \quad r \geq a \end{cases}$$

$$D_r(r) = \begin{cases} \dfrac{\rho_0 r}{3} & \text{for} \quad r \leq a \\ \dfrac{\rho_0 a^3}{3r^2} & \text{for} \quad r \geq a \end{cases}$$

$$\boxed{\mathbf{D} = \begin{cases} \dfrac{\rho_0 r}{3}\mathbf{i}_r & \text{for} \quad r \leq a \\ \dfrac{\rho_0 a^3}{3r^2}\mathbf{i}_r & \text{for} \quad r \geq a \end{cases}} \tag{3.43}$$

The variation of D_r with r is shown plotted in Fig. 3.36.

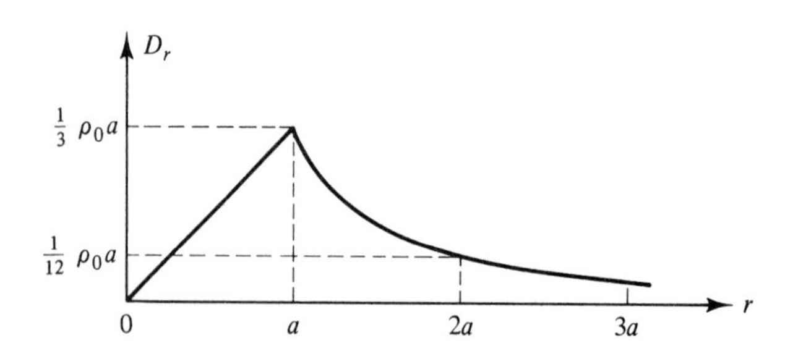

Figure 3.36. Variation of D_r with r for the spherical charge of Fig. 3.35.

Example 3.8

***H** due to a cylindrical wire of current*

Let us consider current flowing with uniform density $\mathbf{J} = J_0\mathbf{i}_z$ A/m² in an infinitely long, solid, cylindrical wire of radius a with its axis along the z-axis, as shown by the cross-sectional view in Fig. 3.37. We wish to find the magnetic field everywhere using (3.36b).

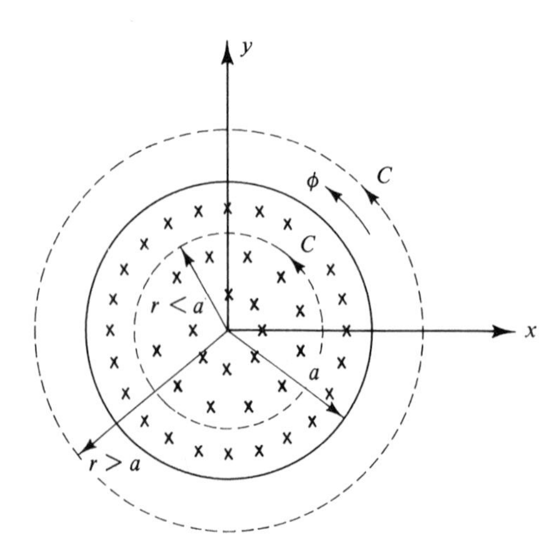

Figure 3.37. For the determination of magnetic field due to an infinitely long, solid, cylindrical wire of uniform current density $J_0\mathbf{i}_z$ A/m².

The current distribution can be thought of as the superposition of infinitely long filamentary wires parallel to the z-axis. Then in view of the symmetry about the z-axis and from the nature of the magnetic field due to an infinitely long wire given by (2.22), we can say that the required $\mathbf{H}$ has only a ϕ component dependent upon r only. Thus

$$\mathbf{H} = H_\phi(r)\mathbf{i}_\phi$$

Choosing then a circular closed path C of radius r lying in the xy-plane and centered at the origin, we obtain

$$\begin{aligned} \oint_C \mathbf{H} \cdot d\mathbf{l} &= \int_{\phi=0}^{2\pi} H_\phi(r)\mathbf{i}_\phi \cdot r\,d\phi\;\mathbf{i}_\phi \\ &= 2\pi r H_\phi(r) \end{aligned} \tag{3.44}$$

Considering the plane surface bounded by C, and noting that the current exists only for $r < a$, we obtain the current enclosed by the closed path to be

$$\begin{aligned} \int_S \mathbf{J} \cdot d\mathbf{S} &= \begin{cases} \displaystyle\int_{r=0}^{r}\int_{\phi=0}^{2\pi} J_0\mathbf{i}_z \cdot r\,dr\,d\phi\;\mathbf{i}_z & \text{for } r \le a \\ \displaystyle\int_{r=0}^{a}\int_{\phi=0}^{2\pi} J_0\mathbf{i}_z \cdot r\,dr\,d\phi\;\mathbf{i}_z & \text{for } r \ge a \end{cases} \\ &= \begin{cases} J_0\pi r^2 & \text{for } r \le a \\ J_0\pi a^2 & \text{for } r \ge a \end{cases} \end{aligned} \tag{3.45}$$

Substituting (3.44) and (3.45) into (3.36b), we get

$$2\pi r H_\phi = \begin{cases} J_0\pi r^2 & \text{for } r \le a \\ J_0\pi a^2 & \text{for } r \ge a \end{cases}$$

$$H_\phi = \begin{cases} \dfrac{J_0 r}{2} & \text{for } r \le a \\[2ex] \dfrac{J_0 a^2}{2r} & \text{for } r \ge a \end{cases}$$

$$\boxed{\mathbf{H} = \begin{cases} \dfrac{J_0 r}{2}\mathbf{i}_\phi & \text{for } r \le a \\[2ex] \dfrac{J_0 a^2}{2r}\mathbf{i}_\phi & \text{for } r \ge a \end{cases}} \tag{3.46}$$

The variation of H_ϕ with r is shown plotted in Fig. 3.38.

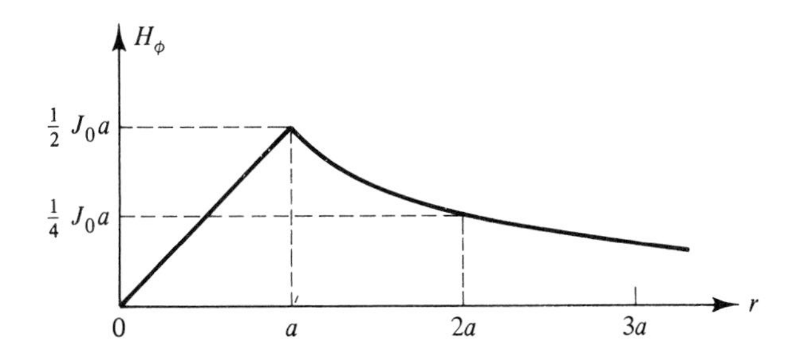

Figure 3.38. Variation of H_ϕ with r for the cylindrical wire of current of Fig. 3.37.

K3.5. Maxwell's equations in integral form for static fields; Uniqueness of current enclosed by a closed path; **D** due to symmetrical charge distributions; **H** due to symmetrical current distributions.

D3.13. Charge is distributed with uniform density ρ_0 C/m^3 inside a regular solid of edges a. Find the displacement flux emanating from one side of the solid for each of the following shapes of the solid: **(a)** tetrahedron; **(b)** cube; and **(c)** octahedron.
Ans. **(a)** $0.0295\rho_0 a^3$ C; **(b)** $0.1667\rho_0 a^3$ C; **(c)** $0.0589\rho_0 a^3$ C

D3.14. The cross section of an infinitely long, solid wire having the z-axis as its axis is a regular polygon of sides a. Current flows in the wire with uniform density $J_0\mathbf{i}_z$ A/m^2. Find the line integral of **H** along one side of the polygon and traversed in the sense of increasing ϕ for each of the following shapes of the polygon: **(a)** equilateral triangle; **(b)** square; and **(c)** octagon.
Ans. **(a)** $0.1433a^2J_0$ A; **(b)** $0.25a^2J_0$ A; **(c)** $0.6036a^2J_0$ A

3.6 BOUNDARY CONDITIONS

Boundary condition explained

In our study of electromagnetics, we will be considering many problems involving more than one medium. Examples are reflections of waves at an air–dielectric interface, determination of capacitance for a multiple-dielectric capacitor, and guiding of waves in a metallic waveguide. To solve a problem involving a boundary surface between different media, we need to know the conditions satisfied by the field components at the boundary. These are known as the *boundary conditions*. They are a set of relationships relating the field components at a point adjacent to and on one side of the boundary to the field components at a corresponding point adjacent to and on the other side of the boundary. These relationships arise from the fact that Maxwell's equations in integral form involve closed paths and surfaces and they must be satisfied for all possible closed paths and surfaces whether they lie entirely in one medium or encompass a portion of the boundary between two different media. In the latter case, Maxwell's equations in integral form must be satisfied collectively by the fields on either side of the boundary, thereby resulting in the boundary conditions.

We shall derive the boundary conditions by considering the Maxwell's equations

$$\oint_C \mathbf{E}\cdot d\mathbf{l} = -\frac{d}{dt}\int_S \mathbf{B}\cdot d\mathbf{S} \tag{3.47a}$$

$$\oint_C \mathbf{H}\cdot d\mathbf{l} = \int_S \mathbf{J}\cdot d\mathbf{S} + \frac{d}{dt}\int_S \mathbf{D}\cdot d\mathbf{S} \tag{3.47b}$$

$$\oint_S \mathbf{D}\cdot d\mathbf{S} = \int_V \rho\, dv \tag{3.47c}$$

$$\oint_S \mathbf{B}\cdot d\mathbf{S} = 0 \tag{3.47d}$$

and applying them one at a time to a closed path or a closed surface encompassing the boundary, and in the limit that the area enclosed by the closed path, or the volume bounded by the closed surface goes to zero. Thus let us consider two

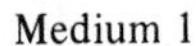

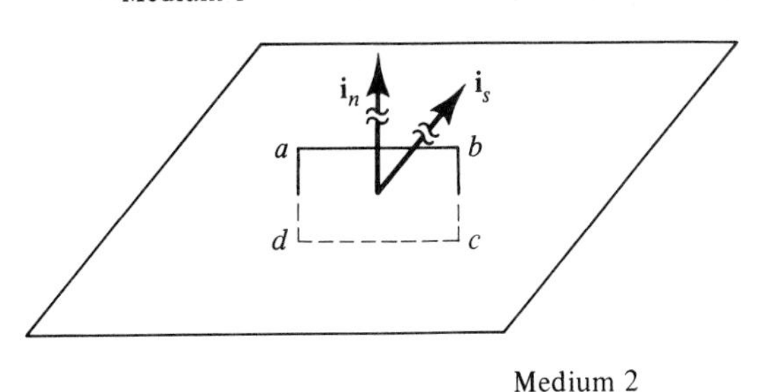

Figure 3.39. For deriving the boundary conditions resulting from Faraday's law and Ampère's circuital law.

semi-infinite media separated by a plane boundary, as shown in Fig. 3.39. Let us denote the quantities pertinent to medium 1 by subscript 1 and the quantities pertinent to medium 2 by subscript 2. Let $\mathbf{i}_n$ be the unit normal vector to the surface and directed into medium 1, as shown in Fig. 3.39, and let all normal components of fields at the boundary in both media denoted by an additional subscript n be directed along $\mathbf{i}_n$. Let the surface charge density (C/m^2) and the surface current density (A/m) on the boundary be ρ_S and $\mathbf{J}_S$, respectively. Note that, in general, the fields at the boundary in both media and the surface charge and current densities are functions of position on the boundary.

Boundary condition for $E_{tangential}$

First, we consider a rectangular closed path *abcda* of infinitesimal area in the plane normal to the boundary and with its sides *ab* and *cd* parallel to and on either side of the boundary, as shown in Fig. 3.39. Applying Faraday's law (3.47a) to this path in the limit that *ad* and *bc* → 0 by making the area *abcd* tend to zero but with *ab* and *cd* remaining on either side of the boundary, we have

$$\lim_{\substack{ad\to 0\\ bc\to 0}} \oint_{abcda} \mathbf{E}\cdot d\mathbf{l} = -\lim_{\substack{ad\to 0\\ bc\to 0}} \frac{d}{dt}\int_{\substack{\text{area}\\ abcd}} \mathbf{B}\cdot d\mathbf{S} \tag{3.48}$$

In this limit, the contributions from *ad* and *bc* to the integral on the left side of (3.48) approach zero. Since *ab* and *cd* are infinitesimal, the sum of the contributions from *ab* and *cd* becomes $[E_{ab}(ab) + E_{cd}(cd)]$, where E_{ab} and E_{cd} are the components of $\mathbf{E}_1$ and $\mathbf{E}_2$ along *ab* and *cd*, respectively. The right side of (3.48) is equal to zero since the magnetic flux crossing the area *abcd* approaches zero as the area *abcd* tends to zero. Thus Eq. (3.48) gives

$$E_{ab}(ab) + E_{cd}(cd) = 0$$

or, since *ab* and *cd* are equal, and $E_{dc} = -E_{cd}$,

$$E_{ab} - E_{dc} = 0 \tag{3.49}$$

Let us now define $\mathbf{i}_s$ to be the unit vector normal to the area *abcd* and in the direction of advance of a right-hand screw as it is turned in the sense of the closed path *abcda*. Noting then that $\mathbf{i}_s \times \mathbf{i}_n$ is the unit vector along *ab*, we can write (3.49) as

$$\mathbf{i}_s \times \mathbf{i}_n \cdot (\mathbf{E}_1 - \mathbf{E}_2) = 0$$

Rearranging the order of the scalar triple product, we obtain

$$\mathbf{i}_s \cdot \mathbf{i}_n \times (\mathbf{E}_1 - \mathbf{E}_2) = 0 \tag{3.50}$$

Since we can choose the rectangle *abcd* to be in any plane normal to the boundary, (3.50) must be true for all orientations of $\mathbf{i}_s$. It then follows that

$$\boxed{\mathbf{i}_n \times (\mathbf{E}_1 - \mathbf{E}_2) = \mathbf{0}} \tag{3.51a}$$

or, in scalar form,

$$\boxed{E_{t1} - E_{t2} = 0} \tag{3.51b}$$

where E_{t1} and E_{t2} are the components of $\mathbf{E}_1$ and $\mathbf{E}_2$, respectively, tangential to the boundary. In words, (3.51a) and (3.51b) state that *at any point on the boundary, the components of* $\mathbf{E}_1$ *and* $\mathbf{E}_2$ *tangential to the boundary are equal.*

Boundary condition for $\mathbf{H}_{tangential}$

Similarly, applying Ampère's circuital law (3.47b) to the closed path *abcda* in the limit that *ad* and *bc* → 0, we have

$$\lim_{\substack{ad \to 0 \\ bc \to 0}} \oint_{abcda} \mathbf{H} \cdot d\mathbf{l} = \lim_{\substack{ad \to 0 \\ bc \to 0}} \int_{\substack{\text{area} \\ abcd}} \mathbf{J} \cdot d\mathbf{S} + \lim_{\substack{ad \to 0 \\ bc \to 0}} \frac{d}{dt} \int_{\substack{\text{area} \\ abcd}} \mathbf{D} \cdot d\mathbf{S} \tag{3.52}$$

Using the same argument as for the left side of (3.48), we obtain the quantity on the left side of (3.52) to be equal to $[H_{ab}(ab) + H_{cd}(cd)]$, where H_{ab} and H_{cd} are the components of $\mathbf{H}_1$ and $\mathbf{H}_2$ along *ab* and *cd*, respectively. The second integral on the right side of (3.52) is zero since the displacement flux crossing the area *abcd* approaches zero as the area *abcd* tends to zero. The first integral on the right side of (3.52) would also be equal to zero but for a contribution from the surface current on the boundary because letting the area *abcd* tend to zero with *ab* and *cd* on either side of the boundary reduces only the volume current, if any, enclosed by it to zero, keeping the surface current still enclosed by it. This contribution is the surface current flowing normal to the line which *abcd* approaches as it tends to zero, that is, $[\mathbf{J}_S \cdot \mathbf{i}_s](ab)$. Thus Eq. (3.52) gives

$$H_{ab}(ab) + H_{cd}(cd) = (\mathbf{J}_S \cdot \mathbf{i}_s)(ab)$$

or, since *ab* and *cd* are equal and $H_{dc} = -H_{cd}$,

$$H_{ab} - H_{dc} = \mathbf{J}_S \cdot \mathbf{i}_s \tag{3.53}$$

In terms of $\mathbf{H}_1$ and $\mathbf{H}_2$, we have

$$\mathbf{i}_s \times \mathbf{i}_n \cdot (\mathbf{H}_1 - \mathbf{H}_2) = \mathbf{J}_S \cdot \mathbf{i}_s$$

or

$$\mathbf{i}_s \cdot \mathbf{i}_n \times (\mathbf{H}_1 - \mathbf{H}_2) = \mathbf{i}_s \cdot \mathbf{J}_S \tag{3.54}$$

Since (3.54) must be true for all orientations of $\mathbf{i}_s$, that is, for a rectangle *abcd* in any plane normal to the boundary, it follows that

$$\boxed{\mathbf{i}_n \times (\mathbf{H}_1 - \mathbf{H}_2) = \mathbf{J}_S} \tag{3.55a}$$

or, in scalar form,

$$\boxed{H_{t1} - H_{t2} = J_S} \tag{3.55b}$$

where H_{t1} and H_{t2} are the components of $\mathbf{H}_1$ and $\mathbf{H}_2$, respectively, tangential to the boundary. In words, Eqs. (3.55a) and (3.55b) state that *at any point on the boundary, the components of* $\mathbf{H}_1$ *and* $\mathbf{H}_2$ *tangential to the boundary are discontinuous by the amount equal to the surface current density at that point*. It should be noted that the information concerning the direction of $\mathbf{J}_S$ relative to that of $(\mathbf{H}_1 - \mathbf{H}_2)$, which is contained in (3.55a), is not present in (3.55b). Thus in general, (3.55b) is not sufficient, and it is necessary to use (3.55a).

Boundary condition for $\mathbf{D}_{normal}$

Now, we consider a rectangular box *abcdefgh* of infinitesimal volume enclosing an infinitesimal area of the boundary and parallel to it, as shown in Fig. 3.40. Applying Gauss' law for the electric field (3.47c) to this box in the limit that the side surfaces (abbreviated *ss*) tend to zero by making the volume of the box tend to zero but with the sides *abcd* and *efgh* remaining on either side of the boundary, we have

$$\lim_{ss\to 0} \oint_{\substack{\text{surface} \\ \text{of the box}}} \mathbf{D} \cdot d\mathbf{S} = \lim_{ss\to 0} \int_{\substack{\text{volume} \\ \text{of the box}}} \rho\, dv \qquad (3.56)$$

In this limit, the contributions from the side surfaces to the integral on the left side of (3.56) approach zero. The sum of the contributions from the top and bottom surfaces becomes $[D_{n1}(abcd) - D_{n2}(efgh)]$ since *abcd* and *efgh* are infinitesimal. The quantity on the right side of (3.56) would be zero but for the surface charge on the boundary since letting the volume of the box tend to zero with the sides *abcd* and *efgh* on either side of it reduces only the volume charge, if any, enclosed by it to zero, keeping the surface charge still enclosed by it. This surface charge is equal to $\rho_S(abcd)$. Thus (3.56) gives

$$D_{n1}(abcd) - D_{n2}(efgh) = \rho_S(abcd)$$

or, since *abcd* and *efgh* are equal,

$$\boxed{D_{n1} - D_{n2} = \rho_S} \qquad (3.57a)$$

In terms of $\mathbf{D}_1$ and $\mathbf{D}_2$, (3.57a) is given by

$$\boxed{\mathbf{i}_n \cdot (\mathbf{D}_1 - \mathbf{D}_2) = \rho_S} \qquad (3.57b)$$

In words, (3.57a) and (3.57b) state that *at any point on the boundary, the components of* $\mathbf{D}_1$ *and* $\mathbf{D}_2$ *normal to the boundary are discontinuous by the amount of the surface charge density at that point*.

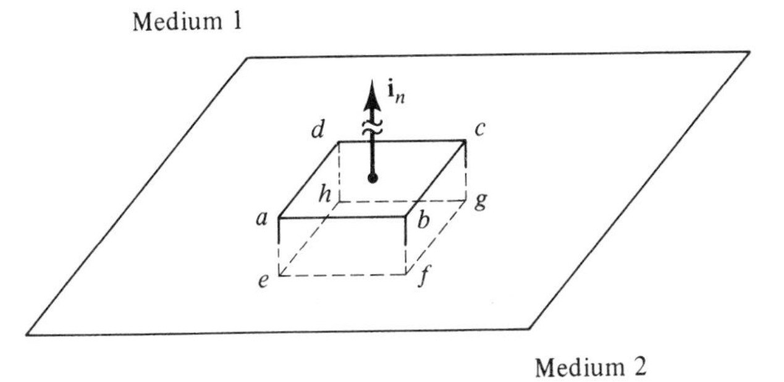

Figure 3.40. For deriving the boundary conditions resulting from the two Gauss' laws.

Boundary condition for $\mathbf{B}_{normal}$

Similarly, applying Gauss' law for the magnetic field (3.47d) to the box *abcdefgh* in the limit that the side surfaces tend to zero, we have

$$\lim_{ss \to 0} \oint_{\substack{\text{surface}\\ \text{of the box}}} \mathbf{B} \cdot d\mathbf{S} = 0 \tag{3.58}$$

Using the same argument as for the left side of (3.56), we obtain the quantity on the left side of (3.58) to be equal to $[B_{n1}(abcd - B_{n2}(efgh)]$. Thus, (3.58) gives

$$B_{n1}(abcd) - B_{n2}(efgh) = 0$$

or, since *abcd* and *efgh* are equal

$$\boxed{B_{n1} - B_{n2} = 0} \tag{3.59a}$$

In terms of $\mathbf{B}_1$ and $\mathbf{B}_2$, (3.59a) is given by

$$\boxed{\mathbf{i}_n \cdot (\mathbf{B}_1 - \mathbf{B}_2) = 0} \tag{3.59b}$$

In words, (3.59a) and (3.59b) state that *at any point on the boundary, the components of* $\mathbf{B}_1$ *and* $\mathbf{B}_2$ *normal to the boundary are equal.*

Summarizing the boundary conditions, we have

$$\mathbf{i}_n \times (\mathbf{E}_1 - \mathbf{E}_2) = \mathbf{0} \tag{3.60a}$$

$$\mathbf{i}_n \times (\mathbf{H}_1 - \mathbf{H}_2) = \mathbf{J}_S \tag{3.60b}$$

$$\mathbf{i}_n \cdot (\mathbf{D}_1 - \mathbf{D}_2) = \rho_S \tag{3.60c}$$

$$\mathbf{i}_n \cdot (\mathbf{B}_1 - \mathbf{B}_2) = 0 \tag{3.60d}$$

or, in scalar form,

$$E_{t1} - E_{t2} = 0 \tag{3.61a}$$

$$H_{t1} - H_{t2} = J_S \tag{3.61b}$$

$$D_{n1} - D_{n2} = \rho_S \tag{3.61c}$$

$$B_{n1} - B_{n2} = 0 \tag{3.61d}$$

as illustrated in Fig. 3.41. Although we have derived these boundary conditions by considering a plane interface between the two media, it should be obvious that we can consider any arbitrary-shaped boundary and obtain the same results by letting the sides *ab* and *cd* of the rectangle and the top and bottom surfaces of the box tend to zero, in addition to the limits that the sides *ad* and *bc* of the rectangle and the side surfaces of the box tend to zero. The boundary conditions given by (3.60a)–(3.60d) are general. When they are applied to particular cases, the special properties of the pertinent media come into play. Two such properties are that (1) no time-varying fields can exist inside a perfect conductor ($\sigma = \infty$) and (2) there can be no free charge and no current due to motion of free charges asso-

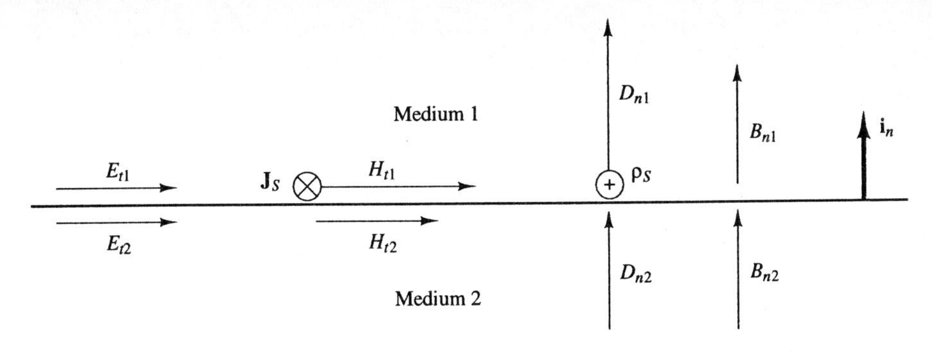

Figure 3.41. For illustrating the boundary conditions at an interface between two different media.

ciated with the surface of a perfect dielectric ($\sigma = 0$). We shall now consider an example.

Example 3.9

In Fig. 3.42, the region $x < 0$ is a perfect conductor, the region $0 < x < d$ is a perfect dielectric of $\varepsilon = 2\varepsilon_0$ and $\mu = \mu_0$, and the region $x > d$ is free space. The electric and magnetic fields in the region $0 < x < d$ are given at a particular instant of time by

$$\mathbf{E} = E_1 \cos \pi x \sin 2\pi z \, \mathbf{i}_x + E_2 \sin \pi x \cos 2\pi z \, \mathbf{i}_z$$

$$\mathbf{H} = H_1 \cos \pi x \sin 2\pi z \, \mathbf{i}_y$$

We wish to find (a) ρ_S and $\mathbf{J}_S$ on the surface $x = 0$ and (b) $\mathbf{E}$ and $\mathbf{H}$ for $x = d+$, that is, immediately adjacent to the $x = d$-plane and on the free-space side, at that instant of time.

(a) Denoting the perfect dielectric medium ($0 < x < d$) to be medium 1 and the perfect conductor medium ($x < 0$) to be medium 2, we have $\mathbf{i}_n = \mathbf{i}_x$, and all fields with subscript 2 are equal to zero. Then from (3.60c) and (3.60b), we obtain

$$[\rho_S]_{x=0} = \mathbf{i}_n \cdot [\mathbf{D}_1]_{x=0} = \mathbf{i}_x \cdot 2\varepsilon_0 E_1 \sin 2\pi z \, \mathbf{i}_x$$
$$= 2\varepsilon_0 E_1 \sin 2\pi z$$
$$[\mathbf{J}_S]_{x=0} = \mathbf{i}_n \times [\mathbf{H}_1]_{x=0} = \mathbf{i}_x \times H_1 \sin 2\pi z \, \mathbf{i}_y$$
$$= H_1 \sin 2\pi z \, \mathbf{i}_z$$

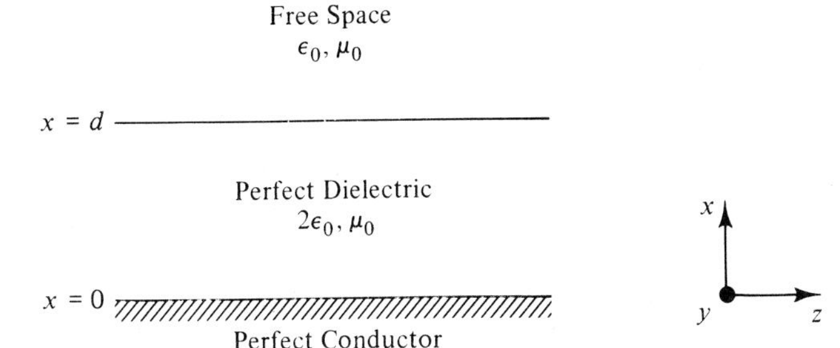

Figure 3.42. For illustrating the application of boundary conditions.

Note that the remaining two boundary conditions (3.60a) and (3.60d) are already satisfied by the given fields since E_y and B_x do not exist and for $x = 0$, $E_z = 0$.

(b) Denoting the perfect dielectric medium $(0 < x < d)$ to be medium 1 and the free-space medium $(x > d)$ to be medium 2 and setting $\rho_S = 0$, we obtain from (3.61a) and (3.61c)

$$[E_y]_{x=d+} = [E_y]_{x=d-} = 0$$

$$[E_z]_{x=d+} = [E_z]_{x=d-} = E_2 \sin \pi d \cos 2\pi z$$

$$[D_x]_{x=d+} = [D_x]_{x=d-} = 2\varepsilon_0[E_x]_{x=d-}$$

$$= 2\varepsilon_0 E_1 \cos \pi d \sin 2\pi z$$

$$[E_x]_{x=d+} = \frac{1}{\varepsilon_0}[D_x]_{x=d+}$$

$$= 2E_1 \cos \pi d \sin 2\pi z$$

Thus

$$[\mathbf{E}]_{x=d+} = 2E_1 \cos \pi d \sin 2\pi z \, \mathbf{i}_x + E_2 \sin \pi d \cos 2\pi z \, \mathbf{i}_z$$

Setting $\mathbf{J}_S = \mathbf{0}$ and using (3.61b) and (3.61d), we obtain

$$[H_y]_{x=d+} = [H_y]_{x=d-} = H_1 \cos \pi d \sin 2\pi z$$

$$[H_z]_{x=d+} = [H_z]_{x=d-} = 0$$

$$[B_x]_{x=d+} = [B_x]_{x=d-} = 0$$

Thus

$$[\mathbf{H}]_{x=d+} = H_1 \cos \pi d \sin 2\pi z \, \mathbf{i}_y$$

K3.6. Boundary conditions; Tangential component of **E**; Tangential component of **H**; Normal component of **D**; Normal component of **B**.

D3.15. For each of the following values of the displacement flux density at a point on the surface of a perfect conductor (no electric field inside and hence $E_t = 0$ on the surface), find the surface charge density at that point: **(a)** $\mathbf{D} = D_0(\mathbf{i}_x - 2\mathbf{i}_y + 2\mathbf{i}_z)$ and pointing away from the surface; **(b)** $\mathbf{D} = D_0(\mathbf{i}_x + \sqrt{3}\mathbf{i}_z)$ and pointing toward the surface; and **(c)** $\mathbf{D} = D_0(0.8\mathbf{i}_x + 0.6\mathbf{i}_y)$ and pointing away from the surface. Assume D_0 to be positive for all cases.
Ans. **(a)** $3D_0$; **(b)** $-2D_0$; **(c)** D_0

D3.16. The region $x > 0$ is a perfect dielectric of permittivity $2\varepsilon_0$ and the region $x < 0$ is a perfect dielectric of permittivity $3\varepsilon_0$. Consider the field components at point 1 on the $+x$-side of the boundary to be denoted by subscript 1 and the field components at the adjacent point 2 on the $-x$-side of the boundary to be denoted by subscript 2. If $\mathbf{E}_1 = E_0(2\mathbf{i}_x + \mathbf{i}_y)$, find the following: **(a)** E_{x1}/E_{x2}; **(b)** E_1/E_2; and **(c)** D_1/D_2.
Ans. **(a)** 1.5; **(b)** $3/\sqrt{5}$; **(c)** $2/\sqrt{5}$

D3.17. The plane $z = 0$ forms the boundary between free space $(z > 0)$ and another medium. Find the following: **(a)** $\mathbf{J}_S(0, 0, 0)$ at $t = 0$ if $z < 0$ is a perfect conductor and $\mathbf{H}(0, 0, 0+) = H_0(3\mathbf{i}_x - 4\mathbf{i}_y) \cos \omega t$; **(b)** $\mathbf{H}(0, 0, 0+)$ if $z < 0$ is a magnetic material of $\mu = 20\mu_0$ and $\mathbf{H}(0, 0, 0-) = H_0(10\mathbf{i}_x + \mathbf{i}_z)$; and **(c)** the ratio of $B(0, 0, 0-)$ to $B(0, 0, 0+)$ for the case of **(b)**.
Ans. **(a)** $H_0(4\mathbf{i}_x + 3\mathbf{i}_y)$; **(b)** $10H_0(\mathbf{i}_x + 2\mathbf{i}_z)$; **(c)** 8.989

3.7 SUMMARY

We first learned in this chapter how to evaluate line and surface integrals of vector quantities and then we introduced Maxwell's equations in integral form. These equations, which form the basis of electromagnetic field theory, are given as follows in words and in mathematical form:

Faraday's law. The electromotive force around a closed path C is equal to the negative of the time rate of change of the magnetic flux enclosed by that path; that is,

$$\oint_C \mathbf{E} \cdot d\mathbf{l} = -\frac{d}{dt}\int_S \mathbf{B} \cdot d\mathbf{S} \tag{3.62}$$

Ampère's circuital law. The magnetomotive force around a closed path C is equal to the sum of the current enclosed by that path due to the actual flow of charges and the displacement current due to the time rate of change of the displacement flux enclosed by that path; that is,

$$\oint_C \mathbf{H} \cdot d\mathbf{l} = \int_S \mathbf{J} \cdot d\mathbf{S} + \frac{d}{dt}\int_S \mathbf{D} \cdot d\mathbf{S} \tag{3.63}$$

Gauss' law for the electric field. The displacement flux emanating from a closed surface S is equal to the charge enclosed by that surface; that is,

$$\oint_S \mathbf{D} \cdot d\mathbf{S} = \int_V \rho\, dv \tag{3.64}$$

Gauss' law for the magnetic field. The magnetic flux emanating from a closed surface S is equal to zero; that is,

$$\oint_S \mathbf{B} \cdot d\mathbf{S} = 0 \tag{3.65}$$

An auxiliary equation, the law of conservation of charge, is given by

$$\oint_S \mathbf{J} \cdot d\mathbf{S} = -\frac{d}{dt}\int_V \rho\, dv \tag{3.66}$$

In words, (3.66) states that the current due to flow of charges emanating from a closed surface is equal to the time rate of decrease of the charge enclosed by that surface.

In using (3.62)–(3.66), we recall that

$$\mathbf{D} = \varepsilon \mathbf{E} \tag{3.67}$$

$$\mathbf{H} = \frac{\mathbf{B}}{\mu} \tag{3.68}$$

where ε and μ are the permittivity and permeability, respectively, of the medium. In addition, if the current density $\mathbf{J}$ is due to conduction, then

$$\mathbf{J} = \mathbf{J}_c = \sigma \mathbf{E} \tag{3.69}$$

where σ is the conductivity of the medium. In evaluating the right sides of (3.62) and (3.63), the normal vectors to the surfaces must be chosen such that they are directed in the right-hand sense, that is, toward the side of advance of a right-hand screw as it is turned around C. In (3.64), (3.65), and (3.66), it is understood that the surface integrals are evaluated so as to find the flux outward from the volume bounded by the surface. We also learned that (3.65) is not independent of (3.62) and that (3.64) follows from (3.63) with the aid of (3.66).

We discussed several applications of Maxwell's equations, including the computation of static electric and magnetic fields due to symmetrical charge and current distributions, respectively. Finally, we derived the boundary conditions resulting from the application of Maxwell's equations to closed paths and closed surfaces encompassing the boundary between two media, and in the limits that the areas enclosed by the closed paths and the volumes bounded by the closed surfaces go to zero. These boundary conditions are given by

$$\mathbf{i}_n \times (\mathbf{E}_1 - \mathbf{E}_2) = \mathbf{0} \tag{3.70a}$$

$$\mathbf{i}_n \times (\mathbf{H}_1 - \mathbf{H}_2) = \mathbf{J}_S \tag{3.70b}$$

$$\mathbf{i}_n \cdot (\mathbf{D}_1 - \mathbf{D}_2) = \rho_S \tag{3.70c}$$

$$\mathbf{i}_n \cdot (\mathbf{B}_1 - \mathbf{B}_2) = 0 \tag{3.70d}$$

where the subscripts 1 and 2 refer to media 1 and 2, respectively, and $\mathbf{i}_n$ is unit vector normal to the boundary at the point under consideration and directed into medium 1. In words, the boundary conditions state that at a point on the boundary, the tangential components of **E** and the normal components of **B** are continuous, whereas the tangential components of **H** are discontinuous by the amount equal to J_S at that point, and the normal components of **D** are discontinuous by the amount equal to ρ_S at that point.

REVIEW QUESTIONS

R3.1. How do you find the work done in moving a test charge by an infinitesimal distance in an electric field? What is the amount of work involved in moving the test charge normal to the electric field?

R3.2. What is the physical interpretation of the line integral of **E** between two points A and B?

R3.3. How do you find the approximate value of the line integral of a vector field along a given path? How do you find the exact value of the line integral?

R3.4. Discuss conservative versus nonconservative fields, giving examples.

R3.5. How do you find the magnetic flux crossing an infinitesimal surface?

R3.6. What is the magnetic flux crossing an infinitesimal surface oriented parallel to the magnetic flux density vector? For what orientation of the infinitesimal surface relative to the magnetic flux density vector is the magnetic flux crossing the surface a maximum?

R3.7. How do you find the approximate value of the surface integral of a vector field over a given surface? How do you find the exact value of the surface integral?

R3.8. Provide physical interpretations for the closed surface integrals of any two vectors of your choice.

R3.9. State Faraday's law.

R3.10. What are the different ways in which an emf is induced around a loop?

R3.11. Discuss the right-hand screw rule convention associated with the application of Faraday's law.

R3.12. To find the induced emf around a planar loop, is it necessary to consider the magnetic flux crossing the plane surface bounded by the loop? Explain.

R3.13. What is Lenz's law?

R3.14. Discuss briefly the motional emf concept.

R3.15. How would you orient a loop antenna to obtain maximum signal from an incident electromagnetic wave which has its magnetic field linearly polarized in the north-south direction?

R3.16. State three applications of Faraday's law.

R3.17. State Ampère's circuital law.

R3.18. What is displacement current? Compare and contrast displacement current with current due to flow of charges.

R3.19. Is it meaningful to consider two different surfaces bounded by a closed path to compute the two different currents on the right side of Ampère's circuital law to find $\oint \mathbf{H} \cdot d\mathbf{l}$ around the closed path?

R3.20. Discuss the relationship between the displacement current emanating from a closed surface and the current due to flow of charges emanating from the same closed surface.

R3.21. Give an example involving displacement current.

R3.22. Discuss briefly the principle of radiation from a wire carrying a time-varying current.

R3.23. State Gauss' law for the electric field.

R3.24. How do you evaluate a volume integral?

R3.25. State Gauss' law for the magnetic field.

R3.26. What is the physical interpretation of Gauss' law for the magnetic field?

R3.27. Discuss the dependence of Gauss' law for the magnetic field on Faraday's law.

R3.28. State the law of conservation of charge.

R3.29. How is Gauss' law for the electric field dependent on Ampère's circuital law?

R3.30. Summarize Maxwell's equations in integral form for time-varying fields.

R3.31. Summarize Maxwell's equations in integral form for static fields.

R3.32. Are static electric and magnetic fields interdependent? Explain.

R3.33. Discuss briefly the application of Gauss' law for the electric field to determine the electric field due to charge distributions.

R3.34. When can you say that the current in a wire enclosed by a closed path is uniquely defined? Give two examples.

R3.35. Give an example in which the current in a wire enclosed by a closed path is not uniquely defined. Is it correct to apply Ampère's circuital law for the static case in such a situation? Explain.

R3.36. Discuss briefly the application of Ampère's circuital law to determine the magnetic field due to current distributions.

R3.37. What is a boundary condition? How do boundary conditions arise?

R3.38. Summarize the boundary conditions for the general case of a boundary between two arbitrary media, indicating correspondingly the Maxwell's equations in integral form from which they are derived.

R3.39. Discuss the boundary conditions on the surface of a perfect conductor.

R3.40. Discuss the boundary conditions at the interface between two perfect dielectric media.

PROBLEMS

P3.1. For the vector field $\mathbf{F} = y^2\mathbf{i}_x + z\mathbf{i}_y + xy\mathbf{i}_z$, find $\int_{(0,0,0)}^{(1,1,1)} \mathbf{F} \cdot d\mathbf{l}$ for each of the following paths from (0, 0, 0) to (1, 1, 1): **(a)** $x = y = z$; and **(b)** $x = y = z^2$.

P3.2. Given $\mathbf{F} = xy\mathbf{i}_x + yz\mathbf{i}_y + zx\mathbf{i}_z$, find $\oint_C \mathbf{F} \cdot d\mathbf{l}$ where C is the closed path comprising the straight lines from (0, 0, 0) to (1, 1, 1), from (1, 1, 1) to (1, 1, 0), and from (1, 1, 0) to (0, 0, 0).

P3.3. For the vector field $\mathbf{F} = xyz(yz\mathbf{i}_x + zx\mathbf{i}_y + xy\mathbf{i}_z)$, find $\int_{(0,0,0)}^{(1,2,3)} \mathbf{F} \cdot d\mathbf{l}$ in each of the following ways: **(a)** along the straight line path from (0, 0, 0) to (1, 2, 3); **(b)** along the straight line paths from (0, 0, 0), to (1, 0, 0), from (1, 0, 0) to (1, 2, 0), and then from (1, 2, 0) to (1, 2, 3); and **(c)** without choosing any particular path. Is the vector field conservative or nonconservative? Explain.

P3.4. Given $\mathbf{A} = 2r \cos \phi \, \mathbf{i}_r + r\mathbf{i}_\phi$ in cylindrical coordinates, evaluate $\oint_C \mathbf{A} \cdot d\mathbf{l}$, where C is the closed path comprising the straight line from (0, 0, 0) to (1, 0, 0), the circular arc from (1, 0, 0) to $(1, 2\pi/3, 0)$ through the point $(1, \pi/2, 0)$, and the straight line from $(1, 2\pi/3, 0)$ to (0, 0, 0).

P3.5. Given $\mathbf{A} = r^2\mathbf{i}_r - r \cos \theta \, \mathbf{i}_\theta + r \sin \theta \, \mathbf{i}_\phi$ in spherical coordinates, evaluate $\oint_C \mathbf{A} \cdot d\mathbf{l}$, where C is the closed path comprising the straight line from (0, 0, 0) to $(1, \pi/3, 0)$, the circular arc from $(1, \pi/3, 0)$ to $(1, \pi/3, \pi/2)$ through the point $(1, \pi/3, \pi/4)$, the circular arc from $(1, \pi/3, \pi/2)$ to $(1, 0, \pi/2)$ through the point $(1, \pi/4, \pi/2)$, and the straight line from $(1, 0, \pi/2)$ to (0, 0, 0).

P3.6. Given $\mathbf{A} = xyz(x\mathbf{i}_x + y\mathbf{i}_y + z\mathbf{i}_z)$, evaluate $\oint_S \mathbf{A} \cdot d\mathbf{S}$, where S is the surface of the cubical box bounded by the planes $x = 0$, $x = 1$, $y = 0$, $y = 1$, $z = 0$, and $z = 1$.

P3.7. Given $\mathbf{A} = (x^2y + 2)\mathbf{i}_x + 3\mathbf{i}_y - 2xyz\mathbf{i}_z$, evaluate $\oint_S \mathbf{A} \cdot d\mathbf{S}$, where S is the surface of the rectangular box bounded by the planes $x = 0$, $x = 1$, $y = 0$, $y = 2$, $z = 0$, and $z = 3$.

P3.8. Given $\mathbf{A} = r \cos \phi \, \mathbf{i}_r - r \sin \phi \, \mathbf{i}_\phi$ in cylindrical coordinates, find $\oint_S \mathbf{A} \cdot d\mathbf{S}$, where S is the surface of the box bounded by the plane surfaces $\phi = 0$, $\phi = \pi$, $z = 0$, and $z = 1$, and the cylindrical surface $r = 2$, $0 < \phi < \pi$.

P3.9. Given $\mathbf{A} = r^2\mathbf{i}_r + r \sin \theta \, \mathbf{i}_\theta$ in spherical coordinates, find $\oint_S \mathbf{A} \cdot d\mathbf{S}$, where S is the surface of that part of the spherical volume of radius unity and lying in the first octant.

P3.10. A magnetic field is given in the xz-plane by $\mathbf{B} = (B_0/x)\mathbf{i}_y$ Wb/m^2, where B_0 is a constant. A rigid rectangular loop is situated in the xz-plane and with its corners at the points (x_0, z_0), $(x_0, z_0 + b)$, $(x_0 + a, z_0 + b)$, and $(x_0 + a, z_0)$. If the loop is moving in that plane with a velocity $\mathbf{v} = v_0\mathbf{i}_x$ m/s, where v_0 is a constant, find by using Faraday's law the induced emf around the loop in the sense defined by connecting the above points in succession. Discuss your result by using the motional emf concept.

P3.11. A magnetic field is given in the xz-plane by $\mathbf{B} = B_0 \cos \pi(x - v_0t) \, \mathbf{i}_y$ Wb/m^2. Consider a rigid square loop situated in the xz-plane with its vertices at $(x, 0, 1)$, $(x, 0, 2)$, $(x + 1, 0, 2)$, and $(x + 1, 0, 1)$. **(a)** Find the expression for the emf induced around the loop in the sense defined by connecting the above points in succession. **(b)** What would be the induced emf if the loop is moving with the velocity $\mathbf{v} = v_0\mathbf{i}_x$ m/s instead of being stationary?

P3.12. A rigid vertical rectangular loop of metallic wire falls under the influence of gravity, as shown in Fig. 3.43, and in the presence of a magnetic field $\mathbf{B} = B_0 z \mathbf{i}_y$ where B_0 is a constant. **(a)** Show that the emf induced around the closed path C of the loop is $B_0 abv$, where v is the downward velocity of the loop. **(b)** If the mass of the loop is m and its electrical resistance is R, obtain the differential equation for v. Assuming the loop starts from rest at $t = 0$, find the solution for v versus t. **(c)** Does the loop fall faster or slower than in the absence of the magnetic field? Explain.

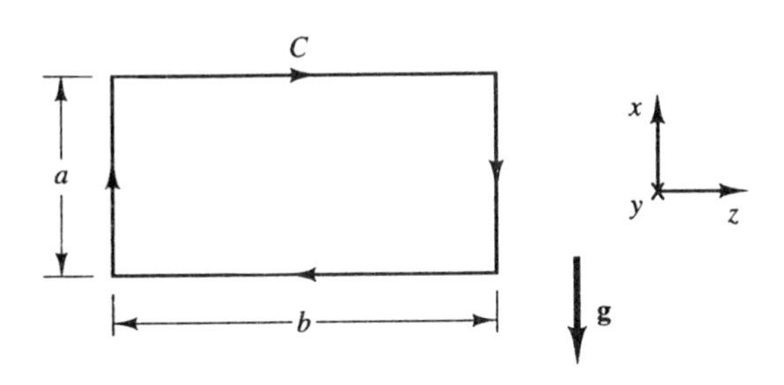

Figure 3.43. For Problem P3.12.

P3.13. A rigid rectangular loop of metallic wire is hung by pivoting one side along the x-axis, as shown in Fig. 3.44. The loop is free to swing about the pivoted side without friction under the influence of gravity and in the presence of a uniform magnetic field $\mathbf{B} = B_0 \mathbf{i}_z$. If the loop is given a slight angular displacement and released, show that the emf induced around the closed path C of the loop is approximately equal to $B_0 ab\omega$ where ω is the angular velocity of swing of the loop toward the vertical. Does the loop swing faster or slower than in the absence of the magnetic field? Explain.

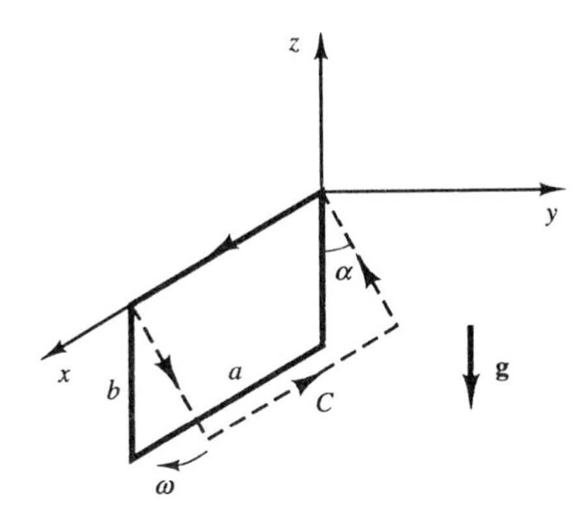

Figure 3.44. For Problem P3.13.

P3.14. A rigid rectangular loop of base b and height h situated normal to the xy-plane and with its sides pivoted to the z-axis revolves about the z-axis with angular velocity ω rad/s in the sense of increasing ϕ, as shown in Fig. 3.45. Find the induced emf around the closed path C of the loop for each of the following magnetic fields: **(a)** $\mathbf{B} = B_0 \mathbf{i}_y$ Wb/m^2 and **(b)** $\mathbf{B} = B_0(y\mathbf{i}_x - x\mathbf{i}_y)$ Wb/m^2. Assume the loop to be in the xz-plane at $t = 0$.

P3.15. A rigid rectangular loop of area A is situated normal to the xy-plane and symmetrically about the z-axis in a region of magnetic field $\mathbf{B}$. It revolves around the z-axis at an angular velocity of ω_1 rad/s in the sense of increasing ϕ, as shown in Fig. 3.46. Find the emf induced around the closed path C of the loop if $\mathbf{B} =$

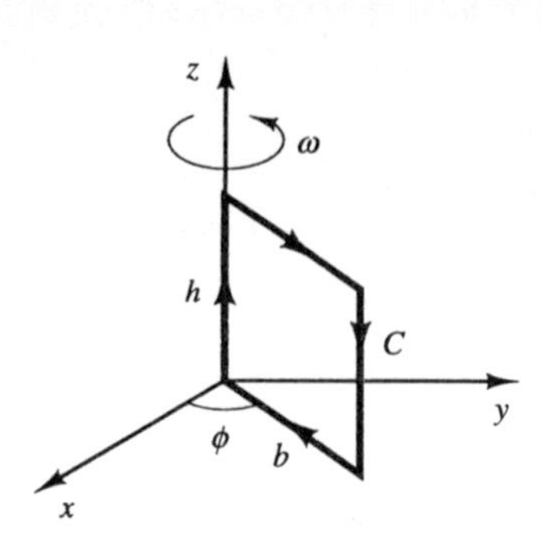

Figure 3.45. For Problem P3.14.

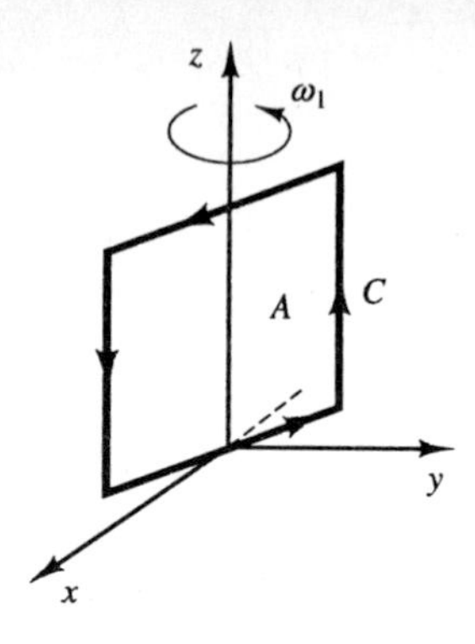

Figure 3.46. For Problem P3.15.

$B_0 \cos \omega_2 t\, \mathbf{i}_x$ Wb/m^2, where B_0 is a constant, and show that the induced emf contains two frequency components $(\omega_1 + \omega_2)$ and $|\omega_1 - \omega_2|$.

P3.16. The magnetic field is given in cylindrical coordinates by $\mathbf{B} = B_0 f(r, t)\mathbf{i}_z$, where B_0 is a constant. If the resulting electric field is given by $\mathbf{E} = E_0 \sin \omega t\, \mathbf{i}_\phi$, where E_0 is a constant, find $f(r, t)$. (*Hint:* Consider a circular path of radius r in the xy-plane and centered at the origin and apply Faraday's law in integral form.)

P3.17. Given that $\mathbf{H} = \pm H_0(t \mp \sqrt{\mu_0\varepsilon_0}\, z)^2\mathbf{i}_y$ and $\mathbf{D} = \sqrt{\mu_0\varepsilon_0}\, H_0(t \mp \sqrt{\mu_0\varepsilon_0}\, z)^2\mathbf{i}_x$ for $z \gtrless 0$, find the current due to flow of charges enclosed by the rectangular closed path from (0, 0, 1) to (0, 1, 1) to (0, 1, −1) to (0, 0, −1) to (0, 0, 1).

P3.18. A current density due to flow of charges is given by $\mathbf{J} = -(x\mathbf{i}_x + y\mathbf{i}_y + z^2\mathbf{i}_z)$ A/m^2. Find the displacement current emanating from each of the following closed surfaces: **(a)** the surface of the cubical box bounded by the planes $x = \pm 2$, $y = \pm 2$, and $z = \pm 2$ and **(b)** the surface of the cylindrical box bounded by the surfaces $r = 1$, $z = 0$, and $z = 2$.

P3.19. A voltage source connected to a parallel-plate capacitor by means of wires sets up a uniform electric field of $\mathbf{E} = 180 \sin 2\pi \times 10^6 t \sin 4\pi \times 10^6 t$ V/m between the plates of the capacitor and normal to the plates. Assume that no field exists outside the region between the plates. If the area of each plate is 0.1 m^2 and the medium between the plates is free space, find the root-mean-square value of the current drawn from the voltage source.

P3.20. Assume that the time-variation of the electric field in Problem P3.19 is as shown in Fig 3.47. Find and plot versus time the current drawn from the voltage source. What is the root-mean-square value of the current?

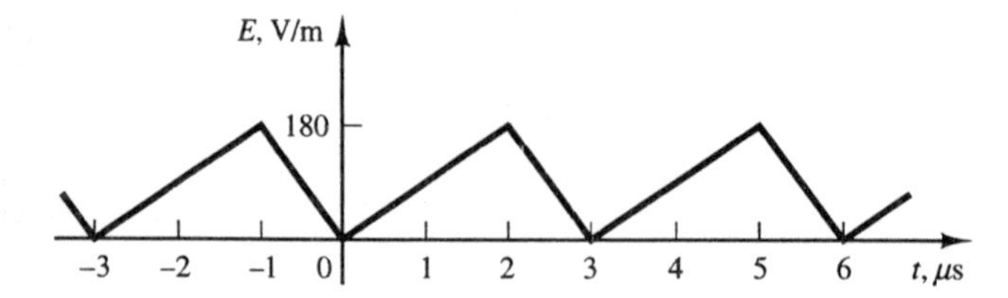

Figure 3.47. For Problem P3.20.

P3.21. A voltage source is connected by means of wires to a parallel-plate capacitor made up of circular plates of radius a in the $z = 0$ and $z = d$ planes and having their centers on the z-axis. The electric field between the plates is given by

$$\mathbf{E} = E_0 \sin \frac{\pi r}{2a} \cos \omega t \, \mathbf{i}_z \qquad \text{for} \quad r < a$$

Find the amplitude of the current drawn from the voltage source, assuming the region between the plates to be free space and that no field exists outside this region.

P3.22. For each of the following charge distributions, find the displacement flux emanating from the surface enclosing the charge: **(a)** $\rho(x, y, z) = \rho_0(x + y + z)^2$ for $0 < x < 1$, $0 < y < 1$, and $0 < z < 1$; **(b)** $\rho(r, \phi, z) = \rho_0 \cos^2 \phi$ for $r < 1$, $0 < \phi < \pi/2$, and $0 < z < 1$ in cylindrical coordinates; and **(c)** $\rho(r, \theta, \phi) = (\rho_0/r) \sin \theta$ for $r < 1$, $0 < \theta < \pi/2$ in spherical coordinates.

P3.23. Using the property that $\oint_S \mathbf{B} \cdot d\mathbf{S} = 0$, find the absolute value of the magnetic flux crossing that portion of the surface $y = \sin x$ bounded by $x = 0$, $x = \pi$, $z = 0$, and $z = 1$ for $\mathbf{B} = B_0(y\mathbf{i}_x - x\mathbf{i}_y)$ Wb/m^2.

P3.24. Given $\mathbf{J} = (x\mathbf{i}_x + y\mathbf{i}_y + z\mathbf{i}_z)$ A/m^2, find the time rate of decrease of the charge contained within each of the following volumes: **(a)** volume bounded by the planes $x = 0$, $x = 1$, $y = 0$, $y = 1$, $z = 0$, and $z = 1$; **(b)** volume bounded by the cylinders $r = 1$ and $r = 2$ and the planes $z = 0$ and $z = 1$; and **(c)** volume bounded by the spherical surfaces $r = 1$ and $r = 2$ and the conical surface $\theta = \pi/3$.

P3.25. Current I flows along a straight wire from a point charge $Q_1(t)$ located at the origin to a point charge $Q_2(t)$ located at (0, 0, 1). Find the line integral of $\mathbf{H}$ along the square closed path having the vertices at (1, 1, 0), (−1, 1, 0), (−1, −1, 0), and (1, −1, 0) and traversed in that order.

P3.26. Current I flows along a straight wire from a point charge $Q_1(t)$ at the origin to a point charge $Q_2(t)$ at the point (2, 2, 2). Find the line integral of $\mathbf{H}$ around the triangular closed path having the vertices at (3, 0, 0), (0, 3, 0), and (0, 0, 3) and traversed in that order.

P3.27. Charge is distributed with density $\rho(x, y, z)$ in a cubical box bounded by the planes $x = \pm 1$ m, $y = \pm 1$ m, and $z = \pm 1$ m. Find the displacement flux emanating from one side of the box for each of the following cases: **(a)** $\rho(x, y, z) = (3 - x^2 - y^2 - z^2)$ C/m^3 and **(b)** $\rho(x, y, z) = |xyz|$ C/m^3.

P3.28. A point charge Q_1 lies at the center of a spherical conducting shell of inner radius a and outer radius b. Find the charge densities on the two surfaces of the conducting shell if it has a net charge Q_2.

P3.29. Charge is distributed with density $\rho = \rho_0 r/a$, where ρ_0 is a constant, in the cylindrical region $r < a$. Find $\mathbf{D}$ everywhere and plot D_r versus r.

P3.30. Charge is distributed with uniform density ρ_0 C/m^3 in the region $a < r < 2a$ in spherical coordinates. Find $\mathbf{D}$ everywhere and plot D_r versus r.

P3.31. Current flows with density $\mathbf{J}(x, y)$ in an infinitely long, thick wire, having the z-axis as its axis. The cross section of the wire in the xy-plane is the square bounded by $x = \pm 1$ m and $y = \pm 1$ m. Find the line integral of $\mathbf{H}$ along one side of the square and traversed in the sense of increasing ϕ for each of the following cases: **(a)** $\mathbf{J}(x, y) = (|x| + |y|)\mathbf{i}_z$ A/m^2 and **(b)** $\mathbf{J}(x, y) = x^2y^2\mathbf{i}_z$ A/m^2.

P3.32. Current flows with density $\mathbf{J} = J_0(r/a)\mathbf{i}_z$ A/m^2 along an infinitely long, solid cylindrical wire of radius a, having the z-axis as its axis. Find $\mathbf{H}$ everywhere and plot H_ϕ versus r.

P3.33. A coaxial cable consists of an inner conductor of radius $3a$ and an outer conductor of inner radius $4a$ and outer radius $5a$. Assume the cable to be infinitely long and its axis to be along the z-axis. Current I flows with uniform density in the $+z$-direction in the inner conductor and returns with uniform density in the $-z$-direction in the outer conductor. Find $\mathbf{H}$ everywhere and plot H_ϕ versus r.

P3.34. Show that the results obtained for the electric field due to the sheet of charge in Example 2.3 and for the magentic field due to the sheet of current in Example 2.6 are consistent with the boundary conditions.

P3.35. The electric field in free space outside a conducting sphere of radius a and centered at the origin resulting from an applied field is given by

$$\mathbf{E} = E_0\left(1 + \frac{2a^3}{r^3}\right)\cos\theta\,\mathbf{i}_r - E_0\left(1 - \frac{a^3}{r^3}\right)\sin\theta\,\mathbf{i}_\theta$$

(a) Show that the tangential component of **E** is zero on the conductor surface. **(b)** Find the charge density on the conductor surface. **(c)** Find the applied and secondary fields. (*Hint*: Consider $a \to 0$.)

P3.36. In Problem P2.25, show that the applied and secondary fields together satisfy the boundary condition of zero tangential component of electric field on the conductor surface. From the boundary condition for the normal component of **D**, find the charge density on the conductor surface and show that the total induced surface charge is $-Q$.

P3.37. Two infinite plane perfectly conducting sheets occupy the planes $x = 0$ and $x = 0.1$ m. Electric and magnetic fields given by

$$\mathbf{E} = E_0 \sin 10\pi x \cos 3\pi \times 10^9 t\,\mathbf{i}_z$$

$$\mathbf{H} = \frac{E_0}{120\pi}\cos 10\pi x \sin 3\pi \times 10^9 t\,\mathbf{i}_y$$

respectively, where E_0 is a constant, exist in the region between the plates which is free space. **(a)** Show that **E** satisfies the boundary condition of zero tangential component on the sheets. **(b)** Find the surface current densities on the two sheets.

P3.38. The rectangular cavity resonator is a box comprising the region $0 < x < a$, $0 < y < b$, and $0 < z < d$, and bounded by perfectly conducting walls on all of its six sides. The time-varying electric and magnetic fields inside the resonator are given by

$$\mathbf{E} = E_0 \sin\frac{\pi x}{a}\sin\frac{\pi z}{d}\cos\omega t\,\mathbf{i}_y$$

$$\mathbf{H} = H_{01}\sin\frac{\pi x}{a}\cos\frac{\pi z}{d}\sin\omega t\,\mathbf{i}_x - H_{02}\cos\frac{\pi x}{a}\sin\frac{\pi z}{d}\sin\omega t\,\mathbf{i}_z$$

where E_0, H_{01}, and H_{02} are constants. Find ρ_S and $\mathbf{J}_S$ on all six walls assuming the medium inside the box to be a perfect dielectric of $\varepsilon = 4\,\varepsilon_0$.

P3.39. Medium 1, comprising the region $r < a$ in spherical coordinates, is a perfect dielectric of permittivity ε_1, whereas medium 2, comprising the region $r > a$ in spherical coordinates, is free space. The electric field intensities in the two media are given by

$$\mathbf{E}_1 = E_{01}(\cos\theta\,\mathbf{i}_r - \sin\theta\,\mathbf{i}_\theta)$$

$$\mathbf{E}_2 = E_{02}\left[\left(1 + \frac{a^3}{2r^3}\right)\cos\theta\,\mathbf{i}_r - \left(1 - \frac{a^3}{4r^3}\right)\sin\theta\,\mathbf{i}_\theta\right]$$

respectively. Find ε_1.

P3.40. Medium 1, comprising the region $r < a$ in spherical coordinates, is a magnetic material of permeability μ_1, whereas medium 2, comprising the region $r > a$ in spherical coordinates, is free space. The magnetic flux densities in the two media are given by

$$\mathbf{B}_1 = B_{01}(\cos\theta\, \mathbf{i}_r - \sin\theta\, \mathbf{i}_\theta)$$

$$\mathbf{B}_2 = B_{02}\left[\left(1 + 1.94\frac{a^3}{r^3}\right)\cos\theta\, \mathbf{i}_r - \left(1 - 0.97\frac{a^3}{r^3}\right)\sin\theta\, \mathbf{i}_\theta\right]$$

respectively. Find μ_1.

PC EXERCISES

PC3.1. Consider the evaluation of the line integral of a vector field of the form

$$\mathbf{A} = A_x(y)\mathbf{i}_x + A_y(x)\mathbf{i}_y$$

along the circular path of radius unity, lying in the xy-plane with its center at the origin and traversed in the sense of increasing ϕ, by dividing the path into n segments of equal length and expressing the line integral as a summation. Write a program that has provision for specifying $A_x(y)$ and $A_y(x)$ through defined function statements and computes for a specified value of n the approximate value of the line integral.

PC3.2. Consider the solution of Problem P3.25 by applying Biot–Savart law to find the magnetic field due to the current carrying wire and evaluating the line integral of **H** around the closed path. Note that it is sufficient to consider one-half of one side of the closed path from symmetry considerations. Write a program to carry out the solution by dividing the wire and the closed path into segments of equal lengths and expressing the line integral as a double summation over the two sets of segments.

PC3.3. Repeat PC3.2 for Problem P3.26.

4

Maxwell's Equations in Differential Form, Potential Functions, and Energy Storage

In Chapter 3 we introduced Maxwell's equations in integral form. We learned that the quantities involved in the formulation of these equations are the scalar quantities, electromotive force, magnetomotive force, magnetic flux, displacement flux, charge, and current, which are related to the field vectors and source densities through line, surface, and volume integrals. Thus the integral forms of Maxwell's equations, while containing all the information pertinent to the interdependence of the field and source quantities over a given region in space, do not permit us to study directly the interaction between the field vectors and their relationships with the source densities at individual points. It is our goal in this chapter to derive the differential forms of Maxwell's equations that apply directly to the field vectors and source densities at a given point.

We shall derive Maxwell's equations in differential form by applying Maxwell's equations in integral form to infinitesimal closed paths, surfaces, and volumes, in the limit that they shrink to points. We will find that the differential equations relate the spatial variations of the field vectors at a given point to their temporal variations and to the charge and current densities at that point. Using Maxwell's equations in differential form, we introduce the electromagnetic potential functions, derive differential equations for the potential functions, and consider the potential functions for the static field case. Finally, we consider the topic of energy storage in electric and magnetic fields.

In the process of deriving the Maxwell's equations in differential form, we shall become familiar with the operations of curl and divergence and then learn the Stokes and divergence theorems. In introducing the potential functions and the associated differential equations, we make use of the operations of gradient and Laplacian.

4.1 FARADAY'S LAW AND AMPÈRE'S CIRCUITAL LAW

Faraday's law, special case

We recall from Chapter 3 that Faraday's law is given in integral form by

$$\boxed{\oint_C \mathbf{E} \cdot d\mathbf{l} = -\frac{d}{dt}\int_S \mathbf{B} \cdot d\mathbf{S}} \tag{4.1}$$

where S is any surface bounded by the closed path C. In the most general case, the electric and magnetic fields have all three components (x, y, and z) and are dependent on all three coordinates (x, y, and z) in addition to time (t). For simplicity, we shall, however, first consider the case in which the electric field has an x component only, which is dependent only on the z coordinate, in addition to time. Thus

$$\boxed{\mathbf{E} = E_x(z, t)\mathbf{i}_x} \tag{4.2}$$

In other words, this simple form of time-varying electric field is everywhere directed in the x-direction and it is uniform in planes parallel to the xy-plane.

Let us now consider a rectangular path C of infinitesimal size lying in a plane parallel to the xz-plane and defined by the points (x, z), $(x, z + \Delta z)$, $(x + \Delta x, z + \Delta z)$, and $(x + \Delta x, z)$, as shown in Fig. 4.1. According to Faraday's law, the emf around the closed path C is equal to the negative of the time rate of change of the magnetic flux enclosed by C. The emf is given by the line integral of $\mathbf{E}$ around C. Thus evaluating the line integrals of $\mathbf{E}$ along the four sides of the rectangular path, we obtain

$$\int_{(x,z)}^{(x,z+\Delta z)} \mathbf{E} \cdot d\mathbf{l} = 0 \qquad \text{since } E_z = 0 \tag{4.3a}$$

$$\int_{(x,z+\Delta z)}^{(x+\Delta x,z+\Delta z)} \mathbf{E} \cdot d\mathbf{l} = [E_x]_{z+\Delta z}\, \Delta x \tag{4.3b}$$

$$\int_{(x+\Delta x,z+\Delta z)}^{(x+\Delta x,z)} \mathbf{E} \cdot d\mathbf{l} = 0 \qquad \text{since } E_z = 0 \tag{4.3c}$$

$$\int_{(x+\Delta x,z)}^{(x,z)} \mathbf{E} \cdot d\mathbf{l} = -[E_x]_z\, \Delta x \tag{4.3d}$$

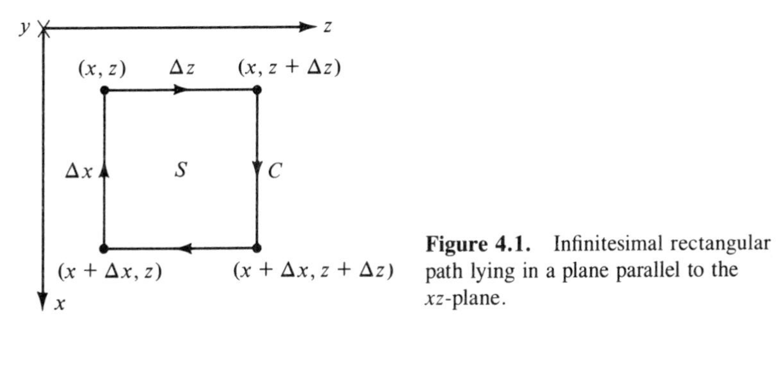

Figure 4.1. Infinitesimal rectangular path lying in a plane parallel to the xz-plane.

Adding up (4.3a)–(4.3d), we obtain

$$\oint_C \mathbf{E} \cdot d\mathbf{l} = [E_x]_{z+\Delta z}\,\Delta x - [E_x]_z\,\Delta x$$
$$= \{[E_x]_{z+\Delta z} - [E_x]_z\}\,\Delta x \tag{4.4}$$

In (4.3a)–(4.3d) and (4.4), $[E_x]_z$ and $[E_x]_{z+\Delta z}$ denote values of E_x evaluated along the sides of the path for which $z = z$ and $z = z + \Delta z$, respectively.

To find the magnetic flux enclosed by C, let us consider the plane surface S bounded by C. According to the right-hand screw rule, we must use the magnetic flux crossing S toward the positive y-direction, that is, into the page, since the path C is traversed in the clockwise sense. The only component of $\mathbf{B}$ normal to the area S is the y-component. Also since the area is infinitesimal in size, we can assume B_y to be uniform over the area and equal to its value at (x, z). The required magnetic flux is then given by

$$\int_S \mathbf{B} \cdot d\mathbf{S} = [B_y]_{(x,z)}\,\Delta x\,\Delta z \tag{4.5}$$

Substituting (4.4) and (4.5) into (4.1) to apply Faraday's law to the rectangular path C under consideration, we get

$$\{[E_x]_{z+\Delta z} - [E_x]_z\}\,\Delta x = -\frac{d}{dt}\{[B_y]_{(x,z)}\,\Delta x\,\Delta z\}$$

or

$$\frac{[E_x]_{z+\Delta z} - [E_x]_z}{\Delta z} = -\frac{\partial [B_y]_{(x,z)}}{\partial t} \tag{4.6}$$

If we now let the rectangular path shrink to the point (x, z) by letting Δx and Δz tend to zero, we obtain

$$\lim_{\substack{\Delta x \to 0 \\ \Delta z \to 0}} \frac{[E_x]_{z+\Delta z} - [E_x]_z}{\Delta z} = -\lim_{\substack{\Delta x \to 0 \\ \Delta z \to 0}} \frac{\partial [B_y]_{(x,z)}}{\partial t}$$

or

$$\boxed{\frac{\partial E_x}{\partial z} = -\frac{\partial B_y}{\partial t}} \tag{4.7}$$

Equation (4.7) is Faraday's law in differential form for the simple case of $\mathbf{E}$ given by (4.2). It relates the variation of E_x with z (space) at a point to the variation of B_y with t (time) at that point. Since this derivation can be carried out for any arbitrary point (x, y, z), it is valid for all points. It tells us in particular that an E_x associated with a time-varying B_y has a differential in the z-direction. This is to be expected since if this is not the case, $\oint \mathbf{E} \cdot d\mathbf{l}$ around the infinitesimal rectangular path would be zero.

Example 4.1

Given $\mathbf{E} = 10 \cos (6\pi \times 10^8 t - 2\pi z)\, \mathbf{i}_x$ V/m, let us find $\mathbf{B}$ that satisfies (4.7).

From (4.7), we have

$$\frac{\partial B_y}{\partial t} = -\frac{\partial E_x}{\partial z}$$

$$= -\frac{\partial}{\partial z}[10 \cos (6\pi \times 10^8 t - 2\pi z)]$$

$$= -20\pi \sin (6\pi \times 10^8 t - 2\pi z)$$

$$B_y = \frac{10^{-7}}{3} \cos (6\pi \times 10^8 t - 2\pi z)$$

$$\mathbf{B} = \frac{10^{-7}}{3} \cos (6\pi \times 10^8 t - 2\pi z)\, \mathbf{i}_y$$

Faraday's law, general case

We shall now proceed to derive the differential form of (4.1) for the general case of the electric field having all three components (x, y, z), each of them depending on all three coordinates (x, y, and z), in addition to time (t); that is,

$$\boxed{\mathbf{E} = E_x(x, y, z, t)\mathbf{i}_x + E_y(x, y, z, t)\mathbf{i}_y + E_z(x, y, z, t)\mathbf{i}_z} \tag{4.8}$$

To do this, let us consider the three infinitesimal rectangular paths in planes parallel to the three mutually orthogonal planes of the Cartesian coordinate system, as shown in Fig. 4.2. Evaluating $\oint \mathbf{E} \cdot d\mathbf{l}$ around the closed paths *abcda*, *adefa*, and *afgba*, we get

$$\oint_{abcda} \mathbf{E} \cdot d\mathbf{l} = [E_y]_{(x,z)}\,\Delta y + [E_z]_{(x,y+\Delta y)}\,\Delta z - [E_y]_{(x,z+\Delta z)}\,\Delta y - [E_z]_{(x,y)}\,\Delta z \tag{4.9a}$$

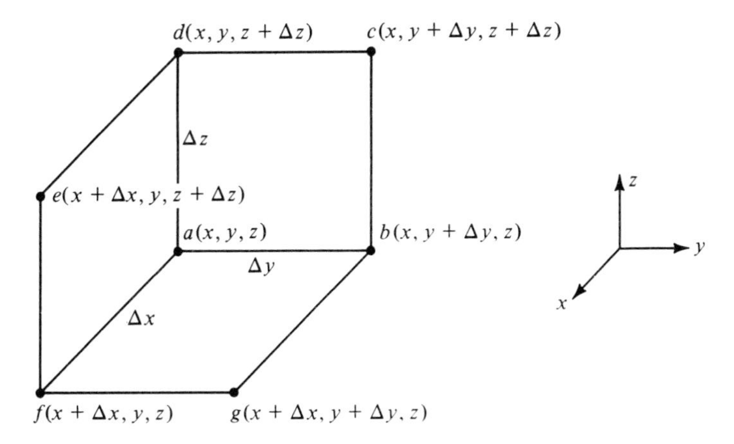

Figure 4.2. Infinitesimal rectangular paths in three mutually orthogonal planes.

$$\oint_{adefa} \mathbf{E} \cdot d\mathbf{l} = [E_z]_{(x,y)} \, \Delta z + [E_x]_{(y,z+\Delta z)} \, \Delta x - [E_z]_{(x+\Delta x,y)} \, \Delta z - [E_x]_{(y,z)} \, \Delta x \tag{4.9b}$$

$$\oint_{afgba} \mathbf{E} \cdot d\mathbf{l} = [E_x]_{(y,z)} \, \Delta x + [E_y]_{(x+\Delta x,z)} \, \Delta y - [E_x]_{(y+\Delta y,z)} \, \Delta x - [E_y]_{(x,z)} \, \Delta y \tag{4.9c}$$

In (4.9a)–(4.9c) the subscripts associated with the field components in the various terms on the right sides of the equations denote the value of the coordinates that remain constant along the sides of the closed paths corresponding to the terms. Now, evaluating $\int \mathbf{B} \cdot d\mathbf{S}$ over the surfaces *abcd*, *adef*, and *afgb*, keeping in mind the right-hand screw rule, we have

$$\int_{abcd} \mathbf{B} \cdot d\mathbf{S} = [B_x]_{(x,y,z)} \, \Delta y \, \Delta z \tag{4.10a}$$

$$\int_{adef} \mathbf{B} \cdot d\mathbf{S} = [B_y]_{(x,y,z)} \, \Delta z \, \Delta x \tag{4.10b}$$

$$\int_{afgb} \mathbf{B} \cdot d\mathbf{S} = [B_z]_{(x,y,z)} \, \Delta x \, \Delta y \tag{4.10c}$$

Applying Faraday's law to each of the three paths by making use of (4.9a)–(4.9c) and (4.10a)–(4.10c) and simplifying, we obtain

$$\frac{[E_z]_{(x,y+\Delta y)} - [E_z]_{(x,y)}}{\Delta y} - \frac{[E_y]_{(x,z+\Delta z)} - [E_y]_{(x,z)}}{\Delta z} = -\frac{\partial [B_x]_{(x,y,z)}}{\partial t} \tag{4.11a}$$

$$\frac{[E_x]_{(y,z+\Delta z)} - [E_x]_{(y,z)}}{\Delta z} - \frac{[E_z]_{(x+\Delta x,y)} - [E_z]_{(x,y)}}{\Delta x} = -\frac{\partial [B_y]_{(x,y,z)}}{\partial t} \tag{4.11b}$$

$$\frac{[E_y]_{(x+\Delta x,z)} - [E_y]_{(x,z)}}{\Delta x} - \frac{[E_x]_{(y+\Delta y,z)} - [E_x]_{(y,z)}}{\Delta y} = -\frac{\partial [B_z]_{(x,y,z)}}{\partial t} \tag{4.11c}$$

If we now let all three paths shrink to the point a by letting Δx, Δy, and Δz tend to zero, (4.11a)–(4.11c) reduce to

$$\frac{\partial E_z}{\partial y} - \frac{\partial E_y}{\partial z} = -\frac{\partial B_x}{\partial t} \tag{4.12a}$$

$$\frac{\partial E_x}{\partial z} - \frac{\partial E_z}{\partial x} = -\frac{\partial B_y}{\partial t} \tag{4.12b}$$

$$\frac{\partial E_y}{\partial x} - \frac{\partial E_x}{\partial y} = -\frac{\partial B_z}{\partial t} \tag{4.12c}$$

Equations (4.12a)–(4.12c) are the differential equations governing the relationships between the space variations of the electric field components and the time variations of the magnetic field components at a point. In particular, we note that the space derivatives are all lateral derivatives, that is, derivatives evaluated along directions lateral to the directions of the field components and not along the directions of the field components. An examination of one of the three equations is sufficient to reveal the physical meaning of these relationships. For example, (4.12a) tells us that a time-varying B_x at a point results in an electric field at that point having y- and z-components such that their net right-lateral differential normal to the x-direction is nonzero. The right-lateral differential of E_y normal to the x-direction is its derivative in the $\mathbf{i}_y \times \mathbf{i}_x$, or $-\mathbf{i}_z$-direction, that is, $\partial E_y/\partial(-z)$ or $-\partial E_y/\partial z$. The right-lateral differential of E_z normal to the x-direction is its derivative in the $\mathbf{i}_z \times \mathbf{i}_x$, or $\mathbf{i}_y$-direction, that is, $\partial E_z/\partial y$. Thus the net right-lateral differential of the y- and z-components of the electric field normal to the x-direction is $(-\partial E_y/\partial z) + (\partial E_z/\partial y)$, or $(\partial E_z/\partial y - \partial E_y/\partial z)$. Figure 4.3(a) shows an example in which the net right-lateral differential is zero although the individual derivatives are nonzero. This is because $\partial E_z/\partial y$ and $\partial E_y/\partial z$ are both positive and equal so that their difference is zero. On the other hand, for the example in Fig. 4.3(b), $\partial E_z/\partial y$ is positive and $\partial E_y/\partial z$ is negative so that their difference, that is, the net right-lateral differential, is nonzero.

Curl (del cross)

Equations (4.12a)–(4.12c) can be combined into a single vector equation as given by

$$\left(\frac{\partial E_z}{\partial y} - \frac{\partial E_y}{\partial z}\right)\mathbf{i}_x + \left(\frac{\partial E_x}{\partial z} - \frac{\partial E_z}{\partial x}\right)\mathbf{i}_y + \left(\frac{\partial E_y}{\partial x} - \frac{\partial E_x}{\partial y}\right)\mathbf{i}_z = -\frac{\partial B_x}{\partial t}\mathbf{i}_x - \frac{\partial B_y}{\partial t}\mathbf{i}_y - \frac{\partial B_z}{\partial t}\mathbf{i}_z \tag{4.13}$$

This can be expressed in determinant form as

$$\boxed{\begin{vmatrix} \mathbf{i}_x & \mathbf{i}_y & \mathbf{i}_z \\ \dfrac{\partial}{\partial x} & \dfrac{\partial}{\partial y} & \dfrac{\partial}{\partial z} \\ E_x & E_y & E_z \end{vmatrix} = -\frac{\partial \mathbf{B}}{\partial t}} \tag{4.14}$$

or as

$$\left(\mathbf{i}_x\frac{\partial}{\partial x} + \mathbf{i}_y\frac{\partial}{\partial y} + \mathbf{i}_z\frac{\partial}{\partial z}\right) \times (E_x\mathbf{i}_x + E_y\mathbf{i}_y + E_z\mathbf{i}_z) = -\frac{\partial \mathbf{B}}{\partial t} \tag{4.15}$$

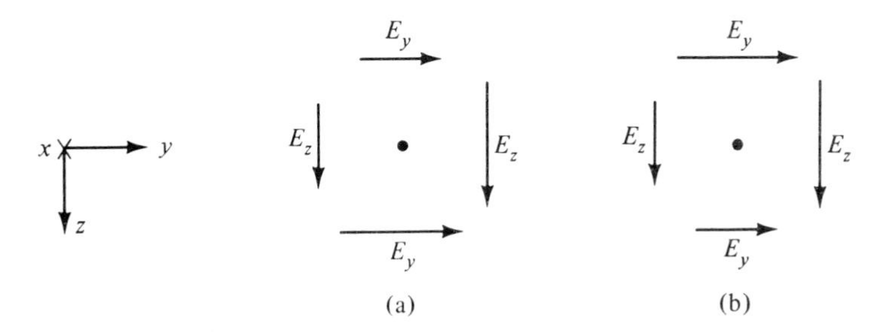

Figure 4.3. For illustrating (a) zero and (b) nonzero net right-lateral differential of E_y and E_z normal to the x-direction.

The left side of (4.14) or (4.15) is known as the *curl of* **E**, denoted as $\nabla \times \mathbf{E}$ (del cross **E**) where ∇ (del) is the vector operator given by

$$\boxed{\nabla = \mathbf{i}_x \frac{\partial}{\partial x} + \mathbf{i}_y \frac{\partial}{\partial y} + \mathbf{i}_z \frac{\partial}{\partial z}} \tag{4.16}$$

Thus we have

$$\boxed{\nabla \times \mathbf{E} = -\frac{\partial \mathbf{B}}{\partial t}} \tag{4.17}$$

Equation (4.17) is Maxwell's equation in differential form corresponding to Faraday's law. It tells us that at a point in an electromagnetic field, the curl of the electric field intensity is equal to the time rate of decrease of the magnetic flux density. We shall discuss curl further in Sec. 4.3, but note that for static fields, $\nabla \times \mathbf{E}$ is equal to the null vector. Thus for a static vector field to be realized as an electric field, the components of its curl must all be zero.

Although we have deduced (4.17) from (4.1) by considering the Cartesian coordinate system, it is independent of the coordinate system since (4.1) is independent of the coordinate system. The expressions for the curl of a vector in cylindrical and spherical coordinate systems are derived in Appendix A. They are reproduced here together with that in (4.14) for the Cartesian coordinate system.

Cartesian:

$$\nabla \times \mathbf{A} = \begin{vmatrix} \mathbf{i}_x & \mathbf{i}_y & \mathbf{i}_z \\ \frac{\partial}{\partial x} & \frac{\partial}{\partial y} & \frac{\partial}{\partial z} \\ A_x & A_y & A_z \end{vmatrix} \tag{4.18a}$$

Cylindrical:

$$\nabla \times \mathbf{A} = \begin{vmatrix} \frac{\mathbf{i}_r}{r} & \mathbf{i}_\phi & \frac{\mathbf{i}_z}{r} \\ \frac{\partial}{\partial r} & \frac{\partial}{\partial \phi} & \frac{\partial}{\partial z} \\ A_r & rA_\phi & A_z \end{vmatrix} \tag{4.18b}$$

Spherical:

$$\nabla \times \mathbf{A} = \begin{vmatrix} \frac{\mathbf{i}_r}{r^2 \sin\theta} & \frac{\mathbf{i}_\theta}{r \sin\theta} & \frac{\mathbf{i}_\phi}{r} \\ \frac{\partial}{\partial r} & \frac{\partial}{\partial \theta} & \frac{\partial}{\partial \phi} \\ A_r & rA_\theta & r\sin\theta\, A_\phi \end{vmatrix} \tag{4.18c}$$

Example 4.2

Find the curls of the following vector fields: (a) $y\mathbf{i}_x - x\mathbf{i}_y$ and (b) $\mathbf{i}_\phi$ in cylindrical coordinates.

(a) Using (4.18a), we have

$$\nabla \times (y\mathbf{i}_x - x\mathbf{i}_y) = \begin{vmatrix} \mathbf{i}_x & \mathbf{i}_y & \mathbf{i}_z \\ \dfrac{\partial}{\partial x} & \dfrac{\partial}{\partial y} & \dfrac{\partial}{\partial z} \\ y & -x & 0 \end{vmatrix}$$

$$= \mathbf{i}_x\left[-\frac{\partial}{\partial z}(-x)\right] + \mathbf{i}_y\left[\frac{\partial}{\partial z}(y)\right] + \mathbf{i}_z\left[\frac{\partial}{\partial x}(-x) - \frac{\partial}{\partial y}(y)\right]$$

$$= -2\mathbf{i}_z$$

(b) Using (4.18b), we obtain

$$\nabla \times \mathbf{i}_\phi = \begin{vmatrix} \dfrac{\mathbf{i}_r}{r} & \mathbf{i}_\phi & \dfrac{\mathbf{i}_z}{r} \\ \dfrac{\partial}{\partial r} & \dfrac{\partial}{\partial \phi} & \dfrac{\partial}{\partial z} \\ 0 & r & 0 \end{vmatrix} = \frac{\mathbf{i}_r}{r}\left[\frac{\partial}{\partial z}(r)\right] + \frac{\mathbf{i}_z}{r}\left[\frac{\partial}{\partial r}(r)\right] = \frac{1}{r}\mathbf{i}_z$$

Ampère's circuital law, general case

We shall now consider the derivation of the differential form of Ampère's circuital law given in integral form by

$$\oint_C \mathbf{H} \cdot d\mathbf{l} = \int_S \mathbf{J} \cdot d\mathbf{S} + \frac{d}{dt}\int_S \mathbf{D} \cdot d\mathbf{S} \tag{4.19}$$

where S is any surface bounded by the closed path C. To do this, we need not repeat the procedure employed in the case of Faraday's law. Instead, we note from (4.1) and (4.17) that in converting to the differential form from integral form, the line integral of $\mathbf{E}$ around the closed path C is replaced by the curl of $\mathbf{E}$, the surface integral of $\mathbf{B}$ over the surface S bounded by C is replaced by $\mathbf{B}$ itself, and the total time derivative is replaced by partial derivative, as shown:

$$\oint_C \mathbf{E} \cdot d\mathbf{l} = -\frac{d}{dt}\int_S \mathbf{B} \cdot d\mathbf{S}$$

$$\downarrow$$

$$\nabla \times \mathbf{E} = -\frac{\partial}{\partial t}(\mathbf{B})$$

Then using the analogy between Ampère's circuital law and Faraday's law, we can write the following:

$$\oint_C \mathbf{H} \cdot d\mathbf{l} = \int_S \mathbf{J} \cdot d\mathbf{S} + \frac{d}{dt}\int_S \mathbf{D} \cdot d\mathbf{S}$$

$$\downarrow$$

$$\nabla \times \mathbf{H} = \mathbf{J} + \frac{\partial}{\partial t}(\mathbf{D})$$

Thus for the general case of the magnetic field having all three components (x, y, and z), each of them depending on all three coordinates (x, y, and z), in addition to time (t), that is, for

$$\mathbf{H} = H_x(x, y, z, t)\mathbf{i}_x + H_y(x, y, z, t)\mathbf{i}_y + H_z(x, y, z, t)\mathbf{i}_z \tag{4.20}$$

the differential form of Ampère's circuital law is given by

$$\nabla \times \mathbf{H} = \mathbf{J} + \frac{\partial \mathbf{D}}{\partial t} \tag{4.21}$$

The quantity $\partial\mathbf{D}/\partial t$ is known as the *displacement current density*. Equation (4.21) tells us that at a point in an electromagnetic field, the curl of the magnetic field intensity is equal to the sum of the current density due to flow of charges and the displacement current density. In Cartesian coordinates, (4.21) becomes

$$\begin{vmatrix} \mathbf{i}_x & \mathbf{i}_y & \mathbf{i}_z \\ \frac{\partial}{\partial x} & \frac{\partial}{\partial y} & \frac{\partial}{\partial z} \\ H_x & H_y & H_z \end{vmatrix} = \mathbf{J} + \frac{\partial \mathbf{D}}{\partial t} \tag{4.22}$$

This is equivalent to three scalar equations relating the lateral space derivatives of the components of $\mathbf{H}$ to the components of the current density and the time derivatives of the electric field components. These scalar equations can be interpreted in a manner similar to the interpretation of (4.12a)–(4.12c) in the case of Faraday's law. Also, expressions similar to (4.22) can be written in the cylindrical and spherical coordinate systems by using the determinant expansions for the curl in those coordinate systems, given by (4.18b) and (4.18c), respectively.

Ampere's circuital law, special case

Having obtained the differential form of Ampère's circuital law for the general case, we can now simplify it for any particular case. Let us consider the particular case of

$$\mathbf{H} = H_y(z, t)\mathbf{i}_y \tag{4.23}$$

that is, a magnetic field directed everywhere in the y-direction and uniform in planes parallel to the xy-plane. Then since **H** does not depend on x and y, we can replace $\partial/\partial x$ and $\partial/\partial y$ in the determinant expansion for $\nabla \times \mathbf{H}$ by zeroes. In addition, setting $H_x = H_z = 0$, we have

$$\begin{vmatrix} \mathbf{i}_x & \mathbf{i}_y & \mathbf{i}_z \\ 0 & 0 & \dfrac{\partial}{\partial z} \\ 0 & H_y & 0 \end{vmatrix} = \mathbf{J} + \frac{\partial \mathbf{D}}{\partial t} \tag{4.24}$$

Equating like components on the two sides and noting that the y- and z-components on the left side are zero, we obtain

$$-\frac{\partial H_y}{\partial z} = J_x + \frac{\partial D_x}{\partial t}$$

or

$$\boxed{\frac{\partial H_y}{\partial z} = -J_x - \frac{\partial D_x}{\partial t}} \tag{4.25}$$

Equation (4.25) is Ampère's circuital law in differential form for the simple case of **H** given by (4.23). It relates the variation of H_y with z (space) at a point to the current density J_x and to the variation of D_x with t (time) at that point. It tells us in particular that an H_y associated with a current density J_x or a time-varying D_x or a nonzero combination of the two quantities has a differential in the z-direction.

Example 4.3

Given $\mathbf{E} = E_0 z^2 e^{-t}\mathbf{i}_x$ in free space ($\mathbf{J} = \mathbf{0}$). We wish to determine if there exists a magnetic field such that both Faraday's law and Ampère's circuital law are satisfied simultaneously.

Using Faraday's law and Ampère's circuital law in succession, we have

$$\frac{\partial B_y}{\partial t} = -\frac{\partial E_x}{\partial z} = -2E_0 z e^{-t}$$

$$B_y = 2E_0 z e^{-t}$$

$$H_y = \frac{2E_0}{\mu_0} z e^{-t}$$

$$\frac{\partial D_x}{\partial t} = -\frac{\partial H_y}{\partial z} = -\frac{2E_0}{\mu_0} e^{-t}$$

$$D_x = \frac{2E_0}{\mu_0} e^{-t}$$

$$E_x = \frac{2E_0}{\mu_0 \varepsilon_0} e^{-t}$$

$$\mathbf{E} = \frac{2E_0}{\mu_0 \varepsilon_0} e^{-t}\mathbf{i}_x$$

which is not the same as the original $\mathbf{E}$. Hence, a magnetic field does not exist which together with the given $\mathbf{E}$ satisfies both laws simultaneously. The pair of fields $\mathbf{E} = E_0 z^2 e^{-t}\mathbf{i}_x$ and $\mathbf{B} = 2E_0 z e^{-t}\mathbf{i}_y$ satisfies only Faraday's law, whereas the pair of fields $\mathbf{B} = 2E_0 z e^{-t}\mathbf{i}_y$ and $\mathbf{E} = (2E_0/\mu_0 \varepsilon_0)e^{-t}\mathbf{i}_x$ satisfies only Ampère's circuital law.

Lumped circuit theory approximations

To generalize the observation made in the example just discussed, there are certain pairs of time-varying electric and magnetic fields which satisfy only Faraday's law as given by (4.17) and certain other pairs which satisfy only Ampère's circuital law as given by (4.21). In the strictest sense, every physically realizable pair of time-varying electric and magnetic fields must satisfy simultaneously both laws as given by (4.17) and (4.21). However, under the low-frequency approximation, it is valid for the fields to satisfy the laws with certain terms neglected in one or both laws. Lumped-circuit theory is based on such approximations. Thus the terminal voltage-to-current relationship $V(t) = d[LI(t)]/dt$ for an inductor is obtained by ignoring the effect of the time-varying electric field, that is, $\partial\mathbf{D}/\partial t$ term in Ampère's circuital law. The terminal current-to-voltage relationship $I(t) = d[CV(t)]/dt$ for a capacitor is obtained by ignoring the effect of the time-varying magnetic field, that is, $\partial\mathbf{B}/\partial t$ term in Faraday's law. The terminal voltage-to-current relationship $V(t) = RI(t)$ for a resistor is obtained by ignoring the effects of both time-varying electric field and time-varying magnetic field, that is, both $\partial\mathbf{D}/\partial t$ term in Ampère's circuital law and $\partial\mathbf{B}/\partial t$ term in Faraday's law. In contrast to these approximations, electromagnetic wave propagation phenomena and transmission-line (distributed circuit) theory are based upon the simultaneous application of the two laws with all terms included, that is, as given by (4.17) and (4.21), as we shall learn in Chapters 6 and 7.

K4.1. Faraday's law in differential form; Ampère's circuital law in differential form; Curl of a vector.

D4.1. Given $\mathbf{E} = E_0 \cos(6\pi \times 10^8 t - 2\pi z)\, \mathbf{i}_x$ V/m, find the time rate of increase of B_y at $t = 10^{-8}$ s for each of the following values of z: **(a)** 0; **(b)** $\frac{1}{4}$ m; and **(c)** $\frac{2}{3}$ m.
Ans. **(a)** 0; **(b)** $2\pi E_0$; **(c)** $-\sqrt{3}\,\pi E_0$

D4.2. For the vector field $\mathbf{A} = xy^2\mathbf{i}_x + xz\mathbf{i}_y + x^2yz\mathbf{i}_z$, find the following: **(a)** the net right-lateral differential of A_x and A_y normal to the z-direction at the point (1, 1, 1); **(b)** the net right-lateral differential of A_y and A_z normal to the x-direction at the point (1, 2, 1); and **(c)** the net right-lateral differential of A_z and A_x normal to the y-direction at the point (1, 1, −1).
Ans. **(a)** −1; **(b)** 0; **(c)** 2

D4.3. Given $\mathbf{J} = \mathbf{0}$ and $\mathbf{H} = H_0 e^{-(3\times 10^8 t - z)^2}\mathbf{i}_y$ A/m, find the time rate of increase of D_x for each of the following cases: **(a)** $z = 2$ m, $t = 10^{-8}$ s; **(b)** $z = 3$ m, $t = \frac{1}{3} \times 10^{-8}$ s; and **(c)** $z = 3$ m, $t = 10^{-8}$ s.
Ans. **(a)** $-0.7358H_0$; **(b)** $0.0733H_0$; **(c)** 0

4.2 GAUSS' LAWS AND THE CONTINUITY EQUATION

Thus far we have derived Maxwell's equations in differential form corresponding to the two Maxwell's equations in integral form involving the line integrals of $\mathbf{E}$ and $\mathbf{H}$ around the closed path, that is, Faraday's law and Ampere's circuital law, respectively. The remaining two Maxwell's equations in integral form, namely,

Gauss' law for the electric field and Gauss' law for the magnetic field, are concerned with the closed surface integrals of **D** and **B**, respectively. We shall in this section derive the differential forms of these two equations.

Gauss' law for the electric field

We recall from Section 3.4 that Gauss' law for the electric field is given by

$$\boxed{\oint_S \mathbf{D} \cdot d\mathbf{S} = \int_V \rho \, dv} \tag{4.26}$$

where V is the volume enclosed by the closed surface S. To derive the differential form of this equation, let us consider a rectangular box of edges of infinitesimal lengths Δx, Δy, and Δz and defined by the six surfaces $x = x$, $x = x + \Delta x$, $y = y$, $y = y + \Delta y$, $z = z$, and $z = z + \Delta z$, as shown in Fig. 4.4, in a region of electric field

$$\boxed{\mathbf{D} = \mathbf{D}_x(x, y, z, t)\mathbf{i}_x + D_y(x, y, z, t)\mathbf{i}_y + D_z(x, y, z, t)\mathbf{i}_z} \tag{4.27}$$

and charge of density $\rho(x, y, z, t)$. According to Gauss' law for the electric field, the displacement flux emanating from the box is equal to the charge enclosed by the box. The displacement flux is given by the surface integral of **D** over the surface of the box, which is comprised of six plane surfaces. Thus evaluating the displacement flux emanating from the box through each of the six plane surfaces of the box, we have

$$\int \mathbf{D} \cdot d\mathbf{S} = -[D_x]_x \, \Delta y \, \Delta z \qquad \text{for the surface } x = x \tag{4.28a}$$

$$\int \mathbf{D} \cdot d\mathbf{S} = [D_x]_{x+\Delta x} \, \Delta y \, \Delta z \qquad \text{for the surface } x = x + \Delta x \tag{4.28b}$$

$$\int \mathbf{D} \cdot d\mathbf{S} = -[D_y]_y \, \Delta z \, \Delta x \qquad \text{for the surface } y = y \tag{4.28c}$$

$$\int \mathbf{D} \cdot d\mathbf{S} = [D_y]_{y+\Delta y} \, \Delta z \, \Delta x \qquad \text{for the surface } y = y + \Delta y \tag{4.28d}$$

$$\int \mathbf{D} \cdot d\mathbf{S} = -[D_z]_z \, \Delta x \, \Delta y \qquad \text{for the surface } z = z \tag{4.28e}$$

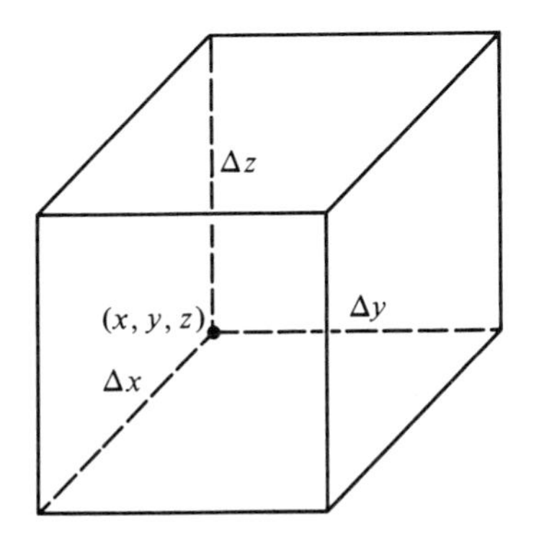

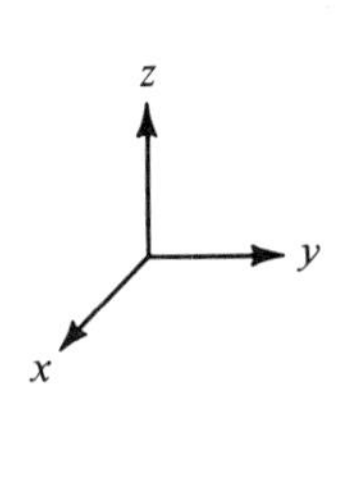

Figure 4.4. Infinitesimal rectangular box.

$$\int \mathbf{D} \cdot d\mathbf{S} = [D_z]_{z+\Delta z}\,\Delta x\,\Delta y \qquad \text{for the surface } z = z + \Delta z \tag{4.28f}$$

Adding up (4.28a)–(4.28f), we obtain the total displacement flux emanating from the box to be

$$\begin{aligned}\oint_S \mathbf{D} \cdot d\mathbf{S} = \{[D_x]_{x+\Delta x} - [D_x]_x\}\,\Delta y\,\Delta z \\ + \{[D_y]_{y+\Delta y} - [D_y]_y\}\,\Delta z\,\Delta x \\ + \{[D_z]_{z+\Delta z} - [D_z]_z\}\,\Delta x\,\Delta y\end{aligned} \tag{4.29}$$

Now the charge enclosed by the rectangular box is given by

$$\int_V \rho\,dv = \rho(x, y, z, t) \cdot \Delta x\,\Delta y\,\Delta z = \rho\,\Delta x\,\Delta y\,\Delta z \tag{4.30}$$

where we have assumed ρ to be uniform throughout the volume of the box and equal to its value at (x, y, z) since the box is infinitesimal in volume.

Substituting (4.29) and (4.30) into (4.26), we get

$$\{[D_x]_{x+\Delta x} - [D_x]_x\}\,\Delta y\,\Delta z + \{[D_y]_{y+\Delta y} - [D_y]_y\}\,\Delta z\,\Delta x + \{[D_z]_{z+\Delta z} - [D_z]_z\}\,\Delta x\,\Delta y = \rho\,\Delta x\,\Delta y\,\Delta z$$

or

$$\frac{[D_x]_{x+\Delta x} - [D_x]_x}{\Delta x} + \frac{[D_y]_{y+\Delta y} - [D_y]_y}{\Delta y} + \frac{[D_z]_{z+\Delta z} - [D_z]_z}{\Delta z} = \rho \tag{4.31}$$

If we now let the box shrink to the point (x, y, z) by letting Δx, Δy, and Δz tend to zero, we obtain

$$\lim_{\Delta x \to 0} \frac{[D_x]_{x+\Delta x} - [D_x]_x}{\Delta x} + \lim_{\Delta y \to 0} \frac{[D_y]_{y+\Delta y} - [D_y]_y}{\Delta y} + \lim_{\Delta z \to 0} \frac{[D_z]_{z+\Delta z} - [D_z]_z}{\Delta z} = \lim_{\substack{\Delta x \to 0 \\ \Delta y \to 0 \\ \Delta z \to 0}} \rho$$

or

$$\boxed{\frac{\partial D_x}{\partial x} + \frac{\partial D_y}{\partial y} + \frac{\partial D_z}{\partial z} = \rho} \tag{4.32}$$

Equation (4.32) is the differential equation governing the relationship between the space variations of the components of **D** to the charge density. In particular, we note that the derivatives are all longitudinal derivatives, that is, derivatives evaluated along the directions of the field components, in contrast to the lateral derivatives encountered in Section 4.1. Thus (4.32) tells us that the net longitudinal differential, that is, the algebraic sum of the longitudinal derivatives, of the components of **D** at a point in space is equal to the charge density at that point. Conversely, a charge density at a point results in an electric field having components of **D** such that their net longitudinal differential is nonzero. Figure 4.5(a) shows an example in which the net longitudinal differential is zero. This is because $\partial D_x/\partial x$ and $\partial D_y/\partial y$ are equal in magnitude but opposite in sign, whereas

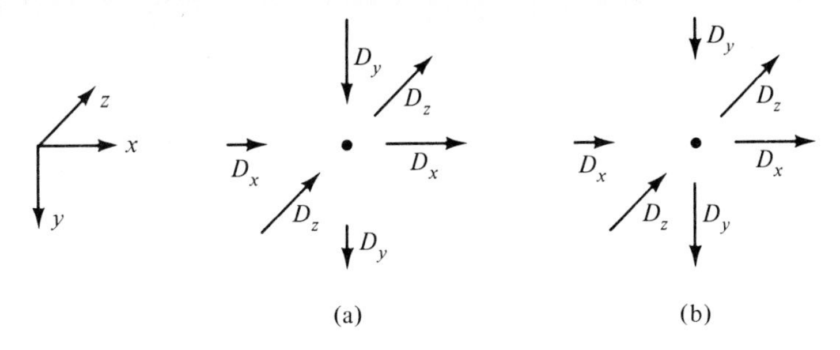

Figure 4.5. For illustrating (a) zero and (b) nonzero net longitudinal differential of the components of **D**.

$\partial D_z/\partial z$ is zero. On the other hand, for the example in Fig. 4.5(b), both $\partial D_x/\partial x$ and $\partial D_y/\partial y$ are positive and $\partial D_z/\partial z$ is zero, so that the net longitudinal differential is nonzero.

Divergence (del dot)

Equation (4.32) can be written in vector notation as

$$\left(\mathbf{i}_x\frac{\partial}{\partial x} + \mathbf{i}_y\frac{\partial}{\partial y} + \mathbf{i}_z\frac{\partial}{\partial z}\right) \cdot (D_x\mathbf{i}_x + D_y\mathbf{i}_y + D_z\mathbf{i}_z) = \rho \tag{4.33}$$

The left side of (4.33) is known as the *divergence of* **D**, denoted as $\nabla \cdot \mathbf{D}$ (del dot **D**). Thus we have

$$\boxed{\nabla \cdot \mathbf{D} = \rho} \tag{4.34}$$

Equation (4.34) is Maxwell's equation in differential form corresponding to Gauss' law for the electric field. It tells us that the divergence of the displacement flux density at a point is equal to the charge density at that point. We shall discuss divergence further in Section 4.3.

Although we have deduced (4.34) from (4.26) by considering the Cartesian coordinate system, it is independent of the coordinate system since (4.26) is independent of the coordinate system. The expressions for the divergence of a vector in cylindrical and spherical coordinate systems are derived in Appendix A. They are reproduced here together with that in (4.32) for the Cartesian coordinate system.

Cartesian:

$$\nabla \cdot \mathbf{A} = \frac{\partial A_x}{\partial x} + \frac{\partial A_y}{\partial y} + \frac{\partial A_z}{\partial z} \tag{4.35a}$$

Cylindrical:

$$\nabla \cdot \mathbf{A} = \frac{1}{r}\frac{\partial}{\partial r}(rA_r) + \frac{1}{r}\frac{\partial A_\phi}{\partial \phi} + \frac{\partial A_z}{\partial z} \tag{4.35b}$$

Spherical:

$$\nabla \cdot \mathbf{A} = \frac{1}{r^2}\frac{\partial}{\partial r}(r^2 A_r) + \frac{1}{r\sin\theta}\frac{\partial}{\partial \theta}(A_\theta \sin\theta) + \frac{1}{r\sin\theta}\frac{\partial A_\phi}{\partial \phi} \tag{4.35c}$$

Example 4.4

Find the divergences of the following vector fields: (a) $3x\mathbf{i}_x + (y - 3)\mathbf{i}_y + (2 - z)\mathbf{i}_z$ and (b) $r^2 \sin\theta\, \mathbf{i}_\theta$ in spherical coordinates.

(a) Using (4.35a), we have

$$\begin{aligned}\nabla \cdot [3x\mathbf{i}_x + (y - 3)\mathbf{i}_y + (2 - z)\mathbf{i}_z] &= \frac{\partial}{\partial x}(3x) + \frac{\partial}{\partial y}(y - 3) + \frac{\partial}{\partial z}(2 - z)\\ &= 3 + 1 - 1 = 3\end{aligned}$$

(b) Using (4.35c), we obtain

$$\begin{aligned}\nabla \cdot r^2 \sin\theta\, \mathbf{i}_\theta &= \frac{1}{r\sin\theta}\frac{\partial}{\partial\theta}(r^2\sin^2\theta)\\ &= \frac{1}{r\sin\theta}(2r^2\sin\theta\cos\theta)\\ &= 2r\cos\theta\end{aligned}$$

Gauss' law for the magnetic field

We shall now consider the derivation of the differential form of Gauss' law for the magnetic field given in integral form by

$$\oint_S \mathbf{B} \cdot d\mathbf{S} = 0 \tag{4.36}$$

where S is any closed surface. To do this, we need not repeat the procedure employed in the case of Gauss' law for the electric field. Instead, we note from (4.26) and (4.34) that in converting to the differential form from integral form, the surface integral of $\mathbf{D}$ over the closed surface S is replaced by the divergence of $\mathbf{D}$ and the volume integral of ρ is replaced by ρ itself, as shown:

$$\begin{array}{ccc}\oint_S \mathbf{D} \cdot d\mathbf{S} & = & \int_V \rho\, dv\\ \downarrow & & \downarrow\\ \nabla \cdot \mathbf{D} & = & \rho\end{array}$$

Then using the analogy between the two Gauss' laws, we can write the following:

$$\begin{array}{ccccc}\oint_S \mathbf{B} \cdot d\mathbf{S} & = & 0 & = & \int_V 0\, dv\\ \downarrow & & & & \downarrow\\ \nabla \cdot \mathbf{B} & = & 0 & & \end{array}$$

Thus Gauss' law in differential form for the magnetic field

$$\mathbf{B} = B_x(x, y, z, t)\mathbf{i}_x + B_y(x, y, z, t)\mathbf{i}_y + B_z(x, y, z, t)\mathbf{i}_z \tag{4.37}$$

is given by

$$\boxed{\nabla \cdot \mathbf{B} = 0} \tag{4.38}$$

which tells us that the divergence of the magnetic flux density at a point is equal to zero. Conversely, for a vector field to be realized as a magnetic field, its divergence must be zero. In Cartesian coordinates, (4.38) becomes

$$\boxed{\frac{\partial B_x}{\partial x} + \frac{\partial B_y}{\partial y} + \frac{\partial B_z}{\partial z} = 0} \tag{4.39}$$

pointing out that the net longitudinal differential of the components of $\mathbf{B}$ is zero. Also, expressions similar to (4.39) can be written in cylindrical and spherical coordinate systems by using the expressions for the divergence in those coordinate systems, given by (4.35b) and (4.35c), respectively.

Example 4.5

Determine if the vector $\mathbf{A} = (1/r^2)(\cos\phi\,\mathbf{i}_r + \sin\phi\,\mathbf{i}_\phi)$ in cylindrical coordinates can represent a magnetic field $\mathbf{B}$.

Noting that

$$\begin{aligned}\nabla \cdot \mathbf{A} &= \frac{1}{r}\frac{\partial}{\partial r}\left(\frac{\cos\phi}{r}\right) + \frac{1}{r}\frac{\partial}{\partial \phi}\left(\frac{\sin\phi}{r^2}\right) \\ &= -\frac{\cos\phi}{r^3} + \frac{\cos\phi}{r^3} = 0\end{aligned}$$

we conclude that the given vector can represent a $\mathbf{B}$.

Continuity equation

We shall conclude this section by deriving the differential form of the law of conservation of charge given in integral form by

$$\boxed{\oint_S \mathbf{J} \cdot d\mathbf{S} = -\frac{d}{dt}\int_V \rho\, dv} \tag{4.40}$$

Using analogy with Gauss' law for the electric field, we can write the following:

$$\oint_S \mathbf{J} \cdot d\mathbf{S} = -\frac{d}{dt}\int_V \rho\, dv$$

$$\downarrow \qquad\qquad \downarrow$$

$$\nabla \cdot \mathbf{J} = -\frac{\partial}{\partial t}(\rho)$$

Thus the differential form of the law of conservation of charge is given by

$$\boxed{\nabla \cdot \mathbf{J} = -\frac{\partial \rho}{\partial t}} \tag{4.41}$$

Equation (4.41) is familiarly known as the *continuity equation*. It tells us that the divergence of the current density due to flow of charges at a point is equal to the time rate of decrease of the charge density at that point. It can be expanded in a given coordinate system by using the expression for the divergence in that coordinate system.

K4.2. Gauss' law for the electric field in differential form; Gauss' law for the magnetic field in differential form; Divergence of a vector; Continuity equation.

D4.4. For the vector field $\mathbf{A} = yz\mathbf{i}_x + xy\mathbf{i}_y + xyz^2\mathbf{i}_z$, find the net longitudinal differential of the components of $\mathbf{A}$ at the following points: **(a)** $(1, 1, -1)$; **(b)** $(1, 1, -\frac{1}{2})$; and **(c)** $(1, 1, 1)$.
Ans. **(a)** -1; **(b)** 0; **(c)** 3

D4.5. The following hold at a point in a charge-free region: (i) the sum of the longitudinal differentials of D_x and D_y is D_0 and (ii) the longitudinal differential of D_y is three times the longitudinal differential of D_z. Find: **(a)** $\frac{\partial D_x}{\partial x}$; **(b)** $\frac{\partial D_y}{\partial y}$; and **(c)** $\frac{\partial D_z}{\partial z}$.
Ans. **(a)** $4D_0$; **(b)** $-3D_0$; **(c)** $-D_0$

D4.6. In a small region around the origin, the current density due to flow of charges is given by $\mathbf{J} = J_0(x^2\mathbf{i}_x + y^2\mathbf{i}_y + z^2\mathbf{i}_z)$ A/m², where J_0 is a constant. Find the time rate of increase of the charge density at each of the following points: **(a)** (0.02, 0.01, 0.01); **(b)** (0.02, −0.01, −0.01); and **(c)** (−0.02, −0.01, 0.01).
Ans. **(a)** $-0.08J_0$ (C/m³)/s; **(b)** 0; **(c)** $0.04J_0$ (C/m³)/s

4.3 CURL AND DIVERGENCE

In Sections 4.1 and 4.2 we derived the differential forms of Maxwell's equations and the law of conservation of charge from their integral forms. Maxwell's equations are given by

$$\nabla \times \mathbf{E} = -\frac{\partial \mathbf{B}}{\partial t}$$

$$\nabla \times \mathbf{H} = \mathbf{J} + \frac{\partial \mathbf{D}}{\partial t}$$

$$\nabla \cdot \mathbf{D} = \rho$$

$$\nabla \cdot \mathbf{B} = 0$$

whereas the continuity equation is given by

$$\nabla \cdot \mathbf{J} = -\frac{\partial \rho}{\partial t}$$

These equations contain two new vector (differential) operations, namely, the curl and the divergence. The curl of a vector is a vector quantity whereas the divergence of a vector is a scalar quantity. In this section we shall introduce the basic definitions of curl and divergence and then discuss physical interpretations of these quantities.

Curl, basic definition

To discuss curl first, let us consider Ampère's circuital law without the displacement current density term; that is,

$$\nabla \times \mathbf{H} = \mathbf{J} \tag{4.42}$$

We wish to express $\nabla \times \mathbf{H}$ at a point in the current region in terms of $\mathbf{H}$ at that point. If we consider an infinitesimal surface $\Delta\mathbf{S}$ at the point and take the dot product of both sides of (4.42) with $\Delta\mathbf{S}$, we get

$$(\nabla \times \mathbf{H}) \cdot \Delta\mathbf{S} = \mathbf{J} \cdot \Delta\mathbf{S} \tag{4.43}$$

But $\mathbf{J} \cdot \Delta\mathbf{S}$ is simply the current crossing the surface $\Delta\mathbf{S}$, and according to Ampère's circuital law in integral form without the displacement current term,

$$\oint_C \mathbf{H} \cdot d\mathbf{l} = \mathbf{J} \cdot \Delta\mathbf{S} \tag{4.44}$$

where C is the closed path bounding $\Delta\mathbf{S}$. Comparing (4.43) and (4.44), we have

$$(\nabla \times \mathbf{H}) \cdot \Delta\mathbf{S} = \oint_C \mathbf{H} \cdot d\mathbf{l}$$

or

$$(\nabla \times \mathbf{H}) \cdot \Delta S\, \mathbf{i}_n = \oint_C \mathbf{H} \cdot d\mathbf{l} \tag{4.45}$$

where $\mathbf{i}_n$ is the unit vector normal to ΔS and directed toward the side of advance of a right-hand screw as it is turned around C. Dividing both sides of (4.45) by ΔS, we obtain

$$(\nabla \times \mathbf{H}) \cdot \mathbf{i}_n = \frac{\oint_C \mathbf{H} \cdot d\mathbf{l}}{\Delta S} \tag{4.46}$$

The maximum value of $(\nabla \times \mathbf{H}) \cdot \mathbf{i}_n$, and hence that of the right side of (4.46), occurs when $\mathbf{i}_n$ is oriented parallel to $\nabla \times \mathbf{H}$, that is, when the surface ΔS is oriented normal to the current density vector $\mathbf{J}$. This maximum value is simply $|\nabla \times \mathbf{H}|$. Thus

$$|\nabla \times \mathbf{H}| = \left[\frac{\oint_C \mathbf{H} \cdot d\mathbf{l}}{\Delta S}\right]_{\max} \tag{4.47}$$

Since the direction of $\nabla \times \mathbf{H}$ is the direction of $\mathbf{J}$, or that of the unit vector normal to ΔS, we can then write

$$\nabla \times \mathbf{H} = \left[\frac{\oint_C \mathbf{H} \cdot d\mathbf{l}}{\Delta S}\right]_{\max} \mathbf{i}_n \tag{4.48}$$

This result is however approximate since (4.45) is exact only in the limit that ΔS tends to zero. Thus

$$\nabla \times \mathbf{H} = \lim_{\Delta S \to 0} \left[\frac{\oint_C \mathbf{H} \cdot d\mathbf{l}}{\Delta S}\right]_{\max} \mathbf{i}_n \tag{4.49}$$

which is the expression for $\nabla \times \mathbf{H}$ at a point in terms of $\mathbf{H}$ at that point. Although we have derived this for the **H** vector, it is a general result and, in fact, is often the starting point for the introduction of curl. Thus for any vector field **A**,

$$\boxed{\nabla \times \mathbf{A} = \lim_{\Delta S \to 0} \left[\frac{\oint_C \mathbf{A} \cdot d\mathbf{l}}{\Delta S} \right]_{\max} \mathbf{i}_n} \tag{4.50}$$

Equation (4.50) tells us that to find the curl of a vector at a point in that vector field, we first consider an infinitesimal surface at that point and compute the closed line integral or circulation of the vector around the periphery of this surface by orienting the surface such that the circulation is maximum. We then divide the circulation by the area of the surface to obtain the maximum value of the circulation per unit area. Since we need this maximum value of the circulation per unit area in the limit that the area tends to zero, we do this by gradually shrinking the area and making sure that each time we compute the circulation per unit area an orientation for the area that maximizes this quantity is maintained. The limiting value to which the maximum circulation per unit area approaches is the magnitude of the curl. The limiting direction to which the normal vector to the surface approaches is the direction of the curl. The task of computing the curl is simplified if we consider one component of the field at a time and compute the curl corresponding to that component since then it is sufficient if we always maintain the orientation of the surface normal to that component axis. In fact, this is what we did in Section 4.1, which led us to the determinant form of curl.

Physical interpretation of curl

We are now ready to discuss the physical interpretation of the curl. We do this with the aid of a simple device known as the *curl meter* which responds to the circulation of the vector field. Although the curl meter may take several forms, we shall consider one consisting of a circular disk that floats in water with a paddle wheel attached to the bottom of the disk, as shown in Fig. 4.6. A dot at the periphery on top of the disk serves to indicate any rotational motion of the curl meter about its axis (i.e., the axis of the paddle wheel). Let us now consider a stream of rectangular cross section carrying water in the z-direction, as shown in Fig. 4.6(a). Let us assume the velocity **v** of the water to be independent of height but increasing sinusoidally from a value of zero at the banks to a maximum value v_0 at the center, as shown in Fig. 4.6(b), and investigate the behavior of the curl meter when it is placed vertically at different points in the stream. We assume that the size of the curl meter is vanishingly small so that it does not disturb the flow of water as we probe its behavior at different points.

Since exactly in midstream the blades of the paddle wheel lying on either side of the centerline are hit by the same velocities, the paddle wheel does not rotate. The curl meter simply slides down the stream without any rotational motion, that is, with the dot on top of the disk maintaining the same position relative to the center of the disk, as shown in Fig. 4.6(c). At a point to the left of the midstream the blades of the paddle wheel are hit by a greater velocity on the right side than on the left side so that the paddle wheel rotates in the counterclockwise sense, as seen looking along the positive y-axis. The curl meter rotates in the counterclockwise direction about its axis as it slides down the stream, as indicated by the changing position of the dot on top of the disk relative to the center of the disk, as shown in Fig. 4.6(d). At a point to the right of midstream, the blades of

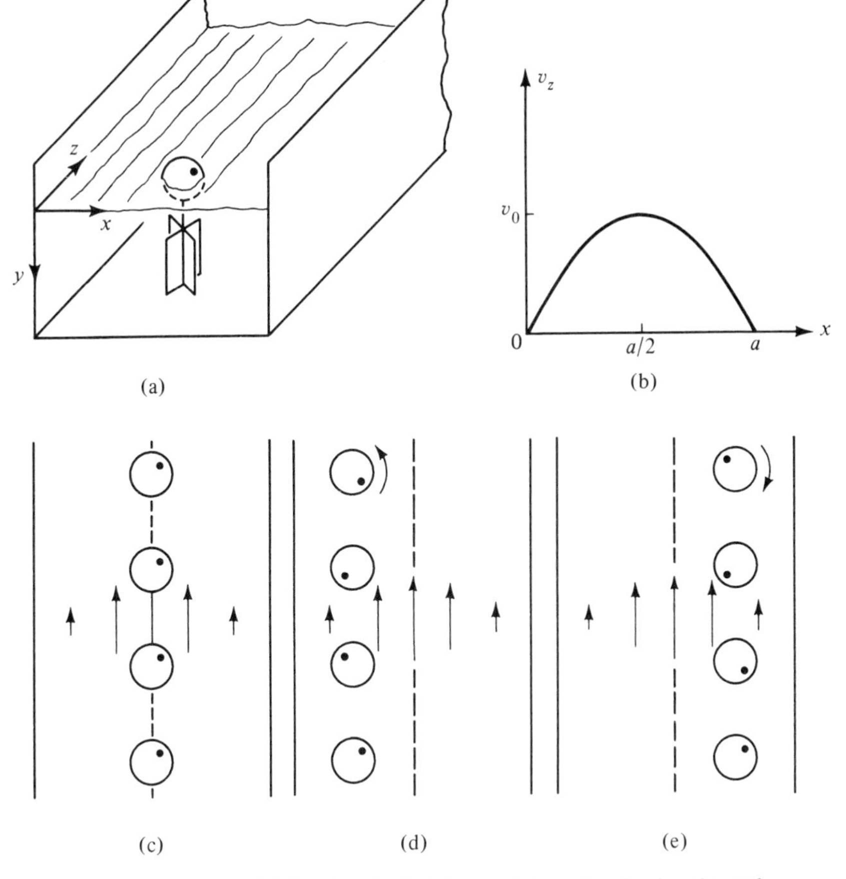

Figure 4.6. For explaining the physical interpretation of curl using the curl meter.

the paddle wheel are hit by a greater velocity on the left side than on the right side so that the paddle wheel rotates in the clockwise sense, as seen looking along the positive y-axis. The curl meter rotates in the clockwise direction about its axis as it slides down the stream, as indicated by the changing position of the dot on top of the disk relative to the center of the disk, as shown in Fig 4.6(e).

If we now pick up the curl meter and insert it in the water with its axis parallel to the x-axis, the curl meter does not rotate because its blades are hit with the same force above and below its axis. If the curl meter is inserted in the water with its axis parallel to the z-axis, it does not rotate since the water flow is then parallel to the blades.

To relate the behavior of the curl meter with the curl of the velocity vector field of the water flow, we note that since the velocity vector is given by

$$\mathbf{v} = v_z(x)\mathbf{i}_z = v_0 \sin \frac{\pi x}{a} \mathbf{i}_z$$

its curl is given by

$$\nabla \times \mathbf{v} = \begin{vmatrix} \mathbf{i}_x & \mathbf{i}_y & \mathbf{i}_z \\ \frac{\partial}{\partial x} & \frac{\partial}{\partial y} & \frac{\partial}{\partial z} \\ 0 & 0 & v_z \end{vmatrix}$$

$$= -\frac{\partial v_z}{\partial x}\mathbf{i}_y$$

$$= -\frac{\pi v_0}{a}\cos\frac{\pi x}{a}\,\mathbf{i}_y$$

Therefore the x- and z-components of the curl are zero, whereas the y-component is nonzero varying with x in a cosinusoidal manner, from negative values left of midstream, to zero at midstream, to positive values right of midstream. Thus no rotation of the curl meter corresponds to zero value for the component of the curl along its axis. Rotation of the curl meter in the counterclockwise or left-hand sense as seen looking along its axis corresponds to a nonzero negative value, and rotation in the clockwise or right-hand sense corresponds to a nonzero positive value for the component of the curl. It can further be visualized that the rate of rotation of the curl meter is a measure of the magnitude of the pertinent nonzero component of the curl.

The foregoing illustration of the physical interpretation of the curl of a vector field can be used to visualize the behavior of electric and magnetic fields. Thus, from

$$\nabla \times \mathbf{E} = -\frac{\partial \mathbf{B}}{\partial t}$$

we know that at a point in an electromagnetic field, the circulation of the electric field per unit area in a given plane is equal to the component of $-\partial \mathbf{B}/\partial t$ along the unit vector normal to that plane and directed in the right-hand sense. Similarly, from

$$\nabla \times \mathbf{H} = \mathbf{J} + \frac{\partial \mathbf{D}}{\partial t}$$

we know that at a point in an electromagnetic field, the circulation of the magnetic field per unit area in a given plane is equal to the component of $\mathbf{J} + \partial \mathbf{D}/\partial t$ along the unit vector normal to that plane and directed in the right-hand sense.

Divergence, basic definition

Turning now to the discussion of divergence, let us consider Gauss' law for the electric field in differential form; that is,

$$\nabla \cdot \mathbf{D} = \rho \tag{4.51}$$

We wish to express $\nabla \cdot \mathbf{D}$ at a point in the charge region in terms of $\mathbf{D}$ at that point. If we consider an infinitesimal volume Δv at that point and multiply both sides of (4.51) by Δv, we get

$$(\nabla \cdot \mathbf{D})\,\Delta v = \rho\,\Delta v \tag{4.52}$$

But $\rho\,\Delta v$ is simply the charge contained in the volume Δv, and according to

Gauss' law for the electric field in integral form,

$$\oint_S \mathbf{D} \cdot d\mathbf{S} = \rho \, \Delta v \tag{4.53}$$

where S is the closed surface bounding Δv. Comparing (4.52) and (4.53), we have

$$(\nabla \cdot \mathbf{D}) \, \Delta v = \oint_S \mathbf{D} \cdot d\mathbf{S} \tag{4.54}$$

Dividing both sides of (4.54) by Δv, we obtain

$$\nabla \cdot \mathbf{D} = \frac{\oint_S \mathbf{D} \cdot d\mathbf{S}}{\Delta v} \tag{4.55}$$

This result is however approximate since (4.54) is exact only in the limit that Δv tends to zero. Thus

$$\nabla \cdot \mathbf{D} = \lim_{\Delta v \to 0} \frac{\oint_S \mathbf{D} \cdot d\mathbf{S}}{\Delta v} \tag{4.56}$$

which is the expression for $\nabla \cdot \mathbf{D}$ at a point in terms of $\mathbf{D}$ at that point. Although we have derived this for the $\mathbf{D}$ vector, it is a general result and, in fact, is often the starting point for the introduction of divergence. Thus for any vector field, $\mathbf{A}$,

$$\boxed{\nabla \cdot \mathbf{A} = \lim_{\Delta v \to 0} \frac{\oint_S \mathbf{A} \cdot d\mathbf{S}}{\Delta v}} \tag{4.57}$$

Equation (4.57) tells us that to find the divergence of a vector at a point in that vector field, we first consider an infinitesimal volume at that point and compute the surface integral of the vector over the surface bounding that volume, that is, the outward flux of the vector field from that volume. We then divide the flux by the volume to obtain the flux per unit volume. Since we need this flux per unit volume in the limit that the volume tends to zero, we do this by gradually shrinking the volume. The limiting value to which the flux per unit volume approaches is the value of the divergence of the vector field at the point to which the volume is shrunk.

Physical interpretation of divergence

We are now ready to discuss the physical interpretation of the divergence. To simplify this task, we shall consider the continuity equation given by

$$\nabla \cdot \mathbf{J} = -\frac{\partial \rho}{\partial t} \tag{4.58}$$

Let us investigate three different cases: (1) positive value, (2) negative value, and (3) zero value of the time rate of decrease of the charge density at a point, that is, the divergence of the current density vector at that point. We shall do this with the aid of a simple device which we shall call the *divergence meter*. The divergence meter can be imagined to be a tiny, elastic balloon enclosing the point and that expands when hit by charges streaming outward from the point and contracts when acted upon by charges streaming inward toward the point. For case 1, that is, when the time rate of decrease of the charge density at the point is positive, there is a net amount of charge streaming out of the point in a given time, resulting in a net current flow outward from the point that will make the imaginary balloon expand. For case 2, that is, when the time rate of decrease of the charge density at the point is negative or the time rate of increase of the charge density is positive, there is a net amount of charge streaming toward the point in a given time, resulting in a net current flow toward the point and the imaginary balloon will contract. For case 3, that is, when the time rate of decrease of the charge density at the point is zero, the balloon will remain unaffected since the charge is streaming out of the point at exactly the same rate as it is streaming into the point. The situation corresponding to case 1 is illustrated in Fig. 4.7(a) and (b), whereas that corresponding to case 2 is illustrated in Fig. 4.7(c) and (d), and that corresponding to case 3 is illustrated in Fig. 4.7(e). Note that in Fig. 4.7(a), (c), and (e), the imaginary balloon slides along the lines of current flow while responding to the divergence by expanding, contracting, or remaining unaffected.

Generalizing the foregoing discussion to the physical interpretation of the divergence of any vector field at a point, we can imagine the vector field to be a velocity field of streaming charges acting upon the divergence meter and obtain in most cases a qualitative picture of the divergence of the vector field. If the divergence meter expands, the divergence is positive and a source of the flux of the vector field exists at that point. If the divergence meter contracts, the divergence is negative and a sink of the flux of the vector field exists at that point. It can be further visualized that the rate of expansion or contraction of the divergence meter is a measure of the magnitude of the divergence. If the divergence meter remains unaffected, the divergence is zero, and neither a source nor a sink of the

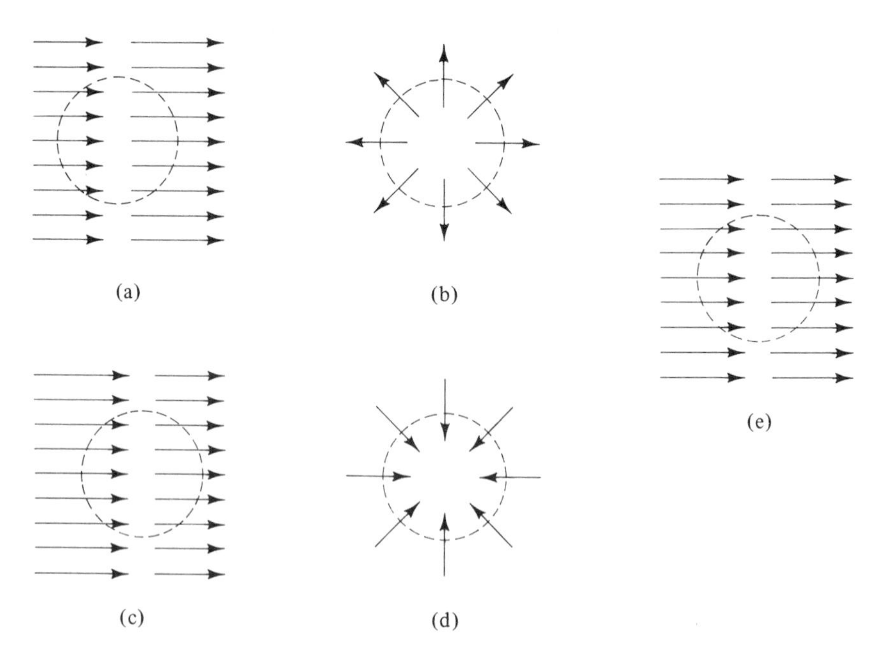

Figure 4.7. For explaining the physical interpretation of divergence using the divergence meter.

flux of the vector field exists at that point; alternatively, there can exist at the point pairs of sources and sinks of equal strengths.

We shall now derive two useful theorems in vector calculus, *Stokes' theorem* and the *divergence theorem*. Stokes' theorem relates the closed line integral of a vector field to the surface integral of the curl of that vector field, whereas the divergence theorem relates the closed surface integral of a vector field to the volume integral of the divergence of that vector field.

Stokes' theorem

To derive Stokes' theorem, let us consider an arbitrary surface S in a magnetic field region and divide this surface into a number of infinitesimal surfaces $\Delta S_1, \Delta S_2, \Delta S_3, \ldots$, bounded by the contours $C_1, C_2, C_3, \ldots$, respectively. Then, applying (4.45) to each one of these infinitesimal surfaces and adding up, we get

$$\sum_j (\nabla \times \mathbf{H})_j \cdot \Delta S_j \, \mathbf{i}_{nj} = \oint_{C_1} \mathbf{H} \cdot d\mathbf{l} + \oint_{C_2} \mathbf{H} \cdot d\mathbf{l} + \cdots \qquad (4.59)$$

where $\mathbf{i}_{nj}$ are unit vectors normal to the surfaces ΔS_j chosen in accordance with the right-hand screw rule. In the limit that the number of infinitesimal surfaces tends to infinity, the left side of (4.59) approaches to the surface integral of $\nabla \times \mathbf{H}$ over the surface S. The right side of (4.59) is simply the closed line integral of $\mathbf{H}$ around the contour C since the contributions to the line integrals from the portions of the contours interior to C cancel, as shown in Fig. 4.8. Thus we get

$$\int_S (\nabla \times \mathbf{H}) \cdot d\mathbf{S} = \oint_C \mathbf{H} \cdot d\mathbf{l} \qquad (4.60)$$

Equation (4.60) is Stokes' theorem. Although we have derived it by considering the $\mathbf{H}$ field, it is general and can be derived from the application of (4.50) to a geometry such as that in Fig. 4.8. Thus for any vector field $\mathbf{A}$,

$$\boxed{\oint_C \mathbf{A} \cdot d\mathbf{l} = \int_S (\nabla \times \mathbf{A}) \cdot d\mathbf{S}} \qquad (4.61)$$

where S is any surface bounded by C.

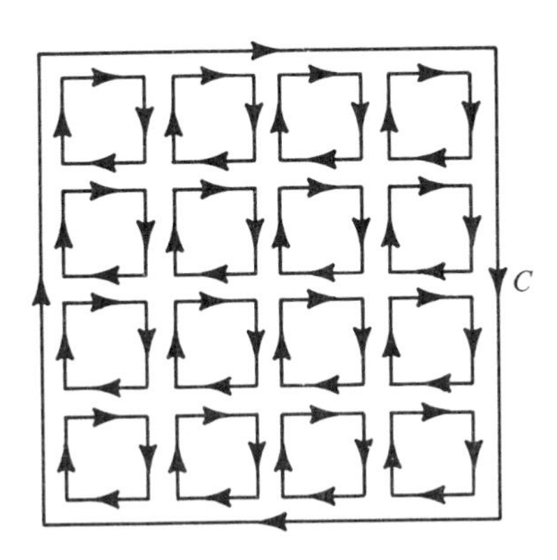

Figure 4.8. For deriving Stokes' theorem.

Example 4.6

Let us evaluate the line integral of Example 3.1 by using Stokes' theorem.

For $\mathbf{F} = x\mathbf{i}_y$,

$$\nabla \times \mathbf{F} = \begin{vmatrix} \mathbf{i}_x & \mathbf{i}_y & \mathbf{i}_z \\ \dfrac{\partial}{\partial x} & \dfrac{\partial}{\partial y} & \dfrac{\partial}{\partial z} \\ 0 & x & 0 \end{vmatrix} = \mathbf{i}_z$$

With reference to Fig. 3.4, we then have

$$\begin{aligned} \oint_{ABCDA} \mathbf{F} \cdot d\mathbf{l} &= \int_{\substack{\text{area} \\ ABCDA}} (\nabla \times \mathbf{F}) \cdot d\mathbf{S} \\ &= \int_{\substack{\text{area} \\ ABCDA}} \mathbf{i}_z \cdot dx\,dy\,\mathbf{i}_z \\ &= \int_{\substack{\text{area} \\ ABCDA}} dx\,dy \\ &= \text{area } ABCDA \\ &= 6 \end{aligned}$$

which agrees with the result obtained in Example 3.1.

Divergence theorem

To derive the divergence theorem, let us consider an arbitrary volume V in an electric field region and divide this volume into a number of infinitesimal volumes $\Delta v_1, \Delta v_2, \Delta v_3, \ldots$, bounded by the surfaces $S_1, S_2, S_3, \ldots$, respectively. Then, applying (4.54) to each one of these infinitesimal volumes and adding up, we get

$$\sum_j (\nabla \cdot \mathbf{D})_j \, \Delta v_j = \oint_{S_1} \mathbf{D} \cdot d\mathbf{S} + \oint_{S_2} \mathbf{D} \cdot d\mathbf{S} + \cdots \tag{4.62}$$

In the limit that the number of the infinitesimal volumes tends to infinity, the left side of (4.62) approaches to the volume integral of $\nabla \cdot \mathbf{D}$ over the volume V. The right side of (4.62) is simply the closed surface integral of $\mathbf{D}$ over S since the contribution to the surface integrals from the portions of the surfaces interior to S cancel, as shown in Fig. 4.9. Thus we get

$$\int_V (\nabla \cdot \mathbf{D})\, dv = \oint_S \mathbf{D} \cdot d\mathbf{S} \tag{4.63}$$

Equation (4.63) is the divergence theorem. Although we have derived it by considering the $\mathbf{D}$ field, it is general and can be derived from the application of (4.57) to a geometry such as that in Fig. 4.9. Thus for any vector field $\mathbf{A}$,

$$\boxed{\oint_S \mathbf{A} \cdot d\mathbf{S} = \int_V (\nabla \cdot \mathbf{A})\, dv} \tag{4.64}$$

where V is the volume bounded by S.

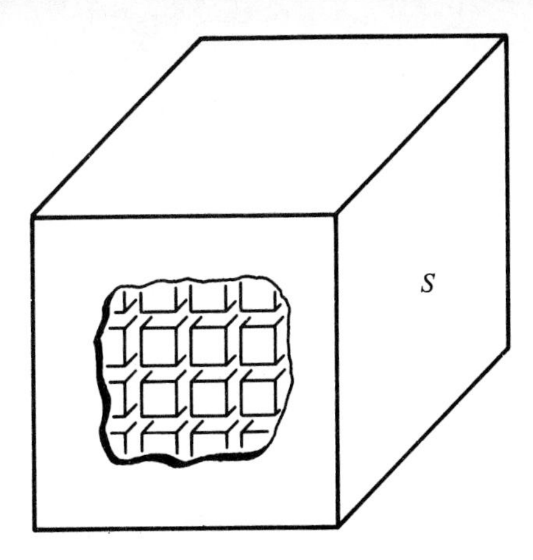

Figure 4.9. For deriving the divergence theorem.

Example 4.7

Divergence of the curl of a vector

By using the Stokes and divergence theorems, show that for any vector field $\mathbf{A}$, $\nabla \cdot \nabla \times \mathbf{A} = 0$.

Let us consider volume V bounded by the closed surface $S_1 + S_2$, where S_1 and S_2 are bounded by the closed paths C_1 and C_2, respectively, as shown in Fig. 4.10. Note that C_1 and C_2 touch each other and are traversed in opposite senses and that $d\mathbf{S}_1$ and $d\mathbf{S}_2$ are directed in the right-hand sense relative to C_1 and C_2, respectively. Then, using divergence and Stokes' theorems in succession, we obtain

$$\begin{aligned}\int_V (\nabla \cdot \nabla \times \mathbf{A})\, dv &= \oint_{S_1+S_2} (\nabla \times \mathbf{A}) \cdot d\mathbf{S} \\ &= \int_{S_1} (\nabla \times \mathbf{A}) \cdot d\mathbf{S}_1 + \int_{S_2} (\nabla \times \mathbf{A}) \cdot d\mathbf{S}_2 \\ &= \oint_{C_1} \mathbf{A} \cdot d\mathbf{l} + \oint_{C_2} \mathbf{A} \cdot d\mathbf{l} \\ &= 0\end{aligned}$$

Since this result holds for any arbitrary volume V, it follows that

$$\boxed{\nabla \cdot \nabla \times \mathbf{A} = 0} \tag{4.65}$$

We shall make use of (4.65) in the following section.

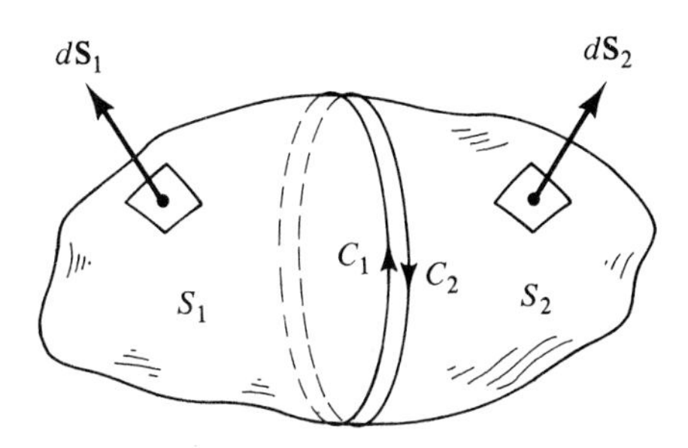

Figure 4.10. For proving the identity $\nabla \cdot \nabla \times \mathbf{A} = 0$.

K4.3. Basic definition of curl; Physical interpretation of curl; Basic definition of divergence; Physical interpretation of divergence; Stokes' theorem; Divergence theorem; Divergence of the curl of a vector.

D4.7. With the aid of the curl meter, determine if the z-component of the curl of the vector field $\mathbf{A} = (x^2 - 4)\mathbf{i}_y$ is positive, zero, or negative at each of the following points: **(a)** (2, −3, 1); **(b)** (0, 2, 4); and **(c)** (−1, 2, −1).
Ans. **(a)** positive; **(b)** zero; **(c)** negative

D4.8. With the aid of the divergence meter, determine if the divergence of the vector field $\mathbf{A} = (x - 2)^2\mathbf{i}_x$ is positive, zero, or negative at each of the following points: **(a)** (2, 4, 3); **(b)** (1, 1, −1); and **(c)** (3, −1, 4).
Ans. **(a)** zero; **(b)** negative; **(c)** positive

D4.9. Using Stokes' theorem, find the absolute value of the line integral of the vector field $(x\mathbf{i}_y + \sqrt{3}y\mathbf{i}_z)$ around each of the following closed paths: **(a)** the perimeter of a square of sides 2 m lying in the xy-plane; **(b)** a circular path of radius $1/\sqrt{\pi}$ m lying in the xy-plane; and **(c)** the perimeter of an equilateral triangle of sides 2 m lying in the yz-plane.
Ans. **(a)** 4; **(b)** 1; **(c)** 3

D4.10. Using the divergence theorem, find the surface integral of the vector field $(x\mathbf{i}_x + y\mathbf{i}_y + z\mathbf{i}_z)$ over each of the following closed surfaces: **(a)** the surface of a cube of sides 1 m; **(b)** the surface of a cylinder of radius $1/\sqrt{\pi}$ m and length 2 m; and **(c)** the surface of a sphere of radius $1/(\pi)^{1/3}$ m.
Ans. **(a)** 3; **(b)** 6; **(c)** 4

4.4 GRADIENT, LAPLACIAN, AND THE POTENTIAL FUNCTIONS

Magnetic vector potential

In Example 4.7 we showed that for any vector $\mathbf{A}$, $\nabla \cdot \nabla \times \mathbf{A} = 0$ (for alternate proof, see Problem P4.16). It then follows from Gauss' law for the magnetic field in differential form, $\nabla \cdot \mathbf{B} = 0$, that the magnetic flux density vector $\mathbf{B}$ can be expressed as the curl of another vector $\mathbf{A}$; that is,

$$\boxed{\mathbf{B} = \nabla \times \mathbf{A}} \tag{4.66}$$

The vector $\mathbf{A}$ in (4.66) is known as the *magnetic vector potential.*

Gradient

Substituting (4.66) into Faraday's law in differential form, $\nabla \times \mathbf{E} = -\partial\mathbf{B}/\partial t$, and rearranging, we then obtain

$$\nabla \times \mathbf{E} + \frac{\partial}{\partial t}(\nabla \times \mathbf{A}) = \mathbf{0}$$

or

$$\nabla \times \left(\mathbf{E} + \frac{\partial \mathbf{A}}{\partial t}\right) = \mathbf{0} \tag{4.67}$$

If the curl of a vector is equal to the null vector, that vector can be expressed as the *gradient* of a scalar, since the curl of the gradient of a scalar function is identically equal to the null vector. The gradient of a scalar, say, Φ, denoted $\nabla\Phi$ (del Φ) is defined in such a manner that the increment $d\Phi$ in Φ from a point P to a neighboring point Q is given by

$$d\Phi = \nabla\Phi \cdot d\mathbf{l} \tag{4.68}$$

where $d\mathbf{l}$ is the differential length vector from P to Q. Applying Stokes' theorem to the vector $\nabla \times \nabla\Phi$ and a surface S bounded by closed path C, we then have

$$\begin{aligned}\int_S (\nabla \times \nabla\Phi) \cdot d\mathbf{S} &= \oint_C \nabla\Phi \cdot d\mathbf{l} \\ &= \oint_C d\Phi \\ &= 0\end{aligned} \tag{4.69}$$

for any single-valued function Φ. Since (4.69) holds for an arbitrary S, it follows that

$$\boxed{\nabla \times \nabla\Phi = \mathbf{0}} \tag{4.70}$$

To obtain the expression for the gradient in the Cartesian coordinate system, we write

$$\begin{aligned}d\Phi &= \frac{\partial\Phi}{\partial x}dx + \frac{\partial\Phi}{\partial y}dy + \frac{\partial\Phi}{\partial z}dz \\ &= \left(\frac{\partial\Phi}{\partial x}\mathbf{i}_x + \frac{\partial\Phi}{\partial y}\mathbf{i}_y + \frac{\partial\Phi}{\partial z}\mathbf{i}_z\right) \cdot (dx\,\mathbf{i}_x + dy\,\mathbf{i}_y + dz\,\mathbf{i}_z)\end{aligned} \tag{4.71}$$

Then comparing with (4.68), we observe that

$$\boxed{\nabla\Phi = \frac{\partial\Phi}{\partial x}\mathbf{i}_x + \frac{\partial\Phi}{\partial y}\mathbf{i}_y + \frac{\partial\Phi}{\partial z}\mathbf{i}_z} \tag{4.72}$$

Note that the right side of (4.72) is simply the vector obtained by applying the del operator to the scalar function Φ. It is for this reason that the gradient of Φ is written as $\nabla\Phi$. Expressions for the gradient in cylindrical and spherical coordinate systems are derived in Appendix A. These are as follows:

Cylindrical:

$$\nabla\Phi = \frac{\partial\Phi}{\partial r}\mathbf{i}_r + \frac{1}{r}\frac{\partial\Phi}{\partial\phi}\mathbf{i}_\phi + \frac{\partial\Phi}{\partial z}\mathbf{i}_z \tag{4.73a}$$

Spherical:

$$\nabla\Phi = \frac{\partial\Phi}{\partial r}\mathbf{i}_r + \frac{1}{r}\frac{\partial\Phi}{\partial\theta}\mathbf{i}_\theta + \frac{1}{r\sin\theta}\frac{\partial\Phi}{\partial\phi}\mathbf{i}_\phi \tag{4.73b}$$

Physical interpretation of gradient

To discuss the physical interpretation of the gradient, let us consider a surface on which Φ is equal to a constant, say, Φ_0, and a point P on that surface, as shown in Fig. 4.11(a). If we now consider another point Q_1 on the same surface and an infinitesimal distance away from P, $d\Phi$ between these two points is zero since Φ is constant on the surface. Thus for the vector $d\mathbf{l}_1$ drawn from P to Q_1,

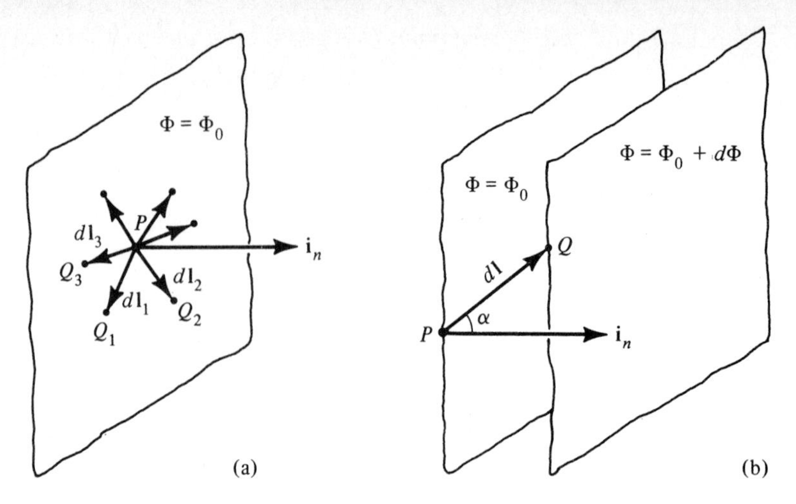

Figure 4.11. For discussing the physical interpretation of the gradient of a scalar function.

$[\nabla\Phi]_P \cdot d\mathbf{l}_1 = 0$ and hence $[\nabla\Phi]_P$ is perpendicular to $d\mathbf{l}_1$. Since this is true for all points $Q_1, Q_2, Q_3, \ldots$ on the constant Φ surface, it follows that $[\nabla\Phi]_P$ must be normal to all possible infinitesimal length vectors $d\mathbf{l}_1, d\mathbf{l}_2, d\mathbf{l}_3, \ldots$ drawn at P and hence is normal to the surface. Denoting $\mathbf{i}_n$ to be the unit normal vector to the surface at P, we then have

$$[\nabla\Phi]_P = |\nabla\Phi|_P \mathbf{i}_n \tag{4.74}$$

Let us now consider two surfaces on which Φ is constant, having values Φ_0 and $\Phi_0 + d\Phi$, as shown in Fig. 4.11(b). Let P and Q be points on the $\Phi = \Phi_0$ and $\Phi = \Phi_0 + d\Phi$ surfaces, respectively, and $d\mathbf{l}$ be the vector drawn from P to Q. Then from (4.68) and (4.74),

$$\begin{aligned} d\Phi &= [\nabla\Phi]_P \cdot d\mathbf{l} \\ &= |\nabla\Phi|_P \mathbf{i}_n \cdot d\mathbf{l} \\ &= |\nabla\Phi|_P \, dl \cos\alpha \end{aligned}$$

where α is the angle between $\mathbf{i}_n$ at P and $d\mathbf{l}$. Thus

$$|\nabla\Phi|_P = \frac{d\Phi}{dl \cos\alpha} \tag{4.75}$$

Since $dl \cos\alpha$ is the distance between the two surfaces along $\mathbf{i}_n$ and hence is the shortest distance between them, it follows that $|\nabla\Phi|_P$ is the maximum rate of increase of Φ at the point P. Thus the gradient of a scalar function Φ at a point is a vector having magnitude equal to the maximum rate of increase of Φ at that point and is directed along the direction of the maximum rate of increase, which is normal to the constant Φ surface passing through that point; that is,

$$\boxed{\nabla\Phi = \frac{d\Phi}{dn}\mathbf{i}_n} \tag{4.76}$$

where dn is a differential length along $\mathbf{i}_n$. The concept of the gradient of a scalar function we just discussed is often utilized to find a unit vector normal to a given surface. We shall illustrate this by means of an example.

Example 4.8

Let us find the unit vector normal to the surface $y = x^2$ at the point (2, 4, 1) by using the concept of the gradient of a scalar.

Writing the equation for the surface as

$$x^2 - y = 0$$

we note that the scalar function that is constant on the surface is given by

$$\Phi(x, y, z) = x^2 - y$$

The gradient of the scalar function is then given by

$$\begin{aligned}\nabla\Phi &= \nabla(x^2 - y)\\ &= \frac{\partial(x^2 - y)}{\partial x}\mathbf{i}_x + \frac{\partial(x^2 - y)}{\partial y}\mathbf{i}_y + \frac{\partial(x^2 - y)}{\partial z}\mathbf{i}_z\\ &= 2x\mathbf{i}_x - \mathbf{i}_y\end{aligned}$$

The value of the gradient at the point (2, 4, 1) is $[2(2)\mathbf{i}_x - \mathbf{i}_y] = (4\mathbf{i}_x - \mathbf{i}_y)$. Thus the required unit vector is

$$\mathbf{i}_n = \pm\frac{4\mathbf{i}_x - \mathbf{i}_y}{|4\mathbf{i}_x - \mathbf{i}_y|} = \pm\left(\frac{4}{\sqrt{17}}\mathbf{i}_x - \frac{1}{\sqrt{17}}\mathbf{i}_y\right)$$

Electric scalar potential

Returning now to (4.67) we write

$$\mathbf{E} + \frac{\partial\mathbf{A}}{\partial t} = -\nabla\Phi \tag{4.77}$$

where we have chosen the scalar to be $-\Phi$, the reason for the minus sign to be explained in Section 4.5. Rearranging (4.77), we obtain

$$\boxed{\mathbf{E} = -\nabla\Phi - \frac{\partial\mathbf{A}}{\partial t}} \tag{4.78}$$

The quantity Φ in (4.78) is known as the electric scalar potential.

Electromagnetic potentials

The electric scalar potential Φ and the magnetic vector potential $\mathbf{A}$ are known as the electromagnetic potentials. As we shall show later in this section, the electric scalar potential is related to the source charge density ρ, whereas the magnetic vector potential is related to the source current density $\mathbf{J}$. For the time-varying case, the two are not independent since the charge and current densities are related through the continuity equation. For a given $\mathbf{J}$, it is sufficient to determine $\mathbf{A}$, since $\mathbf{B}$ can be found from (4.66) and then $\mathbf{E}$ can be found by using Ampère's circuital law $\nabla \times \mathbf{H} = \mathbf{J} + \partial\mathbf{D}/\partial t$. For static fields, that is, for $\partial/\partial t = 0$, the two potentials are independent. Equation (4.66) remains unaltered, whereas (4.78) reduces to $\mathbf{E} = -\nabla\Phi$. We shall consider the static field case in Section 4.5.

To proceed further, we recall that the Maxwell's equations in differential form are given by

$$\nabla \times \mathbf{E} = -\frac{\partial \mathbf{B}}{\partial t} \tag{4.79a}$$

$$\nabla \times \mathbf{H} = \mathbf{J} + \frac{\partial \mathbf{D}}{\partial t} \tag{4.79b}$$

$$\nabla \cdot \mathbf{D} = \rho \tag{4.79c}$$

$$\nabla \cdot \mathbf{B} = 0 \tag{4.79d}$$

From (4.79d), we expressed **B** in the manner

$$\mathbf{B} = \nabla \times \mathbf{A} \tag{4.80}$$

and then from (4.79a), we obtained

$$\mathbf{E} = -\nabla\Phi - \frac{\partial \mathbf{A}}{\partial t} \tag{4.81}$$

We now substitute (4.81) and (4.80) into (4.79c) and (4.79b), respectively, to obtain

$$\nabla \cdot \left(-\nabla\Phi - \frac{\partial \mathbf{A}}{\partial t}\right) = \frac{\rho}{\varepsilon} \tag{4.82a}$$

$$\nabla \times \nabla \times \mathbf{A} - \mu\varepsilon\frac{\partial}{\partial t}\left(-\nabla\Phi - \frac{\partial \mathbf{A}}{\partial t}\right) = \mu\mathbf{J} \tag{4.82b}$$

Laplacian of a scalar

We now define the Laplacian of a scalar quantity Φ, denoted $\nabla^2\Phi$ (del squared Φ) as

$$\boxed{\nabla^2\Phi = \nabla \cdot \nabla\Phi} \tag{4.83}$$

In Cartesian coordinates,

$$\nabla\Phi = \frac{\partial\Phi}{\partial x}\mathbf{i}_x + \frac{\partial\Phi}{\partial y}\mathbf{i}_y + \frac{\partial\Phi}{\partial z}\mathbf{i}_z$$

$$\nabla \cdot \mathbf{A} = \frac{\partial A_x}{\partial x} + \frac{\partial A_y}{\partial y} + \frac{\partial A_z}{\partial z}$$

so that

$$\nabla^2\Phi = \frac{\partial}{\partial x}\left(\frac{\partial\Phi}{\partial x}\right) + \frac{\partial}{\partial y}\left(\frac{\partial\Phi}{\partial y}\right) + \frac{\partial}{\partial z}\left(\frac{\partial\Phi}{\partial z}\right)$$

or

$$\boxed{\nabla^2\Phi = \frac{\partial^2\Phi}{\partial x^2} + \frac{\partial^2\Phi}{\partial y^2} + \frac{\partial^2\Phi}{\partial z^2}} \tag{4.84}$$

Note that the Laplacian of a scalar is a scalar quantity. Expressions for the Laplacian of a scalar in cylindrical and spherical coordinates are derived in Appendix A. These are as follows:

Cylindrical:

$$\nabla^2\Phi = \frac{1}{r}\frac{\partial}{\partial r}\left(r\frac{\partial\Phi}{\partial r}\right) + \frac{1}{r^2}\frac{\partial^2\Phi}{\partial\phi^2} + \frac{\partial^2\Phi}{\partial z^2} \tag{4.85a}$$

Spherical:

$$\nabla^2\Phi = \frac{1}{r^2}\frac{\partial}{\partial r}\left(r^2\frac{\partial\Phi}{\partial r}\right) + \frac{1}{r^2\sin\theta}\frac{\partial}{\partial\theta}\left(\sin\theta\frac{\partial\Phi}{\partial\theta}\right) + \frac{1}{r^2\sin^2\theta}\frac{\partial^2\Phi}{\partial\phi^2} \tag{4.85b}$$

Before proceeding further, it is interesting to note that the four vector differential operations which we have learned thus far in this chapter are such that

The *curl* of a *vector* is a *vector*.
The *divergence* of a *vector* is a *scalar*.
The *gradient* of a *scalar* is a *vector*.
The *Laplacian* of a *scalar* is a *scalar*.

Thus all four combinations of vector and scalar are involved in the four operations.

Laplacian of a vector

Next, we define the Laplacian of a vector, denoted $\nabla^2\mathbf{A}$ as

$$\boxed{\nabla^2\mathbf{A} = \nabla(\nabla\cdot\mathbf{A}) - \nabla\times\nabla\times\mathbf{A}} \tag{4.86}$$

Expanding the right side of (4.86) in Cartesian coordinates and simplifying (see Problem P4.21), we obtain in the Cartesian coordinate system,

$$\boxed{\nabla^2\mathbf{A} = (\nabla^2A_x)\mathbf{i}_x + (\nabla^2A_y)\mathbf{i}_y + (\nabla^2A_z)\mathbf{i}_z} \tag{4.87}$$

Thus in the Cartesian coordinate system, the Laplacian of a vector is a vector whose components are the Laplacians of the corresponding components of $\mathbf{A}$. It should however be cautioned that this simple observation does not hold in the cylindrical and spherical coordinate systems (see, e.g., Problem P4.22).

Using (4.83) and (4.86), we now write (4.82a) and (4.82b) as

$$\nabla^2\Phi + \frac{\partial}{\partial t}(\nabla\cdot\mathbf{A}) = -\frac{\rho}{\varepsilon} \tag{4.88a}$$

$$\nabla^2\mathbf{A} - \nabla\left(\nabla\cdot\mathbf{A} + \mu\varepsilon\frac{\partial\Phi}{\partial t}\right) - \mu\varepsilon\frac{\partial^2\mathbf{A}}{\partial t^2} = -\mu\mathbf{J} \tag{4.88b}$$

Potential function equations

Equations (4.88a) and (4.88b) are a pair of coupled differential equations for Φ and $\mathbf{A}$. To uncouple the equations, we make use of a theorem known as Helmholtz's theorem which states that a vector field is completely specified by its curl and divergence. Therefore, since the curl of $\mathbf{A}$ is given by (4.80), we are at liberty to specify the divergence of $\mathbf{A}$. We do this by setting

$$\boxed{\nabla \cdot \mathbf{A} = -\mu\varepsilon \frac{\partial \Phi}{\partial t}} \tag{4.89}$$

which is known as the Lorentz condition. This uncouples (4.88a) and (4.88b) to give us

$$\nabla^2\Phi - \mu\varepsilon \frac{\partial^2 \Phi}{\partial t^2} = -\frac{\rho}{\varepsilon} \tag{4.90}$$

$$\nabla^2\mathbf{A} - \mu\varepsilon \frac{\partial^2 \mathbf{A}}{\partial t^2} = -\mu \mathbf{J} \tag{4.91}$$

These are the differential equations relating the electromagnetic potentials Φ and $\mathbf{A}$ to the source charge and current densities ρ and $\mathbf{J}$, respectively.

Before proceeding further, we shall show that the continuity equation is implied by the Lorentz condition. To do this, we take the Laplacian of both sides of (4.89). We then have

$$\nabla^2(\nabla \cdot \mathbf{A}) = -\mu\varepsilon\, \nabla^2 \frac{\partial \Phi}{\partial t}$$

or

$$\nabla \cdot \nabla^2\mathbf{A} = -\mu\varepsilon \frac{\partial}{\partial t} \nabla^2\Phi \tag{4.92}$$

Substituting for $\nabla^2\mathbf{A}$ and $\nabla^2\Phi$ in (4.92) from (4.91) and (4.90), respectively, we get

$$\nabla \cdot \left(\mu\varepsilon \frac{\partial^2 \mathbf{A}}{\partial t^2} - \mu \mathbf{J}\right) = -\mu\varepsilon \frac{\partial}{\partial t}\left(\mu\varepsilon \frac{\partial^2 \Phi}{\partial t^2} - \frac{\rho}{\varepsilon}\right)$$

or

$$\mu\varepsilon \frac{\partial^2}{\partial t^2}\left(\nabla \cdot \mathbf{A} + \mu\varepsilon \frac{\partial \Phi}{\partial t}\right) = \mu\left(\nabla \cdot \mathbf{J} + \frac{\partial \rho}{\partial t}\right) \tag{4.93}$$

Thus by assuming the Lorentz condition (4.89), we imply $\nabla \cdot \mathbf{J} + \partial\rho/\partial t = 0$, which is the continuity equation.

As pointed out in the previous section, it is sufficient to determine $\mathbf{A}$ for the time-varying case for a given $\mathbf{J}$. Hence we shall be concerned only with (4.91), which we shall refer to in Section 10.1 in connection with obtaining the electromagnetic field due to an elemental antenna.

K4.4. Magnetic vector potential; Gradient of a scalar; Physical interpretation of gradient; Electric scalar potential; Laplacian of a scalar; Potential function equations.

D4.11. Find the outward pointing unit vectors normal to the closed surface $2x^2 + 2y^2 + z^2 = 8$ at the following points: **(a)** $(\sqrt{2}, \sqrt{2}, 0)$; **(b)** $(1, 1, 2)$; and **(c)** $(1, \sqrt{2}, \sqrt{2})$.
Ans. **(a)** $\dfrac{\mathbf{i}_x + \mathbf{i}_y}{\sqrt{2}}$; **(b)** $\dfrac{\mathbf{i}_x + \mathbf{i}_y + \mathbf{i}_z}{\sqrt{3}}$ **(c)** $\dfrac{\sqrt{2}\mathbf{i}_x + 2\mathbf{i}_y + \mathbf{i}_z}{\sqrt{7}}$

D4.12. Two scalar functions are given by

$$\Phi_1(x, y, z) = x^2 + y^2 + z^2$$

$$\Phi_2(x, y, z) = x + 2y + 2z$$

Find the following at the point (3, 4, 12): **(a)** the maximum rate of increase of Φ_1; **(b)** the maximum rate of increase of Φ_2; and **(c)** the rate of increase of Φ_1 along the direction of the maximum rate of increase of Φ_2.
Ans. **(a)** 26; **(b)** 3; **(c)** $23\frac{1}{3}$

D4.13. Find the Laplacians of the following functions: **(a)** x^2yz^3; **(b)** $\frac{1}{r}\sin\phi$ in cylindrical coordinates; and **(c)** $r^2\cos\theta$ in spherical coordinates.
Ans. **(a)** $2yz^3 + 6x^2yz$; **(b)** 0; **(c)** $4\cos\theta$

4.5 POTENTIAL FUNCTIONS FOR STATIC FIELDS

Potential difference

As already pointed out in the preceding section, Eq. (4.78) reduces to

$$\boxed{\mathbf{E} = -\nabla\Phi} \tag{4.94}$$

for the static field case. We observe from (4.94) that the potential function Φ then is such that the electric field lines are orthogonal to the equipotential surfaces, that is, to the surfaces on which the potential remains constant, as shown in Fig. 4.12. If we consider two such equipotential surfaces corresponding to $\Phi = \Phi_A$ and $\Phi = \Phi_B$, as shown in the figure, the potential difference $\Phi_A - \Phi_B$ is given, according to the definition of the gradient, by

$$\begin{aligned}\Phi_A - \Phi_B &= \int_B^A d\Phi = \int_B^A \nabla\Phi \cdot d\mathbf{l} \\ &= -\int_A^B \nabla\Phi \cdot d\mathbf{l}\end{aligned} \tag{4.95}$$

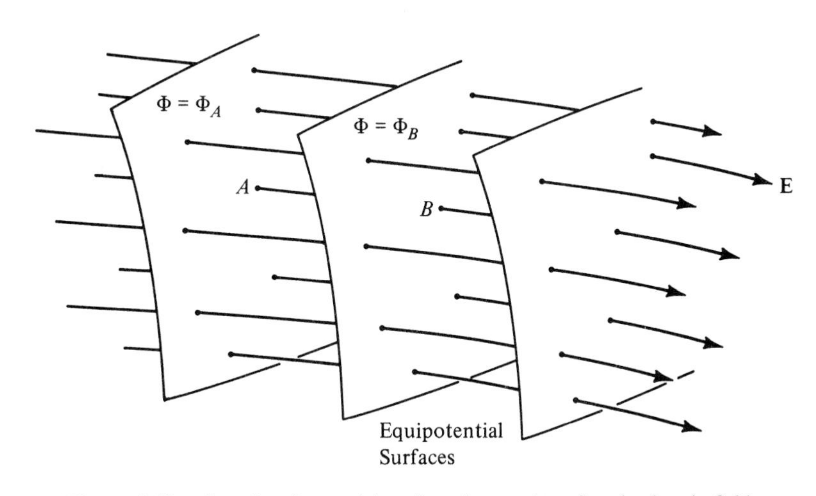

Figure 4.12. Set of equipotential surfaces in a region of static electric field.

Using (4.94), we obtain

$$\boxed{\Phi_A - \Phi_B = \int_A^B \mathbf{E} \cdot d\mathbf{l}} \tag{4.96}$$

We now recall from Section 3.1 that $\int_A^B \mathbf{E} \cdot d\mathbf{l}$ is the voltage between points A and B. Thus the potential difference the static field case has the same meaning as the voltage. The reason for the minus sign in (4.78) and hence in (4.94) is now evident, since without it the voltage between A and B would be the negative of the potential difference between A and B.

Potential difference versus voltage

Before proceeding further we recall that the voltage between two points A and B in a time-varying electric field is in general dependent on the path followed from A to B to evaluate $\int_A^B \mathbf{E} \cdot d\mathbf{l}$ since according to Faraday's law

$$\oint_C \mathbf{E} \cdot d\mathbf{l} = -\frac{d}{dt}\int_S \mathbf{B} \cdot d\mathbf{S}$$

is not in general equal to zero. On the other hand, the potential difference (or voltage) between two points A and B in a static electric field is independent of the path followed from A to B to evaluate $\int_A^B \mathbf{E} \cdot d\mathbf{l}$ since for static fields,

$$\oint_C \mathbf{E} \cdot d\mathbf{l} = 0$$

Thus the potential difference (or voltage) between two points in a static electric field has a unique value. Since the potential difference and voltage have the same meaning for static fields, we shall hereafter replace Φ in (4.94) by V, thereby writing

$$\boxed{\mathbf{E} = -\nabla V} \tag{4.97}$$

Electric potential due to a point charge

Let us now consider the electric field of a point charge and investigate the electric potential due to the point charge. To do this, we recall that the electric field intensity due to a point charge Q is directed radially away from the point charge and its magnitude is $Q/4\pi\varepsilon R^2$ where R is the radial distance from the point charge. Since the equipotential surfaces are everywhere orthogonal to the field lines, it then follows that they are spherical surfaces centered at the point charge, as shown by the cross-sectional view in Fig. 4.13. If we now consider

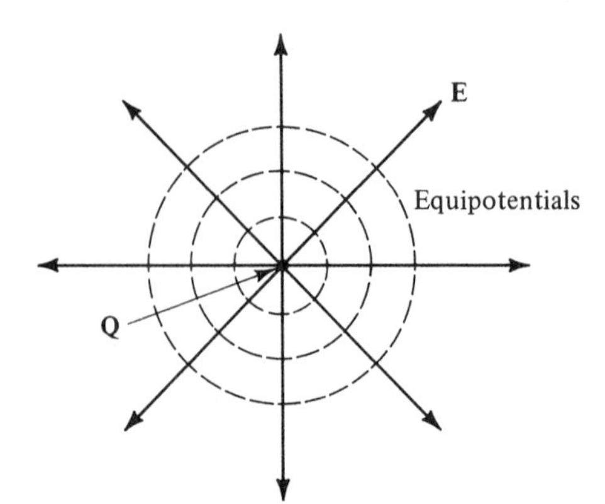

Figure 4.13. Cross-sectional view of equipotential surfaces and electric field lines for a point charge.

two equipotential surfaces of radii R and $R + dR$, the potential drop from the surface of radius R to the surface of radius $R + dR$ is $(Q/4\pi\varepsilon R^2)\,dR$ or, the incremental potential rise dV is given by

$$\begin{aligned} dV &= -\frac{Q}{4\pi\varepsilon R^2}\,dR \\ &= d\left(\frac{Q}{4\pi\varepsilon R} + C\right) \end{aligned} \tag{4.98}$$

where C is a constant. Thus

$$V(R) = \frac{Q}{4\pi\varepsilon R} + C \tag{4.99}$$

Since the potential difference between two points does not depend upon the value of C, we can choose C such that V is zero at some arbitrary reference point. Here we can conveniently set C equal to zero by noting that it is equal to $V(\infty)$ and by choosing $R = \infty$ for the reference point. Thus we obtain the electric potential due to a point charge Q to be

$$\boxed{V = \frac{Q}{4\pi\varepsilon R}} \tag{4.100}$$

We note that the potential drops off inversely with the radial distance away from the point charge.

Equation (4.100) is often the starting point for the computation of the potential field due to static charge distributions and the subsequent determination of the electric field by using (4.97). We shall illustrate this by considering the case of the electric dipole in the following example.

Example 4.9

Electric dipole

As we have learned in Section 2.5, the electric dipole consists of two equal and opposite point charges. Let us consider a static electric dipole consisting of point charges Q and $-Q$ situated on the z-axis at $z = d/2$ and $z = -d/2$, respectively, as shown in Fig. 4.14(a) and find the potential and hence the electric field at a point P far from the dipole.

First we note that in view of the symmetry associated with the dipole around the z-axis, it is convenient to use the spherical coordinate system. Denoting the distance from the point charge Q to P to be r_1 and the distance from the point charge $-Q$ to P to be r_2, we write the expression for the electric potential at P due to the electric dipole as

$$V = \frac{Q}{4\pi\varepsilon r_1} + \frac{-Q}{4\pi\varepsilon r_2} = \frac{Q}{4\pi\varepsilon}\left(\frac{1}{r_1} - \frac{1}{r_2}\right)$$

For a point P far from the dipole, that is, for $r \gg d$, the lines drawn from the two charges to the point are almost parallel. Hence

$$r_1 \approx r - \frac{d}{2}\cos\theta$$

$$r_2 \approx r + \frac{d}{2}\cos\theta$$

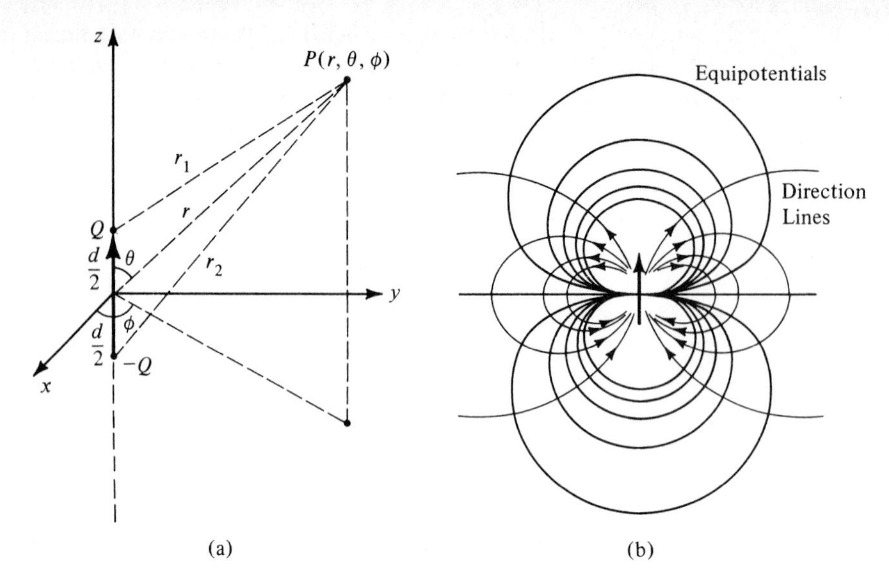

Figure 4.14. (a) Geometry pertinent to the determination of the electric field due to an electric dipole. (b) Cross sections of equipotential surfaces and direction lines of the electric field for the electric dipole.

and

$$\frac{1}{r_1} - \frac{1}{r_2} = \frac{r_2 - r_1}{r_1 r_2} \approx \frac{d \cos \theta}{r^2}$$

so that

$$V \approx \frac{Qd \cos \theta}{4\pi \varepsilon r^2} = \frac{\mathbf{p} \cdot \mathbf{i}_r}{4\pi \varepsilon r^2} \tag{4.101}$$

where $\mathbf{p} = Qd\mathbf{i}_z$ is the dipole moment of the electric dipole. Thus the potential field of the electric dipole drops off inversely with the square of the distance from the dipole. Proceeding further, we obtain the electric field intensity due to the dipole to be

$$\begin{aligned} \mathbf{E} = -\nabla V &= \frac{\partial}{\partial r}\left(\frac{Qd \cos \theta}{4\pi \varepsilon r^2}\right)\mathbf{i}_r - \frac{1}{r}\frac{\partial}{\partial \theta}\left(\frac{Qd \cos \theta}{4\pi \varepsilon r^2}\right)\mathbf{i}_\theta \\ &= \frac{Qd}{4\pi \varepsilon r^3}(2 \cos \theta\, \mathbf{i}_r + \sin \theta\, \mathbf{i}_\theta) \end{aligned} \tag{4.102}$$

Equation (4.101) shows that the equipotential surfaces are given by $r^2 \sec \theta = \text{constant}$, whereas from (4.102), it can be shown that the direction lines of the electric field are given by $r \operatorname{cosec}^2 \theta = \text{constant}$ and $\phi = \text{constant}$. These are shown sketched in Fig. 4.14(b). Alternative to using the equation for the direction lines, they can be sketched by recognizing that (1) they must originate from the positive charge and end on the negative charge and (2) they must be everywhere perpendicular to the equipotential surfaces.

Electrocardiography

A technique in everyday life in which the potential field of an electric dipole is relevant is electrocardiography. This technique is based upon the characterization of the electrical activity of the heart by using a dipole model.[1] The dipole moment, **p**, referred to in medical literature as the *electric force vector* or the *activity* of the heart, sets up an electric potential within the chest cavity and a characteristic pattern of equipotentials on the body surface. The potential differences between various points on the body are measured as a function of time and are used to deduce the temporal evolution of the dipole moment during the cardiac cycle, thereby monitoring changes in the electrical activity of the heart.

We shall now consider an example for illustrating a method of computer plotting of equipotentials when a closed form expression such as that for the electric dipole of Example 4.9 is not available.

Example 4.10

Computer plotting of equipotentials

Let us consider two point charges $Q_1 = 8\pi\varepsilon_0$ C and $Q_2 = -4\pi\varepsilon_0$ C situated at $(-1, 0, 0)$ and $(1, 0, 0)$, respectively, as shown in Fig. 4.15. We wish to discuss the computer plotting of the equipotentials due to the two point charges.

First we recognize that since the equipotential surfaces are surfaces of revolution about the axis of the two charges, it is sufficient to consider the equipotential lines in any plane containing the two charges. Here we shall consider the xz-plane. The equipotential lines are also symmetrical about the x-axis, and hence we shall plot them only on one side of the x-axis and inside the rectangular region having corners at $(-4, 0)$, $(4, 0)$ $(4, 5)$, and $(-4, 5)$.

As we go from Q_1 to Q_2 along the x-axis, the potential varies from $+\infty$ to $-\infty$ and is given by

$$V = \frac{8\pi\varepsilon_0}{4\pi\varepsilon_0(1 + x)} - \frac{4\pi\varepsilon_0}{4\pi\varepsilon_0(1 - x)}$$

$$= \frac{1 - 3x}{1 - x^2}$$

The value of x lying between -1 and 1 for a given potential V_0 is then given by

$$V_0 = \frac{1 - 3x}{1 - x^2}$$

or

$$x = \begin{cases} \dfrac{3 - \sqrt{9 - 4V_0(1 - V_0)}}{2V_0} & \text{for } V_0 \neq 0 \\ \frac{1}{3} & \text{for } V_0 = 0 \end{cases}$$

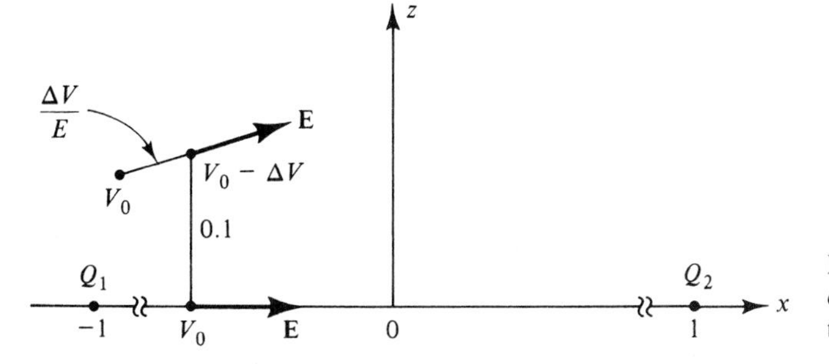

Figure 4.15. For illustrating the procedure for the computer plotting of equipotentials due to two point charges.

[1] See, for example, R. K. Hobbie, "The Electrocardiogram as an Example in Electrostatics," *American Journal of Physics*, June 1973, pp. 824–831.

We shall begin the equipotential line at this value of x on the x-axis for a given value of V_0. To plot the line, we make use of the property that the equipotential lines are orthogonal to the direction lines of $\mathbf{E}$ so that they are tangential to the unit vector $(E_z\mathbf{i}_x - E_x\mathbf{i}_z)/E$. We shall step along this unit vector by a small distance (chosen here to be 0.1) and if necessary correct the position by repeatedly moving along the electric field until the potential is within a specified value (chosen here to be 0.001 V) of that for which the line is being plotted. To correct the position, we make use of the fact that $\nabla V = -\mathbf{E}$. Thus the incremental distance required to be moved op-

PL 4.1. Program listing for plotting the equipotential lines for two point charges.

```
100 '*********************************************************
110 '* PLOTTING OF EQUIPOTENTIAL LINES FOR TWO POINT CHARGES *
120 '*********************************************************
130 BL$="                                                     "
140 K1=2:K2=1:'* VALUES OF POINT CHARGES IN MULTIPLES OF
150 '   4*PI*PERMITTIVITY *
160 DL=.1:XI=139:ZI=144:SX=30:SZ=25
170 CLS:SCREEN 1:COLOR 0,1
180 '* DRAW BOUNDARY AND SCALE MARKS *
190 LINE (19,144)-(19,19):LINE -(259,19):LINE -(259,144)
200 LINE (108,143)-(110,145),3,B
210 LINE (168,143)-(170,145),3,B
220 FOR I=0 TO 6:LINE (49+30*I,19)-(49+30*I,22):NEXT
230 FOR I=0 TO 4:LINE (19,44+25*I)-(22,44+25*I):NEXT
240 FOR I=0 TO 4:LINE (257,44+25*I)-(259,44+25*I):NEXT
250 LOCATE 21,1:INPUT "ENTER VALUE OF POTENTIAL:",VO
260 '* COMPUTE VALUE OF X FOR Z=0 *
270 IF VO=0 THEN X=(K1-K2)/(K1+K2):GOTO 300
280 X=SQR((K1+K2)^2-4*VO*(K1-K2-VO)):X=.5*(K1+K2-X)/VO:Z=0
290 '* PLOT EQUIPOTENTIAL LINE FOR VO *
300 LOCATE 21,1:PRINT "EQUIPOTENTIAL LINE BEING PLOTTED IS FOR
    "
310 PRINT "VALUE OF V =";VO;"V"
320 PSET (XI+X*SX,ZI-Z*SZ):'* PLOT POINT ALONG EQUIPOTENTIAL
330 '   LINE *
340 GOSUB 500:X=X-DL*UZ:Z=Z+DL*UX:'* COMPUTE ELECTRIC FIELD
350 '   DIRECTION AND INCREMENT NORMAL TO IT *
360 V=K1/SQR((X+1)^2+Z*Z)-K2/SQR((X-1)^2+Z*Z):'* COMPUTE
370 '   POTENTIAL *
380 IF ABS(V-VO)<.001 THEN 430:'* CHECK IF CORRECTION OF
390 '   POSITION IS REQUIRED *
400 GOSUB 500:X=X+(V-VO)*UX/E:Z=Z+(V-VO)*UZ/E:'* CORRECT
410 '   POSITION BY MOVING ALONG THE ELECTRIC FIELD *
420 GOTO 360
430 IF ABS(X)>4 OR Z<0 OR Z>5 THEN 460:'*  CHECK IF POINT
440 '   OUTSIDE THE BOUNDARY *
450 GOTO 320
460 LOCATE 23,1:PRINT "STRIKE ANY KEY TO CONTINUE";
470 C$=INPUT$(1)
480 LOCATE 21,1:PRINT BL$:PRINT BL$:PRINT BL$:GOTO 250
490 END
500 '* SUBPROGRAM TO COMPUTE COMPONENTS OF E AND OF UNIT
510 '   VECTOR ALONG E *
520 D1=((X+1)^2+Z*Z)^1.5
530 D2=((X-1)^2+Z*Z)^1.5
540 EX=K1*(X+1)/D1-K2*(X-1)/D2
550 EZ=(K1/D1-K2/D2)*Z
560 E=SQR(EX*EX+EZ*EZ):'* MAGNITUDE OF ELECTRIC FIELD *
570 UX=EX/E:UZ=EZ/E:'* COMPONENTS OF UNIT VECTOR ALONG THE
580 '   FIELD *
590 RETURN
```

posite to the electric field to increase the potential by ΔV is $\Delta V/E$, and hence the distances required to be moved opposite to the x- and z-directions are $(\Delta V/E)(E_x/E)$ and $(\Delta V/E)(E_z/E)$, respectively. The plotting of the line is terminated when the point goes out of the rectangular region.

The listing of a PC program that carries out this procedure is included as PL 4.1. The computer plot obtained from a run of the program for values of potentials ranging from -2 V to 4 V is shown in Fig. 4.16. It should, however, be pointed out that for a complete plot, those equipotential lines which surround both point charges should also be considered.

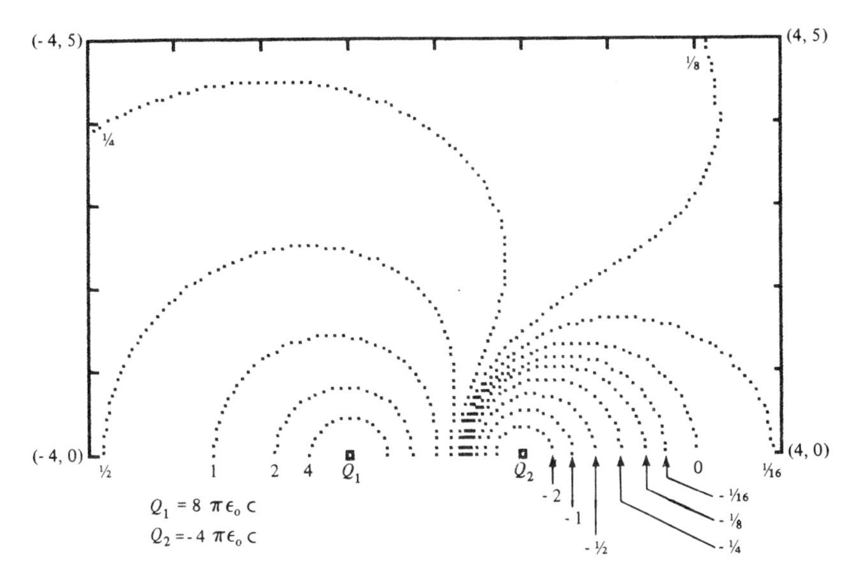

Figure 4.16. Personal computer-generated plot of equipotentials for two point charges, using the program of PL 4.1. The values of potentials are in volts.

The computation of potential can be extended to continuous charge distributions by using superposition in conjunction with the expression for the potential due to a point charge, as in the case of electric field computation in Section 2.1. We shall illustrate by means of an example.

Example 4.11

Potential due to a line charge

An infinitely long line charge of uniform density ρ_{L0} C/m is situated along the z-axis. It is desired to obtain the potential field due to this charge.

First we divide the line into a number of infinitesimal segments each of length dz as shown in Fig. 4.17, such that the charge $\rho_{L0}\,dz$ in each segment can be considered as a point charge. Let us consider a point P at a distance r from the z-axis, with the projection of P onto the z-axis being O. For the sake of generality, we consider the point P_0 at a distance r_0 from O along OP as the reference point for zero potential and write the potential dV at P due to the infinitesimal charge $\rho_{L0}\,dz$ at A as

$$dV = \frac{\rho_{L0}\,dz}{4\pi\varepsilon_0(AP)} - \frac{\rho_{L0}\,dz}{4\pi\varepsilon_0(AP_0)}$$

$$= \frac{\rho_{L0}\,dz}{4\pi\varepsilon_0\sqrt{r^2+z^2}} - \frac{\rho_{L0}\,dz}{4\pi\varepsilon_0\sqrt{r_0^2+z^2}} \tag{4.103}$$

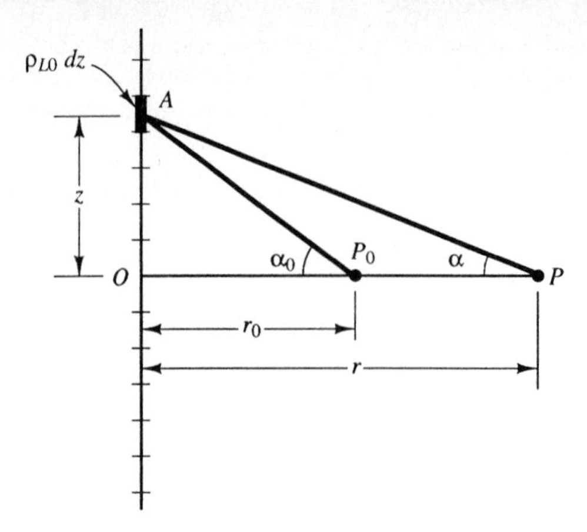

Figure 4.17. Geometry for the computation of the potential field of an infinitely long line charge of uniform density ρ_{L0} C/m.

We will, however, find later that we have to choose the reference point for zero potential at a finite value of r, in contrast to the case of the point charge for which the reference point can be chosen to be infinity. The potential V at P due to the entire line charge is now given by the integral of (4.103), where the integration is to be performed between the limits $z = -\infty$ and $z = \infty$. Thus

$$\begin{aligned} V &= \int_{z=-\infty}^{\infty} dV = \int_{z=-\infty}^{\infty} \left(\frac{\rho_{L0}\,dz}{4\pi\varepsilon_0\sqrt{r^2+z^2}} - \frac{\rho_{L0}\,dz}{4\pi\varepsilon_0\sqrt{r_0^2+z^2}} \right) \\ &= \frac{\rho_{L0}}{2\pi\varepsilon_0}\int_{z=0}^{\infty} \left(\frac{dz}{\sqrt{r^2+z^2}} - \frac{dz}{\sqrt{r_0^2+z^2}} \right) \end{aligned} \tag{4.104}$$

Introducing $z = r \tan \alpha$ and $z = r_0 \tan \alpha_0$ in the first and second terms, respectively, in the integrand on the right side of (4.104), we have

$$\begin{aligned} V &= \frac{\rho_{L0}}{2\pi\varepsilon_0}\left(\int_{\alpha=0}^{\pi/2} \sec\alpha\, d\alpha - \int_{\alpha_0=0}^{\pi/2} \sec\alpha_0\, d\alpha_0 \right) \\ &= \frac{\rho_{L0}}{2\pi\varepsilon_0}\{[\ln(\sec\alpha + \tan\alpha)]_{\alpha=0}^{\pi/2} - [\ln(\sec\alpha_0 + \tan\alpha_0)]_{\alpha_0=0}^{\pi/2}\} \\ &= \frac{\rho_{L0}}{2\pi\varepsilon_0}\left[\ln\frac{(\sqrt{r^2+z^2}+z)r_0}{(\sqrt{r_0^2+z^2}+z)r} \right]_{z=0}^{\infty} \\ &= -\frac{\rho_{L0}}{2\pi\varepsilon_0}\ln\frac{r}{r_0} \end{aligned} \tag{4.105}$$

In view of the cylindrical symmetry about the line charge, (4.105) is the general expression in cylindrical coordinates for the potential field of the infinitely long line charge of uniform density. It can be seen from (4.105) that a choice of $r_0 = \infty$ is not a good choice, since then the potential would be infinity at all points. The difficulty lies in the fact that infinity plus a finite number is still infinity. We also note from (4.105) that the equipotential surfaces are $\ln r/r_0 =$ constant or $r =$ constant, that is, surfaces of cylinders with the line charge as their axis. The result of Example 3.6 shows that the electric field due to the line charge is directed radially away from the line charge. Thus the direction lines of **E** and the equipotential surfaces are indeed orthogonal to each other.

Magnetic vector potential due to a current element

We shall now turn our attention to the magnetic vector potential for the static field case. Thus let us consider a current element of length $d\mathbf{l}$ situated at the origin, as shown in Fig. 4.18, and carrying current I amperes. We shall obtain the magnetic vector potential due to this current element. To do this, we recall from Section 2.2 that the magnetic field due to it at a point $P(r, \theta, \phi)$ is given by

$$\mathbf{B} = \frac{\mu}{4\pi}\frac{I\,d\mathbf{l} \times \mathbf{i}_r}{r^2} \tag{4.106}$$

Expressing $\mathbf{B}$ as

$$\mathbf{B} = \frac{\mu}{4\pi}I\,d\mathbf{l} \times \left(-\boldsymbol{\nabla}\frac{1}{r}\right) \tag{4.107}$$

and using the vector identity

$$\mathbf{A} \times \boldsymbol{\nabla}\Phi = \Phi\boldsymbol{\nabla} \times \mathbf{A} - \boldsymbol{\nabla} \times \Phi\mathbf{A} \tag{4.108}$$

we obtain

$$\mathbf{B} = -\frac{\mu I}{4\pi r}\boldsymbol{\nabla} \times d\mathbf{l} + \boldsymbol{\nabla} \times \frac{\mu I\,d\mathbf{l}}{4\pi r} \tag{4.109}$$

Since $d\mathbf{l}$ is a constant, $\boldsymbol{\nabla} \times d\mathbf{l} = \mathbf{0}$, and (4.109) reduces to

$$\mathbf{B} = \boldsymbol{\nabla} \times \frac{\mu I\,d\mathbf{l}}{4\pi r} \tag{4.110}$$

Comparing (4.110) with (4.66), we see that the magnetic vector potential due to the current element situated at the origin is given by

$$\mathbf{A} = \frac{\mu I\,d\mathbf{l}}{4\pi r} \tag{4.111}$$

It follows from (4.111) that for a current element $I\,d\mathbf{l}$ situated at an arbitrary point, the magnetic vector potential is given by

$$\boxed{\mathbf{A} = \frac{\mu I\,d\mathbf{l}}{4\pi R}} \tag{4.112}$$

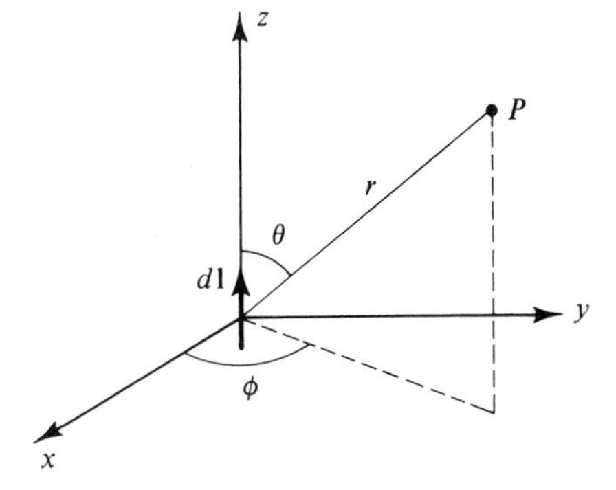

Figure 4.18. For finding the magnetic vector potential due to a current element.

where R is the distance from the current element. Thus it has a magnitude inversely proportional to the radial distance from the element (similar to the inverse distance dependence of the electric scalar potential due to a point charge) and direction parallel to the element. We shall make use of this result in Section 10.1.

K4.5. Potential difference; Potential due to a point charge; Computation of potential due to charge distributions; Electric dipole; Plotting of equipotential lines; Magnetic vector potential due to a current element.

D4.14. In a region of static electric field $\mathbf{E} = yz\mathbf{i}_x + zx\mathbf{i}_y + xy\mathbf{i}_z$, find the potential difference $V_A - V_B$ for each of the following pairs of points: **(a)** $A(2, 1, 1)$ and $B(1, 4, 0.5)$; **(b)** $A(2, 2, 2)$ and $B(1, 1, 1)$; and **(c)** $A(5, 1, 0.2)$ and $B(1, 2, 3)$.
Ans. **(a)** 0 V; **(b)** −7 V; **(c)** 5 V

D4.15. Three point charges are located as follows: $30\pi\varepsilon_0$ C at (3, 4, 0), $10\pi\varepsilon_0$ C at (3, −4, 0), and $-40\pi\varepsilon_0$ C at (−5, 0, 0). Find the following: **(a)** the potential at the point (0, 0, 3.2); **(b)** the coordinate x to three decimal places of the point on the x-axis at which the potential is a maximum; and **(c)** the potential at the point found in **(b)**.
Ans. **(b)** 0 V; **(b)** 3.872 m; **(c)** 1.3155 V

D4.16. For each of the following arrangements of point charges, find the first significant term in the expression for the electric potential at distances far from the origin ($r \gg d$): **(a)** Q at $(0, 0, d)$, $2Q$ at $(0, 0, 0)$, and Q at $(0, 0, -d)$ and **(b)** Q at $(0, 0, d)$, $-2Q$ at $(0, 0, 0)$, and Q at $(0, 0, -d)$.
Ans. **(a)** $Q/\pi\varepsilon_0 r$; **(b)** $(Qd^2/4\pi\varepsilon_0 r^3)(3\cos^2\theta - 1)$

4.6 ENERGY STORAGE IN ELECTRIC AND MAGNETIC FIELDS

In the preceding section, we learned that the potential difference between two points A and B in a static electric field has the meaning of work done per unit charge in moving a test charge from point A to point B. If we transfer a test charge from a point of higher potential to a point of lower potential, the field does the work and hence there is loss in potential energy of the system, which is supplied to the test charge. Where in the system does this energy come from? Alternatively, if we transfer the test charge from a point of lower potential to a point of higher potential, an external agent moving the charge has to do work, thus increasing the potential energy of the system. Where in the system does this energy expended by the external agent reside? Wherever in the system the energy may reside, a convenient way is to think of the energy as being stored in the electric field. In the first case, part of the stored energy in the field is expended in moving the test charge, whereas in the second case the energy expended by the external agent increases the stored energy.

Work required to assemble a system of n point charges

Let us then consider a system of two point charges Q_1 and Q_2 situated an infinite distance apart so that no forces are exerted on either charge and hence the charges are in equilibrium. Then recalling that the potential at a point is the potential difference between that point and a reference point, which for a point charge is conveniently chosen to be at infinity, we note that to bring Q_2 close to Q_1 as shown in Fig. 4.19(a), an amount of work equal to Q_2 times the potential of Q_1 at Q_2 must be expended by an external agent. Thus the potential energy of the

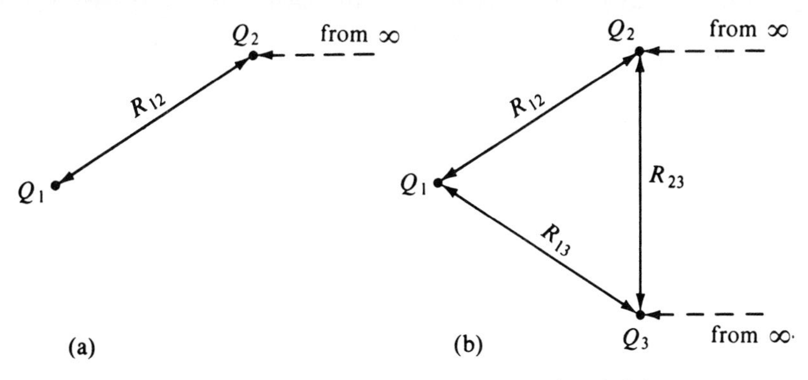

Figure 4.19. Bringing point charges closer from infinity.

system is increased by the amount

$$W_2 = Q_2 V_2^1 \tag{4.113}$$

where V_2^1 is the potential of Q_1 at the location of Q_2. If we start with a system of three charges Q_1, Q_2, Q_3 situated an infinite distance apart from each other, then the amount of work required to bring Q_2 and Q_3 close to Q_1 can be determined in two steps. First we bring Q_2 close to Q_1, for which the work required is given by (4.113). Then we bring Q_3 close to Q_1 as shown in Fig. 4.19(b). But, this time, we have to overcome not only the force exerted on Q_3 by Q_1 but also the force exerted by Q_2. Hence the required work is given by

$$W_3 = Q_3 V_3^1 + Q_3 V_3^2 \tag{4.114}$$

Thus the total work required to bring Q_2 and Q_3 close to Q_1 is

$$W_e = W_2 + W_3 = Q_2 V_2^1 + (Q_3 V_3^1 + Q_3 V_3^2) \tag{4.115}$$

The potential energy of the system is increased by the amount given by (4.115).

We can proceed in this manner and consider a system of n point charges Q_1, Q_2, Q_3, . . . , Q_n initially located infinitely far apart from each other. The total work required in bringing the charges close to each other is given by

$$\begin{aligned} W_e &= W_2 + W_3 + \cdots + W_n \\ &= Q_2 V_2^1 + (Q_3 V_3^1 + Q_3 V_3^2) + (Q_4 V_4^1 + Q_4 V_4^2 + Q_4 V_4^3) + \cdots \\ &= \sum_{i=2}^{n} \sum_{j=1}^{i-1} Q_i V_i^j \end{aligned} \tag{4.116}$$

where V_i^j is the potential of Q_j at the location of Q_i. However, we note that

$$Q_i V_i^j = Q_i \frac{Q_j}{4\pi\varepsilon_0 R_{ji}} = Q_j \frac{Q_i}{4\pi\varepsilon_0 R_{ij}} = Q_j V_j^i \tag{4.117}$$

Hence (4.116) may be written as

$$\begin{aligned} W_e &= Q_1 V_1^2 + (Q_1 V_1^3 + Q_2 V_2^3) + (Q_1 V_1^4 + Q_2 V_2^4 + Q_3 V_3^4) + \cdots \\ &= \sum_{i=2}^{n} \sum_{j=1}^{i-1} Q_j V_j^i \end{aligned} \tag{4.118}$$

Adding (4.116) and (4.117), we have

$$\begin{aligned}
2W_e &= Q_1(V_1^2 + V_1^3 + V_1^4 + \cdots) \\
&\quad + Q_2(V_2^1 + V_2^3 + V_2^4 + \cdots) \\
&\quad + Q_3(V_3^1 + V_3^2 + V_3^4 + \cdots) \\
&\quad + \cdots \\
&= Q_1(\text{potential at } Q_1 \text{ due to all other charges}) \\
&\quad + Q_2(\text{potential at } Q_2 \text{ due to all other charges}) \\
&\quad + Q_3(\text{potential at } Q_3 \text{ due to all other charges}) \\
&\quad + \cdots \\
&= Q_1V_1 + Q_2V_2 + Q_3V_3 + \cdots \\
&= \sum_{i=1}^{n} Q_iV_i
\end{aligned} \tag{4.119}$$

where V_i is the potential at Q_i due to all other charges. Dividing both sides of (4.119) by 2, we have

$$\boxed{W_e = \frac{1}{2}\sum_{i=1}^{n} Q_iV_i} \tag{4.120}$$

Thus the potential energy stored in the system of n point charges is given by (4.120).

Example 4.12

Three point charges of values 1, 2, and 3 C are situated at the corners of an equilateral triangle of sides 1 m. It is desired to find the work required to move these charges to the corners of an equilateral triangle of shorter sides $\frac{1}{2}$ m as shown in Fig. 4.20.

The potential energy stored in the system of three charges at the corners of the larger equilateral triangle is given by

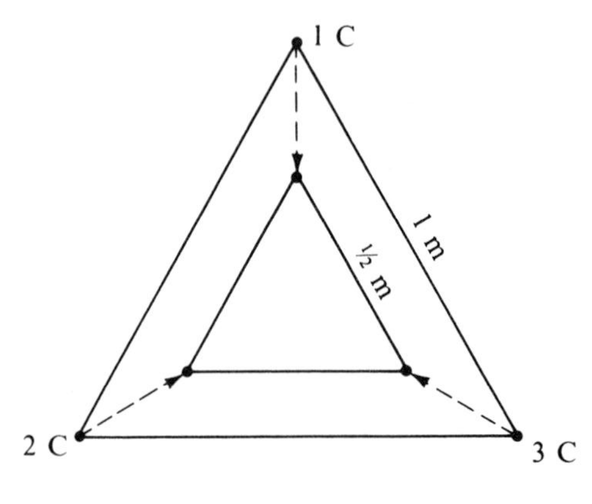

Figure 4.20. Bringing three point charges from the corners of a larger equilaterial triangle to the corners of a smaller equilateral triangle.

$$\frac{1}{2}\sum_{i=1}^{3} Q_i V_i = \frac{1}{2}\left[1\left(\frac{2}{4\pi\varepsilon_0} + \frac{3}{4\pi\varepsilon_0}\right) + 2\left(\frac{1}{4\pi\varepsilon_0} + \frac{3}{4\pi\varepsilon_0}\right) + 3\left(\frac{1}{4\pi\varepsilon_0} + \frac{2}{4\pi\varepsilon_0}\right)\right]$$

$$= \frac{1}{2}\left[\frac{5 + 8 + 9}{4\pi\varepsilon_0}\right] = \frac{11}{4\pi\varepsilon_0} \text{ N-m}$$

The potential energy stored in the system of three charges at the corners of the smaller equilateral triangle is equal to twice this value since all distances are halved. The increase in potential energy of the system in going from the larger to the smaller equilateral triangle is equal to $11/4\pi\varepsilon_0$ N-m. Obviously, this increase in energy must be supplied by an external agent, and hence the work required to move the charges to the corners of the equilateral triangle of sides $\frac{1}{2}$ m from the corners of the equilateral triangle of sides 1 m is equal to $11/4\pi\varepsilon_0$ N-m.

Potential energy in a continuous charge distribution

If we have a continuous distribution of charge with density $\rho(x, y, z)$ instead of an assembly of discrete charges, we can treat it as a continuous collection of infinitesimal charges of value $\rho(x, y, z)\,\Delta v$, each of which can be considered as a point charge, and obtain the potential energy of the system as

$$\begin{aligned} W_e &= \tfrac{1}{2} \lim_{\Delta v \to 0} \sum [\rho(x, y, z)\,\Delta v] V(x, y, z) \\ &= \tfrac{1}{2} \int_{\substack{\text{volume} \\ \text{containing } \rho}} \rho V\, dv \end{aligned} \tag{4.121}$$

Thus far, we have found the potential energy of the charge distribution by considering the work done in assembling the system. We stated at the beginning of this section that the potential energy can be thought of as being stored in the electric field set up by the system of charges. If so, we should be able to express the energy in terms of the electric field. To do this, we substitute for ρ in (4.121) from (4.34) and obtain

$$W_e = \tfrac{1}{2} \int_{\substack{\text{volume} \\ \text{containing } \rho}} (\nabla \cdot \mathbf{D}) V\, dv \tag{4.122}$$

Since $\nabla \cdot \mathbf{D} = 0$ in the region not containing ρ, the value of the integral on the right side of (4.122) is not altered if we change the volume of integration from the volume containing ρ to the entire space. Thus

$$W_e = \tfrac{1}{2} \int_{\text{all space}} (\varepsilon \nabla \cdot \mathbf{E}) V\, dv \tag{4.123}$$

where we have also replaced $\mathbf{D}$ by $\varepsilon\mathbf{E}$. We now use the vector identity

$$\nabla \cdot \Phi\mathbf{A} = \Phi\nabla \cdot \mathbf{A} + \mathbf{A} \cdot \nabla\Phi \tag{4.124}$$

to replace $V\nabla \cdot \mathbf{E}$ on the right side of (4.123) by $\nabla \cdot V\mathbf{E} - \mathbf{E} \cdot \nabla V$ and obtain

$$\begin{aligned} W_e &= \tfrac{1}{2}\varepsilon \int_{\text{all space}} (\nabla \cdot V\mathbf{E} - \mathbf{E} \cdot \nabla V)\, dv \\ &= \tfrac{1}{2}\varepsilon \int_{\text{all space}} \nabla \cdot V\mathbf{E}\, dv + \tfrac{1}{2}\varepsilon \int_{\text{all space}} \mathbf{E} \cdot \mathbf{E}\, dv \end{aligned} \tag{4.125}$$

where we have replaced ∇V by $-\mathbf{E}$ in accordance with (4.97). Using the divergence theorem, we equate the first integral on the right side of (4.125) to a surface integral; thus

$$\int_{\text{all space}} \nabla \cdot V\mathbf{E}\, dv = \int_{\substack{\text{surface}\\\text{bounding}\\\text{all space}}} V\mathbf{E} \cdot \mathbf{i}_n\, dS \tag{4.126}$$

However, as viewed from a surface bounding all space, a charge distribution of finite volume appears as a point charge, say, Q. Hence, as $r \to \infty$, we can write

$$\mathbf{E} \longrightarrow \frac{Q}{4\pi\varepsilon r^2}\mathbf{i}_r$$

$$V \longrightarrow \frac{Q}{4\pi\varepsilon r}$$

$$\begin{aligned}\int_{\substack{\text{surface}\\\text{bounding}\\\text{all space}}} V\mathbf{E} \cdot \mathbf{i}_n\, dS &= \lim_{r\to\infty} \int_{\theta=0}^{\pi}\int_{\phi=0}^{2\pi} \frac{Q}{4\pi\varepsilon r}\,\frac{Q}{4\pi\varepsilon r^2}\mathbf{i}_r \cdot r^2 \sin\theta\, d\theta\, d\phi\, \mathbf{i}_r \\ &= \int_{\theta=0}^{\pi}\int_{\phi=0}^{2\pi} \lim_{r\to\infty} \frac{Q^2}{(4\pi\varepsilon)^2 r} \sin\theta\, d\theta\, d\phi = 0\end{aligned} \tag{4.127}$$

Equation (4.127) also holds for a charge distribution of infinite extent, provided the electric field due to the charge distribution falls off at least as $(1/r^2)\mathbf{i}_r$ and hence the potential falls off at least as $1/r$. Thus (4.125) reduces to

$$W_e = \tfrac{1}{2}\varepsilon \int_{\text{all space}} \mathbf{E} \cdot \mathbf{E}\, dv = \int_{\text{all space}} (\tfrac{1}{2}\varepsilon E^2)\, dv \tag{4.128}$$

Electric stored energy density

Equation (4.128) indicates clearly that the idea of energy residing in the electric field is a valid one provided we integrate $\frac{1}{2}\varepsilon E^2$ throughout the entire space. The quantity $\frac{1}{2}\varepsilon E^2$ is evidently the energy density associated with the electric field; that is,

$$\boxed{w_e = \tfrac{1}{2}\varepsilon E^2} \tag{4.129}$$

Example 4.13

A volume charge is distributed throughout a sphere of radius a m, and centered at the origin, with uniform density ρ_0 C/m³. We wish to find the energy stored in the electric field of this charge distribution.

From Example 3.7, the electric field of the uniformly distributed spherical charge, having its center at the origin, is given by

$$\mathbf{E} = \frac{\mathbf{D}}{\varepsilon} = \begin{cases} \dfrac{\rho_0 r}{3\varepsilon_0}\mathbf{i}_r & \text{for } r < a \\[2ex] \dfrac{\rho_0 a^3}{3\varepsilon_0 r^2}\mathbf{i}_r & \text{for } r > a \end{cases}$$

Hence the energy density in the electric field is given by

$$\frac{1}{2}\varepsilon E_r^2 = \begin{cases} \dfrac{\rho_0^2 r^2}{18\varepsilon_0} & \text{for } r < a \\[2ex] \dfrac{\rho_0^2 a^6}{18\varepsilon_0 r^4} & \text{for } r > a \end{cases}$$

The energy stored in the electric field is

$$W_e = \int_{r=0}^{a} \int_{\theta=0}^{\pi} \int_{\phi=0}^{2\pi} \frac{\rho_0^2 r^2}{18\varepsilon_0} r^2 \sin\theta \, dr \, d\theta \, d\phi$$

$$+ \int_{r=a}^{\infty} \int_{\theta=0}^{\pi} \int_{\phi=0}^{2\pi} \frac{\rho_0^2 a^6}{18\varepsilon_0 r^4} r^2 \sin\theta \, dr \, d\theta \, d\phi$$

$$= \frac{4\pi\rho_0^2 a^5}{15\varepsilon_0}$$

Magnetic stored energy density

Just as energy is stored in an electric field, there is energy storage associated with magnetic field. The expression for the magnetic stored energy density, w_m, can be derived by considering the building up of a current distribution.[2] We shall however simply present the result here. It is given by

$$\boxed{w_m = \tfrac{1}{2}\mu H^2} \qquad (4.130)$$

Let us now consider an example.

Example 4.14

Current I flows in the $+z$-direction with uniform density on the cylindrical surface $r = a$ and returns in the $-z$-direction with uniform density on a second cylindrical surface $r = b$ so that the surface current distribution is given by

$$\mathbf{J}_S = \begin{cases} \dfrac{I}{2\pi a}\mathbf{i}_z & \text{for } r = a \\[2ex] -\dfrac{I}{2\pi b}\mathbf{i}_z & \text{for } r = b \end{cases}$$

We wish to find the energy stored in the magnetic field per unit length of the current distribution.

From application of Ampère's circuital law in integral form, we obtain the magnetic field due to the given current distribution as

$$\mathbf{H} = \begin{cases} \mathbf{0} & \text{for } r < a \\[1ex] \dfrac{I}{2\pi r}\mathbf{i}_\phi & \text{for } a < r < b \\[1ex] \mathbf{0} & \text{for } r > b \end{cases}$$

Hence the energy density in the magnetic field is given by

$$\frac{1}{2}\mu H_\phi^2 = \begin{cases} 0 & \text{for } r < a \\[1ex] \dfrac{\mu_0 I^2}{8\pi^2 r^2} & \text{for } a < r < b \\[1ex] 0 & \text{for } r > b \end{cases}$$

[2] See, for example, N. N. Rao, *Basic Electromagnetics with Applications* (Englewood Cliffs, N.J.: Prentice-Hall, Inc., 1972), pp. 226–230.

The energy stored in the magnetic field per unit length of the current distribution is

$$W_m = \int_{r=a}^{b} \int_{\phi=0}^{2\pi} \int_{z=0}^{1} \frac{\mu_0 I^2}{8\pi^2 r^2} r\, dr\, d\phi\, dz$$

$$= \frac{\mu_0 I^2}{4\pi} \ln \frac{b}{a}$$

K4.6. Potential energy associated with a charge distribution; Energy density in electric field; Energy density in magnetic field.

D4.17. Four point charges, each of value $4\pi\varepsilon_0$ C are situated at the vertices A, B, C, and D (located in that sequence) of a square of sides 1 m. Find the following: **(a)** the work required to assemble the charge distribution; **(b)** the additional work required to move the charge at D to the center of the square; and **(c)** the additional work required to then move the charge at C to the center of the side AB.
Ans. **(a)** $68.037\varepsilon_0$ J; **(b)** $19.2961\varepsilon_0$ J; **(c)** $36.1746\varepsilon_0$ J

D4.18. Two spherical charges, each of the same radius a and the same uniform charge density ρ_0 C/m^3, are situated infinitely apart. The two spherical charges are brought together and made into a single spherical charge of uniform density. Find the following: **(a)** the work required if the radius of the new spherical charge is a; **(b)** the work required if the charge density of the new spherical charge is ρ_0; and **(c)** the radius and the charge density of the new spherical charge for which the work required is zero.
Ans. **(a)** $1.6755\rho_0^2 a^5/\varepsilon_0$ J; **(b)** $0.9842\rho_0^2 a^5/\varepsilon_0$ J; **(c)** $2a$, $0.25\rho_0$

4.7 SUMMARY

We have in this chapter derived the differential forms of Maxwell's equations from their integral forms, which we introduced in Chapter 3. For the general case of electric and magnetic fields having all three components, each of them dependent on all coordinates and time, Maxwell's equations in differential form are given as follows in words and in mathematical form.

Faraday's law. The curl of the electric field intensity is equal to the negative of the time derivative of the magnetic flux density; that is,

$$\nabla \times \mathbf{E} = -\frac{\partial \mathbf{B}}{\partial t} \qquad (4.131)$$

Ampère's circuital law. The curl of the magnetic field intensity is equal to the sum of the current density due to flow of charges and the displacement current density, which is the time derivative of the displacement flux density; that is,

$$\nabla \times \mathbf{H} = \mathbf{J} + \frac{\partial \mathbf{D}}{\partial t} \qquad (4.132)$$

Gauss' law for the electric field. The divergence of the displacement flux density is equal to the charge density; that is,

$$\nabla \cdot \mathbf{D} = \rho \qquad (4.133)$$

Gauss' law for the magnetic field. The divergence of the magnetic flux density is equal to zero; that is,

$$\nabla \cdot \mathbf{B} = 0 \tag{4.134}$$

Auxiliary to (4.131)–(4.134), the continuity equation is given by

$$\nabla \cdot \mathbf{J} = -\frac{\partial \rho}{\partial t} \tag{4.135}$$

This equation, which is the differential form of the law of conservation of charge, states that the sum of the divergence of the current density due to flow of charges and the time derivative of the charge density is equal to zero.

In using (4.131)–(4.135), we recall that

$$\mathbf{D} = \varepsilon \mathbf{E}$$

$$\mathbf{H} = \frac{\mathbf{B}}{\mu}$$

where ε and μ are the permittivity and the permeability, respectively, of the medium. In addition, if the current density $\mathbf{J}$ is due to conduction, then

$$\mathbf{J} = \mathbf{J}_c = \sigma \mathbf{E}$$

where σ is the conductivity of the medium.

We have learned that the basic definitions of curl and divergence, which have enabled us to discuss their physical interpretations with the aid of the curl and divergence meters, are

$$\nabla \times \mathbf{A} = \lim_{\Delta S \to 0} \left[\frac{\oint_C \mathbf{A} \cdot d\mathbf{l}}{\Delta S} \right]_{\max} \mathbf{i}_n$$

$$\nabla \cdot \mathbf{A} = \lim_{\Delta v \to 0} \frac{\oint_S \mathbf{A} \cdot d\mathbf{S}}{\Delta v}$$

Thus the curl of a vector field at a point is a vector whose magnitude is the circulation of that vector field per unit area with the area oriented so as to maximize this quantity and in the limit that the area shrinks to the point. The direction of the vector is normal to the area in the aforementioned limit and in the right-hand sense. The divergence of a vector field at a point is a scalar quantity equal to the net outward flux of that vector field per unit volume in the limit that the volume shrinks to the point. In Cartesian coordinates the expansions for curl and divergence are

$$\nabla \times \mathbf{A} = \begin{vmatrix} \mathbf{i}_x & \mathbf{i}_y & \mathbf{i}_z \\ \frac{\partial}{\partial x} & \frac{\partial}{\partial y} & \frac{\partial}{\partial z} \\ A_x & A_y & A_z \end{vmatrix}$$

$$= \left(\frac{\partial A_z}{\partial y} - \frac{\partial A_y}{\partial z} \right) \mathbf{i}_x + \left(\frac{\partial A_x}{\partial z} - \frac{\partial A_z}{\partial x} \right) \mathbf{i}_y + \left(\frac{\partial A_y}{\partial x} - \frac{\partial A_x}{\partial y} \right) \mathbf{i}_z$$

$$\nabla \cdot \mathbf{A} = \frac{\partial A_x}{\partial x} + \frac{\partial A_y}{\partial y} + \frac{\partial A_z}{\partial z}$$

Thus Maxwell's equations in differential form relate the spatial variations of the field vectors at a point to their temporal variations and to the charge and current densities at that point.

We have also learned two theorems associated with curl and divergence. These are the Stokes' theorem and the divergence theorem given, respectively, by

$$\oint_C \mathbf{A} \cdot d\mathbf{l} = \int_S (\nabla \times \mathbf{A}) \cdot d\mathbf{S}$$

$$\oint_S \mathbf{A} \cdot d\mathbf{S} = \int_V (\nabla \cdot \mathbf{A})\, dv$$

Stokes' theorem enables us to replace the line integral of a vector around a closed path by the surface integral of the curl of that vector over any surface bounded by that closed path, and vice versa. The divergence theorem enables us to replace the surface integral of a vector over a closed surface by the volume integral of the divergence of that vector over the volume bounded by the closed surface and vice versa.

We then introduced the electric scalar and magnetic vector potential functions, Φ and $\mathbf{A}$, respectively. In view of (4.134), we have

$$\mathbf{B} = \nabla \times \mathbf{A} \tag{4.136}$$

and then in view of (4.131), we obtain

$$\mathbf{E} = -\nabla\Phi - \frac{\partial \mathbf{A}}{\partial t} \tag{4.137}$$

In (4.137), $\nabla\Phi$ is the gradient of the scalar function Φ. We learned that the gradient of a scalar Φ is a vector having magnitude equal to the maximum rate of increase of Φ at that point, and its direction is the direction in which the maximum rate of increase occurs, that is, normal to the constant Φ surface passing through that point; that is,

$$\nabla\Phi = \frac{\partial \Phi}{\partial n}\mathbf{i}_n$$

In Cartesian coordinates, the expansion for the gradient is

$$\nabla\Phi = \frac{\partial \Phi}{\partial x}\mathbf{i}_x + \frac{\partial \Phi}{\partial y}\mathbf{i}_y + \frac{\partial \Phi}{\partial z}\mathbf{i}_z$$

Next, we derived two differential equations for the potential functions. These are given by

$$\nabla^2\Phi - \mu\varepsilon\frac{\partial^2 \Phi}{\partial t^2} = -\frac{\rho}{\varepsilon} \tag{4.138a}$$

$$\nabla^2\mathbf{A} - \mu\varepsilon\frac{\partial^2 \mathbf{A}}{\partial t^2} = -\mu\mathbf{J} \tag{4.138b}$$

where $\nabla^2\Phi$ is the Laplacian of the scalar Φ and $\nabla^2\mathbf{A}$ is the Laplacian of the vector $\mathbf{A}$. In Cartesian coordinates,

$$\nabla^2\Phi = \frac{\partial^2 \Phi}{\partial x^2} + \frac{\partial^2 \Phi}{\partial y^2} + \frac{\partial^2 \Phi}{\partial z^2}$$

and

$$\nabla^2\mathbf{A} = (\nabla^2 A_x)\mathbf{i}_x + (\nabla^2 A_y)\mathbf{i}_y + (\nabla^2 A_z)\mathbf{i}_z$$

In deriving (4.138a) and (4.138b), we made use of the Lorentz condition

$$\nabla \cdot \mathbf{A} = -\mu\varepsilon\frac{\partial \Phi}{\partial t}$$

which is consistent with the continuity equation.

We then considered the potential functions for the static field case, for which (4.137) reduces to

$$\mathbf{E} = -\nabla\Phi = -\nabla V \tag{4.139}$$

whereas (4.136) remains unaltered. In (4.139), the symbol Φ is replaced by the symbol V since the electric potential difference between two points in a static electric field has the same meaning as the voltage between the two points. We considered the potential field of a point charge and found that for the point charge

$$V = \frac{Q}{4\pi\varepsilon R} \tag{4.140}$$

where R is the radial distance away from the point charge. The equipotential surfaces for the point charge are thus spherical surfaces centered at the point charge. We illustrated the application of the potential concept in the determination of electric field due to charge distributions by considering the examples of an electric dipole and a line charge. We also discussed a procedure for computer plotting of equipotentials. We then derived the expression for the magnetic vector potential due to a current element. For a current element $I\,d\mathbf{l}$, the magnetic vector potential is given by

$$\mathbf{A} = \frac{\mu I\,d\mathbf{l}}{4\pi R} \tag{4.141}$$

where R is the distance from the current element.

Finally, we discussed energy storage in electric and magnetic fields. We found that the work required to assemble a system of n point charges is given by

$$W_e = \frac{1}{2}\sum_{i=1}^{n} Q_i V_i \tag{4.142}$$

where V_i is the electric potential at the point charge Q_i due to all the other charges. For a continuous charge distribution of density ρ,

$$W_e = \frac{1}{2}\int_{\substack{\text{volume}\\ \text{containing } \rho}} \rho V\,dv$$

which reduces to

$$W_e = \int_{\substack{\text{all}\\ \text{space}}} \left(\frac{1}{2}\varepsilon E^2\right) dv$$

so that the energy density associated with the electric field can be identified to be

$$w_e = \frac{1}{2}\varepsilon E^2 \tag{4.143}$$

Similarly, the energy density associated with the magnetic field is given by

$$w_m = \frac{1}{2}\mu H^2 \tag{4.144}$$

We illustrated by means of examples the determination of energy stored in the electric field of a charge distribution and the magnetic field of a current distribution.

REVIEW QUESTIONS

R4.1. State Faraday's law in differential form for the simple case of $\mathbf{E} = E_x(z, t)\mathbf{i}_x$. How is it derived from Faraday's law in integral form?

R4.2. State Faraday's law in differential form for the general case of an arbitrary electric field. How is it derived from its integral form?

R4.3. What is meant by the net right-lateral differential of the x- and y-components of a vector normal to the z direction? Give an example in which the net right-lateral differential of E_x and E_y normal to the z-direction is zero although the individual derivatives are nonzero.

R4.4. What is the determinant expansion for the curl of a vector in Cartesian coordinates?

R4.5. State Ampère's circuital law in differential form for the general case of an arbitrary magnetic field. How is it derived from its integral form?

R4.6. State Ampère's circuital law in differential form for the simple case of $\mathbf{H} = H_y(z, t)\mathbf{i}_y$. How is it derived from Ampère's circuital law in differential form for the general case?

R4.7. If a pair of **E** and **B** at a point satisfies Faraday's law in differential form, does it necessarily follow that it also satisfies Ampère's circuital law in differential form, and vice versa?

R4.8. State Gauss' law for the electric field in differential form. How is it derived from its integral form?

R4.9. What is meant by the net longitudinal differential of the components of a vector field? Give an example in which the net longitudinal differential of the components of a vector is zero, although the individual derivatives are nonzero.

R4.10. What is the expansion for the divergence of a vector in Cartesian coordinates?

R4.11. State Gauss' law for the magnetic field in differential form. How is it derived from its integral form?

R4.12. How can you determine if a given vector can represent a magnetic field?

R4.13. State the continuity equation and discuss its physical interpretation.

R4.14. Summarize Maxwell's equations in differential form.

R4.15. State and briefly discuss the basic definition of the curl of a vector.

R4.16. What is a curl meter? How does it help visualize the behavior of the curl of a vector field?

R4.17. Provide two examples of physical phenomena in which the curl of a vector field is nonzero.

R4.18. State and briefly discuss the basic definition of the divergence of a vector.

R4.19. What is a divergence meter? How does it help visualize the behavior of the divergence of a vector field?

R4.20. Provide two examples of physical phenomena in which the divergence of a vector field is nonzero.

R4.21. State Stokes' theorem and discuss its application.

R4.22. State the divergence theorem and discuss its application.

R4.23. What is the divergence of the curl of a vector?

R4.24. What are electromagnetic potentials? How do they arise?

R4.25. What is the expansion for the gradient of a scalar in Cartesian coordinates? When can a vector be expressed as the gradient of a scalar?

R4.26. Discuss the physical interpretation for the gradient of a scalar function and the application of the gradient concept for the determination of unit vector normal to a surface.

R4.27. How is the Laplacian of a scalar defined? What is its expansion in Cartesian coordinates?

R4.28. Compare and contrast the operations of curl of a vector, divergence of a vector, gradient of a scalar, and Laplacian of a scalar.

R4.29. How is the Laplacian of a vector defined? What is its expansion in Cartesian coordinates?

R4.30. Outline the derivation of the differential equations for the electromagnetic potentials.

R4.31. What is the relationship between the static electric field intensity and the electric scalar potential?

R4.32. Distinguish between voltage, as applied to time-varying fields, and the potential difference in a static electric field.

R4.33. Describe the electric potential field of a point charge.

R4.34. Discuss the determination of the electric field intensity due to a charge distribution by using the potential concept.

R4.35. Discuss the procedure for the computer plotting of equipotentials due to two (or more) point charges.

R4.36. Compare the magnetic vector potential field due to a current element to the electric scalar potential due to a point charge.

R4.37. Discuss the concept of energy storage in electric and magnetic fields.

R4.38. Outline the derivation of the expression for the work required to assemble a system of n point charges.

R4.39. State and discuss the expression for the energy density associated with the electric field of a charge distribution.

R4.40. State and discuss the expression for the energy density associated with the magnetic field of a current distribution.

PROBLEMS

P4.1. Find the curls of the following vector fields:

(a) $xy\,\mathbf{i}_x + yz\,\mathbf{i}_y + zx\,\mathbf{i}_z$

(b) $\cos y\,\mathbf{i}_x - x \sin y\,\mathbf{i}_y$

(c) $2r \cos \phi\,\mathbf{i}_r + r\,\mathbf{i}_\phi$ in cylindrical coordinates

(d) $(e^{-r}/r)\mathbf{i}_\theta$ in spherical coordinates

P4.2. For each of the following electric fields, find $\mathbf{B}$ that satisfies Faraday's law in differential form:

(a) $\mathbf{E} = E_0 \cos 3\pi z \cos 9\pi \times 10^8 t\ \mathbf{i}_x$

(b) $\mathbf{E} = E_0 \mathbf{i}_y \cos [3\pi \times 10^8 t + 0.2\pi(4x + 3z)]$

P4.3. Obtain the simplified differential equations for the following cases: **(a)** Ampere's circuital law for $\mathbf{H} = H_x(z, t)\mathbf{i}_x$ and **(b)** Faraday's law for $\mathbf{E} = E_\phi(r, t)\mathbf{i}_\phi$ in cylindrical coordinates.

P4.4. For the electric field $\mathbf{E} = E_0 e^{-\alpha z} \cos \omega t\, \mathbf{i}_x$ in free space ($\mathbf{J} = \mathbf{0}$), find $\mathbf{B}$ that satisfies Faraday's law in differential form and then determine if the pair of $\mathbf{E}$ and $\mathbf{B}$ satisfy Ampere's circuital law in differential form.

P4.5. For the electric field $\mathbf{E} = E_0 \cos (\omega t - \alpha y - \beta z)\ \mathbf{i}_x$ in free space ($\mathbf{J} = \mathbf{0}$), find the necessary condition relating α, β, ω, μ_0, and ε_0 for the field to satisfy both of Maxwell's curl equations.

P4.6. For the electric field $\mathbf{E} = E_0 e^{-\alpha y} \cos (\omega t - \beta z)\ \mathbf{i}_x$ in free space ($\mathbf{J} = \mathbf{0}$), find the necessary condition relating α, β, ω, μ_0, and ε_0 for the field to satisfy both of Maxwell's curl equations.

P4.7. Find the divergences of the following vector fields:

(a) $x^2yz\mathbf{i}_x + y^2zx\mathbf{i}_y + z^2xy\mathbf{i}_z$

(b) $3x\mathbf{i}_x + (2 - y)\mathbf{i}_y - (2z + 1)\mathbf{i}_z$

(c) $r \cos \phi\ \mathbf{i}_r - r \sin \phi\ \mathbf{i}_\phi$ in cylindrical coordinates

(d) $r^2\mathbf{i}_r + r \sin \theta\ \mathbf{i}_\theta$ in spherical coordinates

P4.8. For each of the following charge distributions $\rho(x)$, find the corresponding displacement flux density component $D_x(x)$ by using Gauss' law for the electric field in differential form and plot both ρ and D_x versus x.

(a) $$\rho(x) = \begin{cases} \rho_0 \dfrac{x}{a} & \text{for } -a < x < a \\ 0 & \text{otherwise} \end{cases}$$

(b) $$\rho(x) = \begin{cases} \rho_0\left(1 - \dfrac{|x|}{a}\right) & \text{for } -a < x < a \\ 0 & \text{otherwise} \end{cases}$$

P4.9. For each of the following vecctor fields, find the value of the constant k for which the vetor field can be realized as a magnetic field:

(a) $(1/y^k)(2x\mathbf{i}_x + y\mathbf{i}_y)$

(b) $r(\sin k\phi\ \mathbf{i}_r + \cos k\phi\ \mathbf{i}_\phi)$ in cylindrical coordinates

(c) $\left(1 + \dfrac{2}{r^3}\right) \cos \theta\ \mathbf{i}_r + k\left(1 - \dfrac{1}{r^3}\right) \sin \theta\ \mathbf{i}_\theta$ in spherical coordinates

P4.10. Determine which of the following static fields can be realized both as an electric field in a charge-free region and a magnetic field in a current-free region:

(a) $y\mathbf{i}_x + x\mathbf{i}_y$

(b) $\left(1 + \dfrac{1}{r^2}\right) \cos \phi\ \mathbf{i}_r - \left(1 - \dfrac{1}{r^2}\right) \sin \phi\ \mathbf{i}_\phi$ in cylindrical coordinates

(c) $r \sin \theta\ \mathbf{i}_\phi$ in spherical coordinates

P4.11. Discuss with the aid of the curl meter and also by expansion in the Cartesian coordinate system the curl of the velocity vector field associated with the flow of water in the stream of Fig. 4.6(a), except that the velocity v_z varies in a linear manner from zero at the banks to a maximum of v_0 at the center.

P4.12. Discuss with the aid of the curl meter and also by expansion in the Cartesian coordinate system, the curl of the linear velocity vector field associated with points inside the earth due to its spin motion.

P4.13. Discuss the divergences of the following vector fields with the aid of the divergence meter and also by expansion in the appropriate coordinate system: **(a)** the position vector field associated with points in three-dimensional space; **(b)** the linear velocity vector field associated with points inside the earth due to its spin motion; and **(c)** the gravitational field of the earth associated with points outside the earth.

P4.14. Verify Stokes' theorem for the following cases: **(a)** the vector field $\mathbf{A} = xy\mathbf{i}_x + yz\mathbf{i}_y + zx\mathbf{i}_z$ and the closed path comprising the straight lines from (0, 0, 0) to (1, 1, 1), from (1, 1, 1) to (1, 1, 0), and from (1, 1, 0) to (0, 0, 0) and **(b)** the vector field $\mathbf{A} = xyz(yz\mathbf{i}_x + zx\mathbf{i}_y + xy\mathbf{i}_z)$ independent of a closed path.

P4.15. Verify the divergence theorem for the following cases: **(a)** the vector field $\mathbf{A} = xyz(x\mathbf{i}_x + y\mathbf{i}_y + z\mathbf{i}_z)$ and the surface of the cubical box bounded by the planes $x = 0$, $x = 1$, $y = 0$, $y = 1$, $z = 0$, and $z = 1$ and **(b)** the vector field $\mathbf{A} = (x^2y + 2)\mathbf{i}_x + 3\mathbf{i}_y - 2xyz\mathbf{i}_z$ and the surface of the box bounded by the planes $x = 0$, $x = 1$, $y = 0$, $y = 2$, $z = 0$, and $z = 3$.

P4.16. Show by expansion in the Cartesian coordinate system that **(a)** $\boldsymbol{\nabla} \cdot \boldsymbol{\nabla} \times \mathbf{A} = 0$ for any $\mathbf{A}$ and **(b)** $\boldsymbol{\nabla} \times \boldsymbol{\nabla}\Phi = \mathbf{0}$ for any Φ.

P4.17. Determine which of the following vectors can be expressed as the curl of another vector and which of them can be expressed as the gradient of a scalar:

(a) $y^2z\mathbf{i}_x + z^2x\mathbf{i}_y + x^2y\mathbf{i}_z$

(b) $(\sin\phi\,\mathbf{i}_r + \cos\phi\,\mathbf{i}_\phi)$ in cylindrical coordinates

(c) $\dfrac{1}{r^2}(\cos\theta\,\mathbf{i}_r + \sin\theta\,\mathbf{i}_\theta)$ in spherical coordinates

P4.18. Find the scalar functions whose gradients are given by the following vector functions:

(a) $e^{-y}\mathbf{i}_x - xe^{-y}\mathbf{i}_y$

(b) $\dfrac{1}{r^2}(\cos\phi\,\mathbf{i}_r + \sin\phi\,\mathbf{i}_\phi)$ in cylindrical coordinates

(c) $(\cos\theta\,\mathbf{i}_r - \sin\theta\,\mathbf{i}_\theta)$ in spherical coordinates

P4.19. By using the gradient concept, show that the unit vector along the line of intersection of two planes

$$a_1x + a_2y + a_3z = c_1$$

$$b_1x + b_2y + b_3z = c_2$$

which are not parallel is given by

$$\pm\frac{(a_2b_3 - a_3b_2)\mathbf{i}_x + (a_3b_1 - a_1b_3)\mathbf{i}_y + (a_1b_2 - a_2b_1)\mathbf{i}_z}{\sqrt{(a_2b_3 - a_3b_2)^2 + (a_3b_1 - a_1b_3)^2 + (a_1b_2 - a_2b_1)^2}}$$

Then find the unit vector along the intersection of the planes $x + y + z = 3$ and $y = x$.

P4.20. By using the gradient concept, find the equation for the plane which is tangential to the surface $x^2 + 4y^2 + 2z^2 = 10$ at the point (2, 1, 1).

P4.21. Derive the expansion for the Laplacian of a vector in Cartesian coordinates given by (4.87).

P4.22. Show that the Laplacian of a vector in cylindrical coordinates is given by

$$\nabla^2\mathbf{A} = \left(\nabla^2 A_r - \frac{A_r}{r^2} - \frac{2}{r^2}\frac{\partial A_\phi}{\partial \phi}\right)\mathbf{i}_r + \left(\nabla^2 A_\phi - \frac{A_\phi}{r^2} + \frac{2}{r^2}\frac{\partial A_r}{\partial \phi}\right)\mathbf{i}_\phi + (\nabla^2 A_z)\,\mathbf{i}_z.$$

P4.23. Show that the direction lines of the electric field of (4.102) are given by $r \csc^2\theta = \text{constant}$ and $\phi = \text{constant}$.

P4.24. An arrangement of point charges known as the rectangular quadrupole consists of the point charges Q, $-Q$, Q, and $-Q$, at the points (0, 0, 0), (Δx, 0, 0), (Δx, 0, Δz), and (0, 0, Δz), respectively. Obtain the approximate expression for the electric potential and hence for the electric field intensity due to the rectangular quadrupole at distances r from the origin large compared to Δx and Δz.

P4.25. For a point charge Q situated at a vector distance $\mathbf{r}'$ from the origin, show that the electric potential at distances $\mathbf{r}$ from the origin large in magnitude compared to $\mathbf{r}'$ can be expressed as a power series in r in the manner

$$V(\mathbf{r}) = \frac{Q}{4\pi\varepsilon_0 r} + \frac{Q\mathbf{r}'\cdot\mathbf{r}}{4\pi\varepsilon_0 r^3} + \frac{Q}{8\pi\varepsilon_0 r^5}[3(\mathbf{r}'\cdot\mathbf{r})^2 - r^2 r'^2] + \cdots$$

Then find the first two significant terms in the potential for the arrangement of point charges shown in Fig. 4.21 at large distances from it.

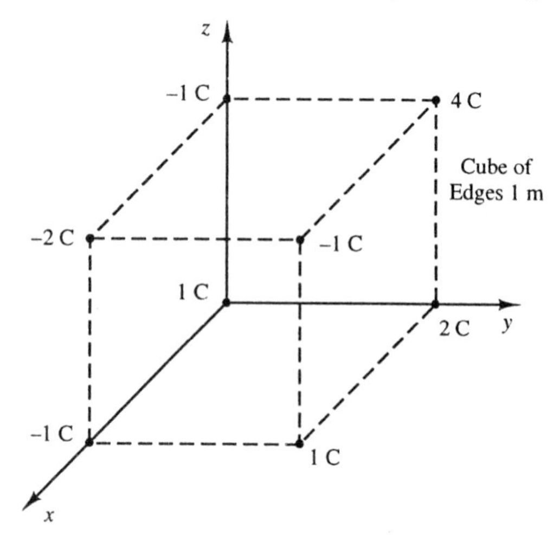

Figure 4.21. For Problem P4.25.

P4.26. For a finitely long line charge of uniform density ρ_{L0} C/m situated along the line between (0, 0, $-a$) and (0, 0, a), obtain the expression for the electric potential at an arbitrary point (r, ϕ, z) in cylindrical coordinates. Further show that the equipotential surfaces are ellipsoids with the ends of the line as their focii.

P4.27. Show that for two infinitely long line charges parallel to the z-axis, having uniform densities $\rho_{L1} = 2k\pi\varepsilon_0$ C/m and $\rho_{L2} = -2\pi\varepsilon_0$ C/m and passing through (-1, 0, 0) and (1, 0, 0), respectively, the potential is given by $V = \ln(r_2/r_1^k)$, where r_l and r_2 are the distances to the point from the line charges 1 and 2, respectively.

P4.28. Consider the surface charge distributed uniformly with density ρ_{S0} C/m^2 on a rectangular-shaped surface of sides a and b. Show that the electric potential at the center of the rectangle is

$$\frac{\rho_{S0}}{2\pi\varepsilon_0}\left(a \ln\frac{\sqrt{a^2+b^2}+b}{a} + b\ln\frac{\sqrt{a^2+b^2}+a}{b}\right)$$

Further show that for a square-shaped surface of sides a, the potential at the center is $(\rho_{S0}a/\pi\varepsilon_0) \ln (1 + \sqrt{2})$.

P4.29. Consider surface charge of uniform density ρ_{S0} C/m² distributed on an infinitely long strip lying between the straight lines $x = -a$, $y = 0$ and $x = a$, $y = 0$. Noting that the electric potential is independent of z, show that the potential difference between two points in the first quadrant of the xy-plane $(x_1, y_1, 0)$ and $(x_2, y_2, 0)$ is given by

$$\frac{\rho_{S0}}{4\pi\varepsilon_0}\left\{(a - x_2) \ln [(a - x_2)^2 + y_2^2] + (a + x_2) \ln [(a + x_2)^2 + y_2^2]\right.$$

$$- (a - x_1) \ln [(a - x_1)^2 + y_1^2] - (a + x_1) \ln [(a + x_1)^2 + y_1^2]$$

$$+ 2y_2\left(\tan^{-1}\frac{a - x_2}{y_2} + \tan^{-1}\frac{a + x_2}{y_2}\right)$$

$$\left.- 2y_1\left(\tan^{-1}\frac{a - x_1}{y_1} + \tan^{-1}\frac{a + x_1}{y_1}\right)\right\}$$

P4.30. Consider a circular current loop of radius a lying in the xy-plane with its center at the origin and with current I flowing in the sense of increasing ϕ, so that the magnetic dipole moment $\mathbf{m}$ is $I\pi a^2\mathbf{i}_z$. Show that far from the dipole such that $r \gg a$, the magnetic vector potential is given by

$$\mathbf{A} \approx \frac{\mu\mathbf{m} \times \mathbf{i}_r}{4\pi r^2}$$

and hence the magnetic flux density is given by

$$\mathbf{B} = \frac{\mu m}{4\pi r^3}(2 \cos \theta \, \mathbf{i}_r + \sin \theta \, \mathbf{i}_\theta)$$

P4.31. By expansion in Cartesian coordinates, show that

$$\mathbf{A} \times \nabla\Phi = \Phi\nabla \times \mathbf{A} - \nabla \times (\Phi\mathbf{A})$$

P4.32. Three point charges Q, Q, and kQ $(k > 0)$ are constrained to move on a circle of radius a. Let α be the angle subtended at the center of the circle by the two equal charges. Show that for the work required to assemble the charge distribution to be a minimum, the relationship between α and k is given by

$$4k \sin^3 \frac{\alpha}{4} = \cos \frac{\alpha}{2}$$

Find the values of α for values of k equal to 1, 2, and 1/2.

P4.33. Charges Q and $-Q$ are distributed uniformly on the surfaces of two concentric spheres of radii a and b $(> a)$, respectively. The outer sphere is made up of two separable hemispheres. Find the work required to carry the two hemispheres to an infinite distance from the inner sphere and join them to form a sphere of uniformly distributed charge $-Q$.

P4.34. A volume charge distribution is given in spherical coordinates by

$$\rho = \begin{cases} \rho_0(r/a) & \text{for } r < a \\ 0 & \text{for } r > a \end{cases}$$

(a) Find the work required to rearrange the charge with uniform density within the region $r < a$.

(b) Find the radius of the spherical region and the charge density for which neither additional work is required nor work is made available in redistributing the charge uniformly within the region.

P4.35. A surface current distribution is given in cylindrical coordinates by

$$\mathbf{J}_S = \begin{cases} (I_1/2\pi a)\mathbf{i}_z & \text{for } r = a \\ -[(I_1 + I_2)/2\pi b]\mathbf{i}_z & \text{for } r = b \\ (I_2/2\pi c)\mathbf{i}_z & \text{for } r = c \end{cases}$$

where $a < b < c$. Find the energy stored in the magnetic field of the current distribution, per unit length along the z-axis.

P4.36. Current I_0 A flows with uniform density $(I_0/\pi a^2)\mathbf{i}_z$ A/m^2 in an infinitely long, solid, cylindrical wire of radius a with its axis along the z-axis and returns with uniform density $-(I_0/4\pi a)\mathbf{i}_z$ A/m on a cylindrical surface of radius $2a$ and coaxial with the solid wire. Find the energy stored in the magnetic field of the current distribution per unit length in the z-direction.

PC EXERCISES

PC4.1. Consider the plotting of equipotential lines for two parallel, infinitely long, line charges of uniform charge densities. Let the charges be parallel to the z-axis and passing through $(-1, 0, 0)$ and $(1, 0, 0)$ with densities $\rho_{L1} = 2k\pi\varepsilon_0$ C/m and $\rho_{L2} = -2\pi\varepsilon_0$ C/m, respectively, where $k > 0$. Using the result of Problem P4.27 and the method of Example 4.10, write a program to plot an equipotential line, passing between the line charges, for a specified value of k and a specified value of the potential. Use the rectangular region in the xy-plane having the corners at the points $(-4, 0)$, $(-4, 5)$ $(4, 5)$, and $(4, 0)$.

PC4.2. Consider the extension of Exercise PC4.1 to plot the equipotential lines which surround both line charges for the case of $k > 1$. Show that as we go from $x = -\infty$ to $x = +\infty$ along the x-axis, the potential V varies in the following manner: (1) increases monotonically from $-\infty$ at $x = -\infty$ to $+\infty$ at $x = -1$; (2) decreases monotonically from $+\infty$ at $x = -1$ to $-\infty$ at $x = 1$; and (3) increases from $-\infty$ at $x = 1$ to a maximum value

$$V = V_m = \left(\ln \frac{2}{k-1} - k \ln \frac{2k}{k-1}\right) \qquad \text{at} \quad x = x_m = \frac{k+1}{k-1}$$

and then decreases to $-\infty$ at $x = +\infty$. Thus all equipotential lines passing through points on the x-axis to the right of x_m surround both line charges. Extend the program of Exercise PC4.1 to plot these equipotential lines and generate a sample plot for value of k specified by your instructor.

PC4.3. Using the procedure employed in Example 4.10, write a program for plotting the equipotential line through a specified point in the xz-plane (instead of plotting an equipotential line for a specified value of potential) for a given set of n point charges in that plane. Use the program to plot a set of equipotentials for values and locations of point charges and boundary of the region in the xz-plane, specified by your instructor.

5

Topics in Static and Quasistatic Fields

In Chapters 3 and 4 we learned Maxwell's equations in integral form and in differential form, respectively. In Chapter 3 we also discussed the computation of static electric and magnetic fields due to symmetrical charge and current distributions, respectively, by using Gauss' law in integral form and Ampère's circuital law in integral form without the displacement current term, respectively. In Chapter 4 we also introduced the electromagnetic potentials and their specializations to static fields. In this chapter we consider several topics in static fields and also extend our study to quasistatic fields, which are low-frequency extensions of static fields.

We begin the chapter with two important differential equations involving the electric potential and discuss several applications based on the solution of these equations. Next we introduce a numerical technique involving the inversion of an integral equation for solving a class of problems for which exact analytical solutions are not in general possible. We then introduce an important relationship between the circuit parameters capacitance, conductance, and inductance for infinitely long, parallel conductor arrangements and consider their determination. In the remainder of the chapter, we extend our study to the quasistatic case, illustrating the determination of the low-frequency behavior of physical structures via the quasistatic field approach, and finally consider two topics pertinent to the study of electromechanical systems.

5.1 POISSON'S AND LAPLACE'S EQUATIONS

Poisson's equation

In Section 4.5 we introduced the static electric potential as related to the static electric field in the manner

$$\boxed{\mathbf{E} = -\nabla V} \tag{5.1}$$

Substituting (5.1) into Maxwell's divergence equation for $\mathbf{D}$ given by

$$\nabla \cdot \mathbf{D} = \rho \tag{5.2}$$

we obtain

$$-\nabla \cdot \varepsilon \nabla V = \rho \tag{5.3}$$

where ε is the permittivity of the medium. Using the vector identity

$$\nabla \cdot \Phi \mathbf{A} = \Phi \nabla \cdot \mathbf{A} + \mathbf{A} \cdot \nabla \Phi \tag{5.4}$$

we can write (5.3) as

$$\varepsilon \nabla \cdot \nabla V + \nabla \varepsilon \cdot \nabla V = -\rho$$

or

$$\boxed{\varepsilon \nabla^2 V + \nabla \varepsilon \cdot \nabla V = -\rho} \tag{5.5}$$

If we assume ε to be uniform in the region of interest, then $\nabla \varepsilon = \mathbf{0}$ and (5.5) becomes

$$\boxed{\nabla^2 V = \frac{-\rho}{\varepsilon}} \tag{5.6}$$

This equation is known as Poisson's equation. It governs the relationship between the volume charge density ρ in a region of uniform permittivity ε to the electric scalar potential V in that region. Note that (5.6) also follows from (4.90) for $\partial/\partial t = 0$ and $\Phi = V$. In Cartesian coordinates, (5.6) becomes

$$\frac{\partial^2 V}{\partial x^2} + \frac{\partial^2 V}{\partial y^2} + \frac{\partial^2 V}{\partial z^2} = -\frac{\rho}{\varepsilon} \tag{5.7}$$

which is a three-dimensional, second-order partial differential equation. For the one-dimensional case in which V varies with x only, $\partial^2 V/\partial y^2$ and $\partial^2 V/\partial z^2$ are both equal to zero, and (5.7) reduces to

$$\boxed{\frac{\partial^2 V}{\partial x^2} = \frac{d^2 V}{dx^2} = -\frac{\rho}{\varepsilon}} \tag{5.8}$$

We shall illustrate the application of (5.8) by means of an example.

Example 5.1

p-n *junction semiconductor*

Let us consider the space charge layer in a p-n junction semiconductor with zero bias, as shown in Fig. 5.1(a), in which the region $x < 0$ is doped p-type and the region $x > 0$ is doped n-type. To review briefly the formation of the space charge layer, we note that since the density of the holes on the p side is larger than that on the n side, there is a tendency for the holes to diffuse to the n side and recombine with the electrons. Similarly, there is a tendency for the electrons on the n side to diffuse to the p side and recombine with the holes. The diffusion of holes leaves be-

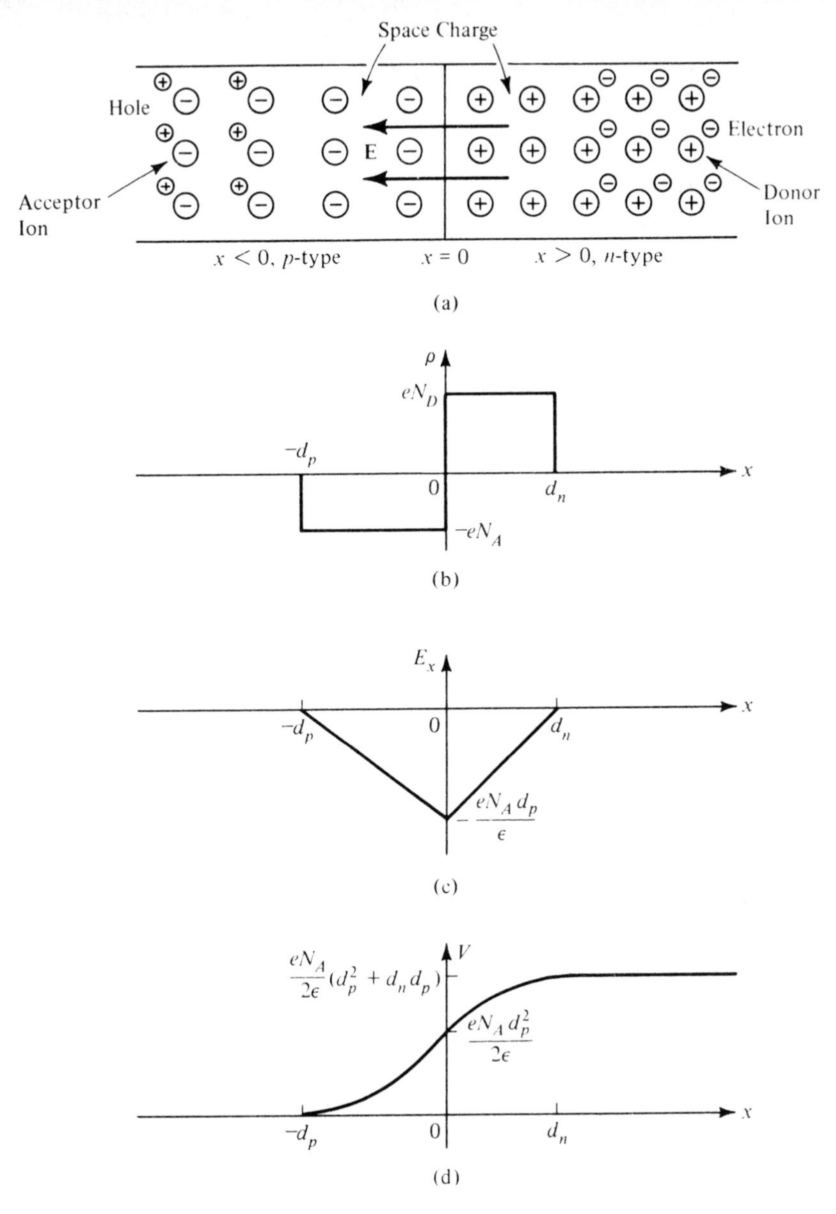

Figure 5.1. For illustrating the application of Poisson's equation for the determination of the potential distribution for a p-n junction semiconductor.

hind negatively charged acceptor atoms, and the diffusion of electrons leaves behind positively charged donor atoms. Since these acceptor and donor atoms are immobile, a space charge layer, also known as the *depletion layer*, is formed in the region of the junction with negative charges on the p side and positive charges on the n side. This space charge gives rise to an electric field directed from the n side of the junction to the p side so that it opposes diffusion of the mobile carriers across the junction thereby resulting in an equilibrium. For simplicity, let us consider an abrupt junction, that is, a junction in which the impurity concentration is constant on either side of the junction. Let N_A and N_D be the acceptor and donor ion concentrations, respectively, and d_p and d_n be the widths in the p and n regions, respectively, of the depletion layer. The space charge density ρ is then given by

$$\rho = \begin{cases} -eN_A & \text{for} \quad -d_p < x < 0 \\ eN_D & \text{for} \quad 0 < x < d_n \end{cases} \tag{5.9}$$

as shown in Fig. 5.1(b), where e is the magnitude of the electronic charge. Since the semiconductor is electrically neutral, the total acceptor charge must be equal to the total donor charge; that is,

$$eN_A d_p = eN_D d_n \tag{5.10}$$

We wish to find the potential distribution in the depletion layer and the depletion layer width in terms of the potential difference across the depletion layer and the acceptor and donor ion concentrations.

Substituting (5.9) into (5.8) we obtain the equation governing the potential distribution to be

$$\frac{d^2V}{dx^2} = \begin{cases} \dfrac{eN_A}{\varepsilon} & \text{for} \quad -d_p < x < 0 \\ -\dfrac{eN_D}{\varepsilon} & \text{for} \quad 0 < x < d_n \end{cases} \tag{5.11}$$

To solve (5.11) for V, we integrate it once and obtain

$$\frac{dV}{dx} = \begin{cases} \dfrac{eN_A}{\varepsilon}x + C_1 & \text{for} \quad -d_p < x < 0 \\ -\dfrac{eN_D}{\varepsilon}x + C_2 & \text{for} \quad 0 < x < d_n \end{cases}$$

where C_1 and C_2 are constants of integration. To evaluate C_1 and C_2, we note that since $\mathbf{E} = -\nabla V = -(\partial V/\partial x)\mathbf{i}_x$, $\partial V/\partial x$ is simply equal to $-E_x$. Since the electric field lines begin on the positive charges and end on the negative charges, and in view of (5.10) the field and hence $\partial V/\partial x$ must vanish at $x = -d_p$ and $x = d_n$, giving us

$$\frac{dV}{dx} = \begin{cases} \dfrac{eN_A}{\varepsilon}(x + d_p) & \text{for} \quad -d_p < x < 0 \\ -\dfrac{eN_D}{\varepsilon}(x - d_n) & \text{for} \quad 0 < x < d_n \end{cases} \tag{5.12}$$

The field intensity, that is, $-dV/dx$, may now be sketched as a function of x as shown in Fig. 5.1(c).

Proceeding further, we integrate (5.12) and obtain

$$V = \begin{cases} \dfrac{eN_A}{2\varepsilon}(x + d_p)^2 + C_3 & \text{for} \quad -d_p < x < 0 \\ -\dfrac{eN_D}{2\varepsilon}(x - d_n)^2 + C_4 & \text{for} \quad 0 < x < d_n \end{cases}$$

where C_3 and C_4 are constants of integration. To evaluate C_3 and C_4, we first set the potential at $x = -d_p$ arbitrarily equal to zero to obtain C_3 equal to zero. Then we make use of the condition that the potential be continuous at $x = 0$, since the discontinuity in dV/dx at $x = 0$ is finite, to obtain

$$\frac{eN_A}{2\varepsilon}d_p^2 = -\frac{eN_D}{2\varepsilon}d_n^2 + C_4$$

or

$$C_4 = \frac{e}{2\varepsilon}(N_A d_p^2 + N_D d_n^2)$$

Substituting this value for C_4 and setting C_3 equal to zero in the expression for V, we get the required solution

$$V = \begin{cases} \frac{eN_A}{2\varepsilon}(x + d_p)^2 & \text{for} \quad -d_p < x < 0 \\ -\frac{eN_D}{2\varepsilon}(x^2 - 2xd_n) + \frac{eN_A}{2\varepsilon}d_p^2 & \text{for} \quad 0 < x < d_n \end{cases} \tag{5.13}$$

The variation of potential with x as given by (5.13) is shown in Fig. 5.1(d).

We can proceed further and find the width $d = d_p + d_n$ of the depletion layer by setting $V(d_n)$ equal to the contact potential, V_0, that is, the potential difference across the depletion layer resulting from the electric field in the layer. Thus

$$\begin{aligned} V_0 = V(d_n) &= \frac{eN_D}{2\varepsilon}d_n^2 + \frac{eN_A}{2\varepsilon}d_p^2 \\ &= \frac{e}{2\varepsilon}\frac{N_D(N_A + N_D)}{N_A + N_D}d_n^2 + \frac{e}{2\varepsilon}\frac{N_A(N_A + N_D)}{N_A + N_D}d_p^2 \\ &= \frac{e}{2\varepsilon}\frac{N_A N_D}{N_A + N_D}(d_n^2 + d_p^2 + 2d_n d_p) \\ &= \frac{e}{2\varepsilon}\frac{N_A N_D}{N_A + N_D}d^2 \end{aligned}$$

where we have made use of (5.10). Finally, we obtain the result that

$$d = \sqrt{\frac{2\varepsilon V_0}{e}\left(\frac{1}{N_A} + \frac{1}{N_D}\right)}$$

which tells us that the depletion layer width is smaller the heavier the doping is. This property is used in tunnel diodes to achieve layer widths on the order of 10^{-6} cm by heavy doping as compared to widths on the order of 10^{-4} cm in ordinary *p-n* junctions.

We have just illustrated an example of the application of Poisson's equation involving the solution for the potential distribution for a given charge distribution. Poisson's equation is even more useful for the solution of problems in which the charge distribution is the quantity to be determined given the functional dependence of the charge density on the potential. We shall, however, proceed to the discussion of Laplace's equation.

Laplace's equation

If the charge density in a region is zero, then Poisson's equation (5.6) reduces to

$$\boxed{\nabla^2 V = 0} \tag{5.14}$$

This equation is known as *Laplace's equation*. It governs the behavior of the potential in a charge-free region characterized by uniform permittivity. In Cartesian coordinates, it is given by

$$\frac{\partial^2 V}{\partial x^2} + \frac{\partial^2 V}{\partial y^2} + \frac{\partial^2 V}{\partial z^2} = 0 \tag{5.15}$$

Laplace's equation is also satisfied by the potential in conductors under steady current condition. For the steady current condition, $\partial\rho/\partial t = 0$, and the continuity equation given for the time-varying case by

$$\nabla \cdot \mathbf{J}_c + \frac{\partial \rho}{\partial t} = 0$$

reduces to

$$\nabla \cdot \mathbf{J}_c = 0 \tag{5.16}$$

Replacing $\mathbf{J}_c$ by $\sigma\mathbf{E} = -\sigma\nabla V$ where σ is the conductivity of the conductor and assuming σ to be constant, we obtain

$$\nabla \cdot \sigma\mathbf{E} = \sigma\nabla \cdot \mathbf{E} = -\sigma\nabla \cdot \nabla V = -\sigma\nabla^2 V = 0$$

or

$$\nabla^2 V = 0$$

The problems for which Laplace's equation is applicable consist of finding the potential distribution in the region between two conductors given the charge distribution on the surfaces of the conductors or the potentials of the conductors or a combination of the two. The procedure involves the solving of Laplace's equation subject to the boundary conditions on the surfaces of the conductors. We shall illustrate this by means of an example involving variation of V in one dimension.

Example 5.2

Parallel-plate arrangement, capacitance

Let us consider two infinite, plane, parallel, perfectly conducting plates occupying the planes $x = 0$ and $x = d$ and kept at potentials $V = 0$ and $V = V_0$, respectively, as shown by the cross-sectional view in Fig. 5.2, and find the solution for Laplace's equation in the region between the plates. The arrangement may be considered an idealization of a parallel-plate capacitor with its plates having dimensions very large compared to the spacing between them.

The potential is obviously a function of x only and hence (5.15) reduces to

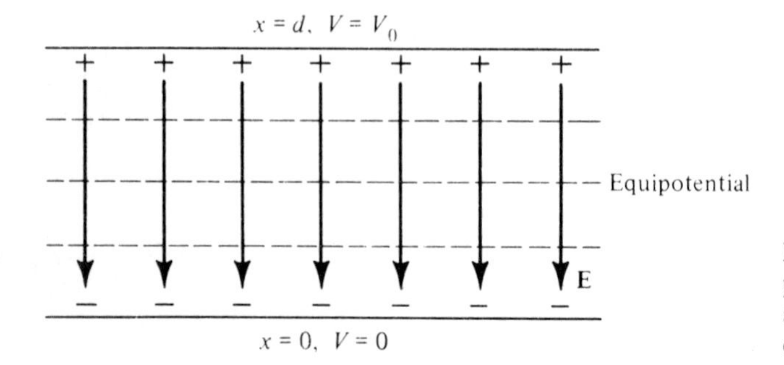

Figure 5.2. Cross-sectional view of parallel-plate capacitor for illustrating the solution of Laplace's equation in one dimension.

$$\frac{\partial^2 V}{\partial x^2} = \frac{d^2 V}{dx^2} = 0$$

Integrating this equation twice, we obtain

$$\boxed{V(x) = Ax + B} \tag{5.17}$$

where A and B are constants of integration. To determine the values of A and B, we make use of the boundary conditions for V; that is,

$$V = 0 \qquad \text{for} \quad x = 0$$

$$V = V_0 \qquad \text{for} \quad x = d$$

giving us

$$0 = A(0) + B \qquad \text{or} \qquad B = 0$$

$$V_0 = A(d) + B = Ad \qquad \text{or} \qquad A = \frac{V_0}{d}$$

Thus the particular solution for the potential here is given by

$$\boxed{V = \frac{V_0}{d}x \qquad \text{for} \quad 0 < x < d} \tag{5.18}$$

which tells us that the equipotentials are planes parallel to the conductors, as shown in Fig. 5.2.

Proceeding further, we obtain

$$\boxed{\mathbf{E} = -\boldsymbol{\nabla} V = -\frac{\partial V}{\partial x}\mathbf{i}_x = -\frac{V_0}{d}\mathbf{i}_x \qquad \text{for} \quad 0 < x < d} \tag{5.19}$$

This field is uniform and directed from the higher potential plate to the lower potential plate, as shown in Fig. 5.2. The surface charge densities on the two plates are given by

$$[\rho_S]_{x=0} = [\mathbf{D}]_{x=0} \cdot \mathbf{i}_x = -\frac{\varepsilon V_0}{d}\mathbf{i}_x \cdot \mathbf{i}_x = -\frac{\varepsilon V_0}{d}$$

$$[\rho_S]_{x=d} = [\mathbf{D}]_{x=d} \cdot (-\mathbf{i}_x) = -\frac{\varepsilon V_0}{d}\mathbf{i}_x \cdot (-\mathbf{i}_x) = \frac{\varepsilon V_0}{d}$$

The magnitude of the surface charge per unit area on either plate is

$$Q = |\rho_S|(1) = \frac{\varepsilon V_0}{d}$$

Finally, we can find the capacitance C per unit area of the plates, defined to be the ratio of Q to V_0. Thus

$$\boxed{C = \frac{Q}{V_0} = \frac{\varepsilon}{d} \text{ per unit area of the plates}} \tag{5.20}$$

The units of capacitance are farads (F).

Parallel-plate arrangement, conductance

If the medium between the plates in Fig. 5.2 is a conductor, then the conduction current density is given by

$$\mathbf{J}_c = \sigma \mathbf{E} = -\frac{\sigma V_0}{d}\mathbf{i}_x$$

The conduction current from the higher potential plate to the lower potential plate per unit area of the plates is

$$I_c = |\mathbf{J}_c|(1) = \frac{\sigma V_0}{d}$$

The ratio of this current to the potential difference is the conductance G (reciprocal of resistance) per unit area of the plates. Thus

$$\boxed{G = \frac{I_c}{V_0} = \frac{\sigma}{d} \text{ per unit area of the plates}} \tag{5.21}$$

The units of conductance are siemens (S).

Cylindrical and spherical capacitors

We have just illustrated the solution of Laplace's equation in one dimension by considering an example involving the variation of V with one Cartesian coordinate. In a similar manner, solutions for one-dimensional Laplace's equations involving variations of V with single coordinates in the other two coordinate systems can be obtained. Of particular interest are the case in which V is a function of the cylindrical coordinate r only, pertinent to the geometry of a capacitor made up of coaxial cylindrical conductors and the case in which V is a function of the spherical coordinate r only, pertinent to the geometry of a capacitor made up of concentric spherical conductors. These two geometries are shown in Fig. 5.3(a) and (b), respectively. The various steps in the solution of Laplace's equation and subsequent determination of capacitance for these two cases are summarized in Table 5.1, which also includes the parallel plane case of Fig. 5.2.

Two-dimensional Laplace's equation

Returning now to (5.15), let us consider the case of variation of V with two dimensions, say, x and y. Then, we obtain the two-dimensional Laplace's equation

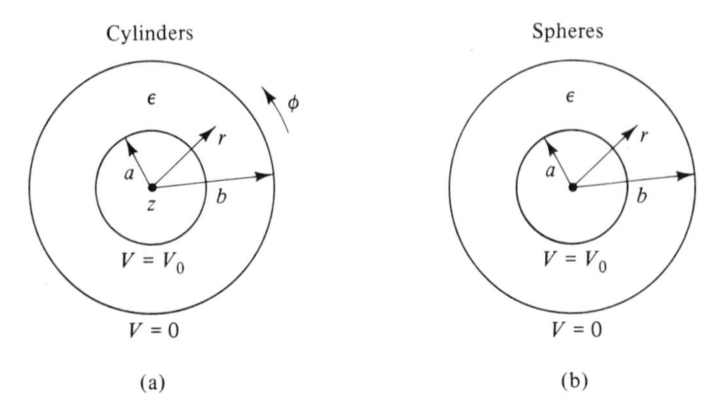

Figure 5.3. Cross-sectional views of capacitors made up of (a) coaxial cylindrical conductors and (b) concentric spherical conductors.

TABLE 5.1 SUMMARY OF VARIOUS STEPS IN THE SOLUTION OF LAPLACE'S EQUATION AND DETERMINATION OF CAPACITANCE FOR THREE ONE-DIMENSIONAL CASES

Geometry	Parallel planes	Coaxial cylinders	Concentric spheres
Figure	5.2	5.3(a)	5.3(b)
Boundary conditions	$V = 0, \quad x = 0$ $V = V_0, \quad x = d$	$V = V_0, \quad r = a$ $V = 0, \quad r = b$	$V = V_0, \quad r = a$ $V = 0, \quad r = b$
Laplace's equation	$\frac{\partial^2 V}{\partial x^2} = 0$	$\frac{1}{r}\frac{\partial}{\partial r}\left(r\frac{\partial V}{\partial r}\right) = 0$	$\frac{1}{r^2}\frac{\partial}{\partial r}\left(r^2\frac{\partial V}{\partial r}\right) = 0$
General solution	$V = Ax + B$	$V = A \ln r + B$	$V = \frac{A}{r} + B$
Particular solution	$V = V_0 \frac{x}{d}$	$V = V_0 \frac{\ln (r/b)}{\ln (a/b)}$	$V = V_0 \frac{1/r - 1/b}{1/a - 1/b}$
Electric field	$-\frac{V_0}{d}\mathbf{i}_x$	$\frac{V_0}{r \ln (b/a)}\mathbf{i}_r$	$\frac{V_0}{r^2(1/a - 1/b)}\mathbf{i}_r$
Surface charge densities	$-\frac{\varepsilon V_0}{d}, \quad x = 0$ $\frac{\varepsilon V_0}{d}, \quad x = d$	$\frac{\varepsilon V_0}{a \ln (b/a)}, \quad r = a$ $\frac{-\varepsilon V_0}{b \ln (b/a)}, \quad r = b$	$\frac{\varepsilon V_0}{a^2(1/a - 1/b)}, \quad r = a$ $\frac{-\varepsilon V_0}{b^2(1/a - 1/b)}, \quad r = b$
Capacitance	$\frac{\varepsilon}{d}$ per unit area	$\frac{2\pi\varepsilon}{\ln (b/a)}$ per unit length	$\frac{4\pi\varepsilon}{1/a - 1/b}$

$$\boxed{\frac{\partial^2 V}{\partial x^2} + \frac{\partial^2 V}{\partial y^2} = 0} \tag{5.22}$$

The analytical technique by means of which an equation of this type is solved is known as the *separation of variables* technique. We shall, however, consider a numerical technique which forms the basis for computer solution and defer the discussion of the separation of variables technique to Section 9.4, where we will encounter a similar partial differential equation.

Numerical solution of two-dimensional Laplace's equation

To introduce the principle behind the numerical solution, let us suppose that we know the potentials V_1, V_2, V_3, and V_4 at four points equidistant from a point $P(0, 0, 0)$ and lying on mutually perpendicular axes, which we call x and y, passing through P as shown in Fig. 5.4, and that we wish to find the potential V_0 at P in terms of V_1, V_2, V_3, and V_4. Then assuming no variation of V in the z-direction, we require that

$$[\nabla^2 V]_P = \left[\frac{\partial^2 V}{\partial x^2} + \frac{\partial^2 V}{\partial y^2}\right]_{(0,0,0)} = 0 \tag{5.23}$$

To solve this equation approximately for V_0, we note that

$$\left[\frac{\partial^2 V}{\partial x^2}\right]_{(0,0,0)} = \left[\frac{\partial}{\partial x}\left(\frac{\partial V}{\partial x}\right)\right]_{(0,0,0)}$$
$$\approx \frac{1}{a}\left\{\left[\frac{\partial V}{\partial x}\right]_{(a/2,0,0)} - \left[\frac{\partial V}{\partial x}\right]_{(-a/2,0,0)}\right\}$$

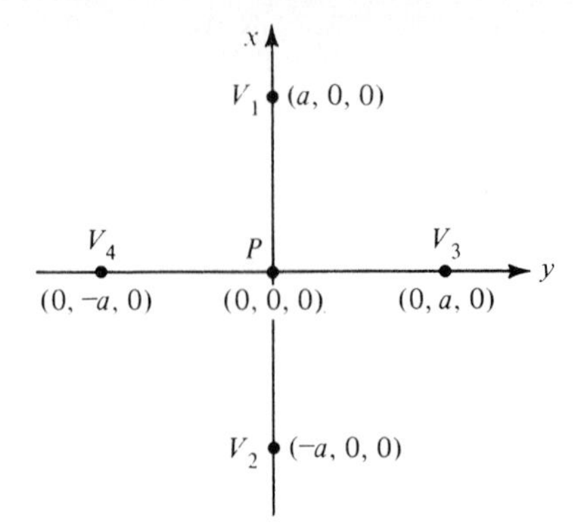

Figure 5.4. For illustrating the principle behind the numerical solution of Laplace's equation in two dimensions.

$$\approx \frac{1}{a}\left\{\frac{[V]_{(a,0,0)} - [V]_{(0,0,0)}}{a} - \frac{[V]_{(0,0,0)} - [V]_{(-a,0,0)}}{a}\right\}$$

$$= \frac{1}{a^2}[(V_1 - V_0) - (V_0 - V_2)]$$

$$= \frac{1}{a^2}(V_1 + V_2 - 2V_0) \tag{5.24a}$$

Similarly,

$$\left[\frac{\partial^2 V}{\partial y^2}\right]_{(0,0,0)} \approx \frac{1}{a^2}(V_3 + V_4 - 2V_0) \tag{5.24b}$$

Substituting (5.24a) and (5.24b) into (5.23) and rearranging, we obtain

$$\boxed{V_0 \approx \tfrac{1}{4}(V_1 + V_2 + V_3 + V_4)} \tag{5.25}$$

Thus the potential at P is approximately equal to the average of the potentials at the four equidistant points lying along mutually perpendicular axes through P. The result becomes more and more accurate as the spacing a becomes less and less. Equation (5.25) is the finite-difference approximation to (5.22) and forms the basis for its computer solution by the *finite-difference method*. We shall illustrate this by means of an example.

Example 5.3

Finite-difference method

Let us consider four infinitely long conducting strips of equal widths, situated such that the cross section of the arrangement is a square and held at potentials V_a, V_b, V_l, and V_r, as shown in Fig. 5.5. Note that the corners are insulated so that the plates do not touch. By dividing the area between the conductors into a 6×6 grid of squares, and using (5.25), we wish to find the approximate values of the potentials at the grid points by the finite-difference method.

The solution consists of obtaining a set of values for the potentials at the grid points such that the potential at each grid point is the average of the potentials at the neighboring four grid points to within a specified tolerance. Thus if we denote the potentials to be V_{11}, V_{12}, V_{13}, V_{14}, V_{15}, V_{21}, V_{22}, . . . , V_{55}, and if the specified tolerance is denoted to be Δ, then the values of the potentials must be such that

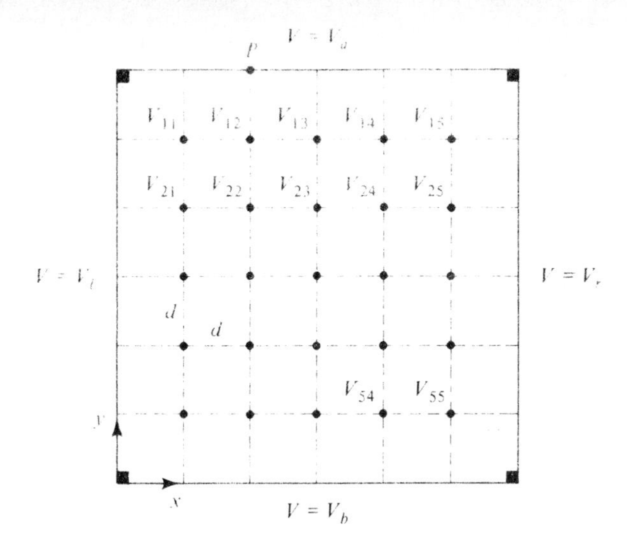

Figure 5.5. Cross-sectional view of an arrangement of four infinitely long conducting strips, with the region inside divided into a 6 × 6 grid of squares.

$$\left| V_{11} - \tfrac{1}{4}(V_a + V_{12} + V_{21} + V_l) \right| < \Delta \tag{5.26a}$$

$$\left| V_{12} - \tfrac{1}{4}(V_a + V_{13} + V_{22} + V_{11}) \right| < \Delta \tag{5.26b}$$

and so on. The simplest technique adaptable to computer solution is to begin with values of zero for all unknown potentials. By traversing the grid in a systematic manner, the average of the four neighboring potentials is computed for each grid point and is used to replace the potential at that grid point if that value differs from the computed average by more than Δ. This procedure is repeated until a final set of values for the unknown potentials consistent with (5.26a), (5.26b), . . . is obtained.

Let us consider some numerical values: $V_a = 100$ V, $V_b = 0$ V, $V_l = 40$ V, $V_r = 0$ V, and $\Delta = 0.01$ V. Then we first set all unknown potentials equal to zero. Beginning at the grid point 11 and traversing the grid rowwise, we replace the zero value for V_{11} by $\frac{1}{4}(100 + 40 + 0 + 0)$ or 35 V, then replace the zero value for V_{12} by $\frac{1}{4}(100 + 35 + 0 + 0)$ or 33.75 V, and so on. After one traversal is completed, we come back to the grid point 11 and traverse the grid again replacing the potential value at each grid point by the average of the then existing values of the four neighboring potentials, as necessary. This procedure is repeated until the desired set of values is obtained.

A PC program for carrying out the procedure just discussed for given set of values of V_a, V_b, V_l, V_r, and Δ is included as PL 5.1. The final set of values for the potentials resulting from a run of the program for $V_a = 100$ V, $V_b = 0$ V, $V_l = 40$ V, $V_r = 0$ V, and $\Delta = 0.01$ V is shown in Fig. 5.6, which also shows the residuals, where a residual at a grid point is the absolute value of the difference between the potential at that grid point and the average of the four neighboring potentials. The residuals are shown below the potential values. It can be seen that all residuals are less than 0.01 V.

PL 5.1. Program listing for numerical solution of Laplace's equation in two dimensions for the potentials at the grid points in the arrangement of Fig. 5.5.

```
100 '***************************************************
110 '* NUMERICAL SOLUTION OF LAPLACE'S EQUATION IN TWO *
120 '* DIMENSIONS                                      *
130 '***************************************************
```

PL 5.1. (continued)

```
140 DIM V(7,7)
150 CLS:SCREEN 0
160 PRINT "ENTER INTEGER VALUES (NOT EXCEEDING 100"
170 PRINT "IN ABSOLUTE VALUE) FOR THE POTENTIALS"
180 PRINT "ON THE FOUR SIDES:":PRINT
190 INPUT "VTOP = ",VTOP
200 INPUT "VBOTTOM = ",VBOT
210 INPUT "VLEFT = ",VLEF
220 INPUT "VRIGHT = ",VRIG
230 PRINT:PRINT "ENTER VALUE OF DELTA:"
240 PRINT:INPUT "VDELTA = ",DV
250 LOCATE 23,1:PRINT "STRIKE ANY KEY TO CONTINUE"
260 C$=INPUT$(1):CLS:NT=0
270 '* PRINT POTENTIAL VALUES ON THE FOUR SIDES AND
280 '  ZEROES AT THE INTERIOR GRID POINTS *
290 FOR J=2 TO 6:JT=6*J-6
300 V(1,J)=VTOP:LOCATE 1,JT:PRINT VTOP
310 V(7,J)=VBOT:LOCATE 19,JT:PRINT VBOT
320 NEXT
330 FOR I=2 TO 6
340 V(I,1)=VLEF:V(I,7)=VRIG:IT=3*I-2
350 FOR J=2 TO 6:V(I,J)=0:NEXT
360 LOCATE IT,1:PRINT V(I,1)
370 FOR J=2 TO 7:JT=6*J-6:LOCATE IT,JT:PRINT V(I,J):NEXT
380 NEXT
390 '* CARRY OUT ITERATION PROCEDURE *
400 IC=0:NT=NT+1:DM=0
410 LOCATE 21,1:PRINT "ITERATION NO. =";NT
420 FOR I=2 TO 6:FOR J=2 TO 6
430 IT=3*I-2:JT=6*J-6:IV=IT+1
440 VIJ=(V(I-1,J)+V(I+1,J)+V(I,J-1)+V(I,J+1))/4
450 DC=ABS(VIJ-V(I,J)):'* RESIDUAL *
460 IF DC>DM THEN DM=DC:'* HIGHEST VALUE OF RESIDUAL IN
470 '  A GIVEN ITERATION *
480 IF DC>=DV THEN V(I,J)=VIJ:IC=1:LOCATE IT,JT:PRINT "
        ":LOCATE IT,JT:PRINT USING "##.##";V(I,J)
490 LOCATE IV,JT:PRINT "      ":LOCATE IV,JT
500 IF DC>= 10 THEN PRINT USING "##.##";DC:GOTO 520
510 PRINT USING "#.###";DC
520 NEXT:NEXT
530 IF IC=1 THEN 400:'* CHECK IF SOLUTION COMPLETED *
540 LOCATE 22,1:PRINT "SOLUTION COMPLETED"
550 LOCATE 23,1:PRINT "VALUE OF DELTA ACHIEVED <=";DM;
560 LOCATE 24,1:PRINT "STRIKE ANY KEY TO CONTINUE";
570 C$=INPUT$(1):GOTO 150
580 END
```

The method we just discussed is known as the *iteration* technique since it involves the iterative process of converging an initially assumed solution to a final one consistent with Laplace's equation in the approximate sense given by (5.25). There are several variations of the iteration technique. For example, by employing an initial guess other than zeros, a faster convergence may be achieved. The end result will, however, still be only to within the specified accuracy. Alternative to the iteration technique, one can write a set of simultaneous equations by applying (5.25) to each grid point and then solve the equations for the unknown potentials (see Problem D5.3).

The solution obtained for the potentials at the grid points by any method can be used to plot approximately the equipotential lines by interpolating between

```
        100   100   100   100   100

  40   65.60 72.71 72.30 65.76 48.10   0
       0.006 0.006 0.004 0.006 0.006

  40   49.69 52.99 50.73 42.68 26.66   0
       0.006 0.007 0.004 0.007 0.006

  40   40.21 38.84 34.95 27.61 15.89   0
       0.004 0.004 0.000 0.004 0.004

  40   32.32 27.23 22.65 16.92  9.29   0
       0.006 0.007 0.004 0.007 0.006

  40   21.86 15.14 11.49  8.19  4.36   0
       0.006 0.006 0.004 0.006 0.006

        0     0     0     0     0

ITERATION NO. = 25
SOLUTION COMPLETED
VALUE OF DELTA ACHIEVED <= 7.423401E-03
```

Figure 5.6. Final set of values of potentials and residuals from a run of the program of PL 5.1, for $V_a = 100$ V, $V_b = 0$ V, $V_l = 40$ V, $V_r = 0$ V, and $\Delta = 0.01$ V.

grid points. An example of such plotting by using an extension of PL 5.1 is shown in Fig. 5.7, which corresponds to that of $V_a = V_l = V_b = 0$ V and $V_r = 100$ V in Fig. 5.5, and an 8×8 grid of squares. Figure 5.7(a) shows the computed potential values at a 4×4 set of grid points (with the remaining grid points omitted for the sake of clarity) and the 25-V equipotential line being plotted. Figure 5.7(b) shows a complete set of equipotential lines from 0 to 100 V in steps of 10 V. Note that in Fig. 5.7(b) the 0-V equipotential line does not follow the boundary at the upper- and lower-left corners. This is because in view of the division of the region into a finite grid of squares (8×8 here) the solution is not influenced by the corner points; that is, the solution for the case of the 0-V conductor following the plotted 0-V line is the same as that for which it follows the original boundary.

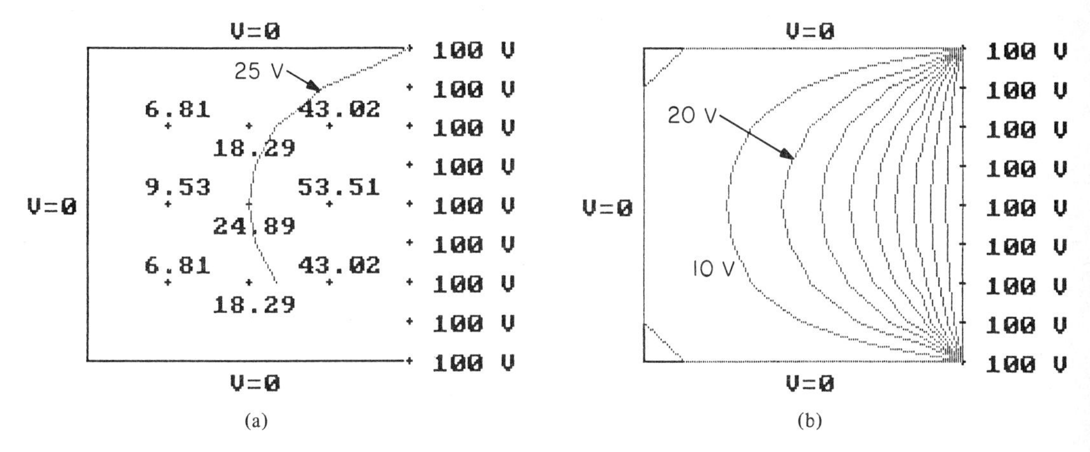

Figure 5.7. (a) Plotting of an equipotential line by interpolation between grid points. (b) Set of equipotential lines from 0 to 100 V in steps of 10 V.

Finally, the solution for the potentials can also be used to find approximate electric field intensities at the grid points by using the potential values to obtain approximate values of $\partial V/\partial x$ and $\partial V/\partial y$. For example, in Fig. 5.5, the electric field intensity at the grid point 12 is given approximately by

$$[\mathbf{E}]_{12} \approx \frac{V_{11} - V_{13}}{2d}\mathbf{i}_x + \frac{V_{22} - V_a}{2d}\mathbf{i}_y$$

where d is the spacing between two adjacent grid points. Similarly, the electric field intensities at points on the conductors can be found and used to obtain the surface charge densities. For example, the surface charge density at the point P on the conductor of potential V_a and adjacent to the grid point 12 is given approximately by

$$[\rho_S]_P \approx -\mathbf{i}_y \cdot \varepsilon\frac{V_{12} - V_a}{d}\mathbf{i}_y$$

$$= \varepsilon\frac{V_a - V_{12}}{d}$$

where ε is the permittivity of the medium between the conductors.

K5.1. Poisson's equation; *p-n* junction; Laplace's equation in one dimension; Parallel-plate arrangement; Capacitance; Conductance; Cylindrical and spherical capacitors; Laplace's equation in two dimensions; Numerical solution; Finite-difference method.

D5.1. The potential distribution in a simplified model of a vacuum diode consisting of cathode in the plane $x = 0$ and anode in the plane $x = d$ and held at a potential V_0 relative to the cathode is given by $V = V_0(x/d)^{4/3}$ for $0 < x < d$. Find the following: **(a)** V at $x = d/8$; **(b)** $\mathbf{E}$ at $x = d/8$; **(c)** ρ at $x = d/8$; and **(d)** ρ_S on the anode.
Ans. **(a)** $V_0/16$; **(b)** $-(2V_0/3d)\mathbf{i}_x$; **(c)** $-16\varepsilon_0 V_0/9d^2$; **(d)** $4\varepsilon_0 V_0/3d$

D5.2. Find the following: **(a)** the spacing between the plates of a parallel-plate capacitor with a dielectric of $\varepsilon = 2.25\varepsilon_0$ and having capacitance per unit area equal to 1000 pF; **(b)** the ratio of the outer radius to the inner radius for a coaxial cylindrical capacitor with a dielectric of $\varepsilon = 2.25\varepsilon_0$, and having capacitance per unit length equal to 100 pF; and **(c)** the radius of an isolated spherical conductor in free space for which the capacitance is 10 pF.
Ans. **(a)** 1.99 cm; **(b)** 3.4903; **(c)** 9 cm

D5.3. Three infinitely long conductor strips are arranged such that the cross section is an isosceles triangle, as shown in Fig. 5.8. The region between the conductors is divided into a grid of points as shown in the figure, where the spacing between adjacent pairs of points is d. By writing equations consistent with (5.39) for the potentials at the grid points A, B, and C and solving them, find the following: **(a)** the approximate potential at the grid point C; **(b)** the approximate electric field intensity at the grid point C; and **(c)** the approximate surface charge density at the point P, assuming the medium between the conductors to be free space.
Ans. **(a)** 8 V; **(b)** $-(5\mathbf{i}_x + 7\mathbf{i}_y)/d$ V/m; **(c)** $-4\varepsilon_0/d$ C/m^2

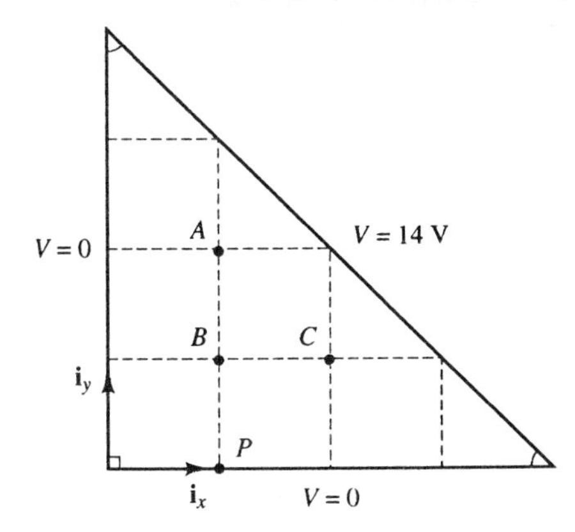

Figure 5.8. For Problem D5.3.

5.2 METHOD OF MOMENTS

In Example 5.2 we discussed the solution of Laplace's equation for the one-dimensional case of two infinite, plane, parallel, perfectly conducting plates, which may be considered an idealization of a parallel-plate capacitor with its plates having large dimensions compared to the spacing between them. We then obtained the expression for the capacitance of the arrangement per unit area of the plates. Because of the idealization, this expression is only approximate for a capacitor with finite-sized plates. It becomes less and less accurate as the size of the plates becomes less and less large for a given spacing between them, since the fringing of the field at the edges of the plates becomes more and more severe. Thus the problem is that in the nonideal case the field distribution between the capacitor plates and the charge distribution on the capacitor plates are not uniform, whereas for the ideal case they are uniform. Hence it is not in general possible to obtain an analytical expression for the capacitance; one has to resort to numerical or graphical techniques.

In this section, we shall consider a numerical technique, known as the method of moments, which is useful for solving a class of problems such as that just discussed. The technique consists of finding the charge distribution on the conductors held at known constant potentials. Thus the problem is the inverse of the problem of finding the potential for a known charge distribution. To cast the technique in general terms, let us consider a surface charge distribution $\rho_S(x, y, z)$ on a given surface. Then applying superposition in conjunction with the expression for the potential due to a point charge given by (4.100), the potential due to the charge distribution can be expressed as

$$V(x, y, z) = \frac{1}{4\pi\varepsilon_0} \int_{\substack{\text{surface of} \\ \text{the charge} \\ \text{distribution}}} \frac{\rho_S(x', y', z')}{R}\, dS' \tag{5.27}$$

where the primes denote source point coordinates. The procedure consists of dividing the surface into a finite number of subsections to approximate the integral in (5.27) by a summation and apply the equation to points on the subsections to obtain a set of linear algebraic equations. The set of equations is then inverted to obtain the desired solution. We shall first illustrate the method by means of an example.

Example 5.4

Thin, straight wire held at known potential

Let us consider a thin, straight wire of length l and radius a ($\ll l$), as shown in Fig. 5.9(a) and held at a potential of 1 V. We wish to obtain the resulting (surface) charge distribution on the wire by the method of moments.

The determination of the charge distribution by the method of moments consists of dividing the wire into a number of segments, assuming the charge density in each segment to be uniform and setting up and solving a set of algebraic equations. For simplicity of illustration, we shall divide the wire into five equal segments numbered 1 through 5 and having surface charge densities $\rho_{S1}, \rho_{S2}, \ldots, \rho_{S5}$. From considerations of symmetry, there are then only three unknowns, since $\rho_{S4} = \rho_{S2}$ and $\rho_{S5} = \rho_{S1}$. Hence we need three independent equations.

An equation is obtained by writing the potential at the center point of a given segment to be the superposition of the potentials at that point due to the charges in the five segments. To obtain the contribution due to a segment, we consider the cylindrical surface charge of uniform density ρ_{S0} coaxial with the z-axis and located symmetrically about the origin, as shown in Fig. 5.9(b), and compute the potential due to it at two points: (1) at the origin and (2) at a point (0, 0, z) where $z > d$, using the approximation $a \ll d$. Case 1 is appropriate to finding the potential due to the charge in a given segment in Fig. 5.9(a) at its own center point, whereas case 2 is appropriate to finding the potential due to the charge in a given segment in Fig. 5.9(a) at the center point of another segment.

Dividing the cylindrical surface charge in Fig. 5.9(b) into a number of ring charges, one of which is shown in the figure, and using superposition, we obtain

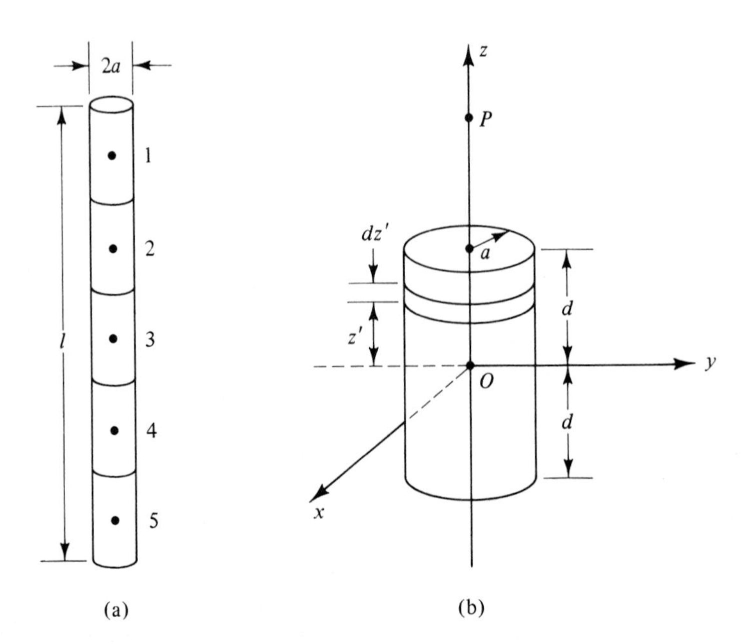

Figure 5.9. (a) Thin wire divided into five equal segments. (b) For the determination of the potential due to a cylindrical surface charge.

$$[V]_{(0,0,0)} = \int_{z'=-d}^{d} \int_{\phi=0}^{2\pi} \frac{\rho_{S0} a \, d\phi \, dz'}{4\pi\varepsilon_0 \sqrt{a^2 + (z')^2}}$$

$$= \frac{\rho_{S0} a}{2\varepsilon_0} \{\ln [z' + \sqrt{a^2 + (z')^2}]\}_{z'=-d}^{d}$$

$$= \frac{\rho_{S0} a}{2\varepsilon_0} \ln \frac{d + \sqrt{a^2 + d^2}}{-d + \sqrt{a^2 + d^2}}$$

which for $a \ll d$ reduces to

$$[V]_{(0,0,0)} \approx \frac{\rho_{S0} a}{2\varepsilon_0} \ln \frac{2d}{-d + d(1 + a^2/2d^2)}$$
$$= \frac{\rho_{S0} a}{\varepsilon_0} \ln \frac{2d}{a} \tag{5.28a}$$

For a point $P(0, 0, z)$ where $z > d$, we can consider the cylindrical surface charge to be a line charge of density $2\pi a \rho_{S0}$ and write

$$[V]_P = \int_{z'=-d}^{d} \frac{2\pi a \rho_{S0} \, dz'}{4\pi\varepsilon_0 (z - z')}$$

$$= \frac{\rho_{S0} a}{2\varepsilon_0} [-\ln (z - z')]_{z'=-d}^{d} \tag{5.28b}$$

$$= \frac{\rho_{S0} a}{2\varepsilon_0} \ln \frac{z + d}{z - d}$$

Applying (5.28a) and (5.28b) to write the equation for the potential at the center of segment 1 in Fig. 5.11(a), we obtain

$$\frac{\rho_{S1} a}{\varepsilon_0} \ln \frac{l}{5a} + \frac{\rho_{S2} a}{2\varepsilon_0} \ln 3 + \frac{\rho_{S3} a}{2\varepsilon_0} \ln \frac{5}{3} + \frac{\rho_{S4} a}{2\varepsilon_0} \ln \frac{7}{5} + \frac{\rho_{S5} a}{2\varepsilon_0} \ln \frac{9}{7} = 1$$

or

$$\rho_{S1}\left(2 \ln \frac{l}{5a} + \ln \frac{9}{7}\right) + \rho_{S2}(\ln 3 + \ln 1.4) + \rho_{S3} \ln \frac{5}{3} = \frac{2\varepsilon_0}{a} \tag{5.29}$$

where we have substituted $\rho_{S5} = \rho_{S1}$ and $\rho_{S4} = \rho_{S2}$. Similarly, writing the equations for the potentials at the center points of segments 2 and 3 and arranging the three equations in matrix form, we get

$$\begin{bmatrix} 2 \ln \frac{l}{5a} + \ln \frac{9}{7} & \ln 3 + \ln 1.4 & \ln \frac{5}{3} \\ \ln 3 + \ln 1.4 & 2 \ln \frac{l}{5a} + \ln \frac{5}{3} & \ln 3 \\ 2 \ln \frac{5}{3} & 2 \ln 3 & 2 \ln \frac{l}{5a} \end{bmatrix} \begin{bmatrix} \rho_{S1} \\ \rho_{S2} \\ \rho_{S3} \end{bmatrix} = \begin{bmatrix} \frac{2\varepsilon_0}{a} \\ \frac{2\varepsilon_0}{a} \\ \frac{2\varepsilon_0}{a} \end{bmatrix} \tag{5.30}$$

By inverting (5.30), the solutions for ρ_{S1}, ρ_{S2}, and ρ_{S3} can be obtained. For a numerical example, if $l = 1$ m and $a = 1$ mm, the values of ρ_{S1}, ρ_{S2}, and ρ_{S3} are $158.38\varepsilon_0$, $145.42\varepsilon_0$, and $143.32\varepsilon_0$, respectively. When a larger number of segments are used, a more accurate solution is obtained for the charge distribution on the wire. The listing of a PC program which computes the values of ρ_S/ε_0 for specified values of l, a, and the number of segments n is included as PL 5.2. The output from a run of the program for $l = 1$ m, $a = 1$ mm, and $n = 21$ is also included with the listing.

PL 5.2. Program listing for computing the charge distribution on a thin, straight wire held at a constant potential of 1 V and the output from a run of the program.

```
100 '*************************************************************
110 '* METHOD OF MOMENTS ANALYSIS FOR THE CHARGE DISTRIBUTION *
120 '* ON A THIN STRAIGHT WIRE HELD AT CONSTANT POTENTIAL OF  *
130 '* 1 V      ,                                             *
140 '*************************************************************
150 DIM A(21,21),Y(11),RHOS(21)
160 CLS:SCREEN 0
170 PRINT "ENTER VALUES OF INPUT PARAMETERS:"
180 INPUT "LENGTH OF WIRE IN M = ",L
190 INPUT "RADIUS OF WIRE IN MM = ",R
200 INPUT "NUMBER OF SEGMENTS = ",N
210 M=INT((N+1)/2):MK=INT(N/2):'* M IS THE NUMBER OF UNKNOWN
220 '  CHARGE DENSITIES TO BE COMPUTED, TAKING SYMMETRY INTO
230 '  ACCOUNT *
240 '* COMPUTE ELEMENTS OF N x N MATRIX *
250 A(1,1)=2*LOG(1000*L/(N*R))
260 FOR J=2 TO N:A(1,J)=LOG((2*J-1)/(2*J-3)):NEXT
270 FOR I=2 TO M
280 A(I,I)=A(1,1)
290 FOR J=1 TO I-1:A(I,J)=A(J,I):NEXT
300 FOR J=I+1 TO N:A(I,J)=A(I-1,J-1):NEXT
310 NEXT
320 '* COMPUTE ELEMENTS FOR THE M x M MATRIX EQUATION *
330 FOR I=1 TO M
340 Y(I)=2000/R
350 IF N=1 THEN 380
360 FOR J=1 TO MK:A(I,J)=A(I,J)+A(I,N-J+1):NEXT:'* ELEMENTS OF
370 '  M x M MATRIX *
380 NEXT
390 '* SOLUTION OF THE M x M MATRIX EQUATION FOR THE M
400 '  UNKNOWNS *
410 FOR K=2 TO M:FOR I=K TO M
420 MF=A(I,K-1)/A(K-1,K-1)
430 FOR J=K TO M:A(I,J)=A(I,J)-A(K-1,J)*MF:NEXT
440 Y(I)=Y(I)-Y(K-1)*MF
450 NEXT:NEXT
460 RHOS(M)=Y(M)/A(M,M)
470 IF M=1 THEN 550
480 FOR I=M-1 TO 1 STEP -1
490 SUM=0
500 FOR J=I+1 TO M:SUM=SUM+A(I,J)*RHOS(J):NEXT
510 RHOS(I)=(Y(I)-SUM)/A(I,I)
520 NEXT
530 '* COMPLETION OF SOLUTION AND PRINTING OF THE CHARGE
540 '  DENSITIES FOR THE N SEGMENTS *
550 JK=0:FOR I=N TO M+1 STEP -1:JK=JK+1:RHOS(I)=RHOS(JK):NEXT
560 PRINT:PRINT "COMPUTED VALUES OF CHARGE DENSITY IN"
570 PRINT "C/M**2 DIVIDED BY PERMITTIVITY ARE:"
580 FOR I=1 TO N:PRINT "SEGMENT";I;":";RHOS(I):NEXT
590 LOCATE 24,1:PRINT "STRIKE ANY KEY TO CONTINUE";:C$=INPUT$(1)
600 GOTO 160
610 END

RUN
ENTER VALUES OF INPUT PARAMETERS:
```

PL 5.2. (continued)

```
LENGTH OF WIRE IN M = 1
RADIUS OF WIRE IN MM = 1
NUMBER OF SEGMENTS = 21

COMPUTED VALUES OF CHARGE DENSITY IN
C/M**2 DIVIDED BY PERMITTIVITY ARE:
SEGMENT 1 : 185.4438
SEGMENT 2 : 161.4599
SEGMENT 3 : 154.4277
SEGMENT 4 : 150.5006
SEGMENT 5 : 147.9689
SEGMENT 6 : 146.2244
SEGMENT 7 : 144.9929
SEGMENT 8 : 144.1315
SEGMENT 9 : 143.5599
SEGMENT 10 : 143.2325
SEGMENT 11 : 143.1259
SEGMENT 12 : 143.2325
SEGMENT 13 : 143.5599
SEGMENT 14 : 144.1315
SEGMENT 15 : 144.9929
SEGMENT 16 : 146.2244
SEGMENT 17 : 147.9689
SEGMENT 18 : 150.5006
SEGMENT 19 : 154.4277
SEGMENT 20 : 161.4599
SEGMENT 21 : 185.4438
STRIKE ANY KEY TO CONTINUE
```

Capacitance of a parallel-plate capacitor

Returning now to the problem of finding the capacitance of a parallel-plate capacitor by the method of moments, let us consider an arrangement in which the spacing between the plates is a, the dimensions of the plates are $2a \times 3a$, and from symmetry considerations the upper plate is held at a potential of 1 V and the lower plate is held at a potential of -1 V. For the purposes of illustration of the method, we shall divide each plate into a 2×3 set of squares, as shown in Fig. 5.10, and assume that within each square, the (surface) charge density is uniform. From symmetry considerations, we then have only two unknown charge densities ρ_{S1} and ρ_{S2}, as shown in the figure. Therefore, it is sufficient to write two independent equations. We shall do this by considering squares 1 and 2 and equating the potentials at the center points of these squares to 1 V.

To write the expression for the potential at the center point of a square due to the charge in a different square, we shall consider that charge to be a point charge at the center of the square. Thus the potential at point 1 due to the charge in square 4 is $\rho_{S1}a^2/4\pi\varepsilon_0 a$, the potential at point 2 due to the charge in square 12 is $-\rho_{S1}a^2/4\pi\varepsilon_0(\sqrt{3}a)$, and so on. To write the expression for the potential at the center point of a square due to the charge in that square, we shall use the result given in Problem P4.32. For example, the potential at point 1 due to the charge in square 1 is $(\rho_{S1}a/\pi\varepsilon_0)\ln(1+\sqrt{2})$. Proceeding in this manner, we obtain the two equations to be

$$\frac{\rho_{S1}a}{\pi\varepsilon_0}\ln(1+\sqrt{2}) + \frac{\rho_{S1}a^2}{4\pi\varepsilon_0}\left(\frac{1}{2a}+\frac{1}{a}+\frac{1}{\sqrt{5}a}-\frac{1}{a}-\frac{1}{\sqrt{5}a}-\frac{1}{\sqrt{2}a}-\frac{1}{\sqrt{6}a}\right) + \frac{\rho_{S2}a^2}{4\pi\varepsilon_0}\left(\frac{1}{a}+\frac{1}{\sqrt{2}a}-\frac{1}{\sqrt{2}a}-\frac{1}{\sqrt{3}a}\right) = 1 \qquad (5.31\text{a})$$

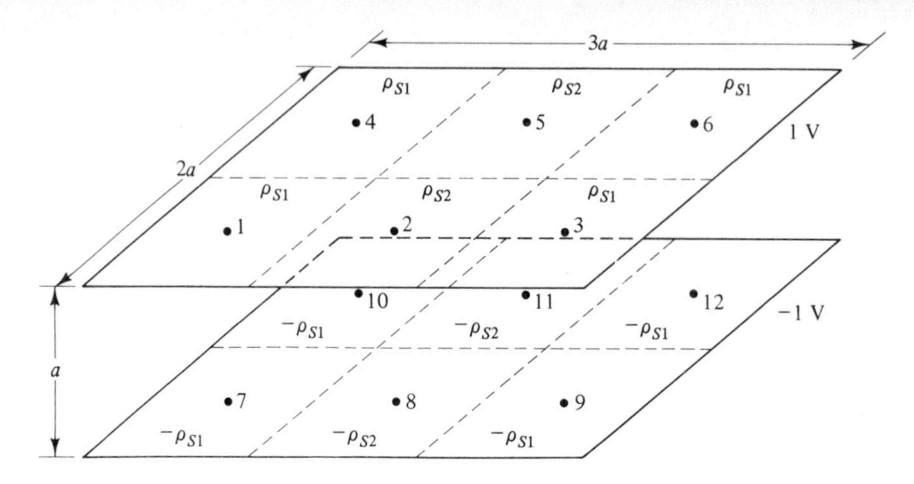

Figure 5.10. For finding the capacitance of a parallel-plate capacitor by the method of moments.

$$\frac{\rho_{S2}a}{\pi\varepsilon_0}\ln(1+\sqrt{2}) + \frac{\rho_{S1}a^2}{4\pi\varepsilon_0}\left(\frac{2}{a} + \frac{2}{\sqrt{2}a} - \frac{2}{\sqrt{2}a} - \frac{2}{\sqrt{3}a}\right) + \frac{\rho_{S2}a^2}{4\pi\varepsilon_0}\left(\frac{1}{a} - \frac{1}{a} - \frac{1}{\sqrt{2}a}\right) = 1 \tag{5.31b}$$

or

$$2.9101\rho_{S1} + 0.4226\rho_{S2} = \frac{4\pi\varepsilon_0}{a} \tag{5.32a}$$

$$0.8453\rho_{S1} + 2.8184\rho_{S2} = \frac{4\pi\varepsilon_0}{a} \tag{5.32b}$$

Solving (5.32a) and (5.32b) for ρ_{S1} and ρ_{S2}, we obtain $\rho_{S1} = 3.8378\varepsilon_0/a$ and $\rho_{S2} = 3.3075\varepsilon_0/a$. The magnitude of charge on either plate is then equal to $(4a^2 \times 3.8378\varepsilon_0/a + 2a^2 \times 3.3075\varepsilon_0/a)$, or $21.9662\varepsilon_0 a$. Finally, noting that the potential difference between the plates is 2 V, the capacitance can be computed to be $10.983\varepsilon_0 a$. A more accurate result can of course be obtained by dividing each plate into a larger number of squares, but it is instructive to compare the value just obtained with the value of $6\varepsilon_0 a$, which follows from the application of $C = \varepsilon_0 A/d$, where A is the area of the plates and d is the spacing between the plates.

K5.2. Method of moments; Thin, straight wire held at known potential; Determination of charge distribution; Parallel-plate capacitor; Determination of capacitance.

D5.4. For the problem of Example 5.4, consider that to compute the potential at the center of a given segment due to the charge in another segment, the charge in that segment can be assumed to be a point charge at the center of that segment. Modify the formulation to obtain the new matrix equation in the place of (5.30) and find the values of ρ_{S1}, ρ_{S2}, and ρ_{S3}, for $l = 1$ m and $a = 1$ mm.
Ans. $159.48\varepsilon_0$ C/m^2; $147.94\varepsilon_0$ C/m^2; $145.77\varepsilon_0$ C/m^2

D5.5. Consider a parallel-plate capacitor having square-shaped plates of sides a and spacing a between the plates. Find the following: **(a)** the capacitance of the capacitor if fringing of fields at the edges of the plates is neglected; **(b)** the capacitance by using the method of moments, considering each plate as one square; and **(c)** the capacitance by dividing each plate into a 2×2 set of squares and using the method of moments. Assume free space for the dielectric.
Ans. **(a)** $\varepsilon_0 a$; **(b)** $2.488\varepsilon_0 a$; **(c)** $2.8367\varepsilon_0 a$

5.3 CAPACITANCE, CONDUCTANCE, AND INDUCTANCE

In Section 5.1 we introduced the capacitance and conductance by considering the solution of Laplace's equation in one dimension. Specifically, we derived the expressions for the capacitance per unit area and the conductance per unit area of a parallel-plate arrangement, the capacitance per unit length of a coaxial cylindrical arrangement, and the capacitance of a concentric spherical arrangement.

Coaxial cylindrical arrangement

Let us now consider the three arrangments shown in Fig. 5.11, each of which is a cross-sectional view of a pair of infinitely long coaxial perfectly conducting cylinders with a material medium between them. In Fig. 5.11(a) the material medium is a dielectric of uniform permittivity ε; in Fig. 5.11(b) it is a conductor of uniform conductivity σ; and in Fig. 5.11(c) it is a magnetic material of uniform permeability μ. In (a) and (b), a potential differrence of V_0 is applied between the conductors, whereas in (c) a current I flows with uniform density in the $+z$-direction on the inner cylinder and returns with uniform density in the $-z$-direction on the outer cylinder.

Capacitance per unit length, $\mathscr{C}$

We know from the discussion in Section 5.1 that the arrangement of Fig. 5.11(a) is that of a coaxial cylindrical capacitor and from Table 5.1 that its capacitance (C) per unit length, defined as the magnitude of the charge per unit length on either conductor to the potential difference between the conductors, is given by

$$\boxed{\mathscr{C} = \frac{C}{l} = \frac{2\pi\varepsilon}{\ln (b/a)}} \tag{5.33}$$

the units of C being farads (F) and hence those of $\mathscr{C}$ being F/m.

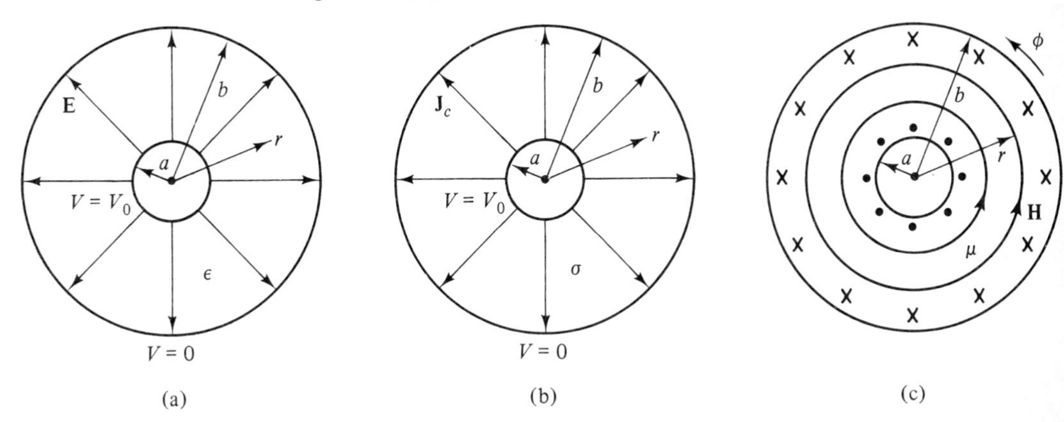

Figure 5.11. Cross sections of three arrangements, each consisting of two infinitely long, coaxial perfectly conducting cylinders. The medium between the cylinders is a perfect dielectric for (a), a conductor for (b), and a magnetic material for (c).

Conductance per unit length, $\mathcal{G}$

Just as in the case of the parallel-plate arrangement of Example 5.2, replacing the dielectric in Fig. 5.11(a) by a conductor as in Fig. 5.11(b) would result in a conduction current of density

$$\mathbf{J}_c = \sigma \mathbf{E} = \frac{\sigma V_0}{r \ln (b/a)} \mathbf{i}_r$$

in the medium and hence a current per unit length

$$I_c = \int_{\phi=0}^{2\pi} \mathbf{J}_c \cdot r \, d\phi \, \mathbf{i}_r = \int_{\phi=0}^{2\pi} \frac{\sigma V_0}{r \ln (b/a)} r \, d\phi$$

$$= \frac{2\pi\sigma V_0}{\ln (b/a)}$$

from the inner cylinder to the outer cylinder. Thus the ratio of the current per unit length from the inner to the outer cylinder to the potential difference between the cylinders, that is, the conductance (G) per unit length of the arrangement, is given by

$$\boxed{\mathcal{G} = \frac{G}{l} = \frac{2\pi\sigma}{\ln (b/a)}} \qquad (5.34)$$

the units of G being siemens (S) and hence those of $\mathcal{G}$ being S/m.

Inductance per unit length, $\mathcal{L}$

Turning now to Fig. 5.11(c), we note from Example 4.14 that the current flow on the cylinders results in a magnetic field between them as given by

$$\mathbf{H} = \frac{I}{2\pi r} \mathbf{i}_\phi \qquad \text{for} \quad a < r < b$$

The magnetic flux density is then given by

$$\mathbf{B} = \mu \mathbf{H} = \frac{\mu I}{2\pi r} \mathbf{i}_\phi \qquad \text{for} \quad a < r < b$$

The magnetic flux linking the current per unit length of the conductors is

$$\psi = \int_{r=a}^{b} \mathbf{B} \cdot dr \, \mathbf{i}_\phi = \int_{r=a}^{b} \frac{\mu I}{2\pi r} dr$$

$$= \frac{\mu I}{2\pi} \ln \frac{b}{a}$$

We now define the inductance (L) per unit length of the arrangement to be the ratio of the magnetic flux linking the current per unit length of the arrangement to the current. Thus

$$\boxed{\mathcal{L} = \frac{L}{l} = \frac{\mu}{2\pi} \ln \frac{b}{a}} \qquad (5.35)$$

The units of L are henrys (H) and hence those of $\mathcal{L}$ are H/m.

An examination of (5.33), (5.34), and (5.35) reveals that

TABLE 5.2 CONDUCTANCE, CAPACITANCE, AND INDUCTANCE PER UNIT LENGTH FOR SOME STRUCTURES CONSISTING OF INFINITELY LONG CONDUCTORS HAVING THE CROSS SECTIONS SHOWN IN FIG. 5.12

Description	Capacitance per unit length, $\mathscr{C}$	Conductance per unit length, $\mathscr{G}$	Inductance per unit length, $\mathscr{L}$
Parallel-plane conductors, Fig. 5.12(a)	$\varepsilon \dfrac{w}{d}$	$\sigma \dfrac{w}{d}$	$\mu \dfrac{d}{w}$
Coaxial cylindrical conductors, Fig. 5.12(b)	$\dfrac{2\pi\varepsilon}{\ln (b/a)}$	$\dfrac{2\pi\sigma}{\ln (b/a)}$	$\dfrac{\mu}{2\pi} \ln \dfrac{b}{a}$
Parallel cylindrical wires, Fig. 5.12(c)	$\dfrac{\pi\varepsilon}{\cosh^{-1} (d/a)}$	$\dfrac{\pi\sigma}{\cosh^{-1} (d/a)}$	$\dfrac{\mu}{\pi} \cosh^{-1} \dfrac{d}{a}$
Eccentric inner conductor, Fig. 5.12(d)	$\dfrac{2\pi\varepsilon}{\cosh^{-1}\left(\dfrac{a^2+b^2-d^2}{2ab}\right)}$	$\dfrac{2\pi\sigma}{\cosh^{-1}\left(\dfrac{a^2+b^2-d^2}{2ab}\right)}$	$\dfrac{\mu}{2\pi} \cosh^{-1} \dfrac{a^2+b^2-d^2}{2ab}$
Shielded parallel cylindrical wires, Fig. 5.12(e)	$\dfrac{\pi\varepsilon}{\ln \dfrac{d(b^2 - d^2/4)}{a(b^2 + d^2/4)}}$	$\dfrac{\pi\sigma}{\ln \dfrac{d(b^2 - d^2/4)}{a(b^2 + d^2/4)}}$	$\dfrac{\mu}{\pi} \ln \dfrac{d(b^2 - d^2/4)}{a(b^2 + d^2/4)}$

$$\boxed{\frac{\mathscr{G}}{\mathscr{C}} = \frac{\sigma}{\varepsilon}} \tag{5.36}$$

and

$$\boxed{\mathscr{L}\mathscr{C} = \mu\varepsilon} \tag{5.37}$$

Thus only one of the three parameters $\mathscr{C}$, $\mathscr{G}$, and $\mathscr{L}$ is independent, with the other two obtainable from it and the material parameters. Although this result is deduced here for the coaxial cylindrical arrangement, it is a general result valid for all arrangements involving two infinitely long, parallel perfect conductors embedded in a homogeneous medium (a medium of uniform material parameters). Expressions for the three quantities $\mathscr{C}$, $\mathscr{G}$, and $\mathscr{L}$ are listed in Table 5.2 for some common configurations of conductors having cross-sectional views shown in Fig. 5.12. The coaxial cylindrical arrangement is repeated for the sake of completion.

Example 5.5

$\mathscr{C}$, $\mathscr{G}$, and $\mathscr{L}$ of parallel-cylindrical wire arrangement

It is desired to obtain the capacitance, conductance, and inductance per unit length of the parallel-cylindrical wire arrangement of Fig. 5.12(c).

In view of (5.36) and (5.37), it is sufficient to find one of the three quantities. Hence we choose to find the capacitance per unit length. Here we shall do this by considering the electric potential field of two parallel, infinitely long, straight line charges of equal and opposite uniform charge densities and showing that the equipotential surfaces are cylinders having their axes parallel to the line charges. By placing conductors in two equipotential surfaces, thereby forming a parallel-wire line, we shall obtain the expression for the capacitance per unit length of the line.

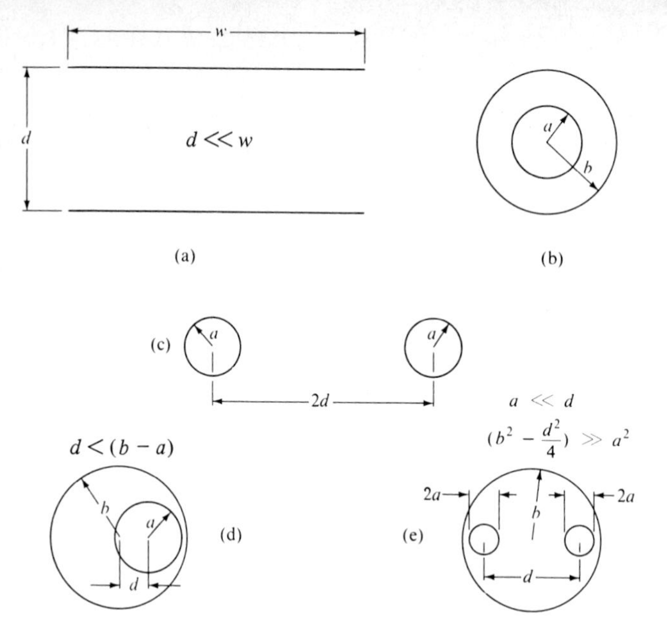

Figure 5.12. Cross sections of some common configurations of parallel, infinitely long conductors.

Let us first consider an infinitely long, straight line charge of uniform density ρ_{L0} C/m situated along the z-axis, as shown in Fig. 5.13(a), and obtain the electric potential due to the line charge. The symmetry associated with the problem indicates that the potential is dependent on the cylindrical coordinate r. Thus we have

$$\nabla^2 V = \frac{1}{r}\frac{\partial}{\partial r}\left(r\frac{\partial V}{\partial r}\right) = 0 \qquad \text{for} \quad r \neq 0$$

$$V = A \ln r + B \tag{5.38}$$

where A and B are constants to be determined. We can arbitrarily set the potential to be zero at a reference value $r = r_0$, giving us $B = -A \ln r_0$ and

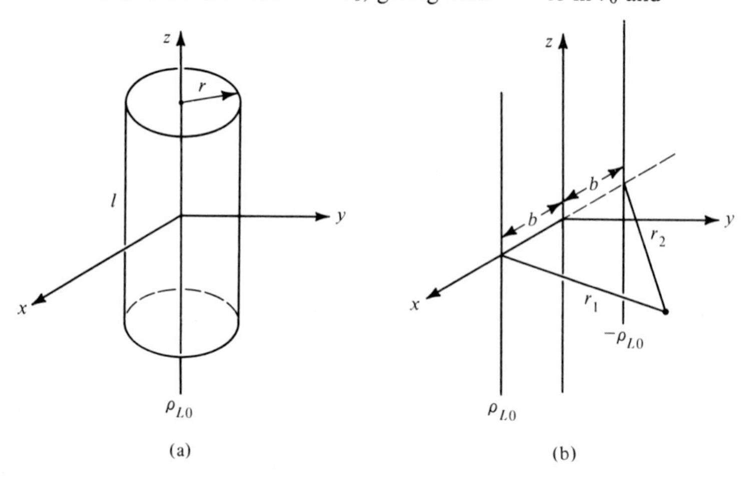

Figure 5.13. (a) Infinitely long line charge of uniform density along the z-axis. (b) Pair of parallel, infinitely long line charges of equal and opposite uniform densities.

$$V = A \ln r - A \ln r_0 = A \ln \frac{r}{r_0} \tag{5.39}$$

To evaluate the arbitrary constant A in (5.39), we find the electric field intensity due to the line charge is given by

$$\mathbf{E} = -\nabla V = -\frac{\partial V}{\partial r}\mathbf{i}_r = -\frac{A}{r}\mathbf{i}_r$$

The electric field is thus directed radial to the line charge. Let us now consider a cylindrical box of radius r and length l coaxial with the line charge, as shown in Fig. 5.13(a), and apply Gauss' law for the electric field in integral form to the surface of the box. For the cylindrical surface,

$$\int \mathbf{D} \cdot d\mathbf{S} = -\frac{\varepsilon A}{r}(2\pi r l)$$

For the top and bottom surfaces, $\int \mathbf{D} \cdot d\mathbf{S} = 0$ since the field is parallel to the surfaces. The charge enclosed by the box is $\rho_{L0} l$. Thus we have

$$-\frac{\varepsilon A}{r}(2\pi r l) = \rho_{L0} l \qquad \text{or} \qquad A = -\frac{\rho_{L0}}{2\pi\varepsilon}$$

Substituting this result in (5.39) we obtain the potential field due to the line charge to be

$$V = -\frac{\rho_{L0}}{2\pi\varepsilon} \ln \frac{r}{r_0} = \frac{\rho_{L0}}{2\pi\varepsilon} \ln \frac{r_0}{r} \tag{5.40}$$

which is consistent with (4.105).

Let us now consider two infinitely long, straight line charges of equal and opposite uniform charge densities ρ_{L0} C/m and $-\rho_{L0}$ C/m, parallel to the z-axis and passing through $x = b$ and $x = -b$, respectively, as shown in Fig. 5.13(b). Applying superposition and using (5.40), we write the potential due to the two line charges as

$$V = \frac{\rho_{L0}}{2\pi\varepsilon} \ln \frac{r_{01}}{r_1} - \frac{\rho_{L0}}{2\pi\varepsilon} \ln \frac{r_{02}}{r_2}$$

where r_1 and r_2 are the distances of the point of interest from the line charges and r_{01} and r_{02} are the distances to the reference point at which the potential is zero. By choosing the reference point to be equidistant from the two line charges, that is, $r_{01} = r_{02}$, we get

$$V = \frac{\rho_{L0}}{2\pi\varepsilon} \ln \frac{r_2}{r_1} \tag{5.41}$$

From (5.41), we note that the equipotential surfaces for the potential field of the line-charge pair are given by

$$\frac{r_2}{r_1} = \text{constant, say, } k \tag{5.42}$$

where k lies between 0 and ∞. In terms of Cartesian coordinates, (5.42) can be written as

$$\frac{(x + b)^2 + y^2}{(x - b)^2 + y^2} = k^2$$

Rearranging, we obtain

$$x^2 - 2b\frac{k^2 + 1}{k^2 - 1}x + y^2 + b^2 = 0$$

or

$$\left(x - b\frac{k^2 + 1}{k^2 - 1}\right)^2 + y^2 = \left(b\frac{2k}{k^2 - 1}\right)^2$$

This equation represents cylinders having their axes along

$$x = b\frac{k^2 + 1}{k^2 - 1}, \qquad y = 0$$

and radii equal to $b[2k/(k^2 - 1)]$. The corresponding potentials are $(\rho_{L0}/2\pi\varepsilon) \ln k$. The cross sections of the equipotential surfaces are shown in Fig. 5.14.

We can now place perfectly conducting cylinders in any two equipotential surfaces without disturbing the field configuration, as shown, for example, by the thick circles in Fig. 5.14, thereby obtaining a parallel-wire line. Letting the distance between their centers be $2d$ and their radii be a, we have

$$\pm d = b\frac{k^2 + 1}{k^2 - 1}$$

$$a = b\frac{2k}{k^2 - 1}$$

Solving these two equations for k and accepting only those solutions lying between 0 and ∞, we obtain

$$k = \frac{d \pm \sqrt{d^2 - a^2}}{a}$$

The potentials of the right ($k > 1$) and left ($k < 1$) conductors are then given, respectively, by

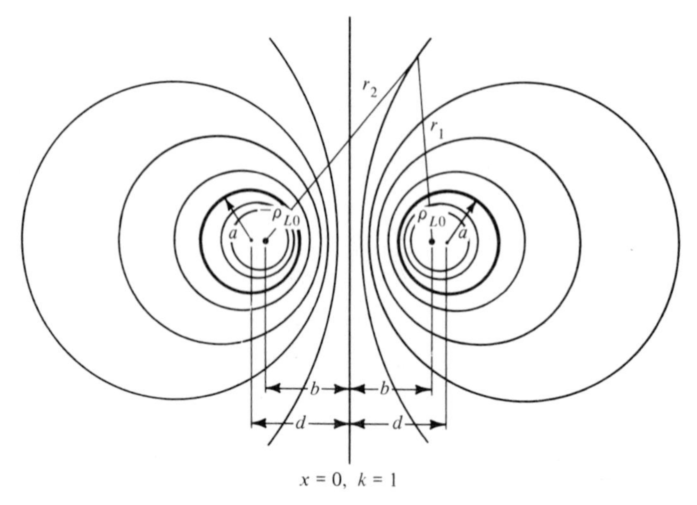

Figure 5.14. Cross sections of equipotential surfaces for the line-charge pair of Fig. 5.13(b). Thick circles represent a cross section of parallel-wire line.

$$V_+ = \frac{\rho_{L0}}{2\pi\varepsilon} \ln \frac{d + \sqrt{d^2 - a^2}}{a}$$

$$V_- = \frac{\rho_{L0}}{2\pi\varepsilon} \ln \frac{d - \sqrt{d^2 - a^2}}{a}$$

$$= -\frac{\rho_{L0}}{2\pi\varepsilon} \ln \frac{d + \sqrt{d^2 - a^2}}{a}$$

The potential difference between the two conductors is

$$V_0 = V_+ - V_- = \frac{\rho_{L0}}{\pi\varepsilon} \ln \frac{d + \sqrt{d^2 - a^2}}{a}$$

Finally, to find the capacitance, we note that since the electric field lines begin on the positive charge and end on the negative charge orthogonal to the equipotentials, the magnitude of the charge on either conductor, which produces the same field as the line-charge pair, must be the same as the line charge itself. Thus considering unit length of the line, we obtain the capacitance per unit length of the parallel-wire line to be

$$\boxed{\begin{aligned} \mathscr{C} = \frac{\rho_{L0}}{V_0} &= \frac{\pi\varepsilon}{\ln [(d + \sqrt{d^2 - a^2})/a]} \\ &= \frac{\pi\varepsilon}{\cosh^{-1} (d/a)} \end{aligned}} \tag{5.43}$$

and hence the expressions for $\mathscr{G}$ and $\mathscr{L}$, as given in Table 5.2.

Internal inductance

If the conductors in a given configuration are not perfect, then the currents flow in the volumes of the conductors instead of being confined to the surfaces. We then have to consider the magnetic field internal to the current distribution in addition to the magnetic field external to it. The inductance associated with the internal field is known as the *internal inductance* as compared to the *external inductance* associated with the external field. The expressions for the inductance per unit length given in Table 5.2 are for the external inductance. To obtain the internal inductance, we have to take into account the fact that different flux lines in the volume occupied by the current distribution link different partial amounts of the total current. We shall illustrate this by means of an example.

Example 5.6

A current I A flows with uniform volume density $\mathbf{J} = J_0\mathbf{i}_z$ A/m^2 along an infinitely long, solid cylindrical conductor of radius a and returns with uniform surface density in the opposite direction along the surface of an infinitely long, perfectly conducting cylinder of radius b ($> a$) and coaxial with the inner conductor. It is desired to find the internal inductance per unit length of the inner conductor.

The cross-sectional view of the conductor arrangement is shown in Fig. 5.15(a). From symmetry considerations, the magnetic field is entirely in the ϕ direction and independent of ϕ. Applying Ampère's circuital law to a circular contour of radius r ($< a$) as shown in Fig. 5.15(a), we have

$$2\pi r H_\phi = \pi r^2 J_0$$

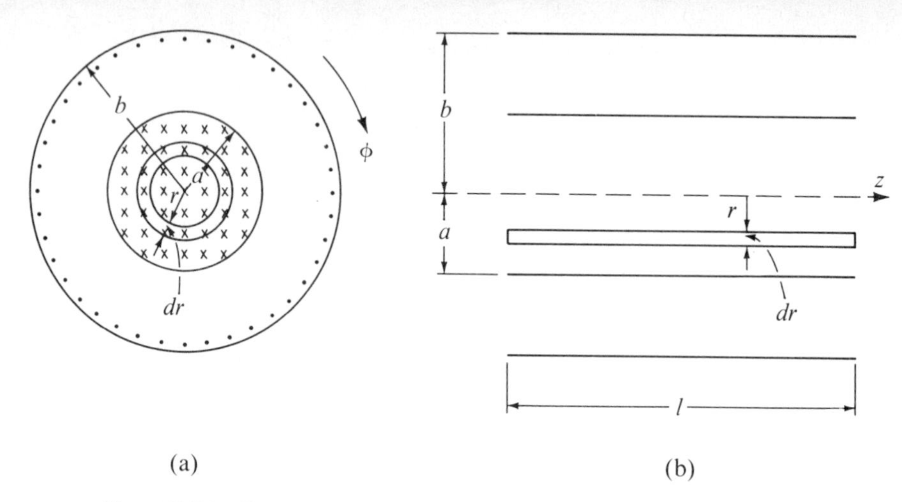

Figure 5.15. For evaluating the internal inductance per unit length associated with a volume current of uniform density along an infinitely long cylindrical conductor.

or

$$\mathbf{H} = H_\phi \mathbf{i}_\phi = \frac{J_0 r}{2}\mathbf{i}_\phi \qquad r < a$$

The corresponding magnetic flux density is given by

$$\mathbf{B} = \mu \mathbf{H} = \frac{\mu J_0 r}{2}\mathbf{i}_\phi \qquad r < a$$

where μ is the permeability of the conductor. Let us now consider a rectangle of infinitesimal width dr in the r direction and length l in the z direction at a distance r from the axis as shown in Fig. 5.15(b). The magnetic flux $d\psi_i$ crossing this rectangular surface is given by

$$\begin{aligned} d\psi_i &= B_\phi(\text{area of the rectangle}) \\ &= \frac{\mu J_0 rl\, dr}{2} \end{aligned}$$

where the subscript i denotes flux internal to the conductor. This flux surrounds only the current flowing within the radius r, as can be seen from Fig. 5.15(a). Let N be the fraction of the total current I linked by this flux. Then

$$\begin{aligned} N &= \frac{\text{current flowing within radius } r\ (< a)}{\text{total current } I} \\ &= \frac{J_0 \pi r^2}{J_0 \pi a^2} = \left(\frac{r}{a}\right)^2 \end{aligned}$$

The contribution from the flux $d\psi_i$ to the internal flux linkage associated with the current I is the product of N and the flux itself, that is, $N\, d\psi_i$. To obtain the internal flux linkage associated with I, we integrate $N\, d\psi_i$ between the limits $r = 0$ and $r = a$, taking into account the dependence of N upon $d\psi_i$. Thus

$$\psi_i = \int_{r=0}^{a} N\, d\psi_i = \int_{r=0}^{a} \left(\frac{r}{a}\right)^2 \frac{\mu J_0 lr}{2}\, dr = \frac{\mu J_0 l a^2}{8}$$

Finally, the required internal inductance per unit length is

$$\mathcal{L}_i = \frac{\psi_i}{lI} = \frac{(\mu J_0 a^2/8)}{(J_0 \pi a^2)} = \frac{\mu}{8\pi} \tag{5.44}$$

From the steps involved in the solution of Example 5.6, we observe that the general expression for the internal inductance is

$$L_{\text{int}} = \frac{1}{I}\int_S N\, d\psi \tag{5.45a}$$

where S is any surface through which the internal magnetic flux associated with I passes. We note that (5.45a) is also good for computing the external inductance since for external inductance N is independent of $d\psi$. Hence

$$L_{\text{ext}} = \frac{N}{I}\int_S d\psi = N\frac{\psi}{I} \tag{5.45b}$$

In (5.45b) the value of N is unity if I is a surface current as for the arrangement of Fig. 5.11(c). On the other hand, for a filamentary wire wound on a core, N is equal to the number of turns of the winding in which case ψ represents the flux through the core, that is, the flux crossing the surface formed by one turn, according to the consideration same as that in conjunction with the discussion of Faraday's law for an N-turn coil (see Fig. 3.13).

Mutual inductance

The discussion pertaining to inductance thus far has been concerned with *self-inductance*, that is, inductance associated with a current distribution by virtue of its own flux linking it. On the other hand, if we have two independent currents I_1 and I_2, we can talk of the flux due to one current linking the second current. This leads to the concept of *mutual inductance*. The mutual inductance denoted as L_{12} is defined as

$$L_{12} = N_1\frac{\psi_{12}}{I_2} \tag{5.46a}$$

where ψ_{12} is the magnetic flux produced by I_2 but linking one turn of the N_1-turn winding carrying current I_1. Similarly,

$$L_{21} = N_2\frac{\psi_{21}}{I_1} \tag{5.46b}$$

where ψ_{21} is the magnetic flux produced by I_1 but linking one turn of the N_2-turn winding carrying current I_2. It can be shown that $L_{21} = L_{12}$. We shall now consider a simple example illustrating the computation of mutual inductance.

Example 5.7

A single straight wire, infinitely long and carrying current I_1, lies below to the left and parallel to a two-wire telephone line carrying current I_2, as shown by the cross-sectional and plan views in Fig. 5.16(a) and (b), respectively. It is desired to obtain

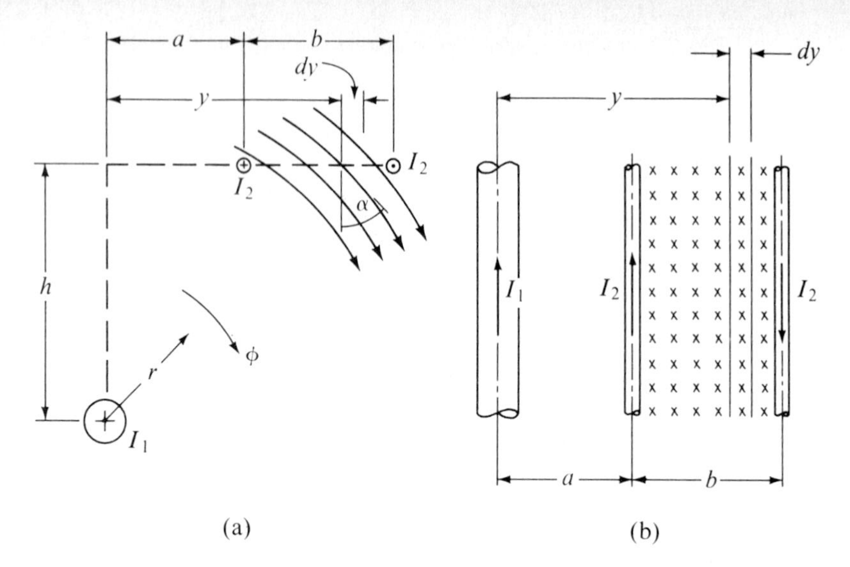

Figure 5.16. For the computation of mutual inductance per unit length between a two-wire telephone line and a single wire parallel to it.

the mutual inductance between the single wire and the telephone line per unit length of the wires. The thickness of the telephone wire is assumed to be negligible.

Choosing a coordinate system with the axis of the single wire as the z-axis and applying Ampère's circuital law to a circular path around the single wire, we obtain the magnetic flux density due to the single wire as

$$\mathbf{B} = \frac{\mu_0 I_1}{2\pi r}\mathbf{i}_\phi$$

The flux $d\psi_{21}$ crossing a rectangular surface of length unity and width dy lying between the telephone wires as shown in Fig. 5.16(b) is then given by

$$d\psi_{21} = B\, dy \cos\alpha = \frac{\mu_0 I_1 y}{2\pi(h^2 + y^2)}\, dy$$

where α is the angle between the flux lines and the normal to the rectangular surface as shown in Fig. 5.16(a). The total flux ψ_{21} crossing the rectangular surface of length unity and extending from one telephone wire to the other is

$$\psi_{21} = \int_{y=a}^{a+b} d\psi_{21} = \int_{y=a}^{a+b} \frac{\mu_0 I_1 y}{2\pi(h^2 + y^2)}\, dy$$

$$= \frac{\mu_0 I_1}{4\pi} \ln \frac{h^2 + (a+b)^2}{h^2 + a^2}$$

This is the flux due to I_1 linking I_2 per unit length along the wires. Thus the required mutual inductance per unit length of the wires is given by

$$\mathcal{L}_{21} = \frac{\psi_{21}}{I_1} = \frac{\mu_0}{4\pi} \ln \frac{h^2 + (a+b)^2}{h^2 + a^2} \text{ H/m}$$

K5.3. Infinitely long, coaxial cylindrical arrangement; Capacitance per unit length ($\mathscr{C}$); Conductance per unit length ($\mathscr{G}$); Inductance per unit length ($\mathscr{L}$); Relationship between $\mathscr{C}$, $\mathscr{G}$, and $\mathscr{L}$; Parallel cylindrical wire arrangement; Internal inductance; Mutual inductance.

D5.6. A coaxial cylindrical conductor arrangement [see Fig. 5.12(b)] has the dimensions $a = 1$ cm and $b = 3$ cm. **(a)** By what value of the distance d should the inner conductor be displaced parallel to the outer conductor [see Fig. 5.12(d)] to increase the capacitance per unit length of the arrangement by 25 percent? **(b)** By what percentage is the inductance per unit length of the arrangement then changed from the original value?
Ans. **(a)** 1.2368 cm; **(b)** -20

D5.7. Figure 5.17 is the cross-sectional view of the coaxial cylindrical conductor arrangement in which a solid conductor of radius a is enclosed by a hollow conductor of inner radius $4a$ and outer radius $5a$. Current I_0 flows in the inner conductor in the $+z$-direction and returns on the outer conductor in the $-z$-direction with densities given by

$$\mathbf{J} = \begin{cases} \dfrac{I_0 e}{\pi a^2}(1 - e^{-r^2/a^2})\mathbf{i}_z & \text{for } 0 < r < a \\ -\dfrac{I_0}{9\pi a^2}\mathbf{i}_z & \text{for } 4a < r < 5a \end{cases}$$

Find the value of N, the fraction of the current I_0 linked by the magnetic flux $d\psi_i$ at a given radius r, for each of the following values of r: **(a)** $0.8a$; **(b)** $3a$; and **(c)** $4.5a$.
Ans. **(a)** 0.4547; **(b)** 1; **(c)** 0.5278

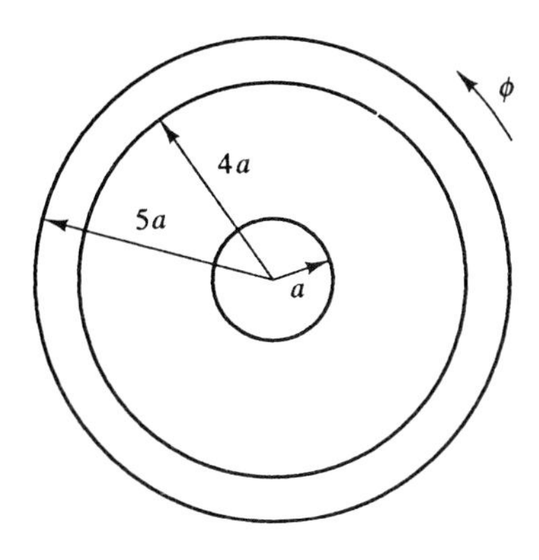

Figure 5.17. For Problem D5.7.

5.4 LOW-FREQUENCY BEHAVIOR VIA QUASISTATICS

Quasistatic extension explained

In Section 4.1 we discussed briefly that lumped circuit theory is based upon approximations resulting from the neglect of certain terms in one or both of Maxwell's curl equations. In this section we elaborate upon this by illustrating the determination of the low-frequency terminal behavior of a physical structure via a quasistatic extension of the static field existing in the structure when the frequency of the source driving the structure is zero. The quasistatic extension consists of starting with a time-varying field having the same spatial characteristics as that of the static field and obtaining the field solutions containing terms up to and including the first power in ω, the radian frequency.

Quasistatic field analysis for an inductor

To introduce the quasistatic field approach, we consider the case of an inductor, as represented by the structure shown in Fig. 5.18(a), in which an arrangement of two parallel plane conductors joined at one end by another conducting sheet is excited by a current source at the other end. We neglect fringing of the fields by assuming that the spacing d between the plates is very small compared to the dimensions of the plates or that the structure is part of a structure of much larger extent in the y- and z-directions. For a constant current source of value I_0 driving the structure at the end $z = -l$, as shown in the figure, such that the surface current densities on the two plates are given by

$$\mathbf{J}_S = \begin{cases} \dfrac{I_0}{w}\mathbf{i}_z & \text{for} \quad x = 0 \\[2ex] -\dfrac{I_0}{w}\mathbf{i}_z & \text{for} \quad x = d \end{cases} \tag{5.47}$$

the medium between the plates is characterized by a uniform y-directed magnetic field as shown by the cross-sectional view in Fig. 5.18(b). From the boundary condition for the tangential magnetic field intensity at the surface of a perfect conductor, the magnitude of this field is I_0/w. Thus we obtain the static magnetic field intensity between the plates to be

$$\mathbf{H} = \frac{I_0}{w}\mathbf{i}_y \qquad \text{for} \quad 0 < x < d \tag{5.48}$$

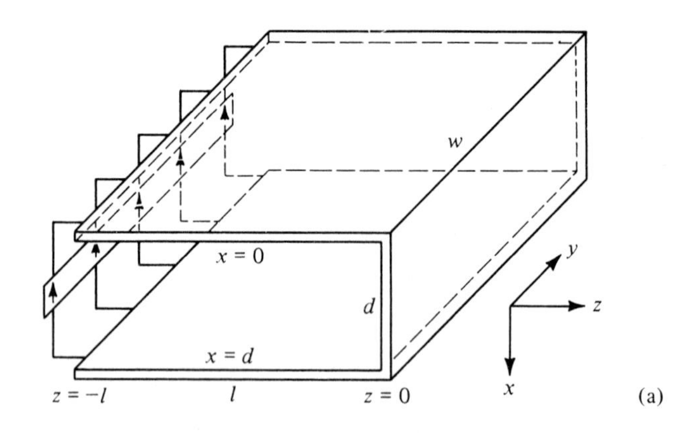

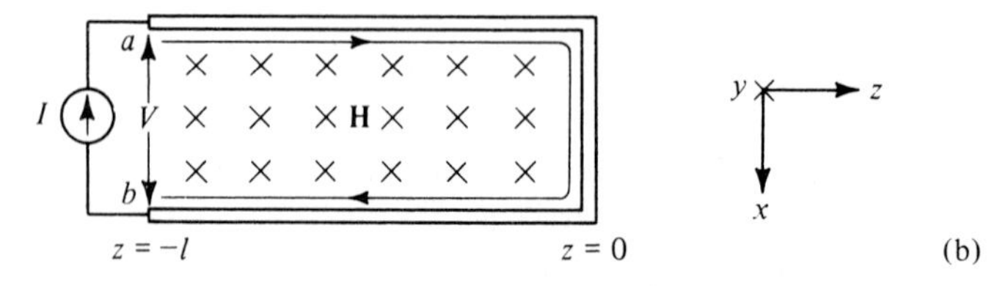

Figure 5.18. (a) Parallel-plate structure short circuited at one end and driven by a current source at the other end. (b) Magnetic field between the plates for a constant current source.

The field is zero outside the plates.

The corresponding magnetic flux density is given by

$$\mathbf{B} = \mu\mathbf{H} = \frac{\mu I_0}{w}\mathbf{i}_y \qquad \text{for} \quad 0 < x < d \tag{5.49}$$

The magnetic flux ψ linking the current is simply the flux crossing the cross-sectional plane of the structure. Since $\mathbf{B}$ is uniform in the cross-sectional plane and normal to it,

$$\psi = B_y(dl) = \frac{\mu dl}{w} I_0 \tag{5.50}$$

The ratio of this magnetic flux to the current, that is, the inductance of the structure, is given by

$$\boxed{L = \frac{\psi}{I_0} = \frac{\mu dl}{w}} \tag{5.51}$$

To discuss the quasistatic behavior of the structure, we now let the current source be varying sinusoidally with time at a frequency ω and assume that the magnetic field between the plates varies accordingly. Thus for

$$I(t) = I_0 \cos \omega t \tag{5.52}$$

we have

$$\mathbf{H}_0 = \frac{I_0}{w} \cos \omega t \; \mathbf{i}_y \tag{5.53}$$

where the subscript 0 denotes that the field is of the zeroth power in ω. In terms of phasor notation, we have

$$\bar{I} = I_0 \tag{5.54}$$

$$\bar{H}_{y0} = \frac{I_0}{w} \tag{5.55}$$

The time-varying magnetic field (5.53) gives rise to an electric field in accordance with Maxwell's curl equation for $\mathbf{E}$. Expansion of the curl equation for the case under consideration gives

$$\frac{\partial E_x}{\partial z} = -\frac{\partial B_{y0}}{\partial t} = -\mu \frac{\partial H_{y0}}{\partial t}$$

or, in phasor form,

$$\frac{\partial \bar{E}_x}{\partial z} = -j\omega\mu \bar{H}_{y0} \tag{5.56}$$

Substituting for $\bar{H}_{y0}$ from (5.55), we have

$$\frac{\partial \bar{E}_x}{\partial z} = -j\omega\mu \frac{I_0}{w}$$

or

$$\bar{E}_x = -j\omega\mu\frac{I_0}{w}z + \bar{C} \tag{5.57}$$

The constant $\bar{C}$ is, however, equal to zero since $[\bar{E}_x]_{z=0} = 0$ to satisfy the boundary condition of zero tangential electric field on the perfect conductor surface. Thus we obtain the quasistatic electric field in the structure to be

$$\boxed{\bar{E}_{x1} = -j\omega\frac{\mu z}{w}I_0} \tag{5.58}$$

where the subscript 1 denotes that the field is of the first power in ω. The value of this field at the input of the structure is given by

$$\boxed{[\bar{E}_{x1}]_{z=-l} = j\omega\mu l\frac{I_0}{w}} \tag{5.59}$$

The voltage developed across the current source is now given by

$$\bar{V} = \int_a^b [\bar{E}_{x1}]_{z=-l}\,dx$$

$$= j\omega\frac{\mu dl}{w}I_0$$

or

$$\boxed{\bar{V} = j\omega L I_0} \tag{5.60}$$

Thus the quasistatic extension of the static field in the structure of Fig. 5.18 illustrates that its input behavior for low frequencies is essentially that of a single inductor of value, the same as that found from static field considerations.

Condition for validity of quasistatic approximation for an inductor

We shall now determine the condition under which the quasistatic approximation is valid, that is, the condition under which the field of the first power in ω is the predominant part of the total field. To do this, we proceed in the following manner. The electric field $\bar{E}_{x1}$ gives rise to a magnetic field in accordance with Maxwell's curl equation for **H**, which for the case under consideration is given by

$$\frac{\partial H_y}{\partial z} = -\varepsilon\frac{\partial E_x}{\partial t}$$

or in phasor form by

$$\frac{\partial \bar{H}_y}{\partial z} = -j\omega\varepsilon\bar{E}_x \tag{5.61}$$

Substituting $\bar{E}_{x1}$ from (5.58) for $\bar{E}_x$ in (5.61), we have

$$\frac{\partial \bar{H}_y}{\partial z} = -\omega^2\mu\varepsilon\frac{z}{w}I_0$$

or

$$\overline{H}_{y2} = -\frac{\omega^2\mu\varepsilon z^2}{2w} I_0 + \overline{C}'' \tag{5.62}$$

where the subscript 2 denotes that the field is of power two in ω. The constant $\overline{C}''$ can be evaluated by noting that at $z = -l$, $\overline{H}_{y2}$ must be zero since $\overline{H}_{y0}$ by itself satisfies the boundary condition $\mathbf{J}_s = \mathbf{i}_n \times \mathbf{H}$. Thus we get

$$\overline{H}_{y2} = -\frac{\omega^2\mu\varepsilon(z^2 - l^2)}{2w} I_0 \tag{5.63}$$

This magnetic field gives rise to an electric field in accordance with Maxwell's curl equation for **E**. Hence we have

$$\begin{aligned}\frac{\partial \overline{E}_x}{\partial z} &= -j\omega\mu\overline{H}_{y2}\\ &= \frac{j\omega^3\mu^2\varepsilon(z^2 - l^2)}{2w} I_0\end{aligned}$$

or

$$\overline{E}_{x3} = \frac{j\omega^3\mu^2\varepsilon(z^3 - 3l^2 z)}{6w} I_0 + \overline{C}''' \tag{5.64}$$

where the subscript 3 denotes that the field is of power 3 in ω. The contant $\overline{C}'''$ has to be equal to zero to satisfy the boundary condition of zero tangential electric field on the conductor surface $z = 0$. Thus we obtain

$$\overline{E}_{x3} = \frac{j\omega^3\mu^2\varepsilon(z^3 - 3l^2 z)}{6w} I_0 \tag{5.65}$$

and hence

$$[\overline{E}_{x3}]_{z=-l} = j\frac{\omega^3\mu^2\varepsilon l^3}{3}\frac{I_0}{w} \tag{5.66}$$

Continuing in this manner, we would obtain

$$[\overline{E}_{x5}]_{z=-l} = j\frac{2\omega^5\mu^3\varepsilon^2 l^5}{15}\frac{I_0}{w} \tag{5.67}$$

and so on. The total electric field at $z = -l$ can then be written as

$$\begin{aligned}[\overline{E}_x]_{z=-l} &= [\overline{E}_{x1}]_{z=-l} + [\overline{E}_{x3}]_{z=-l} + [\overline{E}_{x5}]_{z=-l} + \cdots\\ &= j\omega\mu l\frac{I_0}{w} + j\frac{\omega^3\mu^2\varepsilon l^2}{3}\frac{I_0}{w} + j\frac{2\omega^5\mu^3\varepsilon^2 l^3}{15}\frac{I_0}{w} + \cdots\\ &= j\sqrt{\frac{\mu}{\varepsilon}}\frac{I_0}{w}\left[\omega\sqrt{\mu\varepsilon}l + \frac{1}{3}(\omega\sqrt{\mu\varepsilon}l)^3 + \frac{2}{15}(\omega\sqrt{\mu\varepsilon}l)^5 + \cdots\right]\end{aligned}$$

or

$$\boxed{[\overline{E}_x]_{z=-l} = j\sqrt{\frac{\mu}{\varepsilon}}\frac{I_0}{w}\tan\omega\sqrt{\mu\varepsilon}l} \tag{5.68}$$

From (5.68), it can be seen that for $\omega\sqrt{\mu\varepsilon}l \ll 1$,

$$[\bar{E}_x]_{z=-l} \approx j\sqrt{\frac{\mu}{\varepsilon}}\frac{I_0}{w}\omega\sqrt{\mu\varepsilon}l$$

$$= j\omega\mu l\frac{I_0}{w}$$

which is the same as $[\bar{E}_{x1}]_{z=-l}$. Thus the condition under which the quasistatic approximation is valid is

$$\omega\sqrt{\mu\varepsilon}l \ll 1$$

or

$$\boxed{f \ll \frac{1}{2\pi\sqrt{\mu\varepsilon}l}} \tag{5.69}$$

For frequencies beyond which (5.69) is valid, the input behavior of the structure of Fig. 5.18 is no longer essentially that of a single inductor.

Example 5.8

Low-frequency behavior of a resistor

Let us consider the case of two parallel perfectly conducting plates separated by a lossy medium characterized by conductivity σ, permittivity ε, and permeability μ

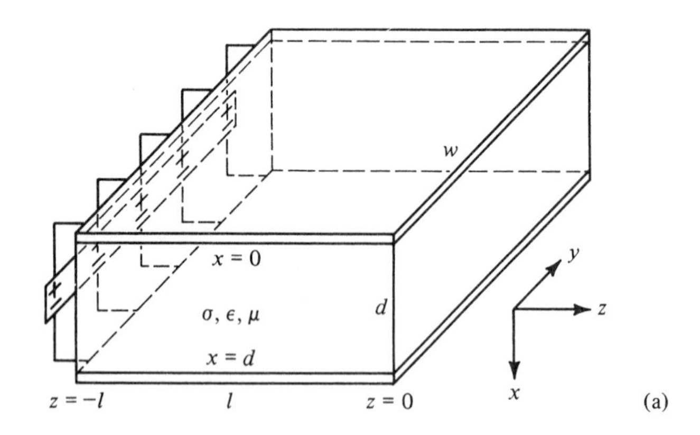

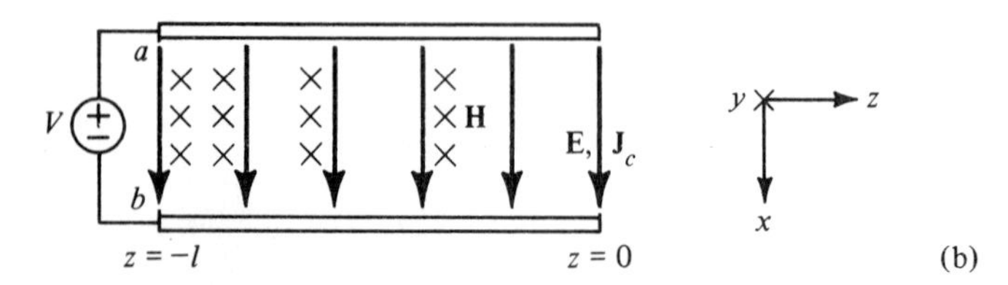

Figure 5.19. (a) Parallel-plate structure with lossy medium between the plates and driven by a voltage source. (b) Electric and magnetic fields between the plates for a constant voltage source.

and driven by a voltage source at one end, as shown in Fig. 5.19(a). We wish to determine its low-frequency behavior by using the quasistatic field approach.

Assuming the voltage source to be a constant voltage source, we first obtain the static electric field in the medium between the plates to be

$$\mathbf{E} = \frac{V_0}{d}\mathbf{i}_x$$

following the procedure of Example 5.2. The conduction current density in the medium is then given by

$$\mathbf{J}_c = \sigma\mathbf{E} = \frac{\sigma V_0}{d}\mathbf{i}_x$$

The conduction current gives rise to a static magnetic field in accordance with Maxwell's curl equation for $\mathbf{H}$ given for static fields by

$$\nabla \times \mathbf{H} = \mathbf{J}_c = \sigma\mathbf{E}$$

For the case under consideration, this reduces to

$$\frac{\partial H_y}{\partial z} = -\sigma E_x = -\frac{\sigma V_0}{d}$$

giving us

$$H_y = -\frac{\sigma V_0 z}{d} + C_1$$

The constant C_1 is, however, equal to zero since $[H_y]_{z=0} = 0$ in view of the boundary condition that the surface current density on the plates must be zero at $z = 0$. Thus the static magnetic field in the medium between the plates is given by

$$\mathbf{H} = -\frac{\sigma V_0 z}{d}\mathbf{i}_y$$

The static electric and magnetic field distributions are shown by the cross-sectional view of the structure in Fig. 5.19(b).

To determine the quasistatic behavior of the structure, we now let the voltage source be varying sinusoidally with time at a frequency ω and assume that the electric and magnetic fields vary with time accordingly. Thus for

$$V = V_0 \cos \omega t$$

we have

$$\mathbf{E}_0 = \frac{V_0}{d}\cos \omega t \ \mathbf{i}_x \qquad (5.70a)$$

$$\mathbf{H}_0 = -\frac{\sigma V_0 z}{d}\cos \omega t \ \mathbf{i}_y \qquad (5.70b)$$

where the subscript 0 denotes that the fields are of the zeroth power in ω. In terms of phasor notation, we have for $\bar{V} = V_0$,

$$\bar{E}_{x0} = \frac{V_0}{d} \qquad (5.71a)$$

$$\bar{H}_{y0} = -\frac{\sigma V_0 z}{d} \qquad (5.71b)$$

The time-varying electric field (5.70a) gives rise to a magnetic field in accordance with

$$\nabla \times \mathbf{H} = \frac{\partial \mathbf{D}_0}{\partial t} = \varepsilon \frac{\partial \mathbf{E}_0}{\partial t}$$

and the time-varying magnetic field (5.70b) gives rise to an electric field in accordance with

$$\nabla \times \mathbf{E} = -\frac{\partial \mathbf{B}_0}{\partial t} = -\mu \frac{\partial \mathbf{H}_0}{\partial t}$$

For the case under consideration and using phasor notation, these equations reduce to

$$\frac{\partial \bar{H}_y}{\partial z} = -j\omega\varepsilon\bar{E}_{x0} = -j\omega\frac{\varepsilon V_0}{d}$$

$$\frac{\partial \bar{E}_x}{\partial z} = -j\omega\mu\bar{H}_{y0} = j\omega\frac{\mu\sigma V_0 z}{d}$$

giving us

$$\bar{H}_{y1} = -j\omega\frac{\varepsilon V_0 z}{d} + \bar{C}_2$$

$$\bar{E}_{x1} = j\omega\frac{\mu\sigma V_0 z^2}{2d} + \bar{C}_3$$

where the subscript 1 denotes that the fields are of the first power in ω. The constant $\bar{C}_2$ is, however, equal to zero in view of the boundary condition that the surface current density on the plates must be zero at $z = 0$. To evaluate the constant $\bar{C}_3$, we note that $[\bar{E}_{x1}]_{z=-l} = 0$ since the boundary condition at the source end, that is,

$$\bar{V} = \int_a^b [\bar{E}_x]_{z=-l}\, dx$$

is satisfied by $\bar{E}_{x0}$ alone. Thus we have

$$j\omega\frac{\mu\sigma V_0(-l)^2}{2d} + \bar{C}_3 = 0$$

or

$$\bar{C}_3 = -j\omega\frac{\mu\sigma V_0 l^2}{2d}$$

Substituting for $\bar{C}_3$ and $\bar{C}_2$ in the expressions for $\bar{E}_{x1}$ and $\bar{H}_{y1}$, respectively, we get

$$\bar{E}_{x1} = j\omega\frac{\mu\sigma V_0(z^2 - l^2)}{2d} \tag{5.72a}$$

$$\bar{H}_{y1} = -j\omega\frac{\varepsilon V_0 z}{d}$$

The result for $\bar{H}_{y1}$ is, however, not complete since $\bar{E}_{x1}$ gives rise to a conduction current of density proportional to ω which in turn provides an additional contribution to $\bar{H}_{y1}$. Denoting this contribution to be $\bar{H}_{y1}^c$, we have

$$\frac{\partial \bar{H}_{y1}^c}{\partial z} = -\sigma \bar{E}_{x1} = -j\omega \frac{\mu\sigma^2 V_0(z^2 - l^2)}{2d}$$

$$\bar{H}_{y1}^c = -j\omega \frac{\mu\sigma^2 V_0(z^3 - 3zl^2)}{6d} + \bar{C}_4$$

The constant $\bar{C}_4$ is zero for the same reason that $\bar{C}_2$ is zero. Hence setting $\bar{C}_4$ equal to zero and adding the resulting expression for $\bar{H}_{y1}^c$ to the right side of the expression for $\bar{H}_{y1}$, we obtain the complete expression for $\bar{H}_{y1}$ as

$$\bar{H}_{y1} = -j\omega \frac{\varepsilon V_0 z}{d} - j\omega \frac{\mu\sigma^2 V_0(z^3 - 3zl^2)}{6d} \tag{5.72b}$$

The total field components correct to the first power in ω are then given by

$$\bar{E}_x = \bar{E}_{x0} + \bar{E}_{x1}$$

$$= \frac{V_0}{d} + j\omega \frac{\mu\sigma V_0(z^2 - l^2)}{2d} \tag{5.73a}$$

$$\bar{H}_y = \bar{H}_{y0} + \bar{H}_{y1}$$

$$= -\frac{\sigma V_0 z}{d} - j\omega \frac{\varepsilon V_0 z}{d} - j\omega \frac{\mu\sigma^2 V_0(z^3 - 3zl^2)}{6d} \tag{5.73b}$$

The current drawn from the voltage source is

$$\bar{I} = w[\bar{H}_y]_{z=-l}$$

$$= \left(\frac{\sigma wl}{d} + j\omega \frac{\varepsilon wl}{d} - j\omega \frac{\mu\sigma^2 wl^3}{3d}\right)\bar{V} \tag{5.74}$$

Finally, the input admittance of the structure is given by

$$\bar{Y} = \frac{\bar{I}}{\bar{V}} = j\omega \frac{\varepsilon wl}{d} + \frac{\sigma wl}{d}\left(1 - j\omega \frac{\mu\sigma l^2}{3}\right)$$

$$\approx j\omega \frac{\varepsilon wl}{d} + \frac{1}{\dfrac{d}{\sigma wl}\left(1 + j\omega \dfrac{\mu\sigma l^2}{3}\right)} \tag{5.75}$$

where we have approximated $\left(1 - j\omega \frac{\mu\sigma l^2}{3}\right)$ by $\left(1 + j\omega \frac{\mu\sigma l^2}{3}\right)^{-1}$. Proceeding further, we have

$$\bar{Y} = j\omega \frac{\varepsilon wl}{d} + \frac{1}{\dfrac{d}{\sigma wl} + j\omega \dfrac{\mu dl}{3w}}$$

$$= j\omega C + \frac{1}{R + (j\omega L/3)} \tag{5.76}$$

where $C = \varepsilon wl/d$ is the capacitance of the structure if the material is a perfect dielectric, $R = d/\sigma wl$ is the dc resistance (reciprocal of the conductance) of the structure, and $L = \mu dl/w$ is the inductance of the structure if the material is lossless and the two plates are short circuited at $z = 0$. The equivalent circuit corresponding to (5.76) consists of capacitance C in parallel with the series combination of resistance R and inductance $L/3$, as shown in Fig. 5.20. Thus the low-frequency

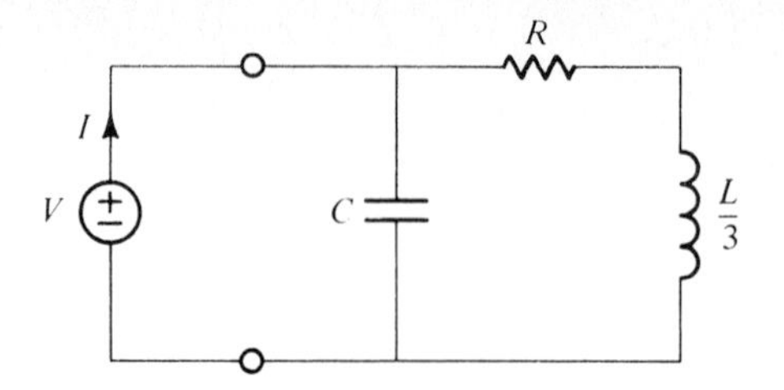

Figure 5.20. Equivalent circuit for the low-frequency input behavior of the structure of Fig. 5.19.

input behavior of the structure of Fig. 5.19 (which acts like a pure resistor at dc) can be represented by the circuit Fig. 5.20, with the understanding of the approximation used in (5.75).

Note that for $\sigma = 0$, (5.74) reduces to

$$\bar{I} = j\omega \frac{\varepsilon \omega l}{d} \bar{V}$$

$$= j\omega C\bar{V}$$

and the input behavior of the structure is essentially that of a single capacitor of the same value as that found from static field considerations.

K5.4. Quasistatic extension of static field; Low-frequency terminal behavior; Inductor; Resistor.

D5.8. For the structure of Fig. 5.18, assume that $l = 10$ cm, $d = 1$ cm, $w = 10$ cm, and that the medium between the conductors is free space. Assuming that the condition for quasistatic approximation given by (5.69) is valid for $f < 1/20\pi\sqrt{\mu\varepsilon}\, l$, find the following: **(a)** the maximum frequency for which the input behavior of the structure is essentially that of a single inductor; **(b)** the value of this inductor; and **(c)** the ratio of the amplitude of the electric field at the input if the structure behaves exactly like a single inductor to the amplitude of the actual electric field at the input for the frequency found in (a).
Ans. **(a)** 47.746 MHz; **(b)** $4\pi \times 10^{-9}$ H; **(c)** 0.9967

5.5 MAGNETIC CIRCUITS

Toroidal conductor versus toroidal magnetic core

Let us consider the two structures shown in Fig. 5.21. The structure of Fig. 5.21(a) is a toroidal conductor of uniform conductivity σ and has a cross-sectional area A and mean circumference l. There is an infinitesimal gap a-b across which a potential difference of V_0 volts is maintained by connecting an appropriate voltage source. Because of the potential difference, an electric field is established in the toroid and a conduction current I_c results from the higher potential surface a to the lower potential surface b as shown in the figure. The structure of Fig. 5.21(b) is a toroidal magnetic core of uniform permeability μ and has a cross-sectional area A and mean circumference l. A current I A is passed through a filamentary wire of N turns wound around the toroid by connecting an appropriate current source. Because of the current through the winding, a magnetic field is established in the toroid and a magnetic flux ψ results in the direction of advance of a right-hand screw as it is turned in the sense of the current.

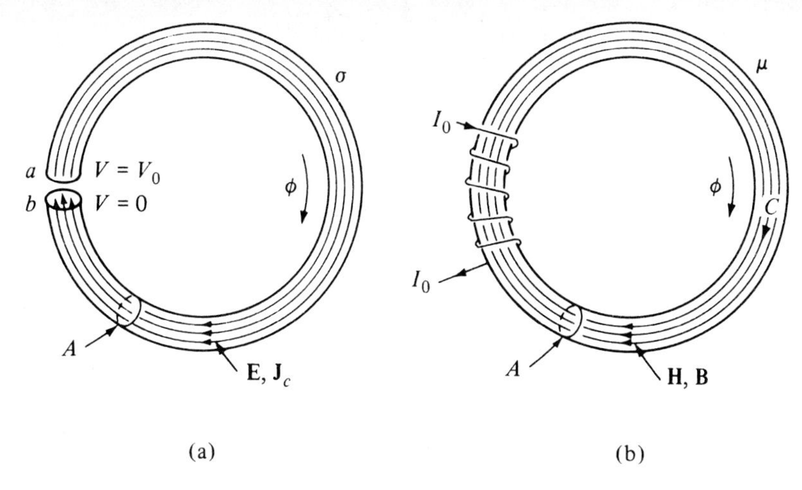

Figure 5.21. (a) Toroidal conductor. (b) Toroidal magnetic core.

Since the conduction current cannot leak into the free space surrounding the conductor, it is confined entirely to the conductor. On the other hand, the magnetic flux can leak into the free space surrounding the magnetic core and hence is not confined completely to the core. However, let us consider the case for which $\mu \gg \mu_0$. Applying the boundary conditions at the boundary between a magnetic material of $\mu \gg \mu_0$ and free space as shown in Fig. 5.22, we have

$$B_1 \sin \alpha_1 = B_2 \sin \alpha_2$$

$$H_1 \cos \alpha_1 = H_2 \cos \alpha_2$$

or

$$\frac{B_1}{H_1} \tan \alpha_1 = \frac{B_2}{H_2} \tan \alpha_2$$

$$\frac{\tan \alpha_1}{\tan \alpha_2} = \frac{\mu_2}{\mu_1} \ll 1$$

Thus $\alpha_1 \ll \alpha_2$, and

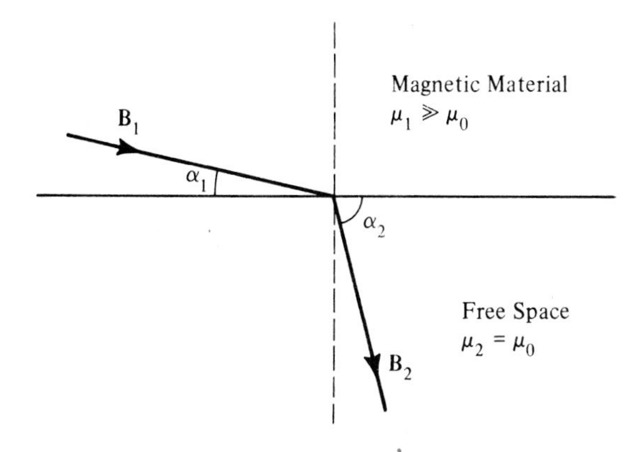

Figure 5.22. Lines of magnetic flux density at the boundary between free space and a magnetic material of $\mu \gg \mu_0$.

$$\frac{B_2}{B_1} = \frac{\sin \alpha_1}{\sin \alpha_2} \ll 1$$

For example, if the values of μ_1 and α_2 are $1000\mu_0$ and 89°, respectively, then $\alpha_1 = 3°16'$ and $\sin \alpha_1/\sin \alpha_2 = 0.057$. We can assume for all practical purposes that the magnetic flux is confined to the magnetic core just as the conduction current is confined to the conductor. The structure of Fig. 5.21(b) is then known as a *magnetic circuit* similar to the *electric circuit* of Fig. 5.21(a). Magnetic circuits are encountered in applications involving electromechanical systems, typical examples of which are electromagnets, transformers, and rotating machines.

For the toroidal conductor of Fig. 5.21(a), we have

$$\int_a^b \mathbf{E} \cdot d\mathbf{l} = V_0 \tag{5.77}$$

Assuming E_ϕ to be uniform over the cross-sectional area and equal to its value E_m at the mean radius of the toroid and proceeding, we obtain

$$lE_m = V_0$$

$$E_m = \frac{V_0}{l}$$

$$J_c = \sigma E_m = \frac{\sigma V_0}{l}$$

$$I_c = J_c A = \frac{\sigma V_0 A}{l}$$

Thus the resistance of the circuit is given by

$$R = \frac{V_0}{I_c} = \frac{l}{\sigma A} \tag{5.78}$$

Similarly, for the toroidal magnetic core of Fig. 5.21(b),

$$\oint_C \mathbf{H} \cdot d\mathbf{l} = NI_0 \tag{5.79}$$

Assuming H_ϕ to be uniform over the cross-sectional area and equal to its value H_m at the mean radius of the toroid, we obtain

$$lH_m = NI_0$$

$$H_m = \frac{NI_0}{l}$$

$$B_m = \mu H_m = \frac{\mu NI_0}{l}$$

$$\psi = B_m A = \frac{\mu NI_0 A}{l}$$

Reluctance defined

We now define the *reluctance* of the magnetic circuit, denoted by the symbol $\mathcal{R}$, as the ratio of the ampere turns NI_0 applied to the magnetic circuit to the magnetic flux ψ. Thus

$$\boxed{\mathcal{R} = \frac{NI_0}{\psi} = \frac{l}{\mu A}} \tag{5.80}$$

The reluctance of the magnetic circuit is analogous to the resistance of an electric circuit and has the units of ampere turns per weber (A-t/Wb). In fact, the complete analogy between the toroidal conductor and the toroidal magnetic core can be seen as follows.

$$V_0 \leftrightarrow NI_0$$

$$\mathbf{E} \leftrightarrow \mathbf{H}$$

$$\mathbf{J}_c \leftrightarrow \mathbf{B}$$

$$\sigma \leftrightarrow \mu$$

$$I_c \leftrightarrow \psi$$

$$R \leftrightarrow \mathcal{R}$$

The equivalent-circuit representations of the two arrangements are shown in Fig. 5.23(a) and (b), respectively. We note from (5.80) that for a given magnetic material, the reluctance appears to be purely a function of the dimensions of the circuit. This is, however, not true since for the ferromagnetic materials used for the cores, μ is a function of the magnetic flux density in the material, as we learned in Section 2.6.

As a numerical example of computations involving the magnetic circuit of Fig. 5.21(b), let us consider a core of cross-sectional area 2 cm^2 and mean circumference 20 cm. Let the material of the core be annealed sheet steel for which the B versus H relationship is shown by the curve of Fig. 5.24. Then to establish a magnetic flux of 3×10^{-4} Wb in the core, the mean flux density must be $(3 \times 10^{-4})/(2 \times 10^{-4})$ or 1.5 Wb/m^2. From Fig. 5.24, the corresponding value of H is 1000 A/m. The number of ampere turns required to establish the flux is then equal to $1000 \times 20 \times 10^{-2}$, or, 200, and the reluctance of the core is $200/(3 \times 10^{-4})$, or $(2/3) \times 10^6$ A-t/Wb. We shall now consider a more detailed example.

Example 5.9

Magnetic circuit with three legs and an air gap

A magnetic circuit containing three legs and with an air gap in the right leg is shown in Fig. 5.25(a). A filamentary wire of N turns carrying current I is wound around the center leg. The core material is annealed sheet steel, for which the B

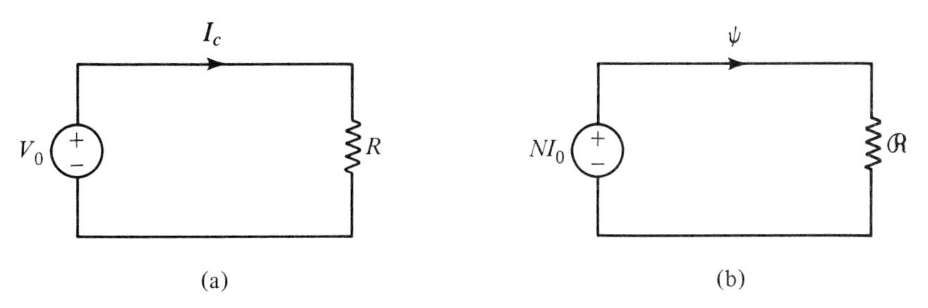

Figure 5.23. Equivalent-circuit representations for the structures of Fig. 5.21(a) and (b), respectively.

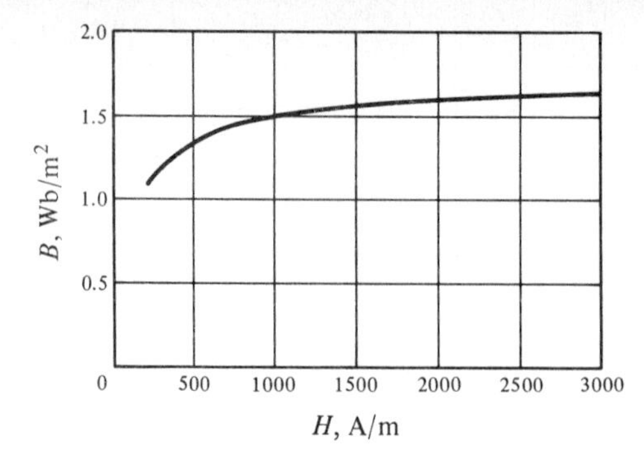

Figure 5.24. *B* versus *H* curve for annealed sheet steel.

versus *H* relationship is shown in Fig. 5.24. The dimensions of the magnetic circuit are

$$A_1 = A_3 = 3 \text{ cm}^2, \qquad A_2 = 6 \text{ cm}^2$$

$$l_1 = l_3 = 20 \text{ cm}, \qquad l_2 = 10 \text{ cm}, \qquad l_g = 0.2 \text{ mm}$$

Let us determine the value of NI required to establish a magnetic flux of 4×10^{-4} Wb in the air gap.

The current in the winding establishes a magnetic flux in the center leg which divides between the right and left legs. Fringing of the flux occurs in the air gap, as shown in Fig. 5.25(b). This is taken into account by using an effective cross section larger than the actual cross section, as shown in Fig. 5.25(c). Using subscripts 1, 2, 3, and *g* for the quantities associated with the left, center, and right legs, and the air gap, respectively, we can write

$$\psi_3 = \psi_g$$

$$\psi_2 = \psi_1 + \psi_3$$

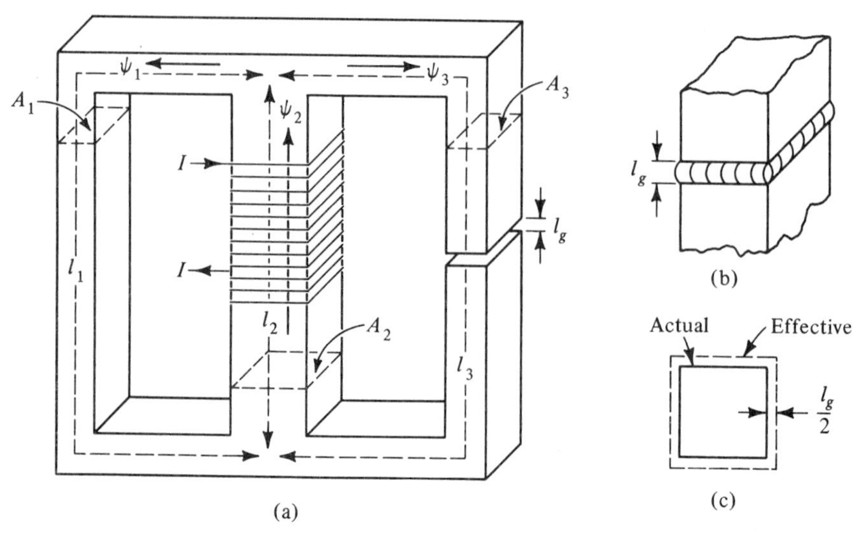

Figure 5.25. (a) Magnetic circuit. (b) Fringing of magnetic flux in the air gap of the magnetic circuit. (c) Effective and actual cross sections for the air gap.

Also, applying Ampère's circuital law to the right and left loops of the magnetic circuit, we obtain, respectively,

$$NI = H_2 l_2 + H_3 l_3 + H_g l_g$$

$$NI = H_2 l_2 + H_1 l_1$$

It follows from these two equations that

$$H_1 l_1 = H_3 l_3 + H_g l_g$$

which can also be written directly from a consideration of the outer loop of the magnetic circuit.

Noting from Fig. 5.25(c) that the effective cross section of the air gap is $(\sqrt{3} + l_g)^2 = 3.07 \text{ cm}^2$, we find the required magnetic flux density in the air gap to be

$$B_g = \frac{\psi_g}{(A_g)_{\text{eff}}} = \frac{4 \times 10^{-4}}{3.07 \times 10^{-4}} = 1.303 \text{ Wb/m}^2$$

The magnetic field intensity in the air gap is

$$H_g = \frac{B_g}{\mu_0} = \frac{1.303}{4\pi \times 10^{-7}} = 0.1037 \times 10^7 \text{ A/m}$$

The flux density in leg 3 is

$$B_3 = \frac{\psi_3}{A_3} = \frac{\psi_g}{A_3} = \frac{4 \times 10^{-4}}{3 \times 10^{-4}} = 1.333 \text{ Wb/m}^2$$

From Fig. 5.24, the value of H_3 is 475 A/m.

Knowing the values of H_g and H_3, we then obtain

$$\begin{aligned} H_1 l_1 &= H_3 l_3 + H_g l_g \\ &= 475 \times 0.2 + 0.1037 \times 10^7 \times 0.2 \times 10^{-3} \\ &= 302.4 \text{ A} \end{aligned}$$

$$H_1 = \frac{302.4}{0.2} = 1512 \text{ A/m}$$

From Fig. 5.24, the value of B_1 is 1.56 Wb/m^2 and hence the flux in leg 1 is

$$\psi_1 = B_1 A_1 = 1.56 \times 3 \times 10^{-4} = 4.68 \times 10^{-4} \text{ Wb}$$

Thus

$$\begin{aligned} \psi_2 &= \psi_1 + \psi_3 \\ &= 4.68 \times 10^{-4} + 4 \times 10^{-4} = 8.68 \times 10^{-4} \text{ Wb} \end{aligned}$$

$$B_2 = \frac{\psi_2}{A_2} = \frac{8.68 \times 10^{-4}}{6 \times 10^{-4}} = 1.447 \text{ Wb/m}^2$$

From Fig. 5.24, the value of H_2 is 750 A/m. Finally, we obtain the required number of ampere turns to be

$$\begin{aligned} NI &= H_2 l_2 + H_1 l_1 \\ &= 750 \times 0.1 + 302.4 \\ &= 377.4 \end{aligned}$$

Note that the equivalent circuit corresponding to the magnetic circuit is as shown in Fig. 5.26, where the reluctances are given by

$$\mathcal{R}_1 = \frac{l_1}{\mu_1 A_1} = \frac{H_1 l_1}{B_1 A_1} = \frac{1512 \times 0.2}{1.56 \times 3 \times 10^{-4}} = 646{,}154 \text{ A-t/Wb}$$

$$\mathcal{R}_2 = \frac{l_2}{\mu_2 A_2} = \frac{H_2 l_2}{B_2 A_2} = \frac{750 \times 0.1}{1.447 \times 6 \times 10^{-4}} = 86{,}386 \text{ A-t/Wb}$$

$$\mathcal{R}_3 = \frac{l_3}{\mu_3 A_3} = \frac{H_3 l_3}{B_3 A_3} = \frac{475 \times 0.2}{1.333 \times 3 \times 10^{-4}} = 237{,}559 \text{ A-t/Wb}$$

$$\mathcal{R}_g = \frac{l_g}{\mu_0 (A_g)_{\text{eff}}} = \frac{0.2 \times 10^{-3}}{4\pi \times 10^{-7} \times 3.07 \times 10^{-4}} = 518{,}420 \text{ A-t/Wb}$$

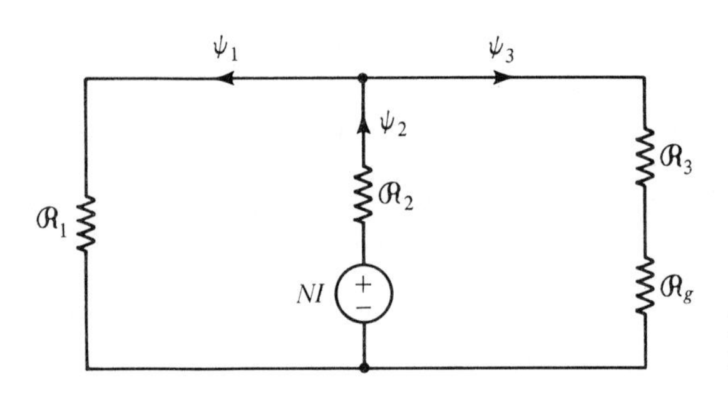

Figure 5.26. Equivalent circuit for the magnetic circuit of Fig. 5.25.

K5.5. Magnetic circuit; Analogy with electric circuit; Reluctance; Air gap.

D5.9. Assume that the portion of B versus H curve of Fig. 5.24 in the range $1500 \leq H \leq 3000$ can be approximated by the straight line

$$B = 1.5 + 5 \times 10^{-5} H$$

For a toroidal magnetic circuit made of annealed sheet steel, find the reluctance for each of the following cases: **(a)** $A = 4$ cm^2, $l = 30$ cm, $H = 1800$ A/m; **(b)** $A = 2$ cm^2, $l = 20$ cm, $NI = 500$ A-t; and **(c)** $A = 5$ cm^2, $l = 25$ cm, $\psi = 8 \times 10^{-4}$ Wb.
Ans. **(a)** 849,057 A-t/Wb; **(b)** 1,538,462 A-t/Wb; **(c)** 625,000 A-t/Wb

D5.10. For the magnetic circuit of Fig. 5.25, assume that the region of operation on the B–H curve of the material is such that μ_r of the material is equal to 4000 for all three legs. Find the reluctance as viewed by the excitation for each of the following cases: **(a)** winding in leg 1; **(b)** winding in leg 2; and **(c)** winding in leg 3.
Ans. **(a)** 164,207 A-t/Wb; **(b)** 143,681 A-t/Wb; **(c)** 689,671 A-t/Wb

5.6 ELECTROMECHANICAL ENERGY CONVERSION

Parallel-plate capacitor with a movable plate

Let us consider a parallel-plate capacitor with one plate fixed and the other plate free to move, as shown by a cross-sectional view in Fig. 5.27. If we assume a positive charge Q on the movable plate and a negative charge $-Q$ on the fixed plate, resulting from the application of a voltage V between the plates, then a force $\mathbf{F}_e$ directed toward the fixed plate is exerted on the movable plate. If this

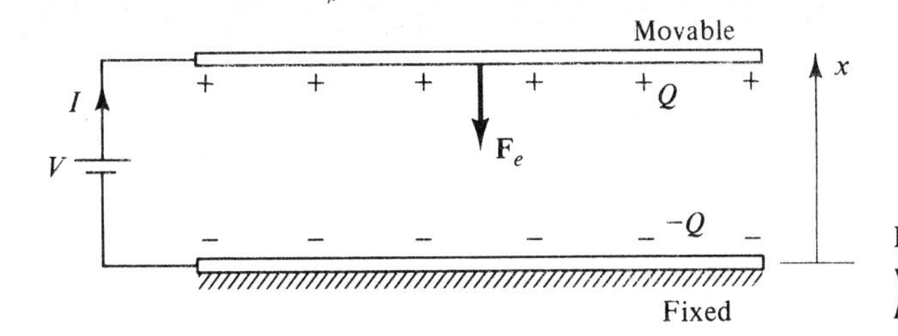

Figure 5.27. Parallel-plate capacitor with a movable plate, depicting the force F_e on the movable plate.

force is allowed to produce a displacement of the movable plate, mechanical work results thereby converting electrical energy in the system into mechanical energy. Conversely, an externally applied mechanical force can be made to act on the movable plate so as to increase the stored electrical energy in the system. Thus energy can be converted from electrical to mechanical or vice versa. A familiar example of the former is in the case of an electrical motor, whereas that of the latter is in the case of an electrical generator. To determine the amount of energy converted from one form to another, we first need to know how to compute the force $\mathbf{F}_e$. In this section we illustrate this computation and discuss the determination of energy converted from one form to another.

Computation of mechanical force of electric origin

The computation of the mechanical force $\mathbf{F}_e$ of electric origin follows from considerations of energy balance associated with the electromechanical system. The energy balance can be expressed as

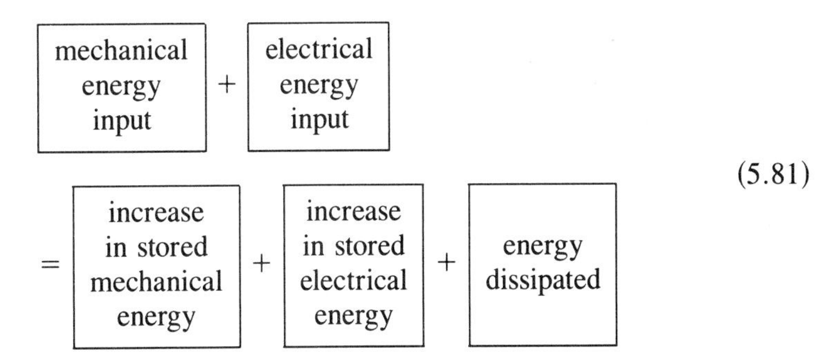

(5.81)

For simplicity, we shall consider the system to be lossless so that the last term on the right side of (5.81) is zero. In using (5.81) to find $\mathbf{F}_e$, we shall apply to the movable element of the system an external force equal to $-\mathbf{F}_e$ and displace the element by an infinitesimal distance in the direction of the external force, so that no change in stored mechanical energy occurs. This eliminates the first term on the right side of (5.81). Thus with reference to the system of Fig. 5.27, we have

$$-F_{ex}\,dx + VI\,dt = dW_e \tag{5.82}$$

where dx is the displacement of the movable plate, I is the current drawn from the voltage source, and W_e is the electric stored energy in the capacitor. Substituting $I = dQ/dt$ from the law of conservation of charge, we obtain

$$-F_{ex}\,dx + V\,dQ = dW_e$$

or

$$\boxed{F_{ex} = -\frac{dW_e}{dx} + V\frac{dQ}{dx}} \tag{5.83}$$

To proceed further, we shall neglect fringing of the electric field at the edges of the capacitor plates so that the charges on the plates and the electric field between the plates are uniformly distributed. Then if A is the area of each plate, we can write the following:

$$W_e = \frac{1}{2}\varepsilon_0 E^2 Ax = \frac{1}{2}\varepsilon_0\left(\frac{V}{x}\right)^2 Ax$$

$$= \frac{\varepsilon_0 V^2 A}{2x}$$

$$\frac{dW_e}{dx} = -\frac{\varepsilon_0 V^2 A}{2x^2}$$

$$Q = CV = \frac{\varepsilon_0 AV}{x}$$

$$\frac{dQ}{dx} = -\frac{\varepsilon_0 AV}{x^2}$$

Thus we obtain

$$F_{ex} = \frac{\varepsilon_0 V^2 A}{2x^2} - \frac{\varepsilon_0 AV^2}{x^2}$$

$$= -\frac{1}{2}\frac{\varepsilon_0 AV^2}{x^2} \tag{5.84}$$

Note that in this procedure, V was held constant since the voltage source was kept connected to the capacitor plates in the process of displacing the plate. If on the other hand, the voltage source is not connected to the capacitor plates in the process of displacing the plate, then Q remains constant, and we can write the following:

$$\frac{dQ}{dx} = 0$$

$$\frac{dW_e}{dx} = \frac{d}{dx}\left(\frac{1}{2}\varepsilon_0 E^2 Ax\right)$$

$$= \frac{d}{dx}\left[\frac{1}{2}\varepsilon_0\left(\frac{Q}{A\varepsilon_0}\right)^2 Ax\right] \tag{5.85}$$

$$= \frac{1}{2}\frac{Q^2}{A\varepsilon_0}$$

$$F_{ex} = -\frac{1}{2}\frac{Q^2}{A\varepsilon_0}$$

The results obtained for F_{ex} in (5.84) and (5.85) appear to be different, but they are not. This can be seen by expressing (5.84) in term of Q or by expressing (5.85) in terms of V. Choosing the first option, we can write (5.84) as

$$F_{ex} = -\frac{1}{2}\frac{\varepsilon_0 AV^2}{x^2}$$
$$= -\frac{1}{2}\varepsilon_0 AE^2$$
$$= -\frac{1}{2}\varepsilon_0 A\left(\frac{Q}{A\varepsilon_0}\right)^2$$
$$= -\frac{1}{2}\frac{Q^2}{A\varepsilon_0}$$

which is the same as that given by (5.85). This is to be expected since Q and V are not independent of each other; they are related through the capacitance of the capacitor. Thus the force $\mathbf{F}_e$ is given by

$$\boxed{\mathbf{F}_e = -\frac{1}{2}\frac{\varepsilon_0 AV^2}{x^2}\mathbf{i}_x = -\frac{1}{2}\frac{Q^2}{A\varepsilon_0}\mathbf{i}_x} \tag{5.86}$$

We shall now illustrate by means of an example the application of the result we obtained for $\mathbf{F}_e$ in the computation of energy converted from electrical to mechanical, or vice versa, in the energy conversion process.

Example 5.10

Energy conversion computation

Assume that in the parallel-plate capacitor of Fig. 5.27 a source of mechanical force $\mathbf{F}$ is applied to the movable plate such that $\mathbf{F}$ is always maintained equal to $-\mathbf{F}_e$. By appropriately varying V and $\mathbf{F}$, the system is made to traverse the closed cycle in the Q–x-plane, shown in Fig. 5.28. We wish to calculate the energy converted per cycle and determine whether the conversion is from electrical to mechanical or vice versa.

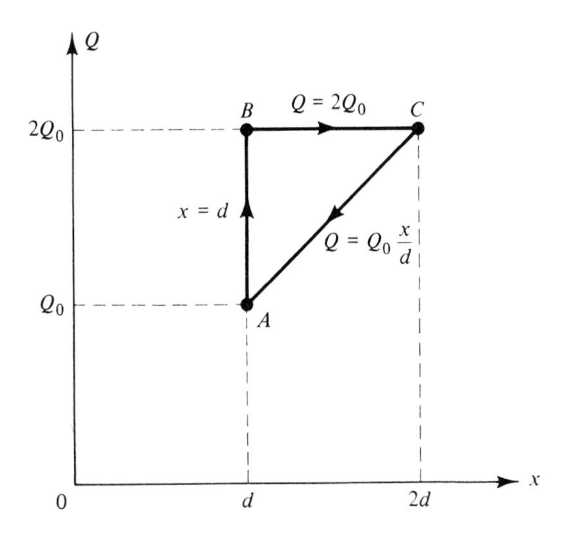

Figure 5.28. Closed cycle traversed by the capacitor system of Fig. 5.27.

Since the system is made to traverse a closed cycle in the Q–x-plane, there is no change in the electrical stored energy from the initial state to the final state. Hence the sum of the mechanical and electrical energy inputs to the system must be zero, or the electrical energy output is equal to the mechanical energy input. The mechanical energy input is given by

$$\begin{aligned}
W_{\substack{\text{mechanical}\\ \text{input}}} &= \oint_{ABCA} F_x\, dx = -\oint_{ABCA} F_{ex}\, dx \\
&= -\int_A^B F_{ex}\, dx - \int_B^C F_{ex}\, dx - \int_C^A F_{ex}\, dx
\end{aligned}$$

From A to B, x remains constant; hence $\int_A^B F_{ex}\, dx$ is zero. From (5.85),

$$F_{ex} = \begin{cases} -\dfrac{2Q_0^2}{A\varepsilon_0} & \text{from } B \text{ to } C \\[2ex] -\dfrac{Q_0^2 x^2}{2A\varepsilon_0 d^2} & \text{from } C \text{ to } A \end{cases}$$

Hence

$$\begin{aligned}
W_{\substack{\text{mechanical}\\ \text{input}}} &= \int_{x=d}^{2d} \frac{2Q_0^2}{A\varepsilon_0}\, dx + \int_{x=2d}^{d} \frac{Q_0^2 x^2}{2A\varepsilon_0 d^2}\, dx \\
&= \frac{2Q_0^2 d}{\varepsilon_0 A} - \frac{7Q_0^2 d}{6\varepsilon_0 A} \\
&= \frac{5}{6}\frac{Q_0^2 d}{\varepsilon_0 A}
\end{aligned}$$

Thus an amount of energy equal to $5Q_0^2 d/6\varepsilon_0 A$ is converted from mechanical to electrical form.

Electromagnet We have thus far considered an electric field electromechanical system, that is, one in which conversion takes place between energy stored in electric field and mechanical energy. For an example of a magnetic field electromechanical system, that is, one in which conversion takes place between energy stored in magnetic field and mechanical energy, let us consider the arrangement shown in Fig. 5.29,

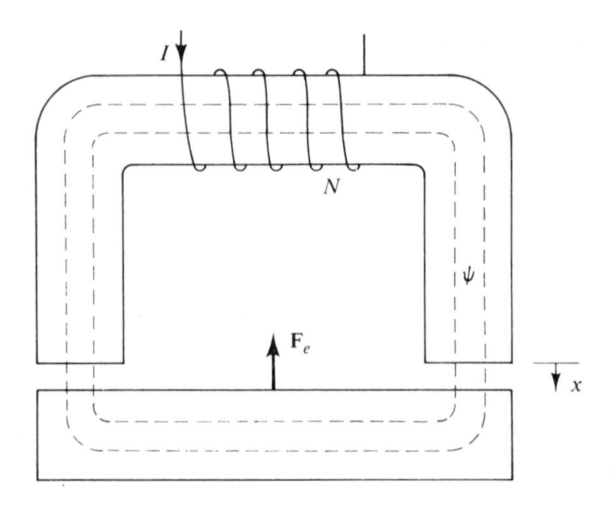

Figure 5.29. Electromagnet.

which is the cross section of an electromagnet. When current is passed through the coil, the armature is pulled upward to close the air gap. The mechanical force $\mathbf{F}_e$ of electric origin can once again be found from energy balance.

In the case of the parallel-plate capacitor of Fig. 5.27, we found $\mathbf{F}_e$ in two ways: by keeping the voltage across the plates constant and by keeping the charge on the plates constant. We found that the two approaches resulted in equivalent expressions for the force. In the present case, we can find $\mathbf{F}_e$ by keeping the current I in the exciting coil to be a constant or by keeping the magnetic flux ψ in the core (and hence in the air gap) to be a constant. The two approaches should result in equivalent expressions for $\mathbf{F}_e$. We shall therefore take advantage of this to simplify the task of finding $\mathbf{F}_e$ by keeping ψ constant, since then no voltage is induced in the coil and hence the electrical energy input term in (5.81) can be set to zero. Also, we shall once again assume a lossless system and apply to the armature an external force equal to $-\mathbf{F}_e$ and displace it by an infinitesimal distance in the direction of the external force. Thus we obtain

$$-F_{ex}\,dx = dW_m$$

$$F_{ex} = -\frac{dW_m}{dx}$$

where W_m is the magnetic stored energy in the system.

Neglecting fringing of flux across the air gap and noting that the displacement of the armature changes only the magnetic energy stored in the air gap, we write the following:

$$H_{\text{gap}} = \frac{\psi}{A\mu_0}$$

$$\begin{aligned}(W_m)_{\text{gap}} &= 2\left[\frac{1}{2}\mu_0(H_{\text{gap}})^2 Ax\right]\\ &= \mu_0\left(\frac{\psi}{A\mu_0}\right)^2 Ax\\ &= \frac{\psi^2 x}{A\mu_0}\end{aligned}$$

where A is the cross-sectional area of each gap and the factor 2 takes into account two gaps. Proceeding further, we have

$$\begin{aligned}\frac{dW_m}{dx} &= \frac{d}{dx}[(W_m)_{\text{gap}}]\\ &= \frac{\psi^2}{A\mu_0}\end{aligned}$$

$$F_{ex} = -\frac{\psi^2}{A\mu_0}$$

$$\boxed{\mathbf{F}_e = -\frac{\psi^2}{A\mu_0}\mathbf{i}_x} \tag{5.87}$$

The expression for $\mathbf{F}_e$ in terms of the current I in the coil which would result from considerations of constant I may now be found by simply expressing ψ in (5.87) in terms of I. Thus if we assume for simplicity that the permeability of the magnetic core material is so high that

$$H_{\text{core}} l_{\text{core}} \ll H_{\text{gap}} l_{\text{gap}}$$

where l_{core} and l_{gap} are the lengths of the core and air gap, respectively, then

$$NI \approx 2H_{\text{gap}}\, x$$

$$H_{\text{gap}} \approx \frac{NI}{2x}$$

$$B_{\text{gap}} \approx \frac{\mu_0 NI}{2x}$$

$$\psi \approx \frac{\mu_0 NIA}{2x}$$

$$\boxed{\mathbf{F}_e \approx -\frac{\mu_0 N^2 I^2 A}{4x^2}\mathbf{i}_x} \qquad (5.88)$$

Finally, the computation of energy converted from electrical to mechanical, or vice versa, in a magnetic field electromechanical system can be performed in a manner similar to that illustrated in Example 5.10 for an electric field system.

K5.6. Mechanical force of electric origin; Energy conversion; Parallel-plate capacitor with movable plate; Electromagnet.

D5.11. For the parallel-plate capacitor of Fig. 5.27, assume $V = 10$ V, $x = 1$ cm, and $A = 0.01$ m^2 and compute $\mathbf{F}_e$ for each of the following cases: **(a)** the dielectric between the plates is free space; **(b)** the dielectric between the plates is a material of permittivity $4\varepsilon_0$; and **(c)** the lower half of the region between the plates is a dielectric of permittivity $4\varepsilon_0$, whereas the upper half is free space.
Ans. **(a)** $-5000\varepsilon_0\mathbf{i}_x$ N; **(b)** $-20{,}000\varepsilon_0\mathbf{i}_x$ N; **(c)** $-12{,}800\varepsilon_0\mathbf{i}_x$ N

5.7 SUMMARY

In this chapter we first introduced Poisson's and Laplace's equations. Poisson's equation given by

$$\nabla^2 V = -\frac{\rho}{\varepsilon}$$

is a differential equation governing the behavior of the electric scalar potential in a region of charge, whereas Laplace's equation

$$\nabla^2 V = 0$$

holds in a charge-free region. We discussed the application of Poisson's and Laplace's equations for the solution of problems involving the variation of V with one dimension only. In particular, we illustrated the solution of Poisson's equation by considering the example of a p-n junction diode and the solution of Laplace's equation by considering the determination of capacitance for several cases. We then discussed and illustrated by means of an example the numerical solution of Laplace's equation in two dimensions

$$\frac{\partial^2 V}{\partial x^2} + \frac{\partial^2 V}{\partial y^2} = 0 \tag{5.89}$$

The numerical solution is based on the finite-difference approximation to (5.89) that the potential V_0 at a point P in the charge-free region is given by

$$V_0 \approx \tfrac{1}{4}(V_1 + V_2 + V_3 + V_4) \tag{5.90}$$

where V_1, V_2, V_3, and V_4 are the potentials at four equidistant points lying along mutually perpendicular axes through P. By using an iterative technique, a set of values for the potentials at appropriately chosen grid points is obtained such that the potential at each grid point satisfies (5.90) to within a specified tolerance.

We then turned our attention to the method of moments, which is a numerical technique useful for solving a class of problems for which exact analytical solutions are in general not possible. Considering for example a surface charge distribution $\rho_S(x, y, z)$ on a given surface, the method of moments technique consists of inverting the integral equation

$$V(x, y, z) = \frac{1}{4\pi\varepsilon_0} \int_{\substack{\text{surface of} \\ \text{the charge} \\ \text{distribution}}} \frac{\rho_S(x', y', z')}{R}\, dS'$$

by approximating the integral as a summation. We illustrated the method of moments technique by means of two examples: (1) finding the charge distribution on a thin, straight wire held at a known potential and (2) finding the capacitance of a parallel-plate capacitor, taking into account fringing of field at the edges of the plates.

We then considered the determination of circuit parameters for infinitely long, parallel conductor arrangements. Specifically, (1) we derived the expressions for the capacitance per unit length ($\mathscr{C}$), the conductance per unit length ($\mathscr{G}$), and the inductance per unit length ($\mathscr{L}$) for a coaxial cylindrical arrangement; (2) we showed that the three circuit parameters are related through the material parameters σ, ε, and μ, as given by

$$\frac{\mathscr{G}}{\mathscr{C}} = \frac{\sigma}{\varepsilon}$$

$$\mathscr{L}\mathscr{C} = \mu\varepsilon$$

and (3) we used these relationships for other geometries of the conductors. We then extended our discussions to internal inductance and mutual inductance.

Next we discussed the quasistatic extension of the static field solution as a means of obtaining the low-frequency behavior of a physical structure. The quasistatic field approach involves starting with a time-varying field having the same spatial characteristics as the static field in the physical structure and then obtaining field solutions containing terms up to and including the first power in frequency by using Maxwell's curl equations for time-varying fields. We illustrated this approach by considering two examples, one of them involving a lossy medium.

We then introduced the magnetic circuit, which is essentially an arrangement of closed paths for magnetic flux to flow around just as current in electric circuits. The closed paths are provided by ferromagnetic cores which because of their high permeability relative to that of the surrounding medium confine the flux almost entirely to within the core regions. We illustrated the analysis of magnetic circuits by considering two examples, one of them including an air gap in one of the legs.

Finally, we introduced the topic of electromechanical energy conversion. By considering examples of a parallel-plate capacitor with one movable plate and an electromagnet, we discussed the determination of mechanical forces of electric origin. We also illustrated energy conversion computation for the parallel-plate capacitor example.

REVIEW QUESTIONS

R5.1. State Poisson's equation. How is it derived?

R5.2. Discuss the application of Poisson's equation for the determination of potential due to the space charge layer in a *p-n* junction semiconductor.

R5.3. State Laplace's equation. In what regions is it valid?

R5.4. Discuss the application of Laplace's equation for a conducting medium.

R5.5. Outline the solution of Laplace's equation in one dimension by considering the variation of potential with x only.

R5.6. Outline the steps in the derivation of the expression for the capacitance of an arrangement of two conductors.

R5.7. Discuss the basis behind the numerical solution of Laplace's equation in two dimensions.

R5.8. Describe the iteration technique for the computer solution of Laplace's equation in two dimensions by the finite-difference method.

R5.9. How would you apply the iteration technique for the computer solution of Laplace's equation in three dimensions?

R5.10. Why is the expression for the capacitance of a parallel-plate capacitor obtained by using Laplace's equation in one dimension approximate?

R5.11. Discuss the formulation behind the problem of finding the charge distribution on a conductor of known potential by the method of moments.

R5.12. Outline by means of an example the procedure for obtaining the charge distribution on a conductor of known potential by the method of moments technique.

R5.13. Discuss the determination of the capacitance of a parallel-plate capacitor by the method of moments technique.

R5.14. Discuss the relationship between the capacitance, conductance, and inductance per unit length for an infinitely long, parallel conductor arrangement.

R5.15. Outline the steps in the derivation of the expressions for the capacitance, conductance, and inductance per unit length of an infinitely long parallel cylindrical-wire arrangement.

R5.16. Distinguish between internal inductance and external inductance. Discuss the concept of flux linkage pertinent to the determination of the internal inductance.

R5.17. Explain the concept of mutual inductance and discuss an example of its computation.

R5.18. What is meant by the quasistatic extension of the static field in a physical structure?

R5.19. Outline the steps involved in the quasistatic extension of the static field in a parallel-plate structure short circuited at one end.

R5.20. Discuss the condition for the validity of the quasistatic approximation for the parallel-plate structure short circuited at one end.

R5.21. Discuss the low-frequency behavior of a parallel-plate structure with a lossy medium between the plates.

R5.22. Discuss the quasistatic behavior of the structure of Fig. 5.19 for $\sigma \approx 0$.

R5.23. What is a magnetic circuit? Why is the magnetic flux in a magnetic circuit confined almost entirely to the core?

R5.24. Define the reluctance of a magnetic circuit. What is the analogous electric circuit quantity? Why is the reluctance for a given set of dimensions of a magnetic circuit not a constant?

R5.25. Discuss the complete analogy between a magnetic circuit and an electric circuit using the example of the toroidal magnetic core versus the toroidal conductor.

R5.26. How is the fringing of the magnetic flux in an air gap in a magnetic circuit taken into account?

R5.27. Discuss by means of an example the analysis of a magnetic circuit with three legs and its equivalent-circuit representation.

R5.28. Discuss by means of an example the phenomenon of electromechanical energy conversion.

R5.29. Outline the computation of mechanical force of electric origin from considerations of energy balance associated with an electromechanical system.

R5.30. Discuss by means of an example the computation of energy converted from electrical to mechanical, or vice versa, in an electromechanical system.

PROBLEMS

P5.1. Assume that the impurity concentration of the *p-n* junction diode of Fig. 5.1(a) is a linear function of distance across the junction. The space charge distribution is then given by

$$\rho = kx \text{ for } -d/2 < x < d/2$$

where d is the width of the space charge region and k is the proportionality constant. Find the solution for the potential inside and outside the space charge region, assuming the potential to be zero at $x = 0$.

P5.2. A space charge density distribution is given by

$$\rho = \begin{cases} \rho_0 \sin x & \text{for } -\pi < x < \pi \\ 0 & \text{otherwise} \end{cases}$$

where ρ_0 is a constant. Find and sketch the potential V versus x for all x. Assume $V = 0$ for $x = 0$.

P5.3. A space charge density distribution is given in spherical coordinates by

$$\rho = \begin{cases} \rho_0 & \text{for } a < r < 2a \\ 0 & \text{otherwise} \end{cases}$$

where ρ_0 is a constant. Find and sketch the potential V versus r for all r.

P5.4. The region between the two plates in Fig. 5.2 is filled with two perfect dielectric media having permittivities ε_1 for $0 < x < t$ (region 1) and ε_2 for $t < x < d$ (region 2). **(a)** Find the solutions for the potentials in the two regions $0 < x < t$ and $t < x < d$. **(b)** Find the capacitance per unit area of the plates.

P5.5. Assume that the two media in Problem P5.4 are imperfect dielectrics having conductivities σ_1 and σ_2 for $0 < x < t$ and $t < x < d$, respectively. **(a)** What are the boundary conditions to be satisfied at $x = t$? **(b)** Find the solutions for the potentials in the two regions. **(c)** Find the potential at $x = t$.

P5.6. Assume that the region between the two plates of Fig. 5.2 is filled with a perfect dielectric of nonuniform permittivity

$$\varepsilon = \varepsilon_1 + (\varepsilon_2 - \varepsilon_1)\frac{x}{d}$$

Find the solution for the potential between the plates and obtain the expression for the capacitance per unit area of the plates.

P5.7. Assume that the region between the coaxial cylindrical conductors of Fig. 5.3(a) is filled with a dielectric of nonuniform permittivity $\varepsilon = \varepsilon_0 b/r$. Obtain the solution for the potential between the conductors and the expression for the capacitance per unit length of the cylinders.

P5.8. Consider the spherical capacitor of Fig. 5.3(b). Obtain the expression for the energy stored in the electric field of the charge distribution and show that it is equal to $\frac{1}{2}CV_0^2$.

P5.9. The cross section of an infinitely long arrangement of conductors normal to the page and that repeats endlessly in the plane of the page is shown in Fig. 5.30. For the grid points shown, find the values of V_1, V_2, V_3, V_4, and V_5, by writing equations consistent with (5.25) and solving them. Then find the approximate magnitude of the field intensity at grid point 2 and the approximate value of the surface charge density at point P, assuming that the spacing between the grid points is d and the medium between the conductors is free space.

P5.10. The cross section of an arrangement of conductors, infinitely long normal to the page, is square, as shown in Fig. 5.31. Three sides are kept at 0 V and the fourth side is kept at 28 V. The region between the conductors is divided into a 4×4 grid of squares. Although there are nine grid points, there are only six unknown potentials $V_A, V_B, \ldots, V_F$, because of symmetry. **(a)** By writing equations consistent with (5.25) for these six potentials and solving the equations, find the values of the potentials. **(b)** Find the approximate magnitude of the electric field intensity at grid point B, assuming that the spacing between grid points is d. **(c)** Find the approximate surface charges per unit length of the arrangement on the 28 V conductor and the 0 V conductor.

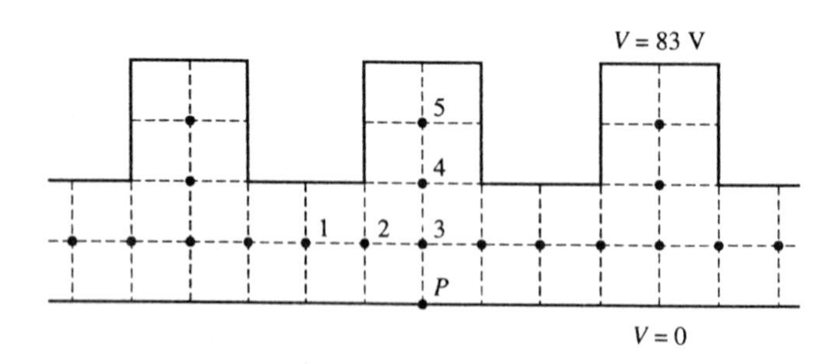

Figure 5.30. For Problem P5.9.

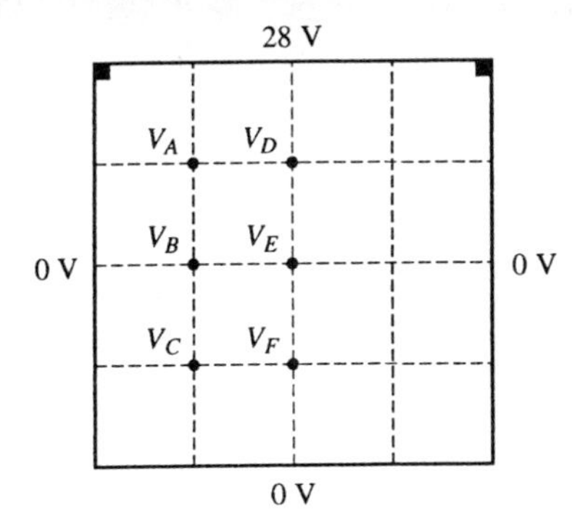

Figure 5.31. For Problem P5.10.

P5.11. In Fig. 5.4, assume that the region below the y-axis ($x < 0$) is a perfect dielectric of relative permittivity ε_r, whereas the region above the y-axis ($x > 0$) is free space. Show that the modified form of (5.25) is given by

$$V_0 \approx \frac{V_1 + \varepsilon_r V_2}{2(1 + \varepsilon_r)} + \frac{V_3 + V_4}{4}$$

P5.12. Consider a thin, straight cylindrical wire of length l and radius a ($\ll l$) bent in the middle to make a 90° angle and held at a potential of 1 V. By dividing the wire into four equal segments and assuming the charge density in each segment to be uniform, and using the method of moments, find the total charge on the wire if $l = 1$ m and $a = 1$ mm. To compute the potential at the center of a given segment due to the charge in another segment, assume the charge to be a point charge at the center of the segment.

P5.13. Consider a thin wire of radius 1 mm in the form of a circular ring of radius 1 m and held at a potential of 1 V. By approximating the shape of the wire to be a $2n$-sided regular polygon inscribing the circle and using the method of moments, obtain the expression for the line charge density along the wire. Use the result of (5.28) for the potential at the center point of a given side of the polygon due to the charge in that side, but to compute the potential at the center point of a given side due to the charge in a different side, consider the charge in that side to be a point charge. Find the value of the line charge density for $n = 10$.

P5.14. Consider two thin wires which are circular, as in Problem P5.13 and arranged such that one wire is directly above and parallel to the second wire at a spacing of 10 cm so as to form a capacitor. Using the method of moments as in Problem P5.13, obtain the expression for the capacitance of the arrangement. Find its value for $n = 10$.

P5.15. A square-shaped conductor of area $3a \times 3a$, with a square-shaped hole of area $a \times a$ in the middle, as shown in Fig. 5.32, is held at a potential of 1 V. By dividing the conductor into eight squares, as shown in the figure, and using the method of moments, find the total charge on the conductor. To find the potential at the center point of a square due to the charge in another square, consider the charge in that square to be a point charge at the center of that square.

P5.16. Assume that a capacitor is made up of two parallel conductors, each having the shape shown in Fig. 5.32. If the spacing between the plates is a, find the capacitance of the arrangement by dividing each conductor into squares as shown in Fig. 5.32 and applying the method of moments.

P5.17. Assume that a parallel-plate capacitor is made up of two unequal square-shaped plates of sides a and $2a$ spaced a apart and with their centers along a line normal

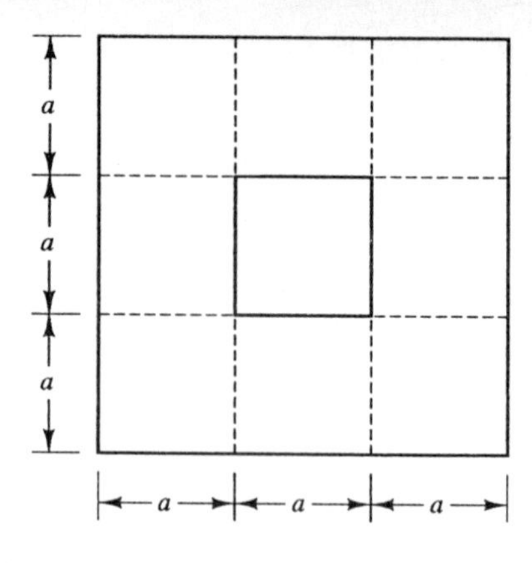

Figure 5.32. For Problem P5.15.

to the plates, as shown in Fig. 5.33. It is desired to find the capacitance of the arrangement by using the method of moments. To do this, assume a potential difference of V_0 applied between the plates and divide each plate into four equal parts. From symmetry considerations on each plate and considering the potential difference between the center point of one square on the top plate and the center point of one square on the bottom plate (since the two plates cannot be considered to be at equal and opposite potentials as for the case of the parallel-plate capacitor with equal-sized plates), write one equation for the unknown surface charge densities on the two plates. Write a second equation by using the equality of magnitudes of the charges on the two plates.

P5.18. For the parallel-wire arrangement of Fig. 5.12(c), show that for $d \gg a$ the capacitance per unit length of the line is $\pi\varepsilon/\ln(2d/a)$. Find the value of d/a for which the exact value of the capacitance per unit length is 1.05 times the value given by the approximate expression for $d \gg a$.

P5.19. For the line-charge pair of Fig. 5.14, show that the direction lines of the electric field are arcs of circles emanating from the positively charged line and terminating on the negatively charged line.

P5.20. Derive the expression for the capacitance per unit length of the arrangement of Fig. 5.12(d). (*Hint:* Consider two conductors placed in two equipotential surfaces surrounding the line charge $-\rho_{L0}$ in Fig. 5.14.)

P5.21. A filamentary wire carrying current I is closely wound around a toroidal magnetic core of rectangular cross section, as shown in Fig. 5.34. The mean radius of the toroidal core is a and the number of turns per unit length along the mean circum-

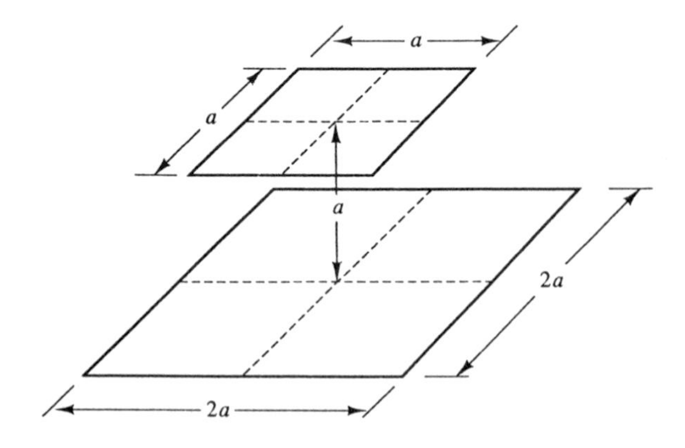

Figure 5.33. For Problem P5.17.

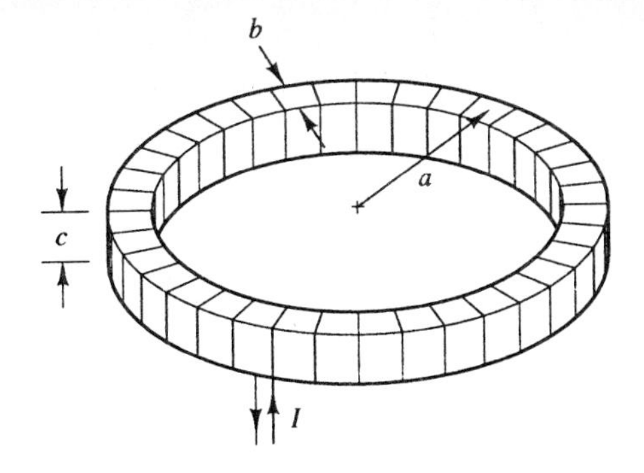

Figure 5.34. For Problem P5.21.

ference of the toroid is N. Find the inductance of the toroid.

P5.22. An infinitely long, uniformly wound solenoid of radius a and having N turns per unit length carries a current I. Find the inductance per unit length of the solenoid. Assume air core ($\mu = \mu_0$).

P5.23. A current I flows with nonuniform volume density given by

$$\mathbf{J} = J_0\left(\frac{r}{a}\right)^2 \mathbf{i}_z$$

along an infinitely long cylindrical conductor of radius a having the z-axis as its axis. The current returns with uniform surface density in the opposite direction along the surface of an infinitely long, perfectly conducting cylinder of radius b ($> a$) and coaxial with the inner conductor. Find the internal inductance per unit length of the inner conductor.

P5.24. Consider the infinitely long solid cylindrical conductor of Fig. 5.15. Obtain the expression for the energy stored per unit length in the magnetic field internal to the current distribution and show that it is equal to $\frac{1}{2}\mathcal{L}_i I^2$, where I is the total current.

P5.25. An infinitely long, uniformly wound solenoid of radius a and having N_1 turns per unit length is coaxial with another infinitely long, uniformly wound solenoid of radius b ($> a$) and having N_2 turns per unit length. Find the mutual inductance per unit length of the solenoids. Assume air core ($\mu = \mu_0$).

P5.26. For the structure of Fig. 5.18, assume $l = 10$ cm, $d = 1$ cm, $w = 10$ cm, and free space for the medium between the plates. **(a)** For a current source $I(t) = 1 \cos 10^6 \pi t$ A, find the voltage developed across the source. **(b)** Repeat part (a) for $I(t) = 1 \cos 10^9 \pi t$ A.

P5.27. For the structure of Fig. 5.18, show that the input behavior for frequencies slightly beyond those for which the quasistatic approximation is valid is equivalent to the parallel combination of L ($= \mu dl/w$) and $\frac{1}{3}C$, where $C = \varepsilon wl/d$ is the capacitance of the structure obtained from static field considerations with the two plates not joined by another conductor at $z = 0$.

P5.28. For the structure of Fig. 5.19 with $\sigma = 0$, show that the input behavior for frequencies slightly beyond those for which the quasistatic approximation is valid is equivalent to the series combination of C ($= \varepsilon wl/d$) and $\frac{1}{3}L$, where $L = \mu dl/w$ is the inductance of the structure obtained from static field considerations with the two plates joined by another conductor at $z = 0$, as in Fig. 5.18.

P5.29. Find the conditions under which the quasistatic input behavior of the structure of

Fig. 5.19 is essentially equivalent to that of **(a)** a single resistor; **(b)** a capacitor C $(= \varepsilon wl/d)$ in parallel with a resistor; and **(c)** a resistor in series with an inductor.

P5.30. For the structure of Fig. 5.18, assume that the medium has nonzero conductivity σ. **(a)** Show that the input behavior correct to the first power in ω is the same as if σ were zero. **(b)** Investigate the input behavior correct to the second power in ω and obtain the equivalent circuit.

P5.31. A toroidal magnetic core has the dimensions $A = 4$ cm^2 and $l = 20$ cm. **(a)** If it is found that for NI equal to 200 A-t a magnetic flux ψ equal to 6×10^{-4} Wb is established in the core, find the permeability μ of the core material. **(b)** If now an air gap of width $l_g = 0.1$ mm is introduced, find the new value of NI required to maintain the flux of 6×10^{-4} Wb, neglecting fringing of flux in the air gap.

P5.32. For the magnetic circuit of Fig. 5.25, assume the air gap to be in the center leg. Find the NI required to establish a magnetic flux of 8.4×10^{-4} Wb in the air gap.

P5.33. For the magnetic circuit of Fig. 5.25, assume that there is an air gap of length 0.2 mm in the left leg in addition to that in the right leg. Find the NI required to establish a magnetic flux of 4×10^{-4} Wb in the air gap in the right leg.

P5.34. For the magnetic circuit of Fig. 5.25, assume that there is no air gap. Find the magnetic flux established in the center leg for an applied NI equal to 300 A-t.

P5.35. For the magnetic circuit of Fig. 5.25, assume that there is no air gap and that $A_1 = 5$ cm^2 with all other dimensions remaining as specified in Example 5.9. Find the magnetic flux density in the center leg for an applied NI equal to 150 A-t.

P5.36. In Fig. 5.35 a dielectric slab of permittivity ε sliding between the plates of a parallel-plate capacitor experiences a mechanical force $\mathbf{F}_e$ of electrical origin. Assuming width w for the plates normal to the page and neglecting fringing of fields at the edges of the plates, find the expression for $\mathbf{F}_e$.

P5.37. In Fig. 5.36, a dielectric material of permittivity ε sliding freely in a cylindrical capacitor experiences a mechanical force $\mathbf{F}_e$ of electrical origin in the axial direction. Show that

$$\mathbf{F}_e = \frac{V_0^2 \pi (\varepsilon - \varepsilon_0)}{\ln \dfrac{b}{a}} \mathbf{i}_x$$

P5.38. Assume that in Example 5.10 the parallel-plate capacitor system of Fig. 5.27 is made to traverse the closed cycle in the V–x-plane shown in Fig. 5.37 instead of the closed cycle in the Q–x-plane shown in Fig. 5.28. Calculate the energy con-

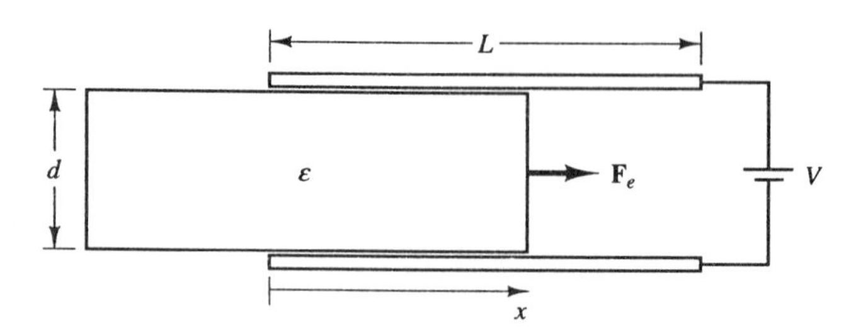

Figure 5.35. For Problem P5.36.

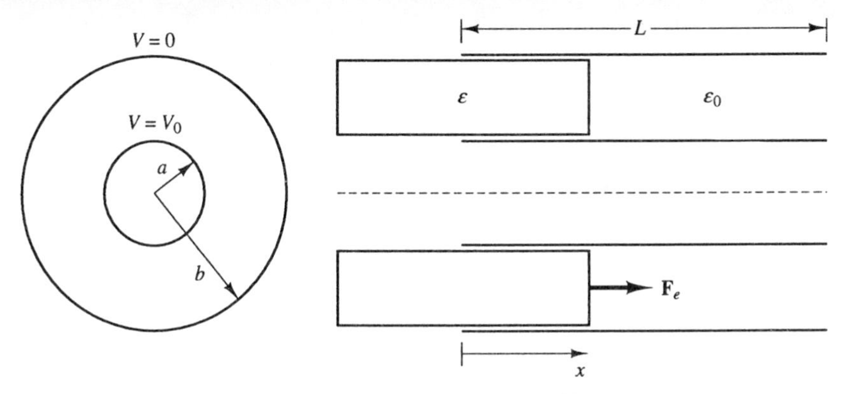

Figure 5.36. For Problem P5.37.

verted per cycle and determine whether the conversion is from mechanical to electrical, or vice versa.

P5.39. Figure 5.38 shows a magnetic field electromechanical device in which the plunger is free to move in the x-direction between two nonmagnetic sleeves. The areas of cross section of all three legs are equal. Using the same assumption as for the case of the electromagnet of Fig. 5.29 for the permeability of the magnetic core, show that the expression for the mechanical force $\mathbf{F}_e$ of electric origin on the plunger, in terms of the current I in the winding, is given by

$$\mathbf{F}_e = -\frac{2N^2I^2A\mu_0}{(2x + d)^2}\mathbf{i}_x$$

P5.40. Assume that the electromechanical device of Fig. 5.38 is made to traverse the closed cycle in the I–x-plane, as shown in Fig. 5.39. Find the energy converted per cycle and determine whether the conversion is from mechanical to electrical, or vice versa. Use $\mathbf{F}_e$ given in Problem P5.39.

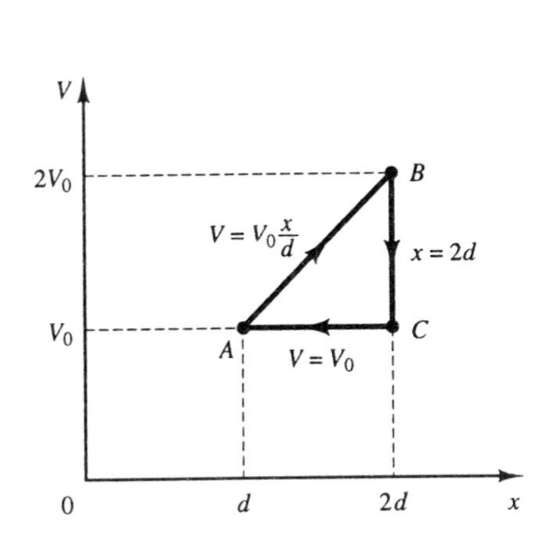

Figure 5.37. For Problem P5.38.

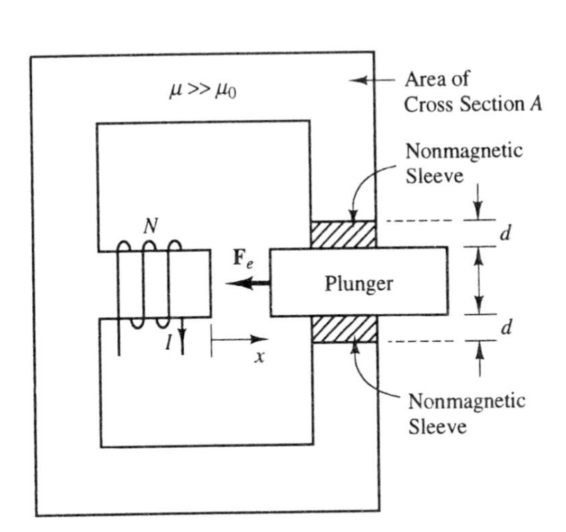

Figure 5.38. For Problem P5.39.

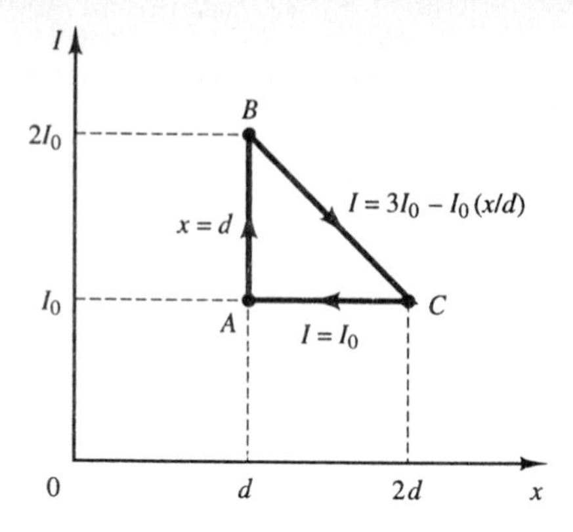

Figure 5.39. For Problem P5.40.

PC EXERCISES

PC5.1. Consider the arrangement of two infinitely long coaxial conductors of square cross sections and with the region between the conductors divided into a grid of squares, as shown in Fig. 5.40. Assume the outer conductor to be at a potential of 100 V and the inner conductor to be at 0 V. Write a program to compute the potentials at the grid points marked by dots to within 0.01 V of the averages of the potentials of the four neighboring points (i.e., use $\Delta = 0.01$ V), and then use these potential values to compute the approximate total surface charge per unit length on the outer conductor and hence the approximate capacitance per unit length of the arrangement.

PC5.2. It can be shown by considering Laplace's equation in three dimensions that the potential at a given point P in a charge-free region is approximately equal to the average of the potentials at the six equidistant points lying along mutually perpendicular axes through P. Using this result in conjunction with the iteration technique, write a program to compute the potential at the center of the cubical box shown in Fig. 5.41 in which the top face is kept at 100 V relative to the five

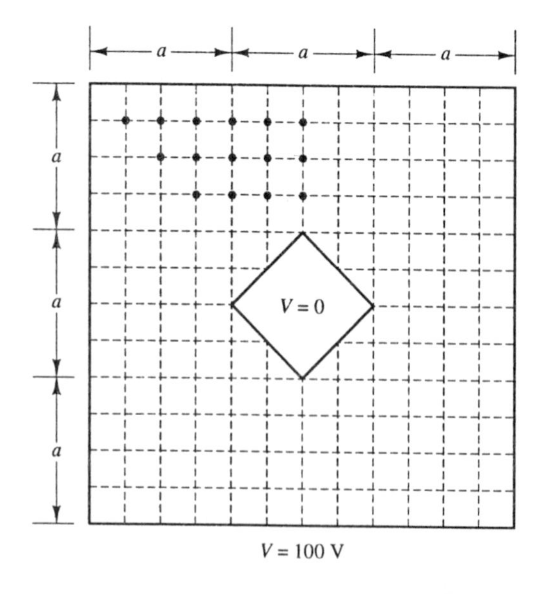

Figure 5.40. For Exercise PC5.1.

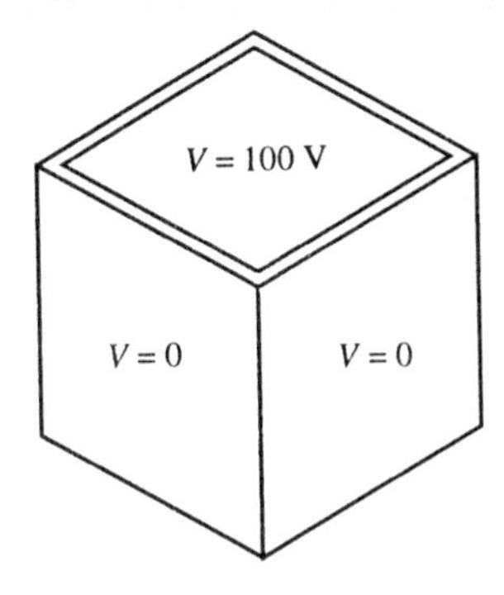

Figure 5.41. For Exercise PC5.2.

other faces. Use an $n \times n \times n$ grid of cubes, where n is a specified even number, and a value of 0.01 V for Δ.

PC5.3. Extend the program of PL5.2 to compute the capacitance of the arrangement of two thin, straight, parallel wires, each having the specification as in Example 5.4 and spacing between the wires equal to kl. Run the program for values of a, l, and k, and the value of n, the number of segments for each wire, specified by your instructor.

PC5.4. For the arrangement of two thin circular wires of Problem P5.14, assume the radius of cross section of each wire to be Δ, the radius of each ring to be a, and the spacing between the rings to be ka. Write a program that computes the capacitance C of the arrangement for specified values of Δ, a, and k, and the value of n, where $2n$ is the number of sides of the regular polygon approximating each ring, as described in Problem P5.13. Test the convergence of C by computing its value for several values of n for a fixed set of values of Δ, a, and k.

PC5.5. Write a program for computing the capacitance of a parallel-plate capacitor of square plates of sides a and spacing ka by dividing each plate into an $n \times n$ set of squares and using the method of moments. The quantities k and n are to be the input parameters. The output is to be the computed capacitance normalized with respect to $\varepsilon_0 a/k$, which is the capacitance of the arrangement if fringing of fields is neglected. Extend the program to generate a plot of the normalized capacitance versus k for the range of values of k specified by your instructor.

PC5.6. Consider the computation of the capacitance of an isolated spherical conductor of radius 1 m by using the method of moments. To do this, let the center of the sphere be at the origin, divide the surface into $2m$ equal intervals in θ and $4n$ equal intervals in ϕ, consider the sphere to be raised to a potential of 1 V, and assume the charge density in each region to be uniform. From symmetry considerations, there are only m unknown charge densities to be determined, namely, those associated with the regions in the θ-direction in the interval $0 < \theta < \pi/2$ (or $\pi/2 < \theta < \pi$) and lying in any of the intervals in ϕ. Write a program which for a given set of integer values for m and n computes the m unknown charge densities and the capacitance of the spherical conductor. Assume that the charge in a given region is a point charge at the center of that region to write the potential due to it at the center point of another region. But to write the potential due to the charge in a given region at the center point of the same region, assume that the shape of that area is a rectangle having the sides equal to the lengths of the arcs in the θ- and ϕ-directions passing through the center point of that region and bounded by that region, and use the result of Problem P4.28. Note that the analytical solution gives a uniform charge density of ε_0 C/m^2 on the spherical surface and a value of $4\pi\varepsilon_0$ F for the capacitance.

6

Uniform Plane Waves

In Chapters 3, 4, and 5 we learned Maxwell's equations in integral form and in differential form and discussed several applications to static and quasistatic fields. We shall now turn our attention to the applications of Maxwell's equations to time-varying fields. Many of these applications are based on electromagnetic wave phenomena, and hence it is necessary to gain an understanding of the basic principles of wave propagation, which is our goal in this chapter. We first consider wave propagation in free space and then extend the discussion to material media, thereby learning how the characteristics of wave propagation are modified from those associated with free space.

We shall employ an approach in this chapter that will enable us not only to learn how the coupling between space variations and time variations of the electric and magnetic fields, as indicated by Maxwell's equations, results in wave motion but also to illustrate the basic principle of radiation of waves from an antenna, which is treated in detail in Chapter 10. We augment our discussion of radiation and propagation of waves by considering such examples as the principle of an antenna array and the Doppler effect. We discuss power flow and energy storage associated with the wave motion and introduce the Poynting vector. Finally, we consider the topic of reflection of waves at a boundary between two different media.

6.1 UNIFORM PLANE WAVES IN TIME DOMAIN IN FREE SPACE

Uniform plane wave defined

In Chapter 4 we learned that the space variations of the electric and magnetic field components are related to the time variations of the magnetic and electric field components, respectively, through Maxwell's equations. This interdependence gives rise to the phenomenon of electromagnetic wave propagation. In the general case, electromagnetic wave propagation involves electric and magnetic fields having more than one component, each dependent on all three coordinates, in addition to time. However, a simple and very useful type of wave that serves as a building block in the study of electromagnetic waves consists of electric and magnetic fields that are perpendicular to each other and to the direction of propagation and are uniform in planes perpendicular to the direction of propagation. These waves are known as *uniform plane waves*. By orienting the coordinate axes such that the electric field is in the x-direction, the magnetic field is in the y-direction, and the direction of propagation is in the z-direction, as shown in Fig. 6.1, we have

$$\mathbf{E} = E_x(z, t)\mathbf{i}_x \tag{6.1a}$$

$$\mathbf{H} = H_y(z, t)\mathbf{i}_y \tag{6.1b}$$

Uniform plane waves do not exist in practice because they cannot be produced by finite-sized antennas. At large distances from physical antennas and ground, however, the waves can be approximated as uniform plane waves. Furthermore, the principles of guiding of electromagnetic waves along transmission lines and waveguides and the principles of many other wave phenomena can be studied basically in terms of uniform plane waves. Hence it is very important that we understand the principles of uniform plane wave propagation.

Infinite plane current sheet source

To illustrate the phenomenon of interaction of electric and magnetic fields giving rise to uniform plane electromagnetic wave propagation and the principle of radiation of electromagnetic waves from an antenna, we shall consider a simple, idealized, hypothetical source. This source consists of an infinite sheet lying in the xy-plane, as shown in Fig. 6.2. On this infinite plane sheet a uniformly distributed current flows in the negative x-direction, as given by

$$\boxed{\mathbf{J}_S = -J_S(t)\mathbf{i}_x \qquad \text{for} \quad z = 0} \tag{6.2}$$

where $J_S(t)$ is a given function of time. Because of the uniformity of the surface current density on the infinite sheet, if we consider any line of width w parallel to the y-axis, as shown in Fig. 6.2, the current crossing that line is simply given by

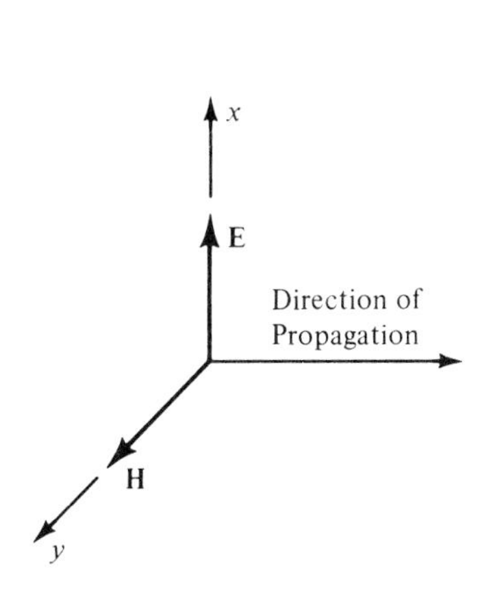

Figure 6.1. Directions of electric and magnetic fields and direction of propagation for a simple case of uniform plane wave.

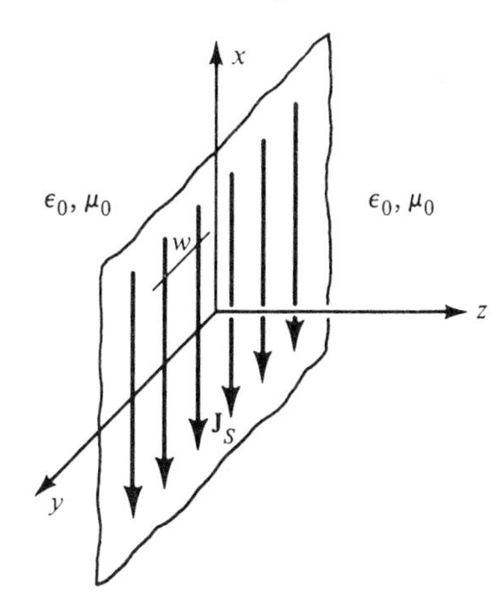

Figure 6.2. Infinite plane sheet in the xy-plane carrying surface current of uniform density.

w times the current density, that is, $wJ_S(t)$. If $J_S(t) = J_{S0} \cos \omega t$, then the current $wJ_{S0} \cos \omega t$, crossing the width w, actually alternates between negative x- and positive x-directions, that is, downward and upward. The time history of this current flow for one period of the sinusoidal variation is illustrated in Fig. 6.3, with the lengths of the lines indicating the magnitudes of the current. We shall consider the medium on either side of the current sheet to be free space.

To find the electromagnetic field due to the time-varying current sheet, we shall begin with Faraday's law and Ampère's circuital law given, respectively, by

$$\nabla \times \mathbf{E} = -\frac{\partial \mathbf{B}}{\partial t} \tag{6.3a}$$

$$\nabla \times \mathbf{H} = \mathbf{J} + \frac{\partial \mathbf{D}}{\partial t} \tag{6.3b}$$

and use a procedure that consists of the following steps:

1. Obtain the particular differential equations for the case under consideration.
2. Derive the general solution to the differential equations of step 1 without regard to the current on the sheet.
3. Show that the solution obtained in step 2 is a superposition of traveling waves propagating in the $+z$- and $-z$-directions.
4. Extend the general solution of step 2 to take into account the current on the sheet, thereby obtaining the required solution.

Although the procedure may be somewhat lengthy, we shall in the process learn several useful concepts and techniques.

1. To obtain the particular differential equations for the case under consideration, we first note that since (6.2) can be thought of as a current *distribution* having only an x-component of the current density which varies only with z, we can set J_y, J_z, and all derivatives with respect to x and y in (6.3a) and (6.3b) equal to zero. Hence (6.3a) and (6.3b) reduce to

$$-\frac{\partial E_y}{\partial z} = -\frac{\partial B_x}{\partial t} \tag{6.4a}$$

$$\frac{\partial E_x}{\partial z} = -\frac{\partial B_y}{\partial t} \tag{6.4b}$$

$$0 = -\frac{\partial B_z}{\partial t} \tag{6.4c}$$

$$-\frac{\partial H_y}{\partial z} = J_x + \frac{\partial D_x}{\partial t} \tag{6.5a}$$

$$\frac{\partial H_x}{\partial z} = \frac{\partial D_y}{\partial t} \tag{6.5b}$$

$$0 = \frac{\partial D_z}{\partial t} \tag{6.5c}$$

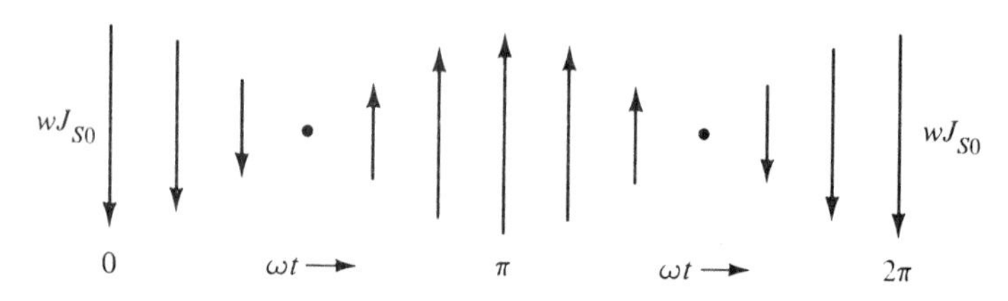

Figure 6.3. Time history of current flow across a line of width w parallel to the y-axis for the current sheet of Fig. 6.2, for $\mathbf{J}_S = -J_{S0} \cos \omega t \, \mathbf{i}_x$.

In these six equations, there are only two equations involving J_x and the pertinent electric and magnetic field components, namely, the simultaneous pair (6.4b) and (6.5a). Thus the equations of interest are

$$\boxed{\begin{aligned}\frac{\partial E_x}{\partial z} &= -\frac{\partial B_y}{\partial t} \quad (6.6a)\\ \frac{\partial H_y}{\partial z} &= -J_x - \frac{\partial D_x}{\partial t} \quad (6.6b)\end{aligned}}$$

which are the same as (4.7) and (4.25), the simplified forms of Faraday's law and Ampère's circuital law, respectively, for the special case of electric and magnetic fields characterized by (6.1a) and (6.1b), respectively.

Derivation of wave equation

2. In applying (6.6a) and (6.6b) to (6.2), we note that J_x in (6.6b) is a volume current density, whereas (6.2) represents a surface current density. Hence we shall solve (6.6a) and (6.6b) by setting $J_x = 0$ and then extend the solution to take into account the current on the sheet. For $J_x = 0$, (6.6a) and (6.6b) become

$$\boxed{\begin{aligned}\frac{\partial E_x}{\partial z} &= -\frac{\partial B_y}{\partial t} = -\mu_0\frac{\partial H_y}{\partial t} \quad (6.7a)\\ \frac{\partial H_y}{\partial z} &= -\frac{\partial D_x}{\partial t} = -\varepsilon_0\frac{\partial E_x}{\partial t} \quad (6.7b)\end{aligned}}$$

Differentiating (6.7a) with respect to z and then substituting for $\partial H_y/\partial z$ from (6.7b), we obtain

$$\frac{\partial^2 E_x}{\partial z^2} = -\mu_0\frac{\partial}{\partial z}\left(\frac{\partial H_y}{\partial t}\right) = -\mu_0\frac{\partial}{\partial t}\left(\frac{\partial H_y}{\partial z}\right) = -\mu_0\frac{\partial}{\partial t}\left(-\varepsilon_0\frac{\partial E_x}{\partial t}\right)$$

or

$$\boxed{\frac{\partial^2 E_x}{\partial z^2} = \mu_0\varepsilon_0\frac{\partial^2 E_x}{\partial t^2}} \quad (6.8)$$

We have thus eliminated H_y from (6.7a) and (6.7b) and obtained a single second-order partial differential equation involving E_x only. Equation (6.8) is known as the *wave equation*. In particular, it is a one-dimensional wave equation in time-domain form, that is, for arbitrary time dependence of E_x.

Solution of wave equation

To obtain the solution for (6.8), we introduce a change of variable by defining $\tau = z\sqrt{\mu_0\varepsilon_0}$. Substituting for z in (6.8) in terms of τ, we then have

$$\frac{\partial^2 E_x}{\partial \tau^2} = \frac{\partial^2 E_x}{\partial t^2} \quad (6.9)$$

or

$$\frac{\partial^2 E_x}{\partial \tau^2} - \frac{\partial^2 E_x}{\partial t^2} = 0$$

$$\left(\frac{\partial}{\partial \tau} + \frac{\partial}{\partial t}\right)\left(\frac{\partial}{\partial \tau} - \frac{\partial}{\partial t}\right)E_x = 0 \quad (6.10)$$

where the quantities in parentheses are operators operating on one another and on E_x. Equation (6.10) is satisfied if

$$\left(\frac{\partial}{\partial \tau} \pm \frac{\partial}{\partial t}\right) E_x = 0$$

or

$$\frac{\partial E_x}{\partial \tau} = \mp \frac{\partial E_x}{\partial t} \tag{6.11}$$

Let us first consider the equation corresponding to the upper sign in (6.11); that is,

$$\frac{\partial E_x}{\partial \tau} = -\frac{\partial E_x}{\partial t}$$

This equation says that the partial derivative of $E_x(\tau, t)$ with respect to τ is equal to the negative of the partial derivative of $E_x(\tau, t)$ with respect to t. The simplest function that satisfies this requirement is the function $(t - \tau)$. It then follows that any arbitrary function of $(t - \tau)$, say, $f(t - \tau)$ satisfies the requirement since

$$\frac{\partial}{\partial t}[f(t - \tau)] = f'(t - \tau)\frac{\partial}{\partial t}(t - \tau) = f'(t - \tau)$$

and

$$\frac{\partial}{\partial \tau}[f(t - \tau)] = f'(t - \tau)\frac{\partial}{\partial \tau}(t - \tau) = -f'(t - \tau) = -\frac{\partial}{\partial t}[f(t - \tau)]$$

where the prime associated with $f'(t - \tau)$ denotes differentiation of f with respect to $(t - \tau)$. In a similar manner, the solution for the equation corresponding to the lower sign in (6.11), that is, for

$$\frac{\partial E_x}{\partial \tau} = \frac{\partial E_x}{\partial t}$$

can be seen to be any arbitrary function of $(t + \tau)$, say, $g(t + \tau)$. Combining the two solutions, we write the solution for (6.11) to be

$$E_x(\tau, t) = Af(t - \tau) + Bg(t + \tau) \tag{6.12}$$

where A and B are arbitrary constants.

Substituting now for τ in (6.12) in terms of z, we obtain the solution for (6.8) to be

$$\boxed{E_x(z, t) = Af(t - z\sqrt{\mu_0 \varepsilon_0}) + Bg(t + z\sqrt{\mu_0 \varepsilon_0})} \tag{6.13}$$

The corresponding solution for $H_y(z, t)$ can be obtained by substituting (6.13) into (6.7a) or (6.76b). Thus using (6.7a),

$$\frac{\partial H_y}{\partial t} = \sqrt{\frac{\varepsilon_0}{\mu_0}}[Af'(t - z\sqrt{\mu_0 \varepsilon_0}) - Bg'(t + z\sqrt{\mu_0 \varepsilon_0})]$$

$$H_y(z, t) = \frac{1}{\sqrt{\mu_0/\varepsilon_0}}[Af(t - z\sqrt{\mu_0\varepsilon_0}) - Bg(t + z\sqrt{\mu_0\varepsilon_0})] \tag{6.14}$$

The fields given by (6.13) and (6.14) are the general solutions to the differential equations (6.7a) and (6.7b).

Traveling wave functions explained

3. To proceed further, we need to know the meanings of the functions f and g in (6.13) and (6.14). To discuss the meaning of f, let us consider a specific example

$$f(t - z\sqrt{\mu_0\varepsilon_0}) = (t - z\sqrt{\mu_0\varepsilon_0})^2$$

Plots of this function versus z for two values of t, $t = 0$ and $t = \sqrt{\mu_0\varepsilon_0}$, are shown in Fig. 6.4(a). An examination of these plots reveals that as time increases from 0 to $\sqrt{\mu_0\varepsilon_0}$, every point on the plot for $t = 0$ moves by one unit in the $+z$-direction, thereby making the plot for $t = \sqrt{\mu_0\varepsilon_0}$ an exact replica of the plot for $t = 0$, except displaced by one unit in the $+z$-direction. The function f is therefore said to represent a *traveling wave propagating in the* $+z$*-direction*, or simply a *(+) wave*. In particular, it is a *uniform plane wave* since its value does not vary with position in a given constant z-plane. By dividing the distance traveled by the time taken, the velocity of propagation of the wave can be obtained to be

$$v_p = \frac{1}{\sqrt{\mu_0\varepsilon_0}} = 3 \times 10^8 \text{ m/s} \tag{6.15}$$

which is equal to c, the velocity of light in free space. Similarly, to discuss the meaning of g, we shall consider

$$g(t + z\sqrt{\mu_0\varepsilon_0}) = (t + z\sqrt{\mu_0\varepsilon_0})^2$$

Then plotting the function versus z for $t = 0$ and $t = \sqrt{\mu_0\varepsilon_0}$, as shown in Fig. 6.4(b), we can see that the plot for $t = \sqrt{\mu_0\varepsilon_0}$ is an exact replica of the plot for $t = 0$, except displaced by one unit in the $-z$-direction. The function g is therefore said to represent a *traveling wave propagating in the* $-z$*-direction*, or simply a *(−) wave*. Once again, it is a uniform plane wave with the velocity of propagation equal to $1/\sqrt{\mu_0\varepsilon_0}$.

To generalize the foregoing discussion of the functions f and g, let us consider two pairs of t and z, say, t_1 and z_1, and $t_1 + \Delta t$ and $z_1 + \Delta z$. Then for the function f to maintain the same value for these two pairs of z and t, we must have

$$t_1 - z_1\sqrt{\mu_0\varepsilon_0} = (t_1 + \Delta t) - (z_1 + \Delta z)\sqrt{\mu_0\varepsilon_0}$$

or

$$\Delta z = \frac{1}{\sqrt{\mu_0\varepsilon_0}}\Delta t$$

Since $\sqrt{\mu_0\varepsilon_0}$ is a positive quantity, this indicates that as time progresses, a given value of the function moves forward in z with the velocity $1/\sqrt{\mu_0\varepsilon_0}$, thereby giving the characteristic of a (+) wave for f. Similarly, for the function g to maintain the same value for the two pairs of t and z, we must have

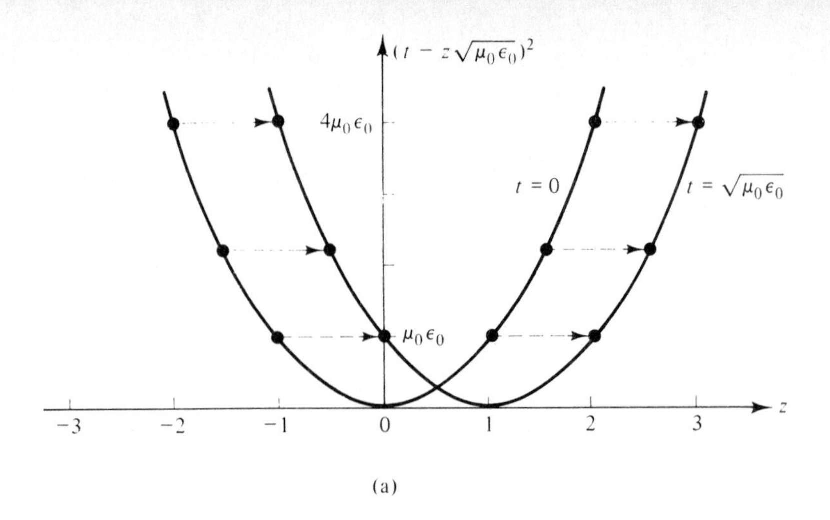

(a)

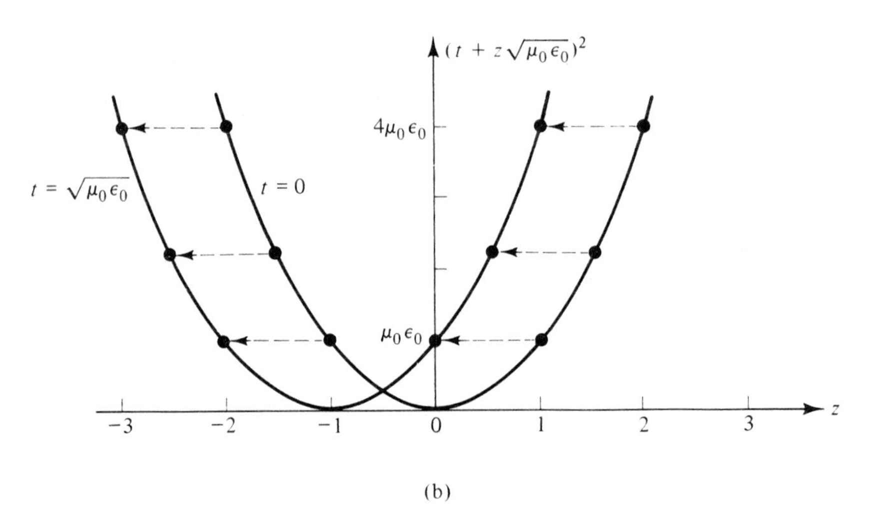

(b)

Figure 6.4. (a) Plots of the function $(t - z\sqrt{\mu_0\varepsilon_0})^2$ versus z for $t = 0$ and $t = \sqrt{\mu_0\varepsilon_0}$. (b) Plots of the function $(t + z\sqrt{\mu_0\varepsilon_0})^2$ versus z for $t = 0$ and $t = \sqrt{\mu_0\varepsilon_0}$.

$$t_1 + z_1\sqrt{\mu_0\varepsilon_0} = (t_1 + \Delta t) + (z_1 + \Delta z)\sqrt{\mu_0\varepsilon_0}$$

or

$$\Delta z = -\frac{1}{\sqrt{\mu_0\varepsilon_0}}\Delta t$$

The minus sign associated with $1/\sqrt{\mu_0\varepsilon_0}$ indicates that as time progresses, a given value of the function moves backward in z with the velocity $1/\sqrt{\mu_0\varepsilon_0}$, giving the characteristic of a $(-)$ wave for g.

We shall now define the intrinsic impedance of free space, η_0, to be

$$\eta_0 = \sqrt{\frac{\mu_0}{\varepsilon_0}} \approx 120\pi\ \Omega = 377\ \Omega \tag{6.16}$$

From (6.13) and (6.14), we see that η_0 is the ratio of E_x to H_y for the $(+)$ wave or the negative of the same ratio for the $(-)$ wave. Since the units of E_x are volts per meter and the units of H_y are amperes per meter, the units of E_x/H_y are volts per ampere or ohms, thereby giving the character of impedance for η_0. Replacing $\sqrt{\mu_0/\varepsilon_0}$ in (6.14) by η_0 and substituting v_p *for* $1/\sqrt{\mu_0\varepsilon_0}$ in the arguments of the functions f and$g$ in both (6.13) and (6.14), we can now write (6.13) and (6.14) as

$$E_x(z, t) = Af\left(t - \frac{z}{v_p}\right) + Bg\left(t + \frac{z}{v_p}\right) \tag{6.17a}$$

$$H_y(z, t) = \frac{1}{\eta_0}\left[Af\left(t - \frac{z}{v_p}\right) - Bg\left(t + \frac{z}{v_p}\right)\right] \tag{6.17b}$$

Electromagnetic field due to the current sheet

4. Having learned that the solution to (6.7a) and (6.7b) consists of superposition of traveling waves propagating in the $+z$- and $-z$-directions, we now make use of this solution together with other considerations to find the electromagnetic field due to the infinite plane current sheet of Fig. 6.2, and with the current density given by (6.2). To do this, we observe the following:

(a) Since the current sheet, which is the source of waves, is in the $z = 0$ plane, there can be only a $(+)$ wave in the region $z > 0$ and only a $(-)$ wave in the region $z < 0$. Thus

$$\mathbf{E}(z, t) = \begin{cases} Af\left(t - \dfrac{z}{v_p}\right)\mathbf{i}_x & \text{for } z > 0 \\ Bg\left(t + \dfrac{z}{v_p}\right)\mathbf{i}_x & \text{for } z < 0 \end{cases} \tag{6.18a}$$

$$\mathbf{H}(z, t) = \begin{cases} \dfrac{A}{\eta_0} f\left(t - \dfrac{z}{v_p}\right)\mathbf{i}_y & \text{for } z > 0 \\ -\dfrac{B}{\eta_0} g\left(t + \dfrac{z}{v_p}\right)\mathbf{i}_y & \text{for } z < 0 \end{cases} \tag{6.18b}$$

(b) From the boundary condition (3.51b) applied to the surface $z = 0$, we have

$$[E_x]_{z=0+} - [E_x]_{z=0-} = 0 \tag{6.19}$$

or $Af(t) = Bg(t)$. Thus (6.18a) and (6.18b) reduce to

$$\mathbf{E}(z, t) = F\left(t \mp \frac{z}{v_p}\right)\mathbf{i}_x \qquad \text{for } z \gtrless 0 \tag{6.20a}$$

$$\mathbf{H}(z, t) = \pm\frac{1}{\eta_0}F\left(t \mp \frac{z}{v_p}\right)\mathbf{i}_y \qquad \text{for } z \gtrless 0 \tag{6.20b}$$

where we have used $Af(t) = Bg(t) = F(t)$.

(c) From the boundary condition (3.55a) applied to the surface $z = 0$, we have

$$\mathbf{i}_z \times \{[\mathbf{H}]_{z=0+} - [\mathbf{H}]_{z=0-}\} = -J_S(t)\mathbf{i}_x \tag{6.21}$$

or $(2/\eta_0)F(t) = J_S(t)$. Thus $F(t) = (\eta_0/2)J_S(t)$, and (6.20a) and (6.20b) become

$$\mathbf{E}(z, t) = \frac{\eta_0}{2} J_S\left(t \mp \frac{z}{v_p}\right)\mathbf{i}_x \quad \text{for} \quad z \gtrless 0 \tag{6.22a}$$

$$\mathbf{H}(z, t) = \pm\frac{1}{2} J_S\left(t \mp \frac{z}{v_p}\right)\mathbf{i}_y \quad \text{for} \quad z \gtrless 0 \tag{6.22b}$$

Equations (6.22a) and (6.22b) represent the complete solution for the electromagnetic field due to the infinite plane current sheet of surface current density given by

$$\mathbf{J}_S(t) = -J_S(t)\mathbf{i}_x \quad \text{for} \quad z = 0 \tag{6.23}$$

The solution corresponds to uniform plane waves having their field components uniform in planes parallel to the current sheet and propagating to either side of the current sheet with the velocity v_p $(= c)$. The time variation of the electric field component E_x in a given $z =$ constant plane is the same as the current density variation delayed by the time $|z|/v_p$ and multiplied by $\eta_0/2$. The time variation of the magnetic field component in a given $z =$ constant plane is the same as the current density variation delayed by $|z|/v_p$ and multiplied by $\pm\frac{1}{2}$ depending on $z \gtrless 0$. Using these properties one can construct plots of the field components versus time for fixed values of z and versus z for fixed values of t. We shall illustrate by means of an example.

Example 6.1

Let us consider the function $J_S(t)$ in (6.23) to be that given in Fig. 6.5. We wish to find and sketch (a) E_x versus t for $z = 300$ m, (b) H_y versus t for $z = -450$ m, (c) E_x versus z for $t = 1$ μs, and (d) H_y versus z for $t = 2.5$ μs.

(a) Since $v_p = c = 3 \times 10^8$ m/s, the time delay corresponding to 300 m is 1 μs. Thus the plot of E_x versus t for $z = 300$ m is the same as that of $J_S(t)$ multiplied by $\eta_0/2$ or 188.5 and delayed by 1 μs, as shown in Fig. 6.6(a).

(b) The time delay corresponding to 450 m is 1.5 μs. Thus the plot of H_y versus t for $z = -450$ m is the same as that of $J_S(t)$ multiplied by $-1/2$ and delayed by 1.5 μs, as shown in Fig. 6.6(b).

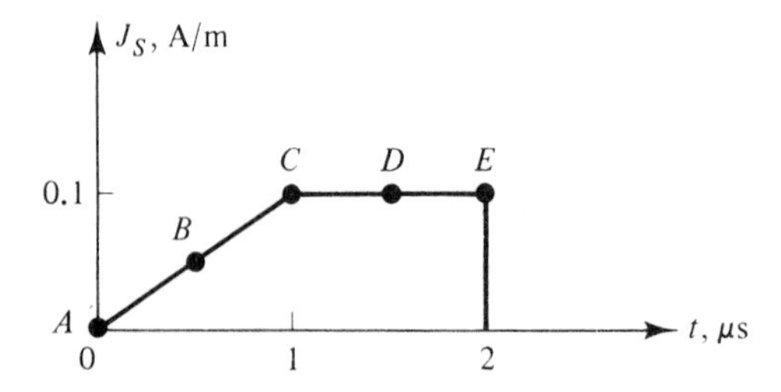

Figure 6.5. Plot of J_S versus t for Example 6.1.

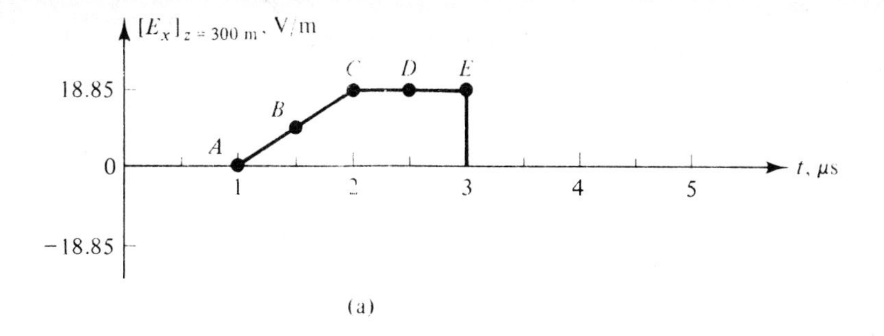

(a)

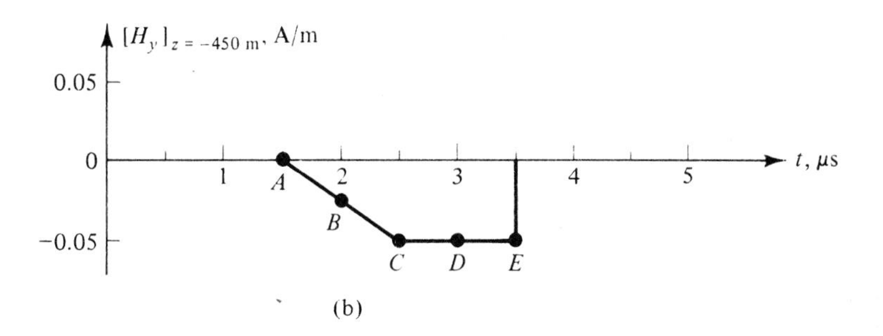

(b)

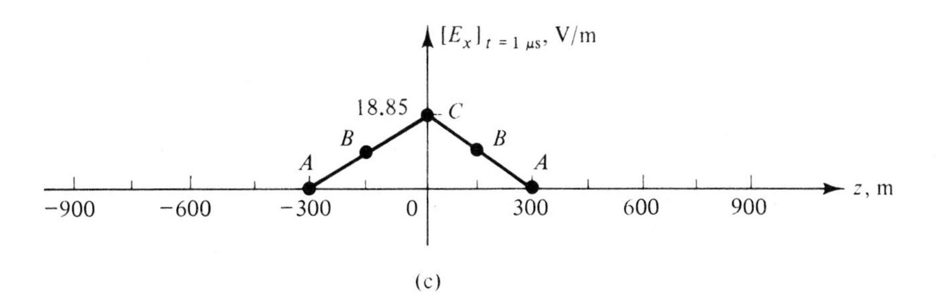

(c)

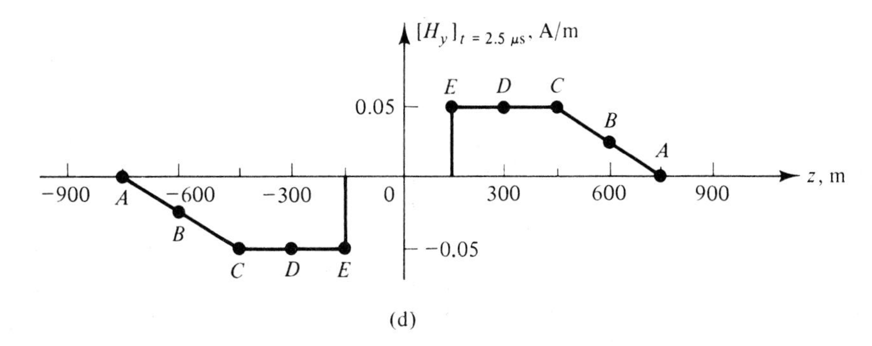

(d)

Figure 6.6. Plots of field components versus t for fixed values of z and versus z for fixed values of t for Example 6.1.

(c) To sketch E_x versus z for a fixed value of t, say, t_1, we use the argument that a given value of E_x existing at the source at an earlier value of time, say, t_2, travels away from the source by the distance equal to $(t_1 - t_2)$ times v_p. Thus at $t = 1\ \mu s$, the values of E_x corresponding to points A and B in Fig. 6.5 move to the locations $z = \pm 300$ m and $z = \pm 150$ m, respectively, and the value of E_x corresponding to point C exists right at the source. Hence the plot of E_x versus z for $t = 1\ \mu s$ is as shown in Fig. 6.6(c). Note that points beyond C in Fig. 6.5 correspond to $t > 1\ \mu s$, and therefore they do not appear in the plot of Fig. 6.6(c).

(d) Using arguments as in part (c), we see that at $t = 2.5\ \mu s$, the values of H_y corresponding to points A, B, C, D, and E in Fig. 6.5 move to the locations $z = \pm 750$ m, ± 600 m, ± 450 m, ± 300 m, and ± 150 m, respectively, as shown in Fig. 6.6(d). Note that the plot is an odd function of z since the factor by which J_{S0} is multiplied to obtain H_y is $\pm \frac{1}{2}$ depending on $z \gtrless 0$.

K6.1. Infinite plane current sheet; Uniform plane wave; Wave equation; Time domain; Traveling wave functions; Velocity of propagation; Intrinsic impedance of free space; Time delay.

D6.1. For each of the following traveling wave functions, find the velocity of propagation both in magnitude and direction: **(a)** $(0.05y - t)^2$; **(b)** $u(t + 0.02x)$; and **(c)** $\cos(2\pi \times 10^8 t - 2\pi z)$.
Ans. **(a)** $20\mathbf{i}_y$ m/s; **(b)** $-50\mathbf{i}_x$ m/s; **(c)** $10^8\mathbf{i}_z$ m/s

D6.2. The time-variation for $z = 0$ of a function $f(z, t)$ representing a traveling wave propagating in the $+z$-direction with velocity 200 m/s is shown in Fig. 6.7. Find the value of the function for each of the following cases: **(a)** $z = 300$ m, $t = 2.0$ s; **(b)** $z = -200$ m, $t = 0.4$ s; and **(c)** $z = 100$ m, $t = 0.5$ s.
Ans. **(a)** $0.25A$; **(b)** $0.6A$; **(c)** 0

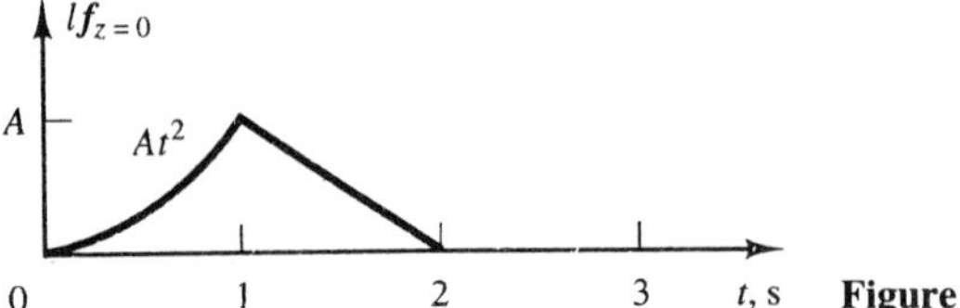

Figure 6.7. For Problem D6.2.

D6.3. The time variation for $z = 0$ of a function $g(z, t)$ representing a traveling wave propagating in the $-z$-direction with velocity 100 m/s is shown in Fig. 6.8. Find the value of the function for each of the following cases: **(a)** $z = 200$ m, $t = 0.2$ s; **(b)** $z = -300$ m, $t = 3.4$ s; and **(c)** $z = 100$ m, $t = 0.6$ s.
Ans. **(a)** $0.9A$; **(b)** $0.4A$; **(c)** A

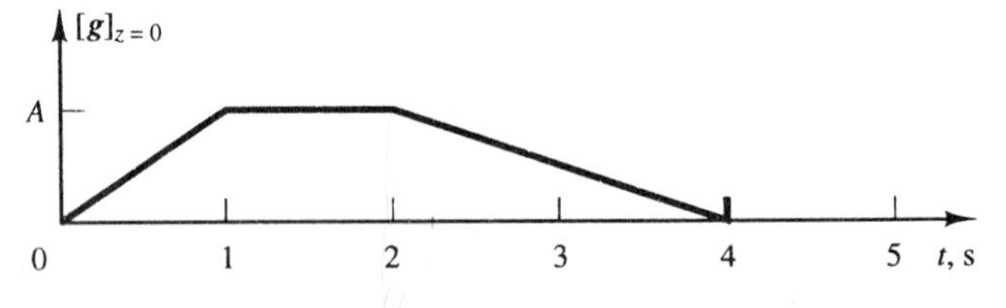

Figure 6.8. For Problem D6.3.

6.2 SINUSOIDALLY TIME-VARYING UNIFORM PLANE WAVES IN FREE SPACE

Solution for the electromagnetic field for the sinusoidal case

In the previous section, we considered the current density on the infinite plane current sheet to be an arbitrary function of time and obtained the solution for the electromagnetic field. As already pointed out in Section 1.5, of particular interest are fields varying sinusoidally with time. These are produced by a source whose current density varies sinusoidally with time. Thus assuming the current density on the infinite plane sheet of Fig. 6.2 to be

$$\mathbf{J}_S = -J_{S0} \cos \omega t \, \mathbf{i}_x \qquad \text{for} \quad z = 0 \tag{6.24}$$

where J_{S0} is the amplitude and ω is the radian frequency, we obtain the corresponding solution for the electromagnetic field by substituting $J_S(t) = J_{S0} \cos \omega t$ in (6.22a) and (6.22b):

$$\mathbf{E} = \frac{\eta_0 J_{S0}}{2} \cos (\omega t \mp \beta z) \, \mathbf{i}_x \qquad \text{for} \quad z \gtrless 0 \tag{6.25a}$$

$$\mathbf{H} = \pm \frac{J_{S0}}{2} \cos (\omega t \mp \beta z) \, \mathbf{i}_y \qquad \text{for} \quad z \gtrless 0 \tag{6.25b}$$

where

$$\beta = \frac{\omega}{v_p} \tag{6.26}$$

Properties and parameters of sinusoidal waves

Equations (6.25a) and (6.25b) represent sinusoidally time-varying uniform plane waves propagating away from the current sheet. The phenomenon is illustrated in Fig. 6.9, which shows sketches of the current density on the sheet and the distance variation of the electric and magnetic fields on either side of the current sheet for three values of t. It should be understood that in these sketches the field variations depicted along the z-axis hold also for any other line parallel to the z-axis. We shall now discuss in detail several important parameters and properties associated with the sinusoidal waves:

1. The argument $(\omega t \mp \beta z)$ of the cosine functions is the phase of the fields. We shall denote the phase by the symbol ϕ. Thus

$$\phi = \omega t \mp \beta z \tag{6.27}$$

Note that ϕ is a function of t and z.

Frequency

2. Since

$$\frac{\partial \phi}{\partial t} = \omega \tag{6.28}$$

the rate of change of phase with time for a fixed value of z is equal to ω, the radian frequency of the wave. The linear frequency given by

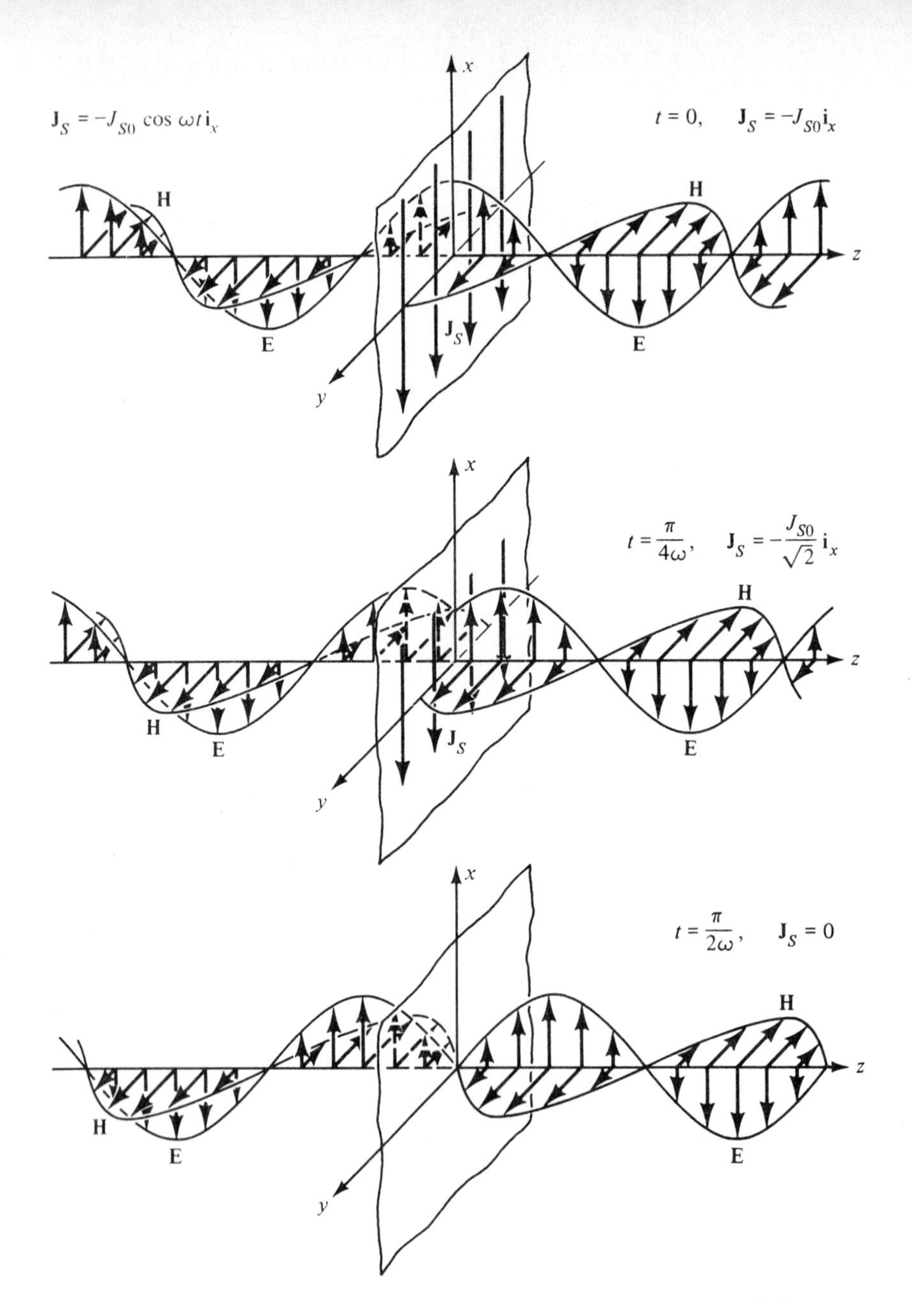

Figure 6.9. Time history of uniform plane electromagnetic wave radiating away from an infinite plane current sheet in free space.

$$\boxed{f = \frac{\omega}{2\pi}} \tag{6.29}$$

is the number of times the phase changes by 2π radians in one second for a fixed value of z. The situation is pertinent to an observer at a point in the field region watching a *movie* of the field variations with time and counting the number of times in one second the field goes through a certain phase point, say, the positive maximum.

Phase constant

3. Since

$$\boxed{\frac{\partial \phi}{\partial z} = \mp \beta} \tag{6.30}$$

the magnitude of the rate of change of phase with distance z for a fixed value of time is equal to β, known as the *phase constant*. The situation is pertinent to taking a *still photograph* of the phenomenon at any given time along the z-axis, counting the number of radians of phase change in one meter.

Wavelength

4. It follows from property 3 that the distance, along the z-direction, in which the phase changes by 2π radians for a fixed value of time is equal to $2\pi/\beta$. This distance is known as the *wavelength*, denoted by the symbol λ. Thus

$$\boxed{\lambda = \frac{2\pi}{\beta}} \tag{6.31}$$

It is the distance between two consecutive positive maximum points on the sinusoid, or between any other two points which are displaced from these two positive maximum points by the same distance and to the same side, as shown in Fig. 6.10.

Phase velocity

5. From (6.26), we note that the velocity of propagation of the wave is given by

$$\boxed{v_p = \frac{\omega}{\beta}} \tag{6.32}$$

Here, it is known as the phase velocity, since a constant value of phase progresses with that velocity along the z-direction. It is the velocity with which an observer has to move along the direction of propagation of the wave to be associated with a particular phase point on the moving sinusoid. Thus it follows from (6.27) that

$$d(\omega t \mp \beta dz) = 0$$

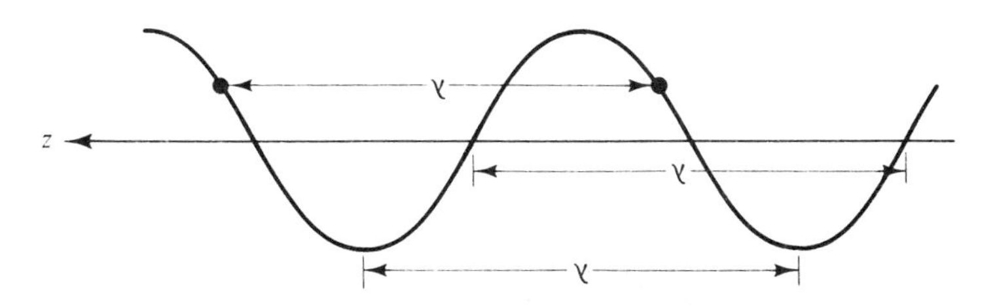

Figure 6.10. For explaining wavelength.

which gives

$$\omega\, dt \mp \beta\, dz = 0$$

$$\frac{dz}{dt} = \pm\frac{\omega}{\beta}$$

where the + and − signs correspond to (+) and (−) waves, respectively. We recall that for free space, $v_p = 1/\sqrt{\mu_0\varepsilon_0} = c = 3 \times 10^8$ m/s.

Classification of waves

6. From (6.31), (6.29), and (6.32), we note that

$$\lambda f = \left(\frac{2\pi}{\beta}\right)\left(\frac{\omega}{2\pi}\right) = \frac{\omega}{\beta}$$

or

$$\boxed{\lambda f = v_p} \tag{6.33}$$

Thus the wavelength and frequency of a wave are not independent of each other but are related through the phase velocity. This is not surprising because λ is a parameter governing the variation of the field with distance for a fixed time, f is a parameter governing the variation of the field with time for a fixed value of z, and we know from Maxwell's equations that the space and time variations of the fields are interdependent. For free space (6.33) gives

$$\boxed{\lambda \text{ in meters} \times f \text{ in hertz} = 3 \times 10^8} \tag{6.34a}$$

or

$$\boxed{\lambda \text{ in meters} \times f \text{ in megahertz} = 300} \tag{6.34b}$$

It can be seen from these relationships that the higher the frequency, the shorter the wavelength. Waves are classified according to frequency or wavelength. Table 6.1 lists the commonly used designations for the various bands up to 300 GHz, where 1 GHz is 10^9 Hz. The corresponding frequency

TABLE 6.1 COMMONLY USED DESIGNATIONS FOR THE VARIOUS FREQUENCY RANGES

Designation	Frequency range	Wavelength range
ELF (extremely low frequency)	30–3000 Hz	10,000–100 km
VLF (very low frequency)	3–30 kHz	100–10 km
LF (low frequency) or long waves	30–300 kHz	10–1 km
MF (medium frequency) or medium waves	300–3000 kHz	1000–100 m
HF (high frequency) or short waves	3–30 MHz	100–10 m
VHF (very high frequency)	30–300 MHz	10–1 m
UHF (ultrahigh frequency)	300–3000 MHz	100–10 cm
Microwaves	1–30 GHz	30–1 cm
Millimeter waves	30–300 GHz	10–1 mm

ranges and wavelength ranges are also given. The frequencies above about 300 GHz fall into regions for infrared and beyond. The AM radio (550–1650 kHz) falls in the medium wave band, whereas the FM radio makes use of 88–108 MHz in the VHF band. The VHF TV channels 2–6 use 54–88 MHz, and 7–13 employ 174–216 MHz. The UHF TV channels are in the 470–890 MHz range. Microwave ovens operate at 2450 MHz. Police traffic radars operate at about 10.5 and 24.1 GHz. Various other ranges in Table 6.1 are used for various other applications too numerous to mention here.

Intrinsic impedance

7. The electric and magnetic fields are such that

$$\boxed{\frac{\text{amplitude of } \mathbf{E}}{\text{amplitude of } \mathbf{H}} = \eta_0} \tag{6.35}$$

We recall that η_0, the intrinsic impedance of free space, has a value approximately equal to 120π or 377 Ω.

8. The electric and magnetic fields have components lying in the planes of constant phase (z = constant planes) and perpendicular to each other and to the direction of propagation. In fact, the cross product of **E** and **H** results in a vector that is directed along the direction of propagation, as can be seen by noting that

$$\boxed{\mathbf{E} \times \mathbf{H} = \pm\frac{\eta_0 J_{S0}^2}{4} \cos^2 (\omega t \mp \beta z)\, \mathbf{i}_z \qquad \text{for} \quad z \gtrless 0} \tag{6.36}$$

Polarization

9. From the discussion of polarization of sinusoidally time-varying fields in Section 1.5, we note that the fields given by (6.25a) and (6.25b) are linearly polarized. Hence the wave is said to be linearly polarized. Two linearly polarized waves of the same frequency propagating in the same direction, with their electric fields (and hence their magnetic fields) perpendicular to each other and out of phase by 90°, add up to give a circularly polarized wave. Since a circle can be traversed in one of two opposite senses, we talk of right-handed or clockwise circular polarization and left-handed or counterclockwise circular polarization. The convention is that if in a given constant phase plane, the tip of the field vector rotates with time in the clockwise sense as seen looking along the direction of propagation of the wave, the wave is said to be right circularly polarized. If the tip of the field vector rotates in the counterclockwise sense, the wave is said to be left circularly polarized. Similar considerations hold for elliptically polarized waves, which arise due to the superposition of two linearly polarized waves in the general case.

We shall now consider two examples of the application of the properties we have learned thus far in this section.

Example 6.2

The electric field of a uniform plane wave is given by

$$\mathbf{E} = 10 \sin (3\pi \times 10^8 t - \pi z)\, \mathbf{i}_x + 10 \cos (3\pi \times 10^8 t - \pi z)\, \mathbf{i}_y \text{ V/m}$$

Let us find (a) the various parameters associated with the wave, (b) the corresponding magnetic field **H**, and (c) the polarization of the wave.

(a) From the argument of the sine and cosine functions, we can identify the following:

$$\omega = 3\pi \times 10^8 \text{ rad/s}$$

$$\beta = \pi \text{ rad/m}$$

Then

$$f = \frac{\omega}{2\pi} = 1.5 \times 10^8 \text{ Hz}$$

$$\lambda = \frac{2\pi}{\beta} = 2 \text{ m}$$

$$v_p = \frac{\omega}{\beta} = 3 \times 10^8 \text{ m/s}$$

Note also that $\lambda f = 3 \times 10^8 = v_p$. In view of the minus sign associated with πz, the direction of propagation of the wave is the $+z$-direction.

(b) The unit vectors $\mathbf{i}_x$ and $\mathbf{i}_y$ associated with the first and second terms, respectively, tell us that the electric field contains components directed along the x- and y-directions. Using the properties 6 and 7 discussed earlier, we obtain the magnetic field of the wave to be

$$\begin{aligned}\mathbf{H} &= \frac{10}{377} \sin (3\pi \times 10^8 t - \pi z)\, \mathbf{i}_y + \frac{10}{377} \cos (3\pi \times 10^8 t - \pi z)\, (-\mathbf{i}_x) \\ &= -\frac{10}{377} \cos (3\pi \times 10^8 t - \pi z)\, \mathbf{i}_x + \frac{10}{377} \sin (3\pi \times 10^8 t - \pi z)\, \mathbf{i}_y \text{ A/m}\end{aligned}$$

(c) The two components of **E** are equal in amplitude, perpendicular, and out of phase by 90°. Therefore, the wave is circularly polarized. To determine if the polarization is right-handed or left-handed, we look at the electric field vectors in the $z = 0$ plane for two values of time, $t = 0$ and $t = \frac{1}{6} \times 10^{-8}$ s $(3\pi \times 10^8 t = \pi/2)$. These are shown in Fig. 6.11. As time progresses, the tip of the vector rotates in the counterclockwise sense, as seen looking in the $+z$-direction. Hence the wave is left circularly polarized.

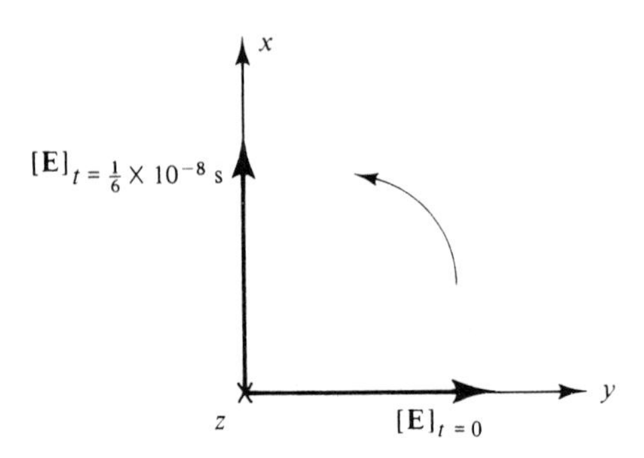

Figure 6.11. For Example 6.2.

Example 6.3

Principle of antenna array

An antenna array consists of two or more antenna elements spaced appropriately and excited with currents having the appropriate amplitudes and phases in order to obtain a desired radiation characteristic. To illustrate the principle of an antenna array, let us consider two infinite plane parallel current sheets, spaced $\lambda/4$ apart and carrying currents of equal amplitudes but out of phase by $\pi/2$ as given by the densities

$$\mathbf{J}_{S1} = -J_{S0} \cos \omega t \ \mathbf{i}_x \qquad \text{for} \quad z = 0$$

$$\mathbf{J}_{S2} = -J_{S0} \sin \omega t \ \mathbf{i}_x \qquad \text{for} \quad z = \frac{\lambda}{4}$$

and find the electric field due to the array of the two current sheets.

We apply the result given by (6.25a) to each current sheet separately and then use superposition to find the required total electric field due to the array of the two current sheets. Thus for the current sheet in the $z = 0$ plane, we have

$$\mathbf{E}_1 = \begin{cases} \dfrac{\eta_0 J_{S0}}{2} \cos(\omega t - \beta z) \ \mathbf{i}_x & \text{for} \quad z > 0 \\[2ex] \dfrac{\eta_0 J_{S0}}{2} \cos(\omega t + \beta z) \ \mathbf{i}_x & \text{for} \quad z < 0 \end{cases}$$

For the current sheet in the $z = \lambda/4$ plane, we have

$$\mathbf{E}_2 = \begin{cases} \dfrac{\eta_0 J_{S0}}{2} \sin\left[\omega t - \beta\left(z - \dfrac{\lambda}{4}\right)\right] \mathbf{i}_x & \text{for} \quad z > \dfrac{\lambda}{4} \\[2ex] \dfrac{\eta_0 J_{S0}}{2} \sin\left[\omega t + \beta\left(z - \dfrac{\lambda}{4}\right)\right] \mathbf{i}_x & \text{for} \quad z < \dfrac{\lambda}{4} \end{cases}$$

$$= \begin{cases} \dfrac{\eta_0 J_{S0}}{2} \sin\left(\omega t - \beta z + \dfrac{\pi}{2}\right) \mathbf{i}_x & \text{for} \quad z > \dfrac{\lambda}{4} \\[2ex] \dfrac{\eta_0 J_{S0}}{2} \sin\left(\omega t + \beta z - \dfrac{\pi}{2}\right) \mathbf{i}_x & \text{for} \quad z < \dfrac{\lambda}{4} \end{cases}$$

$$= \begin{cases} \dfrac{\eta_0 J_{S0}}{2} \cos(\omega t - \beta z) \ \mathbf{i}_x & \text{for} \quad z > \dfrac{\lambda}{4} \\[2ex] -\dfrac{\eta_0 J_{S0}}{2} \cos(\omega t + \beta z) \ \mathbf{i}_x & \text{for} \quad z < \dfrac{\lambda}{4} \end{cases}$$

Now, using superposition, we find the total electric field due to the two current sheets to be

$$\mathbf{E} = \mathbf{E}_1 + \mathbf{E}_2$$

$$= \begin{cases} \eta_0 J_{S0} \cos(\omega t - \beta z) \ \mathbf{i}_x & \text{for} \quad z > \dfrac{\lambda}{4} \\[2ex] \eta_0 J_{S0} \sin \omega t \sin \beta z \ \mathbf{i}_x & \text{for} \quad 0 < z < \dfrac{\lambda}{4} \\[2ex] \mathbf{0} & \text{for} \quad z < 0 \end{cases}$$

Thus the total field is zero in the region $z < 0$, and hence there is no radiation toward that side of the array. In the region $z > \lambda/4$ the total field is twice that of the field due to a single sheet. The phenomenon is illustrated in Fig. 6.12, which shows sketches of the individual fields E_{x1} and E_{x2} and the total field $E_x = E_{x1} + E_{x2}$ for a few values of t. The result that we have obtained here for the total field due to the array of two current sheets, spaced $\lambda/4$ apart and fed with currents of equal amplitudes but out of phase by $\pi/2$, is said to correspond to an "endfire" radiation pattern.

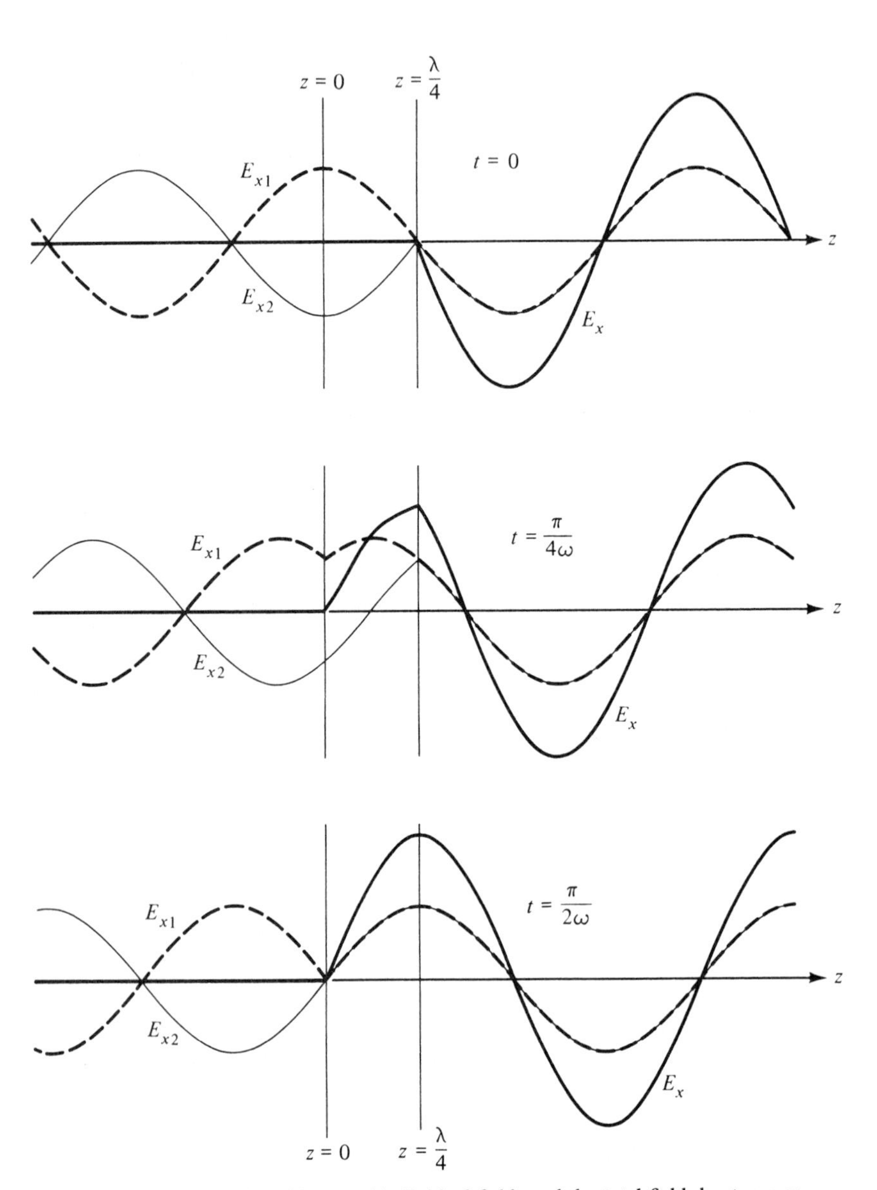

Figure 6.12. Time history of individual fields and the total field due to an array of two infinite plane parallel current sheets.

Doppler effect

Returning now to the solution for the electromagnetic field given by (6.25a) and (6.25b), let us ask ourselves the question, "How does the phase associated with the wave change with time as viewed by a moving observer?" To answer this question, let us consider the (+) wave and an observer moving along the positive z-direction with a velocity v_0 m/s, starting at $z = z_0$ at $t = 0$. Then the position of the observer as a function of time is given by $z = z_0 + v_0 t$ and the phase of the wave at that position is given by

$$\begin{aligned}\phi_{\text{obs}} &= \omega t - \beta(z_0 + v_0 t) \\ &= (\omega - \beta v_0)t - \beta z_0\end{aligned}$$

Ignoring relativistic effects, the rate of change of phase with time or the radian frequency of the wave viewed by the moving observer is

$$\begin{aligned}\omega_{\text{obs}} &= \frac{d}{dt}[(\omega - \beta v_0)t - \beta z_0] \\ &= \omega - \beta v_0 = \omega - \frac{\omega}{v_p} v_0 \\ &= \omega\left(1 - \frac{v_0}{v_p}\right)\end{aligned}$$

or

$$\boxed{f_{\text{obs}} = f\left(1 - \frac{v_0}{v_p}\right)} \tag{6.37}$$

Thus the moving observer views a frequency that is different from that of the source of the wave. This phenomenon of a shift in the frequency of the wave is known as the *Doppler effect*. For an observer moving along the direction of propagation, the Doppler-shifted frequency is less than the actual frequency by the amount fv_0/v_p or v_0/λ. For an observer moving opposite to the direction of propagation of the wave, the Doppler-shifted frequency is higher than the actual frequency by the same amount. The situation is illustrated in Fig. 6.13, which depicts the wave motion as viewed by a stationary observer (A) and two moving observers (O and R), one moving along and the other moving opposite to the direction of propagation of the wave with a velocity which, for simplicity of illustration, is assumed to be one-half the phase velocity of the wave. From the series of sketches for one period of the wave, it can be seen that observer A views a complete cycle of the wave whereas observer O veiws only one-half cycle of the wave and observer R views one and one-half cycles of the wave during that period. Thus the stationary observer A views the same frequency as that of the wave, but moving observer O views a frequency that is one-half that of the wave and moving observer R views a frequency that is one and one-half that of the wave.

Doppler shift also occurs when the current sheet is in motion and the observer is at rest. Assuming that the Doppler-shifted radian frequency as viewed by the stationary observer in the field of a moving current sheet is ω_{obs}, we can write the phase of the (+) wave to be

$$\phi_{\text{obs}} = \omega_{\text{obs}}\left(t - \frac{z}{v_p}\right)$$

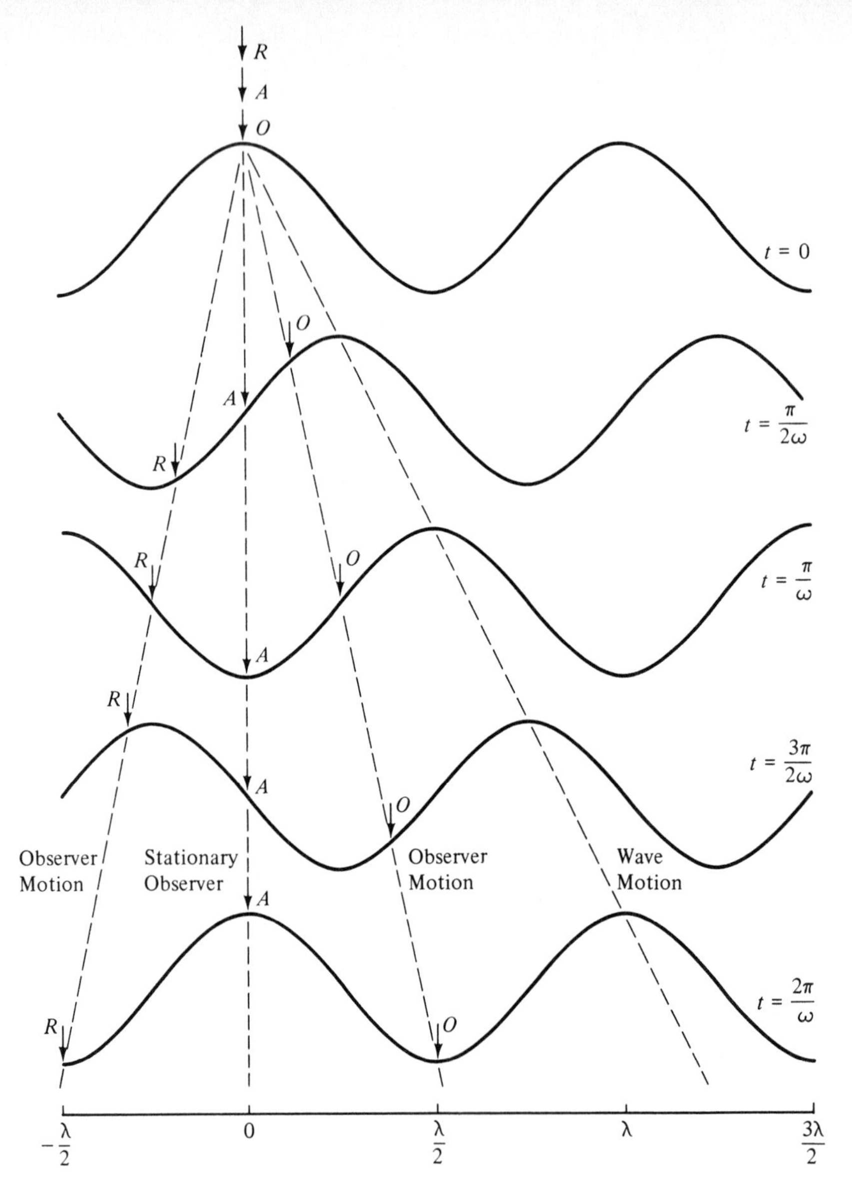

Figure 6.13. Wave motion as viewed by a stationary observer (A) and two moving observers (O and R).

If the current sheet is moving with velocity v_s in the $+z$-direction, starting at $z = 0$ at $t = 0$, then the phase on the current sheet is $\omega_{\text{obs}}(t - v_s t/v_p)$. Since the radian frequency on the current sheet is ω, we have

$$\frac{d}{dt}\left[\omega_{\text{obs}}\left(t - \frac{v_s t}{v_p}\right)\right] = \omega$$

$$\omega_{obs}\left(1 - \frac{v_s}{v_p}\right) = \omega$$

$$\omega_{\text{obs}} = \frac{\omega}{1 - v_s/v_p}$$

or

$$\boxed{f_{\text{obs}} = \frac{f}{1 - v_s/v_p}} \tag{6.38}$$

For $v_s \ll v_p$, (6.38) reduces to

$$\boxed{f_{\text{obs}} \approx f\left(1 + \frac{v_s}{v_p}\right)} \tag{6.39}$$

the right side of which is the same as that in (6.37), with v_0 replaced by $-v_s$. For $v_s > 0$, that is, source moving toward the observer, $f_{\text{obs}} > f$. For $v_s < 0$, that is, source receding from the observer, $f_{\text{obs}} < f$.

Once again, we have ignored relativistic effects in deriving (6.38). When relativistic effects are taken into account, (6.37) and (6.38) are modified to give an expression such that the two values of f_{obs} are exactly equal for the same relative velocity between the source and the observer. This expression is given by

$$f_{\text{obs}} = f\sqrt{\frac{1 - v_r/v_p}{1 + v_r/v_p}}$$

where v_r is the difference between v_0 and v_s. For $v_r \ll v_p$, this expression approximates to the form of (6.37).

Returning to the derivation leading to (6.37), we note that for an observer moving in an arbitrary direction with velocity $\mathbf{v}_o$, only the z-component of $\mathbf{v}_o$ contributes to the Doppler shift since the value of z corresponding to the position of the observer is governed only by the z-component of $\mathbf{v}_o$. The component of $\mathbf{v}_o$ perpendicular to the z-direction does not contribute to the Doppler shift. To generalize this for a wave propagating in an arbitrary direction given by the unit vector $\mathbf{i}_p$ and an observer moving with velocity $\mathbf{v}_o$ in the field of that wave, we can say that only the component of $\mathbf{v}_o$ along $\mathbf{i}_p$ contributes to the Doppler shift. Thus for this general case

$$\boxed{\text{Doppler shift in } f = -f\frac{\mathbf{v}_o \cdot \mathbf{i}_p}{v_p}} \tag{6.40}$$

In the application of (6.40), it should be kept in mind that if the source of waves is a point source instead of an infinite plane sheet of current, the direction of propagation of the wave is along the line from the point source to the observer.

Police radar

The phenomenon of Doppler effect has many applications. An example in everyday life with which some of us might have had an experience is the police traffic radar. In this application, a microwave signal is generated in the radar unit and is transmitted toward the moving car, the speed of which is to be monitored. Part of the incident signal is reflected from the moving car back to the radar unit, Doppler shifted in frequency by an amount proportional to the speed of the car. In the radar unit, the Doppler-shifted signal is received and mixed with the original signal to generate a signal of the difference frequency, which is the Doppler shift. A display in the radar unit is calibrated to indicate the speed of the car, which is the difference frequency multiplied by a proportionality constant. Some other applications of Doppler effect, just to mention a few, are in aircraft navigation, sea-state monitoring, and radio astronomy.

Example 6.4

Let us consider a police radar operating at a frequency $f = 24.1$ GHz and determine the Doppler shift due to an automobile directly approaching the radar at a speed of $v_o = 100$ km/h.

Since the radar operates on the signal reflected from the moving automobile, the Doppler shift is due to the combination of (6.37) and (6.39). Hence it is approximately twice that for one-way transmission. Thus the required Doppler shift in the radar operating frequency is given by

$$\begin{aligned}\Delta f &\approx 2f\frac{v_o}{v_p} \\ &= 2 \times 24.1 \times 10^9 \times \frac{10^5}{3600 \times 3 \times 10^8} \\ &= 4463 \text{ Hz} = 4.463 \text{ kHz}\end{aligned}$$

which is in the audio frequency range.

Experimental setup for Doppler effect

An experimental setup for demonstrating the Doppler effect is shown by the schematic diagram in Fig. 6.14. It consists of a horn radiating a UHF signal generated by a klystron oscillator and connected to it through microwave components as shown in the figure. The signal from the oscillator goes into port 1 of the circu-

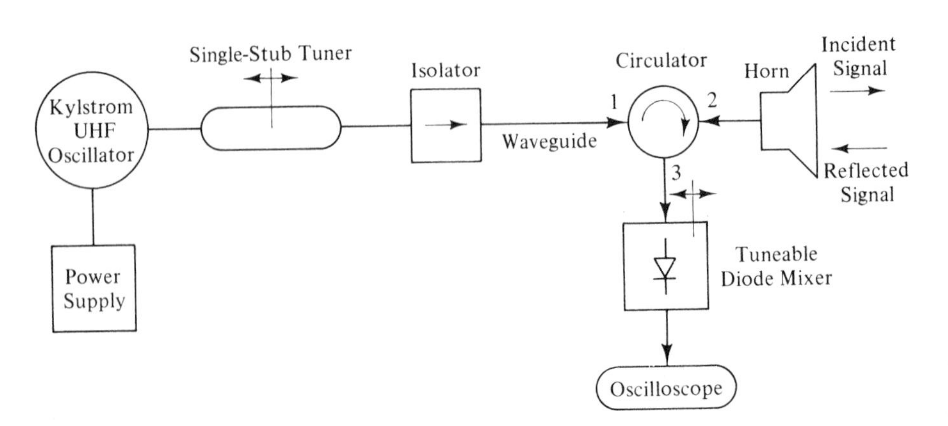

Figure 6.14. Schematic diagram of experimental setup for demonstrating Doppler effect.

lator and comes out of port 2 and is radiated from the horn. A reflective object placed in front of the horn provides a reflected signal, which is picked up by the horn and fed into port 2 of the circulator and comes out of port 3 into the mixer. Part of the original klystron signal going into port 1 also comes out of port 3. Thus signals proportional to the incident and reflected signals are mixed in the mixer, thereby generating the beat frequency signal, which is displayed on the oscilloscope. Since the reflected signal is Doppler shifted by an amount depending upon the motion of the reflective object, the beat frequency is equal to the Doppler shift. The isolator, a one-way device, prevents any signal coming out of port 1 of the circulator from going back to the klystron oscillator and affecting its performance. By placing a metallic sheet in front of the horn and walking toward and away from the horn, the Doppler effect can be demonstrated. Walking sideways, that is, perpendicular to the direction of propagation of the wave, will of course produce very little or no effect.

K6.2. Sinusoidal waves; Phase; Frequency; Wavelength; Phase velocity; Frequency times wavelength; Intrinsic impedance; Polarization; Antenna array; Moving observer; Doppler shift.

D6.4. For a sinusoidally time-varying uniform plane wave propagating in free space, find the following: **(a)** the frequency f, if the phase of the field at a point is observed to change by 3π rad in 0.1 μs; **(b)** the wavelength λ, if the phase of the field at a particular value of time is observed to change by 0.04π in a distance of 1 m along the direction of propagation of the wave; **(c)** the frequency f, if the wavelength is 25 m; and **(d)** the wavelength λ, if the frequency is 5 MHz.
Ans. **(a)** 15 MHz; **(b)** 50 m; **(c)** 12 MHz; **(d)** 60 m

D6.5. The magnetic field of a uniform plane wave in free space is given by

$$\mathbf{H} = H_0 \cos(6\pi \times 10^8 t + 2\pi y)\, \mathbf{i}_x \text{ A/m}$$

Find unit vectors along the following: **(a)** the direction of propagation of the wave; **(b)** the direction of the magnetic field at $t = 0$, $y = 0$; and **(c)** the direction of the electric field at $t = 0$, $y = 0$.
Ans. **(a)** $-\mathbf{i}_y$; **(b)** $\mathbf{i}_x$; **(c)** $-\mathbf{i}_z$

D6.6. For the array of two infinite plane current sheets of Example 6.3, assume that

$$\mathbf{J}_{S2} = -kJ_{S0} \sin \omega t\, \mathbf{i}_x \qquad \text{for } z = \lambda/4$$

where $|k| \leq 1$. Find the value of k for each of the following values of the ratio of the amplitude of the electric field intensity for $z > \lambda/4$ to the amplitude of the electric field intensity for $z < 0$: **(a)** 1/3; **(b)** 3; and **(c)** 7.
Ans. **(a)** $-1/2$; **(b)** 1/2; **(c)** 3/4

D6.7. The electric field intensity of a uniform plane wave propagating in a direction not along one of the coordinate axes is given by

$$\mathbf{E} = E_0(\mathbf{i}_x + 2\mathbf{i}_y + 2\mathbf{i}_z) \cos[9\pi \times 10^8 t - \pi(2x - 2y + z)]$$

Ignoring relativistic effects, find the Doppler shift in the frequency f of the wave viewed by a moving observer for each of the following velocities of the observer: **(a)** $10^3\mathbf{i}_x$ m/s; **(b)** $10^3\mathbf{i}_y$ m/s; **(c)** $10^3\mathbf{i}_z$ m/s; and **(d)** $10^3(\frac{2}{3}\mathbf{i}_x + \frac{1}{3}\mathbf{i}_y - \frac{2}{3}\mathbf{i}_z)$ m/s.
Ans. **(a)** -1000 Hz; **(b)** 1000 Hz; **(c)** -500 Hz; **(d)** 0 Hz

6.3 WAVE EQUATION AND SOLUTION FOR MATERIAL MEDIUM

In Sections 6.1 and 6.2 we discussed uniform plane wave propagation in free space. In this section we extend the treatment to a material medium characterized by conductivity σ, permittivity ε, and permeability μ. We recall that the constitutive relations are

$$\mathbf{J}_c = \sigma \mathbf{E}$$

$$\mathbf{D} = \varepsilon \mathbf{E}$$

$$\mathbf{H} = \frac{\mathbf{B}}{\mu}$$

so that the Maxwell's equations for the material medium are

$$\nabla \times \mathbf{E} = -\frac{\partial \mathbf{B}}{\partial t} = -\mu \frac{\partial \mathbf{H}}{\partial t} \tag{6.41a}$$

$$\nabla \times \mathbf{H} = \mathbf{J} + \frac{\partial \mathbf{D}}{\partial t} = \mathbf{J}_c + \frac{\partial \mathbf{D}}{\partial t} = \sigma \mathbf{E} + \varepsilon \frac{\partial \mathbf{E}}{\partial t} \tag{6.41b}$$

To discuss electromagnetic wave propagation in the material medium, let us consider the infinite plane current sheet of Fig. 6.2, except that now the medium on either side of the sheet is a material instead of free space, as shown in Fig. 6.15.

The electric and magnetic fields for the simple case of the infinite plane current sheet in the $z = 0$ plane and carrying uniformly distributed current in the negative x-direction as given by

$$\mathbf{J}_S = -J_{S0} \cos \omega t \; \mathbf{i}_x \tag{6.42}$$

are of the form

$$\mathbf{E} = E_x(z, t)\mathbf{i}_x \tag{6.43a}$$

$$\mathbf{H} = H_y(z, t)\mathbf{i}_y \tag{6.43b}$$

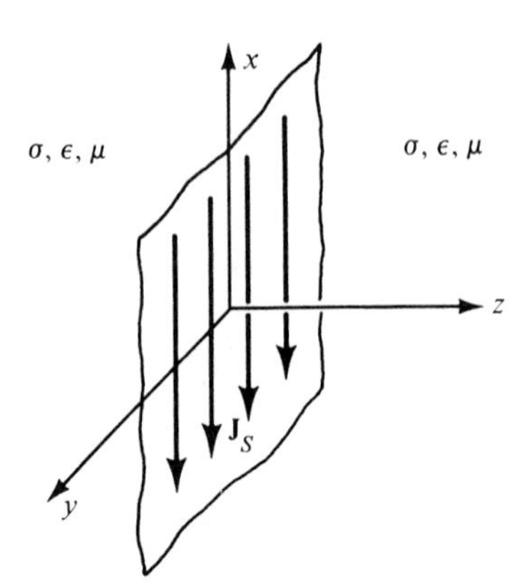

Figure 6.15. Infinite plane current sheet embedded in a material medium.

The corresponding simplified forms of the Maxwell's curl equations are

$$\frac{\partial E_x}{\partial z} = -\mu \frac{\partial H_y}{\partial t} \tag{6.44a}$$

$$\frac{\partial H_y}{\partial z} = -\sigma E_x - \varepsilon \frac{\partial E_x}{\partial t} \tag{6.44b}$$

Without the σE_x term on the right side of (6.44b), these two equations would be the same as (6.7a) and (6.7b) with μ_0 replaced by μ and ε_0 replaced by ε. The addition of the σE_x term complicates the solution in time domain. Hence, it is convenient to consider the solution for the sinusoidally time-varying case by using the phasor technique.

Wave equation

Thus letting

$$E_x(z, t) = \text{Re}[\bar{E}_x(z)e^{j\omega t}] \tag{6.45a}$$

$$H_y(z, t) = \text{Re}[\bar{H}_y(z)e^{j\omega t}] \tag{6.45b}$$

and replacing E_x and H_y in (6.44a) and (6.44b) by their phasors $\bar{E}_x$ and $\bar{H}_y$, respectively, and $\partial/\partial t$ by $j\omega$, we obtain the corresponding differential equations for the phasors $\bar{E}_x$ and $\bar{H}_y$ as

$$\frac{\partial \bar{E}_x}{\partial z} = -j\omega\mu\bar{H}_y \tag{6.46a}$$

$$\frac{\partial \bar{H}_y}{\partial z} = -\sigma\bar{E}_x - j\omega\varepsilon\bar{E}_x = -(\sigma + j\omega\varepsilon)\bar{E}_x \tag{6.46b}$$

Differentiating (6.46a) with respect to z and using (6.46b), we obtain

$$\frac{\partial^2 \bar{E}_x}{\partial z^2} = -j\omega\mu\frac{\partial \bar{H}_y}{\partial z} = j\omega\mu(\sigma + j\omega\varepsilon)\bar{E}_x \tag{6.47}$$

Defining

$$\bar{\gamma} = \sqrt{j\omega\mu(\sigma + j\omega\varepsilon)} \tag{6.48}$$

and substituting in (6.47), we have

$$\frac{\partial^2 \bar{E}_x}{\partial z^2} = \bar{\gamma}^2 \bar{E}_x \tag{6.49}$$

which is the wave equation for $\bar{E}_x$ in the material medium.

The solution to the wave equation (6.49) is given by

$$\bar{E}_x(z) = \bar{A}e^{-\bar{\gamma}z} + \bar{B}e^{\bar{\gamma}z} \tag{6.50}$$

where $\bar{A}$ and $\bar{B}$ are arbitrary constants. Noting that $\bar{\gamma}$ is a complex number and hence can be written as

$$\boxed{\bar{\gamma} = \alpha + j\beta} \tag{6.51}$$

and also writing $\bar{A}$ and $\bar{B}$ in exponential form as $Ae^{j\theta}$ and $Be^{j\phi}$, respectively, we have

$$\bar{E}_x(z) = Ae^{j\theta}e^{-\alpha z}e^{-j\beta z} + Be^{j\phi}e^{\alpha z}e^{j\beta z}$$

or

$$\begin{aligned} E_x(z, t) &= \text{Re}[\bar{E}_x(z)e^{j\omega t}] \\ &= \text{Re}[Ae^{j\theta}e^{-\alpha z}e^{-j\beta z}e^{j\omega t} + Be^{j\phi}e^{\alpha z}e^{j\beta z}e^{j\omega t}] \\ &= Ae^{-\alpha z}\cos(\omega t - \beta z + \theta) + Be^{\alpha z}\cos(\omega t + \beta z + \phi) \end{aligned} \tag{6.52}$$

We now recognize the two terms on the right side of (6.52) as representing uniform plane waves propagating in the positive z- and negative z-directions, respectively, with phase constant β, in view of the factors $\cos(\omega t - \beta z + \theta)$ and $\cos(\omega t + \beta z + \phi)$, respectively. They are, however, multiplied by the factors $e^{-\alpha z}$ and $e^{\alpha z}$, respectively. Hence the amplitude of the field differs from one constant phase surface to another. Since there cannot be a (+) wave in the region $z < 0$, that is, to the left of the current sheet, and since there cannot be a (−) wave in the region $z > 0$, that is, to the right of the current sheet, the solution for the electric field is given by

$$\mathbf{E}(z, t) = \begin{cases} Ae^{-\alpha z}\cos(\omega t - \beta z + \theta)\,\mathbf{i}_x & \text{for} \quad z > 0 \\ Be^{\alpha z}\cos(\omega t + \beta z + \phi)\,\mathbf{i}_x & \text{for} \quad z < 0 \end{cases} \tag{6.53}$$

Attenuation constant

To discuss how the amplitude of E_x varies with z on either side of the current sheet, we note that since σ, ε, and μ are all positive, the phase angle of $j\omega\mu(\sigma + j\omega\varepsilon)$ lies between 90° and 180°, and hence the phase angle of $\bar{\gamma}$ lies between 45° and 90°, making α and β positive quantities. This means that $e^{-\alpha z}$ decreases with increasing value of z, that is, in the positive z-direction, and $e^{\alpha z}$ decreases with decreasing value of z, that is, in the negative z-direction. Thus the exponential factors $e^{-\alpha z}$ and $e^{\alpha z}$ associated with the solutions for E_x in (6.53) have the effect of decreasing the amplitude of the field, that is, attenuating it as it propagates away from the sheet to either side of it. For this reason, the quantity α is known as the *attenuation constant*. The attenuation per unit length is equal to e^{α}. In terms of decibels, this is equal to $20\log_{10} e^{\alpha}$ or 8.686α dB. The units of α are nepers per meter. The quantity $\bar{\gamma}$ is known as the *propagation constant* since its real and imaginary parts, α and β, together determine the propagation characteristics, that is, attenuation and phase shift of the wave.

Having found the solution for the electric field of the wave and discussed its general properties, we now turn to the solution for the corresponding magnetic field by substituting for $\bar{E}_x$ in (6.46a). Thus

$$\begin{aligned} \bar{H}_y &= -\frac{1}{j\omega\mu}\frac{\partial \bar{E}_x}{\partial z} = \frac{\bar{\gamma}}{j\omega\mu}(\bar{A}e^{-\bar{\gamma}z} - \bar{B}e^{\bar{\gamma}z}) \\ &= \sqrt{\frac{\sigma + j\omega\varepsilon}{j\omega\mu}}(\bar{A}e^{-\bar{\gamma}z} - \bar{B}e^{\bar{\gamma}z}) \\ &= \frac{1}{\bar{\eta}}(\bar{A}e^{-\bar{\gamma}z} - \bar{B}e^{\bar{\gamma}z}) \end{aligned} \tag{6.54}$$

where

$$\boxed{\bar{\eta} = \sqrt{\frac{j\omega\mu}{\sigma + j\omega\varepsilon}}} \tag{6.55}$$

is the intrinsic impedance of the medium, which is now complex. Writing

$$\boxed{\bar{\eta} = |\bar{\eta}|e^{j\tau}} \tag{6.56}$$

we obtain the solution for $H_y(z, t)$ as

$$\begin{aligned} H_y(z, t) &= \text{Re}[\bar{H}_y(z)e^{j\omega t}] \\ &= \text{Re}\left[\frac{1}{|\bar{\eta}|e^{j\tau}}Ae^{j\theta}e^{-\alpha z}e^{-j\beta z}e^{j\omega t} - \frac{1}{|\bar{\eta}|e^{j\tau}}Be^{j\phi}e^{\alpha z}e^{j\beta z}e^{j\omega t}\right] \\ &= \frac{A}{|\bar{\eta}|}e^{-\alpha z}\cos(\omega t - \beta z + \theta - \tau) - \frac{B}{|\bar{\eta}|}e^{\alpha z}\cos(\omega t + \beta z + \phi - \tau) \end{aligned} \tag{6.57}$$

Remembering that the first and second terms on the right side of (6.57) correspond to (+) and (−) waves, respectively, and hence represent the solutions for the magnetic field in the regions $z > 0$ and $z < 0$, respectively, we write

$$\mathbf{H}(z, t) = \begin{cases} \dfrac{A}{|\bar{\eta}|}e^{-\alpha z}\cos(\omega t - \beta z + \theta - \tau)\,\mathbf{i}_y & \text{for} \quad z > 0 \qquad (6.58a) \\ -\dfrac{B}{|\bar{\eta}|}e^{\alpha z}\cos(\omega t + \beta z + \phi - \tau)\,\mathbf{i}_y & \text{for} \quad z < 0 \qquad (6.58b) \end{cases}$$

Electromagnetic field due to the current sheet

To complete the solution for the electromagnetic field due to the current sheet embedded in the material medium, we need to find the values of the constants A, B, θ, and ϕ. This is done by using the boundary conditions, as in Section 6.1. Thus from the boundary condition (3.51b) applied to the surface $z = 0$, we have

$$[E_x]_{z=0+} - [E_x]_{z=0-} = 0 \tag{6.59}$$

or $A\cos(\omega t + \theta) - B\cos(\omega t + \phi) = 0$, giving us $A = B$ and $\theta = \phi$. The solutions for **E** and **H** reduce to

$$\mathbf{E}(z, t) = Ae^{\mp\alpha z}\cos(\omega t \mp \beta z + \theta)\,\mathbf{i}_x \qquad \text{for} \quad z \gtrless 0 \tag{6.60a}$$

$$\mathbf{H}(z, t) = \pm\frac{A}{|\bar{\eta}|}e^{\mp\alpha z}\cos(\omega t \mp \beta z + \theta - \tau)\,\mathbf{i}_y \qquad \text{for} \quad z \gtrless 0 \tag{6.60b}$$

Then from the boundary condition (3.55a) applied to the surface $z = 0$, we have

$$\mathbf{i}_z \times \{[\mathbf{H}]_{z=0+} - [\mathbf{H}]_{z=0-}\} = -J_{S0}\cos\omega t\,\mathbf{i}_x \tag{6.61}$$

or

$$\frac{2A}{|\bar{\eta}|}\cos(\omega t + \theta - \tau) = J_{S0}\cos\omega t$$

$$A = \frac{|\bar{\eta}|J_{S0}}{2} \qquad \text{and} \qquad \theta = \tau$$

Thus the electromagnetic field due to the infinite plane currrent sheet of surface current density

$$\boxed{\mathbf{J}_S = -J_{S0} \cos \omega t \, \mathbf{i}_x \qquad \text{for} \quad z = 0}$$

and with a material medium characterized by σ, ε, and μ on either side of it is given by

$$\mathbf{E}(z, t) = \frac{|\bar{\eta}|J_{S0}}{2} e^{\mp \alpha z} \cos (\omega t \mp \beta z + \tau) \, \mathbf{i}_x \qquad \text{for} \quad z \gtrless 0 \tag{6.62a}$$

$$\mathbf{H}(z, t) = \pm \frac{J_{S0}}{2} e^{\mp \alpha z} \cos (\omega t \mp \beta z) \, \mathbf{i}_y \qquad \text{for} \quad z \gtrless 0 \tag{6.62b}$$

Propagation characteristics

As we have already discussed, (6.62a) and (6.62b) represent sinusoidally time-varying uniform plane waves, getting attenuated as they propagate away from the current sheet. The phenomenon is illustrated in Fig. 6.16, which shows sketches of current density on the sheet and the distance variation of the electric and magnetic fields on either side of the current sheet for three values of t. As in Fig. 6.9, it should be understood that in these sketches the field variations depicted along the z-axis hold also for any other line parallel to the z-axis. We shall now discuss further the propagation characteristics associated with these waves:

1. From (6.48) and (6.51) we have

$$\bar{\gamma}^2 = (\alpha + j\beta)^2 = j\omega\mu(\sigma + j\omega\varepsilon)$$

or

$$\alpha^2 - \beta^2 = -\omega^2 \mu\varepsilon \tag{6.63a}$$

$$2\alpha\beta = \omega\mu\sigma \tag{6.63b}$$

Squaring (6.63a) and (6.63b) and adding and then taking the square root, we obtain

$$\alpha^2 + \beta^2 = \omega^2\mu\varepsilon\sqrt{1 + \left(\frac{\sigma}{\omega\varepsilon}\right)^2} \tag{6.64}$$

From (6.63a) and (6.64), we then have

$$\alpha^2 = \frac{1}{2}\left[-\omega^2\mu\varepsilon + \omega^2\mu\varepsilon\sqrt{1 + \left(\frac{\sigma}{\omega\varepsilon}\right)^2}\right]$$

$$\beta^2 = \frac{1}{2}\left[\omega^2\mu\varepsilon + \omega^2\mu\varepsilon\sqrt{1 + \left(\frac{\sigma}{\omega\varepsilon}\right)^2}\right]$$

Since α and β are both positive, we finally get

$$\alpha = \frac{\omega\sqrt{\mu\varepsilon}}{\sqrt{2}}\left[\sqrt{1 + \left(\frac{\sigma}{\omega\varepsilon}\right)^2} - 1\right]^{1/2} \tag{6.65}$$

$$\beta = \frac{\omega\sqrt{\mu\varepsilon}}{\sqrt{2}}\left[\sqrt{1 + \left(\frac{\sigma}{\omega\varepsilon}\right)^2} + 1\right]^{1/2} \tag{6.66}$$

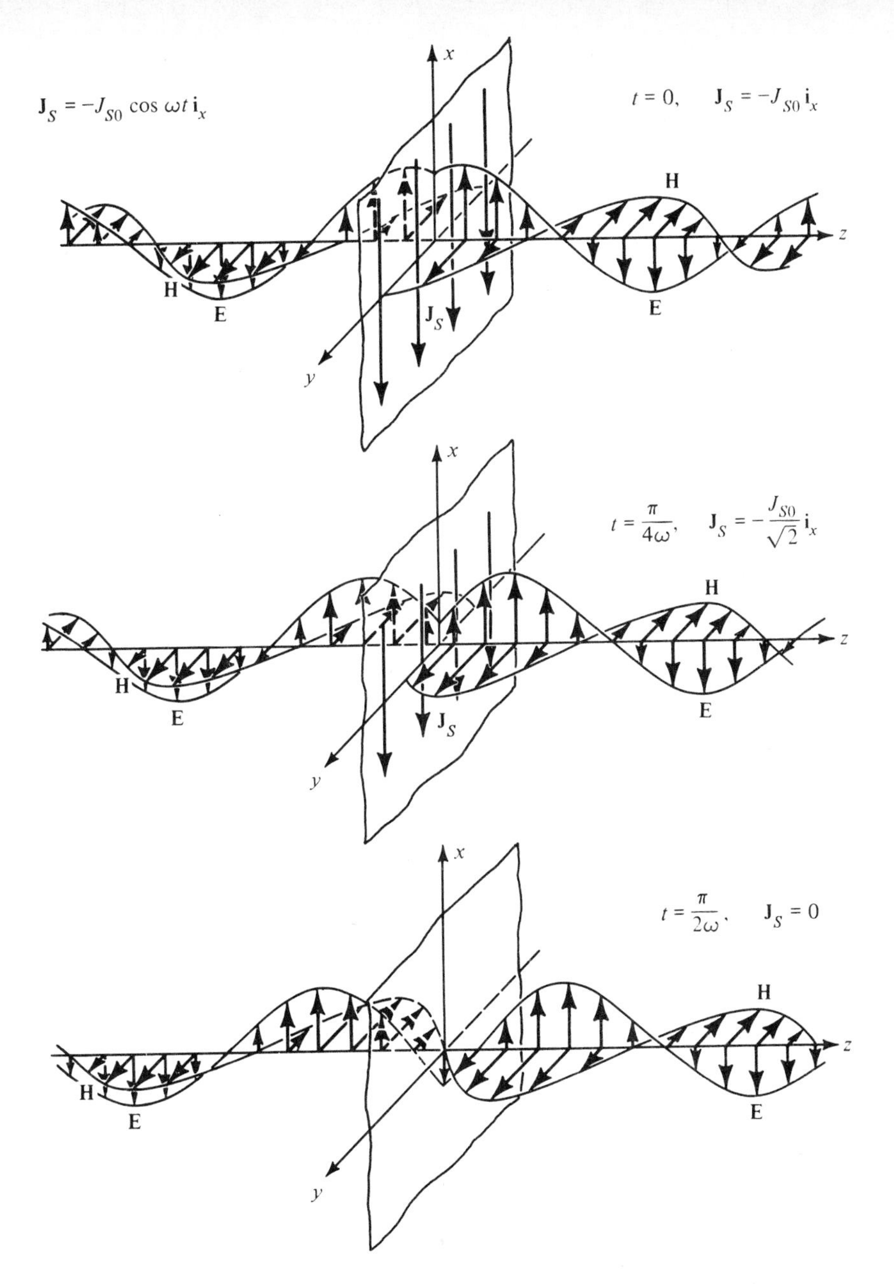

Figure 6.16. Time history of uniform plane electromagnetic wave radiating away from an infinite plane current sheet embedded in a material medium.

We note from (6.65) and (6.66) that α and β are both dependent on σ through the factor $\sigma/\omega\varepsilon$. This factor, known as the *loss tangent*, is the ratio of the magnitude of the conduction current density $\sigma\bar{E}_x$ to the magnitude of the displacement current density $j\omega\varepsilon\bar{E}_x$ in the material medium. In practice, the loss tangent is, however, not simply inversely proportional to ω since both σ and ε are generally functions of frequency. In fact, for many materials, the dependence of $\sigma/\omega\varepsilon$ on ω is more toward constant over wide frequency ranges.

2. The phase velocity of the wave along the direction of propagation is given by

$$v_p = \frac{\omega}{\beta} = \frac{\sqrt{2}}{\sqrt{\mu\varepsilon}}\left[\sqrt{1 + \left(\frac{\sigma}{\omega\varepsilon}\right)^2} + 1\right]^{-1/2} \tag{6.67}$$

   We note that the phase velocity is dependent on the frequency of the wave. Thus waves of different frequencies travel with different phase velocities. Consequently, for a signal comprising a band of frequencies, the different frequency components do not maintain the same phase relationships as they propagate in the medium. This phenomenon is known as *dispersion*. We discuss dispersion in detail in Chapter 9.

3. The wavelength in the medium is given by

$$\lambda = \frac{2\pi}{\beta} = \frac{\sqrt{2}}{f\sqrt{\mu\varepsilon}}\left[\sqrt{1 + \left(\frac{\sigma}{\omega\varepsilon}\right)^2} + 1\right]^{-1/2} \tag{6.68}$$

   In view of the attenuation of the wave with distance, the field variation with distance is not sinusoidal. Hence the wavelength is not exactly equal to the distance between two consecutive positive maxima as in Fig. 6.10. It is, however, still exactly equal to the distance between two alternate zero crossings.

4. The ratio of the amplitude of the electric field to the amplitude of the magnetic field is equal to $|\bar{\eta}|$, the magnitude of the complex intrinsic impedance of the medium. The electric and magnetic fields are out of phase by τ, the phase angle of the intrinsic impedance. In terms of the phasor or complex field components, we have

$$\frac{\bar{E}_x}{\bar{H}_y} = \begin{cases} \bar{\eta} & \text{for the } (+) \text{ wave} \\ -\bar{\eta} & \text{for the } (-) \text{ wave} \end{cases} \tag{6.69}$$

5. From (6.48) and (6.55), we note that

$$\bar{\gamma}\bar{\eta} = j\omega\mu \tag{6.70a}$$

$$\frac{\bar{\gamma}}{\bar{\eta}} = \sigma + j\omega\varepsilon \tag{6.70b}$$

   so that

$$\boxed{\sigma = \text{Re}\left(\frac{\bar{\gamma}}{\bar{\eta}}\right)} \tag{6.71a}$$

$$\boxed{\varepsilon = \frac{1}{\omega}\,\text{Im}\left(\frac{\bar{\gamma}}{\bar{\eta}}\right)} \tag{6.71b}$$

$$\boxed{\mu = \frac{1}{j\omega}\,\bar{\gamma}\bar{\eta}} \tag{6.71c}$$

Using (6.71a)–(6.71c), we can compute the material parameters σ, ε, and μ from a knowledge of the propagation parameters $\bar{\gamma}$ and $\bar{\eta}$ at the frequency of interest.

6. To obtain the electromagnetic field due to a nonsinusoidal source, it is necessary to consider its frequency components and apply superposition, since waves of different frequencies are attenuated by different amounts and travel with different phase velocities. The nonsinusoidal signal changes shape as it propagates in the material medium, unlike in the case of free space.

We shall now consider an example of the computation of $\bar{\gamma}$ and $\bar{\eta}$ given σ, ε, μ, and f.

Example 6.5

The material parameters of a certain food item are given by $\sigma = 2.17$ S/m, $\varepsilon = 47\varepsilon_0$, and $\mu = \mu_0$ at the operating frequency $f = 2.45$ GHz of a microwave oven. We wish to find the propagation parameters α, β, λ, v_p, and $\bar{\eta}$.

Although explicit expressions for α and β in terms of ω, σ, ε, and μ are given by (6.65) and (6.66), it is instructive to compute their values by using complex algebra in conjunction with the expression for $\bar{\gamma}$ given by (6.48). Thus we have

$$
\begin{aligned}
\bar{\gamma} &= \sqrt{j\omega\mu(\sigma + j\omega\varepsilon)} \\
&= \sqrt{j\omega\mu \cdot j\omega\varepsilon\left(1 - j\frac{\sigma}{\omega\varepsilon}\right)} \\
&= j\frac{\omega\sqrt{\varepsilon_r}}{c}\sqrt{1 - j\frac{\sigma}{\omega\varepsilon_r\varepsilon_0}} \\
&= j\frac{2\pi \times 2.45 \times 10^9 \times \sqrt{47}}{3 \times 10^8}\sqrt{1 - j\frac{2.17 \times 36\pi}{2\pi \times 2.45 \times 10^9 \times 47 \times 10^{-9}}} \\
&= j351.782\sqrt{1 - j0.3392} \\
&= j351.782\sqrt{1.0560\angle{-18.7369^\circ}} \\
&= 351.782\angle{90^\circ} \times 1.0276\angle{-9.3685^\circ} \\
&= 361.4912\angle{80.6315^\circ} \\
&= 58.85 + j356.67
\end{aligned}
$$

so that

$$
\begin{aligned}
\alpha &= 58.85 \text{ Np/m} \\
\beta &= 356.67 \text{ rad/m} \\
\lambda &= \frac{2\pi}{\beta} = 0.0176 \text{ m} \\
v_p &= \frac{\omega}{\beta} = 0.4316 \times 10^8 \text{ m/s}
\end{aligned}
$$

Proceeding in a similar manner with (6.55), we obtain

$$
\bar{\eta} = \sqrt{\frac{j\omega\mu}{\sigma + j\omega\varepsilon}}
$$

$$\bar{\eta} = \sqrt{\frac{j\omega\mu}{j\omega\varepsilon[1 - j(\sigma/\omega\varepsilon)]}}$$

$$= \frac{\eta_0}{\sqrt{\varepsilon_r}} \frac{1}{\sqrt{1 - j(\sigma/\omega\varepsilon)}}$$

$$= \frac{120\pi}{\sqrt{47}} \frac{1}{\sqrt{1 - j0.3392}}$$

$$= \frac{54.9898}{1.0276\underline{/-9.3685^\circ}}$$

$$= 53.51\underline{/9.37^\circ}\ \Omega$$

The listing of a PC program to perform these computations and the output from a run of the program are given in PL 6.1.

PL 6.1. Program listing and sample output for the computation of propagation parameters from material parameters at a given frequency.

```
100 '*******************************************************
110 '* COMPUTATION OF PROPAGATION PARAMETERS FROM MATERIAL *
120 '* PARAMETERS                                          *
130 '*******************************************************
140 CLS:SCREEN 0
150 PRINT "ENTER VALUES OF MATERIAL PARAMETERS:":PRINT
160 LOCATE 3,1:INPUT "CONDUCTIVITY IN S/M = ",SI
170 INPUT "RELATIVE PERMITTIVITY = ",ER
180 INPUT "RELATIVE PERMEABILITY = ",MR
190 INPUT "FREQUENCY IN MHZ = ",FR
200 PI=3.141593:EO=10^-9/(36*PI)
210 EP=ER*EO:OM=FR*PI*2*10^6:LT=SI/(OM*EP)
220 BD=OM*SQR(ER*MR)/3E+08:ND=120*PI*SQR(MR/ER)
230 REAL=1:IMAG=-LT:GOSUB 420
240 GM=BD*SQR(MAG):GP=(PANG+PI)/2:'* MAGNITUDE AND PHASE
250 '  ANGLE OF PROPAGATION CONSTANT *
260 AL=GM*COS(GP):BE=GM*SIN(GP):'* ATTENUATION AND PHASE
270 '  CONSTANTS *
280 WL=2*PI/BE:VP=OM/BE:'* WAVELENGTH AND PHASE VELOCITY *
290 NM=ND/SQR(MAG):NP=-90*PANG/PI:'* MAGNITUDE AND PHASE
300 '  ANGLE OF INTRINSIC IMPEDANCE *
310 PRINT:PRINT "COMPUTED VALUES:"
320 PRINT:PRINT "ATTENUATION CONSTANT =";AL;" NP/M"
330 PRINT "PHASE CONSTANT =";BE;" RAD/M"
340 PRINT "WAVELENGTH =";WL;" M"
350 PRINT "PHASE VELOCITY =";VP;" M/S"
360 PRINT "INTRINSIC IMPEDANCE:"
370 PRINT "      MAGNITUDE =";NM;" OHMS"
380 PRINT "      PHASE ANGLE =";NP;" DEG"
390 PRINT:PRINT "STRIKE ANY KEY TO CONTINUE":C$=INPUT$(1)
400 GOTO 140
410 END
420 '* SUBPROGRAM TO CONVERT COMPLEX NUMBER IN RECTANGULAR
430 '  FORM TO POLAR FORM *
440 MAG=SQR(REAL*REAL+IMAG*IMAG)
450 IF REAL=0 THEN PANG=SGN(IMAG)*PI/2:RETURN
460 PANG=ATN(IMAG/REAL)
470 IF REAL<0 THEN PANG=PANG+PI
480 RETURN
```

PL6.1. (continued)

```
RUN
ENTER VALUES OF MATERIAL PARAMETERS:

CONDUCTIVITY IN S/M = 2.17
RELATIVE PERMITTIVITY = 47
RELATIVE PERMEABILITY = 1
FREQUENCY IN MHZ = 2450

COMPUTED VALUES:

ATTENUATION CONSTANT = 58.84623  NP/M
PHASE CONSTANT = 356.67  RAD/M
WAVELENGTH = 1.761624E-02  M
PHASE VELOCITY = 4.31598E+07  M/S
INTRINSIC IMPEDANCE:
     MAGNITUDE = 53.51276  OHMS
     PHASE ANGLE = 9.368718  DEG

STRIKE ANY KEY TO CONTINUE
```

K6.3. Material medium; Sinusoidal waves; Material parameters; Propagation parameters; Attenuation and phase constants; Complex propagation constant; Complex intrinsic impedance; Nonsinusoidal source; Superposition.

D6.8. Compute the propagation constant and intrinsic impedance for the following cases: **(a)** $\sigma = 10^{-5}$ S/m, $\varepsilon = 5\varepsilon_0$, $\mu = \mu_0$, and $f = 10^5$ Hz and **(b)** $\sigma = 4$ S/m, $\varepsilon = 80\varepsilon_0$, $\mu = \mu_0$, and $f = 10^9$ Hz.
Ans. **(a)** $(0.00083 + j0.00476)$ m^{-1}, $163.54\underline{/9.9^\circ}$ Ω; **(b)** $(77.84 + j202.86)$ m^{-1}, $36.34\underline{/20.99^\circ}$ Ω

D6.9. For a uniform plane wave of frequency 10^6 Hz propagating in a nonmagnetic ($\mu = \mu_0$) material medium, the propagation constant is known to be $(0.05 + j0.1)$ m^{-1}. Find the following: **(a)** the distance in which the fields are attenuated by e^{-1}; **(b)** the distance in which the fields undergo a change of phase by 1 rad; **(c)** the distance in which a constant phase of the wave travels in 1 μs; **(d)** the ratio of the amplitudes of the electric and magnetic fields; and **(e)** the phase difference between the electric and magnetic fields.
Ans. **(a)** 20 m; **(b)** 10 m; **(c)** 62.83 m; **(d)** 70.62 Ω; **(e)** 0.1476π

6.4 UNIFORM PLANE WAVES IN DIELECTRICS AND CONDUCTORS

In the preceding section we discussed uniform plane electromagnetic wave propagation in a material medium for the general case. In this section we consider special cases as follows:

Case 1: Perfect dielectrics. Perfect dielectrics are characterized by $\sigma = 0$. Then

$$\bar{\gamma} = \sqrt{j\omega\mu \cdot j\omega\varepsilon} = j\omega\sqrt{\mu\varepsilon} \tag{6.72}$$

is purely imaginary, so that

$$\alpha = 0 \tag{6.73a}$$

$$\beta = \omega\sqrt{\mu\varepsilon} \tag{6.73b}$$

$$v_p = \frac{\omega}{\beta} = \frac{1}{\sqrt{\mu\varepsilon}} \tag{6.73c}$$

$$\lambda = \frac{2\pi}{\beta} = \frac{1}{f\sqrt{\mu\varepsilon}} \tag{6.73d}$$

Further,

$$\bar{\eta} = \sqrt{\frac{j\omega\mu}{j\omega\varepsilon}} = \sqrt{\frac{\mu}{\varepsilon}} \tag{6.74}$$

is purely real. Thus the waves propagate without attenuation and with the electric and magnetic fields in phase, as in free space but with ε_0 replaced by ε and μ_0 replaced by μ. In terms of the relative permittivity ε_r and the relative permeability μ_r of the perfect dielectric medium, the propagation parameters are

$$\beta = \beta_0\sqrt{\mu_r\varepsilon_r} \tag{6.75a}$$

$$v_p = \frac{c}{\sqrt{\mu_r\varepsilon_r}} \tag{6.75b}$$

$$\lambda = \frac{\lambda_0}{\sqrt{\mu_r\varepsilon_r}} \tag{6.75c}$$

$$\eta = \eta_0\sqrt{\frac{\mu_r}{\varepsilon_r}} \tag{6.75d}$$

where the quantities with subscripts "0" refer to free space.

Boundary conditions at interface between perfect dielectrics

Since $\sigma = 0$, $\mathbf{J}_c = \sigma\mathbf{E} = \mathbf{0}$. Thus there cannot be any conduction current in a perfect dielectric, which in turn rules out any accumulation of free charge on the surface of a perfect dielectric. Hence in applying the boundary conditions (3.60a)–(3.60d) to an interface between two perfect dielectric media, we set ρ_S and $\mathbf{J}_S$ equal to zero, thereby obtaining

$$\mathbf{i}_n \times (\mathbf{E}_1 - \mathbf{E}_2) = \mathbf{0} \tag{6.76a}$$

$$\mathbf{i}_n \times (\mathbf{H}_1 - \mathbf{H}_2) = \mathbf{0} \tag{6.76b}$$

$$\mathbf{i}_n \cdot (\mathbf{D}_1 - \mathbf{D}_2) = 0 \tag{6.76c}$$

$$\mathbf{i}_n \cdot (\mathbf{B}_1 - \mathbf{B}_2) = 0 \tag{6.76d}$$

These boundary conditions tell us that the tangential components of **E** and **H**, and the normal components of **D** and **B** are continuous at the boundary.

Case 2: Imperfect dielectrics. Imperfect dielectrics are characterized by $\sigma \neq 0$ but $\sigma/\omega\varepsilon \ll 1$. Recalling that $\sigma\bar{E}_x$ is the conduction current density and $\omega\varepsilon\bar{E}_x$ is the displacement current density, we note that this condition is equivalent to stating that the magnitude of the conduction current density is small compared to the magnitude of the displacement current density. Using the binomial expansion

$$(1 + x)^n = 1 + nx + \frac{n(n-1)}{2!}x^2 + \cdots$$

we can then write

$$\begin{aligned}\bar{\gamma} &= \sqrt{j\omega\mu(\sigma + j\omega\varepsilon)} \\ &= \sqrt{j\omega\mu \cdot j\omega\varepsilon\left(1 - j\frac{\sigma}{\omega\varepsilon}\right)} \\ &= j\omega\sqrt{\mu\varepsilon}\left(1 - j\frac{\sigma}{\omega\varepsilon}\right)^{1/2} \\ &= \frac{\sigma}{2}\sqrt{\frac{\mu}{\varepsilon}}\left(1 - \frac{\sigma^2}{8\omega^2\varepsilon^2}\right) + j\omega\sqrt{\mu\varepsilon}\left(1 + \frac{\sigma^2}{8\omega^2\varepsilon^2}\right)\end{aligned} \tag{6.77}$$

so that

$$\alpha \approx \frac{\sigma}{2}\sqrt{\frac{\mu}{\varepsilon}}\left(1 - \frac{\sigma^2}{8\omega^2\varepsilon^2}\right) \tag{6.78a}$$

$$\beta \approx \omega\sqrt{\mu\varepsilon}\left(1 + \frac{\sigma^2}{8\omega^2\varepsilon^2}\right) \tag{6.78b}$$

$$v_p = \frac{\omega}{\beta} \approx \frac{1}{\sqrt{\mu\varepsilon}}\left(1 - \frac{\sigma^2}{8\omega^2\varepsilon^2}\right) \tag{6.78c}$$

$$\lambda = \frac{2\pi}{\beta} \approx \frac{1}{f\sqrt{\mu\varepsilon}}\left(1 - \frac{\sigma^2}{8\omega^2\varepsilon^2}\right) \tag{6.78d}$$

Further,

$$\begin{aligned}\bar{\eta} &= \sqrt{\frac{j\omega\mu}{\sigma + j\omega\varepsilon}} \\ &= \sqrt{\frac{j\omega\mu}{j\omega\varepsilon}}\left(1 - j\frac{\sigma}{\omega\varepsilon}\right)^{-1/2}\end{aligned}$$

so that

$$\bar{\eta} \approx \sqrt{\frac{\mu}{\varepsilon}}\left[\left(1 - \frac{3}{8}\frac{\sigma^2}{\omega^2\varepsilon^2}\right) + j\frac{\sigma}{2\omega\varepsilon}\right] \tag{6.79}$$

In (6.77)–(6.79) we have retained all terms up to and including the second power in $\sigma/\omega\varepsilon$ and have neglected all higher-order terms, since $\sigma/\omega\varepsilon \ll 1$. For a value of $\sigma/\omega\varepsilon$ equal to 0.1, the quantities β, v_p, and λ are different from those for the corresponding perfect dielectric case by a factor of only 1/800, whereas the intrinsic impedance has a real part differing from the intrinsic impedance of the perfect dielectric medium by a factor of 3/800 and an imaginary part, which is 1/20 of the intrinsic impedance of the perfect dielectric medium. Thus for all practical purposes the only significant feature different from the perfect dielectric case is the attenuation.

Case 3: Good conductors. Good conductors are characterized by $\sigma/\omega\varepsilon >> 1$, just the opposite of imperfect dielectrics. This condition is equivalent to stating that the magnitude of the conduction current density is large compared to the magnitude of the displacement current density. Then

$$\begin{aligned}\bar{\gamma} &= \sqrt{j\omega\mu(\sigma + j\omega\varepsilon)} \\ &\approx \sqrt{j\omega\mu\sigma} \\ &= \sqrt{\omega\mu\sigma}\, e^{j\pi/4} \\ &= \sqrt{\pi f\mu\sigma}(1 + j)\end{aligned} \tag{6.80}$$

so that

$$\alpha \approx \sqrt{\pi f\mu\sigma} \tag{6.81a}$$

$$\beta \approx \sqrt{\pi f\mu\sigma} \tag{6.81b}$$

$$v_p = \frac{\omega}{\beta} \approx \sqrt{\frac{4\pi f}{\mu\sigma}} \tag{6.81c}$$

$$\lambda = \frac{2\pi}{\beta} \approx \sqrt{\frac{4\pi}{f\mu\sigma}} \tag{6.81d}$$

Further,

$$\begin{aligned}\bar{\eta} &= \sqrt{\frac{j\omega\mu}{\sigma + j\omega\varepsilon}} \\ &\approx \sqrt{\frac{j\omega\mu}{\sigma}}\end{aligned}$$

or

$$\begin{aligned}\bar{\eta} &\approx \sqrt{\frac{\omega\mu}{\sigma}}\, e^{j\pi/4} \\ &= \sqrt{\frac{\pi f\mu}{\sigma}}(1 + j)\end{aligned} \tag{6.82}$$

We note that α, β, v_p, and $\bar{\eta}$ are proportional to $\sqrt{f}$, provided that σ and μ are constants. This behavior is much different from the imperfect dielectric case.

Skin effect

To discuss the propagation characteristics of a wave inside a good conductor, let us consider the case of copper. The constants for copper are $\sigma = 5.80 \times 10^7$ S/m, $\varepsilon = \varepsilon_0$, and $\mu = \mu_0$. Hence the frequency at which σ is equal to $\omega\varepsilon$ for copper is equal to $5.8 \times 10^7/2\pi\varepsilon_0$ or 1.04×10^{18} Hz. Thus at frequencies of even several gigahertz, copper behaves like an excellent conductor. To obtain an idea of the attenuation of the wave inside the conductor, we note that the attenuation undergone in a distance of one wavelength is equal to $e^{-\alpha\lambda}$ or $e^{-2\pi}$. In terms of decibels, this is equal to $20 \log_{10} e^{2\pi} = 54.58$ dB. In fact, the field is attenuated by a factor e^{-1} or 0.368 in a distance equal to $1/\alpha$. This distance is known as the *skin depth* and is denoted by the symbol δ. From (6.81a), we obtain

$$\boxed{\delta = \frac{1}{\sqrt{\pi f \mu \sigma}}} \tag{6.83}$$

The skin depth for copper is equal to

$$\frac{1}{\sqrt{\pi f \times 4\pi \times 10^{-7} \times 5.8 \times 10^{7}}} = \frac{0.066}{\sqrt{f}} \text{ m}$$

Thus in copper the fields are attenuated by a factor e^{-1} in a distance of 0.066 mm even at the low frequency of 1 MHz, thereby resulting in the concentration of the fields near to the skin of the conductor. This phenomenon is known as the *skin effect*. It also explains *shielding* by conductors.

Underwater communication

To discuss further the characteristics of wave propagation in a good conductor, we note that the ratio of the wavelength in the conducting medium to the wavelength in a dielectric medium having the same ε and μ as those of the conductor is given by

$$\frac{\lambda_{\text{conductor}}}{\lambda_{\text{dielectric}}} \approx \frac{\sqrt{4\pi/f\mu\sigma}}{1/f\sqrt{\mu\varepsilon}} = \sqrt{\frac{4\pi f \varepsilon}{\sigma}} = \sqrt{\frac{2\omega\varepsilon}{\sigma}} \tag{6.84}$$

Since $\sigma/\omega\varepsilon \gg 1$, $\lambda_{\text{conductor}} \ll \lambda_{\text{dielectric}}$. For example, for seawater, $\sigma = 4$ S/m, $\varepsilon = 80\varepsilon_0$, and $\mu = \mu_0$ so that the ratio of the two wavelengths for $f = 25$ kHz ($\sigma/\omega\varepsilon = 36{,}000$) is equal to 0.00745. Thus for $f = 25$ kHz, the wavelength in seawater is 1/134 of the wavelength in a dielectric having the same ε and μ as those of seawater and a still smaller fraction of the wavelength in free space. Furthermore, the lower the frequency, the smaller is this fraction. Since it is the electrical length, that is, the length in terms of the wavelength, instead of the physical length that determines the radiation characteristics of an antenna, this means that antennas of much shorter length can be used in seawater than in free space. Together with the property that $\alpha \propto \sqrt{f}$, this illustrates that the lower the frequency, the more suitable it is for underwater communication.

For a given frequency, the higher the value of σ, the greater is the value of the attenuation constant, the smaller is the value of the skin depth, and hence the less deep the waves can penetrate. For example, in the heating of malignant tissues (hyperthermia) by RF (radio-frequency) radiation, the waves penetrate much deeper into fat (low water content) than into muscle (high water content).[1]

Equation (6.82) tells us that the intrinsic impedance of a good conductor has a phase angle of 45°. Hence the electric and magnetic fields in the medium are out of phase by 45°. The magnitude of the intrinsic impedance is given by

$$|\bar{\eta}| = \left|(1 + j)\sqrt{\frac{\pi f \mu}{\sigma}}\right| = \sqrt{\frac{2\pi f \mu}{\sigma}} \tag{6.85}$$

As a numerical example, for copper, this quantity is equal to

$$\sqrt{\frac{2\pi f \times 4\pi \times 10^{-7}}{5.8 \times 10^{7}}} = 3.69 \times 10^{-7}\sqrt{f}\ \Omega$$

[1] F. Sterzer et al., "RF Therapy for Malignancy," *IEEE Spectrum*, December 1980, pp. 32–37.

Thus the intrinsic impedance of copper has as low a magnitude as 0.369 Ω even at a frequency of 10^{12} Hz. In fact, by recognizing that

$$|\bar{\eta}| = \sqrt{\frac{2\pi f\mu}{\sigma}} = \sqrt{\frac{\omega\varepsilon}{\sigma}}\sqrt{\frac{\mu}{\varepsilon}} \tag{6.86}$$

we note that the magnitude of the intrinsic impedance of a good conductor medium is a small fraction of the intrinsic impedance of a dielectric medium having the same ε and μ. It follows that for the same electric field, the magnetic field inside a good conductor is much larger than the magnetic field inside a dielectric having the same ε and μ as those of the conductor.

Boundary conditions on a perfect conductor surface

Case 4: Perfect conductors. Perfect conductors are idealizations of good conductors in the limit that $\sigma \to \infty$. From (6.83), we note that the skin depth is equal to zero and hence there is no penetration of fields into the material. Thus no time-varying fields can exist inside a perfect conductor. In view of this, the boundary conditions on a perfect conductor surface are obtained by setting the fields with subscript 2 in (3.60a)–(3.60d) equal to zero. Thus we obtain

$$\mathbf{i}_n \times \mathbf{E} = \mathbf{0} \tag{6.87a}$$

$$\mathbf{i}_n \times \mathbf{H} = \mathbf{J}_S \tag{6.87b}$$

$$\mathbf{i}_n \cdot \mathbf{D} = \rho_S \tag{6.87c}$$

$$\mathbf{i}_n \cdot \mathbf{B} = 0 \tag{6.87d}$$

where we have also omitted subscripts 1 so that **E, H, D**, and **B** are the fields on the perfect conductor surface. The boundary conditions (6.87a) and (6.87d) tell us that on a perfect conductor surface, the tangential component of the electric field intensity and the normal component of the magnetic field intensity are zero. Hence the electric field must be completely normal, and the magnetic field must be completely tangential to the surface. The remaining two boundary conditions (6.87c) and (6.87b) tell us that the (normal) displacement flux density is equal to the surface charge density and the (tangential) magnetic field intensity is equal in magnitude to the surface current density.

Summarizing the discussion of the special cases, we observe that as σ varies from 0 to ∞, a material is classified as a perfect dielectric for $\sigma = 0$, an imperfect dielectric for $\sigma \neq 0$ but $\ll \omega\varepsilon$, a good conductor for $\sigma \gg \omega\varepsilon$, and finally a perfect conductor in the limit that $\sigma \to \infty$. This implies that a material of nonzero σ behaves as an imperfect dielectric for $f \gg f_q$ but as a good conductor for $f \ll f_q$, where f_q, the transition frequency, is equal to $\sigma/2\pi\varepsilon$. In practice, however, the situation is not so simple because, as was already mentioned in Section 6.3, σ and ε are in general functions of frequency.

K6.4. Perfect dielectric; Imperfect dielectric; Good conductor; Conduction current versus displacement current; Skin effect; Perfect conductor; Boundary conditions.

D6.10. For a nonmagnetic ($\mu = \mu_0$) perfect dielectric material, find the relative permittivity for each of the following cases: **(a)** the phase velocity in the dielectric is one-third of its value in free space; **(b)** the rate of change of phase with distance at a fixed time in the dielectric for a wave of frequency f_0 is the same as the rate of

change of phase with distance at a fixed time in free space for a wave of frequency $2f_0$; **(c)** for the same frequency, the wavelength in the dielectric is two-thirds of its value in free space; and **(d)** for the same electric field amplitude, the magnetic field amplitude in the dielectric is 4 times its value in free space.
Ans. **(a)** 9; **(b)** 4; **(c)** 2.25; **(d)** 16

D6.11. For a uniform plane wave of frequency $f = 10^5$ Hz propagating in a good conductor medium, the fields undergo attenuation by the factor $e^{-\pi}$ in a distance of 2.5 m. Find the following: **(a)** the distance in which the fields undergo a change of phase by 2π rad for $f = 10^5$ Hz; **(b)** the distance by which a constant phase travels in 1 μs for $f = 10^5$ Hz; and **(c)** the distance by which a constant phase travels in 1 μs for $f = 10^4$ Hz, assuming the material parameters to be the same as at $f = 10^5$ Hz.
Ans. **(a)** 5 m; (b) 0.5 m; **(c)** 0.1581 m

D6.12. The electric fields of uniform plane waves of the same frequency propagating in three different materials 1, 2, and 3 are given, respectively, by
(a) $\mathbf{E}_1 = E_0 e^{-0.4\pi z} \cos (2\pi \times 10^5 t - 0.4\pi z)\, \mathbf{i}_x$
(b) $\mathbf{E}_2 = E_0 e^{-2\pi \times 10^{-5} z} \cos (2\pi \times 10^5 t - 2\pi \times 10^{-3} z)\, \mathbf{i}_x$
(c) $\mathbf{E}_3 = E_0 e^{-0.004z} \cos (2\pi \times 10^5 t - 0.01 z)\, \mathbf{i}_x$
For each material, determine if at the frequency of operation, it can be classified as an imperfect dielectric or a good conductor or neither of the two.
Ans. **(a)** Good conductor; **(b)** Imperfect dielectric; **(c)** Neither

6.5 POYNTING VECTOR, POWER DISSIPATION, AND ENERGY STORAGE

In the preceding section we found the solution for the electromagnetic field due to an infinite plane current sheet situated in the $z = 0$ plane. For a surface current flowing in the negative x-direction, we found the electric field on the sheet to be directed in the positive x-direction. Since the current is flowing against the force due to the electric field, a certain amount of work must be done by the source of the current to maintain the current flow on the sheet. Let us consider a rectangular area of length Δx and width Δy on the current sheet as shown in Fig. 6.17. Since the current density is $J_{S0} \cos \omega t$, the charge crossing the width Δy in time dt is $dq = J_{S0}\, \Delta y \cos \omega t\, dt$. The force exerted on this charge by the electric field is given by

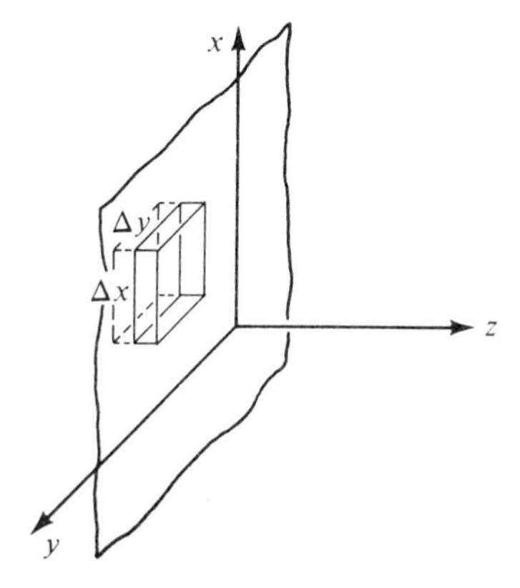

Figure 6.17. For the determination of power flow density associated with the electromagnetic field.

$$\mathbf{F} = dq\,\mathbf{E} = J_{S0}\,\Delta y \cos \omega t\, dt\, E_x \mathbf{i}_x \tag{6.88}$$

The amount of work required to be done against the electric field in displacing this charge by the distance Δx is

$$dw = F_x\,\Delta x = J_{S0}\,E_x \cos \omega t\, dt\, \Delta x\, \Delta y \tag{6.89}$$

Thus the power supplied by the source of the current in maintaining the surface current over the area $\Delta x\,\Delta y$ is

$$\frac{dw}{dt} = J_{S0}E_x \cos \omega t\, \Delta x\, \Delta y \tag{6.90}$$

Recalling that E_x on the sheet is $|\overline{\eta}|\dfrac{J_{S0}}{2}\cos(\omega t + \tau)$, we obtain

$$\frac{dw}{dt} = |\overline{\eta}|\frac{J_{S0}^2}{2}\cos \omega t \cos(\omega t + \tau)\, \Delta x\, \Delta y \tag{6.91}$$

We would expect the power given by (6.91) to be carried by the electromagnetic wave, half of it to either side of the current sheet. To investigate this, we note that the quantity $\mathbf{E} \times \mathbf{H}$ has the units of

$$\frac{\text{newtons}}{\text{coulomb}} \times \frac{\text{amperes}}{\text{meter}} = \frac{\text{newtons}}{\text{coulomb}} \times \frac{\text{coulomb}}{\text{second-meter}} \times \frac{\text{meter}}{\text{meter}}$$

$$= \frac{\text{newton-meters}}{\text{second}} \times \frac{1}{(\text{meter})^2} = \frac{\text{watts}}{(\text{meter})^2}$$

which represents power density. Let us then consider the rectangular box enclosing the area $\Delta x\,\Delta y$ on the current sheet and with its sides almost touching the current sheet on either side of it, as shown in Fig. 6.17. Evaluating the surface integral of $\mathbf{E} \times \mathbf{H}$ over the surface of the rectangular box, we obtain the power flow out of the box to be

$$\begin{aligned}\oint \mathbf{E} \times \mathbf{H} \cdot d\mathbf{S} &= |\overline{\eta}|\frac{J_{S0}^2}{4}\cos(\omega t + \tau)\cos \omega t\, \mathbf{i}_z \cdot \Delta x\, \Delta y\, \mathbf{i}_z \\ &\quad + \left[-|\overline{\eta}|\frac{J_{S0}^2}{4}\cos(\omega t + \tau)\cos \omega t\, \mathbf{i}_z\right] \cdot (-\Delta x\, \Delta y\, \mathbf{i}_z) \\ &= |\overline{\eta}|\frac{J_{S0}^2}{2}\cos(\omega t + \tau)\cos \omega t\, \Delta x\, \Delta y\end{aligned} \tag{6.92}$$

This result is exactly equal to the power supplied by the current source as given by (6.91).

Instantaneous Poynting vector

We now interpret the quantity $\mathbf{E} \times \mathbf{H}$ as the power flow density vector associated with the electromagnetic field. It is known as the *Poynting vector* after J. H. Poynting and is denoted by the symbol $\mathbf{P}$. Thus

$$\boxed{\mathbf{P} = \mathbf{E} \times \mathbf{H}} \tag{6.93}$$

In particular, it is the instantaneous Poynting vector, since $\mathbf{E}$ and $\mathbf{H}$ are instantaneous field vectors. Although we have here introduced the Poynting vector by considering the specific case of the electromagnetic field due to the infinite plane current sheet, the interpretation that $\oint_S \mathbf{E} \times \mathbf{H} \cdot d\mathbf{S}$ is equal to the power flow out of the closed surface S is applicable in the general case.

Let us now consider the region $z > 0$. The magnitude of the Poynting vector in this region is given by

$$\begin{aligned} P_z &= E_x H_y \\ &= |\bar{\eta}| \frac{J_{S0}^2}{4} e^{-2\alpha z} \cos(\omega t - \beta z + \tau) \cos(\omega t - \beta z) \end{aligned} \tag{6.94}$$

The form of variation of P_z with z for $t = 0$ is shown in Fig. 6.18. If we now consider a rectangular box lying between $z = z$ and $z = z + \Delta z$ planes and having dimensions Δx and Δy in the x- and y-directions, respectively, we would in general obtain a nonzero result for the power flowing out of the box, since $\partial P_z/\partial z$ is not everywhere zero. This power flow is given by

$$\begin{aligned} \oint_S \mathbf{P} \cdot d\mathbf{S} &= [P_z]_{z+\Delta z}\, \Delta x\, \Delta y - [P_z]_z\, \Delta x\, \Delta y \\ &= \frac{[P_z]_{z+\Delta z} - [P_z]_z}{\Delta z}\, \Delta x\, \Delta y\, \Delta z \\ &= \frac{\partial P_z}{\partial z} \Delta v \end{aligned} \tag{6.95}$$

where Δv is the volume of the box. Letting P_z equal $E_x H_y$ and using (6.44a) and (6.44b), we obtain

$$\begin{aligned} \oint_S \mathbf{P} \cdot d\mathbf{S} &= \frac{\partial}{\partial z}[E_x H_y]\, \Delta v \\ &= \left(E_x \frac{\partial H_y}{\partial z} + H_y \frac{\partial E_x}{\partial z}\right) \Delta v \\ &= \left[E_x\left(-\sigma E_x - \varepsilon \frac{\partial E_x}{\partial t}\right) + H_y\left(-\mu \frac{\partial H_y}{\partial t}\right)\right] \Delta v \\ &= -\sigma E_x^2\, \Delta v - \frac{\partial}{\partial t}\left(\tfrac{1}{2}\varepsilon E_x^2\, \Delta v\right) - \frac{\partial}{\partial t}\left(\tfrac{1}{2}\mu H_y^2\, \Delta v\right) \end{aligned} \tag{6.96}$$

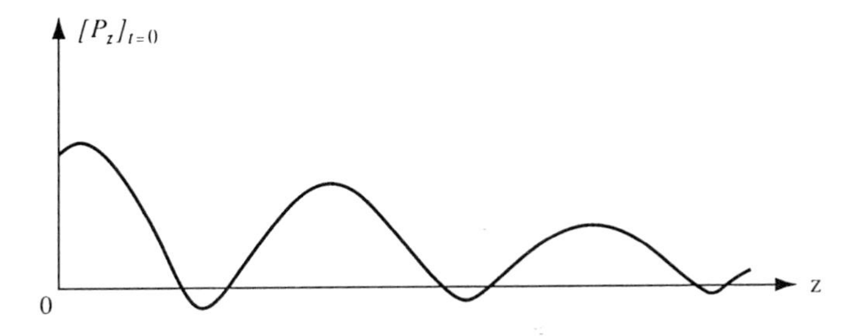

Figure 6.18. For the discussion of power flow associated with the electromagnetic field.

If we ignore the negative signs associated with the terms on the right side of (6.96), then each of these terms represents power flow into the box. Since the property of attenuation is associated with the parameter σ, the quantity $\sigma E_x^2 \Delta v$ is the power dissipated in the box due to flow of conduction current. This can also be seen by considering volume charge of density ρ moving with velocity $\mathbf{v}$ in the volume Δv under the influence of electric field $\mathbf{E}$. The volume charge experiences a force $(\rho \Delta v)\mathbf{E}$ and hence the work done in displacing it by a distance $d\mathbf{l}$ in time dt is $(\rho \Delta v)\mathbf{E} \cdot d\mathbf{l}$, or $(\rho \Delta v)\mathbf{E} \cdot \mathbf{v}\, dt$. The time rate at which the work is done or the power supplied by the field is $(\rho \Delta v)\mathbf{E} \cdot \mathbf{v}$ or $\mathbf{E} \cdot \mathbf{J}$. In the case of conduction current, $\mathbf{J} = \sigma \mathbf{E}$ and the power supplied by the field, which is dissipated in the volume Δv, is $\sigma E^2 \Delta v$. Thus the quantity σE_x^2 represents the power dissipation density (W/m^3) associated with the electromagnetic field due to the conductive property of the medium. The quantities $\frac{1}{2}\epsilon E_x^2$ and $\frac{1}{2} \mu H_y^2$ represent the electric stored energy density (J/m^3) and the magnetic stored energy density (J/m^3), respectively, associated with the electromagnetic field due to the dielectric and magnetic properties, respectively, of the medium.

Poynting's theorem

Equation (6.96) is a special case of a theorem known as *Poynting's theorem*. To derive Poynting's theorem for the general case, we make use of the vector identity

$$\nabla \cdot (\mathbf{E} \times \mathbf{H}) = \mathbf{H} \cdot \nabla \times \mathbf{E} - \mathbf{E} \cdot \nabla \times \mathbf{H} \tag{6.97}$$

and Maxwell's curl equations

$$\nabla \times \mathbf{E} = -\frac{\partial \mathbf{B}}{\partial t} = -\mu \frac{\partial \mathbf{H}}{\partial t}$$

$$\nabla \times \mathbf{H} = \mathbf{J}_c + \frac{\partial \mathbf{D}}{\partial t} = \sigma \mathbf{E} + \varepsilon \frac{\partial \mathbf{E}}{\partial t}$$

to obtain

$$\begin{aligned}
\nabla \cdot (\mathbf{E} \times \mathbf{H}) &= \mathbf{H} \cdot \left(-\mu \frac{\partial \mathbf{H}}{\partial t}\right) - \mathbf{E} \cdot \left(\sigma \mathbf{E} + \varepsilon \frac{\partial \mathbf{E}}{\partial t}\right) \\
&= -\mu \mathbf{H} \cdot \frac{\partial \mathbf{H}}{\partial t} - \sigma \mathbf{E} \cdot \mathbf{E} - \varepsilon \mathbf{E} \cdot \frac{\partial \mathbf{E}}{\partial t} \\
&= -\mu \frac{\partial}{\partial t}(\tfrac{1}{2}\mathbf{H} \cdot \mathbf{H}) - \sigma \mathbf{E} \cdot \mathbf{E} - \varepsilon \frac{\partial}{\partial t}(\tfrac{1}{2}\mathbf{E} \cdot \mathbf{E}) \\
&= -\sigma E^2 - \frac{\partial}{\partial t}(\tfrac{1}{2}\varepsilon E^2) - \frac{\partial}{\partial t}(\tfrac{1}{2}\mu H^2)
\end{aligned} \tag{6.98}$$

Substituting $\mathbf{P}$ for $\mathbf{E} \times \mathbf{H}$ and taking the volume integral of both sides of (6.98) over the volume V, we obtain

$$\begin{aligned}
\int_V (\nabla \cdot \mathbf{P})\, dv = &-\int_V \sigma E^2\, dv - \int_V \frac{\partial}{\partial t}(\tfrac{1}{2}\varepsilon E^2)\, dv \\
&- \int_V \frac{\partial}{\partial t}(\tfrac{1}{2}\mu H^2)\, dv
\end{aligned} \tag{6.99}$$

Interchanging the differentiation operation with time and integration operation over volume in the second and third terms on the right side and replacing the volume integral on the left side by a closed surface integral in accordance with the divergence theorem, we get

$$\oint_S \mathbf{P} \cdot d\mathbf{S} = -\int_V \sigma E^2 \, dv - \frac{\partial}{\partial t}\int_V \tfrac{1}{2}\varepsilon E^2 \, dv - \frac{\partial}{\partial t}\int_V \tfrac{1}{2}\mu H^2 \, dv \qquad (6.100)$$

where S is the surface bounding the volume V. Equation (6.100) is the Poynting theorem for the general case. Since it should hold for any size V, it follows that the power dissipation density, the electric stored energy density, and the magnetic stored energy density are given by

$$p_d = \sigma E^2 \qquad (6.101a)$$

$$w_e = \tfrac{1}{2}\varepsilon E^2 \qquad (6.101b)$$

$$w_m = \tfrac{1}{2}\mu H^2 \qquad (6.101c)$$

respectively.

Time-average power flow

Returning now to Fig. 6.18, we note that there are certain intervals in z for which P_z is negative, although the wave propagation is in the $+z$-direction. This is because of the phase difference between the electric and magnetic fields. There is no inconsistency here since the plot corresponds to only one value of time, namely, $t = 0$. On the other hand, the time-average value of P_z is positive everywhere, as we shall show now. The time-average value of P_z, denoted $\langle P_z \rangle$, is P_z averaged over one period of the sinusoidal time variation of the source; that is,

$$\langle P_z \rangle = \frac{1}{T}\int_0^T P_z(t) \, dt \qquad (6.102)$$

where T $(= 1/f)$ is the period. From (6.94), we have

$$\begin{aligned} \langle P_z \rangle &= \left\langle |\bar{\eta}| \frac{J_{S0}^2}{4} e^{-2\alpha z} \cos(\omega t - \beta z + \tau)\cos(\omega t - \beta z) \right\rangle \\ &= |\bar{\eta}| \frac{J_{S0}^2}{8} e^{-2\alpha z} \langle \cos\tau + \cos(2\omega t - 2\beta z + \tau) \rangle \qquad (6.103) \\ &= |\bar{\eta}| \frac{J_{S0}^2}{8} e^{-2\alpha z} [\langle \cos\tau \rangle + \langle \cos(2\omega t - 2\beta z + \tau) \rangle] \end{aligned}$$

Since $\cos\tau$ is independent of time, $\langle \cos\tau \rangle$ is equal to $\cos\tau$. The quantity $\langle \cos(2\omega t - 2\beta z + \tau) \rangle$ is equal to zero since the integral of a cosine or sine function over each period is zero. Thus (6.103) reduces to

$$\langle P_z \rangle = |\bar{\eta}| \frac{J_{S0}^2}{8} e^{-2\alpha z} \cos\tau \qquad (6.104)$$

which is everywhere positive.

Example 6.6

Let us consider the electric field of a uniform plane wave propagating in seawater ($\sigma = 4$ S/m, $\varepsilon = 80\varepsilon_0$, and $\mu = \mu_0$) in the positive z-direction and having the electric field

$$\mathbf{E} = 1 \cos 5 \times 10^4 \pi t \; \mathbf{i}_x \text{ V/m}$$

at $z = 0$. We wish to find the instantaneous power flow per unit area normal to the z-direction as a function of z and the time-average power flow per unit area normal to the z-direction as a function of z.

From the expression for $\mathbf{E}$, we note that the frequency of the wave is 25 kHz. At this frequency in seawater, the propagation parameters can be computed to be $\alpha = \beta \approx 0.628$ and $\overline{\eta} = 0.222\underline{/45^\circ}$. The expressions for the instantaneous electric and magnetic fields are therefore given by

$$\mathbf{E} = 1e^{-0.628z} \cos (5 \times 10^4 \pi t - 0.628z) \; \mathbf{i}_x \text{ V/m}$$

$$\mathbf{H} = 4.502e^{-0.628z} \cos (5 \times 10^4 \pi t - 0.628z - \pi/4) \; \mathbf{i}_y \text{ A/m}$$

The instantaneous Poynting vector is then given by

$$\begin{aligned}\mathbf{P} &= \mathbf{E} \times \mathbf{H} \\ &= 4.502e^{-1.256z} \cos (5 \times 10^4 \pi t - 0.628z) \\ &\quad \cos (5 \times 10^4 \pi t - 0.628z - \pi/4) \; \mathbf{i}_z \text{ W/m}^2\end{aligned}$$

Thus the instantaneous power flow per unit area normal to the z-direction, which is simply the z-component of the instantaneous Poynting vector, is

$$P_z = 2.251e^{-1.256z}[\cos \pi/4 + \cos (10^5 \pi t - 1.256z - \pi/4)] \text{ W/m}^2$$

Finally, the time-average power flow per unit area normal to the z-direction is

$$\begin{aligned}<P_z> &= 2.251e^{-1.256z} \cos \pi/4 \\ &= 1.592e^{-1.256z} \text{ W/m}^2\end{aligned}$$

Time-average Poynting vector

Returning now to (6.104), we note that $<P_z>$ can be expressed in the manner

$$\begin{aligned}<P_z> &= |\overline{\eta}| \frac{J_{S0}^2}{8} e^{-2\alpha z} \cos \tau \\ &= \frac{1}{2}\text{Re}\left[\left(\frac{|\overline{\eta}|J_{S0}}{2} e^{-\alpha z} e^{j(\quad -\beta z + \tau)}\right)\left(\frac{J_{S0}}{2} e^{-\alpha z} e^{j\beta z}\right)\right] \\ &= \text{Re}\left[\frac{1}{2}\overline{E}_x \overline{H}_y^*\right]\end{aligned} \tag{6.105}$$

where $\overline{E}_x$ and $\overline{H}_y$ are the phasor electric and magnetic field components, respectively. In terms of vector quantities, we have

$$<\mathbf{P}> = \text{Re}\left[\frac{1}{2}\overline{E}_x \mathbf{i}_x \times (\overline{H}_y \mathbf{i}_y)^*\right]$$

$$= \text{Re}\left[\frac{1}{2}\bar{\mathbf{E}} \times \bar{\mathbf{H}}^*\right]$$

$$= \text{Re}\,[\bar{\mathbf{P}}] \tag{6.106}$$

which is the general form of the time-average Poynting vector (see Problem P6.36), where

$$\bar{\mathbf{P}} = \frac{1}{2}\bar{\mathbf{E}} \times \bar{\mathbf{H}}^* \tag{6.107}$$

is the complex Poynting vector.

K6.5. Power flow; Poynting vector; Poynting's theorem; Power dissipation density; Electric stored energy density; Magnetic stored energy density; Time-average power flow; Time-average Poynting vector.

D6.13. The magnetic field associated with a uniform plane wave propagating in the $+z$-direction in a nonmagnetic ($\mu = \mu_0$) material medium is given by

$$\mathbf{H} = H_0 e^{-z} \cos\,(6\pi \times 10^7 t - \sqrt{3}z)\,\mathbf{i}_y \text{ V/m}$$

Find the following: **(a)** the instantaneous power flow across a surface of area 1 m^2 in the $z = 0$ plane at $t = 0$; **(b)** the time-average power flow across a surface of area 1 m^2 in the $z = 0$ plane; and **(c)** the time-average power flow across a surface of area 1 m^2 in the $z = 1$ m plane.
Ans. **(a)** $102.57H_0^2$ W; **(b)** $51.28H_0^2$ W; **(c)** $6.94H_0^2$ W

D6.14. Find the time-average values of the following: **(a)** $A \sin \omega t \sin 3\omega t$; **(b)** $A(\cos^2 \omega t - 0.5 \sin^2 2\omega t)$; and **(c)** $A \sin^6 \omega t$.
Ans. **(a)** 0; **(b)** $0.25A$; **(c)** $0.3125A$

6.6 REFLECTION OF UNIFORM PLANE WAVES

Thus far we have considered uniform plane wave propagation in unbounded media. Practical situations are characterized by propagation involving several different media. When a wave is incident on a boundary between two different media, a reflected wave is produced. In addition, if the second medium is not a perfect conductor, a transmitted wave is set up. Together, these waves satisfy the boundary conditions at the interface between the two media. In this section, we shall consider these phenomena for waves incident normally on plane boundaries.

Normal incidence on a plane interface

To do this, let us consider the situation shown in Fig. 6.19 in which steady-state conditions are established by uniform plane waves of radian frequency ω propagating normal to the plane interface $z = 0$ between two media characterized by two different sets of values of σ, ε, and μ where $\sigma \neq \infty$. We shall assume that a (+) wave is incident from medium 1 ($z < 0$) onto the interface, thereby setting up a reflected (−) wave in that medium, and a transmitted (+) wave in medium 2 ($z > 0$). For convenience, we shall work with the phasor or complex field components. Thus considering the electric fields to be in the x-direction and the magnetic fields to be in the y-direction, we can write the solution for the complex field components in medium 1 to be

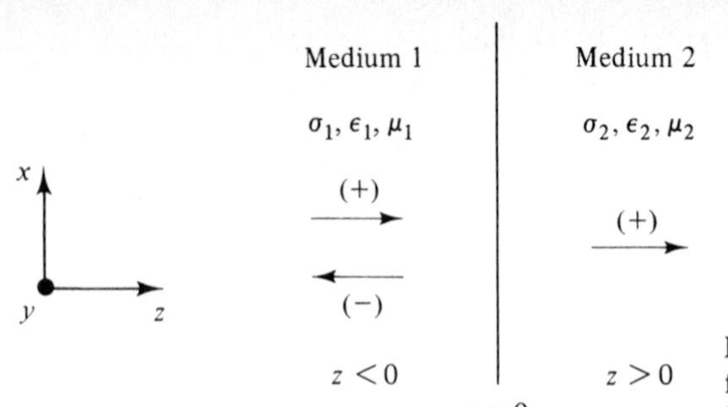

Figure 6.19. Normal incidence of uniform plane waves on a plane interface between two different media.

$$\bar{E}_{1x}(z) = \bar{E}_1^+ e^{-\bar{\gamma}_1 z} + \bar{E}_1^- e^{\bar{\gamma}_1 z} \tag{6.108a}$$

$$\begin{aligned}\bar{H}_{1y}(z) &= \bar{H}_1^+ e^{-\bar{\gamma}_1 z} + \bar{H}_1^- e^{\bar{\gamma}_1 z} \\ &= \frac{1}{\bar{\eta}_1}(\bar{E}_1^+ e^{-\bar{\gamma}_1 z} - \bar{E}_1^- e^{\bar{\gamma}_1 z})\end{aligned} \tag{6.108b}$$

where $\bar{E}_1^+$, $\bar{E}_1^-$, $\bar{H}_1^+$, and $\bar{H}_1^-$ are the incident and reflected wave electric and magnetic field components, respectively, at $z = 0-$ in medium 1 and

$$\bar{\gamma}_1 = \sqrt{j\omega\mu_1(\sigma_1 + j\omega\varepsilon_1)} \tag{6.109a}$$

$$\bar{\eta}_1 = \sqrt{\frac{j\omega\mu_1}{\sigma_1 + j\omega\varepsilon_1}} \tag{6.109b}$$

Recall that the real field corresponding to a complex field component is obtained by multiplying the complex field component by $e^{j\omega t}$ and taking the real part of the product. The complex field components in medium 2 are given by

$$\bar{E}_{2x}(z) = \bar{E}_2^+ e^{-\bar{\gamma}_2 z} \tag{6.110a}$$

$$\begin{aligned}\bar{H}_{2y}(z) &= \bar{H}_2^+ e^{-\bar{\gamma}_2 z} \\ &= \frac{\bar{E}_2^+}{\bar{\eta}_2} e^{-\bar{\gamma}_2 z}\end{aligned} \tag{6.110b}$$

where $\bar{E}_2^+$ and $\bar{H}_2^+$ are the transmitted wave electric and magnetic field components at $z = 0+$ in medium 2 and

$$\bar{\gamma}_2 = \sqrt{j\omega\mu_2(\sigma_2 + j\omega\varepsilon_2)} \tag{6.111a}$$

$$\bar{\eta}_2 = \sqrt{\frac{j\omega\mu_2}{\sigma_2 + j\omega\varepsilon_2}} \tag{6.111b}$$

Reflection and transmission coefficients

To satisfy the boundary conditions at $z = 0$, we note that (1) the components of both electric and magnetic fields are entirely tangential to the interface and (2) in view of the finite conductivities of the media, no surface current exists on the interface (currents flow in the volumes of the media). Hence from the phasor forms of the boundary conditions (3.61a) and (3.61b), we have

$$[\bar{E}_{1x}]_{z=0} = [\bar{E}_{2x}]_{z=0} \tag{6.112a}$$

$$[\bar{H}_{1y}]_{z=0} = [\bar{H}_{2y}]_{z=0} \tag{6.112b}$$

Applying these to the solution pairs given by (6.108a,b) and (6.110a,b), we have

$$\bar{E}_1^+ + \bar{E}_1^- = \bar{E}_2^+ \tag{6.113a}$$

$$\frac{1}{\bar{\eta}_1}(\bar{E}_1^+ - \bar{E}_1^-) = \frac{1}{\bar{\eta}_2}\bar{E}_2^+ \tag{6.113b}$$

We now define the *reflection coefficient* at the boundary, denoted by the symbol $\bar{\Gamma}$, to be the ratio of the reflected wave electric field at the boundary to the incident wave electric field at the boundary. From (6.113a) and (6.113b), we obtain

$$\boxed{\bar{\Gamma} = \frac{\bar{E}_1^-}{\bar{E}_1^+} = \frac{\bar{\eta}_2 - \bar{\eta}_1}{\bar{\eta}_2 + \bar{\eta}_1}} \tag{6.114}$$

Note that the ratio of the reflected wave magnetic field at the boundary to the incident wave magnetic field at the boundary is given by

$$\frac{\bar{H}_1^-}{\bar{H}_1^+} = \frac{-\bar{E}_1^-/\bar{\eta}_1}{\bar{E}_1^+/\bar{\eta}_1} = -\frac{\bar{E}_1^-}{\bar{E}_1^+} = -\bar{\Gamma} \tag{6.115}$$

The ratio of the transmitted wave electric field at the boundary to the incident wave electric field at the boundary, known as the *transmission coefficient* and denoted by the symbol $\bar{\tau}$, is given by

$$\boxed{\bar{\tau} = \frac{\bar{E}_2^+}{\bar{E}_1^+} = \frac{\bar{E}_1^+ + \bar{E}_1^-}{\bar{E}_1^+} = 1 + \bar{\Gamma}} \tag{6.116}$$

where we have used (6.113a). The ratio of the transmitted wave magnetic field at the boundary to the incident wave magnetic field at the boundary is given by

$$\frac{\bar{H}_2^+}{\bar{H}_1^+} = \frac{\bar{H}_1^+ + \bar{H}_1^-}{\bar{H}_1^+} = 1 - \bar{\Gamma} \tag{6.117}$$

The reflection and transmission coefficients given by (6.114) and (6.116), respectively, enable us to find the reflected and transmitted wave fields for a given incident wave field. We observe the following properties of $\bar{\Gamma}$ and $\bar{\tau}$:

1. For $\bar{\eta}_2 = \bar{\eta}_1$, $\bar{\Gamma} = 0$ and $\bar{\tau} = 1$. The incident wave is entirely transmitted. The situation then corresponds to a "matched" condition. A trivial case occurs when the two media have identical values of the material parameters.
2. For $\sigma_1 = \sigma_2 = 0$, that is, when both media are perfect dielectrics, $\bar{\eta}_1$ and $\bar{\eta}_2$ are real. Hence $\bar{\Gamma}$ and $\bar{\tau}$ are real. In particular, if the two media have the same permeability μ but different permittivities ε_1 and ε_2, then

$$\bar{\Gamma} = \frac{\sqrt{\mu/\varepsilon_2} - \sqrt{\mu/\varepsilon_1}}{\sqrt{\mu/\varepsilon_2} + \sqrt{\mu/\varepsilon_1}} = \frac{1 - \sqrt{\varepsilon_2/\varepsilon_1}}{1 + \sqrt{\varepsilon_2/\varepsilon_1}} \tag{6.118}$$

$$\bar{\tau} = \frac{2}{1 + \sqrt{\varepsilon_2/\varepsilon_1}} \tag{6.119}$$

3. For $\sigma_2 \to \infty$, $\bar{\eta}_2 \to 0$, $\bar{\Gamma} \to -1$, and $\bar{\tau} \to 0$. Thus if medium 2 is a perfect conductor, the incident wave is entirely reflected, as it should be since there cannot be any time-varying fields inside a perfect conductor. The superposition of the reflected and incident waves would then give rise to the so-called complete standing waves in medium 1. We shall discuss complete standing waves as well as partial standing waves when we study the topic of sinusoidal steady-state analysis of waves on transmission lines in Chapter 8.

Example 6.7

Region 1 ($z < 0$) is free space, whereas region 2 ($z > 0$) is a material medium characterized by $\sigma = 10^{-4}$ S/m, $\varepsilon = 5\varepsilon_0$, and $\mu = \mu_0$. For a uniform plane wave having the electric field

$$\mathbf{E}_i = E_0 \cos (3\pi \times 10^5 t - 10^{-3}\pi z) \text{ V/m}$$

incident on the interface $z = 0$ from region 1, we wish to obtain the expressions for the reflected and transmitted wave electric and magnetic fields.

From a run of PL6.1 for $\sigma = 10^{-4}$ S/m, $\varepsilon = 5\varepsilon_0$, $\mu = \mu_0$, and $f = \dfrac{3\pi \times 10^5}{2\pi} = 1.5 \times 10^5$ Hz,

$$\bar{\gamma} = (6.283 + j9.425) \times 10^{-3}$$

$$\bar{\eta} = 104.559\underline{/33.69^\circ} = 104.559\underline{/0.1872\pi}$$

Then

$$\bar{\Gamma} = \frac{\bar{\eta} - \eta_0}{\bar{\eta} + \eta_0} = \frac{104.559\underline{/33.69^\circ} - 377}{104.559\underline{/33.69^\circ} + 377}$$

$$= 0.6325\underline{/161.565^\circ} = 0.6325\underline{/0.8976\pi}$$

$$\bar{\tau} = 1 + \bar{\Gamma} = 1 + 0.6325\underline{/161.565^\circ}$$

$$= 0.4472\underline{/26.565^\circ} = 0.4472\underline{/0.1476\pi}$$

Thus the reflected and transmitted wave electric and magnetic fields are given by

$$\mathbf{E}_r = 0.6325 E_0 \cos (3\pi \times 10^5 t + 10^{-3}\pi z + 0.8976\pi)\, \mathbf{i}_x \text{ V/m}$$

$$\mathbf{H}_r = -\frac{0.6325 E_0}{377} \cos (3\pi \times 10^5 t + 10^{-3}\pi z + 0.8976\pi)\, \mathbf{i}_y \text{ A/m}$$

$$= -1.678 \times 10^{-3} E_0 \cos (3\pi \times 10^5 t + 10^{-3}\pi z + 0.8976\pi)\, \mathbf{i}_y \text{ A/m}$$

$$\mathbf{E}_t = 0.4472 E_0 e^{-6.283 \times 10^{-3} z}$$

$$\cdot \cos (3\pi \times 10^5 t - 9.425 \times 10^{-3} z + 0.1476\pi)\, \mathbf{i}_x \text{ V/m}$$

$$\mathbf{H}_t = \frac{0.4472 E_0}{104.559} e^{-6.283 \times 10^{-3} z}$$

$$\cdot \cos (3\pi \times 10^5 t - 9.425 \times 10^{-3} z + 0.1476\pi - 0.1872\pi)\, \mathbf{i}_y \text{ A/m}$$

$$= 4.277 \times 10^{-3} E_0 e^{-6.283 \times 10^{-3} z}$$

$$\cdot \cos (3\pi \times 10^5 t - 9.425 \times 10^{-3} z - 0.0396\pi)\, \mathbf{i}_y \text{ A/m}$$

Note that at $z = 0$, the boundary conditions of $\mathbf{E}_i + \mathbf{E}_r = \mathbf{E}_t$ and $\mathbf{H}_i + \mathbf{H}_r = \mathbf{H}_t$ are satisfied, since

$$E_0 + 0.6325E_0 \cos 0.8976\pi = 0.4472E_0 \cos 0.1476\pi$$

and

$$\frac{E_0}{377} - 1.678 \times 10^{-3} E_0 \cos 0.8976\pi = 4.277 \times 10^{-3} E_0 \cos(-0.0396\pi)$$

K6.6. Plane interface between two material media; Normal incidence of uniform plane waves; Reflection; Transmission; Reflection and transmission coefficients.

D6.15. For each of the following cases of uniform plane waves of frequencey $f = 1$ MHz incident normally from medium 1 ($z < 0$) onto the interface ($z = 0$) with medium 2 ($z > 0$), find the values of $\bar{\Gamma}$ and $\bar{\tau}$. **(a)** Medium 1 is free space and the parameters of medium 2 are $\sigma = 10^{-3}$ S/m, $\varepsilon = 6\varepsilon_0$, and $\mu = \mu_0$ and **(b)** the parameters of medium 1 are $\sigma = 4$ S/m, $\varepsilon = 80\varepsilon_0$, and $\mu = \mu_0$, and the parameters of medium 2 are $\sigma = 10^{-3}$ S/m, $\varepsilon = 80\varepsilon_0$, and $\mu = \mu_0$.
Ans. **(a)** $0.6909\angle 164.177°$, $0.3846\angle 29.331°$; **(b)** $0.9486\angle -2.4155°$, $1.948\angle -1.177°$

D6.16. The regions $z < 0$ and $z > 0$ are nonmagnetic ($\mu = \mu_0$) perfect dielectrics of permittivities ε_1 and ε_2, respectively. For a uniform plane wave incident from the region $z < 0$ normally onto the boundary $z = 0$, find $\varepsilon_2/\varepsilon_1$ for each of the following to hold at $z = 0$: **(a)** the electric field of the reflected wave is $-1/3$ times the electric field of the incident wave; **(b)** the electric field of the transmitted wave is 0.4 times the electric field of the incident wave; and **(c)** the electric field of the transmitted wave is 6 times the electric field of the reflected wave.
Ans. **(a)** 4; **(b)** 16; **(c)** 4/9

6.7 SUMMARY

In this chapter we studied the principles of uniform plane waves. Uniform plane waves are a building block in the study of electromagnetic wave propagation. They are the simplest type of solutions resulting from the coupling of the electric and magnetic fields in Maxwell's curl equations. Their electric and magnetic fields are perpendicular to each other and to the direction of propagation. The fields are *uniform* in the *planes* perpendicular to the direction of propagation.

We first obtained the uniform plane wave solution to Maxwell's equations in time domain in free space by considering an infinite plane current sheet in the xy-plane with uniform surface current density given by

$$\mathbf{J}_S = -J_S(t)\mathbf{i}_x \text{ A/m}$$

and deriving the electromagnetic field due to the current sheet to be

$$\mathbf{E} = \frac{\eta_0}{2} J_S\left(t \mp \frac{z}{v_p}\right)\mathbf{i}_x \quad \text{for} \quad z \gtrless 0 \tag{6.120a}$$

$$\mathbf{H} = \pm\frac{1}{2} J_S\left(t \mp \frac{z}{v_p}\right)\mathbf{i}_y \quad \text{for} \quad z \gtrless 0 \tag{6.120b}$$

where

$$v_p = \frac{1}{\sqrt{\mu_0\varepsilon_0}}$$

and

$$\eta_0 = \sqrt{\frac{\mu_0}{\varepsilon_0}}$$

are the velocity of propagation and intrinsic impedance, respectively. In (6.120a) and (6.120b), the arguments $(t - z/v_p)$ and $(t + z/v_p)$ represent wave motion in the positive z-direction and the negative z-direction, respectively, with the velocity v_p. Thus (6.120a) and (6.120b) correspond to waves propagating away from the current sheet to either side of it. Since the fields are uniform in z-constant planes, they represent uniform plane waves. We discussed how to plot the variations of the field components versus t for fixed values of z and versus z for fixed values of t, for a given function $J_S(t)$.

We then extended the solution to sinusoidally time-varying uniform plane waves by considering the current density on the infinite plane sheet to be

$$\mathbf{J}_S = -J_{S0} \cos \omega t \; \mathbf{i}_x \text{ A/m}$$

and obtaining the corresponding field to be

$$\mathbf{E} = \frac{\eta_0 J_{S0}}{2} \cos(\omega t \mp \beta z)\, \mathbf{i}_x \qquad \text{for} \quad z \lessgtr 0 \tag{6.121a}$$

$$\mathbf{H} = \pm \frac{J_{S0}}{2} \cos(\omega t \mp \beta z)\, \mathbf{i}_y \qquad \text{for} \quad z \gtrless 0 \tag{6.121b}$$

where

$$\beta = \frac{\omega}{v_p} = \omega \sqrt{\mu_0 \varepsilon_0} \tag{6.122}$$

We discussed several important parameters and properties associated with these waves, including polarization. The quantity β is the phase constant, that is, the magnitude of the rate of change of phase with distance along the direction of propagation, for a fixed time. The velocity v_p which from (6.122) is given by

$$v_p = \frac{\omega}{\beta} \tag{6.123}$$

is known as the phase velocity, because it is the velocity with which a particular constant phase progresses along the direction of propagation. The wavelength λ, that is, the distance along the direction of propagation in which the phase changes by 2π radians, for a fixed time, is given by

$$\lambda = \frac{2\pi}{\beta} \tag{6.124}$$

The wavelength is related to the frequency f in a simple manner as given by

$$v_p = \lambda f \tag{6.125}$$

which follows from (6.123) and (6.124) and is a result of the fact that the time and space variations of the fields are interdependent. We discussed the principle of antenna array and the Doppler effect.

Next we extended the treatment of uniform plane waves to a material medium. Starting with the Maxwell's equations for a material medium given by

$$\nabla \times \mathbf{E} = -\frac{\partial \mathbf{B}}{\partial t} = -\mu \frac{\partial \mathbf{H}}{\partial t}$$

$$\nabla \times \mathbf{H} = \mathbf{J}_c + \frac{\partial \mathbf{D}}{\partial t} = \sigma \mathbf{E} + \varepsilon \frac{\partial \mathbf{E}}{\partial t}$$

and using the phasor technique, we considered the infinite plane current sheet of uniform surface current density

$$\mathbf{J}_S = -J_{S0} \cos \omega t \, \mathbf{i}_x \text{ A/m}$$

in the xy-plane and embedded in the material medium and obtained the electromagnetic field due to it to be

$$\mathbf{E} = \frac{|\bar{\eta}| J_{S0}}{2} e^{\mp \alpha z} \cos (\omega t \mp \beta z + \tau) \, \mathbf{i}_x \qquad \text{for} \quad z \gtrless 0 \qquad (6.126a)$$

$$\mathbf{H} = \pm \frac{J_{S0}}{2} e^{\mp \alpha z} \cos (\omega t \mp \beta z) \, \mathbf{i}_y \qquad \text{for} \quad z \gtrless 0 \qquad (6.126b)$$

In (6.126a,b) α and β are the attenuation and phase constants given, respectively, by the real and imaginary parts of the propagation constant, $\bar{\gamma}$. Thus

$$\bar{\gamma} = \alpha + j\beta = \sqrt{j\omega\mu(\sigma + j\omega\varepsilon)}$$

The quantities $|\bar{\eta}|$ and τ are the magnitude and phase angle, respectively, of the intrinsic impedance, $\bar{\eta}$, of the medium. Thus

$$\bar{\eta} = |\bar{\eta}| e^{j\tau} = \sqrt{\frac{j\omega\mu}{\sigma + j\omega\varepsilon}}$$

The solution given by (6.126a) and (6.126b) tells us that the wave propagation in the material medium is characterized by attenuation as indicated by $e^{\mp \alpha z}$ and phase difference between **E** and **H** by the amount τ. We also learned that these properties as well as the phase velocity are frequency dependent.

Having discussed uniform plane wave propagation for the general case of a medium characterized by σ, ε, and μ, we then considered several special cases. These are summarized in the following:

Perfect dielectrics. For these materials, $\sigma = 0$. Wave propagation occurs without attenuation as in free space but with the propagation parameters governed by ε and μ instead of ε_0 and μ_0, respectively. The tangential components of **E** and **H** and the normal components of **D** and **B** are continuous at a boundary between two perfect dielectrics.

Imperfect dielectrics. A material is classified as an imperfect dielectric for $\sigma \ll \omega\varepsilon$, that is, conduction current density small in magnitude compared to the displacement current density. The only significant feature of wave propagation in an imperfect dielectric as compared to that in a perfect dielectric is the attenuation undergone by the wave.

Good conductors. A material is classified as a good conductor for $\sigma \gg \omega\varepsilon$, that is, conduction current density large in magnitude compared to the displacement current density. Wave propagation in a good conductor medium is characterized by attenuation and phase constants both equal to $\sqrt{\pi f \mu \sigma}$. Thus for large values of f and/or σ, the fields do not penetrate very deep into the conductor. This phenomenon is known as the skin effect. From considerations of the frequency dependence of the attenuation and wavelength for a fixed σ, we learned that low frequencies are more suitable for communication with underwater objects. We also learned that the intrinsic impedance of a good conductor medium is very low in magnitude compared to that of a dielectric medium having the same ε and μ.

Perfect conductors. These are idealizations of good conductors in the limit $\sigma \to \infty$. For $\sigma = \infty$, the skin depth, that is, the distance in which the fields inside a conductor are attenuated by a factor e^{-1}, is zero. Hence there can be no penetration of fields into a perfect conductor, so that on a perfect conductor surface, the tangential component of **E** and the normal component of **B** are zero, the normal component of **D** is equal to the surface charge density, and the tangential component of **H** is equal in magnitude to the surface current density.

We then learned that there is power flow, power dissipation, and energy storage associated with the wave propagation. The power flow density is given by the Poynting vector

$$\mathbf{P} = \mathbf{E} \times \mathbf{H}$$

The power dissipation density and the electric and magnetic stored energy densities are given, respectively, by

$$p_d = \sigma E^2$$

$$w_e = \tfrac{1}{2}\varepsilon E^2$$

$$w_m = \tfrac{1}{2}\mu H^2$$

The power flow out of a closed surface S plus the power dissipated in the volume V bounded by S is always equal to the sum of the time rates of decrease of electric and magnetic stored energies in the volume V as given by Poynting's theorem

$$\oint_S \mathbf{P} \cdot d\mathbf{S} = -\int_V \sigma E^2\, dv - \frac{\partial}{\partial t}\int_V \tfrac{1}{2}\varepsilon E^2\, dv - \frac{\partial}{\partial t}\int_V \tfrac{1}{2}\mu H^2\, dv$$

Finally, we considered uniform plane waves incident normally onto a plane boundary between two material media and learned how to compute the reflected and transmitted wave fields for a given incident wave field.

REVIEW QUESTIONS

R6.1. What is a uniform plane wave? Why is the study of uniform plane waves important?

R6.2. Outline the procedure for obtaining from the two Maxwell's curl equations the particular differential equation for the special case of $\mathbf{J} = J_x(z)\mathbf{i}_x$.

R6.3. State the wave equation for the case of $\mathbf{E} = E_x(z, t)\mathbf{i}_x$. Describe the procedure for its solution.

R6.4. Discuss by means of an example how a function $f(t - z\sqrt{\mu_0\varepsilon_0})$ represents a traveling wave propagating in the positive z-direction.

R6.5. Discuss by means of an example how a function $g(t + z\sqrt{\mu_0\varepsilon_0})$ represents a traveling wave propagating in the negative z-direction.

R6.6. What is the significance of the intrinsic impedance of free space? What is its value?

R6.7. Summarize the procedure for obtaining the solution for the electromagnetic field due to the infinite plane sheet of uniform time-varying current density.

R6.8. State and discuss the solution for the electromagnetic field due to the infinite plane sheet of current density $\mathbf{J}_S(t) = -J_S(t)\mathbf{i}_x$ for $z = 0$.

R6.9. Discuss the parameters ω, β, and v_p associated with sinusoidally time-varying uniform plane waves.

R6.10. Define wavelength. What is the relationship among wavelength, frequency, and phase velocity?

R6.11. Discuss the classification of waves according to frequency, giving examples of their application in the different frequency ranges.

R6.12. How is the direction of propagation of a uniform plane wave related to the directions of its fields?

R6.13. Discuss right-handed and left-handed circular polarizations associated with sinusoidally time-varying uniform plane waves.

R6.14. Discuss the principle of an antenna array with the aid of an example.

R6.15. What is Doppler effect? How do you compute the Doppler shift in the frequency of a wave as viewed by an observer moving in an arbitrary direction?

R6.16. Give some examples of the application of Doppler effect.

R6.17. Discuss how the determination of the electromagnetic field due to an infinite plane current sheet of sinusoidally time-varying current density imbedded in a material medium is made convenient by using the phasor technique.

R6.18. What is the propagation constant for a material medium? Discuss the significance of its real and imaginary parts.

R6.19. What is the intrinsic impedance for a material medium? What is the consequence of its complex nature?

R6.20. What is loss tangent? Discuss its significance.

R6.21. Discuss the consequence of the frequency dependence of the phase velocity of a wave in a material medium.

R6.22. How would you obtain the electromagnetic field due to a current sheet of nonsinusoidally time-varying current density embedded in a material medium?

R6.23. What is the condition for a medium to be a perfect dielectric? How do the characteristics of wave propagation in a perfect dielectric medium differ from those of wave propagation in free space?

R6.24. State the boundary conditions at the interface between two perfect dielectrics.

R6.25. What is the criterion for a material to be an imperfect dielectric? What is the significant feature of wave propagation in an imperfect dielectric as compared to that in a perfect dielectric?

R6.26. What is the criterion for a material to be a good conductor? Give two examples of materials that behave as good conductors for frequencies of up to several gigahertz.

R6.27. What is skin effect? Discuss skin depth, giving some numerical values.

R6.28. Why are low-frequency waves more suitable than high-frequency waves for communication with underwater objects?

R6.29. Discuss the consequence of the low intrinsic impedance of a good conductor as compared to that of a dielectric medium having the same ε and μ.

R6.30. Why can there be no fields inside a perfect conductor? What are the boundary conditions at the surface of a perfect conductor?

R6.31. What is the Poynting vector? What is the physical interpretation of the Poynting vector over a closed surface?

R6.32. State Poynting's theorem. How is it derived from Maxwell's curl equations?

R6.33. What are the power dissipation density, the electric stored energy density, and the magnetic stored energy density associated with an electromagnetic field in a material medium?

R6.34. What is time-average power flow? Discuss the time-average power flow versus instantaneous power flow associated with a uniform plane wave in a material medium.

R6.35. Discuss the determination of the reflected and transitted wave fields from the fields of a wave incident normally onto a plane boundary between two material media.

R6.36. What is the consequence of a wave incident on a perfect conductor?

PROBLEMS

P6.1. From Maxwell's curl equations, obtain the particular differential equations for the case of $\mathbf{J} = J_x(y, t)\mathbf{i}_x$.

P6.2. For each of the following functions, plot the value of the function versus z for the two specified values of time and discuss the traveling wave nature of the function:
(a) $e^{-(t-z)^2}$; $t = 0$, $t = 1$ s
(b) $(10^8 t + z)[u(t + 10^{-8}z) - u(t + 10^{-8}z - 2 \times 10^{-8})]$; $t = 0$, $t = 10^{-8}$ s

P6.3. Write expressions for traveling wave functions corresponding to the following cases: **(a)** time-variation at $x = 0$ in the manner $10u(t)$ and propagating in the $-x$-direction with velocity 0.5 m/s; **(b)** time-variation at $y = 0$ in the manner $t \sin 20t$ and propagating in the $+y$-direction with velocity 4 m/s; and **(c)** distance variation at $t = 0$ in the manner $5z^3 e^{-z^2}$ and propagating in the $-z$-direction with velocity 2 m/s.

P6.4. An infinite plane sheet lying in the $z = 0$ plane in free space carries a surface current of density $\mathbf{J}_S = -J_S(t)\mathbf{i}_x$, where $J_S(t)$ is as shown in Fig. 6.20. Find and sketch **(a)** E_x versus t for $z = 300$ m; **(b)** H_y versus t for $z = -600$ m; **(c)** E_x versus z for $t = 3$ μs; and **(d)** H_y versus z for $t = 4$ μs.

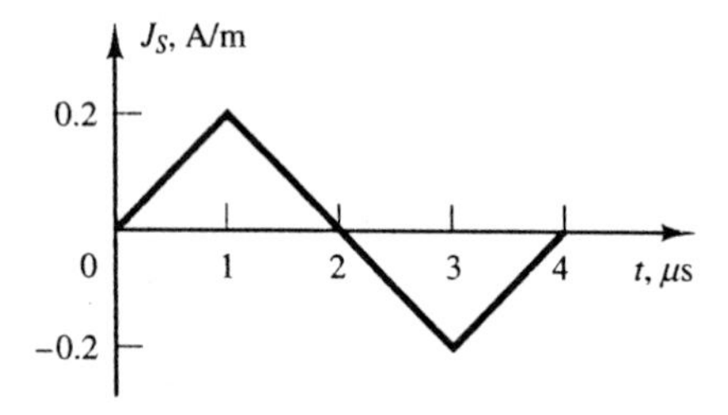

Figure 6.20. For Problem P6.4.

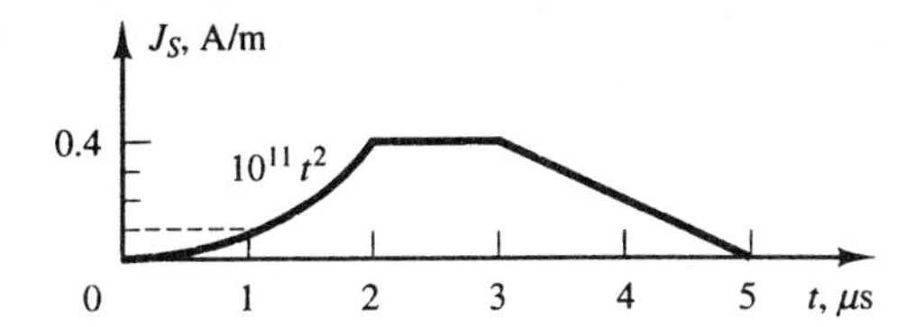

Figure 6.21. For Problems P6.5.

P6.5. An infinite plane sheet lying in the $z = 0$ plane in free space carries a surface current of density $\mathbf{J}_S = -J_S(t)\mathbf{i}_x$, where $J_S(t)$ is as shown in Fig. 6.21. Find and sketch **(a)** E_x versus t in the $z = 300$ m plane; **(b)** E_x versus z for $t = 2$ μs; and **(c)** H_y versus z for $t = 4$ μs.

P6.6. The time-variation of the electric field intensity E_x in the $z = 300$ m plane of a uniform plane wave propagating away from an infinite plane current sheet of current density $\mathbf{J}_S(t) = -J_S(t)\mathbf{i}_x$ lying in the $z = 0$ plane in free space is given by the periodic function shown in Fig. 6.22. Find and sketch **(a)** J_S versus t; **(b)** E_x versus t in $z = -600$ m plane; **(c)** E_x versus z for $t = 2$ μs; and **(d)** H_y versus z for $t = 3$ μs.

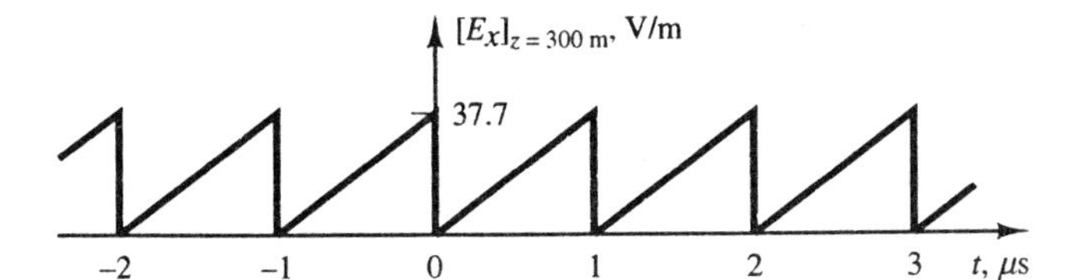

Figure 6.22. For Problem P6.6.

P6.7. The electric field of a uniform plane wave propagating in free space is given by

$$\mathbf{E} = 37.7 \cos (9\pi \times 10^8 t + 3\pi y)\ \mathbf{i}_z \text{ V/m}$$

Find **(a)** the frequency, **(b)** the wavelength, **(c)** the direction of propagation of the wave, and **(d)** the associated magnetic field vector $\mathbf{H}$.

P6.8. Given $\mathbf{J}_S = 0.1(\mathbf{i}_x - \sqrt{3}\mathbf{i}_y) \cos 3\pi \times 10^8 t$ A/m in the $z = 0$ plane in free space, find $\mathbf{E}$ and $\mathbf{H}$ for $z \gtrless 0$. Use the following three steps, which are generalizations of the solution to the electromagnetic field due to the infinite plane current sheet in the $z = 0$ plane:

1. Write the expression for $\mathbf{H}$ on the sheet and on either side of it, by noting that $[\mathbf{H}]_{z=0\pm} = \frac{1}{2}\mathbf{J}_S \times (\pm\mathbf{i}_z) = \frac{1}{2}\mathbf{J}_S \times \mathbf{i}_n$, where $\mathbf{i}_n$ is the unit vector normal to the sheet and directed toward the side of interest.
2. Extend the result of step 1 to write the expression for $\mathbf{H}$ everywhere, that is, for $z \gtrless 0$, considering the traveling wave character of the fields.
3. Write the solution for $\mathbf{E}$ everywhere by noting that (a) the amplitude of $\mathbf{E} = \eta_0 \times$ the amplitude of $\mathbf{H}$ and (b) the direction of $\mathbf{E}$, the direction of $\mathbf{H}$, and the direction of propagation constitute a right-handed orthogonal set, so that $\mathbf{E} = \eta_0 \mathbf{H} \times \mathbf{i}_n$.

P6.9. Given $\mathbf{J}_S = 0.2 \sin 6\pi \times 10^9 t\ \mathbf{i}_z$ A/m in the $y = 0$ plane in free space, find $\mathbf{E}$ and $\mathbf{H}$ for $y \gtrless 0$. Use the three steps outlined in Problem P6.8, except that the current sheet is in the $y = 0$ plane.

P6.10. For each of the following fields, determine if the polarization is right- or left-circular or right- or left-elliptical.

(a) $E_0 \cos (\omega t - \beta x)\ \mathbf{i}_y + E_0 \sin (\omega t - \beta x)\ \mathbf{i}_z$
(b) $E_0 \sin (\omega t + \beta y)\ \mathbf{i}_x + E_0 \cos (\omega t + \beta y)\ \mathbf{i}_z$

(c) $E_0 \cos(\omega t + \beta x)\,\mathbf{i}_y - 2E_0 \sin(\omega t + \beta x)\,\mathbf{i}_z$
(d) $E_0 \cos(\omega t - \beta z)\,\mathbf{i}_x - E_0 \cos(\omega t - \beta z - \pi/3)\,\mathbf{i}_y$

P6.11. Show that a linearly polarized vector can be expressed as the superposition of right- and left-circularly polarized vectors of equal amplitudes by expressing each of the following vectors in terms of right- and left-circularly polarized vectors:
(a) $E_0\mathbf{i}_x \cos(\omega t - \beta z)$
(b) $E_0(\mathbf{i}_x - 2\mathbf{i}_y) \cos(\omega t - \beta z)$

P6.12. Show that an elliptically polarized vector can be expressed as the superposition of right- and left-circularly polarized vectors of unequal amplitudes by expressing each of the following vectors in terms of right- and left-circularly polarized vectors:
(a) $E_0\mathbf{i}_x \cos(\omega t - \beta z) + 3E_0\mathbf{i}_y \sin(\omega t - \beta z)$
(b) $E_0\mathbf{i}_x \cos(\omega t - \beta z + \pi/3) + E_0\mathbf{i}_y \cos(\omega t - \beta z + \pi/6)$

P6.13. The current densities of two infinite plane parallel current sheets are given by

$$\mathbf{J}_{S1} = -J_{S0} \cos \omega t\,\mathbf{i}_x \text{ in the } z = 0 \text{ plane}$$

$$\mathbf{J}_{S2} = -kJ_{S0} \cos \omega t\,\mathbf{i}_x \text{ in the } z = \lambda/2 \text{ plane}$$

Find the electric field intensities in the three regions: **(a)** $z < 0$; **(b)** $0 < z < \lambda/2$; and **(c)** $z > \lambda/2$.

P6.14. For the array of two infinite plane current sheets of Example 6.3, assume that

$$\mathbf{J}_{S2} = -J_{S0} \sin(\omega t + \alpha)\,\mathbf{i}_x \text{ for } z = \lambda/4$$

Obtain the expression for the ratio of the amplitude of the electric field in the region $z > \lambda/4$ to the amplitude of the electric field in the region $z < 0$. Then find the value of the ratio for each of the following values of α: **(a)** 0; **(b)** $\pi/4$; and **(c)** $\pi/2$.

P6.15. The electric field intensity of a uniform plane wave is given by

$$\mathbf{E} = E_0 \cos(3\pi \times 10^9 t - 10\pi x)\,\mathbf{i}_z$$

Ignoring relativistic effects, find the Doppler shift in f for each of the following cases: **(a)** observer moving in the positive x-direction with velocity 10^3 m/s; **(b)** observer moving along the line $x = y = -z$ in the sense of increasing z with velocity 10^3 m/s; and **(c)** observer moving along the line $x = y = 2z$ in the sense of increasing z with velocity 10^3 m/s.

P6.16. Consider an observer moving in free space along the curve $x = y = z^2$ in the direction of increasing z with velocity v_0. For a point source of frequency f located at the point (0, 0, 5), obtain the expression for the Doppler shift in f as a function of z, ignoring relativistic effects. For what location(s) of the observer is the Doppler shift zero?

P6.17. Consider an observer moving on the circumference of a circle of radius a in the xy-plane and centered at the origin with an angular velocity $\omega_0\mathbf{i}_z$ rad/s in free space. Assuming the position of the observer to be $(a, 0)$ at $t = 0$, find and sketch the Doppler shift in f viewed by the observer as a function of time, for each of the following cases: **(a)** the observer is in the field of a uniform plane wave of frequency f propagating in the $+x$-direction; and **(b)** the observer is in the field of a transmitter located at the point $(-2a, 0)$. Ignore relativistic effects. Indicate clearly the points of maxima and zeros in Doppler shift and the corresponding times.

P6.18. An infinite plane sheet in the $z = 0$ plane carries a surface current of density

$$\mathbf{J}_S = -0.2 \cos 2\pi \times 10^6 t \, \mathbf{i}_x \text{ A/m}$$

The medium on either side of the sheet is characterized by $\sigma = 10^{-3}$ S/m, $\varepsilon = 6\varepsilon_0$ and $\mu = \mu_0$. Find **E** and **H** on either side of the current sheet.

P6.19. Consider an array of two infinite plane parallel current sheets of uniform densities given by

$$\mathbf{J}_{S1} = -J_{S0} \cos 2\pi \times 10^6 t \, \mathbf{i}_x \quad \text{in the } z = 0 \text{ plane}$$

$$\mathbf{J}_{S2} = -kJ_{S0} \sin 2\pi \times 10^6 t \, \mathbf{i}_x \quad \text{in the } z = d \text{ plane}$$

situated in a medium characterized by $\sigma = 10^{-3}$ S/m, $\varepsilon = 6\varepsilon_0$, and $\mu = \mu_0$. **(a)** Find the minimum value of d (> 0) and the corresponding value of k for which the fields in the region $z < 0$ are zero. **(b)** For the values of d and k found in (a), obtain the electric field intensity in the region $z > d$.

P6.20. A uniform plane wave of frequency 5×10^5 Hz propagating in a material medium has the following characteristics. (i) The fields are attenuated by the factor e^{-1} in a distance of 28.65 m. (ii) The fields undergo a change in phase by 2π in a distance of 111.2 m. (iii) The ratio of the amplitudes of the electric and magnetic field intensities at a point in the medium is 59.4. **(a)** What is the value of $\bar{\gamma}$? **(b)** What is the value of $\bar{\eta}$? **(c)** Find σ, ϵ, and μ of the medium.

P6.21. For a uniform plane wave propagating in the $+z$-direction in a material medium characterized by $\sigma = 10^{-3}$ S/m, $\varepsilon = 80\varepsilon_0$ and $\mu = \mu_0$, find the electric field intensity as a function of z and t for each of the following magnetic field intensities in the $z = 0$ plane:

(a) $0.1 \cos 2\pi \times 10^5 t \, \mathbf{i}_y$ A/m
(b) $0.1 \cos 6\pi \times 10^5 t \, \mathbf{i}_y$ A/m
(c) $0.1 \cos 2\pi \times 10^5 t \cos 4\pi \times 10^5 t \, \mathbf{i}_y$ A/m
(d) $0.1 \cos^3 2\pi \times 10^5 t \, \mathbf{i}_y$ A/m

P6.22. In Problem P6.18, assume that the region $z > 0$ is free space, whereas the region $z < 0$ is a material medium characterized by $\sigma = 10^{-3}$ S/m, $\varepsilon = 6\varepsilon_0$, and $\mu = \mu_0$. Find **E** and **H** on either side of the current sheet. (*Hint:* Make use of the complex electric and magnetic fields to satisfy the boundary conditions at $z = 0$.)

P6.23. The electric field of a uniform plane wave propagating in a perfect dielectric medium having $\mu = \mu_0$ is given by

$$\mathbf{E} = 10 \cos (6\pi \times 10^7 t - 0.8\pi y) \, \mathbf{i}_x \text{ V/m}$$

Find **(a)** the frequency; **(b)** the wavelength; **(c)** the phase velocity; **(d)** the relative permittivity of the medium; and **(e)** the associated magnetic field vector **H**.

P6.24. An infinite plane sheet lying in the $z = 0$ plane carries a surface current of density $\mathbf{J}_S = -J_S(t)\mathbf{i}_x$ A/m, where $J_S(t)$ is as shown in Fig. 6.23. The medium on either side of the current sheet is a perfect dielectric of $\varepsilon = 2.25\varepsilon_0$ and $\mu = \mu_0$. Find

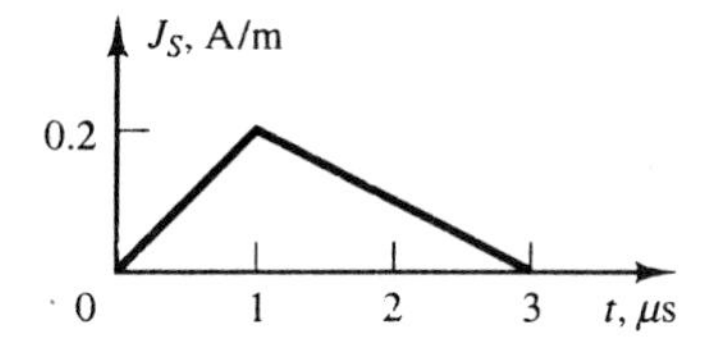

Figure 6.23. For Problem P6.24.

and sketch **(a)** E_x versus t for $z = 200$ m; **(b)** H_y versus t for $z = -300$ m; **(c)** E_x versus z for $t = 2$ μs; and **(d)** H_y versus z for $t = 3$ μs.

P6.25. For a uniform plane wave having $\mathbf{E} = E_x(z, t)\mathbf{i}_x$ and $\mathbf{H} = H_y(z, t)\mathbf{i}_y$ and propagating in the $+z$-direction in a perfect dielectric medium, the time variation of H_y in the $z = 0$ plane and the variation with z of E_x for $t = 5$ μs are shown in Fig. 6.24(a) and (b), respectively. Write the expressions for **E** and **H** if the wave is sinusoidal such that $\mathbf{E}(0, t) = E_0 \cos 3\pi \times 10^9 t\, \mathbf{i}_x$ instead of being a pulse of finite duration.

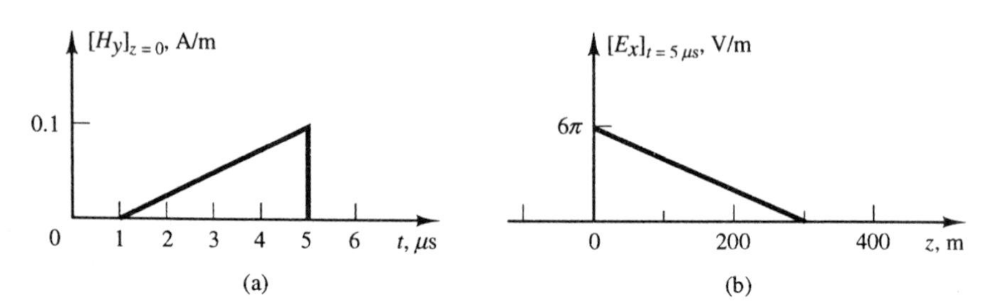

Figure 6.24. For Problem P6.25.

P6.26. An infinite plane sheet lying in the $z = 0$ plane carries a surface current of density

$$\mathbf{J}_S = -0.2 \cos 6\pi \times 10^8 t\, \mathbf{i}_y \text{ A/m}$$

The region $z > 0$ is a perfect dielectric of $\varepsilon = 2.25\varepsilon_0$ and $\mu = \mu_0$, whereas the region $z < 0$ is a perfect dielectric of $\varepsilon = 4\varepsilon_0$ and $\mu = \mu_0$. Find **E** and **H** on both sides of the sheet.

P6.27. For uniform plane wave propagation in ice ($\sigma \approx 10^{-6}$ S/m, $\varepsilon = 3\varepsilon_0$, and $\mu = \mu_0$), compute α, β, v_p, λ, and $\overline{\eta}$ for $f = 1$ MHz. What is the distance in which the fields are attenuated by the factor e^{-1}?

P6.28. Find the following: **(a)** the lowest frequency for which the thickness 2 mm of an aluminum sheet ($\sigma = 3.5 \times 10^7$ S/m) is at least three skin depths; **(b)** the minimum thickness of a copper sheet such that it is at least four skin depths thick in the frequency range 1 MHz to 1 GHz; and **(c)** the minimum conductivity of a material of thickness 3 mm for which it is at least five skin depths thick at $f = 25$ kHz.

P6.29. For uniform plane wave propagation in sea water ($\sigma = 4$ S/m, $\varepsilon = 80\varepsilon_0$, and $\mu = \mu_0$), compute α, δ, β, λ, v_p, and $\overline{\eta}$ for two frequencies: **(a)** $f = 10$ GHz; and **(b)** $f = 100$ kHz.

P6.30. For uniform plane wave propagation in a material medium, the magnetic field intensity in the $z = 0$ plane is given by

$$[\mathbf{H}]_{z=0} = 0.1 \cos^3 2\pi \times 10^8 t\, \mathbf{i}_y \text{ A/m}$$

Find $\mathbf{E}(z, t)$ for each of the following cases: **(a)** the medium is characterized by $\sigma = 0$, $\varepsilon = 9\varepsilon_0$, and $\mu = \mu_0$; **(b)** the medium is characterized by $\sigma = 10^{-3}$ S/m, $\varepsilon = 9\varepsilon_0$, and $\mu = \mu_0$; and **(c)** the medium is characterized by $\sigma = 10$ S/m, $\varepsilon = 9\varepsilon_0$, and $\mu = \mu_0$.

P6.31. Find the magnitude of the Doppler shift in f viewed by an observer moving with velocity 10 m/s along the direction of propagation of a wave propagating in sea water, for each of the following frequencies of the wave: **(a)** $f = 25$ kHz and **(b)** $f = 100$ kHz. Ignore relativistic effects.

P6.32. For each of the following electric field intensities in free space, find the instantaneous and time-average Poynting vectors:

(a) $\mathbf{E} = E_0 \cos(\omega t - \beta z)\,\mathbf{i}_x + E_0 \cos(\omega t - \beta z)\,\mathbf{i}_y$

(b) $\mathbf{E} = E_0 \cos(\omega t - \beta z)\,\mathbf{i}_x + E_0 \sin(\omega t - \beta z)\,\mathbf{i}_y$

(c) $\mathbf{E} = E_0 \cos(\omega t - \beta z)\,\mathbf{i}_x + 2E_0 \sin(\omega t - \beta z)\,\mathbf{i}_y$

P6.33. The electric and magnetic field intensities in the radiation field of an antenna located at the origin are given in spherical coordinates by

$$\mathbf{E} = E_0 \frac{\sin\theta}{r} \cos\omega(t - r\sqrt{\mu_0\varepsilon_0})\,\mathbf{i}_\theta \text{ V/m}$$

$$\mathbf{H} = \frac{E_0}{\sqrt{\mu_0/\varepsilon_0}} \frac{\sin\theta}{r} \cos\omega(t - r\sqrt{\mu_0\varepsilon_0})\,\mathbf{i}_\phi \text{ A/m}$$

Find the following: **(a)** the instantaneous Poynting vector; **(b)** the time-average Poynting vector; and **(c)** the time-average power radiated by the antenna by evaluating the surface integral of the time-average Poynting vector over a spherical surface of radius r centered at the antenna and enclosing the antenna.

P6.34. The electric and magnetic fields in a parallel-plate resonator made up of perfect conductors in the $z = 0$ and $z = l$ planes and with free space between the conductors are given by

$$\mathbf{E} = E_0 \sin\frac{\pi z}{l} \sin\frac{\pi t}{\sqrt{\mu_0\varepsilon_0}\,l}\,\mathbf{i}_x \qquad \text{for } 0 < z < l$$

$$\mathbf{H} = \frac{E_0}{\sqrt{\mu_0/\varepsilon_0}} \cos\frac{\pi z}{l} \cos\frac{\pi t}{\sqrt{\mu_0\varepsilon_0}\,l}\,\mathbf{i}_y \qquad \text{for } 0 < z < l$$

Find the following: **(a)** the electric stored energy in the resonator per unit area of the plates at an instant of time when the magnetic field is zero; **(b)** the magnetic stored energy in the resonator per unit area of the plates at an instant of time when the electric field is zero; and **(c)** the total stored energy in the resonator per unit area of the plates at an arbitrary instant of time.

P6.35. The electric field of a uniform plane wave propagating in a nonmagnetic ($\mu = \mu_0$) material medium is given by

$$\mathbf{E} = E_0\, e^{-z} \cos(2\pi \times 10^6 t - 2z)\,\mathbf{i}_x \text{ V/m}$$

Find the following: **(a)** the time-average power flow per unit area normal to the z-direction and **(b)** the time-average power dissipated in the volume bounded by the planes $x = 0$, $x = 1$, $y = 0$, $y = 1$, $z = 0$, and $z = 1$.

P6.36. Consider the general case of complex electric and magnetic fields

$$\bar{\mathbf{E}}(x, y, z) = \mathbf{E}_0(x, y, z)e^{j\phi(x,y,z)}$$
$$\bar{\mathbf{H}}(x, y, z) = \mathbf{H}_0(x, y, z)e^{j\theta(x,y,z)}$$

Show that the time-average Poynting vector is given by

$$\langle\mathbf{P}\rangle = \text{Re}\left(\frac{1}{2}\bar{\mathbf{E}} \times \bar{\mathbf{H}}^*\right)$$

P6.37. Region 1 ($z < 0$) is free space and region 2 ($z > 0$) is a material medium characterized by $\sigma = 10^{-3}$ S/m, $\varepsilon = 12\varepsilon_0$, and $\mu = \mu_0$. For a uniform plane wave having the electric field

$$\mathbf{E}_i = E_0 \cos(3\pi \times 10^6 t - 0.01\pi z)\,\mathbf{i}_x \text{ V/m}$$

incident on the interface $z = 0$ from region 1, obtain the expressions for the reflected and transmitted wave electric and magnetic fields.

P6.38. Repeat Problem P6.37 for the incident wave electric field given by

$$E_i = E_0 \cos^3 (3\pi \times 10^6 t - 0.01\pi z)\, \mathbf{i}_x \text{ V/m}$$

P6.39. Consider normal incidence of a uniform plane wave of frequency 100 kHz on the plane interface between free space ($z < 0$) and sea water ($z > 0$). Determine the following: **(a)** the amplitude of the incident wave electric field for which the transmitted wave electric field at $z = 1$ m is 1 mV/m and **(b)** the depth at which the transmitted wave electric field is 1 μV/m for the amplitude of the incident wave electric field found in (a).

P6.40. In Fig. 6.25, medium 3 extends to infinity so that no reflected ($-$) wave exists in the medium. For a uniform plane wave having the electric field

$$\mathbf{E}_i = E_0 \cos (3\pi \times 10^8 t - \pi z)\, \mathbf{i}_x \text{ V/m}$$

incident onto the interface $z = 0$, obtain the expressions for the phasor electric and magnetic field components in all three media.

P6.41. A uniform plane wave propagating in the $+z$-direction and having the electric field $\mathbf{E}_i = E_{xi}(t)\mathbf{i}_x$, where $E_{xi}(t)$ in the $z = 0$ plane is as shown in Fig. 6.26, is incident normally from free space ($z < 0$) onto a nonmagnetic ($\mu = \mu_0$), perfect dielectric ($z > 0$) of permittivity $4\varepsilon_0$. Find and sketch the following: **(a)** E_x versus z for $t = 1$ μs and **(b)** H_y versus z for $t = 1$ μs.

P6.42. The region $z < 0$ is a perfect dielectric, whereas the region $z > 0$ is a perfect conductor, as shown in Fig. 6.27. For a uniform plane wave having the electric and magnetic fields

$$\mathbf{E}_i = E_0 \cos (\omega t - \beta z)\, \mathbf{i}_x$$

$$\mathbf{H}_i = \frac{E_0}{\eta} \cos (\omega t - \beta z)\, \mathbf{i}_y$$

where $\beta = \omega\sqrt{\mu\varepsilon}$ and $\eta = \sqrt{\mu/\varepsilon}$, obtain the expressions for the reflected wave electric and magnetic fields and hence the expressions for the total (incident + reflected) electric and magnetic fields in the dielectric, and the current density on the surface of the perfect conductor.

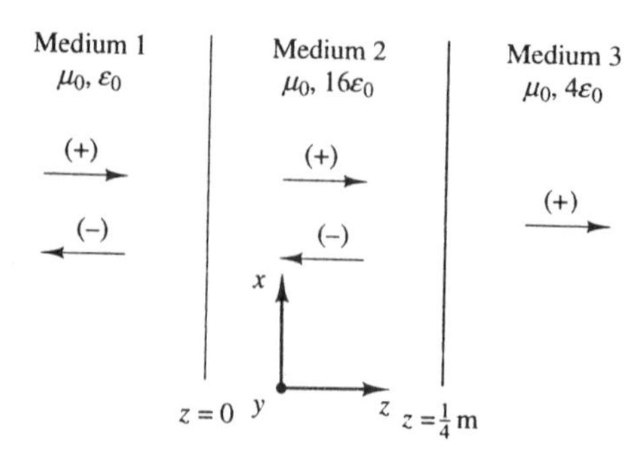

Figure 6.25. For Problem P6.40.

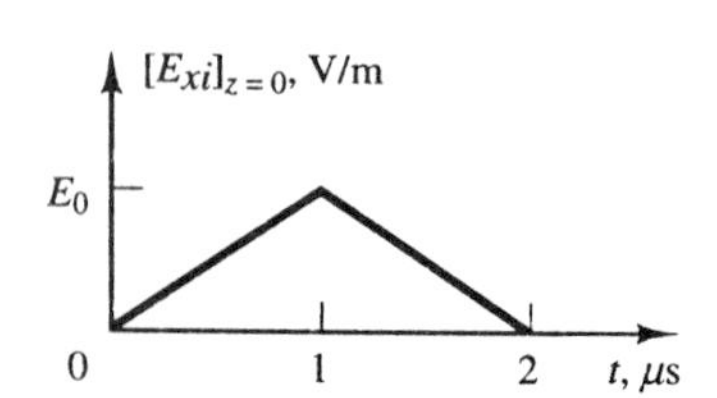

Figure 6.26. For Problem P6.41.

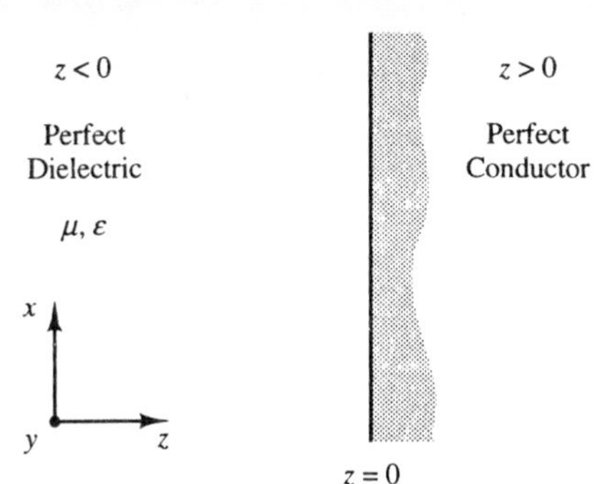

Figure 6.27. For Problem P6.42.

PC EXERCISES

PC6.1. Consider an array of two infinite plane current sheets with current densities given by

$$\mathbf{J}_{S1} = -J_{S0} \cos \omega t \; \mathbf{i}_x \qquad \text{in the } z = 0 \text{ plane}$$

$$\mathbf{J}_{S2} = -kJ_{S0} \cos (\omega t - \phi) \; \mathbf{i}_x \qquad \text{in the } z = a\lambda \text{ plane}$$

in free space where k and a are positive. Write a program for computing the ratio of the amplitude of the electric field in the region $z > a\lambda$ to the amplitude of the electric field in the region $z < 0$, using as input the values of k, ϕ, and a.

PC6.2. Consider the computation of material parameters from propagation parameters. Write a program for computing σ, ε_r, and μ_r given α, β, $|\bar{\eta}|$, and f and making use of the condition that the phase angle of $\bar{\gamma}\bar{\eta}$ must be $\pi/2$ for μ_r to be real.

PC6.3. Consider a uniform plane wave having the magnetic field $\mathbf{H} = 1 \cos^n 2\pi f_0 t \; \mathbf{i}_y$ A/m at $z = 0$ and propagating in the positive z-direction in a material medium. Write a program to compute for a specified value of $z > 0$ the root-mean-square values of the electric and magnetic field intensities and the time-average power density associated with the wave. The input quantities are to be the frequency f_0, the material parameters σ, ε_r, and μ_r, and an odd-integer value for n.

PC6.4. Consider normal incidence of a uniform plane wave from medium 1 ($z < 0$) characterized by σ_1, ε_{r1}, and μ_{r1} onto medium 2 ($z > 0$) characterized by σ_2, ε_{r2}, and μ_{r2}. Assuming the phasor electric field intensity of the incident wave at the interface $z = 0$ to be $1\underline{/0^\circ}$ V/m, write a program that computes the phasor electric and magnetic field intensities for the reflected and transmitted waves at $z = -1$ m and $z = 1$ m, respectively, for given values of the two sets of material parameters and the frequency f of the waves.

7

Transmission Lines 1. Time-Domain Analysis

In Chapter 6 we studied the principles of uniform plane wave propagation first in free space and then in material media. In both cases we were concerned with propagation in unbounded media. In this and the next two chapters we consider guided wave propagation, that is, propagation of waves between boundaries. The boundaries are generally provided by conductors, whereas the media between the boundaries are generally dielectrics. There are two kinds of waveguiding systems: transmission lines and waveguides. A transmission line consists of two or more parallel conductors, whereas a waveguide is generally made up of one conductor. We devote this and the next chapter to transmission lines and their analysis and consider waveguides in Chapter 9.

We introduce the transmission line by considering a uniform plane wave and placing two parallel plane, perfect conductors such that the fields remain unaltered by satisfying the boundary conditions on the perfect conductor surfaces. The wave is then guided between and parallel to the conductors, thus leading to the parallel-plate line. We shall learn to represent a line by the *distributed* parameter equivalent circuit and discuss wave propagation on the line in terms of voltage and current. We introduce the circuit parameters for the parallel-plate line and then discuss several techniques for the computation of line parameters. We then turn our attention to time-domain analysis of transmission-line systems, which is our primary goal in this chapter.

7.1 TRANSMISSION-LINE EQUATIONS AND SOLUTION

Parallel-plate line

In Section 6.4 we learned that the tangential component of the electric field intensity and the normal component of the magnetic field intensity are zero on a perfect conductor surface. Let us now consider the uniform plane electromagnetic wave propagating in the z-direction and having an x-component only of the electric field and a y-component only of the magnetic field, that is,

$$\mathbf{E} = E_x(z, t)\mathbf{i}_x$$

$$\mathbf{H} = H_y(z, t)\mathbf{i}_y$$

and place perfectly conducting sheets in two planes $x = 0$ and $x = d$, as shown in Fig. 7.1. Since the electric field is completely normal and the magnetic field is completely tangential to the sheets, the two boundary conditions just referred to are satisfied, and hence the wave will simply propagate, as though the sheets were not present, being guided by the sheets. We then have a simple case of transmission line, namely, the parallel-plate transmission line. We shall assume the medium between the plates to be a perfect dielectric so that the waves propagate without attenuation and hence the line is lossless.

According to the remaining two boundary conditions, there must be charges and currents on the conductors. The charge densities on the two plates are

$$[\rho_S]_{x=0} = [\mathbf{i}_n \cdot \mathbf{D}]_{x=0} = \mathbf{i}_x \cdot \varepsilon E_x \mathbf{i}_x = \varepsilon E_x \tag{7.1a}$$

$$[\rho_S]_{x=d} = [\mathbf{i}_n \cdot \mathbf{D}]_{x=d} = -\mathbf{i}_x \cdot \varepsilon E_x \mathbf{i}_x = -\varepsilon E_x \tag{7.1b}$$

where ε is the permittivity of the medium between the two plates. The current densities on the two plates are

$$[\mathbf{J}_S]_{x=0} = [\mathbf{i}_n \times \mathbf{H}]_{x=0} = \mathbf{i}_x \times H_y \mathbf{i}_y = H_y \mathbf{i}_z \tag{7.2a}$$

$$[\mathbf{J}_S]_{x=d} = [\mathbf{i}_n \times \mathbf{H}]_{x=d} = -\mathbf{i}_x \times H_y \mathbf{i}_y = -H_y \mathbf{i}_z \tag{7.2b}$$

In (7.1a)–(7.2b) it is understood that the charge and current densities are functions of z and t as E_x and H_y are. Thus the wave propagation along the transmission line is supported by charges and currents on the plates, varying with time and distance along the line, as shown in Fig. 7.1.

Let us now consider finitely sized plates having width w in the y-direction, as shown in Fig. 7.2(a), and neglect fringing of the fields at the edges or assume that the structure is part of a much larger-sized configuration. By considering a constant z-plane, that is, a plane *transverse* to the direction of propagation of the wave, as shown in Fig. 7.2(b), we can find the voltage between the two conductors in terms of the line integral of the electric field intensity evaluated along any path in that plane between the two conductors. Since the electric field is directed

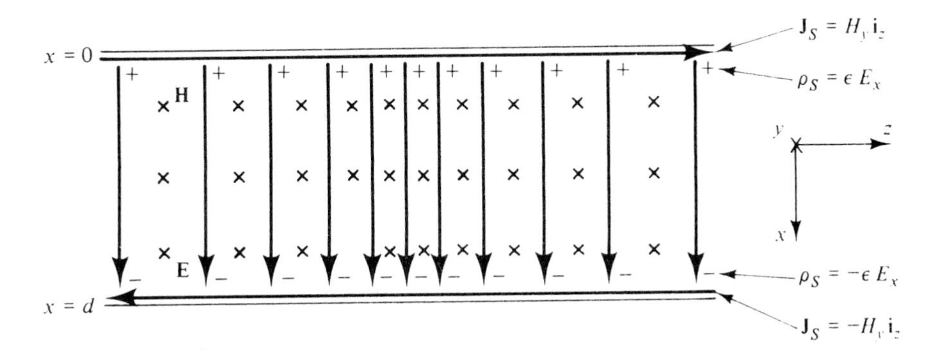

Figure 7.1. Uniform plane electromagnetic wave propagating between two perfectly conducting sheets, supported by charges and currents on the sheets.

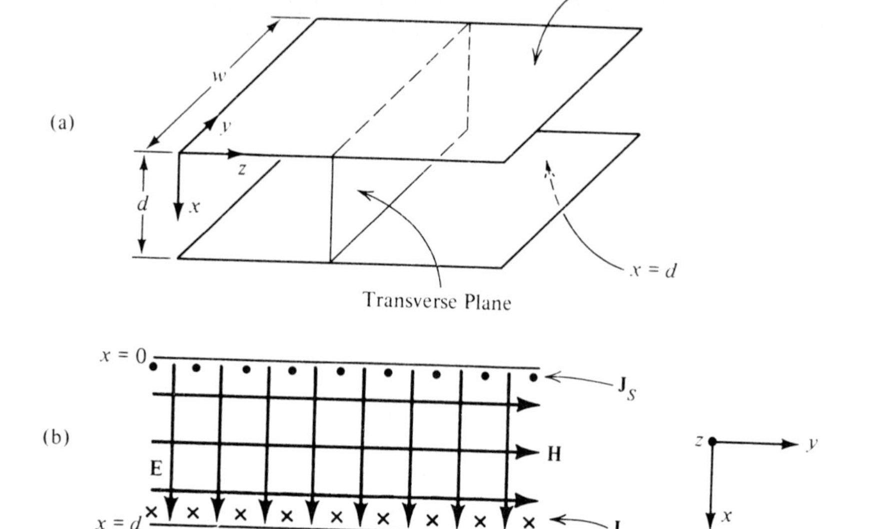

Figure 7.2. (a) Parallel-plate transmission line. (b) Transverse plane of the parallel-plate transmission line.

in the x-direction and since it is uniform in that plane, this voltage is given by

$$V(z, t) = \int_{x=0}^{d} E_x(z, t)\, dx = E_x(z, t) \int_{x=0}^{d} dx = dE_x(z, t) \tag{7.3}$$

Thus each transverse plane is characterized by a voltage between the two conductors which is related simply to the electric field, as given by (7.3). Each transverse plane is also characterized by a current I flowing in the positive z-direction on the upper conductor and in the negative z-direction on the lower conductor. From Fig. 7.2(b) we can see that this current is given by

$$\begin{aligned} I(z, t) &= \int_{y=0}^{w} J_S(z, t)\, dy = \int_{y=0}^{w} H_y(z, t)\, dy = H_y(z, t) \int_{y=0}^{w} dy \\ &= wH_y(z, t) \end{aligned} \tag{7.4}$$

since H_y is uniform in the cross-sectional plane. Thus the current crossing a given transverse plane is related simply to the magnetic field in that plane, as given by (7.4).

Proceeding further, we can find the power flow down the line by evaluating the surface integral of the Poynting vector over a given transverse plane. Thus

$$\begin{aligned} P(z, t) &= \int_{\substack{\text{transverse} \\ \text{plane}}} (\mathbf{E} \times \mathbf{H}) \cdot d\mathbf{S} \\ &= \int_{x=0}^{d} \int_{y=0}^{w} E_x(z, t) H_y(z, t) \mathbf{i}_z \cdot dx\, dy\, \mathbf{i}_z \end{aligned}$$

$$= \int_{x=0}^{d} \int_{y=0}^{w} \frac{V(z, t)}{d} \frac{I(z, t)}{w} \, dx \, dy$$

$$= V(z, t) I(z, t) \tag{7.5}$$

which is the familiar relationship employed in circuit theory.

Transmission-line equations

We now recall from Section 6.3 that E_x and H_y satisfy the two differential equations

$$\frac{\partial E_x}{\partial z} = -\frac{\partial B_y}{\partial t} = -\mu \frac{\partial Hy}{\partial t} \tag{7.6a}$$

$$\frac{\partial H_y}{\partial z} = -\sigma E_x - \varepsilon \frac{\partial E_x}{\partial t} = -\varepsilon \frac{\partial E_x}{\partial t} \tag{7.6b}$$

where we have set $\sigma = 0$ in view of the perfect dielectric medium. From (7.3) and (7.4), however, we have

$$E_x = \frac{V}{d} \tag{7.7a}$$

$$H_y = \frac{I}{w} \tag{7.7b}$$

Substituting for E_x and H_y in (7.6a) and (7.6b) from (7.7a) and (7.7b), respectively, we now obtain two differential equations for voltage and current along the line as

$$\frac{\partial}{\partial z}\left(\frac{V}{d}\right) = -\mu \frac{\partial}{\partial t}\left(\frac{I}{w}\right)$$

$$\frac{\partial}{\partial z}\left(\frac{I}{w}\right) = -\varepsilon \frac{\partial}{\partial t}\left(\frac{V}{d}\right)$$

or

$$\boxed{\frac{\partial V}{\partial z} = -\left(\frac{\mu d}{w}\right) \frac{\partial I}{\partial t}} \tag{7.8a}$$

$$\boxed{\frac{\partial I}{\partial z} = -\left(\frac{\varepsilon w}{d}\right) \frac{\partial V}{\partial t}} \tag{7.8b}$$

These equations are known as the *transmission-line equations*. They characterize the wave propagation along the line in terms of line voltage and line current instead of in terms of the fields.

We now denote two quantities familiarly known as the *circuit parameters*, the inductance and the capacitance per unit length of the transmission line in the z-direction by the symbols $\mathscr{L}$ and $\mathscr{C}$, respectively. We observe from Section 5.3 that the inductance per unit length, having the units henrys per meter (H/m), is the ratio of the magnetic flux per unit length at any value of z to the line current at that value of z. Noting from Fig. 7.2 that the cross-sectional area normal to the magnetic field lines and per unit length in the z-direction is $(d)(1)$ or d, we find the magnetic flux per unit length to be $B_y d$ or $\mu H_y d$. Since the line current is $H_y w$, we then have

$$\boxed{\mathcal{L} = \frac{\mu H_y d}{H_y w} = \frac{\mu d}{w}} \tag{7.9}$$

We also observe that the capacitance per unit length, having the units farads per meter (F/m), is the ratio of the magnitude of the charge per unit length on either plate at any value of z to the line voltage at that value of z. Noting from Fig. 7.2 that the cross-sectional area normal to the electric field lines and per unit length in the z-direction is $(w)(1)$ or w, we find the charge per unit length to be $\rho_S w$ or $\varepsilon E_x w$. Since the line voltage is $E_x d$, we then have

$$\boxed{\mathcal{C} = \frac{\varepsilon E_x w}{E_x d} = \frac{\varepsilon w}{d}} \tag{7.10}$$

We note that $\mathcal{L}$ and $\mathcal{C}$ are purely dependent on the dimensions of the line and are independent of E_x and H_y. We further note that

$$\boxed{\mathcal{L}\mathcal{C} = \mu\varepsilon} \tag{7.11}$$

so that only one of the two parameters $\mathcal{L}$ and $\mathcal{C}$ is independent and the other can be obtained from the knowledge of ε and μ. The results given by (7.9) and (7.10) are the same as those listed in Table 5.2 for the parallel-plane conductor arrangement, whereas (7.11) is the same as given by (5.37).

Replacing now the quantities in parentheses in (7.8a) and (7.8b) by $\mathcal{L}$ and $\mathcal{C}$, respectively, we obtain the transmission-line equations in terms of these parameters as

$$\frac{\partial V}{\partial z} = -\mathcal{L}\frac{\partial I}{\partial t} \tag{7.12a}$$

$$\frac{\partial I}{\partial z} = -\mathcal{C}\frac{\partial V}{\partial t} \tag{7.12b}$$

These equations permit us to discuss wave propagation along the line in terms of circuit quantities instead of in terms of field quantities. It should, however, not be forgotten that the actual phenomenon is one of electromagnetic waves guided by the conductors of the line.

Distributed equivalent circuit

It is customary to represent a transmission line by means of its circuit equivalent, derived from the transmission-line equations (7.12a) and (7.12b). To do this, let us consider a section of infinitesimal length Δz along the line between z and $z + \Delta z$. From (7.12a), we then have

$$\lim_{\Delta z \to 0} \frac{V(z + \Delta z, t) - V(z, t)}{\Delta z} = -\mathcal{L}\frac{\partial I(z, t)}{\partial t}$$

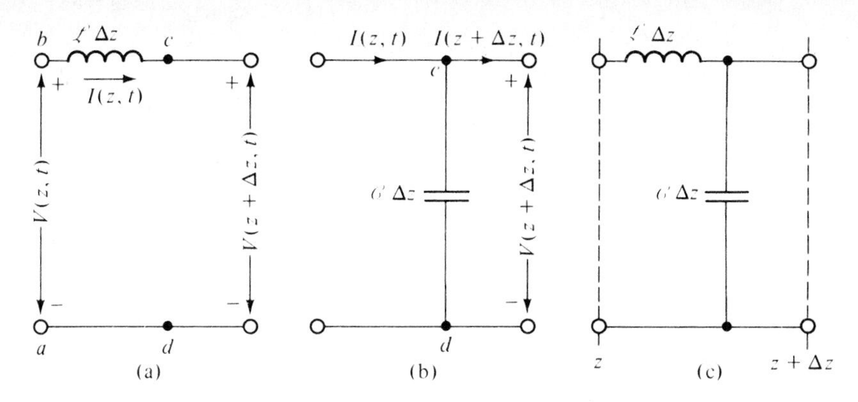

Figure 7.3. Development of circuit equivalent for an infinitesimal length Δz of a transmission line.

or, for $\Delta z \to 0$,

$$V(z + \Delta z, t) - V(z, t) = -\mathcal{L}\, \Delta z \frac{\partial I(z, t)}{\partial t} \tag{7.13a}$$

This equation can be represented by the circuit equivalent shown in Fig. 7.3(a) since it satisfies Kirchhoff's voltage law written around the loop *abcda*. Similarly, from (7.12b), we have

$$\lim_{\Delta z \to 0} \frac{I(z + \Delta z, t) - I(z, t)}{\Delta z} = \lim_{\Delta z \to 0} \left[-\mathcal{C} \frac{\partial V(z + \Delta z, t)}{\partial t} \right]$$

or, for $\Delta z \to 0$,

$$I(z + \Delta z, t) - I(z, t) = -\mathcal{C}\, \Delta z \frac{\partial V(z + \Delta z, t)}{\partial t} \tag{7.13b}$$

This equation can be represented by the circuit equivalent shown in Fig. 7.3(b) since it satisfies Kirchhoff's current law written for node c. Combining the two equations, we then obtain the equivalent circuit shown in Fig. 7.3(c) for a section Δz of the line. It then follows that the circuit representation for a portion of length l of the line consists of an infinite number of such sections in cascade, as shown in Fig. 7.4. Such a circuit is known as a *distributed circuit* as opposed to the *lumped*

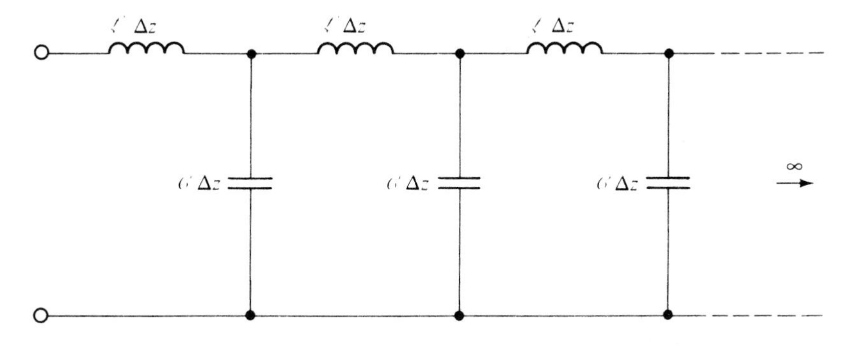

Figure 7.4. Distributed circuit representation of a transmission line.

circuits that are familiar in circuit theory. The distributed circuit notion arises from the fact that the inductance and capacitance are distributed uniformly and overlappingly along the line.

A more physical interpretation of the distributed circuit concept follows from energy considerations. We know that the uniform plane wave propagation between the conductors of the line is characterized by energy storage in the electric and magnetic fields. If we consider a section Δz of the line, the energy stored in the electric field in this section is given by

$$\begin{aligned} W_e &= \frac{1}{2}\varepsilon E_x^2(\text{volume}) = \frac{1}{2}\varepsilon E_x^2(dw\,\Delta z) \\ &= \frac{1}{2}\frac{\varepsilon w}{d}(E_x d)^2\,\Delta z = \frac{1}{2}\mathscr{C}\,\Delta z\,V^2 \end{aligned} \tag{7.14a}$$

The energy stored in the magnetic field in that section is given by

$$\begin{aligned} W_m &= \frac{1}{2}\mu H_y^2(\text{volume}) = \frac{1}{2}\mu H_y^2(dw\,\Delta z) \\ &= \frac{1}{2}\frac{\mu d}{w}(H_y w)^2\,\Delta z = \frac{1}{2}\mathscr{L}\,\Delta z\,I^2 \end{aligned} \tag{7.14b}$$

Thus we note that $\mathscr{L}$ and $\mathscr{C}$ are elements associated with energy storage in the magnetic field and energy storage in the electric field, respectively, for a given infinitesimal section of the line. Since these phenomena occur continuously and since they overlap, the inductance and capacitance must be distributed uniformly and overlappingly along the line.

TEM waves

Thus far we have introduced the transmission-line equations and the distributed circuit concept by considering the parallel-plate line in which the waves are uniform plane waves. In the general case of a line having conductors with arbitrary cross sections, the fields consist of both x- and y-components and are dependent on x- and y-coordinates in addition to the z-coordinate. Thus the fields between the conductors are given by

$$\mathbf{E} = E_x(x, y, z, t)\mathbf{i}_x + \mathbf{E}_y(x, y, z, t)\mathbf{i}_y$$

$$\mathbf{H} = H_x(x, y, z, t)\mathbf{i}_x + \mathbf{H}_y(x, y, z, t)\mathbf{i}_y$$

These fields are no longer uniform in x and y but are directed entirely transverse to the direction of propagation, that is, the z-axis, which is the axis of the transmission line. Hence they are known as *transverse electromagnetic waves*, or *TEM waves*. The uniform plane waves are simply a special case of the transverse electromagnetic waves. The transmission-line equations (7.12a) and (7.12b) and the distributed equivalent circuit of Fig. 7.4 hold for all transmission lines made of perfect conductors and perfect dielectric, that is, for all lossless transmission lines. The quantities that differ from one line to another are the line parameters $\mathscr{L}$ and $\mathscr{C}$, which depend on the geometry of the line. Since there is no z-component of $\mathbf{H}$, the electric field distribution in any given transverse plane at any given instant of time is the same as the static electric field distribution resulting from the application of a potential difference between the conductors equal to the line voltage in that plane at that instant of time. Similarly, since there is no z-component of **E**, the magnetic field distribution in any given transverse plane at any given instant of time is the same as the static magnetic field distribution resulting from current flow on the conductors equal to the line current in that plane at that instant of time. Thus, the values of $\mathscr{L}$ and $\mathscr{C}$ are the same as those obtainable from static field considerations.

General solution

Before we consider several common types of lines, we shall show that the relation (7.11) is valid in general by obtaining the general solution for the transmission-line equations (7.12a) and (7.12b). To do this, we note their analogy with the field equations (6.7a) and (6.7b) in Section 6.1, as follows:

$$\frac{\partial E_x}{\partial z} = -\mu_0 \frac{\partial H_y}{\partial t} \longleftrightarrow \frac{\partial V}{\partial z} = -\mathscr{L}\frac{\partial I}{\partial t}$$

$$\frac{\partial H_y}{\partial z} = -\varepsilon_0 \frac{\partial E_x}{\partial t} \longleftrightarrow \frac{\partial I}{\partial z} = -\mathscr{C}\frac{\partial V}{\partial t}$$

The solutions to (7.12a) and (7.12b) can therefore be written by letting

$$E_x \longrightarrow V$$

$$H_y \longrightarrow I$$

$$\mu_0 \longrightarrow \mathscr{L}$$

$$\varepsilon_0 \longrightarrow \mathscr{C}$$

in the solutions (6.13) and (6.14) to the field equations. Thus we obtain

$$V(z, t) = Af(t - z\sqrt{\mathscr{L}\mathscr{C}}) + Bg(t + z\sqrt{\mathscr{L}\mathscr{C}}) \tag{7.15a}$$

$$I(z, t) = \frac{1}{\sqrt{\mathscr{L}/\mathscr{C}}}[Af(t - z\sqrt{\mathscr{L}\mathscr{C}}) - Bg(t + z\sqrt{\mathscr{L}\mathscr{C}})] \tag{7.15b}$$

These solutions represent voltage and current traveling waves propagating along the $+z$- and $-z$-directions with velocity

$$v_p = \frac{1}{\sqrt{\mathscr{L}\mathscr{C}}} \tag{7.16}$$

in view of the arguments $(t \mp z\sqrt{\mathscr{L}\mathscr{C}})$ for the functions f and g. We however know that the velocity of propagation in terms of the dielectric parameters is given by

$$v_p = \frac{1}{\sqrt{\mu\varepsilon}} \tag{7.17}$$

Therefore it follows that

$$\mathscr{L}\mathscr{C} = \mu\varepsilon \tag{7.18}$$

Characteristic impedance

We now define the *characteristic impedance* of the line to be

$$\boxed{Z_0 = \sqrt{\frac{\mathcal{L}}{\mathcal{C}}}} \tag{7.19}$$

so that (7.15a) and (7.15b) become

$$V(z, t) = Af\left(t - \frac{z}{v_p}\right) + Bg\left(t + \frac{z}{v_p}\right) \tag{7.20a}$$

$$I(z, t) = \frac{1}{Z_0}\left[Af\left(t - \frac{z}{v_p}\right) - Bg\left(t + \frac{z}{v_p}\right)\right] \tag{7.20b}$$

where we have also substituted v_p for $1/\sqrt{\mathcal{L}\mathcal{C}}$. From (7.20a) and (7.20b), it can be seen that *the characteristic impedance is the ratio of the voltage to current in the* (+) *wave or the negative of the same ratio for the* (−) *wave*. It is analogous to the intrinsic impedance of the dielectric medium but not necessarily equal to it. For example, for the parallel-plate line,

$$Z_0 = \sqrt{\frac{\mu d}{w}\bigg/\frac{\varepsilon w}{d}}$$

$$= \eta\frac{d}{w} \tag{7.21}$$

is not equal to η unless d/w is equal to 1. In fact for d/w equal to 1, (7.21) is strictly not valid because fringing of the fields cannot be neglected. Note also that the characteristic impedance of a lossless line is a purely real quantity. We shall find in Section 8.6 that for a lossy line, the characteristic impedance is complex just as the intrinsic impedance of a lossy medium is complex.

K7.1. Parallel-plate line; Transmission-line equations; Circuit parameters; Distributed equivalent circuit; TEM waves; Characteristic impedance; Velocity of propagation.

D7.1. A parallel-plate transmission line is made up of perfect conductors of width $w = 0.2$ m and separation $d = 0.01$ m. The medium between the plates is a dielectric of $\varepsilon = 2.25\varepsilon_0$ and $\mu = \mu_0$. For a uniform plane wave propagating down the line, find the power crossing a given transverse plane for each of the following cases at a given time in that plane: **(a)** the electric field between the plates is 300π V/m; **(b)** the magnetic field between the plates is 7.5 A/m; **(c)** the voltage across the plates is 4π V; and **(d)** the current along the plates is 0.5 A.
Ans. **(a)** 2.25π W; **(b)** 9π W; **(c)** 4π W; **(d)** π W

7.2 DETERMINATION OF LINE PARAMETERS

Equations (7.20a) and (7.20b) are the general solutions for the voltage and current along a lossless line in terms of v_p and Z_0, the parameters that characterize the propagation along the line. While v_p is dependent on the dielectric as given by

(7.17), Z_0 is dependent on the dielectric as well as the geometry associated with the line, in view of (7.19). Combining (7.18) and (7.19), we note that

$$Z_0 = \frac{\sqrt{\mu\varepsilon}}{\mathscr{C}} \tag{7.22}$$

Thus the determination of Z_0 for a given line involving a given homogeneous dielectric medium requires simply the determination of $\mathscr{C}$ of the line and then using (7.22). Since the dielectrics of common transmission lines are generally nonmagnetic, we can further express the propagation parameters in the manner

$$Z_0 = \frac{\sqrt{\varepsilon_r}}{c\mathscr{C}} \tag{7.23a}$$

$$v_p = \frac{c}{\sqrt{\varepsilon_r}} \tag{7.23a}$$

where ε_r is the relative permittivity of the dielectric and $c = 1/\sqrt{\mu_0\varepsilon_0}$ is the velocity of light in free space.

Microstrip line

If the cross section of a transmission line involves more than one dielectric, the situation corresponds to inhomogeneity. An example of this type of line is the microstrip line, used extensively in microwave integrated circuitry and digital systems. The basic microstrip line consists of a high-permittivity substrate material with a conductor strip applied to one side and a conducting ground plane applied to the other side, as shown by the cross-sectional view in Fig. 7.5(a). The approximate electric field distribution is shown in Fig. 7.5(b). Since it is not possible to satisfy the boundary condition of equal phase velocities parallel to the air-dielectric interface with pure TEM waves, the situation for the microstrip line does not correspond exactly to TEM wave propagation, as is the case with any other line involving multiple dielectrics.

The determination of Z_0 and v_p for the case of a line with multiple dielectrics involves a modified procedure, assuming that the inhomogeneity has no effect on $\mathscr{L}$ and the propagation is TEM. Thus if $\mathscr{C}_0$ is the capacitance per unit length of the line with all the dielectrics replaced by free space and $\mathscr{C}$ is the ca-

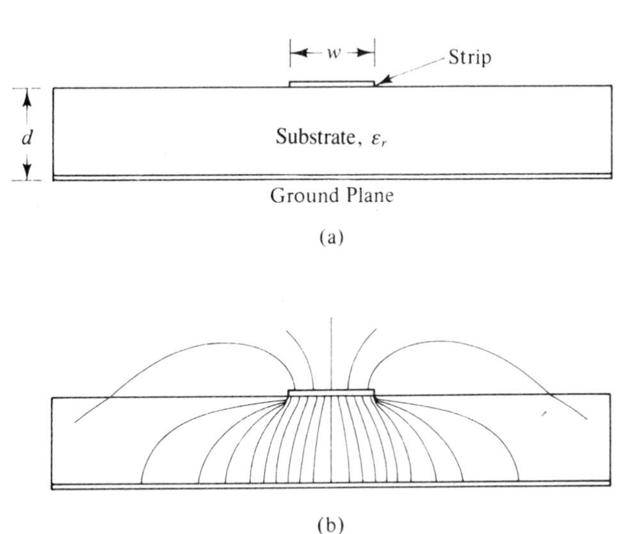

Figure 7.5. (a) Transverse cross-sectional view of a microstrip line. (b) Approximate electric field distribution in the transverse plane.

pacitance per unit length of the line with the dielectrics in place and computed from static field considerations, we can write

$$Z_0 = \sqrt{\frac{\mathcal{L}}{\mathcal{C}}} = \sqrt{\frac{\mathcal{L}\mathcal{C}_0}{\mathcal{C}\mathcal{C}_0}} = \frac{1}{c\sqrt{\mathcal{C}\mathcal{C}_0}} \tag{7.24a}$$

$$v_p = \frac{1}{\sqrt{\mathcal{L}\mathcal{C}}} = \frac{1}{\sqrt{\mathcal{L}\mathcal{C}_0}}\sqrt{\frac{\mathcal{C}_0}{\mathcal{C}}} = c\sqrt{\frac{\mathcal{C}_0}{\mathcal{C}}} \tag{7.24b}$$

where we have assumed nonmagnetic dielectrics. To express (7.24a) and (7.24b) in the form of (7.23a) and (7.23b), respectively, we define an effective relative permittivity $\varepsilon_{r_{\text{eff}}} = \mathcal{C}/\mathcal{C}_0$ so that

$$Z_0 = \frac{\sqrt{\varepsilon_{r_{\text{eff}}}}}{c\mathcal{C}} \tag{7.25a}$$

$$v_p = \frac{c}{\sqrt{\varepsilon_{r_{\text{eff}}}}} \tag{7.25b}$$

Thus the determination of Z_0 and v_p requires the knowledge of both $\mathcal{C}$ and $\mathcal{C}_0$.

The techniques for finding $\mathcal{C}$ (and $\mathcal{C}_0$) and hence the propagation parameters can be broadly divided into three categories: (1) analytical, (2) numerical, and (3) graphical. We discuss these techniques next.

A. Analytical Techniques

Several common types of lines

The analytical techniques are based on the closed-form solution of Laplace's equation subject to the boundary conditions or the equivalent of such a solution. We have already discussed these techniques in Sections 5.1 and 5.3 for several configurations. Hence without further discussion, we shall simply list in Table 7.1 the expressions for Z_0 for some common types of lines, shown by cross-sectional views in Fig. 7.6. Note that in Table 7.1, $\eta = \sqrt{\mu/\varepsilon}$ is the intrinsic impedance of the dielectric medium associated with the line.

B. Numerical Techniques

When a closed-form solution is not possible or when the approximation permitting a closed form solution breaks down, numerical techniques can be employed. We shall consider two examples to illustrate the application of two such techniques.

TABLE 7.1 EXPRESSIONS FOR CHARACTERISTIC IMPEDANCE FOR THE LINES OF FIG. 7.6

Description	Figure	Z_0
Coaxial cable	7.6(a)	$\frac{\eta}{2\pi}\ln\frac{b}{a}$
Parallel-wire line	7.6(b)	$\frac{\eta}{\pi}\cosh^{-1}\frac{d}{a}$
Single wire above ground plane	7.6(c)	$\frac{\eta}{2\pi}\cosh^{-1}\frac{h}{a}$
Shielded parallel-wire line	7.6(d)	$\frac{\eta}{\pi}\ln\frac{d(b^2 - d^2/4)}{a(b^2 + d^2/4)}$

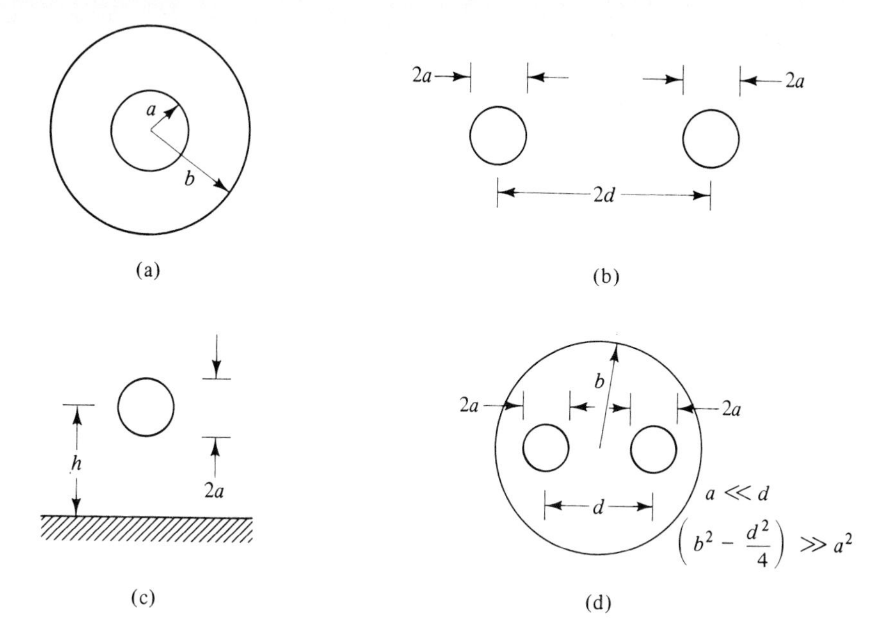

Figure 7.6. Cross sections of some common types of transmission lines.

Example 7.1

Parallel-strip line

The parallel-strip line is the same as the parallel-plate line (see Fig. 7.2) without the imposition of the approximation $d/w \ll 1$, such that fringing of fields cannot be neglected. We wish to find the capacitance per unit length and hence the characteristic impedance of the parallel-strip line embedded in a homogeneous medium (which we shall assume here to be free space) for the case of $d = w$, by using the method of moments.

The procedure for the application of method of moments to find the capacitance per unit length of a parallel-strip line is similar to that used for finding the capacitance of a parallel-plate capacitor in Section 5.2. Thus let us consider the cross-sectional view of the parallel-strip line and divide each conductor into $2n$ substrips, as shown in Fig. 7.7 for $n = 3$, and assume the charge density in each substrip to be uniform. From symmetry consideration, we can apply a potential of 1 V to one of the conductors and -1 V to the other conductor. Also from symmetry considerations, there are only n $(=3)$ unknown charge densities to be determined, namely, the charge densities associated with the substrips in one half of one of the conductors. Thus we need to write a set of n $(=3)$ independent equations for the n $(=3)$

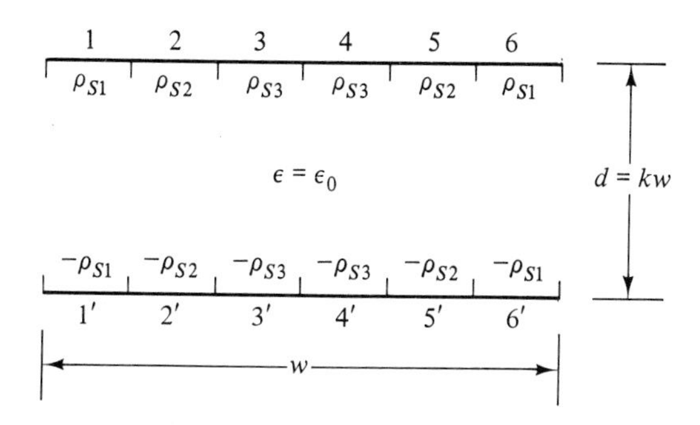

Figure 7.7. Division of the conductors of a parallel-strip line into substrips.

unknown charge densities. To do this, we consider pairs of substrips 11′, 22′, . . . , situated opposite to each other and write the expression for the potential difference between the center points of each pair and set that equal to 2 V. The expression for the potential difference between the center points of a given pair is the sum of the contributions to the potential difference from all $2n$ (=6) pairs. To obtain the contribution from a given pair, we make use of the result given in Problem P4.29 for the potential difference between two points due to an infinitely long strip of uniform surface charge density. For example, let us consider the potential difference between the center points of 1 and 1′. Then the contribution to it from the pair of substrips 1 and 1′ is

$$\frac{\rho_{S1}}{2\pi\varepsilon_0}\left[\frac{w}{6}\ln\frac{(w/12)^2 + d^2}{(w/12)^2} + 4d\tan^{-1}\frac{w}{12d}\right]$$

whereas the contribution from the pair of substrips 2 and 2′ is

$$\frac{\rho_{S2}}{2\pi\varepsilon_0}\left[\frac{w}{4}\ln\frac{(w/4)^2 + d^2}{(w/4)^2} - \frac{w}{12}\ln\frac{(w/12)^2 + d^2}{(w/12)^2} + 2d\left(\tan^{-1}\frac{w}{4d} - \tan^{-1}\frac{w}{12d}\right)\right]$$

Writing contributions in this manner and adding appropriately, we obtain the matrix equation for the three unknown charge densities ρ_{S1}, ρ_{S2}, and ρ_{S3} for the case of $d = w$, that is, $k = 1$, as given by

$$\begin{bmatrix} 1.311 & 0.815 & 0.658 \\ 0.815 & 1.432 & 1.005 \\ 0.658 & 1.005 & 1.779 \end{bmatrix}\begin{bmatrix} \rho_{S1} \\ \rho_{S2} \\ \rho_{S3} \end{bmatrix} = \begin{matrix} 4\pi\varepsilon_0/w \\ 4\pi\varepsilon_0/w \\ 4\pi\varepsilon_0/w \end{matrix}$$

so that

$$\rho_{S1} = \frac{6.0854\varepsilon_0}{w}\ \text{C/m}^2$$

$$\rho_{S2} = \frac{3.2062\varepsilon_0}{w}\ \text{C/m}^2$$

$$\rho_{S3} = \frac{3.003\varepsilon_0}{w}\ \text{C/m}^2$$

The magnitude of the charge per unit length on either conductor is $(w/3) \times (\rho_{S1} + \rho_{S2} + \rho_{S3})(1) = 4.098\varepsilon_0$ C. Thus the capacitance per unit length is given by

$$\mathscr{C} = \frac{4.0982\varepsilon_0}{2} = 2.0491\varepsilon_0\ \text{F/m}$$

Finally, the characteristic impedance of the parallel-strip line for the case of $d = w$ is $\sqrt{\mu_0\varepsilon_0}/2.0491\varepsilon_0$, or 183.98 Ω. The listing of a PC program that computes the values of ρ_S and the values of $\mathscr{C}$ and Z_0 for specified values of d, w, and n is included as PL 7.1. The output from a run of the program for $d = 1$ cm, $w = 1$ cm, and $n = 10$ is also included with the listing.

PL 7.1. Program listing for computing the characteristic inpedance of a parallel-strip line by the method of moments and the output from a run of the program.

```
100 '****************************************************
110 '* COMPUTATION OF THE CHARACTERISTIC IMPEDANCE OF A *
120 '* PARALLEL-SRTRIP TRANSMISSION LINE BY USING THE   *
130 '* METHOD OF MOMENTS                                *
140 '****************************************************
150 DIM A(20,20),Y(20),RHOS(20)
160 PI=3.141593:PF=1E-09/(36*PI)
170 CLS:SCREEN 0
180 PRINT"ENTER VALUES OF INPUT PARAMETERS:"
190 INPUT"W IN CM = ",W
200 INPUT"D IN CM = ",D
210 INPUT"N = ",N
220 W=W/100:D=D/100:WN=W/(2*N):Q=0
230 '* COMPUTE ELEMENTS OF N X N MATRIX *
240 FOR I=1 TO N
250 FOR J=I TO N
260 X1=ABS(I-J)*WN-WN/2:X2=X1+WN
270 X3=(2*N-I-J)*WN+WN/2:X4=X3+WN
280 A(I,J)=X2*LOG(1+(D/X2)^2)-X1*LOG(1+(D/X1)^2)+2*D*(ATN
    (X2/D)-ATN(X1/D))+X4*LOG(1+(D/X4)^2)-X3*LOG(1+(D/X3)^
    2)+2*D*(ATN(X4/D)-ATN(X3/D))
290 NEXT
300 IF I=1 THEN 320
310 FOR J=1 TO I-1:A(I,J)=A(J,I):NEXT
320 Y(I)=4*PI
330 NEXT
340 '* SOLUTION OF THE N X N MATRIX EQUATION *
350 FOR K=2 TO N:FOR I=K TO N
360 MF=A(I,K-1)/A(K-1,K-1)
370 FOR J=K TO N:A(I,J)=A(I,J)-A(K-1,J)*MF:NEXT
380 Y(I)=Y(I)-Y(K-1)*MF
390 NEXT:NEXT
400 RHOS(N)=Y(N)/A(N,N)
410 IF N=1 THEN 490
420 FOR I=N-1 TO 1 STEP -1
430 SUM=0
440 FOR J=I+1 TO N:SUM=SUM+A(I,J)*RHOS(J):NEXT
450 RHOS(I)=(Y(I)-SUM)/A(I,I)
460 NEXT
470 '* COMPLETION OF SOLUTION AND PRINTING OF COMPUTED
480 '  QUANTITIES *
490 FOR I=1 TO N:Q=Q+RHOS(I)*WN:NEXT
500 PRINT:PRINT"COMPUTED VALUES:":PRINT
510 PRINT"CHARGE DENSITIES IN C/M^2 ARE:"
520 FOR I=1 TO N
530 PRINT"SUBSTRIP NO.";I;":";RHOS(I)*PF
540 NEXT
550 PRINT:PRINT"C =";Q*PF;"F/M"
560 PRINT"Z0 =";120*PI/Q;"OHMS"
570 LOCATE 24,1:PRINT"STRIKE ANY KEY TO CONTINUE";
580 C$=INPUT$(1):GOTO 170
590 END

RUN
ENTER VALUES OF INPUT PARAMETERS:
W IN CM = 1
D IN CM = 1
N = 10
```

PL 7.1. (continued)

```
COMPUTED VALUES:

CHARGE DENSITIES IN C/M^2 ARE:
SUBSTRIP NO. 1 : 9.464841E-09
SUBSTRIP NO. 2 : 4.311437E-09
SUBSTRIP NO. 3 : 3.633501E-09
SUBSTRIP NO. 4 : 3.213683E-09
SUBSTRIP NO. 5 : 2.975794E-09
SUBSTRIP NO. 6 : 2.825488E-09
SUBSTRIP NO. 7 : 2.727535E-09
SUBSTRIP NO. 8 : 2.664162E-09
SUBSTRIP NO. 9 : 2.625977E-09
SUBSTRIP NO. 10 : 2.607957E-09

C = 1.852519E-11 F/M
Z0 = 179.9352 OHMS
STRIKE ANY KEY TO CONTINUE
```

Example 7.2

Enclosed-microstrip line

When the bottom conductor of the microstrip line of Fig. 7.6 is extended so as to surround the top conductor, we get the enclosed-microstrip line, as shown by the cross-sectional view in Fig. 7.8. Here we assume a square cross section for the outer conductor and wish to determine the propagation parameters for the line by using the method of finite differences to find the values of capacitance per unit length with and without the dielectric substrate in place, as required by (7.24a) and (7.24b), in view of the inhomogeneity.

For purposes of illustration, we divide the region inside the outer conductor into a 6×6 set of squares with the grid points identified as (i, j), where i is the row number (1 to 5 from top to bottom) and j is the column number (1 to 5 from left to right). We place the inner conductor along the line from grid point (4, 2) to grid point (4, 4) so that the region below row 4 is dielectric substrate (relative permittivity ε_R), and the region above row 4 is free space. We further assume the inner conductor to be kept at 10 V and the outer conductor at 0 V, and apply the iteration procedure, illustrated in Example 5.3 to compute the potentials at the grid points not on the conductors. We note, however, that in view of the inhomogeneity when the

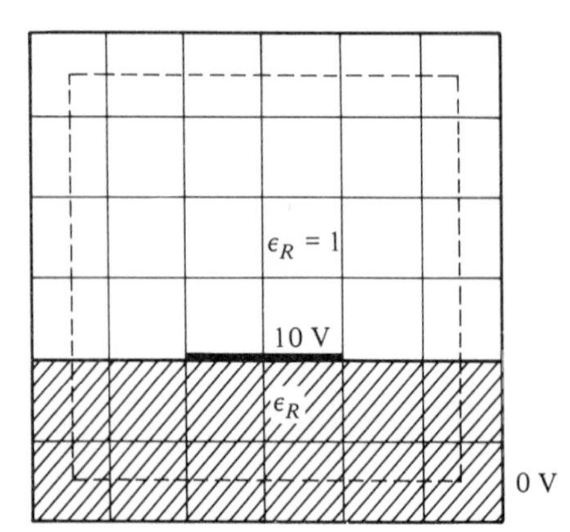

Figure 7.8. Division of the region between the conductors of an enclosed-microstrip line into a set of squares.

dielectric substrate is in place, the modified form of (5.25) given by (see Problem P5.12)

$$V_0 \approx \frac{V_1 + \varepsilon_r V_2}{2(1 + \varepsilon_r)} + \frac{V_3 + V_4}{4} \tag{7.26}$$

needs to be used. Thus the procedure consists of the following steps:

(a) With the dielectric substrate in place, find the solution for the potentials at the grid points not on the conductors consistent with (7.26) to within a specified tolerance (Δ) assumed here to be 0.01 V. Find the magnitude of the charge per unit length along the conductors by applying Gauss' law in integral form to a surface having the cross section the contour that passes through the center points of the squares adjacent to the outer conductor, as shown in Fig. 7.8. Find the capacitance per unit length ($\mathscr{C}$).

(b) With the dielectric replaced by free space, repeat step (a) to obtain the capacitance per unit length ($\mathscr{C}_0$).

(c) Find Z_0 and v_p by using (7.24a) and (7.24b), respectively.

The listing of a PC program for carrying out these steps for the arrangement of Fig. 7.8, except that the inner conductor can be placed between any two specified grid points in a specified row and with the dielectric of specified ε_R below that row, is included as PL 7.2. The sets of values for the potentials obtained from a run of the program for the specific arrangement of Fig. 7.8 for $\varepsilon_R = 10$, and the results for $\mathscr{C}$, $\mathscr{C}_0$, Z_0, and v_p are shown in Fig. 7.9. The upper rows of potential values at the interior grid points correspond to the case of the dielectric substrate in place and the lower rows correspond to the case of the dielectric replaced by free space.

PL 7.2. Program listing for computing the characteristic impedance and velocity of propagation for the enclosed-microstrip line of Fig. 7.8.

```
100 '*****************************************************
110 '* DETERMINATION OF THE CHARACTERISTIC IMPEDANCE AND *
120 '* VELOCITY OF PROPAGATION FOR AN ENCLOSED-MICROSRIP *
130 '* TRANSMISSION LINE                                 *
140 '*****************************************************
150 DIM V(10,10)
160 PI=3.141593:PF=1E-09/(36*PI):DV=.01
170 SB$="STRIKE ANY KEY TO CONTINUE"
180 BL$="                                        "
190 CLS:SCREEN 0
200 PRINT"ENTER VALUES OF INPUT PARAMETERS:"
210 PRINT:PRINT"BEGINNING POINT OF INNER CONDUCTOR:"
220 INPUT"     ROW NO. = ",ROW:ICA=ROW
230 INPUT"     COLUMN NO. = ",COL:JCA=COL
240 PRINT:PRINT"ENDING POINT OF INNER CONDUCTOR:"
250 PRINT"     ROW NO. = ";ICA
260 INPUT"     COLUMN NO. = ",COL:JCB=COL
270 PRINT:INPUT"RELATIVE PERMITTIVITY = ",ER
280 LOCATE 24,1:PRINT SB$;:C$=INPUT$(1)
290 CLS:SCREEN 1
300 IR=1:GOSUB 500:CD=C:' * CAPACITANCE PER UNIT LENGTH
310 ' WITH THE DIELECTRIC SUBSTRATE IN PLACE *
320 IF ER=1 THEN CO=C:LOCATE 21,1:PRINT"C = CO =";C*1E+12;
    "PF/m":GOTO 390
```

PL 7.2. (continued)

```
330 LOCATE 21,1:PRINT"C =";C*1E+12;"PF/m"
340 LOCATE 24,1:PRINT SB$;C$=INPUT$(1);
350 LOCATE 24,1:PRINT BL$;
360 ER=1:IR=2:GOSUB 500:C0=C:' * CAPACITANCE PER UNIT
370 ' LENGTH WITH THE DIELECTRIC REPLACED BY FREE SPACE *
380 LOCATE 21,21:PRINT"C0 =";C*1E+12;"PF/m"
390 LOCATE 24,1:PRINT SB$;C$=INPUT$(1);
400 LOCATE 24,1:PRINT BL$;
410 Z0=1/(SQR(CD*C0)*3E+08)
420 VP=SQR(C0/CD)*3E+08
430 LOCATE 22,1:PRINT "Z0 =";Z0;"OHMS"
440 LOCATE 23,1:PRINT"VP =";VP;"M/S"
450 LOCATE 24,1:PRINT SB$;C$=INPUT$(1)
460 GOTO 190
470 END
500 '* SUBPROGRAM FOR COMPUTING CAPACITANCE PER UNIT
510 '  LENGTH OF THE LINE *
520 FOR I=0 TO 6:FOR J=0 TO 6:V(I,J)=0:NEXT:NEXT
530 FOR J=JCA TO JCB:V(ICA,J)=10:NEXT
540 IF IR=2 THEN 570
550 FOR I=0 TO 6:FOR J=0 TO 6:LOCATE 3*I+1,6*J+1:PRINT V(I
    ,J):NEXT:NEXT
560 GOTO 580
570 FOR I=1 TO 5:FOR J=1 TO 5:LOCATE 3*I+2,6*J+1:PRINT V(I
    ,J):NEXT:NEXT
580 IN=0
590 IC=0:DM=0
600 IF JCA=1 THEN 670
610 FOR J=JCA-1 TO 1 STEP -1
620 VIJ=(V(ICA,J-1)+V(ICA,J+1))/4+(V(ICA-1,J)+V(ICA+1,J)*E
    R)/(2*(1+ER))
630 DC=ABS(VIJ-V(ICA,J))
640 IF DC>DM THEN DM=DC
650 IF DC>=DV THEN V(ICA,J)=VIJ:IC=1
660 NEXT
670 IF JCB=5 THEN 740
680 FOR J=JCB+1 TO 5
690 VIJ=(V(ICA,J-1)+V(ICA,J+1))/4+(V(ICA-1,J)+V(ICA+1,J)*E
    R)/(2*(1+ER))
700 DC=ABS(VIJ-V(ICA,J))
710 IF DC>DM THEN DM=DC
720 IF DC>=DV THEN V(ICA,J)=VIJ:IC=1
730 NEXT
740 IF ICA=1 THEN 810
750 FOR I=ICA-1 TO 1 STEP -1:FOR J=1 TO 5
760 VIJ=(V(I,J-1)+V(I-1,J)+V(I+1,J)+V(I,J+1))/4
770 DC=ABS(VIJ-V(I,J))
780 IF DC>DM THEN DM=DC
790 IF DC>=DV THEN V(I,J)=VIJ:IC=1
800 NEXT:NEXT
810 IF ICA=5 THEN 880
820 FOR I=ICA+1 TO 5:FOR J=1 TO 5
830 VIJ=(V(I,J-1)+V(I-1,J)+V(I+1,J)+V(I,J+1))/4
840 DC=ABS(VIJ-V(I,J))
850 IF DC>DM THEN DM=DC
860 IF DC>=DV THEN V(I,J)=VIJ:IC=1
870 NEXT:NEXT
880 IN=IN+1
890 FOR I=1 TO 5:IT=3*I+IR
900 FOR J=1 TO 5:JT=6*J+1:LOCATE IT,JT:PRINT"      ":LOCATE
     IT,JT:PRINT USING"##.##";V(I,J):NEXT
```

PL 7.2. (continued)

```
910 NEXT
920 IF IC=1 THEN 590
930 BEEP:QO=0
940 FOR J=1 TO 5:QO=QO+V(1,J)+V(5,J)*ER:NEXT
950 FOR I=1 TO ICA-1:QO=QO+V(I,1)+V(I,5):NEXT
960 QO=QO+(V(ICA,1)+V(ICA,5))/2
970 FOR I=ICA+1 TO 5:QO=QO+(V(I,1)+V(I,5))*ER:NEXT
980 QO=QO+(V(ICA,1)+V(ICA,5))*ER/2
990 C=QO*PF/10
1000 RETURN
```

```
0     0     0     0     0     0     0

0     0.69  1.23  1.43  1.23  0.69  0
      0.69  1.24  1.44  1.24  0.69

0     1.53  2.84  3.27  2.84  1.53  0
      1.54  2.85  3.28  2.85  1.54

0     2.58  5.34  5.99  5.34  2.58  0
      2.63  5.36  6.00  5.36  2.63

0     3.47 10.00 10.00 10.00  3.47  0
      3.64 10.00 10.00 10.00  3.64

0     1.89  4.10  4.55  4.10  1.89  0
      1.93  4.12  4.55  4.12  1.94

0     0     0     0     0     0     0

C = 226.4795 PF/m    CO = 37.85039 PF/m
ZO = 36.00221 OHMS
VP = 1.226428E+08 M/S
```

Figure 7.9. Finat set of values for the potentials and the results for $\mathscr{C}$, $\mathscr{C}_0$, Z_0, and v_p for the enclosed-microstrip line of Fig. 7.8 for $\varepsilon_R = 10$, from a run of the program of PL 7.2.

C. Graphical Technique

Field mapping

For a line with arbitrary cross section and involving a homogeneous dielectric, an approximate value of $\mathscr{C}$ and hence of Z_0 can be determined by constructing a *field map*, that is, a graphical sketch of the direction lines of the electric field and associated equipotential lines between the conductors. To illustrate this, let us consider the cross section shown in Fig. 7.10. Assuming that the inner conductor is positive with respect to the outer conductor, we can draw the field map from the following considerations. (1) The electric field lines originate on the inner conductor and normal to it and terminate on the outer conductor and normal to it, since the tangential component of the electric field on the conductor surface must be zero. (2) The equipotential lines must be everywhere perpendicular to the electric field lines. Thus suppose that we start with the inner conductor and draw several lines normal to it at several points on the surface, as shown in Fig. 7.10(b). We can then draw a curved line displaced from the conductor surface and such that it is perpendicular everywhere to the electric field lines of Fig. 7.10(b), as shown in Fig. 7.10(c). This contour represents an equipotential line and forms the basis for further extension of the electric field lines, as shown in Fig. 7.10(d). A second equipotential line can then be drawn so that it is everywhere perpendic-

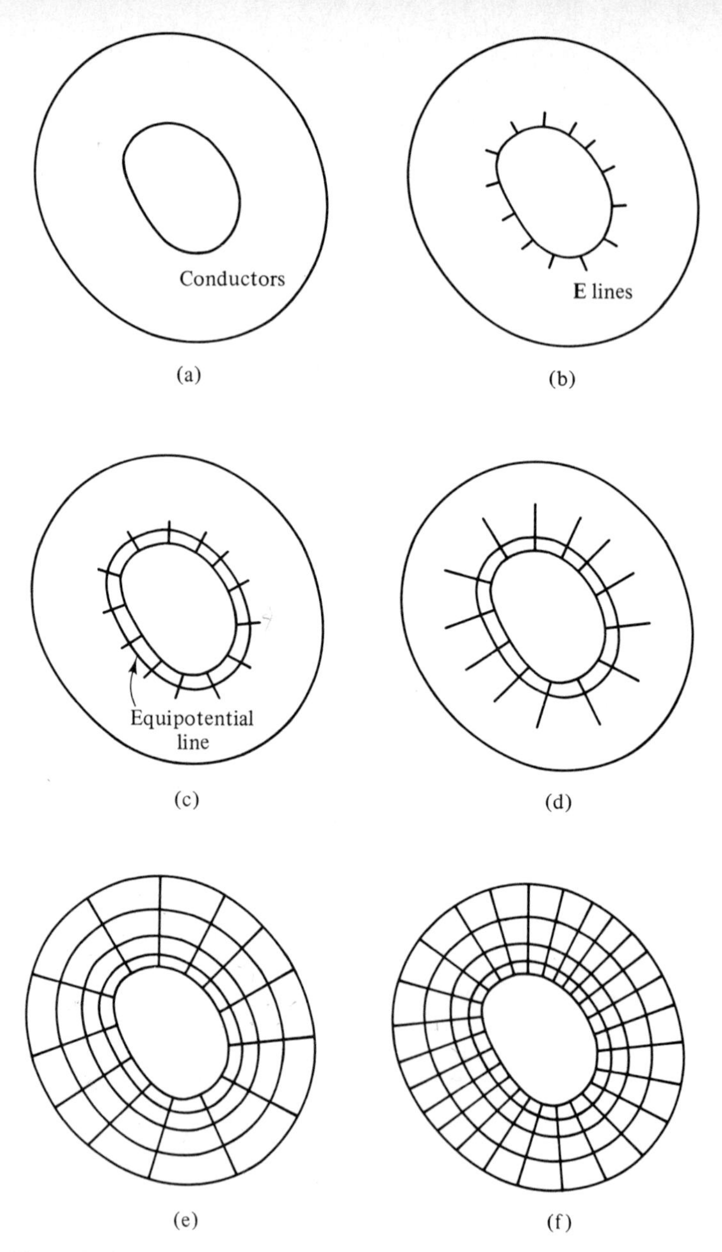

Figure 7.10. For illustrating the construction of a field map for a transmission line of arbitrary cross section.

ular to the extended electric field lines and the procedure continued until the entire cross section between the conductors is filled with two sets of orthogonal contours, as shown in Fig 7.10(e), thereby resulting in a field map made up of curvilinear rectangles. For the actual, time-varying case, the magnetic field lines are the same as the equipotential lines and the field map represents a sketch of the direction lines of electric and magnetic fields between the conductors.

By drawing the field lines with very small spacings, we can make the rectangles so small that each of them can be considered to be the cross section of a parallel-plate line. If we now replace the equipotential lines by perfect conductors, since it does not violate any boundary condition, it can be seen that the arrangement can be viewed as the parallel combination, in the angular direction, of m number of series combinations of n number of parallel-plate lines in the radial direction, where m is the number of rectangles in the angular direction, that is, along a magnetic field line, and n is the number of rectangles in the radial direction, that is, along an electric field line. If $Q_1, Q_2, \ldots, Q_m$ are the charges per unit length associated with the angular direction and $V_1, V_2, \ldots, V_n$ are the potential differences associated with the radial direction, the capacitance per unit length of the line is given by

$$\begin{aligned}
\mathscr{C} &= \frac{Q}{V} = \frac{Q_1 + Q_2 + \cdots + Q_m}{V_1 + V_2 + \cdots + V_n} \\
&= \frac{1}{\dfrac{V_1}{Q_1} + \dfrac{V_2}{Q_1} + \cdots + \dfrac{V_n}{Q_1}} \\
&\quad + \frac{1}{\dfrac{V_1}{Q_2} + \dfrac{V_2}{Q_2} + \cdots + \dfrac{V_n}{Q_2}} \\
&\quad + \cdots + \frac{1}{\dfrac{V_1}{Q_m} + \dfrac{V_2}{Q_m} + \cdots + \dfrac{V_n}{Q_m}} \\
&= \sum_{i=1}^{m} \frac{1}{\displaystyle\sum_{j=1}^{n} \frac{V_j}{Q_i}} \\
&= \sum_{i=1}^{m} \frac{1}{\displaystyle\sum_{j=1}^{n} \frac{1}{\mathscr{C}_{ij}}}
\end{aligned}$$

where $\mathscr{C}_{ij} = Q_i/V_j$ is the capacitance per unit length corresponding to the rectangle ij. The simplicity of the field mapping technique lies in the fact that if the map consists entirely of curvilinear squares (a curvilinear rectangle becomes a curvilinear square if a circle can be inscribed in it), all $\mathscr{C}_{ij}$ are approximately equal to ε, and we obtain the simple formula

$$\mathscr{C} \approx \varepsilon \frac{m}{n} \tag{7.27}$$

and hence

$$\begin{aligned}
Z_0 &= \frac{\sqrt{\mu\varepsilon}}{\mathscr{C}} \\
&\approx \eta \frac{n}{m}
\end{aligned} \tag{7.28}$$

Thus the determination of Z_0 consists of sketching a field map consisting of curvilinear squares, as shown in Fig. 7.10(f), counting the number of squares in each direction and substituting these values in (7.28). For the rough sketch of Fig. 7.10(f), $m = 26$ and $n = 4$ so that $Z_0 \approx 0.154\eta$.

K7.2. Z_0 and v_p for line with homogeneous dielectric; Z_0 and v_p for line with multiple dielectrics; Microstrip line; Parallel-strip line; Method of moments; Enclosed-microstrip line; Finite-difference method; Field mapping; Curvilinear squares.

D7.2. Find the following: **(a)** the ratio b/a of a coaxial cable of $Z_0 = 50\ \Omega$ if $\varepsilon = 2.56\varepsilon_0$; **(b)** the ratio b/a of a coaxial cable of $Z_0 = 75\ \Omega$ if $\varepsilon = 2.25\varepsilon_0$; and **(c)** the ratio d/a of a parallel-wire line of $Z_0 = 300\ \Omega$ if $\varepsilon = \varepsilon_0$.
Ans. **(a)** 3.794; **(b)** 6.521; **(c)** 6.132

D7.3. For the parallel-strip line of Fig. 7.7, find the following for $d = w$: **(a)** the contribution to the potential difference between the center points of substrips 2 and 2′ from the pair of substrips 2 and 2′; **(b)** the contribution to the potential difference between the center points of substrips 2 and 2′ from the pair of substrips 5 and 5′; and **(c)** the contribution to the potential difference between the center points of substrips 1 and 1′ from the pair of substrips 6 and 6′.
Ans. **(a)** $0.1849\rho_{S2}w/\varepsilon_0$; **(b)** $0.043\rho_{S2}w/\varepsilon_0$; **(c)** $0.0238\rho_{S1}w/\varepsilon_0$

D7.4. Two lossless transmission lines 1 and 2 have nonmagnetic ($\mu = \mu_0$), homogeneous perfect dielectrics of $\varepsilon_1 = 2.25\varepsilon_0$ and $\varepsilon_2 = 4\varepsilon_0$, respectively. The values of the ratio m/n corresponding to their curvilinear square field maps are 4 and 5 for lines 1 and 2, respectively. Find **(a)** v_{p1}/v_{p2}, **(b)** $\mathscr{C}_1/\mathscr{C}_2$, and **(c)** Z_{01}/Z_{02}, where the subscripts 1 and 2 denote lines 1 and 2, respectively.
Ans. **(a)** 4/3; **(b)** 0.45; **(c)** 5/3

7.3 LINE TERMINATED BY RESISTIVE LOAD

Notation

In Section 7.1, we obtained the general solutions to the transmission-line equations for the lossless line, as given by (7.20a) and (7.20b). Since these solutions represent superpositions of (+) and (−) wave voltages and (+) and (−) wave currents, we now rewrite them as

$$V(z, t) = V^+\left(t - \frac{z}{v_p}\right) + V^-\left(t + \frac{z}{v_p}\right) \tag{7.29a}$$

$$I(z, t) = \frac{1}{Z_0}\left[V^+\left(t - \frac{z}{v_p}\right) - V^-\left(t + \frac{z}{v_p}\right)\right] \tag{7.29b}$$

or, more concisely,

$$\boxed{\begin{aligned} V &= V^+ + V^- \\ I &= \frac{1}{Z_0}(V^+ - V^-) \end{aligned}} \qquad \begin{aligned}(7.30\text{a})\\(7.30\text{b})\end{aligned}$$

with the understanding that V^+ is a function of $(t - z/v_p)$ and V^- is a function of $(t + z/v_p)$. In terms of (+) and (−) wave currents, the solution for the current may also be written as

$$\boxed{I = I^+ + I^-} \tag{7.31}$$

Comparing (7.30b) and (7.31), we see that

$$I^+ = \frac{V^+}{Z_0} \tag{7.32a}$$

$$I^- = -\frac{V^-}{Z_0} \tag{7.32b}$$

The minus sign in (7.32b) can be understood if we recognize that in writing (7.30a) and (7.31), we follow the notation that both V^+ and V^- have the same polarities with one conductor (say, a) positive with respect to the other conductor (say, b) and that both I^+ and I^- flow in the positive z-direction along conductor a and return in the negative z-direction along conductor b, as shown in Fig. 7.11. The power flow associated with either wave, as given by the product of the corresponding voltage and current, is then directed in the positive z-direction, as shown in Fig. 7.11. Thus

$$P^+ = V^+I^+ = V^+\left(\frac{V^+}{Z_0}\right) = \frac{(V^+)^2}{Z_0} \tag{7.33a}$$

Since $(V^+)^2$ is always positive, regardless of whether V^+ is numerically positive or negative, (7.33a) indicates that the (+) wave power does actually flow in the positive z-direction, as it should. On the other hand,

$$P^- = V^-I^- = V^-\left(-\frac{V^-}{Z_0}\right) = -\frac{(V^-)^2}{Z_0} \tag{7.33b}$$

Since $(V^-)^2$ is always positive, regardless of whether V^- is numerically positive or negative, the minus sign in (7.33b) indicates that P^- is negative, and hence the (−) wave power actually flows in the negative z-direction, as it should.

Excitation by constant voltage source

Let us now consider a line of length l terminated by a load resistance R_L and driven by a constant voltage source V_0 in series with internal resistance R_g, as shown in Fig. 7.12. Note that the conductors of the transmission line are represented by double ruled lines, whereas the connections to the conductors are single ruled lines, to be treated as lumped circuits. We assume that no voltage and cur-

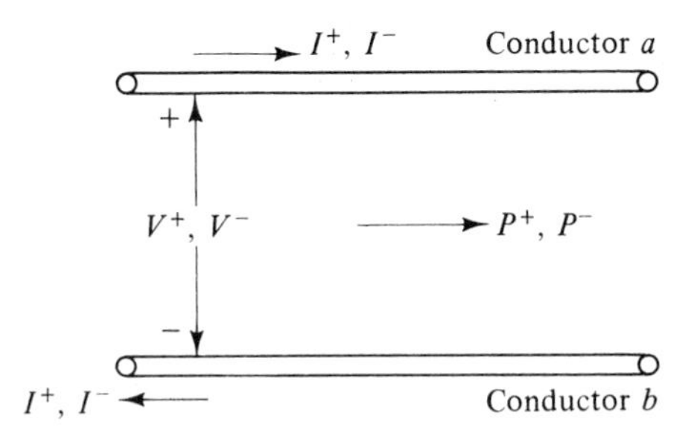

Figure 7.11. Polarities for voltages and currents associated with (+) and (−) waves.

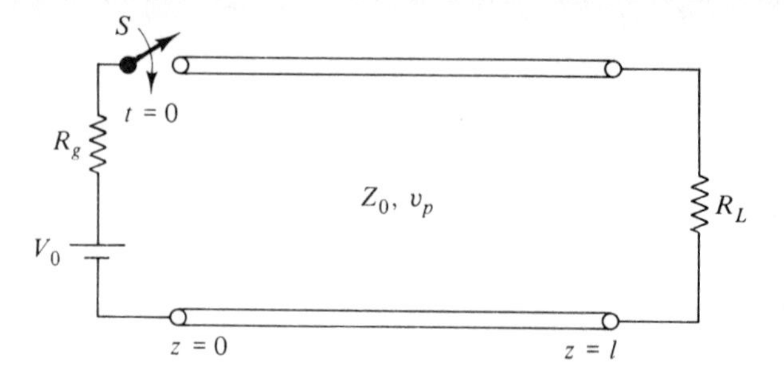

Figure 7.12. Transmission line terminated by a load resistance R_L and driven by a constant voltage source in series with an internal resistance R_g.

rent exist on the line for $t < 0$ and the switch S is closed at $t = 0$. We wish to discuss the transient wave phenomena on the line for $t > 0$. The characteristic impedance of the line and the velocity of propagation are Z_0 and v_p, respectively.

When the switch S is closed at $t = 0$, a (+) wave originates at $z = 0$ and travels toward the load. Let the voltage and current of this (+) wave be V^+ and I^+, respectively. Then we have the situation at $z = 0$, as shown in Fig. 7.13(a). Note that the load resistor does not come into play here since the phenomenon is one of wave propagation; hence, until the (+) wave goes to the load, sets up a reflection, and the reflected wave arrives back at the source, the source does not even know of the existence of R_L. This is a fundamental distinction between ordinary (lumped) circuit theory and transmission-line (distributed circuit) theory. In ordinary circuit theory, no time delay is involved; the effect of a transient in one part of the circuit is felt in all branches of the circuit instantaneously. In a transmission-line system, the effect of a transient at one location on the line is felt at a different location on the line only after an interval of time that the wave takes to travel from the first location to the second. Returning now to the circuit in Fig. 7.13(a), the various quantities must satisfy the boundary condition, that is, Kirchhoff's voltage law around the loop. Thus we have

$$V_0 - I^+ R_g - V^+ = 0 \tag{7.34a}$$

We however know from (7.32a) that $I^+ = V^+/Z_0$. Hence we get

$$V_0 - \frac{V^+}{Z_0} R_g - V^+ = 0 \tag{7.34b}$$

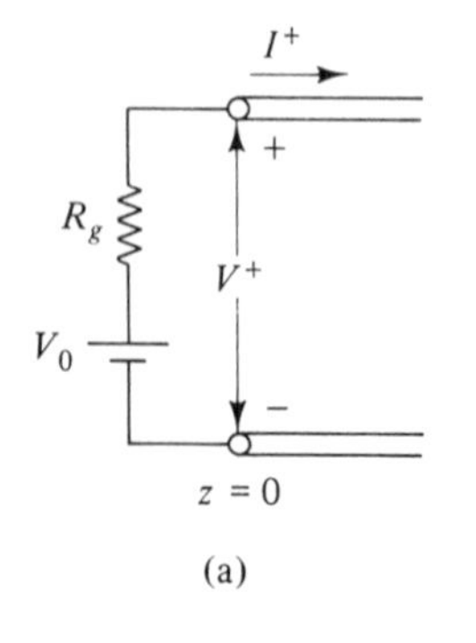

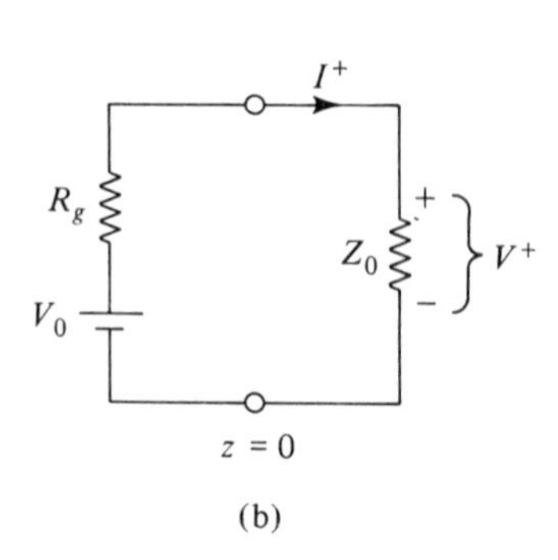

Figure 7.13. (a) For obtaining the (+) wave voltage and current at $z = 0$ for the line of Fig. 7.12. (b) Equivalent circuit for (a).

or

$$V^+ = V_0 \frac{Z_0}{R_g + Z_0} \tag{7.35a}$$

$$I^+ = \frac{V^+}{Z_0} = \frac{V_0}{R_g + Z_0} \tag{7.35b}$$

Thus, we note that the situation in Fig. 7.13(a) is equivalent to the circuit shown in Fig. 7.13(b); that is, the voltage source views a resistance equal to the characteristic impedance of the line, across $z = 0$. This is to be expected since only a (+) wave exists at $z = 0$ and the ratio of the voltage to current in the (+) wave is equal to Z_0.

Reflection coefficient

The (+) wave travels toward the load and reaches the termination at $t = l/v_p$. It does not, however, satisfy the boundary condition there since this condition requires the voltage across the load resistance to be equal to the current through it times its value, R_L. But the voltage-to-current ratio in the (+) wave is equal to Z_0. To resolve this inconsistency, there is only one possibility, which is the setting up of a (−) wave, or a reflected wave. Let the voltage and current in this reflected wave be V^- and I^-, respectively. Then the total voltage across R_L is $V^+ + V^-$, and the total current through it is $I^+ + I^-$, as shown in Fig. 7.14(a). To satisfy the boundary condition, we have

$$V^+ + V^- = R_L(I^+ + I^-) \tag{7.36a}$$

But from (7.32a) and (7.32b), we know that $I^+ = V^+/Z_0$ and $I^- = -V^-/Z_0$, respectively. Hence

$$V^+ + V^- = R_L\left(\frac{V^+}{Z_0} - \frac{V^-}{Z_0}\right) \tag{7.36b}$$

or

$$V^- = V^+ \frac{R_L - Z_0}{R_L + Z_0} \tag{7.37}$$

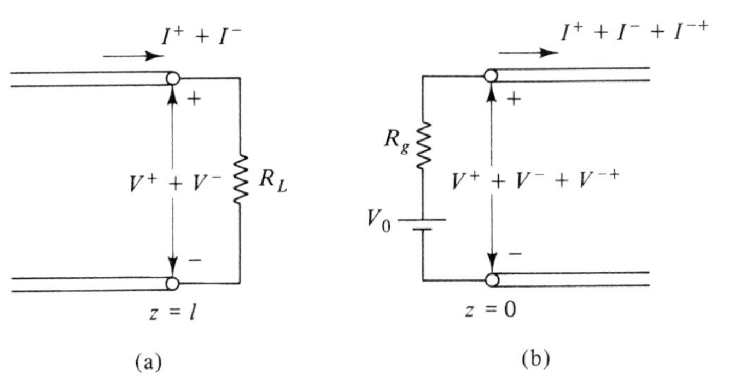

Figure 7.14. For obtaining the voltages and currents associated with (a) the (−) wave and (b) the (−+) wave, for the line of Fig. 7.12.

We now define the *voltage reflection coefficient*, denoted by the symbol Γ, as the ratio of the reflected voltage to the incident voltage. Thus

$$\boxed{\Gamma = \frac{V^-}{V^+} = \frac{R_L - Z_0}{R_L + Z_0}} \tag{7.38}$$

We then note that the *current reflection coefficient* is

$$\boxed{\frac{I^-}{I^+} = \frac{-V^-/Z_0}{V^+/Z_0} = -\frac{V^-}{V^+} = -\Gamma} \tag{7.39}$$

Now, returning to the reflected wave, we observe that this wave travels back toward the source and it reaches there at $t = 2l/v_p$. Since the boundary condition at $z = 0$, which was satisfied by the original (+) wave alone, is then violated, a reflection of the reflection, or a re-reflection, will be set up and it travels toward the load. Let us assume the voltage and current in this re-reflected wave, which is a (+) wave, to be V^{-+} and I^{-+}, respectively, with the superscripts denoting that the (+) wave is a consequence of the (−) wave. Then the total line voltage and the line current at $z = 0$ are $V^+ + V^- + V^{-+}$ and $I^+ + I^- + I^{-+}$, respectively, as shown in Fig. 7.14(b). To satisfy the boundary condition, we have

$$V^+ + V^- + V^{-+} = V_0 - R_g(I^+ + I^- + I^{-+}) \tag{7.40a}$$

But we know that $I^+ = V^+/Z_0$, $I^- = -V^-/Z_0$, and $I^{-+} = V^{-+}/Z_0$. Hence

$$V^+ + V^- + V^{-+} = V_0 - \frac{R_g}{Z_0}(V^+ - V^- + V^{-+}) \tag{7.40b}$$

Furthermore, substituting for V^+ from (7.35a), simplifying, and rearranging, we get

$$V^{-+}\left(1 + \frac{R_g}{Z_0}\right) = V^-\left(\frac{R_g}{Z_0} - 1\right)$$

or

$$\boxed{V^{-+} = V^- \frac{R_g - Z_0}{R_g + Z_0}} \tag{7.41}$$

Reflection coefficients for some special cases

Comparing (7.41) with (7.37), we note that the reflected wave views the source with internal resistance as the internal resistance alone; that is, the voltage source is equivalent to a short circuit insofar as the (−) wave is concerned. A moment's thought will reveal that superposition is at work here. The effect of the voltage source is taken into account by the constant outflow of the original (+) wave from the source. Hence, for the reflection of the reflection, that is, for the (−+) wave, we need only consider the internal resistance R_g. Thus the voltage reflection coefficient formula (7.38) is a general formula and will be used repeatedly. In view of its importance, a brief discussion of the values of Γ for some special cases is in order as follows:

1. $R_L = 0$, or short-circuited line.

$$\Gamma = \frac{0 - Z_0}{0 + Z_0} = -1$$

The reflected voltage is exactly the negative of the incident voltage, thereby keeping the voltage across R_L (short circuit) always zero.

2. $R_L = \infty$, or open-circuited line.

$$\Gamma = \frac{\infty - Z_0}{\infty + Z_0} = 1$$

and the current reflection coefficient $= -\Gamma = -1$. Thus, the reflected current is exactly the negative of the incident current, thereby keeping the current through R_L (open circuit) always zero.

3. $R_L = Z_0$, or line terminated by its characteristic impedance.

$$\Gamma = \frac{Z_0 - Z_0}{Z_0 + Z_0} = 0$$

This corresponds to no reflection, which is to be expected since $R_L(= Z_0)$ is consistent with the voltage to current ratio in the (+) wave alone, and hence there is no violation of boundary condition and no need for the setting up of a reflected wave. Thus, a line terminated by its characteristic impedance is equivalent to an infinitely long line insofar as the source is concerned.

Bounce diagram

Returning to the discussion of the re-reflected wave, we note that this wave reaches the load at time $t = 3l/v_p$ and sets up another reflected wave. This process of bouncing back and forth of waves goes on indefinitely until the steady state is reached. To keep track of this transient phenomenon, we make use of the *bounce diagram* technique. Some other names given for this diagram are *reflection diagram* and *space-time diagram*. We shall introduce the bounce diagram through a numerical example.

Example 7.3

Let us consider the system shown in Fig. 7.15. Note that we have introduced a new quantity T, which is the one-way travel time along the line from $z = 0$ to $z = l$; that is, instead of specifying two quantities l and v_p, we specify $T(=l/v_p)$. Using the bounce diagram technique, we wish to obtain and plot line voltage and current versus t for fixed values z and line voltage and current versus z for fixed values t.

Before we construct the bounce diagram, we need to compute the following quantities:

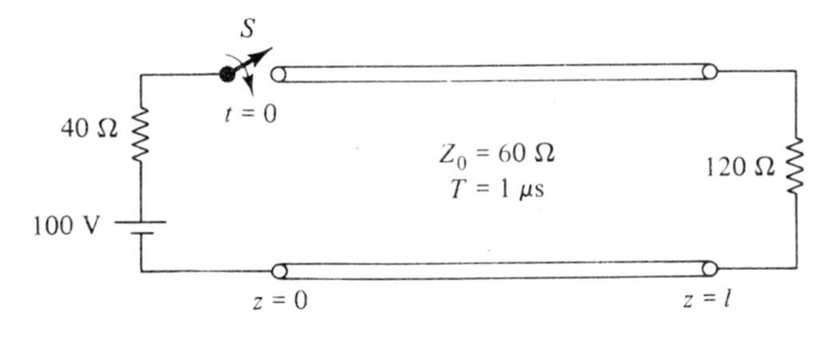

Figure 7.15. Transmission-line system for illustrating the bounce diagram technique of keeping track of the transient phenomenon.

$$\text{voltage carried by the initial (+) wave} = 100\frac{60}{40 + 60} = 60 \text{ V}$$

$$\text{current carried by the initial (+) wave} = \frac{60}{60} = 1 \text{ A}$$

$$\text{voltage reflection coefficient at load, } \Gamma_R = \frac{120 - 60}{120 + 60} = \frac{1}{3}$$

$$\text{voltage reflection coefficient at source, } \Gamma_S = \frac{40 - 60}{40 + 60} = -\frac{1}{5}$$

Construction of bounce diagrams

The bounce diagram is essentially a two-dimensional representation of the transient waves bouncing back and forth on the line. Separate bounce diagrams are drawn for voltage and current as shown in Fig. 7.16(a) and (b), respectively. Position (z) on the line is represented horizontally and the time (t) vertically. Reflection coefficient values for the two ends are shown at the top of the diagrams for quick reference. Note that current reflection coefficients are $-\Gamma_R = -\frac{1}{3}$ and $-\Gamma_S = \frac{1}{5}$, respectively, at the load and at the source. Crisscross lines are drawn as shown in the figures to indicate the progress of the wave as a function of both z and t, with the numerical value for each leg of travel shown beside the line corresponding to that leg and approximately at the center of the line. The arrows indicate the directions of travel. Thus, for example, the first line on the voltage bounce diagram indicates that the intial (+) wave of 60 V takes a time of 1 μs to reach the load end of the line. It sets up a reflected wave of 20 V, which travels back to the source, reaching there at a time of 2 μs, which then gives rise to a (+) wave of voltage -4 V, and so on, with the process continuing indefinitely.

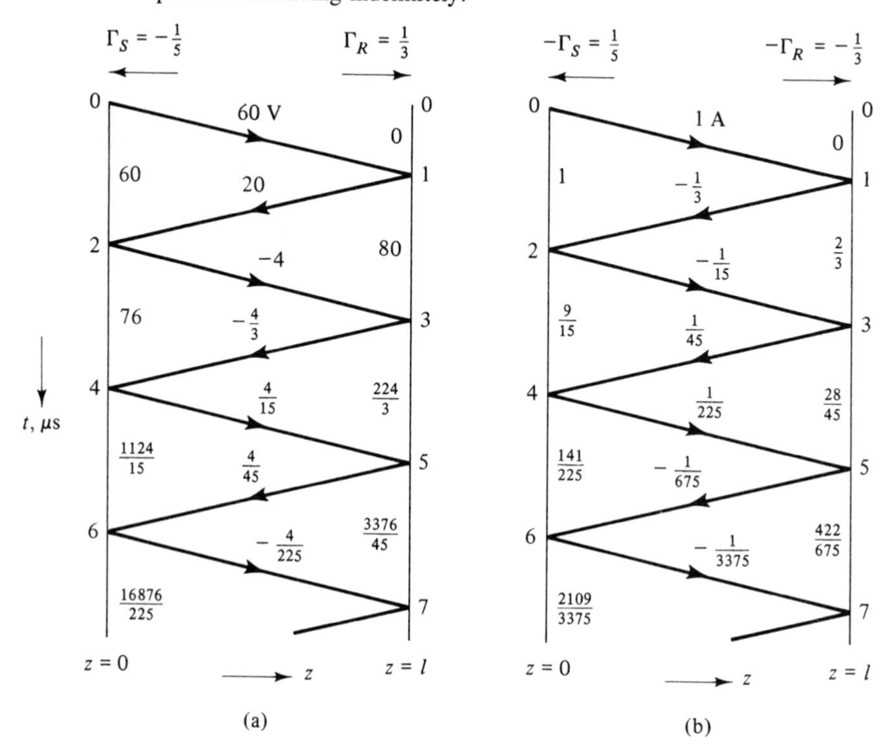

Figure 7.16. (a) Voltage and (b) current bounce diagrams, depicting the bouncing back and forth of the transient waves for the system of Fig. 7.15.

Plots of line voltage and current versus t

Now, to sketch the line voltage and/or current versus time at any value of z, we note that since the voltage source is a constant voltage source, each individual wave voltage and current, once the wave is set up at that value of z, continues to exist there forever. Thus, at any particular time, the voltage (or current) at that value of z is a superposition of all the voltages (or currents) corresponding to the crisscross lines preceding that value of time. These values are marked on the bounce diagrams for $z = 0$ and $z = l$. Noting that each value corresponds to the 2 μs time interval between adjacent crisscross lines, we now sketch the time variations of line voltage and current at $z = 0$ and $z = l$, as shown in Fig. 7.17(a) and (b), respectively. Similarly, by observing that the numbers written along the time axis for $z = 0$ are actually valid for any pair of z and t within the triangle (▷) inside which

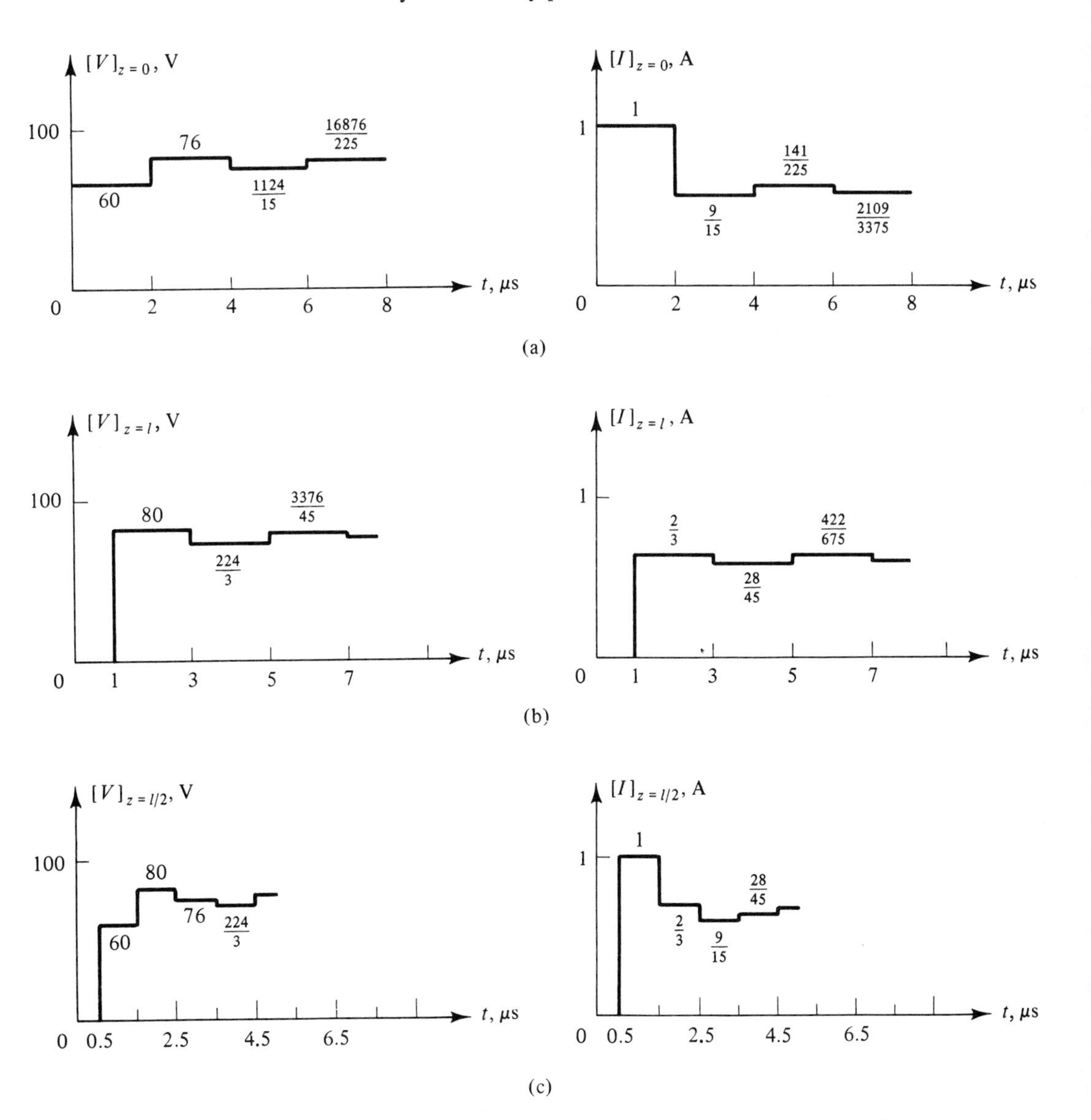

Figure 7.17. Time variations of line voltage and line current at (a) $z = 0$, (b) $z = l$, and (c) $z = l/2$ for the system of Fig. 7.15.

they lie and that the numbers written along the time axis for $z = l$ are actually valid for any pair of z and t within the triangle ($\triangleleft$) inside which they lie, we can draw the sketches of line voltage and current versus time for any other value of z. This is done for $z = l/2$ in Fig. 7.17(c).

Steady-state situation

It can be seen from the sketches of Fig. 7.17 that as time progresses, the line voltage and current tend to converge to certain values, which we can expect to be the steady-state values. In the steady state, the situation consists of a single (+) wave, which is actually a superposition of the infinite number of transient (+) waves, and a single (−) wave, which is actually a superposition of the infinite number of transient (−) waves. Denoting the steady-state (+) wave voltage and current to be V_{SS}^+ and I_{SS}^+, respectively, and the steady-state (−) wave voltage and current to be V_{SS}^- and I_{SS}^-, respectively, we obtain from the bounce diagrams

$$V_{SS}^+ = 60 - 4 + \frac{4}{15} - \cdots = 60\left(1 - \frac{1}{15} + \frac{1}{15^2} - \cdots\right) = 56.25 \text{ V}$$

$$I_{SS}^+ = 1 - \frac{1}{15} + \frac{1}{225} - \cdots = 1 - \frac{1}{15} + \frac{1}{15^2} - \cdots = 0.9375 \text{ A}$$

$$V_{SS}^- = 20 - \frac{4}{3} + \frac{4}{45} - \cdots = 20\left(1 - \frac{1}{15} + \frac{1}{15^2} - \cdots\right) = 18.75 \text{ V}$$

$$I_{SS}^- = -\frac{1}{3} + \frac{1}{45} - \frac{1}{675} + \cdots = -\frac{1}{3}\left(1 - \frac{1}{15} + \frac{1}{15^2} - \cdots\right) = -0.3125 \text{ A}$$

Note that $I_{SS}^+ = V_{SS}^+/Z_0$ and $I_{SS}^- = -V_{SS}^-/Z_0$, as they should be. The steady-state line voltage and current can now be obtained to be

$$V_{SS} = V_{SS}^+ + V_{SS}^- = 75 \text{ V}$$

$$I_{SS} = I_{SS}^+ + I_{SS}^- = 0.625 \text{ A}$$

These are the same as the voltage across R_L and current through R_L if the source and its internal resistance were connected directly to R_L, as shown in Fig. 7.18. This is to be expected since the series inductors and shunt capacitors of the distributed equivalent circuit behave like short circuits and open circuits, respectively, for the constant voltage source in the steady state.

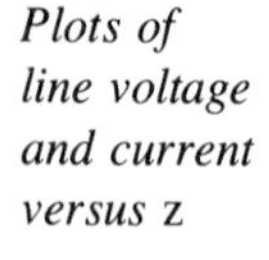

Sketches of line voltage and current as functions of distance (z) along the line for any particular time can also be drawn from considerations similar to those employed for the sketches of Fig. 7.17. For example, suppose we wish to draw the sketch of line voltage versus z for $t = 2.5$ μs. Then we note from the voltage bounce diagram that for $t = 2.5$ μs the line voltage is 76 V from $z = 0$ to $z = l/2$ and 80 V from $z = l/2$ to $z = l$. This is shown sketched in Fig. 7.19(a). Similarly, Fig. 7.19(b) shows the variation of line current versus z for $t = 1\frac{1}{3}$ μs.

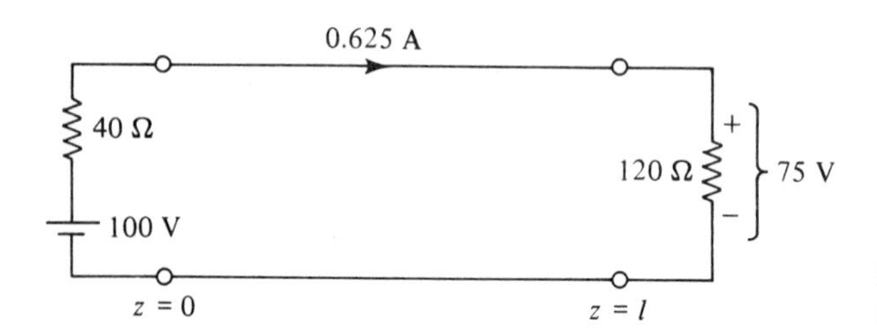

Figure 7.18. Steady-state equivalent for the system of Fig. 7.15.

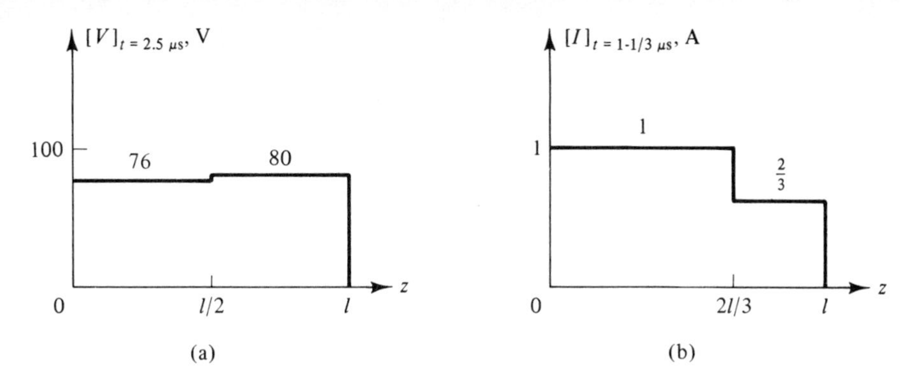

Figure 7.19. Variations with z of (a) line voltage for $t = 2.5\ \mu s$ and (b) line current for $t = 1\frac{1}{3}\ \mu s$, for the system of Fig. 7.15.

All the computations associated with the bounce diagram technique can be carried out conveniently by using a computer program. The listing of a PC program which computes all the voltages and currents in the bounce diagrams and gives the line voltage and current versus z for specified values of time and the output from a run of the program using values for the parameters as in Fig. 7.15 are shown as PL 7.3.

PL 7.3. Program listing and sample output for obtaining the voltage and current variations with distance, for fixed values of time, along a line terminated by a resistance and driven by a constant voltage source in series with internal resistance.

```
100 '***************************************************
110 '* TIME DOMAIN ANALYSIS FOR A LOSSLESS TRANSMISSION *
120 '* LINE TERMINATED BY A RESISTANCE AND DRIVEN BY A  *
130 '* CONSTANT VOLTAGE SOURCE IN SERIES WITH INTERNAL  *
140 '* RESISTANCE                                       *
150 '***************************************************
160 DEF FN TRD(ARG)=INT(ARG*100000!+.5)/100000!:'* ROUNDS
170 '  ARG TO FIVE DECIMAL PLACES *
180 DIM V(20),C(20),VL(20),CL(20)
190 SB$="STRIKE ANY KEY TO CONTINUE"
200 CLS:SCREEN 0:PRINT "ENTER VALUES OF VO IN V, AND RG,R
    L,":PRINT "AND ZO IN OHMS:"
210 PRINT:INPUT "VO = ",VO
220 PRINT:INPUT "RG = ",RG:IF RG<0 THEN 200
230 PRINT:INPUT "RL = ",RL:IF RL<0 THEN 200
240 PRINT:INPUT "ZO = ",ZO:IF ZO<=0 THEN 200
250 LOCATE 23,1:PRINT SB$:C$=INPUT$(1)
260 '* COMPUTE VALUES OF VOLTAGES AND CURRENTS IN THE
270 '  BOUNCE DIAGRAMS *
280 V(0)=VO*ZO/(RG+ZO):C(0)=V(0)/ZO
290 GL=(RL-ZO)/(RL+ZO):GG=(RG-ZO)/(RG+ZO)
300 FOR I=1 TO 15 STEP 2:V(I)=V(I-1)*GL:C(I)=-V(I)/ZO:V(I
    +1)=V(I)*GG:C(I+1)=V(I+1)/ZO:NEXT
310 VL(0)=0:CL(0)=0
320 FOR I=1 TO 16:VL(I)=VL(I-1)+V(I-1):CL(I)=CL(I-1)+C(I-
    1):NEXT
330 '* DETERMINE LINE VOLTAGE AND CURRENT VERSUS DISTANCE
340 '  ALONG THE LINE FOR SPECIFIED VALUE OF TIME *
350 CLS:PRINT "ENTER ANY VALUE FROM 0 TO 16 FOR T/TO"
```

PL 7.3. (continued)

```
360 PRINT "FOR WHICH LINE VOLTAGE AND LINE CURRENT"
370 INPUT "VERSUS Z/L ARE DESIRED: ",TIME
380 IF TIME<0 OR TIME>16 THEN 350
390 CLS:PRINT "LINE VOLTAGE VS. Z/L AT T/TO =";TIME;"IS:"
400 PRINT:J=INT(TIME)
410 IF (J-INT(J/2)*2) THEN ZM=FN TRD(1-TIME+J):J1=J:J2=J+
    1:GOTO 430
420 ZM=FN TRD(TIME-J):J1=J+1:J2=J
430 IF VL(J1)=VL(J2) THEN ZM=0
440 IF ZM=0 THEN 460
450 PRINT FN TRD(VL(J1));"V FROM Z/L = 0 TO Z/L =";ZM
460 IF ZM=1 THEN 480
470 PRINT FN TRD(VL(J2));"V FROM Z/L =";ZM;" TO Z/L = 1"
480 PRINT:PRINT "LINE CURRENT VS. Z/L AT T/TO =";TIME;"IS
    :"
490 PRINT:IF ZM=0 THEN 510
500 PRINT FN TRD(CL(J1));"A FROM Z/L = 0 TO Z/L =";ZM
510 IF ZM=1 THEN 530
520 PRINT FN TRD(CL(J2));"A FROM Z/L =";ZM;" TO Z/L = 1"
530 LOCATE 23,1:PRINT SB$:C$=INPUT$(1)
540 CLS:PRINT "IF YOU WISH TO ENTER ANOTHER VALUE OF"
550 INPUT "T/TO, TYPE Y, OTHERWISE TYPE N: ",W$
560 IF LEFT$(W$,1)="Y" THEN 350
570 CLS:PRINT "IF YOU WISH TO TRY ANOTHER EXAMPLE,"
580 INPUT "TYPE Y, OTHERWISE TYPE N: ",W$
590 IF LEFT$(W$,1)="Y" THEN 200
600 CLS:PRINT "THE END"
610 END
```

```
RUN
ENTER VALUES OF VO IN V, AND RG,RL,
AND ZO IN OHMS:

VO = 100

RG = 40

RL = 120

ZO = 60
STRIKE ANY KEY TO CONTINUE
ENTER ANY VALUE FROM 0 TO 16 FOR T/TO
FOR WHICH LINE VOLTAGE AND LINE CURRENT
VERSUS Z/L ARE DESIRED: 2.5
LINE VOLTAGE VS. Z/L AT T/TO = 2.5 IS:

 76 V FROM Z/L = 0 TO Z/L = .5
 80 V FROM Z/L = .5  TO Z/L = 1

LINE CURRENT VS. Z/L AT T/TO = 2.5 IS:

 .6 A FROM Z/L = 0 TO Z/L = .5
 .66667 A FROM Z/L = .5  TO Z/L = 1
STRIKE ANY KEY TO CONTINUE
IF YOU WISH TO ENTER ANOTHER VALUE OF
T/TO, TYPE Y, OTHERWISE TYPE N: Y
ENTER ANY VALUE FROM 0 TO 16 FOR T/TO
FOR WHICH LINE VOLTAGE AND LINE CURRENT
VERSUS Z/L ARE DESIRED: 1.33333
LINE VOLTAGE VS. Z/L AT T/TO = 1.33333 IS:
```

PL 7.3. (continued)

```
 60 V FROM Z/L = 0 TO Z/L = .66667
 80 V FROM Z/L = .66667  TO Z/L = 1

LINE CURRENT VS. Z/L AT T/TO = 1.33333 IS:

 1 A FROM Z/L = 0 TO Z/L = .66667
 .66667 A FROM Z/L = .66667  TO Z/L = 1
STRIKE ANY KEY TO CONTINUE
IF YOU WISH TO ENTER ANOTHER VALUE OF
T/TO, TYPE Y, OTHERWISE TYPE N: N
IF YOU WISH TO TRY ANOTHER EXAMPLE,
TYPE Y, OTHERWISE TYPE N: N
THE END
```

Excitation by pulse voltage source

In Example 7.3 we introduced the bounce diagram technique for a constant-voltage source. The technique can also be applied if the voltage source is a pulse. In the case of a rectangular pulse, this can be done by representing the pulse as the superposition of two step functions, as shown in Fig. 7.20, and superimposing the bounce diagrams for the two sources one upon another. In doing so, we should note that the bounce diagram for one source begins at a value of time greater than zero. Alternatively, the time variation for each wave can be drawn alongside the time axes beginning at the time of start of the wave. These can then be used to plot the required sketches. An example is in order to illustrate this technique, which can also be used for a pulse of arbitrary shape.

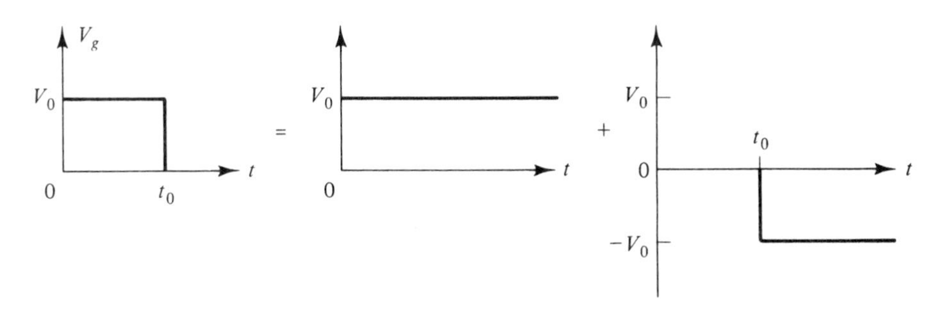

Figure 7.20. Representation of a rectangular pulse as the superposition of two step functions.

Example 7.4

Let us assume that the voltage source in the system of Fig. 7.15 is a 100 V rectangular pulse extending from $t = 0$ to $t = 1$ μs and extend the bounce diagram technique.

Considering, for example, the voltage bounce diagram, we reproduce in Fig. 7.21 part of the voltage bounce diagram of Fig. 7.16(a) and draw the time-variations of the individual pulses alongside the time axes, as shown in the figure. Note that voltage axes are chosen such that positive values are to the left at the left end ($z = 0$) of the diagram, but to the right at the right end ($z = l$) of the diagram.

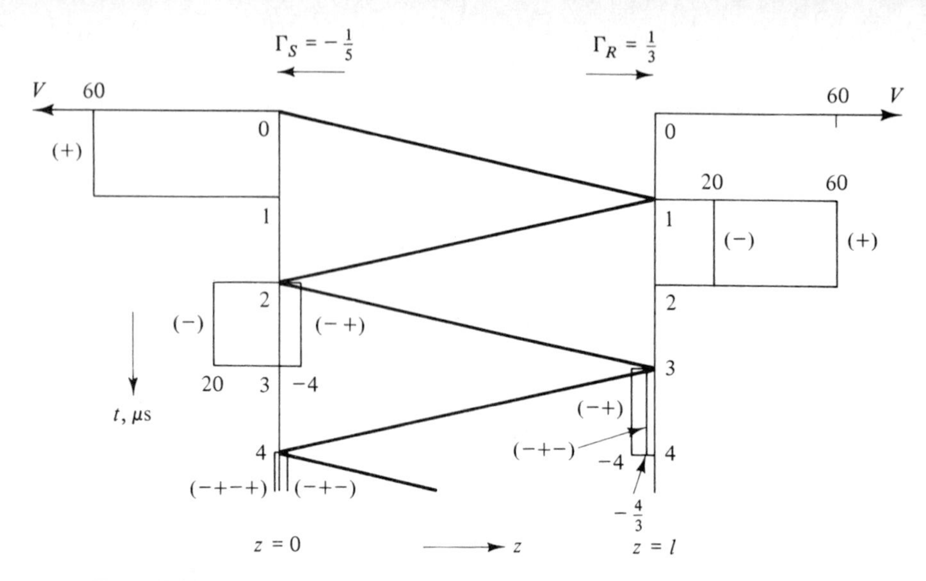

Figure 7.21. Voltage bounce diagram for the system of Fig. 7.15 except that the voltage source is a rectangular pulse of 1-μs duration from $t = 0$ to $t = 1$ μs.

From the voltage bounce diagram, sketches of line voltage versus time at $z = 0$ and $z = l$ can be drawn as shown in Fig. 7.22(a) and (b), respectively. To draw the sketch of line voltage versus time for any other value of z, we note that as time progresses, the (+) wave pulses slide down the crisscross lines from left to right, whereas the (−) wave pulses slide down from right to left. Thus to draw the sketch for $z = l/2$, we displace the time plots of the (+) waves at $z = 0$ and of the (−) waves at $z = l$ forward in time by 0.5 μs, that is, delay them by 0.5 μs, and add them to obtain the plot shown in Fig. 7.22(c).

Sketches of line voltage versus distance (z) along the line for fixed values of time can also be drawn from the bounce diagram based on the phenomenon of the individual pulses sliding down the crisscross lines. Thus, if we wish to sketch $V(z)$ for $t = 2.25$ μs, then we take the portion from $t = 2.25$ μs back to $t = 2.25 - 1 = 1.25$ μs (since the one-way travel time on the line is 1 μs) of all the (+) wave pulses at $z = 0$ and lay them on the line from $z = 0$ to $z = l$, and we take the portion from $t = 2.25$ μs back to $t = 2.25 - 1 = 1.25$ μs of all the (−) wave pulses at $z = l$ and lay them on the line from $z = l$ back to $z = 0$. In this case, we have only one (+) wave pulse—that of the (− +) wave—and only one (−) wave pulse—that of the (−) wave—as shown in Fig. 7.23(a) and (b). The line voltage is then the superposition of these two waveforms, as shown in Fig. 7.23(c).

Similar considerations apply for the current bounce diagram and plots of line current versus t for fixed values of z and line current versus z for fixed values of t.

Relevance of transient bouncing of waves

In this section we have discussed the transient bouncing back and forth of waves along a transmission line. An example in which this phenomenon is of concern is in the interconnection of microelectronic silicon chips in a high-speed digital computer.[1] We shall consider this topic in more detail in Section 7.6. On the

[1]See, for example, Albert J. Blodgett, Jr., "Microelectronic Packaging," *Scientific American*, July 1983, pp. 86–96.

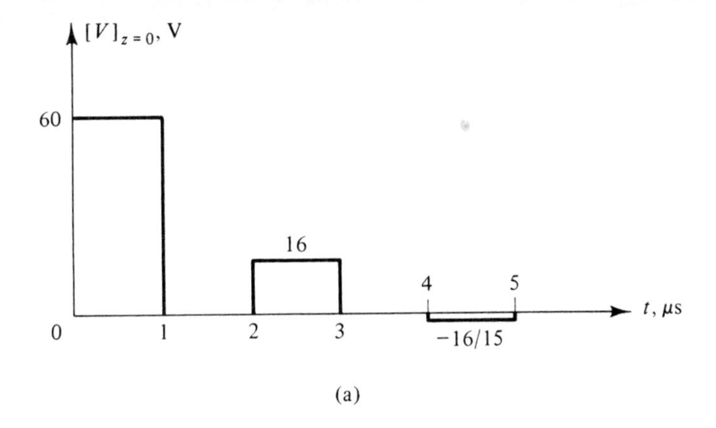

(a)

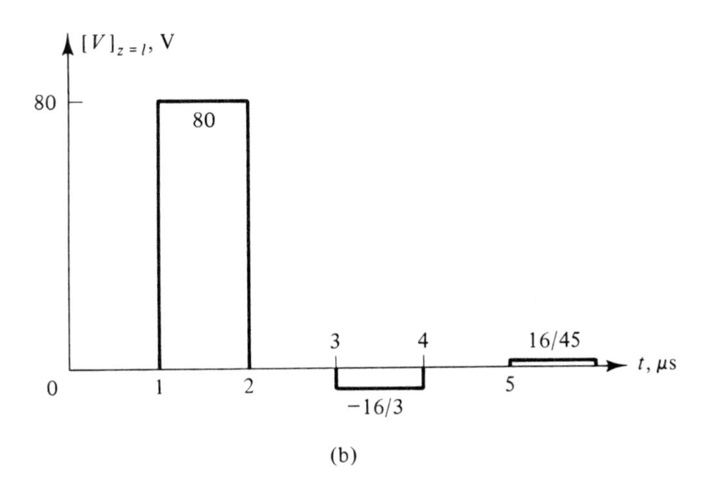

(b)

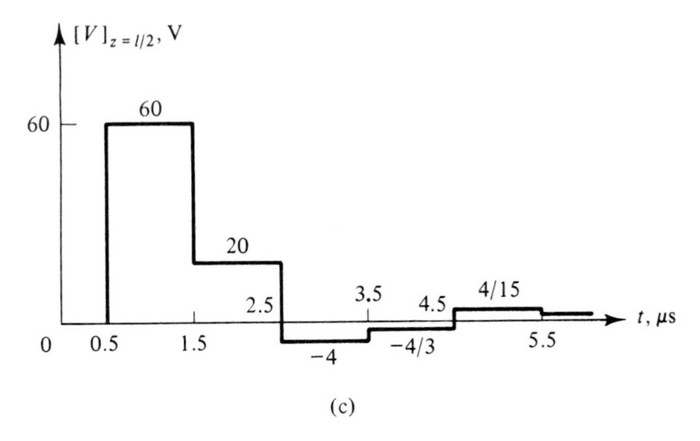

(c)

Figure 7.22. Time variations of line voltage at (a) $z = 0$, (b) $z = l$, and (c) $z = l/2$ for the system of Fig. 7.15, except that the voltage source is a rectangular pulse of 1-μs duration from $t = 0$ to $t = 1$ μs.

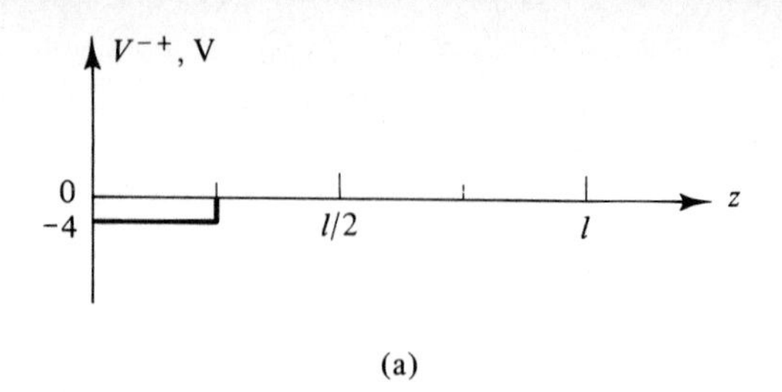

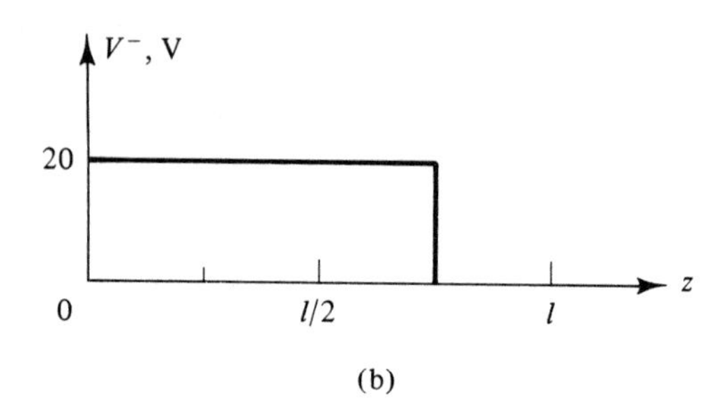

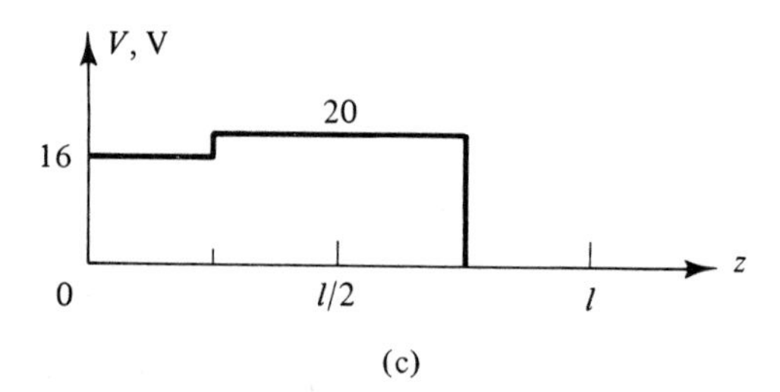

Figure 7.23. Variations with z of (a) the $(-+)$ wave voltage, (b) the $(-)$ wave voltage, and (c) the total line voltage, at $t = 2.25\ \mu s$ for the system of Fig. 7.15, except that the voltage source is a rectangular pulse of 1-μs duration from $t = 0$ to $t = 1 \mu s$.

other hand, the same phenomenon can be used to advantage as a diagnostic tool in the location of faults in transmission-line systems, as we discuss in the following section.

K7.3. (+) wave; (−) wave; Voltage reflection coefficient; Current reflection coefficient; Transient bouncing of waves; Voltage bounce diagram; Current bounce diagram; Steady-state situation; Rectangular pulse voltage source; Superposition.

D7.5. In the system shown in Fig. 7.24, the switch S is closed at $t = 0$. Find the value of R_L for each of the following cases: **(a)** $V(0.5l, 1.7\ \mu s) = 48$ V; **(b)** $V(0.6l, 2.8\ \mu s) = 76$ V; **(c)** $I(0.3l, 4.4\ \mu s) = 1$ A; and **(d)** $I(0.4l, \infty) = 2.5$ A.
Ans. **(a)** 40 Ω; **(b)** 120 Ω; **(c)** 60 Ω; **(d)** 0 Ω

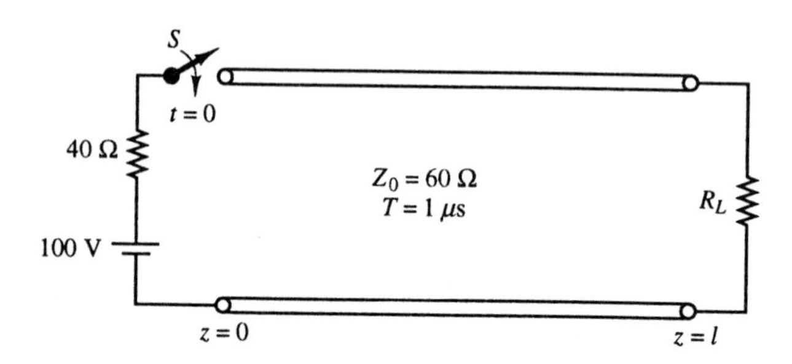

Figure 7.24. For Problem D7.5

D7.6. For a line of characteristic impedance 75 Ω terminated by a resistance and driven by a constant voltage source in series with an internal resistance, the line voltage and current in the steady state are known to be 30 V and 1.2 A, respectively. Find **(a)** the (+) wave voltage; **(b)** the (−) wave voltage; **(c)** the (+) wave current; and **(d)** the (−) wave current, in the steady state.
Ans. **(a)** 60 V; **(b)** −30 V; **(c)** 0.8 A; **(d)** 0.4 A

D7.7 In Fig 7.25, a line of characteristic impedance 50 Ω is terminated by a *passive* nonlinear element. A (+) wave of constant voltage V_0 is incident on the termination. If the volt-ampere characteristic of the nonlinear element in the region of interest is $V_L = 50\ I_L^2$, find the (−) wave voltage for each of the following values of V_0: **(a)** 36 V; **(b)** 50 V; and **(c)** 66 V.
Ans. **(a)** −4 V; **(b)** 0 V; **(c)** 6 V

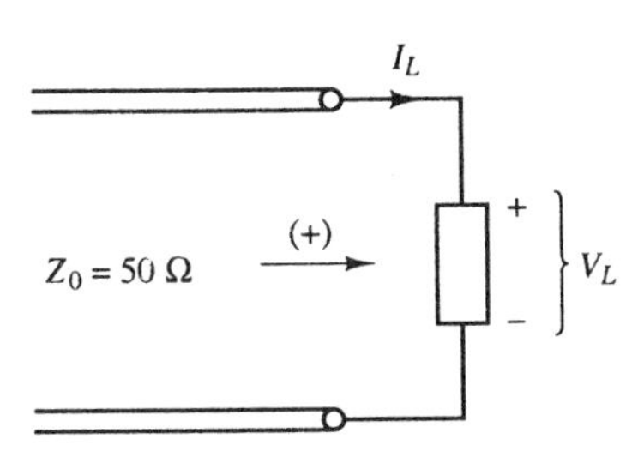

Figure 7.25. For Problem D7.7.

7.4 TRANSMISSION-LINE DISCONTINUITY

Junction between two lines

We now consider the case of a junction between two lines having different values for the parameters Z_0 and v_p, as shown in Fig. 7.26. Assuming that a (+) wave of voltage V^+ and a current I^+ is incident on the junction from line 1, we find that the (+) wave alone cannot satisfy the boundary condition at the junction since the voltage-to-current ratio for that wave is Z_{01}, whereas the characteristic impedance of line 2 is $Z_{02} \neq Z_{01}$. Hence, a reflected wave and a transmitted wave are set up such that the boundary conditions are satisfied. Let the voltages and currents in these waves be V^-, I^-, and V^{++}, I^{++}, respectively, where the superscript ++ denotes that the transmitted wave is a (+) wave resulting from the incident (+) wave. We then have the situation shown in Fig. 7.27(a).

Using the boundary conditions at the junction, we then write

$$V^+ + V^- = V^{++} \tag{7.42a}$$

$$I^+ + I^- = I^{++} \tag{7.42b}$$

But we know that $I^+ = V^+/Z_{01}$, $I^- = -V^-/Z_{01}$, and $I^{++} = V^{++}/Z_{02}$. Hence

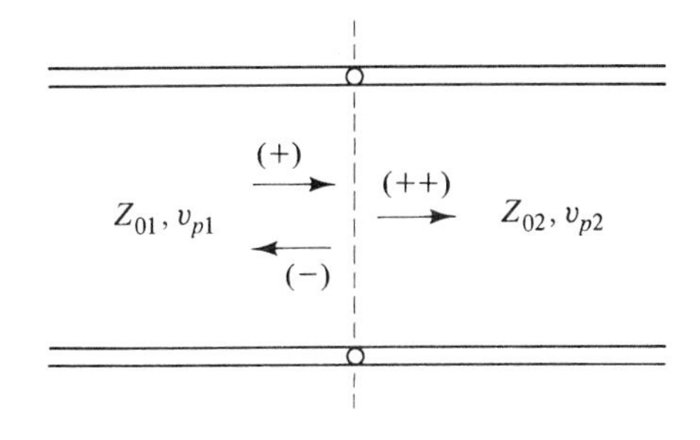

Figure 7.26. Transmission-line junction for illustrating reflection (−) and transmission (++) resulting from an incident (+) wave.

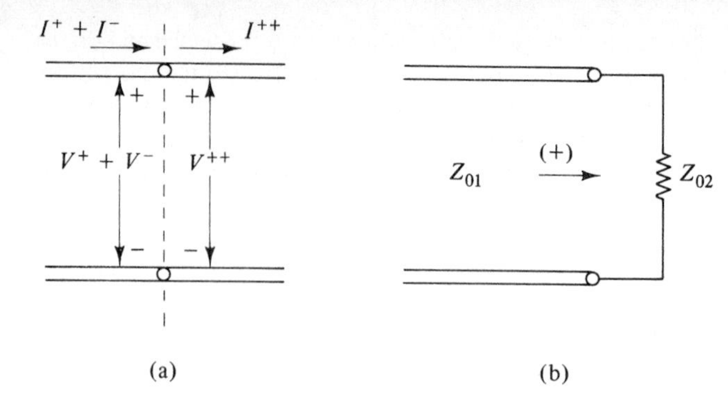

Figure 7.27. (a) For obtaining the reflected (−) wave and transmitted (+ +) wave voltages and currents for the system of Fig. 7.26. (b) Equivalent to (a) for using the reflection coefficient concept.

(7.42b) becomes

$$\frac{V^+}{Z_{01}} - \frac{V^-}{Z_{01}} = \frac{V^{++}}{Z_{02}} \tag{7.43}$$

Combining (7.42a) and (7.43), we have

$$V^+ + V^- = \frac{Z_{02}}{Z_{01}}(V^+ - V^-)$$

$$V^-\left(1 + \frac{Z_{02}}{Z_{01}}\right) = V^+\left(\frac{Z_{02}}{Z_{01}} - 1\right)$$

or

$$\boxed{\Gamma = \frac{V^-}{V^+} = \frac{Z_{02} - Z_{01}}{Z_{02} + Z_{01}}} \tag{7.44}$$

Thus to the incident wave, the transmission line to the right looks like its characteristic impedance Z_{02}, as shown in Fig. 7.27(b). The difference between a resistive load of Z_{02} and a line of characteristic impedance Z_{02} is that, in the first case, power is dissipated in the load, whereas in the second case, the power is transmitted into the line.

Transmission coefficient

We now define the *voltage transmission coefficient,* denoted by the symbol τ_V, as the ratio of the transmitted wave voltage to the incident wave voltage. Thus

$$\boxed{\tau_V = \frac{V^{++}}{V^+} = \frac{V^+ + V^-}{V^+} = 1 + \frac{V^-}{V^+} = 1 + \Gamma} \tag{7.45}$$

The *current transmission coefficient,* τ_C, which is the ratio of the transmitted wave current to the incident wave current, is given by

$$\boxed{\tau_C = \frac{I^{++}}{I^+} = \frac{I^+ + I^-}{I^+} = 1 + \frac{I^-}{I^+} = 1 - \Gamma} \tag{7.46}$$

At this point, one may be puzzled to note that the transmitted voltage can be greater than the incident voltage if Γ is positive. However, this is not of concern, since then the transmitted current will be less than the incident current. Similarly, the transmitted current is greater than the incident current when Γ is negative, but then the transmitted voltage is less than the incident voltage. In fact, what is important is that the transmitted power P^{++} is always less than (or equal to) the incident power P^+, since

$$\begin{aligned} P^{++} &= V^{++}I^{++} = V^+(1+\Gamma)I^+(1-\Gamma) \\ &= V^+I^+(1-\Gamma^2) = (1-\Gamma^2)P^+ \end{aligned} \tag{7.47}$$

and $(1-\Gamma^2) \leq 1$, irrespective of whether Γ is positive or negative.

We shall illustrate the application of reflection and transmission at a junction between lines by means of an example.

Example 7.5

System of three lines

Let us consider the system of three lines in cascade, driven by a unit impulse voltage source $\delta(t)$, as shown in Fig. 7.28(a). We wish to find the output voltage V_o, thereby obtaining the unit impulse response.

To find the output voltage, we draw the voltage bounce diagram, as shown in Fig. 7.28(b). In drawing the bounce diagram, we note that since the internal resistance of the voltage source is 50 Ω, which is equal to Z_{01}, the strength of the impulse that the generator supplies to line 1 is $\frac{1}{2}$. The strengths of the various impulses propagating in the lines are then governed by the reflection and transmission coefficients indicated on the diagram. Also note that the numbers indicated beside the crisscross lines are simply the strengths of the impulses and do not represent constant voltages.

Unit impulse response

From the bounce diagram, we note that the output voltage is a series of impulses. In fact, the phenomenon can be visualized without even drawing the bounce diagram, and the strengths of the impulses can be computed. Thus the strength of the first impulse, which occurs at $t = T_1 + T_2 + T_3 = 6\ \mu$s, is

$$1 \times \frac{50}{50+50} \times \left(1 + \frac{100-50}{100+50}\right) \times \left(1 + \frac{50-100}{50+100}\right) = 1 \times \frac{1}{2} \times \frac{4}{3} \times \frac{2}{3} = \frac{4}{9}$$

Each succeeding impulse is due to the additional reflection and re-reflection of the previous impulse at the right and left end, respectively, of line 2. Hence each succeeding impulse occurs $2T_2$ or 4 μs later than the preceding one, and its strength is

$$\left(\frac{50-100}{50+100}\right)\left(\frac{50-100}{50+100}\right) = \left(-\frac{1}{3}\right)\left(-\frac{1}{3}\right) = \frac{1}{9}$$

times the strength of the previous impulse. We can now write the output voltage as

$$\begin{aligned} V_o(t) &= \frac{4}{9}\delta(t - 6\times 10^{-6}) + \frac{4}{9^2}\delta(t - 10\times 10^{-6}) \\ &\quad + \frac{4}{9^3}\delta(t - 14\times 10^{-6}) + \cdots \\ &= \frac{4}{9}\sum_{n=0}^{\infty}\left(\frac{1}{9}\right)^n \delta(t - 4n\times 10^{-6} - 6\times 10^{-6}) \end{aligned} \tag{7.48}$$

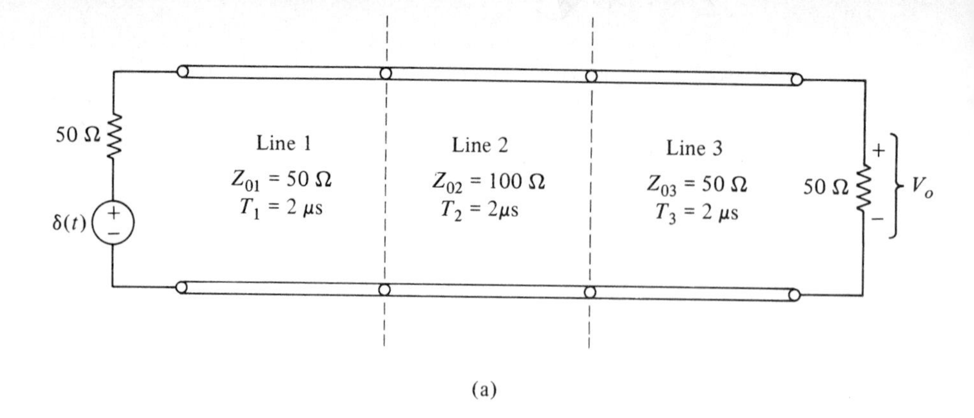

(a)

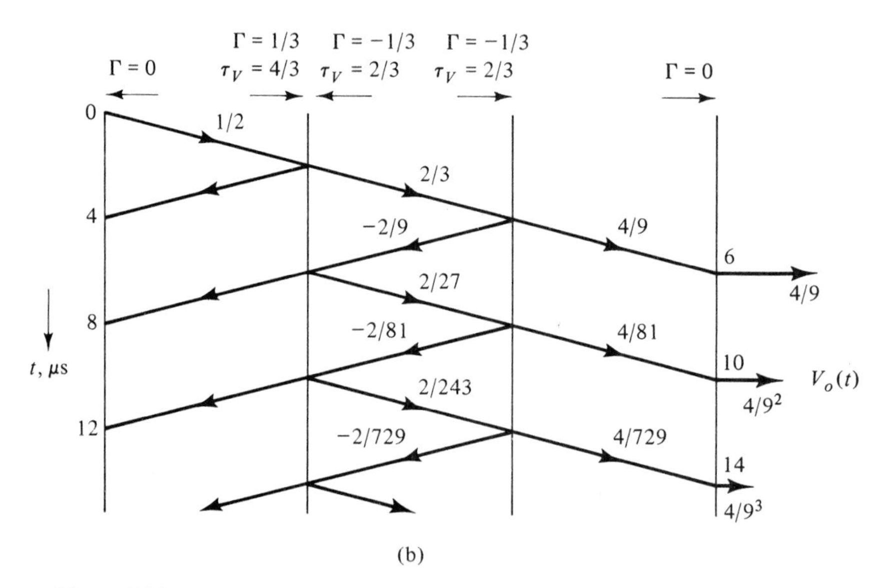

(b)

Figure 7.28. (a) System of three lines in cascade driven by a unit impulse voltage source. (b) Voltage bounce diagram for finding the output voltage $V_o(t)$ for the system of (a).

Note that $\frac{4}{9}$ is the strength of the first impulse and $\frac{1}{9}$ is the multiplication factor for each succeeding impulse. In terms of T_1, T_2, and T_3, we have

$$\begin{aligned} V_o(t) &= \frac{4}{9}\sum_{n=0}^{\infty}\left(\frac{1}{9}\right)^n \delta[t - 2nT_2 - (T_1 + T_2 + T_3)] \\ &= \frac{4}{9}\sum_{n=0}^{\infty}\left(\frac{1}{9}\right)^n \delta(t - 2nT_2 - T_0) \end{aligned} \tag{7.49}$$

where we have replaced $T_1 + T_2 + T_3$ by T_0.

Frequency response

Proceeding further, since the unit impulse response of the system is a series of impulses delayed in time, the response to any other excitation can be found by the superposition of time functions obtained by delaying the exciting function and multiplying by appropriate constants. In particular, by considering $V_g(t) = \cos \omega t$, we can find the frequency response of the system. Thus, assuming $V_g(t) = \cos \omega t$ and substituting the cosine function for the impulse function in (7.49), we obtain the corresponding output voltage to be

$$V_o(t) = \frac{4}{9}\sum_{n=0}^{\infty}\left(\frac{1}{9}\right)^n \cos\omega(t - 2nT_2 - T_0) \tag{7.50}$$

The complex voltage $\bar{V}_o(\omega)$ is then given by

$$\begin{aligned}\bar{V}_o(\omega) &= \frac{4}{9}\sum_{n=0}^{\infty}\left(\frac{1}{9}\right)^n e^{-j\omega(2nT_2+T_0)} \\ &= \frac{4}{9}e^{-j\omega T_0}\sum_{n=0}^{\infty}\left(\frac{1}{9}e^{-j2\omega T_2}\right)^n \\ &= \frac{(4/9)e^{-j\omega T_0}}{1 - (1/9)e^{-j2\omega T_2}}\end{aligned} \tag{7.51}$$

Without going into a detailed discussion of the result given by (7.51), we can conclude the following: maximum amplitude occurs for $2\omega T_2 = 2m\pi$, $m = 0, 1, 2, \ldots$; that is, for $\omega = m\pi/T_2$, $m = 0, 1, 2, \ldots$, and its value is $\frac{4}{9}/(1 - \frac{1}{9}) = 0.5$. Minimum amplitude occurs for $2\omega T_2 = (2m + 1)\pi$, $m = 0, 1, 2, \ldots$; that is, for $\omega = (2m + 1)\pi/2T_2$, $m = 0, 1, 2, \ldots$, and its value is $\frac{4}{9}/(1 + \frac{1}{9}) = 0.4$. The amplitude response can therefore be roughly sketched, as shown in Fig. 7.29.

Radome

A practical situation in which the discussion of this example is applicable is in the design of a radome, which is an enclosure for protecting an antenna from the weather while allowing transparency for electromagnetic waves. A simple, idealized, planar version of the radome is a dielectric slab with free space on either side of it, as shown in Fig. 7.30. For reflection and transmission of uniform plane waves incident normally onto the dielectric slab, the arrangement is equivalent to three lines in cascade, with the characteristic impedances equal to the intrinsic impedances of the media and the velocities of propagation equal to those in the media. Thus the amplitude versus frequency response is of the same form as that in Fig. 7.29, where T_2 is the one-way travel time in the dielectric slab and the maximum is 1 instead of 0.5 (the factor of 0.5 in Fig. 7.29 is due to voltage drop across the internal resistance of the source in the transmission-line system). Hence the lowest frequency for which the dielectric slab is completely transparent is $\omega = \pi/T_2 = \pi c/l\sqrt{\varepsilon_r\mu_r}$ or $f = c/2l\sqrt{\varepsilon_r\mu_r}$. Conversely, for a given frequency f, the minimum thickness for which the slab is transparent is $l = c/2f\sqrt{\varepsilon_r\mu_r} = \lambda/2$, where λ is the wavelength in the dielectric, corresponding to f.

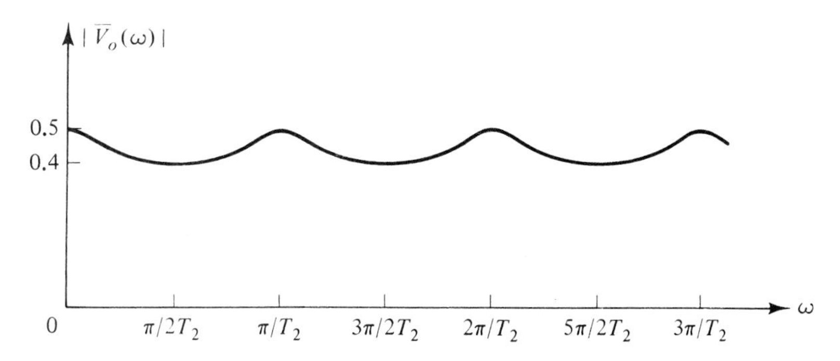

Figure 7.29. Rough sketch of amplitude response versus frequency for the system of Fig. 7.28(a) for sinusoidal excitation.

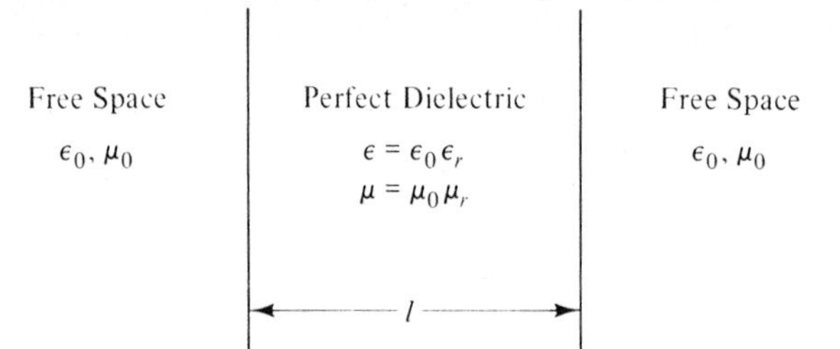

Figure 7.30. Perfect dielectric slab with free space on either side.

Time-domain reflectometry

We shall now discuss *time-domain reflectometry,* abbreviated TDR, a technique by means of which discontinuities in transmission-line systems can be located by making measurements with pulses. The block diagram of a typical TDR system is shown in Fig. 7.31. It consists of a pulse generator whose output is connected to the system under test through a matched attenuator. Voltage pulses are propagated down the transmission-line system, and the incident and reflected pulses are monitored by the display scope using a high-impedance probe. The matched attenuator serves the purpose of absorbing the pulses arriving back from the system so that reflections of those pulses are not produced. We shall illustrate the application of a TDR system by means of an example.

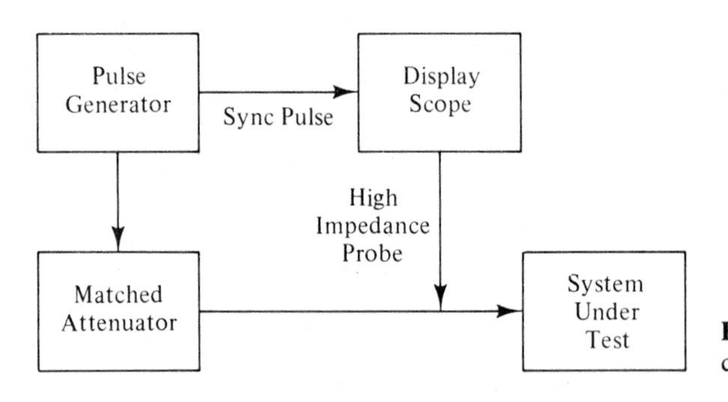

Figure 7.31. Block diagram of a typical time-domain reflectometer.

Example 7.6

Let us consider a transmission line under test as shown in Fig. 7.32, in which a discontinuity exists at $z = 4$ m and the line is short circuited at the far end. We shall first analyze the system to obtain the waveform measured by a TDR system connected at the input end $z = 0$, assuming the TDR pulses to be of amplitude 1 V, duration 10 ns, and repetition rate 10^5 Hz. We shall then discuss how one can deduce the information about the discontinuity from the TDR measurement.

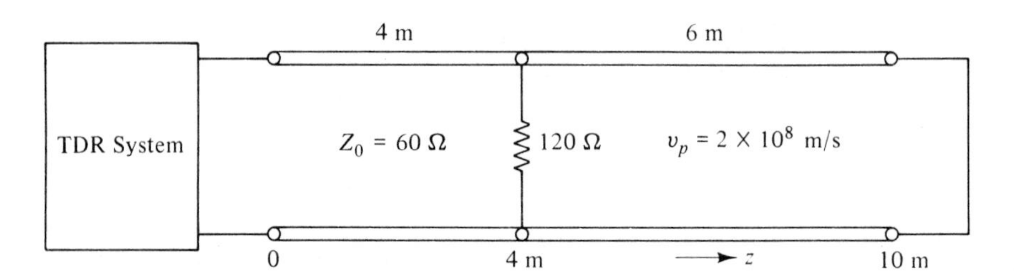

Figure 7.32. Transmission line with discontinuity under test by a TDR system.

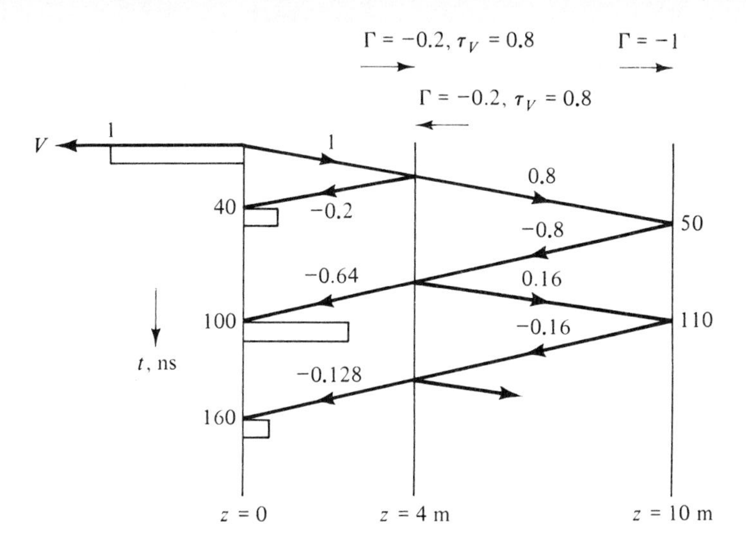

Figure 7.33. Voltage bounce diagram for the system of Fig. 7.32, for TDR pulses of amplitude 1 V.

Assuming that a pulse from the TDR system is incident on the input of the system under test at $t = 0$, we draw the voltage bounce diagram as shown in Fig. 7.33. Note that for a pulse incident on the discontinuity from either side, the resistance viewed is the parallel combination of 120 Ω and Z_0 (= 60 Ω) of the line, or 40 Ω. Hence the reflection and transmission coefficients for the voltage are given, respectively, by

$$\Gamma = \frac{40 - 60}{40 + 60} = -0.2$$

$$\tau_V = 1 + \Gamma = 0.8$$

From the bounce diagram, the voltage pulses which would be viewed on the display scope of the TDR system up to $t = 200$ ns are shown in Fig. 7.34. Subsequent pulses become smaller and smaller in amplitude as time progresses and diminish to insignificant values well before $t = 10\ \mu$s, which is the period of the TDR pulses.

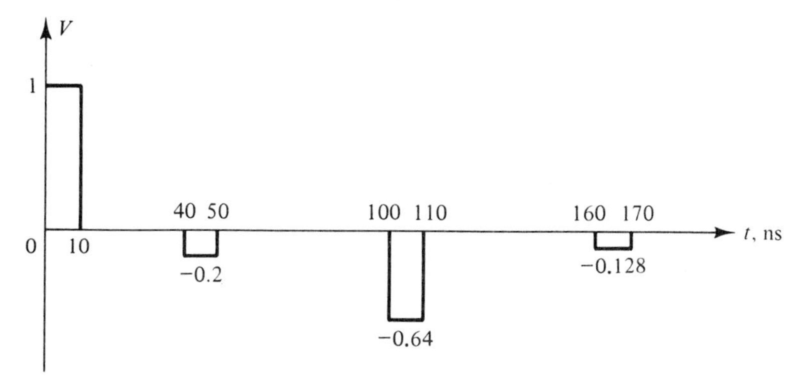

Figure 7.34. Voltage versus time at the input of the transmission line of Fig. 7.32, as displayed by the TDR system.

Now, to discuss how one can deduce information about the discontinuity from the TDR display of Fig. 7.34, without a priori knowledge of the discontinuity but knowing the values of Z_0 and v_p of the line and that the line is short circuited at the far end of unknown distance from the input, we proceed in the following manner:

The first pulse is the outgoing pulse from the TDR system. The second pulse arriving at the input of the system under test at $t = 40$ ns is due to reflection from a discontinuity, since if there is no discontinuity, the voltage of the second pulse should be -1 V and there should be no subsequent pulses. From the voltage of the second pulse, we know that the reflection coefficient at the discontinuity is -0.2. The effective resistance R_L seen by the incident pulse is therefore given by the solution of

$$\frac{R_L - 60}{R_L + 60} = 0.2$$

which is 40 Ω. Since this value is less than the Z_0 of the line, the discontinuity must be due entirely to a resistance in parallel with the line or due to a combination of series and parallel resistors; it cannot be due entirely to a resistance in series with the line. Let us proceed with the assumption of a parallel resistor alone. Then the value of this resistance R must be such that

$$\frac{60R}{60 + R} = 40$$

solving which we obtain $R = 120\ \Omega$. The location of the discontinuity can be deduced by multiplying v_p by 20 ns, which is one-half of the time interval between the first and second pulses. Thus the location is $2 \times 10^8 \times 20 \times 10^{-9} = 4$ m.

Continuing, let us postulate that the third pulse of -0.64 V at $t = 100$ ns is due to reflection occurring at a second discontinuity located at $z = 4 + 2 \times 10^8 \times (60/2) \times 10^{-9} = 10$ m. In terms of the reflection coefficient at the second discontinuity, denoted Γ_2, the voltage of the third pulse would be $\tau_{VR}\Gamma_2\tau_{VL}$, where τ_{VR} and τ_{VL} are the voltage transmission coefficients at $z = 4$ m, for pulses incident from the right and from the left, respectively. Since τ_{VR} and τ_{VL} are both equal to 0.8, we then have $0.64\Gamma_2 = -0.64$ or $\Gamma_2 = -1$, which corresponds to a short circuit, which would then give a fourth pulse of -0.128 V at $t = 160$ ns, and so on. From these reasonings, we confirm the assumption of a parallel resistor of 120 Ω for the discontinuity at $z = 4$ m and also conclude that the short circuit is at $z = 10$ m and that no discontinuities exist between $z = 4$ m and the short circuit. If the value of Γ_2 comes out to be different from -1, then further reasonings are necessary to deduce the information. It should also be noted that the line of reasoning depends on which of the line parameters are known.

K7.4. Voltage transmission coefficient; Current transmission coefficient; Unit impulse response; Frequency response; Time-domain reflectometry.

D7.8. Consider a (+) wave incident from line 1 onto the junction between lines 1 and 2 having characteristic impedances Z_{01} and Z_{02}, respectively. Find the value of Z_{02}/Z_{01} for each of the following cases: **(a)** the reflected wave voltage is $\frac{1}{5}$ times the incident wave voltage: **(b)** the transmitted wave voltage is $\frac{1}{5}$ times the incident wave voltage; **(c)** the reflected wave voltage is $\frac{1}{5}$ times the transmitted wave voltage; and **(d)** the reflected wave current is $\frac{1}{5}$ times the transmitted wave current.
Ans. **(a)** 1.5; **(b)** $\frac{1}{9}$; **(c)** $\frac{5}{3}$; **(d)** $\frac{3}{5}$

D7.9. The output voltage $V_o(t)$ for a system of three lines in cascade is given by

$$V_o(t) = \frac{1}{4}\sum_{n=0}^{\infty}\left(\frac{1}{3}\right)^n \delta(t - 2 \times 10^{-6}n - 3 \times 10^{-6})$$

when the input voltage $V_i(t) = \delta(t)$. If $V_i(t) = \cos \omega t$, find the amplitude of $V_o(t)$ for each of the following values of ω: **(a)** $10^6\pi$; **(b)** $1.25 \times 10^6\pi$; and **(c)** $1.5 \times 10^6\pi$.
Ans. **(a)** 0.375; **(b)** 0.2372; **(c)** 0.1875

D7.10. Consider $(n + 1)$ lines, each of characteristic impedance Z_0, emanating from a common junction, as shown in Fig. 7.35 for $n = 2$. For a wave carrying power P incident on the junction from one of the lines, find the power reflected into that line and the power transmitted into each of the remaining n lines for the following cases: **(a)** $n = 2$; **(b)** $n = 3$; and **(c)** $n = 9$.
Ans. **(a)** $\frac{1}{9}P, \frac{4}{9}P$; **(b)** $\frac{1}{4}P, \frac{1}{4}P$; **(c)** $0.64P, 0.04P$

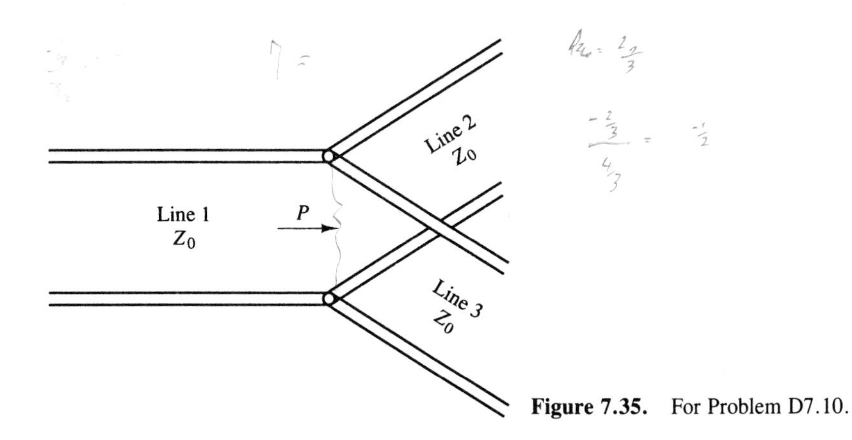

Figure 7.35. For Problem D7.10.

7.5 REACTIVE AND NONLINEAR-RESISTIVE ELEMENTS

Inductive termination

In this section we consider situations in which the concept of reflection coefficient as applicable to linear resistive terminations and discontinuities is not useful. These situations involve reactive elements and nonlinear-resistive elements. Let us first consider the system shown in Fig. 7.36 in which a line of length l is terminated by an inductor L with zero initial current and a constant voltage source

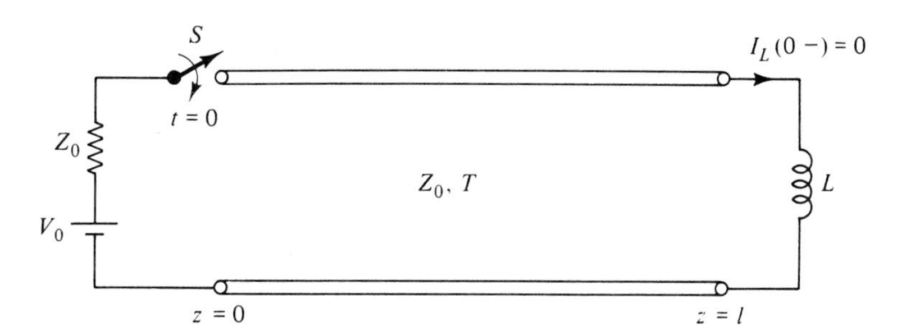

Figure 7.36. Line terminated by an inductor with zero initial current and driven by a constant voltage source in series with internal resistance equal to Z_0 of the line.

with internal resistance equal to the characteristic impedance of the line is connected to the line at $t = 0$. The internal resistance is chosen to be equal to Z_0 so that no reflection takes place at the source end. The moment the switch S is closed at $t = 0$, a $(+)$ wave originates at $z = 0$ with voltage V^+ $(= V_0/2)$ and current I^+ $(= V_0/2Z_0)$ and travels down the line to reach the load end at time T. Since the inductor current cannot change instantaneously from zero to $V_0/2Z_0$, the boundary condition at $z = l$ is violated, and hence a $(-)$ wave is set up. Let the voltage and current in this $(-)$ wave be $V^-(t)$ and $I^-(t)$, respectively. Then the total voltage across L and the total current through L are $(V_0/2) + V^-$ and $(V_0/2Z_0) - (V^-/Z_0)$, respectively, as shown in Fig. 7.37. To satisfy the boundary condition at $z = l$, we then have

$$\frac{V_0}{2} + V^- = L\frac{d}{dt}\left(\frac{V_0}{2Z_0} - \frac{V^-}{Z_0}\right) \tag{7.52}$$

Noting that V_0 is a constant and hence that dV_0/dt is zero, and rearranging we obtain

$$\boxed{\frac{L}{Z_0}\frac{dV^-}{dt} + V^- = -\frac{V_0}{2}} \tag{7.53}$$

This differential equation for $[V^-]_{z=l}$ has to be solved, subject to the intial condition. This initial condition is that the current through the inductor is zero at $t = T$; that is, the inductor behaves initially like an open circuit. Thus at $z = l$,

$$\boxed{\left[\frac{V_0}{2Z_0} - \frac{V^-}{Z_0}\right]_{t=T} = 0}$$

or

$$\boxed{[V^-]_{t=T} = \frac{V_0}{2}} \tag{7.54}$$

The general solution for the differential equation can be written as

$$V^- = -\frac{V_0}{2} + Ae^{-(Z_0/L)t} \tag{7.55}$$

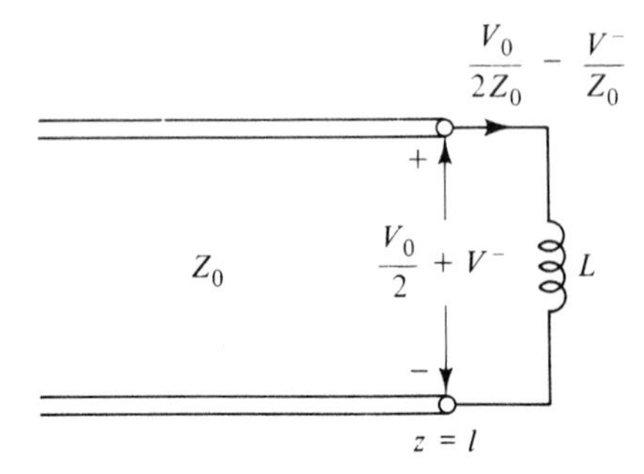

Figure 7.37. For obtaining the $(-)$ wave voltage and current for the system of Fig. 7.36.

where A is an arbitrary constant to be evaluated using (7.54). Thus we have

$$\frac{V_0}{2} = -\frac{V_0}{2} + Ae^{-(Z_0/L)T}$$

or

$$A = V_0 e^{(Z_0/L)T} \tag{7.56}$$

Substituting this result in (7.55), we obtain the solution for $[V^-]_{z=l}$ as

$$\boxed{V^-(l, t) = -\frac{V_0}{2} + V_0 e^{-(Z_0/L)(t-T)} \qquad \text{for} \quad t > T} \tag{7.57}$$

The corresponding solution for the $(-)$ wave current is given by

$$\boxed{I^-(l, t) = -\frac{V^-(l, t)}{Z_0} = \frac{V_0}{2Z_0} - \frac{V_0}{Z_0} e^{-(Z_0/L)(t-T)} \qquad \text{for} \quad t > T} \tag{7.58}$$

The $(-)$ wave, characterized by V^- and I^- as given by (7.57) and (7.58), respectively, travels back toward the source, and it does not set up a reflected wave, since the reflection coefficient at that end is zero. At this point, it can be seen that unlike in the case of linear-resistive terminations and discontinuities, the concept of reflection coefficient is not useful for studying transient behavior when reactive elements are involved. In fact, we note from (7.57) and (7.58) that the ratios of reflected voltage and current to the incident voltage and current, respectively, are no longer constants as in the resistive case.

We may now write the expressions for the total voltage across the inductor and the total current through the inductor as follows:

$$V(l, t) = \frac{V_0}{2} + V^-(l, t) = \begin{cases} 0 & \text{for} \quad t < T \\ V_0\, e^{-(Z_0/L)(t-T)} & \text{for} \quad t > T \end{cases} \tag{7.59}$$

$$I(l, t) = \frac{V_0}{2Z_0} + I^-(l, t) = \begin{cases} 0 & \text{for} \quad t < T \\ (V_0/Z_0)\left[1 - e^{-(Z_0/L)(t-T)}\right] & \text{for} \quad t > T \end{cases} \tag{7.60}$$

These quantities are shown sketched in Fig. 7.38 (a) and (b), respectively. It may be seen from these sketches that in the steady state, the voltage goes to zero and the current goes to V_0/Z_0. This is consistent with the fact that the inductor behaves like a short circuit for the dc voltage in the steady state, and hence the situation in the steady state is the same as that for a short-circuited line. Note also that the variations of the voltage and current from $t = T$ to $t = \infty$ are governed by the time constant L/Z_0, which is that of the inductor L in series with Z_0 of the line. In

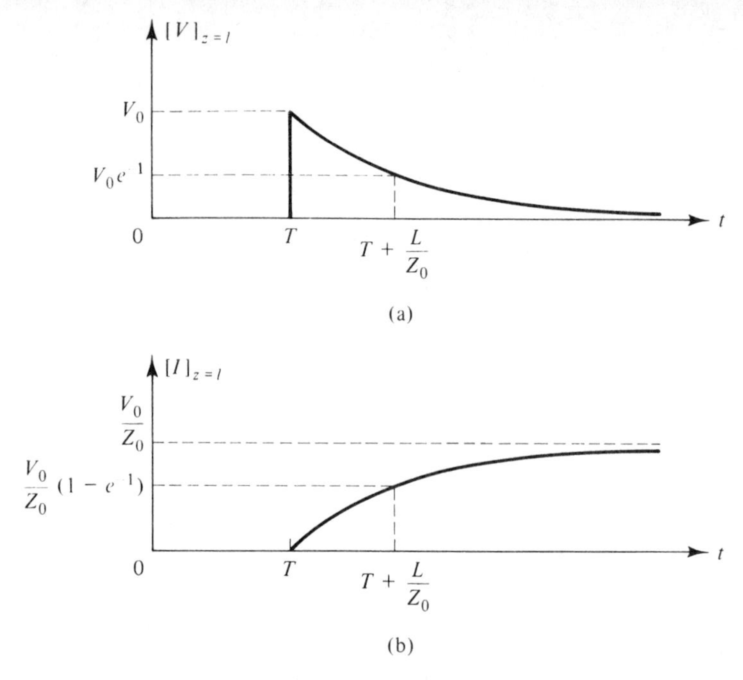

Figure 7.38. Time variations of (a) voltage across the inductor and (b) current through the inductor, for the system of Fig. 7.36.

fact, we can obtain the voltage and current sketches from considerations of initial and final behaviors of the reactive element and the time constant without formally going through the process of setting up the differential equation and solving it. We shall illustrate this procedure by means of an example.

Example 7.7

Capacitive discontinuity

Let us consider the system shown in Fig. 7.39 consisting of a series capacitor of value 10 pF at the junction between the two lines. Note that line 2 is terminated by its own characteristic impedance whereas the internal resistance of the voltage source is equal to the characteristic impedance of line 1 so that no reflections occur at the two ends of the system. We shall assume that the capacitor is initially uncharged and obtain the plots of line voltage and line current at the input $z = 0$ from considerations of initial and final behaviors of the capacitor.

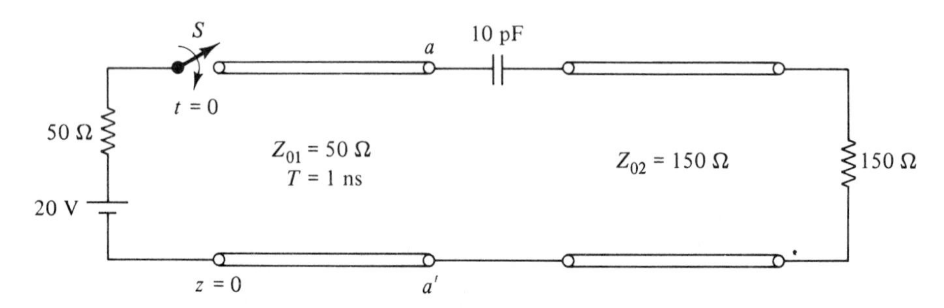

Figure 7.39. Transmission-line system with a capacitive discontinuity.

Plots of line voltage and line current at $z = 0$ versus time are shown in Fig. 7.40(a) and (b), respectively. We shall explain the several features in these plots as follows: When the switch S is closed at $t = 0$, a (+) wave of voltage 10 V and current 0.2 A goes down the line. Since the voltage across a capacitor cannot change instantaneously, the initially uncharged capacitor behaves like a short circuit when the (+) wave impinges on the junction aa' at $t = 1$ ns. Therefore the (+) wave then sees a resistance of Z_{02} (= 150 Ω) across aa' and produces a (−) wave of initial voltage 5 V and initial current −0.1 A. The (−) wave arrives initially at $z = 0$ at $t = 2$ ns, thereby changing the line voltage and line current there to 15 V and 0.1 A, as shown in Fig. 7.40(a) and (b), respectively. In the steady state, the capacitor behaves like an open circuit, which explains the steady-state values of 20 V and 0 A in these plots. Between $t = 2$ ns and $t = \infty$, the voltage and current vary exponentially with a time constant of $10^{-11} \times 200 = 2 \times 10^{-9}$ s = 2 ns, which is that of C (= 10 pF) in series with $(Z_{01} + Z_{02})$ or 200 Ω. Hence, the voltage and current values at $t = 4$ ns are $15 + 5(1 - e^{-1}) = 18.16$ V and $0.1 - 0.1(1 - e^{-1}) = 0.037$ A, respectively.

Finally, the arguments which we have employed to explain the features in Fig. 7.40 can be used to deduce information about the nature of the discontinuity if the plots represent measurements by a time-domain reflectometer.

Nonlinear-resistive termination

Another type of situation in which the concept of reflection coefficient is not useful is that involving nonlinear-resistive elements. The analysis is then made convenient by a grapical technique known as the *load-line technique*. We shall introduce this technique by means of an example.

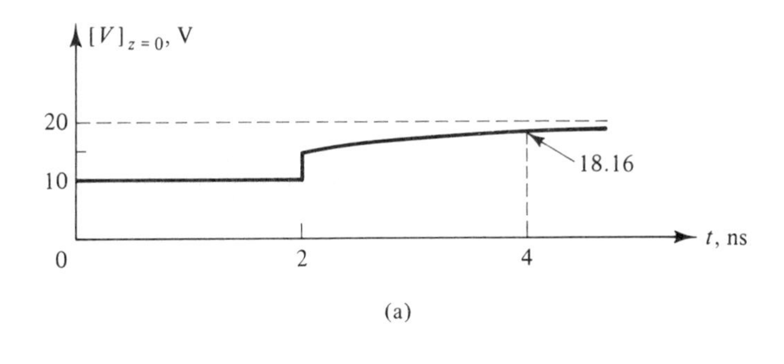

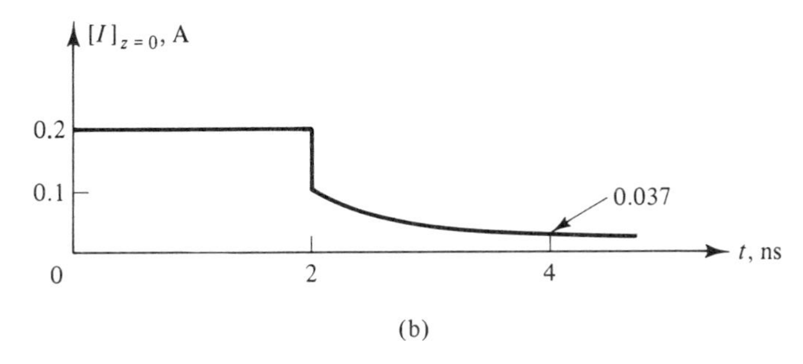

Figure 7.40. Plots of (a) line voltage and (b) line current at $z = 0$ for the system of Fig. 7.39.

Example 7.8

Load-line technique

Let us consider the transmission-line system shown in Fig. 7.41 in which the line is terminated by a passive nonlinear element having the indicated $V-I$ relationship. We wish to obtain the time variations of the voltages V_S and V_L at the source and load ends, respectively, following the closure of the switch S at $t = 0$, using the load-line technique.

With reference to the notation shown in Fig. 7.41, we can write the following equations pertinent to $t = 0+$ at $z = 0$:

$$50 = 200I_S + V_S \tag{7.61a}$$

$$V_S = V^+$$

$$I_S = I^+ = \frac{V^+}{Z_0} = \frac{V_S}{50} \tag{7.61b}$$

where V^+ and I^+ are the voltage and current, respectively, of the (+) wave set up immediately after closure of the switch. The two equations (7.61a) and (7.61b) can be solved graphically by constructing the straight lines representing them, as shown in Fig. 7.42, and obtaining the point of intersection A, which gives the values of V_S and I_S. Note in particular that (7.61b) is a straight line of slope $\frac{1}{50}$ and passing through the origin.

When the (+) wave reaches the load end $z = l$ at $t = T$, a ($-$) wave is set up. We can then write the following equations pertinent to $t = T+$ at $z = l$:

$$V_L = 50I_L|I_L| \tag{7.62a}$$

$$V_L = V^+ + V^-$$

$$I_L = I^+ + I^- = \frac{V^+ - V^-}{Z_0}$$

$$= \frac{V^+ - (V_L - V^+)}{50} = \frac{2V^+ - V_L}{50} \tag{7.62b}$$

where V^- and I^- are the ($-$) wave voltage and current, respectively. The solution for V_L and I_L is then given by the intersection of the nonlinear curve representing (7.62a) and the straight line of slope $-\frac{1}{50}$ corrresponding to (7.26b). Noting from (7.62b) that for $V_L = V^+$, $I_L = V^+/50$, we see that the straight line passes through point A. Thus the solution of (7.62a) and (7.62b) is given by point B in Fig. 7.42.

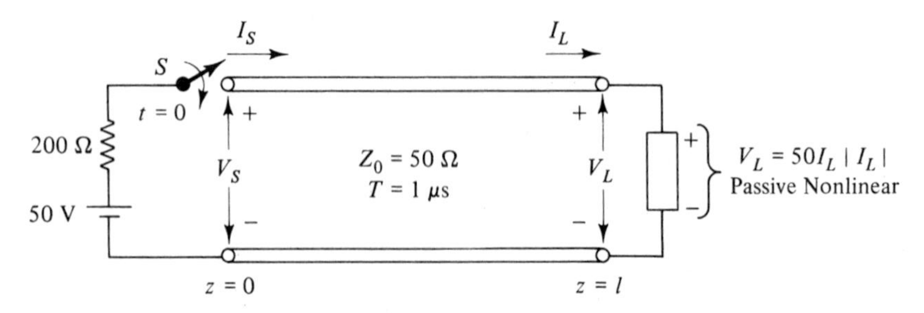

Figure 7.41. Line terminated by a passive nonlinear element and driven by a constant voltage source in series with internal resistance.

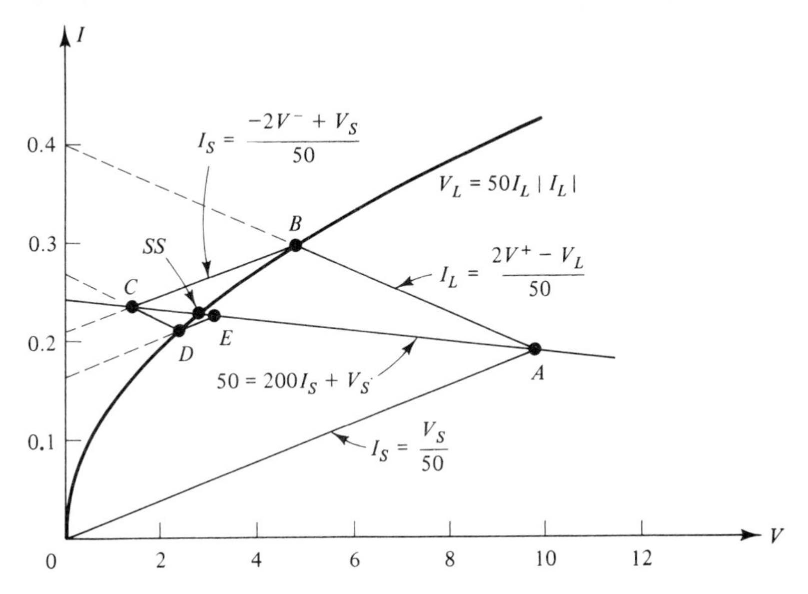

Figure 7.42. Graphical solution for obtaining time variations of V_S and V_L for $t > 0$ in the transmission-line system of Fig. 7.41.

When the $(-)$ wave reaches the source end $z = 0$ at $t = 2T$, it sets up a reflection. Denoting this to be the $(-+)$ wave, we can then write the following equations pertinent to $t = 2T+$ at $z = 0$:

$$50 = 200 I_S + V_S \tag{7.63a}$$

$$V_S = V^+ + V^- + V^{-+}$$

$$I_S = I^+ + I^- + I^{-+} = \frac{V^+ - V^- + V^{-+}}{Z_0}$$

$$= \frac{V^+ - V^- + (V_S - V^+ - V^-)}{50} = \frac{-2V^- + V_S}{50} \tag{7.63b}$$

where V^{-+} and I^{-+} are the $(-+)$ wave voltage and current, respectively. Noting from (7.63b) that for $V_S = V^+ + V^-$, $I_S = (V^+ - V^-)/50$, we see that (7.63b) represents a straight line of slope $\frac{1}{50}$ passing through B. Thus the solution of (7.63a) and (7.63b) is given by point C in Fig. 7.42.

Continuing in this manner, we observe that the solution consists of obtaining the points of intersection on the source and load $V-I$ characteristics by drawing successively straight lines of slope $1/Z_0$ and $-1/Z_0$, beginning at the origin (the initial state) and with each straight line originating at the previous point of intersection, as shown in Fig. 7.42. The points $A, C, E, \ldots,$ give the voltage and current at the source end for $0 < t < 2T$, $2T < t < 4T$, $4T < t < 6T$, $\ldots,$ whereas the points $B, D, \ldots,$ give the voltage and current at the load end for $T < t < 3T$, $3T < t < 5T$, $\ldots$. Thus for example, the time-variations of V_S and V_L are shown in Fig. 7.43(a) and (b), respectively. Finally, it can be seen from Fig. 7.42 that the steady-state values of line voltage and current are reached at the point of intersection (denoted SS) of the source and load $V-I$ characteristics.

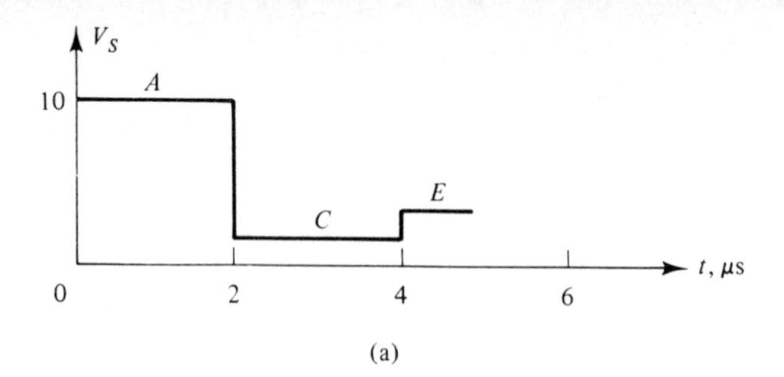

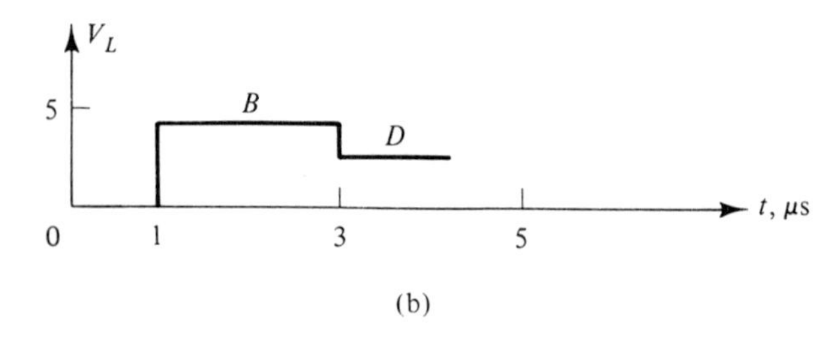

Figure 7.43. Time variations of (a) V_S and (b) V_L, for the transmission-line system of Fig. 7.41. The voltage levels A, B, C, . . . correspond to those in Fig. 7.42.

K7.5. Inductive termination; Capacitive discontinuity; Nonlinear-resistive termination; Load-line technique.

D7.11. In the system of Fig. 7.36, assume that $V_0 = 20$ V, $Z_0 = 50\ \Omega$, and $T = 1\ \mu$s. Find the value of the voltage across the inductor at $t = 2\ \mu$s for each of the following cases: **(a)** $L = 0.1$ mH, $I_L(0-) = 0$ A; **(b)** $L = 0.1$ mH, $I_L(0-) = 0.05$ A; **(c)** $L = 0.05$ mH, $I_L(0-) = 0.1$ A.
Ans. **(a)** 12.13 V; **(b)** 10.61 V; **(c)** 5.52 V

D7.12. In the system shown in Fig. 7.44, the capacitor is initially uncharged. Find the values of the line voltage at $z = 0$ at the following times: **(a)** $t = 2$ ns+; **(b)** $t = \infty$; and **(c)** $t = 3$ ns.
Ans. **(a)** 0 V; **(b)** 15 V; **(c)** 7.2987 V

D7.13. Assume that in the system of Fig. 7.41 the values of the voltage source and its internal resistance are 12 V and 10 Ω, respectively, and that Z_0 of the line is 100 Ω. By using the load-line technique, find the approximate values of: **(a)** V_L at $t = 2\ \mu$s; **(b)** V_S at $t = 3\ \mu$s; **(c)** V_L at $t = 4\ \mu$s; and **(d)** V_L at $t = \infty$.
Ans. **(a)** 2 V; **(b)** 9.3 V; **(c)** 5 V; **(d)** 8 V

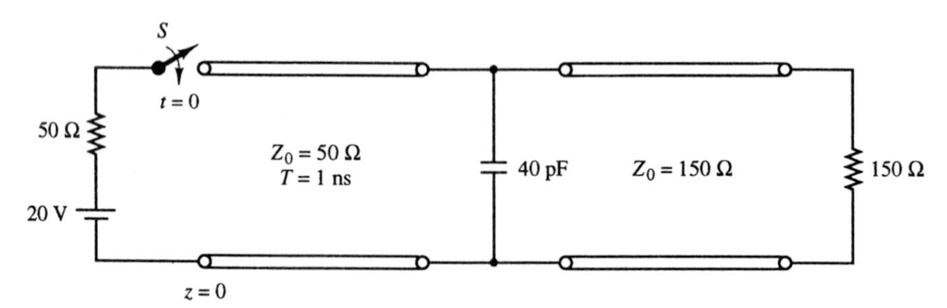

Figure 7.44. For Problem D7.12.

7.6 LINES WITH INITIAL CONDITIONS AND LOGIC GATES

Arbitrary initial distribution

Thus far we have considered lines with quiescent initial conditions, that is, with no initial voltages and currents on them. As a prelude to the discussion of analysis of interconnections between logic gates, we shall now consider lines with nonzero initial conditions. We discuss first the general case of arbitrary initial voltage and current distributions by decomposing them into (+) and (−) wave voltages and currents. To do this, we consider the example shown in Fig. 7.45, in which a line open circuited at both ends is charged initially, say, at $t = 0$, to the voltage and current distributions shown in the figure.

Writing the line voltage and current distributions as sums of (+) and (−) wave voltages and currents, we have

$$V^+(z, 0) + V^-(z, 0) = V(z, 0) \tag{7.64a}$$

$$I^+(z, 0) + I^-(z, 0) = I(z, 0) \tag{7.64b}$$

But we know that $I^+ = V^+/Z_0$ and $I^- = -V^-/Z_0$. Substituting these into (7.64b) and multiplying by Z_0, we get

$$V^+(z, 0) - V^-(z, 0) = Z_0 I(z, 0) \tag{7.65}$$

Solving (7.64a) and (7.65), we obtain

$$V^+(z, 0) = \frac{1}{2}[V(z, 0) + Z_0 I(z, 0)] \tag{7.66a}$$

$$V^-(z, 0) = \frac{1}{2}[V(z, 0) - Z_0 I(z, 0)] \tag{7.66b}$$

Thus, for the distributions $V(z, 0)$ and $I(z, 0)$ given in Fig. 7.45, we obtain the distributions of $V^+(z, 0)$ and $V^-(z, 0)$ as shown by the sketches in Fig. 7.46(a), and hence of $I^+(z, 0)$ and $I^-(z, 0)$, as shown by the sketches in Fig. 7.46(b).

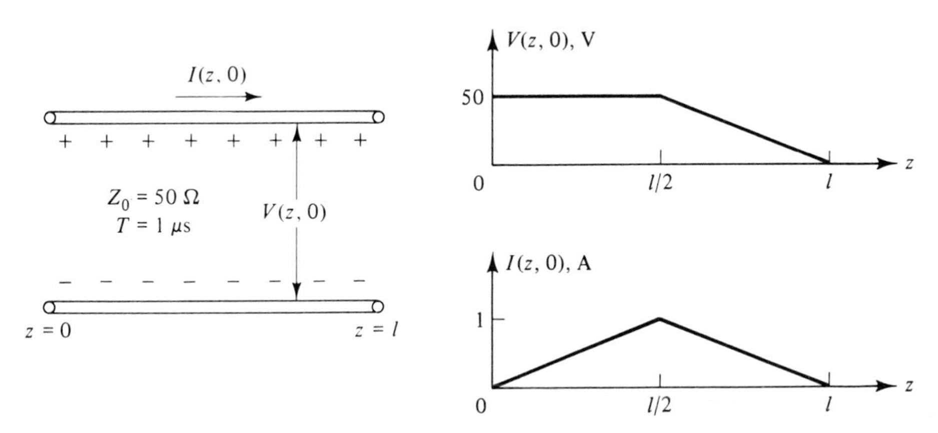

Figure 7.45. Line open circuited at both ends and initially charged to the voltage and current distributions $V(z, 0)$ and $I(z, 0)$, respectively.

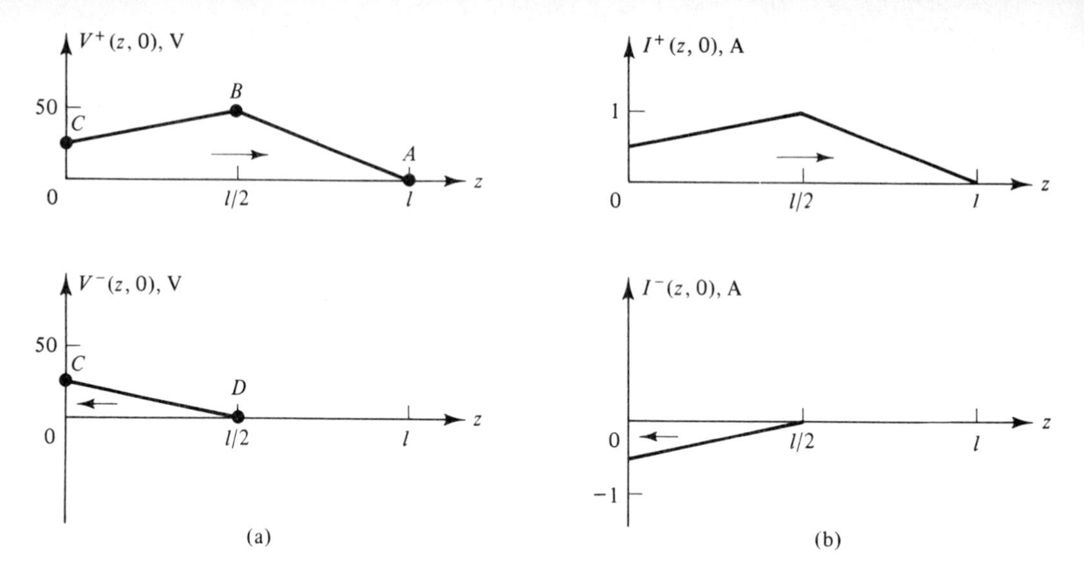

Figure 7.46. Distributions of (a) voltage and (b) current in the (+) and (−) waves obtained by decomposing the voltage and current distributions of Fig. 7.45.

Suppose that we wish to find the voltage and current distributions at some later value of time, say, $t = 0.5\ \mu s$. Then, we note that as the (+) and (−) waves propagate and impinge on the open circuits at $z = l$ and $z = 0$, respectively, they produce the (−) and (+) waves, respectively, consistent with a voltage reflection coefficient of 1 and current reflection coefficient of −1 at both ends. Hence at $t = 0.5\ \mu s$, the (+) and (−) wave voltage and current distributions and their sum distributions are as shown in Fig. 7.47, in which the points A, B, C, and D correspond to the points A, B, C, and D, respectively, in Fig. 7.46. Proceeding in this manner, one can obtain the voltage and current distributions for any value of time.

Suppose that we connect a resistor of value Z_0 at the end $z = l$ at $t = 0$ instead of keeping it open circuited. Then the reflection coefficient at that end becomes zero thereafter and the (+) wave, as it impinges on the resistor, gets absorbed in it instead of producing the (−) wave. The line therefore completely discharges into the resistor by the time $t = 1.5\ \mu s$, with the resulting time variation of voltage across R_L as shown in Fig. 7.48, where the points A, B, C, and D correspond to the points A, B, C, and D, respectively, in Fig. 7.46.

Uniform initial distribution

For a line with uniform initial voltage and current distributions, the analysis can be performed in the same manner as for arbitrary initial voltage and current distributions. Alternatively and more conveniently, the analysis can be carried out with the aid of superposition and bounce diagrams, or by the load-line technique. The basis behind this method lies in the fact that the uniform distribution corresponds to a situation in which the line voltage and current remain constant with time at all points on the line until a change is made at some point on the line. The boundary condition is then violated at that point, and a transient wave of constant voltage and current is set up, to be superimposed on the initial distribution. We shall illustrate this technique of analysis by means of an example.

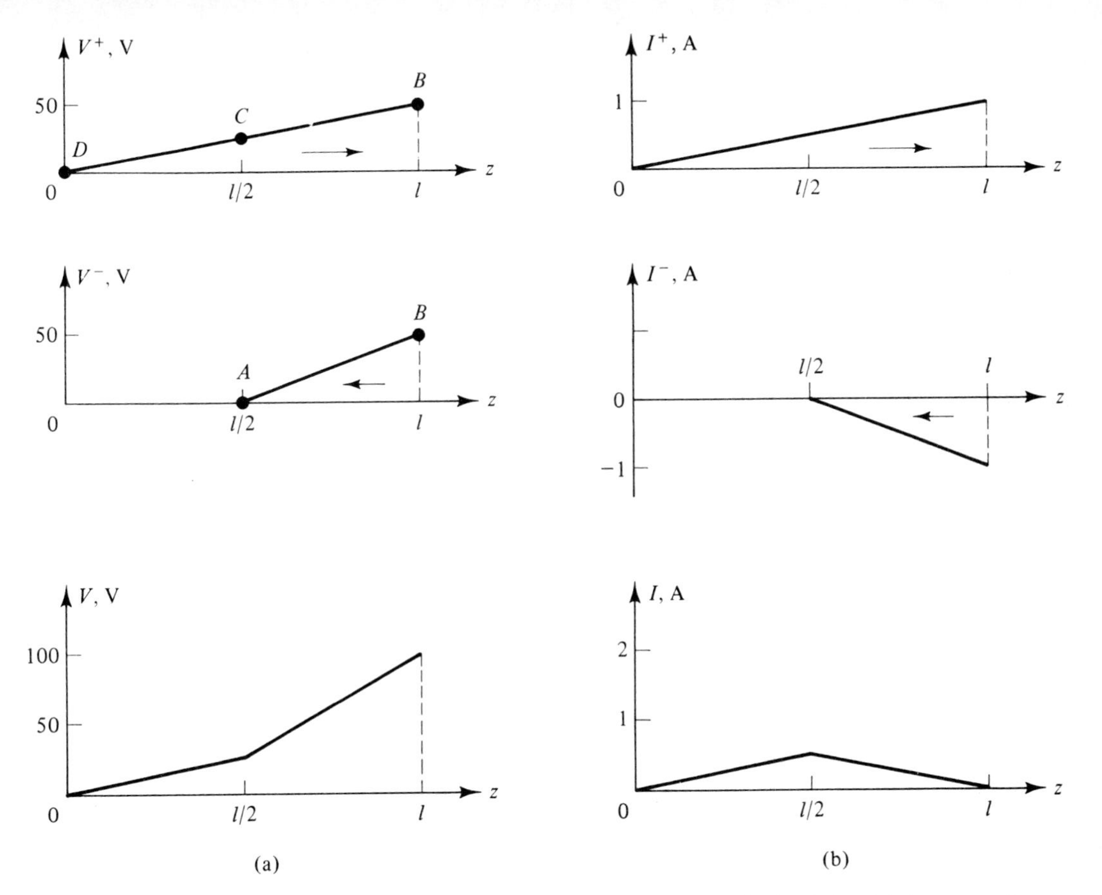

Figure 7.47. Distributions of (a) voltage and (b) current in the (+) and (−) waves and their sum for $t = 0.5\ \mu s$ for the initially charged line of Fig. 7.45.

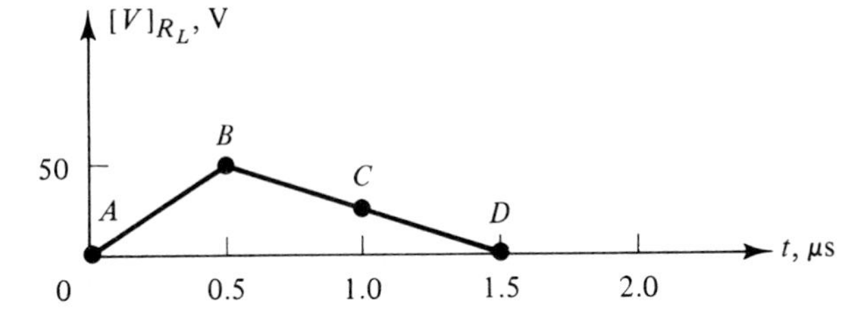

Figure 7.48. Voltage across R_L ($= Z_0 = 50\ \Omega$) resulting from connecting it at $t = 0$ to the end $z = l$ of the line of Fig. 7.45.

Example 7.9

Let us consider a line of $Z_0 = 50\ \Omega$ and $T = 1\ \mu s$ initially charged to uniform voltage $V_0 = 100$ V and zero current. A resistor $R_L = 150\ \Omega$ is connected at $t = 0$ to the end $z = 0$ of the line, as shown in Fig. 7.49(a). We wish to obtain the time variation of the voltage across R_L for $t > 0$.

Since the change is made at $z = 0$ by connecting R_L to the line, a (+) wave originates at $z = 0$ so that the total line voltage at that point is $V_0 + V^+$ and the total

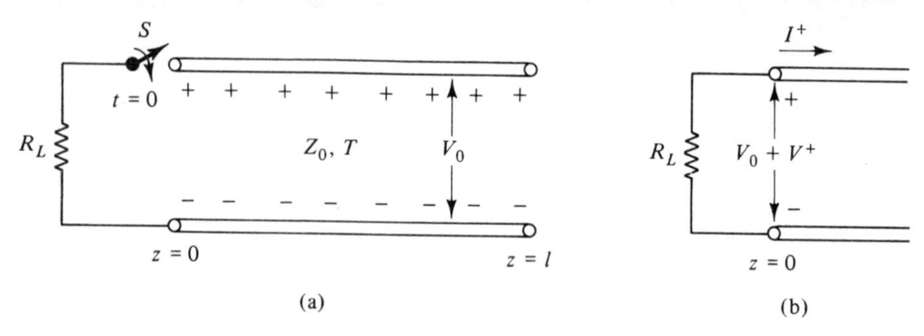

Figure 7.49. (a) Transmission line charged initially to uniform voltage V_0. (b) For obtaining the voltage and current associated with the transient (+) wave resulting from the closure of the switch in (a).

line current is $0 + I^+$, or I^+, as shown in Fig. 7.49(b). To satisfy the boundary condition at $z = 0$, we then write

$$V_0 + V^+ = -R_L I^+ \tag{7.67}$$

But we know that $I^+ = V^+/Z_0$. Hence we have

$$V_0 + V^+ = -\frac{R_L}{Z_0} V^+ \tag{7.68}$$

or

$$V^+ = -V_0 \frac{Z_0}{R_L + Z_0} \tag{7.69a}$$

$$I^+ = -V_0 \frac{1}{R_L + Z_0} \tag{7.69b}$$

For $V_0 = 100$ V, $Z_0 = 50\ \Omega$, and $R_L = 150\ \Omega$, we obtain $V^+ = -25$ V and $I^+ = -0.5$ A.

We may now draw the voltage and current bounce diagrams, as shown in Fig. 7.50. We note that in these bounce diagrams, the initial conditions are accounted for by the horizontal lines drawn at the top, with the numerical values of voltage and current indicated on them. Sketches of line voltage and current versus z for fixed values of t can be drawn from these bounce diagrams in the usual manner. Sketches of line voltage and current versus t for any fixed value of z can also be drawn from the bounce diagrams in the usual manner. Of particular interest is the voltage across R_L, which illustrates how the line discharges into the resistor. The time variation of this voltage is shown in Fig. 7.51.

Alternative to the bounce diagram technique, the behavior of the system can be analyzed by using the load-line technique. Thus noting that the terminal voltage–current characteristics at the ends $z = 0$ and $z = l$ are given by $V = -IR_L = -150I$ and $I = 0$, respectively, and that the characteristic impedance of the line is 50 Ω, we can carry out the load-line construction, as shown in Fig. 7.52, beginning at the point A (100 V, 0 A) and drawing alternately straight lines of slopes $\frac{1}{50}$ and $-\frac{1}{50}$ to obtain the points of intersection $B, C, D, \ldots$. The points $B, D, F, \ldots$ give the line voltage and current values at the end $z = 0$ for intervals of 2 μs beginning at $t = 0$ μs, 2 μs, 6 μs, . . . , whereas the points $C, E, \ldots$ give the line voltage and current values at the end $z = l$ for intervals of 2 μs beginning at $t = 1$ μs, 3 μs, ... For example, the time variation of the line voltage at $z = 0$ provided by the load-line construction is the same as in Fig. 7.51.

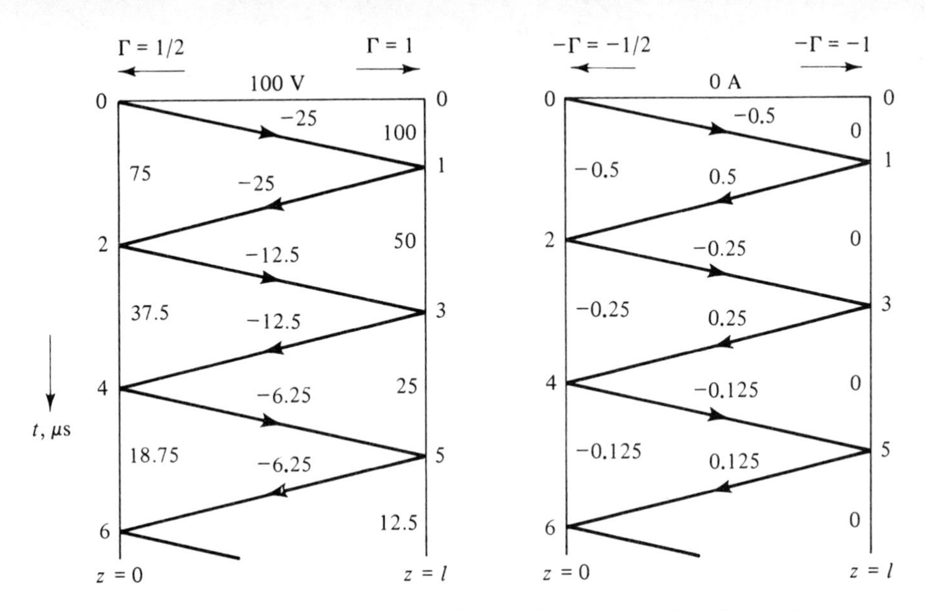

Figure 7.50. Voltage and current bounce diagrams depicting the transient phenomenon for $t > 0$ for the line of Fig. 7.49(a), for $V_0 = 100$ V, $Z_0 = 50\ \Omega$, $R_L = 150\ \Omega$, and $T = 1\ \mu$s.

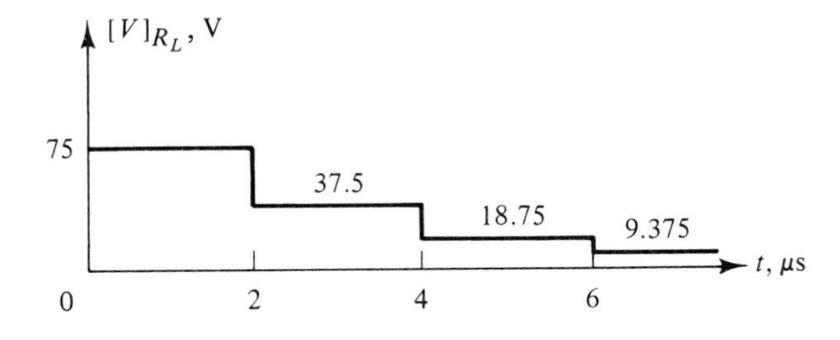

Figure 7.51. Time variation of voltage across R_L for $t > 0$ in Fig. 7.49(a) for $V_0 = 100$ V, $Z_0 = 50\ \Omega$, $R_L = 150\ \Omega$, and $T = 1\ \mu$s.

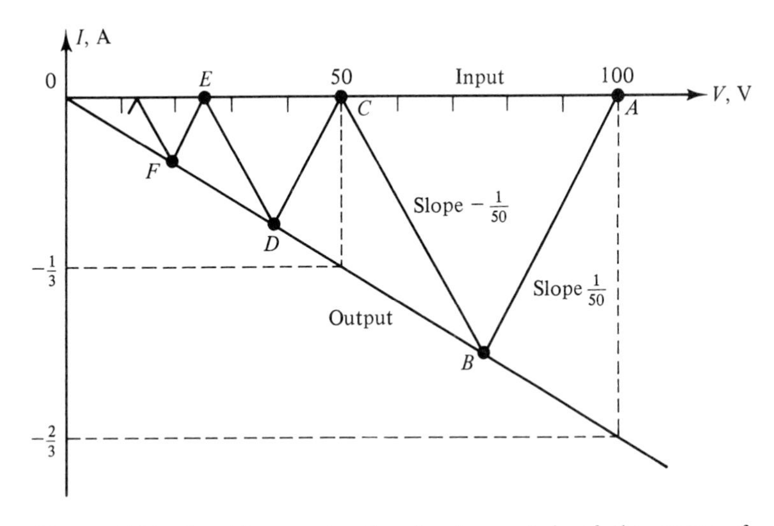

Figure 7.52. Load-line construction for the analysis of the system of Fig. 7.49(a).

Energy balance

Before proceeding further, it is instructive to check the energy balance, that is, to verify that the energy dissipated in the 150-Ω resistor for $t > 0$ is indeed equal to the energy stored in the line at $t = 0-$, since the line is lossless. To do this, we note that, in general, energy is stored in both electric and magnetic fields in the line, with energy densities $\frac{1}{2}\mathscr{C}V^2$ and $\frac{1}{2}\mathscr{L}I^2$, respectively. Thus, for a line charged uniformly to voltage V_0 and current I_0, the total electric and magnetic stored energies are given, respectively, by

$$\begin{aligned} W_e &= \frac{1}{2}\mathscr{C}V_0^2 l = \frac{1}{2}\mathscr{C}V_0^2 v_p T \\ &= \frac{1}{2}\mathscr{C}V_0^2 \frac{1}{\sqrt{\mathscr{L}\mathscr{C}}} T = \frac{1}{2}\frac{V_0^2}{Z_0} T \end{aligned} \tag{7.70a}$$

and

$$\begin{aligned} W_m &= \frac{1}{2}\mathscr{L}I_0^2 l = \frac{1}{2}\mathscr{L}I_0^2 v_p T \\ &= \frac{1}{2}\mathscr{L}I_0^2 \frac{1}{\sqrt{\mathscr{L}\mathscr{C}}} T = \frac{1}{2}I_0^2 Z_0 T \end{aligned} \tag{7.70b}$$

Since for the example under consideration, $V_0 = 100$ V, $I_0 = 0$, and $T = 1\ \mu$s, $W_e = 10^{-4}$ J and $W_m = 0$. Thus, the total initial stored energy in the line is 10^{-4} J. Now, denoting the power dissipated in the resistor to be P_d, we obtain the energy dissipated in the resistor to be

$$\begin{aligned} W_d &= \int_{t=0}^{\infty} P_d\, dt \\ &= \int_0^{2\times10^{-6}} \frac{75^2}{150}\, dt + \int_{2\times10^{-6}}^{4\times10^{-6}} \frac{37.5^2}{150}\, dt + \int_{4\times10^{-6}}^{6\times10^{-6}} \frac{18.75^2}{150}\, dt + \cdots \\ &= \frac{2\times10^{-6}}{150} \times 75^2\left(1 + \frac{1}{4} + \frac{1}{16} + \cdots\right) = 10^{-4}\ \text{J} \end{aligned}$$

which is exactly the same as the initial stored energy in the line, thereby satisfying the energy balance.

Inter-connection between logic gates

We shall now apply the procedure for the use of the load-line technique for a line with uniform initial distribution, just illustrated, to the analysis of the system in Fig. 7.53(a) in which two transistor-transistor logic (TTL) inverters are interconnected by using a transmission line of characteristic impedance Z_0 and one-way travel time T. As the name inverter implies, the gate has an output which is the inverse of the input. Thus if the input is in the HIGH (logic 1) range, the output will be in the LOW (logic 0) range, and vice versa. Typical $V-I$ characteristics for a TTL inverter are shown in Fig. 7.53(b). As shown in this figure, when the system is in the steady state with the output of the first inverter in the 0 state, the voltage and current along the line are given by the intersection of the output 0 characteristic and the input characteristic; when the system is in the steady state with the output of the first inverter in the 1 state, the voltage and current along the line are given by the intersection of the output 1 characteristic and the input characteristic. Thus the line is charged to 0.2 V for the steady-state 0 condition and to 4 V for the steady-state 1 condition. We wish to study the transient phe-

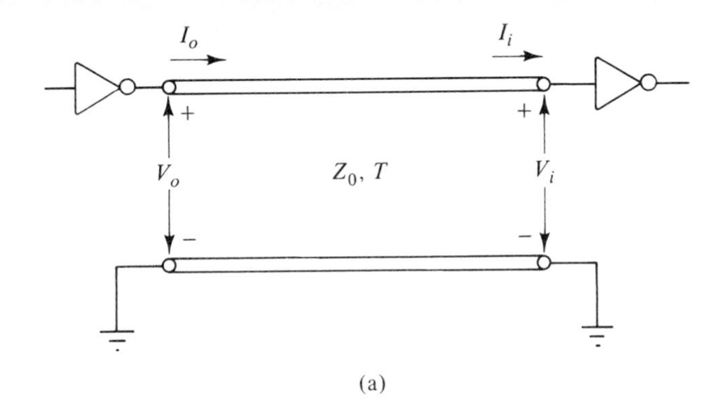

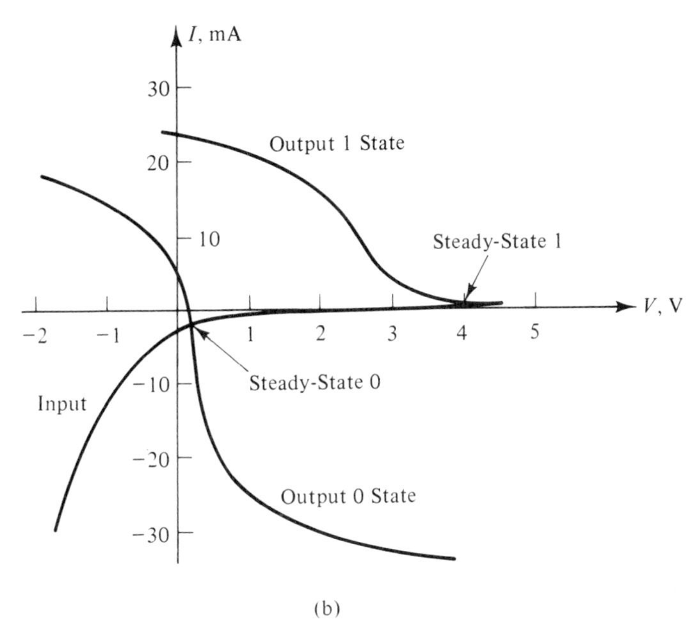

Figure 7.53. (a) Transmission-line interconnection between two logic gates. (b) Typical V–I characteristics for the logic gates.

nomena corresponding to the transition when the output of the first gate switches from the 0 to the 1 state, and vice versa, assuming Z_0 of the line to be 30 Ω.

Analysis of 0-to-1 transition

Considering first the transition from the 0 state to the 1 state, and following the line of argument in Examples 7.8 and 7.9, we carry out the construction shown in Fig. 7.54(a). This construction consists of beginning at the point corresponding to the steady-state 0 (the initial state) and drawing a straight line of slope $\frac{1}{30}$ to intersect with the output 1 characteristic at point A, then drawing from point A a straight line of slope $-\frac{1}{30}$ to intersect the input characteristic at point B, and so on. From this construction, the variation of the voltage V_i at the input of the second gate can be sketched as shown in Fig. 7.54(b), in which the voltage levels correspond to the points 0, B, D, . . . , in Fig. 7.54(a). The effect of the transients on the performance of the system may now be seen by noting from Fig.

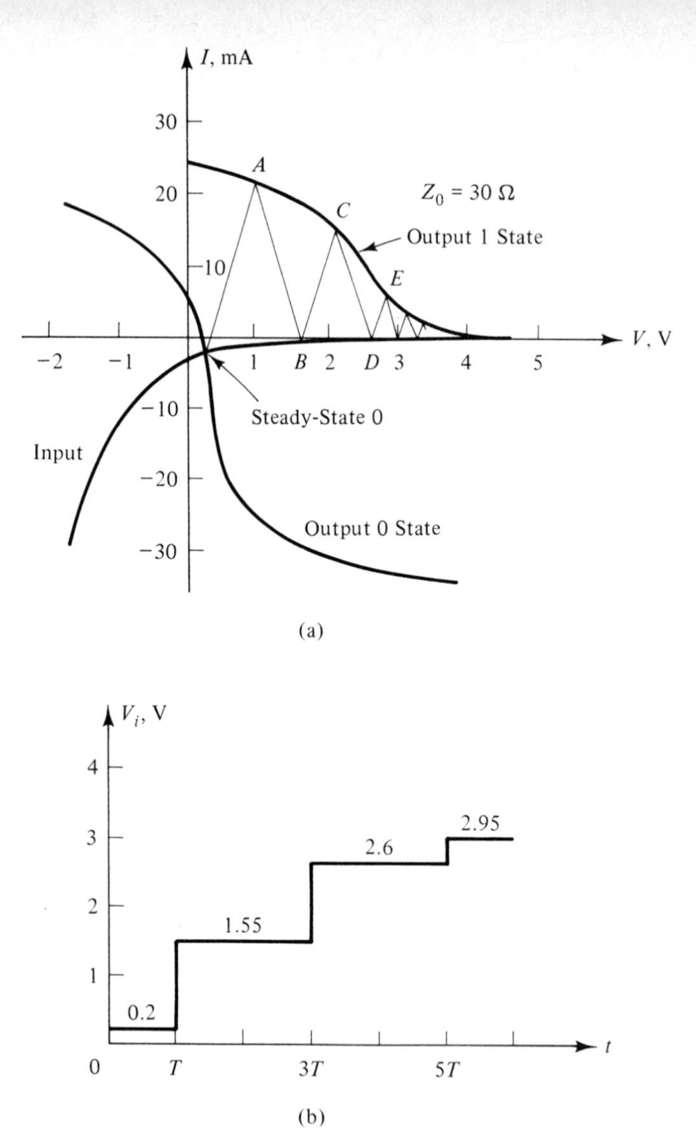

Figure 7.54. (a) Construction based on the load-line technique for analysis of the 0-to-1 transition for the system of Fig. 7.53(a). (b) Plot of V_i versus t obtained from the construction in (a).

7.54(b) that depending on the value of the minimum gate voltage which will reliably be recognized as logic 1, a time delay in excess of T may be involved in the transition from 0 to 1. Thus if this minimum voltage is 2 V, the interconnecting line will result in an extra time delay of $2T$ for the input of the second gate to switch from 0 to 1, since V_i does not exceed 2 V until $t = 3T+$.

Analysis of 1-to-0 transition

Considering next the transition from the 1 state to the 0 state, we carry out the construction shown in Fig. 7.55(a), with the crisscross lines beginning at the point corresponding to the steady-state 1. From this construction, we obtain the plot of V_i versus t, as shown in Fig. 7.55(b), in which the voltage levels corre-

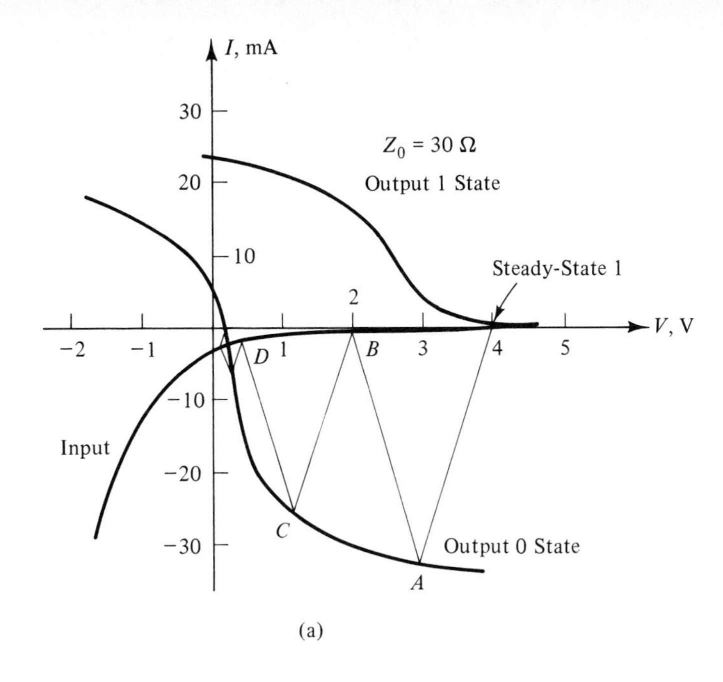

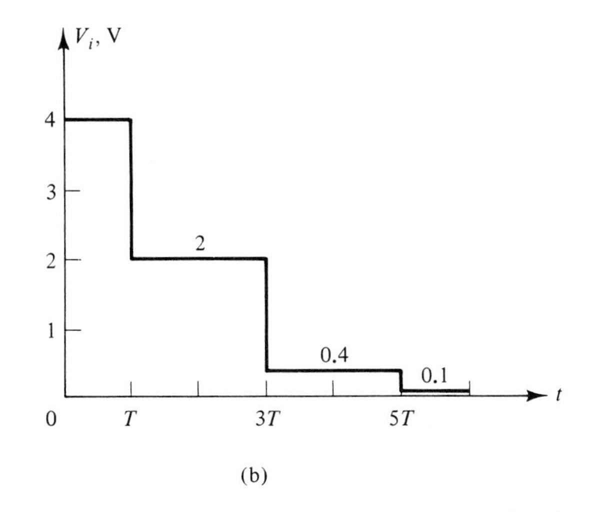

Figure 7.55. (a) Construction based on the load-line technique for analysis of the 1-to-0 transition for the system of Fig. 7.53(a). (b) Plot of V_i versus t obtained from the construction in (a).

spond to the points 1, B, D, . . . , in Fig. 7.55(a). If we assume a maximum gate input voltage which can be readily recognized as logic 0 to be 1 V, it can once again be seen that an extra time delay of $2T$ is involved in the switching of the input of the second gate from 1 to 0, since V_i does not drop below 1 V until $t = 3T+$.

K7.6. Initial conditions; Arbitrary distribution; Uniform distribution; Bounce diagram technique; Load-line technique; Interconnection between logic gates.

D7.14. For the line of Fig. 7.45 with the initial voltage and current distributions as given in the figure, find: **(a)** $V(l/2, 0.25\ \mu\text{s})$; **(b)** $I(l/2, 0.25\ \mu\text{s})$; **(c)** $V(l/4, 1\ \mu\text{s})$; and **(d)** $I(l/4, 1\ \mu\text{s})$.

Ans. **(a)** 37.5 V; **(b)** 0.75 A; **(c)** 25 V; **(d)** −0.5 A

D7.15. In the system shown in Fig. 7.56, a line of characteristic impedance 75 Ω and charged to 10 V is connected at $t = 0$ to another line of characteristic impedance 50 Ω and charged to 5 V. The one-way travel time, T, is equal to 1 μs for both lines. Find **(a)** the value of the voltage at the instant of time when both lines are charged to the same voltage throughout their lengths; **(b)** the value of the current to which the lines are charged at that instant of time; and **(c)** the energy stored in the system at any instant of time.

Ans. **(a)** 7 V; **(b)** 0.4 A; **(c)** $\frac{11}{12}\ \mu\text{J}$

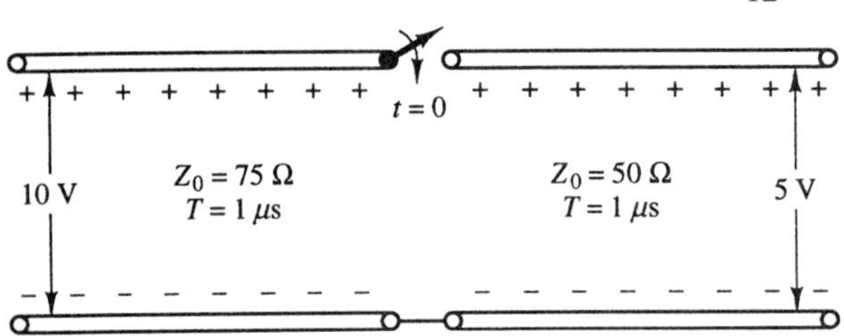

Figure 7.56. For Problem D7.15.

7.7 SUMMARY

In this chapter we first introduced the parallel-plate transmission line by considering a uniform plane wave propagating between two parallel perfectly conducting plates and showed that wave propagation on a transmission line can be discussed in terms of voltage and current, which are related to the electric and magnetic fields, respectively, by deriving the transmission-line equations

$$\frac{\partial V}{\partial z} = -\mathcal{L}\frac{\partial I}{\partial t} \tag{7.71a}$$

$$\frac{\partial I}{\partial z} = -\mathcal{C}\frac{\partial V}{\partial t} \tag{7.71b}$$

which then led us to the concept of the distributed circuit. We learned that propagation along a transmission line in the general case is characterized by transverse electromagnetic waves, with the parameters $\mathcal{L}$ and $\mathcal{C}$ differing from one line to another and derivable from static field considerations. The solutions to the transmission-line equations are

$$V(z, t) = Af\left(t - \frac{z}{v_p}\right) + Bg\left(t + \frac{z}{v_p}\right) \tag{7.72a}$$

$$I(z, t) = \frac{1}{Z_0}\left[Af\left(t - \frac{z}{v_p}\right) - Bg\left(t + \frac{z}{v_p}\right)\right] \tag{7.72b}$$

where $Z_0 = \sqrt{\mathcal{L}/\mathcal{C}}$ is the characteristic impedance of the line and $v_p = 1/\sqrt{\mathcal{L}\mathcal{C}}$ is the velocity of propagation on the line.

We discussed several techniques of determination of Z_0 and v_p for the case of a line with homogeneous dielectric, as well as for the case of a line involving more than one dielectric, an example being the microstrip line. For the former case

$$Z_0 = \frac{\sqrt{\varepsilon_r}}{c\mathscr{C}} \tag{7.73a}$$

$$v_p = \frac{c}{\sqrt{\varepsilon_r}} \tag{7.73b}$$

where ε_r, c, and $\mathscr{C}$ are the relative permittivity of the dielectric, the velocity of light in free space, and the capacitance per unit length of the line computed from static field considerations, respectively. For the latter case, assuming nonmagnetic dielectrics,

$$Z_0 = \frac{1}{c\sqrt{\mathscr{C}\mathscr{C}_0}} \tag{7.74a}$$

$$v_p = c\sqrt{\frac{\mathscr{C}_0}{\mathscr{C}}} \tag{7.74b}$$

where $\mathscr{C}$ is the capacitance per unit length of the line with the dielectrics in place and $\mathscr{C}_0$ is the capacitance per unit length with all dielectrics replaced by free space, both computed from static field considerations. Note that (7.74a) and (7.74b) reduce to (7.73a) and (7.73b), respectively, if all dielectrics are the same, since then $\mathscr{C} = \mathscr{C}_0\varepsilon_r$. Based on these considerations, we presented analytically obtainable expressions for Z_0 for several common types of lines, discussed by means of examples of numerical techniques involving the method of moments and the finite-difference method, and introduced a graphical technique involving field mapping for the case of a line with arbitrary cross section.

We then discussed time-domain analysis of a transmission line terminated by a load resistance R_L and excited by a constant voltage source V_0 in series with internal resistance R_g. Writing the general solutions (7.72a) and (7.72b) concisely in the manner

$$V = V^+ + V^- \tag{7.75a}$$

$$I = I^+ + I^- \tag{7.75b}$$

where

$$I^+ = \frac{V^+}{Z_0} \tag{7.76a}$$

$$I^- = -\frac{V^-}{Z_0} \tag{7.76b}$$

we found that the situation consists of bouncing back and forth of transient (+) and (−) waves between the two ends of the line. The initial (+) wave voltage is $V^+Z_0/(R_g + Z_0)$. All other waves are governed by the reflection coefficients at the two ends of the line, given for the voltage by

$$\Gamma_R = \frac{R_L - Z_0}{R_L + Z_0} \tag{7.77a}$$

and

$$\Gamma_S = \frac{R_g - Z_0}{R_g + Z_0} \tag{7.77b}$$

for the load and source ends, respectively. In the steady state, the situation is the superposition of all the transient waves, equivalent to the sum of a single (+) wave and a single (−) wave. We discussed the bounce diagram technique of keeping track of the transient phenomenon and extended it to a pulse voltage source.

We learned that when a wave is incident from, say, line 1 onto a junction with line 2, reflection occurs just as though line 1 is terminated by a load resistor equal to the characteristic impedance of line 2. A transmitted wave goes into line 2 in accordance with the voltage and current transmission coefficients

$$\tau_V = 1 + \Gamma \tag{7.78a}$$

and

$$\tau_C = 1 - \Gamma \tag{7.78b}$$

respectively, where Γ is the voltage reflection coefficient. Applying this to a system of three lines in cascade, we showed how to obtain the unit impulse response of the system and from it obtain the frequency response. We then extended the analysis to lines with discontinuities to discuss and illustrate by means of an example the application of time-domain reflectometry, an important experimental technique.

We then considered lines with reactive terminations and discontinuities, and nonlinear-resistive terminations, where we learned that the reflection coefficient concept is not useful to study the transient behavior. For the cases of reactive elements, it is necessary to write the differential equations pertinent to the boundary conditions at the terminations and/or discontinuities, and solve them subject to the appropriate initial conditions; alternatively, the required voltages and currents can be obtained from considerations of initial and final behaviors of the reactive element(s), and associated time constant(s). For the nonlinear-resistive termination case, we introduced the load-line technique of time-domain analysis.

As a prelude to the consideration of interconnections between logic gates, we discussed time-domain analysis of lines with nonzero initial conditions. For the general case, the initial voltage and current distributions $V(z, 0)$ and $I(z, 0)$ are decomposed into (+) and (−) wave voltages and currents as given by

$$V^+(z, 0) = \frac{1}{2}[V(z, 0) + Z_0 I(z, 0)]$$

$$V^-(z, 0) = \frac{1}{2}[V(z, 0) - Z_0 I(z, 0)]$$

$$I^+(z, 0) = \frac{1}{Z_0} V^+(z, 0)$$

$$I^-(z, 0) = -\frac{1}{Z_0} V^-(z, 0)$$

The voltage and current distributions for $t > 0$ are then obtained by keeping track of the bouncing of these waves at the two ends of the line. For the special case of uniform distribution, the analysis can be performed more conveniently by considering the situation as one in which a transient wave is superimposed on the initial distribution and using the bounce-diagram technique or the load-line technique. Finally, we applied the load-line technique to the analysis of transmission-line interconnection between logic gates.

REVIEW QUESTIONS

R7.1. Describe the phenomenon of guiding of a uniform plane wave by a pair of parallel, plane, perfectly conducting sheets.

R7.2. Discuss the derivation of the transmission-line equations from the field equations by considering the parallel-plate line.

R7.3. Discuss the concept of the distributed circuit as compared to a lumped circuit.

R7.4. What is a transverse electromagnetic wave? Discuss the electric and magnetic field distributions associated with a transverse electromagnetic wave.

R7.5. Explain why the product of $\mathcal{L}$ and $\mathcal{C}$ of a line is equal to the product of μ and ε of the dielectric of the line.

R7.6. What is the significance of the characteristic impedance of a line? Why is it not in general equal to the intrinsic impedance of the medium between the conductors of the line?

R7.7. Discuss the geometry associated with the microstrip line and the determination of its characteristic impedance and velocity of propagation.

R7.8. Describe the procedure for obtaining Z_0 for a parallel-strip line embedded in a homogeneous medium by using the method of moments.

R7.9. Outline the procedure for obtaining Z_0 and v_p for an enclosed-microstrip line by using the finite-difference method.

R7.10. Describe the procedure for computing the transmission-line parameters by using the field mapping technique.

R7.11. Discuss the general solutions for the line voltage and current and the notation associated with their representation in concise form.

R7.12. What is the fundamental distinction between the occurrence of the response in one branch of a lumped circuit to the application of an excitation in a different branch of the circuit and the occurrence of the response at one location on a transmission line to the application of an excitation at a different location on the line?

R7.13. Describe the phenomenon of the bouncing back and forth of transient waves on a transmission line excited by a constant voltage source in series with internal resistance and terminated by a resistance.

R7.14. What is the nature of the formula for the voltage reflection coefficient? Discuss its values for some special cases.

R7.15. What is the steady-state equivalent of a line excited by a constant voltage source? What is the actual situation in the steady state?

R7.16. Discuss the bounce diagram technique of keeping track of the bouncing back and forth of the transient waves on a transmission line, for a constant voltage source.

R7.17. Discuss the bounce diagram technique of keeping track of the bouncing back and forth of the transient waves on a transmission line for a pulse voltage source.

R7.18. How are the voltage and current transmission coefficients at the junction between two lines related to the voltage reflection coefficient?

R7.19. Explain how it is possible for the transmitted voltage or current at a junction between two lines to exceed the incident voltage or current.

R7.20. Discuss the determination of the unit impulse response of a system of three lines in cascade.

R7.21. Outline the procedure for the determination of the frequency response of a system of three lines in cascade from its unit impulse response.

R7.22. What is a radome? How is it analyzed by using transmission-line equivalent?

R7.23. Describe the technique of locating discontinuities in a transmission-line system by using a time-domain reflectometer.

R7.24. Discuss the transient analysis of a line driven by a constant voltage source in series with a resistance equal to the Z_0 of the line and terminated by an inductor.

R7.25. Why is the concept of reflection coefficient not useful for studying the transient behavior of lines with reactive terminations and discontinuities?

R7.26. Discuss the determination of the transient behavior of lines with reactive terminations and discontinuities without formally setting up the differential equations and solving them.

R7.27. Discuss the load-line technique of obtaining the time-variations of the voltages and currents at the source and load ends of a line from a knowledge of the terminal $V-I$ characteristics.

R7.28. Discuss the determination of the voltage and current distributions on an initially charged line for any given time from the knowledge of the initial voltage and current distributions.

R7.29. Discuss with the aid of an example the discharging of an initially charged line into a resistor.

R7.30. Discuss the bounce diagram technique of transient analysis of a line with uniform initial voltage and current distributions.

R7.31. How do you check the energy balance for the case of a line with initial voltage and/or current distribution(s) and discharged into a resistor?

R7.32. Discuss the analysis of transmission-line interconnection between two logic gates.

PROBLEMS

P7.1. A parallel-plate transmission line is made up of perfect conductors of width $w = 0.1$ m and lying in the planes $x = 0$ and $x = 0.01$ m. The medium between the conductors is a nonmagnetic ($\mu = \mu_0$), perfect dielectric. For a uniform plane wave propagating along the line, the voltage along the line is given by

$$V(z, t) = 10 \cos (3\pi \times 10^8 t - 3\pi z) \text{ V}$$

Neglecting fringing of fields, find **(a)** the electric field intensity $E_x(z, t)$ of the wave; **(b)** the magnetic field intensity $H_y(z, t)$ of the wave; **(c)** the current $I(z, t)$ along the line; and **(d)** the power flow $P(z, t)$ down the line.

P7.2. A parallel-plate transmission line consists of an arrangement of two perfect dielectrics, as shown by the transverse cross section in Fig. 7.57. Note that $\mu_1\varepsilon_1 = \mu_2\varepsilon_2$ so that the TEM waves propagating in the two dielectrics are in phase at all points along the interface between the dielectrics. Neglect fringing of fields and compute the values of $\mathcal{L}$, $\mathcal{C}$, and Z_0 of the line.

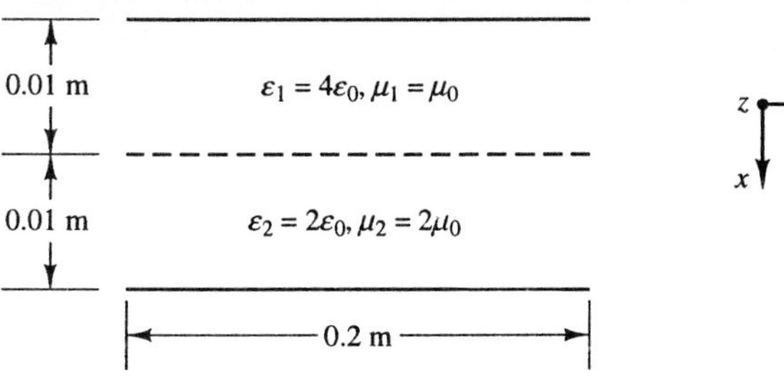

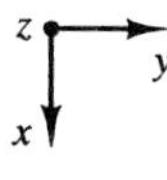

Figure 7.57. For Problem P7.2.

P7.3. Repeat Problem P7.2. for a parallel-plate line having the transverse cross section shown in Fig. 7.58.

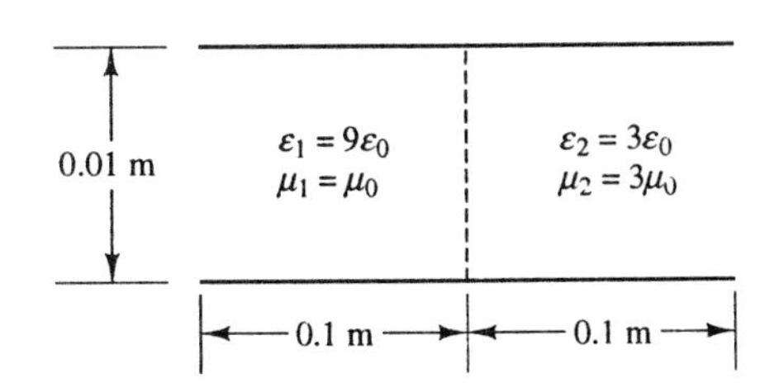

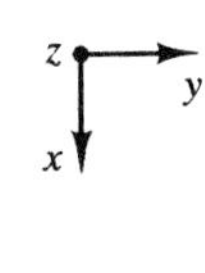

Figure 7.58. For Problem P7.3.

P7.4. Derive the transmission line equations by considering the special case of two infinitely long coaxial cylindrical conductors. Also show that the power flow along the line is equal to the product of the voltage between the conductors and current along the conductors.

P7.5. For the parallel-strip line of Example 7.1, repeat the solution by considering the charges to be line charges along the centerlines of the substrips for writing the contributions to the potential difference between a given pair of substrips due to the charges in a different pair of substrips and using the formula given in Problem P4.27.

P7.6. Consider a parallel-strip line with unequal widths of the conductors, as shown in Fig. 7.59. Obtain the characteristic impedance of the line for the case of $k = 1$ by dividing the conductors into substrips as shown in the figure and using the method of moments. Note that from considerations of symmetry there are only three unknown charge densities ρ_{S1}, ρ_{S2}, and ρ_{S3}. Write two equations by equating the ex-

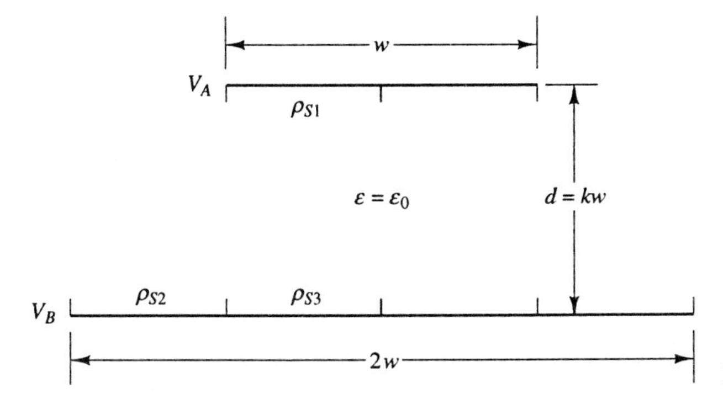

Figure 7.59. For Problem P7.6.

pressions for the potential differences V_{12} and V_{13} to $(V_A - V_B)$ and the third equation from consideration of charge neutrality. Use the result of Problem P4.29 for writing the contributions to the potential differences in all cases.

P7.7. Consider a transmission line having the cross-sectional view shown in Fig. 7.60. With the conductors of the line divided into substrips as shown in the figure, obtain the characteristic impedance by using the method of moments. Note that from considerations of symmetry there are only three unknown charge densities ρ_{S1}, ρ_{S2}, and ρ_{S3}. Write two equations by equating the expressions for the potential differences V_{12} and V_{13} to $(V_A - V_B)$ and the third equation from consideration of charge neutrality. For writing the contribution to the potential difference between a given pair of substrips due to one of those substrips, use the result of Problem P4.29. But for writing the contribution to the potential difference between a given pair of substrips due to a third substrip, consider the charge in that substrip to be a line charge along the centerline of the substrip.

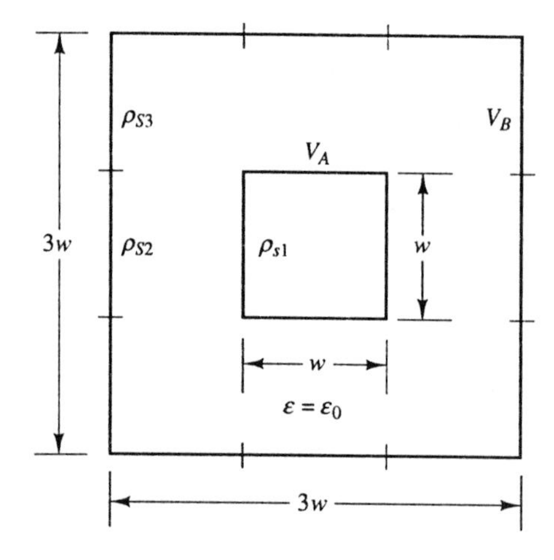

Figure 7.60. For Problem P7.7.

P7.8. For the enclosed-microstrip line of Fig. 7.8, repeat the computations of $\mathscr{C}$, $\mathscr{C}_0$, Z_0, and v_p, by finding the magnitude of the charge per unit length by considering the contour that passes through the center points of the squares adjacent to the center conductor, instead of the one shown in the figure.

P7.9. By applying the curvilinear squares technique to a coaxial cable of inner radius a and outer radius b, show that the characteristic impedance of the cable is $(\eta/2\pi)$ ln b/a, where η is the intrinsic impedance of the dielectric of the cable.

P7.10. The cross section of an eccentric coaxial cable [see Fig. 5.12(d)] consists of an inner circle of radius $a = 2$ cm and an outer circle of radius $b = 5$ cm, with their centers separated by $d = 2$ cm. By constructing a field map consisting of curvilinear squares, obtain the approximate value of Z_0 in terms of η of the dielectric.

P7.11. When one microstrip line is inverted and placed on top of another microstrip line, as shown by the cross-sectional view in Fig. 7.61, a shielded strip line is obtained. Although the sandwich arrangement of this line is more difficult to fabricate than is the microstrip line, it has the advantage that the fields are confined mostly to the substrate region. Assuming for simplicity that the fields are confined to the substrate region, construct a field map consisting of curvilinear squares and compute the approximate value of Z_0 of the line, for the dimensions shown in Fig. 7.61, and considering the substrate to be a perfect dielectric having $\varepsilon = 9\varepsilon_0$ and $\mu = \mu_0$.

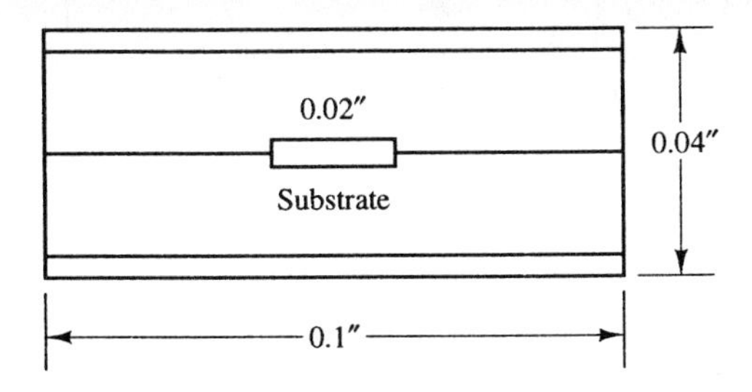

Figure 7.61. For Problem P7.11.

P7.12. In the system shown in Fig. 7.62, the switch S is closed at $t = 0$. Find and sketch **(a)** line voltage versus z for $t = 0.5\ \mu s$; **(b)** line current versus z for $t = 0.5\ \mu s$; **(c)** line voltage versus t for $z = 100$ m; and **(d)** line current versus t for $z = -300$ m.

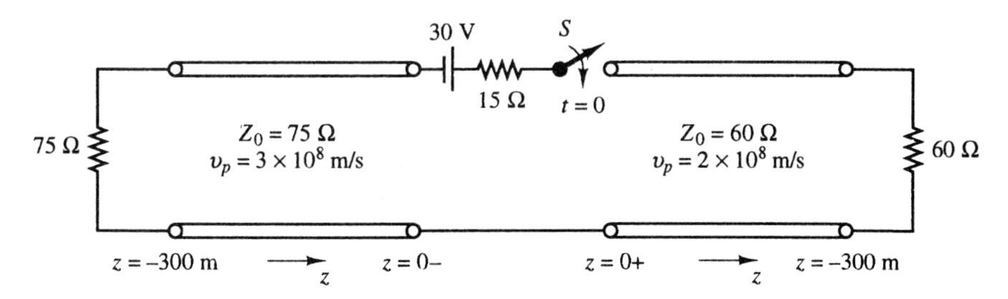

Figure 7.62. For Problem P7.12.

P7.13. In the system shown in Fig. 7.63(a), the switch S is closed at $t = 0$. The line voltage variations with time at $z = 0$ and $z = l$ for the first 5 μs are observed to be as shown in Fig. 7.63(b) and (c), respectively. Find the values of V_0, R_g, R_L, and T.

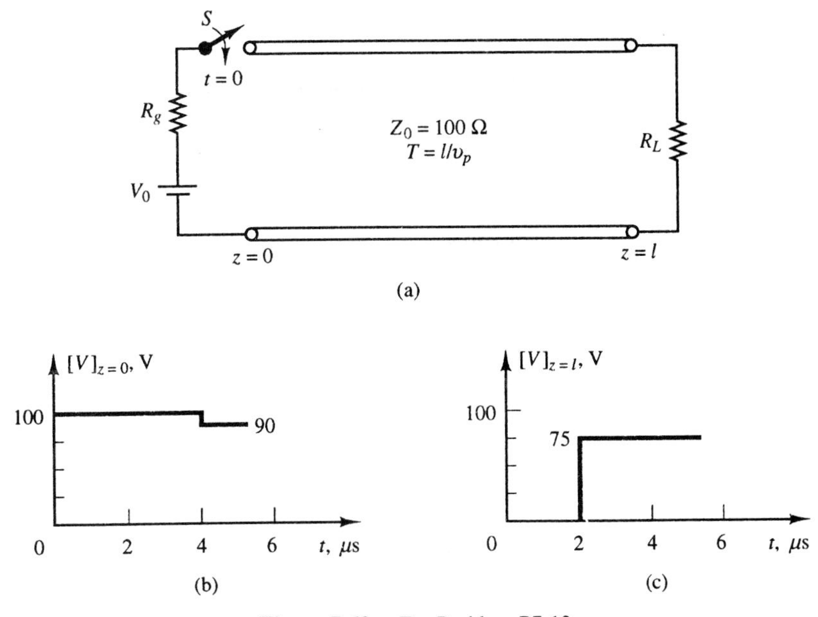

Figure 7.63. For Problem P7.13.

P7.14. In the system shown in Fig. 7.64, the switch S is closed at $t = 0$. Assume $V_g(t)$ to be a direct voltage of 90 V and draw the voltage and current bounce diagrams. From these bounce diagrams, sketch the following: **(a)** line voltage and line current versus t (up to $t = 7.25\ \mu s$) at $z = 0$, $z = l$, and $z = l/2$ and **(b)** line voltage and line current versus z for $t = 1.2\ \mu s$ and $t = 3.5\ \mu s$.

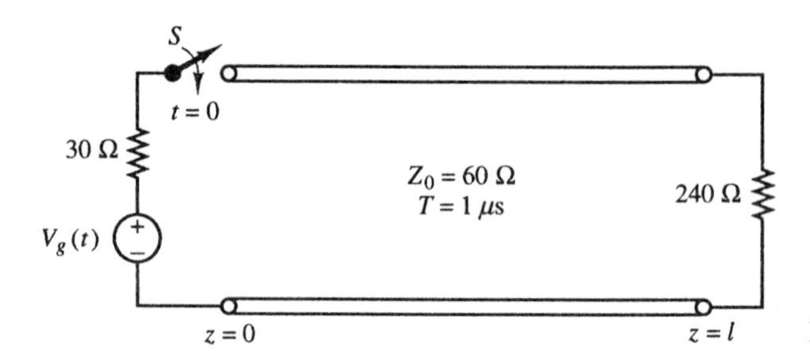

Figure 7.64. For Problem P7.14.

P7.15. For the system of Problem P7.14, assume that the voltage source is of 0.3 μs duration instead of being of infinite duration. Find and sketch the line voltage and line current versus z for $t = 1.2\ \mu s$ and $t = 3.5\ \mu s$.

P7.16. In the system shown in Fig. 7.65, the switch S is closed at $t = 0$. Find and sketch **(a)** the line voltage versus z for $t = 2\frac{1}{2}\mu s$; **(b)** the line current versus z for $t = 2\frac{1}{2}\mu s$; and **(c)** the line voltage at $z = l$ versus t up to $t = 4\ \mu s$.

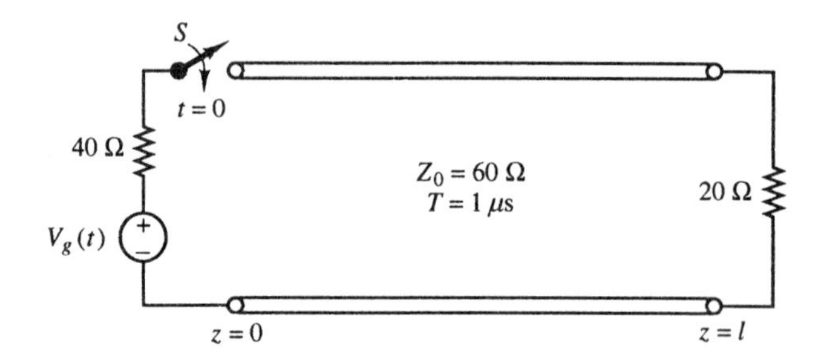

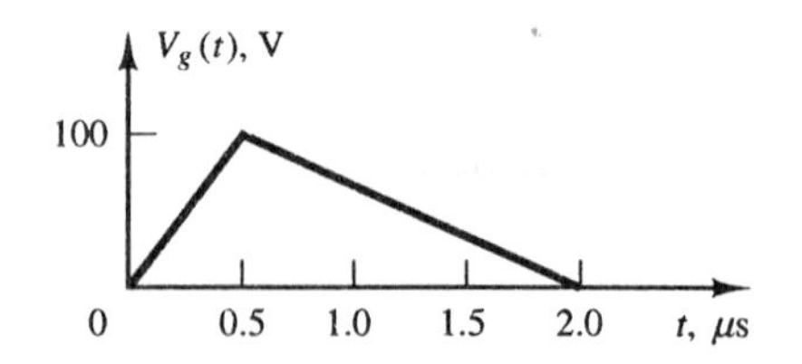

Figure 7.65. For Problem P7.16.

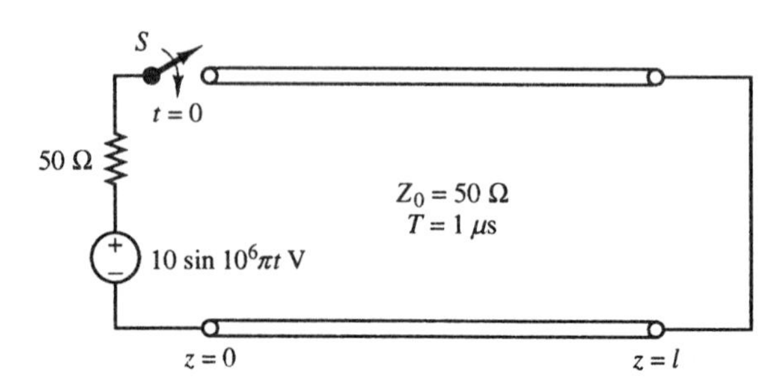

Figure 7.66. For Problem P7.17.

P7.17. In the system shown in Fig. 7.66, the switch S is closed at $t = 0$. Draw the voltage and current bounce diagrams and sketch **(a)** the line voltage and line current versus t for $z = 0$ and $z = l$ and **(b)** the line voltage and line current versus z for $t = 2, \frac{9}{4}, \frac{5}{2}, \frac{11}{4}$, and 3 μs. Note that the period of the source voltage is 2 μs, which is equal to the two-way travel time on the line.

P7.18. In the system shown in Fig. 7.67, an incident wave of voltage V^+ strikes the discontinuity from the left, that is, from line 1. Find the reflected wave voltage and current into line 1 and the transmitted wave voltage and current into line 2.

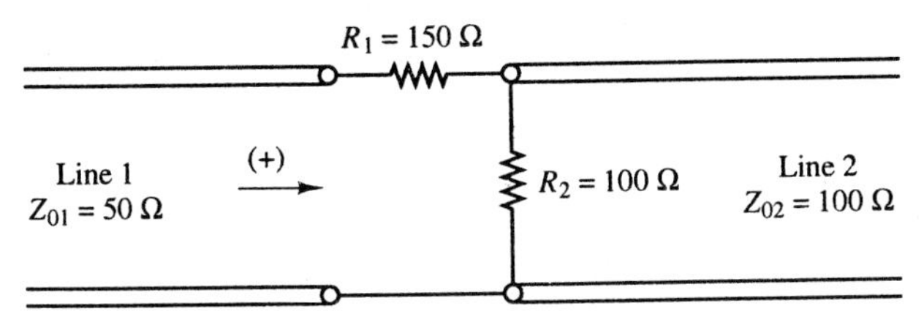

Figure 7.67. For Problem P7.18.

P7.19. In the system shown in Fig. 7.68, **(a)** find the output voltage V_o across the 100 Ω resistor for $V_g(t) = \delta(t)$ and **(b)** find and sketch the amplitude of $V_o(t)$ versus ω for $V_g(t) = \cos \omega t$.

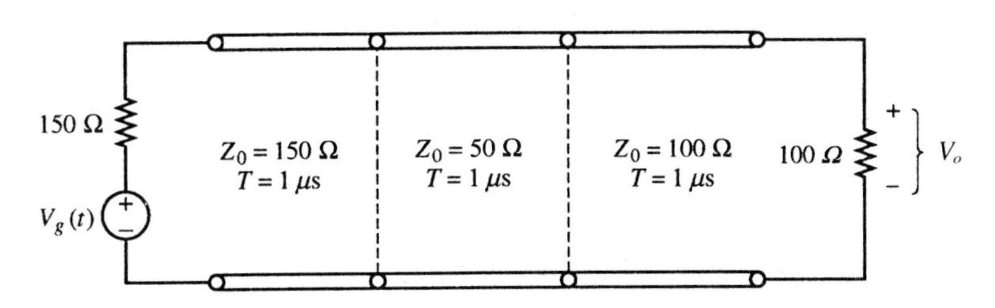

Figure 7.68. For Problem P7.19.

P7.20. In Fig. 7.69(a), the plane I is the input plane from which a uniform plane wave is incident normally on the interface between medium 1 and medium 2, and the plane O is the output plane in which the response of the system is observed. For an incident wave of $E_{xi}(t) = \delta(t)$, find the permittivity ε_2 ($> \varepsilon_0$) and the thickness l of medium 2 required to obtain the electric field $E_{xo}(t)$ in the output plane, as shown in Fig. 7.69(b), in which the interval between successive impulses is 2 μs. Then find the value of A. (*Hint:* Use transmission-line analogy. First find ε_2 and then l.)

P7.21. In Fig. 7.70, a (+) wave carrying power P is incident on the junction a-a' from line 1. Find **(a)** the power reflected into line 1; **(b)** the power transmitted into line 2; and **(c)** the power transmitted into line 3.

P7.22. In the system shown in Fig. 7.71, a (+) wave carrying power P is incident on the junction a-a' from line 1. **(a)** Find the value of R for which there is no reflected wave into line 1. **(b)** For the value of R found in **(a)**, find the power transmitted ainto each of lines 2 and 3.

P7.23. In the system of Fig. 7.32, assume that the discontinuity at $z = 4$ m is a resistor of value 40 Ω in series with the line, instead of the 120-Ω parallel resistor. Find and sketch the waveform that the TDR system would measure up to $t = 200$ ns.

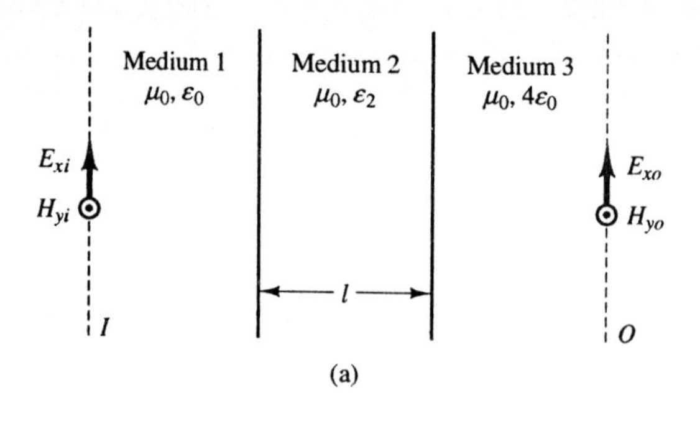

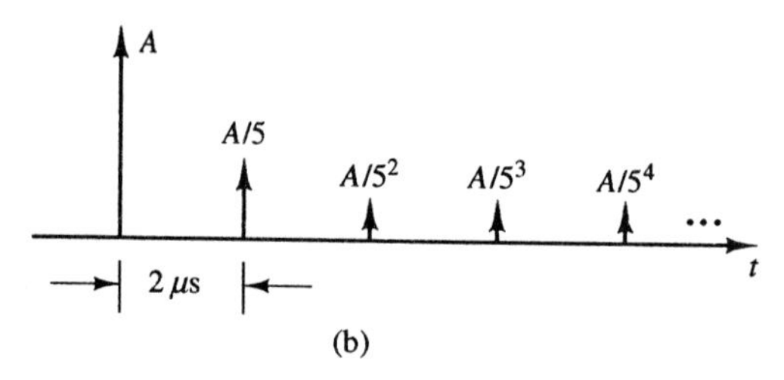

Figure 7.69. For Problem P7.20.

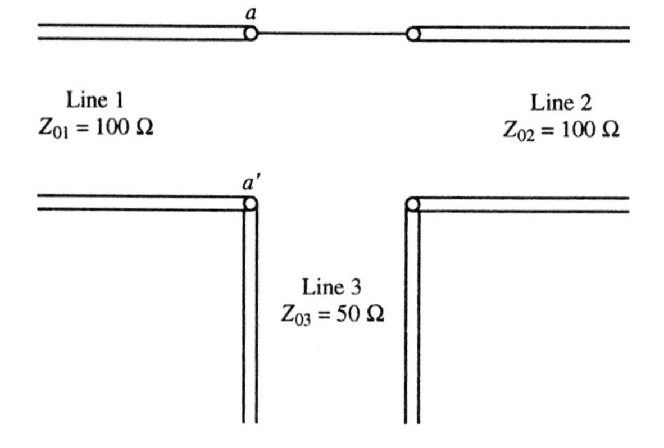

Figure 7.70. For Problem P7.21.

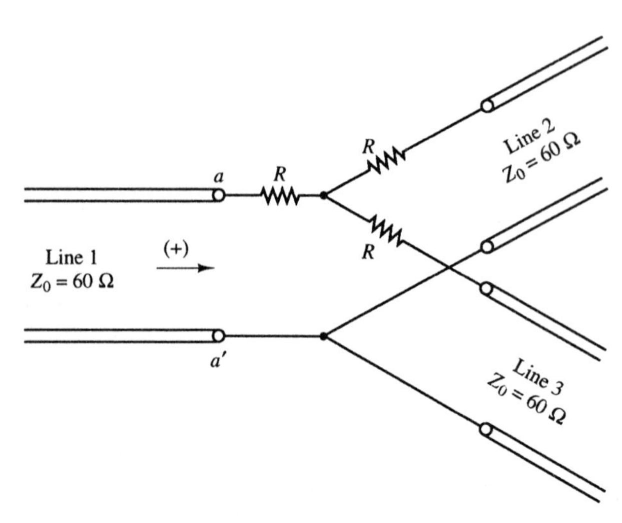

Figure 7.71. For Problem P7.22.

P7.24. In the system shown in Fig. 7.72, the switch S is closed at $t = 0$ with no current in the relay coil and with the relay in position 1. When the relay coil current I_L reaches 1.73 A, the relay switches to position 2; when the current drops to

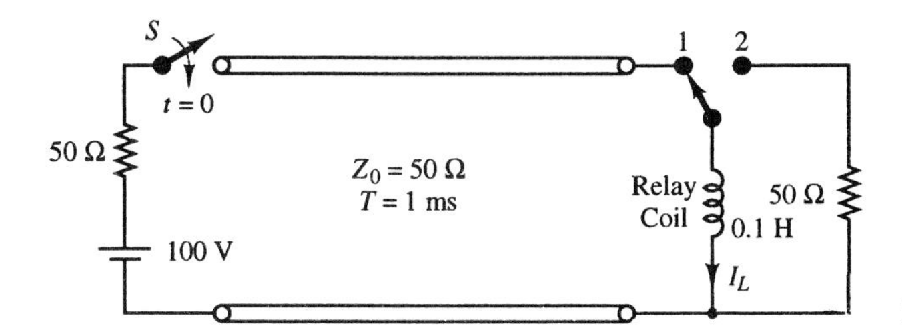

Figure 7.72. For Problem P7.24.

0.636 A, the relay switches back to position 1. **(a)** Find the time t_1 at which the relay switches to position 2. **(b)** Find the time t_2 at which the relay switches back to position 1.

P7.25. In the system shown in Fig. 7.73, the switch S is closed at $t = 0$, with the voltage across the capacitor equal to zero. **(a)** Obtain the differential equation for V^- at $z = l$. **(b)** Find the solution for $V^-(l, t)$.

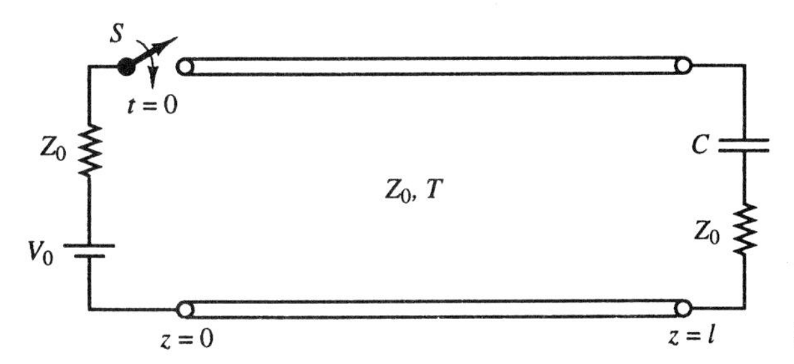

Figure 7.73. For Problem P7.25.

P7.26. In the system shown in Fig. 7.74, the switch S is closed at $t = 0$ with the lines uncharged and with zero current in the inductor. Obtain the solution for the line voltage versus time at $z = l+$.

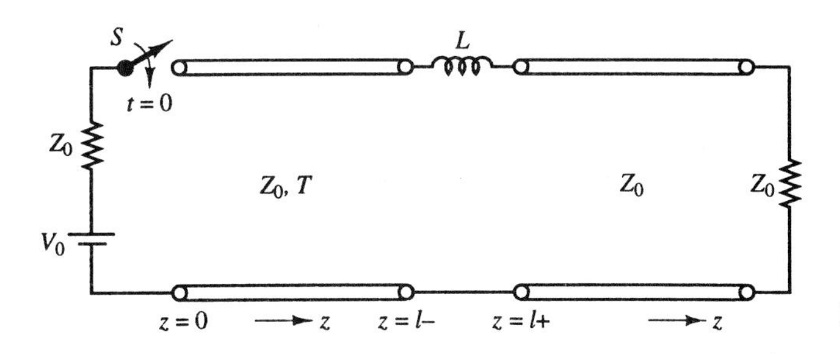

Figure 7.74. For Problem P7.26.

P7.27. In the system shown in Fig. 7.75(a), the network N consists of a single circuit element (R, L, or C). The system is initially uncharged. The switch S is closed at $t = 0$, and the line voltage at $z = 0$ is observed to be as shown in Fig. 7.75(b). **(a)** Determine whether the circuit element is R, L, or C. **(b)** Find the value of Z_{02}/Z_{01}.

P7.28. For the system of Problem P7.14, use the load-line technique to obtain and plot line voltage and line current versus t (up to $t = 5.25\ \mu$s) at $z = 0$ and $z = l$. Also obtain the steady-state values of line voltage and current from the load-line construction.

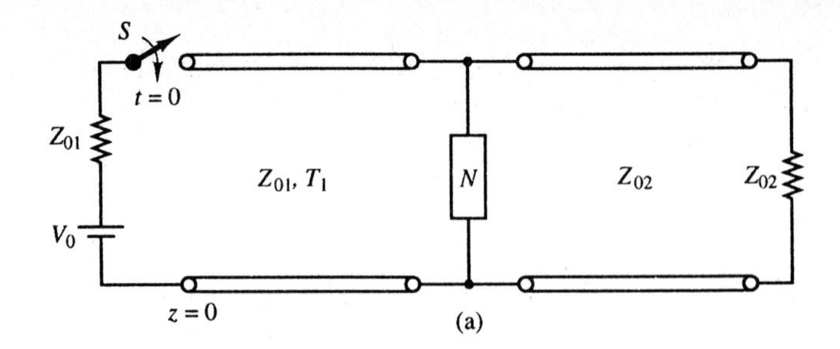

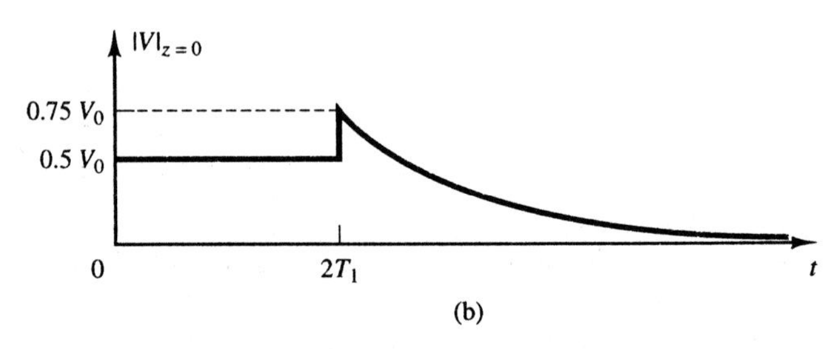

Figure 7.75. For Problem P7.27.

P7.29. In Fig. 7.76(a), the line is short circuited at one end $z = 0$ and open circuited at the other end $z = 100$ m. At $t = 0$, the current is zero throughout the line, and the voltage distribution is given by $V(z, 0) = 10 \sin 0.005\pi z$ V as shown in Fig. 7.76(b). Find and sketch the voltage and current distributions on the line for values of t equal to 0.5 μs and 1 μs.

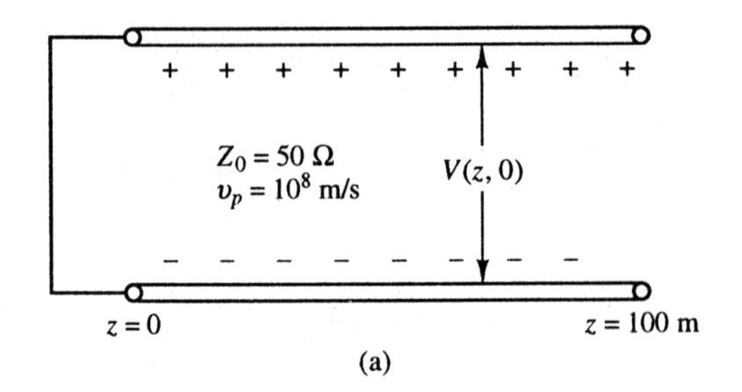

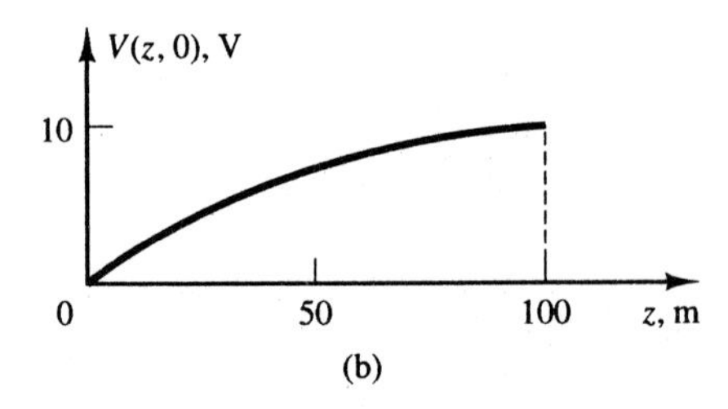

Figure 7.76. For Problem P7.29.

P7.30. In the system of Fig. 7.76(a), assume that a resistor of value 50 Ω is connected at the end $z = 100$ m at $t = 0$. Find and sketch the voltage across the resistor versus t.

P7.31. In the system shown in Fig. 7.77, a passive nonlinear element having the indicated volt-ampere characteristic is connected to an initially charged line at $t = 0$. Find the voltage across the nonlinear element immediately after closure of the switch.

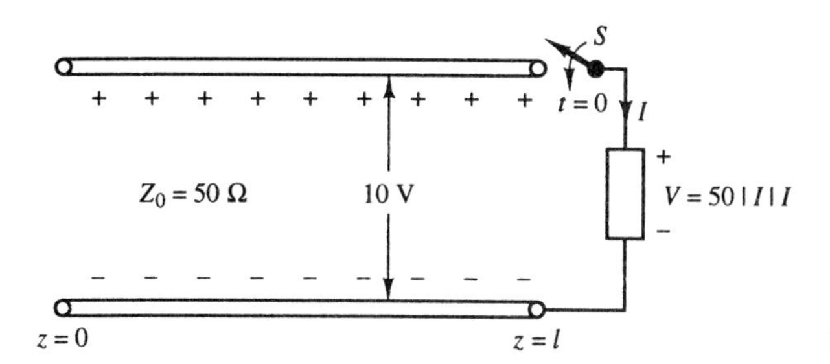

Figure 7.77. For Problem P7.31.

P7.32. In the system shown in Fig. 7.78, steady-state conditions are established with the switch S closed. At $t = 0$, the switch is opened. **(a)** Find and sketch the voltage across the 150 Ω resistor for $t \geq 0$, with the aid of a bounce diagram. **(b)** Show that the total energy dissipated in the 150 Ω resistor after opening the switch is exactly the same as the energy stored in the line before opening the switch.

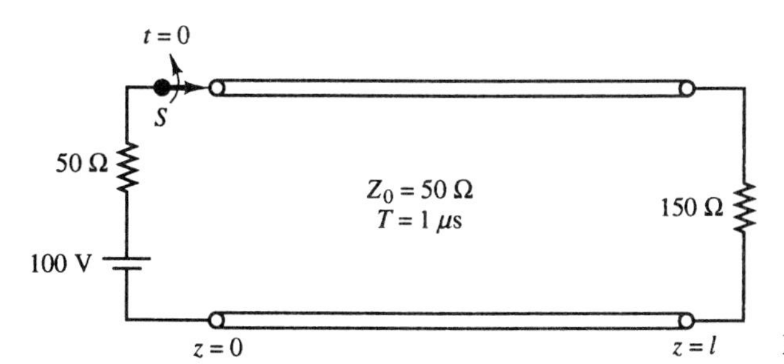

Figure 7.78. For Problem P7.32.

P7.33. In the system shown in Fig. 7.79, steady-state conditions are established with the switch S open and no current in the inductor. At $t = 0$, the switch is closed. **(a)** Obtain the expressions for the line voltage and current versus t at $z = l$. **(b)** Sketch the line voltage and current versus z for $t = T/2$.

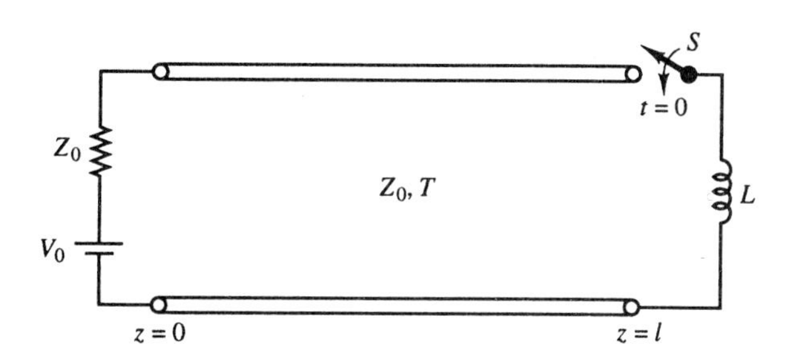

Figure 7.79. For Problem P7.33.

P7.34. For the system of Problem P7.31, use the load-line technique to obtain and plot line voltage versus t from $t = 0$ up to $t = 7l/v_p$ at $z = 0$ and $z = l$.

P7.35. For the example of interconnection between logic gates in Section 7.6, repeat the load-line constructions for $Z_0 = 50\ \Omega$ and draw graphs of V_i versus t for both 0-to-1 and 1-to-0 transitions.

P7.36. For the example of interconnection between logic gates in Section 7.6, find **(a)** the minimum value of Z_0 such that for the transition from 0 to 1, the voltage V_i reaches 2 V at $t = T+$ and **(b)** the minimum value of Z_0 such that for the transition from 1 to 0, the voltage V_i reaches 1 V at $t = T+$.

PC EXERCISES

PC7.1. Consider the determination of the characteristic impedance of a parallel-strip line with unequal widths of the conductors by using the method of moments. To do this, divide each conductor into $2n$ substrips of equal widths, as shown by the cross-sectional view in Fig. 7.80 for $n = 4$. Although there are a total of $4n$ substrips, only $2n$ unknown charge densities need to be determined, from considerations of symmetry. Consider the centerline of one of the substrips (say, 1) to be the reference axis. For each of the remaining $(2n - 1)$ substrips, denoted $2, 3, 4, \ldots, 2n$, write an equation relating the potential difference between the centerline of that substrip and the reference axis to the charge densities, thereby obtaining $(2n - 1)$ independent equations. For writing the contribution to the potential difference between substrip i and substrip 1 due to the charge in the ith substrip, use the result given in Problem P4.29. For writing the contribution to the potential difference between substrip i and substrip 1 due to the charge in a substrip other than the ith substrip, treat the charge in that substrip to be a line charge along its centerline. For the $2n$th equation, use the charge neutrality condition that the total charge per unit length on the two conductors must be zero. Assuming $\varepsilon = \varepsilon_0$ and a potential difference of 1 V between the conductors, write a program that for specified values of k_1, k_2, w, and n computes the $2n$ independent charge densities per unit length, the capacitance per unit length of the line, and the characteristic impedance of the line.

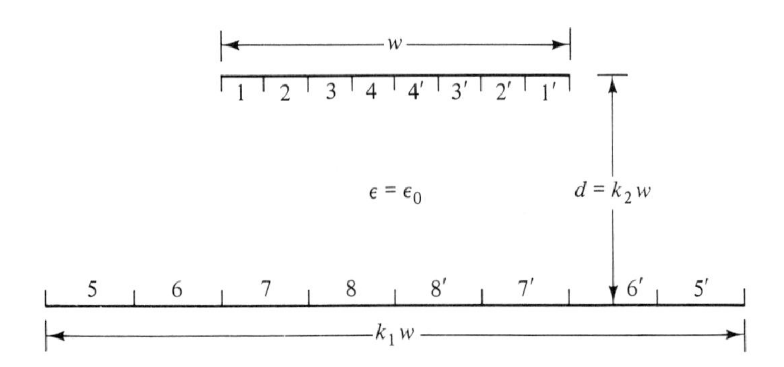

Figure 7.80. For Exercise PC7.1.

PC7.2. Consider the determination of the characteristic impedance of an eccentric coaxial cable [see Fig. 5.12(d)] by using the method of moments. To do this, divide each conductor into $2n$ segments of equal arc lengths, as shown by the cross-sectional view in Fig. 7.81(a) for $n = 4$, and approximate each segment by a plane strip as shown in Fig. 7.81(b). Although there are a total of $4n$ segments, only $2n$ unknown charge densities need to be determined, from considerations of symmetry. Using the method outlined in Exercise PC7.1, write a program that for a specified set of values for a, b, d, and n computes the $2n$ independent charge densities per unit length, the capacitance per unit length of the line, and the characteristic impedance of the line. Assume $\varepsilon = \varepsilon_0$ and a potential difference of 1 V between the conductors. The program is to check to see if $d < (b - a)$ before proceeding further. Note that from the expression for $\mathscr{C}$ of the eccentric coaxial cable given in Table 5.2, the exact analytical value of its characteristic impedance is

$$\frac{\eta}{2\pi} \cosh^{-1} \frac{a^2 + b^2 - d^2}{2ab}$$

and that if $d = 0$, the arrangement reduces to a concentric coaxial cable for which Z_0 is given in Table 7.1.

PC7.3. Modify the program of PL 7.3 for a rectangular pulse voltage source of duration from $t = 0$ to $t = kT$. The value of k is to be considered as additional input to the program. The output from the program is to consist of line voltage and line current versus z for specified values of t/T.

PC7.4. Consider a system of three lines in cascade, as that in Fig. 7.28(a), driven by a voltage source in series with a resistance equal to Z_{01} and terminated by a resistance equal to Z_{03}. Using the values of Z_{01}, Z_{02}, and Z_{03} as input, write a program that computes and plots the frequency response of the amplitude of V_o, as in Fig. 7.29, except in the interval $0 \leq \omega T_2 \leq \pi$ only, since the plot is periodic. An odd number of points n, where n is input to the program, are to be used in drawing the plot.

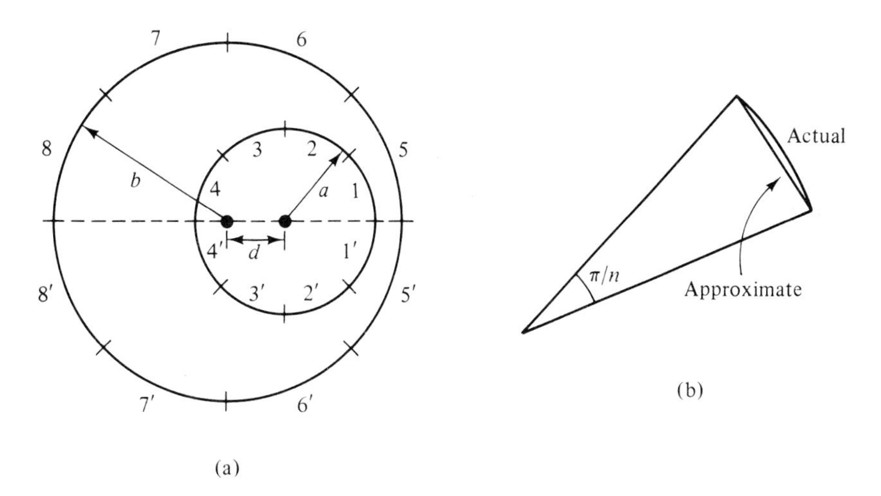

Figure 7.81. For Exercise PC7.2.

PC7.5. Consider n lines in parallel emanating from a junction a–b. For a wave incident on the junction from line i, where $i = 1, 2, 3, \ldots, n$, it is desired to compute the fraction of the incident power reflected into that line and the fraction of the incident power transmitted into each of the remaining $(n - 1)$ lines, assuming that no reflection occurs of the reflected wave and of the transmitted waves. The number n of the lines, the characteristic impedances of the lines, and the number i of the line from which the wave is incident on the junction are to be considered as inputs to the program.

PC7.6. Write a program that applies the load-line technique to the time-domain analysis of a line excited by a constant-voltage source in series with a resistance and terminated by a resistance. The output is to consist of the time variations of the voltages and currents at the two ends of the line, up to a specified value of time, and the steady-state values of the line voltage and line current.

8

Transmission Lines 2. Sinusoidal Steady-State Analysis

In Chapter 7 we introduced transmission lines and discussed time-domain analysis of transmission-line systems. In this chapter we are concerned with the steady-state analysis of transmission-line systems excited by sinusoidally time-varying sources. We recall from Chapter 7 that the phenomenon on a transmission line excited by a source connected to the line at a certain instant of time, say, $t = 0$, consists of the transient bouncing of (+) and (−) waves along the line for $t > 0$. In the steady state, the situation is equivalent to the superposition of one (+) wave, which is the sum of all the transient (+) waves, and one (−) wave, which is the sum of all the transient (−) waves. Thus the general solutions for the line voltage and line current in the sinusoidal steady state are superpositions of voltages and currents, respectively, of sinusoidal (+) and (−) waves. We shall first write these general solutions and then discuss several topics pertinent to sinusoidal steady-state analysis of transmission line systems.

We introduce the standing wave concept by first considering the particular case of a short-circuited line and then the general case of a line terminated by an arbitrary load. We discuss several techniques of transmission-line matching. In this connection, we introduce the Smith chart, a useful graphical aid in the solution of transmission-line problems. Finally we extend our treatment of sinusoidal steady-state analysis to lossy lines.

Although the concepts and techniques discussed in this chapter are based on the analysis of transmission-line systems, many of these are also applicable to the analysis of other, analogous systems. Examples are uniform plane wave propagation involving multiple media, as in Section 6.6, and discontinuities in waveguides, considered in Chapter 9.

8.1 SHORT-CIRCUITED LINE

General solution in the sinusoidal steady state

From (7.20a) and (7.20b), we write the general solutions for the line voltage and line current in the sinusoidal steady state to be

$$V(z, t) = A \cos\left[\omega\left(t - \frac{z}{v_p}\right) + \theta\right] + B \cos\left[\omega\left(t + \frac{z}{v_p}\right) + \phi\right] \tag{8.1a}$$

$$I(z, t) = \frac{1}{Z_0}\left\{A \cos\left[\omega\left(t - \frac{z}{v_p}\right) + \theta\right] - B \cos\left[\omega\left(t + \frac{z}{v_p}\right) + \phi\right]\right\} \tag{8.1b}$$

The corresponding expressions for the phasor line voltage and phasor line current are

$$\bar{V}(z) = \bar{V}^+ e^{-j\beta z} + \bar{V}^- e^{j\beta z} \tag{8.2a}$$

$$\bar{I}(z) = \frac{1}{Z_0}(\bar{V}^+ e^{-j\beta z} - \bar{V}^- e^{j\beta z}) \tag{8.2b}$$

where $\bar{V}^+ = Ae^{j\theta}$ and $\bar{V}^- = Be^{j\phi}$ and we have substituted β for ω/v_p. For sinusoidal steady-state problems, it is convenient to use a distance variable d which increases as we go from the load toward the generator as opposed to z, which increases from the generator toward the load, as shown in Fig. 8.1. The wave that progresses away from the generator is still denoted as the (+) wave, and the wave that progresses toward the generator is still denoted as the (−) wave. In terms of d, the solutions for $\bar{V}$ and $\bar{I}$ are then given by

$$\bar{V}(d) = \bar{V}^+ e^{j\beta d} + \bar{V}^- e^{-j\beta d} \tag{8.3a}$$

$$\bar{I}(d) = \frac{1}{Z_0}(\bar{V}^+ e^{j\beta d} - \bar{V}^- e^{-j\beta d}) \tag{8.3b}$$

Let us now consider a lossless line short circuited at the far end $d = 0$, as shown in Fig. 8.2. We shall assume that sinusoidally time-varying traveling waves exist on the line due to a source which is not shown in the figure and that conditions have reached steady state. We wish to determine the characteristics of the waves satisfying the boundary condition at the short circuit. Since the voltage across a short circuit has to be always equal to zero, this boundary condition is given by

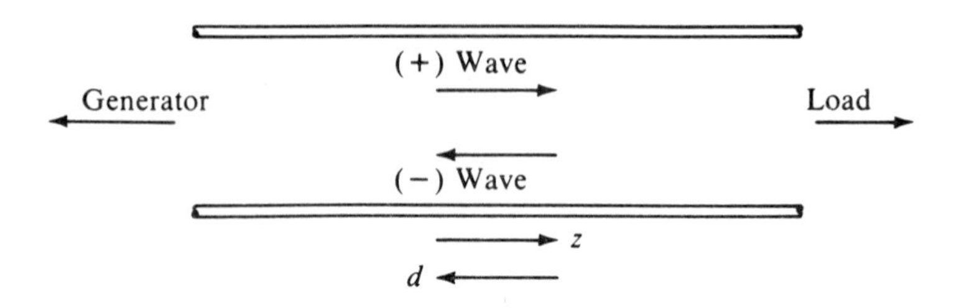

Figure 8.1. For illustrating the distance variable d used for sinusoidal steady-state analysis of traveling waves.

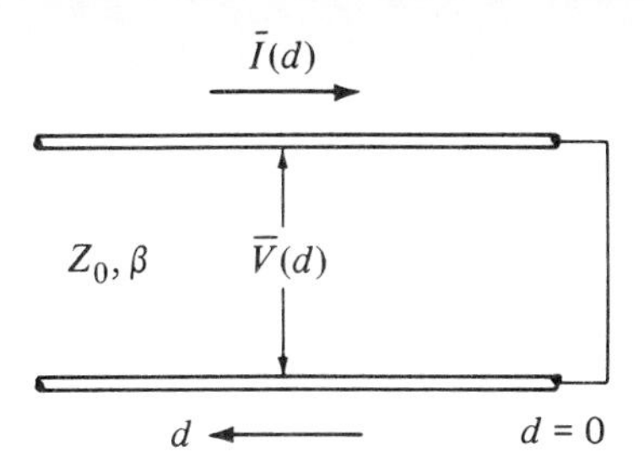

Figure 8.2. Transmission line short circuited at the far end.

$$\bar{V}(0) = 0 \tag{8.4}$$

Applying it to the general solution for $\bar{V}(d)$ given by (8.3a), we obtain

$$\bar{V}(0) = \bar{V}^+ e^{j\beta(0)} + \bar{V}^- e^{-j\beta(0)} = 0$$

or

$$\bar{V}^- = -\bar{V}^+ \tag{8.5}$$

Thus we find that the short circuit gives rise to a (−) or reflected wave whose voltage is exactly the negative of the (+) or incident wave voltage, at the short circuit.

Substituting (8.5) into (8.3a) and (8.3b), we get the particular solutions for the complex voltage and current on the short-circuited line to be

$$\bar{V}(d) = \bar{V}^+ e^{j\beta d} - \bar{V}^+ e^{-j\beta d} = 2j\bar{V}^+ \sin \beta d \tag{8.6a}$$

$$\bar{I}(d) = \frac{1}{Z_0}(\bar{V}^+ e^{j\beta d} + \bar{V}^+ e^{-j\beta d}) = 2\frac{\bar{V}^+}{Z_0} \cos \beta d \tag{8.6b}$$

The real voltage and current are then given by

$$\begin{aligned} V(d, t) &= \text{Re}[\bar{V}(d)e^{j\omega t}] \\ &= \text{Re}(2e^{j\pi/2}|\bar{V}^+|\, e^{j\theta} \sin \beta d\, e^{j\omega t}) \\ &= -2|\bar{V}^+| \sin \beta d \sin (\omega t + \theta) \end{aligned} \tag{8.7a}$$

$$\begin{aligned} I(d, t) &= \text{Re}[\bar{I}(d)e^{j\omega t}] \\ &= \text{Re}\left(2\frac{|\bar{V}^+|}{Z_0} e^{j\theta} \cos \beta d\, e^{j\omega t}\right) \\ &= 2\frac{|\bar{V}^+|}{Z_0} \cos \beta d \cos (\omega t + \theta) \end{aligned} \tag{8.7b}$$

where we have replaced $\bar{V}^+$ by $|\bar{V}^+|e^{j\theta}$ and j by $e^{j\pi/2}$. The instantaneous power flow down the line is given by

$$\begin{aligned} P(d, t) &= V(d, t)I(d, t) \\ &= -\frac{4|\bar{V}^+|^2}{Z_0} \sin \beta d \cos \beta d \sin (\omega t + \theta) \cos (\omega t + \theta) \\ &= -\frac{|\bar{V}^+|^2}{Z_0} \sin 2\beta d \sin 2(\omega t + \theta) \end{aligned} \tag{8.7c}$$

These results for the voltage, current, and power flow on the short-circuited line are illustrated in Fig. 8.3, which shows the variation of each of these quantities with distance from the short circuit for several values of time. The numbers 1, 2, 3, . . . , 9 beside the curves in Fig. 8.3 represent the order of the curves corresponding to values of $(\omega t + \theta)$ equal to 0, $\pi/4$, $\pi/2$, . . . , 2π. From (8.7a), (8.7b), and (8.7c) and from the sketches of Fig. 8.3, we can infer the following:

1. The line voltage is zero for $\sin \beta d = 0$, or $\beta d = 0, \pi, 2\pi, \ldots$, or $d = 0$, $\lambda/2, \lambda, \ldots$, for all values of time. If we short circuit the line at these val-

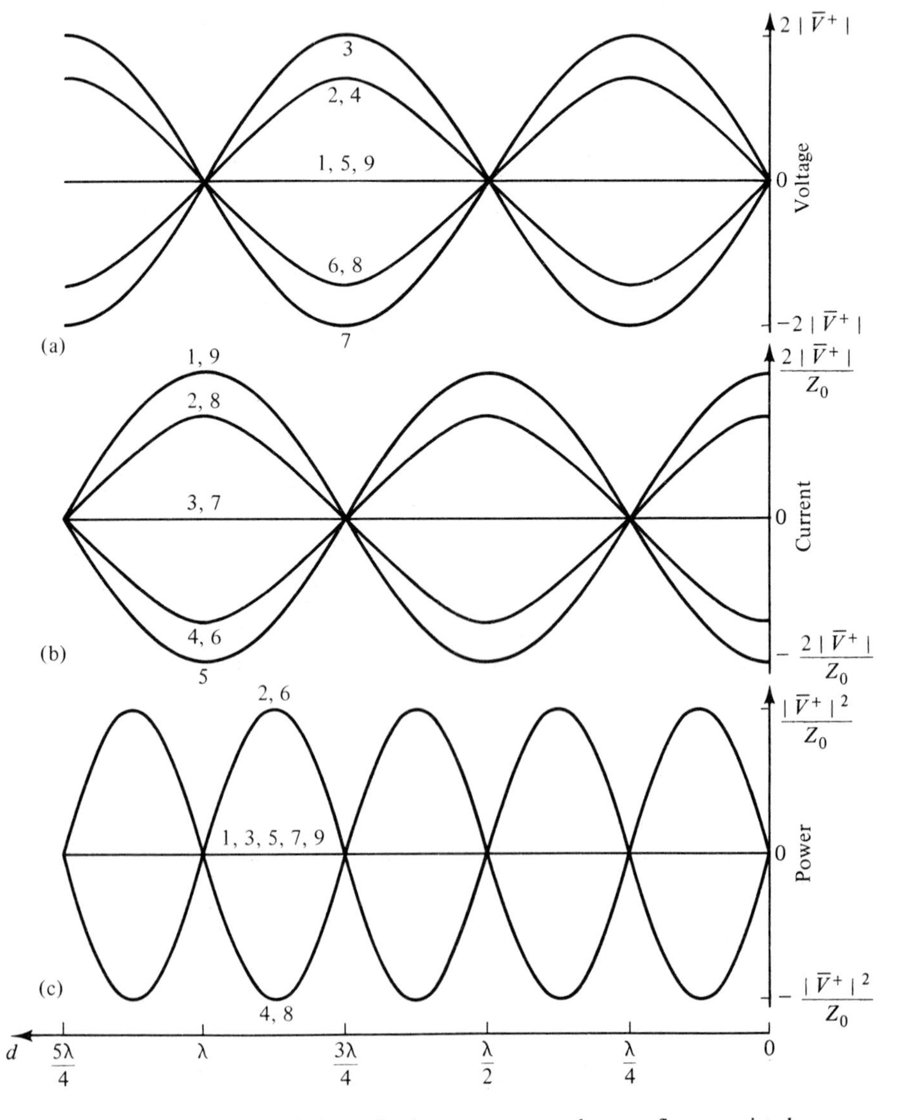

Figure 8.3. Time variations of voltage, current, and power flow associated with standing waves on a short-circuited transmission line.

ues of d, there will be no effect on the voltage and current at any other value of d.

2. The line current is zero for $\cos \beta d = 0$, or $\beta d = \pi/2, 3\pi/2, 5\pi/2, \ldots$, or $d = \lambda/4, 3\lambda/4, 5\lambda/4, \ldots$, for all values of time. If we open circuit the line at these values of d, there will be no effect on the voltage and current at any other value of d.
3. The power flow is zero for $\sin 2\beta d = 0$, or $2\beta d = 0, \pi, 2\pi, \ldots$, or $d = 0, \lambda/4, \lambda/2, \ldots$, for all values of time.

Thus the phenomenon on the short-circuited line is one in which the voltage, current, and power flow oscillate sinusoidally with time with different amplitudes at different locations on the line, unlike in the case of traveling waves in which a given point on the waveform progresses in distance with time. Since there is no feeling of wave motion down the line, these waves are known as *standing waves*. In particular, they represent *complete standing waves* in view of the zero amplitudes of the voltage, current, and power flow at certain locations on the line, as just discussed and as shown in Fig. 8.3. Complete standing waves are the result of (+) and (−) traveling waves of equal amplitudes. Whatever power is incident on the short circuit by the (+) wave is reflected entirely in the form of the (−) wave since the short circuit cannot absorb any power. While there is instantaneous power flow at values of d between the voltage and current nodes, there is no time-average power flow for any value of d, as can be seen from

$$
\begin{aligned}
\langle P \rangle &= \frac{1}{T}\int_{t=0}^{T} P(d, t)\, dt = \frac{\omega}{2\pi}\int_{t=0}^{2\pi/\omega} P(d, t)\, dt \\
&= \frac{\omega}{2\pi}\frac{|\bar{V}^+|^2}{Z_0}\sin 2\beta d \int_{t=0}^{2\pi/\omega} \sin 2(\omega t + \theta)\, dt \\
&= 0
\end{aligned}
$$

Standing wave patterns

From (8.6a) and (8.6b) or (8.7a) and (8.7b), or from Figs. 8.3(a) and 8.3(b), we find that the amplitudes of the sinusoidal time variations of the line voltage and line current as functions of distance along the line are

$$|\bar{V}(d)| = 2|\bar{V}^+||\sin \beta d| = 2|\bar{V}^+|\left|\sin \frac{2\pi}{\lambda} d\right| \tag{8.8a}$$

$$|\bar{I}(d)| = \frac{2|\bar{V}^+|}{Z_0}|\cos \beta d| = \frac{2|\bar{V}^+|}{Z_0}\left|\cos \frac{2\pi}{\lambda} d\right| \tag{8.8b}$$

Sketches of these quantities versus d are shown in Fig. 8.4. These are known as the *standing wave patterns*. They are the patterns of line voltage and line current one would obtain by connecting an ac voltmeter between the conductors of the line and an ac ammeter in series with one of the conductors of the line and observing their readings at various points along the line. Alternatively, one can sample the electric and magnetic fields by means of probes. Standing wave patterns should not be misinterpreted as the voltage and current remaining constant with time at a given point. On the other hand, the voltage and current at every point on the line vary sinusoidally with time, as shown in the insets of Fig. 8.4,

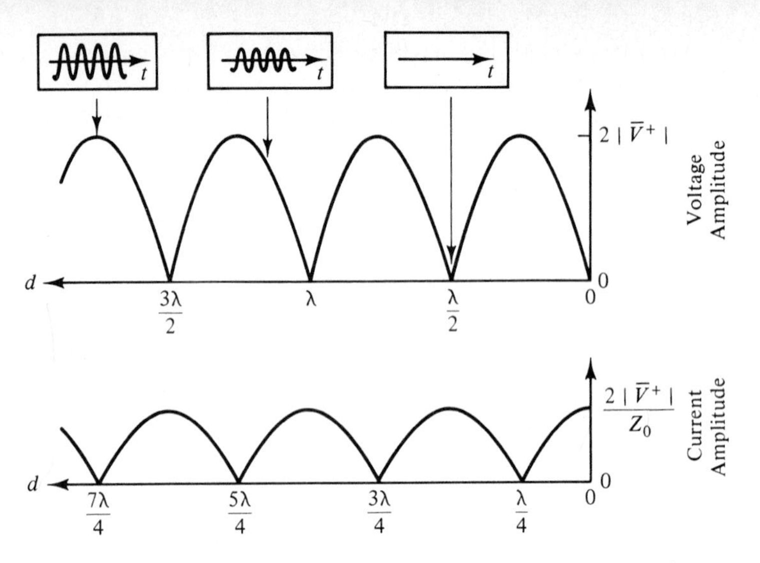

Figure 8.4. Standing wave patterns for voltage and current on a short-circuited line. The insets show time variations of the voltage at points along the line.

with the amplitudes of these sinusoidal variations equal to the magnitudes indicated by the standing wave patterns. Since the distance between successive nodes of voltage or current is equal to $\lambda/2$, a measurement of this distance provides the knowledge of the wavelength. Furthermore, if the phase velocity in the line is known, the frequency of the source can be computed, and vice versa, since $v_p = \lambda f$.

Natural oscillations

Since there is no power flow across a voltage node or a current node of the standing wave patterns, a constant amount of total energy is locked up in every $\lambda/4$ section between two such adjacent nodes with exchange of energy taking place between the electric and magnetic fields. Thus once the line is excited by applying a source of energy, then each $\lambda/4$ section of the line between the voltage and current nodes acts as a resonator entirely independent of the remainder of the line. In fact, the $\lambda/4$ section can be removed from the line by cutting it, that is, open circuiting it, at the current node and short circuiting it at the voltage node, and still be made to maintain forever the oscillations of voltage and current. Such oscillations are called *natural oscillations*. Similarly, sections of lengths equal to multiples of $\lambda/4$ can be removed by always cutting the line at current nodes and short circuiting it at voltage nodes, without disturbing the oscillations.

For a fixed physical length of the line, its electrical length, that is, its length in terms of wavelength, depends upon the frequency. Thus, a line of length equal to one-quarter wavelength at one frequency behaves as a line of length equal to a different multiple of a wavelength at a different frequency. Let us now consider a line of length l, one end of which is open circuited and the other end short circuited, and assume that some energy is stored in this line. Suppose we now pose the question: "What are all the possible standing wave patterns on this line?" To answer this, we note that the voltage across the short circuit must always be zero, and hence the current there must have maximum amplitude. Similarly, the current at the open-circuited end must always be zero, and hence the voltage there must have maximum amplitude. We also know that the standing wave patterns are sinusoidal with the distance between sucessive nodes corresponding to a half sine wave. Thus the least possible variation is a quarter cycle of a sine waveform. This corresponds to a wavelength, say, λ_1, equal to $4l$, and the corresponding standing wave patterns are shown in Fig. 8.5(a).

It is not possible to have a standing wave pattern for which the wavelength is greater than $4l$ since then the pattern on the line of length l will be less than a quarter cycle of a sine wave. On the other hand, it is possible to have a pattern for which the wavelength is less than $4l$ as long as the conditions of zero voltage (maximum current) at the short circuit and zero current (maximum voltage) at the open circuit are satisfied. Obviously, the next largest wavelength λ_2, less than λ_1, for which this condition is satisfied corresponds to the patterns shown in Fig. 8.5(b). For these patterns, $l = 3\lambda_2/4$, or $\lambda_2 = 4l/3$. The next largest wavelength, λ_3, less than λ_2, corresponds to the patterns shown in Fig. 8.5(c). For these patterns, $l = 5\lambda_3/4$, or $\lambda_3 = 4l/5$.

We can continue in this manner and see that any standing wave pattern for which the length of the line is an odd multiple of one-quarter wavelength, that is

$$l = \frac{(2n-1)\lambda_n}{4}, \qquad n = 1, 2, 3, \ldots \tag{8.9}$$

is a valid standing wave pattern. Alternatively, the wavelengths λ_n, corresponding to the valid standing wave patterns, are given by

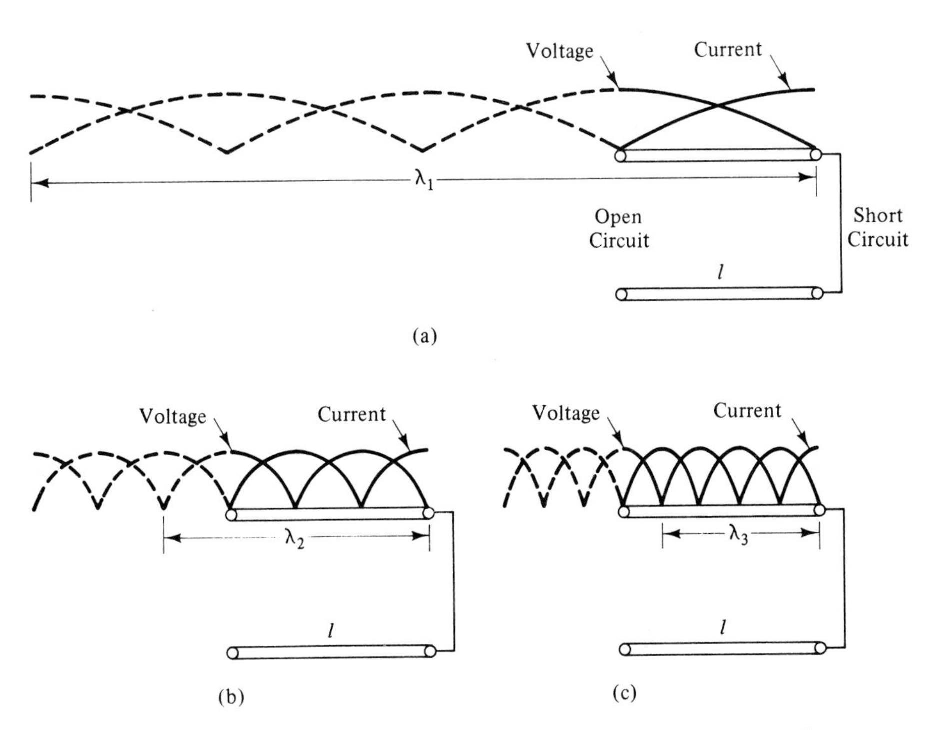

Figure 8.5. Standing wave patterns corresponding to (a) one-quarter cycle, (b) three-quarters cycle, and (c) five-quarters cycle, of a sine wave for the voltage and current amplitude distributions for a line of length l open circuited at one end and short circuited at the other end.

$$\lambda_n = \frac{4l}{2n - 1}, \qquad n = 1, 2, 3, \ldots \tag{8.10}$$

The corresponding frequencies are

$$f_n = \frac{v_p}{\lambda_n} = \frac{(2n - 1)v_p}{4l}, \qquad n = 1, 2, 3, \ldots \tag{8.11}$$

where v_p is the phase velocity. These frequencies are known as the *natural frequencies of oscillation*. The standing wave patterns are said to correspond to the different natural modes of oscillation. The lowest frequency (corresponding to the longest wavelength) is known as the *fundamental frequency of oscillation*, and the corresponding mode is known as the *fundamental mode*. The quantity n is called the *mode number*. In the most general case of nonsinusoidal voltage and current distributions on the line, the situation corresponds to the superposition of some or all of the infinite number of natural modes.

Considerations similar to those for the line open circuited at one end and short circuited at the other end apply to natural oscillations on lines short circuited at both ends or open circuited at both ends.

Input impedance

Returning now to the expressions for the phasor line voltage and the phasor line current given by (8.6a) and (8.6b), respectively, we define the ratio of these two quantities as the line impedance $\bar{Z}(d)$ at that point seen looking toward the short circuit. Thus

$$\bar{Z}(d) = \frac{\bar{V}(d)}{\bar{I}(d)} = \frac{2j\bar{V}^+ \sin \beta d}{2(\bar{V}^+/Z_0) \cos \beta d} = jZ_0 \tan \beta d \tag{8.12}$$

In particular, the input impedance $\bar{Z}_{in}$ of a short-circuited line of length l is given by

$$\boxed{\bar{Z}_{in} = jZ_0 \tan \beta l = jZ_0 \tan \frac{2\pi f}{v_p} l} \tag{8.13}$$

We note from (8.13) that the input impedance of the short-circuited line is purely reactive. As the frequency is varied from a low value upward, the input reactance changes from inductive to capacitive and back to inductive, and so on, as illustrated in Fig. 8.6. The input reactance is zero for values of frequency equal to multiples of $v_p/2l$. These are the frequencies for which l is equal to multiples of $\lambda/2$ so that the line voltage is zero at the input and hence the input sees a short circuit. The input reactance is infinity for values of frequency equal to odd multiples of $v_p/4l$. These are the frequencies for which l is equal to odd multiples of $\lambda/4$ so that the line current is zero at the input and hence the input sees an open circuit.

These properties of the input impedance of a short-circuited line (and, similarly, of an open-circuited line) have several applications. We shall here discuss two of these applications.

Location of short circuit

1. *Determination of the location of a short circuit (or open circuit) in a line.* The principle behind this lies in the fact that as the frequency of a generator connected to the input of a short-circuited (or open-circuited) line is varied continuously upward, the current drawn from it undergoes alternatively maxima and minima corresponding to zero input reactance and infinite input reactance conditions,

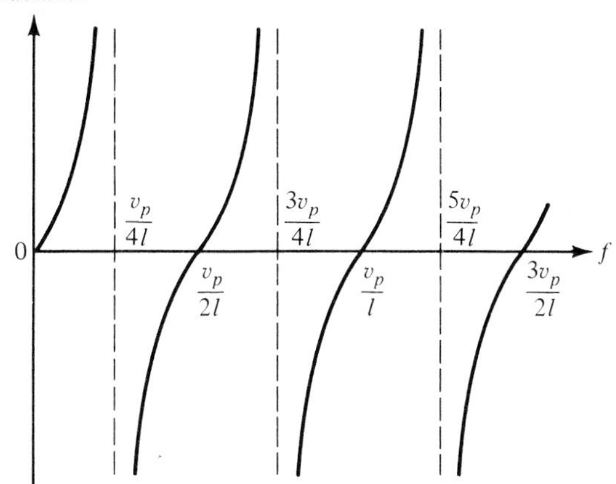

Figure 8.6. Variation of the input reactance of a short-circuited transmission line with frequency.

respectively. Since the difference between a pair of consecutive frequencies for which the input reactance values are zero and infinity is $v_p/4l$, as can be seen from Fig. 8.6, it follows that the difference between successive frequencies for which the currents drawn from the generator are maxima and minima is $v_p/4l$. As a numerical example, if for an air-dielectric line, it is found that as the frequency is varied from 50 MHz upward, the current reaches a minimum for 50.01 MHz and then a maximum for 50.04 MHz, then the distance l of the short circuit from the generator is given by

$$\frac{v_p}{4l} = (50.04 - 50.01) \times 10^6 = 0.03 \times 10^6 = 3 \times 10^4$$

Since $v_p = 3 \times 10^8$ m/s, it follows that

$$l = \frac{3 \times 10^8}{4 \times 3 \times 10^4} = 2500 \text{ m} = 2.5 \text{ km}$$

Alternatively, if the length l is known, we can compute v_p for the dielectric of the line, from which the permittivity of the dielectric can be found, provided that the value of μ (usually equal to μ_0) is known.

Resonant system

2. *Construction of resonant circuits at microwave frequencies.* The principle behind this lies in the fact that the input reactance of a short-circuited line of a given length can be inductive or capacitive, depending upon the frequency, and hence two short-circuited lines connected together form a resonant system. To obtain the characteristic equation for the resonant frequencies of such a system, let us consider the system shown in Fig. 8.7, which is made up of two short-circuited line sections of characteristic impedances Z_{01} and Z_{02}, lengths l_1 and l_2, and phase velocities v_{p1} and v_{p2}. Denoting the voltages and currents just to the left and just to the right of the junction to be $\bar{V}_1$ and $\bar{I}_1$, and $\bar{V}_2$ and $\bar{I}_2$, respectively, as shown in the figure, we write the boundary conditions at the junction as

$$\bar{V}_1 = \bar{V}_2 \tag{8.14a}$$

$$\bar{I}_1 + \bar{I}_2 = 0 \tag{8.14b}$$

Combining the two, we have

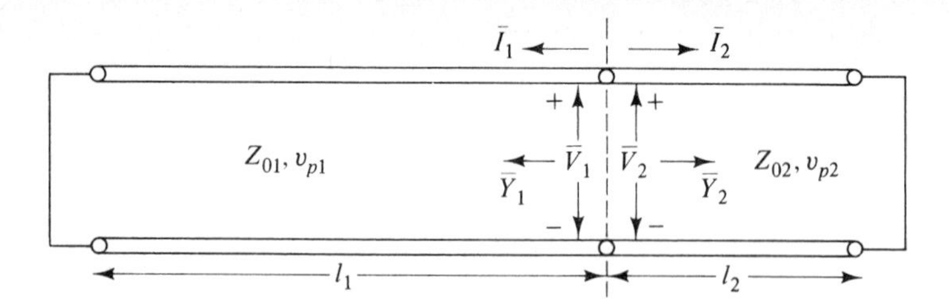

Figure 8.7. Resonant system formed by connecting together two short-circuited line sections.

$$\frac{\bar{I}_1}{\bar{V}_1} + \frac{\bar{I}_2}{\bar{V}_2} = 0$$

or

$$\bar{Y}_1 + \bar{Y}_2 = 0 \tag{8.15}$$

where $\bar{Y}_1$ and $\bar{Y}_2$ are the input admittances of the sections to the left and to the right, respectively, of the junction and seen looking toward the short circuits. Equation (8.15) is the condition for resonance of the system. To express it in terms of the line parameters, we note that

$$\bar{Y}_1 = \frac{1}{\bar{Z}_1} = \frac{1}{jZ_{01} \tan \beta_1 l_1} = \frac{1}{jZ_{01} \tan (2\pi f/v_{p1})l_1} \tag{8.16a}$$

$$\bar{Y}_2 = \frac{1}{\bar{Z}_2} = \frac{1}{jZ_{02} \tan \beta_2 l_2} = \frac{1}{jZ_{02} \tan (2\pi f/v_{p2})l_2} \tag{8.16b}$$

Substituting (8.16a) and (8.16b) into (8.15) and simplifying, we obtain the characteristic equation for the resonant frequencies to be

$$Z_{01} \tan \frac{2\pi f}{v_{p1}} l_1 + Z_{02} \tan \frac{2\pi f}{v_{p2}} l_2 = 0 \tag{8.17}$$

We shall illustrate the computation of the resonant frequencies by means of an example.

Example 8.1

For the system of Fig. 8.7, let us assume $Z_{01} = 2Z_{02} = 60\ \Omega$, $l_1 = 5$ cm, $l_2 = 2$ cm, and $v_{p1} = v_{p2} = c/2$, and obtain the four lowest resonant frequencies of the system.

Substituting the numerical values of the parameters into the characteristic equation (8.17), we obtain

$$\tan \frac{0.2\pi f}{c} + \frac{1}{2} \tan \frac{0.08\pi f}{c} = 0$$

This equation is of the form

$$\tan kx + m \tan x = 0$$

where here $k = 2.5$, $m = 0.5$, and $x = 0.08\pi f/c$. In general, an equation of this type can be solved by plotting $\tan kx$ and $-m \tan x$ to scale versus x and finding the points of intersection. Alternatively, a programmable calculator or a computer can be used. The listing of a PC program that computes the first N nonzero solutions of the equation for specified values of k (> 1.01) and m (> 0) and the output from a run of the program for values of $k = 2.5$, $m = 0.5$, and $N = 4$ are included as PL 8.1.

PL 8.1. Program listing and sample output for computing solutions for the equation $\tan kx + m \tan x = 0$, where $k > 1.01$ and $m > 0$.

```
100 '*******************************************************
110 '* COMPUTATION OF LOWEST N NONZERO SOLUTIONS FOR      *
120 '* F(X)=TAN(K*X)+M*TAN(X)=0, WHERE K IS GREATER THAN *
130 '* UNITY AND M IS POSITIVE                            *
140 '*******************************************************
150 DEF FN TRD(ARG)=INT(ARG*10000+.5)/10000:'* ROUNDS ARG
160 '  TO FOUR DECIMAL PLACES *
170 PI=3.1416:P2=PI/2
180 CLS:LOCATE 1,1:PRINT "ENTER VALUES OF K>1.01 AND M>0:"
190 PRINT:INPUT "K = ",K
200 PRINT:INPUT "M = ",M
210 PRINT:INPUT "ENTER NUMBER OF SOLUTIONS DESIRED: N = ",
    NS:PRINT
220 IF K<1.01 OR M<=0 THEN 180
230 PK=P2/K:I=1:N=0
240 '* COMPUTATION OF X1 AND X2, THE LOWER AND UPPER
250 '  BOUNDS FOR X, FOR A GIVEN SOLUTION *
260 IF (INT(I/2)*2-I) THEN SK=-1:GOTO 280
270 SK=1
280 X1=PK*I:X2=X1+PK
290 IF INT(I/2)=I/2 AND INT(I/(2*K))=I/(2*K) THEN X=X1:GOT
    O 430
300 IF INT(I/K)=I/K THEN S1=SGN(TAN(X1+.0001)):GOTO 320
310 S1=SGN(TAN(X1))
320 IF INT((I+1)/K)=(I+1)/K THEN S2=SGN(TAN(X2-.0001)):GOT
    O 340
330 S2=SGN(TAN(X2))
340 IF S1=SK AND S2=SK THEN I=I+1:GOTO 260
350 IF S1=SK AND S2<>SK THEN X1=(INT(X1/P2)+1)*P2:GOTO 380
360 IF S1<>SK AND S2=SK THEN X2=INT(X2/P2)*P2
370 '* COMPUTATION AND PRINTING OF VALUE OF X *
380 X=(X1+X2)/2
390 F=TAN(K*X)+M*TAN(X)
400 IF F<0 THEN X1=X:GOTO 420
410 X2=X
420 IF ABS(F)>.0001 THEN 380
430 N=N+1:PRINT "SOLUTION NO.";N;" IS: X =";FN TRD(X);"=";
    FN TRD(X/PI);"*PI"
440 IF N=NS THEN PRINT:PRINT "STRIKE ANY KEY TO CONTINUE":
    C$=INPUT$(1):GOTO 180
450 I=I+1:GOTO 260
460 END

RUN
ENTER VALUES OF K>1.01 AND M>0:

K = 2.5

M = .5

ENTER NUMBER OF SOLUTIONS DESIRED: N = 4

SOLUTION NO. 1  IS: X = .9944 = .3165 *PI
SOLUTION NO. 2  IS: X = 1.7476 = .5563 *PI
SOLUTION NO. 3  IS: X = 2.6242 = .8353 *PI
SOLUTION NO. 4  IS: X = 3.659 = 1.1647 *PI

STRIKE ANY KEY TO CONTINUE
```

From the values of x obtained from the computer program, we obtain the lowest four resonant frequencies to be 1.1869×10^9, 2.0861×10^9, 3.1324×10^9, and 4.3676×10^9 Hz, or 1.1869, 2.0861, 3.1324, and 4.3676 GHz.

K8.1. Phasor line voltage and line current; General solutions; Short-circuited line; Complete standing waves; Standing wave patterns; Natural oscillations; Input impedance; Resonant systems.

D8.1. For each of the following characteristics of standing waves on a lossless short-circuited line, find the frequency of the source exciting the line: **(a)** the distance between successive nodes of voltage amplitude is 50 cm and the dielectric is air; **(b)** the distance between successive nodes of current amplitude is 50 cm and the dielectric is nonmagnetic with $\varepsilon = 9\varepsilon_0$; and **(c)** the distance between successive nodes of instantaneous power flow is 50 cm and the dielectric is air.
Ans. **(a)** 300 MHz; **(b)** 100 MHz; **(c)** 150 MHz

D8.2. A lossless coaxial cable of characteristic impedance 50 Ω and having a nonmagnetic ($\mu = \mu_0$), perfect dielectric of permittivity $\varepsilon = 2.25\varepsilon_0$ is short circuited at the far end. Find the minimum length of the line for which the input impedance is equal to the impedance of each of the following at $f = 100$ MHz: **(a)** an inductor of value equal to 0.5 μH; **(b)** an inductor of value equal to the inductance per unit length of the line; and **(c)** an inductor of value equal to the inductance of the line.
Ans. **(a)** 44.98 cm; **(b)** 40.19 cm; **(c)** 143.03 cm

D8.3. A lossless transmission line of length $l = 5$ m, characteristic impedance $Z_0 = 100$ Ω, and having a nonmagnetic ($\mu = \mu_0$), perfect dielectric is short circuited at its far end. A variable frequency voltage source in series with an internal impedance $\bar{Z}_g$ is connected at its input and the line voltage and line current at the input terminals are monitored as the source frequency is varied. It is found that the voltage reaches a maximum amplitude of 10 V at 157.5 MHz and then the current reaches a maximum amplitude of 0.2 A at 165 MHz. Find the following: **(a)** the maximum amplitude of the current in the standing wave pattern on the line at 157.5 MHz; **(b)** the maximum amplitude of the voltage in the standing wave pattern on the line at 165 MHz; **(c)** the magnitude of $\bar{Z}_g$; and **(d)** the permittivity of the dielectric of the line.
Ans. **(a)** 0.1 A; **(b)** 20 V; **(c)** 50 Ω; **(d)** $4\varepsilon_0$

8.2 LINE TERMINATED BY ARBITRARY LOAD

We devoted the preceding section to the short-circuited line. In this section we consider a line terminated by an arbitrary load impedance $\bar{Z}_R$, as shown in Fig. 8.8. Then starting with the general solutions for the complex line voltage and line current given by

$$\bar{V}(d) = \bar{V}^+ e^{j\beta d} + \bar{V}^- e^{-j\beta d} \tag{8.18a}$$

$$\bar{I}(d) = \frac{1}{Z_0}(\bar{V}^+ e^{j\beta d} - \bar{V}^- e^{-j\beta d}) \tag{8.18b}$$

and using the boundary condition at $d = 0$, given by

$$\bar{V}(0) = \bar{Z}_R \bar{I}(0) \tag{8.19}$$

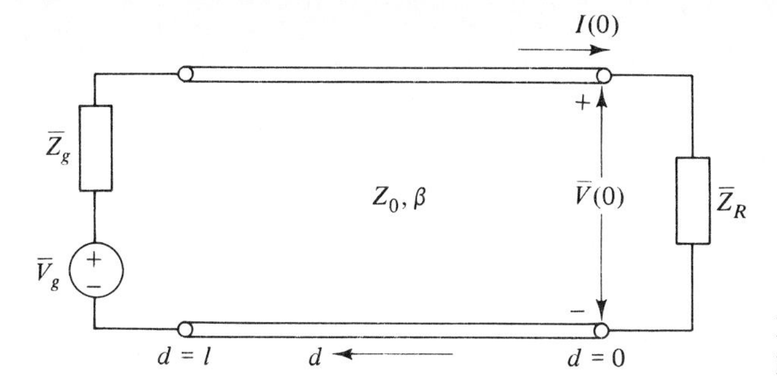

Figure 8.8. Line terminated by a complex load impedance.

we obtain

$$\bar{V}^+ + \bar{V}^- = \frac{\bar{Z}_R}{Z_0}(\bar{V}^+ - \bar{V}^-)$$

or

$$\bar{V}^- = \bar{V}^+ \frac{\bar{Z}_R - Z_0}{\bar{Z}_R + Z_0}$$

Thus the ratio of $\bar{V}^-$, the reflected wave voltage at the load, to $\bar{V}^+$, the incident wave voltage at the load, that is, the voltage reflection coefficient at the load, denoted by $\bar{\Gamma}_R$, is given by

$$\boxed{\bar{\Gamma}_R = \frac{\bar{V}^-}{\bar{V}^+} = \frac{\bar{Z}_R - Z_0}{\bar{Z}_R + Z_0}} \tag{8.20}$$

The solutions for $\bar{V}(d)$ and $\bar{I}(d)$ can then be written as

$$\bar{V}(d) = \bar{V}^+ e^{j\beta d} + \bar{\Gamma}_R \bar{V}^+ e^{-j\beta d} \tag{8.21a}$$

$$\bar{I}(d) = \frac{1}{Z_0}(\bar{V}^+ e^{j\beta d} - \bar{\Gamma}_R \bar{V}^+ e^{-j\beta d}) \tag{8.21b}$$

Generalized reflection coefficient

We now define the generalized voltage reflection coefficient, $\bar{\Gamma}(d)$, that is, the voltage reflection coefficient at any value of d, as the ratio of the reflected wave voltage to the incident wave voltage at that value of d. From (8.21a) we see that

$$\boxed{\bar{\Gamma}(d) = \frac{\bar{\Gamma}_R \bar{V}^+ e^{-j\beta d}}{\bar{V}^+ e^{j\beta d}} = \bar{\Gamma}_R e^{-j2\beta d}} \tag{8.22}$$

so that

$$|\bar{\Gamma}(d)| = |\bar{\Gamma}_R|\,|e^{-j2\beta d}| = |\bar{\Gamma}_R| \tag{8.23a}$$

and

$$\underline{/\bar{\Gamma}(d)} = \underline{/\bar{\Gamma}_R} + \underline{/e^{-j2\beta d}} = \theta - 2\beta d \tag{8.23b}$$

where θ is the phase angle of $\bar{\Gamma}_R$. Thus, the magnitude of the generalized reflection coefficient remains constant along the line and equal to its value at the load, whereas the phase angle varies linearly with d. In terms of $\bar{\Gamma}(d)$, we can write the solutions for $\bar{V}(d)$ and $\bar{I}(d)$ as

$$\begin{aligned}\bar{V}(d) &= \bar{V}^+ e^{j\beta d}(1 + \bar{\Gamma}_R e^{-j2\beta d}) \\ &= \bar{V}^+ e^{j\beta d}[1 + \bar{\Gamma}(d)]\end{aligned} \tag{8.24a}$$

$$\begin{aligned}\bar{I}(d) &= \frac{\bar{V}^+}{Z_0} e^{j\beta d}(1 - \bar{\Gamma}_R e^{-j2\beta d}) \\ &= \frac{\bar{V}^+}{Z_0} e^{j\beta d}[1 - \bar{\Gamma}(d)]\end{aligned} \tag{8.24b}$$

To study the standing wave patterns corresponding to (8.24a) and (8.24b), we look at the magnitudes of $\bar{V}(d)$ and $\bar{I}(d)$. These are given by

$$\begin{aligned}|\bar{V}(d)| &= |\bar{V}^+||e^{j\beta d}||1 + \bar{\Gamma}(d)| \\ &= |\bar{V}^+||1 + \bar{\Gamma}_R e^{-j2\beta d}|\end{aligned} \tag{8.25a}$$

$$\begin{aligned}|\bar{I}(d)| &= \frac{|\bar{V}^+|}{Z_0}|e^{j\beta d}||1 - \bar{\Gamma}(d)| \\ &= \frac{|\bar{V}^+|}{Z_0}|1 - \bar{\Gamma}_R e^{-j2\beta d}|\end{aligned} \tag{8.25b}$$

To sketch $|\bar{V}(d)|$ and $|\bar{I}(d)|$, it is sufficient if we consider the quantities $|1 + \bar{\Gamma}_R e^{-j2\beta d}|$ and $|1 - \bar{\Gamma}_R e^{-j2\beta d}|$ since $|\bar{V}^+|$ is simply a constant, determined by the boundary condition at the source end. Each of these quantities consists of two complex numbers, one of which is a constant equal to $(1 + j0)$ and the other of which has a constant magnitude $|\bar{\Gamma}_R|$ but a variable phase angle $(\theta - 2\beta d)$. To evaluate $|1 + \bar{\Gamma}_R e^{-j2\beta d}|$ and $|1 - \bar{\Gamma}_R e^{-j2\beta d}|$, we make use of the constructions in the complex $\bar{\Gamma}$-plane as shown in Fig. 8.9(a) and (b), respectively. In both diagrams, we draw circles with centers at the origin and having radii equal to $|\bar{\Gamma}_R|$. For $d = 0$, the complex number $\bar{\Gamma}_R e^{-j2\beta d}$ is equal to $\bar{\Gamma}_R$ or $|\bar{\Gamma}_R| e^{j\theta}$, which is repre-

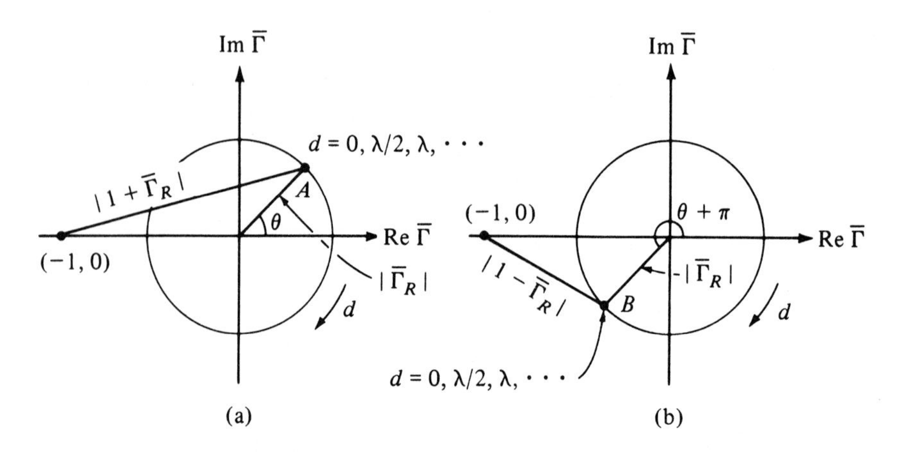

Figure 8.9. $\bar{\Gamma}$-plane diagrams for sketching the voltage and current standing wave patterns for the system of Fig. 8.8.

sented by point A in Fig. 8.9(a). To add $(1 + j0)$ and $\bar{\Gamma}_R$, we simply draw a line from the point $(-1, 0)$ to the point A. The length of this line gives $|1 + \bar{\Gamma}_R|$, which is proportional to the amplitude of the voltage at $d = 0$. As d increases, point A, representing $\bar{\Gamma}_R e^{-j2\beta d}$, moves around the circle in the clockwise direction. The line joining $(-1, 0)$ to the point A whose length is $|1 + \bar{\Gamma}_R e^{-j2\beta d}|$ executes the motion of a crank. To subtract $\bar{\Gamma}_R$ from $(1 + j0)$ we locate point B in Fig. 8.9(b), which is diametrically opposite to point A in Fig. 8.9(a), and draw a line from $(-1, 0)$ to point B. The length of the line gives $|1 - \bar{\Gamma}_R|$, which is proportional to the amplitude of the current at $d = 0$. As d increases, B moves around the circle in the clockwise direction following the movement of A in Fig. 8.9(a). The line joining $(-1, 0)$ to the point B whose length is $|1 - \bar{\Gamma}_R e^{-j2\beta d}|$ executes the motion of a crank. From these constructions and assuming $-\pi \leq \theta < \pi$, we note the following facts:

1. Point A lies along the positive real axis and point B lies along the negative real axis for $(\theta - 2\beta d) = 0, -2\pi, -4\pi, -6\pi, \ldots,$ or $d = (\lambda/4\pi)(\theta + 2n\pi)$, where $n = 0, 1, 2, 3, \ldots$. Hence, at these values of d, the voltage amplitude is maximum and equal to $|\bar{V}^+|(1 + |\bar{\Gamma}_R|)$ whereas the current amplitude is minimum and equal to $(|\bar{V}^+|/Z_0)(1 - |\bar{\Gamma}_R|)$. The voltage and current are in phase.
2. Point A lies along the negative real axis and point B lies along the positive real axis for $(\theta - 2\beta d) = -\pi, -3\pi, -5\pi, -7\pi, \ldots,$ or $d = (\lambda/4\pi)[\theta + (2n - 1)\pi]$, where $n = 1, 2, 3, 4, \ldots$. Hence, at these values of d, the voltage amplitude is minimum and equal to $|\bar{V}^+|(1 - |\bar{\Gamma}_R|)$ whereas the current amplitude is maximum and equal to $(|\bar{V}^+|/Z_0)(1 + |\bar{\Gamma}_R|)$. The voltage and current are in phase.
3. Between maxima and minima, the voltage and current vary in accordance with the lengths of the line joining $(-1, 0)$ to the points A and B, respectively, as they move around the circles. These variations are not sinusoidal with distance. The variations near the minima are sharper than are those near the maxima, and hence the minima can be located more accurately than can the maxima. Also, the voltage and current are not in phase.

Standing wave parameters

From the preceding discussion, we now sketch the standing wave patterns for the line voltage and current, as shown in Fig. 8.10. These patterns correspond to partial standing waves, as compared to complete standing waves in the case of the short-circuited line. There are three parameters associated with the standing wave patterns as follows.

1. *The standing wave ratio, abbreviated as SWR.* This is the ratio of the maximum voltage amplitude V_{max} to the minimum voltage amplitude V_{min} in the standing wave pattern. Thus

$$\boxed{\text{SWR} = \frac{V_{\text{max}}}{V_{\text{min}}} = \frac{|\bar{V}^+|(1 + |\bar{\Gamma}_R|)}{|\bar{V}^+|(1 - |\bar{\Gamma}_R|)} = \frac{1 + |\bar{\Gamma}_R|}{1 - |\bar{\Gamma}_R|}} \tag{8.26}$$

Note also that SWR is equal to the ratio of the maximum current amplitude I_{max} to the minimum current amplitude I_{min} in the standing wave pattern, since

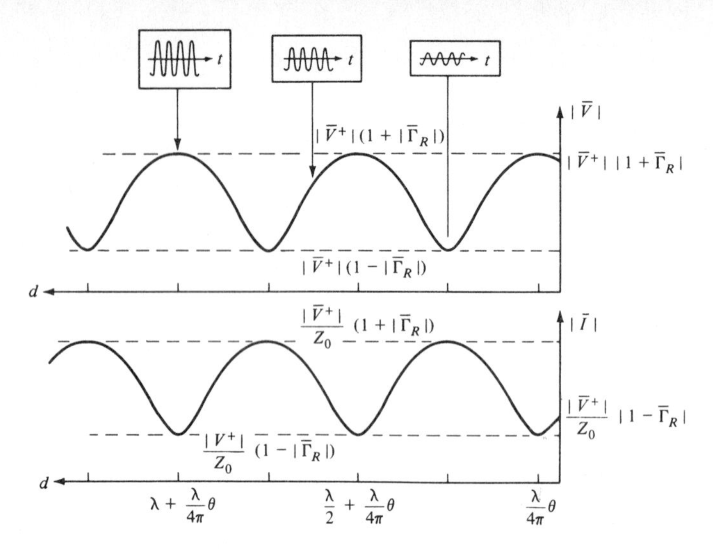

Figure 8.10. Voltage and current standing wave patterns for the system of Fig. 8.8. The insets show time variations of voltage at points along the line.

$$\frac{I_{\text{max}}}{I_{\text{min}}} = \frac{(|\bar{V}^+|/Z_0)(1 + |\bar{\Gamma}_R|)}{(|\bar{V}^+|/Z_0)(1 - |\bar{\Gamma}_R|)} = \frac{1 + |\bar{\Gamma}_R|}{1 - |\bar{\Gamma}_R|}$$

The SWR is a measure of standing waves on the line. It is an easily measurable parameter. We note the following special cases:

(a) For $\bar{\Gamma}_R = 0$, SWR $= 1$ and the standing wave pattern is simply a line representing constant amplitude. This is the case for a semi-infinitely long line or for a line terminated by its characteristic impedance.

(b) For $|\bar{\Gamma}_R| = 1$, SWR $= \infty$ and the standing wave pattern possesses perfect nulls. This is the case for complete standing waves.

2. *The distance of the first voltage minimum from the load, denoted by d_{min}.* The voltage minimum nearest to the load occurs when the phase angle of $\bar{\Gamma}(d) = \bar{\Gamma}_R e^{-j2\beta d}$ is equal to $-\pi$, that is, for $(\theta - 2\beta d)$ equal to $-\pi$. Thus

$$\theta - 2\beta d_{\text{min}} = -\pi \tag{8.27}$$

or

$$\boxed{d_{\text{min}} = \frac{\theta + \pi}{2\beta} = \frac{\lambda}{4\pi}(\theta + \pi)} \tag{8.28}$$

where $-\pi \le \theta < \pi$. If $\theta = 0$, which occurs when $\bar{Z}_R$ is purely real and greater than Z_0, $d_{\text{min}} = \lambda/4$ and a voltage maximum exists right at the load. If $\theta = -\pi$, which occurs when $\bar{Z}_R$ is purely real and less than Z_0, $d_{\text{min}} = 0$ and a voltage minimum exists right at the load.

3. *The wavelength λ.* Since the distance between successive voltage minima is equal to $\lambda/2$, the wavelength is twice the distance between successive voltage minima.

For a numerical example involving a complex $\bar{Z}_R$, let us consider $\bar{Z}_R = (15 - j20)\ \Omega$ and $Z_0 = 50\ \Omega$. Then

$$\bar{\Gamma}_R = \frac{\bar{Z}_R - Z_0}{\bar{Z}_R + Z_0} = \frac{(15 - j20) - 50}{(15 - j20) + 50}$$

$$= \frac{-7 - j4}{13 - j4} = \frac{8.06\underline{/-150.26^\circ}}{13.60\underline{/-17.10^\circ}}$$

$$= 0.593\underline{/-133.16^\circ}$$

$$= 0.593\, e^{-j0.74\pi}$$

$$\text{SWR} = \frac{1 + |\bar{\Gamma}_R|}{1 - |\bar{\Gamma}_R|} = \frac{1 + 0.593}{1 - 0.593} = 3.914$$

$$d_{\text{min}} = \frac{\lambda}{4\pi}(\theta + \pi) = \frac{\lambda}{4\pi}(-0.74\pi + \pi)$$

$$= 0.065\lambda$$

Determination of unknown load impedance

Conversely to the computation of standing wave parameters for a given load impedance, an unknown load impedance can be determined from standing wave measurements on a line of known characteristic impedance. An application in practice is the determination of the input impedance of an antenna by making standing wave measurements on the line feeding the antenna. To outline the basis, we note that by rearranging (8.26) and (8.28), we obtain

$$\boxed{|\bar{\Gamma}_R| = \frac{\text{SWR} - 1}{\text{SWR} + 1}} \tag{8.29}$$

and

$$\boxed{\theta = \frac{4\pi d_{\text{min}}}{\lambda} - \pi} \tag{8.30}$$

Thus the measurement of SWR, d_{min}, and λ provides both the magnitude and phase angle of $\bar{\Gamma}_R$. Then, since from (8.20)

$$\boxed{\bar{Z}_R = Z_0 \frac{1 + \bar{\Gamma}_R}{1 - \bar{\Gamma}_R}} \tag{8.31}$$

we can compute the value of $\bar{Z}_R$.

Slotted line measurements

A method of performing standing wave measurements in the laboratory is by using a *slotted line*. The slotted line is essentially a rigid coaxial line with air dielectric and having a length of about 1 meter (or at least a half-wavelength long). The center conductor is supported by dielectric inserts. A narrow longitudinal slot is cut in the outer conductor, as shown in Fig. 8.11(a). The width of the slot is so small that it has negligible influence on the current flow on the outer conductor, and hence on the field configurations between the conductors. A probe of small length, shown in Fig. 8.11(b), intercepts a portion of the electric field between the inner and outer conductors, and hence a small voltage proportional to the line voltage at the probe's location is developed between the probe and the

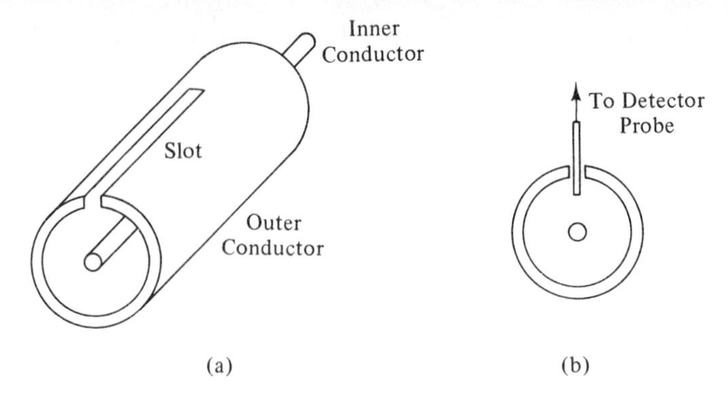

Figure 8.11. (a) Slotted line. (b) Cross-sectional view of the slotted line illustrating the probe arrangement.

outer conductor. The signal frequency voltage thus developed is detected by some sort of detector, and the resulting output is used as an indicator of the amplitude of the line voltage at the probe's location. The amount of energy picked up by the probe is small enough not to disturb appreciably the fields within the line. The probe and the associated detector components are mounted on a carriage arranged to slide mechanically along the longitudinal slot. As the probe is moved along the slot, the detector indication provides a measure of the variation of the voltage as a function of position on the line. Since the SWR is the ratio of V_{max} to V_{min}, the quantity of interest is the ratio of the two readings rather than the absolute values of the readings themselves. Therefore, absolute calibration of the detector is not required, provided that the detector response is uniform in the range of voltages to be measured.

Since it is not always possible to measure the distances of the standing wave pattern minima from the location of the load, the following procedure is employed. First, the line is terminated by a short circuit in the place of the load. One of the nulls in the resulting standing wave pattern is taken as the reference point, as shown in Fig. 8.12(a). This establishes that the location of the load is an inte-

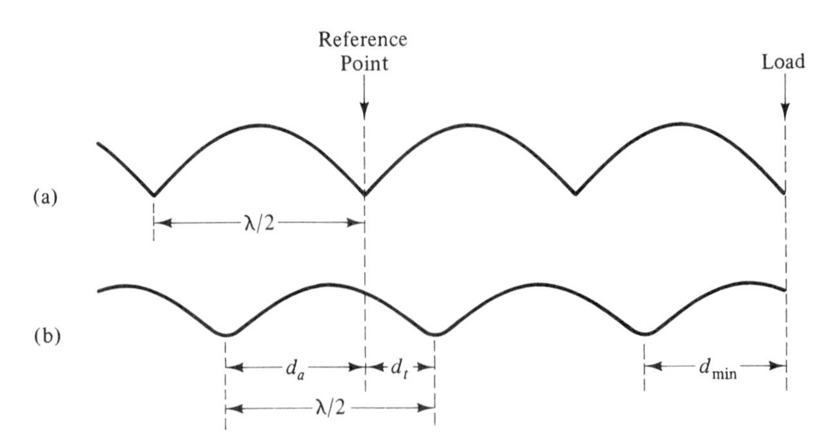

Figure 8.12. For illustrating the procedure employed for the determination of d_{min}, the distance of the first voltage minimum of the standing wave pattern from the load, by making measurements away from the load.

gral multiple of half-wavelengths from the reference point. Next, the short circuit is removed and the load is connected. The voltage minimum then shifts away from the reference point, as shown in Fig. 8.12(b). By measuring this shift, either away from the load or toward the load, the value of $d_{\min}$ can be established. If the shift d_a away from the load is measured, then we can see from Fig. 8.12(b) that $d_{\min}$ is simply equal to d_a. On the other hand, if the shift d_t toward the load is measured, then $d_{\min}$ is equal to $\lambda/2 - d_t$, where $\lambda/2$ is given by the distance between consecutive nulls either in the case of short circuit or with the unknown load as the termination.

We shall illustrate the computation of $\bar{Z}_R$ from standing wave measurements by means of an example.

Example 8.2

Let us assume that measurements performed on a slotted line of characteristic impedance $Z_0 = 50\ \Omega$ provided the following data. First, with the short circuit as the termination, voltage minima were found to be 20 cm apart. Next, with one of the minima marked as the reference point and the short circuit replaced by the unknown load, the SWR was found to be 3.0 and a voltage minimum was found to be at 5.80 cm from the reference point on the side toward the load. We wish to compute the value of the unknown load impedance.

From the value of the SWR, we obtain by using (8.29)

$$|\bar{\Gamma}_R| = \frac{3-1}{3+1} = 0.5$$

Since the distance between successive voltage minima is 20 cm, $\lambda/2$ is equal to 20 cm, or λ is equal to 40 cm. Since the voltage minimum shifted toward the load from the reference point, $d_{\min}$ is equal to $\lambda/2$ minus the shift, or $20 - 5.8 = 14.2$ cm. Then, from (8.30), we get

$$\theta = \frac{4\pi}{40} \times 14.2 - \pi = 0.42\pi$$

Thus

$$\bar{\Gamma}_R = 0.5e^{j0.42\pi}$$

Finally, using (8.31), we compute the value of the load impedance to be

$$\begin{aligned}
\bar{Z}_R &= 50\,\frac{1 + 0.5e^{j0.42\pi}}{1 - 0.5e^{j0.42\pi}} \\
&= 50\,\frac{1.1243 + j0.4843}{0.8757 - j0.4843} \\
&= 50\,\frac{1.2242\underline{/23.303^\circ}}{1.0007\underline{/-28.945^\circ}} \\
&= 61.17\underline{/52.248^\circ} \\
&= (37.45 + j48.365)\ \Omega
\end{aligned}$$

Line impedance

Returning now to the solutions for the complex line voltage and current given by (8.24a) and (8.24b), respectively, we find that the line impedance $\bar{Z}(d)$, that is, the impedance at any value of d seen looking toward the load, is given by

$$\bar{Z}(d) = \frac{\bar{V}(d)}{\bar{I}(d)} = \frac{\bar{V}^+ e^{j\beta d}[1 + \bar{\Gamma}(d)]}{(\bar{V}^+/Z_0)e^{j\beta d}[1 - \bar{\Gamma}(d)]} = Z_0 \frac{1 + \bar{\Gamma}(d)}{1 - \bar{\Gamma}(d)} \tag{8.32}$$

The following properties of the line impedance are of interest:

1. At the location of a voltage maximum of the standing wave pattern, $1 + \bar{\Gamma}(d)$ and $1 - \bar{\Gamma}(d)$ are purely real and equal to their maximum and minimum magnitudes $1 + |\bar{\Gamma}_R|$ and $1 - |\bar{\Gamma}_R|$, respectively. Hence $\bar{Z}(d)$ is purely real and maximum, say, $R_{\max}$, equal to $Z_0[(1 + |\bar{\Gamma}_R|)/(1 - |\bar{\Gamma}_R|)]$, or $Z_0(\text{SWR})$.
2. At the location of a voltage minimum of the standing wave pattern, $1 + \bar{\Gamma}(d)$ and $1 - \bar{\Gamma}(d)$ are purely real and equal to their minimum and maximum magnitudes $1 - |\bar{\Gamma}_R|$ and $1 + |\bar{\Gamma}_R|$, respectively. Hence $\bar{Z}(d)$ is purely real and minimum, say, $R_{\min}$, equal to $Z_0[(1 - |\bar{\Gamma}_R|)/(1 + |\bar{\Gamma}_R|)]$, or $Z_0/(\text{SWR})$.
3. Between voltage maxima and minima, $1 + \bar{\Gamma}(d)$ and $1 - \bar{\Gamma}(d)$ are both complex and out of phase. Hence $\bar{Z}(d)$ is complex, with amplitude lying between $Z_0(\text{SWR})$ and $Z_0/(\text{SWR})$.
4. Since $\bar{\Gamma}(d \pm n\lambda/2) = \bar{\Gamma}(d)e^{\mp j2\beta n\lambda/2} = \bar{\Gamma}(d)e^{\mp j2n\pi} = \bar{\Gamma}(d)$, $n = 1, 2, 3, \ldots$, $\bar{\Gamma}(d)$ repeats at intervals of $\lambda/2$, and hence $\bar{Z}(d)$ repeats at intervals of $\lambda/2$.
5. The product of the line impedances at two values of d separated by $\lambda/4$ is given by

$$\begin{aligned}
[\bar{Z}(d)]\left[\bar{Z}\left(d \pm \frac{\lambda}{4}\right)\right] &= \left[Z_0 \frac{1 + \bar{\Gamma}(d)}{1 - \bar{\Gamma}(d)}\right]\left[Z_0 \frac{1 + \bar{\Gamma}(d \pm \lambda/4)}{1 - \bar{\Gamma}(d \pm \lambda/4)}\right] \\
&= Z_0^2 \left[\frac{1 + \bar{\Gamma}(d)}{1 - \bar{\Gamma}(d)}\right]\left[\frac{1 + \bar{\Gamma}(d)e^{\mp j2\beta\lambda/4}}{1 - \bar{\Gamma}(d)e^{\mp j2\beta\lambda/4}}\right] \\
&= Z_0^2 \left[\frac{1 + \bar{\Gamma}(d)}{1 - \bar{\Gamma}(d)}\right]\left[\frac{1 + \bar{\Gamma}(d)e^{\mp j\pi}}{1 - \bar{\Gamma}(d)e^{\mp j\pi}}\right] \\
&= Z_0^2 \left[\frac{1 + \bar{\Gamma}(d)}{1 - \bar{\Gamma}(d)}\right]\left[\frac{1 - \bar{\Gamma}(d)}{1 + \bar{\Gamma}(d)}\right]
\end{aligned}$$

or

$$[\bar{Z}(d)]\left[\bar{Z}\left(d \pm \frac{\lambda}{4}\right)\right] = Z_0^2 \tag{8.33}$$

This is a useful property, as we shall learn in the following section.

Input impedance

For a line of length l, as in Fig. 8.8, the input impedance is given from (8.32) by

$$\boxed{\bar{Z}_{\text{in}} = \bar{Z}(l) = Z_0 \frac{1 + \bar{\Gamma}(l)}{1 - \bar{\Gamma}(l)}} \tag{8.34}$$

The input impedance is a useful parameter since for a given generator voltage and internal impedance, the power flow down the line can be computed by considering the line voltage and current at any value of d, since the line is lossless; in particular, it is convenient to do this at the input end of the line from input impedance considerations. We shall illustrate this by means of an example.

Example 8.3

Let us consider the system shown in Fig. 8.13, and find the time-average power delivered to the load from input impedance considerations.

We proceed with the solution in the following step-by-step manner:

(a) Compute the reflection coefficient at the load.

$$\bar{\Gamma}_R = \frac{\bar{Z}_R - Z_0}{\bar{Z}_R + Z_0} = \frac{(30 + j40) - 50}{(30 + j40) + 50} = 0.5\underline{/90^\circ}$$

(b) Compute the reflection coefficient $\bar{\Gamma}(l)$ at the input end $d = l$.

$$\begin{aligned}\bar{\Gamma}(l) &= \bar{\Gamma}_R e^{-j2\beta l}\\ &= 0.5\underline{/90^\circ} \times e^{-j(4\pi/\lambda)(0.725\lambda)}\\ &= 0.5\underline{/90^\circ} \times 1\underline{/-162^\circ}\\ &= 0.5\underline{/-72^\circ}\end{aligned}$$

(c) Compute the input impedance.

$$\begin{aligned}\bar{Z}_{\text{in}} = \bar{Z}(l) &= Z_0 \frac{1 + \bar{\Gamma}(l)}{1 - \bar{\Gamma}(l)}\\ &= 50 \frac{1 + 0.5\underline{/-72^\circ}}{1 - 0.5\underline{/-72^\circ}} = 50 \frac{1 + (0.1545 - j0.4755)}{1 - (0.1545 - j0.4755)}\\ &= 50 \frac{1.2486\underline{/-22.385^\circ}}{0.970\underline{/29.353^\circ}} = 64.361\underline{/-51.738^\circ}\\ &= (39.86 - j50.54)\ \Omega\end{aligned}$$

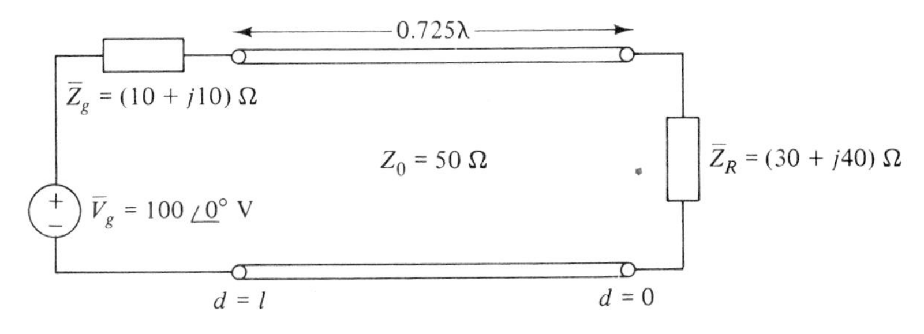

Figure 8.13. Transmission-line system for illustrating the computation of power flow from input impedance considerations.

(d) We now have the equivalent circuit at the input, as shown in Fig. 8.14, from which we compute the current $\bar{I}_g = \bar{I}(l)$, drawn from the generator. Thus we obtain

$$\bar{I}(l) = \bar{I}_g = \frac{\bar{V}_g}{\bar{Z}_g + \bar{Z}_{in}} = \frac{100\underline{/0^\circ}}{(10 + j10) + (39.86 - j50.54)}$$

$$= \frac{100\underline{/0^\circ}}{49.86 - j40.54} = \frac{100\underline{/0^\circ}}{64.261\underline{/-39.114^\circ}}$$

$$= 1.5562\underline{/39.114^\circ}\text{ A}$$

(e) The voltage across the input impedance is then given by

$$\bar{V}(l) = \bar{Z}_{in}\bar{I}(l)$$

$$= 64.361\underline{/-51.738^\circ} \times 1.5562\underline{/39.114^\circ}$$

$$= 100.159\underline{/-12.624^\circ}\text{ V}$$

(f) Finally, the time-average power delivered to the input and hence to the load is given by

$$<P> = \tfrac{1}{2}\text{Re}[\bar{V}(l)\bar{I}^*(l)]$$

$$= \tfrac{1}{2}\text{Re}[100.159\underline{/-12.624^\circ} \times 1.5562\underline{/-39.114^\circ}]$$

$$= \tfrac{1}{2} \times 100.159 \times 1.5562 \times \cos 51.738^\circ$$

$$= 48.26\text{ W}$$

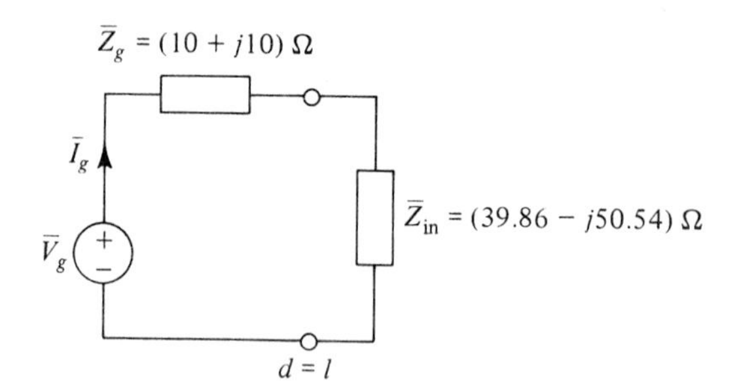

Figure 8.14. Equivalent circuit at the input end $d = l$ for the system of Fig. 8.13.

Normalized impedance and admittance

Returning to (8.32), we now define the normalized line impedance $\bar{z}(d)$ as the ratio of the line impedance to the line characteristic impedance. Thus

$$\boxed{\bar{z}(d) = \frac{\bar{Z}(d)}{Z_0} = \frac{1 + \bar{\Gamma}(d)}{1 - \bar{\Gamma}(d)}} \tag{8.35}$$

Conversely,

$$\boxed{\bar{\Gamma}(d) = \frac{\bar{z}(d) - 1}{\bar{z}(d) + 1}} \tag{8.36}$$

Finally, the line admittance is given by

$$\bar{Y}(d) = \frac{1}{\bar{Z}(d)} = \frac{1}{Z_0}\frac{1 - \bar{\Gamma}(d)}{1 + \bar{\Gamma}(d)}$$

or

$$\bar{Y}(d) = Y_0\frac{1 - \bar{\Gamma}(d)}{1 + \bar{\Gamma}(d)} \tag{8.37}$$

where $Y_0 = 1/Z_0$ is the characteristic admittance of the line. The normalized line admittance is

$$\boxed{\bar{y}(d) = \frac{\bar{Y}(d)}{Y_0} = \frac{1 - \bar{\Gamma}(d)}{1 + \bar{\Gamma}(d)}} \tag{8.38}$$

and conversely,

$$\boxed{\bar{\Gamma}(d) = \frac{1 - \bar{y}(d)}{1 + \bar{y}(d)}} \tag{8.39}$$

We shall use these relationships in the following sections.

K8.2. Arbitrary load; Generalized reflection coefficient; Partial standing waves; Standing wave ratio; Standing wave parameters; Standing wave measurements; Line impedance; Power flow; Normalized line impedance; Normalized line admittance.

D8.4. A line of characteristic impedance 60 Ω is terminated by a load consisting of the series combination of $R = 30\ \Omega$, $L = 1\ \mu$H, and $C = 100\ p$F. Find the values of SWR and d_{min} for each of the following radian frequencies of the source: **(a)** $\omega = 10^8$; **(b)** $\omega = 2 \times 10^8$; and **(c)** $\omega = 0.8 \times 10^8$.
Ans. **(a)** 2, 0; **(b)** 14.94, 0.309λ; **(c)** 3.324, 0.115λ

D8.5. Standing wave measurements are performed on a line of characteristic impedance 60 Ω terminated by a load $\bar{Z}_R$. For each of the following sets of standing wave data, find $\bar{Z}_R$: **(a)** SWR = 1.5, a voltage minimum right at the load; **(b)** SWR = 3.0, two successive voltage minima at 3 cm and 9 cm from the load; and **(c)** SWR = 2.0, two successive voltage minima at 3 cm and 7 cm from the load.
Ans. **(a)** $(40 + j0)$ Ω; **(b)** $(180 + j0)$ Ω; **(c)** $(48 + j36)$ Ω

D8.6. An air dielectric line of characteristic impedance $Z_0 = 75\ \Omega$ is terminated by a load impedance $(45 + j60)$ Ω. Find the input impedance of the line for each of the following pairs of values of the frequency f and the length l of the line: **(a)** $f = 15$ MHz, $l = 5$ m; **(b)** $f = 50$ MHz, $l = 3$ m; and **(c)** $f = 37.5$ MHz, $l = 5$ m.
Ans. **(a)** $(45 - j60)$ Ω; **(b)** $(45 + j60)$ Ω; **(c)** $(225 + j0)$ Ω

8.3 TRANSMISSION-LINE MATCHING

In the preceding section we discussed standing waves on a line terminated by an arbitrary load. In the presence of standing waves, that is, when the load impedance is not equal to the characteristic impedance, it follows from (8.34) that the input impedance of the line will vary with frequency since the electrical length of the line and hence $\bar{\Gamma}(l) = \bar{\Gamma}_R e^{-j2\beta l}$ changes. This sensitivity to frequency increases with the electrical length of the line. To show this, let the length of the line be $l = n\lambda$. If the frequency is changed by an amount Δf, then the change in n is given by

$$\Delta n = \Delta\left(\frac{l}{\lambda}\right) = \Delta\left(\frac{lf}{v_p}\right) = \frac{l}{v_p}\Delta f = \frac{n\lambda}{v_p}\Delta f = n\frac{\Delta f}{f} \tag{8.40}$$

Thus Δn, the change in the number of wavelengths corresponding to the line length, is proportional to n. The variation of the input impedance with frequency puts a limitation on the performance of a transmission-line system from the point of view of communication. For this and other reasons pertaining to power flow, it is desirable to eliminate standing waves on the line by connecting a *matching* device near the load such that the line views an effective impedance equal to its own characteristic impedance on the generator side of the matching device as shown in Fig. 8.15. The matching device should not at the same time absorb any power. It should be noted that *matching*, as referred to here, is not related to maximum power transfer since the condition for maximum power transfer is that the line input impedance must be the complex conjugate of the generator internal impedance. In the following, we discuss three techniques of matching.

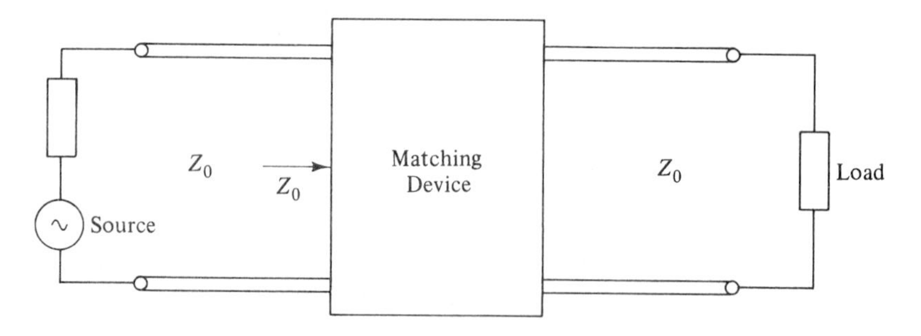

Figure 8.15. For illustrating the principle behind transmission-line "matching."

A. Quarter-Wave Transformer Matching

Quarter-wave transformer matching

The quarter-wave transformer or QWT matching technique makes use of a section of length $\lambda/4$ of a line of characteristic impedance Z_q different from that of the main line, as shown in Fig. 8.16. The principle is based upon the property of line impedance given by (8.33). With reference to the notation of Fig. 8.16, we first note that to achieve a match, $\bar{Z}_1$ must be equal to Z_0. Then since from (8.33) $\bar{Z}_1\bar{Z}_2 = Z_q^2$, $\bar{Z}_2 = Z_q^2/\bar{Z}_1 = Z_q^2/Z_0$ must be purely real. We recall from the discussion of line impedance in Section 8.2 that the line impedance is purely real at locations of voltage maxima and minima of the standing wave pattern. Therefore within the first half-wavelength from the load, there are two solutions for d_q and hence for Z_q.

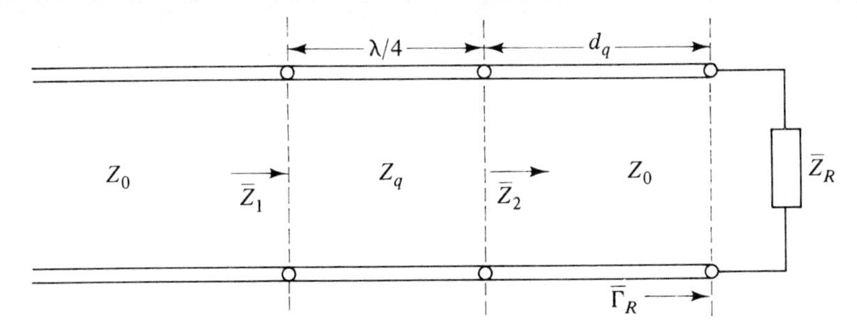

Figure 8.16. For illustrating the quarter-wave transformer matching technique.

If we choose a voltage minimum for the first solution, then from (8.28)

$$\boxed{d_q^{(1)} = \frac{\lambda}{4\pi}(\theta + \pi)} \tag{8.41}$$

where θ is the phase angle of $\bar{\Gamma}_R$ and the superscript (1) refers to solution 1. The value of the line impedance is $Z_0\,(1 - |\bar{\Gamma}_R|)/(1 + |\bar{\Gamma}_R|)$. Hence the value of Z_q is given by

$$Z_0 \cdot Z_0 \frac{1 - |\bar{\Gamma}_R|}{1 + |\bar{\Gamma}_R|} = Z_q^2$$

or

$$\boxed{Z_q^{(1)} = Z_0 \sqrt{\frac{1 - |\bar{\Gamma}_R|}{1 + |\bar{\Gamma}_R|}}} \tag{8.42}$$

For the second solution, the value of d_q corresponds to the location of a voltage maximum which occurs at $\pm\lambda/4$ from the location of the voltage minimum. Thus

$$\boxed{d_q^{(2)} = d_q^{(1)} \pm \frac{\lambda}{4}} \tag{8.43}$$

whichever is positive and less than $\lambda/2$. The corresponding line impedance is Z_0 $(1 + |\bar{\Gamma}_R|)/(1 - |\bar{\Gamma}_R|)$ so that

$$\boxed{Z_q^{(2)} = Z_0 \sqrt{\frac{1 + |\bar{\Gamma}_R|}{1 - |\bar{\Gamma}_R|}}} \tag{8.44}$$

B. Single-Stub Matching

Single-stub matching

Another technique of transmission-line matching known as *stub matching* consists of connecting small sections of short-circuited lines (stubs) of appropriate lengths in parallel with the line, at appropriate distances from the load. In the single-stub

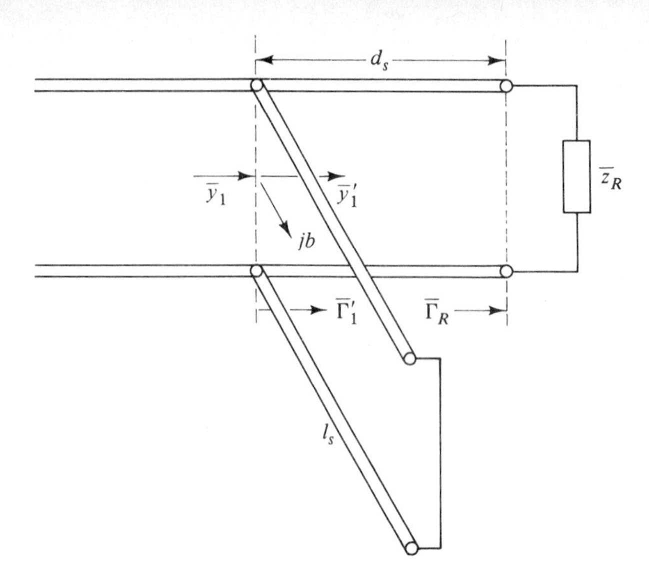

Figure 8.17. For illustrating the single-stub matching technique.

matching technique, one stub is used and a match is achieved by varying the location of the stub and the length of the stub. We shall assume the characteristic impedance of the stub to be the same as that of the line and use the notation shown in Fig. 8.17, in which $\bar{z}_R$ is the normalized load impedance, $\bar{y}_1$ and $\bar{y}'_1$ are the normalized line admittances just to the left and just to the right, respectively, of the stub, and b is the normalized input susceptance of the stub. The solution to the single-stub matching problem then consists of finding the values of d_s and l_s for a given value of $\bar{z}_R$ and hence of $\bar{\Gamma}_R$.

First we observe that to achieve a match, $\bar{y}_1$ must be equal to $(1 + j0)$. Then proceeding to the right of the stub, we can write the following steps:

$$\bar{y}'_1 = 1 - jb \tag{8.45a}$$

$$\bar{\Gamma}'_1 = \frac{1 - \bar{y}'_1}{1 + \bar{y}'_1} = \frac{jb}{2 - jb} \tag{8.45b}$$

$$\begin{aligned} \bar{\Gamma}_R &= \bar{\Gamma}'_1 e^{j2\beta d_s} = \frac{jb}{2 - jb} e^{j2\beta d_s} \\ &= \frac{|b|}{\sqrt{4 + b^2}} e^{j(\pm\pi/2 + \tan^{-1} b/2 + 2\beta d_s + 2n\pi)} \qquad \text{for} \quad b \gtrless 0 \end{aligned} \tag{8.45c}$$

where n is an integer (positive or negative). Thus

$$|\bar{\Gamma}_R| = \frac{|b|}{\sqrt{4 + b^2}} \tag{8.46a}$$

$$\theta = \pm\frac{\pi}{2} + \tan^{-1}\frac{b}{2} + 2\beta d_s + 2n\pi \qquad \text{for} \quad b \gtrless 0 \tag{8.46b}$$

so that

$$b = \pm \frac{2|\overline{\Gamma}_R|}{\sqrt{1 - |\overline{\Gamma}_R|^2}} \tag{8.47}$$

$$d_s = \frac{\lambda}{4\pi}\left(\theta \mp \frac{\pi}{2} - \tan^{-1}\frac{b}{2} - 2n\pi\right) \quad \text{for} \quad b \gtrless 0 \tag{8.48}$$

Thus two solutions are possible for b as given by (8.47) and the corresponding solutions for d_s are given by (8.48), where the integer value for n is chosen such that $0 \leq d_s < \lambda/2$. Finally, to find the solutions for the stub length, we note from (8.13) that the normalized input impedance of a short-circuited line of length l_s is $j \tan \beta l_s$ so that

$$\frac{1}{jb} = j \tan \beta l_s$$

$$\tan \beta l_s = -\frac{1}{b}$$

$$l_s = \begin{cases} \dfrac{\lambda}{2\pi}\left[\tan^{-1}\left(-\dfrac{1}{b}\right)\right] + \dfrac{\lambda}{2} & \text{for} \quad b > 0 \\ \dfrac{\lambda}{2\pi}\left[\tan^{-1}\left(-\dfrac{1}{b}\right)\right] & \text{for} \quad b < 0 \end{cases} \tag{8.49}$$

C. Double-Stub Matching

Double-stub matching

In the single-stub matching technique, it is necessary to vary the distance between the stub and the load as well as the length of the stub to achieve a match for different loads or for different frequencies. This can be inconvenient for some arrangements of lines. When two stubs are used, it is possible to fix their locations and achieve a match for a wide range of loads by adjusting the lengths of the stubs. To discuss the principle behind this *double-stub matching* technique, we make use of the notation shown in Fig. 8.18, in which all admittances and susceptances are normalized quantities with respect to the characteristic admittance of the line. The solution to the double-stub matching problem then consists of finding the values of l_1 and l_2 for a given set of values of $\overline{z}_R$ (and hence of $\overline{\Gamma}_R$), d_1, and d_{12}.

First we observe that to achieve a match, $\overline{y}_2$ must be equal to $(1 + j0)$. Then proceeding to the right in a step-by-step manner, we obtain an expression for $\overline{y}_1'$ in terms of b_1, b_2, and d_{12} as follows:

$$\overline{y}_2' = \overline{y}_2 - jb_2 = 1 - jb_2 \tag{8.50a}$$

$$\overline{\Gamma}_2' = \frac{1 - \overline{y}_2'}{1 + \overline{y}_2'} = \frac{jb_2}{2 - jb_2} \tag{8.50b}$$

$$\overline{\Gamma}_1 = \overline{\Gamma}_2' e^{j2\beta d_{12}} = \frac{jb_2}{2 - jb_2} e^{j2\beta d_{12}} \tag{8.50c}$$

$$\overline{y}_1 = \frac{1 - \overline{\Gamma}_1}{1 + \overline{\Gamma}_1}$$

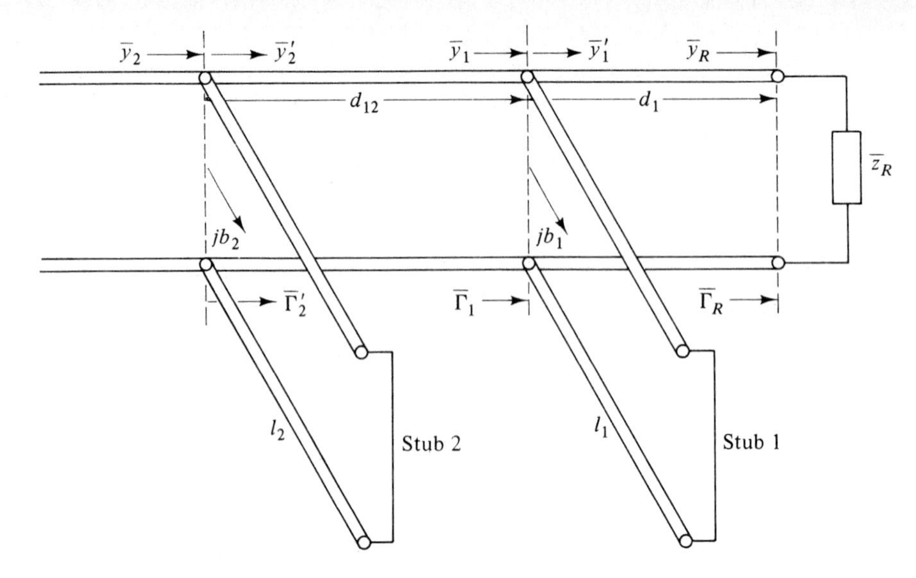

Figure 8.18. For illustrating the double-stub matching technique.

$$= \frac{4 - j(4b_2 \cos 2\beta d_{12} - 2b_2^2 \sin 2\beta d_{12})}{4 - 4b_2 \sin 2\beta d_{12} + 4b_2^2 \sin^2 \beta d_{12}} \tag{8.50d}$$

$$\bar{y}_1' = \bar{y}_1 - jb_1$$

$$= \frac{1}{1 - b_2 \sin 2\beta d_{12} + b_2^2 \sin^2 \beta d_{12}} + j\left(\frac{b_2^2 \sin 2\beta d_{12} - 2b_2 \cos 2\beta d_{12}}{2 - 2b_2 \sin 2\beta d_{12} + 2b_2^2 \sin^2 \beta d_{12}} - b_1\right) \tag{8.50e}$$

For given values of $\bar{z}_R$ and d_1, $\bar{y}_1'$ can be computed in the usual manner, and the real and imaginary parts can be equated to the real and imaginary parts, respectively, on the right side of (8.50e). Noting that b_1 does not appear in the real part expression, we can first compute b_2 by solving the equation for the real parts. Thus letting the real part of $\bar{y}_1'$ as computed from $\bar{z}_R$ and d_1 to be g', we have

$$\frac{1}{1 - b_2 \sin 2\beta d_{12} + b_2^2 \sin^2 \beta d_{12}} = g' \tag{8.51}$$

Rearranging and solving for b_2, we obtain

$$b_2 = \frac{\sin 2\beta d_{12} \pm \sqrt{\sin^2 2\beta d_{12} - 4(1 - 1/g') \sin^2 \beta d_{12}}}{2 \sin^2 \beta d_{12}}$$

or

$$\boxed{b_2 = \frac{\cos \beta d_{12} \pm \sqrt{1/g' - \sin^2 \beta d_{12}}}{\sin \beta d_{12}}} \tag{8.52}$$

We now see that a solution does not exist for b_2 if $g' > 1/\sin^2 \beta d_{12}$, and hence it is not possible to achieve a match for loads which result in real part of $\bar{y}_1'$ greater than $1/\sin^2 \beta d_{12}$. A simple way to get around this problem is to increase d_1 by $\lambda/4$ (see Problem P8.22). Assuming that the condition $g' < 1/\sin^2 \beta d_{12}$ is achieved, we then compute two possible values for b_2 as given by (8.52). From the equation for the imaginary parts of $\bar{y}_1'$, the corresponding values of b_1 are then given by

$$b_1 = \frac{b_2^2 \sin 2\beta d_{12} - 2b_2 \cos 2\beta d_{12}}{2 - 2b_2 \sin 2\beta d_{12} + 2b_2^2 \sin^2 \beta d_{12}} - b' \tag{8.53}$$

where b' is the imaginary part of $\bar{y}_1'$ as computed from $\bar{z}_R$ and d_1. Finally, the lengths of the two stubs are computed from b_1 and b_2 as in the case of the single-stub matching technique.

The listing of a PC program which computes the solutions for all three types of matching techniques for specified values of $\bar{Z}_R$ $(= R_L + jX_L)$ and Z_0 and values of d_1/λ and d_{12}/λ for the double-stub matching case is included as PL 8.2. The output from a run of the program for values of $R_L = 30\ \Omega$, $X_L = -40\ \Omega$, $Z_0 = 50\ \Omega$, $d_1 = 0$, and $d_{12}/\lambda = 0.375$ is also included. Values of odd multiples of $\lambda/8$ are commonly chosen for d_{12}. If the specified value of $\bar{Z}_R$ is such that $g' > 1/\sin^2 \beta d_{12}$, then the value of d_1 is increased by $\lambda/4$ and the double-stub matching solution is continued.

PL 8.2. Program listing and sample output for obtaining the solutions for quarter-wave transformer, single-stub, and double-stub matching techniques.

```
100 '*****************************************************
110 '* SOLUTION OF TRANSMISSION LINE MATCHING PROBLEM FOR *
120 '* QUARTER-WAVE TRANSFORMER, SINGLE STUB, AND DOUBLE  *
130 '* STUB MATCHING TECHNIQUES                           *
140 '*****************************************************
150 PI=3.1416
160 DEF FN TRD(ARG)=INT(ARG*100000!+.5)/100000!:'* ROUNDS
170 '  ARG TO FIVE DECIMAL PLACES *
180 CLS:LOCATE 1,1:PRINT "ENTER VALUES OF ZO, RL, AND XL IN
     OHMS:"
190 INPUT "ZO=",ZO
200 LOCATE 2,15:INPUT "RL=",RL:R=RL/ZO
210 LOCATE 2,30:INPUT "XL=",XL:X=XL/ZO
220 IF RL=ZO AND XL=0 THEN PRINT:PRINT "LINE IS TERMINATED
    BY A MATCHED LOAD":GOTO 550
230 PRINT "ENTER VALUES OF D1 AND D12 FOR DOUBLE"
240 PRINT "STUB MATCHING:"
250 INPUT "D1/WL=",D1
260 LOCATE 5,15:INPUT "D12/WL=",DB
270 '* QUARTER-WAVE TRANSFORMER MATCHING *
280 GOSUB 580:SR=(1+MG)/(1-MG)
290 PRINT:PRINT "SOLUTIONS FOR QWT MATCHING ARE:":PRINT
300 DQ=.25*(PG/PI+1):ZQ=ZO/SQR(SR):GOSUB 320
310 DQ=DQ+.25:ZQ=ZO*SQR(SR):GOSUB 320:GOTO 340
320 IF DQ>=.5 THEN DQ=DQ-.5
330 PRINT "DQ =";FN TRD(DQ);"*WL";"    ZQ =";FN TRD(ZQ);" O
    HMS":RETURN
340 '* SINGLE STUB MATCHING *
350 PRINT:PRINT "SOLUTIONS FOR SINGLE STUB MATCHING ARE:":P
    RINT
```

PL 8.2. (continued)

```
360 B=2*MG/SQR(1-MG*MG):DS=(PG-ATN(B/2))/(4*PI)-.125:GOSUB
    380
370 B=-B:DS=(PG-ATN(B/2))/(4*PI)+.125:GOSUB 380:GOTO 420
380 IF DS<0 THEN DS=DS+.5
390 IF DS>=.5 THEN DS=DS-.5
400 IS=B:GOSUB 710
410 PRINT "DS =";FN TRD(DS);"*WL";"      LS =";FN TRD(LS);"*
    WL":RETURN
420 '* DOUBLE STUB MATCHING *
430 CB=COS(2*PI*DB):SB=SIN(2*PI*DB)
440 PL=PG-4*PI*D1:R=MG*COS(PL):X=MG*SIN(PL):GOSUB 580
450 G=-MG*COS(PG):B=-MG*SIN(PG)
460 IF G<1/(SB*SB) THEN 480
470 D1=D1+.25:PRINT "VALUE OF D1 CHANGED TO";D1;"*WL":GOTO
    440
480 PRINT:PRINT "SOLUTIONS FOR DOUBLE STUB MATCHING ARE:":P
    RINT
490 GI=G:BI=B:B2=CB/SB+SQR(1/(GI*SB*SB)-1):GOSUB 510
500 B2=CB/SB-SQR(1/(GI*SB*SB)-1):GOSUB 510:GOTO 550
510 B1=-BI+(B2*B2*SB*CB-B2*(CB*CB-SB*SB))/(1-2*B2*SB*CB+B2*
    B2*SB*SB)
520 IS=B1:GOSUB 710:L1=LS
530 IS=B2:GOSUB 710:L2=LS
540 PRINT "L1 =";FN TRD(L1);"*WL";"      L2 =";FN TRD(L2);"*
    WL":RETURN
550 LOCATE 23,1:PRINT "STRIKE ANY KEY TO CONTINUE"
560 C$=INPUT$(1):GOTO 180
570 END
580 '* SUBPROGRAM TO COMPUTE REFLECTION COEFFICIENT FROM
590 '  NORMALIZED LINE IMPEDENCE *
600 REAL=R-1:IMAG=X:GOSUB 640:MN=MAG:PN=PANG
610 REAL=R+1:GOSUB 640:MD=MAG:PD=PANG
620 MG=MN/MD:PG=PN-PD
630 RETURN
640 '* SUBPROGRAM TO CONVERT COMPLEX NUMBER IN RECTANGLULAR
650 '  FORM TO ONE IN POLAR FORM *
660 MAG=SQR(REAL*REAL+IMAG*IMAG)
670 IF REAL=0 THEN PANG=SGN(IMAG)*PI/2:RETURN
680 PANG=ATN(IMAG/REAL)
690 IF REAL<0 THEN PANG=PANG+PI
700 RETURN
710 '* SUBPROGRAM TO COMPUTE STUB LENGTH *
720 LS=ATN(-1/IS)/(2*PI):IF LS<0 THEN LS=LS+.5
730 RETURN

RUN
ENTER VALUES OF ZO, RL, AND XL IN OHMS:
ZO=50          RL=30          XL=-40
ENTER VALUES OF D1 AND D12 FOR DOUBLE
STUB MATCHING:
D1/WL=0        D12/WL=.375

SOLUTIONS FOR QWT MATCHING ARE:

DQ = .125 *WL    ZQ = 28.86751  OHMS
DQ = .375 *WL    ZQ = 86.60254  OHMS

SOLUTIONS FOR SINGLE STUB MATCHING ARE:

DS = .20833 *WL     LS = .38641 *WL
DS = .04167 *WL     LS = .11359 *WL
```

PL 8.2. (continued)

```
SOLUTIONS FOR DOUBLE STUB MATCHING ARE:

L1 = .13483 *WL        L2 = .32726 *WL
L1 = .05614 *WL        L2 = .05996 *WL


STRIKE ANY KEY TO CONTINUE
```

Bandwidth

For any transmission-line matched system, the match is disturbed as the frequency is varied from that at which the various electrical lengths and distances are equal to the computed values for achieving the match. For example, in the QWT matched system, the electrical length of the QWT departs from one-quarter wavelength as the frequency is varied from that at which the match is achieved, and the system is no longer matched even if the load does not vary with frequency. A plot of the SWR in the main line to the left of the QWT versus frequency is typically of the shape shown in Fig. 8.19, where f_0 is the design frequency at which the system is matched, and hence the frequency at which the SWR is unity. One can then specify a tolerable value of SWR, say, S, so that there exists an acceptable bandwidth of operation, $f_2 - f_1$. Similar considerations apply to the single-stub and double-stub matched systems.

SWR versus frequency computation

To discuss a procedure by means of which the SWR versus frequency curve can be computed for all three types of matching techniques discussed, let us consider a transmission-line system having n discontinuities, as shown for $n = 2$ in Fig. 8.20. At each discontinuity, there can exist a stub and a change in characteristic impedance. We shall consider a specification of zero for the length of the stub to mean no stub is present instead of a stub of zero length. This does not result in a conflict since for any matched system using short-circuited stubs, values of zero cannot be obtained for stub lengths since then SWR would be infinity. With this understanding, Fig. 8.20 can be used to represent all three types of matching systems by specifying values for the various parameters as shown in Table 8.1.

Then to compute the SWR in the main line at a given frequency, we first note that since $\lambda \propto 1/f$, the electrical length of a line section or of a stub is proportional to f. Thus at a frequency f, the electrical length is equal to f/f_0 times its

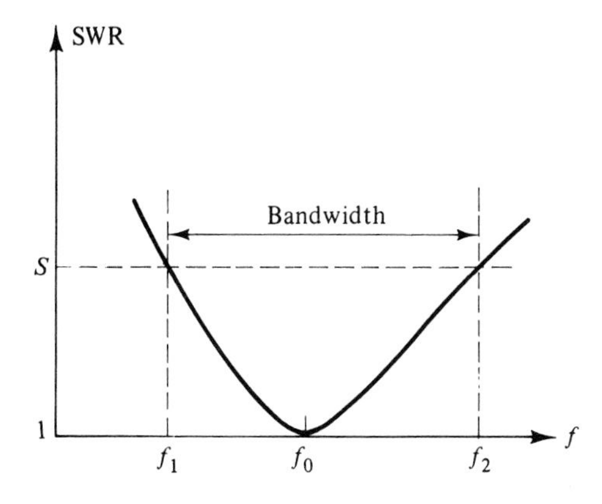

Figure 8.19. The SWR versus frequency curve illustrating the bandwith between the two frequencies, f_1 and f_2, on either side of the design frequency f_0, at which the SWR is a specified value, S (> 1).

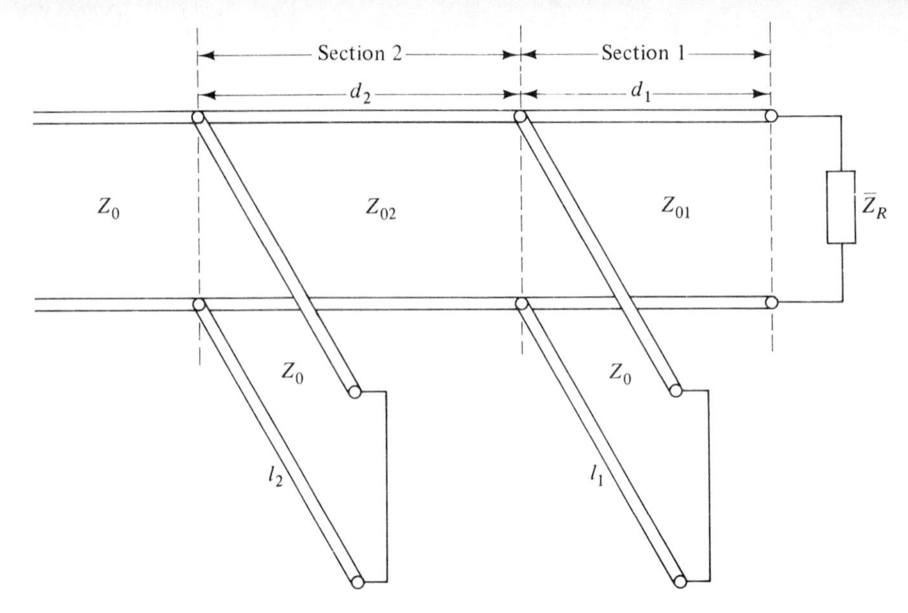

Figure 8.20. Transmission-line system for computing the SWR versus frequency curve for QWT, single-stub, and double-stub matched systems.

TABLE 8.1 VALUES OF PARAMETERS FOR USING THE SYSTEM OF FIG. 8.20 FOR THREE DIFFERENT CASES

System	n	Z_{01}	d_1	l_1	Z_{02}	d_2	l_2
QWT	2	Z_0	d_q	0[a]	Z_q	$\frac{1}{4}$	0[a]
Single stub	1	Z_0	d_s	l_s	—	—	—
Double stub	2	Z_0	d_1	l_1	Z_0	d_{12}	l_2

[a] Value of zero means no stub present.

value at f_0. For a given f/f_0, the procedure consists of starting at the load and computing in succession the line admittance to the left of the stub at each discontinuity from a knowledge of the line admittance at the output of the line section to the right of that stub, until the line admittance to the left of the last discontinuity is found and used to compute the required SWR. In carrying out this procedure, we observe the following:

1. To compute the normalized admittance, say, $\bar{y}_i$, at the input (left) end of a line section of length l from the normalized admittance, say, $\bar{y}_o$, at the output (right) end of that section, we use the formula

$$\bar{y}_i = \frac{1 - \bar{\Gamma}_i}{1 + \bar{\Gamma}_i} = \frac{1 - \bar{\Gamma}_o e^{-j2\beta l}}{1 + \bar{\Gamma}_o e^{-j2\beta l}}$$

$$= \frac{1 - [(1 - \bar{y}_o)/(1 + \bar{y}_o)]e^{-j2\beta l}}{1 + [(1 - \bar{y}_o)/(1 + \bar{y}_o)]e^{-j2\beta l}}$$

or

$$\bar{y}_i = \frac{j \sin \beta l + \bar{y}_o \cos \beta l}{\cos \beta l + j\bar{y}_o \sin \beta l} \tag{8.54}$$

where $\bar{\Gamma}_i$ and $\bar{\Gamma}_o$ are the reflection coefficients at the input and output ends, respectively.

2. To compute the line admittance to the left of a stub, we add the input admittance of the stub to the line admittance to the right of the stub.

The listing of a PC program which computes SWR versus f/f_0 using the procedure just discussed is presented as PL 8.3. The output from a run of the program is also included. It can be seen from PL 8.3 that the values of input parameters used for the run are pertinent to the first of the two solutions for the double-stub matching case in the output of PL 8.2. The frequency variation of $\bar{Z}_R$ is taken into account by assuming $\bar{Z}_R$ to be the series combination of a single resistor and a single reactive element.

PL 8.3. Program listing and sample output for computing SWR in the main line versus frequency for the transmission-line system of Fig. 8.20.

```
100 '*********************************************************
110 '* COMPUTATION OF SWR VERSUS FREQUENCY FOR A TRANSMISSION *
120 '* LINE SYSTEM MATCHED AT F=FO                           *
130 '*********************************************************
140 DIM D(10),ZS(10),L(10)
150 PI=3.1416
160 DEF FN TR(ARG)=INT(ARG*10000+.5)/10000:'* ROUNDS ARG TO
170 '  FOUR DECIMAL PLACES *
180 CLS:LOCATE 1,1:PRINT "ENTER VALUES OF ZO, RL, AND XL IN OHM
    S:"
190 INPUT "ZO=",ZO
200 LOCATE 2,15:INPUT "RL=",RL:R=RL/ZO
210 LOCATE 2,30:INPUT "XL=",XL:X=XL/ZO
220 PRINT:INPUT "ENTER NUMBER OF SECTIONS: ",NS:PRINT
230 PRINT "FOR EACH SECTION, ENTER ITS ELECTRICAL"
240 PRINT "LENGTH, ZO IN OHMS, AND ELECTRICAL"
250 PRINT "LENGTH OF STUB (IF NO STUB, ENTER 0):"
260 I=1:IL=10
270 LOCATE IL,1:PRINT I;":":'* NUMBER OF SECTIONS *
280 LOCATE IL,6:INPUT "DS/WL=",D(I):'* ELECTRICAL LENGTH OF
290 '  SECTION *
300 LOCATE IL,19:INPUT "ZO=",ZS(I):'* CHARACTERISTIC
310 '  IMPEDANCE OF SECTION *
320 LOCATE IL,31:INPUT "LS/WL=",L(I):'* ELECTRICAL LENGTH OF
330 '  STUB *
340 IF I<NS THEN I=I+1:IL=IL+1:GOTO 270
350 PRINT:PRINT "ENTER THE FOLLOWING:":PRINT
360 INPUT "MINIMUM VALUE OF F/FO = ",F1
370 INPUT "MAXIMUM VALUE OF F/FO = ",F2
380 INPUT "STEP IN F/FO = ",FS
390 PRINT:PRINT "STRIKE ANY KEY TO CONTINUE":C$=INPUT$(1)
400 CLS:LOCATE 1,1:PRINT "VARIATION OF SWR WITH FREQUENCY IS:":
    PRINT
410 '* COMPUTE SWR VERSUS FREQUENCY *
420 FOR RT=F1 TO F2 STEP FS
430 IF XL>0 THEN X=XL*RT/ZO:GOTO 450
440 X=XL/(RT*ZO)
```

PL 8.3. (continued)

```
450 R=RL/ZO:MZ2=R*R+X*X:G=R/MZ2:B=-X/MZ2:'* COMPUTATION OF
460 '  NORMALIZED LOAD ADMITTANCE *
470 FOR I=1 TO NS
480 IF D(I)=0 THEN 540
490 DI=D(I)*RT:ZR=ZS(I)/ZO
500 G=G*ZR:B=B*ZR
510 CL=COS(2*PI*DI):SL=SIN(2*PI*DI)
520 RN=G*CL:IN=B*CL+SL:RD=CL-B*SL:ID=G*SL
530 MN2=RD*RD+ID*ID:G=(RN*RD+IN*ID)/(MN2*ZR):B=(IN*RD-RN*ID)/(M
    N2*ZR)
540 IF L(I)=0 THEN 590:'* NO STUB *
550 BX=TAN(2*PI*L(I)*RT)
560 IF BX=0 THEN 620:'* IF INPUT REACTANCE OF STUB IS ZERO,
570 '  THEN SET SWR EQUAL TO INFINITY *
580 B=B-1/BX
590 NEXT
600 MG=SQR(((1-G)^2+B*B)/((1+G)^2+B*B)):SWR=(1+MG)/(1-MG)
610 PRINT USING "F/FO = #.##";RT;:LOCATE ,15:PRINT "SWR =";FN T
    R(SWR):GOTO 630
620 PRINT USING "F/FO = #.##";RT;:LOCATE ,15:PRINT "SWR = INFIN
    ITY"
630 NEXT
640 PRINT:PRINT "STRIKE ANY KEY TO CONTINUE"
650 C$=INPUT$(1):GOTO 180
660 END
```

```
RUN
ENTER VALUES OF ZO, RL, AND XL IN OHMS:
ZO=50           RL=30              XL=-40

ENTER NUMBER OF SECTIONS: 2

FOR EACH SECTION, ENTER ITS ELECTRICAL
LENGTH, ZO IN OHMS, AND ELECTRICAL
LENGTH OF STUB (IF NO STUB, ENTER 0):

 1 : DS/WL=0       ZO=50         LS/WL=.1348
 2 : DS/WL=.375    ZO=50         LS/WL=.3273

ENTER THE FOLLOWING:

MINIMUM VALUE OF F/FO = .9
MAXIMUM VALUE OF F/FO = 1.1
STEP IN F/FO = .02

STRIKE ANY KEY TO CONTINUE

VARIATION OF SWR WITH FREQUENCY IS:

F/FO = 0.90   SWR = 1.9249
F/FO = 0.92   SWR = 1.7124
F/FO = 0.94   SWR = 1.5117
F/FO = 0.96   SWR = 1.325
F/FO = 0.98   SWR = 1.1543
F/FO = 1.00   SWR = 1.0006
F/FO = 1.02   SWR = 1.1583
F/FO = 1.04   SWR = 1.3459
F/FO = 1.06   SWR = 1.5663
F/FO = 1.08   SWR = 1.8236
F/FO = 1.10   SWR = 2.1216

STRIKE ANY KEY TO CONTINUE
```

K8.3. Matching; Quarter-wave transformer; Single stub; Double stub; Bandwidth.

D8.7. For a line of characteristic impedance 75 Ω, find the location nearest to the load and the characteristic impedance of a quarter-wave transformer required to achieve a match for each of the following values of $\bar{\Gamma}_R$: **(a)** 1/9; **(b)** $-j0.5$; and **(c)** $j1/3$.
Ans. **(a)** 83.85 Ω, 0; **(b)** 43.30 Ω, 0.125λ; **(c)** 106.07 Ω, 0.125λ

D8.8. For each of the following values of $\bar{Z}_R$ terminating a line of characteristic impedance 60 Ω, find the lowest value of d_s and the corresponding smallest value of the length l_s of a single short-circuited stub of characteristic impedance 60 Ω required to achieve a match between the line and the load: **(a)** $\bar{Z}_R = 30$ Ω and **(b)** $\bar{Z}_R = (12 - j24)$ Ω.
Ans. **(a)** 0.098λ, 0.348λ; **(b)** 0, 0.074λ

D8.9. For each of the following sets of values of d_1, d_{12}, and $\bar{z}_R$, associated with the double-stub matching technique, determine whether or not it is possible to achieve a match between the line and the load: **(a)** $d_1 = 0$, $d_{12} = 3\lambda/8$, $\bar{z}_R = 0.3 + j0.4$; **(b)** $d_1 = \lambda/8$, $d_{12} = 3\lambda/8$, $\bar{z}_R = 0.5$; and **(c)** $d_1 = \lambda/4$, $d_{12} = 5\lambda/8$, $\bar{z}_R = 2.5 - j5.0$.
Ans. **(a)** Yes; **(b)** yes; **(c)** no

8.4 THE SMITH CHART 1. BASIC PROCEDURES

In the preceding section we considered transmission-line matching techniques and computer solutions of matching problems. In this section we discuss some basic procedures using the Smith chart. Introduced in 1939 by P. H. Smith,[1] the Smith chart continues to be a popular graphical aid in the solution of transmission-line problems, including simulation on personal computers.[2]

Construction

The *Smith chart* is a transformation from the complex $\bar{Z}$-plane (or $\bar{Y}$-plane) to the complex $\bar{\Gamma}$-plane. To discuss the basis behind the construction of the Smith chart, we begin with the relationship for the reflection coefficient in terms of the normalized line impedance as given by

$$\bar{\Gamma}(d) = \frac{\bar{z}(d) - 1}{\bar{z}(d) + 1} \tag{8.55}$$

Letting $\bar{z}(d) = r + jx$, we have

$$\bar{\Gamma}(d) = \frac{r + jx - 1}{r + jx + 1} = \frac{(r - 1) + jx}{(r + 1) + jx}$$

and

$$|\bar{\Gamma}(d)| = \left[\frac{(r - 1)^2 + x^2}{(r + 1)^2 + x^2}\right]^{1/2} \leq 1$$

[1] P. H. Smith, "Transmission-Line Calculator," *Electronics,* January 1939, pp. 29–31.

[2] See, for example, M. Felton, "Moving the Smith Chart to a Low-Cost Computer," *Microwave Journal,* October 1983, pp. 131–133, and N. N. Rao, "PC-Assisted Instruction of Introductory Electromagnetics," *IEEE Transactions on Education,* February 1990, pp. 51–59.

for positive values of r. Thus, for the passive line impedances, the reflection coefficient lies inside or on the circle of unit radius in the $\bar{\Gamma}$-plane. We will hereafter call this circle the *unit circle*. Conversely, each point inside or on the unit circle represents a possible value of reflection coefficient corresponding to a unique value of passive normalized line impedance. Hence all possible values of passive normalized line impedances can be mapped onto the region bounded by the unit circle.

To determine how the normalized line impedance values are mapped onto the region bounded by the unit circle, we note that

$$\bar{\Gamma} = \frac{r + jx - 1}{r + jx + 1} = \frac{r^2 - 1 + x^2}{(r + 1)^2 + x^2} + j\frac{2x}{(r + 1)^2 + x^2}$$

so that

$$\boxed{\begin{aligned} \text{Re}(\bar{\Gamma}) &= \frac{r^2 - 1 + x^2}{(r + 1)^2 + x^2} \\ \text{Im}(\bar{\Gamma}) &= \frac{2x}{(r + 1)^2 + x^2} \end{aligned}} \qquad \begin{aligned} &(8.56\text{a}) \\ &(8.56\text{b}) \end{aligned}$$

Let us now discuss different cases:

1. $\bar{z}$ is purely real; that is, $x = 0$. Then

$$\text{Re}(\bar{\Gamma}) = \frac{r - 1}{r + 1} \qquad \text{and} \qquad \text{Im}(\bar{\Gamma}) = 0$$

   Purely real values of $\bar{z}$ are represented by points on the real axis. For example, $r = 0, \frac{1}{3}, 1, 3$, and ∞ are represented by $\bar{\Gamma} = -1, -\frac{1}{2}, 0, \frac{1}{2}$, and 1, respectively, as shown in Fig. 8.21(a).

2. $\bar{z}$ is purely imaginary; that is, $r = 0$. Then

$$|\bar{\Gamma}| = \left| \frac{x^2 - 1}{x^2 + 1} + j\frac{2x}{x^2 + 1} \right| = 1$$

   and

$$\underline{/\bar{\Gamma}} = \tan^{-1} \frac{2x}{x^2 - 1}$$

   Purely imaginary values of $\bar{z}$ are represented by points on the unit circle. For example, $x = 0, 1, \infty, -1$, and $-\infty$ are represented by $\bar{\Gamma} = 1\underline{/\pi}$, $1\underline{/\pi/2}$, $1\underline{/0^\circ}$, $1\underline{/-\pi/2}$, and $1\underline{/2\pi}$, respectively, as shown in Fig. 8.21(b).

3. $\bar{z}$ is complex, but its real part is constant. Then

$$\left[\text{Re}(\bar{\Gamma}) - \frac{r}{r + 1}\right]^2 + [\text{Im}(\bar{\Gamma})]^2$$

$$= \left[\frac{r^2 - 1 + x^2}{(r + 1)^2 + x^2} - \frac{r}{r + 1}\right]^2 + \left[\frac{2x}{(r + 1)^2 + x^2}\right]^2 = \left(\frac{1}{r + 1}\right)^2$$

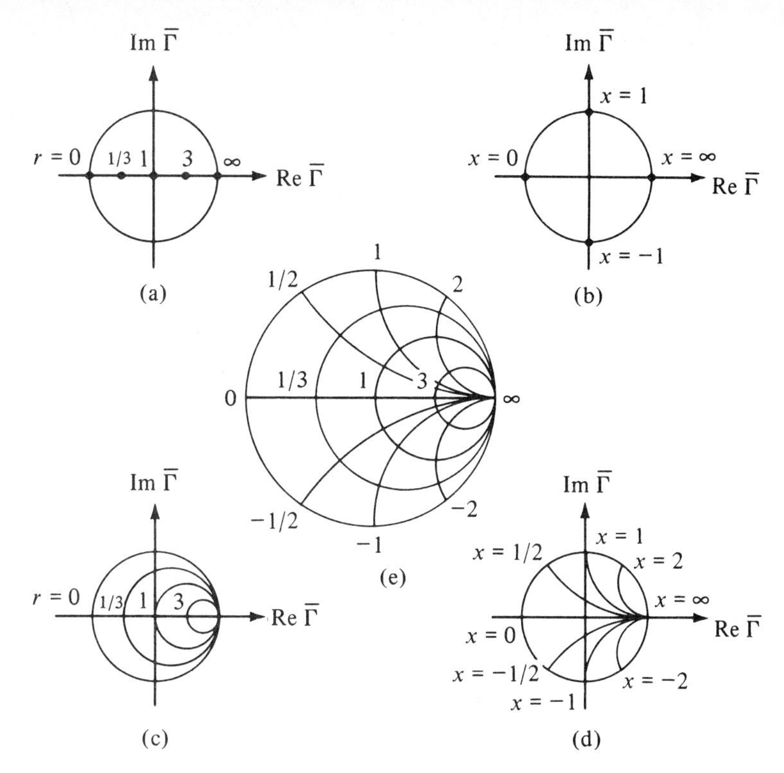

Figure 8.21. Development of the Smith chart by transformation from $\bar{z}$ to $\bar{\Gamma}$.

This is the equation of a circle with center at $\text{Re}(\bar{\Gamma}) = r/(r+1)$ and $\text{Im}(\bar{\Gamma}) = 0$ and radius equal to $1/(r+1)$. Thus loci of constant r are circles in the $\bar{\Gamma}$-plane with centers at $[r/(r+1), 0]$ and radii $1/(r+1)$. For example, for $r = 0, \frac{1}{3}, 1, 3$, and ∞, the centers of the circles are $(0, 0)$, $(\frac{1}{4}, 0)$, $(\frac{1}{2}, 0)$, $(\frac{3}{4}, 0)$, and $(1, 0)$, respectively, and the radii are $1, \frac{3}{4}, \frac{1}{2}, \frac{1}{4}$, and 0, respectively. These circles are shown in Fig. 8.21(c).

4. $\bar{z}$ is complex, but its imaginary part is constant. Then

$$[\text{Re}(\bar{\Gamma}) - 1]^2 + \left[\text{Im}(\bar{\Gamma}) - \frac{1}{x}\right]^2$$
$$= \left[\frac{r^2 - 1 + x^2}{(r+1)^2 + x^2} - 1\right]^2 + \left[\frac{2x}{(r+1)^2 + x^2} - \frac{1}{x}\right]^2 = \left(\frac{1}{x}\right)^2$$

This is the equation of a circle with center at $\text{Re}(\bar{\Gamma}) = 1$ and $\text{Im}(\bar{\Gamma}) = 1/x$ and radius equal to $1/|x|$. Thus locii of constant x are circles in the $\bar{\Gamma}$-plane with centers at $(1, 1/x)$ and radii equal to $1/|x|$. For example, for $x = 0$, $\pm\frac{1}{2}$, ± 1, ± 2, and $\pm\infty$, the centers of the circles are $(1, \infty)$, $(1, \pm 2)$, $(1, \pm 1)$, $(1, \pm\frac{1}{2})$, and $(1, 0)$, respectively, and the radii are $\infty, 2, 1, \frac{1}{2}$, and 0, respectively. Portions of these circles that fall inside the unit circle are shown in Fig. 8.21(d). Portions that fall outside the unit circle represent active impedances.

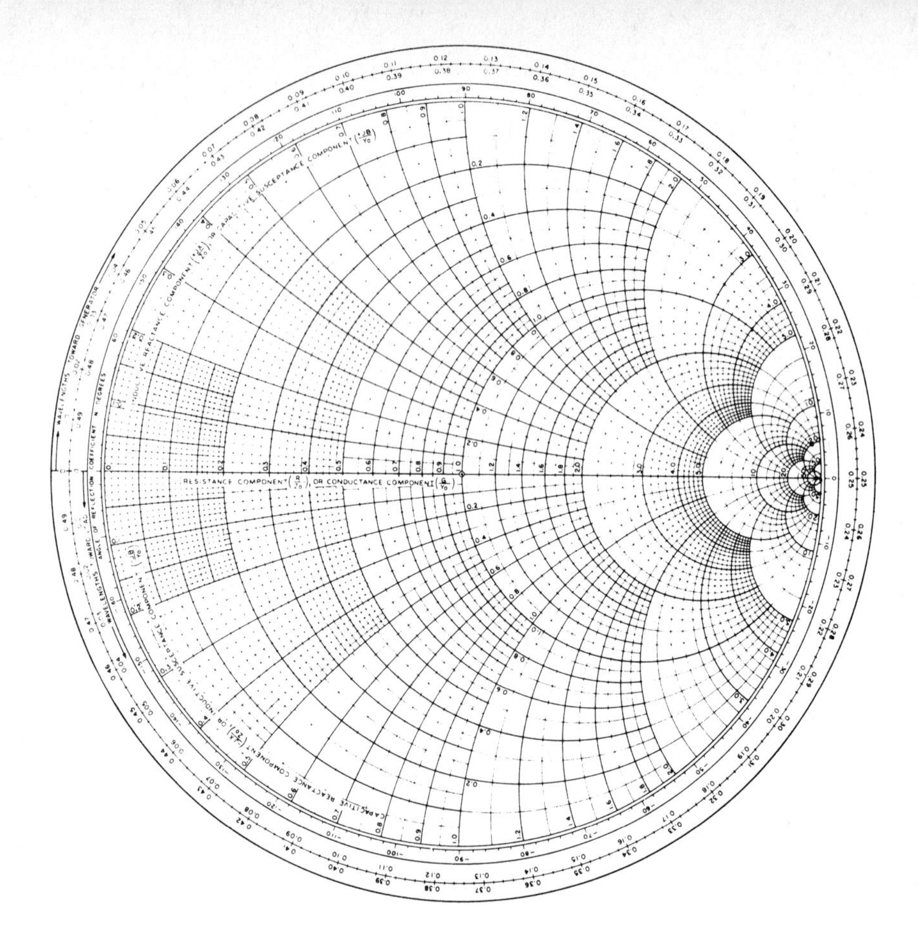

Figure 8.22. The Smith chart. (Copyrighted by and reproduced with the permission of Kay Elemetrics Corp., Pine Brook, N.J.)

Combining (c) and (d), we obtain the chart of Fig. 8.21(e). In a commercially available form shown in Fig. 8.22, the Smith chart contains circles of constant r and constant x for very small increments of r and x, respectively, so that interpolation between the contours can be carried out accurately. We now consider an example to illustrate some basic procedures using the Smith chart.

Example 8.4

Some basic procedures

A transmission line of characteristic impedance 50 Ω is terminated by a load impedance $\bar{Z}_R = (15 - j20)\ \Omega$. It is desired to find the following quantities by using the Smith chart.

(1) Reflection coefficient at the load
(2) SWR on the line

(3) Distance of the first voltage minimum of the standing wave pattern from the load

(4) Line impedance at $d = 0.05\lambda$

(5) Line admittance at $d = 0.05\lambda$

(6) Location nearest to the load at which the real part of the line admittance is equal to the line characteristic admittance

We proceed with the solution of the problem in the following step-by-step manner with reference to Fig. 8.23.

(a) Find the normalized load impedance.

$$\bar{z}_R = \frac{\bar{Z}_R}{Z_0} = \frac{15 - j20}{50} = 0.3 - j0.4$$

(b) Locate the normalized load impedance on the Smith chart at the intersection of the 0.3 constant normalized resistance circle and -0.4 constant normalized reactance circle (point A).

(c) Locating point A actually amounts to computing the reflection coefficient at the load since the Smith chart is a transformation in the $\bar{\Gamma}$-plane. The magnitude of the reflection coefficient is the distance from the center (O) of the Smith chart (origin of the $\bar{\Gamma}$-plane) to the point A based on a radius of unity for the outermost circle. For this example, $|\bar{\Gamma}_R| = 0.6$. The phase angle of $\bar{\Gamma}_R$ is the angle measured from the horizontal axis to the right of O (positive real axis in the $\bar{\Gamma}$-plane) to the line OA in the counterclockwise direction. This an-

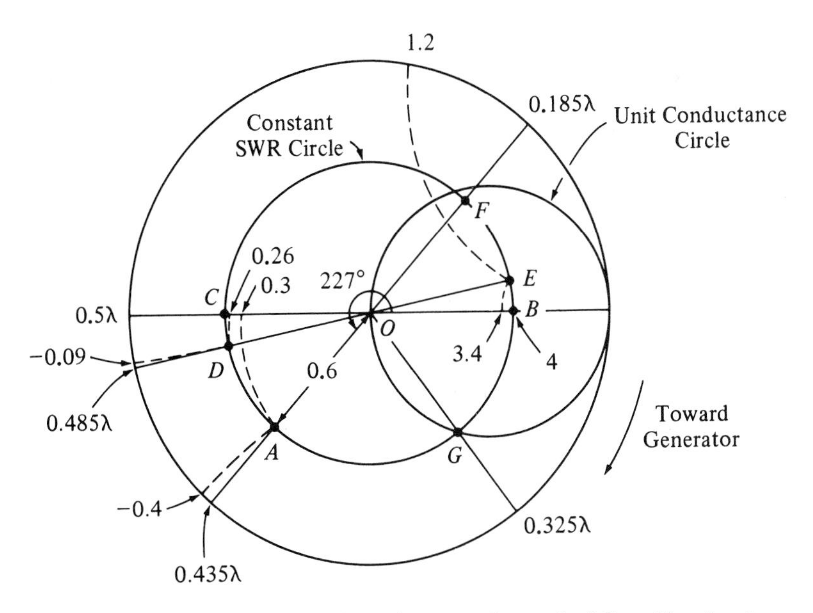

Figure 8.23. For illustrating the various procedures to be followed in using the Smith chart.

gle is indicated on the chart along its circumference. For this example, $\underline{/\bar{\Gamma}_R} = 227°$. Thus

$$\bar{\Gamma}_R = 0.6\, e^{j1.261\pi}$$

(d) To find the SWR, we recall that at the location of a voltage maximum, the line impedance is purely real and given by

$$R_{\text{max}} = Z_0\,(\text{SWR}) \tag{8.57}$$

Thus the normalized value of R_{max} is equal to the SWR. We therefore move along the line to the location of the voltage maximum, which involves going around the constant $|\bar{\Gamma}|$ circle to the point on the positive real axis. To do this on the Smith chart, we draw a circle passing through A and with center at O. This circle is known as the *constant SWR circle* since for points on this circle, $|\bar{\Gamma}|$ and hence SWR $= (1 + |\bar{\Gamma}|)/(1 - |\bar{\Gamma}|)$ is a constant. Impedance values along this circle are normalized line impedances as seen moving along the line. In particular, since point B (the intersection of the constant SWR circle with the horizontal axis to the right of O) corresponds to voltage maximum, the normalized impedance value at point B, which is purely real and maximum, is equal to the SWR. Thus, for this example, SWR $= 4$.

(e) Just as point B represents the position of a voltage maximum on the line, point C (intersection of the constant SWR circle with the horizontal axis to the left of O, i.e., the negative real axis of the $\bar{\Gamma}$-plane) represents the location of a voltage minimum. Hence, to find the distance of the first voltage minimum from the load, we move along the constant SWR circle starting at point A (load impedance) toward the generator (clockwise direction on the chart) to reach point C. Distance moved along the constant SWR circle in this process can be determined by recognizing that one complete revolution around the chart ($\bar{\Gamma}$-plane diagram) constitutes movement on the line by 0.5λ. However, it is not necessary to compute in this manner since distance scales in terms of λ are provided along the periphery of the chart for movement in both directions. For this example, the distance from the load to the first voltage minimum $= (0.5 - 0.435)\lambda = 0.065\lambda$. Conversely, if the SWR and the location of the voltage minimum are specified, we can find the load impedance by following the foregoing procedures in reverse.

(f) To find the line impedance at $d = 0.05\lambda$, we start at point A and move along the constant SWR circle toward the generator (in the clockwise direction) by a distance of 0.05λ to reach point D. This step is equivalent to finding the reflection coefficient at $d = 0.05\lambda$ knowing the reflection coefficient at $d = 0$ and then computing the normalized line impedance by using (8.35). Thus, from the coordinates corresponding to point D, the normalized line impedance at $d = 0.05\lambda$ is $(0.26 - j0.09)$, and hence the line impedance at $d = 0.05\lambda$ is $50(0.26 - j0.09)$ or $(13 - j4.5)\ \Omega$.

(g) To find the line admittance at $d = 0.05\lambda$, we recall that

$$[\bar{Z}(d)]\left[\bar{Z}\left(d + \frac{\lambda}{4}\right)\right] = Z_0^2$$

so that

$$[\bar{z}(d)]\left[\bar{z}\left(d + \frac{\lambda}{4}\right)\right] = 1$$

or

$$\bar{y}(d) = \bar{z}\left(d + \frac{\lambda}{4}\right) \tag{8.58}$$

Thus the normalized line admittance at point D is the same as the normalized line impedance at a distance $\lambda/4$ from it. Hence, to find $\bar{y}(0.05\lambda)$, we start at point D and move along the constant SWR circle by a distance $\lambda/4$ to reach point E (we note that this point is diametrically opposite to point D) and read its coordinates. This gives $\bar{y}(0.05\lambda) = (3.4 + j1.2)$. We then have $\bar{Y}(0.05\lambda) = \bar{y}(0.05\lambda) \times Y_0 = (3.4 + j1.2) \times 1/50 = (0.068 + j0.024)$ S.

(h) Relationship (8.58) permits us to use the Smith chart as an admittance chart instead of an impedance chart. In other words, if we want to find the normalized line admittance $\bar{y}(Q)$ at a point Q on the line, knowing the normalized line admittance $\bar{y}(P)$ at another point P on the line, we can simply locate $\bar{y}(P)$ by entering the chart at coordinates equal to its real and imaginary parts and then moving along the constant SWR circle by the amount of the distance from P to Q in the proper direction to obtain the coordinates equal to the real and imaginary parts of $\bar{y}(Q)$. Thus it is not necessary first to locate $\bar{z}(P)$ diametrically opposite to $\bar{y}(P)$ on the constant SWR circle, then move along the constant SWR circle to locate $\bar{z}(Q)$, and then find $\bar{y}(Q)$ diametrically opposite to $\bar{z}(Q)$. To find the location nearest to the load at which the real part of the line admittance is equal to the line characteristic admittance, we first locate $\bar{y}(0)$ at point F diametrically opposite to point A which corresponds to $\bar{z}(0)$. We then move along the constant SWR circle toward the generator to reach point G on the circle corresponding to constant real part equal to unity (we call this circle the *unit conductance circle*). Distance moved from F to G is read off the chart as $(0.325 - 0.185)\lambda = 0.14\lambda$. This is the distance closest to the load at which the real part of the normalized line admittance is equal to unity and hence the real part of the line admittance is equal to line characteristic admittance.

K8.4. $\bar{\Gamma}$-plane; Unit circle; Transformation from $\bar{z}$ (or $\bar{y}$) to $\bar{\Gamma}$; Smith chart; Constant SWR circle; Unit conductance circle.

D8.10. Find the values of $\bar{\Gamma}$ in polar form onto which the following normalized impedances are mapped: **(a)** $0.25 + j0$; **(b)** $0 - j0.5$; **(c)** $3 + j3$; **(d)** $-1 + j2$.
Ans. **(a)** $0.6\angle 180^\circ$; **(b)** $1\angle 233.13^\circ$; **(c)** $0.721\angle 19.44^\circ$; **(d)** $1.414\angle 45^\circ$

D8.11. Find the following using the Smith chart: **(a)** the normalized input impedance of a line of length 0.1λ and terminated by a normalized load impedance $(2 + j1)$; **(b)** the normalized input admittance of a short-circuited stub of length 0.17λ; and **(c)** the shortest length of an open-circuited stub having the normalized input admittance $j0.4$.
Ans. **(a)** $1.4 - j1.1$; **(b)** $-j0.55$; **(c)** 0.06λ

8.5 THE SMITH CHART 2. APPLICATIONS

In the preceding section we introduced the Smith chart and discussed some basic procedures. In this section we first consider by means of examples graphical solutions of transmission-line matching problems using the Smith chart and then discuss further applications.

Example 8.5

Single-stub matching solution

Let us consider a transmission line of characteristic impedance $Z_0 = 50\ \Omega$ terminated by a load impedance $\bar{Z}_R = (30 - j40)\ \Omega$ and illustrate the solution of the single-stub matching problem by using the Smith chart, assuming Z_0 of the stub to be 50 Ω.

With reference to the notation in Fig. 8.17, we recall that to achieve a match, the stub must be located at a point on the line at which the real part of the normalized line admittance is equal to unity; the imaginary part of the line admittance at that point is then canceled by appropriately choosing the length of the stub. Hence we proceed with the solution in the following step-by-step manner with reference to Fig. 8.24.

(a) Find the normalized load impedance.

$$\bar{z}_R = \frac{\bar{Z}_R}{Z_0} = \frac{30 - j40}{50} = 0.6 - j0.8$$

Locate the normalized load impedance on the Smith chart at point A.

(b) Draw the constant SWR circle passing through point A. This is the locus of the normalized line impedance as well as the normalized line admittance. Starting at point A, go around the constant SWR circle by half a revolution to reach point B diametrically opposite to point A. Point B corresponds to the normalized load admittance.

(c) Starting at point B, go around the constant SWR circle toward the generator until point C on the unit conductance circle is reached. This point corresponds to the normalized line admittance having the real part equal to

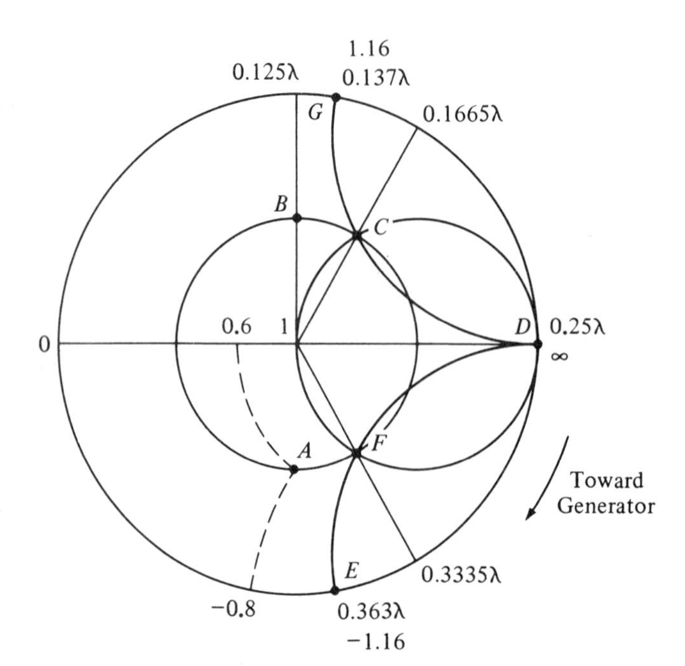

Figure 8.24. Solution of single-stub matching problem by using the Smith chart.

unity, and hence it corresponds to the location of the stub. The distance moved from point B to point C (not from point A to point C) is equal to the distance from the load at which the stub must be located. Thus the location of the stub from the load = $(0.1665 - 0.125)\lambda = 0.0415\lambda$.

(d) Read off the Smith chart the normalized susceptance value corresponding to point C. This value is 1.16, and it is the imaginary part of the normalized line admittance at the location of the stub. The imaginary part of the line admittance is equal to $1.16 \times Y_0 = (1.16/50)$ S. The input susceptance of the stub must therefore be equal to $-(1.16/50)$ S.

(e) This step consists of finding the length of a short-circuited stub having an input susceptance equal to $-(1.16/50)$ S. We can use the Smith chart for this purpose since this simply consists of finding the distance between two points on a line (the stub in this case) at which the admittances (purely imaginary in this case) are known. Thus, since the short circuit corresponds to a susceptance of infinity, we start at point D and move toward the generator along the constant SWR circle through D (the outermost circle) to reach point E corresponding to $-j1.16$, which is the input admittance of the stub normalized with respect to its own characteristic admittance. The distance moved from D to E is the required length of the stub. Thus length of the short-circuited stub = $(0.363 - 0.25)\lambda = 0.113\lambda$.

(f) The results obtained for the location and the length of the stub agree with one of the solutions found analytically in Section 8.3 by using PL 8.2. The second solution can be obtained by noting that in step (c), we can go around the constant SWR circle from point B until point F on the unit conductance circle is reached, instead of stopping at point C. The stub location for this solution is $(0.3335 - 0.125)\lambda = 0.2085\lambda$. The required input susceptance of the stub is $(1.16/50)$ S. The length of the stub is the distance from point D to point G in the clockwise direction. This is $(0.137 + 0.25)\lambda = 0.387\lambda$. These values are the same as the second solution obtained in Section 8.3.

Example 8.6

Double-stub matching solution

For the line of characteristic impedance $Z_0 = 50\ \Omega$ and load impedance $\bar{Z}_R = (30 - j40)\ \Omega$ of Example 8.5, it is desired to solve the double-stub matching problem by using the Smith chart and assuming Z_0 of both stubs to be 50 Ω, the first stub to be located at the load, and distance between stubs equal to 0.375λ.

With reference to the notation of Fig. 8.18, we first note that to achieve a match, $\bar{y}_2' = 1 - jb_2$ must fall on the unit conductance circle. Now since $\bar{y}_2'$ and $\bar{y}_1$ correspond to locations at the end points of the line section between the stubs, for a given $\bar{y}_1$, $\bar{y}_2'$ can be obtained by drawing the constant SWR circle through $\bar{y}_1$ and going toward the generator (clockwise direction) by the distance d_{12} from $\bar{y}_1$. Conversely, to obtain $\bar{y}_1$ for a given $\bar{y}_2'$, we start at $\bar{y}_2'$ and go toward the load (counterclockwise direction) by the distance d_{12} along the constant SWR circle. Hence, for $\bar{y}_2'$ to fall on the unit conductance circle, $\bar{y}_1$ must fall on a circle which is obtained by pivoting the unit conductance circle at the center of the Smith chart (point O) and rotating it toward the load by the distance d_{12}, as shown in Fig. 8.25 for $d_{12} = 3\lambda/8$. We shall call this circle the *auxiliary circle*. Thus the auxiliary circle is the *locus of $\bar{y}_1$ for possible match*.

The matching procedure consists of first locating $\bar{y}_R$ on the Smith chart and then moving along the constant SWR circle through $\bar{y}_R$ toward the generator by the distance d_1 between the load and the first stub, thereby locating $\bar{y}_1'$. The right amount of susceptance is then added to $\bar{y}_1'$ to reach a point on the auxiliary circle.

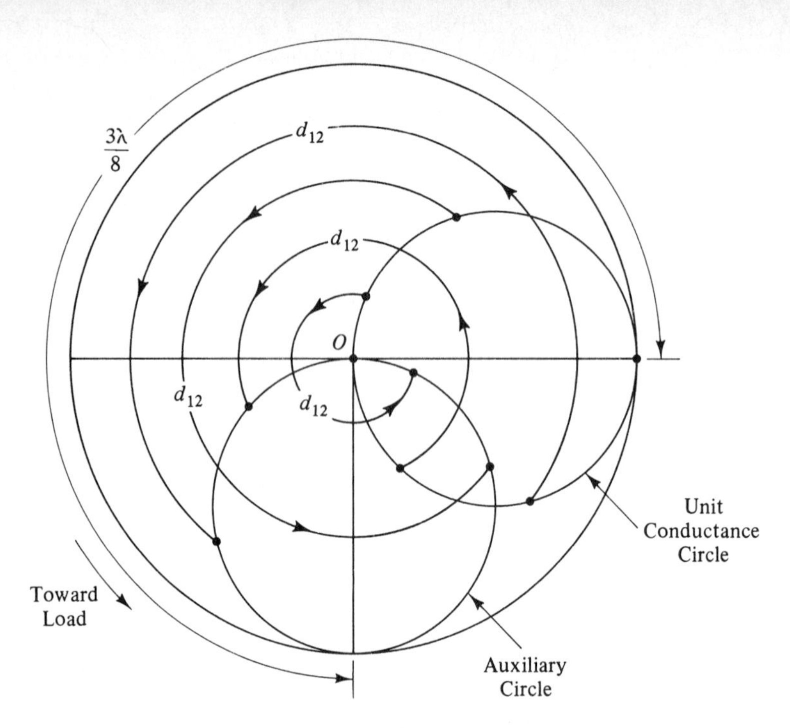

Figure 8.25. Rotation of the unit conductance circle by $d_{12}(= 3\lambda/8)$ toward the load about O for illustrating the construction of the auxiliary circle, that is, the locus of $\bar{y}_1$ for possible match for the double-stub matching arrangement of Fig. 8.18.

This point corresponds to $\bar{y}_1$ and determines a new constant SWR circle. By going along this new constant SWR circle toward the generator by the distance d_{12}, $\bar{y}_2'$ is located on the unit conductance circle. The amount of susceptance added to $\bar{y}_1'$ is the required normalized input susceptance of the first stub, whereas the negative of the imaginary part of $\bar{y}_2'$ is the required normalized input susceptance of the second stub.

Considering now the numerical values of $\bar{z}_R = (30 - j40)/50 = (0.6 - j0.8)$, $d_1 = 0$, and $d_{12} = 0.375\lambda$, we proceed with the solution in the following step-by-step manner with reference to Fig. 8.26.

(a) Locate $\bar{z}_R = (0.6 - j0.8)$ at point A and draw the constant SWR circle through A.

(b) Locate point B on the constant SWR circle and diametrically opposite to point A. This point corresponds to $\bar{y}_R$. Since d_1 is equal to zero, it also corresponds to $\bar{y}_1'$. If d_1 is not equal to zero, then $\bar{y}_1'$ has to be located by going along the constant SWR circle toward the generator by the distance d_1 from point B.

(c) Draw the auxiliary circle which is the circle obtained by pivoting the unit conductance circle at the center of the chart and rotating it by the distance $d_{12} = 0.375\lambda$ toward the load.

(d) This step consists of adding the right amount of susceptance to $\bar{y}_1'$ to get to a point on the auxiliary circle. Hence, starting at point B, go along the constant conductance circle to reach point C on the auxiliary circle. This point corresponds to $\bar{y}_1$. The required normalized input susceptance of the first stub can now be found by noting that $\bar{y}_1 = \bar{y}_1' + jb_1$, and hence

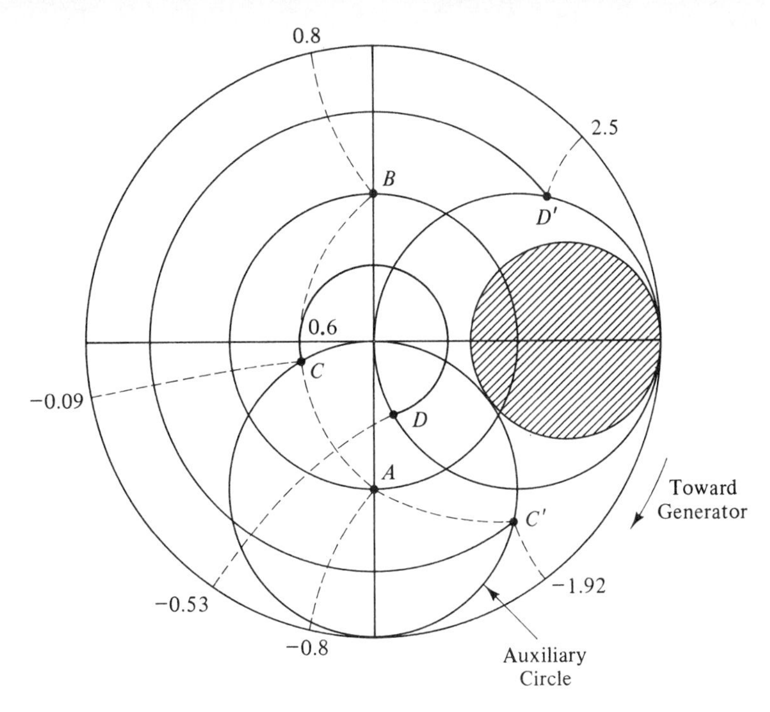

Figure 8.26. Solution of the double-stub matching problem by using the Smith chart.

$$jb_1 = \bar{y}_1 - \bar{y}_1' = (0.6 - j0.09) - (0.6 + j0.8) = -j0.89$$

(e) Starting at point C, go along the constant SWR circle through C toward the generator by $d_{12} = 0.375\lambda$ to reach point D on the unit conductance circle. This point corresponds to $\bar{y}_2'$. Note that the SWR on the portion of the line between the stubs is different from the SWR to the right of the first stub because of the discontinuity introduced by the stub. The required normalized input susceptance of the second stub can now be found by reading the imaginary part of $\bar{y}_2'$ and taking its negative. Thus

$$jb_2 = -j[\text{Im}(\bar{y}_2')] = j0.53$$

(f) This step consists of finding the lengths of the two stubs having the normalized input susceptances found in steps (d) and (e), by using the procedure discussed in Example 8.5. Thus we obtain

$$l_1, \text{ length of first stub} = (0.385 - 0.25)\lambda = 0.135\lambda$$

$$l_2, \text{ length of second stub} = (0.077 + 0.25)\lambda = 0.327\lambda$$

which agree with one of the solutions found analytically in Section 8.3 by using PL 8.2.

(g) Finally, the second solution can be obtained by going from point B to point C' on the auxiliary circle and then to point D' on the unit conductance circle, and computing jb_1 and jb_2 as in steps (d) and (e). Thus we obtain

$$jb_1 = (0.6 - j1.92) - (0.6 + j0.8) = -j2.72$$

$$jb_2 = -j2.5$$

giving us

$$l_1 = (0.306 - 0.25)\lambda = 0.056\lambda$$

$$l_2 = (0.31 - 0.25)\lambda = 0.06\lambda$$

These values are the same as the second solution obtained in Section 8.3.

Before proceeding further, we recall from Section 8.3 that in the double-stub matching technique, it is not possible to achieve a match for loads which result in real part of $\bar{y}_1'$ greater than $1/\sin^2 \beta d_{12}$. For $d_{12} = 3\lambda/8$, $1/\sin^2 \beta d_{12} = 2$, and a match cannot be achieved if the real part of $\bar{y}_1'$ is greater than 2. This is easily evident from the Smith chart construction in Fig. 8.26, since if point B falls inside the crosshatched region (real part > 2), it is not possible to reach a point on the auxiliary circle by moving on the constant conductance circle through B. The crosshatched region is therefore called the *forbidden region of* $\bar{y}_1'$ *for possible match*. As pointed out in Section 8.3, a solution to the problem is to increase d_1 by $\lambda/4$. This effectively rotates the forbidden region by 180° about the center of the chart, thereby making possible a match.

Transformation across a discontinuity

To illustrate the application of the Smith chart further, we shall now discuss a very useful property of the reflection coefficient and hence of the Smith chart. This has to do with the transformation of the reflection coefficient from one side of a discontinuity to the other side of the discontinuity. Let us, for example, consider the system shown in Fig. 8.27, which consists of a junction between two lines of characteristic admittances Y_{01} and Y_{02} and in addition, an admittance $\bar{Y}_d$ connected across the junction. If $\bar{Y}_d = 0$, then the system reduces to a simple junction between two lines. If $Y_{01} = Y_{02}$, then the system reduces to an admittance discontinuity in the same line.

Let $\bar{y}_1 = \bar{Y}_1/Y_{01}$ and $\bar{y}_2 = \bar{Y}_2/Y_{02}$ be the normalized admittances to the left and to the right, respectively, of the junction, and let the corresponding reflection coefficients be $\bar{\Gamma}_1$ and $\bar{\Gamma}_2$, respectively, as shown in Fig. 8.27. Then, since $\bar{Y}_1 = \bar{Y}_2 + \bar{Y}_d$, we have

$$\begin{aligned} \bar{y}_1 &= \frac{\bar{Y}_1}{Y_{01}} = \frac{\bar{Y}_2}{Y_{01}} + \frac{\bar{Y}_d}{Y_{01}} \\ &= \frac{Y_{02}}{Y_{01}}\left(\frac{\bar{Y}_2}{Y_{02}} + \frac{\bar{Y}_d}{Y_{02}}\right) \\ &= a(\bar{y}_2 + \bar{y}_d) \end{aligned} \tag{8.59}$$

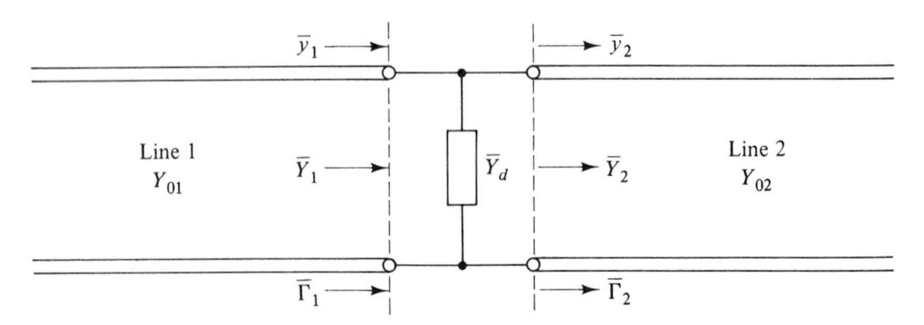

Figure 8.27. Transmission-line system for deriving the transformation of $\bar{\Gamma}$ across a discontinuity.

where $a = Y_{02}/Y_{01}$ is the ratio of the characteristic admittances of the two lines and $\bar{y}_d = \bar{Y}_d/Y_{02}$ is the normalized value of $\bar{Y}_d$ with respect to Y_{02}. Substituting for $\bar{y}_1$ and $\bar{y}_2$ in (8.59) in terms of $\bar{\Gamma}_1$ and $\bar{\Gamma}_2$, respectively, we have

$$\frac{1 - \bar{\Gamma}_1}{1 + \bar{\Gamma}_1} = a\left(\frac{1 - \bar{\Gamma}_2}{1 + \bar{\Gamma}_2} + \bar{y}_d\right) \tag{8.60}$$

Rearranging (8.60), we obtain

$$\boxed{\bar{\Gamma}_1 = \frac{(1 + a - a\bar{y}_d)\bar{\Gamma}_2 + (1 - a - a\bar{y}_d)}{(1 - a + a\bar{y}_d)\bar{\Gamma}_2 + (1 + a + a\bar{y}_d)}} \tag{8.61}$$

Equation (8.61) is of the form of the so-called *bilinear transformation*, between two complex planes, a property of which is that circles in one plane are transformed into circles in the second plane. Consequently, loci of $\bar{\Gamma}_2$ which are circles in the $\bar{\Gamma}$-plane are mapped on to loci of $\bar{\Gamma}_1$ which are also circles in the $\bar{\Gamma}$-plane, and vice versa. Since the Smith chart is a transformation (also bilinear) from $\bar{z}$ or $\bar{y}$ to $\bar{\Gamma}$, this means that loci of $\bar{y}_2$ which are circles are mapped on to loci of $\bar{y}_1$, which are also circles. Since a circle is defined completely by three points, it is therefore sufficient if we use any three points on the locus of $\bar{y}_2$ and find the corresponding three points for $\bar{y}_1$. By locating the center at the intersection of the perpendicular bisectors of lines joining any two pairs of these three points, we can then draw the circle passing through these points, that is, the locus of $\bar{y}_1$. While we have demonstrated this property by considering the discontinuity of the form shown in Fig. 8.27, it can be shown that the property holds for the case of any linear, passive, bilateral network serving as the discontinuity. We shall now consider an example.

Example 8.7

Let us consider the system shown in Fig. 8.28, in which a line is terminated by a normalized admittance $\bar{y}_R = (0.6 + j0.8)$ and a normalized susceptance of value $b = 0.8$ connected between the two conductors of the line forms the discontinuity. We wish to find the locus of the normalized admittance $\bar{y}_1$ to the left of the discontinuity as the susceptance slides along the line, and then determine the location, nearest to the load, of the susceptance for which the SWR to the left of it is minimized.

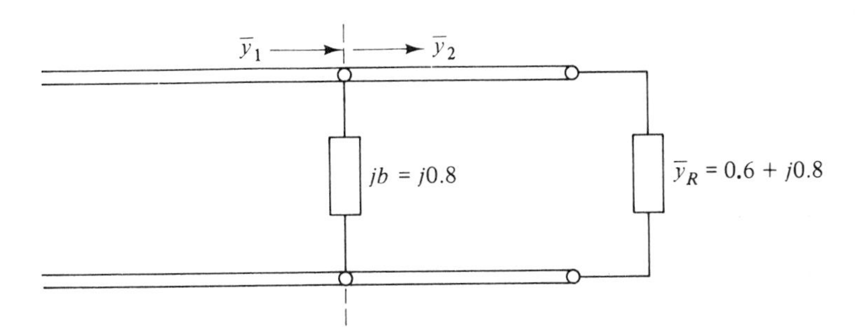

Figure 8.28. Transmission-line system in which a susceptance of fixed value sliding along the line forms a discontinuity.

To construct the locus of $\bar{y}_1$, we first locate $\bar{y}_R = (0.6 + j0.8)$ on the Smith chart at point A and draw the constant SWR circle passing through A, as shown in Fig. 8.29. This circle is the locus of $\bar{y}_2$, the normalized admittance just to the right of the discontinuity as the distance between the load and the discontinuity is varied, that is, as the susceptance slides along the line. We then choose any three points on the locus of $\bar{y}_2$ and locate the corresponding three points for $\bar{y}_1 = \bar{y}_2 + j0.8$. Here, we choose the points A, B, and C. Following the constant conductance circles through these points by the amount of normalized susceptance $+0.8$, we obtain the points D, E, and F, respectively. We then draw the circle passing through these points to obtain the locus of $\bar{y}_1$.

Proceeding further, we note that each point on the locus of $\bar{y}_1$ corresponds to a value of SWR to the left of the susceptance, obtained by following the constant SWR circle through that point to the r value at the V_{max} point. In particular, it can be seen that minimum SWR is achieved to the left of the susceptance for $\bar{y}_1$ lying at point G, which is the closest point to the center of the chart, and the minimum SWR value is 1.35. The distance from the load at which the susceptance must be connected to achieve this minimum SWR can be found by locating the $\bar{y}_2$ corresponding to the $\bar{y}_1$ at G by following the constant conductance circle through G by the amount -0.8 to reach point H. The distance from point A to point H toward the generator is the required distance. It is equal to $(0.346 - 0.125)\lambda$, or 0.221λ.

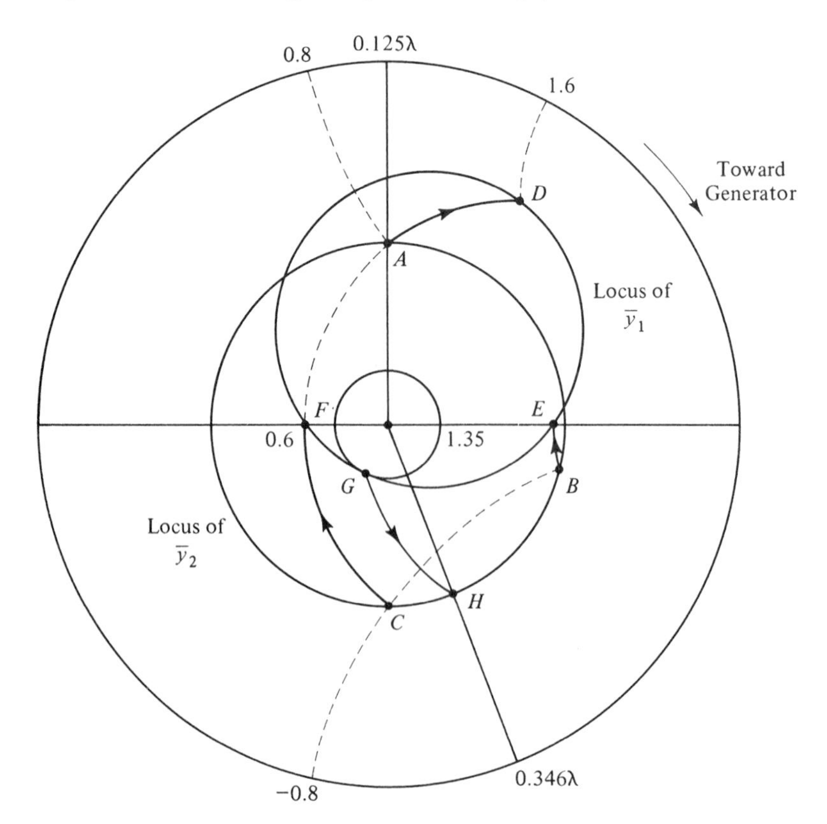

Figure 8.29. Construction of the locus of $\bar{y}_1$ for the system of Fig. 8.28 as the susceptance b slides along the line and determination of the minimum SWR that can be achieved to the left of the susceptance and the location of the susceptance to achieve the minimum SWR.

Together with the basic procedures discussed in the previous section, the methods which we have discussed in this section can be extended to solve many other problems using the Smith chart. We include some of these in the problems.

K8.5. Single-stub matching; Double-stub matching; Auxiliary circle; Forbidden region of $\bar{y}_1'$ for possible match; Transformation of $\bar{\Gamma}$ across a discontinuity.

D8.12. A line of characteristic impedance 100 Ω is terminated by a load of impedance $(50 + j65)$ Ω. Find the following using the Smith chart: **(a)** the SWR on the line; **(b)** the minimum SWR that can be achieved on the line by connecting a stub in parallel with the line at the load; and **(c)** the minimum SWR that can be achieved on the line by connecting a stub in series with the line at the load.
Ans. **(a)** 3.0; **(b)** 1.33; **(c)** 2.0

D8.13. A line of characteristic impedance 50 Ω is terminated by a load of impedance $(100 + j100)$ Ω. Find the following by using the Smith chart: **(a)** the minimum distance at which a reactance of value 50 Ω must be connected in parallel with the line to minimize the SWR to the left of the reactance and the minimum SWR achieved; **(b)** the minimum length of a line section of characteristic impedance 100 Ω between the main line and the load required to minimize the SWR on the main line and the minimum SWR achieved; and **(c)** the characteristic impedance of a $\lambda/4$ section of line inserted between the main line and the load to minimize the SWR on the main line and the minimum SWR achieved.
Ans. **(a)** 0.204λ, 1.63; **(b)** 0.338λ, 1.30; **(c)** 83.7Ω, 2.42

8.6 THE LOSSY LINE

Distributed equivalent circuit

Thus far we have been concerned with lossless lines. We learned in Section 7.1 that the distributed equivalent circuit for a lossless line consists of series inductors and shunt capacitors, representing energy storage in magnetic and electric fields, respectively. A lossy line is characterized by imperfect but good conductors and imperfect dielectric giving rise to power dissipation, thereby modifying the distributed equivalent circuit. The power dissipation in the conductors is taken into account by a resistance in series with the inductor whereas the power dissipation in the dielectric is taken into account by a conductance in parallel with the capacitor. In addition, the magnetic field inside the conductors is taken into account by an additional inductance in the series branch. Thus the distributed equivalent circuit for the lossy line is as shown in Fig. 8.30, where $\mathcal{L}$ includes the additional inductance just mentioned.

Transmission-line equations and solution

To discuss wave propagation on a lossy line, we first obtain the transmission-line equations by applying Kirchhoff's voltage and current laws to the circuit of Fig. 8.30. Thus we have

$$V(z + \Delta z, t) - V(z, t) = -\mathcal{R}\,\Delta z\, I(z, t) - \mathcal{L}\,\Delta z\,\frac{\partial I(z, t)}{\partial t} \tag{8.62a}$$

$$I(z + \Delta z, t) - I(z, t) = -\mathcal{G}\,\Delta z\, V(z, t) - \mathcal{C}\,\Delta z\,\frac{\partial V(z, t)}{\partial t} \tag{8.62b}$$

Dividing both sides of (8.62a) and (8.62b) by Δz and letting $\Delta z \to 0$, we obtain the transmission-line equations

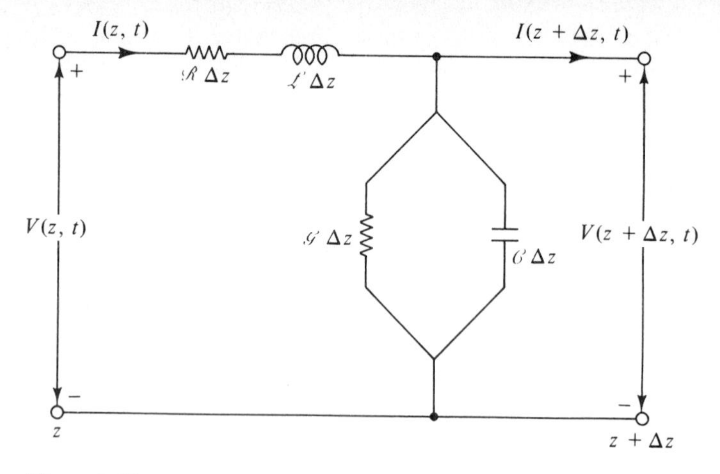

Figure 8.30. Distributed equivalent circuit for a lossy transmission line.

$$\boxed{\begin{aligned} \frac{\partial V(z, t)}{\partial z} &= -\mathcal{R}I(z, t) - \mathcal{L}\frac{\partial I(z, t)}{\partial t} \qquad (8.63a) \\ \frac{\partial I(z, t)}{\partial z} &= -\mathcal{G}V(z, t) - \mathcal{C}\frac{\partial V(z, t)}{\partial t} \qquad (8.63b) \end{aligned}}$$

The corresponding equations in terms of phasor voltage and current are

$$\frac{\partial \bar{V}(z)}{\partial z} = -\mathcal{R}\bar{I}(z) - j\omega\mathcal{L}\bar{I}(z)$$

$$\frac{\partial \bar{I}(z)}{\partial z} = -\mathcal{G}\bar{V}(z) - j\omega\mathcal{C}\bar{V}(z)$$

or

$$\boxed{\begin{aligned} \frac{\partial \bar{V}}{\partial z} &= -(\mathcal{R} + j\omega\mathcal{L})\bar{I} \qquad (8.64a) \\ \frac{\partial \bar{I}}{\partial z} &= -(\mathcal{G} + j\omega\mathcal{C})\bar{V} \qquad (8.64b) \end{aligned}}$$

where $\bar{V}$ and $\bar{I}$ are understood to be functions of z.

Combining the two transmission-line equations (8.64a) and (8.64b) by eliminating $\bar{I}$, we obtain the wave equation

$$\begin{aligned} \frac{\partial^2 \bar{V}}{\partial z^2} &= -(\mathcal{R} + j\omega\mathcal{L})\frac{\partial \bar{I}}{\partial z} \\ &= (\mathcal{R} + j\omega\mathcal{L})(\mathcal{G} + j\omega\mathcal{C})\bar{V} \end{aligned}$$

or

$$\frac{\partial^2 \bar{V}}{\partial z^2} = \bar{\gamma}^2\bar{V} \qquad (8.65)$$

where

$$\boxed{\begin{aligned}\bar{\gamma} &= \alpha + j\beta \\ &= \sqrt{(\mathcal{R} + j\omega\mathcal{L})(\mathcal{G} + j\omega\mathcal{C})}\end{aligned}} \tag{8.66}$$

The solution for $\bar{V}(z)$ is given by

$$\bar{V}(z) = \bar{A}e^{-\bar{\gamma}z} + \bar{B}e^{\bar{\gamma}z} \tag{8.67}$$

where $\bar{A} = Ae^{j\theta}$ and $\bar{B} = Be^{j\phi}$ are arbitrary constants. It then follows that

$$\begin{aligned}V(z, t) &= \text{Re}[Ae^{j\theta}e^{-\alpha z}e^{-j\beta z}e^{j\omega t} + Be^{j\phi}e^{\alpha z}e^{j\beta z}e^{j\omega t}] \\ &= Ae^{-\alpha z}\cos(\omega t - \beta z + \theta) + Be^{\alpha z}\cos(\omega t + \beta z + \phi)\end{aligned}$$

Noting that the first and second terms on the right side correspond to waves propagating in the $+z$ and $-z$-directions, respectively, we write (8.67) as

$$\boxed{\bar{V}(z) = \bar{V}^{+}e^{-\bar{\gamma}z} + \bar{V}^{-}e^{\bar{\gamma}z}} \tag{8.68}$$

where the superscripts + and − denote (+) and (−) waves, respectively. The quantity β, which is the imaginary part of $\bar{\gamma}$ is, of course, the phase constant, that is, the rate of change of phase with z for a fixed time, for either wave. The quantity α, which is the real part of $\bar{\gamma}$, is the attentuation constant, denoting that the waves get attenuated by the factor e^{α} per unit distance as they propagate in their respective directions. Thus the quantity $\bar{\gamma}$ $(= \alpha + j\beta)$ is the propagation constant associated with the wave. We recall that the units of α are nepers per meter. Proceding further, we obtain the corresponding solution for the phasor line current by substituting (8.68) into one of the transmission-line equations. Thus using (8.64a), we obtain

$$\begin{aligned}\bar{I}(z) &= -\frac{1}{\mathcal{R} + j\omega\mathcal{L}}\frac{\partial\bar{V}}{\partial z} \\ &= -\frac{1}{\mathcal{R} + j\omega\mathcal{L}}[-\bar{\gamma}\bar{V}^{+}e^{-\bar{\gamma}z} - \bar{\gamma}\bar{V}^{-}e^{\bar{\gamma}z}]\end{aligned}$$

or

$$\boxed{\bar{I}(z) = \frac{1}{\bar{Z}_0}(\bar{V}^{+}e^{-\bar{\gamma}z} - \bar{V}^{-}e^{\bar{\gamma}z})} \tag{8.69}$$

where

$$\boxed{\bar{Z}_0 = \sqrt{\frac{\mathcal{R} + j\omega\mathcal{L}}{\mathcal{G} + j\omega\mathcal{C}}}} \tag{8.70}$$

is the characteristic impedance of the line, which is now complex.

Low-loss line

Equations (8.68) and (8.69) are the general solutions for the phasor line voltage and current, respectively, with the associated propagation constant and characteristic impedance given by (8.66) and (8.70), respectively. While it is possible to obtain explicit expressions for α and β, as well as for the real and imaginary parts of $\bar{Z}_0$ in terms of ω, $\mathcal{R}$, $\mathcal{L}$, $\mathcal{G}$, and $\mathcal{C}$, such expressions are often not meaningful since $\mathcal{R}$, $\mathcal{L}$, $\mathcal{G}$, and $\mathcal{C}$ are themselves functions of frequency. Hence in practice these quantities are obtained from experimental determination of characteristic impedance and propagation constant. However, for the special case of the low-loss line, that is, for $\omega\mathcal{L} \gg \mathcal{R}$, and $\omega\mathcal{C} \gg \mathcal{G}$, we have

$$\begin{aligned}
\bar{\gamma} &= \sqrt{j\omega\mathcal{L}\left(1 + \frac{\mathcal{R}}{j\omega\mathcal{L}}\right)j\omega\mathcal{C}\left(1 + \frac{\mathcal{G}}{j\omega\mathcal{C}}\right)} \\
&\approx j\omega\sqrt{\mathcal{L}\mathcal{C}}\sqrt{1 + \frac{\mathcal{R}}{j\omega\mathcal{L}} + \frac{\mathcal{G}}{j\omega\mathcal{C}}} \\
&\approx j\omega\sqrt{\mathcal{L}\mathcal{C}}\left[1 + \frac{1}{2}\left(\frac{\mathcal{R}}{j\omega\mathcal{L}} + \frac{\mathcal{G}}{j\omega\mathcal{C}}\right)\right] \\
&\approx \frac{1}{2}\left(\mathcal{R}\sqrt{\frac{\mathcal{C}}{\mathcal{L}}} + \mathcal{G}\sqrt{\frac{\mathcal{L}}{\mathcal{C}}}\right) + j\omega\sqrt{\mathcal{L}\mathcal{C}}
\end{aligned}$$

so that

$$\alpha \approx \frac{1}{2}\left(\mathcal{R}\sqrt{\frac{\mathcal{C}}{\mathcal{L}}} + \mathcal{G}\sqrt{\frac{\mathcal{L}}{\mathcal{C}}}\right) \tag{8.71a}$$

$$\beta = \omega\sqrt{\mathcal{L}\mathcal{C}} \tag{8.71b}$$

$$v_p = \frac{\omega}{\beta} \approx \frac{1}{\sqrt{\mathcal{L}\mathcal{C}}} \tag{8.71c}$$

Similarly,

$$\begin{aligned}
\bar{Z}_0 &= \sqrt{\frac{j\omega\mathcal{L}(1 + \mathcal{R}/j\omega\mathcal{L})}{j\omega\mathcal{C}(1 + \mathcal{G}/j\omega\mathcal{C})}} \\
&\approx \sqrt{\frac{\mathcal{L}}{\mathcal{C}}}\sqrt{\left(1 + \frac{\mathcal{R}}{j\omega\mathcal{L}}\right)\left(1 - \frac{\mathcal{G}}{j\omega\mathcal{C}}\right)} \\
&\approx \sqrt{\frac{\mathcal{L}}{\mathcal{C}}}\sqrt{\left(1 + \frac{\mathcal{R}}{j\omega\mathcal{L}} - \frac{\mathcal{G}}{j\omega\mathcal{C}}\right)} \\
&\approx \sqrt{\frac{\mathcal{L}}{\mathcal{C}}}\left[1 + \frac{1}{2}\left(\frac{\mathcal{R}}{j\omega\mathcal{L}} - \frac{\mathcal{G}}{j\omega\mathcal{C}}\right)\right] \\
&\approx \sqrt{\frac{\mathcal{L}}{\mathcal{C}}}
\end{aligned} \tag{8.71d}$$

Thus for the low-loss line, the expressions for β and $\bar{Z}_0$ are essentially the same as those for a lossless line. Note that the low-loss conditions $\omega\mathcal{L} \gg \mathcal{R}$ and $\omega\mathcal{C} \gg \mathcal{G}$ are valid for very high frequencies or for very small values of $\mathcal{R}$ and $\mathcal{G}$ at lower frequencies.

Experimental determination of $\bar{Z}_0$ and $\bar{\gamma}$

As already pointed out, for the general case, it is more convenient to determine experimentally the values of $\bar{Z}_0$ and $\bar{\gamma}$ than it is to compute them analytically. The experimental technique is based upon the measurements of the input impedance of the line for two values of load impedance. To obtain the expression for the input impedance, we first write the general solutions for the phasor line voltage and current given by (8.68) and (8.69), respectively, in terms of the distance variable d, measured from the load toward the generator, as opposed to z, measured from the generator toward the load. Thus we have

$$\bar{V}(d) = \bar{V}^+ e^{\bar{\gamma}d} + \bar{V}^- e^{-\bar{\gamma}d} \tag{8.72a}$$

$$\bar{I}(d) = \frac{1}{\bar{Z}_0}(\bar{V}^+ e^{\bar{\gamma}d} - \bar{V}^- e^{-\bar{\gamma}d}) \tag{8.72b}$$

or

$$\bar{V}(d) = \bar{V}^+ e^{\bar{\gamma}d}[1 + \bar{\Gamma}(d)] \tag{8.73a}$$

$$\bar{I}(d) = \frac{\bar{V}^+}{\bar{Z}_0} e^{\bar{\gamma}d}[1 - \bar{\Gamma}(d)] \tag{8.73b}$$

where

$$\begin{aligned}\bar{\Gamma}(d) &= \frac{\bar{V}^-(d)}{\bar{V}^+(d)} = \frac{\bar{V}^- e^{-\bar{\gamma}d}}{\bar{V}^+ e^{\bar{\gamma}d}} \\ &= \bar{\Gamma}_R e^{-2\bar{\gamma}d} = \bar{\Gamma}_R e^{-2\alpha d} e^{-j2\beta d}\end{aligned} \tag{8.74}$$

is the voltage reflection coefficient at any value of d, and $\bar{\Gamma}_R$ is the voltage reflection coefficient at the load.

The line impedance is given by

$$\begin{aligned}\bar{Z}(d) &= \frac{\bar{V}(d)}{\bar{I}(d)} = \bar{Z}_0 \frac{1 + \bar{\Gamma}(d)}{1 - \bar{\Gamma}(d)} \\ &= \bar{Z}_0 \frac{1 + \bar{\Gamma}_R e^{-2\bar{\gamma}d}}{1 - \bar{\Gamma}_R e^{-2\bar{\gamma}d}}\end{aligned} \tag{8.75}$$

The input impedance of a line of length l terminated by a load impedance $\bar{Z}_R$, as shown in Fig. 8.31, is then given in terms of $\bar{Z}_R$ by

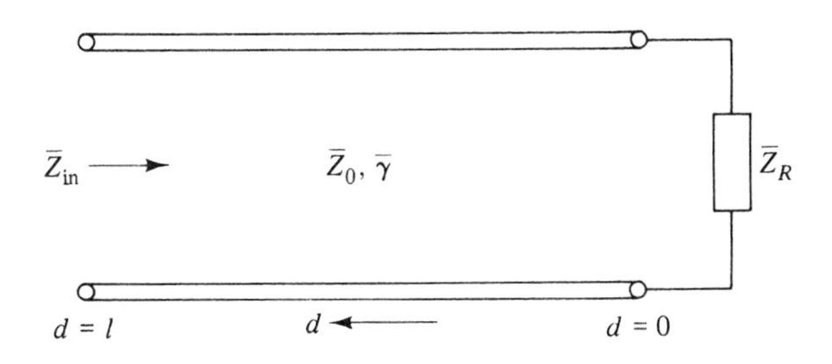

Figure 8.31. Lossy line of length l terminated by $\bar{Z}_R$.

$$\bar{Z}_{in} = \bar{Z}(l) = \bar{Z}_0 \frac{1 + \bar{\Gamma}_R e^{-2\bar{\gamma}l}}{1 - \bar{\Gamma}_R e^{-2\bar{\gamma}l}}$$

$$= \bar{Z}_0 \frac{1 + [(\bar{Z}_R - \bar{Z}_0)/(\bar{Z}_R + \bar{Z}_0)]e^{-2\bar{\gamma}l}}{1 - [(\bar{Z}_R - \bar{Z}_0)/(\bar{Z}_R + \bar{Z}_0)]e^{-2\bar{\gamma}l}}$$

$$= \bar{Z}_0 \frac{(\bar{Z}_R + \bar{Z}_0) + (\bar{Z}_R - \bar{Z}_0)e^{-2\bar{\gamma}l}}{(\bar{Z}_R + \bar{Z}_0) - (\bar{Z}_R - \bar{Z}_0)e^{-2\bar{\gamma}l}}$$

$$= Z_0 \frac{\bar{Z}_R \cosh \bar{\gamma}l + \bar{Z}_0 \sinh \bar{\gamma}l}{\bar{Z}_R \sinh \bar{\gamma}l + \bar{Z}_0 \cosh \bar{\gamma}l}$$

or

$$\boxed{\bar{Z}_{in} = \bar{Z}_0 \frac{\bar{Z}_R + \bar{Z}_0 \tanh \bar{\gamma}l}{\bar{Z}_R \tanh \bar{\gamma}l + \bar{Z}_0}} \tag{8.76}$$

Let us now consider two values of $\bar{Z}_R$; in particular, $\bar{Z}_R = 0$ and $\bar{Z}_R = \infty$, corresponding to short circuit and open circuit, respectively. Then denoting the corresponding input impedances to be $\bar{Z}^s_{in}$ and $\bar{Z}^o_{in}$, respectively, we have from (8.76),

$$\boxed{\bar{Z}^s_{in} = \bar{Z}_0 \tanh \bar{\gamma}l} \tag{8.77a}$$

$$\boxed{\bar{Z}^o_{in} = \bar{Z}_0 \coth \bar{\gamma}l} \tag{8.77b}$$

from which we obtain

$$\boxed{\bar{Z}_0 = \sqrt{\bar{Z}^s_{in}\bar{Z}^o_{in}}} \tag{8.78}$$

and

$$\boxed{\tanh \bar{\gamma}l = \sqrt{\frac{\bar{Z}^s_{in}}{\bar{Z}^o_{in}}}} \tag{8.79}$$

To illustrate the computation of $\bar{Z}_0$ and $\bar{\gamma}$ by means of a numerical example, let us assume that at a certain frequency, measurements indicated

$$\bar{Z}^s_{in} = (30 - j40)\ \Omega$$

$$\bar{Z}^o_{in} = (30 + j40)\ \Omega$$

Then from (8.78)

$$\bar{Z}_0 = \sqrt{(30 - j40)(30 + j40)} = 50\ \Omega$$

From (8.79),

$$\tanh \bar{\gamma}l = \sqrt{\frac{30 - j40}{30 + j40}} = \sqrt{\frac{50\angle -53.13^\circ}{50\angle 53.13^\circ}}$$

$$= 1\angle -53.13^\circ = 0.6 - j0.8$$

$$\bar{\gamma}l = \tanh^{-1}(0.6 - j0.8)$$

Using the identity

$$\tanh^{-1} x = \frac{1}{2} \ln \frac{1 + x}{1 - x}$$

we then have

$$\bar{\gamma}l = \frac{1}{2} \ln \frac{1.6 - j0.8}{0.4 + j0.8} = \frac{1}{2} \ln \frac{1.789 \underline{/-26.565^\circ}}{0.894 \underline{/63.435^\circ}}$$

$$= \frac{1}{2} \ln 2\underline{/-90^\circ} = \frac{1}{2} \ln [2e^{j(2n\pi - \pi/2)}]$$

$$= \frac{1}{2}[\ln 2 + j(2n\pi - \pi/2)]$$

$$= 0.3466 + j(n\pi - \pi/4), \qquad n = 0, 1, 2, \ldots$$

Thus

$$\alpha l = 0.3466$$

$$\alpha = 0.3466/l$$

whereas

$$\beta l = n\pi - \pi/4, \qquad n = 1, 2, \ldots$$

where $n = 0$ is ruled out since it gives negative value for β. Note that β can only be determined to within $n\pi$. However, if the approximate value of β is known, then the correct value of n and hence of β can be determined.

In practice, since a perfect open-circuited termination can often be difficult to achieve, it may be desirable to consider the second value of $\bar{Z}_R$ to be arbitrary instead of being equal to ∞. Denoting the corresponding input impedance to be $\bar{Z}_{in}$, we then have from (8.76) and (8.77a)

$$\bar{Z}_{in} = \bar{Z}_0^2 \frac{\bar{Z}_R + \bar{Z}_{in}^s}{\bar{Z}_R \bar{Z}_{in}^s + \bar{Z}_0^2} \tag{8.80}$$

and hence

$$\bar{Z}_0 = \sqrt{\frac{\bar{Z}_R \bar{Z}_{in}^s \bar{Z}_{in}}{\bar{Z}_R + \bar{Z}_{in}^s - \bar{Z}_{in}}} \tag{8.81}$$

Knowing the value of $\bar{Z}_0$ from (8.81), we can then compute the value of $\bar{\gamma}$ by using (8.77a).

Power flow

We shall conclude this section with a discussion of power flow down the line. From (8.73a) and (8.73b), the time-average power flow down the line is given by

$$\langle P \rangle = \tfrac{1}{2} \text{Re}[\bar{V}(d)\bar{I}^*(d)]$$

$$= \tfrac{1}{2} \text{Re}\left\{\bar{V}^+ e^{\bar{\gamma}d}[1 + \bar{\Gamma}(d)] \frac{(\bar{V}^+)^*}{\bar{Z}_0^*} e^{\bar{\gamma}^* d}[1 - \bar{\Gamma}^*(d)]\right\}$$

$$= \tfrac{1}{2} \text{Re}\left\{\frac{|\bar{V}^+|^2}{\bar{Z}_0^*} e^{2\alpha d}[1 - |\bar{\Gamma}(d)|^2 + \bar{\Gamma}(d) - \bar{\Gamma}^*(d)]\right\}$$

$$= \tfrac{1}{2} \text{Re}\{|\bar{V}^+|^2 \bar{Y}_0^* e^{2\alpha d}[1 - |\bar{\Gamma}(d)|^2 + j2 \,\text{Im}\, \bar{\Gamma}(d)]\}$$

or

$$\boxed{<P> = \tfrac{1}{2}|\bar{V}^+|^2 e^{2\alpha d}\{G_0[1 - |\bar{\Gamma}(d)|^2] + 2B_0 \operatorname{Im} \bar{\Gamma}(d)\}} \tag{8.82}$$

where

$$\bar{Y}_0 = \frac{1}{\bar{Z}_0} = G_0 + jB_0$$

is the characteristic admittance of the line. For a given source voltage and impedance, we can compute the value of $|\bar{V}^+|$ from line impedance and power flow considerations at the input end of the line and use that value for further computations. We shall illustrate this by means of an example.

Example 8.8

Let us consider the low-loss line system shown in Fig. 8.32, and compute the time-average power delivered to the input of the line, the time-average power delivered to the load, and the time-average power dissipated in the line.

We proceed with the solution in a step-by-step manner as follows:

(a) The reflection coefficient at the load end is given by

$$\bar{\Gamma}_R = \frac{\bar{Z}_R - \bar{Z}_0}{\bar{Z}_R + \bar{Z}_0} = \frac{150 - 50}{150 + 50} = 0.5$$

(b) Noting that α is specified in nepers per wavelength, we obtain the reflection coefficient at the input end $d = l$ as

$$\begin{aligned}\bar{\Gamma}(l) &= \bar{\Gamma}_R e^{-2\bar{\gamma}l} = \bar{\Gamma}_R e^{-2\alpha l} e^{-j2\beta l} \\ &= 0.5e^{-0.204} e^{-j40.8\pi} \\ &= 0.4077\underline{/-144^\circ}\end{aligned}$$

(c) The input impedance of the line is given by

$$\begin{aligned}\bar{Z}_{\text{in}} = \bar{Z}(l) &= Z_0 \frac{1 + \bar{\Gamma}(l)}{1 - \bar{\Gamma}(l)} \\ &= 50\,\frac{1 + 0.4077\underline{/-144^\circ}}{1 - 0.4077\underline{/-144^\circ}} = 50\,\frac{1 + (-0.33 - j0.24)}{1 - (-0.33 - j0.24)}\end{aligned}$$

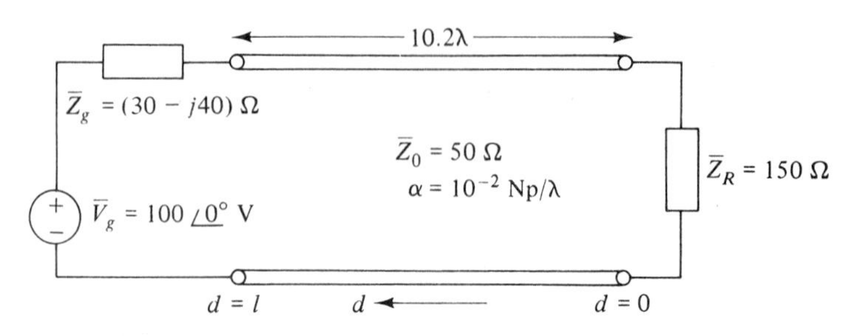

Figure 8.32. Lossy transmission-line system for illustrating the computation of power flow at the two ends of the line and the power dissipated in the line.

$$= 50\,\frac{0.7117\angle{-19.708^\circ}}{1.3515\angle{10.229^\circ}} = 26.33\angle{-29.937^\circ}$$

$$= (22.817 - j13.140)\ \Omega$$

(d) The current $\bar{I}_g = \bar{I}(l)$ drawn from the voltage generator can be obtained as

$$\bar{I}(l) = \frac{\bar{V}_g}{\bar{Z}_g + \bar{Z}_{\text{in}}} = \frac{100\angle{0^\circ}}{(30 - j40) + (22.817 - j13.140)}$$

$$= \frac{100\angle{0^\circ}}{52.817 - j53.140} = \frac{100\angle{0^\circ}}{74.923\angle{-45.175^\circ}}$$

$$= 1.3347\angle{45.175^\circ}\ \text{A}$$

(e) The voltage at the input end of the line is given by

$$\bar{V}(l) = \bar{Z}_{\text{in}}\bar{I}(l)$$

$$= 26.33\angle{-29.937^\circ} \times 1.3347\angle{45.175^\circ}$$

$$= 35.143\angle{15.238^\circ}\ \text{V}$$

(f) The time-average power flow at the input end of the line is given by

$$\langle P(l)\rangle = \tfrac{1}{2}\,\text{Re}[\bar{V}(l)\bar{I}^*(l)]$$

$$= \tfrac{1}{2}\,\text{Re}[35.143\angle{15.238^\circ} \times 1.3347\angle{-45.175^\circ}]$$

$$= \tfrac{1}{2} \times 35.143 \times 1.3347 \times \cos 29.937^\circ$$

$$= 20.32\ \text{W}$$

(g) Noting that $B_0 = 0$, we then obtain the value of $|\bar{V}^+|$ by applying (8.82) to $d = l$. Thus

$$|\bar{V}^+| = \sqrt{\frac{2\,\langle P(l)\rangle\, e^{-2\alpha l}}{G_0[1 - |\bar{\Gamma}(l)|^2]}}$$

$$= \sqrt{\frac{2 \times 20.32 \times e^{-0.204}}{0.02(1 - 0.4077^2)}}$$

$$= 44.58\ \text{V}$$

(h) The time-average power delivered to the load is then given by

$$\langle P(0)\rangle = \tfrac{1}{2}|\bar{V}^+|^2\, G_0(1 - |\bar{\Gamma}_R|^2)$$

$$= \tfrac{1}{2} \times 44.58^2 \times 0.02(1 - 0.25)$$

$$= 14.91\ \text{W}$$

(i) Finally, the time-average power dissipated in the line is

$$\langle P_d\rangle = \langle P(l)\rangle - \langle P(0)\rangle$$

$$= (20.32 - 14.91)$$

$$= 5.41\ \text{W}$$

K8.6. Distributed equivalent circuit; Transmission-line equations; Complex propagation constant; Complex characteristic impedance; Input impedance; $\bar{Z}_0$ and $\bar{\gamma}$ from input impedance considerations; Power flow; Power dissipation.

D8.14. For a lossy line of length $l = 16.3\lambda$ and characterized by $Z_0 = 60\ \Omega$ and $\alpha = 0.02$ Np/λ, find the input impedance for each of the following values of $\bar{Z}_R$: **(a)** $\bar{Z}_R = 0$; **(b)** $\bar{Z}_R = \infty$; and **(c)** $\bar{Z}_R = (36 + j0)\ \Omega$.
Ans. **(a)** $(102.04 - j85.77)\ \Omega$; **(b)** $(20.67 + j17.38)\ \Omega$; **(c)** $(73.17 - j11.39)\ \Omega$

D8.15. A lossy line of length $l = 10\lambda$ and characterized by $\bar{Z}_0 = 100\ \Omega$ and $\alpha = 10^{-2}$ Np/λ is terminated by a load impedance $\bar{Z}_R$. If a power of 10 W is to be delivered to the load, determine how much power should be delivered to the input terminals of the line for each of the following values of $\bar{Z}_R$: **(a)** $\bar{Z}_R = 100\ \Omega$; **(b)** $\bar{Z}_R = 20\ \Omega$; and **(c)** $\bar{Z}_R = 300\ \Omega$.
Ans. **(a)** 12.214 W; **(b)** 15.436 W; **(c)** 13.556 W

8.7 SUMMARY

In this chapter we began our study of sinusoidal steady-state analysis of lossless transmission lines by expressing the general solutions for the phasor line voltage and line current in terms of the distance variable d measured from the load toward the source. These solutions are

$$\bar{V}(d) = \bar{V}^+ e^{j\beta d} + \bar{V}^- e^{-j\beta d}$$

$$\bar{I}(d) = \frac{1}{Z_0}(\bar{V}^+ e^{j\beta d} - \bar{V}^- e^{-j\beta d})$$

By applying these general solutions to the case of a line short circuited at the far end and obtaining the particular solutions for that case, we discussed the standing wave phenomenon resulting from the complete reflection of waves by the short circuit. We introduced the concept of a standing wave pattern and discussed the phenomenon of natural oscillations. We examined the frequency behavior of the input impedance of a short-circuited line of length l, given by

$$\bar{Z}_{\text{in}} = jZ_0 \tan \beta l$$

and illustrated (1) its application in a technique for locating short circuit in a line and (2) the computation of resonant frequencies for a system formed by connecting together short-circuited line sections.

Next we considered the general case of a line terminated by an arbitrary load $\bar{Z}_R$ and introduced the concept of the generalized voltage reflection coefficient, as the ratio of the phasor reflected wave voltage at any value of d to the phasor incident wave voltage at that value of d. It is given by

$$\bar{\Gamma}(d) = \bar{\Gamma}_R e^{-j2\beta d}$$

where

$$\bar{\Gamma}_R = |\bar{\Gamma}_R|e^{j\theta} = \frac{\bar{Z}_R - Z_0}{\bar{Z}_R + Z_0}$$

is the voltage reflection coefficient at the load. We then expressed the solutions for the line voltage and line current in terms of $\bar{\Gamma}(d)$ and discussed the construction of standing wave patterns from the solutions. We learned that together with the property that distance between successive voltage minima of the standing wave patterns is $\lambda/2$, the quantities

$$\text{SWR} = \frac{1 + |\bar{\Gamma}_R|}{1 - |\bar{\Gamma}_R|}$$

and

$$d_{\min} = \frac{\lambda}{4\pi}(\theta + \pi)$$

constitute an important set of parameters associated with the standing waves. The SWR, which is the ratio of the maximum voltage amplitude to the minimum voltage amplitude in the standing wave pattern, and $d_{\min}$, which is the distance of the first voltage minimum of the standing wave pattern from the load, are easily measurable quantities. We then defined the ratio of the complex line voltage to the complex line current at a given value of d to be the line impedance $\bar{Z}(d)$, given by

$$\bar{Z}(d) = Z_0 \frac{1 + \bar{\Gamma}(d)}{1 - \bar{\Gamma}(d)}$$

and discussed its several properties as well as the computation of power flow along the line from considerations of input impedance of the line.

We then turned our attention to the topic of transmission-line matching, which consists of eliminating standing waves by connecting a matching device near the load such that the line views an effective impedance equal to its own characteristic impedance, on the generator side of the matching device. We discussed the need for matching and three techniques of matching: (1) quarter-wave transformer, (2) single stub, and (3) double stub. The quarter-wave transformer technique is based on a property of the line impedance that

$$[\bar{Z}(d)]\left[\bar{Z}\left(d + \frac{\lambda}{4}\right)\right] = Z_0^2$$

whereas the stub-matching techniques make use of the property that the input impedance of a lossless line short circuited (or open circuited) at the far end is purely reactive. We also discussed the departure of SWR from unity as the frequency is varied from that at which the match is achieved and illustrated a procedure for computation of the SWR versus frequency.

Next we introduced the Smith chart, a popular graphical aid in the solution of transmission-line problems. We learned that the Smith chart is based upon the transformation from the $\bar{z}$-plane to the $\bar{\Gamma}$-plane in accordance with the relationship

$$\bar{\Gamma}(d) = \frac{\bar{z}(d) - 1}{\bar{z}(d) + 1}$$

where

$$\bar{z}(d) = \frac{\bar{Z}(d)}{Z_0}$$

is the normalized line impedance. We discussed the construction of the Smith chart, some basic procedures, and the solution of transmission-line matching problems. We also discussed a useful property associated with the transformation of the reflection coefficient across a discontinuity and illustrated its application by means of an example.

Finally, we extended our analysis of lossless lines briefly to lossy lines, with the discussion of (1) the distributed equivalent circuit, (2) computation of characteristic impedance and propagation constant from input impedance measurements, and (3) computation of power flow at the generator and load ends of the line and power dissipated in the line.

REVIEW QUESTIONS

R8.1. Discuss the general solutions for the line voltage and line current in terms of the distance variable d in the sinusoidal steady state

R8.2. State the boundary condition at a short circuit on a line. For an open-circuited line, what is the boundary condition to be satisfied at the open circuit?

R8.3. What is a standing wave? How do complete standing waves arise? Discuss their characteristics.

R8.4. What is a standing wave pattern? Discuss the voltage and current standing wave patterns for a short-circuited line.

R8.5. Explain the phenomenon of natural oscillations and the determination of natural frequencies of oscillation by means of an example.

R8.6. Discuss the variation with frequency of the input reactance of a short-circuited line and its application in the determination of the location of a short circuit.

R8.7. Outline the method of computation of resonant frequencies of a system formed by connecting together two short-circuited line sections.

R8.8. How is the generalized voltage reflection coefficient defined? Discuss its variation along the line.

R8.9. Discuss the sketching of standing wave patterns for line voltage and current on a line terminated by an arbitrary load.

R8.10. Define standing wave ratio (SWR). What are the standing wave ratios for **(a)** a semi-infinitely long line; **(b)** a short-circuited line; **(c)** an open-circuited line; and **(d)** a line terminated by its characteristic impedance?

R8.11. Discuss the slotted line technique for performing standing wave measurements on a line and the determination of an unknown load impedance from the standing wave measurements.

R8.12. How is line impedance defined? Summarize the several properties of the line impedance.

R8.13. Outline the procedure for the determination of time-average power flow down a line from input impedance considerations.

R8.14. Define normalized line impedance and normalized line admittance. How are they related to the voltage reflection coefficient?

R8.15. Discuss the reasons for transmission-line matching and the principle behind matching.

R8.16. Which property of line impedance forms the basis for the quarter-wave transformer (QWT) technique of transmission-line matching? Outline the solution for the QWT matching problem.

R8.17. What is a stub? Outline the solution for the single-stub matching problem.

R8.18. Outline the solution for the double-stub matching problem.

R8.19. Discuss the "bandwidth" associated with a transmission-line matched system and the procedure for obtaining the SWR in the main line versus frequency.

R8.20. What is the basis behind the construction of the Smith chart? Briefly discuss the mapping of the normalized line impedances on to the $\bar{\Gamma}$-plane.

R8.21. Why is a circle with its center at the center of the Smith chart known as a constant SWR circle? Where on the circle is the corresponding SWR value marked?

R8.22. Using the Smith chart, how do you find the normalized line admittance at a point on the line, given the normalized line impedance at that point?

R8.23. Briefly describe the solution to the single-stub matching problem by using the Smith chart.

R8.24. Briefly describe the solution to the double-stub matching problem by using the Smith chart.

R8.25. Discuss the forbidden region of $\bar{y}'_1$ for possible match associated with the double-stub matching technique.

R8.26. Discuss the transformation of the reflection coefficient from one side of a transmission-line discontinuity to the other side of the discontinuity and an application of the property associated with this transformation.

R8.27. Discuss the modification of the distributed equivalent circuit for the lossless line case to the lossy line case.

R8.28. What are the conditions under which a lossy line can be classified as a low-loss line? Compare the propagation parameters of the low-loss line with those for the lossless line.

R8.29. Discuss the computation of $\bar{Z}_0$ and $\bar{\gamma}$ for a lossy line from a knowledge of the input impedances of the line with short-circuit and open-circuit terminations.

R8.30. Briefly outline the procedure for the computation of time-average power flow at the input and the load ends of a lossy line and hence the time-average power dissipated in the line.

PROBLEMS

P8.1. For a line open circuited at the far end, as shown in Fig. 8.33, obtain the solutions for the phasor line voltage and current and sketch the voltage and current standing wave patterns, as in Fig. 8.4.

P8.2. In the system shown in Fig. 8.34, $V_g(t) = V_0 \cos 2\pi f_0 t$. Find and sketch the standing wave patterns for the line voltage and line current, indicating the values of the maximum voltage and current amplitudes for each of the cases **(a)** $l = \lambda/4$ and **(b)** $l = \lambda/2$.

P8.3. In the system shown in Fig. 8.35, the source voltage is

$$V_g(t) = V_0 \cos^3 2\pi f_0 t$$

and $l = \lambda/4$ at $f = f_0$. Find the root-mean-square (rms) values of the line voltage and line current at values of d/l equal to 0, 1/3, 1/2, and 1. (*Note:* The rms value of the sum of the voltages of two harmonically related frequencies is equal to the square root of the sum of the squares of the rms values of the individual voltages.)

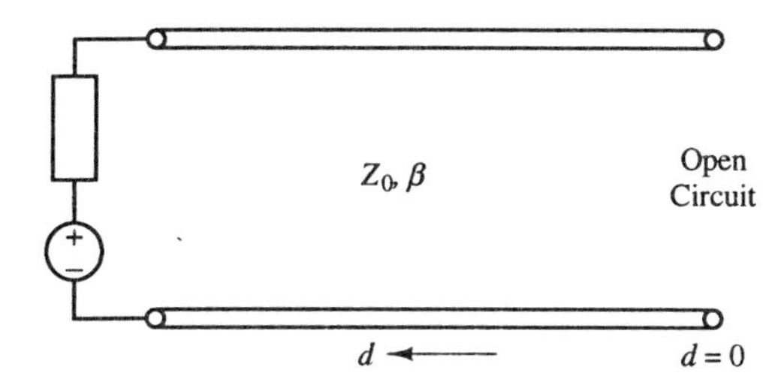

Figure 8.33. For Problem P8.1.

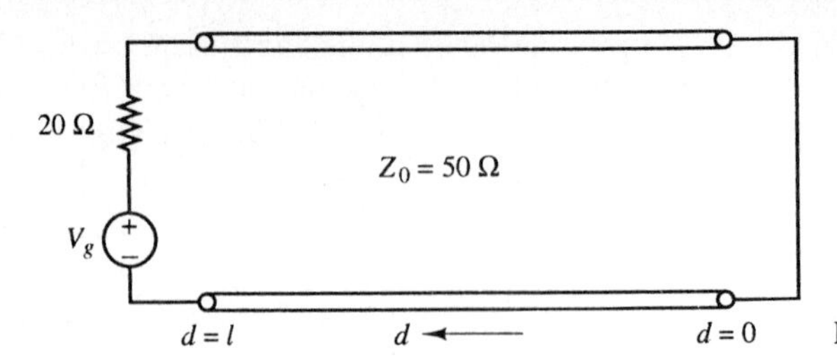

Figure 8.34. For Problem P8.2.

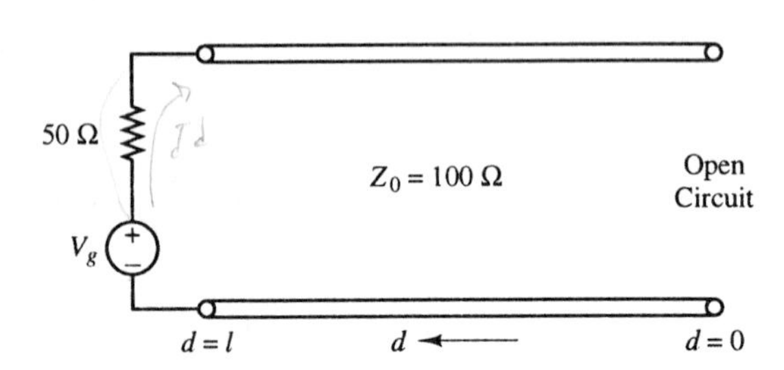

Figure 8.35. For Problem P8.3.

P8.4. In the system shown in Fig. 8.36(a), the source voltage is periodic as shown in Fig. 8.36(b). Find the reading of ammeter A, if it reads root-mean-square values.

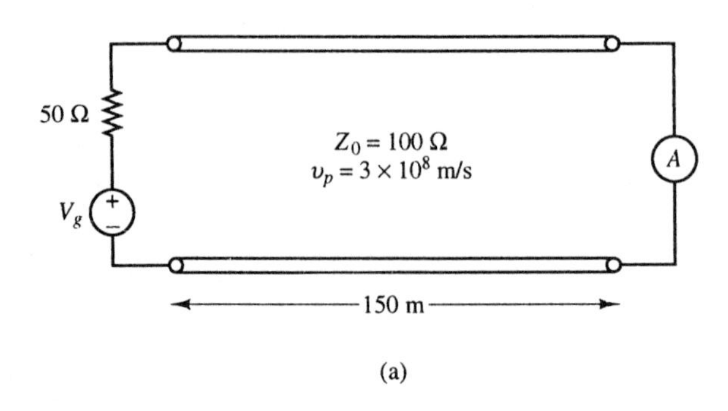

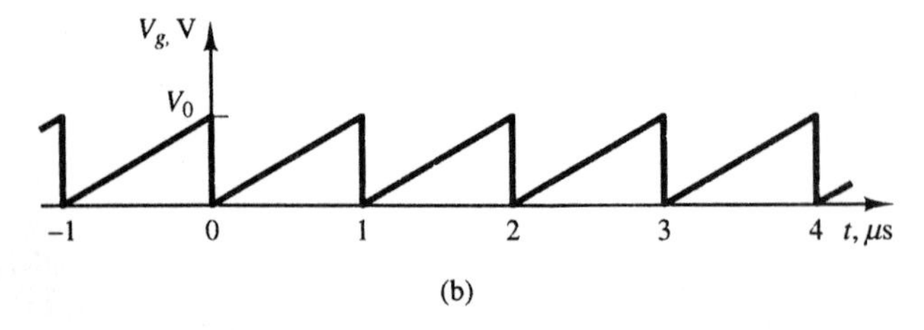

Figure 8.36. For Problem P8.4.

P8.5. In the arrangement shown in Fig. 8.37, a dielectric slab is sandwiched between two parallel, perfect conductors. For uniform plane waves bouncing back and forth normal to the conductors, find the natural frequencies of oscillation.

P8.6. A ring transmission line is formed as shown in Fig. 8.38(a) by connecting the ends a and a' of the conductors of a line of length l [shown in Fig. 8.38(b)] to the ends b and b', respectively, of the same conductors. Find the natural frequencies of oscillation of the system.

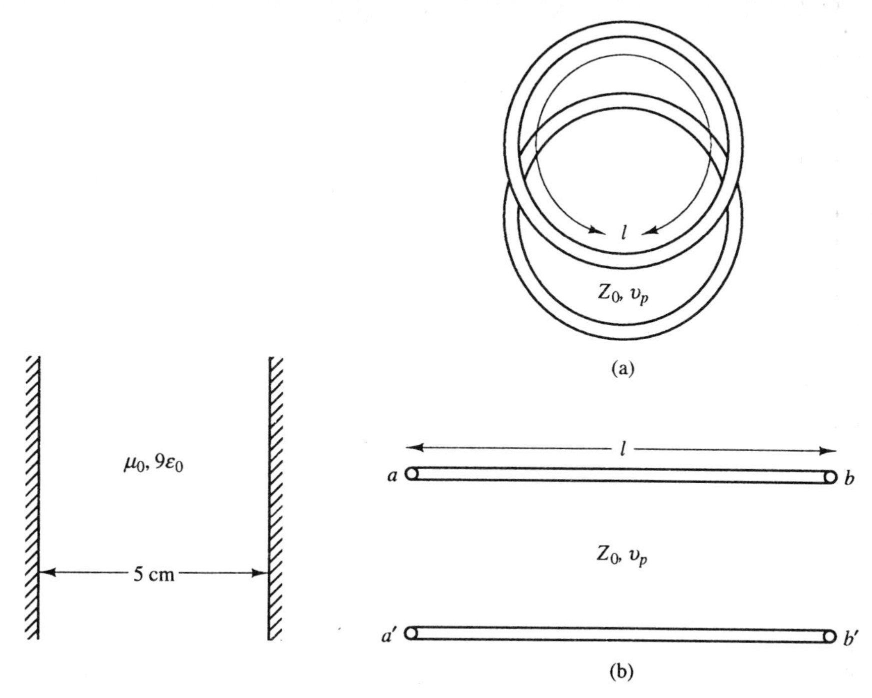

Figure 8.37. For Problem P8.5.

Figure 8.38. For Problem P8.6.

P8.7. Show that for $f \ll v_p/2\pi l$ the input impedance of a short-circuited line of length l and phase velocity v_p is essentially that of a single inductor of value $\mathcal{L}l$, where $\mathcal{L}$ is the inductance per unit length of the line. Assuming that the criterion $f \ll v_p/2\pi l$ is satisfied for frequencies $f \leq 0.1v_p/2\pi l$, compute the maximum length of an air-dielectric short-circuited line for which the input impedance is approximately that of an inductance equal to the total inductance of the line for $f = 100$ MHz.

P8.8. In the example involving the location of a short circuit in a line, solve for the distance of the short circuit from the generator by considering the standing wave patterns for the two frequencies of interest and deducing the number of wavelengths at one of the two frequencies.

P8.9. An air-dielectric transmission line of characteristic impedance 100 Ω and length 15 cm is short circuited at one end and terminated by a capacitor of value 5 pF at the other end. Find the three lowest resonant frequencies of the system.

P8.10. The arrangement shown in Fig. 8.39 is that of a parallel-plate resonator made up of two dielectric slabs sandwiched between perfect conductors and in which uniform plane waves bounce back and forth normal to the conductors. **(a)** Show that the resonant frequencies of the system are given by the roots of the characteristic equation

$$\tan \omega\sqrt{\mu_0\varepsilon_1}t + \sqrt{\frac{\varepsilon_1}{\varepsilon_2}} \tan \omega\sqrt{\mu_0\varepsilon_2}(l - t) = 0$$

(b) Find the five lowest resonant frequencies if $t = l/2$, $l = 5$ cm, $\varepsilon_1 = 4\varepsilon_0$, and $\varepsilon_2 = 16\varepsilon_0$.

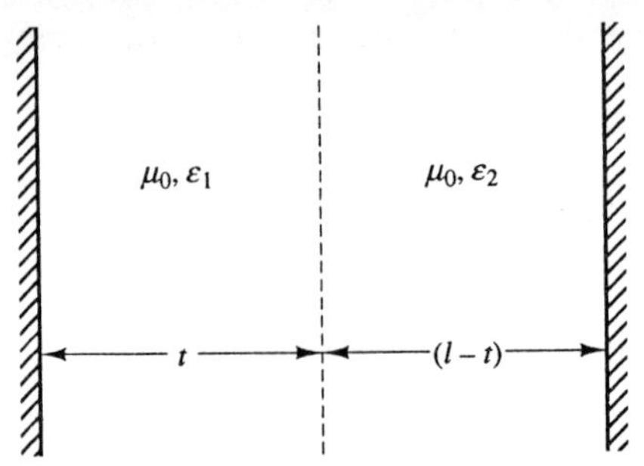

Figure 8.39. For Problem P8.10.

P8.11. For a line of characteristic impedance Z_0 terminated by a purely reactive load jX, show that the SWR is equal to infinity and the value of d_{min} is $(\lambda/2\pi)[\pi - \tan^{-1}(X/Z_0)]$ for $X > 0$ and $(\lambda/2\pi)\tan^{-1}(|X|/Z_0)$ for $X < 0$.

P8.12. In the system shown in Fig. 8.40, a line of characteristic impedance 50 Ω is terminated by a series R, L, C circuit having the values $R = 50\ \Omega$, $L = 1\ \mu\text{H}$, and $C = 100$ pF. **(a)** Find the source frequency f_0 for which there are no standing waves on the line. **(b)** Find the source frequencies f_1 and f_2 on either side of f_0 for which the SWR on the line is 2.0.

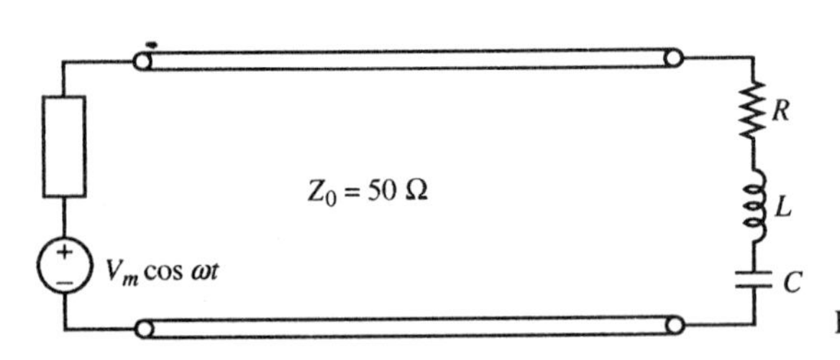

Figure 8.40. For Problem P8.12.

P8.13. A slotted coaxial line of characteristic impedance 50 Ω was used to measure an unknown load impedance. First, the receiving end of the line was short circuited. The voltage minima were found to be 0.6 m apart. One of the minima was marked as the reference point. Next the unknown impedance was connected to the receiving end of the line. The SWR was found to be 5.0 and a voltage minimum was found to be 0.1 m from the reference point toward the load. Find the value of the unknown load impedance.

P8.14. In the system shown in Fig. 8.41, assume uniform plane waves of frequency f incident normally onto the interface from medium 1. **(a)** Find the SWR in medium 1 for $f = 10^9$ Hz, if $l = 2.5$ cm. **(b)** Find the three lowest values of f for which complete transmission occurs if $l = 2.5$ cm. **(c)** Find the three lowest values of l for which complete transmission occurs for $f = 10^9$ Hz.

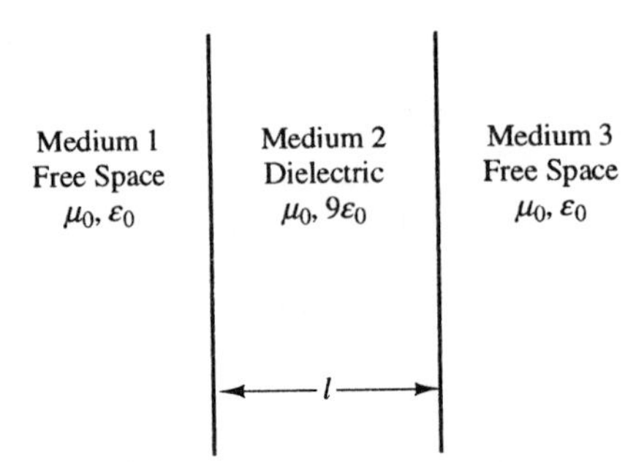

Figure 8.41. For Problem P8.14.

P8.15. For the system shown in Fig. 8.42, find the input impedance of the line and the time-average power delivered to the load.

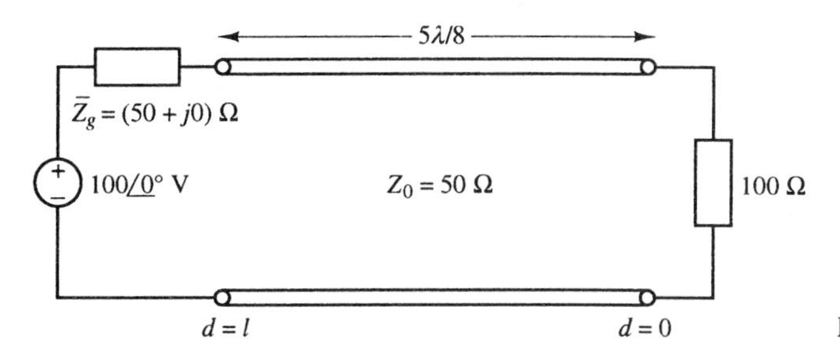

Figure 8.42. For Problem P8.15.

P8.16. In the system shown in Fig. 8.43, find **(a)** the value of the load impedance that enables maximum power transfer from the generator to the load and **(b)** the power transferred to the load for the value found in (a). (*Hint:* Apply maximum power transfer theorem at $d = l$.)

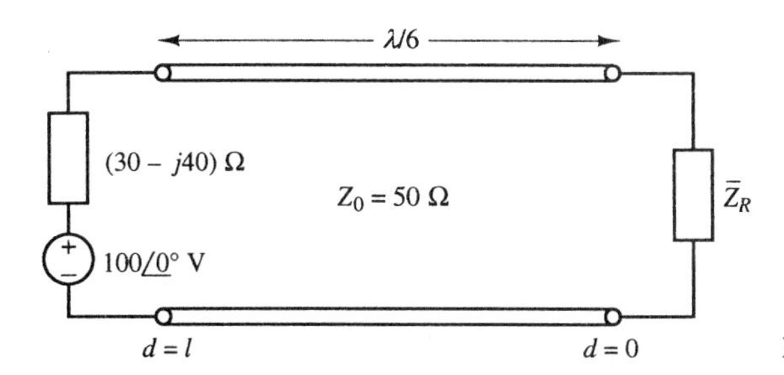

Figure 8.43. For Problem P8.16.

P8.17. The ring transmission line of Fig. 8.38(a) is excited by connecting a voltage source $V_g(t) = V_0 \cos \omega t$ across the conductors at some location on the line [such as aa' (or bb') in Fig. 8.38(b)]. Find the impedance viewed by the voltage source.

P8.18. In the system shown in Fig. 8.44, find the values of the reactance X and the characteristic impedance Z_{02} of line 2 for which the power delivered to the load $\bar{Z}_R$ is a maximum.

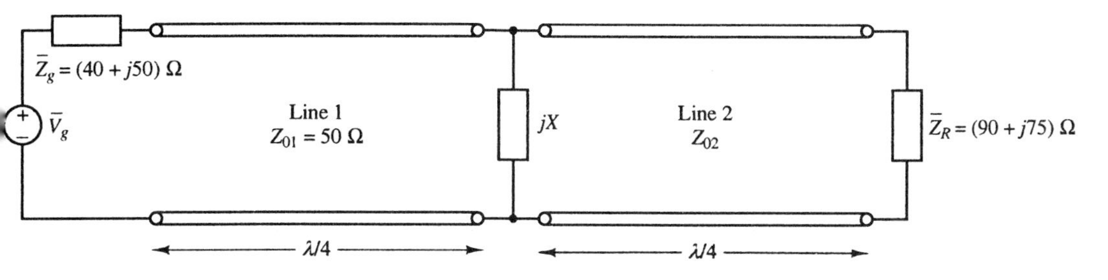

Figure 8.44. For Problem P8.18.

P8.19. In the arrangement shown in Fig. 8.45, a quarter-wave dielectric coating is employed to eliminate reflections of uniform plane waves of frequency 2500 MHz incident normally from free space onto a dielectric of permittivity $9\varepsilon_0$. Assuming $\mu = \mu_0$, find the thickness in centimeters and the permittivity of the dielectric coating.

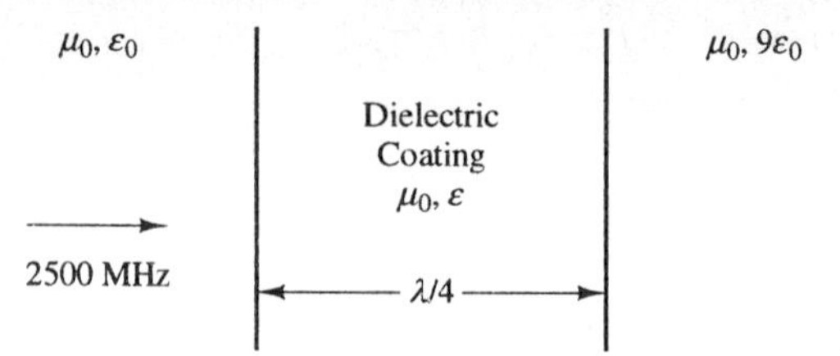

Figure 8.45. For Problem P8.19.

P8.20. In the system shown in Fig. 8.46, the $\lambda/4$ section of characteristic impedance 50 Ω is used to minimize the SWR to the left of the section. Find analytically the minimum value of d_1 that minimizes the SWR and the minimum value of the SWR.

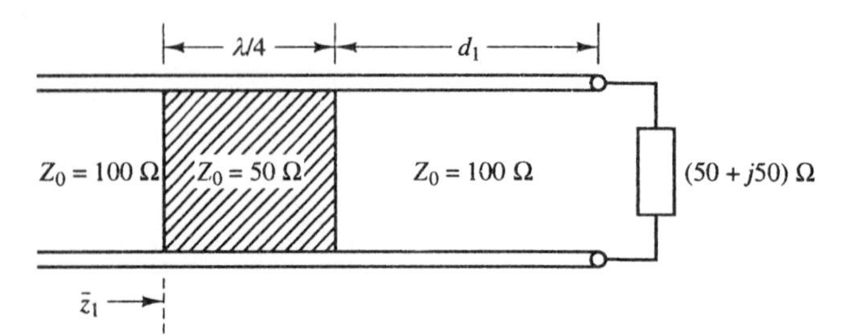

Figure 8.46. For Problem P8.20.

P8.21. Figure 8.47 shows an arrangement, known as the alternated-line transformer, for achieving a matched interconnection between two lines of different characteristic impedances Z_{01} and Z_{02}. It consists of two sections of the same characteristic impedances as those of the lines to be matched but alternated, as shown in the figure. The electrical lengths of the two sections are equal. Show that to achieve a match, the required electrical length of each section is

$$\frac{l}{\lambda} = \frac{1}{2\pi}\tan^{-1}\sqrt{\frac{n}{n^2 + n + 1}}$$

where $n = Z_{02}/Z_{01}$.

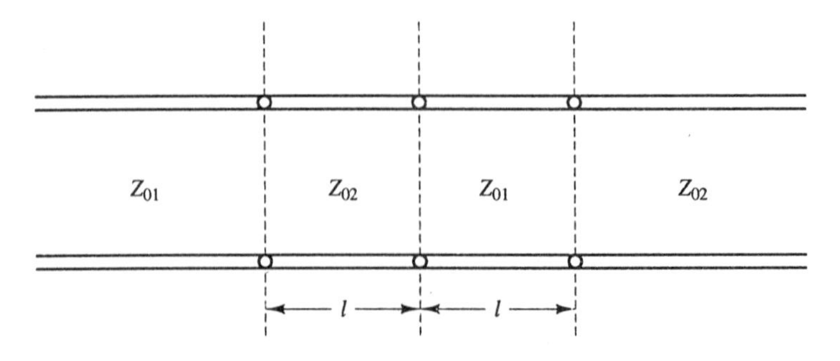

Figure 8.47. For Problem P8.21.

P8.22. We learned that in the double-stub matching technique, a solution does not exist for b_2 *if* $g' > 1/\sin^2 \beta d_{12}$—see (8.52) and associated discussion. Show that one means of resolving this problem is by increasing d_1 by $\lambda/4$.

P8.23. In the arrangement shown in Fig. 8.48, a quarter-wave transformer is employed to eliminate reflections of uniform plane waves of frequency 1500 MHz incident normally from the free space side. Find analytically the bandwidth between frequencies on either side of 1500 MHz at which the SWR in free space is 1.5.

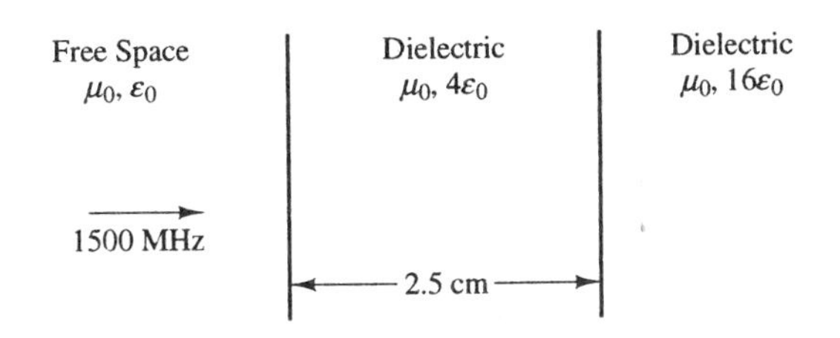

Figure 8.48. For Problem P8.23.

P8.24. The transformation

$$\bar{\Gamma} = \frac{\bar{z} - 1}{\bar{z} + 1}$$

which forms the basis for the construction of the Smith chart maps circles in the complex $\bar{z}$-plane into circles in the complex $\bar{\Gamma}$-plane. For the circle in the $\bar{z}$-plane given by $(r - 2)^2 + x^2 = 1$, find the equation for the circle in the $\bar{\Gamma}$-plane. (*Hint:* Consider three points on the circle in the $\bar{z}$-plane, find the corresponding three points in the $\bar{\Gamma}$-plane, and then find the equation.)

P8.25. Using the inverse of the procedure suggested in Problem P8.24, find the equation of the circle in the $\bar{z}$-plane which maps on to the circle in the $\bar{\Gamma}$-plane given by $(\text{Re } \bar{\Gamma})^2 + (\text{Im } \bar{\Gamma} - 0.5)^2 = 0.25$.

P8.26. For a transmission line of characteristic impedance 100 Ω terminated by a load impedance $(80 + j200)$ Ω, find the following quantities by using the Smith chart: **(a)** reflection coefficient at the load; **(b)** SWR on the line; **(c)** the distance of the first voltage minimum of the standing wave pattern from the load; **(d)** the line impedance at $d = 0.1\lambda$; **(e)** the line admittance at $d = 0.1\lambda$; and **(f)** the location nearest to the load at which the real part of the line admittance is equal to the line characteristic admittance.

P8.27. In the arrangement shown in Fig. 8.49, uniform plane waves of frequency 1.5 GHz are incident normally from medium 1 onto the interface between medium 1 and medium 2. By using the Smith chart find the SWR in **(a)** medium 3; **(b)** medium 2; and **(c)** medium 1.

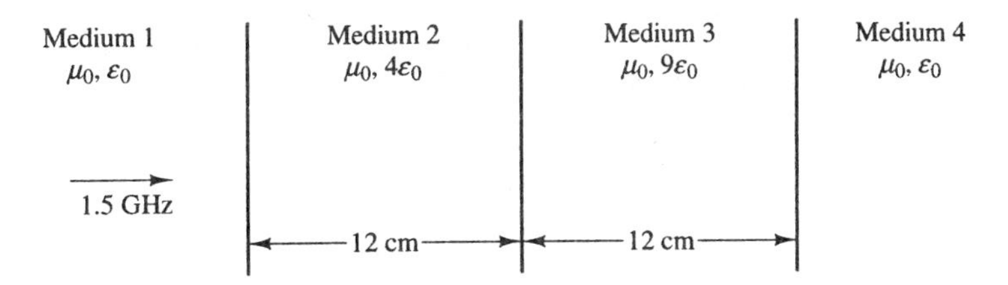

Figure 8.49. For Problem P8.27.

P8.28. Standing wave measurements on a line of characteristic impedance 50 Ω indicate an SWR of 5.0 and a voltage minimum at a distance of $5\lambda/12$ from the load. Determine the value of the load impedance by using the Smith chart.

P8.29. A transmission line of characteristic impedance 50 Ω is terminated by a certain load impedance. It is found that the SWR on the line is equal to 5.0 and that the first voltage minimum of the standing wave pattern is located to be at 0.1λ from the load. Using the Smith chart, determine the location nearest to the load and the length of a short-circuited stub of characteristic impedance 50 Ω connected in parallel with the line required to achieve a match between the line and the load.

P8.30. Standing wave measurements on a line of characteristic impedance 50 Ω indicate SWR on the line to be 3.0 and the location of the first voltage minimum of the standing wave pattern to be 0.16λ from the load. Assuming $d_1 = 0.1\lambda$ and $d_{12} = 0.625\lambda$ and using the Smith chart, find the lengths of the two short-circuited stubs of characteristic impedance 50 Ω required to achieve a match between the line and the load.

P8.31. It is proposed to match a transmission line of characteristic impedance 100 Ω to a load impedance $(7.5 - j30)$ Ω by using a double-stub arrangement with spacing between stubs, d_{12}, equal to $3\lambda/8$. By using the Smith chart determine the forbidden range of values of d_1 within the first half-wavelength to achieve the match.

P8.32. Solve Problem P8.20 by using the Smith chart.

P8.33 In the system shown in Fig. 8.50, two line sections, each of length $\lambda/4$ and characteristic impedance 50 Ω, are employed. By using the Smith chart find the locations of the two $\lambda/4$ sections, that is, the values of l_1 and l_2 to achieve a match between the 100-Ω line and the load. Use the notation shown in the figure.

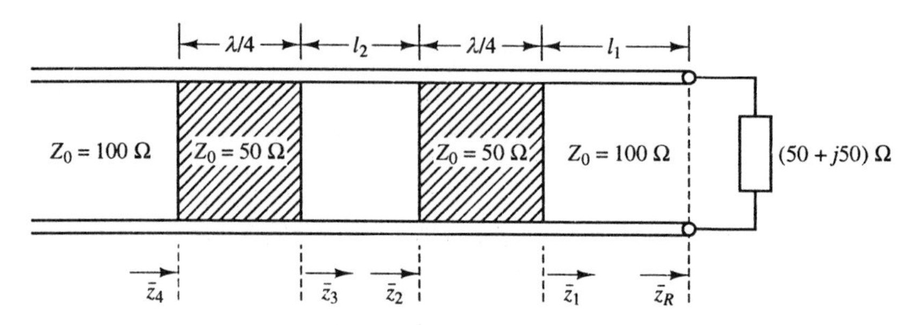

Figure 8.50. For Problem P8.33.

P8.34. In the system shown in Fig. 8.51, it is proposed to achieve a match between line 1 of characteristic impedance 150 Ω and line 2 of characteristic impedance 50 Ω by inserting a line section of characteristic impedance 25 Ω in line 2. By using the Smith chart find the values of l_1 and l_2 to achieve the match. Assume line 2 to be infinitely long. Use the notation shown in the figure.

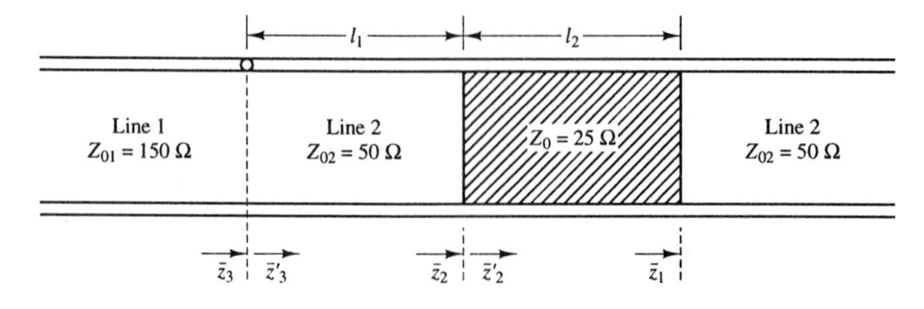

Figure 8.51. For Problem P8.34.

P8.35. Solve Problem P8.23 by using the Smith chart.

P8.36. For a lossy line having the parameters $\mathcal{R} = 0.03\ \Omega/\text{m}$, $\mathcal{L} = 1.0\ \mu\text{H/m}$, $\mathcal{G} = 3 \times 10^{-9}$ S/m, and $\mathcal{C} = 50$ pF/m, compute the values of $\bar{Z}_0$ and $\bar{\gamma}$ for $f = 10$ kHz.

P8.37. Assume that for a lossy line $\mathcal{R}/\mathcal{L} = \mathcal{G}/\mathcal{C}$ (Heaviside's distortionless line). Show that the expressions for v_p and $\bar{Z}_0$ are then the same as those for the lossless line and that α is equal to $\sqrt{\mathcal{R}\mathcal{G}}$.

P8.38. The input impedance of a lossy line of length 30 m is measured at a frequency of 100 MHz for two cases: with the output short circuited, it is $(44 + j90)\ \Omega$, and with the output open circuited, it is $(44 - j90)\ \Omega$. Find **(a)** the characteristic impedance of the line; **(b)** the attenuation constant of the line; and **(c)** the phase velocity in the line, assuming its approximate value to be 2×10^8 m/s.

P8.39. For the lossy transmission line system shown in Fig. 8.52, find the time-average power flow at the input end of the line, the time-average power delivered to the load, and the time-average power dissipated in the line.

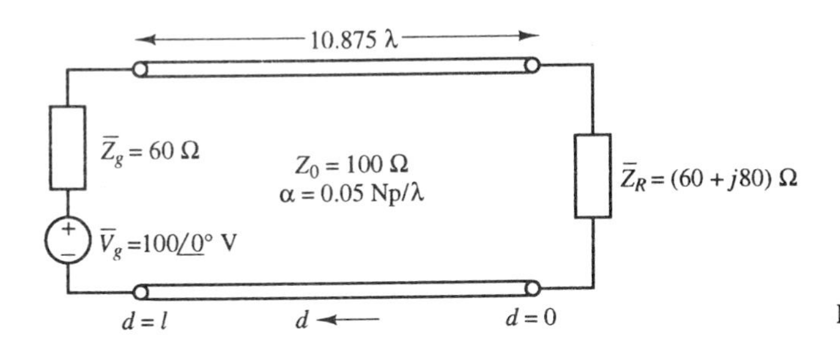

Figure 8.52. For Problem P8.39.

P8.40. In Fig. 8.53, uniform plane waves are incident normally onto a coating of good conductor material of conductivity σ and thickness l on a perfect dielectric slab of thickness $\lambda/4$ and backed by a perfect conductor. Show that no reflection occurs from the coating if $|\bar{\gamma}_c l| \ll 1$ and $\sigma = 1/\eta_0 l$, where $\bar{\gamma}_c$ is the propagation constant in the good conductor material.

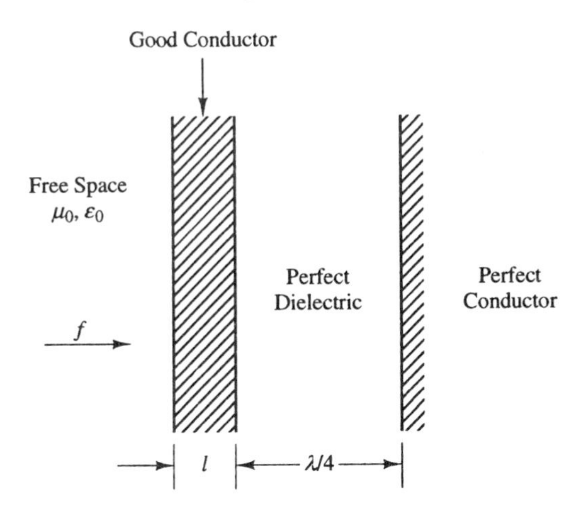

Figure 8.53. For Problem P8.40.

PC EXERCISES

PC8.1. The arrangement shown in Fig. 8.54 is that of a parallel-plate resonator made up of two plane perfect conductors coated with a dielectric and in which uniform plane waves bounce back and forth normal to the plates. It is desired to investi-

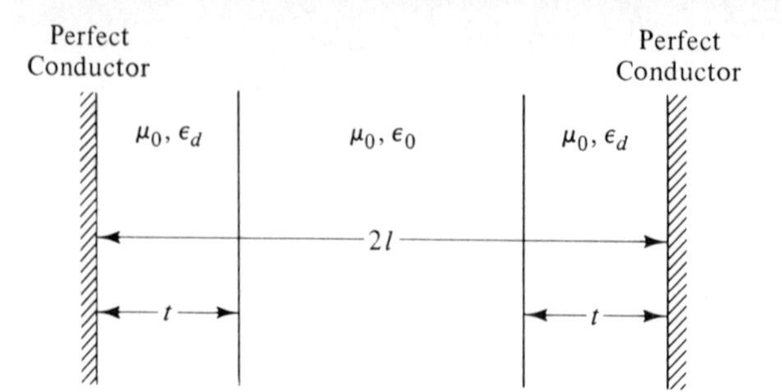

Figure 8.54. For Exercise PC8.1.

gate the variation of the lowest resonant frequency of the system as the thickness t of the dielectric coating varies from 0 to l. Show that the characteristic equation for this resonant frequency is given by

$$\tan \omega \sqrt{\mu_0 \varepsilon_0}(l - t) \tan \omega \sqrt{\mu_0 \varepsilon_d}t = \sqrt{\frac{\varepsilon_d}{\varepsilon_0}}$$

Assuming the input quantity to be the value of $\varepsilon_d/\varepsilon_0$, write a program that computes the ratio of the lowest resonant frequency for $t \neq 0$ to the lowest resonant frequency for $t = 0$, for a specified value of $t/l \leq 1$.

PC8.2. Consider n transmission lines in cascade terminated by a load impedance $\bar{Z}_R$ and driven by a voltage generator $V_g\underline{/0^\circ}$ in series with internal impedance $\bar{Z}_g$. Write a program that computes the SWR in each line and the time-average power delivered to the load. The number n of the lines, their electrical lengths and characteristic impedances, the load impedance, the source voltage, and its internal impedance are to be provided as input quantities to the program.

PC8.3. Consider matching between two dielectric media 1 and 2 having relative permittivities ε_{r1} and ε_{r2}, respectively, by employing a dielectric slab of relative permittivity ε_{rs} sandwiched between media 1 and 2 and having thickness $\lambda/4$ for uniform plane waves of frequency f_0 (see Problem P8.19, for example). Assume the waves to be incident from medium 1. Write a program that computes for specified values of ε_{r1}, ε_{r2} and f_0 in megahertz, the value of ε_{rs}, the thickness of the slab in centimeters, the bandwidth between the two frequencies on either side of f_0 at which the SWR in medium 1 is a specified value S, and the SWR versus frequency variation within this bandwidth. Note that the value of S, to be specified as input to the program, cannot exceed a maximum value that depends upon the values of ε_{r1} and ε_{r2}.

PC8.4. Consider the solution of Problem P8.33 by analytical formulation. Write a program based on such a formulation to compute the values of l_1/λ and l_2/λ for specified values of the characteristic impedance of the line, the characteristic impedance of the $\lambda/4$ sections, and the load impedance, and then compute the SWR versus frequency variation on either side of the frequency at which the match is achieved. For a complex load, assume series combination of a single resistor and a single reactive element for the purpose of SWR versus frequency computation.

PC8.5. Consider the computation of the line voltage and line current amplitudes versus distance along a lossy line of length l, terminated by a load impedance $\bar{Z}_R$ and driven by a voltage source $V_g\underline{/0^\circ}$ in series with internal impedance $\bar{Z}_g$. Note that from (8.73a), (8.73b), and (8.74), these are given by

$$|\bar{V}(d)| = |\bar{V}^+|e^{\alpha d}|1 + \bar{\Gamma}_R e^{-2\alpha d} e^{-j2\beta d}|$$

$$|\bar{I}(d)| = \frac{|\bar{V}^+|}{|\bar{Z}_0|} e^{\alpha d}|1 - \bar{\Gamma}_R e^{-2\alpha d} e^{-j2\beta d}|$$

Assuming the input quantities to be the value of α in nepers per wavelength, the value of l/λ, the value of V_g, and the values of $\bar{Z}_0$, $\bar{Z}_R$, and $\bar{Z}_g$ in rectangular or polar form, write a program which computes $|\bar{V}|$ and $|\bar{I}|$ versus d/λ from d/λ equal to zero to d/λ equal to l/λ in steps of Δ/λ, where Δ/λ is to be specified as input. The program is also to compute the time-average power at the input end of the line, the time-average power delivered to the load, and the time-average power dissipated in the line.

9

Metallic Waveguides and Resonators

In Chapters 7 and 8 we studied the principles of transmission lines, one of the two kinds of waveguiding systems. We learned that transmission lines are made up of two (or more) parallel conductors. The second kind of waveguiding system, namely, waveguide, generally consists of a single conductor. It is our goal in this chapter to learn the principles of metallic waveguides. We first consider the parallel-plate waveguide, that is, a waveguide consisting of two parallel plane conductors, and then extend it to rectangular and cylindrical waveguides.

We will learn that guiding of waves in these waveguides is accomplished by the bouncing of the waves obliquely between the walls of the guide, as compared to the case of a transmission line in which the waves slide parallel to the conductors of the line. We will also learn that waveguides are characterized by cutoff, which is the phenomenon of no propagation in a certain range of frequencies, and dispersion, which is the phenomenon of propagating waves of different frequencies possessing different phase velocities along the waveguide. In connection with the latter characteristic, we shall introduce the concept of group velocity. We also discuss the principles of cavity resonators, the microwave counterparts of resonant circuits, and losses in waveguides and resonators.

To introduce the parallel-plate waveguide, we shall make use of the superposition of two uniform plane waves propagating at an angle to each other. Hence we begin the chapter with a discussion of uniform plane wave propagation in an arbitrary direction relative to the coordinate axes.

9.1 UNIFORM PLANE WAVE PROPAGATION IN AN ARBITRARY DIRECTION

Two dimensions

In Chapter 6 we introduced the uniform plane wave propagating in the z-direction by considering an infinite plane current sheet lying in the xy-plane. If the current sheet lies in a plane making an angle to the xy-plane, the uniform plane wave would then propagate in a direction different from the z-direction. Thus let us first consider the two-dimensional case of a uniform plane wave propagating in a perfect dielectric medium in the z'-direction making an angle θ with the negative x-axis, as shown in Fig. 9.1. Let the electric field of the wave be entirely in the y-direction. The magnetic field would then be directed as shown in the figure so that $\mathbf{E} \times \mathbf{H}$ points in the z'-direction.

We can write the expression for the electric field of the wave as

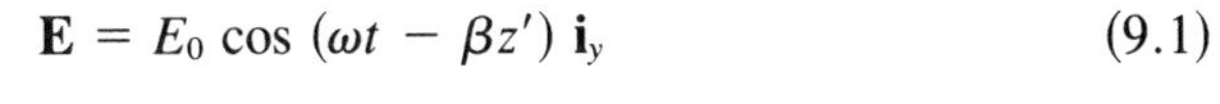

$$\mathbf{E} = E_0 \cos(\omega t - \beta z')\, \mathbf{i}_y \tag{9.1}$$

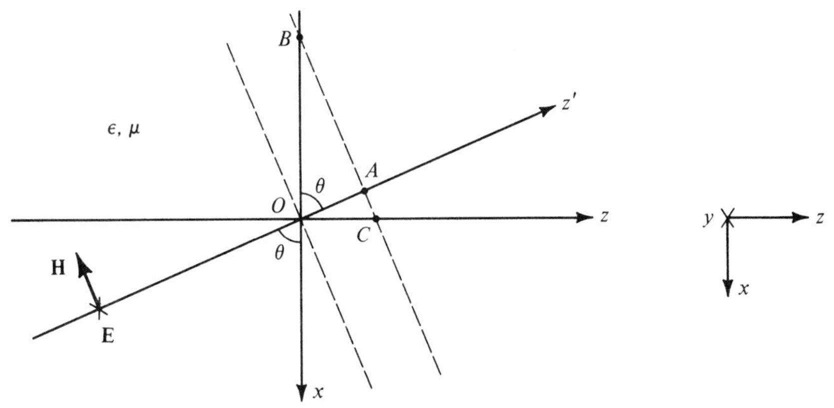

Figure 9.1. Uniform plane wave propagating in the z'-direction lying in the xz-plane and making an angle θ with the negative x-axis.

where $\beta = \omega\sqrt{\mu\varepsilon}$ is the phase constant, that is, the rate of change of phase with distance along the z'-direction for a fixed value of time. From the construction of Fig. 9.2(a), we, however, have

$$z' = -x \cos\theta + z \sin\theta \tag{9.2}$$

so that

$$\begin{aligned} \mathbf{E} &= E_0 \cos[\omega t - \beta(-x\cos\theta + z\sin\theta)]\, \mathbf{i}_y \\ &= E_0 \cos[(\omega t - (-\beta\cos\theta)x - (\beta\sin\theta)z]\, \mathbf{i}_y \\ &= E_0 \cos(\omega t - \beta_x x - \beta_z z)\, \mathbf{i}_y \end{aligned} \tag{9.3}$$

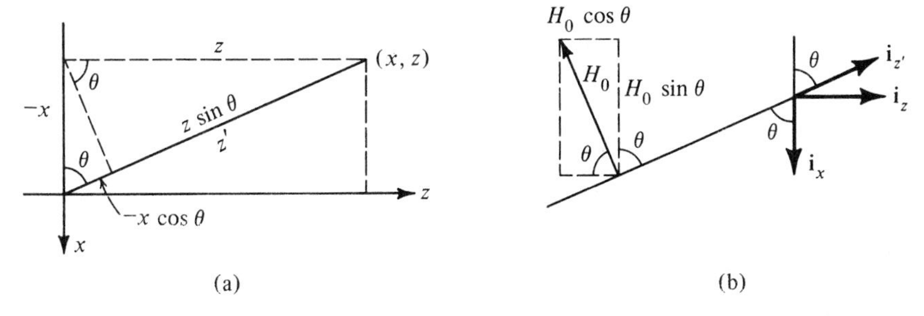

Figure 9.2. Constructions pertinent to the formulation of the expressions for the fields of the uniform plane wave of Fig. 9.1.

where $\beta_x = -\beta \cos \theta$ and $\beta_z = \beta \sin \theta$ are the phase constants in the positive x- and positive z-directions, respectively.

We note that $|\beta_x|$ and $|\beta_z|$ are less than β, the phase constant along the direction of propagation of the wave. This can also be seen from Fig. 9.1 in which two constant phase surfaces are shown by dashed lines passing through the points O and A on the z'-axis. Since the distance along the x-direction between the two constant phase surfaces, that is, the distance OB is equal to $OA/\cos \theta$, the rate of change of phase with distance along the x-direction is equal to

$$\beta \frac{OA}{OB} = \frac{\beta(OA)}{OA/\cos \theta} = \beta \cos \theta$$

The minus sign for β_x simply signifies the fact that insofar as the x-axis is concerned, the wave is progressing in the negative x-direction. Similarly, since the distance along the z-direction between the two constant phase surfaces, that is, the distance OC is equal to $OA/\sin \theta$, the rate of change of phase with distance along the z-direction is equal to

$$\beta \frac{OA}{OC} = \frac{\beta(OA)}{OA/\sin \theta} = \beta \sin \theta$$

Since the wave is progressing along the positive z-direction, β_z is positive. We further note that

$$\beta_x^2 + \beta_z^2 = (-\beta \cos \theta)^2 + (\beta \sin \theta)^2 = \beta^2 \tag{9.4}$$

and that

$$-\cos \theta \, \mathbf{i}_x + \sin \theta \, \mathbf{i}_z = \mathbf{i}_{z'} \tag{9.5}$$

where $\mathbf{i}_{z'}$ is the unit vector directed along z'-direction, as shown in Fig. 9.2(b). Thus the vector

$$\boldsymbol{\beta} = (-\beta \cos \theta)\mathbf{i}_x + (\beta \sin \theta)\mathbf{i}_z = \beta_x \mathbf{i}_x + \beta_z \mathbf{i}_z \tag{9.6}$$

defines completely the direction of propagation and the phase constant along the direction of propagation. Hence the vector $\boldsymbol{\beta}$ is known as the *propagation vector*.

The expression for the magnetic field of the wave can be written as

$$\mathbf{H} = \mathbf{H}_0 \cos (\omega t - \beta z') \tag{9.7}$$

where

$$|\mathbf{H}_0| = \frac{E_0}{\sqrt{\mu/\varepsilon}} = \frac{E_0}{\eta} \tag{9.8}$$

since the ratio of the electric field intensity to the magnetic field intensity of a uniform plane wave is equal to the intrinsic impedance of the medium. From the construction in Fig. 9.2(b), we observe that

$$\mathbf{H}_0 = H_0(-\sin \theta \, \mathbf{i}_x - \cos \theta \, \mathbf{i}_z) \tag{9.9}$$

Thus using (9.9) and substituting for z' from (9.2), we obtain

$$\begin{aligned} \mathbf{H} &= H_0(-\sin \theta \, \mathbf{i}_x - \cos \theta \, \mathbf{i}_z) \cos [\omega t - \beta(-x \cos \theta + z \sin \theta)] \\ &= -\frac{E_0}{\eta}(\sin \theta \, \mathbf{i}_x + \cos \theta \, \mathbf{i}_z) \cos [\omega t - \beta_x x - \beta_z z] \end{aligned} \tag{9.10}$$

Generalization to three dimensions

Generalizing the foregoing treatment to the case of a uniform plane wave propagating in a completely arbitrary direction in three dimensions, as shown in Fig. 9.3, and characterized by phase constants β_x, β_y, and β_z in the x-, y-, and z-directions, respectively, we can write the expression for the electric field as

$$\begin{aligned}\mathbf{E} &= \mathbf{E}_0 \cos(\omega t - \beta_x x - \beta_y y - \beta_z z + \phi_0) \\ &= \mathbf{E}_0 \cos[\omega t - (\beta_x \mathbf{i}_x + \beta_y \mathbf{i}_y + \beta_z \mathbf{i}_z) \cdot (x\mathbf{i}_x + y\mathbf{i}_y + z\mathbf{i}_z) + \phi_0] \\ &= \mathbf{E}_0 \cos(\omega t - \boldsymbol{\beta} \cdot \mathbf{r} + \phi_0)\end{aligned} \tag{9.11}$$

where

$$\boldsymbol{\beta} = \beta_x \mathbf{i}_x + \beta_y \mathbf{i}_y + \beta_z \mathbf{i}_z \tag{9.12}$$

is the propagation vector,

$$\mathbf{r} = x\mathbf{i}_x + y\mathbf{i}_y + z\mathbf{i}_z \tag{9.13}$$

is the position vector, and ϕ_0 is the phase at the origin at $t = 0$. We recall that the position vector is the vector drawn from the origin to the point (x, y, z) and hence has components x, y, and z along the x-, y-, and z-axes, respectively. The expression for the magnetic field of the wave is then given by

$$\mathbf{H} = \mathbf{H}_0 \cos(\omega t - \boldsymbol{\beta} \cdot \mathbf{r} + \phi_0) \tag{9.14}$$

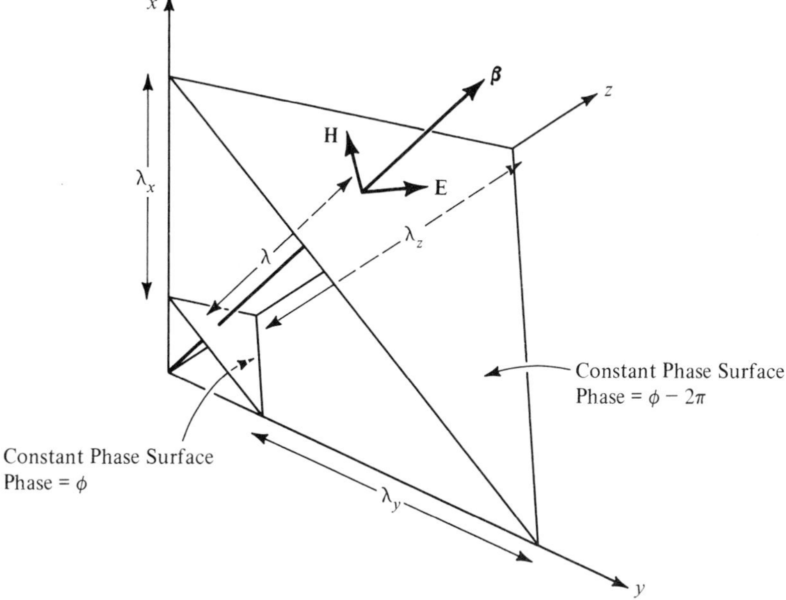

Figure 9.3. The various quantities associated with a uniform plane wave propagating in an arbitrary direction.

where

$$\boxed{|\mathbf{H}_0| = \frac{|\mathbf{E}_0|}{\eta}} \tag{9.15}$$

Since **E**, **H**, and the direction of propagation are mutually perpendicular to each other, it follows that

$$\mathbf{E}_0 \cdot \boldsymbol{\beta} = 0 \tag{9.16a}$$

$$\mathbf{H}_0 \cdot \boldsymbol{\beta} = 0 \tag{9.16b}$$

$$\mathbf{E}_0 \cdot \mathbf{H}_0 = 0 \tag{9.16c}$$

In particular, $\mathbf{E} \times \mathbf{H}$ should be directed along the propagation vector $\boldsymbol{\beta}$, as illustrated in Fig. 9.3, so that $\boldsymbol{\beta} \times \mathbf{E}_0$ is directed along $\mathbf{H}_0$. We can therefore combine the facts (9.16) and (9.15) to obtain

$$\begin{aligned} \mathbf{H}_0 &= \frac{\mathbf{i}_\beta \times \mathbf{E}_0}{\eta} = \frac{\mathbf{i}_\beta \times \mathbf{E}_0}{\sqrt{\mu/\varepsilon}} = \frac{\omega\sqrt{\mu\varepsilon}\,\mathbf{i}_\beta \times \mathbf{E}_0}{\omega\mu} \\ &= \frac{\beta\mathbf{i}_\beta \times \mathbf{E}_0}{\omega\mu} = \frac{\boldsymbol{\beta} \times \mathbf{E}_0}{\omega\mu} \end{aligned} \tag{9.17}$$

where $\mathbf{i}_\beta$ is the unit vector along $\boldsymbol{\beta}$. Thus

$$\boxed{\mathbf{H} = \frac{1}{\omega\mu}\boldsymbol{\beta} \times \mathbf{E}} \tag{9.18}$$

Apparent wavelengths and phase velocities

Returning to Fig. 9.3, we can define several quantities pertinent to the uniform plane wave propagation in an arbitrary direction. The apparent wavelengths λ_x, λ_y, and λ_z along the coordinate axes x, y, and z, respectively, are the distances measured along those respective axes between two consecutive constant phase surfaces between which the phase difference is 2π, as shown in the figure, at a fixed time. From the interpretations of β_x, β_y, and β_z as being the phase constants along the x-, y-, and z-axes, respectively, we have

$$\lambda_x = \frac{2\pi}{\beta_x} \tag{9.19a}$$

$$\lambda_y = \frac{2\pi}{\beta_y} \tag{9.19b}$$

$$\lambda_z = \frac{2\pi}{\beta_z} \tag{9.19c}$$

We note that the wavelength λ along the direction of propagation is related to λ_x, λ_y, and λ_z in the manner

$$\frac{1}{\lambda^2} = \frac{1}{(2\pi/\beta)^2} = \frac{\beta^2}{4\pi^2} = \frac{\beta_x^2 + \beta_y^2 + \beta_z^2}{4\pi^2}$$

$$\frac{1}{\lambda^2} = \frac{1}{\lambda_x^2} + \frac{1}{\lambda_y^2} + \frac{1}{\lambda_z^2} \tag{9.20}$$

The apparent phase velocities v_{px}, v_{py}, and v_{pz} along the x-, y-, and z-axes, respectively, are the velocities with which the phase of the wave progresses with time along the respective axes. Thus

$$v_{px} = \frac{\omega}{\beta_x} \tag{9.21a}$$

$$v_{py} = \frac{\omega}{\beta_y} \tag{9.21b}$$

$$v_{pz} = \frac{\omega}{\beta_z} \tag{9.21c}$$

The phase velocity v_p along the direction of propagation is related to v_{px}, v_{py}, and v_{pz} in the manner

$$\begin{aligned} \frac{1}{v_p^2} &= \frac{1}{(\omega/\beta)^2} = \frac{\beta^2}{\omega^2} = \frac{\beta_x^2 + \beta_y^2 + \beta_z^2}{\omega^2} \\ &= \frac{1}{v_{px}^2} + \frac{1}{v_{py}^2} + \frac{1}{v_{pz}^2} \end{aligned} \tag{9.22}$$

The apparent wavelengths and phase velocities along the coordinate axes are greater than the actual wavelength and phase velocity, respectively, along the direction of propagation of the wave. This fact can be understood physically by considering, for example, water waves in an ocean striking the shore at an angle. The distance along the shoreline between two successive crests is greater than the distance between the same two crests measured along a line normal to the orientation of the crests. Also, to keep pace with a particular crest an observer has to run faster along the shoreline rather than in a direction normal to the orientation of the crests. We shall now consider an example.

Example 9.1

Let us consider a 30-MHz uniform plane wave propagating in free space and given by the electric field vector

$$\mathbf{E} = 5(\mathbf{i}_x + \sqrt{3}\mathbf{i}_y) \cos [6\pi \times 10^7 t - 0.05\pi (3x - \sqrt{3}y + 2z)] \text{ V/m}$$

We wish to verify the properties and find the magnetic field vector **H** and other parameters associated with the wave.

Comparing the given expression for **E** with the general expression (9.11), we have

$$\begin{aligned} \mathbf{E}_0 &= 5(\mathbf{i}_x + \sqrt{3}\mathbf{i}_y) \\ \boldsymbol{\beta} \cdot \mathbf{r} &= 0.05\pi (3x - \sqrt{3}y + 2z) \\ &= 0.05\pi (3\mathbf{i}_x - \sqrt{3}\mathbf{i}_y + 2\mathbf{i}_z) \cdot (x\mathbf{i}_x + y\mathbf{i}_y + z\mathbf{i}_z) \\ \boldsymbol{\beta} &= 0.05\pi (3\mathbf{i}_x - \sqrt{3}\mathbf{i}_y + 2\mathbf{i}_z) \\ \boldsymbol{\beta} \cdot \mathbf{E}_0 &= 0.05\pi (3\mathbf{i}_x - \sqrt{3}\mathbf{i}_y + 2\mathbf{i}_z) \cdot 5(\mathbf{i}_x + \sqrt{3}\mathbf{i}_y) \\ &= 0.25\pi (3 - 3) = 0 \end{aligned}$$

Hence (9.16a) is satisfied; $\mathbf{E}_0$ is perpendicular to $\boldsymbol{\beta}$.

$$\beta = |\boldsymbol{\beta}| = 0.05\pi|3\mathbf{i}_x - \sqrt{3}\mathbf{i}_y + 2\mathbf{i}_z| = 0.05\pi\sqrt{9 + 3 + 4} = 0.2\pi$$

$$\lambda = \frac{2\pi}{\beta} = \frac{2\pi}{0.2\pi} = 10\,\text{m}$$

This does correspond to a frequency of $(3 \times 10^8)/10$ Hz or 30 MHz in free space. The direction of propagation is along the unit vector

$$\mathbf{i}_\beta = \frac{\boldsymbol{\beta}}{|\boldsymbol{\beta}|} = \frac{3\mathbf{i}_x - \sqrt{3}\mathbf{i}_y + 2\mathbf{i}_z}{\sqrt{9 + 3 + 4}} = \frac{3}{4}\mathbf{i}_x - \frac{\sqrt{3}}{4}\mathbf{i}_y + \frac{1}{2}\mathbf{i}_z$$

From (9.17),

$$\begin{aligned}
\mathbf{H}_0 &= \frac{1}{\omega\mu_0}\boldsymbol{\beta} \times \mathbf{E}_0 \\
&= \frac{0.05\pi \times 5}{6\pi \times 10^7 \times 4\pi \times 10^{-7}}(3\mathbf{i}_x - \sqrt{3}\mathbf{i}_y + 2\mathbf{i}_z) \times (\mathbf{i}_x + \sqrt{3}\mathbf{i}_y) \\
&= \frac{1}{96\pi}\begin{vmatrix} \mathbf{i}_x & \mathbf{i}_y & \mathbf{i}_z \\ 3 & -\sqrt{3} & 2 \\ 1 & \sqrt{3} & 0 \end{vmatrix} \\
&= \frac{1}{48\pi}(-\sqrt{3}\mathbf{i}_x + \mathbf{i}_y + 2\sqrt{3}\mathbf{i}_z)
\end{aligned}$$

Thus

$$\mathbf{H} = \frac{1}{48\pi}(-\sqrt{3}\mathbf{i}_x + \mathbf{i}_y + 2\sqrt{3}\mathbf{i}_z) \times \cos\,[6\pi \times 10^7 t - 0.05\pi(3x - \sqrt{3}y + 2z)]\ \text{A/m}$$

To verify the expression for $\mathbf{H}$ just derived, we note that

$$\begin{aligned}
\mathbf{H}_0 \cdot \boldsymbol{\beta} &= \left[\frac{1}{48\pi}(-\sqrt{3}\mathbf{i}_x + \mathbf{i}_y + 2\sqrt{3}\mathbf{i}_z)\right] \cdot [0.05\pi\,(3\mathbf{i}_x - \sqrt{3}\mathbf{i}_y + 2\mathbf{i}_z)] \\
&= \frac{0.05}{48}(-3\sqrt{3} - \sqrt{3} + 4\sqrt{3}) = 0
\end{aligned}$$

$$\begin{aligned}
\mathbf{E}_0 \cdot \mathbf{H}_0 &= 5(\mathbf{i}_x + \sqrt{3}\mathbf{i}_y) \cdot \frac{1}{48\pi}(-\sqrt{3}\mathbf{i}_x + \mathbf{i}_y + 2\sqrt{3}\mathbf{i}_z) \\
&= \frac{5}{48\pi}(-\sqrt{3} + \sqrt{3}) = 0
\end{aligned}$$

$$\begin{aligned}
\frac{|\mathbf{E}_0|}{|\mathbf{H}_0|} &= \frac{5|\mathbf{i}_x + \sqrt{3}\mathbf{i}_y|}{(1/48\pi)|-\sqrt{3}\mathbf{i}_x + \mathbf{i}_y + 2\sqrt{3}\mathbf{i}_z|} = \frac{5\sqrt{1 + 3}}{(1/48\pi)\sqrt{3 + 1 + 12}} \\
&= \frac{10}{1/12\pi} = 120\pi = \eta_0
\end{aligned}$$

Hence (9.16b), (9.16c), and (9.15) are satisfied.

Proceeding further, we find that

$$\beta_x = 0.05\pi \times 3 = 0.15\pi$$
$$\beta_y = -0.05\pi \times \sqrt{3} = -0.05\sqrt{3}\pi$$
$$\beta_z = 0.05\pi \times 2 = 0.1\pi$$

We then obtain

$$\lambda_x = \frac{2\pi}{\beta_x} = \frac{2\pi}{0.15\pi} = \frac{40}{3} \text{ m} = 13.333 \text{ m}$$

$$\lambda_y = \frac{2\pi}{|\beta_y|} = \frac{2\pi}{0.05\sqrt{3}\pi} = \frac{40}{\sqrt{3}} \text{ m} = 23.094 \text{ m}$$

$$\lambda_z = \frac{2\pi}{\beta_z} = \frac{2\pi}{0.1\pi} = 20 \text{ m}$$

$$v_{px} = \frac{\omega}{\beta_x} = \frac{6\pi \times 10^7}{0.15\pi} = 4 \times 10^8 \text{ m/s}$$

$$v_{py} = \frac{\omega}{|\beta_y|} = \frac{6\pi \times 10^7}{0.05\sqrt{3}\pi} = 4\sqrt{3} \times 10^8 \text{ m/s} = 6.928 \times 10^8 \text{ m/s}$$

$$v_{pz} = \frac{\omega}{\beta_z} = \frac{6\pi \times 10^7}{0.1\pi} = 6 \times 10^8 \text{ m/s}$$

Finally, to verify (9.20) and (9.22), we note that

$$\frac{1}{\lambda_x^2} + \frac{1}{\lambda_y^2} + \frac{1}{\lambda_z^2} = \frac{1}{(40/3)^2} + \frac{1}{(40/\sqrt{3})^2} + \frac{1}{20^2}$$
$$= \frac{9}{1600} + \frac{3}{1600} + \frac{4}{1600} = \frac{1}{100} = \frac{1}{10^2} = \frac{1}{\lambda^2}$$

and

$$\frac{1}{v_{px}^2} + \frac{1}{v_{py}^2} + \frac{1}{v_{pz}^2} = \frac{1}{(4 \times 10^8)^2} + \frac{1}{(4\sqrt{3} \times 10^8)^2} + \frac{1}{(6 \times 10^8)^2}$$
$$= \frac{1}{16 \times 10^{16}} + \frac{1}{48 \times 10^{16}} + \frac{1}{36 \times 10^{16}}$$
$$= \frac{1}{9 \times 10^{16}} = \frac{1}{(3 \times 10^8)^2} = \frac{1}{v_p^2}$$

K9.1. Uniform plane wave; Propagation in an arbitrary direction; Propagation vector; Apparent wavelengths; Apparent phase velocities.

D9.1. For each of the following cases of a uniform plane wave propagating in free space, find the frequency f: **(a)** wavelength along the direction of propagation of the wave is 2 m; **(b)** the propagation vector is $\pi(1.2\mathbf{i}_x + 0.9\mathbf{i}_y)$ rad/m; and **(c)** the apparent wavelengths along three mutually perpendicular directions are 1 m, 1 m, and 2 m.
Ans. **(a)** 150 MHz; **(b)** 225 MHz; **(c)** 450 MHz

D9.2. For a uniform plane wave of frequency 150 MHz propagating away from the origin into the first octant in a nonmagnetic ($\mu = \mu_0$), perfect dielectric medium of $\varepsilon = 2\varepsilon_0$, the apparent wavelengths along the x- and y-directions are found to be $2\frac{1}{2}$ m and $3\frac{1}{3}$ m, respectively. Find **(a)** the phase constant along the x-direction; **(b)** the apparent wavelength along the z-direction; **(c)** the apparent phase velocity along the direction of the unit vector $\frac{1}{13}(3\mathbf{i}_x - 4\mathbf{i}_y + 12\mathbf{i}_z)$; and **(d)** the equation of the plane if the source of the wave is an infinite plane sheet of uniform current density passing through the origin.
Ans. **(a)** 0.8π rad/m; **(b)** 2 m; **(c)** 3.25×10^8 m/s; **(d)** $4x + 3y + 5z = 0$

9.2 TE AND TM WAVES IN A PARALLEL-PLATE WAVEGUIDE

TE waves

In the preceding section we introduced uniform plane wave propagation in an arbitrary direction. Let us now consider the superposition of two uniform plane waves propagating symmetrically with respect to the z-axis as shown in Fig. 9.4 and having the electric fields entirely in the y-direction as given by

$$\begin{aligned}\mathbf{E}_1 &= -\frac{E_0}{2}\cos(\omega t - \boldsymbol{\beta}_1 \cdot \mathbf{r})\,\mathbf{i}_y \\ &= -\frac{E_0}{2}\cos(\omega t + \beta x\cos\theta - \beta z\sin\theta)\,\mathbf{i}_y\end{aligned} \tag{9.23a}$$

$$\begin{aligned}\mathbf{E}_2 &= \frac{E_0}{2}\cos(\omega t - \boldsymbol{\beta}_2 \cdot \mathbf{r})\,\mathbf{i}_y \\ &= \frac{E_0}{2}\cos(\omega t - \beta x\cos\theta - \beta z\sin\theta)\,\mathbf{i}_y\end{aligned} \tag{9.23b}$$

where $\beta = \omega\sqrt{\mu\varepsilon}$, with ε and μ being the permittivity and the permeability, respectively, of the medium. The corresponding magnetic fields are given by

$$\mathbf{H}_1 = \frac{E_0}{2\eta}(\sin\theta\,\mathbf{i}_x + \cos\theta\,\mathbf{i}_z)\cos(\omega t + \beta x\cos\theta - \beta z\sin\theta) \tag{9.24a}$$

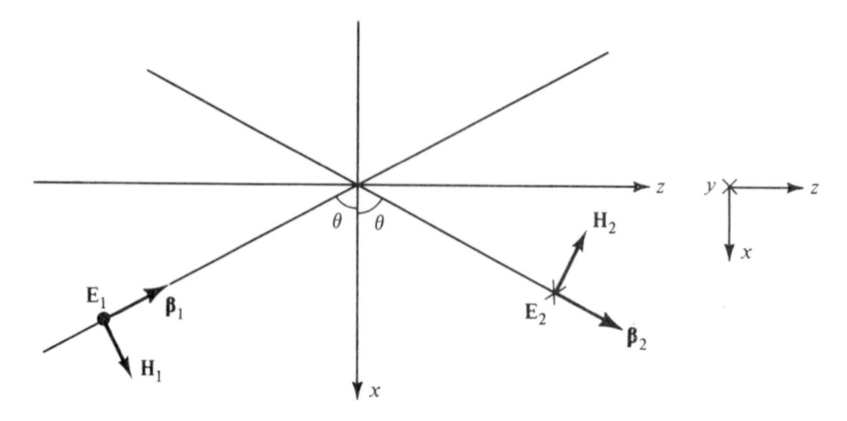

Figure 9.4. Superposition of two uniform plane waves propagating symmetrically with respect to the z-axis.

$$\mathbf{H}_2 = \frac{E_0}{2\eta}(-\sin\theta\,\mathbf{i}_x + \cos\theta\,\mathbf{i}_z)\cos(\omega t - \beta x\cos\theta - \beta z\sin\theta) \tag{9.24b}$$

where $\eta = \sqrt{\mu/\varepsilon}$. The electric and magnetic fields of the superposition of the two waves are given by

$$\begin{aligned}
\mathbf{E} &= \mathbf{E}_1 + \mathbf{E}_2 \\
&= -\frac{E_0}{2}[\cos(\omega t - \beta z\sin\theta + \beta x\cos\theta) \\
&\quad - \cos(\omega t - \beta z\sin\theta - \beta x\cos\theta)]\,\mathbf{i}_y \\
&= E_0\sin(\beta x\cos\theta)\sin(\omega t - \beta z\sin\theta)\,\mathbf{i}_y
\end{aligned} \tag{9.25a}$$

$$\begin{aligned}
\mathbf{H} &= \mathbf{H}_1 + \mathbf{H}_2 \\
&= \frac{E_0}{2\eta}\sin\theta\,[\cos(\omega t - \beta z\sin\theta + \beta x\cos\theta) \\
&\quad - \cos(\omega t - \beta z\sin\theta - \beta x\cos\theta)]\,\mathbf{i}_x \\
&\quad + \frac{E_0}{2\eta}\cos\theta\,[\cos(\omega t - \beta z\sin\theta + \beta x\cos\theta) \\
&\quad + \cos(\omega t - \beta z\sin\theta - \beta x\cos\theta)]\,\mathbf{i}_z \\
&= -\frac{E_0}{\eta}\sin\theta\sin(\beta x\cos\theta)\sin(\omega t - \beta z\sin\theta)\,\mathbf{i}_x \\
&\quad + \frac{E_0}{\eta}\cos\theta\cos(\beta x\cos\theta)\cos(\omega t - \beta z\sin\theta)\,\mathbf{i}_z
\end{aligned} \tag{9.25b}$$

In view of the factors $\sin(\beta x\cos\theta)$ and $\cos(\beta x\cos\theta)$ for the x-dependence and the factors $\sin(\omega t - \beta z\sin\theta)$ and $\cos(\omega t - \beta z\sin\theta)$ for the z-dependence, the composite fields have standing wave character in the x-direction and traveling wave character in the z-direction. Thus we have standing waves in the x-direction moving bodily in the z-direction, as illustrated in Fig. 9.5, by considering the electric field for two different times. In fact, we find that the Poynting vector is given by

$$\begin{aligned}
\mathbf{P} &= \mathbf{E}\times\mathbf{H} = E_y\mathbf{i}_y \times (H_x\mathbf{i}_x + H_z\mathbf{i}_z) \\
&= -E_yH_x\mathbf{i}_z + E_yH_z\mathbf{i}_x \\
&= \frac{E_0^2}{\eta}\sin\theta\sin^2(\beta x\cos\theta)\sin^2(\omega t - \beta z\sin\theta)\,\mathbf{i}_z \\
&\quad + \frac{E_0^2}{4\eta}\cos\theta\sin(2\beta x\cos\theta)\sin 2(\omega t - \beta z\sin\theta)\,\mathbf{i}_x
\end{aligned} \tag{9.26}$$

The time-average Poynting vector is given by

$$\begin{aligned}
\langle\mathbf{P}\rangle &= \frac{E_0^2}{\eta}\sin\theta\sin^2(\beta x\cos\theta)\langle\sin^2(\omega t - \beta z\sin\theta)\rangle\,\mathbf{i}_z \\
&\quad + \frac{E_0^2}{4\eta}\cos\theta\sin(2\beta x\cos\theta)\langle\sin 2(\omega t - \beta z\sin\theta)\rangle\,\mathbf{i}_x
\end{aligned}$$

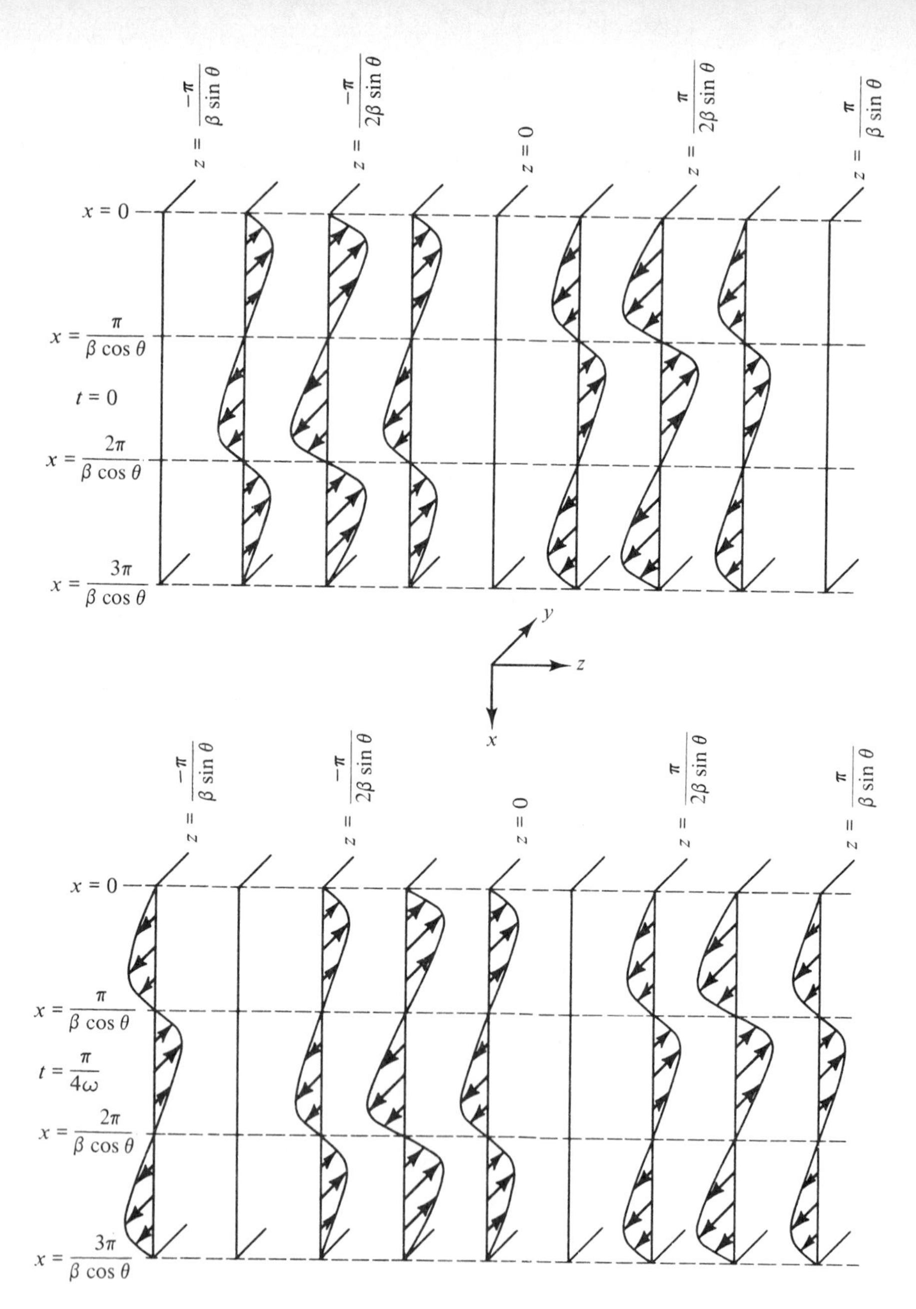

Figure 9.5. Standing waves in the x-direction moving bodily in the z-direction.

$$\langle \mathbf{P} \rangle = \frac{E_0^2}{2\eta} \sin\theta \sin^2 (\beta x \cos\theta)\, \mathbf{i}_z \tag{9.27}$$

Thus the time-average power flow is entirely in the z-direction, thereby verifying our interpretation of the field expressions. Since the composite electric field is directed entirely transverse to the z-direction, that is, the direction of time-average power flow, whereas the composite magnetic field is not, the composite wave is known as the *transverse electric*, or TE wave.

From the expressions for the fields for the TE wave given by (9.25a) and (9.25b), we note that the electric field is zero for sin ($\beta x \cos\theta$) equal to zero, or

$$\beta x \cos\theta = \pm m\pi, \qquad m = 0, 1, 2, 3, \ldots$$

$$x = \pm\frac{m\pi}{\beta\cos\theta} = \pm\frac{m\lambda}{2\cos\theta}, \qquad m = 0, 1, 2, 3, \ldots \tag{9.28}$$

where

$$\lambda = \frac{2\pi}{\beta} = \frac{2\pi}{\omega\sqrt{\mu\varepsilon}} = \frac{1}{f\sqrt{\mu\varepsilon}}$$

Thus if we place perfectly conducting sheets in these planes, the waves will propagate undisturbed, that is, as though the sheets were not present since the boundary condition that the tangential component of the electric field be zero on the surface of a perfect conductor is satisfied in these planes. The boundary condition that the normal component of the magnetic field be zero on the surface of a perfect conductor is also satisfied since H_x is zero in these planes.

Parallel-plate waveguide

If we consider any two adjacent sheets, the situation is actually one of uniform plane waves bouncing obliquely between the sheets, as illustrated in Fig. 9.6 for two sheets in the planes $x = 0$ and $x = \lambda/(2\cos\theta)$, thereby guiding the wave and hence the energy in the z-direction, parallel to the plates. Thus we have a *parallel-plate waveguide*, as compared to the parallel-plate transmission line in which the uniform plane wave slides parallel to the plates. We note from the constant phase surfaces of the obliquely bouncing wave shown in Fig. 9.6 that $\lambda/(2\cos\theta)$ is simply one-half of the apparent wavelength of that wave in the x-direction, that is, normal to the plates. Thus the fields have one-half apparent wavelength in the x-direction. If we place the perfectly conducting sheets in the planes $x = 0$ and $x = m\lambda/(2\cos\theta)$, the fields will then have m number of one-half apparent wavelengths in the x-direction between the plates. The fields have no variations in the y-direction. Thus the fields are said to correspond to *$TE_{m,0}$ modes*, where the subscript m refers to the x-direction, denoting m number of one-half apparent wavelengths in that direction and the subscript 0 refers to the y-direction, denoting zero number of one-half apparent wavelengths in that direction.

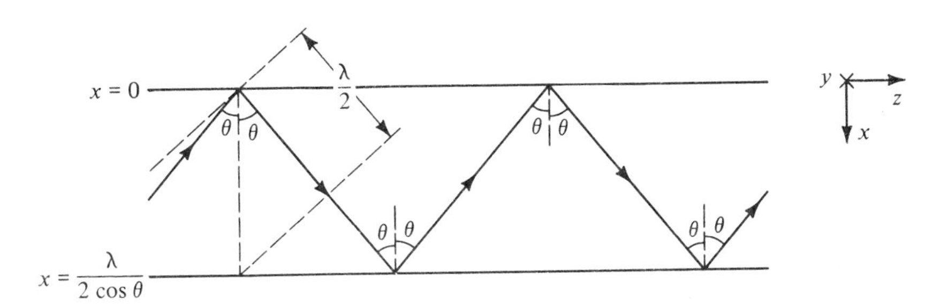

Figure 9.6. Uniform plane waves bouncing obliquely between two parallel plane perfectly conducting sheets.

Cutoff phenomenon

Let us now consider a parallel-plate waveguide with perfectly conducting plates situated in the planes $x = 0$ and $x = a$, that is, having a fixed spacing a between them, as shown in Fig. 9.7(a). Then, for $TE_{m,0}$ waves guided by the plates, we have from (9.28),

$$a = \frac{m\lambda}{2 \cos \theta}$$

or

$$\cos \theta = \frac{m\lambda}{2a} = \frac{m}{2a} \frac{1}{f\sqrt{\mu\varepsilon}} \tag{9.29}$$

Thus waves of different wavelengths (or frequencies) bounce obliquely between the plates at different values of the angle θ. For very small wavelengths (very high frequencies), $m\lambda/2a$ is small, $\cos \theta \approx 0$, $\theta \approx 90°$, and the waves simply slide between the plates as in the case of the transmission line, as shown in Fig. 9.7(b). As λ increases (f decreases), $m\lambda/2a$ increases, θ decreases, and the waves bounce more and more obliquely, as shown in Fig. 9.7(c)–(e), until λ becomes equal to $2a/m$ for which $\cos \theta = 1$, $\theta = 0°$, and the waves simply bounce back and forth normally to the plates, as shown in Fig. 9.7(f), without any feeling of being guided parallel to the plates. For $\lambda > 2a/m$, $m\lambda/2a > 1$, $\cos \theta > 1$, and θ has no real solution, indicating that propagation does not occur for these wavelengths in the waveguide mode. This condition is known as the *cutoff* condition.

The cutoff wavelength, denoted by the symbol λ_c, is given by

$$\boxed{\lambda_c = \frac{2a}{m}} \tag{9.30}$$

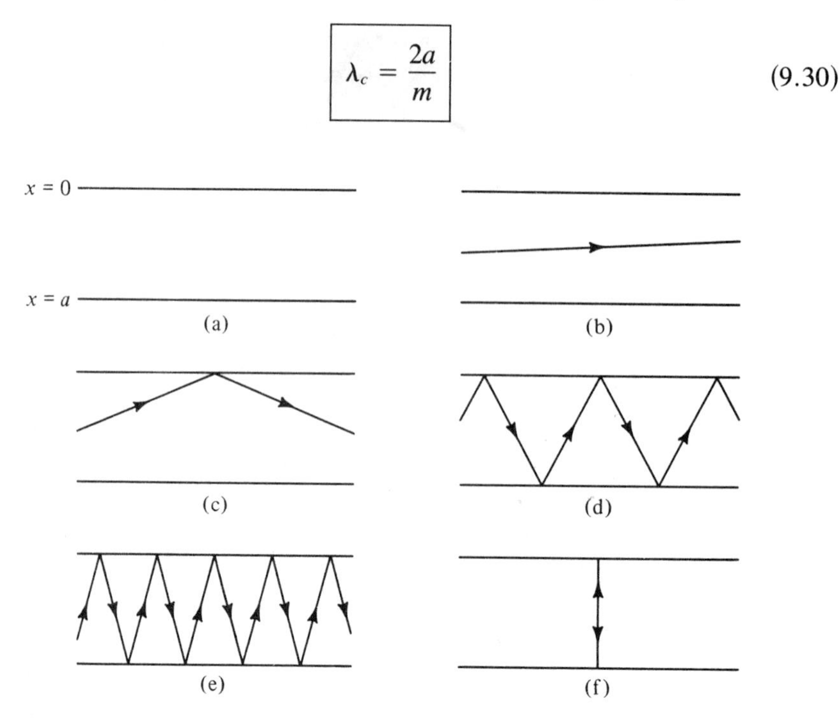

Figure 9.7. For illustrating the phenomenon of cutoff in a parallel-plate wave-guide.

This is simply the wavelength for which the spacing a is equal to m number of one-half wavelengths. Propagation of a particular mode is possible only if λ is less than the value of λ_c for that mode. The cutoff frequency is given by

$$\boxed{f_c = \frac{m}{2a\sqrt{\mu\varepsilon}}} \tag{9.31}$$

Propagation of a particular mode is possible only if f is greater than the value of f_c for that mode. Consequently, waves of a given frequency f can propagate in all modes for which the cutoff wavelengths are greater than the wavelength or the cutoff frequencies are less than the frequency.

Substituting λ_c for $2a/m$ in (9.29), we have

$$\cos\theta = \frac{\lambda}{\lambda_c} = \frac{f_c}{f} \tag{9.32a}$$

$$\sin\theta = \sqrt{1-\cos^2\theta} = \sqrt{1-\left(\frac{\lambda}{\lambda_c}\right)^2} = \sqrt{1-\left(\frac{f_c}{f}\right)^2} \tag{9.32b}$$

$$\beta\cos\theta = \frac{2\pi}{\lambda}\frac{\lambda}{\lambda_c} = \frac{2\pi}{\lambda_c} = \frac{m\pi}{a} \tag{9.32c}$$

$$\beta\sin\theta = \frac{2\pi}{\lambda}\sqrt{1-\left(\frac{\lambda}{\lambda_c}\right)^2} \tag{9.32d}$$

We see from (9.32d) that the phase constant along the z-direction, that is, $\beta_z(=\beta\sin\theta)$, is real for $\lambda < \lambda_c$ and imaginary for $\lambda > \lambda_c$. Since

$$\begin{aligned}\cos(\omega t \mp j|\beta_z|z) &= \text{Re}\, e^{j(\omega t \mp j|\beta_z|z)} \\ &= \text{Re}(e^{\pm|\beta_z|z}e^{j\omega t}) \\ &= e^{\pm|\beta_z|z}\cos\omega t\end{aligned}$$

an imaginary value of the phase constant does not correspond to wave propagation. This once again explains the cutoff phenomenon. We now define the guide wavelength, λ_g, to be the wavelength in the z-direction, that is, along the guide. This is given by

$$\boxed{\lambda_g = \frac{2\pi}{\beta_z} = \frac{2\pi}{\beta\sin\theta} = \frac{\lambda}{\sqrt{1-(\lambda/\lambda_c)^2}} = \frac{\lambda}{\sqrt{1-(f_c/f)^2}}} \tag{9.33}$$

This is simply the apparent wavelength, in the z-direction, of the obliquely bouncing uniform plane waves. The phase velocity along the guide axis, which is simply the apparent phase velocity, in the z-direction, of the obliquely bouncing uniform plane waves, is

$$\boxed{v_{pz} = \frac{\omega}{\beta_z} = \frac{\omega}{\beta\sin\theta} = \frac{v_p}{\sqrt{1-(\lambda/\lambda_c)^2}} = \frac{v_p}{\sqrt{1-(f_c/f)^2}}} \tag{9.34}$$

Field expressions for $TE_{m,0}$ modes

Finally, substituting (9.32a)–(9.32d) in the field expressions (9.25a) and (9.25b), we obtain

$$\mathbf{E} = E_0 \sin\left(\frac{m\pi x}{a}\right) \sin(\omega t - \beta_z z)\, \mathbf{i}_y \tag{9.35a}$$

$$\mathbf{H} = -\frac{E_0}{\eta}\frac{\lambda}{\lambda_g} \sin\left(\frac{m\pi x}{a}\right) \sin(\omega t - \beta_z z)\, \mathbf{i}_x + \frac{E_0}{\eta}\frac{\lambda}{\lambda_c} \cos\left(\frac{m\pi x}{a}\right) \cos(\omega t - \beta_z z)\, \mathbf{i}_z \tag{9.35b}$$

These expressions for the $TE_{m,0}$ mode fields in the parallel-plate waveguide do not contain the angle θ. They clearly indicate the standing wave character of the fields in the x-direction, having m one-half sinusoidal variations between the plates. We shall now consider an example.

Example 9.2

Let us assume the spacing a between the plates of an air-dielectric parallel-plate waveguide to be 5 cm and investigate the propagating $TE_{m,0}$ modes for f = 10,000 MHz.

From (9.30), the cutoff wavelengths for $TE_{m,0}$ modes are given by

$$\lambda_c = \frac{2a}{m} = \frac{10}{m}\ \text{cm} = \frac{0.1}{m}\ \text{m}$$

This result is independent of the dielectric between the plates. Since the medium between the plates is free space, the cutoff frequencies for the $TE_{m,0}$ modes are

$$f_c = \frac{3 \times 10^8}{\lambda_c} = \frac{3 \times 10^8}{0.1/m} = 3m \times 10^9\ \text{Hz}$$

For f = 10,000 MHz = 10^{10} Hz, the propagating modes are $TE_{1,0}(f_c = 3 \times 10^9$ Hz), $TE_{2,0}(f_c = 6 \times 10^9$ Hz), and $TE_{3,0}(f_c = 9 \times 10^9$ Hz).

For each propagating mode, we can find θ, λ_g, and v_{pz} by using (9.32a), (9.33), and (9.34), respectively. Values of these quantities are listed in the following:

Mode	λ_c (cm)	f_c (MHz)	θ (deg)	λ_g (cm)	v_{pz} (m/s)
$TE_{1,0}$	10	3000	72.54	3.145	3.145×10^8
$TE_{2,0}$	5	6000	53.13	3.75	3.75×10^8
$TE_{3,0}$	3.33	9000	25.84	6.882	6.882×10^8

TM waves

We have thus far considered transverse electric or TE waves in a parallel-plate waveguide. In a similar manner, it is possible to have propagation of transverse magnetic or TM waves, so termed because the magnetic field is directed entirely transverse to the direction of time-average power flow whereas the electric field is not. The field expressions for TM waves can be obtained by starting with two uniform plane waves having their magnetic fields entirely in the y-direction and proceeding in a manner similar to the development of TE waves. We shall however not pursue that approach. Instead, we shall, by analogy with (9.35a), write the expression for the magnetic field of the TM wave and then derive the electric field by using one of Maxwell's curl equations.

Field expressions for TM$_{m,0}$ modes

Thus assuming the guide to be made up of parallel plates in the $x = 0$ and $x = a$ planes, and writing the expression for the magnetic field of the TM$_{m,0}$ wave and using

$$\nabla \times \mathbf{H} = \frac{\partial \mathbf{D}}{\partial t}$$

we obtain the fields for the TM modes to be

$$\mathbf{H} = H_0 \cos\left(\frac{m\pi x}{a}\right) \sin(\omega t - \beta_z z)\,\mathbf{i}_y \tag{9.36a}$$

$$\begin{aligned}\mathbf{E} &= \frac{\lambda}{\lambda_g}\eta H_0 \cos\left(\frac{m\pi x}{a}\right) \sin(\omega t - \beta_z z)\,\mathbf{i}_x \\ &\quad + \frac{\lambda}{\lambda_c}\eta H_0 \sin\left(\frac{m\pi x}{a}\right) \cos(\omega t - \beta_z z)\,\mathbf{i}_z\end{aligned} \tag{9.36b}$$

Note that the x-variation of H_y is cosinusoidal, which leads to sinusoidal variation for E_z so that the boundary condition of zero tangential electric field is satisfied on the two plates. The parameters λ_c and λ_g in (9.36a) and (9.36b) and the other parameters f_c and v_{pz} for the TM modes are the same as those for the TE modes, given by (9.30), (9.33), (9.31), and (9.34), respectively.

Parallel-plate waveguide discontinuity

Let us now consider reflection and transmission at a dielectric discontinuity in a parallel-plate guide, as shown in Fig. 9.8. If a TE or TM wave is incident on the junction from section 1, then it will set up a reflected wave into section 1 and a transmitted wave into section 2, provided that mode propagates in that section. The fields corresponding to these incident, reflected, and transmitted waves must satisfy the boundary conditions at the dielectric discontinuity.

Considering first TE waves and denoting the incident, reflected, and transmitted wave fields by the subscripts i, r, and t, respectively, we have from the continuity of the tangential component of **E** at a dielectric discontinuity,

$$E_{yi} + E_{yr} = E_{yt} \qquad \text{at } z = 0 \tag{9.37a}$$

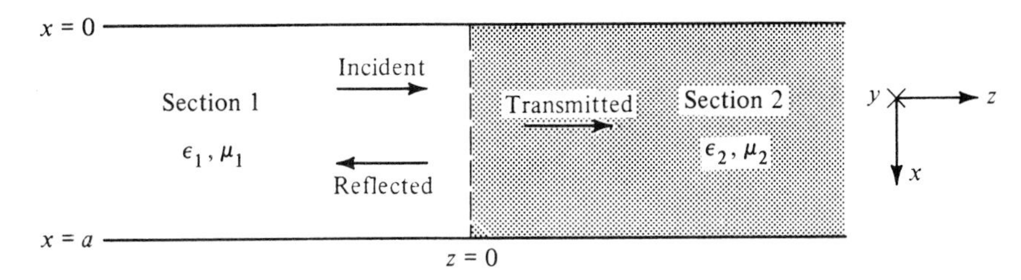

Figure 9.8. For consideration of reflection and transmission at a dielectric discontinuity in a parallel-plate waveguide.

and from the continuity of the tangential component of **H** at a dielectric discontinuity,

$$H_{xi} + H_{xr} = H_{xt} \qquad \text{at } z = 0 \tag{9.37b}$$

We now define the guide characteristic impedance, η_{g1}, of section 1 as

$$\eta_{g1} = \frac{E_{yi}}{-H_{xi}} \tag{9.38}$$

Recognizing that $\mathbf{i}_y \times (-\mathbf{i}_x) = \mathbf{i}_z$, we note that η_{g1} is simply the ratio of the transverse components of the electric and magnetic fields of the $\text{TE}_{m,0}$ wave which give rise to time-average power flow down the guide. From (9.35a) and (9.35b) applied to section 1, we have

$$\eta_{g1} = \eta_1 \frac{\lambda_{g1}}{\lambda_1} = \frac{\eta_1}{\sqrt{1 - (\lambda_1/\lambda_c)^2}} = \frac{\eta_1}{\sqrt{1 - (f_{c1}/f)^2}} \tag{9.39}$$

The guide characteristic impedance is analogous to the characteristic impedance of a transmission line, if we recognize that E_{yi} and $-H_{xi}$ are analogous to V^+ and I^+, respectively. In terms of the reflected wave fields, it then follows that

$$\eta_{g1} = -\left(\frac{E_{yr}}{-H_{xr}}\right) = \frac{E_{yr}}{H_{xr}} \tag{9.40}$$

This result can also be seen from the fact that for the reflected wave, the power flow is in the negative z-direction, and since $\mathbf{i}_y \times \mathbf{i}_x = -\mathbf{i}_z$, η_{g1} is equal to E_{yr}/H_{xr}. For the transmitted wave fields, we have

$$\frac{E_{yt}}{-H_{xt}} = \eta_{g2} \tag{9.41}$$

where

$$\eta_{g2} = \eta_2 \frac{\lambda_{g2}}{\lambda_2} = \frac{\eta_2}{\sqrt{1 - (\lambda_2/\lambda_c)^2}} = \frac{\eta_2}{\sqrt{1 - (f_{c2}/f)^2}} \tag{9.42}$$

is the guide characteristic impedance of section 2.

Using (9.38), (9.40), and (9.41), (9.37b) can be written as

$$\frac{E_{yi}}{\eta_{g1}} - \frac{E_{yr}}{\eta_{g1}} = \frac{E_{yt}}{\eta_{g2}} \tag{9.43}$$

Solving (9.37a) and (9.43), we get

$$E_{yi}\left(1 - \frac{\eta_{g2}}{\eta_{g1}}\right) + E_{yr}\left(1 + \frac{\eta_{g2}}{\eta_{g1}}\right) = 0$$

or the reflection coefficient at the junction is given by

$$\boxed{\Gamma = \frac{E_{yr}}{E_{yi}} = \frac{\eta_{g2} - \eta_{g1}}{\eta_{g2} + \eta_{g1}}} \tag{9.44}$$

This expression for the reflection coefficient is the same as that for the voltage reflection coefficient at the load of a lossless transmission line of characteristic impedance η_{g1} terminated by a resistive load η_{g2}. It is also the same as the voltage reflection coefficient at the junction between two transmission lines 1 and 2 having the characteristic impedances η_{g1} and η_{g2}, respectively, as shown in Fig. 9.9, where line 2 is infinitely long and hence its input impedance is equal to η_{g2}. Thus insofar as reflection and transmission at the discontinuity are concerned, each waveguide section can be replaced by a transmission line of characteristic impedance equal to the guide characteristic impedance given for the TE modes by

$$\boxed{[\eta_g]_{\text{TE}} = \frac{\eta}{\sqrt{1 - (f_c/f)^2}}} \tag{9.45}$$

It should be noted that unlike the characteristic impedance of a lossless line, which is a constant independent of frequency, the guide characteristic impedance of the lossless waveguide is a function of the frequency and the mode of propagation. Before considering TM modes, it should be pointed out that the power reflection coefficient is Γ^2 so that the reflected power is Γ^2 times the incident power and the transmitted power into section 2 is $(1 - \Gamma^2)$ times the incident power.

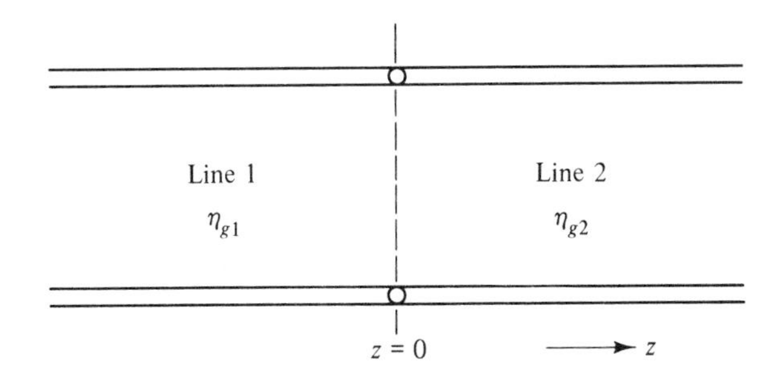

Figure 9.9. Transmission-line equivalent of parallel-plate waveguide discontinuity.

Turning now to TM waves, we observe from (9.36a) and (9.36b) that the ratio of the transverse electric field component E_x to the transverse magnetic field component H_y, which together are responsible for time-average power flow in the z-direction, is equal to $\eta\lambda/\lambda_g$, and hence the guide characteristic impedance for TM waves is given by

$$\boxed{[\eta_g]_{\text{TM}} = \eta\sqrt{1 - (f_c/f)^2}} \tag{9.46}$$

Thus the transmission-line equivalent for reflection and transmission of TM waves at the waveguide discontinuity is the same as in Fig. 9.9 except that η_{g1} and η_{g2} follow from (9.46). We shall now consider an example.

Example 9.3

Let us consider the parallel-plate waveguide discontinuity shown in Fig. 9.10. We wish to find the power reflection coefficients for $\text{TE}_{1,0}$ and $\text{TM}_{1,0}$ waves of frequency $f = 5000$ MHz incident on the junction from the free space side.

For the $\text{TE}_{1,0}$ mode or for the $\text{TM}_{1,0}$ mode, $\lambda_c = 2a = 10$ cm, independent of the dielectric. For $f = 5000$ MHz,

$$\lambda_1 = \text{wavelength on the free space side} = \frac{3 \times 10^8}{5 \times 10^9} = 6 \text{ cm}$$

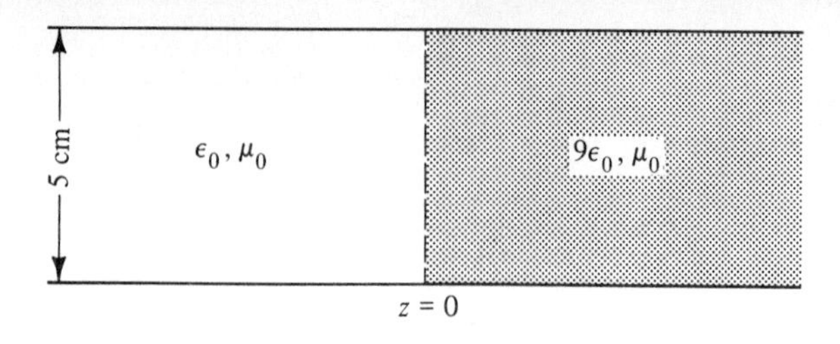

Figure 9.10. For illustrating the computation of reflection and transmission coefficients at a parallel-plate waveguide discontinuity.

$$\lambda_2 = \text{wavelength on the dielectric side} = \frac{3 \times 10^8}{\sqrt{9} \times 5 \times 10^9} = \frac{6}{3} = 2 \text{ cm}$$

Since $\lambda < \lambda_c$ in both sections, $TE_{1,0}$ and $TM_{1,0}$ modes propagate in both sections. Thus for the $TE_{1,0}$ mode,

$$\eta_{g1} = \frac{\eta_1}{\sqrt{1 - (\lambda_1/\lambda_c)^2}} = \frac{120\pi}{\sqrt{1 - (6/10)^2}} = 471.24\,\Omega$$

$$\eta_{g2} = \frac{\eta_2}{\sqrt{1 - (\lambda_2/\lambda_c)^2}} = \frac{120\pi/\sqrt{9}}{\sqrt{1 - (2/10)^2}} = \frac{40\pi}{\sqrt{1 - 0.04}} = 128.25\,\Omega$$

$$\Gamma^2 = \left(\frac{\eta_{g2} - \eta_{g1}}{\eta_{g2} + \eta_{g1}}\right)^2 = \left(\frac{128.25 - 471.24}{128.25 + 471.24}\right)^2 = (-0.572)^2 = 0.327$$

For the $TM_{1,0}$ mode,

$$\eta_{g1} = \eta_1\sqrt{1 - (\lambda_1/\lambda_c)^2} = 301.59\,\Omega$$

$$\eta_{g2} = \eta_2\sqrt{1 - (\lambda_2/\lambda_c)^2} = 123.12\,\Omega$$

$$\Gamma^2 = \left(\frac{\eta_{g2} - \eta_{g1}}{\eta_{g2} + \eta_{g1}}\right)^2 = \left(\frac{123.12 - 301.59}{123.12 + 301.59}\right)^2 = (-0.42)^2 = 0.176$$

We have in this section introduced the principle of waveguides by considering the parallel-plate waveguide. In practice, however, waveguides are generally made up of a single conductor having rectangular or circular cross section. We shall defer the consideration of rectangular waveguides to Section 9.4 and discuss in Section 9.3 the important phenomenon of dispersion, characteristic of propagation in parallel-plate as well as rectangular and circular waveguides and leading to the concept of group velocity.

Earth-ionosphere waveguide

But first we shall conclude this section with a brief description of a naturally occurring waveguide, although of spherical geometry. This is the earth-ionosphere waveguide. The ionosphere is a region of the upper atmosphere extending from approximately 50 km to more than 1000 km above the earth. In this region the constituent gases are ionized, mostly because of ultraviolet radiation from the sun, thereby resulting in the production of positive ions and electrons that are free to move under the influence of the fields of a wave incident upon the medium. The positive ions are, however, heavy compared to the electrons, and hence they are relatively immobile. The electron motion produces a current that influences the wave propagation. The electron density in the ionosphere exists in several layers known as the *D*, *E*, and *F* layers in which the ionization changes with the hour of the day, the season, and the sunspot cycle. However, for the purpose of our discussion, it is sufficient to assume that the electron density increases continuously from zero at the lower boundary, reaching a peak at some height, typically lying between 250 and 350 km, and then decreases continuously, as shown in Fig. 9.11(a). The wave propagation is influenced by the electrons in such a manner that waves of very low frequencies are reflected at the base. As the frequency is increased, the waves penetrate deeper into the region but still return to earth after reflection. When their frequency exceeds certain value, typically between 20 and 40 MHz depending on the angle of incidence, they penetrate through the maximum of the layer and hence do not return to the earth. Thus for frequencies in the VLF range and lower, the lower boundary of the ionosphere and the earth form a waveguide, thereby permitting waveguide mode of propagation.

K9.2. Transverse electric wave; Transverse magnetic wave; Parallel-plate waveguide; Cutoff frequency; Cutoff wavelength; Guide wavelength; Guide characteristic impedance.

D9.3. The dimension a of an air-dielectric parallel-plate waveguide is 3 cm. Find the values of θ and λ_g for each of the following cases: **(a)** $f = 6000$ MHz, $TE_{1,0}$ mode; **(b)** $f = 15{,}000$ MHz, $TE_{1,0}$ mode; and **(c)** $f = 15{,}000$ MHz, $TE_{2,0}$ mode.
Ans. **(a)** 33.56°, 9.045 cm; **(b)** 70.53°, 2.121 cm; **(c)** 48.19°, 2.683 cm

D9.4. TE waves are excited in an air-dielectric parallel-plate waveguide having the plates in the $x = 0$ and $x = 5$ cm planes by setting up at its input $z = 0$ a field distribution having

$$\mathbf{E} = 40 \sin^3 20\pi x \sin 2 \times 10^{10}\pi t \, \mathbf{i}_y \text{ V/m}$$

Noting that the electric field of a propagating $TE_{m,0}$ mode is of the form given by (9.35a), find E_0 for each of the following modes: **(a)** $TE_{1,0}$; **(b)** $TE_{2,0}$; and **(c)** $TE_{3,0}$.
Ans. **(a)** 30 V/m; **(b)** 0 V/m; **(c)** −10 V/m

D9.5. For a parallel-plate waveguide of spacing $a = 3$ cm and filled with a dielectric of $\varepsilon = 6.25\varepsilon_0$ and $\mu = \mu_0$, find the values of the guide characteristic impedance for

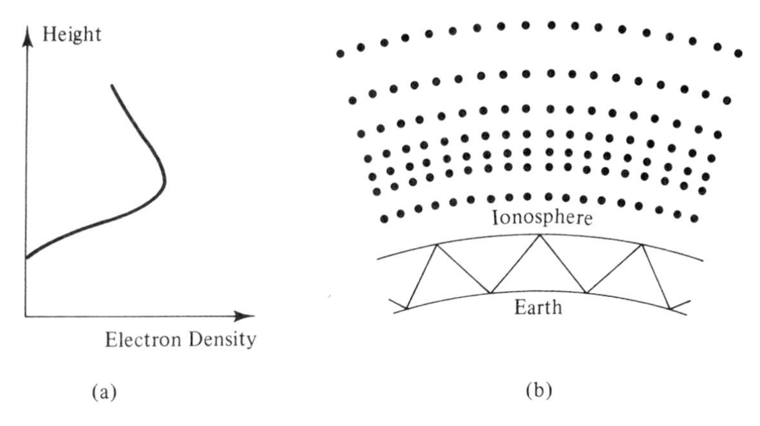

Figure 9.11. (a) Variation of electron density with height for a simplified ionosphere. (b) Depiction of waveguide mode of propagation in the earth-ionosphere waveguide.

each of the following cases: **(a)** $TE_{1,0}$ mode of $f = 3000$ MHz; **(b)** $TM_{1,0}$ mode of $f = 3000$ MHz; and **(c)** $TE_{1,0}$ mode of $f = 6000$ MHz.
Ans. **(a)** 202.3 Ω; **(b)** 112.4 Ω; **(c)** 159.94 Ω

9.3 DISPERSION AND GROUP VELOCITY

In the preceding section we learned that for the propagating range of frequencies, the phase velocity and the wavelength along the axis of the parallel-plate waveguide are given by

$$v_{pz} = \frac{v_p}{\sqrt{1 - (f_c/f)^2}} \tag{9.47}$$

and

$$\lambda_g = \frac{\lambda}{\sqrt{1 - (f_c/f)^2}} \tag{9.48}$$

where $v_p = 1/\sqrt{\mu\varepsilon}$, $\lambda = v_p/f = 1/f\sqrt{\mu\varepsilon}$, and f_c is the cutoff frequency. We note that for a particular mode, the phase velocity of propagation along the guide axis varies with the frequency. As a consequence of this characteristic of the guided wave propagation, the field patterns of the different frequency components of a signal comprising a band of frequencies do not maintain the same phase relationships as they propagate down the guide. This phenomenon is known as *dispersion*, so termed after the phenomenon of dispersion of colors by a prism.

To discuss dispersion, let us consider a simple example of two infinitely long trains A and B traveling in parallel, one below the other, with each train made up of boxcars of identical size and having wavy tops, as shown in Fig. 9.12. Let the spacings between the peaks (centers) of successive boxcars be 50 m and 90 m, and let the speeds of the trains be 20 m/s and 30 m/s, for trains A and B, respectively. Let the peaks of the cars numbered 0 for the two trains be aligned at time $t = 0$, as shown in Fig. 9.12(a). Now, as time progresses, the two peaks get out of alignment as shown, for example, for $t = 1$ s in Fig 9.12(b), since train B is traveling faster than train A. But at the same time, the gap between the peaks of cars numbered -1 decreases. This continues until at $t = 4$ s, the peak of car -1 of train A having moved by a distance of 80 m aligns with the peak of car -1 of train B, which will have moved by a distance of 120 m, as shown in Fig. 9.12(c). For an observer following the movement of the two trains as a group, the group appears to have moved by a distance of 30 m although the individual trains will have moved by 80 m and 120 m, respectively. Thus we can talk of a *group velocity*, that is, the velocity with which the group as a whole is moving. In this case, the group velocity is (30 m)/(4 s) or 7.5 m/s.

The situation in the case of the guided wave propagation of two different frequencies in the parallel-plate waveguide is exactly similar to the two-train example just discussed. The distance between the peaks of two successive cars is analogous to the guide wavelength, and the speed of the train is analogous to the phase velocity along the guide axis. Thus let us consider the field patterns corresponding to two waves of frequencies f_A and f_B propagating in the same mode, having guide wavelengths λ_{gA} and λ_{gB}, and phase velocities along the guide axis v_{pzA} and v_{pzB}, respectively, as shown, for example, for the electric field of the

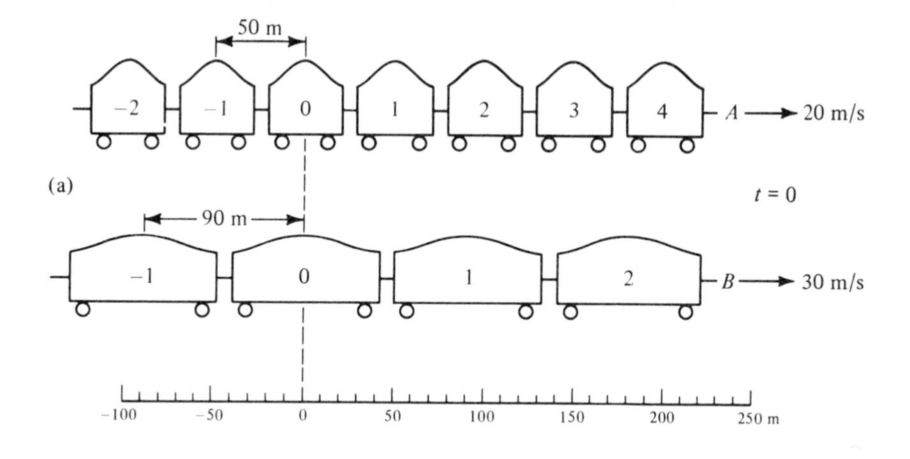

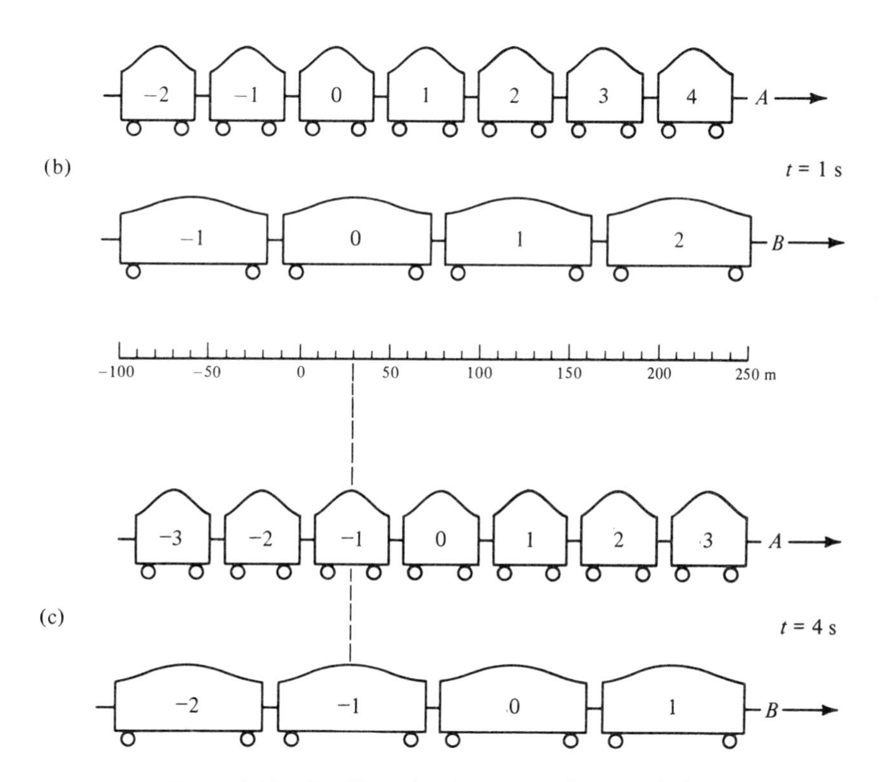

Figure 9.12. For illustrating the concept of group velocity.

$TE_{1,0}$ mode in Fig. 9.13. Let the positive peaks numbered 0 of the two patterns be aligned at $t = 0$, as shown in Fig. 9.13(a). As the individual waves travel with their respective phase velocities along the guide, these two peaks get out of alignment, but some time later, say, Δt, the positive peaks numbered -1 will align at some distance, say, Δz, from the location of the alignment of the 0 peaks, as shown in Fig. 9.13(b). Since the -1th peak of wave A will have traveled a distance $\lambda_{gA} + \Delta z$ with a phase velocity v_{pzA} and the -1th peak of wave B will have traveled a distance $\lambda_{gB} + \Delta z$ with a phase velocity v_{pzB} in this time Δt, we have

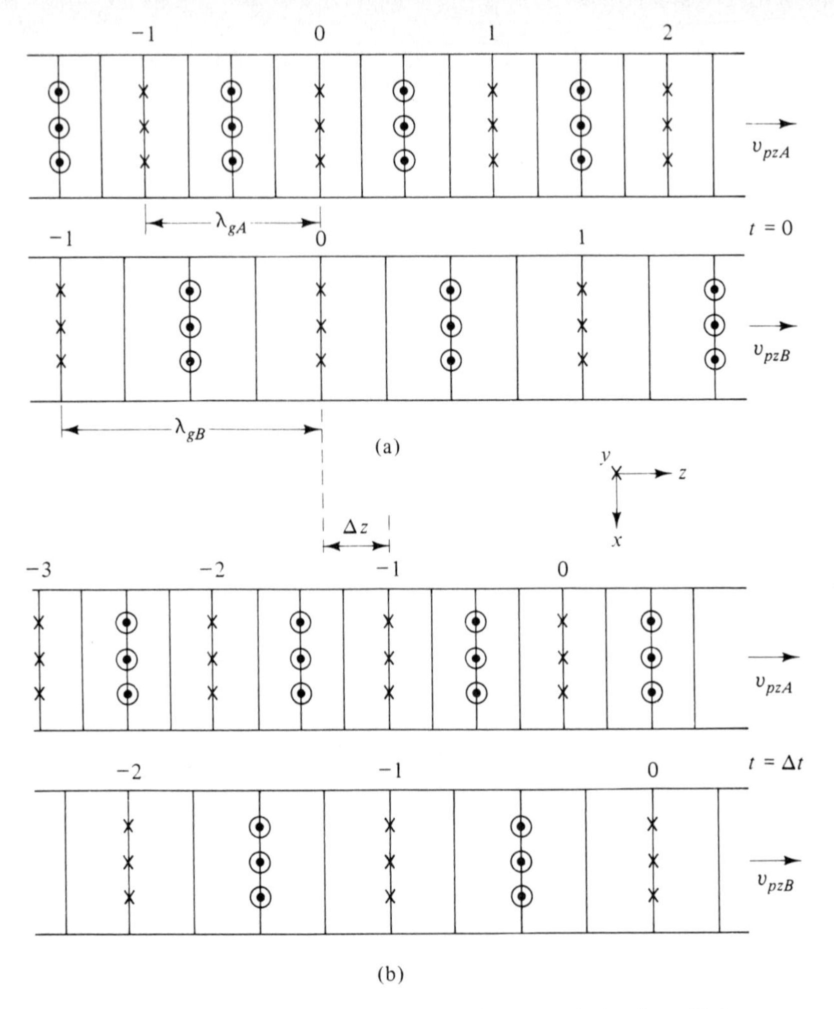

Figure 9.13. For illustrating the concept of group velocity for guided wave propagation.

$$\lambda_{gA} + \Delta z = v_{pzA}\,\Delta t \tag{9.49a}$$

$$\lambda_{gB} + \Delta z = v_{pzB}\,\Delta t \tag{9.49b}$$

Solving (9.49a) and (9.49b) for Δt and Δz, we obtain

$$\Delta t = \frac{\lambda_{gA} - \lambda_{gB}}{v_{pzA} - v_{pzB}} \tag{9.50a}$$

and

$$\Delta z = \frac{\lambda_{gA} v_{pzB} - \lambda_{gB} v_{pzA}}{v_{pzA} - v_{pzB}} \tag{9.50b}$$

The group velocity, v_g, is then given by

$$v_g = \frac{\Delta z}{\Delta t} = \frac{\lambda_{gA} v_{pzB} - \lambda_{gB} v_{pzA}}{\lambda_{gA} - \lambda_{gB}} = \frac{\lambda_{gA}\lambda_{gB} f_B - \lambda_{gB}\lambda_{gA} f_A}{\lambda_{gA}\lambda_{gB}\left(\dfrac{1}{\lambda_{gB}} - \dfrac{1}{\lambda_{gA}}\right)}$$

$$v_g = \frac{f_B - f_A}{\dfrac{1}{\lambda_{gB}} - \dfrac{1}{\lambda_{gA}}}$$

or

$$\boxed{v_g = \frac{\omega_B - \omega_A}{\beta_{zB} - \beta_{zA}}} \tag{9.51}$$

where β_{zA} and β_{zB} are the phase constants along the guide axis, corresponding to f_A and f_B, respectively. Thus the group velocity of a signal comprised of two frequencies is the ratio of the difference between the two radian frequencies to the difference between the corresponding phase constants along the guide axis.

Dispersion diagram

If we now have a signal comprised of a number of frequencies, then a value of group velocity can be obtained for each pair of these frequencies in accordance with (9.51). In general, these values of group velocity will all be different. In fact, this is the case for wave propagation in the parallel-plate guide, as can be seen from Fig. 9.14, which is a plot of ω versus β_z corresponding to the parallel-plate guide for which

$$\beta_z = \frac{2\pi}{\lambda_g} = \frac{2\pi}{\lambda}\sqrt{1 - \left(\frac{\lambda}{\lambda_c}\right)^2} = \omega\sqrt{\mu\varepsilon}\sqrt{1 - \left(\frac{f_c}{f}\right)^2} \tag{9.52}$$

Such a plot known as the ω–β_z *diagram* or *dispersion diagram*. Note that the dispersion diagram begins at $\omega = \omega_c$ on the ω-axis, since for $\omega < \omega_c$ propagation does not occur. The phase velocity along the guide axis given for a particular frequency by

$$\boxed{v_{pz} = \frac{\omega}{\beta_z}} \tag{9.53}$$

is equal to the slope of the line drawn from the origin to the point, on the dispersion curve, corresponding to that frequency, as shown in the figure for the three frequencies ω_1, ω_2, and ω_3. The group velocity for a particular pair of frequencies is given by the slope of the line joining the two points, on the curve, corre-

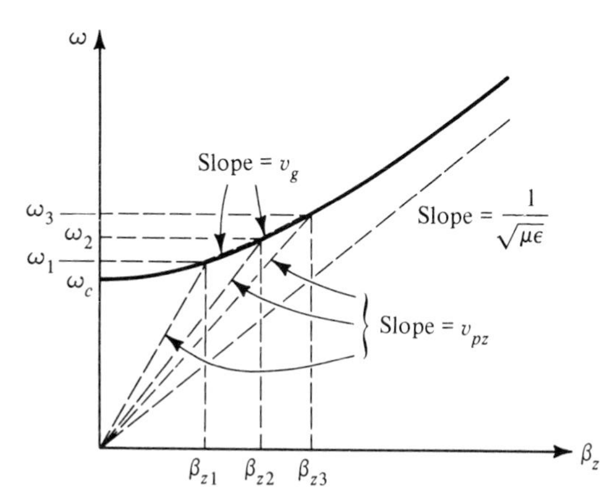

Figure 9.14. Dispersion diagram for the parallel-plate waveguide.

sponding to the two frequencies as shown in the figure for the two pairs ω_1, ω_2 and ω_2, ω_3, Since the curve is nonlinear, it can be seen that the two group velocities are not equal. We cannot then attribute a particular value of group velocity for the group of the three frequencies ω_1, ω_2, and ω_3.

If, however, the three frequencies are very close, as in the case of a narrow-band signal, it is meaningful to assign a group velocity to the entire group having a value equal to the slope of the tangent to the dispersion curve at the center frequency. Thus the group velocity corresponding to a narrow band of frequencies centered around a predominant frequency ω is given by

$$\boxed{v_g = \frac{d\omega}{d\beta_z}} \tag{9.54}$$

For the parallel-plate waveguide under consideration, we have from (9.52)

$$\begin{aligned}
\frac{d\beta_z}{d\omega} &= \sqrt{\mu\varepsilon}\sqrt{1 - \left(\frac{f_c}{f}\right)^2} + \omega\sqrt{\mu\varepsilon}\cdot\frac{1}{2}\left(1 - \frac{f_c^2}{f^2}\right)^{-1/2}\frac{f_c^2}{\pi f^3} \\
&= \sqrt{\mu\varepsilon}\left(1 - \frac{f_c^2}{f^2} + \frac{\omega}{2\pi}\frac{f_c^2}{f^3}\right)\left(1 - \frac{f_c^2}{f^2}\right)^{-1/2} \\
&= \sqrt{\mu\varepsilon}\left(1 - \frac{f_c^2}{f^2}\right)^{-1/2}
\end{aligned}$$

and hence, from (9.54),

$$\boxed{v_g = \frac{d\omega}{d\beta_z} = \frac{1}{\sqrt{\mu\varepsilon}}\sqrt{1 - \frac{f_c^2}{f^2}} = v_p\sqrt{1 - \left(\frac{f_c}{f}\right)^2}} \tag{9.55}$$

From (9.47) and (9.55), we note that $v_{pz} > v_p$, $v_g < v_p$, and

$$v_{pz}v_g = v_p^2$$

For a numerical example, let us consider the air-dielectric parallel-plate waveguide of spacing $a = 5$ cm and a narrow-band signal of center frequency $f =$ 10,000 MHz propagating in the $\text{TE}_{1,0}$ mode. Then from Example 9.2, $f_c =$ 3000 MHz, and from (9.55),

$$\begin{aligned}
v_g &= 3 \times 10^8\sqrt{1 - (3/10)^2} \\
&= 2.862 \times 10^8 \text{ m/s}
\end{aligned}$$

as compared to $v_{pz} = 3.145 \times 10^8$ m/s found in Example 9.2.

Amplitude-modulated signal

An example of a narrow-band signal is an amplitude-modulated signal, having a carrier frequency ω modulated by a low-frequency $\Delta\omega \ll \omega$ as given by

$$E_x(t) = E_{x0}(1 + m\cos\Delta\omega\cdot t)\cos\omega t \tag{9.56}$$

where m is the percentage modulation. Such a signal is actually equivalent to a superposition of unmodulated signals of three frequencies $\omega - \Delta\omega$, ω, and $\omega + \Delta\omega$, as can be seen by expanding the right side of (9.56). Thus

$$E_x(t) = E_{x0} \cos \omega t + mE_{x0} \cos \omega t \cos \Delta\omega \cdot t$$
$$= E_{x0} \cos \omega t + \frac{mE_{x0}}{2} [\cos (\omega - \Delta\omega)t + \cos (\omega + \Delta\omega)t] \tag{9.57}$$

The frequencies $\omega - \Delta\omega$ and $\omega + \Delta\omega$ are the side frequencies. When the amplitude-modulated signal propagates in a dispersive channel such as the parallel-plate waveguide under consideration, the different frequency components undergo phase changes in accordance with their respective phase constants. Thus if $\beta_z - \Delta\beta_z$, β_z, and $\beta_z + \Delta\beta_z$ are the phase constants corresponding to $\omega - \Delta\omega$, ω, and $\omega + \Delta\omega$, respectively, assuming linearity of the dispersion curve within the narrow band, the amplitude-modulated wave is given by

$$E_x(z, t) = E_{x0} \cos (\omega t - \beta_z z)$$
$$+ \frac{mE_{x0}}{2} \{\cos [(\omega - \Delta\omega)t - (\beta_z - \Delta\beta_z)z]$$
$$+ \cos [(\omega + \Delta\omega)t - (\beta_z + \Delta\beta_z)z]\}$$
$$= E_{x0} \cos (\omega t - \beta_z z) \tag{9.58}$$
$$+ \frac{mE_{x0}}{2} \{\cos [(\omega t - \beta_z z) - (\Delta\omega \cdot t - \Delta\beta_z \cdot z)]$$
$$+ \cos [(\omega t - \beta_z z) + (\Delta\omega \cdot t - \Delta\beta_z \cdot z)]\}$$
$$= E_{x0} \cos (\omega t - \beta_z z) + mE_{x0} \cos (\omega t - \beta_z z) \cos (\Delta\omega \cdot t - \Delta\beta_z \cdot z)$$
$$= E_{x0}[1 + m \cos (\Delta\omega \cdot t - \Delta\beta_z \cdot z)] \cos (\omega t - \beta_z z)$$

This indicates that although the carrier frequency phase changes in accordance with the phase constant β_z, the modulation envelope and hence the information travels with the group velocity $\Delta\omega/\Delta\beta_z$, as shown in Fig. 9.15. In view of this and since v_g is less than v_p, the fact that v_{pz} is greater than v_p is not a violation of the theory of relativity. Since it is always necessary to use some modulation technique to convey information from one point to another, the information always takes more time to reach from one point to another in a dispersive channel than in

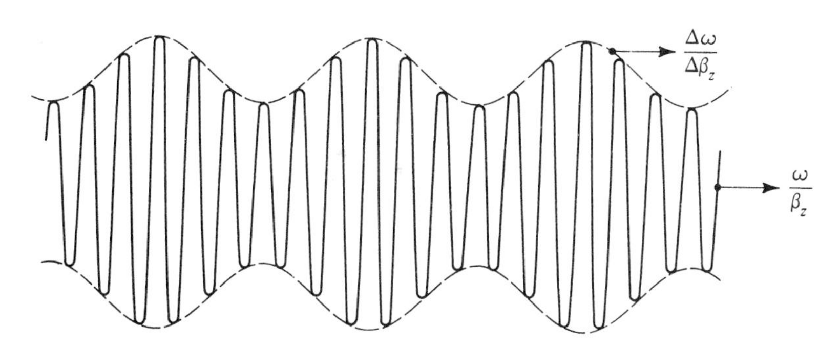

Figure 9.15. For illustrating that the modulation envelope travels with the group velocity.

the corresponding nondispersive medium. For further understanding of the concept of group velocity, the reader is advised to view a movie narrated by Van Duzer.[1]

K9.3. Dispersion; Group velocity; Dispersion diagram; Narrow-band signal.

D9.6. The $\omega - \beta_z$ curve for a dispersive channel can be approximated by

$$\omega = \omega_0 + k\beta_z^2$$

in the vicinity of $\omega = 1.5\omega_0$, where k is a positive constant. Find **(a)** the phase velocity for a signal of $\omega = 1.4\omega_0$; **(b)** the phase velocity for a signal of $\omega = 1.6\omega_0$; **(c)** the group velocity for a signal composed of two radian frequencies $1.4\omega_0$ and $1.6\omega_0$; and **(d)** the group velocity for a narrow-band signal having the center radian frequency $1.5\omega_0$.
Ans. **(a)** $2.2136\sqrt{k\omega_0}$; **(b)** $2.0656\sqrt{k\omega_0}$; **(c)** $1.4071\sqrt{k\omega_0}$; **(d)** $1.4142\sqrt{k\omega_0}$

9.4 RECTANGULAR WAVEGUIDE AND CAVITY RESONATOR

TE modes in rectangular waveguide

In Section 9.2 we mentioned that a common form of waveguide is the rectangular waveguide. To introduce the rectangular waveguide, we begin with TE modes in a parallel-plate waveguide. We recall that the parallel-plate waveguide is made up of two perfectly conducting sheets in the planes $x = 0$ and $x = a$ and that the electric field of the $TE_{m,0}$ mode has only a y-component with m number of one-half sinusoidal variations in the x-direction and no variations in the y-direction. If we now introduce two perfectly conducting sheets in two constant y-planes, say, $y = 0$ and $y = b$, the field distribution will remain unaltered since the electric field is entirely normal to the plates, and hence the boundary condition of zero tangential electric field is satisfied for both sheets. We then have the rectangular waveguide, a metallic pipe with rectangular cross section in the xy-plane, as shown in Fig. 9.16.

Since the $TE_{m,0}$ mode field expressions derived for the parallel-plate waveguide satisfy the boundary conditions for the rectangular waveguide, those expressions as well as the entire discussion of the parallel-plate waveguide case hold also for $TE_{m,0}$ mode propagation in the rectangular waveguide case. We learned that the $TE_{m,0}$ modes can be interpreted as due to uniform plane waves having electric field in the y-direction and bouncing obliquely between the conducting walls $x = 0$ and $x = a$, and with the associated cutoff condition characterized by bouncing of the waves back and forth normally to these walls, as shown in Fig. 9.17(a). For the cutoff condition, the dimension a is equal to m number of one-half wavelengths such that

$$\boxed{[\lambda_c]_{TE_{m,0}} = \frac{2a}{m}} \tag{9.59}$$

[1] T. Van Duzer, *Wave Velocities, Dispersion and the Omega–Beta Diagram* (Newton, Mass.: Educational Development Center).

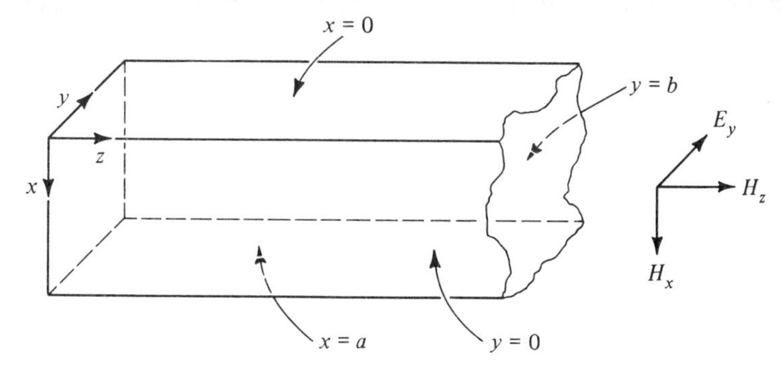

Figure 9.16. Rectangular waveguide.

In a similar manner, we can have uniform plane waves having electric field in the x-direction and bouncing obliquely between the walls $y = 0$ and $y = b$, and with the associated cutoff condition characterized by bouncing of the waves back and forth normally to these walls, as shown in Fig. 9.17(b), thereby resulting in $TE_{0,n}$ modes having no variations in the x-direction and n number of one-half sinusoidal variations in the y-direction. For the cutoff condition, the dimension b is equal to n number of one-half wavelengths such that

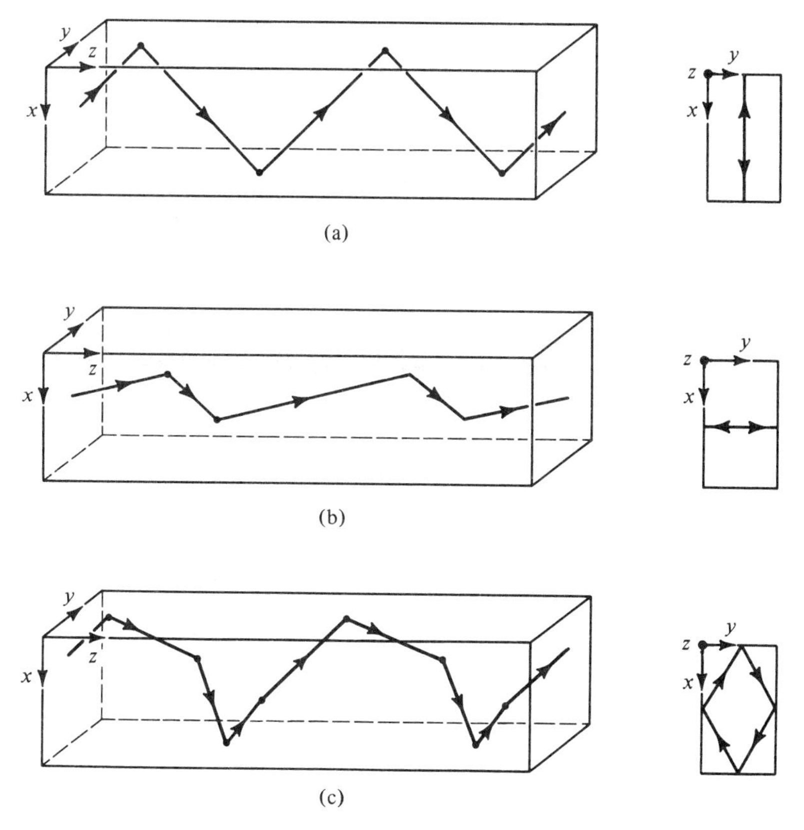

Figure 9.17. Propagation and cutoff of (a) $TE_{m,0}$, (b) $TE_{0,n}$, and (c) $TE_{m,n}$ modes in a rectangular waveguide.

$$[\lambda_c]_{TE_{0,n}} = \frac{2b}{n} \tag{9.60}$$

We can even have $TE_{m,n}$ modes having m number of one-half sinusoidal variations in the x-direction and n number of one-half sinusoidal variations in the y-direction due to uniform plane waves having both x- and y-components of the electric field and bouncing obliquely between all four walls of the guide and with the associated cutoff condition characterized by bouncing of the waves back and forth obliquely between the four walls as shown, for example, in Fig. 9.17(c). For the cutoff condition, the dimension a must be equal to m number of one-half apparent wavelengths in the x-direction, and the dimension b must be equal to n number of one-half apparent wavelengths in the y-direction such that

$$\frac{1}{[\lambda_c]^2_{TE_{m,n}}} = \frac{1}{(2a/m)^2} + \frac{1}{(2b/n)^2}$$

or

$$[\lambda_c]_{TE_{m,n}} = \frac{1}{\sqrt{(m/2a)^2 + (n/2b)^2}} \tag{9.61}$$

Derivation of field expressions for TE modes

The discussion thus far of $TE_{m,n}$ modes in a rectangular waveguide has been based on qualitative reasoning. We shall now derive the field expressions for the TE modes. To do this, we shall first show, by making use of the expansions for the Maxwell's curl equations in Cartesian coordinates, that all transverse (x and y) field components are derivable from the longitudinal field component H_z. It is convenient to use the phasor forms of the field components and differential equations. Since all components of the fields are then dependent on t and z in the manner $e^{j[\omega t - \beta_z z]}$, we can replace $\partial/\partial t$ by $j\omega$ and $\partial/\partial z$ by $-j\beta_z$. Furthermore, $E_z = 0$ in view of TE modes and J_x, J_y, and J_z are all zero since the medium inside the waveguide is a perfect dielectric. Thus the phasor forms of (4.12a)–(4.12c) and of the component equations of (4.22) pertinent to the discussion here are

$$j\beta_z \bar{E}_y = -j\omega\mu \bar{H}_x \tag{9.62a}$$

$$-j\beta_z \bar{E}_x = -j\omega\mu \bar{H}_y \tag{9.62b}$$

$$\frac{\partial \bar{E}_y}{\partial x} - \frac{\partial \bar{E}_x}{\partial y} = -j\omega\mu \bar{H}_z \tag{9.62c}$$

$$\frac{\partial \bar{H}_z}{\partial y} + j\beta_z \bar{H}_y = j\omega\varepsilon \bar{E}_x \tag{9.62d}$$

$$-j\beta_z \bar{H}_x - \frac{\partial \bar{H}_z}{\partial x} = j\omega\varepsilon \bar{E}_y \tag{9.62e}$$

$$\frac{\partial \bar{H}_y}{\partial x} - \frac{\partial \bar{H}_x}{\partial y} = 0 \tag{9.62f}$$

Solving (9.62a), (9.62b), (9.62d), and (9.62e), for $\bar{E}_x$, $\bar{E}_y$, $\bar{H}_x$, and $\bar{H}_y$ in terms of $\bar{H}_z$, we obtain

$$\bar{E}_x = \frac{j\omega\mu}{\beta_z^2 - \beta^2} \frac{\partial \bar{H}_z}{\partial y} \tag{9.63a}$$

$$\bar{E}_y = -\frac{j\omega\mu}{\beta_z^2 - \beta^2} \frac{\partial \bar{H}_z}{\partial x} \tag{9.63b}$$

$$\overline{H}_x = j\frac{\beta_z}{\beta_z^2 - \beta^2}\frac{\partial \overline{H}_z}{\partial x} \tag{9.63c}$$

$$\overline{H}_y = j\frac{\beta_z}{\beta_z^2 - \beta^2}\frac{\partial \overline{H}_z}{\partial y} \tag{9.63d}$$

Furthermore by substituting (9.63a) and (9.63b) into (9.62c) and rearranging, we obtain a differential equation for $\overline{H}_z$ as given by

$$\boxed{\frac{\partial^2 \overline{H}_z}{\partial x^2} + \frac{\partial^2 \overline{H}_z}{\partial y^2} + (\beta_z^2 - \beta^2)\overline{H}_z = 0} \tag{9.64}$$

Recall that $\beta = \omega\sqrt{\mu\varepsilon}$ so that $\beta^2 = \omega^2\mu\varepsilon$.

Separation of variables technique

To solve (9.64) for $\overline{H}_z$, we make use of the *separation of variables* technique. This consists of assuming that the required function of the two variables x and y is the product of two functions, one of which is a function of x only and the second is a function of y only. Thus denoting these functions to be $\overline{X}$ and $\overline{Y}$, we have

$$\boxed{\overline{H}_z(x, y, z) = \overline{X}(x)\overline{Y}(y)e^{-j\beta_z z}} \tag{9.65}$$

Substituting (9.65) into (9.64), we then obtain

$$\overline{X}''(x)\,\overline{Y}(y) + \overline{X}(x)\,\overline{Y}''(y) + (\beta_z^2 - \beta^2)\overline{X}(x)\,\overline{Y}(y) = 0$$

or

$$\frac{\overline{X}''}{\overline{X}} + \frac{\overline{Y}''}{\overline{Y}} = \beta_z^2 - \beta^2 \tag{9.66}$$

where the primes denote differentiation with respect to the respective variables. Equation (9.66) says that a function of x only plus a function of y only is equal to a constant. For this to be satisfied, both functions must be equal to constants. Hence we write

$$\frac{\overline{X}''}{\overline{X}} = -\beta_x^2, \text{ a constant} \tag{9.67a}$$

and

$$\frac{\overline{Y}''}{\overline{Y}} = -\beta_y^2, \text{ a constant} \tag{9.67b}$$

or

$$\frac{d^2\overline{X}}{dx^2} = -\beta_x^2\overline{X} \tag{9.68a}$$

and

$$\frac{d^2\overline{Y}}{dy^2} = -\beta_y^2\overline{Y} \tag{9.68b}$$

We have thus obtained two ordinary differential equations involving separately the two variables x and y; hence the technique is known as the *separation of variables* technique.

The solutions to (9.68a) and (9.68b) are given by

$$\bar{X}(x) = \bar{A}_1 e^{j\beta_x x} + \bar{A}_2 e^{-j\beta_x x}$$

$$\bar{Y}(y) = \bar{B}_1 e^{j\beta_y y} + \bar{B}_2 e^{-j\beta_y y}$$

so that

$$\bar{H}_z = (\bar{A}_1 e^{j\beta_x x} + A_2 e^{-j\beta_x x})(\bar{B}_1 e^{j\beta_y y} + \bar{B}_2 e^{-j\beta_y y}) e^{-j\beta_z z} \tag{9.69}$$

where $\bar{A}_1$, $\bar{A}_2$, $\bar{B}_1$, and $\bar{B}_2$ are constants. We also note from substitution of (9.67a) and (9.67b) into (9.66) that

$$-\beta_x^2 - \beta_y^2 = \beta_z^2 - \beta^2$$

or

$$\beta_z^2 = \beta^2 - \beta_x^2 - \beta_y^2 \tag{9.70}$$

Now, to determine the constants in (9.69), we make use of the boundary conditions which require that the tangential components of the electric field intensity on all four walls of the guide be zero. Thus we have

$$\bar{E}_x = 0 \qquad \text{for } y = 0,\ 0 < x < a$$

$$\bar{E}_x = 0 \qquad \text{for } y = b,\ 0 < x < a$$

$$\bar{E}_y = 0 \qquad \text{for } x = 0,\ 0 < y < b$$

$$\bar{E}_y = 0 \qquad \text{for } x = a,\ 0 < y < b$$

To apply these boundary conditions to (9.69), we have to translate them into boundary conditions involving $\bar{H}_z$. From (9.63a) and (9.63b), these are

$$\frac{\partial \bar{H}_z}{\partial y} = 0 \qquad \text{for } y = 0,\ 0 < x < a \tag{9.71a}$$

$$\frac{\partial \bar{H}_z}{\partial y} = 0 \qquad \text{for } y = b,\ 0 < x < a \tag{9.71b}$$

$$\frac{\partial \bar{H}_z}{\partial x} = 0 \qquad \text{for } x = 0,\ 0 < y < b \tag{9.71c}$$

$$\frac{\partial \bar{H}_z}{\partial x} = 0 \qquad \text{for } x = a,\ 0 < y < b \tag{9.71d}$$

Using (9.71c) and (9.71a) in conjunction with (9.69), we get

$$\bar{A}_1 - \bar{A}_2 = 0 \qquad \text{or} \qquad \bar{A}_2 = \bar{A}_1$$

$$\bar{B}_1 - \bar{B}_2 = 0 \qquad \text{or} \qquad \bar{B}_2 = \bar{B}_1$$

which then simplifies (9.69) to

$$\bar{H}_z = \bar{A} \cos \beta_x x \cos \beta_y y \, e^{-j\beta_z z} \tag{9.72}$$

where $\bar{A}$ is a constant. Using the remaining two boundary conditions (9.71d) and (9.71b), we then obtain

$$\sin \beta_x a = 0 \qquad \text{or} \qquad \beta_x = \frac{m\pi}{a}, \quad m = 0, 1, 2, \ldots \tag{9.73a}$$

$$\sin \beta_y b = 0 \qquad \text{or} \qquad \beta_y = \frac{n\pi}{b}, \quad n = 0, 1, 2, \ldots \tag{9.73b}$$

Thus the solution for $\bar{H}_z$ for the $\text{TE}_{m,n}$ mode is given by

$$\boxed{\bar{H}_z = \bar{A} \cos \frac{m\pi x}{a} \cos \frac{n\pi y}{b} e^{-j\beta_z z}} \tag{9.74}$$

We also note by substituting (9.73a) and (9.73b) in (9.70) that

$$\beta_z^2 = \beta^2 - \left(\frac{m\pi}{a}\right)^2 - \left(\frac{n\pi}{b}\right)^2 \tag{9.75}$$

For propagation to occur, the exponent β_z in (9.74) must be real. Hence the cutoff condition is given by

$$\omega^2 \mu\varepsilon - \left(\frac{m\pi}{a}\right)^2 - \left(\frac{n\pi}{b}\right)^2 = 0 \tag{9.76}$$

or the cutoff frequency is given by

$$\boxed{f_c = \frac{1}{\sqrt{\mu\varepsilon}} \sqrt{\left(\frac{m}{2a}\right)^2 + \left(\frac{n}{2b}\right)^2}} \tag{9.77}$$

and the cutoff wavelength is

$$\boxed{\begin{aligned} \lambda_c &= \frac{1}{\sqrt{\mu\varepsilon} f_c} \\ &= \frac{1}{\sqrt{(m/2a)^2 + (n/2b)^2}} \end{aligned}} \tag{9.78}$$

which is the same as given by (9.61). Now, from (9.75) and (9.78), we have

$$\begin{aligned} \beta_z^2 - \beta^2 &= -\left(\frac{m\pi}{a}\right)^2 - \left(\frac{n\pi}{b}\right)^2 \\ &= -(2\pi)^2\left[\left(\frac{m}{2a}\right)^2 + \left(\frac{n}{2b}\right)^2\right] \\ &= -\left(\frac{2\pi}{\lambda_c}\right)^2 \end{aligned} \tag{9.79}$$

Substituting (9.74) and (9.79) into (9.63a)–(9.63d), we obtain the expressions for the transverse field components:

$$\bar{E}_x = j\frac{\omega\mu\lambda_c^2}{4\pi^2}\frac{n\pi}{b}\bar{A}\cos\frac{m\pi x}{a}\sin\frac{n\pi y}{b}e^{-j\beta_z z} \tag{9.80a}$$

$$\bar{E}_y = -j\frac{\omega\mu\lambda_c^2}{4\pi^2}\frac{m\pi}{a}\bar{A}\sin\frac{m\pi x}{a}\cos\frac{n\pi y}{b}e^{-j\beta_z z} \tag{9.80b}$$

$$\bar{H}_x = j\frac{\lambda_c^2}{2\pi\lambda_g}\frac{m\pi}{a}\bar{A}\sin\frac{m\pi x}{a}\cos\frac{n\pi y}{b}e^{-j\beta_z z} \tag{9.80c}$$

$$\bar{H}_y = j\frac{\lambda_c^2}{2\pi\lambda_g}\frac{n\pi}{b}\bar{A}\cos\frac{m\pi x}{a}\sin\frac{n\pi y}{b}e^{-j\beta_z z} \tag{9.80d}$$

Note that the sine terms in these field expressions satisfy the boundary conditions of zero tangential electric field and zero normal magnetic field at the walls of the waveguide. It can also be seen that if both m and n are equal to zero, then all transverse field components go to zero. Therefore, *for TE modes, either m or n can be zero, but both m and n cannot be zero*.

TM modes in rectangular waveguide

The entire procedure for the derivation of the field expressions can be repeated for TM waves by starting with the longitudinal field component $\bar{E}_z$. We shall however not pursue the derivation here but instead present the final expressions. This is done in Table 9.1 together with the TE mode field expressions. The upper signs of the $\mp$ and $\pm$ signs in these expressions refer to waves propagating in the $+z$-direction, whereas the lower signs refer to waves propagating in the $-z$-direction. Note from the expression for $\bar{E}_z$ in Table 9.1 that the x- and y-variations of $\bar{E}_z$ are sinusoidal so that $\bar{E}_z$ goes to zero on all four walls of the waveguide. This arises because $\bar{E}_z$, being longitudinal, is tangential to all four walls, and the boundary conditions require that the tangential components of **E** be zero on the walls. It can also be seen that if either m or n is equal to zero, then $\bar{E}_z = 0$. Therefore, *for TM modes both m and n must be nonzero*. Also listed in Table 9.1 are the expressions for the cutoff frequency f_c, the cutoff wavelength λ_c, the guide wavelength λ_g, the phase velocity v_{pz} along the guide axis, and the guide characteristic impedance η_g, all of which have the same interpretations as the corresponding quantities for the parallel-plate waveguide case.

Dominant mode

The foregoing discussion of the modes of propagation in a rectangular waveguide points out that a signal of given frequency can propagate in several modes, namely, all modes for which the cutoff frequencies are less than the signal frequency or the cutoff wavelengths are greater than the signal wavelength. Waveguides are, however, designed so that only one mode, the mode with the lowest cutoff frequency (or the largest cutoff wavelength), propagates. This is known as the *dominant mode*. From Table 9.1, we can see that the dominant mode is the $TE_{1,0}$ mode or the $TE_{0,1}$ mode, depending on whether the dimension a or the dimension b is the larger of the two. By convention, the larger dimension is designated to be a, and hence the $TE_{1,0}$ mode is the dominant mode. We shall now consider an example.

Finding propagating modes

Example 9.4

It is desired to determine the lowest four cutoff frequencies referred to the cutoff frequency of the dominant mode for three cases of rectangular waveguide dimensions:

TABLE 9.1 FIELD EXPRESSIONS AND ASSOCIATED PARAMETERS FOR TE AND TM MODES IN A RECTANGULAR WAVEGUIDE

Transverse electric (TE) waves	Transverse magnetic (TM) waves
Field Expressions: $(m, n = 0, 1, 2, \ldots,$ but not both zero$)$	Field Expressions: $(m, n = 1, 2, 3, \ldots)$
$\bar{E}_z = 0$	$\bar{H}_z = 0$
$\bar{H}_z = \bar{A} \cos \frac{m\pi x}{a} \cos \frac{n\pi y}{b} e^{\mp j\beta_z z}$	$\bar{E}_z = \bar{A} \sin \frac{m\pi x}{a} \sin \frac{n\pi y}{b} e^{\mp j\beta_z z}$
$\bar{E}_x = j\frac{\lambda_c^2}{4\pi^2} \omega\mu \frac{n\pi}{b} \bar{A} \cos \frac{m\pi x}{a} \sin \frac{n\pi y}{b} e^{\mp j\beta_z z}$	$\bar{E}_x = \mp j\frac{\lambda_c^2}{2\pi\lambda_g} \frac{m\pi}{a} \bar{A} \cos \frac{m\pi x}{a} \sin \frac{n\pi y}{b} e^{\mp j\beta_z z}$
$\bar{E}_y = -j\frac{\lambda_c^2}{4\pi^2} \omega\mu \frac{m\pi}{a} \bar{A} \sin \frac{m\pi x}{a} \cos \frac{n\pi y}{b} e^{\mp j\beta_z z}$	$\bar{E}_y = \mp j\frac{\lambda_c^2}{2\pi\lambda_g} \frac{n\pi}{b} \bar{A} \sin \frac{m\pi x}{a} \cos \frac{n\pi y}{b} e^{\mp j\beta_z z}$
$\bar{H}_x = \mp \frac{\bar{E}_y}{\eta_g}$	$\bar{H}_x = \mp \frac{\bar{E}_y}{\eta_g}$
$\bar{H}_y = \pm \frac{\bar{E}_x}{\eta_g}$	$\bar{H}_y = \pm \frac{\bar{E}_x}{\eta_g}$
$f_c = \frac{1}{2\sqrt{\mu\varepsilon}} \sqrt{\left(\frac{m}{a}\right)^2 + \left(\frac{n}{b}\right)^2}$	$f_c = \frac{1}{2\sqrt{\mu\varepsilon}} \sqrt{\left(\frac{m}{a}\right)^2 + \left(\frac{n}{b}\right)^2}$
$\lambda_c = \frac{2}{\sqrt{(m/a)^2 + (n/b)^2}}$	$\lambda_c = \frac{2}{\sqrt{(m/a)^2 + (n/b)^2}}$
$\lambda_g = \frac{\lambda}{\sqrt{1 - (\lambda/\lambda_c)^2}} = \frac{\lambda}{\sqrt{1 - (f_c/f)^2}}$	$\lambda_g = \frac{\lambda}{\sqrt{1 - (\lambda/\lambda_c)^2}} = \frac{\lambda}{\sqrt{1 - (f_c/f)^2}}$
$v_{pz} = \frac{1}{\sqrt{\mu\varepsilon}\sqrt{1 - (f_c/f)^2}} = \frac{1}{\sqrt{\mu\varepsilon}\sqrt{1 - (\lambda/\lambda_c)^2}}$	$v_{pz} = \frac{1}{\sqrt{\mu\varepsilon}\sqrt{1 - (f_c/f)^2}} = \frac{1}{\sqrt{\mu\varepsilon}\sqrt{1 - (\lambda/\lambda_c)^2}}$
$\eta_g = \frac{\sqrt{\mu/\varepsilon}}{\sqrt{1 - (f_c/f)^2}} = \frac{\sqrt{\mu/\varepsilon}}{\sqrt{1 - (\lambda/\lambda_c)^2}}$	$\eta_g = \sqrt{\frac{\mu}{\varepsilon}} \sqrt{1 - \left(\frac{f_c}{f}\right)^2} = \sqrt{\frac{\mu}{\varepsilon}} \sqrt{1 - \left(\frac{\lambda}{\lambda_c}\right)^2}$

$b/a = 1$, $b/a = \frac{1}{2}$, and $b/a = \frac{1}{3}$. Given $a = 3$ cm, it is then desired to find the propagating mode(s) for $f = 9000$ MHz for each of the three cases.

From Table 9.1, the expression for the cutoff wavelength for a $\text{TE}_{m,n}$ mode where $m = 0, 1, 2, 3, \ldots$ and $n = 0, 1, 2, 3, \ldots$ but not both m and n equal to zero and for a $\text{TM}_{m,n}$ mode where $m = 1, 2, 3, \ldots$ and $n = 1, 2, 3, \ldots$ is given by

$$\lambda_c = \frac{1}{\sqrt{(m/2a)^2 + (n/2b)^2}}$$

The corresponding expression for the cutoff frequency is

$$f_c = \frac{v_p}{\lambda_c} = \frac{1}{\sqrt{\mu\varepsilon}} \sqrt{\left(\frac{m}{2a}\right)^2 + \left(\frac{n}{2b}\right)^2}$$

$$= \frac{1}{2a\sqrt{\mu\varepsilon}} \sqrt{m^2 + \left(n\frac{a}{b}\right)^2}$$

The cutoff frequency of the dominant mode $\text{TE}_{1,0}$ is $1/2a\sqrt{\mu\varepsilon}$. Hence

$$\frac{f_c}{[f_c]_{\text{TE}_{1,0}}} = \sqrt{m^2 + \left(n\frac{a}{b}\right)^2}$$

By assigning different pairs of values for m and n, the lowest four values of $f_c/[f_c]_{TE_{1,0}}$ can be computed for each of the three specified values of b/a. These computed values and the corresponding modes are shown in Fig. 9.18.

For $a = 3$ cm, and assuming free space for the dielectric in the waveguide,

$$[f_c]_{TE_{1,0}} = \frac{1}{2a\sqrt{\mu\varepsilon}} = \frac{3 \times 10^8}{2 \times 0.03} = 5000 \text{ MHz}$$

Hence for a signal of frequency $f = 9000$ MHz, all the modes for which $f_c/[f_c]_{TE_{1,0}}$ is less than 1.8 propagate. From Fig. 9.18, these are

$$\begin{aligned} &TE_{1,0},\ TE_{0,1},\ TM_{1,1},\ TE_{1,1} && \text{for} \quad b/a = 1 \\ &TE_{1,0} && \text{for} \quad b/a = \tfrac{1}{2} \\ &TE_{1,0} && \text{for} \quad b/a = \tfrac{1}{3} \end{aligned}$$

It can be seen from Fig. 9.18 that for $b/a \le \frac{1}{2}$, the second lowest cutoff frequency which corresponds to that of the $TE_{2,0}$ mode is twice the cutoff frequency of the dominant mode $TE_{1,0}$. For this reason, the dimension b of a rectangular waveguide is generally chosen to be less than or equal to $a/2$ in order to achieve single-mode transmission over a complete octave (factor of two) range of frequencies.

Transmission-line analogy

As in the case of the parallel-plate waveguide, reflection and transmission at discontinuities in rectangular waveguides can be studied by using the transmission-line analogy. We shall illustrate this by means of an example.

Example 9.5

A rectangular waveguide extending in the z-direction and having the dimensions $a = 4$ cm and $b = 2$ cm has a dielectric discontinuity at $z = 0$, as shown in Fig. 9.19. For $TE_{1,0}$ waves of frequency $f = 5000$ MHz incident from section 1, we

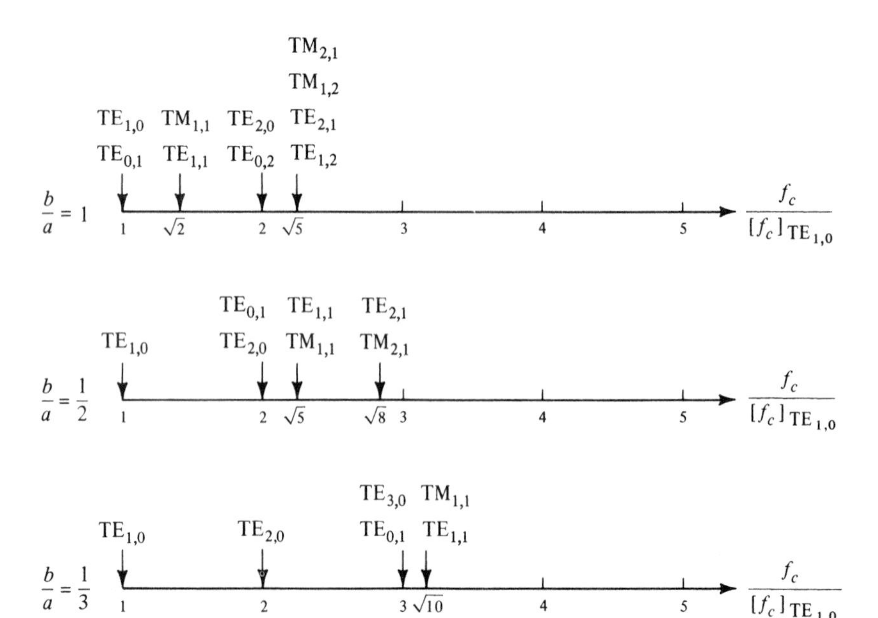

Figure 9.18. Lowest four cutoff frequencies referred to the cutoff frequency of the dominant mode for three cases of rectangular waveguide dimensions.

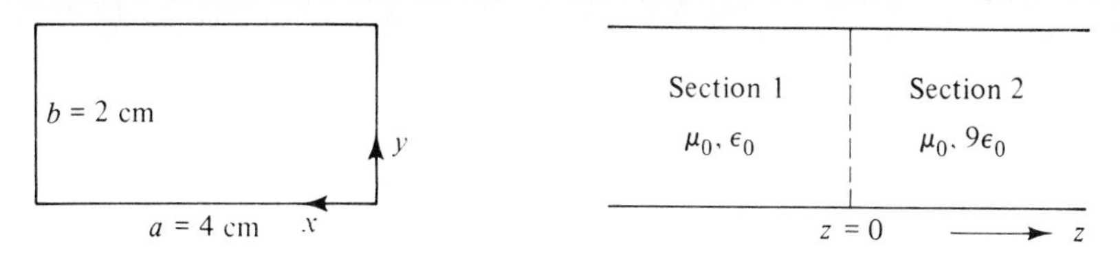

Figure 9.19. Rectangular waveguide discontinuity.

wish to find (a) the transmission-line equivalent and (b) the length and the permittivity of a quarter-wave section required to achieve a match between the two sections.

(a) First we note that for the $TE_{1,0}$ mode, $\lambda_c = 2a = 8$ cm for both sections. For $f = 5000$ MHz, the wavelength in free space is $\lambda_1 = 6$ cm and the wavelength in a dielectric of permittivity $9\varepsilon_0$ is $\lambda_2 = 2$ cm. Since λ_1 and λ_2 are both less than λ_c, the $TE_{1,0}$ mode propagates in both sections. Denoting the guide parameters associated with sections 1 and 2 by subscripts 1 and 2, respectively, we then obtain

$$\eta_{g1} = \frac{\eta_1}{\sqrt{1 - (\lambda_1/\lambda_c)^2}} = \frac{377}{\sqrt{1 - (6/8)^2}} = 570\ \Omega$$

$$\eta_{g2} = \frac{\eta_2}{\sqrt{1 - (\lambda_2/\lambda_c)^2}} = \frac{377/3}{\sqrt{1 - (2/8)^2}} = 129.8\ \Omega$$

Thus the transmission-line equivalent is as shown in Fig. 9.20.

(b) The characteristic impedance of a quarter-wave section required to achieve a match between line 1 and line 2 must be equal to $\sqrt{\eta_{g1}\eta_{g2}}$. Denoting the parameters associated with the quarter-wave section by subscript 3, we then have

$$\eta_{g3} = \frac{\eta_3}{\sqrt{1 - (\lambda_3/\lambda_c)^2}} = \sqrt{\eta_{g1}\eta_{g2}}$$

or

$$\frac{\eta_1\sqrt{\varepsilon_0/\varepsilon_3}}{\sqrt{1 - (\lambda_1/\lambda_c)^2(\varepsilon_0/\varepsilon_3)}} = \sqrt{\eta_{g1}\eta_{g2}}$$

$$\frac{\varepsilon_0/\varepsilon_3}{1 - (6/8)^2(\varepsilon_0/\varepsilon_3)} = \frac{570 \times 129.8}{(377)^2} = 0.5205$$

solving which we obtain $\varepsilon_3 = 2.484\varepsilon_0$. To find the length of the quarter-wave section, we compute

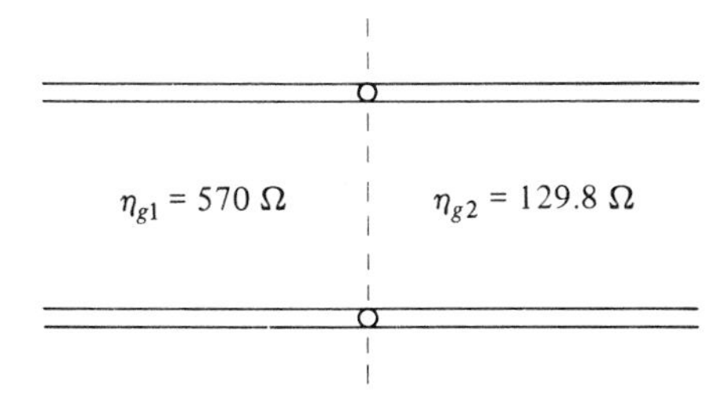

Figure 9.20. Transmission-line equivalent for the rectangular waveguide discontinuity of Fig. 9.19 for $TE_{1,0}$ waves of frequency 5000 MHz.

$$\lambda_{g3} = \frac{\lambda_3}{\sqrt{1 - (\lambda_3/\lambda_c)^2}} = \frac{\lambda_1\sqrt{\varepsilon_0/\varepsilon_3}}{\sqrt{1 - (\lambda_1/\lambda_c)^2(\varepsilon_0/\varepsilon_3)}}$$

$$= \frac{6 \times 0.6345}{\sqrt{1 - (9/16) \times 0.4026}} = 4.33\,\text{cm}$$

Thus the length of the quarter-wave section is $\lambda_{g3}/4$ or 1.0825 cm.

Cavity resonator

Let us now consider guided waves of equal amplitude propagating in the positive z- and negative z-directions in a rectangular waveguide. This can be achieved by terminating the guide by a perfectly conducting sheet in a constant z-plane, that is, a transverse plane of the guide. Due to perfect reflection from the sheet, the fields will then be characterized by standing wave nature along the guide axis, that is, in the z-direction, in addition to the standing wave nature in the x- and y-directions. The standing wave pattern along the guide axis will have nulls of transverse electric field on the terminating sheet and in planes parallel to it at distances of integer multiples of $\lambda_g/2$ from that sheet. Placing of perfect conductors in these planes will not disturb the fields since the boundary condition of zero tangential electric field is satisfied in those planes.

Conversely, if we place two perfectly conducting sheets in two constant z-planes separated by a distance d, then, for the boundary conditions to be satisfied, d must be equal to an integer multiple of $\lambda_g/2$. We then have a rectangular box of dimensions a, b, and d in the x-, y-, and z-directions, respectively, as shown in Fig. 9.21. Such a structure is known as a *cavity resonator* and is the counterpart of the low-frequency lumped parameter resonant circuit at microwave frequencies, since it supports oscillations at frequencies for which the foregoing condition, that is,

$$d = l\frac{\lambda_g}{2}, \qquad l = 1, 2, 3, \ldots \tag{9.81}$$

is satisfied. Substituting for λ_g in (9.81) from Table 9.1 and rearranging, we obtain

$$\frac{2d}{l} = \frac{\lambda}{\sqrt{1 - (\lambda/\lambda_c)^2}}$$

or

$$\frac{1}{\lambda^2} - \frac{1}{\lambda_c^2} = \left(\frac{l}{2d}\right)^2$$

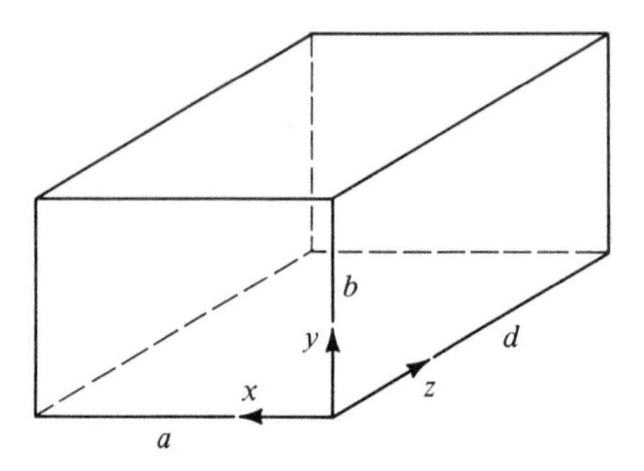

Figure 9.21. Rectangular cavity resonator.

which upon substitution for λ_c gives

$$\frac{1}{\lambda_2} = \left(\frac{m}{2a}\right)^2 + \left(\frac{n}{2b}\right)^2 + \left(\frac{l}{2d}\right)^2$$

$$\lambda = \frac{1}{\sqrt{(m/2a)^2 + (n/2b)^2 + (l/2d)^2}} \tag{9.82}$$

The expression for the frequencies of oscillation is thus given by

$$\boxed{f_{\text{osc}} = \frac{v_p}{\lambda} = \frac{1}{\sqrt{\mu\varepsilon}}\sqrt{\left(\frac{m}{2a}\right)^2 + \left(\frac{n}{2b}\right)^2 + \left(\frac{l}{2d}\right)^2}} \tag{9.83}$$

The modes are designated by three subscripts in the manner $\text{TE}_{m,n,l}$ and $\text{TM}_{m,n,l}$. Since m, n, and l can assume combinations of integer values, an infinite number of frequencies of oscillation are possible for a given set of dimensions of the cavity resonator. Also, a given frequency of oscillation may correspond to more than one mode. We recall that for TE modes $m, n = 0, 1, 2, \ldots$, but not both zero, whereas for TM modes $m, n = 1, 2, 3, \ldots$. For both TE and TM modes $l = 1, 2, 3, \ldots$, as given in (9.81). In addition TM modes at cutoff ($\beta_z = 0$, $\lambda_g = \infty$ and $\eta_g = 0$) satisfy the boundary conditions since then $\bar{E}_x$ and $\bar{E}_y$ both go to zero. Hence for TM modes $l = 0$ is allowed. We shall now consider an example.

Example 9.6

The dimensions of a rectangular cavity resonator with air dielectric are $a = 4$ cm, $b = 2$ cm, and $d = 4$ cm. It is desired to determine the three lowest frequencies of oscillation and specify the mode(s) of oscillation, transverse with respect to the z-direction, for each frequency.

By substituting $\mu = \mu_0$, $\varepsilon = \varepsilon_0$, and the given dimensions for a, b, and d in (9.83), we obtain

$$f_{\text{osc}} = 3 \times 10^8\sqrt{\left(\frac{m}{0.08}\right)^2 + \left(\frac{n}{0.04}\right)^2 + \left(\frac{l}{0.08}\right)^2}$$

$$= 3750\sqrt{m^2 + 4n^2 + l^2}\ \text{MHz}$$

By assigning combinations of integer values for m, n, and l and keeping in mind the restrictions on these values as discussed, we obtain the three lowest frequencies of oscillation and the corresponding modes to be

$3750 \times \sqrt{2} = 5303$ MHz for $\text{TE}_{1,0,1}$ mode

$3750 \times \sqrt{5} = 8385$ MHz for $\text{TE}_{0,1,1}$, $\text{TE}_{2,0,1}$, $\text{TE}_{1,0,2}$, and $\text{TM}_{1,1,0}$ modes

$3750 \times \sqrt{6} = 9186$ MHz for $\text{TE}_{1,1,1}$ and $\text{TM}_{1,1,1}$ modes

K9.4. Rectangular waveguide; TE and TM modes; Dominant mode; Cavity resonator; Frequency of oscillation.

D9.7. A generator of fundamental frequency 2000 MHz and rich in harmonics excites a rectangular waveguide. Find all frequencies that propagate in only TE modes for each of the following cases: **(a)** $a = 5$ cm, $b = 2.5$ cm, $\varepsilon = \varepsilon_0$; **(b)** $a = 4.5$ cm, $b = 1.5$ cm, $\varepsilon = 4\varepsilon_0$; and **(c)** $a = 6$ cm, $b = 6$ cm, $\varepsilon = \varepsilon_0$. Assume that $\mu = \mu_0$ for all cases.

Ans. **(a)** 4000, 6000 MHz; **(b)** 2000, 4000 MHz; **(c)** None

D9.8. For the rectangular waveguide discontinuity of Fig 9.19, find the power reflection coefficient for incidence from section 1 for each of the following cases: **(a)** $TE_{1,0}$ wave of frequency $f = 10{,}000$ MHz; **(b)** $TE_{1,1}$ wave of frequency $f = 10{,}000$ MHz; and **(c)** $TM_{1,1}$ wave of frequency $f = 10{,}000$ MHz.
Ans. **(a)** 0.2756; **(b)** 0.4649; **(c)** 0.0676

D9.9. The frequencies of oscillation for an air-dielectric rectangular cavity resonator of dimensions a, b, and d, in the x-, y-, and z-directions, respectively, are given for three modes as follows:

$$f_{osc} = 3000\sqrt{5} \text{ MHz for } TE_{1,0,1} \text{ mode}$$

$$f_{osc} = 3000\sqrt{26} \text{ MHz for } TE_{0,1,1} \text{ mode}$$

$$f_{osc} = 3000\sqrt{30} \text{ MHz for } TM_{1,1,1} \text{ mode}$$

Find the values of a, b, and d in cm.
Ans. 2.5 cm, 1 cm, 5 cm

9.5 CYLINDRICAL WAVEGUIDE AND CAVITY RESONATOR

Thus far in this chapter we have been concerned with the guiding of waves between metallic boundaries involving rectangular geometries. We shall now extend the treatment to cylindrical geometry by considering the case of a cylindrical waveguide, which is simply a hollow tube of circular cross section of a radius a and extending along the z-direction, as shown in Fig. 9.22. Thus for TM waves in a cylindrical waveguide,

$$H_z = 0, \qquad E_z \neq 0$$

and all other field components can be expressed in terms of E_z, whereas for TE waves,

$$E_z = 0, \qquad H_z \neq 0$$

and all other field components can be expressed in terms of H_z.

Solution of Maxwell's curl equations in cylindrical coordinates

To derive the field expressions for TM and TE modes in a cylindrical waveguide, we begin with Maxwell's curl equations in cylindrical coordinates and pro-

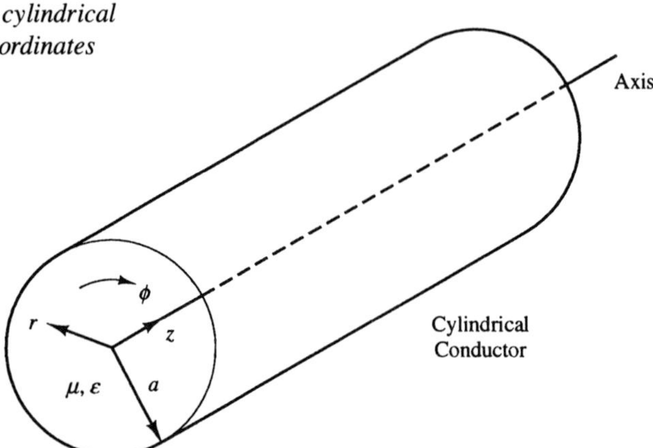

Figure 9.22 A cylindrical waveguide extending in the z-direction.

ceed similarly to the rectangular waveguide case, using the phasor forms of the field components in the manner $e^{j[\omega t \mp \beta_z z]}$, where the upper and lower signs represent wave propagation in the $+z$- and $-z$-directions, respectively. Thus replacing $\partial/\partial t$ by $j\omega$ and $\partial/\partial z$ by $\mp j\beta_z$ in the expansions for the Maxwell's curl equations in cylindrical coordinates, we have for a perfect dielectric medium inside the waveguide,

$$\frac{1}{r}\frac{\partial \bar{E}_z}{\partial \phi} \pm j\beta_z \bar{E}_\phi = -j\omega\mu \bar{H}_r \quad (9.84a) \qquad \frac{1}{r}\frac{\partial \bar{H}_z}{\partial \phi} \pm j\beta_z \bar{H}_\phi = j\omega\varepsilon \bar{E}_r \quad (9.84d)$$

$$\mp j\beta_z \bar{E}_r - \frac{\partial \bar{E}_z}{\partial r} = -j\omega\mu \bar{H}_\phi \quad (9.84b) \qquad \mp j\beta_z \bar{H}_r - \frac{\partial \bar{H}_z}{\partial r} = j\omega\varepsilon \bar{E}_\phi \quad (9.84e)$$

$$\frac{1}{r}\frac{\partial}{\partial r}(r\bar{E}_\phi) - \frac{1}{r}\frac{\partial \bar{E}_r}{\partial \phi} = -j\omega\mu \bar{H}_z \quad (9.84c) \qquad \frac{1}{r}\frac{\partial}{\partial r}(r\bar{H}_\phi) - \frac{1}{r}\frac{\partial \bar{H}_r}{\partial \phi} = j\omega\varepsilon \bar{E}_z \quad (9.84f)$$

Solving (9.84a), (9.84b), (9.84d), and (9.84e), we obtain

$$\bar{E}_r = \frac{1}{\beta_z^2 - \beta^2}\left[\pm j\beta_z \frac{\partial \bar{E}_z}{\partial r} + \frac{j\omega\mu}{r}\frac{\partial \bar{H}_z}{\partial \phi}\right] \tag{9.85a}$$

$$\bar{E}_\phi = -\frac{1}{\beta_z^2 - \beta^2}\left[\mp j\frac{\beta_z}{r}\frac{\partial \bar{E}_z}{\partial \phi} + j\omega\mu \frac{\partial \bar{H}_z}{\partial r}\right] \tag{9.85b}$$

$$\bar{H}_r = -\frac{1}{\beta_z^2 - \beta^2}\left[\frac{j\omega\varepsilon}{r}\frac{\partial \bar{E}_z}{\partial \phi} \mp j\beta_z \frac{\partial \bar{H}_z}{\partial r}\right] \tag{9.85c}$$

$$\bar{H}_\phi = \frac{1}{\beta_z^2 - \beta^2}\left[j\omega\varepsilon \frac{\partial \bar{E}_z}{\partial r} \pm j\frac{\beta_z}{r}\frac{\partial \bar{H}_z}{\partial \phi}\right] \tag{9.85d}$$

thereby expressing the transverse field components in terms of the longitudinal field components. Recall that $\beta^2 = \omega^2\mu\varepsilon$.

Now, setting $\bar{H}_z = 0$ and substituting for $\bar{H}_r$ and $\bar{H}_\phi$ in (9.84f), we obtain the differential equation for $\bar{E}_z$ for the TM-mode case

$$\frac{1}{r}\frac{\partial}{\partial r}\left(r\frac{\partial \bar{E}_z}{\partial r}\right) + \frac{1}{r^2}\frac{\partial^2 \bar{E}_z}{\partial \phi^2} - (\beta_z^2 - \beta^2)\bar{E}_z = 0 \tag{9.86a}$$

Similarly, setting $\bar{E}_z = 0$ and substituting for $\bar{E}_r$ and $\bar{E}_\phi$ in (9.84c), we obtain the differential equation for $\bar{H}_z$ for the TE-mode case

$$\frac{1}{r}\frac{\partial}{\partial r}\left[r\frac{\partial \bar{H}_z}{\partial r}\right] + \frac{1}{r^2}\frac{\partial^2 \bar{H}_z}{\partial \phi^2} - (\beta_z^2 - \beta^2)\bar{H}_z = 0 \tag{9.86b}$$

In view of the similarity of (9.86a) and (9.86b), we let $\bar{\psi}$ stand for $\bar{E}_z$ in the case of TM waves and $\bar{H}_z$ in the case of TE waves and consider the solution of the differential equation

$$\frac{1}{r}\frac{\partial}{\partial r}\left(r\frac{\partial \bar{\psi}}{\partial r}\right) + \frac{1}{r^2}\frac{\partial^2 \bar{\psi}}{\partial \phi^2} - (\beta_z^2 - \beta^2)\bar{\psi} = 0$$

or

$$\boxed{\frac{1}{r}\frac{\partial}{\partial r}\left[r\frac{\partial \bar{\psi}}{\partial r}\right] + \frac{1}{r^2}\frac{\partial^2 \bar{\psi}}{\partial \phi^2} + \beta_c^2 \bar{\psi} = 0} \tag{9.87}$$

where

$$\boxed{\beta_c^2 = \beta^2 - \beta_z^2} \tag{9.88}$$

To solve this equation, we make use of the separation of variables technique as in Section 9.4. Thus letting

$$\bar{\psi} = \bar{R}(r)\bar{\Phi}(\phi)e^{\mp j\beta_z z} \tag{9.89}$$

and substituting into (9.87), we have

$$\bar{\Phi}\frac{1}{r}\frac{\partial}{\partial r}\left(r\frac{\partial \bar{R}}{\partial r}\right) + \frac{\bar{R}}{r^2}\frac{\partial^2 \bar{\Phi}}{\partial \phi^2} + \beta_c^2 \bar{R}\bar{\Phi} = 0$$

or

$$\frac{r}{\bar{R}}(\bar{R}' + r\bar{R}'') + \beta_c^2 r^2 = -\frac{\bar{\Phi}''}{\bar{\Phi}} = n^2, \text{ a constant} \tag{9.90}$$

where the primes denote differentiation with respect to the respective variables, thereby obtaining the two separate differential equations

$$\bar{\Phi}'' + n^2\bar{\Phi} = 0 \tag{9.91a}$$

$$\bar{R}'' + \frac{1}{r}\bar{R}' + \left(\beta_c^2 - \frac{n^2}{r^2}\right)\bar{R} = 0 \tag{9.91b}$$

Bessel functions

The solution for (9.91a) is of the form

$$\bar{\Phi}(\phi) = \bar{A}_n \cos n\phi + \bar{B}_n \sin n\phi \tag{9.92a}$$

where $\bar{A}_n$ and $\bar{B}_n$ are constants. Equation (9.91b) is a Bessel's equation which has the solution

$$\bar{R}(r) = \bar{C}_n J_n(\beta_c r) + \bar{D}_n N_n(\beta_c r) \tag{9.92b}$$

where $\bar{C}_n$ and $\bar{D}_n$ are constants, and

$J_n(\beta_c r)$	= Bessel function of the first kind of order n and argument $\beta_c r$
$N_n(\beta_c r)$	= Bessel function of the second kind of order n and argument $\beta_c r$

The variations of $J_n(x)$ and $N_n(x)$ with their argument x are shown for a few values of n in Fig. 9.23(a) and (b), respectively. The function $N_n(x)$ has the property that $N_n(0) \rightarrow \infty$ for all orders. Since the fields must remain finite inside the guide which includes $r = 0$ (the z-axis), it follows that $\bar{D}_n$ must be zero. Hence the solution for $R(r)$ pertinent to the cylindrical waveguide problem is given by

$$R(r) = \bar{C}_n J_n(\beta_c r)$$

The solution for $\bar{\psi}$ is thus given by

$$\bar{\psi} = J_n(\beta_c r)(\bar{A}_n \cos n\phi + \bar{B}_n \sin n\phi)e^{\mp j\beta_z z} \tag{9.93}$$

where we have absorbed $\bar{C}_n$ into $\bar{A}_n$ and $\bar{B}_n$. This is the general solution for $\bar{\psi}$ which we use for $\bar{E}_z$ in the case of TM waves and the $\bar{H}_z$ in the case of TE waves to obtain their particular solutions satisfying the boundary conditions.

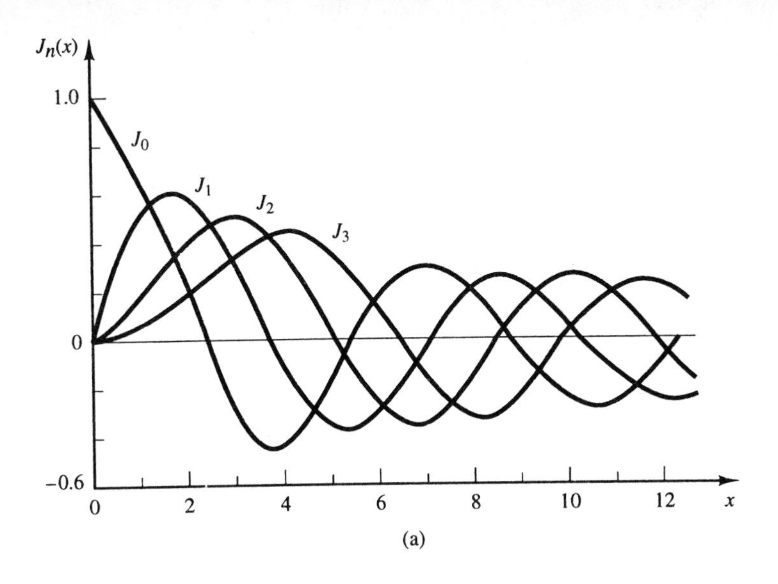

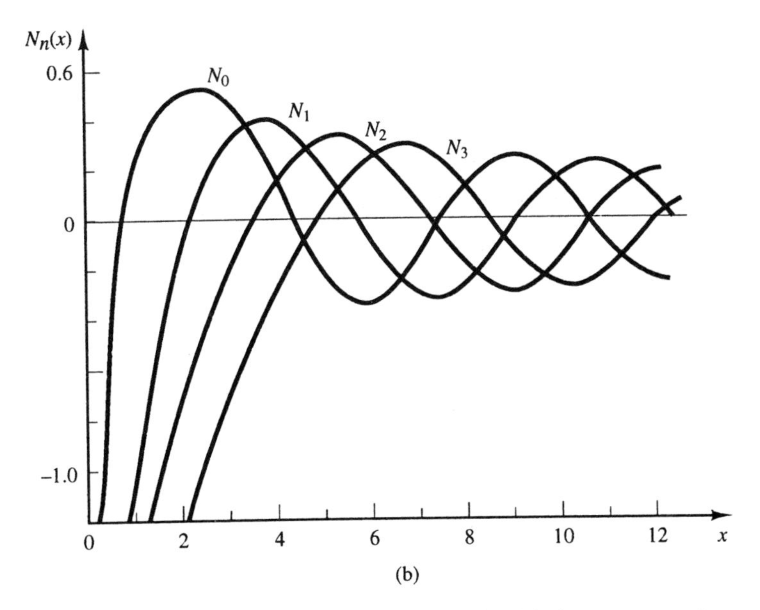

Figure 9.23 Variations of (a) $J_n(x)$ and (b) $N_n(x)$, with the argument x for a few values of n.

TM modes

For TM modes, $\overline{H}_z = 0$ and $\overline{\psi} = \overline{E}_z$. Hence

$$\boxed{\overline{E}_z = J_n(\beta_c r)(\overline{A}_n \cos n\phi + B_n \sin n\phi)e^{\mp j\beta_z z}} \tag{9.94}$$

Since for a cylindrical waveguide, $0 \leq \phi \leq 2\pi$ and $\overline{E}_z$ must be a single-valued function, the solution must be periodic in ϕ with a period 2π. This requires that n be an integer. Thus

$$n = 0, 1, 2, 3, \ldots$$

The boundary condition at $r = a$ requires that the tangential component of $\overline{\mathbf{E}}$ be zero. Since $\overline{E}_z$ is tangential, it follows that $[\overline{E}_z]_{r=a} = 0$ for all ϕ, giving us

$$0 = J_n(\beta_c a)(\overline{A}_n \cos n\phi + \overline{B}_n \sin n\phi)e^{\mp j\beta_z z}$$

or

$$\boxed{J_n(\beta_c a) = 0} \tag{9.95}$$

Note that this also makes the remaining tangential component $\overline{E}_\phi$ equal to zero at $r = a$ since from (9.85b), $\overline{E}_\phi \propto \dfrac{\partial \overline{E}_z}{\partial \phi}$. Equation (9.95) tells us that only certain values of $\beta_c a$ and hence of β_c are allowed. These are the values for which the Bessel function goes to zero, that is, they are the roots of the equation

$$J_n(x) = 0 \tag{9.96}$$

For a given value of n, there are an infinite number of roots as can be seen from Fig. 9.23(a). Denoting the number of the root to be m, we list in Table 9.2 the lowest three ($m = 1$ 2, 3) nonvanishing roots for the first five values of n ($n = 0, 1, 2, 3, 4$).

With the understanding that the values of $\beta_c a$ are given by those in Table 9.2, we can obtain the expressions for the transverse field components by noting that $\overline{H}_z = 0$ and substituting (9.94) into (9.85a)–(9.85d)

$$\overline{E}_r = \mp j\frac{\beta_z}{\beta_c} J_n'(\beta_c r)(\overline{A}_n \cos n\phi + \overline{B}_n \sin n\phi)e^{\mp j\beta_z z} \tag{9.97a}$$

$$\overline{E}_\phi = \pm j\frac{\beta_z n}{\beta_c^2 r} J_n(\beta_c r)(\overline{A}_n \sin n\phi - \overline{B}_n \cos n\phi)e^{\mp j\beta_z z} \tag{9.97b}$$

$$\overline{H}_r = \mp\frac{\omega\varepsilon}{\beta_z}\overline{E}_\phi \tag{9.97c}$$

$$\overline{H}_\phi = \pm\frac{\omega\varepsilon}{\beta_z}\overline{E}_r \tag{9.97d}$$

In (9.97a), the prime associated with J_n denotes differentiation with respect to the argument $\beta_c r$.

The allowed values of $\beta_c a$ may be written as $(\beta_c a)_{n,m}$ where the first subscript refers to the order of the Bessel function and the second subscript denotes

TABLE 9.2 Roots of $J_n(x)$

m \ n	0	1	2	3	4
1	2.40	3.83	5.14	6.38	7.59
2	5.52	7.02	8.42	9.76	11.06
3	8.65	10.17	11.62	13.02	14.37

the mth root of the nth order Bessel function. The corresponding modes are designated as $TM_{n,m}$ modes. In terms of the field configurations, the first subscript refers to the number of complete sinusoidal variations of the field components in the ϕ direction. For example, $n = 0$ means that the field components have no variations in ϕ, $n = 1$ means that they have one sinusoidal variation in ϕ, and so on. The second subscript refers to the number of quasi half-cycle variations (except that in the case of $n = 0$, the first one is a one-quarter cycle) of the field components in the r direction in accordance with the behavior of the Bessel functions, as shown in Fig. 9.23(a), and their derivatives, depending upon the component.

TE modes

For TE modes, $\bar{E}_z = 0$, and $\bar{\psi} = \bar{H}_z$. Hence

$$\boxed{\bar{H}_z = J_n(\beta_c r)(\bar{A}_n \cos n\phi + \bar{B}_n \sin n\phi)e^{\mp j\beta_z z}} \tag{9.98}$$

Again, n must be an integer as in the case of TM modes since the solution must be periodic in ϕ with period 2π in order that $\bar{H}_z$ is single valued. Thus

$$n = 0, 1, 2, 3, \ldots$$

The boundary condition at $r = a$ requires that the tangential component of $\bar{\mathbf{E}}$ be zero. Since $\bar{E}_z \equiv 0$ for TE modes, we need to consider $\bar{E}_\phi$, which is the remaining tangential component of $\bar{\mathbf{E}}$. From (9.85b), we note that $\bar{E}_\phi \propto \dfrac{\partial \bar{H}_z}{\partial r}$. Thus we have

$$\left[\frac{\partial \bar{H}_z}{\partial r}\right]_{r=a} = 0$$

or,

$$0 = J_n'(\beta_c a)(A_n \cos n\phi + B_n \sin n\phi)e^{\mp j\beta_z z}$$

where the prime associated with J_n denotes derivative of the Bessel function. It follows that

$$\boxed{J_n'(\beta_c a) = 0} \tag{9.99}$$

Equation (9.99) tells us that again in this case, only certain values of $\beta_c a$ and hence of β_c are allowed. These are the values for which the derivative of the Bessel function goes to zero, that is, they are the roots of the equation

$$J_n'(x) = 0 \tag{9.100}$$

It can be seen by visualizing the derivatives of the graphs of the Bessel functions in Fig. 9.23(a) that for a given value of n, there are infinite number of roots for (9.100), corresponding to the points at which the slopes of the Bessel functions are zero. Denoting the number of the root to be m, we list in Table 9.3 the lowest three ($m = 1, 2, 3$) nonvanishing roots for the first five values of n ($n = 0, 1, 2, 3, 4$).

With the understanding that the values of $\beta_c a$ are given by those in Table 9.3, we can obtain the expressions for the transverse field components by noting that $\bar{E}_z = 0$ and substituting (9.98) into (9.85a)–(9.85d)

TABLE 9.3 Roots of $J_n'(x)$

m \ n	0	1	2	3	4
1	3.83	1.84	3.05	4.20	5.32
2	7.02	5.33	6.71	8.02	9.28
3	10.17	8.54	9.97	11.35	12.68

$$\overline{E}_r = j\frac{\omega\mu n}{\beta_c^2 r} J_n(\beta_c r)(\overline{A}_n \sin n\phi - \overline{B}_n \cos n\phi)e^{\mp j\beta_z z} \tag{9.101a}$$

$$\overline{E}_\phi = j\frac{\omega\mu}{\beta_c} J_n'(\beta_c r)(\overline{A}_n \cos n\phi + \overline{B}_n \sin n\phi)e^{\mp j\beta_z z} \tag{9.101b}$$

$$\overline{H}_r = \mp\frac{\beta_z}{\omega\mu}\overline{E}_\phi \tag{9.101c}$$

$$\overline{H}_\phi = \pm\frac{\beta_z}{\omega\mu}\overline{E}_r \tag{9.101d}$$

The allowed values of $\beta_c a$ may once again be written as $(\beta_c a)_{n,m}$ and the modes designated as $\text{TE}_{n,m}$ modes. As in the case of $\text{TM}_{n,m}$ modes, the first subscript refers to the number of complete sinusoidal variations of the field components in the ϕ direction. The second subscript refers to the number of quasi half-cycle variations (except that for $n \neq 0$ the first one is a one-quarter cycle) of the field components in the r direction.

Characteristics of TM and TE modes

Let $(\beta_c)_{n,m}$ be the values of β_c for the (n,m)th mode found from the roots of $J_n(\beta_c a)$ for TM modes and of $J_n'(\beta_c a)$ for TE modes. Then recalling that

$$\beta_c^2 = \beta^2 - \beta_z^2 = \omega^2\mu\varepsilon - \beta_z^2$$

and that the cutoff condition occurs for β_z equal to zero, we note that the cutoff frequencies are given by

$$\omega^2\mu\varepsilon = (\beta_c)_{n,m}^2$$

$$(f_c)_{n,m} = \frac{(\beta_c)_{n,m}}{2\pi\sqrt{\mu\varepsilon}} \tag{9.102}$$

and the cutoff wavelengths are given by

$$(\lambda_c)_{n,m} = \frac{1}{\sqrt{\mu\varepsilon}(f_c)_{n,m}} = \frac{2\pi}{(\beta_c)_{n,m}} \tag{9.103}$$

Proceeding further, for the propagating range of frequencies for a given mode we have from (9.88),

$$\beta_z^2 = \beta^2 - \beta_c^2$$
$$= \left(\frac{2\pi}{\lambda}\right)^2\left[1 - \left(\frac{f_c}{f}\right)^2\right]$$
$$= \left(\frac{2\pi}{\lambda}\right)^2\left[1 - \left(\frac{\lambda}{\lambda_c}\right)^2\right]$$

or

$$\lambda_g = \frac{\lambda}{\sqrt{1 - (f_c/f)^2}} = \frac{\lambda}{\sqrt{1 - (\lambda/\lambda_c)^2}} \tag{9.104}$$

The phase velocity along the guide axis is given by

$$v_{pz} = \lambda_g f = \frac{1}{\sqrt{\mu\varepsilon}}\frac{1}{\sqrt{1 - (f_c/f)^2}} = \frac{1}{\sqrt{\mu\varepsilon}}\frac{1}{\sqrt{1 - (\lambda/\lambda_c)^2}} \tag{9.105}$$

Finally, by taking the ratios of appropriate pairs of transverse electric and magnetic field components given by (9.97a)–(9.97d) for TM modes and (9.101a)–(9.101d) for TE modes, we obtain the guide characteristic impedances to be

$$\begin{aligned}[\eta_g]_{\text{TM}} &= \frac{\bar{E}_r}{\pm\bar{H}_\phi} = \frac{\bar{E}_\phi}{\mp\bar{H}_r} = \frac{\beta_z}{\omega\varepsilon} = \frac{2\pi}{\lambda_g\omega\varepsilon} = \sqrt{\frac{\mu}{\varepsilon}}\frac{\lambda}{\lambda_g} \\ &= \sqrt{\frac{\mu}{\varepsilon}}\sqrt{1 - \left(\frac{f_c}{f}\right)^2} = \sqrt{\frac{\mu}{\varepsilon}}\sqrt{1 - \left(\frac{\lambda}{\lambda_c}\right)^2}\end{aligned} \tag{9.106a}$$

and

$$\begin{aligned}[\eta_g]_{\text{TE}} &= \frac{\bar{E}_r}{\pm\bar{H}_\phi} = \frac{\bar{E}_\phi}{\mp\bar{H}_r} = \frac{\omega\mu}{\beta_z} = \frac{\omega\mu\lambda_g}{2\pi} = \sqrt{\frac{\mu}{\varepsilon}}\frac{\lambda_g}{\lambda} \\ &= \frac{\sqrt{\mu/\varepsilon}}{\sqrt{1 - (f_c/f)^2}} = \frac{\sqrt{\mu/\varepsilon}}{\sqrt{1 - (\lambda/\lambda_c)^2}}\end{aligned} \tag{9.106b}$$

Note that the expressions (9.104)–(9.106b) are the same as the corresponding expressions for the rectangular waveguide case. We shall now consider an example.

Example 9.7

Finding propagating modes

The radius of an air-dielectric cylindrical waveguide is given by $a = 5$ cm. It is desired to find the propagating modes and their characteristics for a signal of frequency $f = 3$ GHz.

From Tables 9.2 and 9.3, we note that the lowest root is 1.84, corresponding to the $\text{TE}_{1,1}$ mode. Thus $\text{TE}_{1,1}$ mode is the dominant mode and its cutoff frequency is given by

$$[f_c]_{TE_{1,1}} = \frac{[(\beta_c a)_{1,1}]/a}{2\pi\sqrt{\mu_0\varepsilon_0}} = \frac{1.84}{5 \times 10^{-2}} \times \frac{3 \times 10^8}{2\pi}$$
$$= 1.757 \text{ GHz}$$

In fact, by choosing the roots in increasing order of value from Tables 9.2 and 9.3 and dividing them by a and substituting in (9.102), we can find the cutoff frequencies in the increasing order. The four lowest cutoff frequencies and the corresponding modes found in this manner are listed in Table 9.4. For any given frequency f, propagation occurs in all modes for which $f > f_c$. Thus for a signal of frequency 3 GHz, the propagating modes are $TE_{1,1}$, $TM_{0,1}$, and $TE_{2,1}$. The corresponding values of λ_c, λ_g, v_{pz}, and η_g, computed by using (9.103)–(9.106b), are also listed in Table 9.4.

TABLE 9.4 THE FOUR LOWEST CUTOFF FREQUENCIES AND THE CORRESPONDING MODES FOR AN AIR-DIELECTRIC CYLINDRICAL WAVEGUIDE OF RADIUS a = 5 CM AND THE PARAMETERS FOR THE PROPAGATING MODES FOR f = 3 GHz.

$\beta_c a$	Mode(s)	f_c (GHz)	λ_c (cm)	λ_g (cm)	v_{pz} (m/s)	η_g (ohms)
1.84	$TE_{1,1}$	1.757	17.074	12.337	3.701×10^8	465.10
2.40	$TM_{0,1}$	2.292	13.090	15.500	4.650×10^8	243.24
3.05	$TE_{2,1}$	2.913	10.300	41.827	12.548×10^8	1576.84
3.83	$TE_{0,1}$, $TM_{1,1}$	3.657	8.203	—	—	—

Cylindrical cavity resonator

As in the case of the rectangular waveguide, by placing two perfectly conducting sheets in two constant z-planes separated by a distance d, we can have a cylindrical cavity resonator that supports oscillations at frequencies for which

$$d = l\frac{\lambda_g}{2}, \qquad l = 1, 2, 3, \ldots \tag{9.107}$$

Substituting for λ_g from (9.104), we have

$$\frac{2d}{l} = \frac{\lambda}{\sqrt{1 - (\lambda/\lambda_c)^2}}$$

$$\frac{1}{\lambda^2} = \frac{1}{\lambda_c^2} + \left(\frac{l}{2d}\right)^2$$

$$\boxed{f_{\text{osc}} = \frac{v_p}{\lambda} = \frac{1}{\sqrt{\mu\varepsilon}}\sqrt{\frac{1}{\lambda_c^2} + \left(\frac{l}{2d}\right)^2}} \tag{9.108}$$

The modes are designated $TE_{n,m,l}$ and $TM_{n,m,l}$, where n and m are allowed values as discussed earlier. For both TE and TM modes, $l = 1, 2, 3, \ldots$, as given by (9.107). In addition, as for the rectangular waveguide case, TM modes at cutoff ($\beta_z = 0$, $\lambda_g = \infty$ and $\eta_g = 0$) satisfy the boundary conditions since both $\bar{E}_r$ and $\bar{E}_\phi$ go to zero. Hence for TM modes $l = 0$ is allowed. Let us consider an example.

Example 9.8

A cylindrical cavity resonator is formed by placing two perfectly conducting sheets 5 cm apart in the cross-sectional planes of the cylindrical waveguide of Example 9.7, so that d = 5 cm. It is desired to find the four lowest frequencies of oscillation and identify the mode(s) of oscillation for each frequency.

By substituting $\mu = \mu_0$, $\varepsilon = \varepsilon_0$, and $d = 5$ cm, in (9.108), we obtain

$$f_{osc} = 3 \times 10^8 \sqrt{\frac{1}{\lambda_c^2} + \left(\frac{l}{0.1}\right)^2}$$
$$= 3000\sqrt{\left(\frac{0.1}{\lambda_c}\right)^2 + l^2} \text{ MHz}$$

By using the results of Table 9.4 and assigning values to l, as discussed, we obtain the four lowest frequencies of oscillation and the corresponding modes to be

$$3000 \times \frac{10}{13.09} = 2292 \text{ MHz for } TM_{0,1,0} \text{ mode}$$

$$3000\sqrt{\left(\frac{10}{17.074}\right)^2 + 1} = 3477 \text{ MHz for } TE_{1,1,1} \text{ mode}$$

$$3000 \times \frac{10}{8.203} = 3657 \text{ MHz for } TM_{1,1,0} \text{ mode}$$

$$3000\sqrt{\left(\frac{10}{13.09}\right)^2 + 1} = 3775 \text{ MHz for } TM_{0,1,1} \text{ mode}$$

K9.5. Cylindrical waveguide; Bessel functions; TM and TE modes; Cylindrical cavity resonator.

D9.10. An air-dielectric waveguide with the z-axis as its axis has a semicircular cross section of radius $a = 5$ cm, as shown in Fig. 9.24. Find the mode with the lowest cutoff frequency and the corresponding value of the cutoff frequency for **(a)** TE waves and **(b)** TM waves.
Ans. **(a)** $TE_{1,1}$, 1.757 GHz; **(b)** $TM_{1,1}$, 3.657 GHz

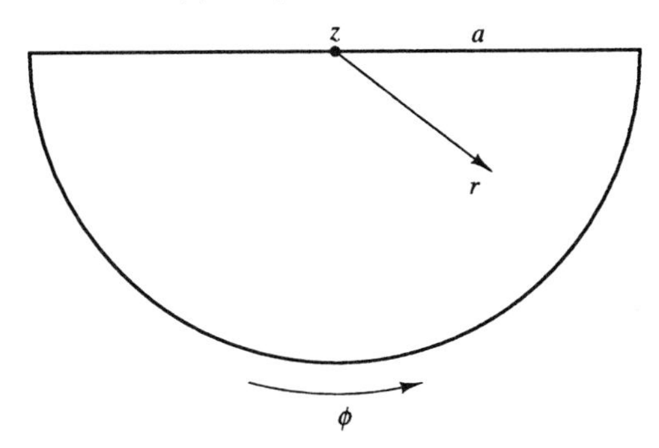

Figure 9.24 For Problem D9.10.

D9.11. The resonant frequencies for the $TM_{0,1,0}$ mode and the $TE_{1,1,1}$ mode of an air-dielectric cylindrical cavity resonator are both known to be 3000 MHz. Find the values of the dimensions a and d of the resonator.
Ans. 3.82 cm, 7.79 cm

9.6 LOSSES IN WAVEGUIDES AND RESONATORS

Loss in dielectric

In this section we shall extend our discussion of waveguides and resonators to consider the effects of lossy materials. Power dissipation in the imperfect dielectric of a guide results in loss that follows simply from the attenuation constant for the case of a uniform plane wave propagating in the dielectric. If for the purpose of illustration we consider the TE or TM wave in a parallel-plate waveguide, then we know that progress of the composite TE or TM wave along the guide by a distance d involves travel of the component uniform plane waves obliquely to the plates by a distance $d/\sqrt{1-(f_c/f)^2}$. Thus if α_d' is the attenuation constant for uniform plane wave propagation in the dielectric, then the attenuation constant α_d for the TE or TM wave along the guide axis is $\alpha_d'/\sqrt{1-(f_c/f)^2}$ and the attenuation $e^{\mp\alpha_d z}$ is equal to $e^{\mp[\alpha_d'/\sqrt{1-(f_c/f)^2}]z}$. From Section 6.4 we recall that for slightly imperfect dielectric ($\sigma/\omega\varepsilon \ll 1$), $\alpha_d' \approx \frac{\sigma}{2}\sqrt{\frac{\mu}{\varepsilon}}\left(1-\frac{\sigma^2}{8\omega^2\varepsilon^2}\right) \approx \frac{\sigma}{2}\sqrt{\frac{\mu}{\varepsilon}}$.

Basis for analysis of loss in conductors

Unlike the case of the imperfect dielectric, attenuation of the wave due to power flow into the imperfect conductors of a guide as the wave propagates down the guide involves an elaborate treatment. Since the conductors are only slightly imperfect ($\sigma/\omega\varepsilon \gg 1$), the procedure is based on considering the situation as though a plane wave having the same magnetic field components as those given by the appropriate tangential magnetic field components on that wall for the perfect conductor case propagates normally into the conductor and then computing the power flow into the wall (assumed to be of infinite depth in view of the rapid attenuation of fields as they propagate into a good conductor). Now, for a tangential magnetic field $\overline{\mathbf{H}}_t$ on a given wall, the electric field vector of a uniform plane wave propagating into the wall (designated to be in the direction $\mathbf{i}_n$) is $\overline{\eta}_c\overline{\mathbf{H}}_t \times \mathbf{i}_n$, where $\overline{\eta}_c$ is the intrinsic impedance of the conductor. The complex Poynting vector is

$$\begin{aligned}\overline{\mathbf{P}} &= \frac{1}{2}\overline{\mathbf{E}} \times \overline{\mathbf{H}}^* = \frac{1}{2}\overline{\eta}_c(\overline{\mathbf{H}}_t \times \mathbf{i}_n) \times \overline{\mathbf{H}}_t^* \\ &= \frac{1}{2}\overline{\eta}_c[\mathbf{i}_n(\overline{\mathbf{H}}_t \cdot \overline{\mathbf{H}}_t^*) - \overline{\mathbf{H}}_t(\mathbf{i}_n \cdot \overline{\mathbf{H}}_t^*)] \\ &= \frac{1}{2}\overline{\eta}_c\overline{\mathbf{H}}_t \cdot \overline{\mathbf{H}}_t^*\,\mathbf{i}_n\end{aligned} \tag{9.109}$$

The time-average power flowing into the conductor of conductivity σ for a length Δz along the guide is given by

$$\begin{aligned}\Delta\langle P_d\rangle &= \int_l (\text{Re}\,\overline{\mathbf{P}}) \cdot dl\,\Delta z\,\mathbf{i}_n \\ &= \int_l \text{Re}\left(\frac{1}{2}\overline{\eta}_c\overline{\mathbf{H}}_t \cdot \overline{\mathbf{H}}_t^*\right) dl\,\Delta z \\ &= \frac{\Delta z}{2\sigma\delta}\int_l \overline{\mathbf{H}}_t \cdot \overline{\mathbf{H}}_t^*\,dl\end{aligned} \tag{9.110}$$

where $\delta(=1/\sqrt{\pi f\mu\sigma})$ is the skin depth at the frequency of operation f, dl is the differential length element along the transverse dimension, and $\int_l$ denotes integration performed along the transverse dimension. We shall illustrate the application of (9.110) by means of an example.

Example 9.9

Attenuation constant for $TE_{1,0}$ mode in a rectangular waveguide

Let us consider the propagation of $TE_{1,0}$ mode in a rectangular waveguide and obtain the expression for the attenuation constant α_c associated with it due to imperfect but good conductors comprising the walls of the guide.

To obtain the attenuation constant α_c, we note that since for a given mode the fields are attenuated in the manner $e^{-\alpha_c z}$ where the z-direction is assumed to be the guide axis, the time-average power flow $<P_f>$ down the guide varies in the manner $e^{-2\alpha_c z}$. The time-average power dissipated over an infinitesimal distance Δz at any value of z along the guide is then given by

$$\begin{aligned}\Delta<P_d> &= \frac{-\partial<P_f>}{\partial z}\Delta z \\ &= 2\alpha_c<P_f>\Delta z\end{aligned}$$

so that

$$\boxed{\alpha_c = \frac{1}{2<P_f>}\frac{\Delta<P_d>}{\Delta z}} \tag{9.111}$$

Thus the attenuation constant is one-half the ratio of the time-average power dissipated per unit distance at any value of z along the guide to the time-average power flow down the guide at that value of z. In fact, this is a general result applicable for any lossy traveling wave. The procedure for computing α_c for a given mode therefore consists of evaluating $\Delta<P_d>$ and $<P_f>$ for that mode.

To find $\Delta<P_d>$, we consider the different walls of the waveguide separately. For each wall, we compute the time-average power flowing into the conductor over a surface made up of distance Δz along the guide axis and the entire transverse dimension of that wall by using (9.110). To proceed further, we substitute $m = 1$ and $n = 0$ in the TE mode field expressions given in Table 9.1, and considering the (+) wave only, we obtain the $TE_{1,0}$ mode field components in a lossless waveguide to be

$$\bar{E}_z = \bar{E}_x = \bar{H}_y = 0 \tag{9.112a}$$

$$\bar{H}_z = \bar{A}\cos\frac{\pi x}{a}e^{-j\beta_z z} \tag{9.112b}$$

$$\bar{E}_y = -j\frac{\lambda_c^2}{4\pi^2}\omega\mu\frac{\pi}{a}\bar{A}\sin\frac{\pi x}{a}e^{-j\beta_z z} \tag{9.112c}$$

$$\bar{H}_x = -\frac{\bar{E}_y}{\eta_g} = j\frac{2a}{\lambda_g}\bar{A}\sin\frac{\pi x}{a}e^{-j\beta_z z} \tag{9.112d}$$

where a and b are the dimensions of the guide in the x- and y-directions, respectively, as shown in Fig. 9.25. For the lossy case, the field components are multiplied by $e^{-\alpha_c z}$. Since we compute the quantities $\Delta<P_d>$ and $<P_f>$ at some particular value of z, say z_0, we can absorb the factors $e^{-\alpha_c z_0}$ and $e^{-j\beta_z z_0}$ into the constant $\bar{A}$. Also, each nonzero tangential component of magnetic field on a given wall will be accompanied by a tangential electric field perpendicular to it so as to produce power flow into the conductor. Since some of these tangential electric field components are longitudinal, the mode is no longer exactly TE mode. However, these components are very small in magnitude; hence, the mode is almost TE mode.

We shall now consider the different walls and compute the corresponding values of $\Delta<P_d>$ with the aid of Fig. 9.25.

Right side wall ($x = 0$):

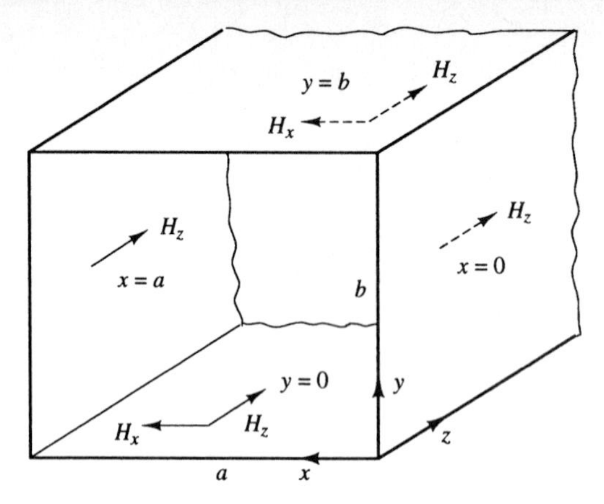

Figure 9.25 Rectangular waveguide with imperfectly conducting walls and showing the tangential magnetic field components on the walls for the $TE_{1,0}$ mode.

$$\overline{\mathbf{H}}_t = \bar{A}\mathbf{i}_z$$

$$\overline{\mathbf{H}}_t \cdot \overline{\mathbf{H}}_t^* = |\bar{A}|^2$$

$$\Delta\langle P_d\rangle = \frac{\Delta z}{2\sigma\delta}\int_{y=0}^{b} |\bar{A}|^2\, dy$$

$$= \frac{|\bar{A}|^2 b}{2\sigma\delta}\Delta z \tag{9.113a}$$

Left side wall ($x = a$):
Same as for right side wall.

$$\Delta\langle P_d\rangle = \frac{|\bar{A}|^2 b}{2\sigma\delta}\,\Delta z \tag{9.113b}$$

Bottom wall ($y = 0$):

$$\overline{\mathbf{H}}_t = \bar{A}\left(j\frac{2a}{\lambda_g}\sin\frac{\pi x}{a}\,\mathbf{i}_x + \cos\frac{\pi x}{a}\,\mathbf{i}_z\right)$$

$$\overline{\mathbf{H}}_t \cdot \overline{\mathbf{H}}_t^* = |\bar{A}|^2\left(\frac{4a^2}{\lambda_g^2}\sin^2\frac{\pi x}{a} + \cos^2\frac{\pi x}{a}\right)$$

$$\Delta\langle P_d\rangle = \frac{\Delta z}{2\sigma\delta}\int_{x=0}^{a} |\bar{A}|^2\left(\frac{4a^2}{\lambda_g^2}\sin^2\frac{\pi x}{a} + \cos^2\frac{\pi x}{a}\right) dx$$

$$= \frac{|\bar{A}|^2\,\Delta z}{4\sigma\delta}\left(\frac{4a^3}{\lambda_g^2} + a\right) \tag{9.113c}$$

Top wall ($y = b$):
Same as for bottom wall.

$$\Delta\langle P_d\rangle = \frac{|\bar{A}|^2\,\Delta z}{4\sigma\delta}\left(\frac{4a^3}{\lambda_g^2} + a\right) \tag{9.113d}$$

Adding up the contributions from all four walls, we obtain the total time-average power dissipated over an infinitesimal length Δz along the guide to be

$$\Delta\langle P_d\rangle = \frac{|\bar{A}|^2}{2\sigma\delta}\left(\frac{4a^3}{\lambda_g^2} + a + 2b\right)\Delta z$$

$$= \frac{|\bar{A}|^2}{2\sigma\delta}\left[\frac{4a^3}{\lambda^2}\left(1 - \frac{\lambda^2}{4a^2}\right) + a + 2b\right]\Delta z$$

$$= \frac{|\bar{A}|^2}{\sigma\delta}\left(\frac{2a^3}{\lambda^2} + b\right)\Delta z \tag{9.114}$$

Now, to find the time-average power transmitted down the guide, we note that the time-average Poynting vector is given by

$$<\mathbf{P}> = \frac{1}{2}\text{Re}\ \bar{\mathbf{E}} \times \bar{\mathbf{H}}^*$$

$$= -\frac{1}{2\eta_g}\bar{E}_y\bar{E}_y^*\mathbf{i}_z$$

$$= \frac{1}{2\eta_g}\frac{\lambda_c^4}{16\pi^4}\omega^2\mu^2\frac{\pi^2}{a^2}|\bar{A}|^2\sin^2\frac{\pi x}{a}\mathbf{i}_z$$

$$= \frac{\omega^2\mu^2a^2}{2\eta_g\pi^2}|\bar{A}|^2\sin^2\frac{\pi x}{a}\mathbf{i}_z \tag{9.115}$$

The time-average power transmitted down the guide is then given by

$$<P_f> = \int_{y=0}^{b}\int_{x=0}^{a} <\mathbf{P}> \cdot\, dx\, dy\, \mathbf{i}_z$$

$$= \frac{\omega^2\mu^2a^2}{2\eta_g\pi^2}|\bar{A}|^2\int_{y=0}^{b}\int_{x=0}^{a}\sin^2\frac{\pi x}{a}\,dx\,dy$$

$$= \frac{\omega^2\mu^2a^2b}{4\eta_g\pi^2}|\bar{A}|^2 \tag{9.116}$$

Finally, the attenuation constant is given by

$$\alpha_c = \frac{\Delta<P_d>}{2<P_f>\Delta z}$$

$$= \frac{|\bar{A}|^2}{2\sigma\delta}\left(\frac{2a^3}{\lambda^2} + b\right)\frac{4\eta_g\pi^2}{\omega^2\mu^2a^3b|\bar{A}|^2}$$

$$= \frac{1}{\sigma\delta}\left(1 + \frac{b\lambda^2}{2a^3}\right)\frac{\eta\varepsilon}{\mu b\sqrt{1 - (f_c/f)^2}}$$

$$= \frac{1}{\sigma\delta}\left[1 + \frac{2b}{a}\left(\frac{\lambda}{\lambda_c}\right)^2\right]\frac{1}{b\eta\sqrt{1 - (f_c/f)^2}}$$

or

$$\boxed{\alpha_c = \frac{1}{\sigma\delta b\eta\sqrt{1 - (f_c/f)^2}}\left[1 + \frac{2b}{a}\left(\frac{f_c}{f}\right)^2\right]} \tag{9.117)]}$$

Note that for $f \to f_c$, $\alpha_c \to \infty$. For $f \gg f_c$, $\alpha_c \approx \dfrac{1}{\sigma\delta b\eta} \propto \sqrt{f}$, so that for $f \to \infty$, $\alpha_c \to \infty$. Thus as f varies from f_c to infinity, α_c varies from infinity to some minimum value and then increases to infinity. The minimum value of α_c occurs for

$$\left(\frac{f}{f_c}\right)^2 = \left(1.5 + \frac{3b}{a}\right) + \sqrt{\left(1.5 + \frac{3b}{a}\right)^2 - \frac{2b}{a}} \tag{9.118}$$

For example, for $\frac{b}{a} = \frac{1}{2}$, the minimum value occurs for $f/f_c \approx 2.4142$.

Q factor of a resonator

To proceed further, let us now consider the walls of a cavity resonator to be imperfect but good conductors. Then we can talk of the quality or Q factor of the resonator and derive the expression for it. The Q factor, which is a measure of the frequency selectivity of the resonator, is defined as

$$\boxed{\begin{aligned} Q &= 2\pi \frac{\text{energy stored in the resonator}}{\text{energy dissipated per cycle}} \\ &= 2\pi f \frac{\text{energy stored in the resonator}}{\text{time average power dissipated}} \end{aligned}} \tag{9.119}$$

Since the conductors are good conductors, the power dissipated in them can be computed by using analysis as in Example 9.9 for the waveguide case. As for the energy stored in the cavity, it is distributed between the electric and magnetic fields at any arbitrary instant of time. But there are particular values of time at which the electric field is maximum and the magnetic field is zero, and vice versa. At these values of time, the entire energy is stored in one of the two fields. This can be taken advantage of to obtain the stored energy. We shall illustrate the determination of the Q factor by means of an example.

Example 9.10

Q factor for $TE_{1,0,1}$ mode of a rectangular cavity resonator

Let us consider the $TE_{1,0,1}$ mode of oscillation in the rectangular cavity resonator of Fig. 9.21 and obtain the expression for the Q factor associated with it due to imperfect but good conductors comprising the walls of the resonator.

First we obtain the expressions for $TE_{1,0,1}$ mode field components by superimposing the (+) and (−) wave field components for the $TE_{1,0}$ waves from Table 9.1 and satisfying the boundary conditions of zero tangential electric fields at the ends $z = 0$ and $z = d$. Thus we have

$$\bar{E}_z = \bar{E}_x = \bar{H}_y = 0 \tag{9.120a}$$

$$\bar{H}_z = \cos\frac{\pi x}{a}[\bar{A}_1 e^{-j\beta_z z} + \bar{A}_2 e^{j\beta_z z}] \tag{9.120b}$$

$$\bar{E}_y = -j\frac{\lambda_c^2}{4\pi^2}\omega\mu\frac{\pi}{a}\sin\frac{\pi x}{a}[\bar{A}_1 e^{-j\beta_z z} + \bar{A}_2 e^{j\beta_z z}] \tag{9.120c}$$

$$\bar{H}_x = j\frac{1}{\eta_g}\frac{\lambda_c^2}{4\pi^2}\omega\mu\frac{\pi}{a}\sin\frac{\pi x}{a}[\bar{A}_1 e^{-j\beta_z z} - \bar{A}_2 e^{j\beta_z z}] \tag{9.120d}$$

$$[\bar{E}_y]_{z=0} = [\bar{E}_y]_{z=d} = 0 \tag{9.121}$$

so that

$$\bar{A}_1 + \bar{A}_2 = 0 \quad \text{or} \quad \bar{A}_2 = -\bar{A}_1 \tag{9.122a}$$

$$\sin\beta_z d = 0 \quad \text{or} \quad \beta_z = \pi/d \tag{9.122b}$$

giving us the required field components

$$\bar{E}_z = \bar{E}_x = \bar{H}_y = 0 \tag{9.123a}$$

$$\bar{H}_z = \bar{A} \cos \frac{\pi x}{a} \sin \frac{\pi z}{d} \tag{9.123b}$$

$$\bar{E}_y = -j\bar{A}\frac{\omega\mu a}{\pi} \sin \frac{\pi x}{a} \sin \frac{\pi z}{d} \tag{9.123c}$$

$$\bar{H}_x = -\bar{A}\frac{a}{d} \sin \frac{\pi x}{a} \cos \frac{\pi z}{d} \tag{9.123d}$$

where $\bar{A} = -2j\bar{A}_1$ and we have also substituted $\lambda_c = 2a$ and $\omega\mu/\eta_g = 2\pi/\lambda_g = \pi/d$.

To find the energy stored in the cavity, we shall make use of the electric field. Noting that the amplitude of the only electric field component E_y, which is the value of E_y at the instant of time the magnetic field throughout the cavity is zero, is given by

$$|\bar{E}_y| = |\bar{A}|\frac{\omega\mu a}{\pi} \sin \frac{\pi x}{a} \sin \frac{\pi z}{d} \tag{9.124}$$

and integrating the energy density throughout the volume of the cavity, we obtain the energy stored in the cavity to be

$$\begin{aligned}
W_{\text{stored}} &= \int_{x=0}^{a} \int_{y=0}^{b} \int_{z=0}^{d} \frac{1}{2}\varepsilon|\bar{E}_y|^2\,dv \\
&= \frac{1}{2}\varepsilon|\bar{A}|^2 \frac{\omega^2\mu^2a^2}{\pi^2} \int_{x=0}^{a} \int_{y=0}^{b} \int_{z=0}^{d} \sin^2 \frac{\pi x}{a} \sin^2 \frac{\pi z}{d}\,dx\,dy\,dz \\
&= |\bar{A}|^2\frac{\omega^2\mu^2\varepsilon}{8\pi^2}a^3bd
\end{aligned} \tag{9.125}$$

To find the time-average power dissipated in the walls of the cavity, we note from the application of (9.109) that for a given wall, the time-average power dissipated is

$$\begin{aligned}
\langle P_d \rangle &= \int_S (\text{Re}\,\bar{\mathbf{P}}) \cdot dS\,\mathbf{i}_n \\
&= \int_S \text{Re}\left(\frac{1}{2}\bar{\eta}_c\bar{\mathbf{H}}_t \cdot \bar{\mathbf{H}}_t^*\right) dS \\
&= \frac{1}{2\sigma\delta}\int_S \bar{\mathbf{H}}_t \cdot \bar{\mathbf{H}}_t^*\,dS
\end{aligned} \tag{9.126}$$

where S is the surface of the wall. Applying this result to the different walls of the cavity, we compute the corresponding values of $\langle P_d \rangle$ with the aid of Fig. 9.26.

Right side wall ($x = 0$):

$$\bar{\mathbf{H}}_t = \bar{A} \sin \frac{\pi z}{d}\mathbf{i}_z$$

$$\bar{\mathbf{H}}_t \cdot \bar{\mathbf{H}}_t^* = |\bar{A}|^2 \sin^2 \frac{\pi z}{d}$$

$$\begin{aligned}
\langle P_d \rangle &= \frac{1}{2\sigma\delta}\int_{y=0}^{b}\int_{z=0}^{d} \bar{\mathbf{H}}_t \cdot \bar{\mathbf{H}}_t^*\,dy\,dz \\
&= \frac{|\bar{A}|^2\,bd}{4\sigma\delta}
\end{aligned} \tag{9.127a}$$

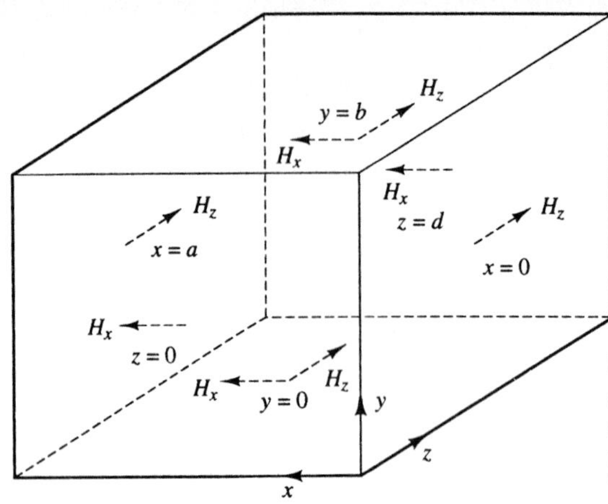

Figure 9.26 Rectangular cavity resonator with imperfectly conducting walls and showing the tangential magnetic field components on the walls for the $TE_{1,0,1}$ mode.

Left side wall ($x = a$):
Same as for right side wall.

$$<P_d> = \frac{|\bar{A}|^2 bd}{4\sigma\delta} \tag{9.127b}$$

Bottom wall ($y = 0$):

$$\overline{\mathbf{H}}_t = \bar{A}\left[-\frac{a}{d}\sin\frac{\pi x}{a}\cos\frac{\pi z}{d}\,\mathbf{i}_x + \cos\frac{\pi x}{a}\sin\frac{\pi z}{d}\,\mathbf{i}_z\right]$$

$$\overline{\mathbf{H}}_t \cdot \overline{\mathbf{H}}_t^* = |\bar{A}|^2\left[\left(\frac{a}{d}\right)^2 \sin^2\frac{\pi x}{a}\cos^2\frac{\pi z}{d} + \cos^2\frac{\pi x}{a}\sin^2\frac{\pi z}{d}\right]$$

$$<P_d> = \frac{1}{2\sigma\delta}\int_{x=0}^{a}\int_{z=0}^{d} \overline{\mathbf{H}}_t \cdot \overline{\mathbf{H}}_t^*\,dx\,dz$$

$$= \frac{|\bar{A}|^2}{8\sigma\delta}\left(\frac{a^3}{d} + ad\right) \tag{9.127c}$$

Top wall ($y = b$):
Same as for bottom wall.

$$<P_d> = \frac{|\bar{A}|^2}{8\sigma\delta}\left(\frac{a^3}{d} + ad\right) \tag{9.127d}$$

Front wall ($z = 0$):

$$\overline{\mathbf{H}}_t = -\bar{A}\frac{a}{d}\sin\frac{\pi x}{a}\,\mathbf{i}_x$$

$$\overline{\mathbf{H}}_t \cdot \overline{\mathbf{H}}_t^* = |\bar{A}|^2\left(\frac{a}{d}\right)^2 \sin^2\frac{\pi x}{a}$$

$$<P_d> = \frac{1}{2\sigma\delta}\int_{x=0}^{a}\int_{y=0}^{b} \overline{\mathbf{H}}_t \cdot \overline{\mathbf{H}}_t^*\,dx\,dy$$

$$= \frac{|\bar{A}|^2}{4\sigma\delta}\frac{a^3 b}{d^2} \tag{9.127e}$$

Back wall ($z = d$):

Same as for front wall.

$$<P_d> = \frac{|\bar{A}|^2}{4\sigma\delta} \frac{a^3 b}{d^2} \tag{9.127f}$$

Adding up the contributions from all six walls, we obtain the total time-average power dissipated to be

$$<P_d> = \frac{|\bar{A}|^2}{2\sigma\delta}\left(bd + \frac{a^3}{2d} + \frac{ad}{2} + \frac{a^3 b}{d^2}\right) \tag{9.128}$$

Substituting (9.125) and (9.128) into (9.119), we obtain

$$Q = \omega \frac{\omega^2 \mu\varepsilon a^3 bd/8\pi^2}{\dfrac{1}{2\sigma\delta}\left(bd + \dfrac{a^3}{2d} + \dfrac{ad}{2} + \dfrac{a^3 b}{d^2}\right)}$$

$$= \frac{\omega^3 \mu^2 \varepsilon\sigma\delta}{4\pi^2} \frac{2a^3 bd^3}{2bd^3 + a^3 d + ad^3 + 2a^3 b} \tag{9.129}$$

From (9.83) the resonant frequency ω for the $TE_{1,0,1}$ mode is however given by

$$\omega = \frac{\pi}{\sqrt{\mu\varepsilon}}\left[\left(\frac{1}{a}\right)^2 + \left(\frac{1}{d}\right)^2\right]^{1/2} \tag{9.130}$$

Thus (9.129) reduces to

$$\boxed{Q = \frac{\pi\sigma\delta\eta}{2} \frac{b(a^2 + d^2)^{3/2}}{ad(a^2 + d^2) + 2b(a^3 + d^3)}} \tag{9.131}$$

For a numerical example for the $TE_{1,0,1}$ mode of Example 9.6 and with the walls of the cavity made of copper ($\sigma = 5.80 \times 10^7$ S/m), $\delta = 9.075 \times 10^{-7}$ m and the value of Q is about 11020.

K9.6. Attenuation constant; $TE_{1,0}$ mode in a rectangular waveguide; Q factor; $TE_{1,0,1}$ mode in a rectangular cavity resonator.

D9.12. For each of the following cases of $TE_{1,0}$ waves propagating in a rectangular waveguide with copper walls, find the frequency of operation for which the attenuation constant α_c is a minimum and the minimum value of α_c: **(a)** $a = 3$ cm, $b = 1.5$ cm, and air-dielectric and **(b)** $a = b = 3$ cm, and dielectric of $\varepsilon = 4\varepsilon_0$ and $\mu = \mu_0$.
Ans. **(a)** 12.0711 GHz, 0.00653; **(b)** 7.4045 GHz, 0.0047

D9.13. Find Q for $TE_{1,0,1}$ mode for each of the following cases of a rectangular cavity resonator with copper walls: **(a)** $a = b = d = 5$ cm, air-dielectric; **(b)** $a = 2.5$ cm, $b = 2$ cm, $d = 5$ cm, air-dielectric; **(c)** $a = b = d = 5$ cm; dielectric with $\varepsilon = 2.25\varepsilon_0$ and $\mu = \mu_0$.
Ans. **(a)** 16,434; **(b)** 10,160; **(c)** 13,417

9.7 SUMMARY

In this chapter we studied the principles of waveguides. To introduce the waveguiding phenomenon, we first learned how to write the expressions for the electric and magnetic fields of a uniform plane wave propagating in an arbitrary direction with respect to the coordinate axes. These expressions are given by

$$\mathbf{E} = \mathbf{E}_0 \cos(\omega t - \boldsymbol{\beta} \cdot \mathbf{r} + \phi_0)$$

$$\mathbf{H} = \mathbf{H}_0 \cos(\omega t - \boldsymbol{\beta} \cdot \mathbf{r} + \phi_0)$$

where $\boldsymbol{\beta}$ and $\mathbf{r}$ are the propagation and position vectors given by

$$\boldsymbol{\beta} = \beta_x \mathbf{i}_x + \beta_y \mathbf{i}_y + \beta_z \mathbf{i}_z$$

$$\mathbf{r} = x\mathbf{i}_x + y\mathbf{i}_y + z\mathbf{i}_z$$

and ϕ_0 is the phase of the wave at the origin at $t = 0$. The magnitude of $\boldsymbol{\beta}$ is equal to $\omega\sqrt{\mu\varepsilon}$, the phase constant along the direction of propagation of the wave. The direction of $\boldsymbol{\beta}$ is the direction of propagation of the wave. We learned that

$$\mathbf{E}_0 \cdot \boldsymbol{\beta} = 0$$

$$\mathbf{H}_0 \cdot \boldsymbol{\beta} = 0$$

$$\mathbf{E}_0 \cdot \mathbf{H}_0 = 0$$

that is, $\mathbf{E}_0$, $\mathbf{H}_0$, and $\boldsymbol{\beta}$ are mutually perpendicular, and that

$$\frac{|\mathbf{E}_0|}{|\mathbf{H}_0|} = \eta = \sqrt{\frac{\mu}{\varepsilon}}$$

Also, since $\mathbf{E} \times \mathbf{H}$ should be directed along the propagation vector $\boldsymbol{\beta}$, it then follows that

$$\mathbf{H} = \frac{1}{\omega\mu}\boldsymbol{\beta} \times \mathbf{E}$$

The quantities β_x, β_y, and β_z are the phase constants along the x-, y-, and z-axes, respectively. The apparent wavelengths and the apparent phase velocities along the coordinate axes are given, respectively, by

$$\lambda_i = \frac{2\pi}{\beta_i}, \qquad i = x, y, z$$

$$v_{pi} = \frac{\omega}{\beta_i}, \qquad i = x, y, z$$

By considering the superposition of two uniform plane waves having only y-components of electric fields and propagating at an angle to each other and placing perfect conductors in two constant x-planes such that the boundary condition of zero tangential electric field is satisfied, we introduced the parallel-plate waveguide. We learned that the composite wave is a transverse electric, or TE wave since the electric field is entirely transverse to the direction of time-average power flow, that is, the guide axis, but the magnetic field is not. In terms of the uniform plane wave propagation, the phenomenon is one of waves bouncing obliquely between the conductors as they progress down the guide. For a fixed spacing a between the conductors of the guide, waves of different frequencies bounce obliquely at different angles such that the spacing a is equal to an integer, say, m number of one-half apparent wavelengths normal to the plates and hence the fields have m number of one-half sinusoidal variations normal to the plates. These are said to correspond to $TE_{m,0}$ modes, where the subscript 0 implies no variations of the fields in the direction parallel to the plates and transverse to the guide axis. When the frequency is such that the spacing a is equal to m one-half wavelengths, the waves bounce normally to the plates without the feeling of being guided along the axis, thereby leading to the cutoff condition. Thus the cutoff wavelengths corresponding to $TE_{m,0}$ modes are given by

$$\lambda_c = \frac{2a}{m}$$

and the cutoff frequencies are given by

$$f_c = \frac{v_p}{\lambda_c} = \frac{m}{2a\sqrt{\mu\varepsilon}}$$

A given frequency signal can propagate in all modes for which $\lambda < \lambda_c$ or $f > f_c$. For the propagating range of frequencies, the wavelength along the guide axis, that is, the guide wavelength, and the phase velocity along the guide axis are given, respectively, by

$$\lambda_g = \frac{\lambda}{\sqrt{1 - (\lambda/\lambda_c)^2}} = \frac{\lambda}{\sqrt{1 - (f_c/f)^2}}$$

$$v_{pz} = \frac{v_p}{\sqrt{1 - (\lambda/\lambda_c)^2}} = \frac{v_p}{\sqrt{1 - (f_c/f)^2}}$$

As compared to TE modes, the transverse magnetic or TM modes have their magnetic fields entirely transverse to the direction of time-average power flow. They are obtained by considering two uniform plane waves having only y-components of magnetic fields and propagating at an angle to each other and placing two perfect conductors in two constant x-planes. The expressions for the propagation parameters λ_c, f_c, λ_g, and v_{pz} for the TM modes are the same as those for the TE modes.

We discussed the solution of problems involving reflection and transmission at a discontinuity in a waveguide by using the transmission-line analogy. This consists of replacing each section of the waveguide by a transmission line whose characteristic impedance is equal to the guide characteristic impedance and then computing the reflection and transmission coefficients as in the transmission-line case. The guide characteristic impedance, η_g, is the ratio of a transverse electric field component to the corresponding transverse magnetic field component, which together with the electric field component gives rise to time-average power flow down the guide. It is given for TE modes by

$$[\eta_g]_{TE} = \frac{\eta}{\sqrt{1 - (\lambda/\lambda_c)^2}} = \frac{\eta}{\sqrt{1 - (f_c/f)^2}}$$

and for TM modes by

$$[\eta_g]_{TM} = \eta\sqrt{1 - \left(\frac{\lambda}{\lambda_c}\right)^2} = \eta\sqrt{1 - \left(\frac{f_c}{f}\right)^2}$$

We then discussed the phenomenon of dispersion arising from the frequency dependence of the phase velocity along the guide axis, and we introduced the concept of group velocity. Group velocity is the velocity with which the envelope of a narrow-band modulated signal travels in the dispersive channel, and hence it is the velocity with which the information is transmitted. It is given by

$$v_g = \frac{d\omega}{d\beta_z}$$

where β_z is the phase constant along the guide axis. On the dispersion diagram or the $\omega - \beta_z$ diagram, the group velocity is equal to the slope of the tangent to the dispersion curve at the center frequency of the narrow-band signal. For the parallel-plate waveguide, it is given by

$$v_g = v_p\sqrt{1 - \left(\frac{f_c}{f}\right)^2}$$

We extended the treatment of the parallel-plate waveguide to the rectangular waveguide, which is a metallic pipe of rectangular cross section. By considering a rectangular waveguide of cross-sectional dimensions a and b, we discussed transverse electric or TE modes, as well as transverse magnetic or TM modes, and learned that while $TE_{m,n}$ modes can include values of m or n equal to zero, $TM_{m,n}$ modes require that both m and n be nonzero, where m and n refer to the number of one-half sinusoidal variations of the fields along the dimensions a and b, respectively. The cutoff wavelengths for the $TE_{m,n}$ or $TM_{m,n}$ modes are given by

$$\lambda_c = \frac{1}{\sqrt{(m/2a)^2 + (n/2b)^2}}$$

The mode that has the largest cutoff wavelength or the lowest cutoff frequency is the dominant mode, which here is the $TE_{1,0}$ mode.

By placing perfect conductors in two transverse planes of a rectangular waveguide separated by an integer multiple of one-half the guide wavelength, we introduced the cavity resonator, which is the microwave counterpart of the lumped parameter resonant circuit encountered in low-frequency circuit theory. For a rectangular cavity resonator having dimensions a, b, and d, the frequencies of oscillation for the $TE_{m,n,l}$ or $TM_{m,n,l}$ modes are given by

$$f_{osc} = \frac{1}{\sqrt{\mu\varepsilon}}\sqrt{\left(\frac{m}{2a}\right)^2 + \left(\frac{n}{2b}\right)^2 + \left(\frac{l}{2d}\right)^2}$$

where l refers to the number of one-half sinusoidal variations of the fields along the dimension d.

Next we introduced the cylindrical waveguide, which is a metallic pipe of cylindrical cross section. We learned that guided modes in the cylindrical waveguide are characterized by field variations in the radial direction in accordance with Bessel functions and sinusoidal variations in the angular direction. The modes are designated as $TE_{n,m}$ and $TM_{n,m}$, where the first subscript refers to the angular variations and the second to the radial variations. The $TE_{1,1}$ mode is the dominant mode. We also discussed the cylindrical cavity resonator formed by placing perfect conductors in two transverse planes of the guide, as in the case of the rectangular cavity resonator.

Finally, we considered losses in waveguides and resonators and discussed by means of examples the determination of attenuation constant for a propagating mode in a waveguide and the Q factor, a measure of frequency selectivity, for an oscillating mode in a resonator.

REVIEW QUESTIONS

R9.1. What is the propagation vector? Interpret the significance of its magnitude and direction.

R9.2. Discuss how the phase constants along the coordinate axes are less than the phase constant along the direction of propagation of a uniform plane wave propagating in an arbitrary direction.

R9.3. Write the expressions for the electric and magnetic fields of a uniform plane wave propagating in an arbitrary direction and list all the conditions to be satisfied by the electric field, magnetic field, and propagation vectors.

R9.4. What are apparent wavelengths? Why are they longer than the wavelength along the direction of propagation?

R9.5. What are apparent phase velocities? Why are they greater than the phase velocity along the direction of propagation?

R9.6. Discuss how the superposition of two uniform plane waves propagating at an angle to each other gives rise to a composite wave consisting of standing waves traveling bodily transverse to the standing waves.

R9.7. What is a transverse electric wave? Discuss the reasoning behind the nomenclature $TE_{m,0}$ modes.

R9.8. Compare the phenomenon of guiding of uniform plane waves in a parallel-plate waveguide with that in a parallel-plate transmission line.

R9.9. Discuss how the cutoff condition arises in a parallel-plate waveguide. Explain the relationship between the cutoff wavelength and the spacing between the plates of a parallel-plate waveguide based on the phenomenon at cutoff.

R9.10. Is the cutoff wavelength dependent on the dielectric in the waveguide? Is the cutoff frequency dependent on the dielectric in the waveguide?

R9.11. What is guide wavelength?

R9.12. Provide a physical explanation for the frequency dependence of the phase velocity along the guide axis.

R9.13. What is a transverse magnetic wave? Compare and contrast TE and TM waves in a parallel-plate waveguide.

R9.14. How is guide characteristic impedance defined? Discuss guide characteristic impedance for both TE and TM modes.

R9.15. Discuss the use of the transmission-line analogy for solving problems involving reflection and transmission at a waveguide discontinuity.

R9.16. Why are the reflection and transmission coefficients for a given mode at a lossless waveguide discontinuity dependent on frequency whereas the reflection and transmission coefficients at the junction of two lossless lines are independent of frequency?

R9.17. Discuss the phenomenon of guiding of waves in the earth-ionosphere waveguide.

R9.18. Discuss the phenomenon of dispersion.

R9.19. Discuss the concept of group velocity with the aid of an example.

R9.20. What is a dispersion diagram? Explain how the phase and group velocities can be determined from a dispersion diagram.

R9.21. When is it meaningful to attribute a group velocity to a signal comprised of more than two frequencies? Why?

R9.22. Discuss the propagation of a narrow-band amplitude-modulated signal in a dispersive channel.

R9.23. Discuss the nomenclature associated with the modes of propagation in a rectangular waveguide.

R9.24. Explain the relationship between the cutoff wavelength and the dimensions of a rectangular waveguide based on the phenomenon at cutoff.

R9.25. Briefly outline the procedure for deriving the expressions for TE mode fields in a rectangular waveguide.

R9.26. Compare and contrast TE and TM modes in a rectangular waveguide.

R9.27. What is the dominant mode? Which one of the rectangular waveguide modes is the dominant mode?

R9.28. Why is the dimension b of a rectangular waveguide generally chosen to be less than or equal to one-half the dimension a?

R9.29. What is a cavity resonator?

R9.30. How do the dimensions of a rectangular cavity resonator determine the frequencies of oscillation of the resonator?

R9.31. Briefly outline the procedure for deriving the expressions for the TE and TM mode fields in a cylindrical waveguide.

R9.32. Compare and contrast TE and TM modes in a cylindrical waveguide.

R9.33. Which one of the cylindrical waveguide modes is the dominant mode?

R9.34. Discuss the basis for the computation of power loss associated with slightly imperfect conductors comprising the walls of a waveguide.

R9.35. Briefly outline the procedure for the determinatin of the attenuation constant for a propagating mode in a waveguide with slightly imperfect conductors.

R9.36. How is the Q factor of a resonator defined? Briefly outline the procedure for the determination of the Q factor of a cavity resonator with slighly imperfect conductors.

PROBLEMS

P9.1. The electric field of a uniform plane wave propagating in a perfect dielectric medium having $\varepsilon = 2.25\varepsilon_0$ and $\mu = \mu_0$ is given by

$$\mathbf{E} = 10(3\mathbf{i}_x + 5\mathbf{i}_y - 4\mathbf{i}_z)\cos\left[2\pi \times 10^7 t - 0.02\pi(3x + 4z)\right]$$

Find **(a)** the frequency, **(b)** the direction of propagation, **(c)** the wavelength along the direction of propagation, **(d)** the apparent wavelengths along the x-, y-, and z-axes, and **(e)** the apparent phase velocities along th x-, y-, and z-axes.

P9.2. Given

$$\mathbf{E} = 10(2\mathbf{i}_x + \sqrt{3}\mathbf{i}_y + \mathbf{i}_z)\cos\left[6\pi \times 10^7 t - 0.1\pi(y - \sqrt{3}z)\right]$$

(a) Determine if the given **E** represents the electric field of a uniform plane wave propagating in free space. **(b)** If the answer to part (a) is "yes," find the corresponding magnetic field vector **H**.

P9.3. Given

$$\mathbf{E} = (2\mathbf{i}_x + \mathbf{i}_y - 2\mathbf{i}_z) \cos [15\pi \times 10^7 t - \pi(2x - 2y + z)]$$

$$\mathbf{H} = \frac{1}{80\pi}(\mathbf{i}_x + 2\mathbf{i}_y + 2\mathbf{i}_z) \cos [15\pi \times 10^7 t - \pi(2x - 2y + z)]$$

(a) Perform all the necessary tests and determine if these fields represent a uniform plane wave propagating in a perfect dielectric medium. **(b)** If the answer is "yes," find the permittivity and the permeability of the medium.

P9.4. The apparent phase velocities of a uniform plane wave propagating in a perfect dielectric medium are measured in three directions as follows: 2×10^8 m/s along the x-direction, 2×10^8 m/s along the direction of the unit vector $\frac{1}{5}(3\mathbf{i}_x + 4\mathbf{i}_y)$, and 3×10^8 m/s along the direction of the unit vector $\frac{1}{3}(\mathbf{i}_x - 2\mathbf{i}_y + 2\mathbf{i}_z)$. Find the direction of propagation of the wave and the phase velocity along the direction of the unit vector $\frac{1}{3}(2\mathbf{i}_x + 2\mathbf{i}_y - \mathbf{i}_z)$.

P9.5. The dimension a of a parallel-plate waveguide with a dielectric of $\varepsilon = 2.25\varepsilon_0$ and $\mu = \mu_0$ is 3 cm. Determine the propagating modes for a wave of frequency 9000 MHz. For each propagating mode, find λ_c, f_c, θ, λ_g, and v_{pz}.

P9.6. TE modes are excited in a parallel-plate waveguide filled with a perfect dielectric of $\varepsilon = 2.25\varepsilon_0$ and $\mu = \mu_0$ and having the plates in the $x = 0$ and $x = 10$ cm planes by setting up at its input $z = 0$ the electric field distribution

$$\mathbf{E} = E_0 \sin 10\pi x \cos 10^9\pi t \cos 4 \times 10^9\pi t \; \mathbf{i}_y$$

Find the expression for the electric field of the propagating wave.

P9.7. TM mode is excited in a parallel-plate waveguide filled with a dielectric of $\varepsilon = 4\varepsilon_0$ and $\mu = \mu_0$ and having the plates in the $x = 0$ and $x = 5$ cm planes by setting up at its input $z = 0$ the magnetic field distribution

$$\mathbf{H} = H_0 \cos 40\pi x \sin 8 \times 10^9\pi t \; \mathbf{i}_y$$

Find the expressions for the electric and magnetic fields of the propagating wave.

P9.8. For the parallel-plate waveguide discontinuity of Example 9.3, find the power reflection coefficients for $f = 7500$ MHz propagating in **(a)** $\text{TE}_{2,0}$ mode and **(b)** $\text{TM}_{1,0}$ mode.

P9.9. The left half of a parallel-plate waveguide of dimension $a = 4$ cm is filled with a dielectric of $\varepsilon = 2.25\varepsilon_0$ and $\mu = \mu_0$. The right half is filled with a dielectric of $\varepsilon = 4\varepsilon_0$ and $\mu = \mu_0$. For waves of frequency 3000 MHz incident on the discontinuity from the left, find the power reflection coefficient for each of the following cases: **(a)** TEM mode; **(b)** $\text{TE}_{1,0}$ mode; and **(c)** $\text{TM}_{1,0}$ mode.

P9.10. For a TM wave incident on a parallel-plate waveguide discontinuity involving two nonmagnetic ($\mu = \mu_0$), perfect dielectric sections, show that complete transmission occurs for a frequency f equal to $\sqrt{f_{c1}^2 + f_{c2}^2}$, where f_{c1} and f_{c2} are the cutoff frequencies corresponding to the two sections. Then compute the value of f at which complete transmission occurs for the case of Problem P9.9(c).

P9.11. For a backward wave (to be distinguished from a backward traveling wave), the group and phase velocities are in opposite directions. Consider the $\omega - \beta_z$ relationship given approximately by

$$\omega = \omega_0 + \frac{k}{\beta_z}$$

in the vicinity of $\omega = 2\omega_0$, where k is a positive constant. Find **(a)** the phase velocity for $\omega = 2\omega_0$; **(b)** the group velocity for a signal composed of the two frequencies $1.9\omega_0$ and $2.1\omega_0$; and **(c)** the group velocity for a narrow-band signal having the center frequency $2\omega_0$.

P9.12. For an air-dielectric parallel-plate waveguide having the dimension $a = 3$ cm, find the group velocity for **(a)** a signal composed of the two frequencies $f_1 = 6000$ MHz and $f_2 = 9000$ MHz and **(b)** a narrow-band signal having the center frequency 6000 MHz. Assume TE or TM mode propagation.

P9.13. For the signal of Problem P9.6, find the group velocity of propagation down the guide.

P9.14. By considering the parallel-plate waveguide, show that a point on the obliquely bouncing wavefront, traveling with the phase velocity along the oblique direction, progresses parallel to the guide axis with the group velocity.

P9.15. For a rectangular waveguide of dimensions $a = 5$ cm and $b = 2.5$ cm and having a dielectric of $\varepsilon = 2.25\varepsilon_0$ and $\mu = \mu_0$, find all propagating modes for $f = 5000$ MHz and for each mode, find the values of β_z, λ_g, v_{pz}, and η_g.

P9.16. For $f = 6000$ MHz, find the ranges of the dimensions a and b of an air-dielectric rectangular waveguide such that $TE_{1,0}$ mode propagates with a 20 percent safety factor ($f = 1.20 f_c$) but also such that the frequency is at least 20 percent below the cutoff frequency of the next-higher-order mode.

P9.17. A rectangular waveguide of dimensions $a = 3$ cm and $b = 1.5$ cm has a dielectric discontinuity, as shown in Fig. 9.27. A $TE_{1,0}$ wave of frequency 6000 MHz is incident on the discontinuity from the free-space side. **(a)** Find the SWR in the free-space section. **(b)** Find the length and the permittivity of a quarter-wave section required to achieve a match between the two media.

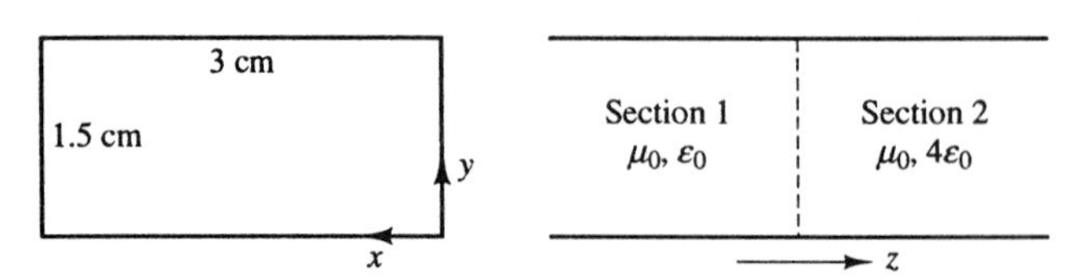

Figure 9.27 For Problem P9.17.

P9.18. A rectangular waveguide of dimensions $a = 4$ cm and $b = 2$ cm has a dielectric discontinuity, as shown in Fig. 9.28. A $TM_{1,1}$ wave of frequency 10,000 MHz is incident on the discontinuity from the free-space side. **(a)** Find the SWR in the free-space section. **(b)** Find the length and the permittivity of a quarter-wave section required to achieve a match between the two media.

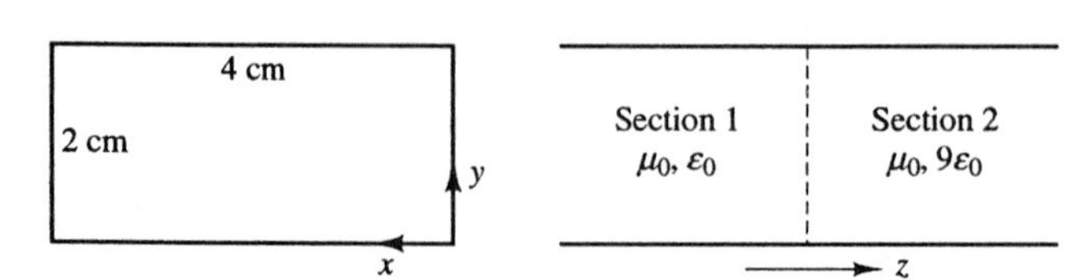

Figure 9.28 For Problem P9.18.

P9.19. A dielectric slab of thickness 4 cm and permittivity $2.25\varepsilon_0$ exists in an air-dielectric rectangular waveguide of dimensions $a = 3$ cm and $b = 1.5$ cm, as shown in Fig. 9.29. Find the lowest frequency for which the dielectric slab is transparent (that is, allows complete transmission) for $TE_{1,0}$ mode propagation in the waveguide.

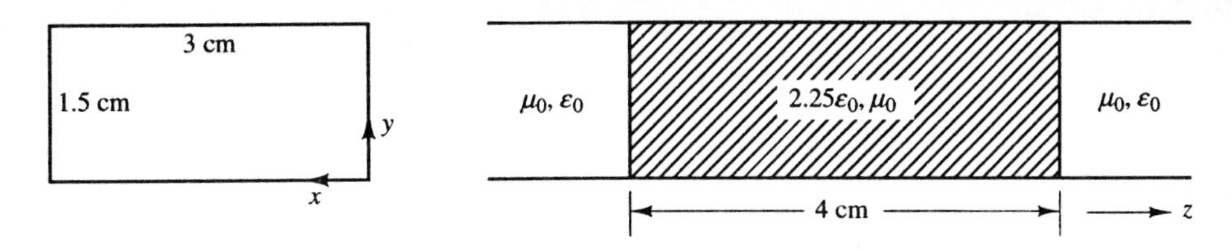

Figure 9.29 For Problem P9.19.

P9.20. A dielectric slab of thickness l and permittivity $4\varepsilon_0$ exists in an air-dielectric waveguide of square cross section $a = b = 3$ cm, as shown in Fig. 9.30. **(a)** Find the value of f for which the dielectric slab is transparent (that is, complete transmission occurs) for $TM_{1,1}$ mode propagation, independent of the value of l. **(b)** For $l = 1$ cm, find the two lowest frequencies for which the dielectric slab is transparent for $TM_{1,1}$ mode propagation.

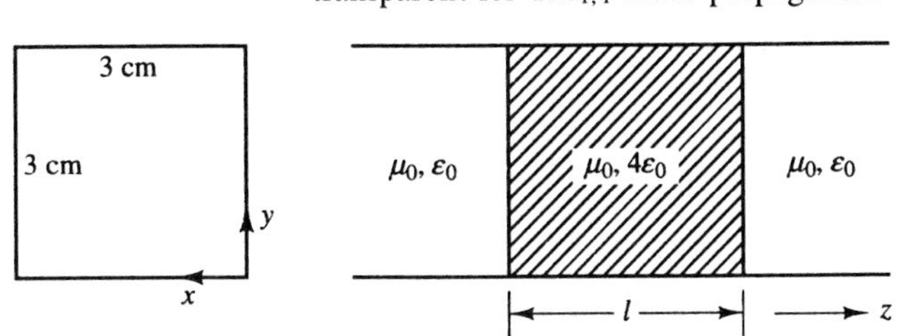

Figure 9.30 For Problem P9.20.

P9.21. For an air-dielectric rectangular cavity resonator having the dimensions $a = 3$ cm, $b = 2$ cm, and $d = 4$ cm, find the four lowest frequencies of oscillation and identify the mode(s) for each frequency.

P9.22. For a cubical cavity resonator having the dimensions $a = b = d = 2.5$ cm and filled with a dielectric of $\varepsilon = 4\varepsilon_0$ and $\mu = \mu_0$, find the three lowest frequencies of oscillation and identify the mode(s) for each frequency.

P9.23. For a cylindrical waveguide of radius $a = 3$ cm and having a dielectric of $\varepsilon = 2.25\varepsilon_0$ and $\mu = \mu_0$, find the propagating modes for a signal of frequency 5.0 GHz. For each of the propagating modes, find and tabulate the values of f_c, λ_c, λ_g, v_{pz}, and η_g, as in Table. 9.4.

P9.24. An air-dielectric waveguide has the cross section shown in Fig. 9.31. For the radius $a = 2$ cm, determine the lowest three cutoff frequencies and identify the corresponding modes.

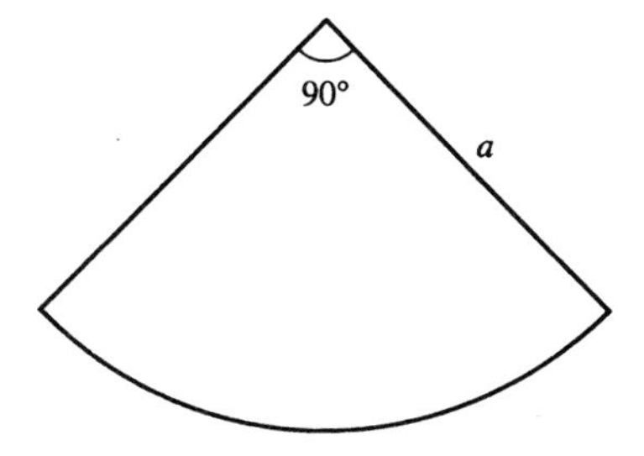

Figure 9.31 For Problem P9.24.

P9.25. Consider the use of the alternated transformer arrangement (see Problem P8.21) to achieve a match between two sections of a cylindrical waveguide, as shown in Fig. 9.32. Find the minimum values of l_2 and l_3 in centimeters to achieve the desired match for $TE_{1,1}$ mode at a frequency $f = 3$ GHz.

P9.26. Repeat Problem P9.25 for $TM_{0,1}$ mode.

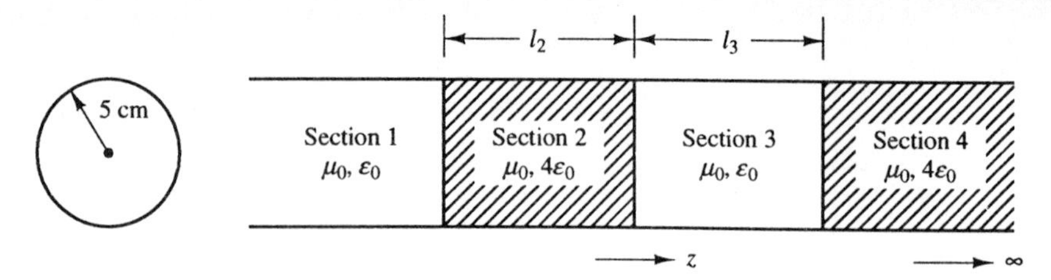

Figure 9.32 For Problem P9.25.

P9.27. A cylindrical cavity resonator is formed by placing two perfectly conducting sheets 4 cm apart in the cross-sectional planes of the cylindrical waveguide of Problem P9.23. Find the five lowest frequencies of oscillation and identify the mode(s) of oscillation for each frequency.

P9.28. Consider $\text{TE}_{n,m}$ modes bouncing between the walls $z = 0$ and $z = d$ of a cylindrical cavity resonator of radius a with one-half $(0 < z < d/2)$ filled with a nonmagnetic $(\mu = \mu_0)$, perfect dielectric of permittivity ε_1 and the other half $(d/2 < z < d)$ filled with a nonmagnetic, perfect dielectric of permittivity ε_2. **(a)** Show that the condition for oscillation is given by

$$\sqrt{\frac{\varepsilon_2}{\varepsilon_1} - \left(\frac{\lambda_1}{\lambda_c}\right)^2} \tan\left[\frac{\pi d}{\lambda_1}\sqrt{1 - \left(\frac{\lambda_1}{\lambda_c}\right)^2}\right] + \sqrt{1 - \left(\frac{\lambda_1}{\lambda_c}\right)^2} \tan\left[\frac{\pi d}{\lambda_1}\sqrt{\frac{\varepsilon_2}{\varepsilon_1} - \left(\frac{\lambda_1}{\lambda_2}\right)^2}\right] = 0$$

where λ_1 is the wavelength in dielectric 1 and λ_c is the cutoff wavelength. **(b)** Compute the value of the lowest resonant frequency for $a = 5$ cm, $d = 5$ cm, $\varepsilon_1 = \varepsilon_0$, and $\varepsilon_2 = 4\varepsilon_0$.

P9.29. For a parallel-plate waveguide with imperfect but good conductors of conductivity σ and spacing a, show that the attenuation constant α_c for TEM wave propagation along the guide is $1/\sigma\delta a\eta$. Compute the value of α_c for $a = 5$ cm, $f =$ 5,000 MHz, copper plates, and air-dielectric.

P9.30. Repeat Problem P9.29 for TE wave propagation to show that α_c is equal to $2(f_c/f)^2/[\sigma\delta a\eta\sqrt{1 - (f_c/f)^2}]$ and compute the value of α_c for the data specified in that problem and for the $\text{TE}_{1,0}$ mode.

P9.31. Repeat Problem P9.29 for TM wave propagation to show that α_c is equal to $2/[\sigma\delta a\eta\sqrt{1 - (f_c/f)^2}]$ and compute the value of α_c for the data specified in that problem and for the $\text{TM}_{1,0}$ mode.

P9.32. Show that the expression for the attenuation constant for the $\text{TM}_{1,1}$ mode in a rectangular waveguide with imperfect but good conductors is given by

$$\alpha_c = \frac{2}{\sigma\delta b\eta\sqrt{1 - (f_c/f)^2}}\left[\frac{(b/a)^3 + 1}{(b/a)^2 + 1}\right]$$

Further show that the minimum value of α_c occurs for $f/f_c = \sqrt{3}$.

P9.33. For a parallel-plate resonator consisting of two infinite, plane, perfectly conducting plates in the $z = 0$ and $z = l$ planes and separated by a perfect dielectric, as shown in Fig. 9.33, the electric and magnetic fields are given by

$$\mathbf{E} = E_0 \sin\frac{n\pi z}{l} \sin\frac{n\pi t}{l\sqrt{\mu\varepsilon}}\,\mathbf{i}_x$$

$$\mathbf{H} = \frac{E_0}{\sqrt{\mu/\varepsilon}} \cos\frac{n\pi z}{l} \cos\frac{n\pi t}{l\sqrt{\mu\varepsilon}}\,\mathbf{i}_y$$

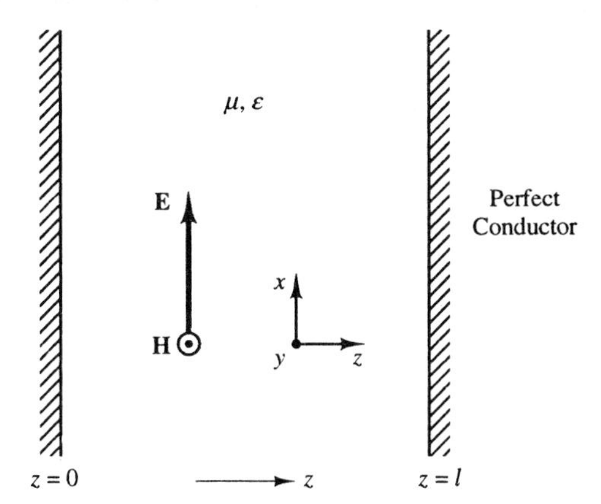

Figure 9.33 For Problem P9.33.

where $n = 1, 2, 3, \ldots$ **(a)** Show that the energy stored in the resonator per unit area of the plates is $\frac{1}{4}\varepsilon E_0^2 l$. **(b)** If the plates are made of imperfect but good conductors, show that the Q of the resonator is $l/2\delta$. **(c)** Compute the value of Q for the fundamental mode of oscillation ($n = 1$), for $l = 1$ cm, assuming air-dielectric and copper plates.

P9.34. For the parallel-plate resonator of Problem P9.33, assume that the dielectric is slightly lossy with conductivity $\sigma_d \ll \omega\varepsilon$. **(a)** Assuming the plates to be perfect conductors, show that the Q of the resonator is given by $Q_1 = \omega\varepsilon/\sigma_d$. **(b)** If in addition to the slightly lossy dielectric the plates are made up of imperfect but good conductors, show that the Q of the resonator is given by

$$\frac{1}{Q} = \frac{1}{Q_1} + \frac{1}{Q_2}$$

where Q_1 is as given in part (a) and Q_2 is equal to $l/2\delta$, as in Problem P9.33.

P9.35. Obtain the expression for the Q factor for $TM_{m,n,l}$ ($l \neq 0$) mode in a cubical cavity resonator of sides a and show that it is equal to $a/4\delta$.

P9.36. Consider a cylindrical cavity resonator of radius a and dimension d in the axial direction and made up of imperfect but good conductors of conductivity σ. Show that the Q factor for the $TM_{0,1,0}$ mode of oscillation is given by $1.2\eta\sigma\delta/(a/d + 1)$. Use the following:

$$J_0'(kr) = -J_1(kr)$$

$$\int rJ_0^2(kr)\,dr = \frac{r^2}{2}[J_0^2(kr) + J_1^2(kr)]$$

$$\int rJ_1^2(kr)\,dr = \frac{r^2}{2}[J_1^2(kr) - J_0(kr)J_2(kr)]$$

PC EXERCISES

PC9.1. Consider a rectangular waveguide containing n dielectric discontinuities, with the first section being free space and the last section extending to infinity. Write a program that computes the SWR in the first section for a $TE_{1,0}$ wave incident on the first discontinuity from that section. The input quantities are to be the dimensions a and b of the guide in centimeters, the frequency of the $TE_{1,0}$ wave in gigahertz, and the lengths in centimeters and relative permittivities of the n sections following the n discontinuities.

PC9.2. Consider a dielectric discontinuity in a rectangular waveguide, one side of which is free space and the other side a dielectric of relative permittivity ε_r, and $TE_{1,0}$ waves of frequency f_0 incident on the discontinuity from the free-space side. Write a program that computes the minimum length and the relative permittivity of a quarter-wave section required to achieve a match between the two sections and then compute the SWR in the free-space section versus frequency on either side of f_0. Assume the input parameters to be the dimensions a and b of the waveguide in centimeters, the frequency f_0 in gigahertz, and the value of ε_r.

PC9.3. Repeat Exercise PC9.2 for $TM_{1,1}$ waves incident on the dielectric discontinuity, noting that a relative permittivity of less than unity for the quarter-wave section is to be ruled out.

PC9.4. To generalize the cylindrical cavity resonator of Problem P9.28, assume that the section $0 < z < kd$, where $0 \leq k \leq 1$ is filled with a nonmagnetic ($\mu = \mu_0$), perfect dielectric of permittivity ε_1 and the section $kd < z < d$ is filled with a nonmagnetic, perfect dielectric of permittivity ε_2. Obtain the condition for oscillation and write a program to compute the value of the lowest resonant frequency for specified values of a, d, k, ε_1, and ε_2. For values of $a = 5$ cm, $d = 5$ cm, $\varepsilon_1 = \varepsilon_0$, and $\varepsilon_2 = 4\varepsilon_0$, compute the values of the lowest resonant frequency for values of $0 \leq k \leq 1$ in steps of 0.1 and plot versus k the ratio of the resonant frequency to its value for $k = 0$.

10

Electromagnetic Principles for Photonics

Electromagnetic wave phenomena cover a wide span of frequencies. In Chapters 7, 8, and 9, we studied transmission lines and metallic waveguides and resonators. Traditionally the treatment of transmission and guiding of waves at an introductory level is limited to these topics which are of principal interest up to the microwave region. With the advent of the photonics era, the movement of the region of interest beyond microwaves into the optical regime is accelerating. While the electromagnetic principles associated with the topics studied thus far are applicable to optical frequencies, directly or indirectly, there are certain other principles that are particularly relevant to the optical region and are also applicable to lower frequencies. This chapter is devoted to a selection of such topics.

We shall first consider the basic topic of reflection and transmission of uniform plane waves incident obliquely on a dielectric interface and learn of the fundamental phenomenon of total internal reflection. This is the phenomenon of the waves from a medium of higher permittivity incident onto one of lower permittivity at an angle from the normal to the interface exceeding a critical value getting entirely reflected, with the boundary condition being satisfied by an evanescent field in the lower permittivity medium. It forms the basis for waveguiding in a dielectric slab or rod surrounded by another dielectric medium of lower permittivity, as compared to the situation in a metallic waveguide in which the guiding is governed by reflections from the conducting walls of the guide. We shall begin with the dielectric slab waveguide and progressively extend the discussion to the optical fiber, learning several concepts in this process. Finally we shall consider the topics of interference and diffraction, and wave propagation in an anisotropic medium.

10.1 REFLECTION AND REFRACTION OF PLANE WAVES

Let us consider a uniform plane wave incident obliquely on a plane boundary between two different perfect dielectric media at an angle of incidence θ_i to the normal to the boundary, as shown in Fig. 10.1. To satisfy the boundary conditions at the interface between the two media, a reflected wave and a transmitted wave will be set up. Let θ_r be the angle of reflection and θ_t be the angle of transmission. Then without writing the expressions for the fields, we can find the relationship among θ_i, θ_r, and θ_t by noting that for the incident, reflected, and transmitted waves to be in step at the boundary, their apparent phase velocities parallel to the boundary must be equal; that is,

$$\boxed{\frac{v_{p1}}{\sin\theta_i} = \frac{v_{p1}}{\sin\theta_r} = \frac{v_{p2}}{\sin\theta_t}} \tag{10.1}$$

where v_{p1} $(= 1/\sqrt{\mu_1\varepsilon_1})$ and v_{p2} $(= 1/\sqrt{\mu_2\varepsilon_2})$ are the phase velocities along the directions of propagation of the waves in medium 1 and medium 2, respectively.

Laws of reflection and refraction

From (10.1), we have

$$\sin\theta_r = \sin\theta_i \tag{10.2a}$$

$$\sin\theta_t = \frac{v_{p2}}{v_{p1}}\sin\theta_i = \sqrt{\frac{\mu_1\varepsilon_1}{\mu_2\varepsilon_2}}\sin\theta_i \tag{10.2b}$$

or

$$\theta_r = \theta_i \tag{10.3a}$$

$$\theta_t = \sin^{-1}\left(\sqrt{\frac{\mu_1\varepsilon_1}{\mu_2\varepsilon_2}}\sin\theta_i\right) \tag{10.3b}$$

Equation (10.3a) is known as the *law of reflection* and (10.3b) is known as the *law of refraction*, or *Snell's law*. Snell's law is commonly cast in terms of the refractive index, denoted by the symbol n and defined as the ratio of the velocity of light in free space to the phase velocity in the medium. Thus if n_1 $(= c/v_{p1})$ and

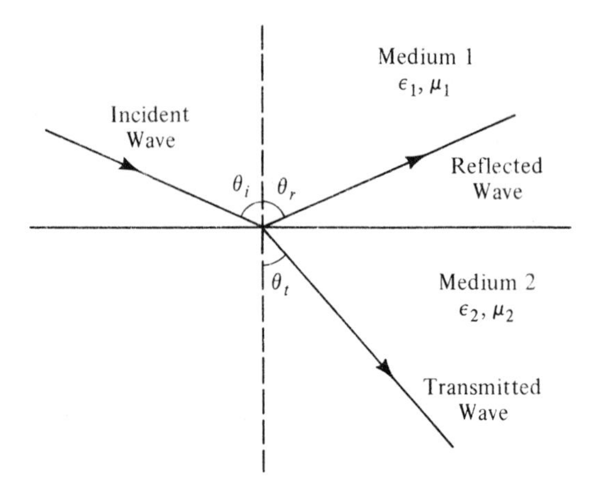

Figure 10.1. Reflection and transmission of an obliquely incident uniform plane wave on a plane boundary between two different perfect dielectric media.

n_2 ($= c/v_{p2}$) are the (phase) refractive indices for media 1 and 2, respectively, then

$$\boxed{\theta_t = \sin^{-1}\left(\frac{n_1}{n_2}\sin\theta_i\right)} \tag{10.4}$$

For two dielectrics having $\mu_1 = \mu_2 = \mu_0$, which is usually the case, (10.4) reduces to

$$\boxed{\theta_t = \sin^{-1}\left(\sqrt{\frac{\varepsilon_1}{\varepsilon_2}}\sin\theta_i\right)} \tag{10.5}$$

We shall now consider the derivation of the expressions for the reflection and transmission coefficients at the boundary. To do this, we distinguish between two cases: (1) the electric field vector of the wave linearly polarized parallel to the interface and (2) the magnetic field vector of the wave linearly polarized parallel to the interface. The law of reflection and Snell's law hold for both cases since they result from the fact that the apparent phase velocities of the incident, reflected, and transmitted waves parallel to the boundary must be equal.

Perpendicular polarization

The geometry pertinent to the case of the electric field vector parallel to the interface is shown in Fig. 10.2 in which the interface is assumed to be in the $x = 0$ plane, and the subscripts i, r, and t associated with the field symbols denote incident, reflected, and transmitted waves, respectively. The plane of incidence, that is, the plane containing the normal to the interface and the propagation vectors, is assumed to be in the xz-plane so that the electric field vectors are

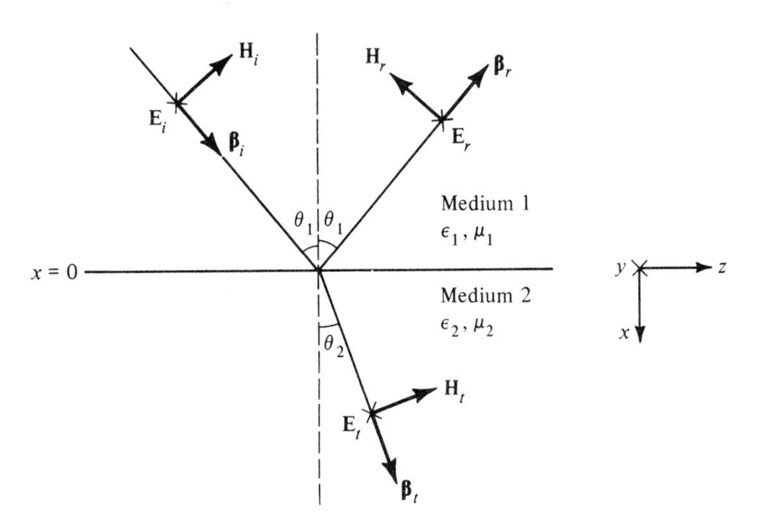

Figure 10.2. For obtaining the reflection and transmission coefficients for an obliquely incident uniform plane wave on a dielectric interface with its electric field perpendicular to the plane of incidence.

entirely in the y-direction. The corresponding magnetic field vectors are then as shown in the figure so as to be consistent with the condition that **E**, **H**, and $\boldsymbol{\beta}$ form a right-handed mutually orthogonal set of vectors. Since the electric field vectors are perpendicular to the plane of incidence, this case is also said to correspond to perpendicular polarization. The angle of incidence is assumed to be θ_1. From the law of reflection (9.86a), the angle of reflection is then also θ_1. The angle of transmission, assumed to be θ_2, is related to θ_1 by Snell's law, given by (10.3b).

The boundary conditions to be satisfied at the interface $x = 0$ are that (1) the tangential component of the electric field intensity be continuous and (2) the tangential component of the magnetic field intensity be continuous. Thus, we have at the interface $x = 0$

$$E_{yi} + E_{yr} = E_{yt} \tag{10.6a}$$

$$H_{zi} + H_{zr} = H_{zt} \tag{10.6b}$$

Expressing the quantities in (10.6a) and (10.6b) in terms of the total fields, we obtain

$$E_i + E_r = E_t \tag{10.7a}$$

$$H_i \cos \theta_1 - H_r \cos \theta_1 = H_t \cos \theta_2 \tag{10.7b}$$

We also know from one of the properties of uniform plane waves that

$$\frac{E_i}{H_i} = \frac{E_r}{H_r} = \eta_1 = \sqrt{\frac{\mu_1}{\varepsilon_1}} \tag{10.8a}$$

$$\frac{E_t}{H_t} = \eta_2 = \sqrt{\frac{\mu_2}{\varepsilon_2}} \tag{10.8b}$$

Substituting (10.8a) and (10.8b) into (10.7b) and rearranging, we get

$$E_i - E_r = E_t \frac{\eta_1 \cos \theta_2}{\eta_2 \cos \theta_1} \tag{10.9}$$

Solving (10.7a) and (10.9) for E_i and E_r, we have

$$E_i = \frac{E_t}{2}\left(1 + \frac{\eta_1 \cos \theta_2}{\eta_2 \cos \theta_1}\right) \tag{10.10a}$$

$$E_r = \frac{E_t}{2}\left(1 - \frac{\eta_1 \cos \theta_2}{\eta_2 \cos \theta_1}\right) \tag{10.10b}$$

We now define the reflection coefficient $\Gamma_\perp$ and the transmission coefficient $\tau_\perp$ as

$$\boxed{\Gamma_\perp = \frac{E_r}{E_i} = \frac{E_{yr}}{E_{yi}}} \tag{10.11a}$$

$$\boxed{\tau_\perp = \frac{E_t}{E_i} = \frac{E_{yt}}{E_{yi}}} \tag{10.11b}$$

where the subscript $\perp$ refers to perpendicular polarization. From (10.10a) and (10.10b), we then obtain

$$\Gamma_\perp = \frac{\eta_2 \cos \theta_1 - \eta_1 \cos \theta_2}{\eta_2 \cos \theta_1 + \eta_1 \cos \theta_2} \tag{10.12a}$$

$$\tau_\perp = \frac{2\eta_2 \cos \theta_1}{\eta_2 \cos \theta_1 + \eta_1 \cos \theta_2} \tag{10.12b}$$

Equations (10.12a) and (10.12b) are known as the Fresnel reflection and transmission coefficients, respectively, for perpendicular polarization.

Parallel polarization

Before we discuss the result given by (10.12a) and (10.12b), we shall derive the corresponding expressions for the case in which the magnetic field of the wave is parallel to the interface. The geometry pertinent to this case is shown in Fig. 10.3. Here again the plane of incidence is chosen to be the *xz*-plane so that the magnetic field vectors are entirely in the *y*-direction. The corresponding electric field vectors are then as shown in the figure so as to be consistent with the condition that **E**, **H**, and $\boldsymbol{\beta}$ form a right-handed mutually orthogonal set of vectors. Since the electric field vectors are parallel to the plane of incidence, this case is also said to correspond to parallel polarization.

Once again the boundary conditions to be satisfied at the interface $x = 0$ are that (1) the tangential component of the electric field intensity be continuous and (2) the tangential component of the magnetic field intensity be continuous. Thus we have at the interface $x = 0$,

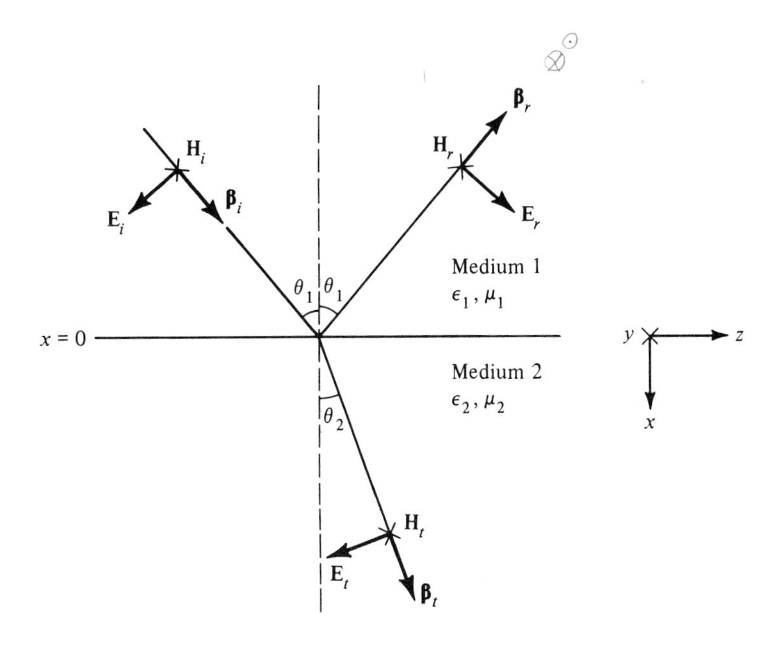

Figure 10.3. For obtaining the reflection and transmission coefficients for an obliquely incident uniform plane wave on a dielectric interface with its electric field parallel to the plane of incidence.

$$E_{zi} + E_{zr} = E_{zt} \tag{10.13a}$$

$$H_{yi} + H_{yr} = H_{yt} \tag{10.13b}$$

Expressing the quantities in (10.13a) and (10.13b) in terms of the total fields and also using (10.8a) and (10.8b), we obtain

$$E_i - E_r = E_t \frac{\cos\theta_2}{\cos\theta_1} \tag{10.14a}$$

$$E_i + E_r = E_t \frac{\eta_1}{\eta_2} \tag{10.14b}$$

Solving (10.14a) and (10.14b) for E_i and E_r, we have

$$E_i = \frac{E_t}{2}\left(\frac{\eta_1}{\eta_2} + \frac{\cos\theta_2}{\cos\theta_1}\right) \tag{10.15a}$$

$$E_r = \frac{E_t}{2}\left(\frac{\eta_1}{\eta_2} - \frac{\cos\theta_2}{\cos\theta_1}\right) \tag{10.15b}$$

We now define the reflection coefficient $\Gamma_\parallel$ and the transmission coefficient $\tau_\parallel$ as

$$\boxed{\Gamma_\parallel = -\frac{E_r}{E_i}} \tag{10.16a}$$

$$\boxed{\tau_\parallel = \frac{E_t}{E_i}} \tag{10.16b}$$

where the subscript $\parallel$ refers to parallel polarization. From (10.15a) and (10.15b), we then obtain

$$\boxed{\Gamma_\parallel = \frac{\eta_2\cos\theta_2 - \eta_1\cos\theta_1}{\eta_2\cos\theta_2 + \eta_1\cos\theta_1}} \tag{10.17a}$$

$$\boxed{\tau_\parallel = \frac{2\eta_2\cos\theta_1}{\eta_2\cos\theta_2 + \eta_1\cos\theta_1}} \tag{10.17b}$$

Note from (10.16a) and (10.16b) that

$$\frac{E_{zr}}{E_{zi}} = \frac{E_r\cos\theta_1}{-E_i\cos\theta_1} = -\frac{E_r}{E_i} = \Gamma_\parallel \tag{10.18a}$$

$$\frac{E_{zt}}{E_{zi}} = \frac{-E_t\cos\theta_2}{-E_i\cos\theta_1} = \tau_\parallel\frac{\cos\theta_2}{\cos\theta_1} \tag{10.18b}$$

Equations (10.17a) and (10.17b) are known as the Fresnel reflection and transmission coefficients, respectively, for parallel polarization.

We shall now discuss the results given by (10.12a), (10.12b), (10.17a), and (10.17b) for the reflection and transmission coefficients for the two cases:

1. For $\theta_1 = 0$, that is, for the case of normal incidence of the uniform plane wave upon the interface, $\theta_2 = 0$ and

$$\Gamma_\perp = \frac{\eta_2 - \eta_1}{\eta_2 + \eta_1}, \qquad \Gamma_\parallel = \frac{\eta_2 - \eta_1}{\eta_2 + \eta_1}$$

$$\tau_\perp = \frac{2\eta_2}{\eta_2 + \eta_1}, \qquad \tau_\parallel = \frac{2\eta_2}{\eta_2 + \eta_1}$$

Thus the reflection coefficients as well as the transmission coefficients for the two cases become equal as they should, since for normal incidence there is no difference between the two polarizations except for rotation by 90° parallel to the interface.

Total internal reflection

2. $\Gamma_\perp = 1$ and $\Gamma_\parallel = -1$ if $\cos\theta_2 = 0$; that is,

$$\sqrt{1 - \sin^2\theta_2} = \sqrt{1 - \frac{\mu_1\varepsilon_1}{\mu_2\varepsilon_2}\sin^2\theta_1} = 0$$

or

$$\sin\theta_1 = \sqrt{\frac{\mu_2\varepsilon_2}{\mu_1\varepsilon_1}} \tag{10.19}$$

where we have used Snell's law given by (10.3b) to express $\sin\theta_2$ in terms of $\sin\theta_1$. If we assume $\mu_2 = \mu_1 = \mu_0$ as is usually the case, (10.19) has real solutions for θ_1 for $\varepsilon_2 < \varepsilon_1$. Thus, for $\varepsilon_2 < \varepsilon_1$, that is, for transmission from a dielectric medium of higher permittivity into a dielectric medium of lower permittivity, there is a critical angle of incidence θ_c given by

$$\boxed{\theta_c = \sin^{-1}\sqrt{\frac{\varepsilon_2}{\varepsilon_1}}} \tag{10.20}$$

for which θ_2 is equal to 90° and $|\Gamma_\perp|$ and $|\Gamma_\parallel| = 1$. For $\theta_1 > \theta_c$, $\sin\theta_2$ becomes greater than 1, $\cos\theta_2$ becomes imaginary, and $\Gamma_\perp$ and $\Gamma_\parallel$ become complex, but with their magnitudes equal to unity, and *total internal reflection* occurs; that is, the incident wave is entirely reflected, the boundary condition being satisfied by an evanescent field in medium 2. To explain the evanescent nature, we note with reference to the geometry of Fig. 10.2 or Fig. 10.3 that

$$\beta_{x2}^2 + \beta_{z2}^2 = \beta_t^2 = \omega^2\mu_2\varepsilon_2$$

or

$$\beta_{x2}^2 = \omega^2\mu_2\varepsilon_2 - \beta_{z2}^2$$

For $\theta_1 = \theta_c$, $\beta_{z2} = \beta_{z1} = \omega^2\mu_1\varepsilon_1 \sin^2\theta_c = \omega^2\mu_2\varepsilon_2$, and $\beta_{x2}^2 = 0$. Therefore, for $\theta_1 > \theta_c$, $\beta_{z2} = \beta_{z1} = \omega^2\mu_1\varepsilon_1\sin^2\theta_1 > \omega^2\mu_2\varepsilon_2$, and $\beta_{x2}^2 < 0$. Thus β_{x2} becomes imaginary, corresponding to exponential decay of field in the x-direction without a propagating wave character. The phenomenon of total internal reflection is the fundamental principle of optical waveguides, since if we have a dielectric slab of permittivity ε_1 sandwiched between two dielectric media of permittivity $\varepsilon_2 < \varepsilon_1$, then by launching waves at an angle of incidence greater than the critical angle, it is possible to achieve guided wave propagation within the slab, as we shall learn in the next section.

3. $\Gamma_\perp = 0$ for $\eta_2 \cos\theta_1 = \eta_1 \cos\theta_2$; that is, for

$$\eta_2\sqrt{1 - \sin^2\theta_1} = \eta_1\sqrt{1 - \frac{\mu_1\varepsilon_1}{\mu_2\varepsilon_2}\sin^2\theta_1}$$

or

$$\sin^2\theta_1 = \frac{\eta_2^2 - \eta_1^2}{\eta_2^2 - \eta_1^2(\mu_1\varepsilon_1/\mu_2\varepsilon_2)} = \mu_2\frac{\mu_2 - \mu_1(\varepsilon_2/\varepsilon_1)}{\mu_2^2 - \mu_1^2} \qquad (10.21)$$

For the usual case of transmission between two dielectric materials, that is, for $\mu_2 = \mu_1$ and $\varepsilon_2 \neq \varepsilon_1$, this equation has no real solution for θ_1, and hence there is no angle of incidence for which the reflection coefficient is zero for the case of perpendicular polarization.

4. $\Gamma_\| = 0$ for $\eta_2 \cos\theta_2 = \eta_1 \cos\theta_1$; that is, for

$$\eta_2\sqrt{1 - \frac{\mu_1\varepsilon_1}{\mu_2\varepsilon_2}\sin^2\theta_1} = \eta_1\sqrt{1 - \sin^2\theta_1}$$

or

$$\sin^2\theta_1 = \frac{\eta_2^2 - \eta_1^2}{\eta_2^2(\mu_1\varepsilon_1/\mu_2\varepsilon_2) - \eta_1^2} = \varepsilon_2\frac{(\mu_2/\mu_1)\varepsilon_1 - \varepsilon_2}{\varepsilon_1^2 - \varepsilon_2^2} \qquad (10.22)$$

If we assume $\mu_2 = \mu_1$, this equation reduces to

$$\sin^2\theta_1 = \frac{\varepsilon_2}{\varepsilon_1 + \varepsilon_2}$$

which then gives

$$\cos^2\theta_1 = 1 - \sin^2\theta_1 = \frac{\varepsilon_1}{\varepsilon_1 + \varepsilon_2}$$

and

$$\tan\theta_1 = \sqrt{\frac{\varepsilon_2}{\varepsilon_1}}$$

Thus there exists a value of the angle of incidence θ_p, given by

$$\boxed{\theta_p = \tan^{-1}\sqrt{\frac{\varepsilon_2}{\varepsilon_1}}} \qquad (10.23)$$

for which the reflection coefficient is zero, and hence there is complete transmission for the case of parallel polarization.

Brewster angle

5. In view of cases 3 and 4, for an elliptically polarized wave incident on the interface at the angle θ_p, the reflected wave will be linearly polarized per-

pendicular to the plane of incidence. For this reason, the angle θ_p is known as the *polarizing angle*. It is also known as the *Brewster angle*. The phenomenon associated with the Brewster angle has several applications. An example is in gas lasers in which the discharge tube lying between the mirrors of a Fabry–Perot resonator is sealed by glass windows placed at the Brewster angle, as shown in Fig. 10.4, to minimize reflections from the ends of the tube so that the laser behavior is governed by the mirrors external to the tube.

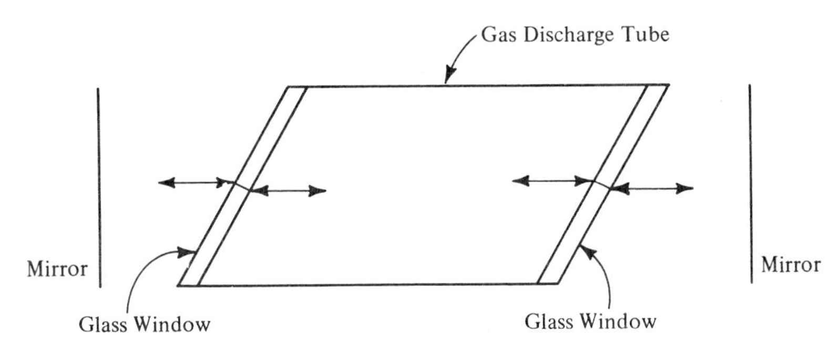

Figure 10.4. For illustrating the application of the Brewster angle effect in gas lasers.

We shall now consider an example.

Example 10.1

A uniform plane wave having the electric field

$$\mathbf{E}_i = E_0\left(\frac{\sqrt{3}}{2}\mathbf{i}_x - \frac{1}{2}\mathbf{i}_z\right)\cos\left[6\pi \times 10^9 t - 10\pi(x + \sqrt{3}z)\right]$$

is incident on the interface between free space and a dielectric medium of $\varepsilon = 1.5\varepsilon_0$ and $\mu = \mu_0$, as shown in Fig. 10.5. We wish to obtain the expressions for the electric fields of the reflected and transmitted waves.

First we note from the given $\mathbf{E}_i$ that the propagation vector of the incident wave is given by

$$\boldsymbol{\beta}_i = 10\pi(\mathbf{i}_x + \sqrt{3}\mathbf{i}_z) = 20\pi\left(\frac{1}{2}\mathbf{i}_x + \frac{\sqrt{3}}{2}\mathbf{i}_z\right)$$

the direction of which is consistent with the angle of incidence of 60°. We also note that the electric field vector (which is perpendicular to $\boldsymbol{\beta}_i$) is entirely in the plane of incidence. Thus the situation corresponds to one of parallel polarization, as in Fig. 10.3.

To obtain the required fields, we first find, by using (10.5) and with reference to the notation of Fig. 10.3, that

$$\sin\theta_2 = \sqrt{\frac{\varepsilon_0}{1.5\varepsilon_0}}\sin 60^\circ = \frac{1}{\sqrt{2}}$$

or $\theta_2 = 45°$. Then from (10.17a)–(10.17b) and (10.18a)–(10.18b), we have

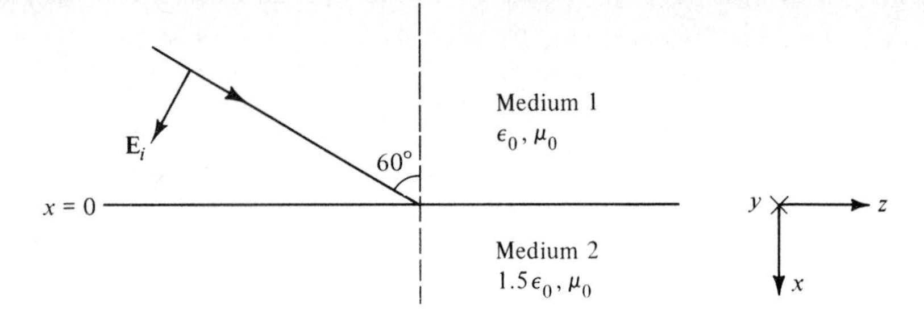

Figure 10.5. For Example 10.1.

$$\Gamma_{\|} = \frac{(\eta_0/\sqrt{1.5}) \cos 45° - \eta_0 \cos 60°}{(\eta_0/\sqrt{1.5}) \cos 45° + \eta_0 \cos 60°}$$

$$= \frac{2 - \sqrt{3}}{2 + \sqrt{3}} = 0.072$$

$$\tau_{\|} = \frac{2(\eta_0/\sqrt{1.5}) \cos 60°}{(\eta_0/\sqrt{1.5}) \cos 45° + \eta_0 \cos 60°}$$

$$= \frac{2\sqrt{2}}{2 + \sqrt{3}} = 0.758$$

$$\frac{E_r}{E_i} = -0.072$$

$$\frac{E_t}{E_i} = 0.758$$

Finally, noting with the aid of Fig. 10.6 that

$$\boldsymbol{\beta}_r = 20\pi\left(-\frac{1}{2}\mathbf{i}_x + \frac{\sqrt{3}}{2}\mathbf{i}_z\right) = 10\pi(-\mathbf{i}_x + \sqrt{3}\,\mathbf{i}_z)$$

and

$$\boldsymbol{\beta}_t = 20\pi\sqrt{1.5}\left(\frac{1}{\sqrt{2}}\mathbf{i}_x + \frac{1}{\sqrt{2}}\mathbf{i}_z\right) = 10\sqrt{3}\pi(\mathbf{i}_x + \mathbf{i}_z)$$

we write the expressions for the reflected and transmitted wave fields to be

$$\mathbf{E}_r = -0.072E_0\left(\frac{\sqrt{3}}{2}\mathbf{i}_x + \frac{1}{2}\mathbf{i}_z\right) \cos\,[6\pi \times 10^9 t + 10\pi(x - \sqrt{3}z)]$$

and

$$\mathbf{E}_t = 0.758E_0\left(\frac{1}{\sqrt{2}}\mathbf{i}_x - \frac{1}{\sqrt{2}}\mathbf{i}_z\right) \cos\,[6\pi \times 10^9 t - 10\sqrt{3}\pi(x + z)]$$

Note that for $x = 0$, $E_{zi} + E_{zr} = E_{zt}$ and $E_{xi} + E_{xr} = 1.5E_{xt}$ so that the fields do indeed satisfy the boundary conditions.

K10.1. Oblique incidence of uniform plane waves; Plane interface; Law of reflection; Snell's law; Perpendicular and parallel polarizations; Total internal reflection; Brewster angle.

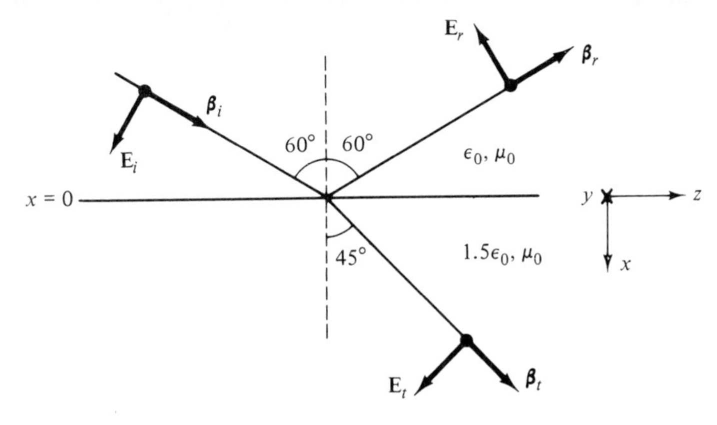

Figure 10.6. For writing the expressions for the reflected and transmitted wave electric fields for Example 10.1.

D10.1. Consider a plane boundary between medium 1 ($\varepsilon = \varepsilon_1$, $\mu = \mu_0$) and medium 2 ($\varepsilon = \varepsilon_2$, $\mu = \mu_0$). Find the value of $\varepsilon_2/\varepsilon_1$ for each of the following cases of uniform plane waves incident on the boundary from medium 1. **(a)** Total internal reflection occurs for $\theta_i \geq 60°$; **(b)** the reflection coefficient for parallel polarization is zero for $\theta_i = 60°$; and **(c)** the critical angle of incidence for total internal reflection is the same as the Brewster angle for incidence from medium 2.
Ans. **(a)** 0.75; **(b)** 3; **(c)** 0.618

D10.2. In Figs. 10.2 and 10.3, assume that $\varepsilon_1 = 3\varepsilon_0$, $\varepsilon_2 = 9\varepsilon_0$, $\mu_1 = \mu_2 = \mu_0$, and $\theta_i = 45°$. Find **(a)** E_r/E_i and E_t/E_i for the case of perpendicular polarization (Fig. 10.2) and **(b)** E_r/E_i and E_t/E_i for the case of parallel polarization (Fig. 10.3).
Ans. **(a)** -0.382, 0.618; **(b)** 0.146, 0.662

10.2 DIELECTRIC SLAB GUIDE

In the preceding section we learned that for a wave incident obliquely from a dielectric medium of permittivity ε_1 onto another dielectric medium of permittivity $\varepsilon_2 < \varepsilon_1$, total internal reflection occurs for angles of incidence θ_i exceeding the critical angle θ_c given by

$$\theta_c = \sin^{-1}\sqrt{\frac{\varepsilon_2}{\varepsilon_1}} \tag{10.24}$$

where it is assumed that $\mu = \mu_0$ everywhere. In this section we shall consider the dielectric slab waveguide, which forms the basis for thin-film waveguides, used extensively in integrated optics.

A. Wave-Bounce Approach

Description

The dielectric slab waveguide consists of a dielectric slab of permittivity ε_1, sandwiched between two dielectric media of permittivities less than ε_1. For simplicity, we shall consider the symmetric waveguide, that is, one for which the permittivities of the dielectrics on either side of the slab are the same and equal to ε_2, as shown in Fig. 10.7. Then by launching waves at an angle of incidence $\theta_i > \theta_c$ where θ_c is given by (10.24), it is possible to achieve guided wave propagation within the slab, as shown in the figure. For a given thickness d of the slab and for a given frequency of the waves, there are only discrete values of θ_i for which the guiding can take place. In other words, guiding of a wave of a given frequency is not ensured simply because the condition for total internal reflection is met.

Self-consistency condition for guidance

The allowed values of θ_i are dictated by the self-consistency condition, which can be explained with the aid of the construction in Fig. 10.8, as follows. If we consider a point A on a given wavefront designated 1 and follow that wavefront as it moves to position 1′ passing through point B, reflects at the interface $x = d/2$ giving rise to wavefront designated 2, then moves to position 2′ passing through point C, reflects at the interface $x = -d/2$ giving rise to wavefront designated 3, and moves to position 3′ passing through A, then the total phase shift undergone must be equal to an integer multiple of 2π. If λ_0 is the wavelength in free space corresponding to the wave frequency, the self-consistency condition is given by

$$\begin{aligned} \frac{2\pi\sqrt{\varepsilon_{r1}}}{\lambda_0}(AB\cos\theta_i) + \underline{/\bar{\Gamma}_B} + \frac{2\pi\sqrt{\varepsilon_{r1}}}{\lambda_0}(BC\cos\theta_i) \\ + \underline{/\bar{\Gamma}_A} + \frac{2\pi\sqrt{\varepsilon_{r1}}}{\lambda_0}(CA\cos\theta_i) = 2m\pi, \qquad m = 0, 1, 2, \ldots \end{aligned} \tag{10.25}$$

where $\bar{\Gamma}_A$ and $\bar{\Gamma}_B$ are the reflection coefficients at the interfaces $x = 0$ and $x = d$, respectively, and $\varepsilon_{r1} = \varepsilon_1/\varepsilon_0$. We recall that under conditions of total internal reflection, the reflection coefficients (10.12a) and (10.17a) become complex with their magnitudes equal to unity. For the symmetric waveguide, $\bar{\Gamma}_A = \bar{\Gamma}_B$. Thus substituting $\bar{\Gamma}$ for $\bar{\Gamma}_A$ and $\bar{\Gamma}_B$ and $2d$ for $(AB + BC + CA)$, we write (10.25) as

$$\frac{4\pi d\sqrt{\varepsilon_{r1}}}{\lambda_0}\cos\theta_i + 2\underline{/\bar{\Gamma}} = 2m\pi, \qquad m = 0, 1, 2, \ldots$$

or

$$\boxed{\frac{2\pi d\sqrt{\varepsilon_{r1}}}{\lambda_0}\cos\theta_i + \underline{/\bar{\Gamma}} = m\pi, \qquad m = 0, 1, 2, \ldots} \tag{10.26}$$

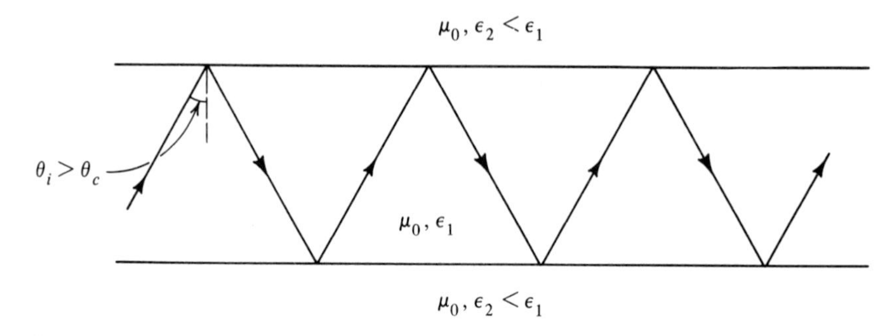

Figure 10.7. Total internal reflection in a dielectric slab waveguide.

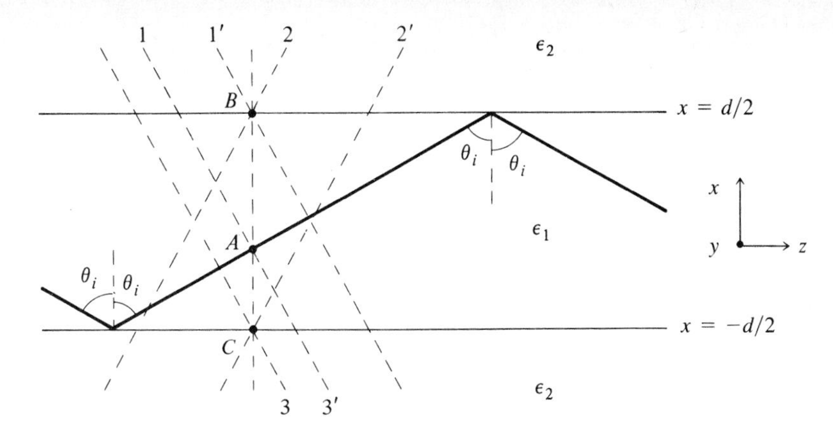

Figure 10.8. For explaining the self-consistency condition for waveguiding in a dielectric slab guide.

Characteristic equation for TE modes and solution

To proceed further, we need to distinguish between the cases of perpendicular and parallel polarizations as defined in the preceding section, since the reflection coefficients for the two cases are different. We shall here consider only the case of perpendicular polarization. The situation then corresponds to TE modes since the electric field has no longitudinal or z-component. Thus substituting

$$\cos\theta_1 = \cos\theta_i$$

and

$$\begin{aligned}\cos\theta_2 &= \sqrt{1-\sin^2\theta_2}\\ &= j\sqrt{\sin^2\theta_2 - 1}\\ &= j\sqrt{\frac{\varepsilon_1}{\varepsilon_2}\sin^2\theta_i - 1}\end{aligned}$$

in (10.12), we obtain

$$\overline{\Gamma}_\perp = \frac{\eta_2\cos\theta_i - j\eta_1\sqrt{(\varepsilon_1/\varepsilon_2)\sin^2\theta_i - 1}}{\eta_2\cos\theta_i + j\eta_1\sqrt{(\varepsilon_1/\varepsilon_2)\sin^2\theta_i - 1}} \tag{10.27}$$

so that

$$\begin{aligned}\underline{/\overline{\Gamma}_\perp} &= -2\tan^{-1}\frac{\eta_1\sqrt{(\varepsilon_1/\varepsilon_2)\sin^2\theta_i - 1}}{\eta_2\cos\theta_i}\\ &= -2\tan^{-1}\frac{\sqrt{\sin^2\theta_i - (\varepsilon_2/\varepsilon_1)}}{\cos\theta_i}\end{aligned} \tag{10.28}$$

Substituting (10.28) into (10.26), we then obtain

$$\frac{2\pi d\sqrt{\varepsilon_{r1}}}{\lambda_0}\cos\theta_i - 2\tan^{-1}\frac{\sqrt{\sin^2\theta_i - (\varepsilon_2/\varepsilon_1)}}{\cos\theta_i} = m\pi, \qquad m = 0, 1, 2, \ldots$$

or

$$\tan\left(\frac{\pi d\sqrt{\varepsilon_{r1}}}{\lambda_0}\cos\theta_i - \frac{m\pi}{2}\right) = \frac{\sqrt{\sin^2\theta_i - (\varepsilon_2/\varepsilon_1)}}{\cos\theta_i}, \qquad m = 0, 1, 2, \ldots$$

or

$$\boxed{\tan[f(\theta_i)] = \begin{cases} g(\theta_i), & m = 0, 2, 4, \ldots \\ -\dfrac{1}{g(\theta_i)}, & m = 1, 3, 5, \ldots \end{cases}} \tag{10.29}$$

where

$$f(\theta_i) = \frac{\pi d\sqrt{\varepsilon_{r1}}}{\lambda_0}\cos\theta_i \tag{10.30a}$$

$$g(\theta_i) = \frac{\sqrt{\sin^2\theta_i - (\varepsilon_2/\varepsilon_1)}}{\cos\theta_i} \tag{10.30b}$$

Equation (10.29) is the characteristic equation for the guiding of TE waves in the dielectric slab. For given values of ε_1, ε_2, d, and λ_0, the solutions for θ_i can be obtained by plotting the two sides of (10.29) versus θ_i and finding the points of intersection. The nature of this construction is shown in Fig. 10.9. Each solution corresponds to one mode. It can be seen from (10.30a) and Fig. 10.9 that for a given set of values of ε_1 and ε_2, fewer solutions are obtained for θ_i as the ratio (d/λ_0) becomes smaller, since the number of branches of the plot of $\tan[f(\theta_i)]$ between $\theta_i = \pi/2$ and $\theta_i = \theta_c$ become fewer. It can also be seen that there is always one solution even for arbitrarily low values of (d/λ_0), that is, for large values of λ_0 or low frequencies, for a given d.

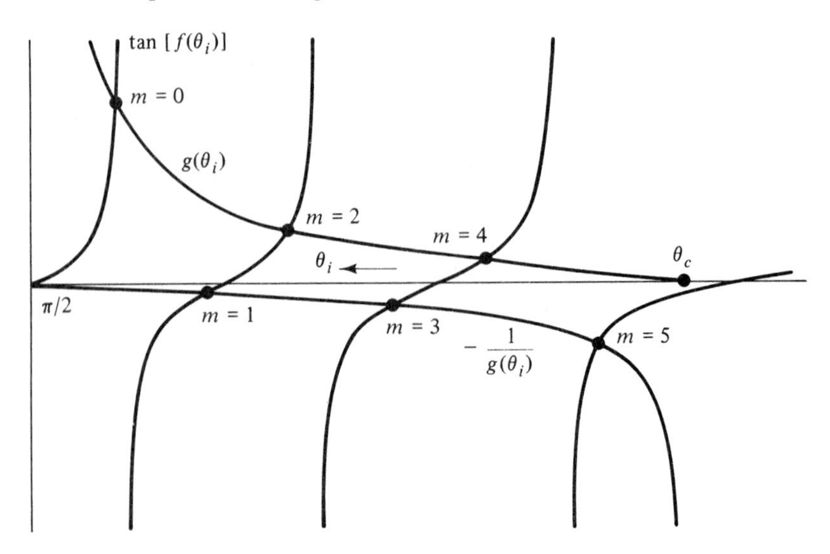

Figure 10.9. Graphical construction pertinent to the solution of Eq. (10.29).

Alternative to the graphical solution, we can use a computer to solve (10.29). The listing of a PC program which computes the allowed values of θ_i for specified values of ε_{r1}, ε_{r2}, d, and λ_0 is included as PL 10.1. Sample output from a run of the program for values of $\varepsilon_{r1} = 4$, $\varepsilon_{r2} = 1$, $d = 10$ mm, and $\lambda_0 = 5$ mm is also included.

PL10.1. Program listing and sample output for computing allowed values of angles of incidence for TE guided modes in a symmetric dielectric slab waveguide.

```
100 '*****************************************************
110 '* COMPUTAION OF ALLOWED VALUES OF ANGLES OF INCIDENCE *
120 '* FOR TE GUIDED MODES IN A SYMMETRIC DIELECTRIC SLAB  *
130 '* WAVEGUIDE.                                          *
140 '*****************************************************
150 CLS:SCREEN 0
160 PRINT "ENTER VALUES OF INPUT PARAMETERS:"
170 PRINT:INPUT "REL. PERMITTIVITY ER1 = ",E1
180 INPUT "REL. PERMITTIVITY ER2 = ",E2
190 INPUT "THICKNESS IN MM = ",D
200 INPUT "WAVELENGTH IN MM = ",WL
210 PI=3.1416:RD=180/PI
220 ER=E2/E1:ES=SQR(ER)
230 TMAX=ATN(ES/SQR(1-ES*ES)):'* CRITICAL ANGLE *
240 CNST=WL/(D*SQR(E1))
250 PRINT:PRINT "COMPUTED VALUES ARE:":PRINT
260 '* COMPUTATION OF T1 AND T2, THE LOWER AND UPPER BOUNDS
270 '  FOR THETA, THE ALLOWED ANGLE OF INCIDENCE, FOR A
280 '  GIVEN VALUE OF M *
290 M=0:T1=PI/2:GOTO 340
300 ARG=M*CNST/2
310 IF ARG>=1 THEN 480:'* SOLUTION COMPLETED *
320 T1=ATN(SQR(1-ARG*ARG)/ARG)
330 IF T1<=TMAX THEN 480:'* SOLUTION COMPLETED *
340 ARG=(M+1)*CNST/2
350 IF ARG>=1 THEN T2=TMAX:GOTO 390
360 T2=ATN(SQR(1-ARG*ARG)/ARG)
370 IF TMAX>T2 THEN T2=TMAX
380 '* COMPUTATION AND PRINTING OF VALUE OF THETA *
390 THETA=(T1+T2)/2
400 LHS=TAN(PI*COS(THETA)/CNST)
410 RHS=SQR((SIN(THETA)^2-ER))/COS(THETA)
420 IF (M-INT(M/2)*2)<>0 THEN RHS=-1/RHS
430 IF ABS(RHS-LHS)<ABS(RHS/10000) THEN 460
440 IF RHS>LHS THEN T1=THETA:GOTO 390
450 T2=THETA:GOTO 390
460 PRINT "M =";M;"  THETA =";THETA*RD;"DEG"
470 M=M+1:GOTO 300
480 LOCATE 23,1:PRINT "STRIKE ANY KEY TO CONTINUE"
490 C$=INPUT$(1)
500 GOTO 150
510 END
RUN
ENTER VALUES OF INPUT PARAMETERS:

REL. PERMITTIVITY ER1 = 4
REL. PERMITTIVITY ER2 = 1
THICKNESS IN MM = 10
WAVELENGTH IN MM = 5
```

PL10.1. (continued)

```
COMPUTED VALUES ARE:

M = 0    THETA = 83.42783 DEG
M = 1    THETA = 76.77756 DEG
M = 2    THETA = 69.96263 DEG
M = 3    THETA = 62.87805 DEG
M = 4    THETA = 55.38428 DEG
M = 5    THETA = 47.28283 DEG
M = 6    THETA = 38.30225 DEG
STRIKE ANY KEY TO CONTINUE
```

Cutoff frequencies

Returning now to Fig. 10.9, we designate the modes associated with the solutions as TE_m modes, where $m = 0, 1, 2,$ correspond to the values of m on the plot. We note from the plot that the solution for a given TE_m mode for $m > 1$ does not exist if $f(\theta_c) < m\pi/2$. Therefore the cutoff condition is given by

$$\frac{\pi d \sqrt{\varepsilon_{r1}}}{\lambda_0} \cos \theta_c < \frac{m\pi}{2}$$

$$\frac{\pi d \sqrt{\varepsilon_{r1}}}{\lambda_0} \sqrt{1 - \frac{\varepsilon_2}{\varepsilon_1}} < \frac{m\pi}{2}$$

$$\lambda_0 > \frac{2d\sqrt{\varepsilon_{r1} - \varepsilon_{r2}}}{m} \tag{10.31}$$

where we have used (10.24). The cutoff frequency is given by

$$f_c = \frac{c}{\lambda_0} = \frac{mc}{2d\sqrt{\varepsilon_{r1} - \varepsilon_{r2}}}$$

The fundamental mode, TE_0, has no cutoff frequency. Thus

$$\boxed{f_c = \frac{mc}{2d\sqrt{\varepsilon_{r1} - \varepsilon_{r2}}}, \qquad m = 0, 1, 2, \ldots} \tag{10.32}$$

Example 10.2

Finding propagating modes

For the symmetric dielectric slab waveguide of Fig. 10.8, let $\varepsilon_1 = 2.56\varepsilon_0$, $\varepsilon_2 = \varepsilon_0$, and $d = 10\lambda_0$. We wish to find the number of TE modes that can propagate by guidance in the slab.

From (10.32),

$$\begin{aligned} f_c &= \frac{mc}{20\lambda_0\sqrt{2.56 - 1}} \\ &= \frac{mf}{24.98}, \qquad m = 0, 1, 2, \ldots \end{aligned}$$

Thus for $m > 24$, $f_c > f$ and the modes are cut off. Therefore, the number of propagating TE modes is 25, corresponding to $m = 0, 1, 2, \ldots, 24$.

TM modes

The entire discussion for guided waves in the dielectric slab guide can be repeated for TM modes by using $\bar{\Gamma}_{\parallel}$ in the place of $\bar{\Gamma}_{\perp}$ in (10.26) to derive the characteristic equation for guidance. We shall include the derivation as Problem P10.10.

B. Wave-Field Approach

Field behavior for guided modes

A formal approach to the investigation of guided modes in the dielectric slab involves the derivation of the field expressions. This is done by recognizing with reference to the geometry in Fig. 10.8 that (a) in the slab the fields have standing wave character in the x-direction and traveling wave character in the z-direction; (b) outside the slab, the fields are evanescent, that is, they decay exponentially away from it in the x-direction and have traveling wave character in the z-direction; and (c) from symmetry considerations, the fields should be even or odd with respect to x.

Field expressions for even TE modes

Let us first consider even TE modes, that is, modes with the transverse field components having even symmetry with respect to x. Then we write the expression for the (only) electric field component $\bar{E}_y$ to be

$$\bar{E}_y = \begin{cases} \bar{A} \cos \beta_{x1} x \, e^{-j\beta_z z} & \text{for } |x| < d/2 \\ \bar{B} e^{-\alpha_{x2} x} \, e^{-j\beta_z z} & \text{for } x > d/2 \\ \bar{B} e^{\alpha_{x2} x} \, e^{-j\beta_z z} & \text{for } x < -d/2 \end{cases} \tag{10.33}$$

where $\bar{A}$ and $\bar{B}$ are constants. Note that subscripts 1 and 2 denote regions of permittivities ε_1 and ε_2, respectively, and that the phase constant β_z does not have a subscript 1 or 2, since it must be the same in all three regions in view of the requirement that the fields be in phase at the boundaries $x = \pm d/2$ for all z. Continuity of $\bar{E}_y$ at $x = \pm d/2$ further requires that

$$\bar{A} \cos \beta_{x1} \frac{d}{2} = \bar{B} e^{-\alpha_{x2} d/2}$$

so that

$$\bar{B} = \bar{A} e^{\alpha_{x2} d/2} \cos \beta_{x1} \frac{d}{2} \tag{10.34}$$

and hence

$$\bar{E}_y = \begin{cases} \bar{A} \cos \beta_{x1} x \, e^{-j\beta_z z} & \text{for } |x| < d/2 \\ \bar{A} \cos \beta_{x1} \frac{d}{2} \, e^{-\alpha_{x2}(x - d/2)} e^{-j\beta_z z} & \text{for } x > d/2 \\ \bar{A} \cos \beta_{x1} \frac{d}{2} \, e^{\alpha_{x2}(x + d/2)} e^{-j\beta_z z} & \text{for } x < -d/2 \end{cases} \tag{10.35}$$

To obtain the corresponding magnetic field components, we use the phasor forms of (4.12a)–(4.12c) with the understanding that $\frac{\partial}{\partial t} \to j\omega$, $\frac{\partial}{\partial y} = 0$, $\frac{\partial}{\partial z} \to -j\beta_z$ and $E_x = E_z = 0$:

$$j\beta_z \bar{E}_y = -j\omega \bar{B}_x \tag{10.36a}$$

$$\frac{\partial \bar{E}_y}{\partial x} = -j\omega \bar{B}_z \tag{10.36b}$$

Thus

$$\bar{H}_x = -\frac{\beta_z}{\omega\mu_0}\bar{E}_y \tag{10.37}$$

$$\bar{H}_z = \begin{cases} -\dfrac{j\beta_{x1}}{\omega\mu_0}\bar{A}\sin\beta_{x1}x\, e^{-j\beta_z z} & \text{for } |x| < d/2 \\ -\dfrac{j\alpha_{x2}}{\omega\mu_0}\bar{A}\cos\beta_{x1}\dfrac{d}{2}\, e^{-\alpha_{x2}(x-d/2)}e^{-j\beta_z z} & \text{for } x > d/2 \\ \dfrac{j\alpha_{x2}}{\omega\mu_0}\bar{A}\cos\beta_{x1}\dfrac{d}{2}\, e^{\alpha_{x2}(x+d/2)}e^{-j\beta_z z} & \text{for } x < -d/2 \end{cases} \tag{10.38}$$

Now, continuity of $\bar{H}_z$ at $x = \pm d/2$ requires that

$$\tan\beta_{x1}\frac{d}{2} = \frac{\alpha_{x2}}{\beta_{x1}} \tag{10.39}$$

We also know that β_{x1}, β_z, and α_{x2} are not independent since together the field components must also satisfy the component equations of (4.22) in phasor form with $\mathbf{J} = 0$:

$$-j\beta_z\bar{H}_x - \frac{\partial\bar{H}_z}{\partial x} = \begin{cases} j\omega\varepsilon_1\bar{E}_y & \text{for } |x| < d/2 \\ j\omega\varepsilon_2\bar{E}_y & \text{for } x > d/2 \\ j\omega\varepsilon_2\bar{E}_y & \text{for } x < -d/2 \end{cases} \tag{10.40}$$

Substitution of (10.35), (10.37), and (10.38) gives us

$$\beta_{x1}^2 + \beta_z^2 = \omega^2\mu_0\varepsilon_1 \tag{10.41a}$$

$$-\alpha_{x2}^2 + \beta_z^2 = \omega^2\mu_0\varepsilon_2 \tag{10.41b}$$

or

$$\frac{\alpha_{x2}}{\beta_{x1}} = \sqrt{\frac{\omega^2\mu_0(\varepsilon_1 - \varepsilon_2)}{\beta_{x1}^2} - 1} \tag{10.42}$$

Combining (10.39) and (10.42), we obtain the characeristic equation for guidance to be

$$\tan\left(\beta_{x1}\frac{d}{2}\right) = \sqrt{\frac{\omega^2\mu_0(\varepsilon_1 - \varepsilon_2)}{\beta_{x1}^2} - 1}$$

$$\tan\left(\frac{\beta_1 d}{2}\cos\theta_i\right) = \sqrt{\frac{\omega^2\mu_0(\varepsilon_1 - \varepsilon_2)}{\omega^2\mu_0\varepsilon_1\cos^2\theta_i} - 1}$$

$$\tan\left(\frac{\pi d\sqrt{\varepsilon_{r1}}}{\lambda_0}\cos\theta_i\right) = \frac{\sqrt{\sin^2\theta_i - (\varepsilon_2/\varepsilon_1)}}{\cos\theta_i}$$

$$\boxed{\tan[f(\theta_i)] = g(\theta_i)} \tag{10.43}$$

which is the same as (10.29) for $m = 0, 2, 4, \ldots$.

Proceeding further, we can interpret the mode number m in terms of the field variations with x in the following manner. For a given value of m, we observe from Fig. 10.9 that $m\pi/2 < \beta_{x1}d/2$ [$= f(\theta_i)$] $< (m + 1)\pi/2$. Thus from $x = 0$ to $x = \pm d/2$, $\cos \beta_{x1}x$ varies from $\cos 0°$ to some value between $\cos m\pi/2$ and $\cos (m + 1)\pi/2$. Near cutoff, $\beta_{x1}d/2 \to m\pi/2$, $\tan (\beta_{x1}d/2) \to 0$, and $\alpha_{x2} \to 0$. The variation of E_y with x is as illustrated in Fig. 10.10(a) for $m = 2$. At high frequencies far from cutoff, $\beta_{x1}d/2 \to (m + 1)\pi/2$, $\tan \beta_{x1}d/2 \to \infty$, and $\alpha_{x2} \to \infty$. The variation of E_y with x is as illustrated in Fig. 10.10(b). Figure 10.10(c) illustrates the situation intermediate to those near cutoff and far from cutoff. Thus within the thickness of the slab, the behavior of the field components varies from m half-sine variations near cutoff toward $(m + 1)$ half-sine variations far from cutoff, with the evanescence outside the dielectric slab dictated by $\alpha_{x2} \to 0$ near cutoff toward $\alpha_{x2} \to \infty$ far from cutoff.

Field expressions for odd TE modes

The field expressions for the odd TE modes, that is, modes with the transverse field components having odd symmetry with respect to x, can be obtained by writing the expression for $\bar{E}_y$ to be

$$\bar{E}_y = \begin{cases} \bar{C} \sin \beta_{x1}x \, e^{-j\beta_z z} & \text{for } |x| < d/2 \\ \bar{D}e^{-\alpha_{x2}x}e^{-j\beta_z z} & \text{for } x > d/2 \\ -\bar{D}e^{\alpha_{x2}x}e^{-j\beta_z z} & \text{for } x < -d/2 \end{cases} \tag{10.44}$$

where $\bar{C}$ and $\bar{D}$ are constants and proceeding in a manner similar to that for the even modes. We shall omit the details and write down the final results:

$$\bar{E}_y = \begin{cases} \bar{C} \sin \beta_{x1}x \, e^{-j\beta_z z} & \text{for } |x| < d/2 \\ \bar{C} \sin \beta_{x1}\dfrac{d}{2} \, e^{-\alpha_{x2}(x-d/2)}e^{-j\beta_z z} & \text{for } x > d/2 \\ -\bar{C} \sin \beta_{x1}\dfrac{d}{2} \, e^{\alpha_{x2}(x+d/2)}e^{-j\beta_z z} & \text{for } x < -d/2 \end{cases} \tag{10.45}$$

$$\bar{H}_x = -\frac{\beta_z}{\omega\mu_0}\bar{E}_y \tag{10.46}$$

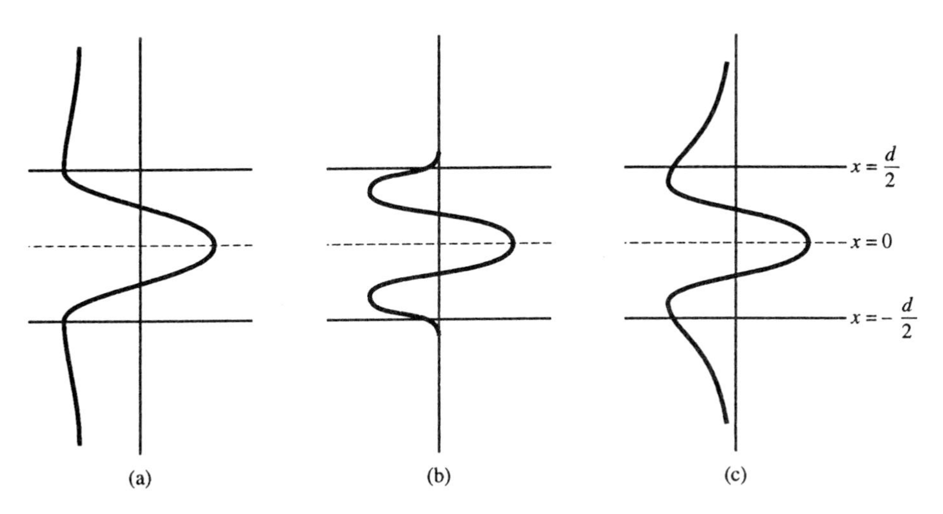

Figure 10.10. Variations of E_y with x for TE_2 mode in the symmetric dielectric slab waveguide for **(a)** near cutoff; **(b)** far from cutoff; and **(c)** intermediate to (a) and (b).

$$\overline{H}_z = \begin{cases} \dfrac{j\beta_{x1}}{\omega\mu_0}\overline{C}\cos\beta_{x1}x\, e^{-j\beta_z z} & \text{for } |x| < d/2 \\ -\dfrac{j\alpha_{x2}}{\omega\mu_0}\overline{C}\sin\beta_{x1}\dfrac{d}{2}\, e^{-\alpha_{x2}(x-d/2)}e^{-j\beta_z z} & \text{for } x > d/2 \\ -\dfrac{j\alpha_{x2}}{\omega\mu_0}\overline{C}\sin\beta_{x1}\dfrac{d}{2}\, e^{\alpha_{x2}(x+d/2)}e^{-j\beta_z z} & \text{for } x < -d/2 \end{cases} \tag{10.47}$$

Continuity of $\overline{H}_z$ at $x = \pm d/2$ requires that

$$\begin{aligned} \cot\beta_{x1}\frac{d}{2} &= -\frac{\alpha_{x2}}{\beta_{x1}} \\ &= -\sqrt{\frac{\omega^2\mu_0(\varepsilon_1 - \varepsilon_2)}{\beta_{x1}^2} - 1} \end{aligned} \tag{10.48}$$

where we have used (10.42). Thus the characteristic equation for guidance is

$$\boxed{\tan f(\theta_i) = -\frac{1}{g(\theta_i)}} \tag{10.49}$$

which is the same as (10.29) for $m = 1, 3, 5, \ldots$.

Proceeding further, we observe from Fig. 10.9 that for a given value of m, $m\pi/2 < \beta_{x1}d/2\ [= f(\theta_i)] < (m + 1)\pi/2$. Thus from $x = 0$ to $x = \pm d/2$, $\sin\beta_{x1}x$ varies from $\sin 0°$ to some value between $\sin m\pi/2$ and $\sin (m + 1)\pi/2$. Near cutoff, $\beta_{x1}d/2 \to m\pi/2$, $\cot(\beta_{x1}d/2) \to 0$, and $\alpha_{x2} \to 0$. At high frequencies far from cutoff, $\beta_{x1}d/2 \to (m + 1)\pi/2$, $\cot(\beta_{x1}d/2) \to \infty$, and $\alpha_{x2} \to \infty$. Thus the variation of E_y with x for $m = 1$ is illustrated in Fig. 10.11 for three situations: (a) near cutoff; (b) far from cutoff; and (c) intermediate to (a) and (b). As in the case of the even modes, the behavior of the field components varies from m half-sine variations near cutoff toward $(m + 1)$ half-sine variations far from cutoff.

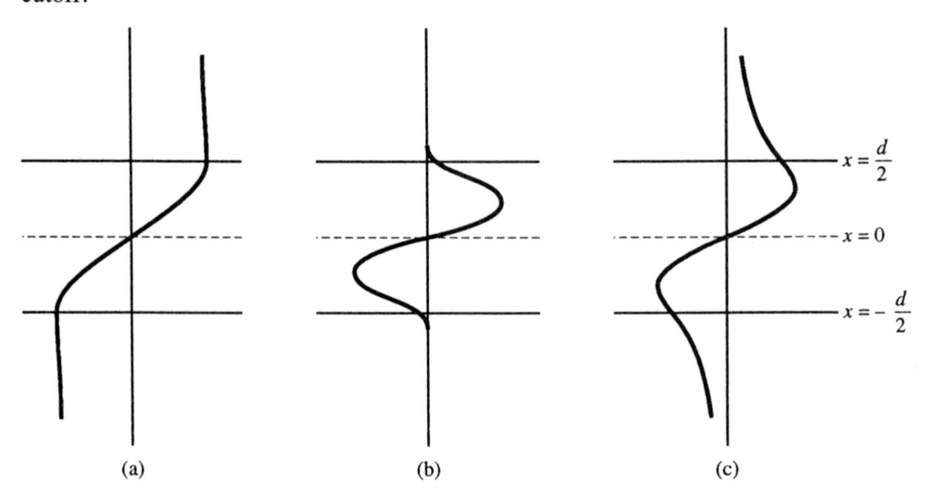

Figure 10.11. Variations of E_y with x for TE_1 mode in the symmetric dielectric slab waveguide for **(a)** near cutoff; **(b)** far from cutoff; and **(c)** intermediate to (a) and (b).

Power flow Let us now investigate the time-average power flow down the symmetric slab waveguide for TE modes. First we write the complex Poynting vector associated with the TE mode fields as given by

$$\begin{aligned}\mathbf{P} &= \frac{1}{2}\overline{\mathbf{E}} \times \overline{\mathbf{H}}^* \\ &= \frac{1}{2}(\bar{E}_y \bar{H}_z^* \mathbf{i}_x - \bar{E}_y \bar{H}_x^* \mathbf{i}_z)\end{aligned} \tag{10.50}$$

Then noting from (10.35), (10.37), and (10.38) that $\bar{E}_y \bar{H}_x^*$ is real, whereas as $\bar{E}_y \bar{H}_z^*$ is imaginary, we obtain the time-average Poynting vector as given by

$$\begin{aligned}\langle\mathbf{P}\rangle &= \mathrm{Re}\,\overline{\mathbf{P}} \\ &= \frac{\beta_z}{2\omega\mu_0}|\bar{E}_y|^2\,\mathbf{i}_z \\ &= \frac{\beta_z|\bar{A}|^2}{2\omega\mu_0}\begin{cases}\cos^2\beta_{x1}x\,\mathbf{i}_z & \text{for } |x| < d/2 \\ \cos^2\beta_{x1}\dfrac{d}{2}\,e^{-2\alpha_{x2}(x-d/2)}\mathbf{i}_z & \text{for } x > d/2 \\ \cos^2\beta_{x1}\dfrac{d}{2}\,e^{2\alpha_{x2}(x+d/2)}\mathbf{i}_z & \text{for } x < -d/2\end{cases}\end{aligned} \tag{10.51}$$

where we have used the even mode field expression. For the odd modes, the $\cos^2$ terms will be replaced by $\sin^2$ terms and the final result will be the same. The time-average power flow along the guide per unit length in the y-direction (because of the independence of the fields with y) is then given by

$$\begin{aligned}\langle P\rangle &= \int_{y=0}^{1}\int_{x=-\infty}^{\infty}\langle\mathbf{P}\rangle\cdot dx\,dy\,\mathbf{i}_z \\ &= \frac{\beta_z|\bar{A}|^2}{\omega\mu_0}\left[\int_0^{d/2}\cos^2\beta_{x1}x\,dx + \int_{d/2}^{\infty}\cos^2\beta_{x1}\frac{d}{2}\,e^{-2\alpha_{x2}(x-d/2)}\,dx\right] \\ &= \frac{\beta_z|\bar{A}|^2}{\omega\mu_0}\left[\frac{d}{4} + \frac{\sin\beta_{x1}d}{4\beta_{x1}} + \frac{\cos^2\beta_{x1}d/2}{2\alpha_{x2}}\right]\end{aligned} \tag{10.52}$$

Using (10.39) and substituting

$$\begin{aligned}\sin\beta_{x1}d &= 2\sin\beta_{x1}\frac{d}{2}\cos\beta_{x1}\frac{d}{2} \\ &= \frac{2\beta_{x1}}{\alpha_{x2}}\sin^2\beta_{x1}\frac{d}{2}\end{aligned}$$

we obtain

$$\begin{aligned}\langle P\rangle &= \frac{\beta_z|\bar{A}|^2}{\omega\mu_0}\left(\frac{d}{4} + \frac{1}{2\alpha_{x2}}\right) \\ &= \frac{\beta_z|\bar{A}|^2}{4\omega\mu_0}\left(d + \frac{2}{\alpha_{x2}}\right)\end{aligned} \tag{10.53}$$

Besides giving the expression for the time-average power flow along the guide, (10.53) leads to the definition of fictitious effective boundaries at $x = \pm d_{\text{eff}}/2$, where

$$\boxed{d_{\text{eff}} = d + \frac{2}{\alpha_{x2}}} \tag{10.54}$$

as shown in Fig. 10.12. The physical interpretation of the placement of these effective boundaries relates to the phase shift that the waves experience upon being total internally reflected at the actual boundaries.

Example 10.3

For the values of ε_{r1}, ε_{r2}, d, and λ_0 used for the running of PL10.1, it is desired to find d_{eff} for the first three modes.

From (10.41b),

$$\begin{aligned}
\alpha_{x2}^2 &= \beta_z^2 - \omega^2\mu_0\varepsilon_2 \\
&= \omega^2\mu_0\varepsilon_1 \sin^2\theta_i - \omega^2\mu_0\varepsilon_2 \\
&= \omega^2\mu_0\varepsilon_0(\varepsilon_{r1}\sin^2\theta_i - \varepsilon_{r2}) \\
\alpha_{x2} &= \frac{2\pi}{\lambda_0}\sqrt{\varepsilon_{r1}\sin^2\theta_i - \varepsilon_{r2}}
\end{aligned}$$

Substituting $\lambda_0 = 5$ mm, $\varepsilon_{r1} = 4$, and $\varepsilon_{r2} = 1$, we have

$$\alpha_{x2} = \frac{2\pi}{5 \times 10^{-3}}\sqrt{4\sin^2\theta_i - 1}$$

From the output of PL10.1, we can then compute α_{x2} and hence d_{eff} for the first three modes, as listed in Table 10.1.

TM modes

The entire solution for the field components can be repeated for TM modes by starting with the expression for the (only) magnetic field component $\overline{H}_y$ and proceeding in a manner similar to that used for the TE mode case. We shall not however pursue these details here but include them as Problems P10.15 and P10.16.

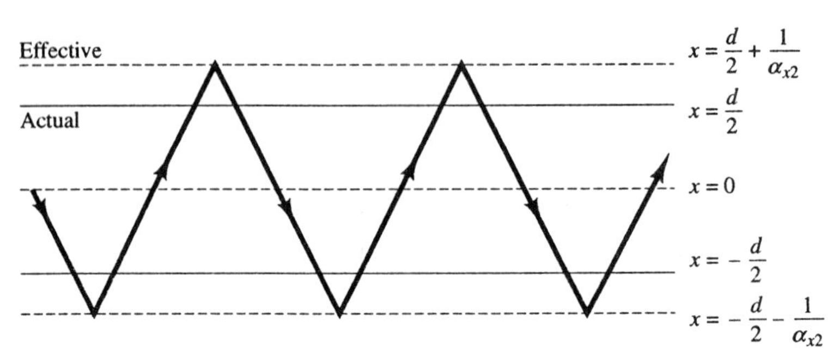

Figure 10.12. For illustrating the effective boundaries for waveguiding along the symmetric dielectric slab guide.

TABLE 10.1 VALUES OF θ_i, α_{x2}, AND d_{eff} FOR EXAMPLE 10.3.

Mode	θ_i (deg)	$\alpha_{x2}(\text{m}^{-1})$	d_{eff} (mm)
TE_0	83.42783	2157.47	10.927
TE_1	76.77756	2099.27	10.953
TE_2	69.96263	1998.97	11.001

Radiation modes

We have thus far discussed modes that are guided within the slab. Another type of mode that is possible for the dielectric slab guide is that for which the field variations with x are sinusoidal not only in the slab but also outside it. The situation can be visualized by locating perfectly conducting plates on either side of the slab and parallel to it as in Fig. 10.13 and displacing the conductors to infinity, thereby obtaining the slab waveguide in the limiting case. Waves that are incident from medium 1 on the interface at angles of incidence less than the critical angle for total internal reflection are transmitted into medium 2 and are reflected at the conductor, giving rise to ray paths such as the one shown. The modes established when the associated self-consistency condition is satisfied are known as the radiation modes. These modes are important in the coupling of energy in and out of the dielectric slab.

K10.2. Dielectric slab waveguide; Guiding by total internal reflection; Self-consistency condition; Characteristic equation for guidance; Propagating modes; Derivation of field expressions; Characteristic equation for guidance; Mode behavior; Power flow; Radiation modes.

D10.3. For a symmetric dielectric slab waveguide, $\varepsilon_1 = 2.25\varepsilon_0$ and $\varepsilon_2 = \varepsilon_0$. Find the following for TE modes: (a) the lowest value of d/λ_0 for which an allowed value of θ_i is 60°; **(b)** the lowest values of d/λ_0 for which an allowed value of θ_i is 75°; and **(c)** the second lowest value of d/λ_0 for which an allowed value of θ_i is 75°.
Ans. **(a)** 0.3545; **(b)** 0.9972; **(c)** 2.2852

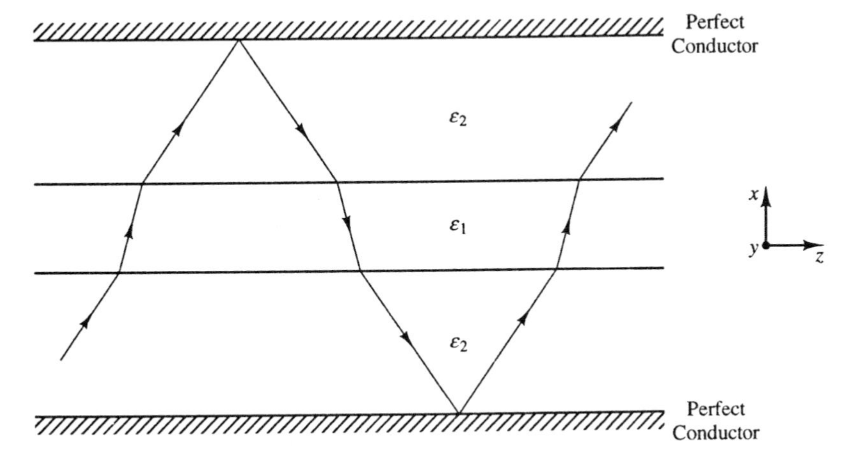

Figure 10.13. For explaining the mechanism pertinent to radiation modes in a dielectric slab guide.

D10.4. For a symmetric dielectric slab waveguide, $\varepsilon_{r1} = 4$ and $\varepsilon_{r2} = 1$. Find the value of d_{eff}/d for the TE_0 mode for each of the following values of d/λ_0: **(a)** 2; **(b)** 5; and **(c)** 0.5.
Ans. **(a)** 1.0927; **(b)** 1.0368; **(c)** 1.4047

10.3 RAY TRACING AND GRADED-INDEX GUIDE

For the dielectric slab waveguide of the preceding section, the permittivity undergoes an abrupt discontinuity from a uniform value of ε_1 in the slab to a uniform value of ε_2 on either side of the slab. When the permittivity varies within the thickness of the slab, the arrangement is known as a graded-index guide, as compared to the step-index guide of the previous section, where the word "index" refers to the refractive index n ($= c/v_p = \sqrt{\varepsilon_r}$ for a nonmagnetic dielectric). To extend our discussion of guided wave propagation to a graded-index slab waveguide, we first introduce the general topic of geometrical optics and ray tracing.

Geometrical optics approximation explained

Geometrical optics is that branch of optics which allows us to study wave phenomena by tracing "rays," which are paths normal to the wavefronts, from the local application of the laws of reflection and refraction (Snell's law). Whenever the wavefront extends and is uniform over many wavelengths and when the boundaries are large compared to the wavelength, ray tracing can be usefully employed. Also, as long as the radii of curvature are large in comparison to the wavelength, the boundaries as well as the wavefronts can be nonplanar. In fact, we have already made use of geometrical optics concepts to introduce the cutoff phenomenon in a parallel-plate waveguide in Section 9.2, to discuss cutoff wavelengths for modes in a metallic rectangular waveguide in Section 9.4, and to derive the characteristic equation for guidance in the dielectric slab guide in Section 10.2. In all of these cases the media were uniform and the boundaries plane and abrupt so that the ray paths were all straight lines. For nonuniform media, the ray paths become curved.

Ray tracing formulation

To formulate the ray tracing procedure, let us consider the arrangement shown in Fig. 10.14 in which a medium of continuously varying refractive index

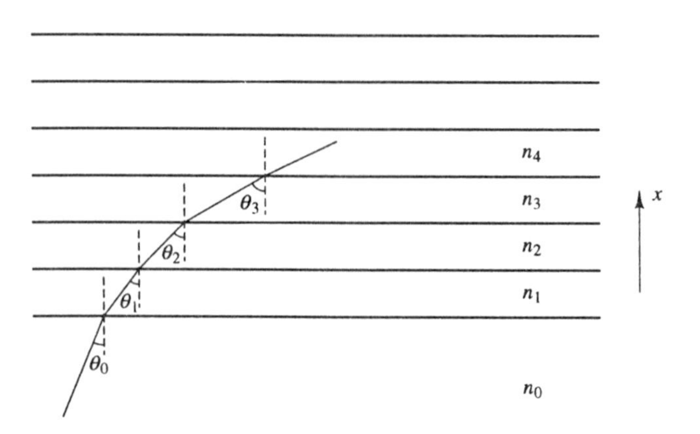

Figure 10.14. The bending of ray paths in a series of plane dielectric slabs of uniform refractive indices.

$n(x)$ is approximated by a series of plane slabs of uniform refractive indices n_1, n_2, n_3, ... Let a wave be incident from the medium of refractive index n_0 at an angle θ_0 from the vertical. Then assuming $n_0 < n_1 < n_2 < n_3 < \ldots$, the ray path bends more and more away from the vertical in accordance with Snell's law applied at the interfaces

$$\boxed{n_0 \sin \theta_0 = n_1 \sin \theta_1 = n_2 \sin \theta_2 = \ldots} \tag{10.55}$$

with the path in each layer being a straight line. In the limit that the thickness of each layer goes to zero, the refractive index varies continuously with x and the ray path becomes curved. To trace the path of the ray, let us consider a differential segment ds along the ray path, having the components dx and dz in the x- and z-directions, respectively, as shown in Fig. 10.15. Then

$$\frac{dz}{dx} = \tan \theta = \frac{\sin \theta}{\sqrt{1 - \sin^2 \theta}} \tag{10.56}$$

From (10.55), $n \sin \theta = n_0 \sin \theta_0$, or $\sin \theta = \dfrac{n_0}{n} \sin \theta_0$. Substituting in (10.56), we obtain

$$\boxed{\frac{dz}{dx} = \frac{n_0 \sin \theta_0}{\sqrt{n^2 - n_0^2 \sin^2 \theta_0}}} \tag{10.57}$$

For a given refractive index profile $n(x)$, the solution of (10.57) gives the ray trajectory in the xz-plane. In general, the solution has to be carried out numerically. For certain functions for $n(x)$, an analytical solution is possible. We shall illustrate by means of an example.

Example 10.4

Ray path for linear profile of permittivity

Let us consider a variation of refractive index as given by

$$n^2 = n_0^2(1 - \alpha x)$$

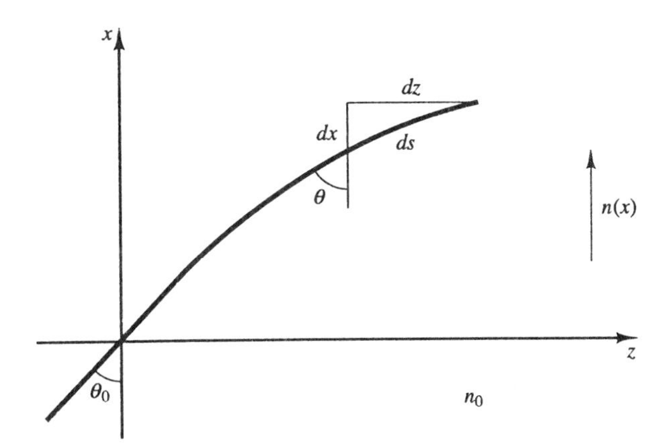

Figure 10.15. For the formulation of ray tracing for continuous variation of refractive index.

where $\alpha > 0$. Note that for a nonmagnetic dielectric medium, this corresponds to a linear profile of permittivity. We wish to find the ray trajectory in the medium in the xz-plane for a wave entering the medium at $x = 0$ and $z = 0$ at an angle θ_0 from the vertical (x-direction).

From (10.57), we have

$$\frac{dz}{dx} = \frac{n_0 \sin\theta_0}{\sqrt{n_0^2 - n_0^2\alpha x - n_0^2 \sin^2\theta_0}}$$

$$= \frac{\sin\theta_0}{\sqrt{\cos^2\theta_0 - \alpha x}}$$

The ray trajectory is given by

$$z(x) = \int_0^x \frac{\sin\theta_0}{\sqrt{\cos^2\theta_0 - \alpha x}}\, dx$$

$$= -\sin\theta_0 \left[\frac{2}{\alpha}\sqrt{\cos^2\theta_0 - \alpha x}\right]_0^x$$

$$= -\frac{2\sin\theta_0}{\alpha}\left[\sqrt{\cos^2\theta_0 - \alpha x} - \cos\theta_0\right]$$

Rearranging, we have

$$\cos^2\theta_0 - \alpha x = \left(-\frac{\alpha z}{2\sin\theta_0} + \cos\theta_0\right)^2$$

or

$$x = \frac{\cos^2\theta_0}{\alpha} - \left(\frac{\cos\theta_0}{\sqrt{\alpha}} - \frac{\sqrt{\alpha}}{2\sin\theta_0} z\right)^2$$

Thus the ray trajectory is parabolic, with the parabola having its apex at $x = \dfrac{\cos^2\theta_0}{\alpha}$ and $z = \dfrac{\sin 2\theta_0}{\alpha}$. Note that at $x = \dfrac{\cos^2\theta_0}{\alpha}$, $n^2 = n_0^2(1 - \cos^2\theta_0)$ or, $n = n_0 \sin\theta_0$, and $\theta = 90°$, consistent with the solution obtained.

Paraxial ray approximation

When the ray trajectories are nearly along the propagation axis, the rays are known as paraxial rays. For paraxial rays, the angle δ ($= 90° - \theta$) that the ray makes with the propagation axis, which here is the z-axis as shown in Fig. 10.16, is small such that approximations of $\sin\delta \approx \delta$ and $\cos\delta \approx 1$ can be used.

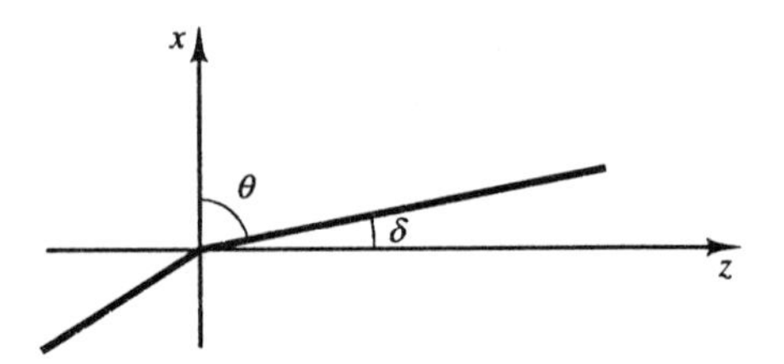

Figure 10.16. Geometry pertinent to the paraxial ray approximation.

Example 10.5

Paraxial rays in parabolic index profile

An important example of graded refractive index profile is given by

$$\boxed{n^2 = n_0^2(1 - \alpha^2 x^2)} \tag{10.58}$$

For α sufficiently small that $\alpha x \ll 1$, which is usually the case,

$$n(x) = n_0\sqrt{1 - \alpha^2 x^2} \approx n_0\left(1 - \frac{1}{2}\alpha^2 x^2\right) \tag{10.59}$$

corresponds to a parabolic variation. We wish to investigate paraxial rays for this profile.

Let us consider a ray making an angle δ_0 with the z-axis at the point $x = 0$. Then from (10.57),

$$\begin{aligned}
\frac{dz}{dx} &= \frac{n_0 \sin \theta_0}{\sqrt{n^2 - n_0^2 \sin^2 \theta_0}} = \frac{n_0 \cos \delta_0}{\sqrt{n^2 - n_0^2 \cos^2 \delta_0}} \\
&= \frac{n_0 \cos \delta_0}{\sqrt{n_0^2(1 - \alpha^2 x^2) - n_0^2 \cos^2 \delta_0}} \\
&= \frac{\cos \delta_0}{\sqrt{\sin^2 \delta_0 - \alpha^2 x^2}} \\
&\approx \frac{1}{\sqrt{\delta_0^2 - \alpha^2 x^2}} \\
z &= \int_0^x \frac{dx}{\sqrt{\delta_0^2 - \alpha^2 x^2}} \\
&= \left[\frac{1}{\alpha} \sin^{-1}\left(\frac{\alpha}{\delta_0} x\right)\right]_0^x \\
&= \frac{1}{\alpha} \sin^{-1} \frac{\alpha x}{\delta_0}
\end{aligned}$$

or

$$\boxed{x = \frac{\delta_0}{\alpha} \sin \alpha z} \tag{10.60}$$

The ray oscillates about the axis with a period of $2\pi/\alpha$, known as the *pitch,* independent of δ_0, and a peak amplitude $\dfrac{\delta_0}{\alpha}$, as shown in Fig. 10.17 for a few values of δ_0.

Guidance condition for graded-index guide

We may now discuss wave guidance in a graded-index guide. To do this, let us consider, for simplicity, the symmetric refractive index profile of the shape shown in Fig. 10.18(a). Then for those waves that are total internally reflected within $-d/2 < x < d/2$, a sketch of the ray path can be drawn, as in Fig.

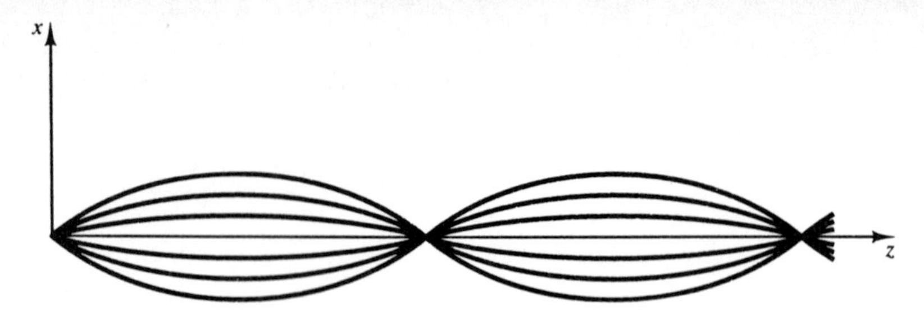

Figure 10.17. Paraxial rays in a parabolic index profile.

10.18(b), with apex points at $x = \pm x_a$, where $x_a < d/2$. From Snell's law, $n(\pm x_a) = n_1(x) \sin\theta(x)$, since $\theta = 90°$ at $x = \pm x_a$. Using the same reasoning as for writing (10.25), we can write the self-consistency condition for guidance to be

$$\int_{x=-x_a}^{x_a} \frac{2\pi}{\lambda_0}\sqrt{\varepsilon_{r1}(x)} \cos\theta(x)\,dx + \underline{/[\bar{\Gamma}]_{x=x_a}} + \int_{x=x_a}^{-x_a} \frac{2\pi}{\lambda_0}\sqrt{\varepsilon_{r1}(x)} \cos\theta(x)\,dx + \underline{/[\bar{\Gamma}]_{x=-x_a}} = 2m\pi, \qquad m = 0, 1, 2, \ldots \tag{10.61}$$

In view of symmetry, (10.61) reduces to

$$\frac{4\pi}{\lambda_0}\int_{x=0}^{x_a} \sqrt{\varepsilon_{r1}(x)} \cos\theta(x)\,dx + \underline{/\bar{\Gamma}_a} = m\pi, \qquad m = 0, 1, 2, \ldots \tag{10.62}$$

where $\bar{\Gamma}_a = [\bar{\Gamma}]_{x=x_a}$.

To proceed further, we consider the TE case and find $\underline{/\bar{\Gamma}_a}$ by using (10.28). We first write $\bar{\Gamma}_a$ as

$$\begin{aligned}\bar{\Gamma}_a &= \frac{\eta_2' \cos\theta_i - j\eta_1'\sqrt{(\varepsilon_1'/\varepsilon_2')\sin^2\theta_i - 1}}{\eta_2' \cos\theta_i + j\eta_1'\sqrt{(\varepsilon_1'/\varepsilon_2')\sin^2\theta_i - 1}} \\ &= \frac{\sqrt{\varepsilon_1'}\cos\theta_i - j\sqrt{\varepsilon_1'\sin^2\theta_i - \varepsilon_2'}}{\sqrt{\varepsilon_1'}\cos\theta_i + j\sqrt{\varepsilon_1'\sin^2\theta_i - \varepsilon_2'}}\end{aligned} \tag{10.63}$$

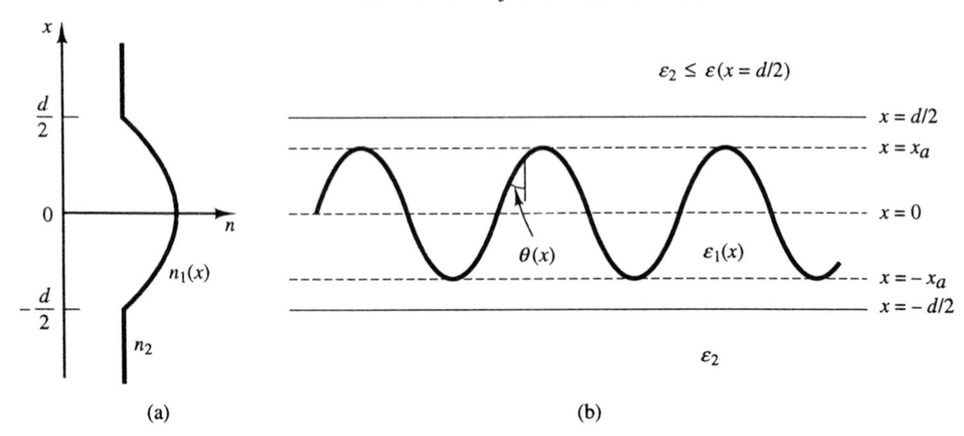

Figure 10.18. (a) Refractive index profile for a symmetrical graded-index guide. (b) Ray path within the graded-index region.

where we have inserted primes so as not to confuse with the notation of Fig. 10.18. Now we note that for the situation under consideration,

$$\varepsilon_1' = [\varepsilon]_{x=x_a-\Delta x} = \varepsilon(x_{a^-})$$

$$\varepsilon_2' = [\varepsilon]_{x=x_a+\Delta x} = \varepsilon(x_{a^+})$$

$$\varepsilon_1' \sin^2 \theta_i = [\varepsilon]_{x=x_a} \sin^2 90^\circ = \varepsilon(x_a)$$

$$\sqrt{\varepsilon_1'} \cos \theta_i = \sqrt{\varepsilon_1' - \varepsilon_1' \sin^2 \theta_i} = \sqrt{\varepsilon(x_{a^-}) - \varepsilon(x_a)}$$

so that

$$\overline{\Gamma}_a = \frac{\sqrt{\varepsilon(x_{a^-}) - \varepsilon(x_a)} - j\sqrt{\varepsilon(x_a) - \varepsilon(x_{a^+})}}{\sqrt{\varepsilon(x_{a^-}) - \varepsilon(x_a)} + j\sqrt{\varepsilon(x_a) - \varepsilon(x_{a^+})}} \tag{10.64}$$

In view of the continuous variation of ε_1, we now have to take the limit of the right side of (10.64) as x_{a^-} and x_{a^+} tend to x_a. We note however that this results in a situation of zero divided by zero. To avoid this, we write that in the vicinity of $x = x_a$,

$$\varepsilon(x) \approx \varepsilon(x_a) + (x - x_a)\frac{d\varepsilon}{dx} = \varepsilon(x_a) + \Delta x \frac{d\varepsilon}{dx}$$

Thus

$$\varepsilon(x_a^-) - \varepsilon(x_a) \approx \varepsilon(x_a) - \Delta x \frac{d\varepsilon}{dx} - \varepsilon(x_a) = -\Delta x \frac{d\varepsilon}{dx}$$

$$\varepsilon(x_a) - \varepsilon(x_a^+) \approx \varepsilon(x_a) - \varepsilon(x_a) - \Delta x \frac{d\varepsilon}{dx} = -\Delta x \frac{d\varepsilon}{dx}$$

$$\overline{\Gamma}_a = \lim_{\Delta x \to 0} \frac{\sqrt{-\Delta x \dfrac{d\varepsilon}{dx}} - j\sqrt{-\Delta x \dfrac{d\varepsilon}{dx}}}{\sqrt{-\Delta x \dfrac{d\varepsilon}{dx}} + j\sqrt{-\Delta x \dfrac{d\varepsilon}{dx}}}$$

$$= \frac{1 - j}{1 + j} = 1\underline{/-\pi/2} \tag{10.65}$$

so that $\underline{/\overline{\Gamma}_a} = -\pi/2$. The same result can be shown to hold for the TM case, which makes use of $\Gamma_\parallel$ (see Problem P10.19).

Finally, substituting $\underline{/\overline{\Gamma}_a} = -\pi/2$ and also

$$\sqrt{\varepsilon_{r1}(x)} \cos \theta = \sqrt{\varepsilon_{r1}(x) - \varepsilon_{r1}(x) \sin^2 \theta}$$

$$= \sqrt{\varepsilon_{r1}(x) - \varepsilon_{r1}(x_a)} \tag{10.66}$$

into (10.62), we obtain the characteristic equation for guidance to be

$$\boxed{\frac{4\pi}{\lambda_0} \int_{x=0}^{x_a} \sqrt{\varepsilon_{r1}(x) - \varepsilon_{r1}(x_a)}\, dx = \left(m + \frac{1}{2}\right)\pi, \qquad m = 0, 1, 2, \ldots} \tag{10.67}$$

or in terms of refractive index

$$\boxed{\frac{4\pi}{\lambda_0}\int_{x=0}^{x_a} \sqrt{n_1^2(x) - n_1^2(x_a)}\, dx = \left(m + \frac{1}{2}\right)\pi, \qquad m = 0, 1, 2, \ldots} \tag{10.68}$$

As in the case of the step-index guide, each value of m corresponds to a mode. For a given value of m and for a given profile $n_1(x)$, (10.68) must in general be solved numerically. There are however certain refractive index profiles that permit analytical solution. An example is in order.

Example 10.6

Modes in parabolic index guide

Let us consider the refractive index profile of Example 10.5 given by

$$\boxed{n_1^2(x) = n_0^2\,[1 - \alpha^2 x^2]} \tag{10.69}$$

where α is such that $[n_1]_{x=\pm d/2} \geq n_2$, and investigate guided waves in the slab.

Substituting for $n_1^2(x)$ in (10.68), we have

$$\frac{4\pi}{\lambda_0}\int_{x=0}^{x_a} \sqrt{n_0^2(1 - \alpha^2 x^2) - n_0^2(1 - \alpha^2 x_a^2)}\, dx = \left(m + \frac{1}{2}\right)\pi, \qquad m = 0, 1, 2, \ldots$$

$$\frac{4\pi n_0 \alpha}{\lambda_0}\int_{x=0}^{x_a} \sqrt{x_a^2 - x^2}\, dx = \left(m + \frac{1}{2}\right)\pi, \qquad m = 0, 1, 2, \ldots$$

$$\frac{2\pi n_0 \alpha}{\lambda_0}\left[x\sqrt{x_a^2 - x^2} + x_a^2 \sin^{-1}\frac{x}{x_a}\right]_{x=0}^{x_a} = \left(m + \frac{1}{2}\right)\pi, \qquad m = 0, 1, 2, \ldots$$

$$\boxed{x_a^2 = \frac{(2m + 1)\lambda_0}{2\pi n_0 \alpha}, \qquad m = 0, 1, 2, \ldots} \tag{10.70}$$

The value of x_a increases with the mode number m, as can also be seen in general from (10.68). Recall that $x_a < d/2$. For paraxial modes, the ray trajectories are given by

$$\begin{aligned} x &= x_a \sin \alpha z \\ &= \sqrt{\frac{(2m + 1)\lambda_0}{2\pi n_0 \alpha}} \sin \alpha z \end{aligned} \tag{10.71}$$

In contrast with the modes of the step-index guide in which the ray paths for all modes extend to the boundaries of the slab with varying values of the pitch, for these modes the ray paths possess amplitudes increasing with m but with a fixed pitch.

The listing of a PC program that carries out the numerical solution of (10.68) by an iterative procedure is included as PL10.2. For a given refractive index profile and specified values of d and λ_0, the iterative procedure consists of starting with $m = 0$ and computing x_a that satisfies (10.68). To do this, x_a is set equal to $d/2$ and the integral on the left side is evaluated numerically. If this results in a value of less than $\pi/2$ for the left side of (10.68), then it means that a solution does not exist for any value of m and the computation is terminated. If the value is greater than $\pi/2$, then an interval-bisection procedure is used iteratively, beginning with the interval from 0 to $d/2$, until a value of x_a that satisfies

(10.68) to a desired accuracy is found. The value of m is then increased in steps of unity and the computation repeated for each value of m, beginning with the search interval extending from the solution for x_a found for the previous value of m to $d/2$. The entire computation is terminated when a value of m is reached for which the left side of (10.68) yields a value of less than $\left(m + \frac{1}{2}\right)\pi$. To illustrate the solution, the function for $n_1(x)$ given by (10.69) is considered by using a defined function statement. The output from a run of the program for values of $n_0 = 1.5$, $\alpha = 1\ (\text{mm})^{-1}$, $d = 1$ mm, and $\lambda_0 = 0.5$ mm is included.

PL10.2. Program listing and sample output for solution of the characteristic equation for guidance in a graded-index dielectric slab waveguide.

```
100 '***********************************************************
110 '* SOLUTION OF CHARACTERISTIC EQUATION FOR GUIDANCE *
120 '* IN A GRADED-INDEX DIELECTRIC SLAB WAVEGUIDE       *
130 '***********************************************************
140 DIM FINT(100)
150 DEF FN N(X)=NO*SQR(1-(ALPHA*X)^2)
160 CLS:SCREEN 0
170 PRINT"ENTER VALUES OF INPUT PARAMETERS:"
180 PRINT:INPUT"NO = ",NO
190 INPUT"ALPHA IN 1/MM = ",ALPHA
200 INPUT"WAVELENGTH IN MM = ",WL
210 INPUT"THICKNESS IN MM = ",D
220 PRINT:PRINT"COMPUTED VALUES ARE:":PRINT
230 M=0:XA=0
240 CONST=(M+.5)*WL/4:NN=100
250 XA1=XA:XA2=D/2:XA=XA2
260 GOSUB 420
270 IF SUM<CONST THEN 370
280 GOTO 310
290 IF SUM<CONST THEN XA1=XA:GOTO 310
300 XA2=XA
310 IF ABS(SUM-CONST)<CONST/100 THEN 350
320 XA=(XA1+XA2)/2
330 GOSUB 420
340 GOTO 290
350 PRINT"M =";M;"     XA =";XA;"MM";"          N =";FN N(XA)
360 M=M+1:GOTO 240
370 PRINT"M >=";M;"   SOLUTION DOES NOT EXIST"
380 PRINT:PRINT "STRIKE ANY KEY TO CONTINUE"
390 C$=INPUT$(1)
400 GOTO 160
410 END
420 '* SUBPROGRAM FOR EVALUATING INTEGRAL *
430 FINT(0)=0:SUM=0
440 FOR I=1 TO NN-1
450 X=XA*I/NN
460 FINT(I)=SQR((FN N(X))^2-(FN N(XA))^2)
470 SUM=SUM+(FINT(I)+FINT(I-1))/2
480 NEXT
490 SUM=SUM+FINT(I)/2
500 SUM=SUM*XA/NN
510 RETURN

RUN
ENTER VALUES OF INPUT PARAMETERS:

NO = 1.5
ALPHA IN 1/MM = 1
```

PL10.2. (continued)

```
WAVELENGTH IN MM = 0.5
THICKNESS IN MM = 1

COMPUTED VALUES ARE:

M = 0       XA = .2304688 MM        N = 1.45962
M = 1       XA = .3989258 MM        N = 1.375475
M >= 2      SOLUTION DOES NOT EXIST

STRIKE ANY KEY TO CONTINUE
```

Intermodal dispersion

Returning now to the result of (10.71), we consider its consequence in intermodal dispersion, the type of dispersion resulting from different travel times of rays corresponding to different modes. Because rays of higher modes travel farther but with greater velocities (lower refractive index), the travel times of the different rays are nearly equalized, thereby almost eliminating intermodal dispersion. To discuss in quantitative terms, we note that the phase constant along the guide axis is given by

$$\begin{aligned}
\beta_z &= \omega\sqrt{\mu[\varepsilon]_{x=x_a}} \\
&= \omega\sqrt{\mu_0\varepsilon_0[n_1]_{x=x_a}} \\
&= \frac{\omega}{c}n_0\sqrt{1-\alpha^2x_a^2} \\
&\approx \frac{\omega n_0}{c}\left(1-\frac{1}{2}\alpha^2x_a^2\right) \qquad \text{for } \alpha^2x_a^2 \ll 1 \\
&= \frac{\omega n_0}{c}\left[1-\frac{(2m+1)\lambda_0\alpha}{4\pi n_0}\right] \\
&= \frac{\omega n_0}{c}-\frac{(2m+1)\alpha}{2} \\
\frac{d\beta_z}{d\omega} &= \frac{n_0}{c}
\end{aligned}$$

Thus the group velocity along the guide axis is given by

$$v_{gz} = \frac{d\omega}{d\beta_z} = \frac{c}{n_0}$$

independent of m.

K10.3. Ray tracing; Paraxial rays; Parabolic-index profile; Graded-index guide; Intermodal dispersion.

D10.5. In Fig. 10.19, a medium of continuously varying refractive index is approximated by a series of plane slabs of uniform refractive indices. If $n_0 = 1$ and $\theta_0 = 60°$, find the following: **(a)** θ_1 if $n_1 = 1.5$; **(b)** n_3 if $\theta_3 = 30°$; and **(c)** θ_t.
Ans. **(a)** 35.26°; **(b)** $\sqrt{3}$; **(c)** 30°

D10.6. In Example 10.6, let $n_0 = 1.5$, $\alpha = 0.05/\lambda_0$, and $d = 10\lambda_0$. Compute the following: **(a)** x_a for $m = 0$; **(b)** the maximum value of m; and **(c)** the paraxial angle

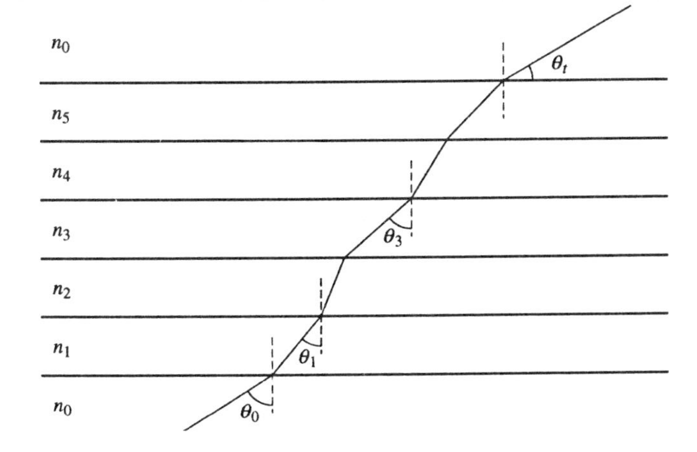

Figure 10.19. For Problem D10.5.

δ_0 for the value of m computed in (b).
Ans. **(a)** $1.4567\lambda_0$; **(b)** 5; **(c)** 13.88°

10.4 OPTICAL FIBER

Description Thus far in this chapter we have discussed the guiding of waves in a dielectric slab waveguide. Another common form of optical waveguide is the optical fiber. An optical fiber, so termed because of its filamentary appearance, consists typically of a core and a cladding, having circular cross sections as shown in Fig. 10.20(a). The core is made up of a material of permittivity greater than that of the cladding so that a critical angle exists for waves inside the core incident on the interface between the core and the cladding, and hence waveguiding is made possible in the core by total internal reflection. The phenomenon of guiding may be visualized by considering a longitudinal cross section of the fiber through its axis, shown in Fig. 10.20(b), and comparing it with that of the slab waveguide shown in Fig. 10.7. While this situation corresponds to meridional rays, skewed rays whose paths lie in planes offset from the fiber axis also explain the guiding mechanism. Although the cladding is not essential for the purpose of waveguiding in the core since the permittivity of the core material is greater than that of free space, the cladding serves two useful purposes: (1) It avoids scattering and field distortion by the supporting structure of the fiber since the field decays exponentially

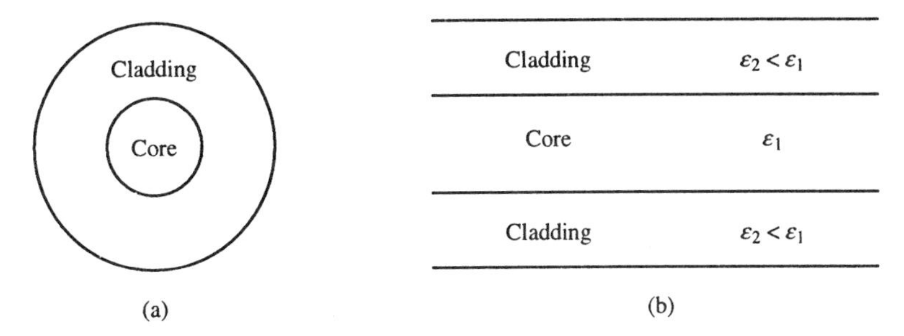

Figure 10.20. (a) Transverse and (b) longitudinal cross sections of an optical fiber.

outside the core and hence is negligible outside the cladding; (2) it allows a single-mode propagation for a larger value of the radius of the core than permitted in the absence of the cladding.

Optical fibers are used predominantly in communication, among other applications. The first commercial light-wave communications system was put into operation in May 1977 in downtown Chicago by interconnecting two switching offices of the Illinois Bell Telephone Company and a large commercial building to carry voice, data, and video signals.[1] Another milestone was reached early in 1983 when American Telephone & Telegraph Company began carrying some telephone calls between New York City and Washington, D.C., by light, thereby signaling the entrance of light-wave communication into the long-distance market.[2] Yet another milestone was reached when the first fiber-optic telephone cable across the Atlantic was put into service on December 14, 1988, vastly increasing the number of calls that can be placed at one time to Europe.[3]

To simplify analysis of waveguiding in an optical fiber, we shall consider the cladding region to extend to infinity, so that the geometry is one of a cylindrical dielectric rod of permittivity greater than that of the surrounding medium, as shown in Fig. 10.21. In addition, we shall consider the permittivity of the core to be uniform, thereby corresponding to the case of a step-index fiber. To carry out the field analysis, we make use of our previous experience with the cylindrical metallic waveguide and the dielectric slab waveguide. First, we know that the transverse field components can be expressed in terms of the longitudinal field components E_z and H_z. We shall however find that the modes do not all separate into TE and TM modes and that the situation leads to the so-called hybrid modes. Second, we know in analogy with the case of the slab guide that radially decaying fields can be expected outside the core.

We recall from Section 9.5 that the differential equation to be satisfied by the fields for the cylindrical geometry is that given by (9.87)

$$\boxed{\frac{1}{r}\frac{\partial}{\partial r}\left(r\frac{\partial\overline{\psi}}{\partial r}\right) + \frac{1}{r^2}\frac{\partial^2\overline{\psi}}{\partial\phi^2} + \beta_c^2\overline{\psi} = 0} \tag{10.72}$$

where $\overline{\psi}$ stands for $\overline{E}_z$ or $\overline{H}_z$ and

$$\boxed{\beta_c^2 = \omega^2\mu\varepsilon - \left(\frac{2\pi}{\lambda_g}\right)^2 = \beta^2 - \beta_z^2} \tag{10.73}$$

We also learned that (10.72) is separable into a differential equation involving r only and one involving ϕ only. The solution for the r-variation is a superposition of the Bessel functions $J_l(\beta_c r)$ and $N_l(\beta_c r)$, whereas the solutions for the ϕ-variation are sinusoidal ($\cos l\phi$ and $\sin l\phi$). The sinusoidal variations in ϕ require l to

[1] See, for example, W. S. Boyle, "Light-Wave Communications," *Scientific American*, August 1977, pp. 40–48.

[2] "Optic Fiber Net Sends AT&T Light Years Ahead," *The New York Times News Service*, published in the *Chicago Tribune*, February 25, 1983.

[3] "Fiber-Optic Phone Cable Links Calls Across Atlantic," *The New York Times News Service*, published in the *Champaign-Urbana* (Illinois) *News-Gazette*, December 14, 1988.

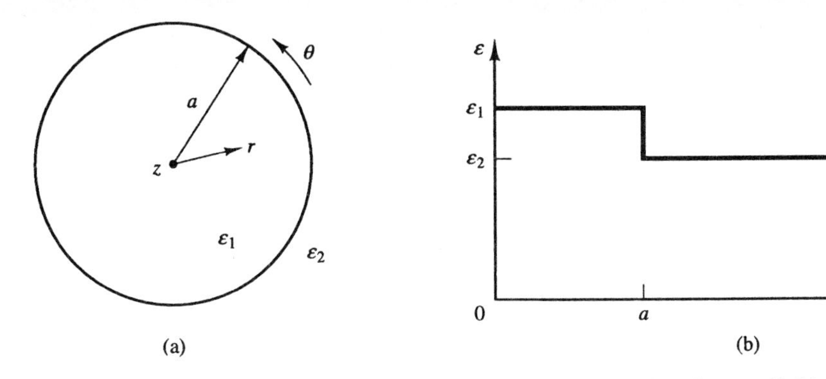

Figure 10.21. (a) Cross section of a cylindrical dielectric rod surrounded by a cladding region extending to infinity. (b) Permittivity profile for the arrangement of (a).

be an integer. The behaviors of the Bessel functions for real arguments are shown in Fig. 9.23.

Field solution in the core

Inside the core, $\beta = \beta_1 = \omega\sqrt{\mu\varepsilon_1}$, $\beta_{c1}^2 = \beta_1^2 - \beta_z^2 > 0$, and β_c is real. The solutions for the r-variation are J_l and N_l with real arguments. But since N_l has the property that $N_l(0) \to \infty$ for all orders, we rule it out. Thus

$$\boxed{\overline{\psi}(r, \phi, z) = \overline{A}J_l(\beta_{c1}r)\begin{Bmatrix}\cos l\phi \\ \sin l\phi\end{Bmatrix}e^{-j\beta_z z} \text{ for } r < a} \tag{10.74}$$

where we consider only the (+) wave, for simplicity. Defining $u = \beta_{c1}a$, we can write the longitudinal components and then the transverse field components by using (9.85a)–(9.85d) as follows:

$$\overline{E}_{z1} = \overline{A}J_l\left(u\frac{r}{a}\right)\begin{Bmatrix}\cos l\phi \\ \sin l\phi\end{Bmatrix}e^{-j\beta_z z} \tag{10.75a}$$

$$\overline{H}_{z1} = \overline{B}J_l\left(u\frac{r}{a}\right)\begin{Bmatrix}\sin l\phi \\ \cos l\phi\end{Bmatrix}e^{-j\beta_z z} \tag{10.75b}$$

$$\overline{E}_{r1} = \left[-\frac{j\beta_z a}{u}\overline{A}J_l'\left(u\frac{r}{a}\right) \mp \frac{j\omega\mu a^2 l}{u^2 r}\overline{B}J_l\left(u\frac{r}{a}\right)\right]\begin{Bmatrix}\cos l\phi \\ \sin l\phi\end{Bmatrix}e^{-j\beta_z z} \tag{10.75c}$$

$$\overline{H}_{r1} = \left[\mp\frac{j\omega\varepsilon_1 a^2 l}{u^2 r}\overline{A}J_l\left(u\frac{r}{a}\right) - \frac{j\beta_z a}{u}\overline{B}J_l'\left(u\frac{r}{a}\right)\right]\begin{Bmatrix}\sin l\phi \\ \cos l\phi\end{Bmatrix}e^{-j\beta_z z} \tag{10.75d}$$

$$\overline{E}_{\phi 1} = \left[\pm\frac{j\beta_z a^2 l}{u^2 r}\overline{A}J_l\left(u\frac{r}{a}\right) + \frac{j\omega\mu a}{u}\overline{B}J_l'\left(u\frac{r}{a}\right)\right]\begin{Bmatrix}\sin l\phi \\ \cos l\phi\end{Bmatrix}e^{-j\beta_z z} \tag{10.75e}$$

$$\overline{H}_{\phi 1} = \left[-\frac{j\omega\varepsilon_1 a}{u}\overline{A}J_l'\left(u\frac{r}{a}\right) \mp \frac{j\beta_z a^2 l}{u^2 r}\overline{B}J_l\left(u\frac{r}{a}\right)\right]\begin{Bmatrix}\cos l\phi \\ \sin l\phi\end{Bmatrix}e^{-j\beta_z z} \tag{10.75f}$$

In (10.75a)–(10.75f), the upper functions and signs go together and the lower functions and signs go together. Note that when $\overline{E}_{z1} \propto \cos l\phi$, $\overline{H}_{z1} \propto \sin l\phi$. This is because when a given transverse component is proportional to $\overline{E}_z$ (or $\overline{H}_z$), it is proportional to $\partial\overline{H}_z/\partial\phi$ (or $\partial\overline{E}_z/\partial\phi$).

Field solution in the cladding

In the cladding, $\beta = \beta_2 = \omega\sqrt{\mu\varepsilon_2}$, $\beta_{c2}^2 = \beta_2^2 - \beta_z^2 < 0$, and β_c is imaginary. The solutions for the r-variation are J_l and N_l with imaginary arguments jwr/a, where $w^2 = -\beta_{c2}^2 a^2$. It is common practice then to represent the solution in terms of the modified Bessel functions I_l and K_l with real arguments wr/a. The behaviors of these functions for a few values of l are shown in Fig. 10.22. Since I_l has the property that $I_l(\infty) \to \infty$ for all orders, we rule it out. We can thus write the solutions for the field components as follows:

$$\bar{E}_{z2} = \bar{C}K_l\left(w\frac{r}{a}\right)\begin{Bmatrix}\cos l\phi \\ \sin l\phi\end{Bmatrix}e^{-j\beta_z z} \tag{10.76a}$$

$$\bar{H}_{z2} = \bar{D}K_l\left(w\frac{r}{a}\right)\begin{Bmatrix}\sin l\phi \\ \cos l\phi\end{Bmatrix}e^{-j\beta_z z} \tag{10.76b}$$

$$\bar{E}_{r2} = \left[\frac{j\beta_z a}{w}\bar{C}K_l'\left(w\frac{r}{a}\right) \pm \frac{j\omega\mu a^2 l}{w^2 r}\bar{D}K_l\left(w\frac{r}{a}\right)\right]\begin{Bmatrix}\cos l\phi \\ \sin l\phi\end{Bmatrix}e^{-j\beta_z z} \tag{10.76c}$$

$$\bar{H}_{r2} = \left[\mp\frac{j\omega\varepsilon_2 a^2 l}{w^2 r}\bar{C}K_l\left(w\frac{r}{a}\right) + \frac{j\beta_z a}{w}\bar{D}K_l'\left(w\frac{r}{a}\right)\right]\begin{Bmatrix}\sin l\phi \\ \cos l\phi\end{Bmatrix}e^{-j\beta_z z} \tag{10.76d}$$

$$\bar{E}_{\phi 2} = \left[\mp\frac{j\beta_z a^2 l}{w^2 r}\bar{C}K_l\left(w\frac{r}{a}\right) - \frac{j\omega\mu a}{w}\bar{D}K_l'\left(w\frac{r}{a}\right)\right]\begin{Bmatrix}\sin l\phi \\ \cos l\phi\end{Bmatrix}e^{-j\beta_z z} \tag{10.76e}$$

$$\bar{H}_{\phi 2} = \left[\frac{j\omega\varepsilon_2 a}{w}\bar{C}K_l'\left(w\frac{r}{a}\right) \pm \frac{j\beta_z a^2 l}{w^2 r}\bar{D}K_l\left(w\frac{r}{a}\right)\right]\begin{Bmatrix}\cos l\phi \\ \sin l\phi\end{Bmatrix}e^{-j\beta_z z} \tag{10.76f}$$

Guidance condition

To obtain the guidance condition, we need to satisfy the boundary conditions for the field components. We now see that for $l = 0$ the set of equations separates into two groups $\bar{E}_{z1,2}$, $\bar{E}_{r1,2}$, and $\bar{H}_{\phi 1,2}$ involving the constants $\bar{A}$ and $\bar{C}$ and $\bar{H}_{z1,2}$, $\bar{H}_{r1,2}$, and $\bar{E}_{\phi 1,2}$ involving the constants $\bar{B}$ and $\bar{D}$. The first group corresponds to the TM case and the characteristic equation obtained by setting $\bar{E}_{z1}/\bar{H}_{\phi 1} = \bar{E}_{z2}/\bar{H}_{\phi 2}$ at $r = a$ is

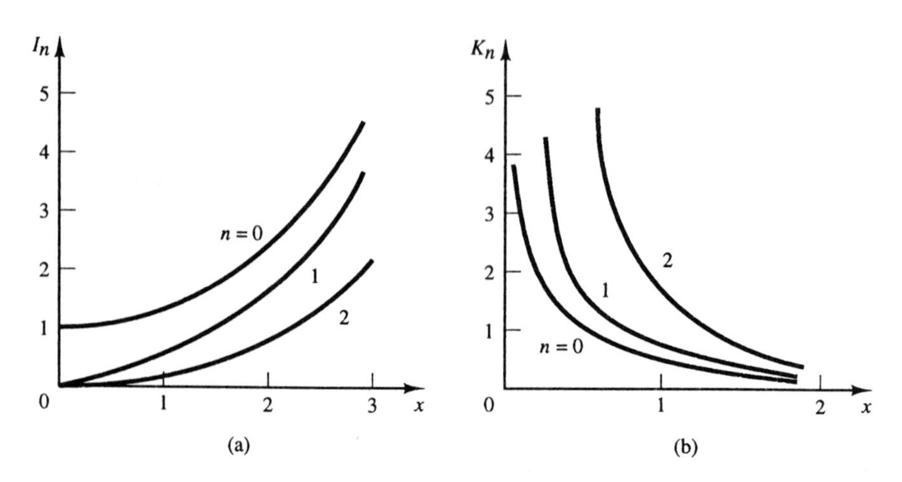

Figure 10.22. Behaviors of the modified Bessel functions **(a)** $I_n(x)$ and **(b)** $K_n(x)$.

$$\frac{J_1(u)}{J_0(u)} = -\frac{\varepsilon_2 u}{\varepsilon_1 w}\frac{K_1(w)}{K_0(w)} \tag{10.77}$$

where we have used the property that $J_0'(u) = -J_1(u)$ and $K_0'(w) = -K_1(w)$. The second group corresponds to the TE case and the characteristic equation obtained by setting $\overline{E}_{\phi 1}/\overline{H}_{z1} = \overline{E}_{\phi 2}/\overline{H}_{z2}$ at $r = a$ is

$$\frac{J_1(u)}{J_0(u)} = -\frac{u}{w}\frac{K_1(w)}{K_0(w)} \tag{10.78}$$

For $l \neq 0$, the boundary conditions cannot be satisfied by the two separate groups and hence the fields can no longer be separated into TM and TE modes but are known as *hybrid modes*. The characteristic equation obtained by setting $\overline{E}_{z1} = \overline{E}_{z2}$, $\overline{H}_{z1} = \overline{H}_{z2}$, $\overline{E}_{z1}/\overline{H}_{\phi 1} = \overline{E}_{z2}/\overline{H}_{\phi 2}$, and $\overline{E}_{\phi 1}/\overline{H}_{z1} = \overline{E}_{\phi 2}/\overline{H}_{z2}$ at $r = a$ is then given by (see Problem P10.22)

$$\left[\frac{J_l'(u)}{uJ_l(u)} + \frac{K_l'(w)}{wK_l(w)}\right]\left[\frac{\beta_1^2 J_l'(u)}{uJ_l(u)} + \frac{\beta_2^2 K_l'(w)}{wK_l(w)}\right] = \frac{\beta_z^2 l^2 V^4}{u^4 w^4} = l^2\left(\frac{1}{u^2} + \frac{1}{w^2}\right)\left(\frac{\beta_1^2}{u^2} + \frac{\beta_2^2}{w^2}\right) \tag{10.79}$$

where

$$V^2 = u^2 + w^2 = (\beta_1^2 - \beta_2^2)a^2 \tag{10.80}$$

Note that

$$V = \sqrt{\beta_1^2 - \beta_2^2}\,a = \frac{2\pi a}{\lambda_0}\sqrt{\varepsilon_{r1} - \varepsilon_{r2}} = \frac{2\pi a}{\lambda_0}\sqrt{n_1^2 - n_2^2} = \frac{2\pi a}{\lambda_0}(\mathrm{NA}) \tag{10.81}$$

where

$$\mathrm{NA} = \sqrt{n_1^2 - n_2^2} \tag{10.82}$$

is known as the *numerical aperture,* an important parameter of physical significance (see Problem P10.24).

Determination of the characteristics of the modes for a given value of l requires the solution of (10.77) or (10.78) for $l = 0$ (TM and TE modes) and (10.79) for $l \neq 0$ (hybrid modes), all with the constraint (10.80), from a knowledge of the values of ω, ε_1, ε_2, and a. It is convenient to replace the derivatives of the Bessel functions by the Bessel functions themselves, using the recursive formulas

$$J_{l-1}(x) + J_{l+1}(x) = \frac{2l}{x} J_l(x) \tag{10.83a}$$

$$J_{l-1}(x) - J_{l+1}(x) = 2J_l'(x) \tag{10.83b}$$

$$K_{l-1}(x) + K_{l+1}(x) = -2K_l'(x) \tag{10.83c}$$

$$K_{l-1}(x) - K_{l+1}(x) = -\frac{2l}{x} K_l(x) \tag{10.83d}$$

and express (10.79) in the manner (see Problem P10.25)

$$\boxed{(J^+ + K^+)(\beta_1^2 J^- - \beta_2^2 K^-) + (J^- - K^-)(\beta_1^2 J^+ + \beta_2^2 K^+) = 0} \tag{10.84}$$

where

$$\boxed{J^\pm = \frac{J_{l\pm 1}(u)}{uJ_l(u)} \quad \text{and} \quad K^\pm = \frac{K_{l\pm 1}(w)}{wK_l(w)}} \tag{10.85}$$

Hybrid modes

Let us first consider the hybrid modes. Then a method of solution consists of plotting (10.80) and (10.84) in the u-w plane and looking for the points of intersection. To do this, we note that (10.80) is simply the equation of a circle in the u-w plane, whereas (10.84) results in many branches. The approximate plot of Fig. 10.23 shows the nature of the first three branches for $l = 1$ and the circle for $V = 2$. The mode designations $\text{HE}_{l,m}$ and $\text{EH}_{l,m}$ for the hybrid modes denote although hybrid, H_z is predominant for HE modes and E_z is predominant for EH modes. The subscript m denotes the number of the solution for the particular value of l. It can be seen that for the situation shown in Fig. 10.23 $\text{HE}_{1,2}$ and $\text{EH}_{1,1}$ are cut off, whereas a solution exists for $\text{HE}_{1,1}$. In fact, since the branch for $\text{HE}_{1,1}$ originates at (0, 0), that mode has no cutoff. Thus $\text{HE}_{1,1}$ mode is the dominant mode. As ω increases, V increases, and more and more points of intersection with the V-circle occur, corresponding to higher order modes.

Single-mode operation

The modes with the lowest nonzero cutoff frequency are the TM and TE modes governed by the characteristic equations (10.77) and (10.78), respectively. Therefore, to determine the condition for single-mode operation, we consider

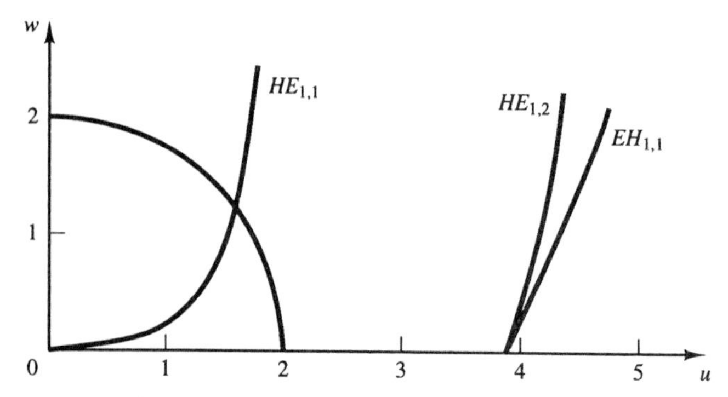

Figure 10.23. For illustrating the nature of solution of the characteristic equation for guided modes in an optical fiber.

these equations. Since the cutoff occurs for $w = 0$ (condition of fields extending to infinity in medium 2) and $K_1/K_0 = \infty$ for $w = 0$, we get $J_1(u) = 0$ or $u = 2.405$ for both modes. Also for $w = 0$, $V = \sqrt{u^2 + w^2} = u$. Thus for single-mode operation, V must be less than 2.405.

Example 10.7

A fiber with core and cladding refractive indices $n_1 = 1.45$ and $n_2 = 1.44$, respectively, has a core radius of $a = 25\ \mu\text{m}$. We wish to find the minimum free-space wavelength, λ_0, for single-mode operation.

For $n_1 = 1.45$ and $n_2 = 1.44$,

$$\text{NA} = \sqrt{(1.45)^2 - (1.44)^2} = 0.17$$

Thus, λ_0 is given by

$$V = \frac{2\pi a}{\lambda_0}(\text{NA}) < 2.405$$

$$\lambda_0 > \frac{2\pi a(\text{NA})}{2.405}$$

$$= \frac{2\pi \times 25 \times 0.17}{2.405}\ \mu\text{m}$$

$$= 11.1033\ \mu\text{m}$$

Weak guidance and LP modes

For most practical fibers, the refractive indices of the core and the cladding are nearly equal ($\varepsilon_1 \approx \varepsilon_2$), corresponding to weak guidance. The rays are then paraxial and the longitudinal components of the fields are much smaller than the transverse components, so that the waves are almost TEM, the simplest such waves being linearly polarized along two orthogonal axes. The associated modes are designated as $\text{LP}_{l,m}$ modes, formed in general by superpositions of the exact modes. The $\text{LP}_{0,1}$ mode, which corresponds to just the $\text{HE}_{1,1}$ mode, is the dominant mode with no cutoff. Thus in terms of LP modes, single-mode operation refers to the $\text{LP}_{0,1}$ mode and it is ensured for $V < 2.405$.

K10.4. Core; Cladding; Field solutions; Hybrid modes; $\text{HE}_{1,1}$ mode; LP modes.

D10.7. A fiber has core radius $a = 10\ \mu\text{m}$. Find the following for single-mode propagation at $\lambda_0 = 10\ \mu\text{m}$: **(a)** maximum allowable value of the numerical aperture; **(b)** minimum allowable value of the cladding refractive index n_2, if the core refractive index $n_1 = 1.5$; and **(c)** maximum allowable value of the core refractive index n_1, if the difference between the core and cladding refractive indices is 0.04. *Ans.* **(a)** 0.3828; **(b)** 1.4503; **(c)** 1.8513

10.5 INTERFERENCE AND DIFFRACTION

Interference explained

In this section, we turn our attention to two related topics which are based on the superposition of waves. When two or more waves are superimposed, the resulting distribution of intensity is in general not merely the sum of the distributions of the intensities of the individual waves; instead, it varies between maxima, which exceed the sum of the individual intensities, and minima, which may go to zero.

This phenomenon is called interference. In the terminology of light, intensity means the time-average power crossing a unit area perpendicular to the direction of power flow, that is, the time-average Poynting vector, which is proportional to the square of the amplitude of the electric field. Thus denoting intensity by the symbol I, we have

$$\boxed{I = k<\mathbf{E}\cdot\mathbf{E}> = k<E^2>} \tag{10.86}$$

where k is the constant of proportionality, dependent on the medium. For free space, k is equal to $1/2\eta_0$.

Let us now consider two uniform plane waves whose electric fields at a point P are $\mathbf{E}_1$ and $\mathbf{E}_2$, respectively. Then the intensities of the individual waves are

$$I_1 = k<E_1^2> \text{ and } I_2 = k<E_2^2> \tag{10.87}$$

The intensity of the superposition of the two waves is

$$\begin{aligned} I &= k<(\mathbf{E}_1 + \mathbf{E}_2)\cdot(\mathbf{E}_1 + \mathbf{E}_2)> \\ &= k<E_1^2> + k<E_2^2> + 2k<\mathbf{E}_1\cdot\mathbf{E}_2> \\ &= I_1 + I_2 + 2k<E_1E_2\cos\alpha> \end{aligned} \tag{10.88}$$

where α is the angle between the directions of E_1 and E_2. The third term on the right side is seen to be the interference term. Depending upon the sign of this term, it can be seen that the intensity of the composite wave can be greater than or less than the sum of the individual intensities. It is equal to the sum of the two intensities only if $\cos\alpha$ is equal to zero, which occurs when $\mathbf{E}_1$ and $\mathbf{E}_2$ are perpendicular. Thus two waves whose fields are polarized perpendicular to each other do not produce interference.

Two-beam interference

An experimental demonstration of interference of light was first performed by Thomas Young. Young's experiment consisted of light from a monochromatic point source S_0 behind a screen containing two pinholes separated by a distance of the order of a millimeter passing through the pinholes, thereby giving rise to two light beams creating an interference pattern on an observation screen, located at a distance R of the order of several meters beyond the pinholes, as shown in Fig. 10.24.

The two waves generated at the pinholes are spherical waves and in phase. At a point P on the observation screen, the waves can be approximated as plane waves. The interference phenomenon can therefore be analyzed by superimposing two traveling waves propagating at the angles θ_1 and θ_2 to the vertical. Because of the large value of R, the difference between these angles is small so that the two lines S_1P and S_2P can be considered to be parallel and $(r_2 - r_1) \approx d\cos\theta$. As the point P moves vertically on the observation screen, the path difference $d\cos\theta$ varies and hence the phase difference between the two waves varies, creating the interference pattern.

Assuming equal amplitudes and zero phase angle at the slits, the individual electric fields of the two beams can be written as

$$E_1 = \frac{A}{r_1}\cos(\omega t - \beta_{r1}) \tag{10.89a}$$

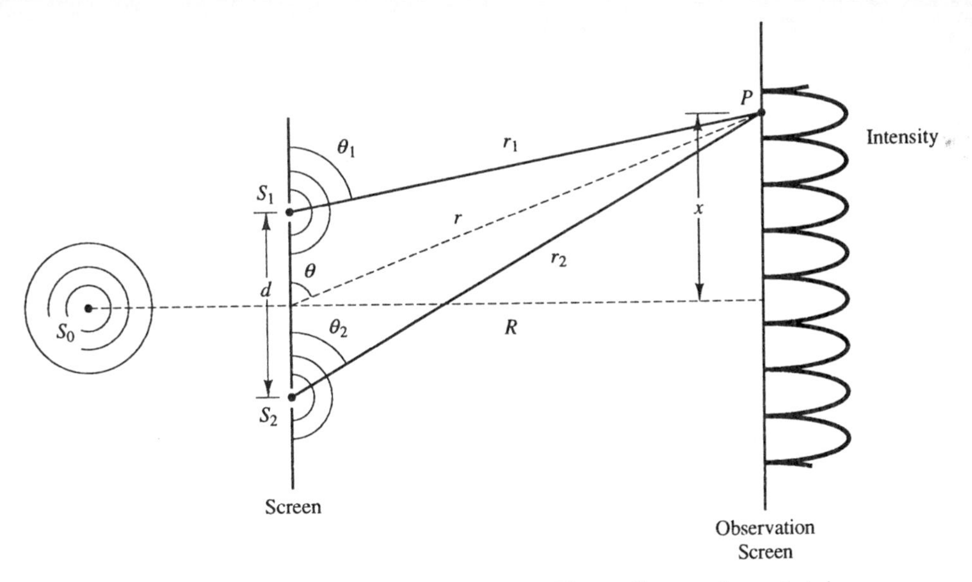

Figure 10.24. Experimental arrangement of Thomas Young to demonstrate interference of light.

$$E_2 = \frac{A}{r_2} \cos (\omega t - \beta_{r2}) \tag{10.89b}$$

Setting $r_1 \approx r_2 \approx r$ in the amplitude factors, we obtain the total electric field to be

$$E = E_1 + E_2 \approx \frac{A}{r} [\cos (\omega t - \beta r_1) + \cos (\omega t - \beta r_2)] \tag{10.90}$$

The intensity of the beam on the screen is therefore given by

$$\begin{aligned} I &= k<E^2> \\ &= \frac{kA^2}{r^2} <\cos^2 (\omega t - \beta r_1) + \cos^2 (\omega t - \beta r_2) \\ &\quad + 2 \cos (\omega t - \beta r_1) \cos (\omega t - \beta r_2)> \\ &= \frac{kA^2}{r^2}\left[\frac{1}{2} + \frac{1}{2} + \cos \beta(r_2 - r_1)\right] \\ &= \frac{2kA^2}{r^2} \cos^2 \frac{\beta(r_2 - r_1)}{2} \\ &= \frac{2kA^2}{r^2} \cos^2 \frac{\beta d \cos \theta}{2} \end{aligned} \tag{10.91}$$

Thus the intensity varies between zero and $\frac{2kA^2}{r^2}$. In terms of the distance x, we can write $\cos \theta = \frac{x}{r} = \frac{x}{\sqrt{R^2 + x^2}}$. Furthermore, for $x \ll R$, we can set $\sqrt{R^2 + x^2} \approx R$, so that we obtain

$$\boxed{I \approx \frac{2kA^2}{R^2}\cos^2\frac{\pi dx}{\lambda R}} \tag{10.92}$$

Thus the maxima of the intensity occur for $x = \dfrac{n\lambda R}{d}$, $n = 0, \pm 1, \pm 2, \ldots$, and the minima occur for $x = \dfrac{(2n+1)\lambda R}{d}$, $n = 0, \pm 1, \pm 2, \ldots$. The separation between the maxima and the minima is a constant, equal to $\Delta x = \lambda R/d$. For a numerical example, for $d = 0.1$ mm, $R = 2$ m, and $\lambda = 0.5\ \mu$m, $\Delta x = 1$ cm.

Conversely to the above, if λ is unknown, it can be computed from a measurement of Δx and given the values of R and d. For two beams, the sharpness of the interference fringes is not sufficient to permit accurate measurement of Δx, but by using a large number of beams the sharpness can be increased. We shall not pursue the multiple-source problem here but consider it in connection with antenna arrays in Section 11.4. We shall however include here an example of a similar phenomenon resulting from multiple reflections and transmissions of waves from a single source.

Example 10.8

Multiple-beam interference

Let us consider a uniform plane wave of electric field $\bar{E}_i$ and wavelength λ_0, incident at an angle θ_i on a plane dielectric slab of thickness d and refractive index n, as shown in Fig. 10.25. For simplicity, we shall consider the medium on either side of the dielectric slab to be free space. We wish to investigate the interference resulting from the (infinite) number of waves produced by reflections and transmissions at the two interfaces, a few of which are shown in Fig. 10.25.

With reference to the notation shown in Fig. 10.25, and denoting Γ and τ to be the reflection and transmission coefficients, respectively, for incidence from free space on to the dielectric and Γ' and τ' to be the reflection and transmission

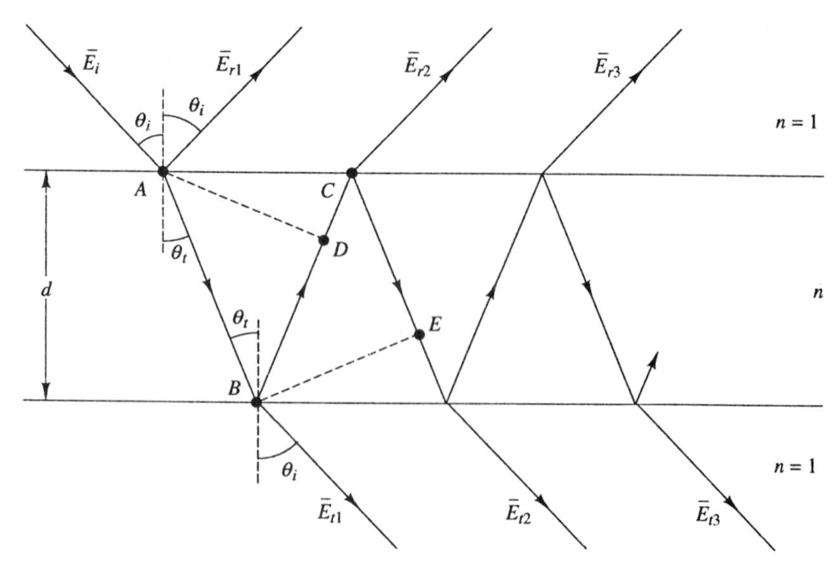

Figure 10.25. Multiple reflections and transmissions for plane wave incidence on a plane dielectric slab.

coefficients, respectively, for incidence from the dielectric on to free space, we can write the expressions for the successively reflected and transmitted wave electric fields as follows:

$$\bar{E}_{r1} = \bar{E}_i \Gamma \qquad (10.93a) \qquad\qquad \bar{E}_{t1} = \bar{E}_i \tau\tau' \qquad (10.93d)$$

$$\bar{E}_{r2} = \bar{E}_i \tau\Gamma'\tau' e^{-j\delta} \qquad (10.93b) \qquad\qquad \bar{E}_{t2} = \bar{E}_i \tau(\Gamma')^2\tau' e^{-j\delta} \qquad (10.93e)$$

$$\bar{E}_{r3} = \bar{E}_i \tau(\Gamma')^3\tau' e^{-j2\delta} \qquad (10.93c) \qquad\qquad \bar{E}_{t3} = \bar{E}_i \tau(\Gamma')^4\tau' e^{-j2\delta} \qquad (10.93f)$$

$$\cdots \qquad\qquad\qquad\qquad \cdots$$

where

$$\boxed{\begin{aligned}\delta &= \frac{2\pi n}{\lambda_0}(AB + BD) = \frac{2\pi n}{\lambda_0}(BC + CE) \\ &= \frac{2\pi n}{\lambda_0}\frac{d}{\cos\theta_t}(1 + \cos 2\theta_t) \\ &= \frac{4\pi n d}{\lambda_0}\cos\theta_t\end{aligned}} \qquad (10.94)$$

is the additional phase shift undergone by successive reflected (or transmitted) waves. Summing up all the reflected wave fields, we obtain the total reflected wave field to be

$$\begin{aligned}\bar{E}_r &= \bar{E}_i\,\{\Gamma + \tau\tau'\bar{\Gamma}'e^{-j\delta}[1 + (\Gamma')^2e^{-j\delta} + \ldots]\} \\ &= \bar{E}_i\left[\Gamma + \frac{\tau\tau'\Gamma'e^{-j\delta}}{1 - (\Gamma')^2e^{-j\delta}}\right]\end{aligned} \qquad (10.95a)$$

Similarly, the total transmitted wave field is given by

$$\begin{aligned}\bar{E}_t &= \bar{E}_i\tau\tau'[1 + (\Gamma')^2e^{-j\delta} + \ldots] \\ &= \bar{E}_i\left[\frac{\tau\tau'}{1 - (\Gamma')^2e^{-j\delta}}\right]\end{aligned} \qquad (10.95b)$$

While the specific expressions for Γ, Γ', τ, and τ' depend upon the polarization of $\mathbf{E}_i$ and are given in Section 10.1 by the Fresnel coefficients, we can write, regardless of polarization (see Problem P10.28), that

$$\boxed{\Gamma' = -\Gamma} \qquad (10.96a)$$

and

$$\boxed{\tau\tau' + \Gamma^2 = 1} \qquad (10.96b)$$

Substituting (10.96a) and (10.96b) into (10.95a) and (10.95b), we obtain

$$\bar{E}_r = \bar{E}_i\left[\frac{\Gamma(1 - e^{-j\delta})}{1 - \Gamma^2e^{-j\delta}}\right] \qquad (10.97a)$$

$$\bar{E}_t = \bar{E}_i\,\frac{1 - \Gamma^2}{1 - \Gamma^2e^{-j\delta}} \qquad (10.97b)$$

The fractions of the incident intensity that are reflected and transmitted are given, respectively, by

$$\boxed{\frac{I_r}{I_i} = \frac{|\bar{E}_r|^2}{|\bar{E}_i|^2} = \frac{4F^2 \sin^2 \dfrac{\delta}{2}}{\pi^2 + 4F^2 \sin^2 \dfrac{\delta}{2}}} \tag{10.98a}$$

$$\boxed{\frac{I_t}{I_i} = \frac{|\bar{E}_t|^2}{|\bar{E}_i|^2} = \frac{\pi^2}{\pi^2 + 4F^2 \sin^2 \dfrac{\delta}{2}}} \tag{10.98b}$$

where

$$\boxed{F = \frac{\pi|\Gamma|}{1 - \Gamma^2}} \tag{10.99}$$

is called the finesse. Note that the sum of right sides of (10.98a) and (10.98b) is equal to unity, as it should be.

Fabry–Perot etalon

The transmission characteristic of the arrangement of Fig. 10.25, a model for the Fabry–Perot etalon, or interferometer, can now be discussed with the aid of the plot of the right side of (10.98b) versus δ, which is shown in Fig. 10.26, for several values of F and hence Γ. It can be seen that maximum transmission of unity occurs for

$$\delta = \frac{4\pi n d}{\lambda_0} \cos \theta_t = 2m\pi, \quad m = 1, 2, 3, \ldots \tag{10.100}$$

with the sharpness of the maxima increasing with F. For given values of λ_0, d, and n, the plot can be thought of as variation of I_t/I_0 with θ_t (and hence θ_i), thereby corresponding to the interference pattern. For fixed values of d, n, and θ_t, the peaks in Fig. 10.26 correspond to two adjacent frequencies at which I_t/I_i is unity. From

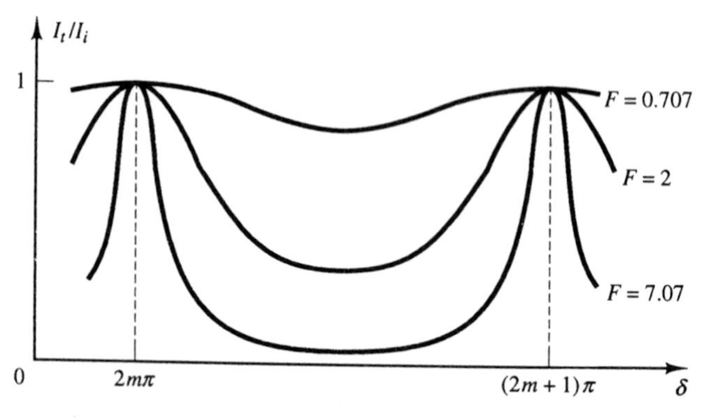

Figure 10.26. Plots of the transmission characteristics of the arrangement of Fig. 10.25 for several values of F.

(10.100), this frequency separation can be seen to be equal to $c/(2nd \cos \theta_t)$. Also, for given values of d and n, different values of λ_0 (and hence f) give rise to interference patterns of different periodicities, thereby allowing resolution of closely spaced frequencies for high values of F (see Problem P10.29).

Diffraction explained

When an object is placed between a source of light and an observation screen, the shadow on the screen contains a fine structure of interference fringes in the vicinity of the boundary separating the dark shadow from the rest of the brightly illuminated screen, as compared to a simple sharp boundary between the dark and bright regions. This phenomenon, which occurs due to the "bending" of a portion of the beam, is known as "diffraction." Just as interference is a manifestation of the superposition of light beams, diffraction is also a manifestation of the superposition of light beams. Interference usually applies to the interaction of only a few beams with one another, whereas diffraction usually pertains to the superposition of a large number, even a continuous distribution of beams, although the distinction is not sharp. The phenomenon of diffraction is in contrast to the principle of geometrical optics, which has to do with light traveling in straight lines except for bending by reflection and refraction, and which is strictly valid under certain conditions: (a) The dimensions of the object in the path of light are very large compared to the wavelength, and (b) the region of importance is not close to the boundary of the shadow.

Huygens–Fresnel principle

While the exact treatment of the phenomenon of diffraction involves solution as a boundary value problem and is very difficult, it can be studied in approximate but sufficiently accurate terms by using the Huygens-Fresnel principle, as long as the distance from the diffracting object to the point of observation is more than about ten wavelengths. To explain this principle, let us consider a plane monochromatic wave incident normally on a screen in the xy-plane with an aperture cut into it, as shown in Fig. 10.27. Then, according to this principle, the in-

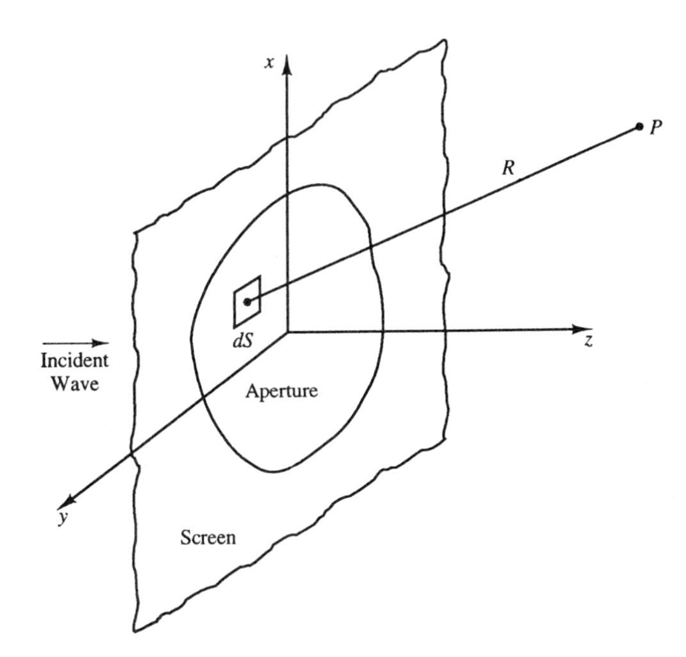

Figure 10.27. Geometry pertinent to diffraction by an aperture on a screen.

cident wave may be thought of as giving rise to secondary (spherical) waves emanating from every point in the aperture and which interfere with one another to the right of the screen. The scalar field at a point P is approximately given by

$$\boxed{\bar{E}(P) \approx \frac{j\beta}{2\pi}\int_S \frac{\bar{E}(x', y', 0)}{R}e^{-j\beta R}\,dS} \tag{10.101}$$

where S is the area of the aperture and $\bar{E}(x', y', 0)$ is the scalar field in the aperture. We shall illustrate the application of (10.101) by means of an example.

Example 10.9

Diffraction by a circular aperture

Let us assume that the aperture of Fig. 10.27 is a circular hole of radius a having its center at the origin and illuminated by a uniform plane wave of electric field intensity E_0 at the aperture, as shown in Fig. 10.28. We wish to investigate the diffracted field along the z-axis.

Applying (10.101) to the geometry in Fig. 10.28 and noting the circular symmetry about the z-axis, we obtain

$$\bar{E}(0, 0, z) = \frac{j\beta}{2\pi}\int_{r=0}^{a}\int_{\phi=0}^{2\pi}\frac{E_0}{\sqrt{r^2 + z^2}}e^{-j\beta\sqrt{r^2+z^2}}r\,dr\,d\phi \tag{10.102}$$

Making the change of variable $R = \sqrt{r^2 + z^2}$, and hence $dR = \dfrac{r\,dr}{R}$, we obtain

$$\begin{aligned}\bar{E}(0, 0, z) &= j\beta E_0\int_z^{\sqrt{a^2+z^2}} e^{-j\beta R}\,dR \\ &= -E_0e^{-j\beta\sqrt{a^2+z^2}} + E_0e^{-j\beta z}\end{aligned} \tag{10.103}$$

Proceeding further, the intensity is given by

$$I = \frac{1}{2\eta_0}|\bar{E}|^2 = \frac{1}{2\eta_0}\bar{E}\bar{E}^*$$

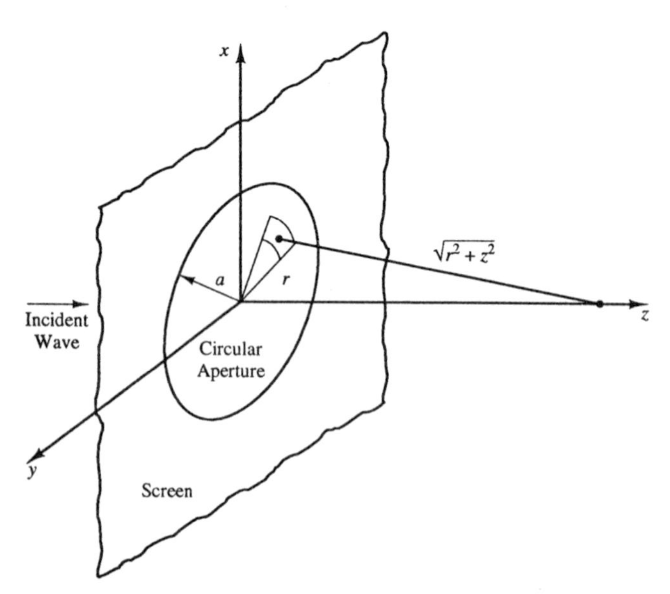

Figure 10.28. Geometry pertinent to diffraction by a circular aperture illuminated by a plane wave.

$$I = \frac{E_0^2}{2\eta_0}[2 - 2\cos\beta(\sqrt{a^2 + z^2} - z)]$$
$$= \frac{E_0^2}{\eta_0}\sin^2\frac{\beta(\sqrt{a^2 + z^2} - z)}{2} \tag{10.104}$$

For $z \gg a$, (10.104) reduces to

$$\boxed{\begin{aligned} I &\approx \frac{2E_0^2}{\eta_0}\sin^2\frac{\beta a^2}{4z} \\ &= \frac{2E_0^2}{\eta_0}\sin^2\frac{\pi a^2}{2\lambda z} \end{aligned}} \tag{10.105}$$

Fresnel vs. Fraunhofer diffraction

A sketch of the result given by (10.105) is shown in Fig. 10.29. It can be seen that for $z < a^2/\lambda$, the intensity fluctuates between maxima of $2E_0^2/\eta_0$ and minima of zero, corresponding to constructive and destructive interference, respectively, of the spherical waves. The situation is said to correspond to Fresnel diffraction. For $z > a^2/\lambda$, the intensity decreases monotonically. For $z \gg a^2/\lambda$, the diffraction is known as Fraunhofer diffraction and the evaluation of the integral in (10.101) becomes easier, because waves arriving at P from the aperture approach plane waves, thereby permitting the simplifying plane wave approximation in the integrand. We shall however not pursue the topic here but defer the consideration to Section 11.5 where the determination of the far field due to an aperture distribution is identical to that of the solution for Fraunhofer diffraction. In practice, the boundary between the Fresnel and Fraunhofer diffraction regions is taken to be $2D^2/\lambda$, where D is the diameter of the circular aperture, or in the case of a noncircular aperture it is its maximum linear dimension (see Problem P10.31).

K10.5. Intensity; Two-beam interference; Multiple-beam interference; Finesse; Fabry–Perot etalon; Diffraction; Huygens–Fresnel principle; Circular aperture; Fresnel diffraction; Fraunhofer diffraction.

D10.8. For a uniform plane wave incident from free space onto a plane dielectric slab of thickness $d = 5.1\lambda_0$ and refractive index $n = 4$ at an angle $\theta_i = 60°$, find F and

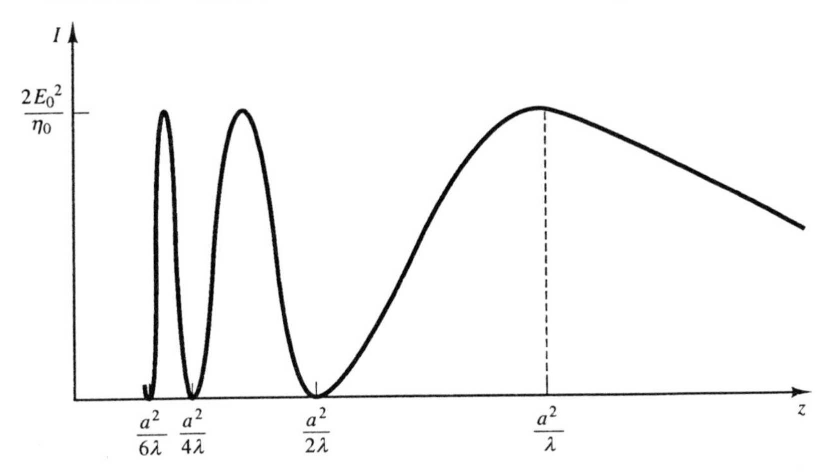

Figure 10.29. Variation of the intensity along the z-axis for the arrangement of Fig. 10.28.

I_t/I_i for each of the two polarizations: **(a)** perpendicular and **(b)** parallel.
Ans. **(a)** 6.034, 0.2113 **(b)** 1.226, 0.8665

10.6 WAVE PROPAGATION IN ANISOTROPIC MEDIUM

Linear characteristic polarizations

In Section 2.5 we learned that for certain dielectric materials known as "anisotropic dielectric materials," **D** is not in general parallel to **E** and the relationship between **D** and **E** is expressed by means of a permittivity tensor consisting of a 3×3 matrix. Let us consider an anisotropic dielectric medium characterized by the simple **D** to **E** relationship given by

$$\begin{bmatrix} D_x \\ D_y \\ D_z \end{bmatrix} = \begin{bmatrix} \varepsilon_x & 0 & 0 \\ 0 & \varepsilon_y & 0 \\ 0 & 0 & \varepsilon_z \end{bmatrix} \begin{bmatrix} E_x \\ E_y \\ E_z \end{bmatrix} \tag{10.106}$$

and having the permeability μ_0. It is easy to see that the characteristic polarizations for this case are all linear directed along the coordinate axes and having the effective permittivities ε_x, ε_y, and ε_z for the x-, y- and z-directed polarizations, respectively. These axes are then known as the principal axes. Let us consider a uniform plane wave propagating along one of the principal axes, say, the z-direction. The wave will then generally contain both x- and y-components of the fields. It can be decomposed into two waves, one having an x-directed electric field and the other having a y-directed electric field. These component waves travel individually in the anisotropic medium as though it is isotropic but with different phase velocities since the effective permittivities are different. In view of this, the phase relationship between the two waves, and hence the polarization of the composite wave, changes with distance along the direction of propagation. Also, when they encounter a discontinuity the component waves undergo reflection and transmission by different amounts. We shall illustrate by means of an example.

Example 10.10

Propagation along a principal axis

Let us consider a uniform plane wave of frequency 1500 MHz incident from free space ($z < 0$) normally onto an anisotropic perfect dielectric medium ($z > 0$), characterized by the permittivity matrix

$$[\varepsilon] = \varepsilon_0 \begin{bmatrix} 4 & 0 & 0 \\ 0 & 9 & 0 \\ 0 & 0 & 4 \end{bmatrix}$$

and $\mu = \mu_0$. We wish to discuss the reflected and transmitted waves for several cases of incident waves.

Case 1. The incident wave has only an x-component of **E** as given by

$$\mathbf{E}_i = E_0 \cos (3 \times 10^9 \pi t - 10\pi z)\, \mathbf{i}_x$$

Then the effective permittivity of the anisotropic medium is $4\varepsilon_0$, and from (6.118) and (6.119), $\bar{\Gamma} = -\frac{1}{3}$ and $\bar{\tau} = \frac{2}{3}$. The reflected and transmitted wave electric fields are

$$\mathbf{E}_r = -\frac{E_0}{3} \cos (3 \times 10^9 \pi t + 10\pi z)\, \mathbf{i}_x$$

$$\mathbf{E}_t = \frac{2E_0}{3} \cos (3 \times 10^9 \pi t - 20\pi z)\, \mathbf{i}_x$$

where we have made use of the fact that for the transmitted wave, the phase constant is $\omega \sqrt{\mu_o \cdot 4\varepsilon_0} = 2\omega \sqrt{\mu_0 \varepsilon_0} = 2 \times 10\pi = 20\pi$.

Case 2. The incident wave has only a y-component of $\mathbf{E}$ as given by

$$\mathbf{E}_i = E_0 \cos (3 \times 10^9 \pi t - 10\pi z)\, \mathbf{i}_y$$

Then the effective permittivity of the anisotropic medium is $9\varepsilon_0$, and from (6.118) and (6.119), $\bar{\Gamma} = -\frac{1}{2}$ and $\bar{\tau} = \frac{1}{2}$. The reflected and transmitted wave electric fields are

$$\mathbf{E}_r = -\frac{E_0}{2} \cos (3 \times 10^9 \pi t + 10\pi z)\, \mathbf{i}_y$$

$$\mathbf{E}_t = \frac{E_0}{2} \cos (3 \times 10^9 \pi t - 30\pi z)\, \mathbf{i}_y$$

where we have made use of the fact that for the transmitted wave, the phase constant is $\omega \sqrt{\mu_0 \cdot 9\varepsilon_0} = 3\omega \sqrt{\mu_0 \varepsilon_0} = 3 \times 10\pi = 30\pi$.

Case 3. The incident wave has both x- and y-components of $\mathbf{E}$ and is linearly polarized as given by

$$\mathbf{E}_i = E_1 \cos (3 \times 10^9 \pi t - 10\pi z)\, \mathbf{i}_x + E_2 \cos (3 \times 10^9 \pi t - 10\pi z)\, \mathbf{i}_y$$

Then from superposition of cases 1 and 2, the reflected and transmitted wave electric fields are given by

$$\mathbf{E}_r = -\frac{E_1}{3} \cos (3 \times 10^9 \pi t + 10\pi z)\, \mathbf{i}_x - \frac{E_2}{2} \cos (3 \times 10^9 \pi t + 10\pi z)\, \mathbf{i}_y$$

$$\mathbf{E}_t = \frac{2E_1}{3} \cos (3 \times 10^9 \pi t - 20\pi z)\, \mathbf{i}_x + \frac{E_2}{2} \cos (3 \times 10^9 \pi t - 30\pi z)\, \mathbf{i}_y$$

Note that $\mathbf{E}_r$ is linearly polarized, although along a direction making an angle to that of the direction of polarization of $\mathbf{E}_i$. The polarization of $\mathbf{E}_t$, on the other hand, varies with z since the phase difference between the x- and y-components of the electric field is $\Delta\phi = 10\pi z$. As the transmitted wave propagates in the z-direction, $\Delta\phi$ changes from zero at $z = 0$ to $\pi/2$ at $z = 0.05$ m to π at $z = 0.1$ m, and so on. Thus the polarization changes from linear at $z = 0$ to elliptical for $z > 0$, becoming linear again at $z = 0.1$ m, but rotated by an angle as shown in Fig. 10.30, and so on.

The simple form of permittivity tensor given by (10.106) can be realized for certain anisotropic crystals by an appropriate choice of the coordinate system. If the permittivities ε_x, ε_y, and ε_z are all different, then the crystal is said to be *biaxial*. If two of the three are equal, then it is said to be *uniaxial*.

Wave plates

To generalize the observation in Example 10.10, the phase difference between the x- and y-components of $\mathbf{E}$ in the anisotropic medium can be expressed as

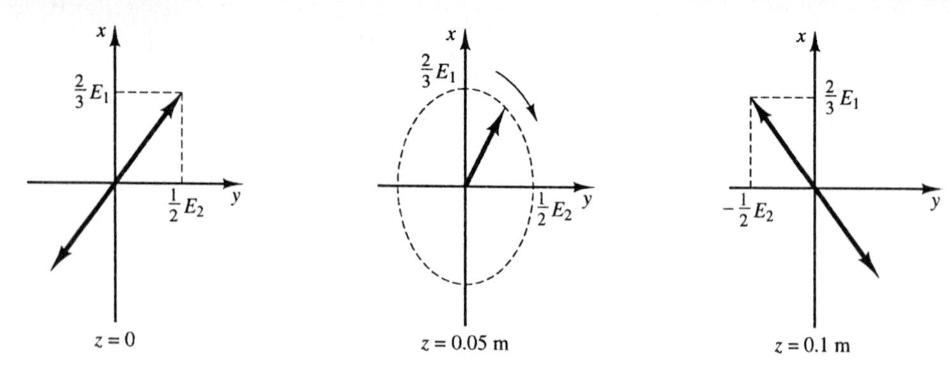

Figure 10.30. Change in polarization versus z of the transmitted wave electric field of Example 10.10.

$$\begin{aligned}\Delta\phi &= (\omega\sqrt{\mu_0\varepsilon_x} - \omega\sqrt{\mu_0\varepsilon_y})z \\ &= \frac{\omega}{c}(\sqrt{\varepsilon_{rx}} - \sqrt{\varepsilon_{ry}})z \\ &= \frac{2\pi}{\lambda_0}(n_x - n_y)z\end{aligned} \tag{10.107}$$

where λ_0 is the free-space wavelength and n_x and n_y are the refractive indices. The result given by (10.107) is the basis behind *wave plates* or *retardation plates*. The word "retardation" refers to the fact that the phase of one of the two components lags that of the second component. If $\Delta\phi = 2m\pi$, where m is an integer, the plate is called a *full-wave plate*. If reflections from the surfaces of the plate are considered to be negligible, as is usually the case, then it can be seen that the state of polarization of the wave at the output plane of the plate is the same as that at the input plane. For $\Delta\phi = (2m + 1)\pi$ and $(2m + 1)\pi/2$, the arrangement corresponds to *half-wave plate* and *quarter-wave plate*, respectively. A half-wave plate results in a rotation of the direction of linear polarization, as illustrated in Fig. 10.30, which corresponds to $\Delta\phi = \pi$. Note that the direction has shifted by twice the angle which it initially makes with the x- (or y-) direction. A quarter-wave plate can transform a linearly polarized wave into a circularly polarized wave.

Circular characteristic polarizations

For a different example of an anisotropic medium than that characterized by (10.106), let us consider the **D** to **E** relationship of the form

$$\bar{\mathbf{D}} = \varepsilon\bar{\mathbf{E}} + j\varepsilon_0\gamma\bar{\mathbf{B}} \times \bar{\mathbf{E}} \tag{10.108}$$

which is exhibited by certain materials when placed in a static magnetic field **B**, where γ is a constant depending on the material. For a uniform plane wave propagating in the z-direction, and $\bar{\mathbf{B}} = B_0\mathbf{i}_z$, we have

$$\begin{aligned}\bar{D}_x &= \varepsilon\bar{E}_x - j\varepsilon_0\gamma B_0\bar{E}_y \\ \bar{D}_y &= j\varepsilon_0\gamma B_0\bar{E}_x + \varepsilon\bar{E}_y\end{aligned}$$

To find the characteristic polarizations, we set $\bar{D}_x/\bar{D}_y = \bar{E}_x/\bar{E}_y$. Thus

$$\frac{\varepsilon \bar{E}_x - j\varepsilon_0 \gamma B_0 \bar{E}_y}{j\varepsilon_0 \gamma B_0 \bar{E}_x + \varepsilon \bar{E}_y} = \frac{\bar{E}_x}{\bar{E}_y}$$

Solving for $\bar{E}_x/\bar{E}_y$, we get

$$\boxed{\frac{\bar{E}_x}{\bar{E}_y} = \pm j}$$

This result corresponds to equal amplitudes of E_x and E_y and phase differences of $\pm 90°$. Thus the characteristic polarizations are both circular, rotating in opposite senses as viewed along the z-direction.

The effective permittivities of the medium corresponding to the characteristic polarizations are

$$\begin{aligned} \frac{\bar{D}_x}{\bar{E}_x} &= \frac{\varepsilon \bar{E}_x - j\varepsilon_0 \gamma B_0 \bar{E}_y}{\bar{E}_x} \\ &= \varepsilon - j\varepsilon_0 \gamma B_0 \frac{\bar{E}_y}{\bar{E}_x} \\ &= \varepsilon \mp \varepsilon_0 \gamma B_0 \qquad \text{for} \quad \frac{\bar{E}_x}{\bar{E}_y} = \pm j \end{aligned} \tag{10.109}$$

The phase constants associated with the propagation of the characteristic waves are

$$\boxed{\beta_\pm = \omega\sqrt{\mu(\varepsilon \mp \varepsilon_0 \gamma B_0)}} \tag{10.110}$$

where the subscripts + and − refer to $\bar{E}_x/\bar{E}_y = +j$ and $\bar{E}_x/\bar{E}_y = -j$, respectively.

Let us now consider the electric field of the wave to be linearly polarized in the x-direction at $z = 0$, that is,

$$\mathbf{E}(0) = E_0 \cos \omega t \, \mathbf{i}_x \tag{10.111}$$

Then we can express (10.111) as the superposition of two circularly polarized fields having opposite senses of rotation in the xy-plane in the manner

$$\mathbf{E}(0) = \left(\frac{E_0}{2} \cos \omega t \, \mathbf{i}_x + \frac{E_0}{2} \sin \omega t \, \mathbf{i}_y\right) + \left(\frac{E_0}{2} \cos \omega t \, \mathbf{i}_x - \frac{E_0}{2} \sin \omega t \, \mathbf{i}_y\right) \tag{10.112}$$

The circularly polarized field inside the first pair of parentheses on the right side of (10.112) corresponds to

$$\frac{\bar{E}_x}{\bar{E}_y} = \frac{E_0/2}{-jE_0/2} = +j$$

whereas that inside the second pair of parentheses corresponds to

$$\frac{\bar{E}_x}{\bar{E}_y} = \frac{E_0/2}{jE_0/2} = -j$$

Assuming propagation in the positive z-direction, the field at an arbitrary value of z is then given by

$$
\begin{aligned}
\mathbf{E}(z) &= \left[\frac{E_0}{2}\cos(\omega t - \beta_+ z)\,\mathbf{i}_x + \frac{E_0}{2}\sin(\omega t - \beta_+ z)\,\mathbf{i}_y\right] \\
&\quad + \left[\frac{E_0}{2}\cos(\omega t - \beta_- z)\,\mathbf{i}_x - \frac{E_0}{2}\sin(\omega t - \beta_- z)\,\mathbf{i}_y\right] \\
&= \left[\frac{E_0}{2}\cos\left(\omega t - \frac{\beta_+ + \beta_-}{2}z - \frac{\beta_+ - \beta_-}{2}z\right)\mathbf{i}_x\right. \\
&\quad \left. + \frac{E_0}{2}\sin\left(\omega t - \frac{\beta_+ + \beta_-}{2}z - \frac{\beta_+ - \beta_-}{2}z\right)\mathbf{i}_y\right] \qquad (10.113) \\
&\quad + \left[\frac{E_0}{2}\cos\left(\omega t - \frac{\beta_+ + \beta_-}{2}z + \frac{\beta_+ - \beta_-}{2}z\right)\mathbf{i}_x\right. \\
&\quad \left. - \frac{E_0}{2}\sin\left(\omega t - \frac{\beta_+ + \beta_-}{2}z + \frac{\beta_+ - \beta_-}{2}z\right)\mathbf{i}_y\right] \\
&= \left[E_0\cos\left(\frac{\beta_- - \beta_+}{2}z\right)\mathbf{i}_x\right. \\
&\quad \left. + E_0\sin\left(\frac{\beta_- - \beta_+}{2}z\right)\mathbf{i}_y\right]\cos\left(\omega t - \frac{\beta_+ + \beta_-}{2}z\right)
\end{aligned}
$$

Faraday rotation

The result given by (10.113) indicates that the x- and y-components of the field are in phase at any given value of z. Hence the field is linearly polarized for all values of z. The direction of polarization is, however, a function of z since $\tan^{-1}(E_y/E_x)$, the angle made by the field vector with the x-axis, is $[(\beta_- - \beta_+)/2]z$. Thus the direction of polarization rotates linearly with z at a rate of $(\beta_- - \beta_+)/2$. This phenomenon is known as "Faraday rotation" and is illustrated with the aid of the sketches in Fig. 10.31. The sketches in any given column correspond to a fixed value of z whereas the sketches in a given row correspond to a fixed value of t. At $z = 0$, the field is linearly polarized in the x-direction and is the superposition of two counter-rotating circularly polarized fields as shown by the time series of sketches in the first column. If the medium is isotropic, the two counter-rotating circularly polarized fields undergo the same amount of phase lag with z and the field remains linearly polarized in the x-direction as shown by the dashed lines in the second and third columns. For the case of the anisotropic medium, the two circularly polarized fields undergo different amounts of phase lag with z. Hence their superposition results in a linear polarization making an angle with the x-direction and increasing linearly with z as shown by the solid lines in the second and third columns.

Magneto-optical switch

The phenomenon of Faraday rotation that we have just discussed forms the basis for a number of devices. A simple example is illustrated by the magneto-optical switch. The magneto-optical switch is a device for modulating a laser beam by switching an electric current on and off. The electric current generates a magnetic field that rotates the magnetization vector in a magnetic iron-garnet film on a substrate of garnet in the plane of the film through which a light wave passes.

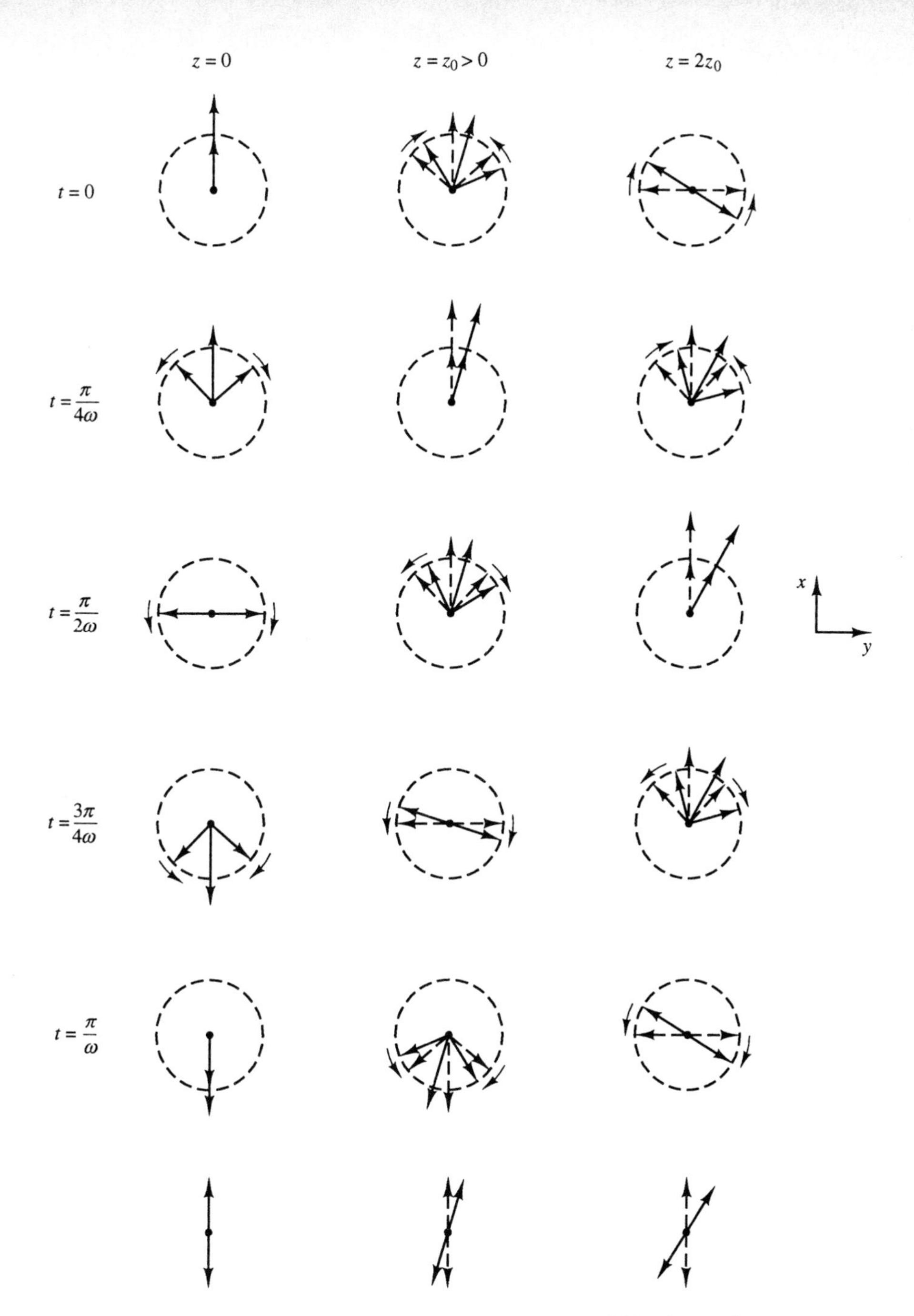

Figure 10.31. For illustrating the phenomenon of Faraday rotation.

When it enters the film, the light wave field is linearly polarized normal to the plane of the film. If the current in the electric circuit is off, the magnetization vector is normal to the direction of propagation of the wave and the wave emerges out of the film without change of polarization, as shown in Fig. 10.32(a). If the current in the electric circuit is on, the magnetization vector is parallel to the di-

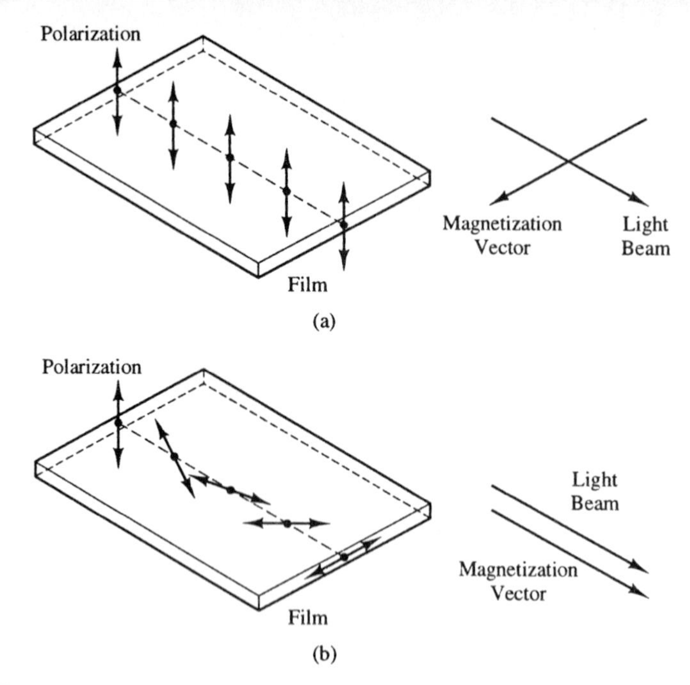

Figure 10.32. For illustrating the principle of operation of a magneto-optical switch.

rection of propagation of the wave, the light wave undergoes Faraday rotation and emerges out of the film with its polarization rotated by 90°, as shown in Fig. 10.32(b). After it emerges out of the film, the light beam is passed through a polarizer which has the property of absorbing light of the original polarization but passing through the light of the 90°-rotated polarization. Thus the beam is made to turn on and off by the switching on and off of the current in the electric circuit. In this manner, any coded message can be made to be carried by the light beam.

K10.6. Anisotropic dielectric materials; Characteristic polarizations; Wave plates; Faraday rotation.

D10.9. At $\lambda_0 = 0.633\ \mu\text{m}$, the refractive indices of mica are given by $n_x = 1.594$ (fast axis) and $n_y = 1.599$ (slow axis). Find the following: **(a)** the minimum thickness of a mica sheet to act as a half-wave plate; **(b)** the number of wavelengths undergone by the wave in the thickness of the plate for the x-polarization; and **(c)** the number of wavelengths undergone by the wave in the thickness of the plate for the y-polarization.
Ans. **(a)** 63.3 μm; **(b)** 159.4; **(c)** 159.9

10.7 SUMMARY

In this chapter, we first considered oblique incidence of a uniform plane wave upon the boundary between two perfect dielectric media. We derived the *laws of reflection and refraction,* given, respectively, by

$$\theta_r = \theta_i$$

$$\theta_t = \sin^{-1}\left(\sqrt{\frac{\mu_1\varepsilon_1}{\mu_2\varepsilon_2}}\sin\theta_i\right)$$

where θ_i, θ_r, and θ_t are the angles of incidence, reflection, and transmission, respectively, of a uniform plane wave incident from medium 1 (ε_1, μ_1) onto medium 2 (ε_2, μ_2). The law of refraction is also known as *Snell's law*. We then derived the expressions for the reflection and transmission coefficients for the cases of perpendicular and parallel polarizations. An examination of these expressions revealed the following, under the assumption of $\mu_1 = \mu_2$: (1) For incidence from a medium of higher permittivity onto one of lower permittivity, there is a critical angle of incidence given by

$$\theta_c = \sin^{-1}\sqrt{\frac{\varepsilon_2}{\varepsilon_1}}$$

beyond which total internal reflection occurs, and (2) for the case of parallel polarization, there is an angle of incidence, known as the Brewster angle and given by

$$\theta_p = \tan^{-1}\sqrt{\frac{\varepsilon_2}{\varepsilon_1}}$$

for which the reflection coefficient is zero.

Next, we introduced the dielectric slab waveguide, consisting of a dielectric slab of permittivity ε_1 sandwiched between two dielectric media of permittivities $\varepsilon_2 < \varepsilon_1$. We learned that by launching waves at an angle of incidence θ_i greater than the critical angle for total internal reflection, it is possible to achieve guided wave propagation within the slab. For a given frequency, several modes are possible corresponding to values of θ_i which satisfy the self-consistency condition associated with the bouncing waves. We derived the characteristic equation for computing these values of θ_i for the TE case and discussed its solution. The modes are designated TE_m modes and their cutoff frequencies are given by

$$f_c = \frac{mc}{2d\sqrt{\varepsilon_{r1} - \varepsilon_{r2}}}, \qquad m = 0, 1, 2, \ldots$$

where d is the thickness of the slab. The fundamental mode, TE_0, has no cutoff frequency. We also discussed the guided modes by using the approach of deriving the field expressions, based on the behavior that (a) in the transverse (x-) direction the fields have standing wave character inside the slab and are evanescent outside the slab, and (b) in the longitudinal (z-) direction the fields have traveling wave character both inside and outside the slab. Dividing the modes into even and odd modes with respect to x from symmetry considerations, we derived the field expressions for the TE modes and (a) obtained the associated characteristic equation for guidance to be the same as that obtained from the wave-bounce approach, (b) discussed the field behavior from near cutoff to far from cutoff, and (c) investigated power flow down the guide.

To extend the treatment of dielectric waveguide to one of graded-index guide, that is, one in which the refractive index varies within the thickness of the slab, we first introduced the topic of ray tracing, making use of the geometrical optics concept. The ray tracing procedure involves the application of Snell's law in conjunction with the geometry associated with the problem. Although in general the solution has to be carried out numerically, for certain functions for the refractive index variation analytical solutions are possible, as illustrated by considering (a) a linear profile of permittivity, and (b) paraxial rays, that is, rays that make small angles to the propagation axis, in a parabolic index profile. For the latter case, we found that the ray oscillates about the axis with a pitch independent of the angle of takeoff from the axis at the starting point. We then considered the graded-index guide having a symmetric refractive index profile and derived the condition for guidance. Applying the guidance condition to investigate modes in a parabolic index guide, we showed that for paraxial rays in the parabolic index guide intermodal dispersion is nearly eliminated.

Proceeding further, we introduced the optical fiber, which consists typically of a core and a cladding having circular cross sections. By assuming the cladding region to extend to infinity so that the situation corresponds to one of a cylindrical dielectric rod, we carried out the field analysis and learned that in addition to TE ($E_z = 0$) and TM ($H_z = 0$) modes, the fields correspond to the so-called hybrid modes, designated HE and EH. For the hybrid modes, both E_z and H_z are not equal to zero; however, for the HE modes H_z is predominant, whereas for the EH modes E_z is predominant. If fact, the dominant mode is the $HE_{1,1}$ mode, having no cutoff. For single-mode operation at a given wavelength λ_0, the condition is given by

$$V = \frac{2\pi a}{\lambda_0}(\text{NA}) < 2.405$$

where a is the radius of the core and NA is the numerical aperture given by

$$\text{NA} = \sqrt{n_1^2 - n_2^2}$$

n_1 and n_2 being the refractive indices of the core and the cladding, respectively. For weak guidance ($n_1 \approx n_2$), the modes are designated LP, with $LP_{0,1}$ mode corresponding to the $HE_{1,1}$ mode.

Next we turned our attention to two related topics, interference and diffraction, which are both based on superposition of waves. While the distinction is not sharp, interference usually applies to the interaction of only a few light beams with one another, whereas diffraction usually pertains to the superposition of a large number, even a continuous distribution, of beams. We discussed interference by considering (a) the two-beam interference experiment of Thomas Young and (b) multiple-beam interference due to plane wave incidence obliquely on a plane dielectric slab, the latter arrangement constituting a model for the Fabry–Perot etalon or interferometer. For diffraction, we introduced the Huygens–Fresnel principle, according to which each point on a wavefront generates a spherical wave, and illustrated its application by considering the example of a plane wave incident on a circular aperture in a screen. By investigating diffraction along the axis of the aperture, we discussed briefly Fresnel versus Fraunhofer diffraction.

Finally, we discussed the topic of wave propagation in an anisotropic medium. By considering the example of a uniform plane wave incident normally along a principal axis of a uniaxial crystal, thereby resulting in the characteristic polarizations to be linear, we illustrated the principle behind wave plates or retardation plates. By means of another example of an anisotropic medium for which the characteristic polarizations are circular, we introduced Faraday rotation, which is the phenomenon of rotation of the direction of polarization of a linearly polarized wave as it propagates in the medium, and discussed the operation of a magneto-optical switch, a device employing Faraday rotation for modulating a light beam.

REVIEW QUESTIONS

R10.1. Discuss the condition required to be satisfied by the incident, reflected, and transmitted waves at the interface between two dielectric media.

R10.2. What is Snell's law?

R10.3. What is meant by the plane of incidence? Distinguish between the two different linear polarizations pertinent to the derivation of the reflection and transmission coefficients for oblique incidence on a dielectric interface.

R10.4. Briefly discuss the determination of the Fresnel reflection and transmission coefficients for an obliquely incident wave on a dielectric interface.

R10.5. What is total internal reflection? Discuss the nature of the reflection coefficient and the manner in which the boundary condition is satisfied for angle of incidence greater than the critical angle for total internal reflection?

R10.6. What is the Brewster angle? What is the polarization of the reflected wave for an elliptically polarized wave incident on a dielectric interface at the Brewster angle? Discuss an application of the Brewster angle effect.

R10.7. Discuss the principle of optical waveguides by considering the dielectric slab waveguide.

R10.8. Explain the self-consistency condition for waveguiding in a dielectric slab waveguide.

R10.9. Discuss the dependence of the number of propagating modes in a dielectric slab waveguide upon the ratio of the thickness d of the dielectric slab to the wavelength λ_0.

R10.10. Considering TE modes in a dielectric slab guide, specify the fundamental mode and discuss the associated cutoff condition.

R10.11. Outline the considerations that come into play in deriving the field expressions for the modes in a dielectric slab guide.

R10.12. Discuss the mode designation for a dielectric slab guide with reference to the field variations in the guide. Further discuss the behavior of the even and odd modes as the situation changes from near cutoff to far from cutoff.

R10.13. Discuss the concept of effective boundary for waveguiding along a dielectric slab guide.

R10.14. Explain radiation modes with reference to a dielectric slab guide.

R10.15. What is geometrical optics approximation? Under what conditions is it valid?

R10.16. Outline the formulation of the procedure for ray tracing in a plane-stratified medium of continuously varying refractive index.

R10.17. What is paraxial ray approximation? Discuss paraxial rays in a medium of parabolic index profile.

R10.18. Outline the derivation of the self-consistency condition for wave guidance in a graded-index dielectric slab guide.

R10.19. Compare and contrast the ray trajectories associated with modes in a graded-index guide with those associated with modes in a step-index guide.

R10.20. What is intermodal dispersion? Why is it minimized for the case of paraxial rays in a parabolic index guide?

R10.21. Provide a brief description of the optical fiber.

R10.22. Outline the steps involved in obtaining the guidance condition for a wave along a cylindrical dielectric rod surrounded by a cladding region extending to infinity.

R10.23. What are hybrid modes? Why do they arise for guided waves in an optical fiber but not for those in a cylindrical metallic waveguide?

R10.24. Discuss the condition for single-mode operation of an optical fiber.

R10.25. What is interference? Under what condition(s) do two waves not produce interference?

R10.26. Describe Young's two-beam interference experiment.

R10.27. Discuss the phenomenon of multiple-beam interference resulting from the incidence of a uniform plane wave obliquely on a plane dielectric slab and its application to the Fabry–Perot etalon.

R10.28. What is diffraction? Compare and contrast the phenomenon of diffraction with the principle of geometrical optics.

R10.29. Describe the Huygens–Fresnel principle for the solution of a diffraction problem.

R10.30. Using the example of diffraction along the axis of a circular aperture in a plane screen, discuss Fresnel versus Fraunhofer diffraction.

R10.31. When does a wave propagate in an anisotropic medium without change in its polarization?

R10.32. Discuss the principle behind wave plates, providing specific examples.

R10.33. What is Faraday rotation? When does Faraday rotation take place in an anisotropic medium?

R10.34. Consult appropriate reference books and list three applications of Faraday rotation.

R10.35. What is a magneto-optical switch? Discuss its operation.

PROBLEMS

P10.1. In Example 10.1, assume that

$$\mathbf{E}_i = E_0(\mathbf{i}_x - \mathbf{i}_z) \cos [6\pi \times 10^8 t - \sqrt{2}\pi(x + z)]$$

and the angle of incidence is 45°. Obtain the expressions for the electric fields of the reflected and transmitted waves.

P10.2. Repeat Problem P10.1 for

$$\mathbf{E}_i = E_0\mathbf{i}_y \cos [6\pi \times 10^8 t - \sqrt{2}\pi(x + z)]$$

P10.3. In Example 10.1, assume that the permittivity ε_2 of medium 2 is unknown and that

$$\mathbf{E}_i = E_0\left(\frac{\sqrt{3}}{2}\mathbf{i}_x - \frac{1}{2}\mathbf{i}_z\right)\cos\,[6\pi \times 10^9 t - 10\pi(x + \sqrt{3}z)]$$
$$+ E_0\mathbf{i}_y \sin\,[6\pi \times 10^9 t - 10\pi(x + \sqrt{3}z)]$$

(a) Find the value of ε_2 for which the reflected wave is linearly polarized.
(b) For the value of ε_2 found in (a), find the expressions for the reflected and transmitted wave electric fields.

P10.4. In Example 10.1, assume that $\varepsilon_2 = 6\varepsilon_0$ and

$$\mathbf{E}_i = \left(\frac{\sqrt{3}}{2}\mathbf{i}_x - \frac{1}{2}\mathbf{i}_z\right)E_0\cos\,[6\pi \times 10^9 t - 10\pi(x + \sqrt{3}z)]$$
$$+ aE_0\mathbf{i}_y \sin\,[6\pi \times 10^9 t - 10\pi(x + \sqrt{3}z)]$$

Find the value(s) of a for each of the following cases: **(a)** the reflected wave is right circularly polarized; **(b)** the transmitted wave is left circularly polarized; and **(c)** the axial ratio (ratio of the major axis to the minor axis) of the polarization ellipse for the reflected wave electric field is the same as that for the transmitted wave electric field.

P10.5. Show that for a wave incident on a plane boundary between two dielectric media at the Brewster angle, the angle of transmission is equal to 90° minus the Brewster angle. Provide a geometric interpretation for this result based on power flow.

P10.6. For oblique incidence of a uniform plane wave on a dielectric interface, show that the Fresnel reflection and transmission coefficients are consistent with the condition that for power flow normal to the interface, the sum of the reflected power and the transmitted power be equal to the incident power, for each of the two cases: **(a)** perpendicular polarization and **(b)** parallel polarization.

P10.7. A thin film waveguide employed in integrated optics consists of a substrate upon which a thin film of refractive index (c/v_p) greater than that of the substrate is deposited. The medium above the film is air. For relative permittivities of the substrate and the film equal to 2.25 and 2.4, respectively, find the minimum bouncing angle of total internally reflected waves in the film. Assume $\mu = \mu_0$ for both substrate and film.

P10.8. For a symmetric dielectric slab waveguide, $\varepsilon_1 = 2.25\varepsilon_0$ and $\varepsilon_2 = \varepsilon_0$. **(a)** Find the number of propagating TE modes for $d/\lambda_0 = 10$. **(b)** Find the maximum value of d/λ_0 for which the waveguide supports only one TE mode.

P10.9. Design a symmetric dielectric slab waveguide, with $\varepsilon_{r1} = 2.25$ and $\varepsilon_{r2} = 2.13$, by finding the value of d/λ_0 such that the TE_1 mode operates at 20 percent above its cutoff frequency.

P10.10. Consider the derivation of the characteristic equation for guiding of waves in the symmetric dielectric slab waveguide for the case of parallel polarization, which corresponds to TM modes. Noting that in Fig. 10.3, $H_r/H_i = E_r/E_i = -\Gamma_\parallel$, where $\Gamma_\parallel$ is given by (10.17a), show that the characteristic equation is given by

$$\tan\,[f(\theta_i)] = \begin{cases} g(\theta_i), & m = 0, 1, 2, \ldots \\ -\dfrac{1}{g(\theta_i)}, & m = 1, 3, 5, \ldots \end{cases}$$

where

$$f(\theta_i) = \frac{\pi d\sqrt{\varepsilon_{r1}}}{\lambda_0}\cos\theta_i$$

$$g(\theta_i) = \frac{\sqrt{\sin^2\theta_i - (\varepsilon_2/\varepsilon_1)}}{(\varepsilon_2/\varepsilon_1)\cos\theta_i}$$

P10.11. For an asymmetric dielectric slab waveguide made up of a dielectric slab of thickness d and permittivity ε_1, sandwiched between two dielectric media of permittivities ε_2 ($< \varepsilon_1$) and ε_3 ($< \varepsilon_1, \varepsilon_2$), show that the characteristic equation for guiding of TE waves is given by

$$\tan[f(\theta_i)] = g(\theta_i), \qquad m = 0, 1, 2, \ldots$$

where

$$f(\theta_i) = \frac{2\pi d\sqrt{\varepsilon_{r1}}}{\lambda_0}\cos\theta_i$$

$$g(\theta_i) = \frac{\cos\theta_i\left[\sqrt{\sin^2\theta_i - (\varepsilon_2/\varepsilon_1)} + \sqrt{\sin^2\theta_i - (\varepsilon_3/\varepsilon_1)}\right]}{\cos^2\theta_i - \sqrt{[\sin^2\theta_i - (\varepsilon_2/\varepsilon_1)][\sin^2\theta_i - (\varepsilon_3/\varepsilon_1)]}}$$

and

$$\sin^{-1}\sqrt{\varepsilon_2/\varepsilon_1} \le \theta_i < \pi/2$$

P10.12. Repeat Problem P10.11 for TM modes to show that the characteristic equation for guiding of TM waves is given by

$$\tan[f(\theta_i)] = g(\theta_i), \quad m = 0, 1, 2, \ldots$$

where

$$f(\theta_i) = \frac{2\pi d\sqrt{\varepsilon_{r1}}}{\lambda_0}\cos\theta_i$$

$$g(\theta_i) = \frac{\cos\theta_i\left[(\varepsilon_3/\varepsilon_1)\sqrt{\sin^2\theta_i - (\varepsilon_2/\varepsilon_1)} + (\varepsilon_2/\varepsilon_1)\sqrt{\sin^2\theta_i - (\varepsilon_3/\varepsilon_1)}\right]}{(\varepsilon_2\varepsilon_3/\varepsilon_1^2)\cos^2\theta_i - \sqrt{[\sin^2\theta_i - (\varepsilon_2/\varepsilon_1)][\sin^2\theta_i - (\varepsilon_3/\varepsilon_1)]}}$$

and

$$\sin^{-1}\sqrt{\varepsilon_2/\varepsilon_1} \le \theta_i < \pi/2$$

P10.13. For an asymmetric dielectric slab waveguide, the field solutions are neither even nor odd. Hence, considering TE modes, the derivation of the field expressions begins with the solution for $\bar{E}_y$ of the form

$$\bar{E}_y = \begin{cases} \bar{A}\cos(\beta_{x1}x + \psi)e^{-j\beta_z z} & \text{for } |x| < d/2 \\ \bar{B}_1 e^{-\alpha_{x2}x}e^{-j\beta_z z} & \text{for } x > d/2 \\ \bar{B}_2 e^{-\alpha_{x3}x}e^{-j\beta_z z} & \text{for } x < -d/2 \end{cases}$$

where $\varepsilon = \varepsilon_1$ for $|x| < d/2$, $\varepsilon = \varepsilon_2$ ($< \varepsilon_1$) for $x > d/2$, and $\varepsilon = \varepsilon_3$ ($< \varepsilon_1, \varepsilon_2$) for $x < -d/2$. Derive the field expressions and show that they lead to the characteristic equation for guidance the same as that given in Problem P10.11.

P10.14. The ratio of the power associated with the region of the slab ($|x| < d/2$) to the total power ($-\infty < x < \infty$) is known as the *power confinement factor*. Show that for a given frequency the power confinement factor is the highest for the dominant mode. Find the power confinement factors for the three modes in Example 10.3.

P10.15. Beginning with an expression for $\bar{H}_y$ analogous to that of $\bar{E}_y$ for even TE modes given by (10.33), derive the field expressions for even TM modes for the symmetric dielectric slab waveguide and obtain the characteristic equation for guidance.

P10.16. Repeat Problem P10.15 for odd TM modes, beginning with an expression for $\overline{H}_y$ analogous to that of $\overline{E}_y$ for odd TE modes given by (10.44).

P10.17. The laws of geometrical optics can be derived from Fermat's principle which states that the optical path length $\int_A^B n\, ds$ of a ray of light from point A to point B is an extremum, so that the variation in the optical path length, $\delta \int_A^B n\, ds$, is equal to zero. Derive from this property the laws of reflection and refraction for oblique incidence on a plane boundary between two different perfect dielectric media.

P10.18. Consider ray tracing in spherical geometry, as shown in Fig. 10.33, in which n is a function of r, the radial distance from the center of a spherical interface (radius r_0). A ray is incident on the interface at an angle θ_0 from a medium of uniform refractive index n_0. Show that the ray path for $r > r_0$ is governed by a modified form of Snell's law given by

$$nr \sin \theta = \text{constant} = n_0 r_0 \sin \theta_0$$

and hence by the solution of the differential equation

$$\frac{d\theta}{dr} = \frac{n_0 r_0 \sin \theta}{r\sqrt{n^2 r^2 - n_0^2 r_0^2 \sin^2 \theta_0}}$$

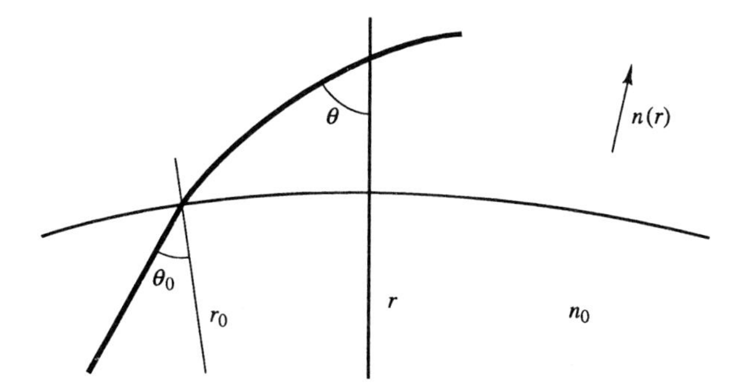

Figure 10.33. For Problem P10.18.

P10.19. In Example 10.5, assume

$$n(x) = n_0(1 - \alpha|x|)$$

Obtain the solution for $z(x)$ for paraxial rays and find the approximate peak amplitude and the approximate pitch in terms of δ_0.

P10.20. Show that the guidance condition for the graded index guide for the TM case is the same as that given by (10.67) or (10.68) for the TE case by showing that $\overline{\Gamma}_a$ for the TM case is $1\underline{/-\pi/2}$. Note that $\overline{\Gamma}$ to be used for the TM case is $-\Gamma_\parallel$ (see Problem P10.10), where $\overline{\Gamma}_\parallel$ is given by (10.17a).

P10.21. In Example 10.6, assume

$$n_1^2(x) = n_0^2(1 - \alpha|x|)$$

Obtain the expression for x_a for the mth mode.

P10.22. Supply the missing steps in the derivation of the characteristic equation (10.79) for the case of $l \neq 0$ from the boundary conditions $\overline{E}_{z1} = \overline{E}_{z2}$, $\overline{H}_{z1} = \overline{H}_{z2}$, $\overline{E}_{z1}/\overline{H}_{\phi 1} = \overline{E}_{z2}/\overline{H}_{\phi 2}$, and $\overline{E}_{\phi 1}/\overline{H}_{z1} = \overline{E}_{\phi 2}/\overline{H}_{z2}$ at $r = a$.

P10.23. Show that the guidance conditions (10.77)–(10.79) are consistent with the boundary conditions for r-components of the fields given by (10.75c), (10.75d), (10.76c), and (10.76d).

P10.24. Assume that a wave is incident from air onto the core of an optical fiber at an angle θ_a, as shown by the cross-sectional view in Fig. 10.34. Show that the maximum allowable value of θ_a for guiding of the wave in the core by total internal reflection is given by

$$\sin\,[\theta_a]_{\max} = \sqrt{n_1^2 - n_2^2}$$

which is defined to be the numerical aperture (NA) of the optical fiber. This provides the physical interpretation for NA.

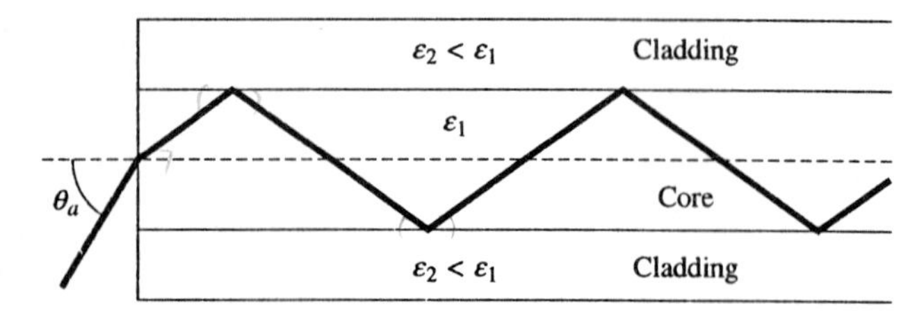

Figure 10.34. For Problem P10.24.

P10.25. Supply the missing steps in the derivation of (10.84) from (10.79) by using (10.83a)–(10.83d). [*Hint:* Use (10.83b) and (10.83c) for the left side of (10.79) and (10.83a) and (10.83d) for the right side of (10.79).]

P10.26. For an optical fiber with core and cladding refractive indices $n_1 = 1.50$ and $n_2 = 1.40$, find the maximum value of a/λ_0 for single-mode operation.

P10.27. An alternate arrangement to Young's experiment of Fig. 10.24 is that of Lloyd's mirror, shown in Fig. 10.35. It consists of a source S_1, placed some distance away from a plane mirror M. Direct light from the source interfering with reflected light, appearing to be coming from an image source S_2, produces fringes as shown in the figure. Since the distance d from the plane of the mirror to the source is very small, the reflection can be thought of as occurring at nearly grazing incidence. Show that the intensity variation on the screen is approximately in accordance with $\sin^2\left(\frac{2\pi xd}{\lambda D}\right)$. (*Note:* A phase change of π occurs at the mirror surface for the reflected light, since the reflection coefficient is -1.)

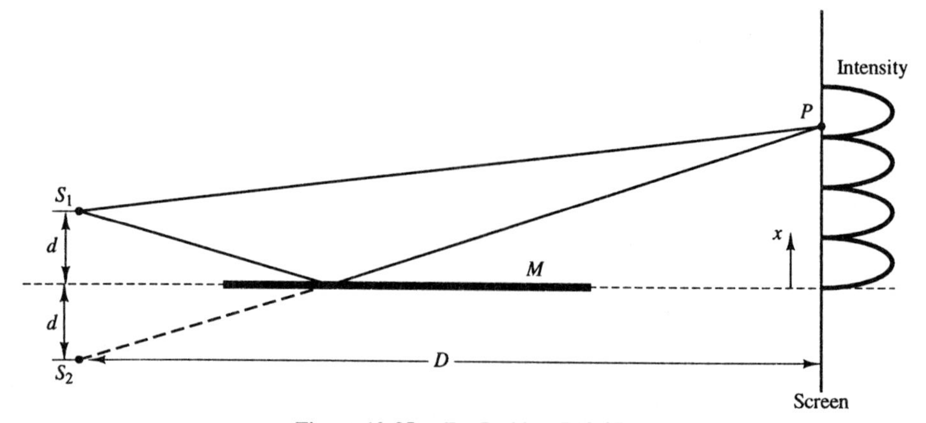

Figure 10.35. For Problem P10.27.

P10.28. Verify that the Fresnel coefficients given in Section 10.1 satisfy (10.96a) and (10.96b) for both cases of the polarization of $\mathbf{E}_i$: **(a)** perpendicular and **(b)** parallel. Further show that (10.96b) is consistent with conservation of power flow normal to the dielectric slab.

P10.29. Assume that the limiting frequency resolution of the Fabry–Perot etalon is defined as the separation between the two frequencies at which I_t/I_i is $\frac{1}{2}$. Then, assuming further that this frequency difference is small compared to the frequency range, $\Delta f = c/(2nd \cos \theta)$, between two adjacent peaks of the interference pattern for fixed d and θ_t, show that it is equal to $\Delta f/F$.

P10.30. Consider the arrangement complementary to that in Example 10.9, that is, a circular disk of radius a having its center at the origin and illuminated by a plane wave of electric field intensity E_0. Obtain the expression for the diffracted field along the axis of the disk and show that the intensity is a constant, independent of distance from the center of the disk. (*Hint:* Use the fact that for two complementary screens placed together in the same plane, no aperture exits so that no diffraction results.)

P10.31. The boundary between the Fresnel and Fraunhofer diffraction regions is determined by the maximum allowable departure of the phase of the waves emanating from the aperture and arriving at the observation point (and hence, vice versa) from that of a plane wave. For a maximum allowable departure of $\pi/8$, which is the value used in practice, show by considering the circular aperture that the boundary between the two regions is $2D^2/\lambda$, where D is the diameter of the hole.

P10.32. For all three cases in Example 10.10, find the expressions for the incident, reflected, and transmitted wave magnetic fields.

P10.33. Medium 1 ($z < 0$) is free space, whereas medium 2 ($z > 0$) is a nonmagnetic ($\mu = \mu_0$), anisotropic perfect dielectric characterized by

$$[\varepsilon] = \varepsilon_0 \begin{bmatrix} 6.25 & 0 & 0 \\ 0 & 2.25 & 0 \\ 0 & 0 & 6.25 \end{bmatrix}$$

For a uniform plane wave having the electric field

$$\mathbf{E}_i = E_0[\cos(6\pi \times 10^8 t - 2\pi z)\,\mathbf{i}_x + \sin(6\pi \times 10^8 t - 2\pi z)\,\mathbf{i}_y]$$

incident on the interface $z = 0$ from medium 1, find the following: **(a)** the reflected wave electric and magnetic fields; and **(b)** the transmitted wave electric and magnetic fields.

P10.34. Medium 1 ($z < 0$) is free space, whereas medium 2 ($z > 0$) is a nonmagnetic ($\mu = \mu_0$), anisotropic perfect dielectric characterized by

$$[\varepsilon] = \varepsilon_0 \begin{bmatrix} 8 & 2 & 0 \\ 2 & 5 & 0 \\ 0 & 0 & 4 \end{bmatrix}$$

A uniform plane wave having the electric field

$$\mathbf{E}_i = E_0(\mathbf{i}_x - 2\mathbf{i}_y)\cos(6\pi \times 10^9 t - 20\pi z)$$

is incident on the interface $z = 0$ from medium 1. **(a)** Write the expression for the incident wave magnetic field. **(b)** Noting that $\mathbf{E}_i$ corresponds to one of the characteristic polarizations for the anisotropic dielectric, obtain the reflected and transmitted wave electric and magnetic fields.

P10.35. Assume that the incident wave electric field in Problem P10.34 is given by

$$\mathbf{E}_i = E_0 \cos(6\pi \times 10^9 t - 20\pi z)\,\mathbf{i}_x$$

Obtain the reflected and transmitted wave electric and magnetic fields. (*Hint:*

Express $\mathbf{E}_i$ in terms of the two characteristic polarizations for the anisotropic dielectric for $\mathbf{E} = E_x \mathbf{i}_x + E_y \mathbf{i}_y$ and use superposition.)

P10.36. Show that for $\gamma B_0 \ll \varepsilon/\varepsilon_0$, the Faraday rotating power $(\beta_- - \beta_+)/2$ of the medium characterized by (10.108) is approximately equal to $-\dfrac{\pi \gamma B_0}{\lambda_0 \sqrt{\varepsilon/\varepsilon_0}}$.

PC EXERCISES

PC10.1. Consider the computation of allowed values of angle of incidence for TE guided modes in an asymmetric dielectric slab waveguide. Using the result of Problem 10.11, write a program to compute these angles for specified values of ε_{r1}, ε_{r2}, ε_{r3}, thickness d of the slab in millimeters, and the free-space wavelength λ_0 of the waves in millimeters.

PC10.2. Run the program of PL10.2 for a refractive index profile and values of d and λ_0 specified by your instructor.

11

Antennas

In the preceding five chapters we studied the principles of propagation and transmission of electromagnetic waves. The remaining important topic pertinent to electromagnetic wave phenomena is radiation of electromagnetic waves. We have, in fact, touched on the principle of radiation of electromagnetic waves in Chapter 6 when we derived the electromagnetic field due to the infinite plane sheet of time-varying, spatially uniform current density. We learned that the current sheet gives rise to uniform plane waves radiating away from the sheet to either side of it. We pointed out at that time that the infinite plane current sheet is, however, an idealized, hypothetical source. With the experience gained thus far in our study of the elements of engineering electromagnetics, we are now in a position to learn the principles of radiation from physical antennas, which is our goal in this chapter.

We begin the chapter with the derivation of the electromagnetic field due to an elemental wire antenna, known as the *Hertzian dipole*. After studying the radiation characteristics of the Hertzian dipole, we consider the example of a half-wave dipole to illustrate the use of superposition to represent an arbitrary wire antenna as a series of Hertzian dipoles to determine its radiation fields. We also discuss the principles of arrays of physical antennas and the concept of image antennas to take into account ground effects. Next we study radiation from aperture antennas. Finally, we consider briefly the receiving properties of antennas and learn of their reciprocity with the radiating properties.

11.1 HERTZIAN DIPOLE

The *Hertzian dipole* is an elemental antenna consisting of an infinitesimally long piece of wire carrying an alternating current $I(t)$, as shown in Fig. 11.1. To maintain the current flow in the wire, we postulate two point charges $Q_1(t)$ and $Q_2(t)$

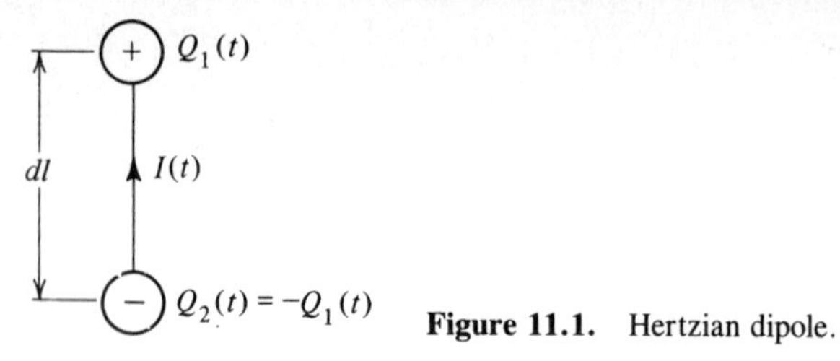

Figure 11.1. Hertzian dipole.

terminating the wire at its two ends, so that the law of conservation of charge is satisfied. Thus if

$$\boxed{I(t) = I_0 \cos \omega t} \tag{11.1}$$

then

$$\frac{dQ_1}{dt} = I(t) = I_0 \cos \omega t \tag{11.2a}$$

$$\frac{dQ_2}{dt} = -I(t) = -I_0 \cos \omega t \tag{11.2b}$$

and

$$Q_1(t) = \frac{I_0}{\omega} \sin \omega t \tag{11.3a}$$

$$Q_2(t) = -\frac{I_0}{\omega} \sin \omega t = -Q_1(t) \tag{11.3b}$$

The time variations of I, Q_1, and Q_2, given by (11.1), (11.3a) and (11.3b), respectively, are illustrated by the curves and the series of sketches for the dipoles in Fig. 11.2, corresponding to one complete period. The different sizes of the arrows associated with the dipoles denote the different strengths of the current whereas the number of the plus or minus signs is indicative of the strength of the charges.

To determine the electromagnetic field due to the Hertzian dipole, we consider the dipole to be situated at the origin and oriented along the z-axis, in a perfect dielectric medium. We shall use an approach based upon the magnetic vector potential and obtain electric and magnetic fields consistent with Maxwell's equations, while fulfilling certain other pertinent requirements. We shall begin with the magnetic vector potential for the static case and then extend it to the time-varying current element. To do this, we recall from Section 4.5 that for a current element of length $d\mathbf{l} = dl\, \mathbf{i}_z$ situated at the origin, as shown in Fig. 11.3 and carrying current I, the magnetic field at a point $P(r, \theta, \phi)$ is given by

$$\boxed{\mathbf{A} = \frac{\mu I\, d\mathbf{l}}{4\pi r} = \frac{\mu I\, dl}{4\pi r}\mathbf{i}_z} \tag{11.4}$$

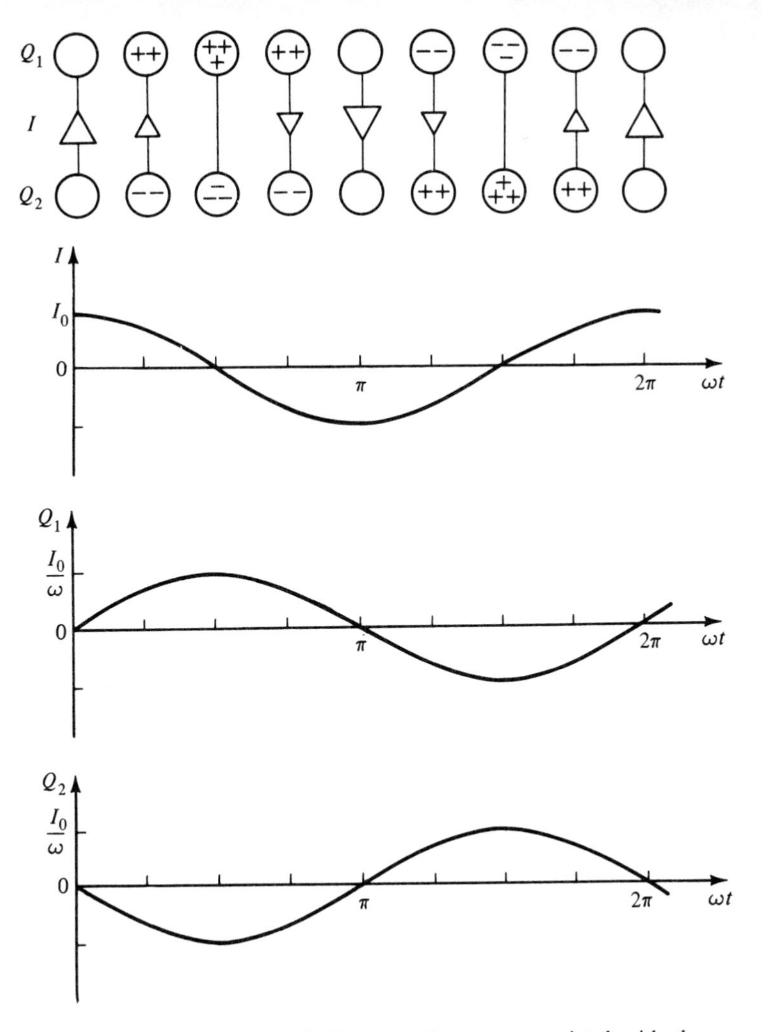

Figure 11.2. Time variations of charges and current associated with the Hertzian dipole.

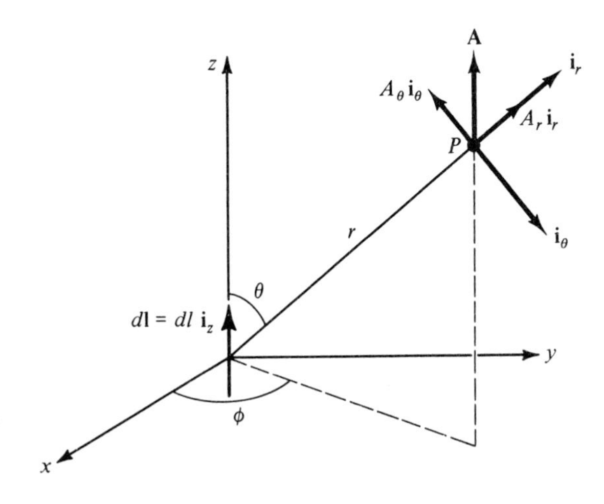

Figure 11.3. For finding the magnetic vector potential due to an infinitesimal current element.

Retarded potential

If the current in the element is now assumed to be time varying in the manner $I = I_0 \cos \omega t$, we might expect the corresponding magnetic vector potential to be that in (11.4) with I replaced by $I_0 \cos \omega t$. Proceeding in this manner would however lead to fields inconsistent with Maxwell's equations. The reason is that time-varying electric and magnetic fields give rise to wave propagation, according to which the effect of the source current at a given value of time is felt at a distance r from the origin after a time delay of r/v_p, where v_p is the velocity of propagation of the wave. Conversely, the effect felt at a distance r from the origin at time t is due to the current which existed at the origin at an earlier time $(t - r/v_p)$. Thus for the time-varying current element $I_0\, dl \cos \omega t\, \mathbf{i}_z$ situated at the origin, the magnetic vector potential is given by

$$\boxed{\begin{aligned}\mathbf{A} &= \frac{\mu I_0\, dl}{4\pi r} \cos \omega\left(t - \frac{r}{v_p}\right) \mathbf{i}_z \\ &= \frac{\mu I_0\, dl}{4\pi r} \cos (\omega t - \beta r)\, \mathbf{i}_z\end{aligned}} \tag{11.5}$$

where we have replaced ω/v_p by β, the phase constant. The result given by (11.5) is known as the *retarded* magnetic vector potential in view of the phase-lag factor βr contained in it.

To augment the reasoning behind the retarded magnetic vector potential, recall that in Section 4.4 we derived differential equations for the electromagnetic potentials. For the magnetic vector potential, we obtained

$$\nabla^2 \mathbf{A} - \mu\varepsilon \frac{\partial^2 \mathbf{A}}{\partial t^2} = -\mu \mathbf{J} \tag{11.6}$$

which reduces to

$$\nabla^2 A_z - \mu\varepsilon \frac{\partial^2 A_z}{\partial t^2} = -\mu J_z \tag{11.7}$$

for $\mathbf{A} = A_z \mathbf{i}_z$ and $\mathbf{J} = J_z \mathbf{i}_z$. Equation (11.7) has the form of the wave equation, except in three dimensions and with the source term on the right side. Thus the solution for A_z must be of the form of a traveling wave while reducing to the static field case for no time variations.

Fields due to Hertzian dipole

Expressing $\mathbf{A}$ in (11.5) in terms of its components in spherical coordinates, as shown in Fig. 11.3, we obtain

$$\mathbf{A} = \frac{\mu I_0\, dl \cos (\omega t - \beta r)}{4\pi r} (\cos \theta\, \mathbf{i}_r - \sin \theta\, \mathbf{i}_\theta) \tag{11.8}$$

The magnetic field due to the Hertzian dipole is then given by

$$\mathbf{H} = \frac{\mathbf{B}}{\mu} = \frac{1}{\mu} \nabla \times \mathbf{A}$$

$$= \frac{1}{\mu} \begin{vmatrix} \dfrac{\mathbf{i}_r}{r^2 \sin \theta} & \dfrac{\mathbf{i}_\theta}{r \sin \theta} & \dfrac{\mathbf{i}_\phi}{r} \\ \dfrac{\partial}{\partial r} & \dfrac{\partial}{\partial \theta} & \dfrac{\partial}{\partial \phi} \\ A_r & rA_\theta & 0 \end{vmatrix} = \frac{1}{\mu r}\left[\frac{\partial}{\partial r}(rA_\theta) - \frac{\partial A_r}{\partial \theta}\right]$$

or

$$\mathbf{H} = \frac{I_0\, dl \sin\theta}{4\pi}\left[\frac{\cos(\omega t - \beta r)}{r^2} - \frac{\beta \sin(\omega t - \beta r)}{r}\right]\mathbf{i}_\phi \tag{11.9}$$

Using Maxwell's curl equation for **H** with **J** set equal to zero in view of perfect dielectric medium, we then have

$$\frac{\partial \mathbf{E}}{\partial t} = \frac{1}{\varepsilon}\nabla \times \mathbf{H}$$

$$= \frac{1}{\varepsilon}\begin{vmatrix} \dfrac{\mathbf{i}_r}{r^2 \sin\theta} & \dfrac{\mathbf{i}_\theta}{r \sin\theta} & \dfrac{\mathbf{i}_\phi}{r} \\ \dfrac{\partial}{\partial r} & \dfrac{\partial}{\partial \theta} & \dfrac{\partial}{\partial \phi} \\ 0 & 0 & r \sin\theta\, H_\phi \end{vmatrix}$$

$$= \frac{1}{\varepsilon r^2 \sin\theta}\frac{\partial}{\partial\theta}(r\sin\theta\, H_\phi)\,\mathbf{i}_r - \frac{1}{\varepsilon r \sin\theta}\frac{\partial}{\partial r}(r \sin\theta\, H_\phi)\,\mathbf{i}_\theta$$

or

$$\begin{aligned}\mathbf{E} = {} & \frac{2 I_0\, dl \cos\theta}{4\pi\varepsilon\omega}\left[\frac{\sin(\omega t - \beta r)}{r^3} + \frac{\beta \cos(\omega t - \beta r)}{r^2}\right]\mathbf{i}_r \\ & + \frac{I_0\, dl \sin\theta}{4\pi\varepsilon\omega}\left[\frac{\sin(\omega t - \beta r)}{r^3} + \frac{\beta \cos(\omega t - \beta r)}{r^2} \right. \\ & \left. - \frac{\beta^2 \sin(\omega t - \beta r)}{r}\right]\mathbf{i}_\theta \end{aligned} \tag{11.10}$$

Equations (11.10) and (11.9) represent the electric and magnetic fields, respectively, due to the Hertzian dipole. The following observations are pertinent to these field expressions:

1. They satisfy the two Maxwell's curl equations. In fact, we have obtained (11.10) from (11.9) by using the curl equation for **H**. The reader is urged to verify that (11.9) follows from (11.10) through the curl equation for **E**.
2. They contain terms involving $1/r^3$, $1/r^2$, and $1/r$. Far from the dipole such that $\beta r \gg 1$, the $1/r^3$ and $1/r^2$ terms are negligible compared to the $1/r$ terms so that the fields vary inversely with r. Furthermore, for any value of r, the time-average value of the θ-component of the Poynting vector due to the fields is zero, and the contribution to the time-average value of the r-component is completely from the $1/r$ terms (see Problem P11.2). Thus the time-average Poynting vector varies proportionately to $1/r^2$ and is directed entirely in the radial direction. This is consistent with the physical requirement that for the time-average power crossing all possible spherical surfaces centered at the dipole to be the same, the power density must be inversely proportional to r^2, since the surface areas of the spherical surfaces are proportional to the squares of their radii.

3. For $\beta r \ll 1$, the $1/r^3$ terms dominate the $1/r^2$ terms which in turn dominate the $1/r$ terms. Also, $\sin(\omega t - \beta r) \approx (\sin \omega t - \beta r \cos \omega t)$ and $\cos(\omega t - \beta r) \approx (\cos \omega t + \beta r \sin \omega t)$, so that

$$\mathbf{E} \approx \frac{I_0\, dl \sin \omega t}{4\pi \varepsilon \omega r^3}(2 \cos \theta\, \mathbf{i}_r + \sin \theta\, \mathbf{i}_\theta) \tag{11.11}$$

$$\mathbf{H} \approx \frac{I_0\, dl \cos \omega t}{4\pi r^2} \sin \theta\, \mathbf{i}_\phi \tag{11.12}$$

Equation (11.11) is the same as (4.102) with Q replaced by $(I_0/\omega) \sin \omega t$, that is, $Q_1(t)$ in Fig. 11.1, and d replaced by dl. Equation (11.12) gives the same $\mathbf{B}$ as the magnetic field given by Biot–Savart law applied to a current element $I\, dl\, \mathbf{i}_z$ at the origin and then I replaced by $I_0 \cos \omega t$, that is, $I(t)$ in Fig. 11.1. Thus electrically close to the dipole, where retardation effects are negligible, the field expressions approach toward the corresponding static field expressions with the static source terms simply replaced by the time-varying source terms.

Example 11.1

Let us consider in free space a Hertzian dipole of length 0.1 m situated at the origin and along the z-axis, carrying the current $10 \cos 2\pi \times 10^7 t$ A. We wish to obtain the electric and magnetic fields at the point $(5, \pi/6, 0)$.

For convenience in computation of the amplitudes and phase angles of the field components, we shall express the field components in phasor form. Thus replacing $\cos(\omega t - \beta r)$ by $e^{-j\beta r}$ and $\sin(\omega t - \beta r)$ by $-je^{-j\beta r}$, we have

$$\begin{aligned} \bar{E}_r &= \frac{2I_0\, dl \cos \theta}{4\pi \varepsilon \omega}\left(-\frac{j}{r^3} + \frac{\beta}{r^2}\right)e^{-j\beta r} \\ &= \frac{2\beta^2 \eta I_0\, dl \cos \theta}{4\pi}\left[-j\frac{1}{(\beta r)^3} + \frac{1}{(\beta r)^2}\right]e^{-j\beta r} \end{aligned} \tag{11.13}$$

$$\begin{aligned} \bar{E}_\theta &= \frac{I_0\, dl \sin \theta}{4\pi \varepsilon \omega}\left(-\frac{j}{r^3} + \frac{\beta}{r^2} + \frac{j\beta^2}{r}\right)e^{-j\beta r} \\ &= \frac{\beta^2 \eta I_0\, dl \sin \theta}{4\pi}\left[-j\frac{1}{(\beta r)^3} + \frac{1}{(\beta r)^2} + j\frac{1}{\beta r}\right]e^{-j\beta r} \end{aligned} \tag{11.14}$$

$$\begin{aligned} \bar{H}_\phi &= \frac{I_0\, dl \sin \theta}{4\pi}\left(\frac{1}{r^2} + \frac{j\beta}{r}\right)e^{-j\beta r} \\ &= \frac{\beta_2 I_0\, dl \sin \theta}{4\pi}\left[\frac{1}{(\beta r)^2} + j\frac{1}{\beta r}\right]e^{-j\beta r} \end{aligned} \tag{11.15}$$

where $\eta = \sqrt{\mu/\varepsilon}$ is the intrinsic impedance of the medium. Using $I_0 = 10$ A, $dl = 0.1$ m, $f = 10^7$ Hz, $\mu = \mu_0$, $\varepsilon = \varepsilon_0$, $r = 5$ m, and $\theta = \pi/6$, and carrying out the computations with the aid of the PC program included as PL 11.1, we obtain

$$\bar{E}_r = 2.8739\underline{/-103.679^\circ} \text{ V/m}$$

$$\bar{E}_\theta = 0.6025\underline{/-54.728^\circ} \text{ V/m}$$

$$\bar{H}_\phi = 0.0023\underline{/-13.679^\circ} \text{ A/m}$$

PL 11.1. Program listing and sample output for computing the fields due to a Hertzian dipole.

```
100 '**************************************************
110 '* COMPUTATION OF FIELDS DUE TO A HERTZIAN DIPOLE *
120 '* ORIENTED ALONG THE Z-AXIS AND LOCATED AT THE   *
130 '* ORIGIN IN FREE SPACE (PHASE ANGLE OF CURRENT   *
140 '* IS ASSUMED TO BE ZERO)                          *
150 '**************************************************
160 PI=3.1416:RD=180/PI
170 SCREEN 0:CLS:PRINT "ENTER VALUES OF INPUT PARAMETER
    S:":PRINT
180 INPUT "CURRENT IN AMPERES = ",I
190 INPUT "LENGTH OF DIPOLE IN METERS = ",L
200 INPUT "FREQUENCY IN MHZ = ",F
210 INPUT "R IN METERS = ",R
220 INPUT "THETA IN DEGREES = ",THETA
230 BETA=PI*F/150:ETA=120*PI:BR=BETA*R
240 C1=BETA*BETA*I*L/(4*PI):CP=C1*SIN(THETA/RD)
250 CT=CP*ETA:CR=2*ETA*C1*COS(THETA/RD)
260 PRINT:PRINT "COMPUTED VALUES ARE:"
270 PRINT:PRINT "R-COMPONENT OF E:"
280 REAL=1/(BR*BR)
290 IMAG=-REAL/BR:C=CR:U$="V/M":GOSUB 370
300 PRINT:PRINT "THETA-COMPONENT OF E:"
310 IMAG=IMAG+1/BR:C=CT:GOSUB 370
320 PRINT:PRINT "PHI-COMPONENT OF H:"
330 IMAG=1/BR:C=CP:U$="A/M":GOSUB 370
340 PRINT:PRINT "STRIKE ANY KEY TO CONTINUE"
350 C$=INPUT$(1):GOTO 170
360 END
370 MAG=SQR(REAL*REAL+IMAG*IMAG)
380 PANG=ATN(IMAG/REAL)-BR
390 IF ABS(PANG)<PI THEN 410
400 PANG=PANG+2*PI:GOTO 390
410 MAG=C*MAG:PANG=PANG*RD
420 PRINT "      MAGNITUDE = ";MAG;U$
430 PRINT "      PHASE ANGLE = ";PANG;"DEG"
440 RETURN
```

```
RUN
ENTER VALUES OF INPUT PARAMETERS:

CURRENT IN AMPERES = 10
LENGTH OF DIPOLE IN METERS = .1
FREQUENCY IN MHZ = 10
R IN METERS = 5
THETA IN DEGREES = 30

COMPUTED VALUES ARE:

R-COMPONENT OF E:
      MAGNITUDE =  2.873907 V/M
      PHASE ANGLE = -103.6791 DEG

THETA-COMPONENT OF E:
      MAGNITUDE =  .6025501 V/M
      PHASE ANGLE = -54.72812 DEG

PHI-COMPONENT OF H:
      MAGNITUDE =  2.304522E-03 A/M
      PHASE ANGLE = -13.67934 DEG

STRIKE ANY KEY TO CONTINUE
```

Thus the required fields are

$$\mathbf{E} = 2.8739 \cos (2\pi \times 10^7 t - 0.576\pi)\, \mathbf{i}_r + 0.6025 \cos (2\pi \times 10^7 t - 0.304\pi)\, \mathbf{i}_\theta \text{ V/m}$$

$$\mathbf{H} = 0.0023 \cos (2\pi \times 10^7 t - 0.076\pi)\, \mathbf{i}_\phi \text{ A/m}$$

K11.1. Hertzian dipole; Retarded magnetic vector potential; Complete electromagnetic field; Behavior far from the dipole ($\beta r \gg 1$); Behavior close to the dipole ($\beta r \ll 1$).

D11.1. Consider a Hertzian dipole of length 0.1λ carrying sinusoidally time-varying current of amplitude 4π A. Find the magnitude of the electric dipole moment for each of the following cases: **(a)** $f = 10$ MHz, medium is free space; **(b)** $f = 100$ kHz, medium is free space; and **(c)** $f = 25$ kHz, medium is sea water ($\sigma = 4$ S/m, $\varepsilon = \varepsilon_0$, and $\mu = \mu_0$).
Ans. **(a)** 6×10^{-7} C-m; **(b)** 6×10^{-3} C-m; **(c)** 8×10^{-5} C-m

D11.2. Three Hertzian dipoles of lengths 1, 1, and 2 m are situated at the origin oriented along the positive x-, y-, and z-axes, respectively, and carrying currents $1 \cos 2\pi \times 10^6 t$, $2 \sin 2\pi \times 10^6 t$, and $2 \cos 2\pi \times 10^6 t$ A, respectively. The medium is free space. Find the following at (0, 0, 50) in Cartesian coordinates: **(a)** E_x; **(b)** E_y; and **(c)** E_z.
Ans. **(a)** $12.051 \cos (2\pi \times 10^6 t + 0.696\pi)$ mV/m; **(b)** $24.102 \cos (2\pi \times 10^6 t + 0.196\pi)$ mV/m; **(c)** $0.1327 \cos (2\pi \times 10^6 t - 0.576\pi)$ V/m

11.2 RADIATION RESISTANCE AND DIRECTIVITY

Radiation fields

In the preceding section we derived the expressions for the complete electromagnetic field due to the Hertzian dipole. These expressions look very complicated. Fortunately, it is seldom necessary to work with the complete field expressions because one is often interested in the field far from the dipole which is governed predominantly by the terms involving $1/r$. Thus from (11.10) and (11.9), we find that for a Hertzian dipole of length dl oriented along the z-axis and carrying current

$$I = I_0 \cos \omega t \tag{11.16}$$

the electric and magnetic fields at values of r *far from the dipole* are given by

$$\mathbf{E} = -\frac{\beta^2 I_0\, dl \sin\theta}{4\pi\varepsilon\omega r} \sin(\omega t - \beta r)\, \mathbf{i}_\theta = -\frac{\eta\beta I_0\, dl \sin\theta}{4\pi r} \sin(\omega t - \beta r)\, \mathbf{i}_\theta \tag{11.17a}$$

$$\mathbf{H} = -\frac{\beta I_0\, dl \sin\theta}{4\pi r} \sin(\omega t - \beta r)\, \mathbf{i}_\phi \tag{11.17b}$$

These fields are known as the *radiation fields*, since they are the components of the total fields that contribute to the time-average radiated power away from the dipole. Before we discuss the nature of these fields, let us find out quantitatively what we mean by *far from the dipole*. To do this, we look at the expression for the complete magnetic field given by (11.9) and note that the ratio of the amplitudes of the $1/r^2$ and $1/r$ terms is equal to $1/\beta r$. Hence for $\beta r \gg 1$, the $1/r^2$ term is negligible compared to the $1/r$ term as already pointed out in the previous section. This means that for $r \gg 1/\beta$, or $r \gg \lambda/2\pi$, that is, even at a distance of a few wavelengths from the dipole, the fields are predominantly radiation fields.

Returning now to the expressions for the radiation fields given by (11.17a) and (11.17b), we note that at any given point, (1) the electric field (E_θ), the magnetic field (H_ϕ), and the direction of propagation (r) are mutually perpendicular and (2) the ratio of E_θ to H_ϕ is equal to η, which are characteristic of uniform plane waves. The phase of the field, however, is uniform over the surfaces $r =$ constant, that is, spherical surfaces centered at the dipole, whereas the amplitude of the field is uniform over surfaces $(\sin\theta)/r =$ constant. Hence the fields are only locally uniform plane waves, that is, over any small area normal to the r-direction at a given point.

The Poynting vector due to the radiation frequency fields is given by

$$\begin{aligned} \mathbf{P} &= \mathbf{E} \times \mathbf{H} \\ &= E_\theta \mathbf{i}_\theta \times H_\phi \mathbf{i}_\phi = E_\theta H_\phi \mathbf{i}_r \\ &= \frac{\eta \beta^2 I_0^2 (dl)^2 \sin^2\theta}{16\pi^2 r^2} \sin^2(\omega t - \beta r)\, \mathbf{i}_r \end{aligned} \tag{11.18}$$

By evaluating the surface integral of the Poynting vector over any surface enclosing the dipole, we can find the power flow out of that surface, that is, the power "radiated" by the dipole. For convenience in evaluating the surface integral, we choose the spherical surface of radius r and centered at the dipole, as shown in Fig. 11.4. Thus noting that the differential surface area on the spherical surface is $(r\,d\theta)(r\sin\theta\,d\phi)\mathbf{i}_r$ or $r^2\sin\theta\,d\theta\,d\phi\,\mathbf{i}_r$, we obtain the instantaneous power radiated to be

$$P_{\text{rad}} = \int_{\theta=0}^{\pi}\int_{\phi=0}^{2\pi} \mathbf{P}\cdot r^2\sin\theta\,d\theta\,d\phi\,\mathbf{i}_r$$

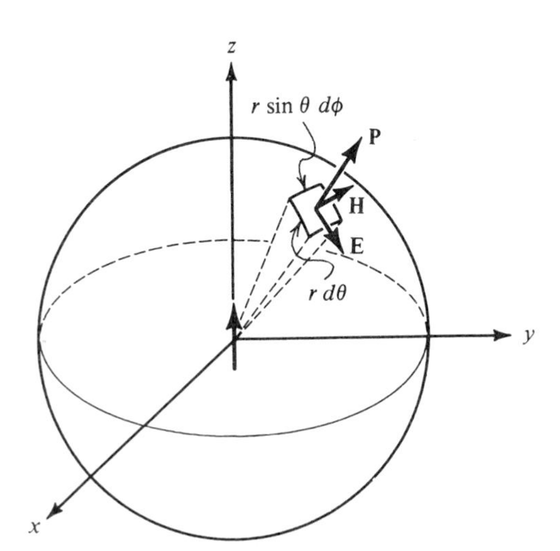

Figure 11.4. For computing the power radiated by the Hertzian dipole.

$$
\begin{aligned}
P_{\text{rad}} &= \int_{\theta=0}^{\pi}\int_{\phi=0}^{2\pi} \frac{\eta\beta^2 I_0^2 (dl)^2 \sin^3\theta}{16\pi^2} \sin^2(\omega t - \beta r)\, d\theta\, d\phi \\
&= \frac{\eta\beta^2 I_0^2 (dl)^2}{8\pi} \sin^2(\omega t - \beta r) \int_{\theta=0}^{\pi} \sin^3\theta\, d\theta \\
&= \frac{\eta\beta^2 I_0^2 (dl)^2}{6\pi} \sin^2(\omega t - \beta r) \\
&= \frac{2\pi\eta I_0^2}{3}\left(\frac{dl}{\lambda}\right)^2 \sin^2(\omega t - \beta r)
\end{aligned} \tag{11.19}
$$

The time-average power radiated by the dipole, that is, the average of P_{rad} over one period of the current variation, is

$$
\begin{aligned}
\langle P_{\text{rad}} \rangle &= \frac{2\pi\eta I_0^2}{3}\left(\frac{dl}{\lambda}\right)^2 \langle \sin^2(\omega t - \beta r) \rangle \\
&= \frac{\pi\eta I_0^2}{3}\left(\frac{dl}{\lambda}\right)^2 \\
&= \frac{1}{2} I_0^2 \left[\frac{2\pi\eta}{3}\left(\frac{dl}{\lambda}\right)^2\right]
\end{aligned} \tag{11.20}
$$

Radiation resistance

We now define a quantity known as the *radiation resistance* of the antenna, denoted by the symbol R_{rad}, as the value of a fictitious resistor that dissipates the same amount of time-average power as that radiated by the antenna when a current of the same peak amplitude as that in the antenna is passed through it. Recalling that the average power dissipated in a resistor R when a current $I_0 \cos \omega t$ is passed through it is $\frac{1}{2}I_0^2 R$, we note from (11.20) that the radiation resistance of the Hertzian dipole is

$$
\boxed{R_{\text{rad}} = \frac{2\pi\eta}{3}\left(\frac{dl}{\lambda}\right)^2 \ \Omega} \tag{11.21}
$$

For free space, $\eta = \eta_0 = 120\pi\ \Omega$, and

$$
\boxed{R_{\text{rad}} = 80\pi^2\left(\frac{dl}{\lambda}\right)^2 \ \Omega} \tag{11.22}
$$

As a numerical example, for (dl/λ) equal to 0.01, $R_{\text{rad}} = 80\pi^2(0.01)^2 = 0.08\ \Omega$. Thus for a current of peak amplitude 1 A, the time-average radiated power is equal to 0.04 W. This indicates that a Hertzian dipole of length 0.01λ is not a very effective radiator.

We note from (11.21) that the radiation resistance and hence the radiated power are proportional to the square of the electrical length, that is, the physical length expressed in terms of wavelength, of the dipole. The result given by (11.21) is, however, valid only for small values of dl/λ since if dl/λ is not small, the amplitude of the current along the antenna can no longer be uniform and its variation must be taken into account in deriving the radiation fields and hence the radiation resistance. We shall do this in the following section for a half-wave dipole, that is, for a dipole of length equal to $\lambda/2$.

Radiation pattern

Let us now examine the directional characteristics of the radiation from the Hertzian dipole. We note from (11.17a) and (11.17b) that, for a constant r, the amplitude of the fields is proportional to $\sin\theta$. Similarly, we note from (11.18) that for a constant r, the power density is proportional to $\sin^2\theta$. Thus an observer wandering on the surface of an imaginary sphere centered at the dipole views different amplitudes of the fields and of the power density at different points on the surface. The situation is illustrated in Fig. 11.5(a) for the power density by attaching to different points on the spherical surface vectors having lengths proportional to the Poynting vectors at those points. It can be seen that the power density is

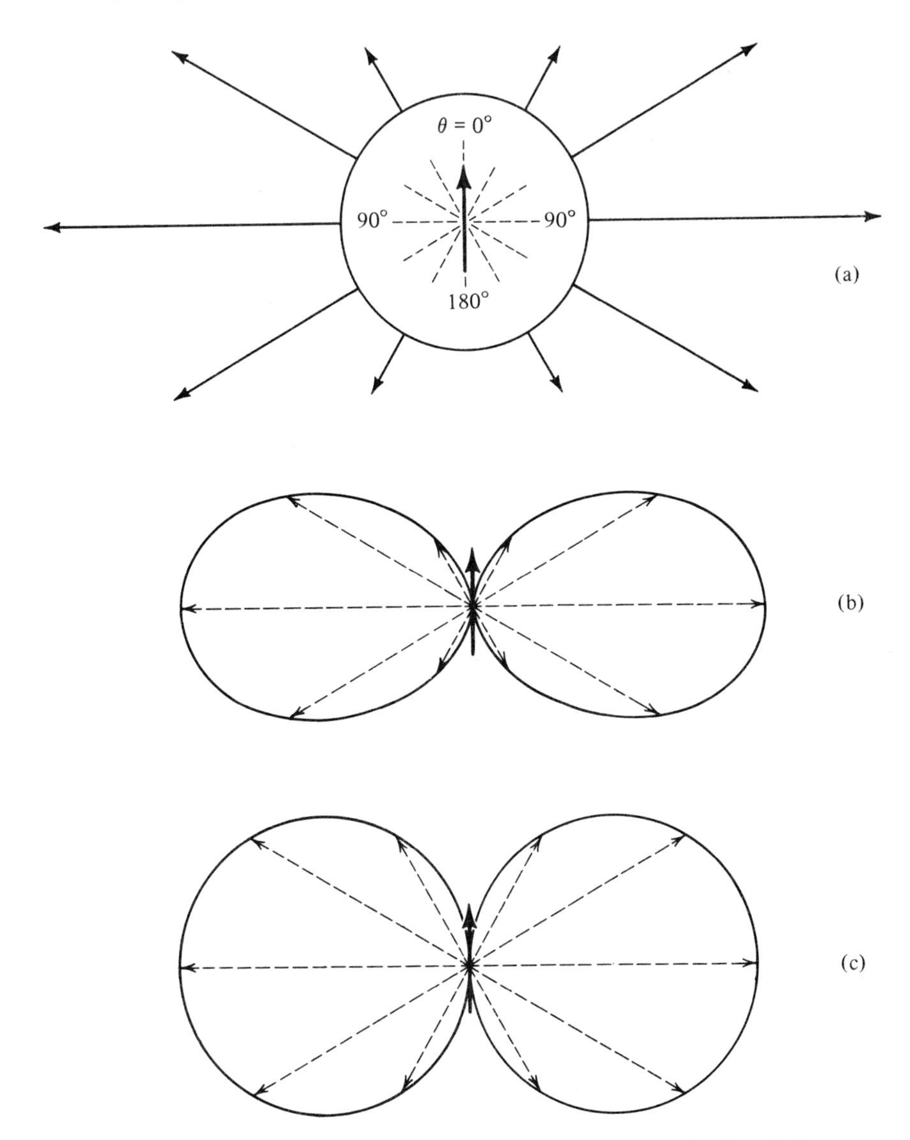

Figure 11.5. Directional characteristics of radiation from the Hertzian dipole.

largest for $\theta = \pi/2$, that is, in the plane normal to the axis of the dipole, and decreases continuously toward the axis of the dipole, becoming zero along the axis.

It is customary to depict the radiation characteristic by means of a *radiation pattern*, as shown in Fig. 11.5(b), which can be imagined to be obtained by shrinking the radius of the spherical surface in Fig. 11.5(a) to zero with the Poynting vectors attached to it and then joining the tips of the Poynting vectors. Thus the distance from the dipole point to a point on the radiation pattern is proportional to the power density in the direction of that point. Similarly, the radiation pattern for the fields can be drawn as shown in Fig. 11.5(c), based upon the $\sin\theta$ dependence of the fields. In view of the independence of the fields from ϕ, the patterns of Fig. 11.5(b) and (c) are valid for any plane containing the axis of the dipole. In fact, the three-dimensional radiation patterns can be imagined to be the figures obtained by revolving these patterns about the dipole axis. For a general case, the radiation may also depend on ϕ, and hence it will be necessary to draw a radiation pattern for the $\theta = \pi/2$ plane. Here, this pattern is merely a circle centered at the dipole.

Directivity

We now define a parameter known as the *directivity* of the antenna, denoted by the symbol D, as the ratio of the maximum power density radiated by the antenna to the average power density. To elaborate on the definition of D, imagine that we take the power radiated by the antenna and distribute it equally in all directions by shortening some of the vectors in Fig. 11.5(a) and lengthening the others so that they all have equal lengths. The pattern then becomes nondirectional, and the power density, which is the same in all directions, will be less than the maximum power density of the original pattern. Obviously, the more directional the radiation pattern of an antenna is, the greater is the directivity.

From (11.18), we obtain the maximum power density radiated by the Hertzian dipole to be

$$\begin{aligned}[P_r]_{\max} &= \frac{\eta\beta^2 I_0^2 (dl)^2 [\sin^2\theta]_{\max}}{16\pi^2 r^2}\sin^2(\omega t - \beta r) \\ &= \frac{\eta\beta^2 I_0^2 (dl)^2}{16\pi^2 r^2}\sin^2(\omega t - \beta r)\end{aligned} \tag{11.23}$$

By dividing the radiated power given by (11.19) by the surface area $4\pi r^2$ of the sphere of radius r, we obtain the average power density to be

$$[P_r]_{\text{av}} = \frac{P_{\text{rad}}}{4\pi r^2} = \frac{\eta\beta^2 I_0^2 (dl)^2}{24\pi^2 r^2}\sin^2(\omega t - \beta r) \tag{11.24}$$

Thus the directivity of the Hertzian dipole is given by

$$\boxed{D = \frac{[P_r]_{\max}}{[P_r]_{\text{av}}} = 1.5} \tag{11.25}$$

To generalize the computation of directivity for an arbitrary radiation pattern, let us consider

$$P_r = \frac{P_0 \sin^2(\omega t - \beta r)}{r^2} f(\theta, \phi) \tag{11.26}$$

where P_0 is a constant and $f(\theta, \phi)$ is the power density pattern. Then

$$[P_r]_{max} = \frac{P_0 \sin^2 (\omega t - \beta r)}{r^2}[f(\theta, \phi)]_{max}$$

$$[P_r]_{av} = \frac{P_{rad}}{4\pi r^2}$$

$$= \frac{1}{4\pi r^2}\int_{\theta=0}^{\pi}\int_{\phi=0}^{2\pi} \frac{P_0 \sin^2 (\omega t - \beta r)}{r^2} f(\theta, \phi)\, \mathbf{i}_r \cdot r^2 \sin\theta \, d\theta \, d\phi \, \mathbf{i}_r$$

$$= \frac{P_0 \sin^2 (\omega t - \beta r)}{4\pi r^2}\int_{\theta=0}^{\pi}\int_{\phi=0}^{2\pi} f(\theta, \phi) \sin\theta \, d\theta \, d\phi$$

$$\boxed{D = 4\pi \frac{[f(\theta, \phi)]_{max}}{\int_{\theta=0}^{\pi}\int_{\phi=0}^{2\pi} f(\theta, \phi) \sin\theta \, d\theta \, d\phi}} \qquad (11.27)$$

Example 11.2

Let us compute the directivity corresponding to the power density pattern function $f(\theta, \phi) = \sin^2\theta \cos^2\theta$.

From (11.27),

$$D = 4\pi \frac{[\sin^2\theta \cos^2\theta]_{max}}{\int_{\theta=0}^{\pi}\int_{\phi=0}^{2\pi} \sin^3\theta \cos^2\theta \, d\theta \, d\phi}$$

$$= 4\pi \frac{[\frac{1}{4}\sin 2\theta]_{max}}{2\pi \int_{\theta=0}^{\pi} (\sin^3\theta - \sin^5\theta) d\theta}$$

$$= \frac{1}{2}\frac{1}{(4/3) - (16/15)}$$

$$= 1\frac{7}{8}$$

The ratio of the power density radiated by the antenna as a function of direction to the average power density is given by $Df(\theta, \phi)$. This quantity is known as the *directive gain of the antenna*. Another useful parameter is the power gain of the antenna, which takes into account the ohmic power losses in the antenna. It is denoted by the symbol G and is proportional to the directive gain, the proportionality factor being the power efficiency of the antenna, which is the ratio of the power radiated by the antenna to the power supplied to it by the source of excitation.

K11.2. Radiation fields; $\beta r \gg 1$; $1/r$ terms; Time-average radiated power; Radiation resistance; Radiation pattern; Power density; Directivity.

D11.3. Three Hertzian dipoles of lengths 1, 2, and 2 m are situated at the origin oriented along the positive x-, y-, and z-axes, respectively, carrying currents $1 \cos 2\pi \times$

$10^6 t$, 2 cos $2\pi \times 10^6 t$, and 2 sin $2\pi \times 10^6 t$ A, respectively. Determine the polarizations (including right-hand or left-hand sense in the case of circular and elliptical) of the radiation field at each of the following points: **(a)** a point on the x-axis; **(b)** a point on the y-axis; and **(c)** a point on the z-axis.

Ans. **(a)** right circular; **(b)** left elliptical; **(c)** linear

D11.4. Compute the directivity responding to each of the following functions $f(\theta, \phi)$ in (11.27):

(a) $f(\theta, \phi) = \begin{cases} 1 & \text{for } 0 < \theta < \pi/2 \\ 0 & \text{otherwise} \end{cases}$

(b) $f(\theta, \phi) = \begin{cases} \sin^2 \theta & \text{for } 0 < \theta < \pi/2 \\ 0 & \text{otherwise} \end{cases}$

(c) $f(\theta, \phi) = \begin{cases} 1 & \text{for } 0 < \theta < \pi/2 \\ \sin^2 \theta & \text{for } \pi/2 < \theta < \pi \end{cases}$

Ans. **(a)** 2; **(b)** 3; **(c)** 1.2

11.3 LINEAR ANTENNAS

In the preceding section we found the radiation fields due to a Hertzian dipole, which is an elemental antenna of infinitesimal length. If we now have an antenna of any length having a specified current distribution, we can divide it into a series of Hertzian dipoles, and by applying superposition we can find the radiation fields for that antenna. We illlustrate this procedure in this section by first considering the half-wave dipole, which is a commonly used form of antenna.

Half-wave dipole

The half-wave dipole is a center fed, straight wire antenna of length L equal to $\lambda/2$ and having the current distribution

$$\boxed{I(z) = I_0 \cos \frac{\pi z}{L} \cos \omega t \qquad \text{for} \quad -L/2 < z < L/2} \tag{11.28}$$

where the dipole is assumed to be oriented along the z-axis with its center at the origin, as shown in Fig. 11.6(a). As can be seen from Fig. 11.6(a), the amplitude of the current distribution varies cosinusoidally along the antenna with zeros at the ends and maximum at the center. To see how this distribution comes about, the half-wave dipole may be imagined to be the evolution of an open-circuited transmission line with the conductors folded perpendicularly to the line at points $\lambda/4$ from the end of the line. The current standing wave pattern for an open-circuited line is shown in Fig. 11.6(b). It consists of zero current at the open circuit and maximum current at $\lambda/4$ from the open circuit, that is, at points a and a'. Hence it can be seen that when the conductors are folded perpendicularly to the line at a and a', the half-wave dipole shown in Fig. 11.6(a) results.

Now to find the radiation field due to the half-wave dipole, we divide it into a number of Hertzian dipoles, each of length dz', as shown in Fig. 11.7. If we consider one of these dipoles situated at distance z' from the origin, then from (11.28) the current in this dipole is $I_0 \cos (\pi z'/L) \cos \omega t$. From (11.17a) and (11.17b), the radiation fields due to this dipole at point P situated at distance r' from it are given by

$$d\mathbf{E} = -\frac{\eta \beta I_0 \cos (\pi z'/L)\, dz' \sin \theta'}{4\pi r'} \sin (\omega t - \beta r')\, \mathbf{i}_{\theta'} \tag{11.29a}$$

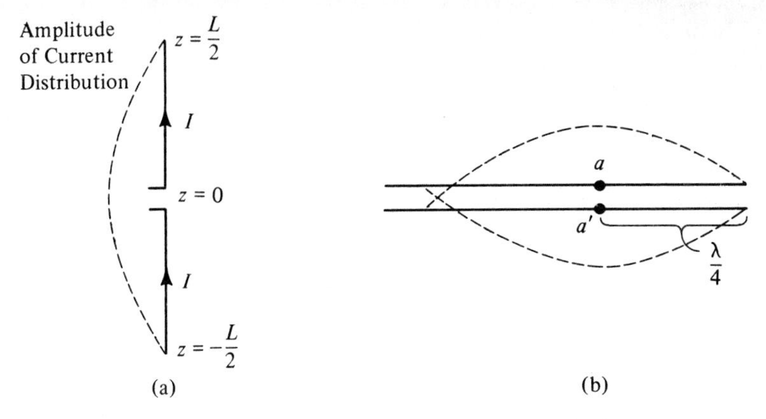

Figure 11.6. (a) Half-wave dipole. (b) Open-circuited transmission line for illustrating the evolution of the half-wave dipole.

$$d\mathbf{H} = -\frac{\beta I_0 \cos(\pi z'/L)\, dz' \sin\theta'}{4\pi r'} \sin(\omega t - \beta r')\, \mathbf{i}_\phi \tag{11.29b}$$

where θ' is the angle between the z-axis and the line from the current element to the point P and $\mathbf{i}_{\theta'}$ is the unit vector perpendicular to that line, as shown in Fig. 11.7. The fields due to the entire current distribution of the half-wave dipole are then given by

$$\begin{aligned}\mathbf{E} &= \int_{z'=-L/2}^{L/2} d\mathbf{E} \\ &= -\int_{z'=-L/2}^{L/2} \frac{\eta\beta I_0 \cos(\pi z'/L) \sin\theta'\, dz'}{4\pi r'} \sin(\omega t - \beta r')\, \mathbf{i}_{\theta'}\end{aligned} \tag{11.30a}$$

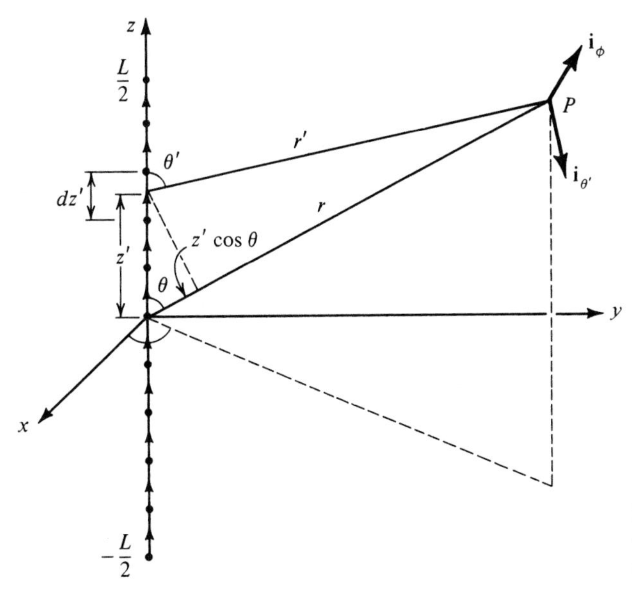

Figure 11.7. For the determination of the radiation field due to the half-wave dipole.

$$\mathbf{H} = \int_{z'=-L/2}^{L/2} d\mathbf{H}$$

$$= -\int_{z'=-L/2}^{L/2} \frac{\beta I_0 \cos(\pi z'/L) \sin\theta' \, dz'}{4\pi r'} \sin(\omega t - \beta r') \, \mathbf{i}_\phi \tag{11.30b}$$

where r', θ', and $\mathbf{i}_{\theta'}$ are functions of z'.

For radiation fields, r' is at least equal to several wavelengths and hence $\gg L$. We can therefore set $\mathbf{i}_{\theta'} \approx \mathbf{i}_\theta$ and $\theta' \approx \theta$ since they do not vary significantly for $-L/2 < z' < L/2$. We can also set $r' \approx r$ in the amplitude factors for the same reason, but for r' in the phase factors, we substitute $r - z' \cos\theta$ since the phase angle in $\sin(\omega t - \beta r') = \sin(\omega t - \pi r'/L)$ can vary appreciably over the range $-L/2 < z' < L/2$. For example, if $L = 2$ m ($\lambda = 4$ m), $\theta = 0$, and $r = 10$, then r' varies from 11 for $z' = -L/2$ to 9 for $z' = L/2$, and $\pi r'/L$ varies from 5.5π for $z' = -L/2$ to 4.5π for $z' = L/2$. Thus we have

$$\mathbf{E} = E_\theta \mathbf{i}_\theta$$

where

$$E_\theta = -\int_{z'=-L/2}^{L/2} \frac{\eta\beta I_0 \cos(\pi z'/L) \sin\theta}{4\pi r} \sin(\omega t - \beta r + \beta z' \cos\theta) \, dz'$$

$$= -\frac{\eta(\pi/L) I_0 \sin\theta}{4\pi r} \int_{z'=-L/2}^{L/2} \cos\frac{\pi z'}{L} \sin\left(\omega t - \frac{\pi}{L} r + \frac{\pi}{L} z' \cos\theta\right) dz'$$

Evaluating the integral, we obtain

$$\boxed{E_\theta = -\frac{\eta I_0}{2\pi r} \frac{\cos[(\pi/2)\cos\theta]}{\sin\theta} \sin\left(\omega t - \frac{\pi}{L} r\right)} \tag{11.31a}$$

Similarly,

$$\mathbf{H} = H_\phi \mathbf{i}_\phi$$

where

$$\boxed{H_\phi = -\frac{I_0}{2\pi r} \frac{\cos[(\pi/2)\cos\theta]}{\sin\theta} \sin\left(\omega t - \frac{\pi}{L} r\right)} \tag{11.31b}$$

The Poynting vector due to the radiation fields of the half-wave dipole is given by

$$\mathbf{P} = \mathbf{E} \times \mathbf{H} = E_\theta H_\phi \mathbf{i}_r$$

$$= \frac{\eta I_0^2}{4\pi^2 r^2} \frac{\cos^2[(\pi/2)\cos\theta]}{\sin^2\theta} \sin^2\left(\omega t - \frac{\pi}{L} r\right) \mathbf{i}_r \tag{11.32}$$

The power radiated by the half-wave dipole is given by

$$P_{\text{rad}} = \int_{\theta=0}^{\pi} \int_{\phi=0}^{2\pi} \mathbf{P} \cdot r^2 \sin\theta \, d\theta \, d\phi \, \mathbf{i}_r$$

$$P_{\text{rad}} = \int_{\theta=0}^{\pi}\int_{\phi=0}^{2\pi} \frac{\eta I_0^2}{4\pi^2} \frac{\cos^2[(\pi/2)\cos\theta]}{\sin\theta} \sin^2\left(\omega t - \frac{\pi}{L}r\right) d\theta\, d\phi$$

$$= \frac{\eta I_0^2}{\pi} \sin^2\left(\omega t - \frac{\pi}{L}r\right) \int_{\theta=0}^{\pi/2} \frac{\cos^2[(\pi/2)\cos\theta]}{\sin\theta} d\theta \tag{11.33}$$

$$= \frac{0.609\eta I_0^2}{\pi} \sin^2\left(\omega t - \frac{\pi}{L}r\right)$$

where we have used the result

$$\int_{\theta=0}^{\pi/2} \frac{\cos^2[(\pi/2)\cos\theta]}{\sin\theta} d\theta = 0.609$$

obtainable by numerical integration. The time-average radiated power is

$$\langle P_{\text{rad}} \rangle = \frac{0.609\eta I_0^2}{\pi} \left\langle \sin^2\left(\omega t - \frac{\pi}{L}r\right)\right\rangle$$

$$= \frac{1}{2} I_0^2 \left(\frac{0.609\eta}{\pi}\right) \tag{11.34}$$

Thus the radiation resistance of the half-wave dipole is

$$\boxed{R_{\text{rad}} = \frac{0.609\eta}{\pi}\ \Omega} \tag{11.35}$$

For free space, $\eta = \eta_0 = 120\pi\ \Omega$, and

$$\boxed{R_{\text{rad}} = 0.609 \times 120 = 73\ \Omega} \tag{11.36}$$

Turning our attention now to the directional characteristics of the half-wave dipole, we note from (11.31a) and (11.31b) that the radiation pattern for the fields is $\{\cos[(\pi/2)\cos\theta]\}/\sin\theta$ whereas for the power density, it is $\{\cos^2[(\pi/2)\cos\theta]\}/\sin^2\theta$. These patterns, shown in Fig. 11.8(a) and (b), are slightly more directional than the corresponding patterns for the Hertzian dipole. The directivity of the half-wave dipole may now be found by using (11.27). Thus

$$D = 4\pi \frac{\{\cos^2[(\pi/2)\cos\theta]/\sin^2\theta\}_{\max}}{\int_{\theta=0}^{\pi}\int_{\phi=0}^{2\pi}\{\cos^2[(\pi/2)\cos\theta]/\sin^2\theta\}\sin\theta\, d\theta\, d\phi}$$

$$= 4\pi \frac{1}{2\pi \times 2 \times 0.609}$$

or

$$\boxed{D = 1.642} \tag{11.37}$$

Arbitrary length

For a center-fed linear antenna of length L equal to an arbitrary number of wavelengths, the current distribution can be written as

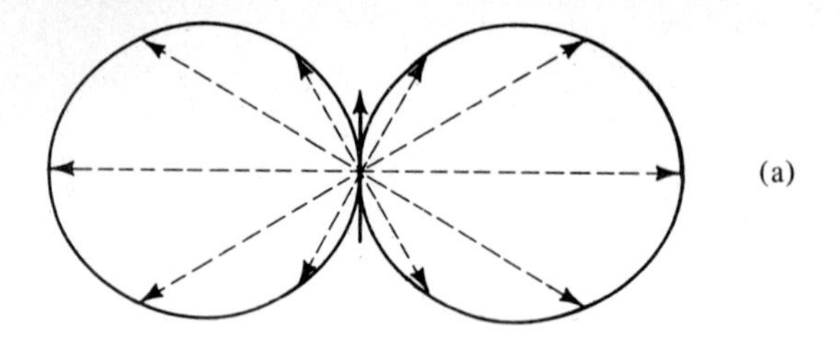

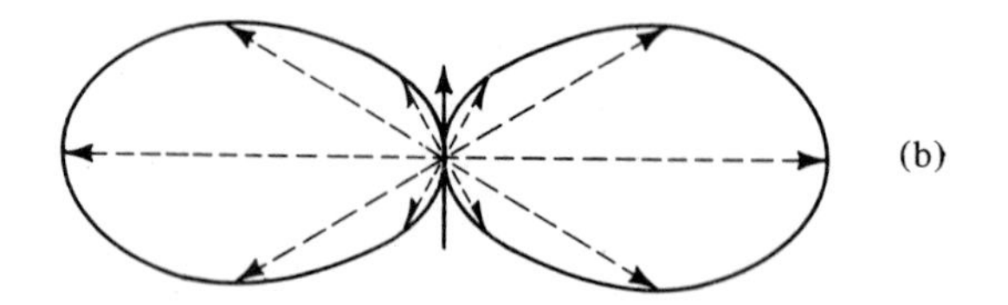

Figure 11.8. Radiation patterns for (a) the fields and (b) the power density due to the half-wave dipole.

$$I(z) = \begin{cases} I_0 \sin \beta\left(\dfrac{L}{2} + z\right) \cos \omega t & \text{for} \quad -\dfrac{L}{2} < z < 0 \\ I_0 \sin \beta\left(\dfrac{L}{2} - z\right) \cos \omega t & \text{for} \quad 0 < z < \dfrac{L}{2} \end{cases} \tag{11.38}$$

where once again the antenna is assumed to be oriented along the z-axis with its center at the origin. Note that the current distribution is such that the amplitude of the current goes to zero at the two ends of the antenna and varies sinusoidally along the antenna with phase reversals every half wavelength from the ends, as shown, for example, for $L = 5\lambda/2$ in Fig. 11.9. Note also that for $L = \lambda/2$, (11.38) reduces to (11.28). Using (11.38) and proceeding in the same manner as for the half-wave dipole, the components of the radiation fields, the radiation resistance, and the directivity for the linear antenna of arbitrary electrical length can be obtained. The results are

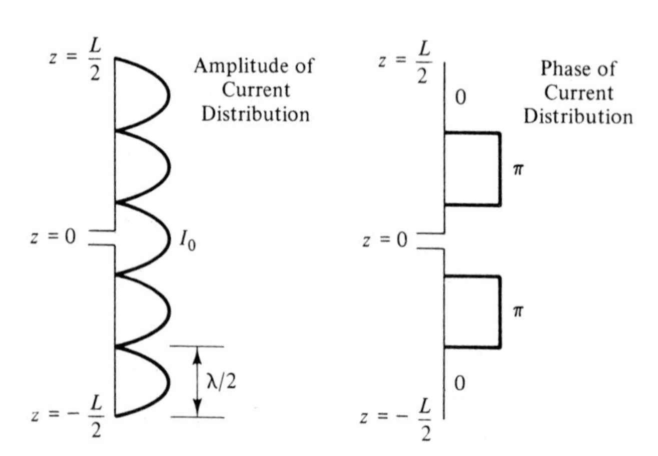

Figure 11.9. Variations of amplitude and phase of current distribution along a linear antenna of length $L = 5\lambda/2$.

$$E_\theta = -\frac{\eta I_0}{2\pi r} F(\theta) \sin(\omega t - \beta r) \tag{11.39a}$$

$$H_\phi = -\frac{I_0}{2\pi r} F(\theta) \sin(\omega t - \beta r) \tag{11.39b}$$

$$R_{\text{rad}} = \frac{\eta}{\pi} \int_{\theta=0}^{\pi/2} F^2(\theta) \sin\theta \, d\theta \tag{11.39c}$$

$$D = \frac{[F^2(\theta)]_{\max}}{\int_{\theta=0}^{\pi/2} F^2(\theta) \sin\theta \, d\theta} \tag{11.39d}$$

where

$$F(\theta) = \frac{\cos\left(\frac{\beta L}{2}\cos\theta\right) - \cos\frac{\beta L}{2}}{\sin\theta} \tag{11.40}$$

is the radiation pattern for the fields. For $L = k\lambda$, (11.40) reduces to

$$F(\theta) = \frac{\cos(k\pi\cos\theta) - \cos(k\pi)}{\sin\theta} \tag{11.41}$$

The listing of a PC program which for a specified value of k computes and plots the radiation pattern given by (11.41), and computes the radiation resistance and directivity by evaluating the integrals in (11.39c) and (11.39d) numerically, is included as PL 11.2.

PL 11.2. Program listing for plotting of radiation pattern and computation of radiation resistance and directivity of a linear antenna of length equal to k wavelengths.

```
100 '*****************************************************
110 '* PLOTTING OF RADIATION PATTERN AND COMPUTATION     *
120 '* OF RADIATION RESISTANCE AND DIRECTIVITY FOR A     *
130 '* LINEAR ANTENNA OF LENGTH EQUAL TO K WAVELENGTHS *
140 '*****************************************************
150 DIM E(100),CT(100),ST(100)
160 PI=3.1416:DR=PI/180
170 SC=1.2:'* SCALE FACTOR TO EQUALIZE VERTICAL AND
180 '  HORIZONTAL SCALES *
190 CLS:SCREEN 1:COLOR 0,1
200 LOCATE 21,1:INPUT "ENTER VALUE OF K: ",K
210 LOCATE 21,1:PRINT "LENGTH OF DIPOLE =";K;"* WAVELENG
    TH"
220 '* COMPUTATION OF RADIATION PATTERN, RADIATION
230 '  RESISTANCE AND DIRECTIVITY *
240 LOCATE 22,1:PRINT "RADIATION PATTERN BEING COMPUTED"
250 C1=COS(K*PI):FR=0:EMAX=0
260 FOR I=1 TO 89
270 THETA=I*DR:CT(I)=COS(THETA):ST(I)=SIN(THETA)
280 FT=ABS(COS(K*PI*CT(I))-C1)
290 E(I)=FT/ST(I)
300 IF E(I)>EMAX THEN EMAX=E(I)
```

PL 11.2. (continued)

```
310 FR=FR+FT*FT/ST(I)
320 NEXT
330 E(90)=1-C1:CT(90)=0:ST(90)=1
340 IF E(90)>EMAX THEN EMAX=E(90)
350 IT=(FR+.5*E(90)*E(90))*DR
360 RAD=IT*120:'* RADIATION RESISTANCE *
370 DIR=(EMAX*EMAX)/IT:'* DIRECTIVITY *
380 C2=80/EMAX:PSET (160,80),3
390 '* PLOTTING OF RADIATION PATTERN *
400 LOCATE 22,1:PRINT "RADIATION PATTERN BEING PLOTTED
    "
410 FOR I=1 TO 90
420 EI=E(I)*C2:VD=EI*CT(I):HD=EI*ST(I)*SC
430 PSET (160+HD,80-VD),3
440 PSET (160+HD,80+VD),3
450 PSET (160-HD,80-VD),3
460 PSET (160-HD,80+VD),3
470 NEXT
480 LOCATE 22,1:PRINT "RADIATION RESISTANCE =";RAD;"OHMS
    "
490 LOCATE 23,1:PRINT "DIRECTIVITY =";DIR
500 LOCATE 24,1:PRINT "STRIKE ANY KEY TO CONTINUE";:C$=I
    NPUT$(1):GOTO 190
510 END
```

The radiation pattern obtained from a run of PL 11.2 for $k = 2.5$ is shown in Fig. 11.10. The radiation resistance and directivity for $k = 2.5$ as computed by the program are 120.768 Ω and 3.058, respectively.

K11.3. Half-wave dipole; Radiation fields; Radiation characteristics; Linear antenna; Arbitrary length.

D11.5. A center-fed linear antenna in free space has the current distribution of the form given by (11.38), where $I_0 = 1$ A. Find the amplitude of E_θ at $r = 100$ m for each of the following cases: **(a)** $L = 2$ m, $f = 75$ MHz, $\theta = 60°$; **(b)** $L = 2$ m, $f = 200$ MHz, $\theta = 60°$; and **(c)** $L = 4$ m, $f = 300$ MHz, $\theta = 30°$.
Ans. **(a)** 0.49 V/m; **(b)** 0 V/m; **(c)** 1.335 V/m

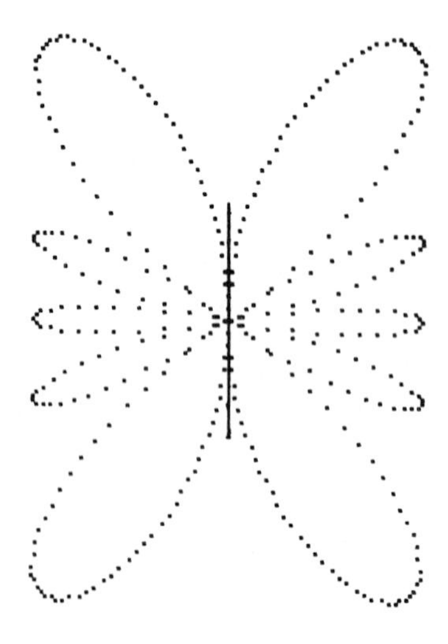

Figure 11.10. Computer-generated plot of radiation pattern for a linear antenna of length 2.5λ, using the program of PL 11.2.

In Section 6.2 we illustrated the principle of an antenna array by considering an array of two parallel, infinite plane current sheets of uniform densities. We learned that by appropriately choosing the spacing between the current sheets and the amplitudes and phases of the current densities, a desired radiation characteristic can be obtained. The infinite plane current sheet is, however, a hypothetical antenna for which the fields are truly uniform plane waves propagating in the one dimension normal to the sheet. Now that we have gained some knowledge of physical antennas, in this section we consider arrays of such antennas.

Array of two Hertzian dipoles

The simplest array we can consider consists of two Hertzian dipoles, oriented parallel to the z-axis and situated at points on the x-axis on either side of and equidistant from the origin, as shown in Fig. 11.11. We shall consider the amplitudes of the currents in the two dipoles to be equal, but we shall allow a phase difference α between them. Thus if $I_1(t)$ and $I_2(t)$ are the currents in the dipoles situated at $(d/2, 0, 0)$ and $(-d/2, 0, 0)$, respectively, then

$$I_1 = I_0 \cos\left(\omega t + \frac{\alpha}{2}\right) \tag{11.42a}$$

$$I_2 = I_0 \cos\left(\omega t - \frac{\alpha}{2}\right) \tag{11.42b}$$

For simplicity, we consider a point P in the xz-plane and compute the radiation field at that point due to the array of the two dipoles. To do this, we note from (11.17a) that the electric field intensities at the point P due to the individual dipoles are given by

$$\mathbf{E}_1 = -\frac{\eta\beta I_0\, dl \sin\theta_1}{4\pi r_1} \sin\left(\omega t - \beta r_1 + \frac{\alpha}{2}\right) \mathbf{i}_{\theta_1} \tag{11.43a}$$

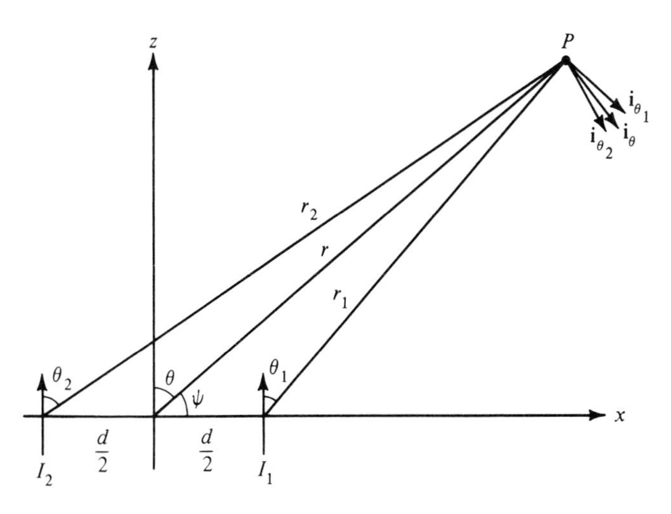

Figure 11.11. For computing the radiation field due to an array of two Hertzian dipoles.

$$\mathbf{E}_2 = -\frac{\eta\beta I_0\, dl \sin\theta_2}{4\pi r_2} \sin\left(\omega t - \beta r_2 - \frac{\alpha}{2}\right)\mathbf{i}_{\theta_2} \tag{11.43b}$$

where θ_1, θ_2, r_1, r_2, $\mathbf{i}_{\theta_1}$, and $\mathbf{i}_{\theta 2}$ are as shown in Fig. 11.11.

For $r \gg d$, that is, for points far from the array, which is the region of interest, we can set $\theta_1 \approx \theta_2 \approx \theta$ and $\mathbf{i}_{\theta_1} \approx \mathbf{i}_{\theta_2} \approx \mathbf{i}_\theta$. Also, we can set $r_1 \approx r_2 \approx r$ in the amplitude factors, but for r_1 and r_2 in the phase factors, we substitute

$$r_1 \approx r - \frac{d}{2}\cos\psi \tag{11.44a}$$

$$r_2 \approx r + \frac{d}{2}\cos\psi \tag{11.44b}$$

where ψ is the angle made by the line from the origin to P with the axis of the array, that is, the x-axis, as shown in Fig. 11.11. Thus we obtain the resultant field to be

$$\begin{aligned}\mathbf{E} &= \mathbf{E}_1 + \mathbf{E}_2 \\ &= -\frac{\eta\beta I_0\, dl \sin\theta}{4\pi r}\left[\sin\left(\omega t - \beta r + \frac{\beta d}{2}\cos\psi + \frac{\alpha}{2}\right)\right. \\ &\quad \left. + \sin\left(\omega t - \beta r - \frac{\beta d}{2}\cos\psi - \frac{\alpha}{2}\right)\right]\mathbf{i}_\theta \\ &= -\frac{2\eta\beta I_0\, dl \sin\theta}{4\pi r}\cos\left(\frac{\beta d\cos\psi + \alpha}{2}\right)\sin(\omega t - \beta r)\,\mathbf{i}_\theta\end{aligned} \tag{11.45}$$

Unit, group, and resultant patterns

Comparing (11.45) with the expression for the electric field at P due to a single dipole situated at the origin, we note that the resultant field of the array is simply equal to the single dipole field multiplied by the factor $2\cos[(\beta d\cos\psi + \alpha)/2]$, known as the *array factor*. Thus the radiation pattern of the resultant field is given by the product of $\sin\theta$, which is the radiation pattern of the single dipole field, and $\cos[(\beta d\cos\psi + \alpha)/2]$, which is the radiation pattern of the array if the antennas were isotropic. We shall call these three patterns the *resultant pattern*, the *unit pattern*, and the *group pattern*, respectively. It is apparent that the group pattern is independent of the nature of the individual antennas as long as they have the same spacing and carry currents having the same relative amplitudes and phase differences. It can also be seen that the group pattern is the same in any plane containing the axis of the array. In other words, the three-dimensional group pattern is simply the pattern obtained by revolving the group pattern in the xz-plane about the x-axis, that is, the axis of the array.

Example 11.3

For the array of two antennas carrying currents having equal amplitudes, let us consider several pairs of d and α and investigate the group patterns.

Case 1: $d = \lambda/2$, $\alpha = 0$. The group pattern is

$$\left|\cos\left(\frac{\beta\lambda}{4}\cos\psi\right)\right| = \cos\left(\frac{\pi}{2}\cos\psi\right)$$

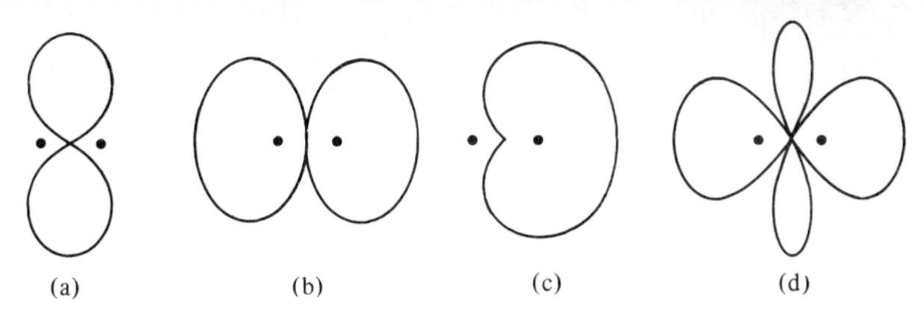

Figure 11.12. Group patterns for an array of two antennas carrying currents of equal amplitude for (a) $d = \lambda/2$, $\alpha = 0$, (b) $d = \lambda/2$, $\alpha = \pi$, (c) $d = \lambda/4$, $\alpha = -\pi/2$, and (d) $d = \lambda$, $\alpha = 0$.

This is shown sketched in Fig. 11.12(a). It has maxima perpendicular to the axis of the array and nulls along the axis of the array. Such a pattern is known as a *broadside pattern*.

Case 2: $d = \lambda/2$, $\alpha = \pi$. The group pattern is

$$\left|\cos\left(\frac{\beta\lambda}{4}\cos\psi + \frac{\pi}{2}\right)\right| = \left|\sin\left(\frac{\pi}{2}\cos\psi\right)\right|$$

This is shown sketched in Fig. 11.12(b). It has maxima along the axis of the array and nulls perpendicular to the axis of the array. Such a pattern is known as an *endfire pattern*.

Case 3: $d = \lambda/4$, $\alpha = -\pi/2$. The group pattern is

$$\left|\cos\left(\frac{\beta\lambda}{8}\cos\psi - \frac{\pi}{4}\right)\right| = \cos\left(\frac{\pi}{4}\cos\psi - \frac{\pi}{4}\right)$$

This is shown sketched in Fig. 11.12(c). It has a maximum along $\psi = 0$ and null along $\psi = \pi$. Again, this is an endfire pattern, but directed to one side. This case is the same as the one considered in Section 6.2.

Case 4: $d = \lambda$, $\alpha = 0$. The group pattern is

$$\left|\cos\left(\frac{\beta\lambda}{2}\cos\psi\right)\right| = |\cos(\pi\cos\psi)|$$

This is shown sketched in Fig. 11.12(d). It has maxima along $\psi = 0°$, $90°$, and $180°$ and nulls along $\psi = 60°$ and $120°$.

Proceeding further, we can obtain the resultant pattern for an array of two Hertzian dipoles by multiplying the unit pattern by the group pattern. Thus recalling that the unit pattern for the Hertzian dipole is $\sin\theta$ in the plane of the dipole and considering values of $\lambda/2$ and 0 for d and α, respectively, for which the group pattern is given in Fig. 11.12(a), we obtain the resultant pattern in the xz-plane, as shown in Fig. 11.13(a). In the xy-plane, that is, the plane normal to the axis of the dipole, the unit pattern is a circle, and hence the resultant pattern is the same as the group pattern, as illustrated in Fig. 11.13(b).

Example 11.4.

Pattern multiplication

The procedure of multiplication of the unit and group patterns to obtain the resultant pattern illustrated in Example 11.3 is known as the *pattern multiplication* technique. Let us consider a linear array of four isotropic antennas spaced $\lambda/2$ apart and fed in

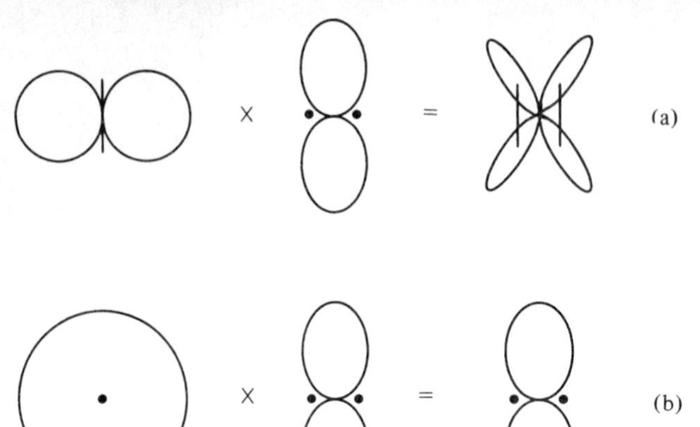

Figure 11.13. Determination of the resultant pattern of an antenna array by multiplication of unit and group patterns.

phase, as shown in Fig. 11.14(a) and obtain the resultant pattern, by using the pattern multiplication technique.

To obtain the resultant pattern of the four-element array, we replace it by a two-element array of spacing λ, as shown in Fig. 11.14(b), in which each element forms a unit representing a two-element array of spacing $\lambda/2$. The unit pattern is then the pattern shown in Fig. 11.12(a). The group pattern, which is the pattern of two isotropic radiators having $d = \lambda$ and $\alpha = 0$, is the pattern given in Fig. 11.12(d). The resultant pattern of the four-element array is the product of these two patterns, as illustrated in Fig. 11.14(c). If the individual elements of the four-element array are not isotropic, then this pattern becomes the group pattern for the determination of the new resultant pattern.

Uniform linear array of n antennas

Let us now consider a uniform linear array of n antennas of spacing d, as shown in Fig. 11.15. Then assuming currents of equal amplitude I_0 and progressive phase shift α, that is, in the manner $I_0 \cos \omega t$, $I_0 \cos (\omega t + \alpha)$, $I_0 \cos (\omega t + 2\alpha)$, . . . for antennas 1, 2, 3, . . . , respectively, we can obtain the far field ($r \gg nd$) as follows. If the complex electric field at the point (r_0, ψ) due to element 1 is assumed to be $1e^{-j\beta r_0}$, then the complex electric fields at that point due to elements 2, 3, . . . are $1e^{j\alpha}e^{j\beta(r_0 - d \cos \psi)}$, $1e^{j2\alpha}e^{j\beta(r_0 - 2d \cos \psi)}$, . . . , so that the field due to the n-element array is

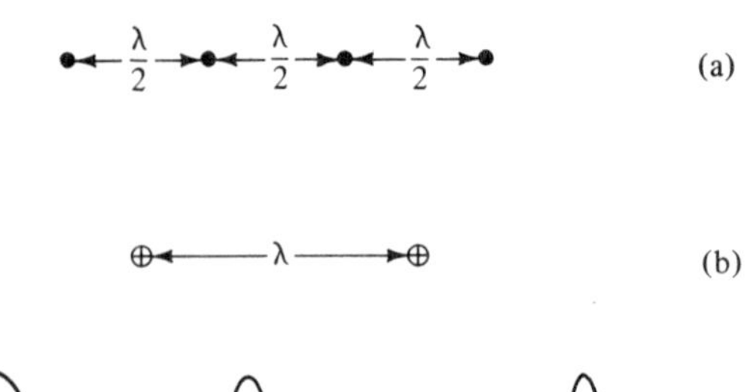

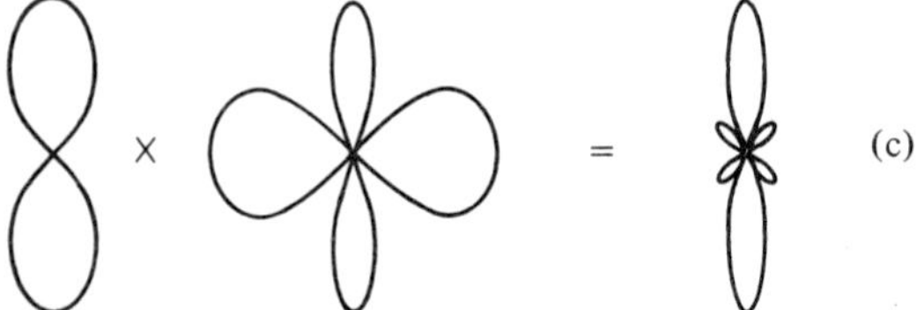

Figure 11.14. Determination of the resultant pattern for a linear array of four isotropic antennas.

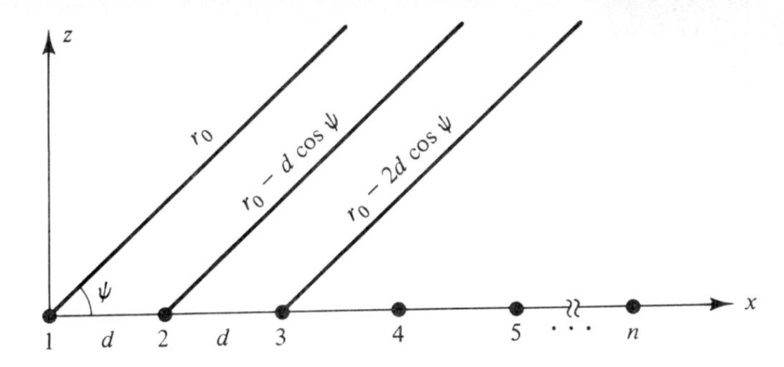

Figure 11.15. For obtaining the group pattern for a uniform linear array of n antennas.

$$
\begin{aligned}
\bar{E}(\psi) &= 1e^{-j\beta r_0} + 1e^{j\alpha}e^{-j\beta(r_0 - d\cos\psi)} \\
&\quad + 1e^{j2\alpha}e^{-j\beta(r_0 - 2d\cos\psi)} + \cdots \\
&\quad + 1e^{j(n-1)\alpha}e^{-j\beta[r_0 - (n-1)d\cos\psi]} \\
&= [1 + e^{j(\beta d\cos\psi + \alpha)} + e^{j2(\beta d\cos\psi + \alpha)} \\
&\quad + \cdots + e^{j(n-1)(\beta d\cos\psi + \alpha)}]e^{-j\beta r_0} \\
&= \frac{1 - e^{jn(\beta d\cos\psi + \alpha)}}{1 - e^{j(\beta d\cos\psi + \alpha)}}e^{-j\beta r_0}
\end{aligned}
\tag{11.46}
$$

The magnitude of $\bar{E}$ is given by

$$
\begin{aligned}
|\bar{E}(\psi)| &= \left| \frac{1 - e^{jn(\beta d\cos\psi + \alpha)}}{1 - e^{j(\beta d\cos\psi + \alpha)}} \right| \\
&= \left| \frac{\sin n[(\beta d\cos\psi + \alpha)/2]}{\sin [(\beta d\cos\psi + \alpha)/2]} \right|
\end{aligned}
\tag{11.47}
$$

which has a maximum value of n for $\beta d \cos \psi + \alpha = 0, 2\pi, 4\pi, \ldots$. Thus the group pattern is

$$
\boxed{F(\psi) = \frac{1}{n}\left| \frac{\sin n[(\beta d\cos\psi + \alpha)/2]}{\sin [(\beta d\cos\psi + \alpha)/2]} \right|}
\tag{11.48}
$$

Note that for $n = 2$, (11.48) reduces to $\cos [(\beta d \cos \psi + \alpha)/2]$, which is the group pattern obtained for the two-element array. The nulls of the pattern occur for $n(\beta d \cos \psi + \alpha) = 2m\pi$, where m is any integer but not equal to $0, n, 2n, \ldots$. For $d = k\lambda$, (11.48) reduces to

$$
F(\psi) = \frac{1}{n}\left| \frac{\sin n(\pi k\cos\psi + \alpha/2)}{\sin (\pi k\cos\psi + \alpha/2)} \right|
\tag{11.49}
$$

The listing of a PC program which for specified values of n and k generates a sequence of plots of F versus ψ ($0 \leq \psi \leq 180°$) for values of α ranging from $-180°$ to $150°$ in steps of $30°$ is included as PL 11.3.

PL 11.3. Program listing for plotting of group patterns for a uniform linear array of n antennas for varying values of progressive phase shift.

```
100 '*****************************************************
110 '* PLOTTING OF GROUP PATTERNS FOR A UNIFORM LINEAR    *
120 '* ARRAY OF N ANTENNAS FOR VALUES OF THE PROGRESSIVE *
130 '* PHASE SHIFT RANGING FROM -180 TO 150 DEG IN STEPS *
140 '* OF 30 DEG                                         *
150 '*****************************************************
160 DIM CP(100),HC(12),VC(12)
170 PI=3.1416:DR=PI/180
180 CLS:SCREEN 1:COLOR 0,1
190 LOCATE 21,1:INPUT "ENTER VALUE OF N: ",N
200 LOCATE 21,1:PRINT "NUMBER OF ELEMENTS =";N
210 LOCATE 22,1:INPUT "ENTER VALUE OF K: ",K:'* ELEMENT
220 '  SPACING IN WAVELENGTHS *
230 LOCATE 22,1:PRINT "ELEMENT SPACING =";K;"* WAVELENGTH"
240 LOCATE 23,1:PRINT "PLEASE WAIT"
250 C1=PI*K:JK=0
260 FOR J=0 TO 90:CP(J)=C1*COS(2*J*DR):NEXT
270 FOR I=1 TO 4:FOR J=0 TO 2
280 JK=JK+1:HC(JK)=94*J:VC(JK)=36*I
290 NEXT:NEXT
300 '* PLOTTING OF GROUP PATTERNS *
310 FOR I=1 TO 12
320 LINE (HC(I),VC(I))-(HC(I)+90,VC(I)-30),3,B
330 AL=(I-7)*15*DR
340 LOCATE 23,1:PRINT "VALUE OF ALPHA =";30*(I-7);"DEG "
350 FOR J=0 TO 90
360 ARG=CP(J)+AL:DEN=SIN(ARG)
370 IF DEN=0 THEN E=30:GOTO 400
380 E=30*ABS(SIN(N*ARG)/(N*DEN))
390 IF E>30 THEN E=30
400 PSET (HC(I)+J,VC(I)-E),3
410 NEXT
420 NEXT
430 '* DRAWING OF HORIZONTAL SCALE MARKS *
440 FOR I=1 TO 3
450 LINE (HC(I),156)-(HC(I)+90,156),3
460 FOR J=0 TO 6
470 LINE (HC(I)+15*J,155)-(HC(I)+15*J,153)
480 NEXT
490 NEXT
500 LOCATE 23,1:PRINT "PLOTTING COMPLETED       "
510 LOCATE 24,1:PRINT "STRIKE ANY KEY TO CONTINUE";
520 C$=INPUT$(1):GOTO 180
530 END
```

The output resulting from a run of the program of PL 11.3 for $n = 6$ and $k = 0.5$ is shown in Fig. 11.16. It can be seen that as the value of α is varied, the value of ψ along which the principal maximum of the group pattern occurs varies in a continuous manner, as to be expected.

Principle of phased array

The behavior illustrated in Fig. 11.16 is the basis for the principle of phased arrays. In a phased array, the phase differences between the elements of the array are varied electronically to scan the radiation pattern over a desired angle without having to move the antenna structure mechanically.

Log-periodic dipole array

A type of array that is commonly seen is the *log-periodic dipole array*, which is an example of a broadband array. To discuss briefly, we first note that the directional properties of antennas and antenna arrays depend on their electri-

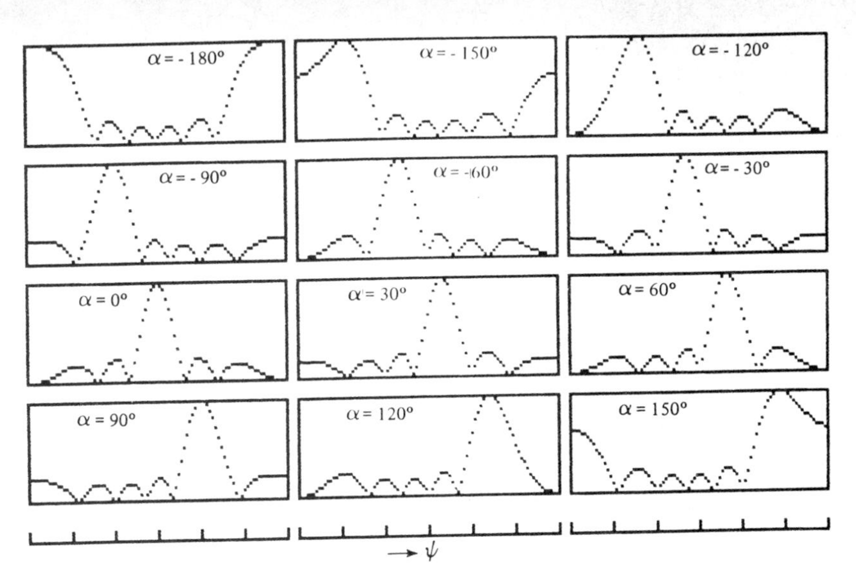

Figure 11.16. Plots of group patterns resulting from a run of the program of PL 11.3 for $n = 6$ and $k = 0.5$. The horizontal scale for ψ for each plot is such that ψ varies for 0 to 180°.

cal dimensions, that is, the dimensions expressed in terms of the wavelength at the operating frequency. Hence an antenna of fixed physical dimensions exhibits frequency-dependent characteristics. This very fact suggests that for an antenna to be frequency independent, its electrical size must remain constant with frequency, and hence its physical size should increase proportionately to the wavelength. Alternatively, for an antenna of fixed physical dimensions, the active region, that is, the region responsible for the predominant radiation should vary with frequency, that is, scale itself in such a manner that its electrical size remains the same. An example in which this is the case is the log-periodic dipole array, shown in Fig. 11.17. As the name implies, it employs a number of dipoles. The dipole lengths and the spacings between consecutive dipoles increase along the array by a constant scale factor such that

$$\frac{l_{i+1}}{l_i} = \frac{d_{i+1}}{d_i} = \tau \tag{11.50}$$

From the principle of scaling, it is evident that for this structure extending from zero to infinity and energized at the apex, the properties repeat at frequencies given by $\tau^n f$, where n takes integer values. When plotted on a logarithmic scale, these frequencies are equally spaced at intervals of $\log \tau$. It is for this reason that the structure is termed *log periodic*.

The log-periodic dipole array is fed by a transmission line, as shown in Fig. 11.17, such that a 180° phase shift is introduced between successive elements in addition to that corresponding to the spacing between the elements. The resulting radiation pattern is directed toward the apex, that is, toward the source. Almost all the radiation takes place from those elements which are in the vicinity of a half wavelength long. The operating band of frequencies is therefore bounded on

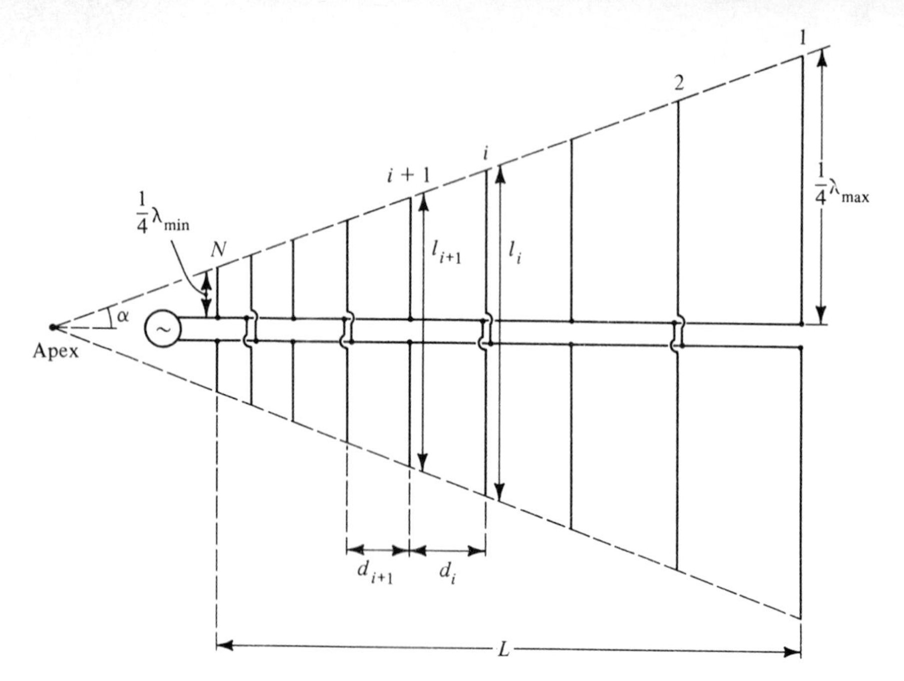

Figure 11.17. Log-periodic dipole array.

the low side by frequencies at which the largest elements are approximately a half wavelength long and on the high side by frequencies corresponding to the size of the smallest elements. As the frequency is varied, the radiating or active region moves back and forth along the array. Since practically all the input power is radiated by the active region, the larger elements to the right of it are not excited. Furthermore, because the radiation is toward the apex, these larger elements are essentially in a field-free region and hence do not significantly influence the operation. Although the shorter elements to the left of the active region are in the antenna beam, they have small influence on the pattern because of their short lengths, close spacings, and the 180° phase shift.

Image antennas

Thus far we have considered the antennas to be situated in an unbounded medium so that the waves radiate in all directions from the antenna without giving rise to reflections from any obstacles. In practice, however, we have to consider the effect of the ground even if no other obstacles are present. To do this, it is reasonable to assume that the ground is a perfect conductor and use the concept of image antennas, which together with the actual antennas form arrays.

To introduce this concept, let us consider a Hertzian dipole oriented vertically and located at a height h above a plane, perfect conductor surface, as shown in Fig. 11.18(a). Since no waves can penetrate into the perfect conductor, as we learned in Section 6.4, the waves radiated from the dipole onto the conductor give rise to reflected waves, as shown in Fig. 11.18(a) for two directions of incidence. For a given incident wave onto the conductor surface, the angle of reflection is equal to the angle of incidence, as can be seen intuitively from the following reasons: (1) the reflected wave must propagate away from the conductor surface, (2) the apparent wavelengths of the incident and reflected waves parallel to the conductor surface must be equal, and (3) the tangential component of the resultant

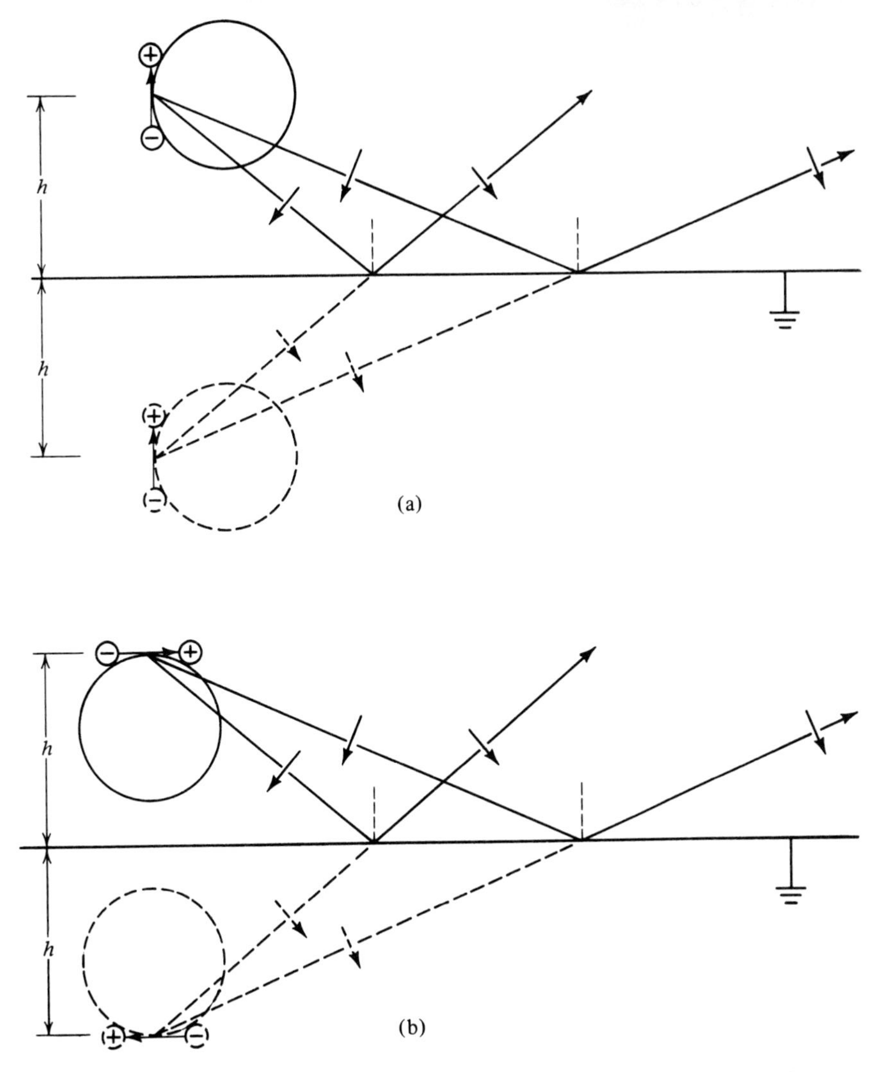

Figure 11.18. For illustrating the concept of image antennas. (a) Vertical Hertzian dipole and (b) horizontal Hertzian dipole above a plane perfect conductor surface.

electric field on the conductor surface must be zero, which also determines the polarity of the reflected wave electric field. Also because of (3), the reflected wave amplitude must equal the incident wave amplitude. If we now produce the directions of propagation of the two reflected waves backward, they meet at a point which is directly beneath the dipole and at the same distance h below the conductor surface as the dipole is above it. Thus the reflected waves appear to be originating from an antenna, which is the *image* of the actual antenna about the conductor surface. This image antenna must also be a vertical antenna since in order for the boundary condition of zero tangential electric field to be satisfied at all points on the conductor surface, the image antenna must have the same radiation pattern as that of the actual antenna, as shown in Fig. 11.18(a). In particular, the current in the image antenna must be directed in the same sense as that in the actual antenna to be consistent with the polarity of the reflected wave electric field. It can therefore be seen that the charges associated with the image dipole have signs opposite to those of the corresponding charges associated with the actual dipole.

A similar reasoning can be applied to the case of a horizontal Hertzian dipole above a perfect conductor surface, as shown in Fig. 11.18(b). Here it can be seen that the current in the image antenna is directed in the opposite sense to that in the actual antenna. This again results in charges associated with the image dipole having signs opposite to those of the corresponding charges associated with the actual dipole. In fact, this is always the case.

From the foregoing discussion it can be seen that the field due to an antenna in the presence of the conductor is the same as the resultant field of the array formed by the actual antenna and the image antenna. There is, of course, no field inside the conductor. The image antenna is only a virtual antenna that serves to simplify the field determination outside the conductor. The simplification results from the fact that we can use the knowledge gained on antenna arrays to determine the radiation pattern.

For example, for a vertical Hertzian dipole at a height of $\lambda/2$ above the conductor surface, the radiation pattern in the vertical plane is the product of the unit pattern, which is the radiation pattern of the single dipole in the plane of its axis, and the group pattern corresponding to an array of two isotropic radiators spaced λ apart and fed in phase. This multiplication and the resultant pattern are illustrated in Fig. 11.19. The radiation patterns for the case of the horizontal dipole can be obtained in a similar manner.

Corner reflector

To discuss another example of the application of the image-antenna concept, we consider the corner reflector, an arrangement of two plane perfect conductors at an angle to each other, as shown by the cross-sectional view in Fig. 11.20 for the case of 90° angle. We shall assume that each conductor is semi-infinite in extent. For a Hertzian dipole situated parallel to both conductors, the locations and polarities of the images can be obtained to be as shown in the figure. Using the pattern multiplication technique, the radiation pattern in the cross-sectional plane can then be obtained.

For an example, let $d_1 = d_2 = \lambda/4$. Then using the notation in Fig. 11.20, we can consider antennas 1 and 2 as constituting a unit for which the pattern is $|\sin[(\pi/2)\sin\psi]|$, which is that of case 2 in Example 11.3, except that ψ is measured from the line which is perpendicular to the axis of the array. Antennas 3 and 4 constitute a similar unit except for opposite polarity so that the group pattern for the two units is $|\sin[(\pi/2)\cos\psi]|$. Thus the required radiation pattern is

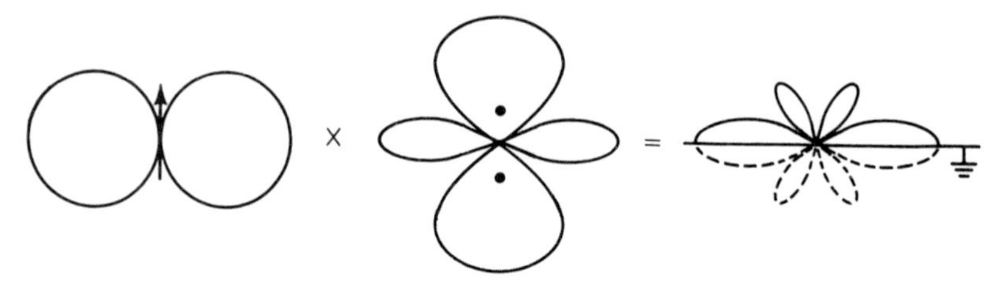

Figure 11.19. Determination of radiation pattern in the vertical plane for a vertical Hertzian dipole above a plane perfect conductor surface.

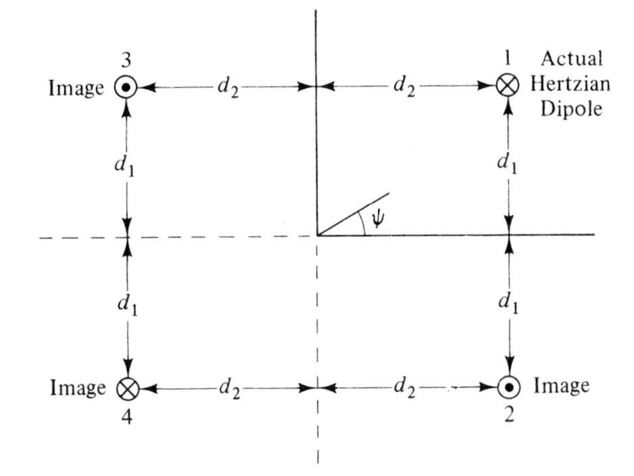

Figure 11.20. Application of image-antenna concept to obtain the radiation pattern for a Hertzian dipole in the presence of a corner reflector.

$$\left| \sin\left(\frac{\pi}{2}\sin\psi\right)\sin\left(\frac{\pi}{2}\cos\psi\right)\right|$$

which is shown plotted in Fig. 11.21.

K11.4. Antenna array; Unit pattern; Group pattern; Resultant pattern; Pattern multiplication; Uniform linear array; Image antenna concept; Corner reflector.

D11.6. For the array of two antennas of Example 11.3, assume that $d = 3\lambda/2$ and $\alpha = \pi/2$. Find the three lowest values of ψ for which the group pattern has nulls.
Ans. 33.56°, 80.41°, 120°

D11.7. Obtain the expression for the resultant pattern for each of the following cases of linear array of isotropic antennas: **(a)** three antennas carrying currents with amplitudes in the ratio 1:2:1, spaced λ apart and fed in phase; **(b)** five antennas carrying currents with amplitudes in the ratio 1:2:2:2:1, spaced $\lambda/2$ apart and with progressive phase shift of 180°; and **(c)** five antennas carrying currents in the ratio 1:2:3:2:1, spaced λ apart and fed in phase.
Ans. **(a)** $\cos^2(\pi\cos\psi)$; **(b)** $\sin^2[(\pi/2)\cos\psi]\,|\cos(\pi\cos\psi)|$;
(c) $[\sin^2(3\pi\cos\psi)]/[9\sin^2(\pi\cos\psi)]$

D11.8. For the Hertzian dipole in the presence of the corner reflector of Fig. 11.20, let r be the ratio of the radiation field at a point in the cross-sectional plane and along the line extending from the corner through the dipole to the radiation field at the

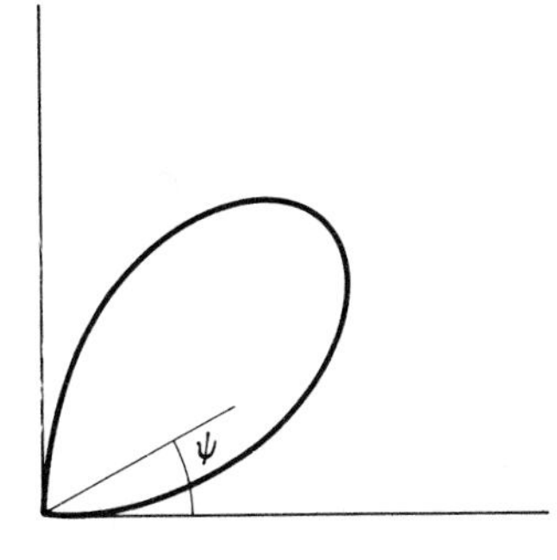

Figure 11.21. Radiation pattern in the cross-sectional plane for the case of $d_1 = d_2 = \lambda/4$ in the arrangement of Fig. 11.20.

same point in the absence of the corner reflector. Find the value of r for each of the following cases: **(a)** $d_1 = d_2 = \sqrt{2}\lambda$; **(b)** $d_1 = d_2 = \lambda/4\sqrt{2}$; and **(c)** $d_1 = 0.3\lambda$, $d_2 = 0.4\lambda$.
Ans. **(a)** 0; **(b)** 2; **(c)** 3.275

11.5 APERTURE ANTENNAS

Description and examples An important class of antennas, termed aperture antennas, is one for which the radiation is computed from a knowledge of the field distribution in an aperture instead of from a current distribution associated with the source of radiation, as has been the case thus far. The corner reflector discussed in the previous section is, in the practical case of finite-sized conductors (and hence defining an aperture), an example of such an antenna. Besides reflectors such as the corner reflector, other examples of aperture antennas are horns extending from waveguides, slots in conducting enclosures, and lenses. Essentially for an aperture antenna the primary source, which is elsewhere, sets up the field distribution in the aperture, which in turn is assumed to give rise to secondary waves in accordance with the Huygens–Fresnel principle, introduced in Section 10.5.

Far-field determination In particular, as mentioned in Section 10.5, the determination of the far field from an aperture antenna is the same as setting up the problem to solve for Fraunhofer diffraction from the aperture. To review briefly, consider a plane monochromatic wave incident normally on a screen in the x-y plane, with an aperture cut into it, as shown in Fig. 11.22. Then, according to the Huygens–Fresnel principle, the incident wave may be thought of as giving rise to secondary (spherical) waves emanating from every point in the aperture and which interfere with one another to produce the field distribution away from the aperture. The scalar

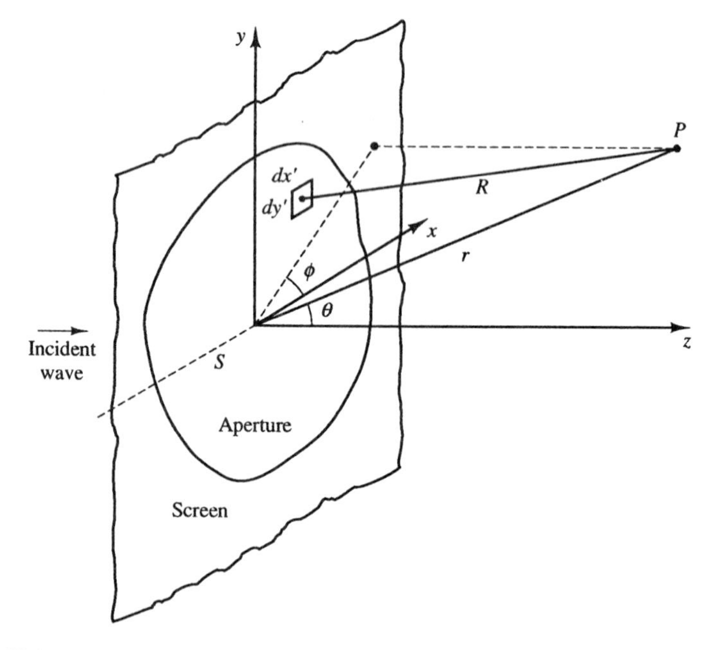

Figure 11.22. Geometry pertinent to the determination of the far field for radiation from an aperture antenna.

field at a point P is approximately given by

$$\bar{E}(P) \approx \frac{j\beta}{2\pi} \int_S \frac{\bar{E}(x', y', 0)}{R} e^{-j\beta R}\, dS \tag{11.51}$$

where S is the area of the aperture and $\bar{E}(x', y', 0)$ is the scalar field in the aperture. For the Fraunhofer approximation, the waves arriving at P approach plane waves, thereby permitting simplification of the integrand in (11.51) by using the plane wave approximation. This consists of assuming that the lines from points in the aperture $(x', y', 0)$ to the observation point $P(x, y, z)$ are all parallel $(x', y' \ll r)$, so that

$$\begin{aligned} R &= \sqrt{(x - x')^2 + (y - y')^2 + z^2} \\ &= \sqrt{r^2 - 2xx' - 2yy' + (x')^2 + (y')^2} \\ &= r\left[1 - \frac{2xx'}{r^2} - \frac{2yy'}{r^2} + \left(\frac{x'}{r}\right)^2 + \left(\frac{y'}{r}\right)^2\right]^{1/2} \\ &\approx r\left(1 - \frac{xx'}{r^2} - \frac{yy'}{r^2}\right) \\ &= r - x' \sin\theta \cos\phi - y' \sin\theta \sin\phi \end{aligned} \tag{11.52}$$

For the R in the denominator in the integrand, further approximation can be made as $R \approx r$. Thus (11.51) reduces to

$$\boxed{\bar{E}(x, y, z) \approx \frac{j\beta}{2\pi r} e^{-j\beta r} \int_S \bar{E}(x', y', 0)\, e^{j\beta \sin\theta\,(x' \cos\phi + y' \sin\phi)}\, dx'\, dy'} \tag{11.53}$$

Equation (11.53) is the starting point for the determination of the far-field distribution for an aperture antenna. We shall illustrate by means of an example.

Example 11.5

Rectangular aperture with uniform excitation

Let us consider a rectangular aperture in the *x-y* plane and centered at the origin with a uniform field distribution $\mathbf{E} = E_0\mathbf{i}_y$ in it, as shown in Fig. 11.23, and investigate the characteristics of the far field due to it.

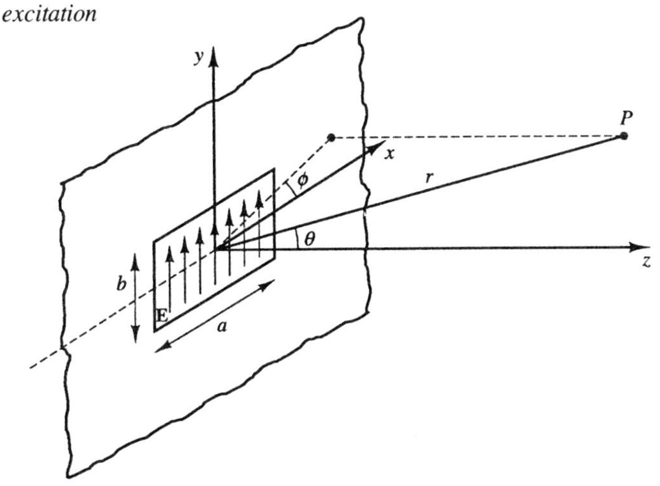

Figure 11.23 Rectangular aperture antenna with a uniform field distribution in the aperture.

Applying (11.53) to the rectangular aperture, we have at a point $P(r, \theta, \phi)$ far from the aperture

$$\bar{E} \approx \frac{j\beta e^{-j\beta r}}{2\pi r} \int_{x'=-a/2}^{a/2} \int_{y'=-b/2}^{b/2} E_0 e^{j\beta \sin\theta\, (x' \cos\phi + y' \sin\phi)}\, dx'\, dy'$$
$$= \frac{j\beta E_0 e^{-j\beta r}}{2\pi r} \int_{-a/2}^{a/2} e^{jx'\beta \sin\theta \cos\phi}\, dx' \int_{-b/2}^{b/2} e^{jy'\beta \sin\theta \sin\phi}\, dy' \tag{11.54}$$

Evaluating the integrals, we obtain

$$\boxed{\bar{E} = \frac{j\beta E_0 ab e^{-j\beta r}}{2\pi r}\left(\frac{\sin \psi_1}{\psi_1}\right)\left(\frac{\sin \psi_2}{\psi_2}\right)} \tag{11.55}$$

where

$$\boxed{\psi_1 = \frac{\beta a \sin\theta \cos\phi}{2}} \tag{11.56a}$$

and

$$\boxed{\psi_2 = \frac{\beta b \sin\theta \sin\phi}{2}} \tag{11.56b}$$

Radiation characteristics

The quantities of interest in (11.55) are the $(\sin \psi)/\psi$ type of terms, which determine the radiation pattern. To discuss this, we consider the two coordinate planes $\phi = 0$ and $\phi = 90°$ and find from (11.55) that the amplitudes of the fields in these two planes are given by

$$\boxed{\begin{aligned} |\bar{E}|_{\phi=0} &= \frac{\beta E_0 ab}{2\pi r}\left|\frac{\sin \psi_1}{\psi_1}\right|_{\phi=0} \\ &= \frac{\beta E_0 ab}{2\pi r}\left|\frac{\sin[(\beta a \sin\theta)/2]}{(\beta a \sin\theta)/2}\right| \end{aligned}} \tag{11.57a}$$

and

$$\boxed{\begin{aligned} |\bar{E}|_{\phi=90°} &= \frac{\beta E_0 ab}{2\pi r}\left|\frac{\sin \psi_2}{\psi_2}\right|_{\phi=90°} \\ &= \frac{\beta E_0 ab}{2\pi r}\left|\frac{\sin[(\beta b \sin\theta)/2]}{(\beta b \sin\theta)/2}\right| \end{aligned}} \tag{11.57b}$$

where we have used the fact that $\lim_{\Delta\to 0} \frac{\sin \Delta}{\Delta}$ is equal to 1. Thus in both planes the behavior is the same except for the appearance of the different dimensions a and b in the $(\sin \psi)/\psi$ factors in (11.57a) and (11.57b), respectively.

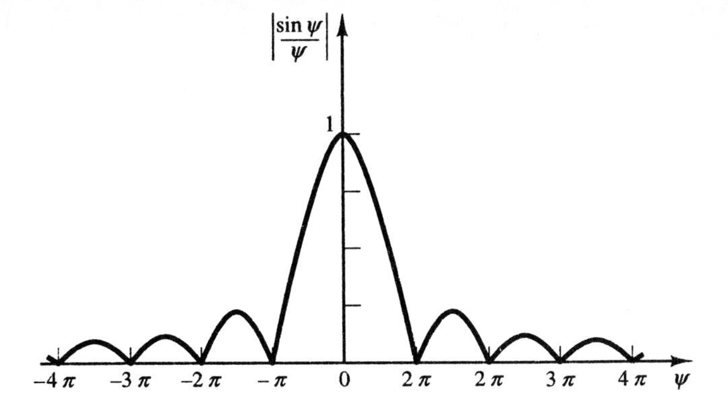

Figure 11.24 Variation of $|(\sin\psi)/\psi|$ with ψ, pertinent to the radiation pattern for the rectangular aperture antenna of Fig. 11.23.

To examine this behavior, we consider the plot of $|(\sin\psi)/\psi|$ versus ψ, which is shown in Fig. 11.24. We note that it indicates a strong central maximum of unity at $\psi = 0$ and a series of secondary (weaker) maxima on either side of it, with nulls occurring at $|\psi| = m\pi$, $m = 1, 2, 3, \ldots$. The secondary maxima, which occur at $|\psi| = 1.4303\pi, 2.459\pi, 3.471\pi, \ldots$, are successively less intense, having values 0.2172, 0.1284, 0.0913, . . . , respectively. If we consider the fact that the power density is proportional to $|\bar{E}|^2$, then the insignificance of these maxima becomes more evident, since the successive maxima of $|(\sin\psi)/\psi|^2$ are 1, 0.0472, 0.0165, 0.0083, ... Thus the quantity of interest is the beam width between the first nulls (BWFN) between which the radiation is concentrated. The BWFN is given by twice the value of θ corresponding to the first null. For the $\phi = 0$ plane this value is given by

$$\frac{\beta a \sin\theta}{2} = \pi \tag{11.58}$$

For narrow beams, which is the case in practice, $\sin\theta \approx \theta$ in this range so that (11.58) can be written as

$$\frac{\beta a\theta}{2} \approx \pi$$

$$\theta \approx \frac{2\pi}{\beta a} = \frac{\lambda}{a} \tag{11.59}$$

or

$$\boxed{[\text{BWFN}]_{\phi=0} \approx \frac{2\lambda}{a}} \tag{11.60a}$$

Similarly,

$$\boxed{[\text{BWFN}]_{\phi=90^\circ} \approx \frac{2\lambda}{b}} \tag{11.60b}$$

Finally, we consider the determination of the directivity of the rectangular aperture antenna. To do this, it is convenient to use the basic definition that

$$D = \frac{[P_r]_{\max}}{[P_r]_{\text{av}}} = \frac{\langle P_r \rangle_{\max}}{\langle P_r \rangle_{\text{av}}} = \frac{4\pi r^2 \langle P_r \rangle_{\max}}{\langle P_{\text{rad}} \rangle} \tag{11.61}$$

instead of using (11.27), since P_{rad}, the power radiated from the antenna, being the same as that passing through the aperture, is much easier to compute from the aperture field distribution as compared to the evaluation of the integral in (11.27). Thus in view of the uniform distribution of $\bar{\mathbf{E}}(x, y, 0) = E_0\mathbf{i}_y$ in the aperture,

$$<P_{rad}> = \frac{1}{2}\frac{E_0^2}{\eta_0}(ab) \tag{11.62}$$

and from (11.55)

$$\bar{E}_{max} = \frac{j\beta E_0 abe^{-j\beta r}}{2\pi r}$$

$$<P_r>_{max} = \frac{1}{2}\frac{|\bar{E}_{max}|^2}{\eta_0}$$

$$= \frac{\beta^2 E_0^2 a^2 b^2}{8\pi^2 r^2 \eta_0} \tag{11.63}$$

Substituting (11.62) and (11.63) into (11.61), we obtain

$$\boxed{D = \frac{\beta^2 ab}{\pi} = \frac{4\pi}{\lambda^2}(ab)} \tag{11.64}$$

This result tells us that the directivity of the rectangular aperture antenna is $4\pi/\lambda^2$ times the physical aperture, ab. Although we have derived it here for the rectangular aperture, it is true for an aperture of any shape with uniform excitation.

K11.5. Aperture antenna; Far field; Rectangular aperture; Uniform excitation; BWFN.

D11.9. For the rectangular aperture antenna of Fig. 11.23, the BWFN in the $\phi = 0$ plane is 0.1 rad and the directivity is 800π. Find the following in degrees: **(a)** the BWFN in the $\phi = 90°$ plane; **(b)** the half-power beamwidth (HPBW), that is, twice the value of θ for which the power density is one-half of the maximum power density in the $\phi = 0$ plane; and **(c)** the beamwidth between the first secondary maxima in the $\phi = 0$ plane.
Ans. **(a)** 11.46; **(b)** 2.54; **(c)** 8.19

11.6 RECEIVING PROPERTIES

Reciprocity

Thus far we have considered the radiating, or transmitting, properties of antennas. Fortunately, it is not necessary to repeat all the derivations for the discussion of the receiving properties of antennas since reciprocity dictates that the receiving pattern of an antenna be the same as its transmitting pattern. To illustrate this in simple terms without going through the general proof of reciprocity, let us consider a Hertzian dipole situated at the origin and directed along the z-axis, as shown in Fig. 11.25. We know that the radiation pattern is then given by $\sin\theta$ and that the polarization of the radiated field is such that the electric field is in the plane of the dipole axis.

To investigate the receiving properties of the Hertzian dipole, we assume that it is situated in the radiation field of a second antenna so that the incoming

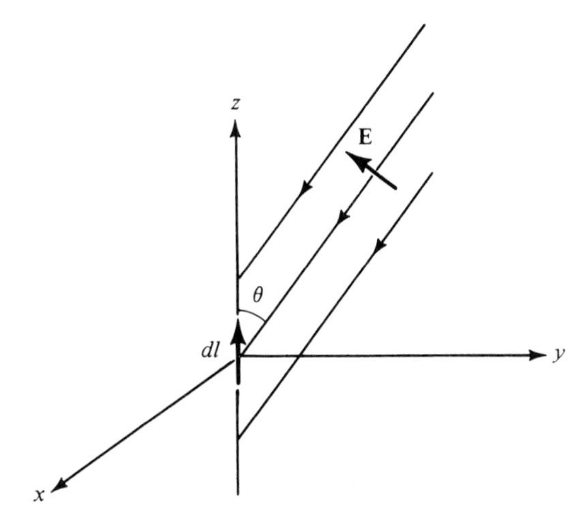

Figure 11.25. For investigating the receiving properties of a Hertzian dipole.

waves are essentially uniform plane waves. Thus let us consider a uniform plane wave with its electric field **E** in the plane of the dipole and incident on the dipole at an angle θ with its axis, as shown in Fig. 11.25. Then the component of the incident electric field parallel to the dipole is $E \sin \theta$. Since the dipole is infinitesimal in length, the voltage induced in the dipole, which is the line integral of the electric field intensity along the length of the dipole, is simply equal to $(E \sin \theta)\, dl$ or to $E\, dl \sin \theta$. This indicates that for a given amplitude of the incident wave field, the induced voltage in the dipole is proportional to $\sin \theta$. Furthermore, for an incident uniform plane wave having its electric field normal to the dipole axis, the voltage induced in the dipole is zero; that is, the dipole does not respond to polarization with electric field normal to the plane of its axis. These properties are reciprocal to the transmitting properties of the dipole. Since an arbitrary antenna can be decomposed into a series of Hertzian dipoles, it then follows that reciprocity holds for an arbitrary antenna. Thus the receiving pattern of an antenna is the same as its transmitting pattern.

Loop antenna

Let us consider the loop antenna, a common type of receiving antenna. A simple form of loop antenna consists of a circular loop of wire with a pair of terminals. We shall orient the circular loop antenna with its axis aligned with the z-axis, as shown in Fig. 11.26, and we shall assume that it is electrically short; that is, its dimensions are small compared to the wavelength of the incident wave, so that the spatial variation of the field over the area of the loop is negligible. For a uniform plane wave incident on the loop, we can find the voltage induced in the loop, that is, the line integral of the electric field intensity around the loop, by using Faraday's law. Thus if **H** is the magnetic field intensity associated with the wave, the magnitude of the induced voltage is given by

$$\begin{aligned} |V| &= \left| -\frac{d}{dt} \int_{\substack{\text{area of} \\ \text{the loop}}} \mathbf{B} \cdot d\mathbf{S} \right| \\ &= \left| -\mu \frac{d}{dt} \int_{\substack{\text{area of} \\ \text{the loop}}} \mathbf{H} \cdot dS\, \mathbf{i}_z \right| \qquad (11.65) \\ &= \mu A \left| \frac{\partial H_z}{\partial t} \right| \end{aligned}$$

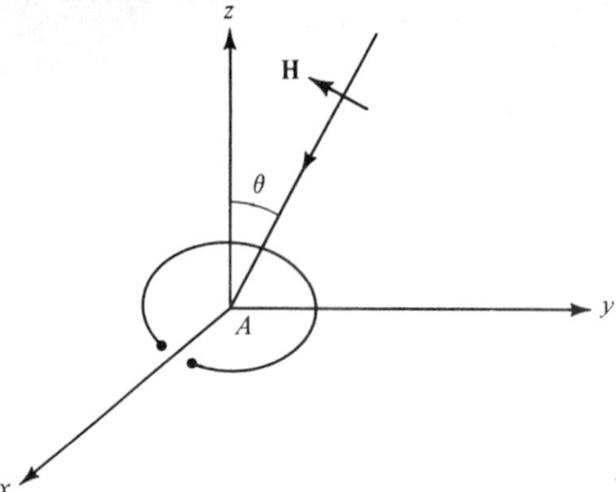

Figure 11.26. Circular loop antenna.

where A is the area of the loop. Hence the loop does not respond to a wave having its magnetic field entirely parallel to the plane of the loop, that is, normal to the axis of the loop.

For a wave having its magnetic field in the plane of the axis of the loop, and incident on the loop at an angle θ with its axis, as shown in Fig. 11.26, $H_z = H \sin \theta$ and hence the induced voltage has a magnitude

$$|V| = \mu A \left| \frac{\partial H}{\partial t} \right| \sin \theta \tag{11.66}$$

Thus the receiving pattern of the loop antenna is given by sin θ, the same as that of a Hertzian dipole aligned with the axis of the loop antenna. The loop antenna, however, responds best to polarization with the magnetic field in the plane of its axis, whereas the Hertzian dipole responds best to polarization with the electric field in the plane of its axis.

Example 11.6

The directional properties of a receiving antenna can be used to locate the source of an incident signal. To illustrate the principle, as already discussed in Section 3.2, let us consider two vertical loop antennas, numbered 1 and 2, situated on the x-axis at $x = 0$ m and $x = 200$ m, respectively. By rotating the loop antennas about the vertical (z-axis), it is found that no (or minimum) signal is induced in antenna 1 when it is in the xz-plane and in antenna 2 when it is in a plane making an angle of 5° with the axis, as shown by the top view in Fig. 11.27. Let us find the location of the source of the signal.

Since the receiving properties of a loop antenna are such that no signal is induced for a wave arriving along its axis, the source of the signal is located at the intersection of the axes of the two loops when they are oriented so as to receive no (or minimum) signal. From simple geometrical considerations, the source of the signal is therefore located on the y-axis at $y = 200/\tan 5°$ or 2.286 km.

Effective area

A useful parameter associated with the receiving properties of an antenna is the effective area, denoted A_e and defined as the ratio of the time-average power delivered to a matched load connected to the antenna to the time-average power

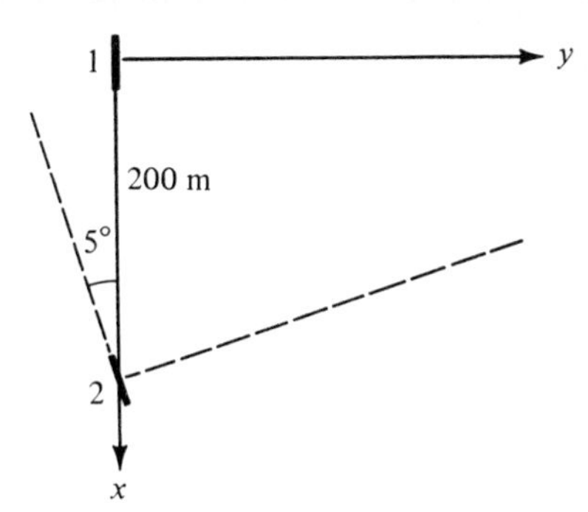

Figure 11.27. Top view of two loop antennas used to locate the source of an incident signal.

density of the appropriately polarized incident wave at the antenna. The matched condition is achieved when the load impedance is equal to the complex conjugate of the antenna impedance.

Let us consider the Hertzian dipole and derive the expression for its effective area. First, with reference to the equivalent circuit shown in Fig. 11.28, where $\bar{V}_{oc}$ is the open-circuit voltage induced between the terminals of the antenna, $\bar{Z}_A = R_A + jX_A$ is the antenna impedance, and $\bar{Z}_L = \bar{Z}_A^*$ is the load impedance, we note that the time-average power delivered to the matched load is

$$\begin{aligned} P_R &= \frac{1}{2}\left(\frac{|\bar{V}_{oc}|}{2R_A}\right)^2 R_A \\ &= \frac{|\bar{V}_{oc}|^2}{8R_A} \end{aligned} \tag{11.67}$$

For a Hertzian dipole of length l, the open-circuit voltage is

$$\bar{V}_{oc} = \bar{E}l \tag{11.68}$$

where $\bar{E}$ is the electric field of an incident wave linearly polarized parallel to the dipole axis. Substituting (11.68) into (11.67), we get

$$P_R = \frac{|\bar{E}|^2 l^2}{8R_A} \tag{11.69}$$

For a lossless dipole, $R_A = R_{rad} = 80\pi^2(l/\lambda)^2$, so that

$$P_R = \frac{|\bar{E}|^2 \lambda^2}{640\pi^2} \tag{11.70}$$

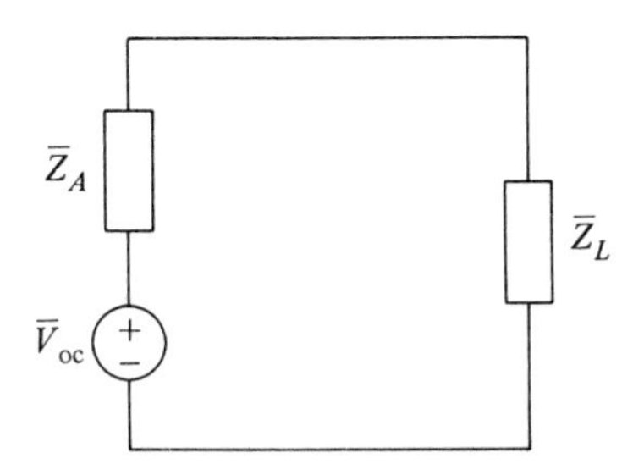

Figure 11.28. Equivalent circuit for a receiving antenna connected to a load.

The time-average power density at the antenna is

$$\frac{|\bar{E}|^2}{2\eta_0} = \frac{|\bar{E}|^2}{240\pi} \tag{11.71}$$

Thus the effective area is

$$\begin{aligned} A_e &= \frac{|\bar{E}|^2\lambda^2/640\pi^2}{|\bar{E}|^2/240\pi} \\ &= \frac{3\lambda^2}{8\pi} \end{aligned} \tag{11.72}$$

or

$$\boxed{A_e = 0.1194\lambda^2} \tag{11.73}$$

In practice, R_A is greater than R_{rad} due to losses in the antenna, and the effective area is less than that given by (11.73). Rewriting (11.72) as

$$A_e = 1.5 \times \frac{\lambda^2}{4\pi}$$

and recalling that the directivity of the Hertzian dipole is 1.5, we observe that

$$\boxed{A_e = \frac{\lambda^2}{4\pi}D} \tag{11.74}$$

Although we have obtained this result for a Hertzian dipole, it can be shown that it holds for any antenna. It is of interest to note from (11.74) and (11.64) that the effective area of a rectangular aperture antenna for uniform field distribution in the aperture is equal to the physical aperture, which is to be expected

Friis transmission formula

We shall now derive the *Friis transmission formula*, an important equation in making communication link calculations. To do this, let us consider two antennas, one transmitting and the other receiving, separated by a distance d. Let us assume that the antennas are oriented and polarization matched so as to maximize the received signal. Then if P_T is the transmitter power radiated by the transmitting antenna, the power density at the receiving antenna is $(P_T/4\pi d^2)D_T$, where D_T is the directivity of the transmitting antenna. The power received by a matched load connected to the terminals of the receiving antenna is then given by

$$P_R = \frac{P_T D_T}{4\pi d^2}A_{eR} \tag{11.75}$$

where A_{eR} is the effective area of the receiving antenna. Thus the ratio of P_R to P_T is given by

$$\frac{P_R}{P_T} = \frac{D_T A_{eR}}{4\pi d^2} \tag{11.76}$$

Denoting A_{eT} to be the effective area of the transmitting antenna if it were receiving and using (11.74), we obtain

$$\boxed{\frac{P_R}{P_T} = \frac{A_{eT} A_{eR}}{\lambda^2 d^2}} \tag{11.77}$$

Equation (11.77) is the Friis transmission formula. It gives the maximum value of P_R/P_T for a given d and for a given pair of transmitting and receiving antennas. If the antennas are not oriented to receive the maximum signal, or if a polarization mismatch exists, or if the receiving antenna is not matched to its load, P_R/P_T would be less than that given by (11.77). Losses in the antennas would also decrease the value of P_R/P_T.

An alternative formula to (11.77) is obtained by substituting for A_{eR} in (11.76) in terms of the directivity D_R of the receiving antenna if it were used for transmitting. Thus we obtain

$$\boxed{\frac{P_R}{P_T} = \frac{D_T D_R \lambda^2}{16\pi^2 d^2}} \tag{11.78}$$

K11.6. Receiving pattern; Reciprocity with transmitting pattern; Effective area; Communication link; Friis transmission formula.

D11.10. A communication link in free space uses two linear antennas of equal lengths L, oriented parallel to each other and normal to the line joining their centers. The antennas are separated by a distance $d = 1$ km. Find the maximum value of P_R/P_T for each of the following cases: **(a)** $L = 1$ m, $f = 10$ MHz; **(b)** $L = 1$ m, $f = 150$ MHz; and **(c)** $L = 2$ m, $f = 75$ MHz.
Ans. **(a)** 12.8×10^{-6}; **(b)** 6.8×10^{-8}; **(c)** 27.25×10^{-8}

11.7 SUMMARY

In this chapter we studied the principles of antennas. We first introduced the Hertzian dipole, which is an elemental wire antenna, and derived the electromagnetic field due to the Hertzian dipole by using the retarded magnetic vector potential. For a Hertzian dipole of length dl, oriented along the z-axis at the origin and carrying current

$$I(t) = I_0 \cos \omega t$$

we found the complete electromagnetic field to be given by

$$\mathbf{E} = \frac{2I_0\, dl \cos\theta}{4\pi\varepsilon\omega}\left[\frac{\sin(\omega t - \beta r)}{r^3} + \frac{\beta\cos(\omega t - \beta r)}{r^2}\right]\mathbf{i}_r$$

$$+ \frac{I_0\, dl \sin\theta}{4\pi\varepsilon\omega}\left[\frac{\sin(\omega t - \beta r)}{r^3} + \frac{\beta\cos(\omega t - \beta r)}{r^2} - \frac{\beta^2 \sin \omega t - \beta r)}{r}\right]\mathbf{i}_\theta$$

$$\mathbf{H} = \frac{I_0\, dl \sin\theta}{4\pi}\left[\frac{\cos(\omega t - \beta r)}{r^2} - \frac{\beta \sin(\omega t - \beta r)}{r}\right]\mathbf{i}_\phi$$

where $\beta = \omega\sqrt{\mu\varepsilon}$ is the phase constant.

For $\beta r \gg 1$ or for $r \gg \lambda/2\pi$, the only important terms in the complete field expressions are the $1/r$ terms since the remaining terms are negligible compared to these terms. Thus for $r \gg \lambda/2\pi$, the Hertzian dipole fields are given by

$$\mathbf{E} = -\frac{\eta\beta I_0\, dl \sin\theta}{4\pi r} \sin(\omega t - \beta r)\, \mathbf{i}_\theta$$

$$\mathbf{H} = -\frac{\beta I_0\, dl \sin\theta}{4\pi r} \sin(\omega t - \beta r)\, \mathbf{i}_\phi$$

where $\eta = \sqrt{\mu/\varepsilon}$ is the intrinsic impedance of the medium. These fields, known as the radiation fields, correspond to locally uniform plane waves radiating away from the dipole and, in fact, are the only components of the complete fields contributing to the time-average radiated power. We found the time-average power radiated by the Hertzian dipole to be given by

$$\langle P_{\text{rad}} \rangle = \frac{1}{2} I_0^2 \left[\frac{2\pi\eta}{3}\left(\frac{dl}{\lambda}\right)^2\right]$$

and identified the quantity inside the brackets to be its radiation resistance. The radiation resistance, R_{rad}, of an antenna is the value of a fictitious resistor that will dissipate the same amount of time-average power as that radiated by the antenna when a current of the same peak amplitude as that in the antenna is passed through it. Thus for the Hertzian dipole,

$$R_{\text{rad}} = \frac{2\pi\eta}{3}\left(\frac{dl}{\lambda}\right)^2$$

We then examined the directional characteristics of the radiation fields of the Hertzian dipole as indicated by the factor $\sin\theta$ in the field expressions and hence by the factor $\sin^2\theta$ for the power density. We discussed the radiation patterns and introduced the concept of the directivity of an antenna. The directivity, D, of an antenna is defined as the ratio of the maximum power density radiated by the antenna to the average power density. For the Hertzian dipole,

$$D = 1.5$$

For the general case of a power density pattern $f(\theta, \phi)$, the directivity is given by

$$D = 4\pi \frac{[f(\theta, \phi)]_{\max}}{\int_{\theta=0}^{\pi}\int_{\phi=0}^{2\pi} f(\theta, \phi) \sin\theta\, d\theta\, d\phi}$$

As an illustration of obtaining the radiation fields due to a wire antenna of arbitrary length and arbitrary current distribution by representing it as a series of Hertzian dipoles and using superposition, we considered the example of a half-wave dipole of length L $(= \lambda/2)$, oriented along the z-axis with its center at the origin and having the current distribution given by

$$I(z) = I_0 \cos\frac{\pi z}{L} \cos\omega t \qquad \text{for} \quad -L/2 < z < L/2$$

the radiation fields are

$$\mathbf{E} = -\frac{\eta I_0}{2\pi r}\frac{\cos[(\pi/2)\cos\theta]}{\sin\theta} \sin\left(\omega t - \frac{\pi}{L}r\right) \mathbf{i}_\theta$$

$$\mathbf{H} = -\frac{I_0}{2\pi r}\frac{\cos[(\pi/2)\cos\theta]}{\sin\theta}\sin\left(\omega t - \frac{\pi}{L}r\right)\mathbf{i}_\phi$$

From these, we sketched the radiation patterns and computed the radiation resistance and the directivity of the half-wave dipole to be

$$R_{rad} = 73\ \Omega \quad \text{for free space}$$

$$D = 1.642$$

We then extended the computation of these quantities to the case of a center-fed linear antenna of length equal to an arbitrary number of wavelengths.

We discussed antenna arrays and introduced the technique of obtaining the resultant radiation pattern of an array by multiplication of the unit and the group patterns. For an array of two antennas having the spacing d and fed with currents of equal amplitude but differing in phase by α, we found the group pattern for the fields to be $|\cos[(\beta d \cos\psi + \alpha)/2]|$, where ψ is the angle measured from the axis of the array, and we investigated the group patterns for several pairs of values of d and α. For example, for $d = \lambda/2$ and $\alpha = 0$, the pattern corresponds to maximum radiation broadside to the axis of the array, whereas for $d = \lambda/2$ and $\alpha = \pi$, the pattern corresponds to maximum radiation endfire to the axis of the array. We generalized the treatment to a uniform linear array of n antennas and briefly discussed the principle of a broadband array.

To take into account the effect of ground on antennas, we introduced the concept of an image antenna in a perfect conductor and discussed the application of the array techniques in conjunction with the actual and the image antennas to obtain the radiation pattern of the actual antenna in the presence of the ground. As another example of the image antenna concept, we considered the corner reflector.

Next we discussed the far-field determination for an aperture antenna by recalling that it is equivalent to setting up the problem to solve for Fraunhofer diffraction from the aperture, which consists of using the plane wave approximation. By considering the example of a rectangular aperture with uniform field distribution in it, we illustrated the solution and studied the resulting radiation pattern and its characteristics.

Finally, we discussed receiving properties of antennas. In particular (1) we discussed the reciprocity between the receiving and radiating properties of an antenna by considering the simple case of a Hertzian dipole, (2) we considered the loop antenna and illustrated the application of its directional properties for locating the source of a radio signal, and (3) we introduced the effective area concept and derived the Friis transmission formula.

REVIEW QUESTIONS

R11.1. What is a Hertzian dipole? Discuss the time variations of the current and charges associated with the Hertzian dipole.

R11.2. Discuss the analogy between the magnetic vector potential due to an infinitesimal current element and the electric scalar potential due to a point charge.

R11.3. To what does the word *retarded* in the terminology *retarded magnetic vector potential* refer? Explain.

R11.4. Outline the derivation of the electromagnetic field due to the Hertzian dipole.

R11.5. Discuss the characteristics of the electromagnetic field due to the Hertzian dipole.

R11.6. What are radiation fields? Why are they important? Discuss their characteristics.

R11.7. Define the radiation resistance of an antenna.

R11.8. Why is the expression for the radiation resistance of a Hertzian dipole not valid for a linear antenna of any length?

R11.9. What is a radiation pattern?

R11.10. Discuss the radiation pattern for the power density due to the Hertzian dipole.

R11.11. Define the directivity of an antenna. What is the directivity of a Hertzian dipole?

R11.12. How do you find the radiation fields due to an antenna of arbitrary length and arbitrary current distribution?

R11.13. Discuss the evolution of the half-wave dipole from an open-circuited transmission line.

R11.14. Justify the approximations involved in evaluating the integrals in the determination of the radiation fields due to the half-wave dipole.

R11.15. What are the values of the radiation resistance and the directivity for a half-wave dipole?

R11.16. What is an antenna array?

R11.17. Justify the approximations involved in the determination of the resultant field of an array of two antennas.

R11.18. What is an array factor? Provide a physical explanation for the array factor.

R11.19. Discuss the concept of unit and group patterns and their multiplication to obtain the resultant pattern of an array.

R11.20. Distinguish between broadside and endfire radiation patterns.

R11.21. Discuss the principle of a phased array.

R11.22. Discuss the principle of a broadband array using as an example the log-periodic dipole array.

R11.23. Discuss the concept of an image antenna to find the field of an antenna in the vicinity of a perfect conductor.

R11.24. What determines the sense of the current flow in an image antenna relative to that in the actual antenna?

R11.25. How does the concept of an image antenna simplify the determination of the radiation pattern of an antenna above a perfect conductor surface?

R11.26. Discuss the application of the image antenna concept to the 90° corner reflector.

R11.27. Explain the distinguishing feature pertinent to the computation of radiation from an aperture antenna.

R11.28. Give examples of aperture antennas.

R11.29. Discuss the determination of the far field for an aperture antenna.

R11.30. Describe the radiation pattern for the far field of a rectangular aperture antenna with uniform field distribution in the aperture and discuss its characteristics.

R11.31. Discuss the reciprocity associated with the transmitting and receiving properties of an antenna. Can you think of a situation in which reciprocity does not hold?

R11.32. What is the receiving pattern of a loop antenna? How should you orient a loop antenna to receive **(a)** a maximum signal and **(b)** a minimum signal?

R11.33. Discuss the application of the directional receiving properties of a loop antenna in the location of the source of a radio signal.

R11.34. How is the effective area of a receiving antenna defined?

R11.35. Outline the derivation of the expression for the effective area of a Hertzian dipole.

R11.36. Discuss the derivation of the Friis transmission formula.

PROBLEMS

P11.1. Show that Eqs. (11.9) and (11.10) satisfy the Maxwell's curl equations for **E**.

P11.2. For the electromagnetic field due to the Hertzian dipole, show that **(a)** the time-average value of the θ-component of the Poynting vector is zero and **(b)** the contribution to the time-average value of the r-component of the Poynting vector is completely from the terms involving $1/r$.

P11.3. Show that the field expressions obtained by replacing ωt in Eqs. (11.11) and (11.12) by $(\omega t - \beta r)$ do not satisfy Maxwell's curl equations.

P11.4. A Hertzian dipole of length 0.1 m situated at the origin and oriented along the positive z-direction carries the current $10 \cos^3 2\pi \times 10^7 t$ A. Find the root-mean-square values of E_r, E_θ, and H_ϕ at the point $(5, \pi/6, 0)$. Assume free space for the medium.

P11.5. Show that the radiation fields given by Eqs. (11.17a) and (11.17b) do not by themselves satisfy simultaneously the Maxwell's curl equations.

P11.6. Find the value of r at which the amplitude of the radiation field in the θ-component of **E** in (11.10) is equal to the resultant amplitude of the remaining two terms.

P11.7. A Hertzian dipole of length 1 m situated in free space carries the current $10 \cos 4\pi \times 10^6 t$ A. Calculate the radiation resistance and the time-average power radiated by the dipole.

P11.8. Find the amplitude I_0 of the current with which a Hertzian dipole of length 0.1 m has to be excited at a frequency of 30 MHz to produce an electric field intensity of amplitude 1 mV/m at a distance of 1 km broadside to the dipole in free space. What is the time-average power radiated for the computed value of I_0?

P11.9. The power density pattern for an antenna located at the origin is given by

$$f(\theta, \phi) = \begin{cases} \csc^2 \theta & \text{for } \pi/6 \le \theta \le \pi/2 \\ 0 & \text{otherwise} \end{cases}$$

Find the directivity of the antenna.

P11.10. Find the ratio of the currents in two antennas having directivities D_1 and D_2 and radiation resistances R_{rad1} and R_{rad2} for which the maximum time-average radiated power densities are equal.

P11.11. For the half-wave dipole of Section 11.3, find the magnetic vector potential for the radiation fields and show that the radiation fields obtained from it are the same as those given by Eqs. (11.31a) and (11.31b).

P11.12. Find the maximum amplitude I_0 of the current with which a linear dipole of length 5 m has to be excited at a frequency of 30 MHz in order to produce an electric field intensity of amplitude 1 mV/m at a distance of 1 km broadside to the dipole in free space. What is the time-average power radiated for the computed value of I_0?

P11.13. Repeat Problem P11.12 for a linear dipole of length 5 m at a frequency of 150 MHz.

P11.14. A short dipole is a center-fed straight-wire antenna having a length small compared to a wavelength. The amplitude of the current distribution can then be approximated as decreasing linearly from a maximum at the center to zero at the ends. Thus for a short dipole of length L lying along the z-axis between $z = -L/2$ and $z = L/2$, the current distribution is given by

$$I(z) = \begin{cases} I_0\left(1 + \dfrac{2z}{L}\right) \cos \omega t & \text{for } -L/2 < z < 0 \\ I_0\left(1 - \dfrac{2z}{L}\right) \cos \omega t & 0 < z < L/2 \end{cases}$$

(a) Obtain the radiation fields of the short dipole. **(b)** Find the radiation resistance and the directivity of the short dipole.

P11.15. Consider a circular loop antenna of radius a such that the circumference is small compared to the wavelength. Assume the loop antenna to be in the xy-plane with its center at the origin and the loop current to be $I = I_0 \cos \omega t$ in the sense of increasing ϕ. Show that for obtaining the radiation fields, the magnetic vector potential due to the loop antenna is given by

$$\mathbf{A} = \frac{\mu_0 I_0 \pi a^2 \beta \sin \theta}{4\pi r} \sin (\omega t - \beta r)\, \mathbf{i}_\phi$$

where $\beta = \omega/v_p$. Then show that the radiation fields are

$$\mathbf{E} = \frac{\eta I_0 \pi a^2 \beta^2 \sin \theta}{4\pi r} \cos (\omega t - \beta r)\, \mathbf{i}_\phi$$

$$\mathbf{H} = -\frac{I_0 \pi a^2 \beta^2 \sin \theta}{4\pi r} \cos (\omega t - \beta r)\, \mathbf{i}_\theta$$

P11.16. Find the radiation resistance and the directivity of the circular loop antenna of Problem P11.15. Compare the dependence of the radiation resistance on the electrical size (circumference/wavelength) to the dependence of the radiation resistance of the Hertzian dipole on its electrical size (length/wavelength).

P11.17. For the array of two antennas of Example 11.3, find and sketch the group patterns for each of the following cases: **(a)** $d = \lambda$, $\alpha = \pi/2$ and **(b)** $d = 2\lambda$, $\alpha = 0$.

P11.18. For the array of two Hertzian dipoles in Fig. 11.11, find and sketch the resultant pattern in the xz-plane for each of the following cases: **(a)** $d = \lambda/4$, $\alpha = -\pi/2$ and **(b)** $d = \lambda$, $\alpha = 0$.

P11.19. For the array of two Hertzian dipoles in Fig. 11.11, assume that $d \ll \lambda$ and $\alpha = \pi$. Obtain an approximate expression for the three-dimensional power density pattern $f(\theta, \phi)$ and find the directivity.

P11.20. For a linear binomial array of n antennas, the amplitudes of the currents in the elements are proportional to the coefficients in the polynomial $(1 + x)^{n-1}$. Show that the group pattern is $|\cos [(\beta d \cos \psi + \alpha)/2]|^{n-1}$, where d is the spacing between the elements and α is the progressive phase shift.

P11.21. The pattern multiplication technique can be used in reverse to synthesize an array for a specified radiation pattern. Find an arrangement of isotropic elements for the group pattern

$$\frac{\sin(2\pi \cos\psi)}{4 \sin[(\pi/2)\cos\psi]} \cos^2\left(\frac{\pi}{2}\cos\psi\right)$$

P11.22. Repeat Problem P11.21 for the group pattern

$$\frac{\cos^2(6\pi \cos\psi)}{9\cos^2(2\pi\cos\psi)}$$

P11.23. For a horizontal half-wave dipole at a height $\lambda/4$ above a plane, perfect conductor surface, find and sketch the radiation pattern in **(a)** the vertical plane perpendicular to the axis of the antenna and **(b)** the vertical plane containing the axis of the antenna.

P11.24. For a vertical antenna of length $\lambda/4$ above a plane, perfect conductor surface, find **(a)** the radiation pattern in the vertical plane and **(b)** the directivity.

P11.25. A Hertzian dipole is situated parallel to one side and perpendicular to the other side of a 90° corner reflector, as shown in Fig. 11.29. Find the expression for the radiation pattern in the plane of the paper as a function of the angle θ and the distances d_1 and d_2.

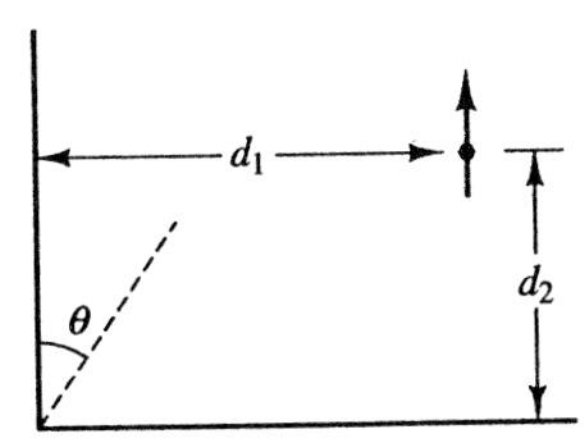

Figure 11.29. For Problem P11.25.

P11.26. A corner reflector is made up of two semi-infinite plane, perfect conductors at an angle of 60°, as shown by the cross-sectional view in Fig. 11.30. A Hertzian dipole is situated parallel to the conductors at a distance of $k\lambda$ from the corner along the bisector of the two conductors. Find the ratio of the radiation field at a point broadside to the dipole and along the bisector of the conductors to the radiation field at the same point in the absence of the corner reflector for the following values of k: **(a)** $\frac{1}{4}$; **(b)** $\frac{1}{2}$; and **(c)** 1.

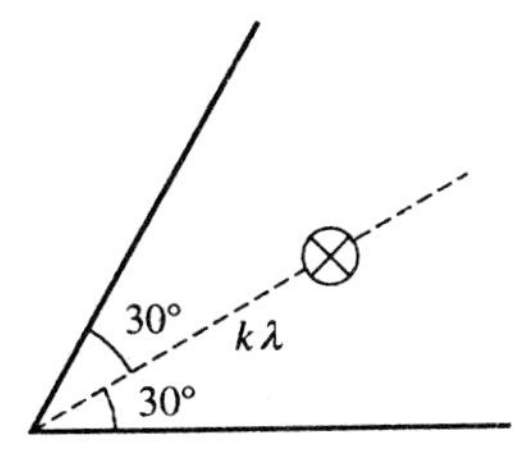

Figure 11.30. For Problem P11.26.

P11.27. A corner reflector occupies the three coordinate planes in the Cartesian coordinate system that bound the first octant. For a Hertzian dipole oriented in the positive z-direction and located at the point $(\lambda, \lambda, \lambda)$, find the ratio of the radiation field along the line $x = y = z$ to the radiation field at the same point in the absence of the corner reflector.

P11.28. For the rectangular aperture antenna of Example 11.5, assume that the field distribution in the aperture is nonuniform as given by,

$$\bar{\mathbf{E}}(x, y, 0) = E_0 \cos \frac{\pi x}{a} \mathbf{i}_y \text{ for } -a/2 < x < a/2, \; -b/2 < y < b/2$$

Obtain the expression for the far field and hence the expressions for the following: **(a)** BWFN in the $\phi = 0$ plane; **(b)** BWFN in the $\phi = 90°$ plane; **(c)** HPBW (see Problem D11.9) in the $\phi = 0$ plane; and **(d)** the directivity.

P11.29. Repeat Problem P11.28 for

$$\bar{\mathbf{E}}(x, y, 0) = E_0\left(1 - \frac{2|x|}{a}\right)\mathbf{i}_y \text{ for } -a/2 < x < a/2, \; -b/2 < y < b/2$$

P11.30. Consider a circular aperture of radius a in the xy-plane and centered at the origin. For uniform field distribution $\bar{\mathbf{E}} = E_0\mathbf{i}_x$ in the aperture, show that the far-field radiation pattern is in accordance with $J_1(\beta a \sin \theta)/(\beta a \sin \theta)$. Further, given that the first nonzero root of $J_1(x) = 0$ is 3.83, show that in any constant-ϕ plane the BWFN is approximately equal to $1.22\lambda/a$.

[*Note:* $\frac{1}{2\pi}\int_0^{2\pi} e^{jx \cos \alpha}\, d\alpha = J_0(x) \quad \int xJ_0(x)\, dx = xJ_1(x)$]

P11.31. Consider the uniform linear array of n isotropic antennas of Fig. 11.15 for the case of $\alpha = 0$ so that the group pattern is a broadside pattern. Show that for large n and for $nd \gg \lambda$, the radiation pattern is the same as that in one of the coordinate planes ($\phi = 0$ or $\phi = 90°$) for the rectangular aperture antenna with uniform field distribution of Example 11.5, and hence the BWFN is approximately equal to $2\lambda/L$, where L is the length of the array.

P11.32. Consider the uniform linear array of n isotropic antennas of Fig. 11.15 for the case of $\alpha = -\beta d$ so that the group pattern is an endfire pattern. Show that for large n and for $nd \gg \lambda$, the BWFN is approximately equal to $\sqrt{8\lambda/L}$.

P11.33. An arrangement of two identical Hertzian dipoles situated at the origin and oriented along the x- and y-axes, known as the turnstile antenna, is used for receiving circularly polarized signals arriving along the z-axis. Determine how you would combine the voltages induced in the two dipoles so that the turnstile antenna is responsive to circular polarization rotating in the clockwise sense as viewed by the antenna but not to that of the counterclockwise sense of rotation.

P11.34. An interferometer consists of an array of two identical antennas with spacing $d = 3\lambda$. For a uniform plane wave incident on the array at an angle ψ to the axis of the array, as shown in Fig. 11.31, the phase difference $\Delta\phi$ between the voltage induced in antenna 1 and the voltage induced in antenna 2 is measured to be 60°. Find all possible values of ψ, taking into account the fact that the phase measurement is ambiguous by the amount $\pm 2n\pi$, where n is an integer. What should be the upper limit for the spacing between the antennas for ψ to be determined unambiguously for the measured value of 60° for the phase difference?

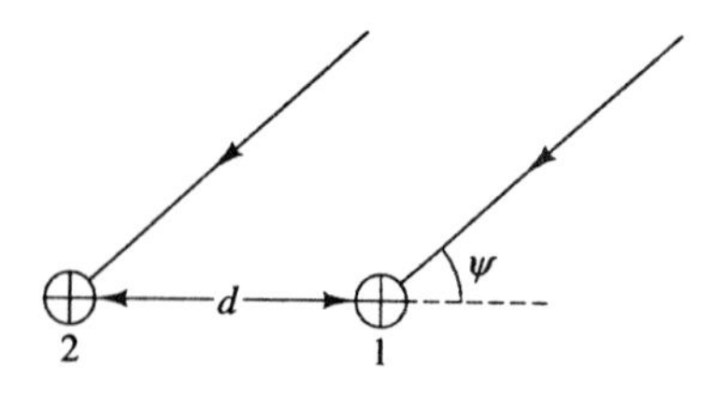

Figure 11.31. For Problem P11.34.

P11.35. A communication link at a frequency of 30 MHz uses a half-wave dipole for the transmitting antenna and a small loop for the receiving antenna. The antennas are oriented so as to receive maximum signal and the receiving antenna is matched to its load. The transmitting antenna is excited by a current of maximum amplitude $I_0 = 10$ A. If the received time-average power is to be 1 μW, find the maximum allowable distance between the two antennas. Assume the antennas to be lossless.

PC EXERCISES

PC11.1. Modify the program of PL 11.3 to plot the group patterns for a linear binomial array of n antennas (see Problem P11.20). The input parameters are to be the same as in PL 11.3.

PC11.2. Write a program to plot the radiation pattern for the arrangement of Problem P11.25 (Fig. 11.29) in the cross-sectional plane containing the Hertzian dipole. The input quantities are to be the values of d_1/λ and d_2/λ.

PC11.3. Consider the two-element interferometer in Problem P11.34. Assuming the input parameters to be the spacing d as a ratio of wavelength and the measured phase difference $\Delta\phi$ in degrees between the voltage induced in antenna 1 and the voltage induced in antenna 2, write a program to compute all possible values of ψ in degrees.

Appendix A

Curl, Divergence, Gradient, and Laplacian in Cylindrical and Spherical Coordinate Systems

In Chapter 4 we introduced the curl, divergence, gradient, and Laplacian and derived the expressions for them in the Cartesian coordinate system. In this appendix we derive the corresponding expressions in the cylindrical and spherical coordinate systems. Considering first the cylindrical coordinate system, we recall from Section 1.3 that the infinitesimal box defined by the three orthogonal surfaces intersecting at point $P(r, \phi, z)$ and the three orthogonal surfaces intersecting at point $Q(r + dr, \phi + d\phi, z + dz)$ is as shown in Fig. A.1.

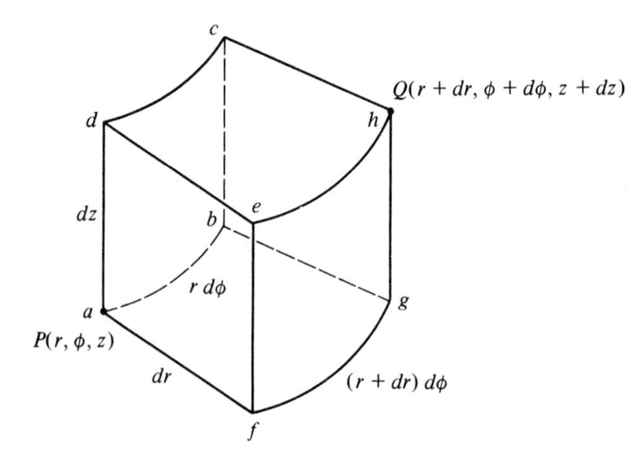

Figure A.1. Infinitesimal box formed by incrementing the coordinates in the cylindrical coordinate system.

From the basic definition of the curl of a vector introduced in Section 4.3 and given by

$$\nabla \times \mathbf{A} = \lim_{\Delta S \to 0} \left[\frac{\oint_C \mathbf{A} \cdot d\mathbf{l}}{\Delta S} \right]_{\max} \mathbf{i}_n \tag{A.1}$$

we find the components of $\nabla \times \mathbf{A}$ as follows with the aid of Fig. A.1:

$$(\nabla \times \mathbf{A})_r = \lim_{\substack{d\phi \to 0 \\ dz \to 0}} \frac{\oint_{abcda} \mathbf{A} \cdot d\mathbf{l}}{\text{area } abcd}$$

$$= \lim_{\substack{d\phi \to 0 \\ dz \to 0}} \frac{\left\{ \begin{matrix} [A_\phi]_{(r,z)} r\, d\phi + [A_z]_{(r,\phi+d\phi)}\, dz \\ -[A_\phi]_{(r,z+dz)} r\, d\phi - [A_z]_{(r,\phi)}\, dz \end{matrix} \right\}}{r\, d\phi\, dz} \tag{A.2a}$$

$$= \lim_{d\phi \to 0} \frac{[A_z]_{(r,\phi+d\phi)} - [A_z]_{(r,\phi)}}{r\, d\phi} + \lim_{dz \to 0} \frac{[A_\phi]_{(r,z)} - [A_\phi]_{(r,z+dz)}}{dz}$$

$$= \frac{1}{r} \frac{\partial A_z}{\partial \phi} - \frac{\partial A_\phi}{\partial z}$$

$$(\nabla \times \mathbf{A})_\phi = \lim_{\substack{dz \to 0 \\ dr \to 0}} \frac{\oint_{adefa} \mathbf{A} \cdot d\mathbf{l}}{\text{area } adef}$$

$$= \lim_{\substack{dz \to 0 \\ dr \to 0}} \frac{\left\{ \begin{matrix} [A_z]_{(r,\phi)}\, dz + [A_r]_{(\phi,z+dz)}\, dr \\ -[A_z]_{(r+dr,\phi)}\, dz - [A_r]_{(\phi,z)}\, dr \end{matrix} \right\}}{dr\, dz} \tag{A.2b}$$

$$= \lim_{dz \to 0} \frac{[A_r]_{(\phi,z+dz)} - [A_r]_{(\phi,z)}}{dz} + \lim_{dr \to 0} \frac{[A_z]_{(r,\phi)} - [A_z]_{(r+dr,\phi)}}{dr}$$

$$= \frac{\partial A_r}{\partial z} - \frac{\partial A_z}{\partial r}$$

$$(\nabla \times \mathbf{A})_z = \lim_{\substack{dr \to 0 \\ d\phi \to 0}} \frac{\oint_{afgba} \mathbf{A} \cdot d\mathbf{l}}{\text{area } afgb}$$

$$= \lim_{\substack{dr \to 0 \\ d\phi \to 0}} \frac{\left\{ \begin{matrix} [A_r]_{(\phi,z)}\, dr + [A_\phi]_{(r+dr,z)}(r + dr)\, d\phi \\ -\, [A_r]_{(\phi+d\phi,z)}\, dr - [A_\phi]_{(r,z)} r\, d\phi \end{matrix} \right\}}{r\, dr\, d\phi} \tag{A.2c}$$

$$= \lim_{dr \to 0} \frac{[rA_\phi]_{(r+dr,z)} - [rA_\phi]_{(r,z)}}{r\, dr} + \lim_{d\phi \to 0} \frac{[A_r]_{(\phi,z)} - [A_r]_{(\phi+d\phi,z)}}{r\, d\phi}$$

$$= \frac{1}{r} \frac{\partial}{\partial r}(rA_\phi) - \frac{1}{r} \frac{\partial A_r}{\partial \phi}$$

Combining (A.2a), (A.2b) and (A.2c), we obtain the expression for the curl of a vector in cylindrical coordinates as

$$\nabla \times \mathbf{A} = \left[\frac{1}{r} \frac{\partial A_z}{\partial \phi} - \frac{\partial A_\phi}{\partial z} \right] \mathbf{i}_r + \left[\frac{\partial A_r}{\partial z} - \frac{\partial A_z}{\partial r} \right] \mathbf{i}_\phi + \frac{1}{r} \left[\frac{\partial}{\partial r}(rA_\phi) - \frac{\partial A_r}{\partial \phi} \right] \mathbf{i}_z$$

$$= \begin{vmatrix} \dfrac{\mathbf{i}_r}{r} & \mathbf{i}_\phi & \dfrac{\mathbf{i}_z}{r} \\ \dfrac{\partial}{\partial r} & \dfrac{\partial}{\partial \phi} & \dfrac{\partial}{\partial z} \\ A_r & rA_\phi & A_z \end{vmatrix} \tag{A.3}$$

To find the expression for the divergence, we make use of the basic definition of the divergence of a vector, introduced in Section 4.3 and given by

$$\mathbf{\nabla} \cdot \mathbf{A} = \lim_{\Delta v \to 0} \frac{\oint_S \mathbf{A} \cdot d\mathbf{S}}{\Delta v} \tag{A.4}$$

Evaluating the right side of (A.4) for the box of Fig. A.1, we obtain

$$\begin{aligned} \mathbf{\nabla} \cdot \mathbf{A} &= \lim_{\substack{dr \to 0 \\ d\phi \to 0 \\ dz \to 0}} \frac{\left\{\begin{array}{l} [A_r]_{r+dr}(r + dr)\, d\phi\, dz - [A_r]_r r\, d\phi\, dz + [A_\phi]_{\phi+d\phi}\, dr\, dz \\ - [A_\phi]_\phi\, dr\, dz + [A_z]_{z+dz} r\, dr\, d\phi - [A_z]_z r\, dr\, d\phi \end{array}\right\}}{r\, dr\, d\phi\, dz} \\ &= \lim_{dr \to 0} \frac{[rA_r]_{r+dr} - [rA_r]_r}{r\, dr} + \lim_{d\phi \to 0} \frac{[A_\phi]_{\phi+d\phi} - [A_\phi]_\phi}{r\, d\phi} \\ &\quad + \lim_{dz \to 0} \frac{[A_z]_{z+dz} - [A_z]_z}{dz} \\ &= \frac{1}{r}\frac{\partial}{\partial r}(rA_r) + \frac{1}{r}\frac{\partial A_\phi}{\partial \phi} + \frac{\partial A_z}{\partial z} \end{aligned} \tag{A.5}$$

To obtain the expression for the gradient of a scalar, we recall from Section 1.3 that in cylindrical coordinates,

$$d\mathbf{l} = dr\, \mathbf{i}_r + r\, d\phi\, \mathbf{i}_\phi + dz\, \mathbf{i}_z \tag{A.6}$$

Therefore

$$\begin{aligned} d\Phi &= \frac{\partial \Phi}{\partial r} dr + \frac{\partial \Phi}{\partial \phi} d\phi + \frac{\partial \Phi}{\partial z} dz \\ &= \left(\frac{\partial \Phi}{\partial r}\mathbf{i}_r + \frac{1}{r}\frac{\partial \Phi}{\partial \phi}\mathbf{i}_\phi + \frac{\partial \Phi}{\partial z}\mathbf{i}_z\right) \cdot (dr\, \mathbf{i}_r + r\, d\phi\, \mathbf{i}_\phi + dz\, \mathbf{i}_z) \\ &= \nabla\Phi \cdot d\mathbf{l} \end{aligned} \tag{A.7}$$

Thus

$$\mathbf{\nabla}\Phi = \frac{\partial \Phi}{\partial r}\mathbf{i}_r + \frac{1}{r}\frac{\partial \Phi}{\partial \phi}\mathbf{i}_\phi + \frac{\partial \Phi}{\partial z}\mathbf{i}_z \tag{A.8}$$

To derive the expression for the Laplacian of a scalar, we recall from Section 4.5 that

$$\nabla^2\Phi = \mathbf{\nabla} \cdot \mathbf{\nabla}\Phi \tag{A.9}$$

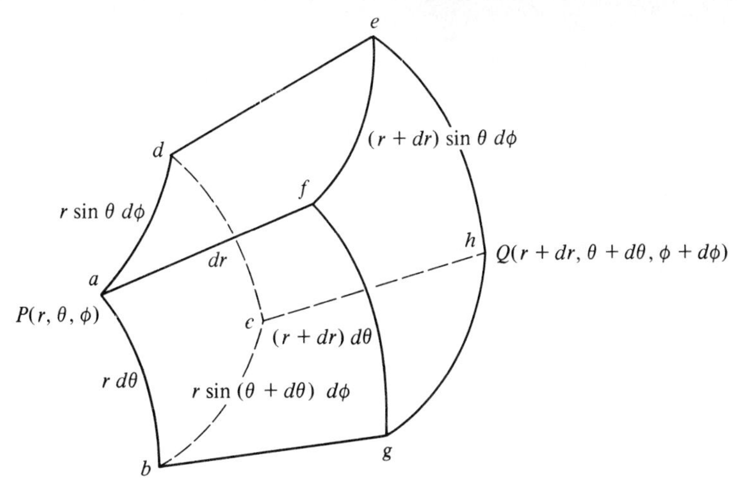

Figure A.2. Infinitesimal box formed by incrementing the coordinates in the spherical coordinate system.

Then using (A.5) and (A.8), we obtain

$$\begin{aligned}\nabla^2\Phi &= \frac{1}{r}\frac{\partial}{\partial r}\left(r\frac{\partial\Phi}{\partial r}\right) + \frac{1}{r}\frac{\partial}{\partial\phi}\left(\frac{1}{r}\frac{\partial\Phi}{\partial\phi}\right) + \frac{\partial}{\partial z}\left(\frac{\partial\Phi}{\partial z}\right) \\ &= \frac{1}{r}\frac{\partial}{\partial r}\left(r\frac{\partial\Phi}{\partial r}\right) + \frac{1}{r^2}\frac{\partial^2\Phi}{\partial\phi^2} + \frac{\partial^2\Phi}{\partial z^2}\end{aligned} \tag{A.10}$$

Turning now to the spherical coordinate system, we recall from Section 1.3 that the infinitesimal box defined by the three orthogonal surfaces intersecting at $P(r, \theta, \phi)$ and the three orthogonal surfaces intersecting at $Q(r + dr, \theta + d\theta, \phi + d\phi)$ is as shown in Fig. A.2. From the basic definition of the curl of a vector given by (A.1), we then find the components of $\nabla \times \mathbf{A}$ as follows with the aid of Fig. A.2:

$$\begin{aligned}(\nabla \times \mathbf{A})_r &= \lim_{\substack{d\theta\to 0\\ d\phi\to 0}} \frac{\oint_{abcda} \mathbf{A}\cdot d\mathbf{l}}{\text{area } abcd} \\ &= \lim_{\substack{d\theta\to 0\\ d\phi\to 0}} \frac{\left\{\begin{aligned}&[A_\theta]_{(r,\phi)} r\, d\theta + [A_\phi]_{(r,\theta+d\theta)} r \sin(\theta + d\theta)\, d\phi \\ &- [A_\theta]_{(r,\phi+d\phi)} r\, d\theta - [A_\phi]_{(r,\theta)} r \sin\theta\, d\phi\end{aligned}\right\}}{r^2 \sin\theta\, d\theta\, d\phi} \\ &= \lim_{d\theta\to 0} \frac{[A_\phi \sin\theta]_{(r,\theta+d\theta)} - [A_\phi \sin\theta]_{(r,\theta)}}{r \sin\theta\, d\theta} \\ &\quad + \lim_{d\phi\to 0} \frac{[A_\theta]_{(r,\phi)} - [A_\theta]_{(r,\phi+d\phi)}}{r \sin\theta\, d\phi} \\ &= \frac{1}{r\sin\theta}\frac{\partial}{\partial\theta}(A_\phi \sin\theta) - \frac{1}{r\sin\theta}\frac{\partial A_\theta}{\partial\phi}\end{aligned} \tag{A.11a}$$

$$(\nabla \times \mathbf{A})_\theta = \lim_{\substack{d\phi \to 0 \\ dr \to 0}} \frac{\oint_{adefa} \mathbf{A} \cdot d\mathbf{l}}{\text{area } adef}$$

$$= \lim_{\substack{d\phi \to 0 \\ dr \to 0}} \frac{\left\{ \begin{array}{l} [A_\phi]_{(r,\theta)} r \sin\theta \, d\phi + [A_r]_{(\theta,\phi+d\phi)} \, dr \\ - [A_\phi]_{(r+dr,\theta)}(r + dr) \sin\theta \, d\phi - [A_r]_{(\theta,\phi)} \, dr \end{array} \right\}}{r \sin\theta \, dr \, d\phi}$$

$$= \lim_{d\phi \to 0} \frac{[A_r]_{(\theta,\phi+d\phi)} - [A_r]_{(\theta,\phi)}}{r \sin\theta \, d\phi} \tag{A.11b}$$

$$+ \lim_{dr \to 0} \frac{[rA_\phi]_{(r,\theta)} - [rA_\phi]_{(r+dr,\theta)}}{r \, dr}$$

$$= \frac{1}{r \sin\theta} \frac{\partial A_r}{\partial \phi} - \frac{1}{r} \frac{\partial}{\partial r}(rA_\phi)$$

$$(\nabla \times \mathbf{A})_\phi = \lim_{\substack{dr \to 0 \\ d\theta \to 0}} \frac{\oint_{afgba} \mathbf{A} \cdot d\mathbf{l}}{\text{area } afgb}$$

$$= \lim_{\substack{dr \to 0 \\ d\theta \to 0}} \frac{\left\{ \begin{array}{l} [A_r]_{(\theta,\phi)} \, dr + [A_\theta]_{(r+dr,\phi)}(r + dr) \, d\theta \\ - [A_r]_{(\theta+d\theta,\phi)} \, dr - [A_\theta]_{(r,\phi)} r \, d\theta \end{array} \right\}}{r \, dr \, d\theta}$$

$$= \lim_{dr \to 0} \frac{[rA_\theta]_{(r+dr,\phi)} - [rA_\theta]_{(r,\phi)}}{r \, dr} \tag{A.11c}$$

$$+ \lim_{d\theta \to 0} \frac{[A_r]_{(\theta,\phi)} \, dr - [A_r]_{(\theta+d\theta,\phi)} \, dr}{r \, d\theta}$$

$$= \frac{1}{r} \frac{\partial}{\partial r}(rA_\theta) - \frac{1}{r} \frac{\partial A_r}{\partial \theta}$$

Combining (A.11a), (A.11b), and (A.11c), we obtain the expression for the curl of a vector in spherical coordinates as

$$\nabla \times \mathbf{A} = \frac{1}{r \sin\theta} \left[\frac{\partial}{\partial \theta}(A_\phi \sin\theta) - \frac{\partial A_\theta}{\partial \phi} \right] \mathbf{i}_r$$

$$+ \frac{1}{r} \left[\frac{1}{\sin\theta} \frac{\partial A_r}{\partial \phi} - \frac{\partial}{\partial r}(rA_\phi) \right] \mathbf{i}_\theta$$

$$+ \frac{1}{r} \left[\frac{\partial}{\partial r}(rA_\theta) - \frac{\partial A_r}{\partial \theta} \right] \mathbf{i}_\phi \tag{A.12}$$

$$= \begin{vmatrix} \dfrac{\mathbf{i}_r}{r^2 \sin\theta} & \dfrac{\mathbf{i}_\theta}{r \sin\theta} & \dfrac{\mathbf{i}_\phi}{r} \\ \dfrac{\partial}{\partial r} & \dfrac{\partial}{\partial \theta} & \dfrac{\partial}{\partial \phi} \\ A_r & rA_\theta & r \sin\theta \, A_\phi \end{vmatrix}$$

To find the expression for the divergence, we make use of the basic definition of the divergence of a vector given by (A.4) and by evaluating its right side for the box of Fig. A.2, we obtain

$$\nabla \cdot \mathbf{A} = \lim_{\substack{dr\to 0\\ d\theta\to 0\\ d\phi\to 0}} \frac{\begin{Bmatrix} [A_r]_{r+dr}(r+dr)^2 \sin\theta\, d\theta\, d\phi - [A_r]_r r^2 \sin\theta\, d\theta\, d\phi \\ + [A_\theta]_{\theta+d\theta} r \sin(\theta + d\theta)\, dr\, d\phi - [A_\theta]_\theta r \sin\theta\, dr\, d\phi \\ + [A_\phi]_{\phi+d\phi} r\, dr\, d\theta - [A_\phi]_\phi r\, dr\, d\theta \end{Bmatrix}}{r^2 \sin\theta\, dr\, d\theta\, d\phi}$$

$$= \lim_{dr\to 0} \frac{[r^2 A_r]_{r+dr} - [r^2 A_r]_r}{r^2\, dr} + \lim_{d\theta\to 0} \frac{[A_\theta \sin\theta]_{\theta+d\theta} - [A_\theta \sin\theta]_\theta}{r \sin\theta\, d\theta}$$

$$+ \lim_{d\phi\to 0} \frac{[A_\phi]_{\phi+d\phi} - [A_\phi]_\phi}{r \sin\theta\, d\phi}$$

$$= \frac{1}{r^2}\frac{\partial}{\partial r}(r^2 A_r) + \frac{1}{r\sin\theta}\frac{\partial}{\partial\theta}(A_\theta \sin\theta) + \frac{1}{r\sin\theta}\frac{\partial A_\phi}{\partial\phi} \tag{A.13}$$

To obtain the expression for the gradient of a scalar, we recall from Section 1.3 that in spherical coordinates,

$$d\mathbf{l} = dr\, \mathbf{i}_r + r\, d\theta\, \mathbf{i}_\theta + r\sin\theta\, d\phi\, \mathbf{i}_\phi \tag{A.14}$$

Therefore

$$\begin{aligned} d\Phi &= \frac{\partial\Phi}{\partial r}dr + \frac{\partial\Phi}{\partial\theta}d\theta + \frac{\partial\Phi}{\partial\phi}d\phi \\ &= \left(\frac{\partial\Phi}{\partial r}\mathbf{i}_r + \frac{1}{r}\frac{\partial\Phi}{\partial\theta}\mathbf{i}_\theta + \frac{1}{r\sin\theta}\frac{\partial\Phi}{\partial\phi}\mathbf{i}_\phi\right) \\ &\quad \cdot (dr\, \mathbf{i}_r + r\, d\theta\, \mathbf{i}_\theta + r\sin\theta\, d\phi\, \mathbf{i}_\phi) \\ &= \nabla\Phi \cdot d\mathbf{l} \end{aligned} \tag{A.15}$$

Thus

$$\nabla\Phi = \frac{\partial\Phi}{\partial r}\mathbf{i}_r + \frac{1}{r}\frac{\partial\Phi}{\partial\theta}\mathbf{i}_\theta + \frac{1}{r\sin\theta}\frac{\partial\Phi}{\partial\phi}\mathbf{i}_\phi \tag{A.16}$$

To derive the expression for the Laplacian of a scalar, we make use of (A.9), in conjunction with (A.13) and (A.16). Thus we obtain

$$\begin{aligned} \nabla^2\Phi &= \frac{1}{r^2}\frac{\partial}{\partial r}\left(r^2\frac{\partial\Phi}{\partial r}\right) + \frac{1}{r\sin\theta}\frac{\partial}{\partial\theta}\left(\frac{1}{r}\frac{\partial\Phi}{\partial\theta}\sin\theta\right) \\ &\quad + \frac{1}{r\sin\theta}\frac{\partial}{\partial\phi}\left(\frac{1}{r\sin\theta}\frac{\partial\Phi}{\partial\phi}\right) \\ &= \frac{1}{r^2}\frac{\partial}{\partial r}\left(r^2\frac{\partial\Phi}{\partial r}\right) + \frac{1}{r^2\sin\theta}\frac{\partial}{\partial\theta}\left(\sin\theta\frac{\partial\Phi}{\partial\theta}\right) \\ &\quad + \frac{1}{r^2\sin^2\theta}\frac{\partial^2\Phi}{\partial\phi^2} \end{aligned} \tag{A.17}$$

Appendix B

Units and Dimensions

In 1960 the International System of Units was given official status at the Eleventh General Conference on weights and measures held in Paris, France. This system of units is an expanded version of the rationalized meter–kilogram–second–ampere (MKSA) system of units and is based on six fundamental or basic units. The six basic units are the units of length, mass, time, current, temperature, and luminous intensity.

The international unit of length is the meter. It is exactly 1,650,763.73 times the wavelength in vacuum of the radiation corresponding to the unperturbed transition between the levels $2p_{10}$ and $5d_5$ of the atom of krypton-86, the orange-red line. The international unit of mass is the kilogram. It is the mass of the International Prototype Kilogram which is a particular cylinder of platinum-iridium alloy preserved in a vault at Sèvres, France, by the International Bureau of Weights and Measures. The international unit of time is the second. It is equal to 9,192,631,770 times the period corresponding to the frequency of the transition between the hyperfine levels $F = 4$, $M = 0$ and $F = 3$, $M = 0$ of the fundamental state $^2S_{1/2}$ of the cesium-133 atom unperturbed by external fields.

To present the definition for the international unit of current, we first define the newton, which is the unit of force, derived from the fundamental units meter, kilogram, and second in the following manner. Since velocity is rate of change of distance with time, its unit is meter per second. Since acceleration is rate of change of velocity with time, its unit is meter per second per second or meter per second squared. Since force is mass times acceleration, its unit is kilogram-meter per second squared, also known as the newton. Thus the newton is that force which imparts an acceleration of 1 meter per second squared to a mass of 1 kilogram. The international unit of current, which is the ampere, can now be defined. It is the constant current which when maintained in two straight, infinitely long, parallel conductors of negligible cross section and placed 1 meter apart in vacuum produces a force of 2×10^{-7} newton per meter length of the conductors.

The international unit of temperature is the kelvin. It is based on the definition of the thermodynamic scale of temperature by means of the triple point of water as a fixed fundamental point to which a temperature of exactly 273.16 kelvin is attributed. The international unit of luminous intensity is the candela. It is defined such that the luminance of a blackbody radiator at the freezing temperature of platinum is 60 candelas per square centimeter.

We have just defined the six basic units of the International System of Units. Two supplementary units are the radian and the steradian for plane angle and solid angle, respectively. All other units are derived units. For example, the unit of charge which is the coulomb is the amount of charge transported in 1 second by a current of 1 ampere; the unit of energy which is the joule is the work done when the point of application of a force of 1 newton is displaced a distance of 1 meter in the direction of the force; the unit of power which is the watt is the power which gives rise to the production of energy at a rate of 1 joule per second; the unit of electric potential difference which is the volt is the difference of electric potential between two points of a conducting wire carrying constant current of 1 ampere when the power dissipated between these points is equal to 1 watt; and so on. The units for the various quantities used in this book are listed in Table B.1, together with the symbols of the quantities and their dimensions.

TABLE B.1. SYMBOLS, UNITS, AND DIMENSIONS OF VARIOUS QUANTITIES

Quantity	Symbol	Unit	Dimensions
Admittance	$\bar{Y}$	siemens	$M^{-1}L^{-2}TQ^2$
Area	A	square meter	L^2
Attenuation constant	α	neper/meter	L^{-1}
Capacitance	C	farad	$M^{-1}L^{-2}T^2Q^2$
Capacitance per unit length	$\mathscr{C}$	farad/meter	$M^{-1}L^{-3}T^2Q^2$
Cartesian coordinates	x	meter	L
	y	meter	L
	z	meter	L
Characteristic admittance	Y_0	siemens	$M^{-1}L^{-2}TQ^2$
Characteristic impedance	Z_0	ohm	$ML^2T^{-1}Q^{-2}$
Charge	Q, q	coulomb	Q
Conductance	G	siemens	$M^{-1}L^{-2}TQ^2$
Conductance per unit length	$\mathscr{G}$	siemens/meter	$M^{-1}L^{-3}TQ^2$
Conduction current density	$\mathbf{J}_c$	ampere/square meter	$L^{-2}T^{-1}Q$
Conductivity	σ	siemens/meter	$M^{-1}L^{-3}TQ^2$
Current	I	ampere	$T^{-1}Q$
Cutoff frequency	f_c	hertz	T^{-1}
Cutoff wavelength	λ_c	meter	L
Cylindrical coordinates	r, r_c	meter	L
	ϕ	radian	—
	z	meter	L
Differential length element	$d\mathbf{l}$	meter	L
Differential surface element	$d\mathbf{S}$	square meter	L^2
Differential volume element	dv	cubic meter	L^3
Directivity	D	—	—
Displacement flux density	$\mathbf{D}$	coulomb/square meter	$L^{-2}Q$
Electric dipole moment	$\mathbf{p}$	coulomb-meter	LQ
Electric field intensity	$\mathbf{E}$	volt/meter	$MLT^{-2}Q^{-1}$
Electric potential	V	volt	$ML^2T^{-2}Q^{-1}$
Electric susceptibility	χ_e	—	—
Electron density	N_e	$(\text{meter})^{-3}$	L^{-3}

TABLE B.1. (continued)

Quantity	Symbol	Unit	Dimensions
Electronic charge	e	coulomb	Q
Energy	W	joule	ML^2T^{-2}
Energy density	w	joule/cubic meter	$ML^{-1}T^{-2}$
Force	$\mathbf{F}$	newton	MLT^{-2}
Frequency	f	hertz	T^{-1}
Group velocity	v_g	meter/second	LT^{-1}
Guide characteristic impedance	η_g	ohm	$ML^2T^{-1}Q^{-2}$
Guide wavelength	λ_g	meter	L
Impedance	$\bar{Z}$	ohm	$ML^2T^{-1}Q^{-2}$
Inductance	L	henry	ML^2Q^{-2}
Inductance per unit length	$\mathcal{L}$	henry/meter	MLQ^{-2}
Intensity	I	watts/square meter	MT^{-3}
Intrinsic impedance	η	ohm	$ML^2T^{-1}Q^{-2}$
Length	l	meter	L
Line charge density	ρ_L	coulomb/meter	$L^{-1}Q$
Magnetic dipole moment	$\mathbf{m}$	ampere-square meter	$L^2T^{-1}Q$
Magnetic field intensity	$\mathbf{H}$	ampere/meter	$L^{-1}T^{-1}Q$
Magnetic flux	ψ	weber	$ML^2T^{-1}Q^{-1}$
Magnetic flux density	$\mathbf{B}$	tesla or weber/square meter	$MT^{-1}Q^{-1}$
Magnetic susceptibility	χ_m	—	—
Magnetic vector potential	$\mathbf{A}$	weber/meter	$MLT^{-1}Q^{-1}$
Magnetization surface current density	$\mathbf{J}_{mS}$	ampere/meter	$L^{-1}T^{-1}Q$
Magnetization vector	$\mathbf{M}$	ampere/meter	$L^{-1}T^{-1}Q$
Mass	m	kilogram	M
Mobility	μ	square meter/volt-second	$M^{-1}TQ$
Permeability	μ	henry/meter	MLQ^{-2}
Permeability of free space	μ_0	henry/meter	MLQ^{-2}
Permittivity	ε	farad/meter	$M^{-1}L^{-3}T^2Q^2$
Permittivity of free space	ε_0	farad/meter	$M^{-1}L^{-3}T^2Q^2$
Phase constant	β	radian/meter	L^{-1}
Phase velocity	v_p	meter/second	LT^{-1}
Polarization surface charge density	ρ_{pS}	coulomb/square meter	$L^{-2}Q$
Polarization vector	$\mathbf{P}$	coulomb/square meter	$L^{-2}Q$
Power	P	watt	ML^2T^{-3}
Power density	p	watt/square meter	MT^{-3}
Poynting vector	$\mathbf{P}$	watt/square meter	MT^{-3}
Propagation constant	$\bar{\gamma}$	complex neper/meter	L^{-1}
Propagation vector	$\boldsymbol{\beta}$	radian/meter	L^{-1}
Q factor	Q	—	—
Radian frequency	ω	radian/second	T^{-1}
Radiation resistance	R_{rad}	ohm	$ML^2T^{-1}Q^{-2}$
Reactance	X	ohm	$ML^2T^{-1}Q^{-2}$
Reflection coefficient	Γ	—	—
Refractive index	n	—	—
Relative permeability	μ_r	—	—
Relative permittivity	ε_r	—	—
Reluctance	$\mathcal{R}$	ampere (turn)/weber	$M^{-1}L^{-2}Q^2$
Resistance	R	ohm	$ML^2T^{-1}Q^{-2}$
Skin depth	δ	meter	L

TABLE B.1. (continued)

Quantity	Symbol	Unit	Dimensions
Spherical coordinates	r, r_s	meter	L
	θ	radian	—
	ϕ	radian	—
Standing wave ratio	SWR	—	—
Surface charge density	ρ_S	coulomb/square meter	$L^{-2}Q$
Surface current density	$\mathbf{J}_S$	ampere/meter	$L^{-1}T^{-1}Q$
Susceptance	B	siemens	$M^{-1}L^{-2}TQ^2$
Time	t	second	T
Transmission coefficient	τ	—	—
Unit normal vector	$\mathbf{i}_n$	—	—
Velocity	v	meter/second	LT^{-1}
Velocity of light in free space	c	meter/second	LT^{-1}
Voltage	V	volt	$ML^2T^{-2}Q^{-1}$
Volume	V	cubic meter	L^3
Volume charge density	ρ	coulomb/cubic meter	$L^{-3}Q$
Volume current density	$\mathbf{J}$	ampere/square meter	$L^{-2}T^{-1}Q$
Wavelength	λ	meter	L
Work	W	joule	ML^2T^{-2}

Dimensions are a convenient means of checking the possible validity of a derived equation. The dimension of a given quantity can be expressed as some combination of a set of fundamental dimensions. These fundamental dimensions are mass (M), length (L), and time (T). In electromagnetics, it is the usual practice to consider the charge (Q), instead of the current, as the additional fundamental dimension. For the quantities listed in Table B.1, these four dimensions are sufficient. Thus, for example, the dimension of velocity is length (L) divided by time (T), that is, LT^{-1}; the dimension of acceleration is length (L) divided by time squared (T^2), that is, LT^{-2}; the dimension of force is mass (M) times acceleration (LT^{-2}), that is, MLT^{-2}; the dimension of ampere is charge (Q) divided by time (T), that is, QT^{-1}; and so on.

To illustrate the application of dimensions for checking the possible validity of a derived equation, let us consider the equation for the phase velocity of an electromagnetic wave in free space, given by

$$v_p = \frac{1}{\sqrt{\mu_0 \varepsilon_0}}$$

We know that the dimension of v_p is LT^{-1}. Hence we have to show that the dimension of $1/\sqrt{\mu_0 \varepsilon_0}$ is also LT^{-1}. To do this, we note from Coulomb's law that

$$\varepsilon_0 = \frac{Q_1 Q_2}{4\pi F R^2}$$

Hence the dimension of ε_0 is $Q^2/[MLT^{-2})(L^2)]$ or $M^{-1}L^{-3}T^2Q^2$. We note from Ampère's law of force applied to two infinitesimal current elements parallel to each other and normal to the line joining them that

$$\mu_0 = \frac{4\pi F R^2}{(I_1\, dl_1)(I_2\, dl_2)}$$

Hence the dimension of μ_0 is $[(MLT^{-2})(L^2)]/(QT^{-1}L)^2]$ or MLQ^{-2}. Now we obtain the dimension of $1/\sqrt{\mu_0\varepsilon_0}$ as $1/\sqrt{(M^{-1}L^{-3}T^2Q^2)(MLQ^{-2})}$ or LT^{-1}, which is the same as the dimension of v_p. It should be noted, however, that the test for the equality of the dimensions of the two sides of a derived equation is not a sufficient test to establish the equality of the two sides since any dimensionless constants associated with the equation may be in error.

It is not always necessary to refer to the table of dimensions for checking the possible validity of a derived equation. For example, let us assume that we have derived the expression for the characteristic impedance of a transmission line (i.e., $\sqrt{\mathscr{L}/\mathscr{C}}$) and we wish to verify that $\sqrt{\mathscr{L}/\mathscr{C}}$ does indeed have the dimension of impedance. To do this, we write

$$\sqrt{\frac{\mathscr{L}}{\mathscr{C}}} = \sqrt{\frac{\omega\mathscr{L}l}{\omega\mathscr{C}l}} = \sqrt{\frac{\omega L}{\omega C}} = \sqrt{(\omega L)\left(\frac{1}{\omega C}\right)}$$

We now recognize from our knowledge of circuit theory that both ωL and $1/\omega C$, being the reactances of L and C, respectively, have the dimension of impedance. Hence we conclude that $\sqrt{\mathscr{L}/\mathscr{C}}$ has the dimension of $\sqrt{(\text{impedance})^2}$ or impedance.

Suggested Collateral and Further Reading

ADLER, R. B., L. J. CHU, and R. M. FANO, *Electromagnetic Energy Transmission and Radiation*. New York: John Wiley & Sons, Inc., 1960.

DAVIDSON, C. W., *Transmission Lines for Communications,* 2nd ed. New York: John Wiley & Sons, Inc., 1989.

DIAMENT, P., *Wave Transmission and Fiber Optics*. New York: Macmillan Publishing Company, 1990.

HAYT, W. H., JR., *Engineering Electromagnetics,* 5th ed. New York: McGraw-Hill Book Company, Inc., 1989.

JORDAN, E. C., and K. G. BALMAIN, *Electromagnetic Waves and Radiating Systems,* 2nd ed. Englewood Cliffs, NJ: Prentice-Hall, Inc., 1968.

KRAUS, J. D., *Electromagnetics,* 4th ed. New York: McGraw-Hill Book Company, Inc., 1992.

LEE, D. L., *Electromagnetic Principles of Integrated Optics*. New York: John Wiley & Sons, Inc., 1986.

RAMO, S., J. R. WHINNERY, and T. VAN DUZER, *Fields and Waves in Communication Electronics,* 2nd ed. New York: John Wiley & Sons, Inc., 1984.

RAO, N. N., *Basic Electromagnetics with Applications*. Englewood Cliffs, NJ: Prentice-Hall, Inc., 1972.

SALEH, B. E. A., and M. C. TEICH, *Fundamentals of Photonics*. New York: John Wiley & Sons, Inc., 1991.

SESHADRI, S. R., *Fundamentals of Transmission Lines and Electromagnetic Fields*. Reading, MA: Addison-Wesley Publishing Company, 1971.

SHEN, L. C., and J. A. KONG, *Applied Electromagneticism,* 2nd ed. Boston: PWS Publishers, 1987.

Answers to Selected Problems

CHAPTER 1

P1.1. **(a)** 2 m **(b)** 0.7143 m northward and 0.2474 m eastward **(c)** 0.7559 m, 19.1° east of north

P1.4. **(a)** $\frac{1}{2}(\mathbf{A} + \mathbf{B})$ **(b)** $\left(\dfrac{\mathbf{B} \times \mathbf{A}}{\mathbf{B} \times \mathbf{C} + \mathbf{C} \times \mathbf{A}}\right)\mathbf{C}$

P1.6. **(b)** (i) neither parallel nor perpendicular (ii) perpendicular (iii) parallel

P1.9. **(a)** 2.132 **(b)** yes

P1.12. $\pm\dfrac{\mathbf{i}_x + \mathbf{i}_y + 4\mathbf{i}_z}{3\sqrt{2}}$

P1.15. **(a)** 7 **(b)** yes

P1.18. **(a)** $-\dfrac{\sqrt{3}}{2}$ **(b)** 0 **(c)** $\frac{1}{2}$

P1.20. **(a)** $\mathbf{i}_\phi$ **(b)** $0.2145\mathbf{i}_r - 0.2145\mathbf{i}_\theta + 0.9529\mathbf{i}_\phi$

P1.22. **(a)** $-mMG\dfrac{x\mathbf{i}_x + y\mathbf{i}_y + z\mathbf{i}_z}{(x^2 + y^2 + z^2)^{3/2}}$ **(b)** $\omega r_s \sin\theta\, \mathbf{i}_\phi$

P1.25. $r \sin^2\theta = C_1$, $\phi = C_2$

P1.27. $f(z, t)$ represents a standing wave

P1.29. **(a)** elliptical **(b)** -1 **(c)** 0, 2

P1.33. $8 + j0$, $0 + j8$, $-8 + j0$, $0 - j8$

P1.35. $0.6 \cos(10^6 t - 6.87°) + 0.1015 \cos(3 \times 10^6 t + 23.96°)$

CHAPTER 2

P2.1. $\dfrac{0.1949Q^2}{\varepsilon_0 L^2}$ away from the center of the tetrahedron

P2.3. **(a)** $\dfrac{d^2z}{dt^2} + \dfrac{Q|e|}{\pi\varepsilon_0 md^3} = 0$ **(b)** $z = z_0 \cos\sqrt{\dfrac{Q|e|}{\pi\varepsilon_0 md^3}}\,t$

P2.5. $-\dfrac{\rho_{L0}a^2}{4\varepsilon_0(a^2+z^2)^{3/2}}\mathbf{i}_y$

P2.7. $2\pi z\left(\ln\dfrac{a+\sqrt{a^2+z^2}}{z} - \dfrac{a}{\sqrt{a^2+z^2}}\right)\mathbf{i}_z$

P2.9. $0.1454\mu_0(I\,dz)^2$ toward center of the circle

P2.11. $\dfrac{\mu_0 I}{2}\dfrac{a^2}{(a^2+z^2)^{3/2}}\mathbf{i}_z$

P2.14. **(a)** $B_0(-2\mathbf{i}_x - 2\mathbf{i}_y - \mathbf{i}_z)$ **(b)** $B_0(-2\mathbf{i}_x - 4\mathbf{i}_y + 5\mathbf{i}_z)$ **(c)** $B_0(2\mathbf{i}_x + 4\mathbf{i}_y - 5\mathbf{i}_z)$

P2.17. $\dfrac{E_0}{3B_0}(2\mathbf{i}_x - \mathbf{i}_y - 2\mathbf{i}_z)$

P2.19. **(a)** $\dfrac{qE_0}{2}\mathbf{i}_z$ **(b)** $-\dfrac{qE_0}{2}\mathbf{i}_z$ **(c)** $-qE_0\mathbf{i}_z$

P2.23. $\rho_{S1} = \frac{1}{2}(\rho_{SA}+\rho_{SB})$, $\rho_{S2} = \frac{1}{2}(\rho_{SA}-\rho_{SB})$, $\rho_{S3} = \frac{1}{2}(\rho_{SB}-\rho_{SA})$, $\rho_{S4} = \frac{1}{2}(\rho_{SA}+\rho_{SB})$

P2.25. $2\times 10^{-6}(2\mathbf{i}_x + \mathbf{i}_y)$ N-m

P2.28. $\dfrac{E_y}{E_x} = \dfrac{(\varepsilon_{yy}-\varepsilon_{xx}) \pm \sqrt{(\varepsilon_{xx}-\varepsilon_{yy})^2 + 4\varepsilon_{xy}\varepsilon_{yx}}}{2\varepsilon_{xy}}$

$\varepsilon_{\text{eff}} = \frac{1}{2}\left[(\varepsilon_{xx}+\varepsilon_{xy}) \pm \sqrt{(\varepsilon_{xx}-\varepsilon_{yy})^2 + 4\varepsilon_{xy}\varepsilon_{yx}}\right]$

P2.30. $2\times 10^{-12}\pi(\mathbf{i}_x + \mathbf{i}_y)$ N-m

CHAPTER 3

P3.1. **(a)** $1\frac{1}{6}$ **(b)** $1\frac{1}{5}$

P3.4. 3.5944

P3.7. 0

P3.9. $2\pi/3$

P3.11. **(a)** $-2B_0v_0 \cos\pi(x - v_0t)$ **(b)** 0

P3.14. **(a)** $B_0 hb\omega \sin\omega t$ **(b)** 0

P3.16. $\dfrac{E_0}{\omega B_0 r}\cos\omega t$

P3.18. **(a)** 32 A **(b)** 8π A

P3.20. 0.1125 mA

P3.23. $-\pi^2/2$ Wb

P3.26. $\frac{3}{4}I$

P3.29. $\dfrac{\rho_0 r^2}{3a}\mathbf{i}_r$ for $r < a$, $\dfrac{\rho_0 a^2}{3r}\mathbf{i}_r$ for $r > a$

P3.32. $\dfrac{J_0 r^2}{3a}\mathbf{i}_\phi$ for $r < a$, $\dfrac{J_0 a^2}{3r}\mathbf{i}_\phi$ for $r > a$

P3.35. **(b)** $3\varepsilon_0 E_0 \cos\theta$

(c) $\mathbf{E}_a = E_0 \cos\theta\, \mathbf{i}_r - E_0 \sin\theta\, \mathbf{i}_\theta$

$[\mathbf{E}_s]_{r<a} = -E_0 \cos\theta\, \mathbf{i}_r + E_0 \sin\theta\, \mathbf{i}_\theta$

$[\mathbf{E}_s]_{r>a} = \dfrac{E_0 a^3}{r^3}(2\cos\theta\, \mathbf{i}_r + \sin\theta\, \mathbf{i}_\theta)$

P3.37. **(b)** $\dfrac{E_0}{120\pi} \sin 3\pi \times 10^9 t\, \mathbf{i}_z$ on both sheets

P3.40. $98\mu_0$

CHAPTER 4

P4.2. **(a)** $\dfrac{E_0}{3 \times 10^8} \sin 3\pi z \sin 9\pi \times 10^8 t\, \mathbf{i}_y$

(b) $\dfrac{E_0(0.6\mathbf{i}_x - 0.8\mathbf{i}_z)}{3 \times 10^8} \cos[3\pi \times 10^8 t + 0.2\pi(4x + 3z)]$

P4.5. $\alpha^2 + \beta^2 = \omega^2 \mu_0 \varepsilon_0$

P4.8. **(a)** $\dfrac{\rho_0}{2a}(x^2 - a^2)$ for $-a < x < a$, 0 otherwise

(b) $-\dfrac{\rho_0 a}{2}$ for $x < -a$, $\rho_0\left(x + \dfrac{x^2}{2a}\right)$ for $-a < x < 0$, $\rho_0\left(x - \dfrac{x^2}{2a}\right)$ for $0 < x < a$, $\dfrac{\rho_0 a}{2}$ for $x > a$

P4.10. **(a)** yes **(b)** yes **(c)** no

P4.12. $\nabla \times \mathbf{v} = 2\omega\, \mathbf{i}_z$

P4.15. **(a)** $\oint_S \mathbf{A} \cdot d\mathbf{S} = \int_V (\nabla \cdot \mathbf{A})\, dv = \dfrac{3}{4}$

(b) $\oint_S \mathbf{A} \cdot d\mathbf{S} = \int_V (\nabla \cdot \mathbf{A})\, dv = 0$

P4.18. **(a)** xe^{-y} **(b)** $-\dfrac{1}{r}\cos\phi$ **(c)** $r\cos\theta$

P4.20. $x + 2y + z = 5$

P4.24. $V \approx \dfrac{3Q\, \Delta x\, \Delta z}{4\pi\varepsilon_0 r^3} \sin\theta \cos\theta \cos\phi$

$\mathbf{E} \approx \dfrac{3Q\, \Delta x\, \Delta z}{4\pi\varepsilon_0 r^4}(3\sin\theta\cos\theta\cos\phi\, \mathbf{i}_r - \cos 2\theta \cos\phi\, \mathbf{i}_\theta + \cos\theta \sin\phi\, \mathbf{i}_\phi)$

P4.26. $V = \dfrac{\rho_{L0}}{4\pi\varepsilon} \ln \dfrac{\sqrt{r^2 + (z+a)^2} + (z+a)}{\sqrt{r^2 + (z-a)^2} + (z-a)}$

P4.32. 120°, 101.6284°, 137.6641°

P4.34. **(a)** $\dfrac{\pi\rho_0^2 a^5}{140\varepsilon_0}$ **(b)** $\dfrac{21}{20}a$, $\dfrac{2000}{3087}\rho_0$

P4.36. $0.0751\mu_0 I_0^2$

CHAPTER 5

P5.1. $-\dfrac{kd^3}{24\varepsilon}$ for $x < -\dfrac{d}{2}$, $-\dfrac{kx^3}{6\varepsilon} + \dfrac{kd^2x}{8\varepsilon}$ for $-\dfrac{d}{2} < x < \dfrac{d}{2}$, $\dfrac{kd^3}{24\varepsilon}$ for $x > \dfrac{d}{2}$

P5.4. **(a)** $\dfrac{\varepsilon_2 x}{\varepsilon_2 t + \varepsilon_1(d - t)} V_0$ for $0 < x < t$, $\dfrac{\varepsilon_2 t + \varepsilon_1(x - t)}{\varepsilon_2 t + \varepsilon_1(d - t)} V_0$ for $t < x < d$

(b) $\dfrac{\varepsilon_1 \varepsilon_2}{\varepsilon_2 t + \varepsilon_1(d - t)}$

P5.7. $\dfrac{V_0(r - b)}{(a - b)}, \dfrac{2\pi\varepsilon_0 b}{b - a}$

P5.10. **(a)** $V_A = 12$ V, $V_B = 5.25$ V, $V_C = 2$ V, $V_D = 14.75$ V, $V_E = 7$ V, $V_F = 2.75$ V

(b) $\dfrac{6.1033}{d}$ V/m **(c)** $61.25\varepsilon_0$ C/m, $-61.25\varepsilon_0$ C/m

P5.13. $2\pi\varepsilon_0\left[\ln\left(2000 \sin \dfrac{\pi}{2n}\right) + \left(\tan \dfrac{\pi}{2n}\right)\left(\dfrac{1}{2} + \sum_{j=1}^{n-1} \operatorname{cosec} \dfrac{j\pi}{2n}\right)\right]^{-1}$, $0.7065\varepsilon_0$ C/m

P5.15. $12.536\varepsilon_0 a$

P5.17. $4.0853\varepsilon_0 a$

P5.21. $2\pi\mu N^2 a^2 c \ln \dfrac{2a + b}{2a - b}$

P5.23. $\dfrac{\mu}{16\pi}$

P5.26. **(a)** $-0.0395 \sin 10^6 \pi t$ V **(b)** $-65.285 \sin 10^9 \pi t$ V

P5.29. **(a)** $\sigma\sqrt{\dfrac{\mu}{\varepsilon}}\, l = \sqrt{3}$ **(b)** $\sigma\sqrt{\dfrac{\mu}{\varepsilon}}\, l \ll 1$ **(c)** $\sqrt{\omega\mu\sigma}\, l \ll 1$ and $\dfrac{\sigma}{\omega\varepsilon} \gg 1$

P5.32. 400 A-t

P5.34. 9×10^{-4} Wb

P5.36. $\dfrac{V^2 w}{2d}(\varepsilon - \varepsilon_0)\mathbf{i}_x$

P5.38. $\dfrac{\varepsilon_0 A V_0^2}{4d}$ from mechanical to electrical

CHAPTER 6

P6.1. $\dfrac{\partial E_x}{\partial y} = \dfrac{\partial B_z}{\partial t}, \dfrac{\partial H_z}{\partial y} = J_x + \dfrac{\partial D_x}{\partial t}$

P6.4. **(b)**

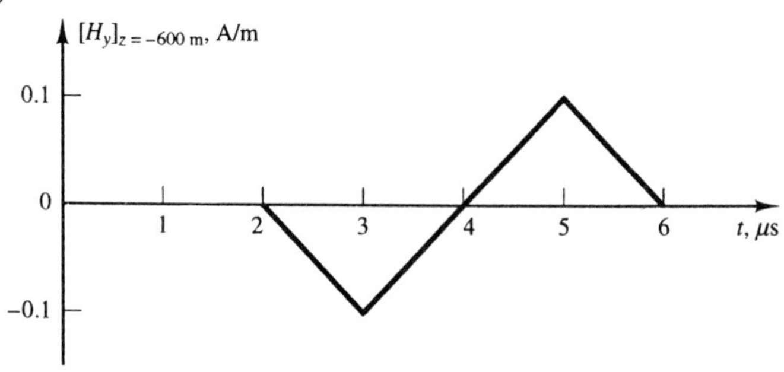

(c)

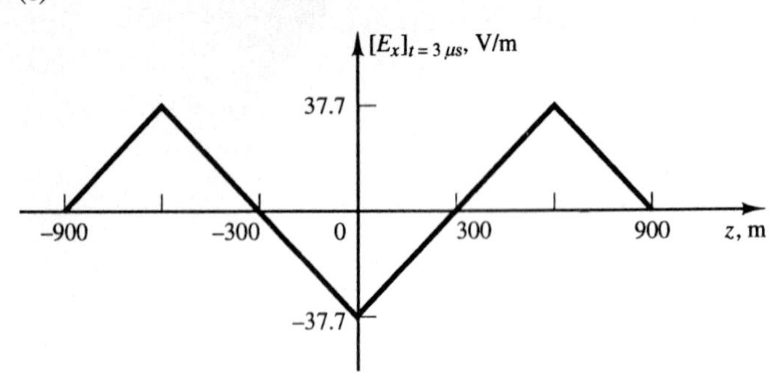

P6.8. $\mathbf{E} = -18.85(\mathbf{i}_x - \sqrt{3}\mathbf{i}_y)\cos(3\pi \times 10^8 t \mp \pi z)$ V/m for $z \gtrless 0$
$\mathbf{H} = \mp 0.05(\sqrt{3}\mathbf{i}_x + \mathbf{i}_y)\cos(3\pi \times 10^8 t \mp \pi z)$ A/m for $z \gtrless 0$

P6.10. **(a)** right circular **(b)** left circular **(c)** right elliptical **(d)** left elliptical

P6.12. **(a)** $2E_0[\cos(\omega t - \beta z)\,\mathbf{i}_x + \sin(\omega t - \beta z)\,\mathbf{i}_y]$
$-E_0[\cos(\omega t - \beta z)\,\mathbf{i}_x - \sin(\omega t - \beta z)\,\mathbf{i}_y]$

(b) $-\dfrac{\sqrt{3}E_0}{2}[\sin(\omega t - \beta z)\,\mathbf{i}_x - \cos(\omega t - \beta z)\,\mathbf{i}_y]$

$+\dfrac{E_0}{2}[\cos(\omega t - \beta z)\,\mathbf{i}_x - \sin(\omega t - \beta z)\,\mathbf{i}_y]$

P6.14. cosec α + cot α, **(a)** ∞ **(b)** 2.4142 **(c)** 1

P6.17. **(a)** $\dfrac{\omega_0 af}{c}\sin\omega_0 t$ **(b)** $\dfrac{2f\omega_0 a}{c}\dfrac{\sin\omega_0 t}{\sqrt{5 + 4\cos\omega_0 t}}$

P6.20. **(a)** $0.0349 + j0.0565$ **(b)** $59.4\angle 31.7^\circ\ \Omega$ **(c)** 10^{-3} S/m, $18\varepsilon_0$, μ_0

P6.23. **(a)** 30 MHz **(b)** 2.5 m **(c)** 0.75×10^8 m/s **(d)** 16

(e) $-\dfrac{1}{3\pi}\cos(6\pi \times 10^7 t - 0.8\pi y)\,\mathbf{i}_z$ A/m

P6.26.

$$\mathbf{E} = \begin{cases} \dfrac{48\pi}{7}\cos(6\pi \times 10^8 t - 3\pi z)\,\mathbf{i}_x \text{ V/m} & \text{for } z > 0 \\ \dfrac{48\pi}{7}\cos(6\pi \times 10^8 t + 4\pi z)\,\mathbf{i}_x \text{ V/m} & \text{for } z < 0 \end{cases}$$

$$\mathbf{H} = \begin{cases} \dfrac{3}{35}\cos(6\pi \times 10^8 t - 3\pi z)\,\mathbf{i}_y \text{ A/m} & \text{for } z > 0 \\ -\dfrac{4}{35}\cos(6\pi \times 10^8 t + 4\pi z)\,\mathbf{i}_y \text{ A/m} & \text{for } z < 0 \end{cases}$$

P6.28. **(a)** 16.2838 kHz **(b)** 0.264 mm **(c)** 2.81×10^7 S/m

P6.31. **(a)** 1 Hz **(b)** 2 Hz

P6.33. **(a)** $\dfrac{E_0^2}{\sqrt{\mu_0/\varepsilon_0}}\dfrac{\sin^2\theta}{r^2}\cos^2\omega(t - r\sqrt{\mu_0\varepsilon_0})\,\mathbf{i}_r$

(b) $\dfrac{E_0^2}{2\sqrt{\mu_0/\varepsilon_0}}\dfrac{\sin^2\theta}{r^2}\,\mathbf{i}_r$ **(c)** $\dfrac{4E_0^2\pi}{3\sqrt{\mu_0/\varepsilon_0}}$

P6.35. **(a)** $0.1266E_0^2 e^{-2z}$ W/m² **(b)** $0.1095E_0^2$ W

P6.37. $\mathbf{E}_r = 0.6363E_0 \cos(3\pi \times 10^6 t + 0.01\pi z + 0.938\pi)\,\mathbf{i}_x$ V/m

$\mathbf{E}_t = 0.3955E_0 e^{-0.0495z} \cos(3\pi \times 10^6 t - 0.1196z + 0.1009\pi)\,\mathbf{i}_x$ V/m

$\mathbf{H}_r = -1.688 \times 10^{-3} E_0 \cos(3\pi \times 10^6 t + 0.01\pi z + 0.938\pi)\,\mathbf{i}_y$ A/m

$\mathbf{H}_t = 4.322 \times 10^{-3} E_0 e^{-0.0495z} \cos(3\pi \times 10^6 t - 0.1196z - 0.0241\pi)\,\mathbf{i}_y$ A/m

P6.39. **(a)** 1.492 V/m **(b)** 6.497 m

P6.42. $\mathbf{E}_r = -E_0 \cos(\omega t + \beta z)\,\mathbf{i}_x$

$$\mathbf{H}_r = \frac{E_0}{\eta} \cos(\omega t + \beta z)\,\mathbf{i}_y$$

$$\mathbf{E}_{\text{tot}} = 2E_0 \sin \omega t \sin \beta z\,\mathbf{i}_x$$

$$\mathbf{H}_{\text{tot}} = \frac{2E_0}{\eta} \cos \omega t \cos \beta z\,\mathbf{i}_y$$

$$\mathbf{J}_S = \frac{2E_0}{\eta} \cos \omega t\,\mathbf{i}_x$$

CHAPTER 7

P7.1. **(a)** $1000 \cos(3\pi \times 10^8 t - 3\pi z)$ V/m

(b) $\frac{25}{\pi} \cos(3\pi \times 10^8 t - 3\pi z)$ A/m

(c) $\frac{2.5}{\pi} \cos(3\pi \times 10^8 t - 3\pi z)$ A

(d) $\frac{25}{\pi} \cos^2(3\pi \times 10^8 t - 3\pi z)$ W

P7.3. $\frac{3}{40}\mu_0$ H/m, $120\varepsilon_0$ F/m, $3\pi\ \Omega$

P7.6. 160.55 Ω

P7.10. $\frac{1}{9}\eta$

P7.13. 125 V, 25 Ω, 60 Ω, 2 μs

P7.15.

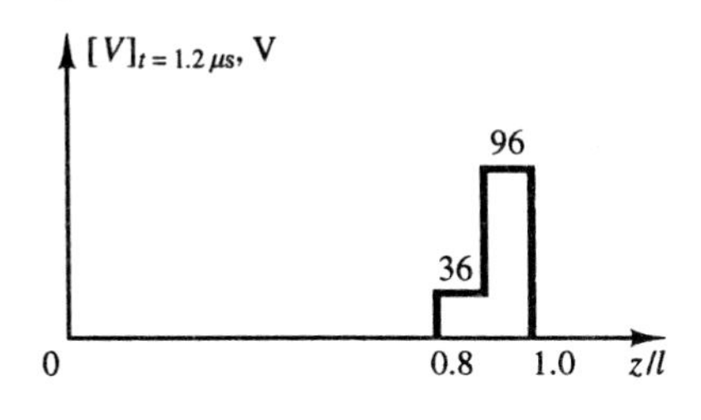

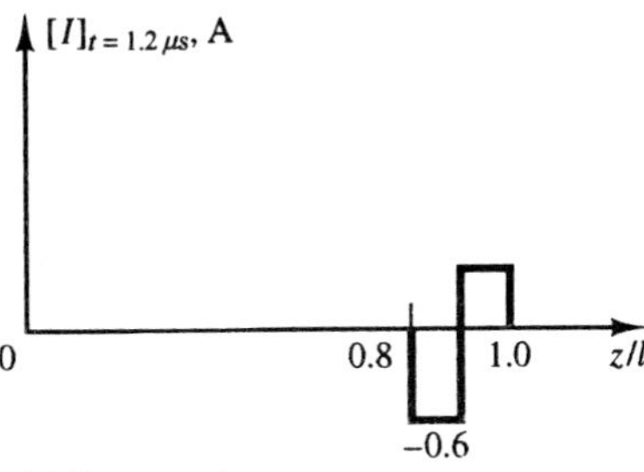

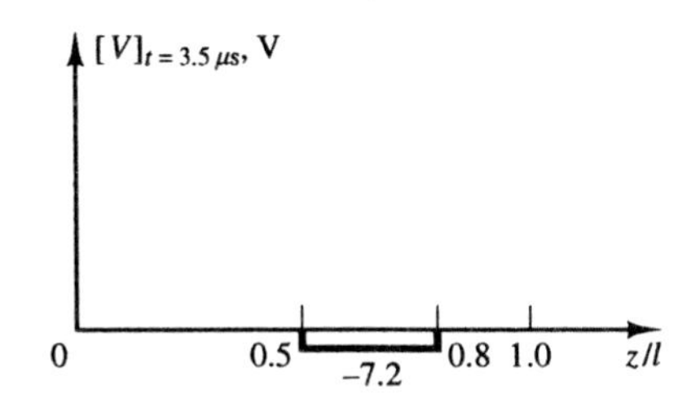

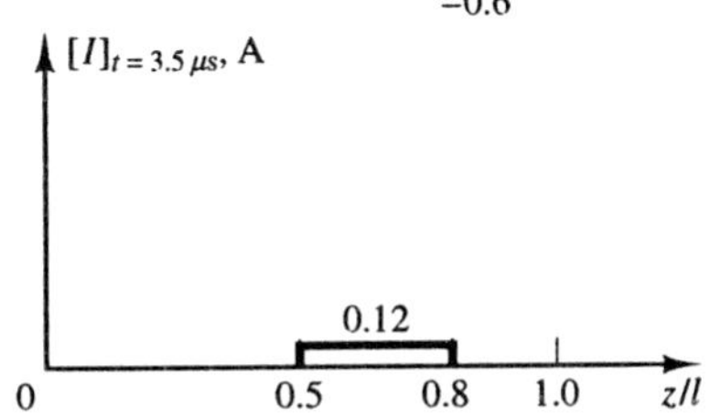

P7.18. $V^- = 0.6V^+$, $I^- = -0.012V^+$, $V^{++} = 0.4V^+$, $I^{++} = 0.004V^+$

P7.20. $16\varepsilon_0$, 75 m, $\frac{8}{15}$

P7.22. **(a)** 20 Ω **(b)** $\frac{1}{4}P$

P7.24. **(a)** 5 ms **(b)** 7 ms

P7.26. $\frac{V_0}{2}[1 - e^{-(2Z_0/L)(t - T)}]$ for $t \geq T$

P7.30.

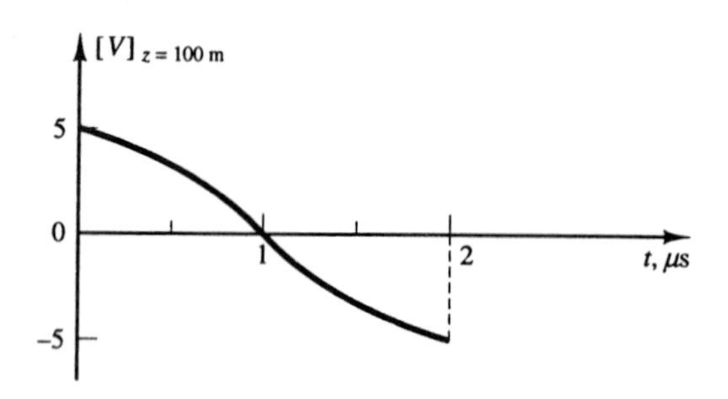

P7.32. **(a)**

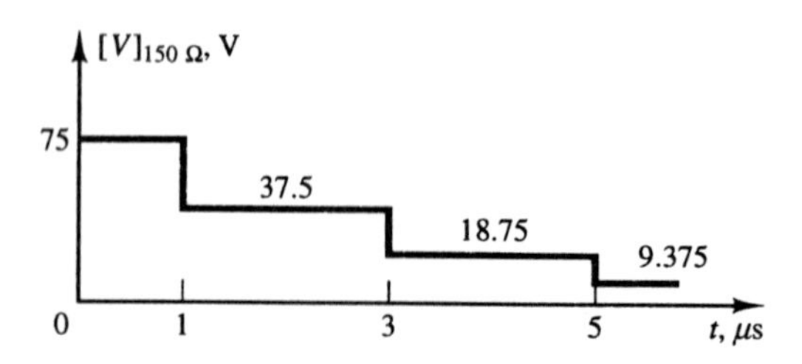

(b) 62.5×10^{-6} J

P7.34.

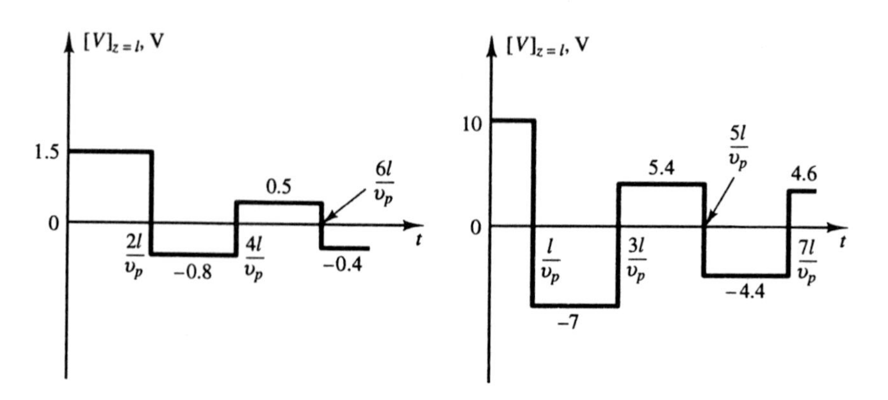

P7.36. **(a)** 38.4 Ω **(b)** 48.4 Ω

CHAPTER 8

P8.2. **(a)**

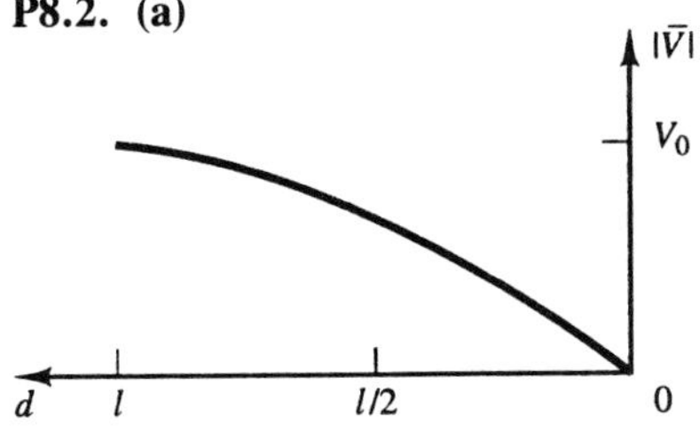

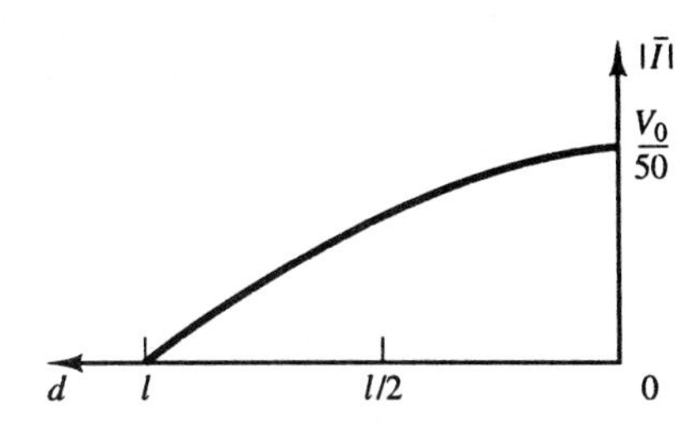

(b)

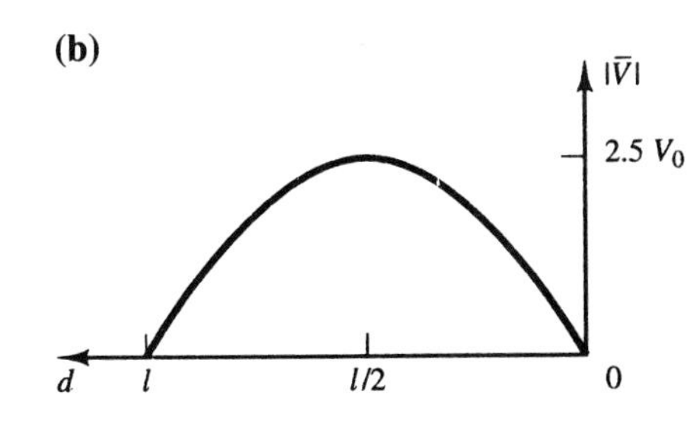

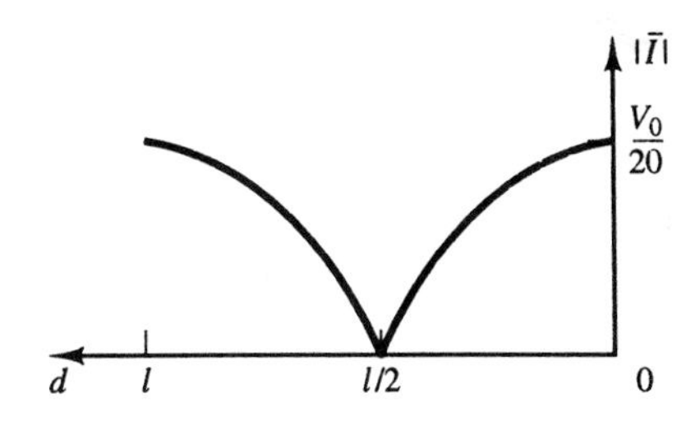

P8.4. $0.01155V_0$

P8.6. nv_p/l, $n = 1, 2, 3, \ldots$

P8.9. 0.2739 GHz, 1.0904 GHz, 2.0491 GHz

P8.12. **(a)** 15.92 MHz **(b)** 13.35 MHz, 18.98 MHz

P8.14. **(a)** 9 **(b)** 2×10^9 Hz, 4×10^9 Hz, 6×10^9 Hz **(c)** 5 cm, 10 cm, 20 cm

P8.17. $-j\dfrac{Z_0}{2} \cot \dfrac{\omega l}{2v_p}$

P8.20. 0.4119λ, 1.53

P8.23. 526.8 MHz

P8.25. $(r - 1)^2 + (x - 1)^2 = 1$

P8.27. **(a)** 3 **(b)** 4.4 **(c)** 2.25

P8.29. 0.032λ, 0.083λ

P8.31. 0.019λ to 0.075λ

P8.33. $l_1 = 0.328\lambda$, $l_2 = 0.458\lambda$
Alternate solution: $l_1 = 0.496\lambda$, $l_2 = 0.042\lambda$

P8.36. $(144.23 - j32.59)\ \Omega$, $(1.035 + j0.975) \times 10^{-4}\ \text{m}^{-1}$

P8.39. 20.659 W, 5.375 W, 15.284 W

CHAPTER 9

P9.2. **(a)** yes **(b)** $\dfrac{1}{12\pi}(2\mathbf{i}_x - \sqrt{3}\mathbf{i}_y - \mathbf{i}_z) \cos [6\pi \times 10^7 t - 0.1\pi(y - \sqrt{3}z)]$

P9.4. $\dfrac{1}{3}(2\mathbf{i}_x + \mathbf{i}_y + 2\mathbf{i}_z)$, 3×10^8 m/s

P9.6. $\dfrac{E_0}{2} \sin 10\pi x\, [\cos (3 \times 10^9 \pi t - 35.1241z) + \cos (5 \times 10^9 \pi t - 71.9829z)]\mathbf{i}_y$

P9.9. **(a)** 0.0204 **(b)** 0.0938 **(c)** 0.0008523

P9.12. **(a)** 2.16×10^8 m/s **(b)** 1.6583×10^8 m/s

P9.16. 3 cm $\leq a \leq$ 4 cm, $b \leq$ 2 cm

P9.18. **(a)** 1.70 **(b)** 0.366 cm, $4.9\varepsilon_0$

P9.20. **(a)** 7.906 GHz **(b)** 7.906 GHz, 8.291 GHz

P9.22. $3\sqrt{2}$ GHz ($TE_{1,0,1}$, $TE_{0,1,1}$, $TM_{1,1,0}$)
$3\sqrt{3}$ GHz ($TE_{1,1,1}$, $TM_{1,1,1}$)
$3\sqrt{5}$ GHz ($TE_{2,0,1}$, $TE_{0,2,1}$, $TE_{1,0,2}$, $TE_{0,1,2}$, $TM_{1,2,0}$, $TM_{2,1,0}$)

P9.24. 7.281 GHz ($TE_{2,1}$)), 9.143 GHz ($TE_{0,1}$), 12.271 GHz ($TM_{2,1}$)

P9.26. $l_2 = 0.4442$ cm, $l_3 = 1.2726$ cm

P9.28. **(b)** 2091.32 MHz

P9.30. 0.882×10^{-3}

P9.33. **(c)** 9,276

CHAPTER 10

P10.2. $\mathbf{E}_r = -0.1716E_0\mathbf{i}_y \cos\left[6\pi \times 10^8 t + \sqrt{2}\pi(x - z)\right]$
$\mathbf{E}_t = 0.8284E_0\mathbf{i}_y \cos\left[6\pi \times 10^8 t - \sqrt{2}\pi(\sqrt{2}x + z)\right]$

P10.4. **(a)** -0.2087 **(b)** -1.2919 **(c)** ± 0.5192

P10.7. 75.52°

P10.9. $\sqrt{3}$

P10.14. 0.9152, 0.9130, 0.9090

P10.19. $z = \dfrac{1}{\alpha} \ln \dfrac{\alpha x - 1 + \sqrt{\alpha^2 x^2 - 2\alpha x + \delta_0^2}}{\delta_0 - 1}; \dfrac{\delta_0^2}{2\alpha}, \dfrac{2\delta_0}{\alpha}$

P10.26. 0.7108

P10.33. **(a)** $\mathbf{E}_r = -E_0\left[\tfrac{3}{7}\cos(6\pi \times 10^8 t + 2\pi z)\,\mathbf{i}_x + \tfrac{1}{5}\sin(6\pi \times 10^8 t + 2\pi z)\,\mathbf{i}_y\right]$

$\mathbf{H}_r = \dfrac{E_0}{\eta_0}\left[-\tfrac{1}{5}\sin(6\pi \times 10^8 t + 2\pi z)\,\mathbf{i}_x + \tfrac{3}{7}\cos(6\pi \times 10^8 t + 2\pi z)\,\mathbf{i}_y\right]$

(b) $\mathbf{E}_t = E_0\left[\tfrac{4}{7}\cos(6\pi \times 10^8 t + 5\pi z)\,\mathbf{i}_x + \tfrac{4}{5}\sin(6\pi \times 10^8 t + 3\pi z)\,\mathbf{i}_y\right]$

$\mathbf{H}_t = \dfrac{E_0}{\eta_0}\left[-\tfrac{6}{5}\sin(6\pi \times 10^8 t + 3\pi z)\,\mathbf{i}_x + \tfrac{10}{7}\cos(6\pi \times 10^8 t + 5\pi z)\,\mathbf{i}_y\right]$

P10.35. $\mathbf{E}_r = -\dfrac{E_0}{15}(7\mathbf{i}_x + \mathbf{i}_y)\cos(6\pi \times 10^9 t + 20\pi z)$

$$\mathbf{E}_t = \frac{E_0}{5}(2\mathbf{i}_x + \mathbf{i}_y)\cos(6\pi \times 10^9 t - 60\pi z) + \frac{2E_0}{15}(\mathbf{i}_x - 2\mathbf{i}_y)\cos(6\pi \times 10^9 t - 40\pi z)$$

$$\mathbf{H}_r = -\frac{E_0}{15\eta_0}(\mathbf{i}_x - 7\mathbf{i}_y)\cos(6\pi \times 10^9 t + 20\pi z)$$

$$\mathbf{H}_t = -\frac{3E_0}{5\eta_0}(\mathbf{i}_x - 2\mathbf{i}_y)\cos(6\pi \times 10^9 t - 60\pi z) + \frac{4E_0}{15\eta_0}(2\mathbf{i}_x + \mathbf{i}_y)\cos(6\pi \times 10^9 t - 40\pi z)$$

CHAPTER 11

P11.4. 1.5721 V/m, 0.4506 V/m, 1.5343×10^{-3} A/m

P11.6. 0.2024λ

P11.8. 0.5305 A, 0.0111 W

P11.10. $\sqrt{\dfrac{D_2}{D_1}\dfrac{R_{\text{rad}\,2}}{R_{\text{rad}\,1}}}$

P11.13. 0.0167 A, 0.0168 W

P11.16. $\dfrac{\pi\eta}{6}\left(\dfrac{2\pi a}{\lambda}\right)^4$, 1.5

P11.19. $\sin^4\theta\cos^2\phi$, 3.75

P11.22. Five elements spaced 2λ apart and fed in the ratio $1:-2:3:-2:1$

P11.24. **(a)** $\dfrac{\cos[(\pi/2)\cos\theta]}{\sin\theta}$ **(b)** 3.284

P11.26. **(a)** 0.8284 **(b)** 4 **(c)** 0

P11.29. $\bar{E} = \dfrac{j\beta E_0 ab e^{-j\beta r}}{4\pi r}\left[\dfrac{\sin(\psi_1/2)}{(\psi_1/2)}\right]^2\left(\dfrac{\sin\psi_2}{\psi_2}\right)$

where $\psi_1 = (\beta a/2)\sin\theta\cos\phi$ and $\psi_2 = (\beta b/2)\sin\theta\sin\phi$

(a) $4\dfrac{\lambda}{a}$ **(b)** $2\dfrac{\lambda}{b}$ **(c)** $1.28\dfrac{\lambda}{a}$ **(d)** $\dfrac{3\pi}{\lambda^2}ab$

P11.34. 160.81°, 127.67°, 106.13°, 86.82°, 67.11°, 43.76°; $(5/6)\lambda$

Index